Beginner's Guide to SOLIDWORKS 2018 – Level II

Sheet Metal, Top Down Design, Weldments, Surfacing and Molds

Alejandro Reyes, MSME, CSWE, CSWI
Certified SOLIDWORKS Expert

SDC Publications
P.O. Box 1334
Mission, KS 66222
913-262-2664
www.SDCpublications.com
Publisher: Stephen Schroff

ISBN-13: 978-1-63057-166-5
ISBN-10: 1-63057-166-0

Printed and bound in the United States of America.

Acknowledgements

Beginner's Guide to SOLIDWORKS 2018 – Level II is dedicated to the love of my life, my wife Patricia, and my kids Liz, Ale and Hector, who have given me the time, support and patience to write this book. To you, all my love.

Also, I wish to thank the hundreds of students, users, professors and engineers whose great ideas and words of encouragement helped me to improve this book.

About the Author

Alejandro Reyes holds a BSME from the Instituto Tecnológico de Ciudad Juárez, Mexico, in electro-mechanical engineering and a Master's Degree from the University of Texas at El Paso in mechanical design, with strong focus in Materials Science and Finite Element Analysis.

Alejandro spent more than 8 years as a SOLIDWORKS Value Added Reseller. During this time, he was a Certified SOLIDWORKS Instructor and Support Technician, CosmosWorks Support Technician, and a Certified SOLIDWORKS Expert, credentials that he still maintains. Alejandro has over 20 years of experience using CAD/CAM/FEA software and is currently the President of MechaniCAD Inc.

His professional interests include finding alternatives and improvements to existing products, FEA analyses and new technologies. On a personal level, he enjoys spending time with his family and friends.

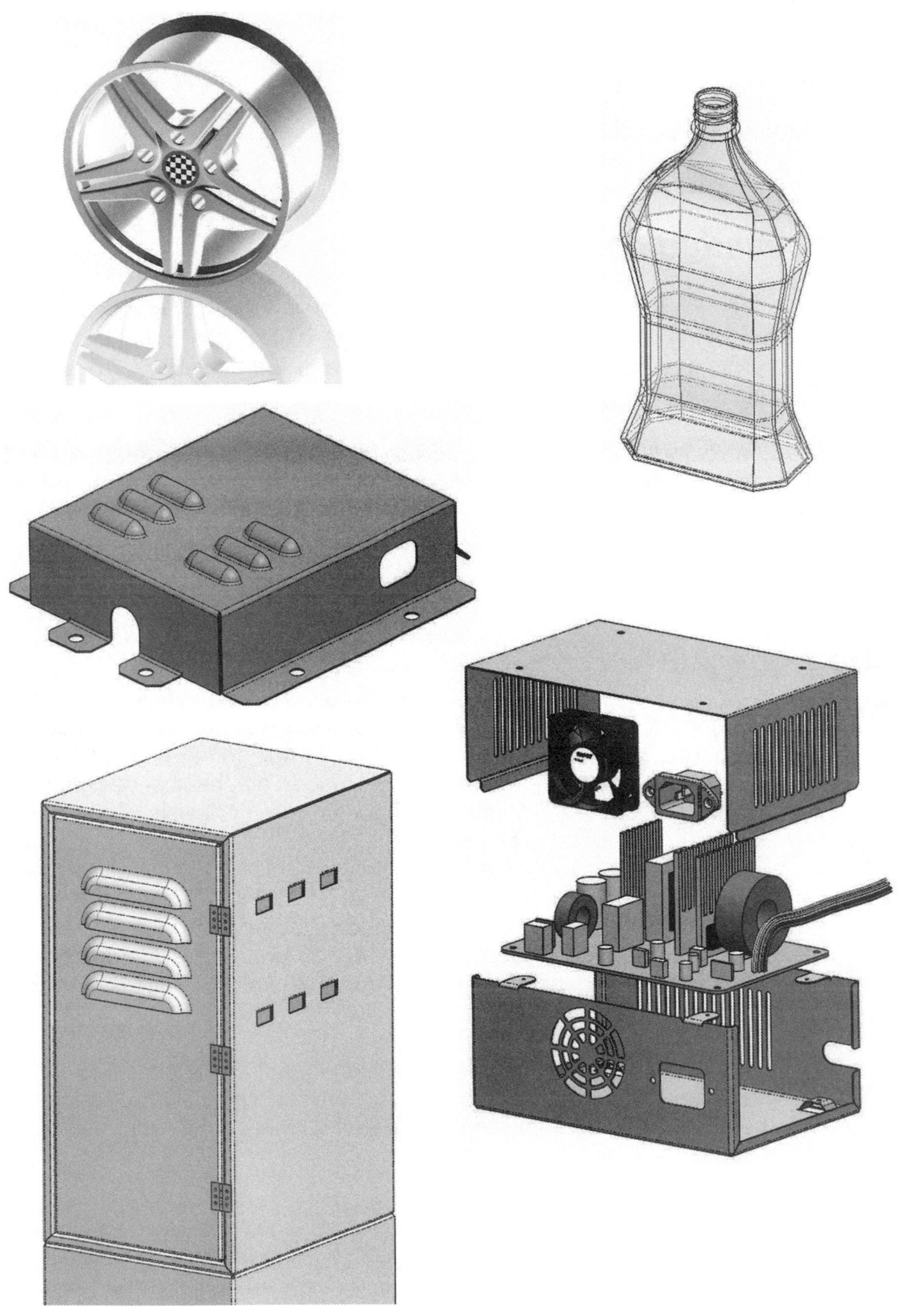

Table of Contents

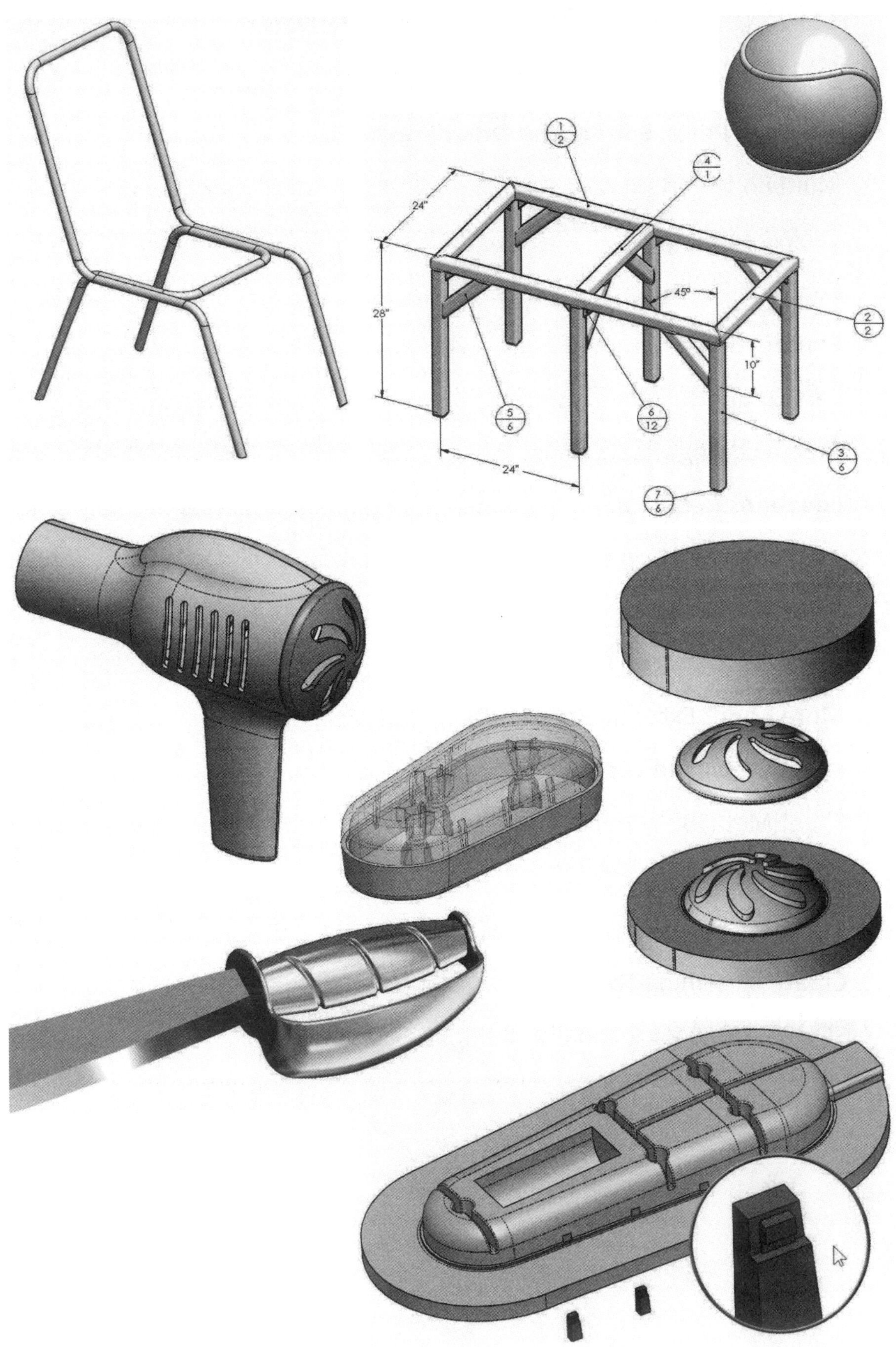

1
2
4
1
24"
28"
45°
2
2
10"
5
6
6
12
3
6
24"
7
6

List of commands introduced in each chapter. Note that many commands are used extensively in following chapters after they have been presented.

Multi Body Parts

Multi Body
Local Operations
Hide/Show Body
Merge
Feature Scope
Delete Body
Body Pattern
Combine Bodies
  Add
  Subtract
  Common
Bodies to keep
Offset from Surface
Sketch Picture

Contour Selection

Regions available
Contour Selection
Shared Sketch
Start From: Condition

Part Editing

What's Wrong
Parent/Child relations
Sketch Editing
Dangling Relations
Delete Absorbed Features
Sketch Relations
Over Defined Sketch
Not Solved Sketch
SketchXpert
View Sketch Relations

Equations

Rename Dimensions
Pattern Seed Only
Add Equations
Edit Equations
Delete Equations
Link Values

Top Down Design

New Part
Edit In Context
Assembly Transparency
Internal Part
Externalize Part
Edit Assembly
External References
  In Context
  Out of Context
  Locked
  Broken
List External References

Sheet Metal and Top Down Design

Base Flange
Sheet Metal Thickness
Bend Radius
Bend Allowance
Bend Deduction
K-Factor
Auto-Relief
  Rectangular
  Obround
  Tear
Flat Pattern
Forming Tools
Modify Sketch
Link to Thickness
Normal Cut
3D Content Central
Vent feature
Miter Flange
Unfold/Fold Bend
Edge Flange
Build Library Features
Library Parts
Mate Reference
Break Corners
Jog bend
Flat Pattern Drawing
Bend Notes
Convert to Sheet Metal
Closed Corners
Selection Filters
Sketch Pattern
Feature Driven Pattern
Hem Feature
Creating Forming Tools
Component Pattern
Collision Detection
Flexible Sub Assemblies
Assembly Features

3D Sketch

3D Sketch
3D Sketch Relations
Derived Sketch
Projected Curve

Weldments

3D Sketch review
Cut list
Weldment feature
Structural Member
Corner Treatment
  End Miter
  End Butt
Locate Profile
Rotate Profile
Trim/Extend
Gusset
End Cap
Weld Beads
Weldment Cut List
Weldment Drawings
Cut List Table
Weld Table
Weld Symbols
Save Bodies to Assembly
Structural Member Libraries

Surfacing

Revolved Surface
Lofted Surface
Extruded Surface
Direction of Extrusion
Extrude with Draft
Trim Surface
Mutual Trim
Planar Surface
Filled Surface
Knit Surface
Constant Width Fillet
Thicken
Body Split
Face Fillet
Extend Surface
Extrude From
Mirror Bodies
Swept Surface
Twist Along Path
Sweep Cut
Master Model
Mounting Boss Feature
Lip/Groove Feature
Snap Hook/Groove Feature

Mold Tools

Draft Analysis
Direction of Pull
Positive Draft
Negative Draft
Draft
Neutral Plane
Rollback/Roll Forward
Scale
Parting Line
Parting Surface
Tooling Split
Composite Curve
Swept Surface
Shut Off Surfaces
Move/Copy body
Delete Face
Face Classification
Manual Parting Line Selection
Select Open Loop
Rib
Side Core

Notes:

Introduction

Beginner's Guide to SOLIDWORKS – Level II starts where *Beginner's Guide – Level I* ends, following the same easy to read style, but this time covering advanced topics and techniques.

The purpose of this book is to teach advanced SOLIDWORKS functionality including sheet metal, surfacing, how to create components in the context of an assembly referencing other components (Top-down design), propagate design changes with SOLIDWORKS' parametric capabilities, mold design, weldments, and more while explaining the basic concepts of each trade to allow the reader to understand the how and why of each operation.

This book uses simple examples to allow the reader to better understand commands and environments, as well as to make it easier to explain the purpose of each step, maximizing the learning time by focusing on one task at a time. Keep in mind that this book is focused on learning SOLIDWORKS, and the specific details of some of the topics that will be covered go far beyond the scope of this book; entire books have been written about sheet metal processes, mold design, welded structures, and such. With this in mind, please remember that this book will teach you how to use SOLIDWORKS' tools for those trades, and at no time will attempt to teach the trade itself, but we'll do our best to explain the general details in order to understand the processes.

At the end of this book, the reader will have acquired enough skills to be competitive when it comes to designing with SOLIDWORKS, and while there are many less frequently used commands and options available that will not be covered, rest assured that those commands that are covered are the most commonly used by SOLIDWORKS designers.

IMPORTANT: The exercise files required to complete some of the exercises and extra content can be accessed using the code included with this book. Exercise files only will be freely available from www.sdcpublications.com/downloads/978-1-63057-166-5. The files include high resolution images of the exercises and the finished files made throughout the book to help you practice and enhance your skills. We hope you'll learn new skills and enjoy reading this book as much as we enjoyed making it. We always love to hear ideas and comments from our readers. If you have any, please share them with us and our readers; we'll try our best to accommodate any suggestions to improve and expand the content of this book, and while we cannot guarantee that they will make it into the next edition, we can assure you that we reply to all email messages.

alejandro@mechanicad.com

Prerequisites

This book has been written assuming the reader has knowledge of the following topics:

- Familiarity with the Windows operating system,
- General knowledge of mechanical design and drafting, and
- Previous experience with basic SOLIDWORKS functionality.
- A general understanding of sheet metal, structural and molding processes will help to better understand the tools used in each section.

Note about the screen images: The images on your screen *may* be slightly different from this book. The images in the book were made using Windows 10 Professional and SOLIDWORKS' 2018 default installation settings. Unless otherwise noted, the only changes made to SOLIDWORKS' default options were adding a white background and in some instances the preview colors were changed to improve clarity in print and/or electronic format.

IMPORTANT INFORMATION ABOUT FILES: High resolution images of the exercises in this book and the files needed for some exercises are included in the exercise files. Copy all the files to a local folder in your PC to have them ready when needed.

Multi Body Parts, Editing and Other Tools

Multi body parts

SOLIDWORKS offers several tools to help us model components easier and faster with powerful options like multi body parts and contour selection. Multi body parts means that we can have more than one "solid" body in a part. Up until now we have been modeling single body parts, meaning that each feature either added or removed material to the part but always resulted in a single solid body. Designing with multi body parts allows us to do things that sometimes are very difficult or maybe even impossible using a single body component, like, among other things, combine bodies, remove one body from another or calculate their common volume. Multi body operations are particularly useful when working with mold design and are the essence of the weldments environment (both are covered later in this book).

Contour Selection

Besides working with single body parts, we have also worked with single and multiple contour open or closed sketches. In this section, we'll cover how to use Contour selection to work with sketches that contain multiple entities sharing an endpoint and/or having intersecting lines. We'll learn how to take advantage of these situations and make the best of them.

Part Editing

Another important topic when modeling in SOLIDWORKS is editing parts to fix an error or make a change to a component. After all, when we are designing, at some point we will most likely have to go back and make changes to our design, and sometimes those changes can cause trouble in subsequent features. In the Editing section, we'll cover different options available when editing parts, as well as fixing and recovering from errors.

Equations

The last topic we'll cover in this section deals with maintaining design intent; we'll learn how to add, edit and delete equations in a part, giving us powerful tools to make our designs more flexible, robust and predictable when making design changes.

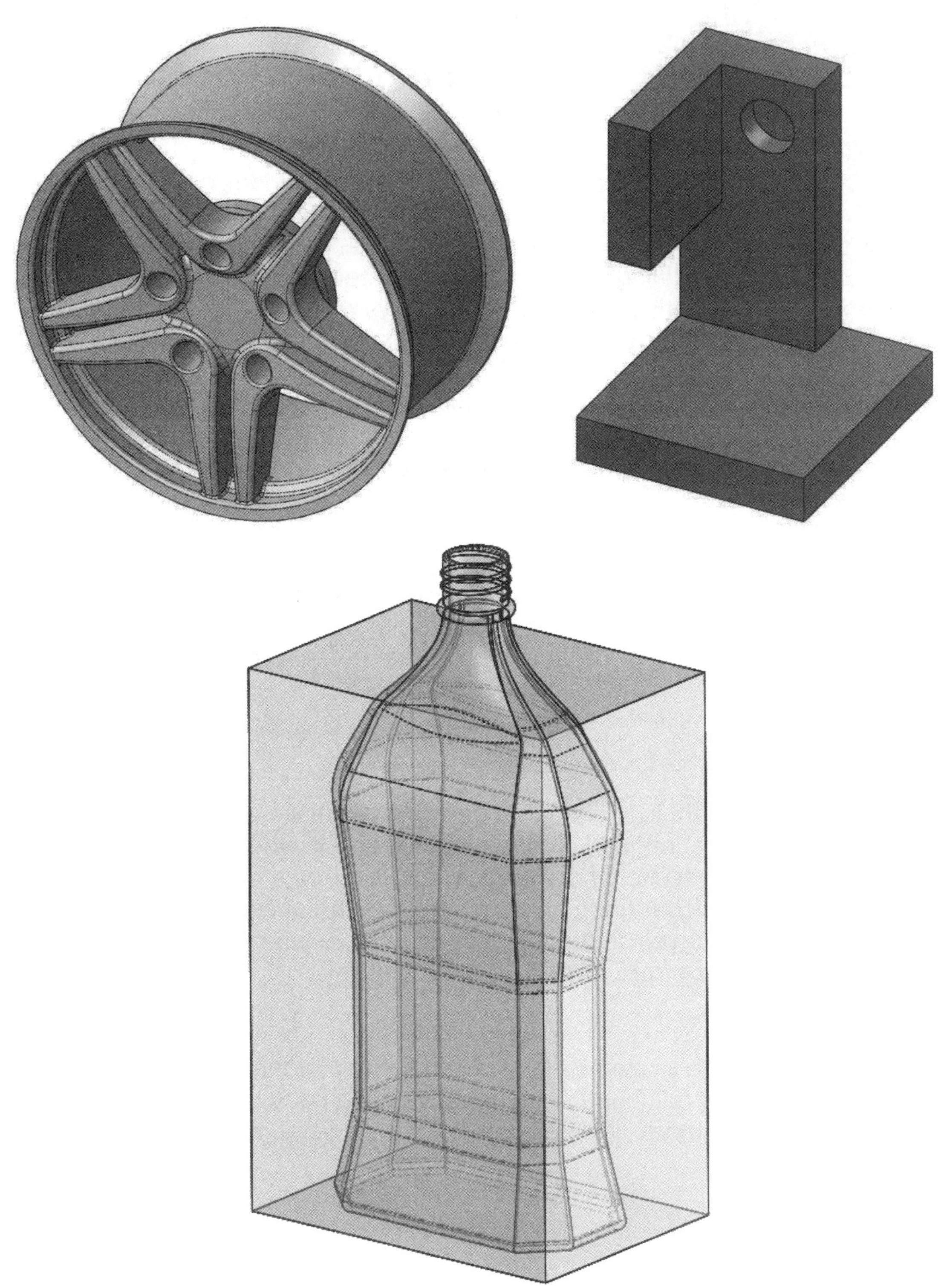

Multi Body Parts

As we stated earlier, multi body parts allow us to do certain operations that would be difficult to accomplish with a single body part. First, we'll talk about the general concept of multi body parts, and then we'll look at different ways to use them.

1.1. - Multi body parts are made by adding material that is not connected to the current solid body, in other words, that is purposely <u>not</u> merged with it. A different way to create a multi body part is by splitting an existing part into multiple bodies (this technique will be shown later). To show how a multi body part works, make a new part, and add a sketch to the *"Front Plane"* as shown. Do not worry about dimensions in this example, we are just illustrating the concept of how multi bodies work.

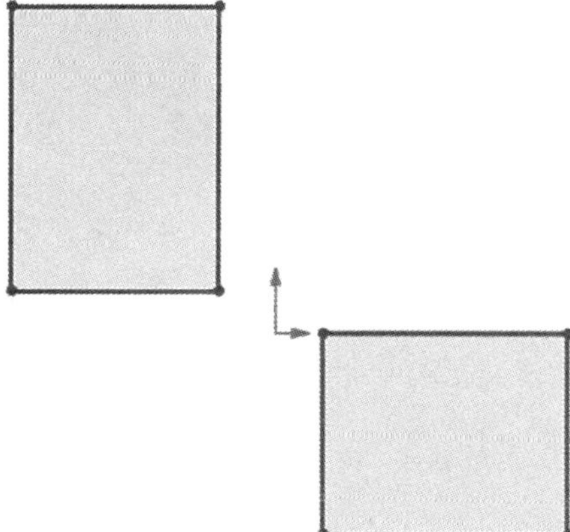

1.2. - Select the "**Boss Extrude**" command. Notice that we are not given a warning about having two separate bodies; it just works. Extrude the sketch any size that looks similar to the following image (dimensions are not important at this time) and click **OK** to finish. Making two or more extruded/revolved/swept features would work just the same.

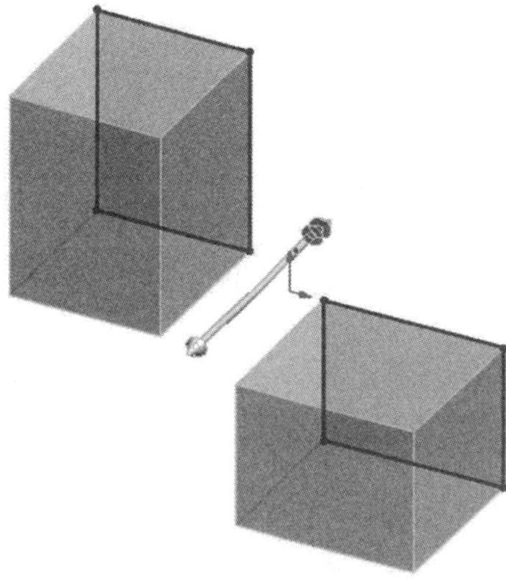

The first thing we notice is a new folder in the FeatureManager called "*Solid Bodies (2).*" This folder is automatically added when SOLIDWORKS detects multiple disjointed bodies in a part and lists the number of bodies found in the part (in this case 2). If we expand the folder, we can see the two bodies in our part listed under it. The important thing to know and remember is that **multi body parts are not to be confused or used in place of an assembly**; parts and assemblies have significant differences and each serves a different purpose. A multi body part is used mostly as a means to an end.

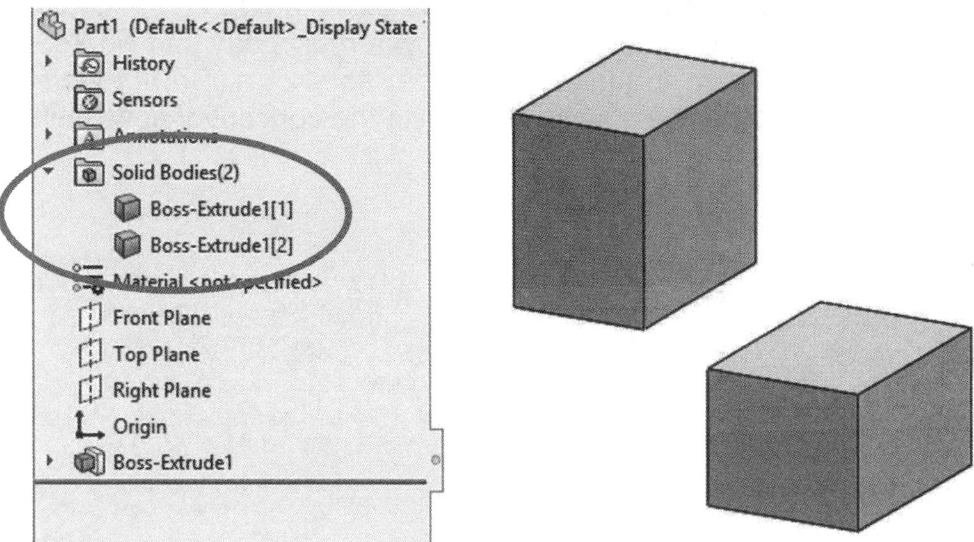

1.3. - The next step is to add a new feature. When working in a multi body part we can make 'local' operations to each body, in other words, a feature affecting only one solid body. Local operations are applied features like fillets, shells, chamfers, etc., which can only be applied to a single body at a time. Select the "Shell" command from the Features toolbar, and shell the bottom body removing the indicated faces.

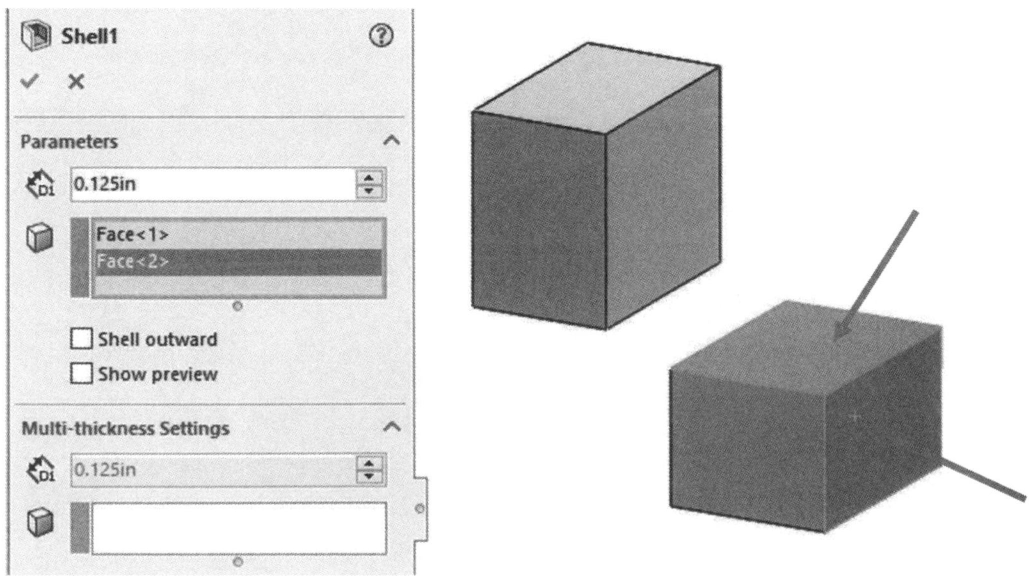

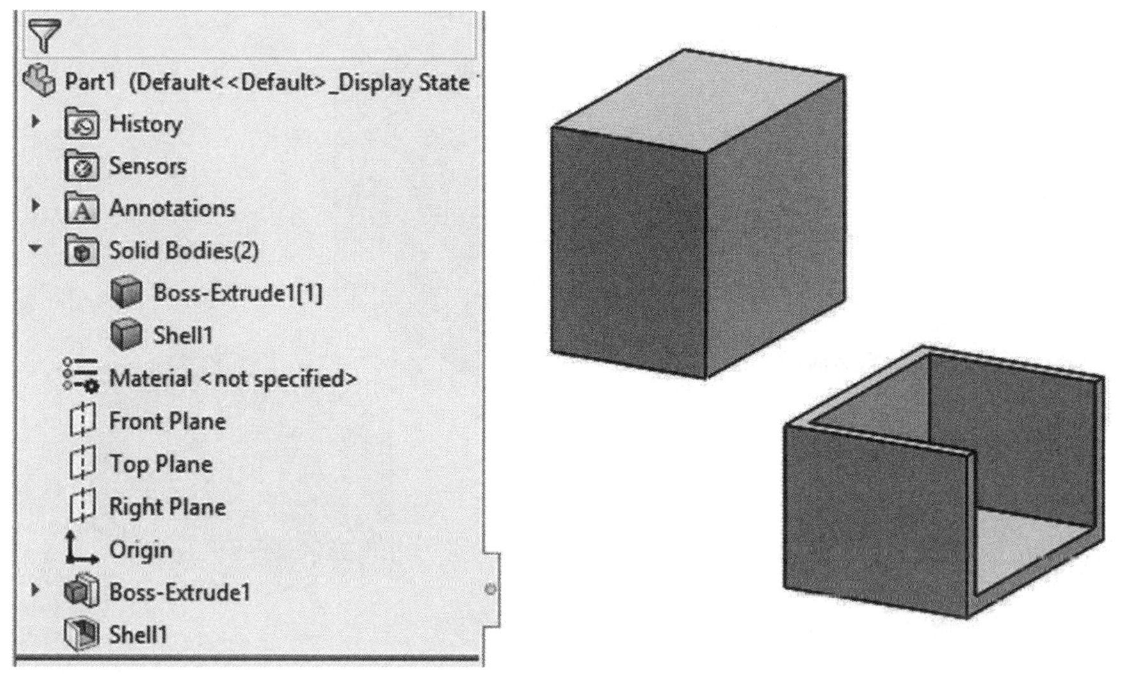

1.4. – Now we'll round a couple of corners in the other body. Select the **Fillet** command and round two corners of the top body as shown.

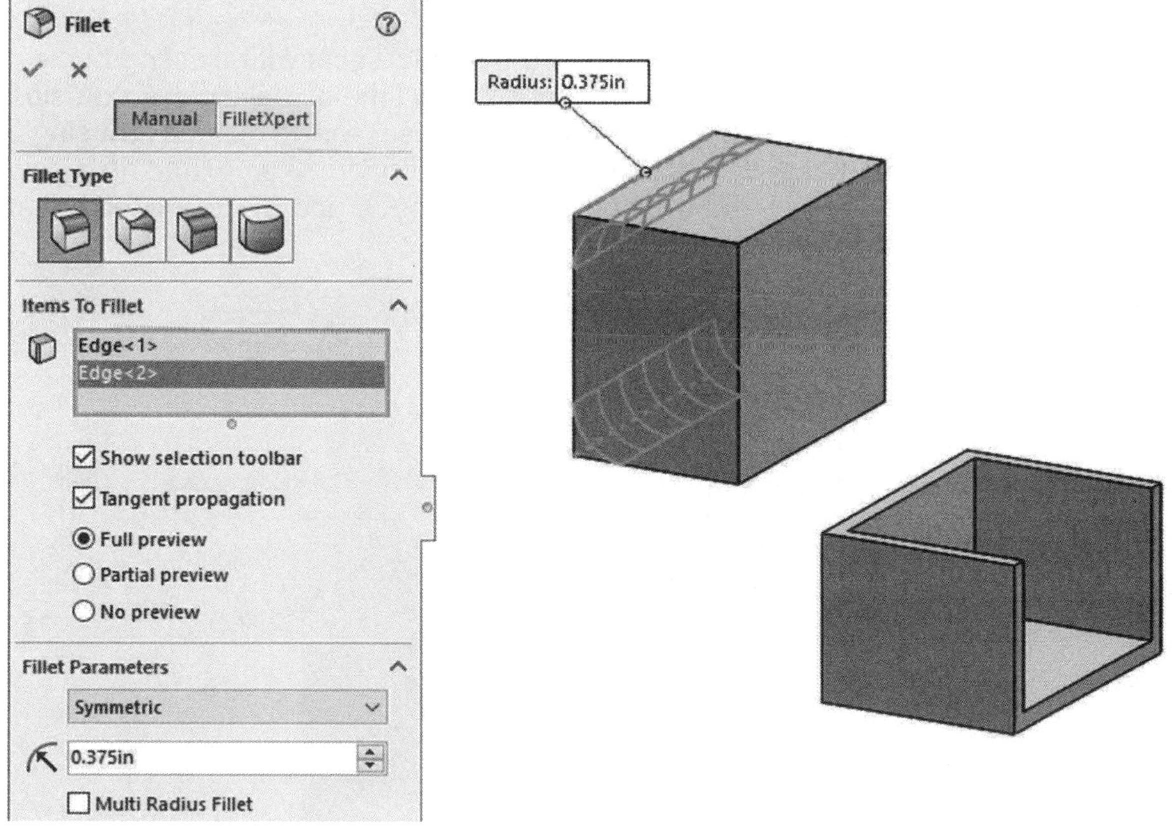

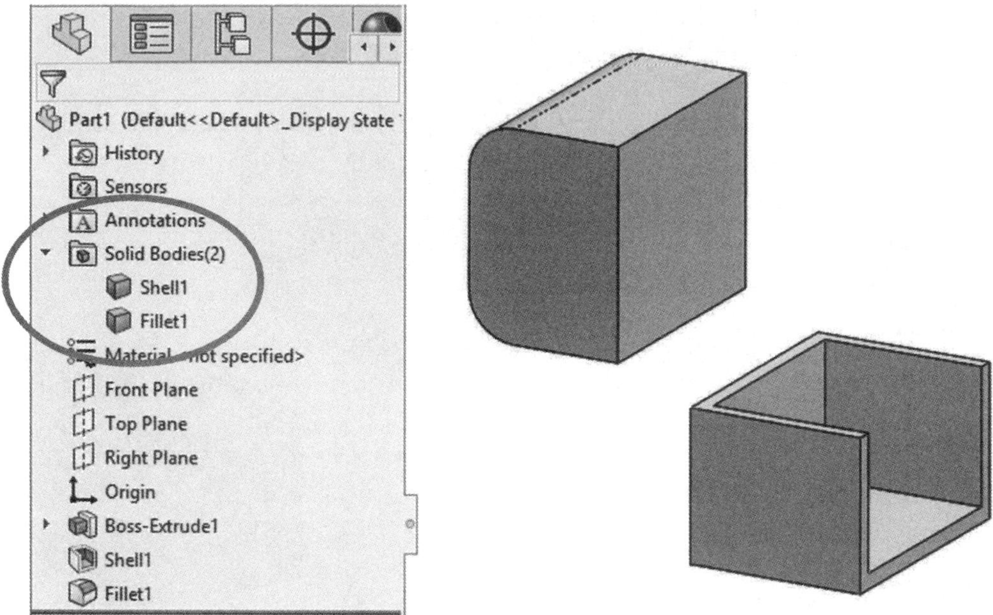

To help the reader understand the concept of local operations better, “local” means that we can add applied features (features that don't require a sketch like Shell, Fillet, Chamfer, Draft, etc.) to one body at a time. Notice the name of a body changes to match the last feature applied to it.

1.5. - If we add another boss extrude feature that intersects with existing bodies, by default it will automatically merge with them. At the time of making the extrusion we can optionally select which bodies to “merge” (or fuse) with to make a new single body. Add the following sketch to explore the merge option when extruding a new boss into existing bodies. Select the front face of a body and make a sketch that intersects both bodies as shown.

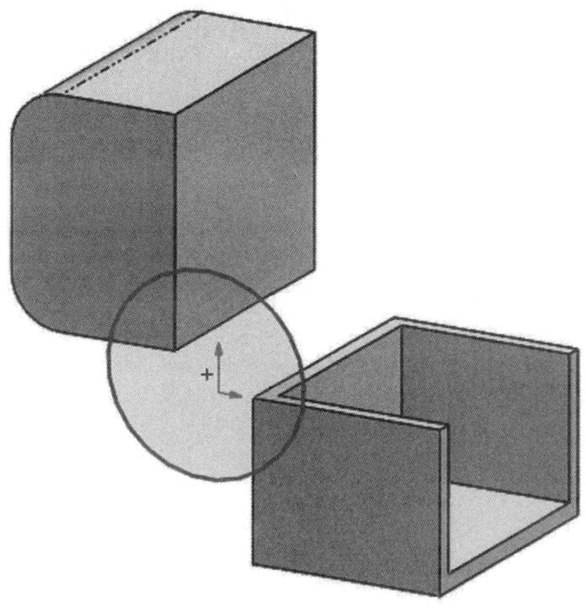

1.6. – Extrude the sketch *into* the existing bodies. (The same would happen if we extrude *away* from the bodies because they are touching the new extrusion.)

Notice the "Merge result" option in the Extrude command. It's always been there (except when there are no solid bodies) and by default is always checked. When a part has multiple bodies, an additional option at the bottom called "Feature Scope" is shown. This is where we can select the existing bodies we want to merge the new extrusion with.

The "Merge result" options for the new extrusion automatically includes "All bodies" or manually "Selected bodies." By default, the "Feature Scope" option is set to "Selected bodies" and "Auto-select." These two options mean that, by default, the new feature will merge with any solid body it intersects or touches. If the "Merge result" option is un-checked, "Feature Scope" is automatically removed, and the new extrusion will not be merged with any solid body.

For this step uncheck the option "Merge Result" and click OK to finish.

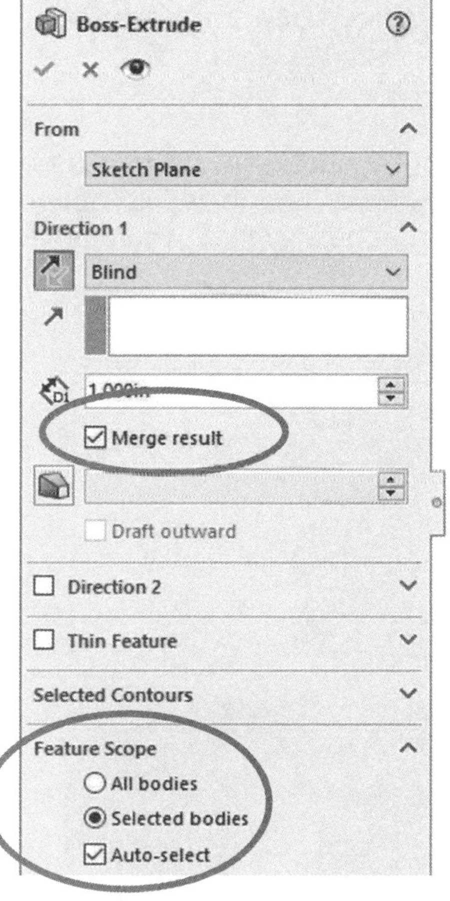

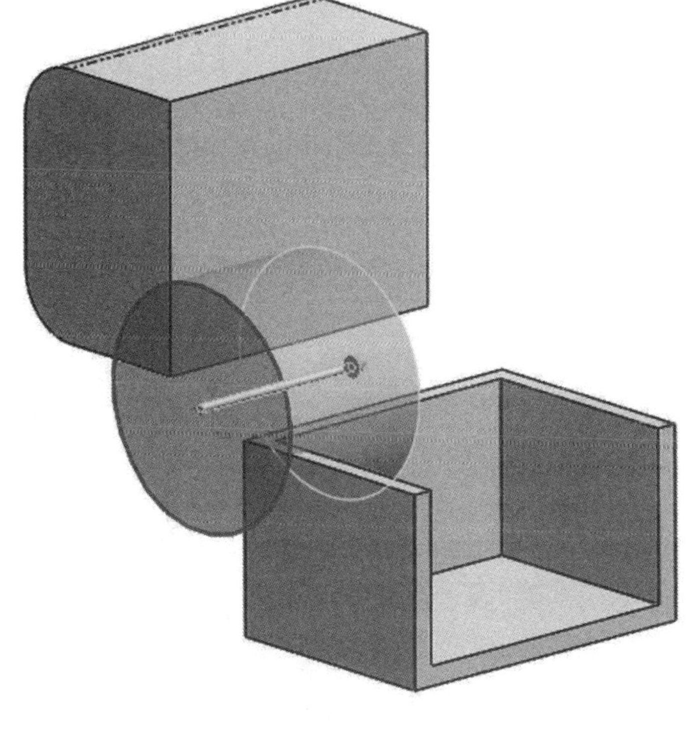

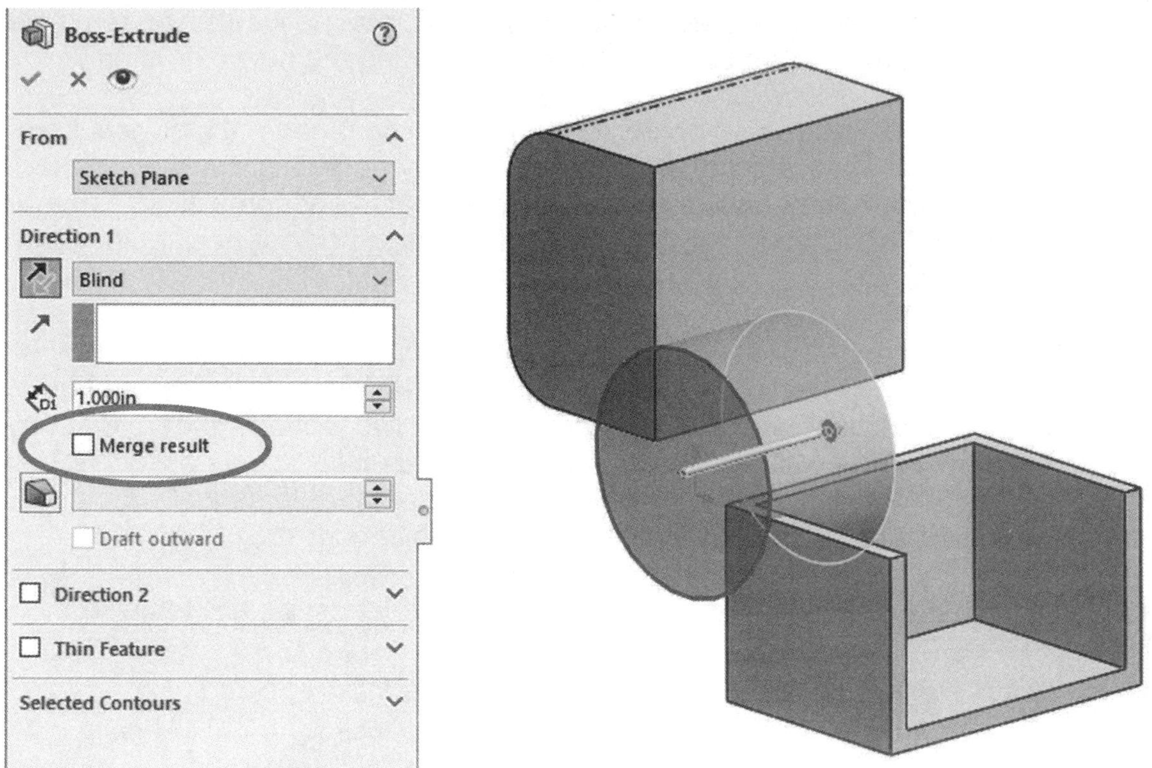

1.7. - After adding this extrusion we have three different bodies in our part. See how the different bodies' edges intersect each other. If the "Merge result" option had been checked, we would not see the overlapping edges as they would have merged into a single body.

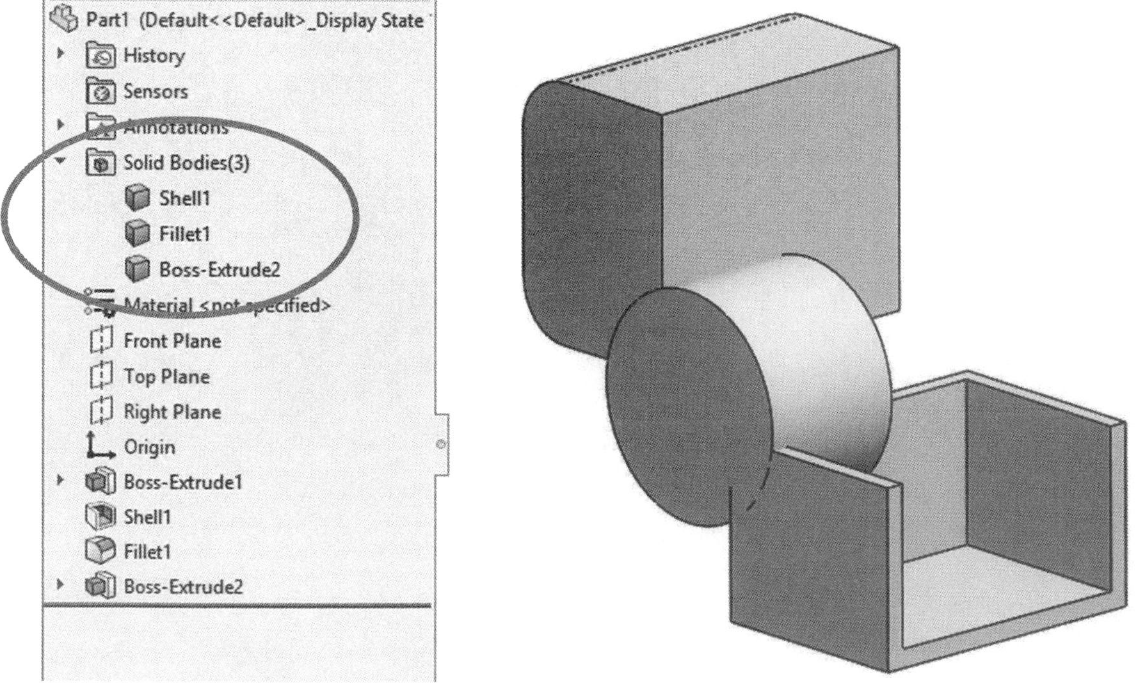

Edit the last extrusion to explore the effect of different "Merge Result" and "Feature Scope" combinations.

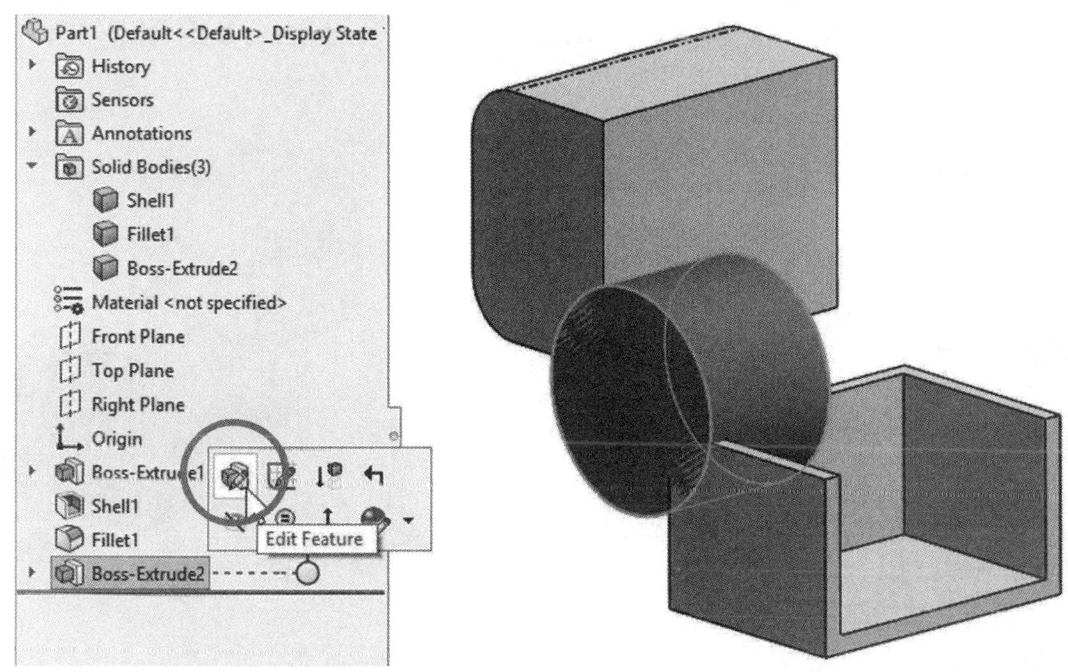

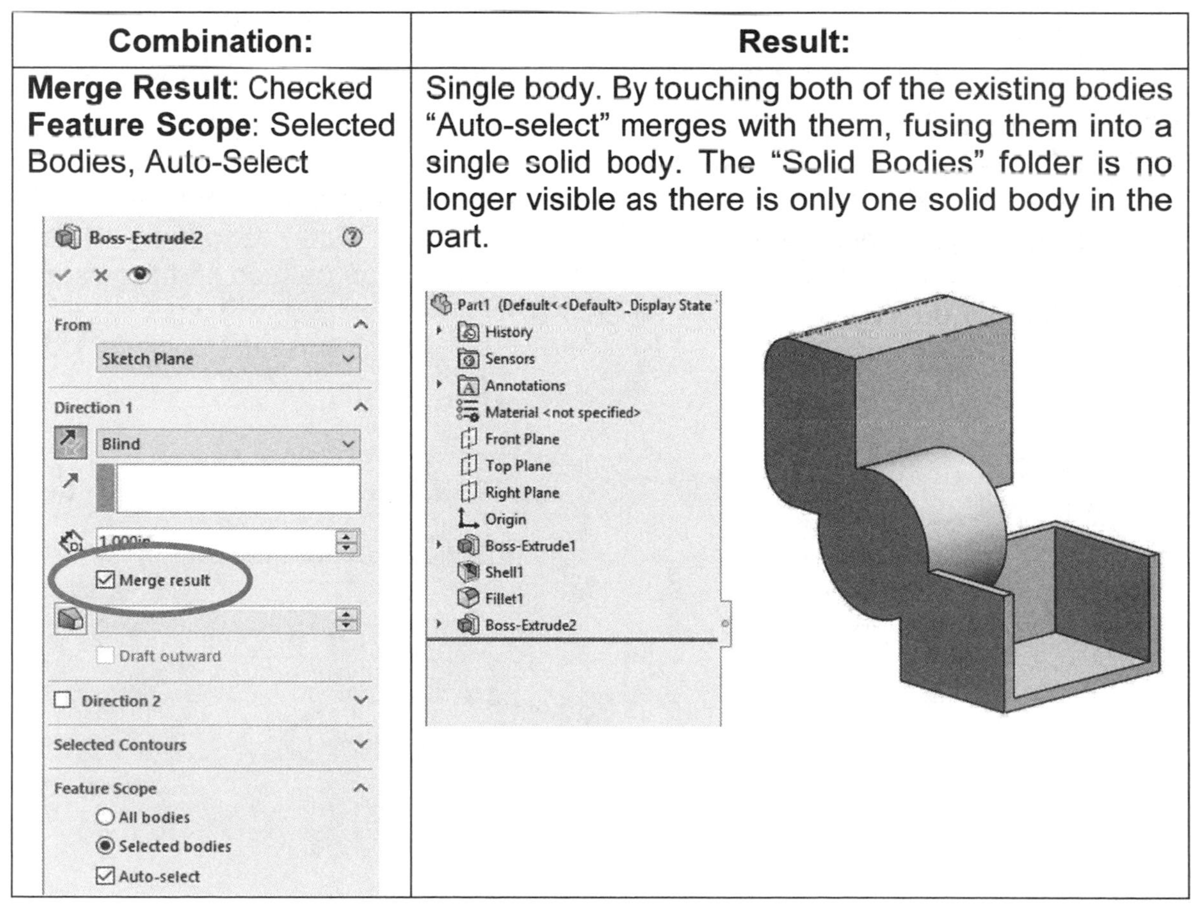

Combination:	Result:
Merge Result: Checked **Feature Scope**: Selected Bodies, Auto-Select	Single body. By touching both of the existing bodies "Auto-select" merges with them, fusing them into a single solid body. The "Solid Bodies" folder is no longer visible as there is only one solid body in the part.

Merge result: Checked **Auto-select**: Unchecked, Add shell body to selection box	Two bodies. We are telling the Boss-Extrude to merge only to the Shell body (or whichever body we pick). Notice the edges of the upper body and the Boss-Extrude overlap, as they are different bodies.

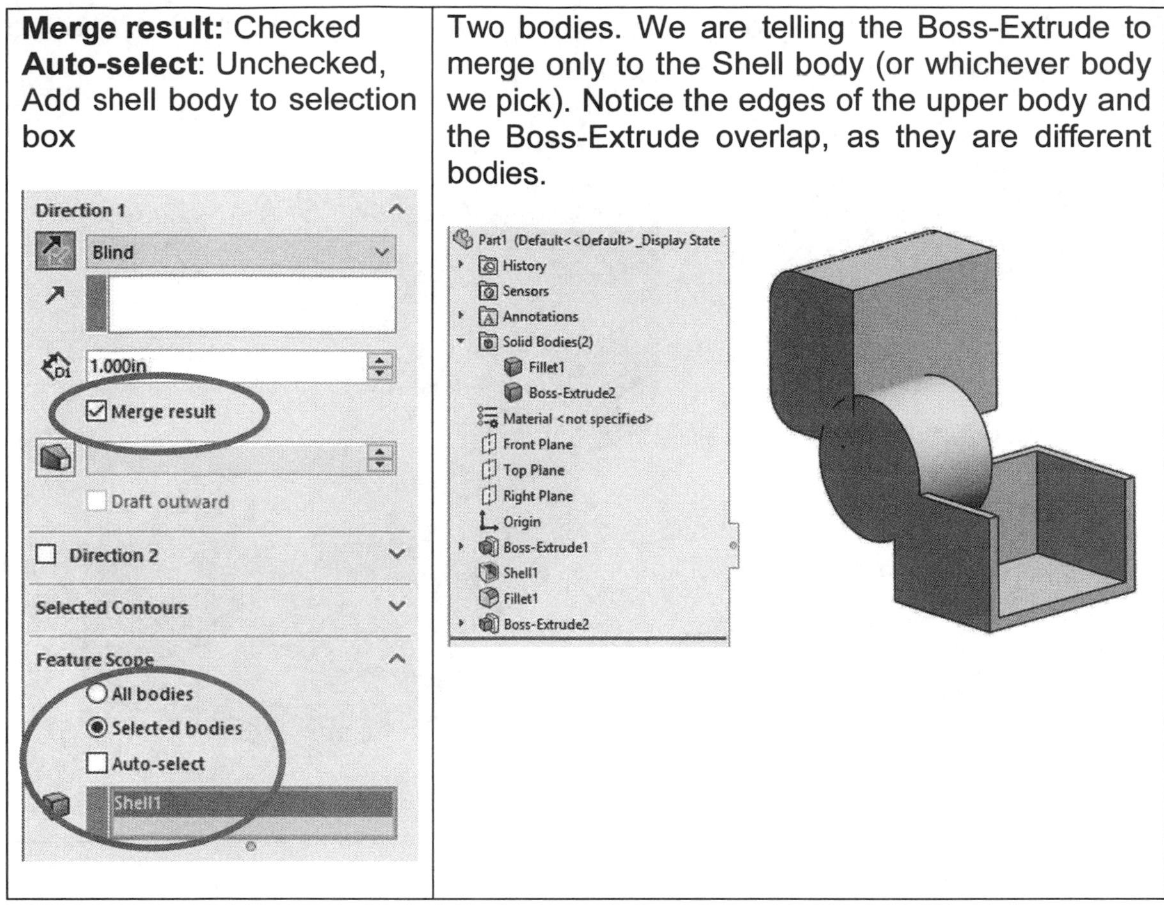

1.8. - The “Merge result“ option works the same way with any feature that adds material to the part, including revolved boss, sweep, loft, etc. Now we'll see how it works when material is removed. Delete the last Boss-Extrude feature and keep the sketch. When deleting the “Boss-Extrude,” make sure the option “Delete absorbed features” is unchecked; this way the boss will be deleted, but the sketch will be left behind.

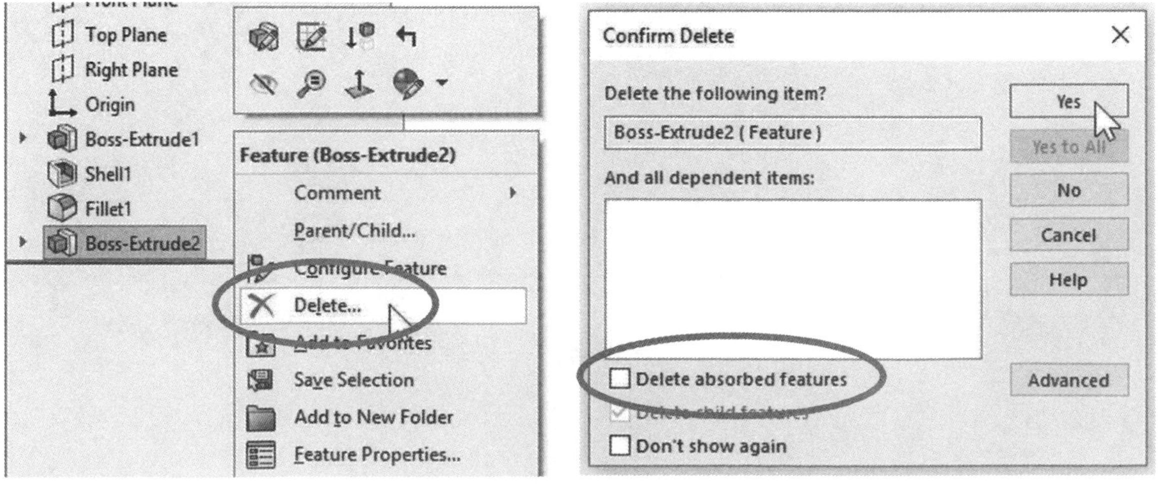

Select the sketch in the FeatureManager and make a Cut-Extrude feature using the "Through All" end condition. In this case the "Merge Result" option is replaced by "**Feature Scope**" with the same selection options: "All bodies" or "Selected Bodies" with/without "Auto-select." Leave the "Auto-select" option checked and click **OK** to finish.

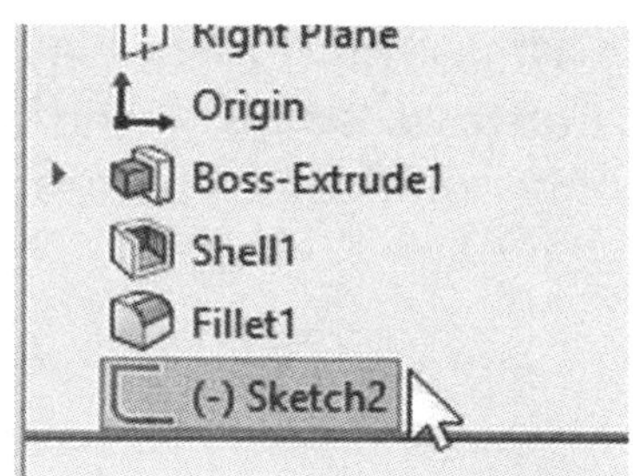

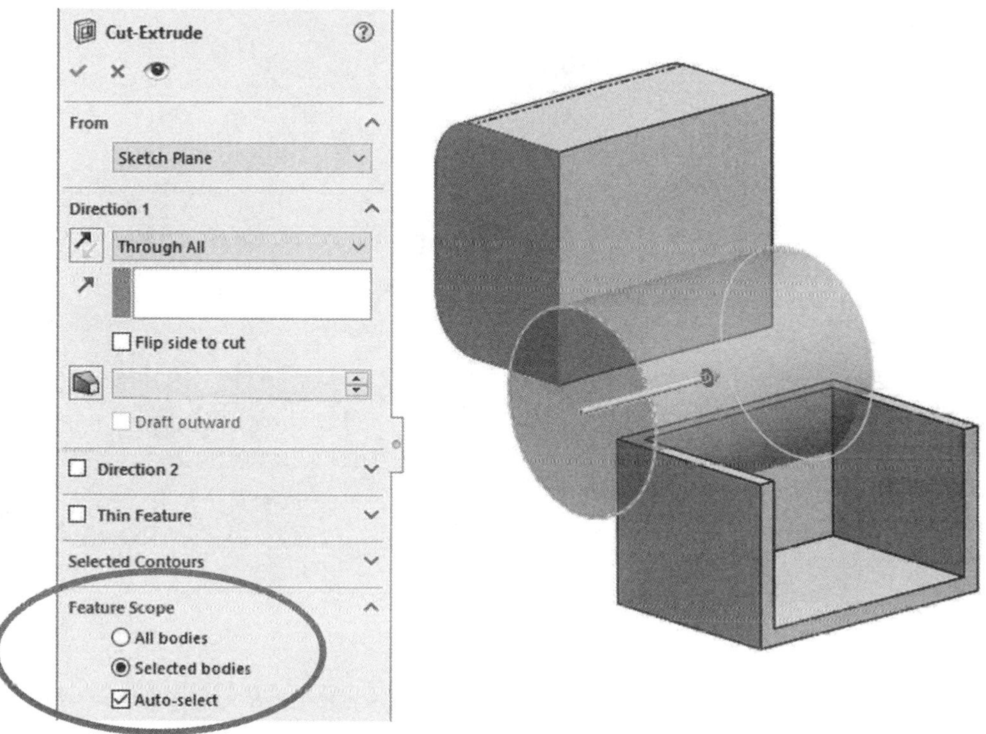

1.9. - What we end up with is the same two bodies we had before, but now they have a cut through them because the cut automatically selected them.

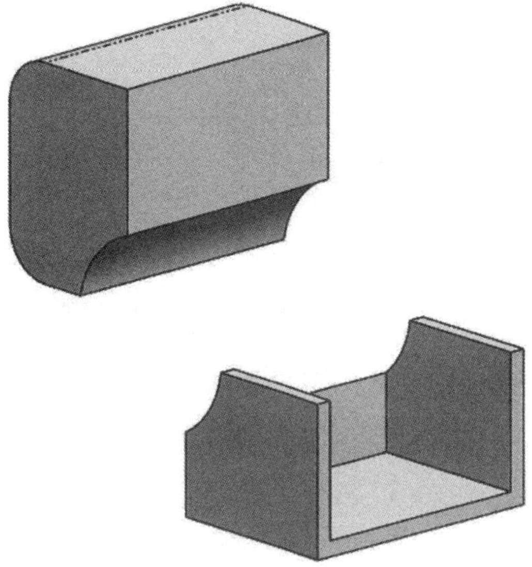

1.10. - Edit the "*Cut-Extrude1*" feature, turn the "Auto-select" option OFF in the "**Feature Scope**," and select the top body only. Click **OK** to finish. Now we are only cutting into the top solid, even when the cut overlaps the lower body.

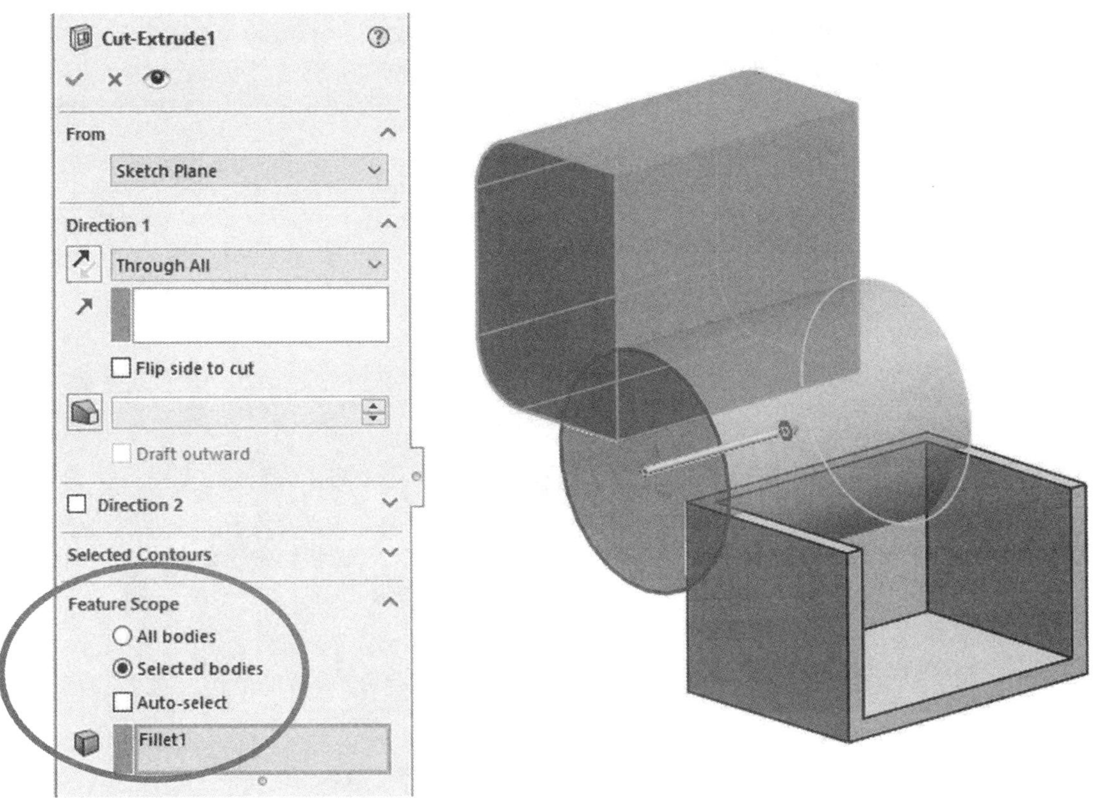

And just as with the features that add material, this technique works the same way with all the features that remove material including revolved cut, sweep cut, loft cut, etc.

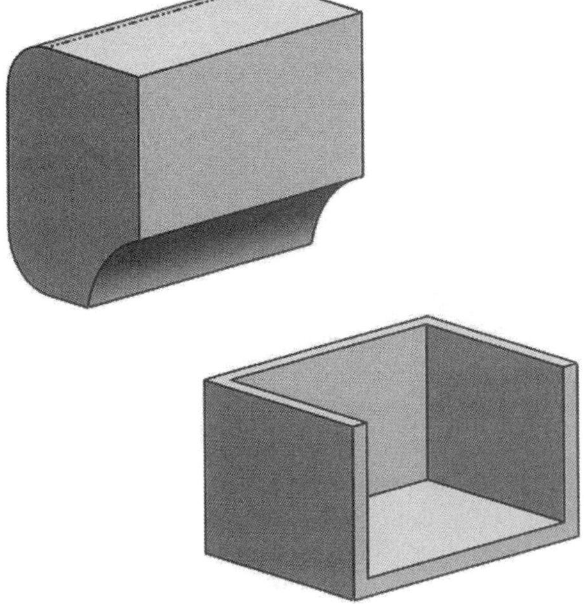

Car Wheel

2.1. - Multi bodies is a powerful technique to model parts that would otherwise be difficult to complete. A frequently used technique is called "bridging"; this means to connect two or more bodies by adding material between them to merge into a single solid body. Reasons to use this technique may include a model where we know what opposite sides/ends of a part look like, but we may not know what the middle (the "*bridge*") should be like.

In this example we'll make a car's wheel using multi-body techniques to complete it, including obtaining the common volume between bodies and adding (merging) two different bodies. We know what the actual tire and hub dimensions should be, but we don't know yet what the spokes will look like; we just know it must look great.

Make a new part; we'll assume the dimensions for the wheel are as shown in the following sketches. The first feature for the wheel will be the hub or mounting pad. Draw the following sketch in the *"Right Plane"* of the part and make a 360° Boss-Revolve feature.

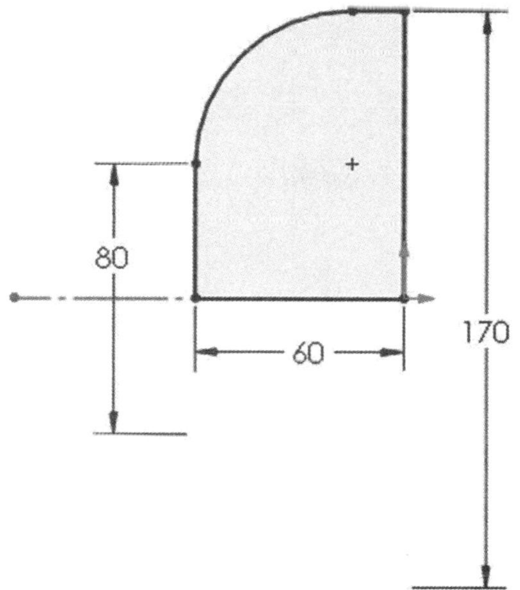

Our model is dimensioned in millimeters, so be sure to change your model's dimensions accordingly (**Tools, Options, Document Options, Units,** or in the status bar's **Unit System**). Pay attention to the diameter dimensions (doubled about the horizontal centerline).

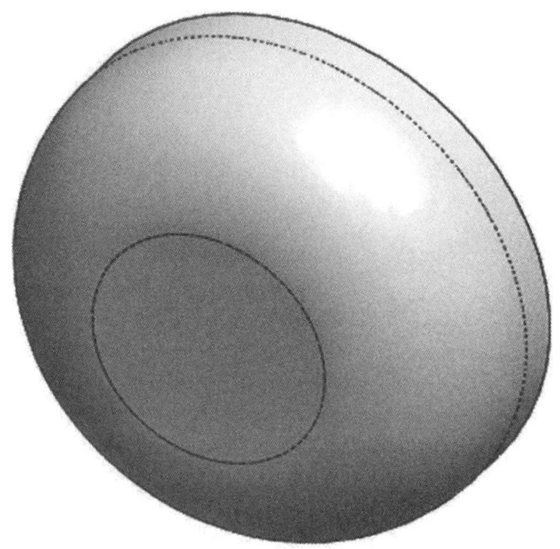

2.2. - Add a new sketch in the front flat face of the part, and add the following sketch. Draw the arcs and make them tangent to the circular edges of the first extrusion as indicated.

To draw a sketch and make a mirror copy about the centerline at the same time, make a centerline and, while it's selected, turn on the "**Dynamic Mirror**" tool in the menu "**Tools, Sketch Tools, Dynamic Mirror**." After it is activated, the centerline will show an equal sign at both ends letting us know that whatever is drawn on one side will be automatically copied in the other side. At the same time, symmetric geometric relations will be automatically added between pairs of geometric elements. To turn it OFF, deactivate it in the menu.

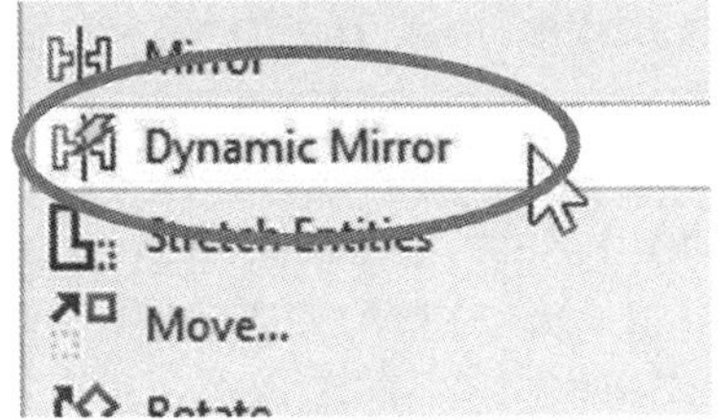

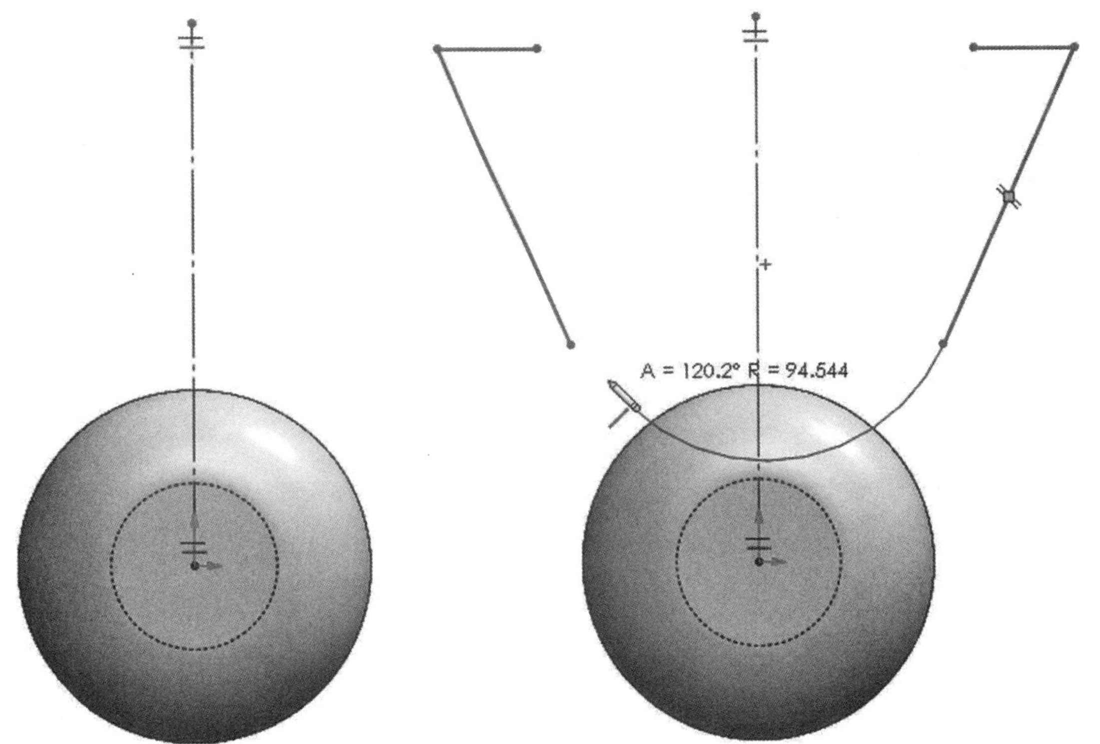

When using the "Dynamic Mirror" command, if we draw a line, arc, etc. crossing the centerline so that the mirror element would overlap the original, we get a warning letting us know, and the mirror copy is not created.

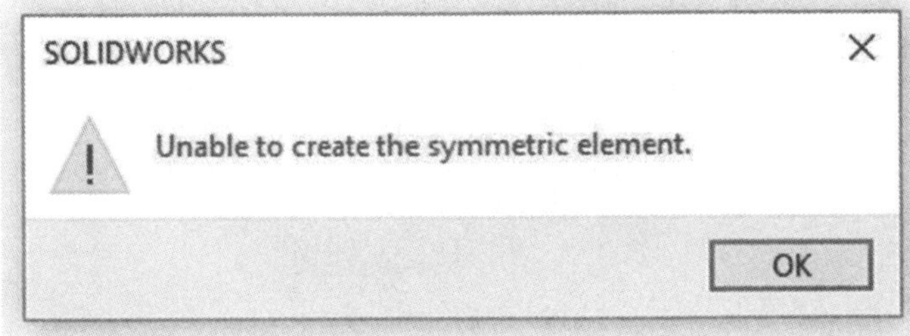

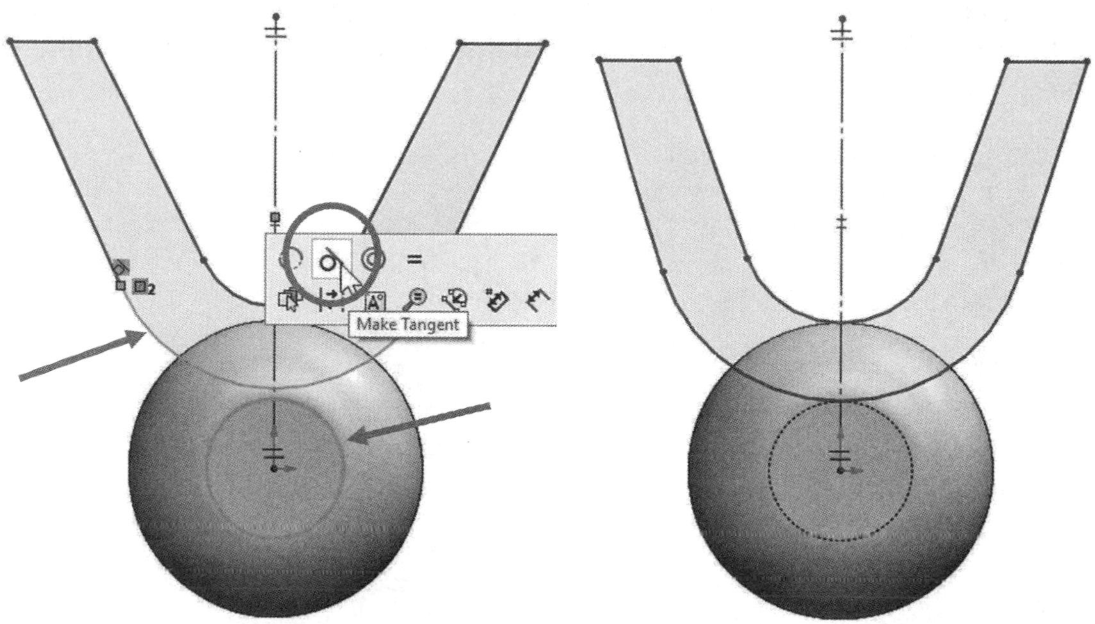

After drawing the sketch dimension it.

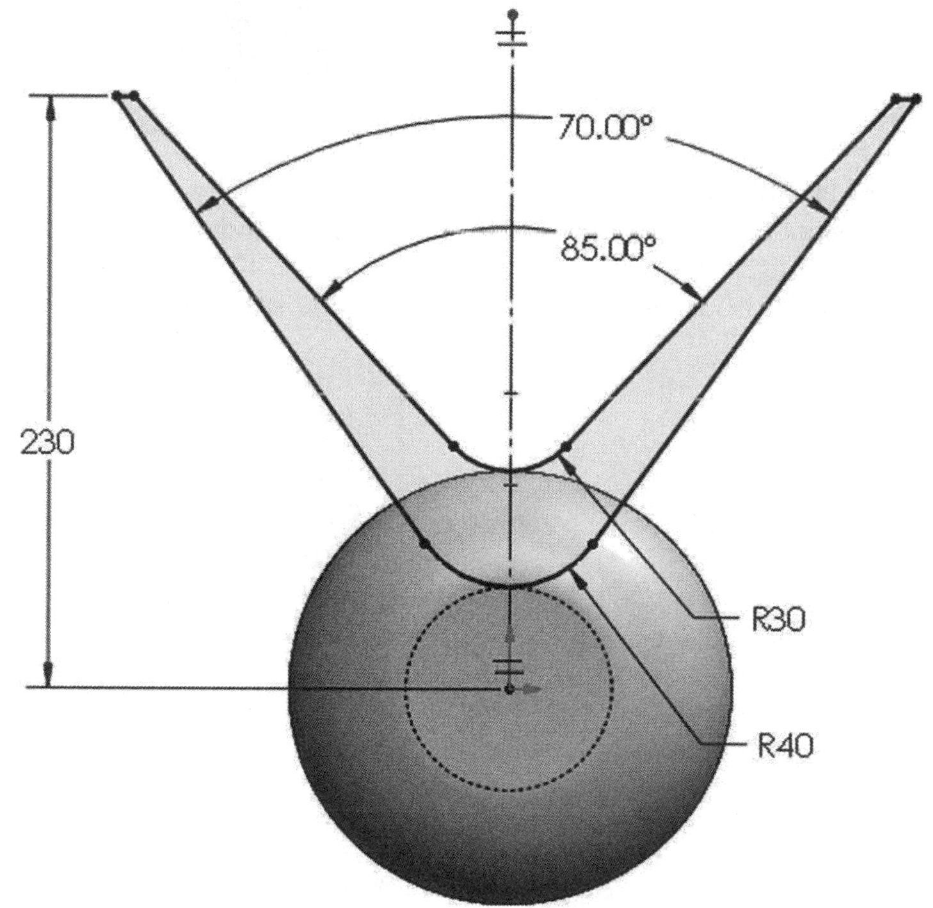

2.3. - Make the extrusion 20mm towards the front (Direction 1) and "Through All" going to the back (Direction 2). Uncheck "Merge result" OFF to have two different bodies.

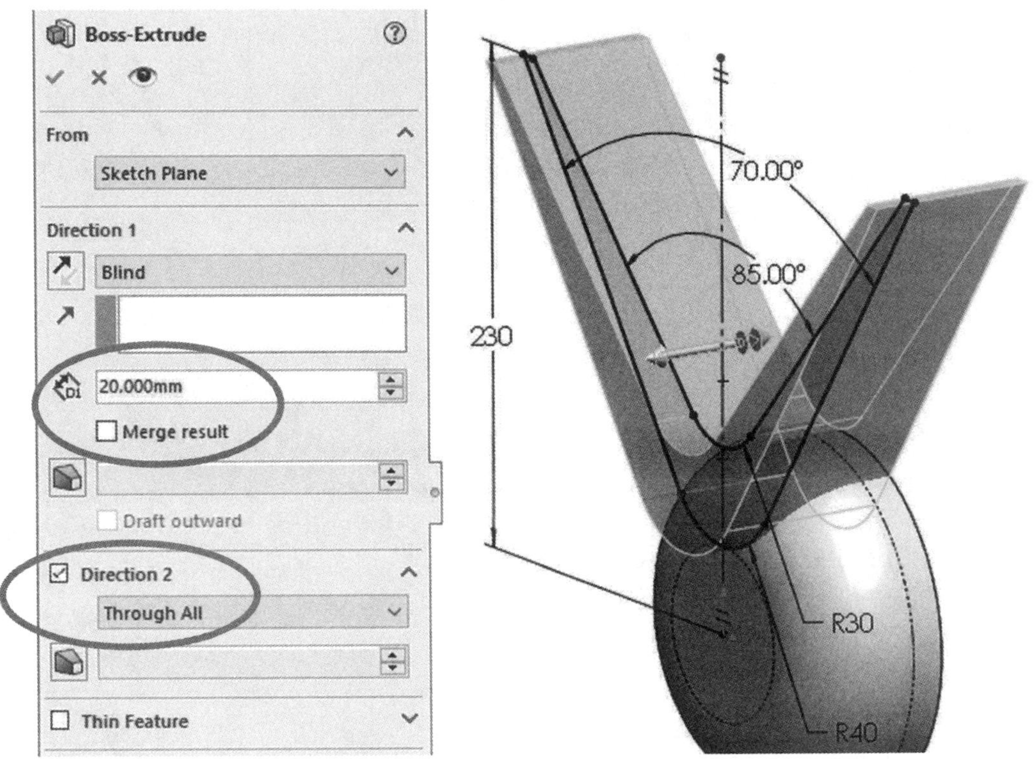

2.4. - In the next step we'll make the outside rim of the wheel. To better visualize this step, the second body will be hidden from view. Just like we can hide a part in an assembly, we can hide a solid body in a part. In the "*Solid Bodies*" folder, select the *"Boss-Extrude1"* body and select "**Hide**" from the pop-up toolbar.

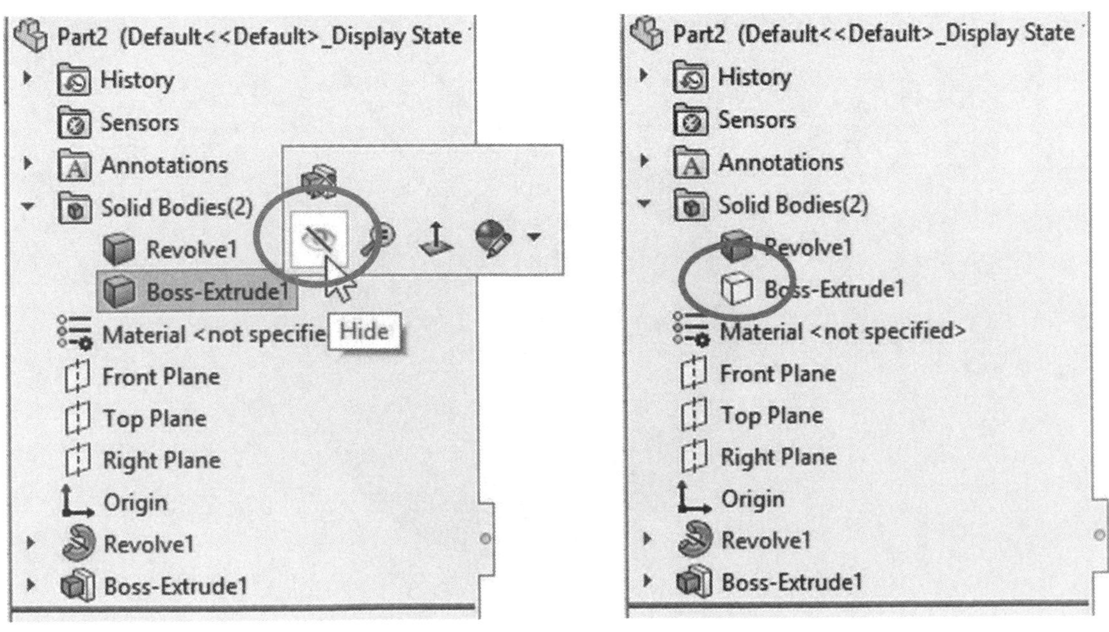

2.5. – Switch to a Right view and add the following sketch in the "*Right Plane*"; we'll use it to create a thin revolved feature to build the wheel's rim. The 360mm and 380mm dimensions are doubled about the horizontal centerline. The endpoints at the ends of the sketch are at the same height; be sure to add a horizontal geometric relation between them.

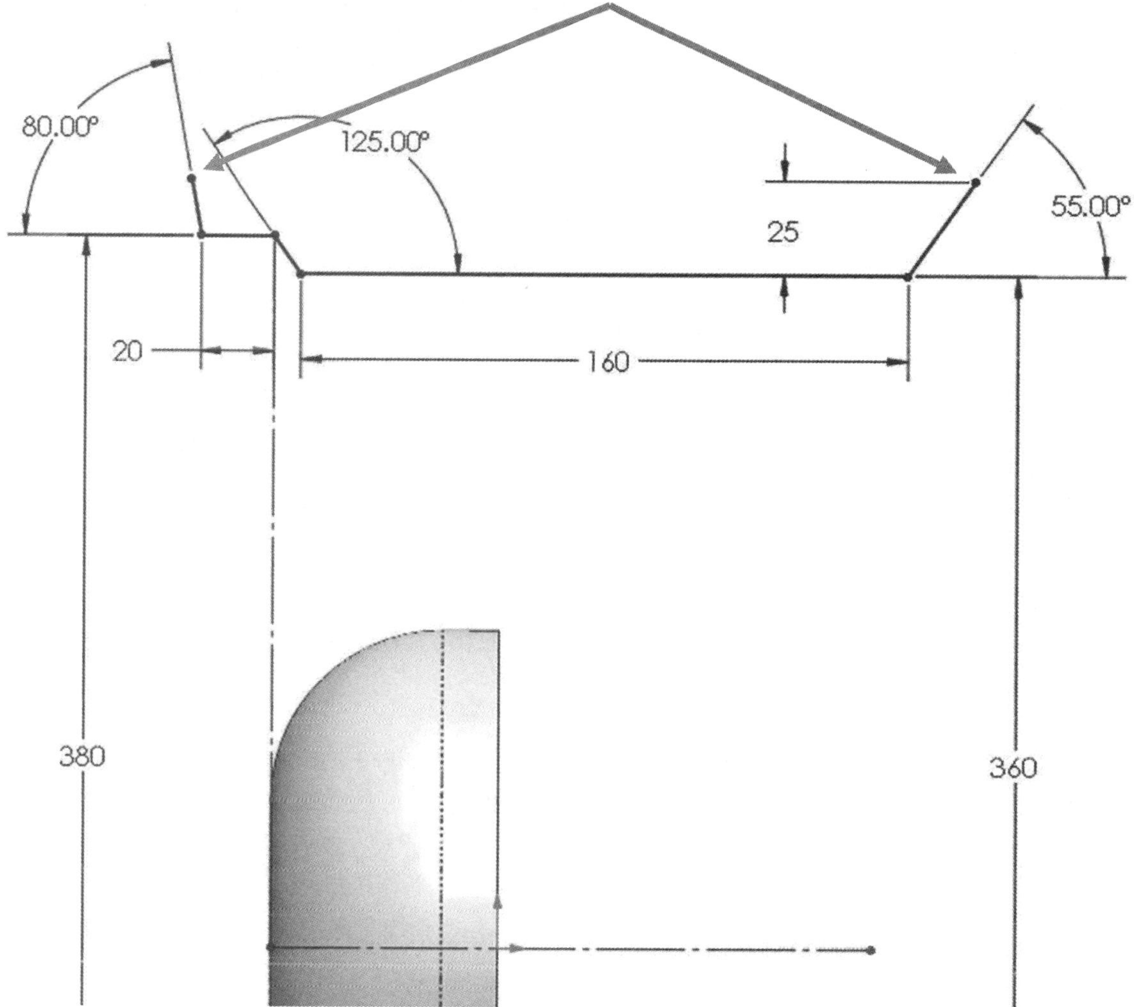

2.6. - Select the "**Revolved Boss/Base**" command and make a 360° thin revolve feature. SOLIDWORKS will ask if we want to close the sketch; select "No" to continue. Use the horizontal centerline as the "Axis of Revolution," turn off the "Merge result" checkbox and make the "Thin Feature" 5mm thick, going away from the wheel.

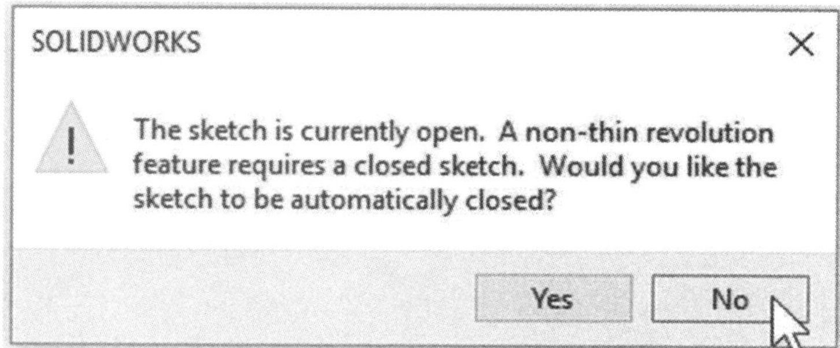

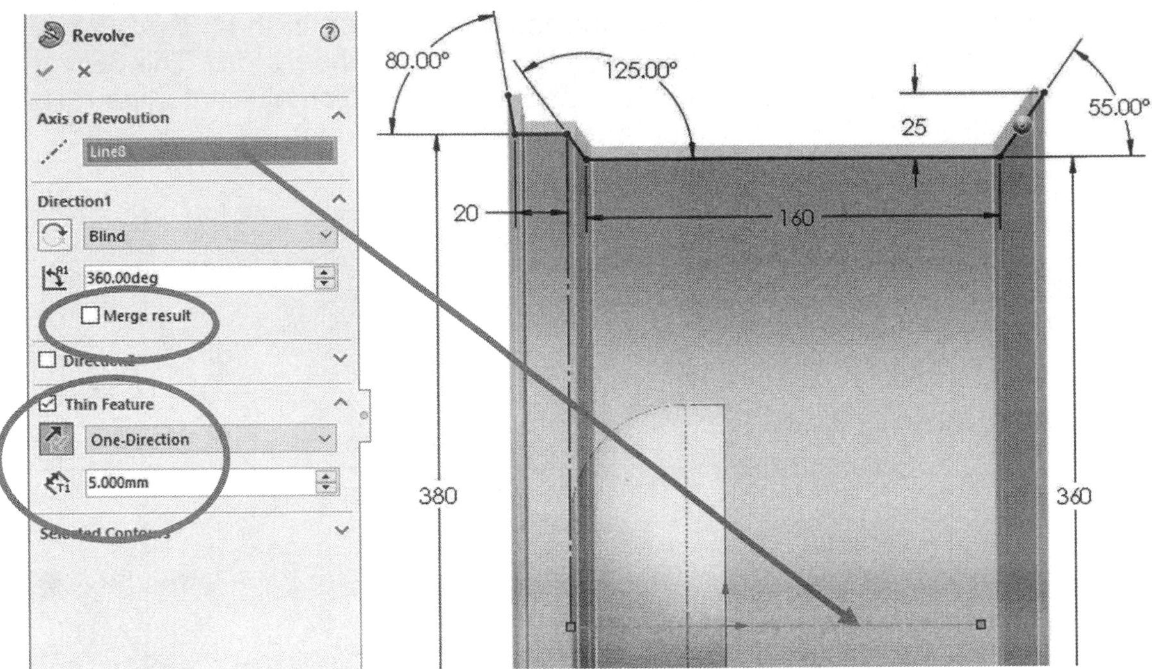

2.7. - Now we have three different solid bodies in our part. The next feature will be another revolved feature that will be used to build the wheel's spokes.

Before adding the new solid body, hide the *"Revolve-Thin1"* body and show the *"Boss-Extrude1"* body.

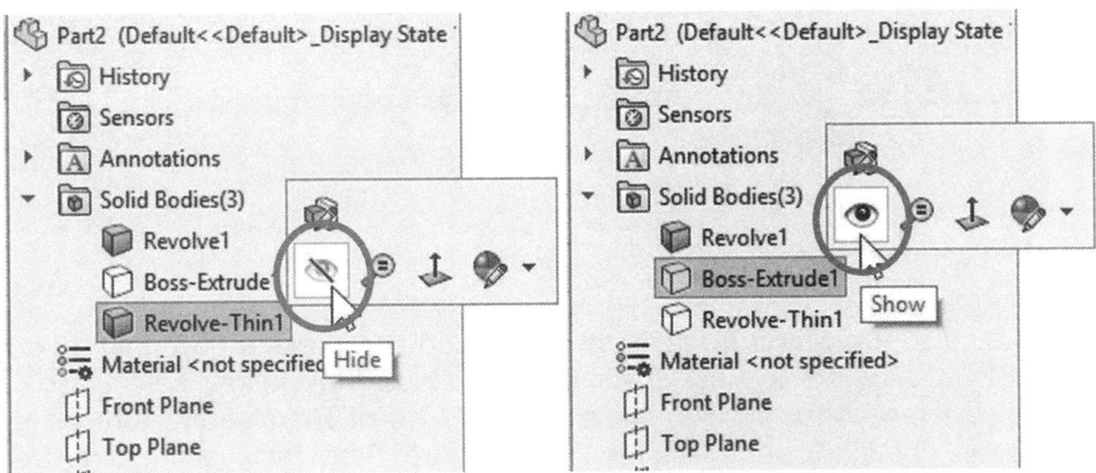

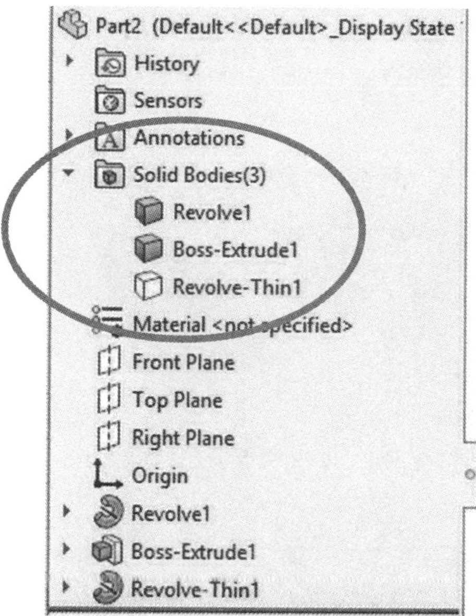

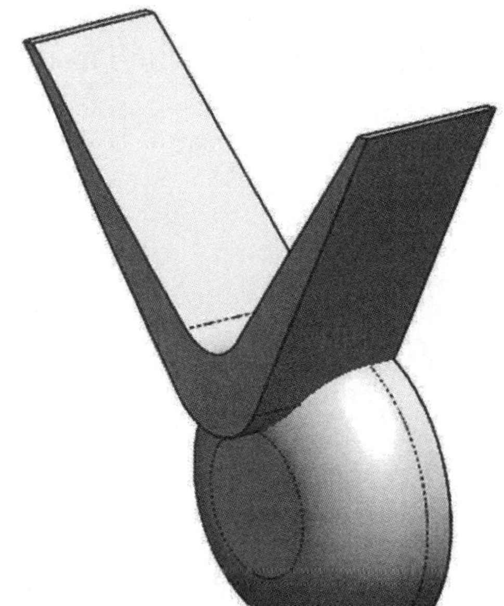

2.8. - After hiding the revolved thin feature, add a new sketch in the *"Right Plane."* Note: The arc on the left is tangent to the solid body's edge, and the arc on the right is tangent to the vertical centerline. Don't forget the horizontal centerline at the origin to make the revolved feature about it. We can use a "**Three Point Arc**" for this sketch. The part is shown in both shaded and wireframe modes for clarity.

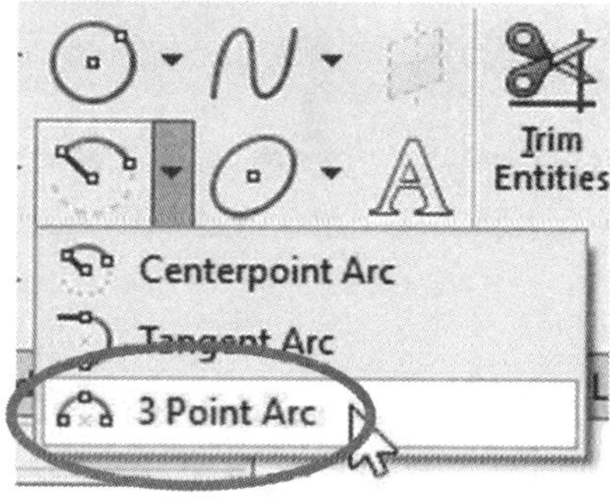

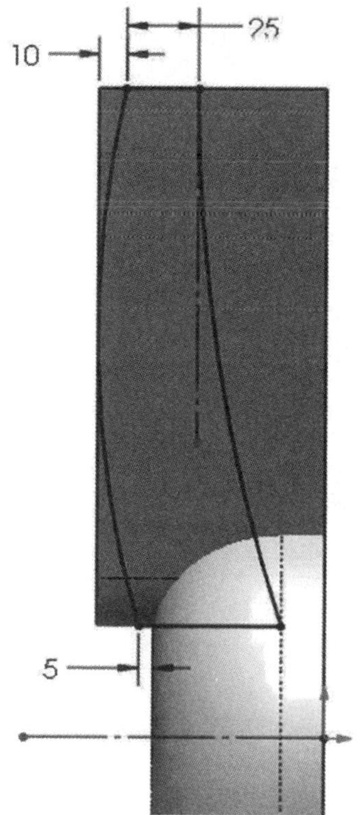

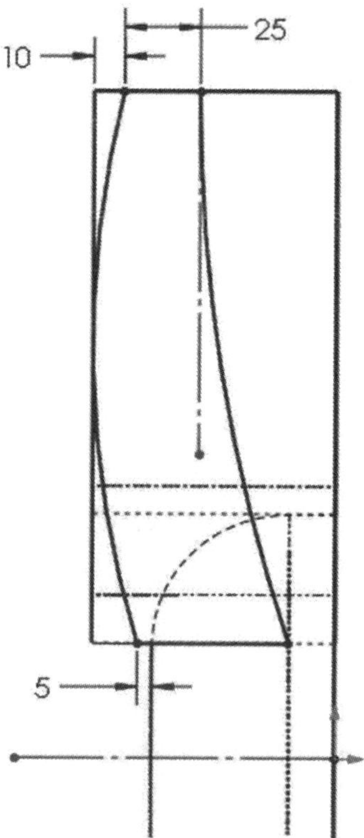

2.9. - Make a 360° revolved boss using the horizontal centerline, and uncheck the "Merge result" option to create the fourth solid body.

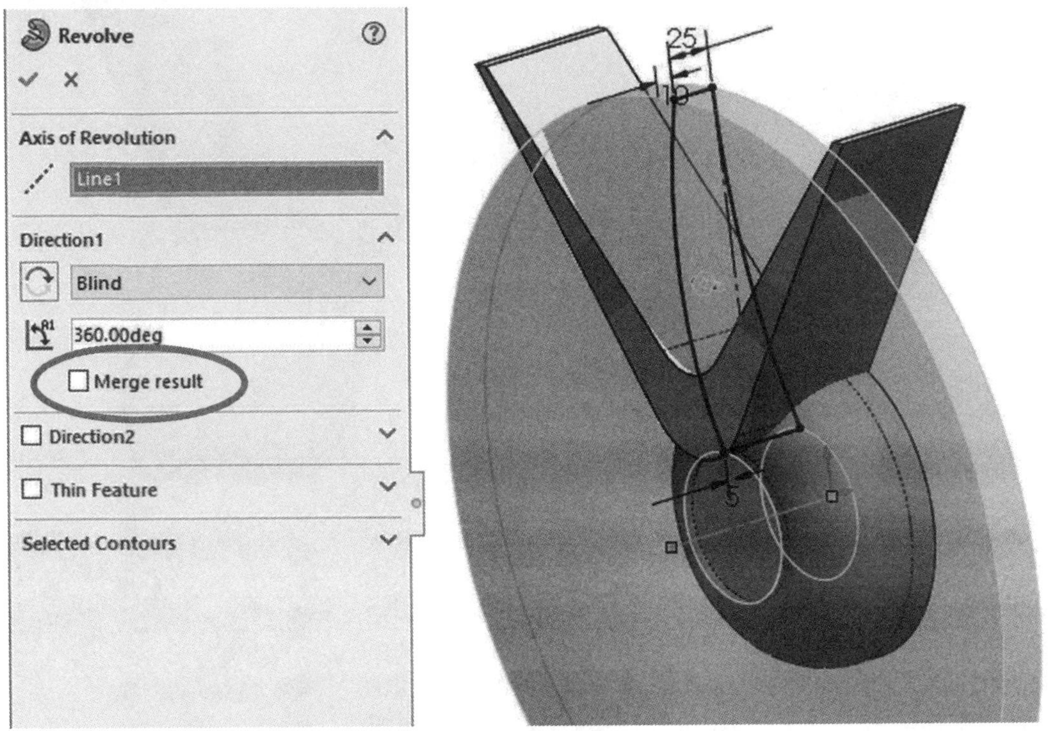

2.10. - In the next step we will obtain the common volume between two of the bodies to get the wheel's spokes. While holding the "Ctrl" key, select the *"Boss-Extrude1"* and the *"Revolve2"* bodies from the *"Solid Bodies"* folder, right-mouse-click in either one and select "**Combine**" from the pop-up menu, or the menu "**Insert, Features, Combine.**"

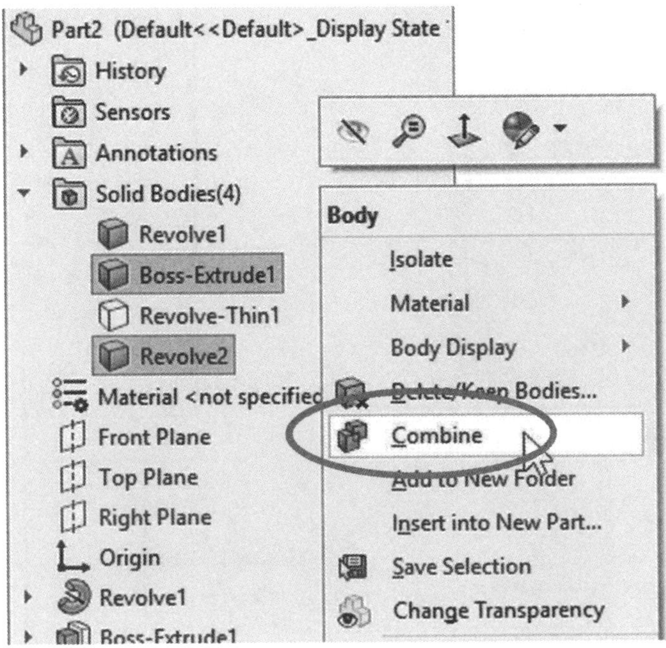

In the "Operation Type" options select "Common" and then press the "Show Preview" button (it will change to "Hide Preview"). The "**Combine**" command is used to Add bodies to form a new one, subtract one or more bodies from another, or get the Common volume between two or more bodies, as in this step. After clicking OK, the bodies selected in the "Bodies to combine" list are consumed by the operation and replaced by the new body.

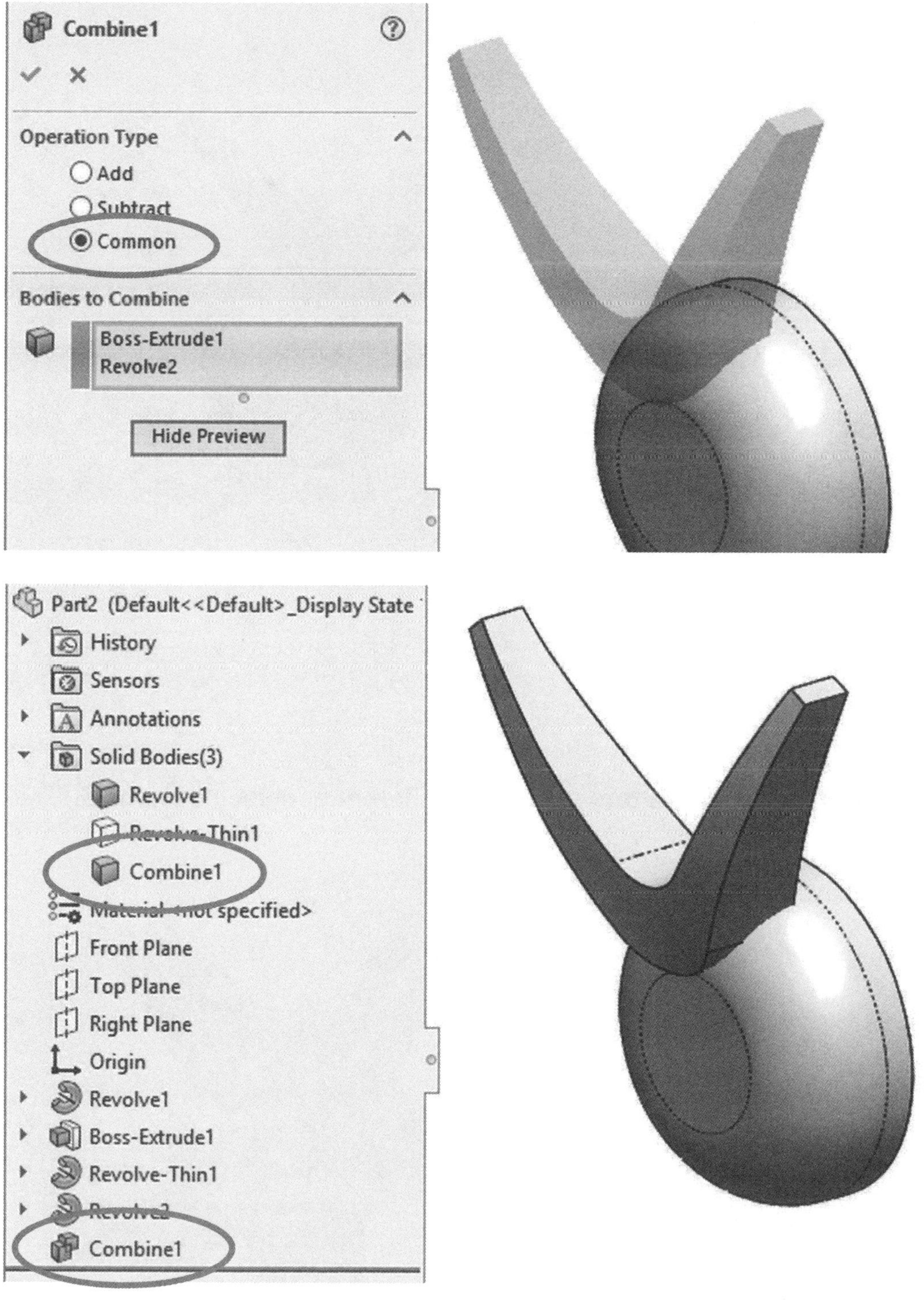

2.11. - Now that we have the body we are interested in to create the spokes, we need to make a circular pattern using the combined body as a seed feature. Select the "Circular Pattern" command using a circular edge of the "*Revolve1*" body for the "Pattern Axis." For this step, instead of selecting the *"Combine1"* feature to make our pattern, select the "Bodies to Pattern" selection box and select the combined body for the pattern. Change the number of instances to 5 equally spaced in 360 degrees. Click OK to continue.

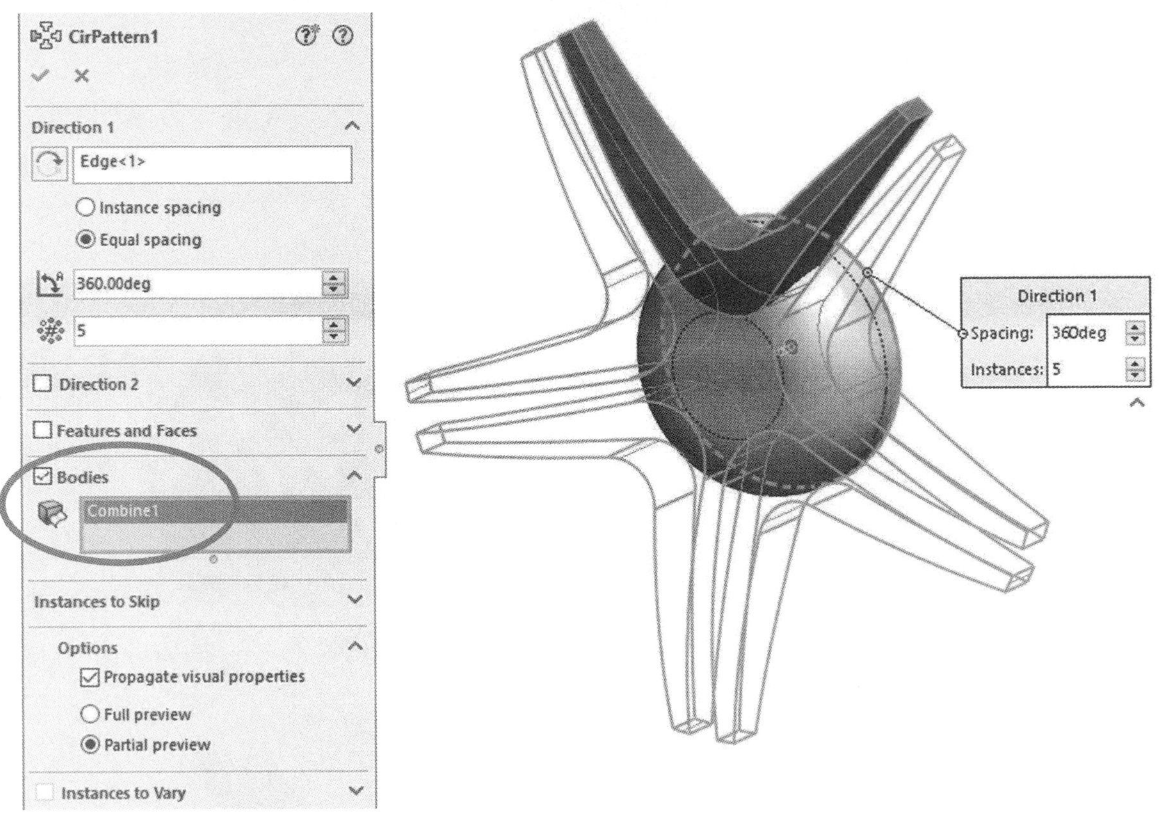

After adding the circular pattern we have 7 different bodies in our wheel.

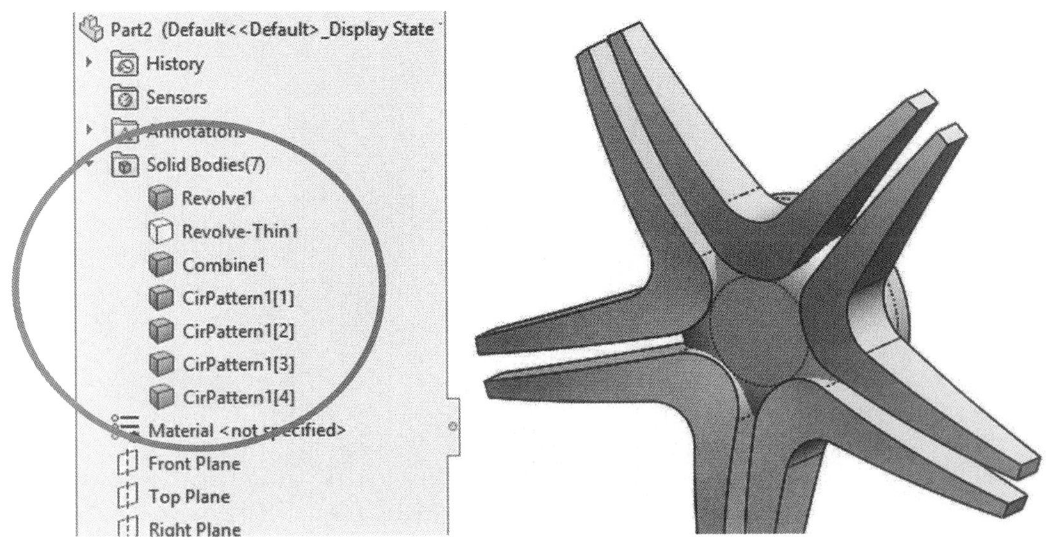

2.12. - The next step could be done after we combine the bodies into a single body, but we chose to do it now to show more multi-body functionality. In this step, we'll use the following sketch to make the holes for the lug nuts. Add the sketch in the *"Right Plane."* Note the 25mm and 16mm dimensions are doubled about the top centerline; the 125mm is doubled about the centerline in the middle of the part. Make the sketch big enough to extend well past the left side of the part.

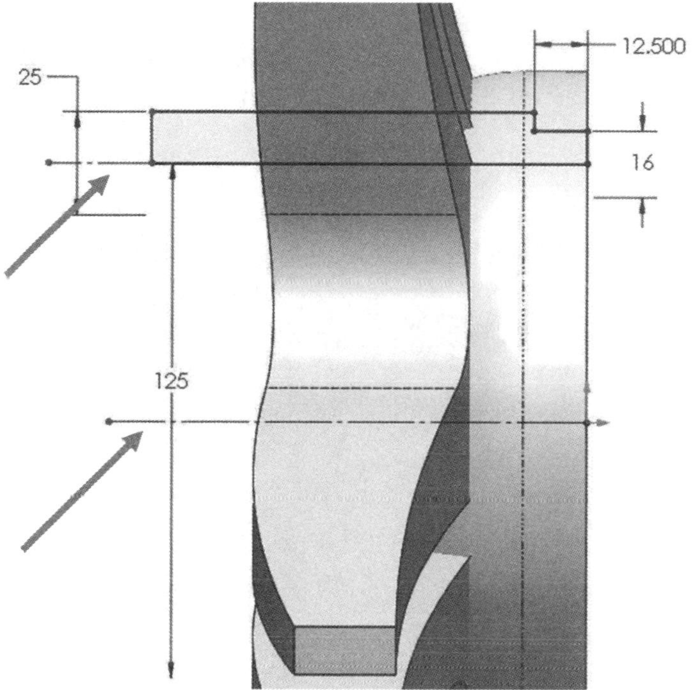

2.13. - Make a "Revolved Cut" using the top centerline as the axis of revolution, and set "Feature Scope" to "Selected Bodies, Auto-select."

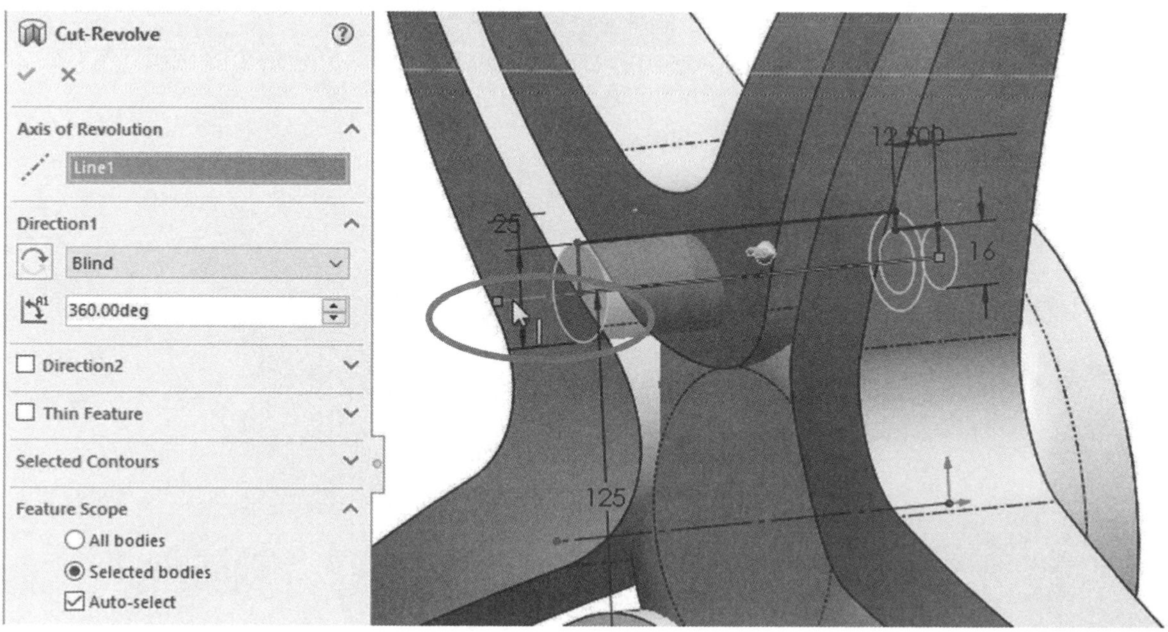

2.14. - Make a circular pattern with 5 copies of the hole just added. Under "Feature Scope" use the "Selected Bodies" option, uncheck "Auto-select" and add the remaining four bodies made with the circular pattern and the center body (the first revolved feature we did). Click OK to finish the pattern. All the bodies affected now have the name of "*CirPattern2*".

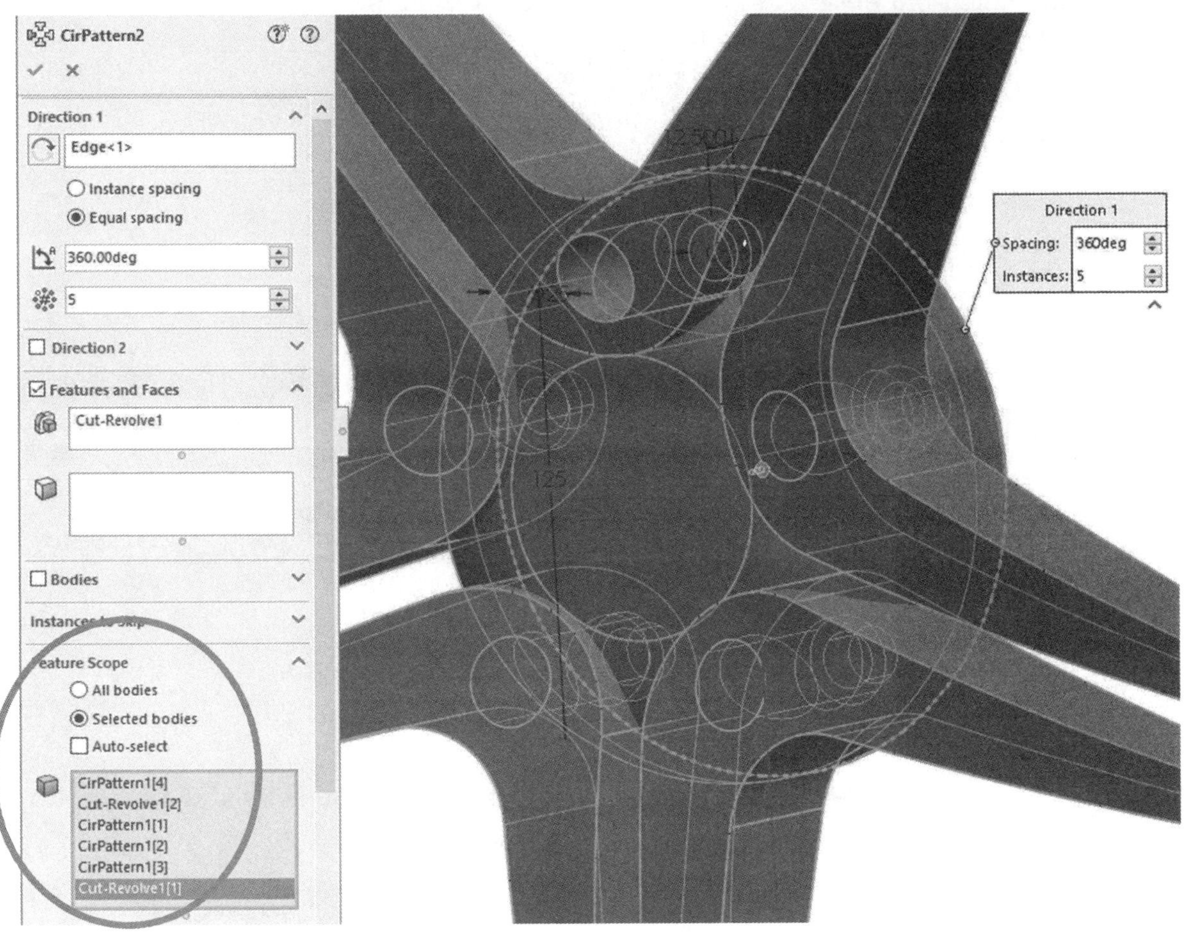

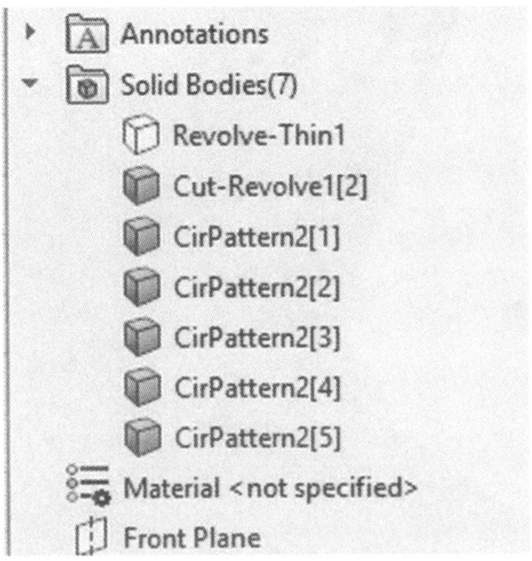

2.15. - Now that the holes for the bolts are complete we can merge the spokes and the hub into a single solid body. In the "Solid Bodies" folder select the patterned bodies and the hub, right-mouse-click in any of them and select "**Combine**" from the pop-up menu, or use the menu "**Insert, Features, Combine**." Select "Add" in the "Operation Type" and click OK to finish.

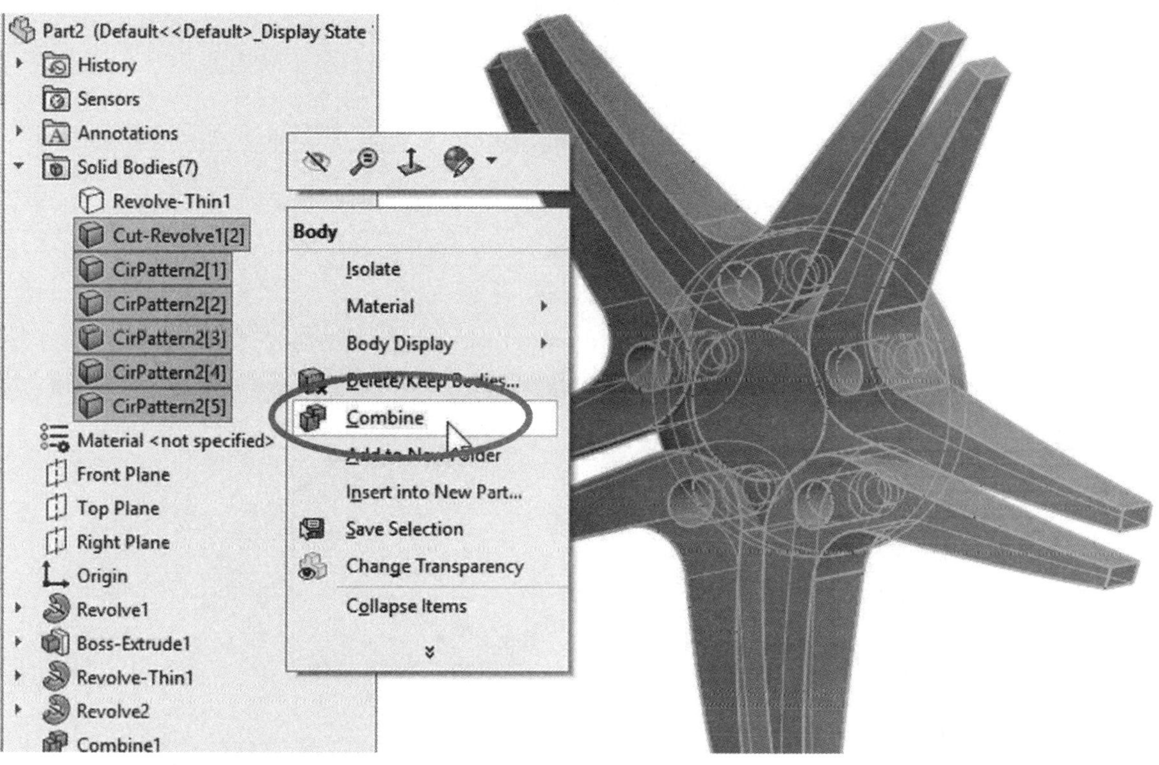

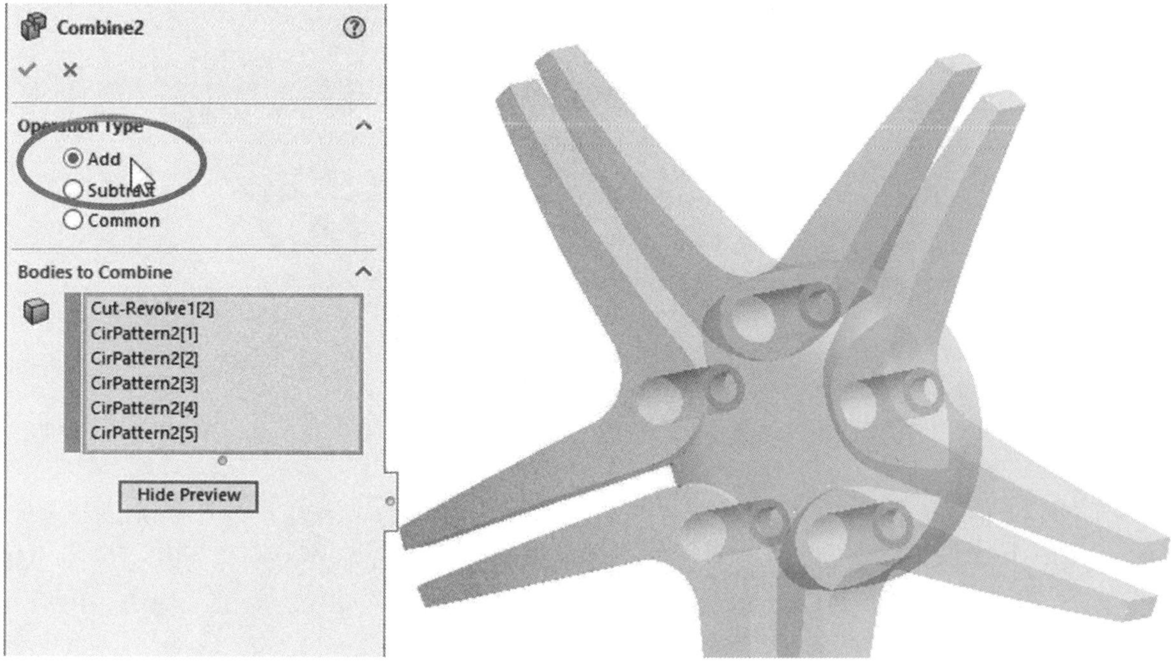

2.16. - After adding the bodies in the previous step we have two bodies in the part. One is the hub with the spokes and the other (hidden up to this point) is the wheel itself. In the "Solid Bodies" folder select the hidden body and click "Show" from the pop-up menu.

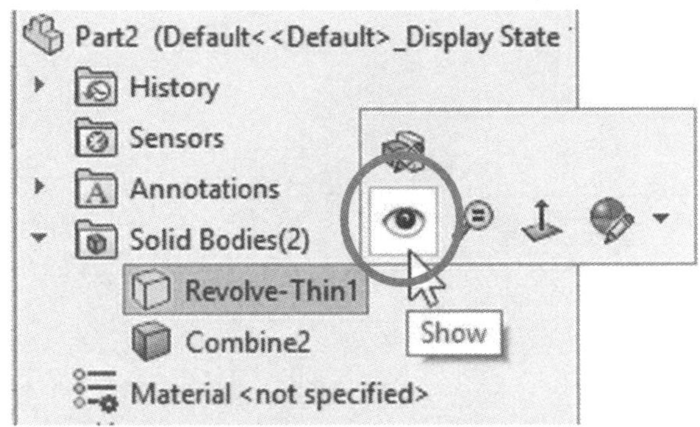

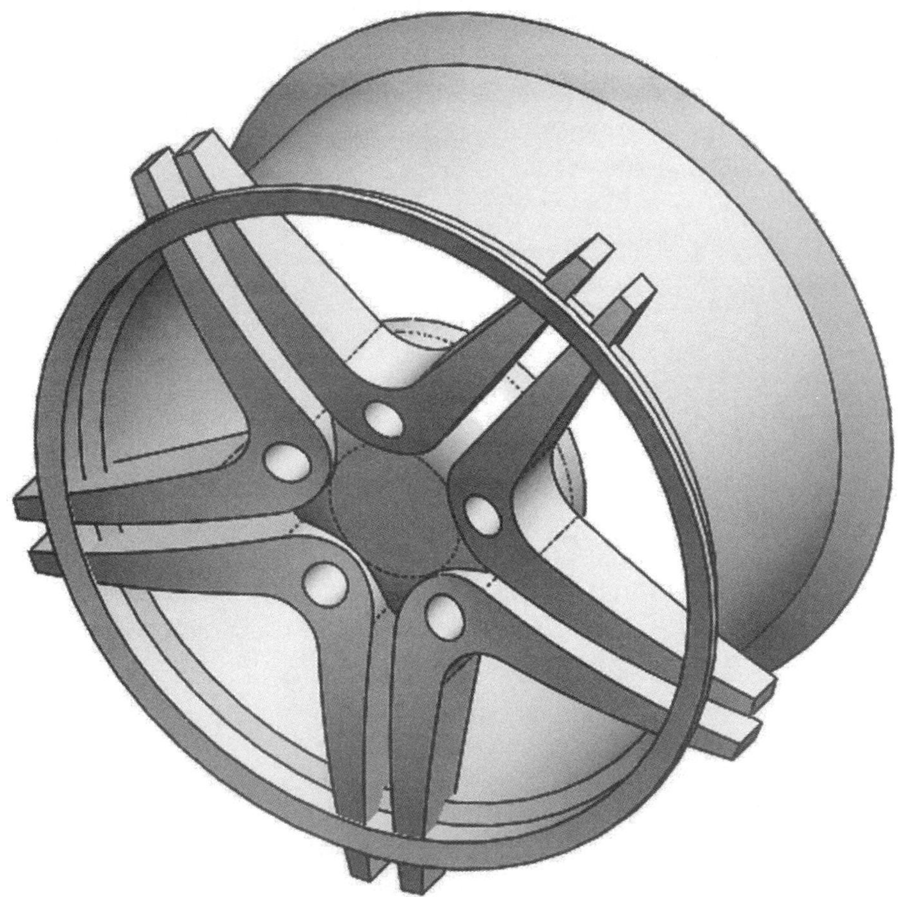

To hide or show a solid body we can select any feature that modified it in the FeatureManager and use the same "Hide/Show" command from the pop-up toolbar, not just in the "*Solid Bodies*" folder.

2.17. - To make the rim better looking add a fillet to round off the sharp edges of the "*Revolve-Thin1*" feature. Add a 5mm to all of the inside and outside sharp edges in the rim. All the edges can be rounded by selecting four faces.

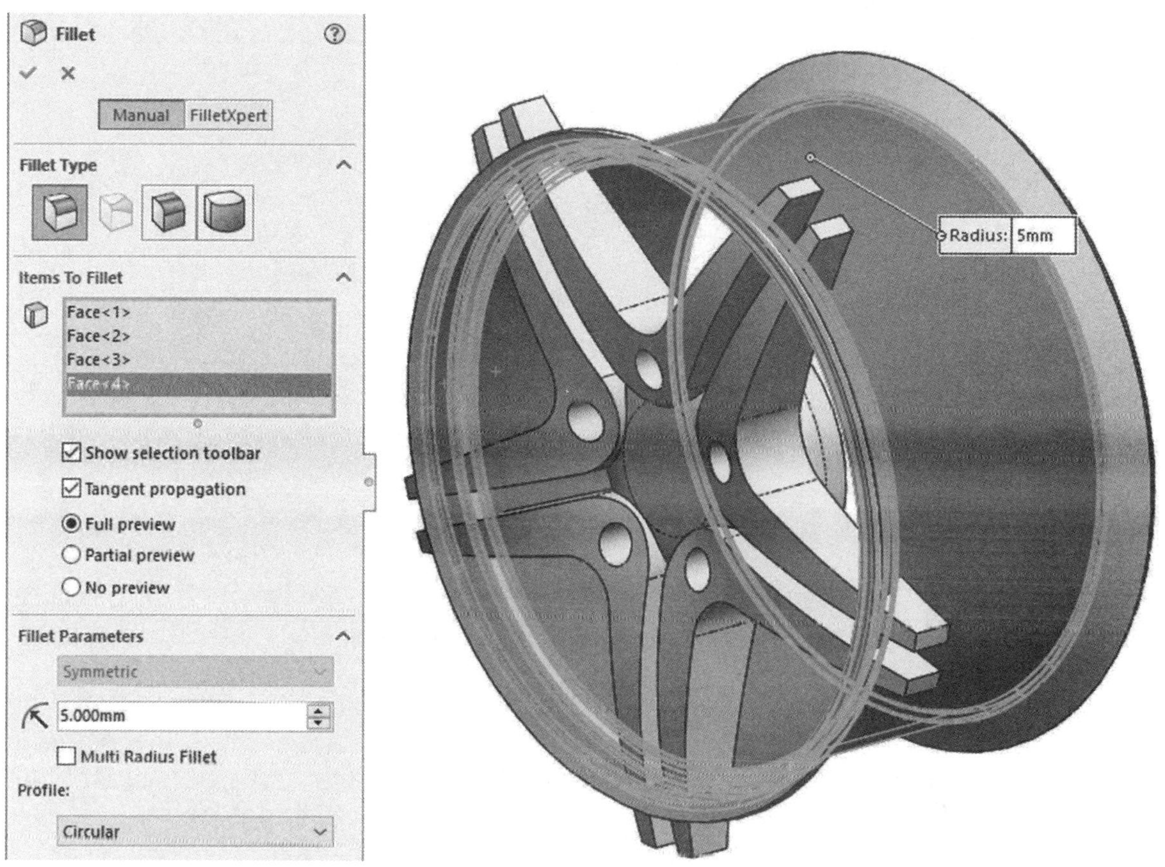

2.18. - The next step is to trim the excess material from the spokes that protrude beyond the rim. We'll use a "Revolved Cut" to trim it. Using the "**Intersection Curve**" command we can create a sketch that matches the profile of the rim to cut the spokes. Add a new sketch in the *"Right Plane"* and select the "**Intersection Curve**" command from the "**Convert Entities**" drop down menu, or the menu "**Tools, Sketch Tools, Intersection Curve.**"

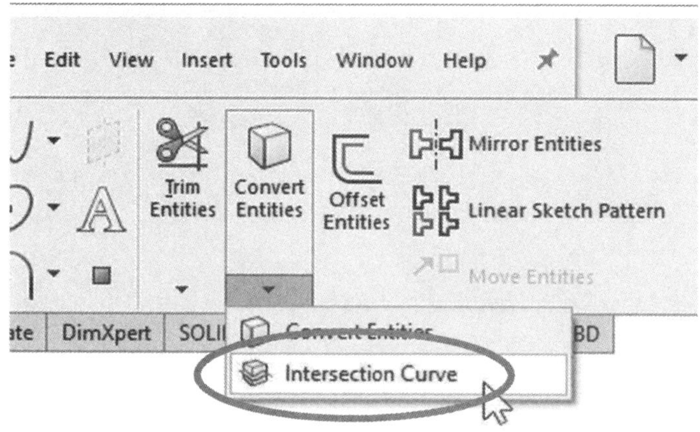

Select all the faces outside of the rim (or inside, in this case selecting either side will give us the same result); right-mouse-click in any face and select "**Select Tangency**" from the pop-up menu. This way the selection will propagate through the faces until no tangency is found. (The fillet we added in the previous step created the tangent faces to propagate.) After selecting the faces click OK to close the command and continue.

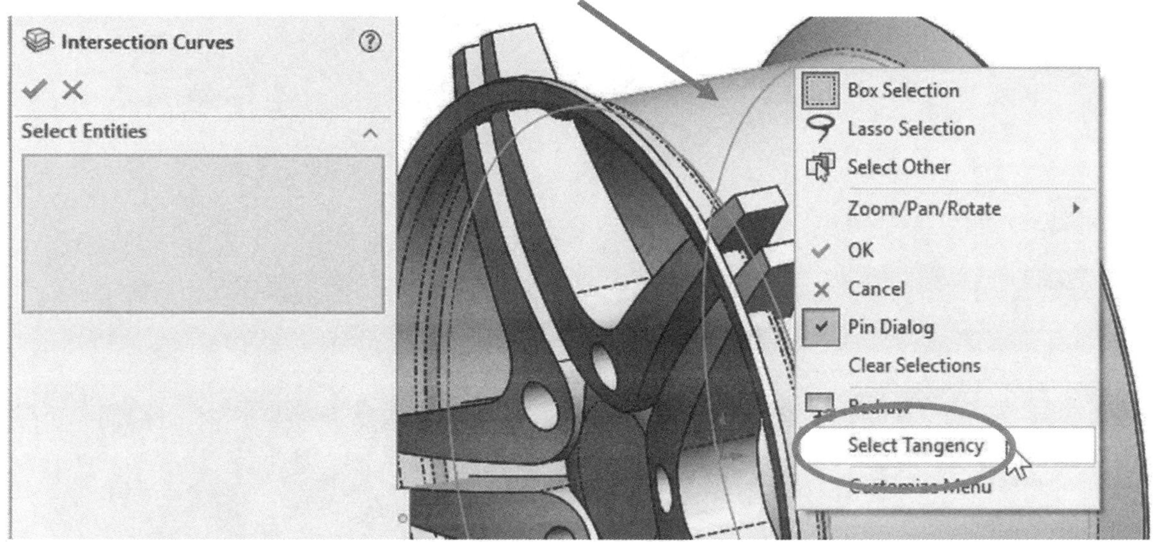

2.19. - The "**Intersection Curve**" command creates sketch entities where the selected faces intersect the sketch plane. Because the faces cross the plane in two different places (top and bottom) we get two profiles. Window-select the bottom profile and delete it; we only need the top side for this feature.

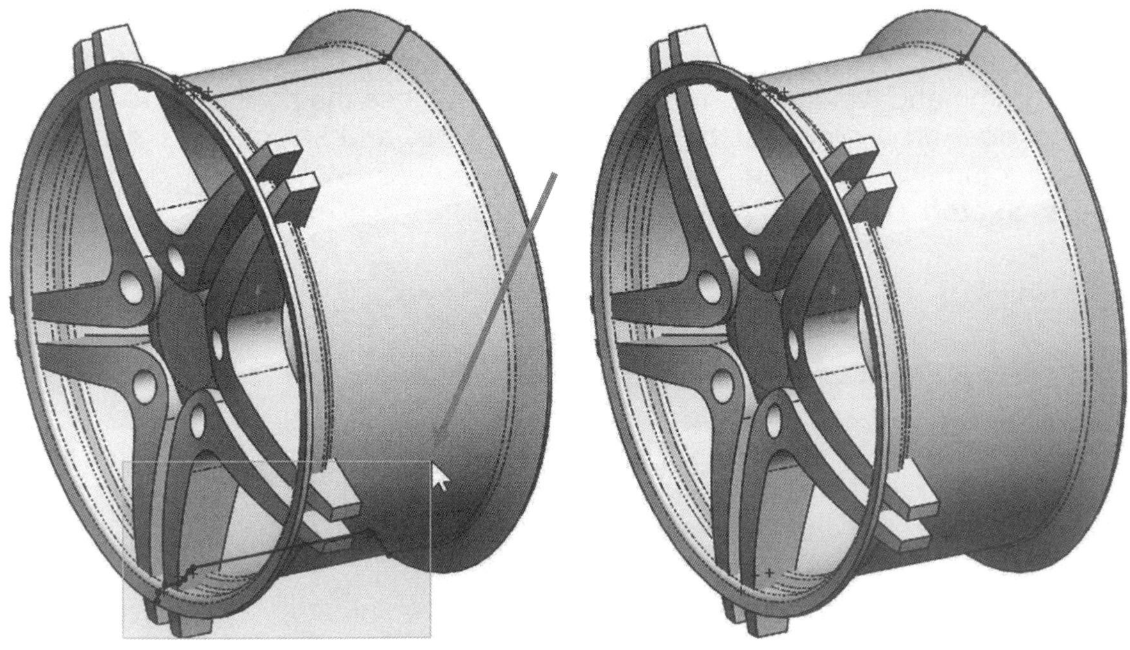

2.20. - Add a horizontal centerline starting at the origin and make a "**Revolved Cut**." We'll get the message alerting us that the sketch is open asking if we want to close it. In this case, since we want to create a closed profile, select "Yes" when asked. A new line will be automatically added connecting the open ends of the sketch. The next image is displayed using the "Shaded" view style (No edges) for clarity.

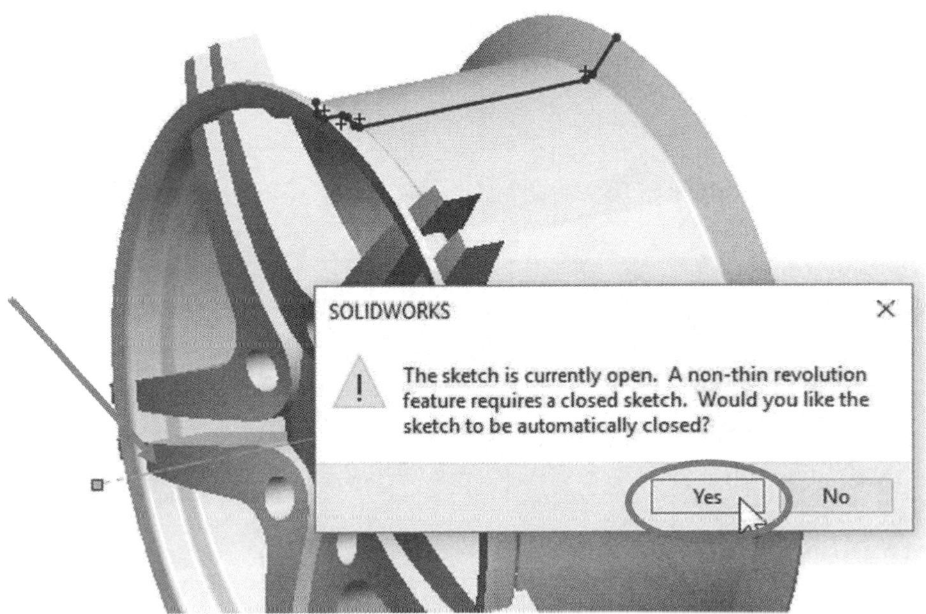

2.21. – In the "Feature Scope" section uncheck "Auto-select"; since we only want to cut the spoke's body (*"Combine2"*), add it to the selection list and click OK to continue.

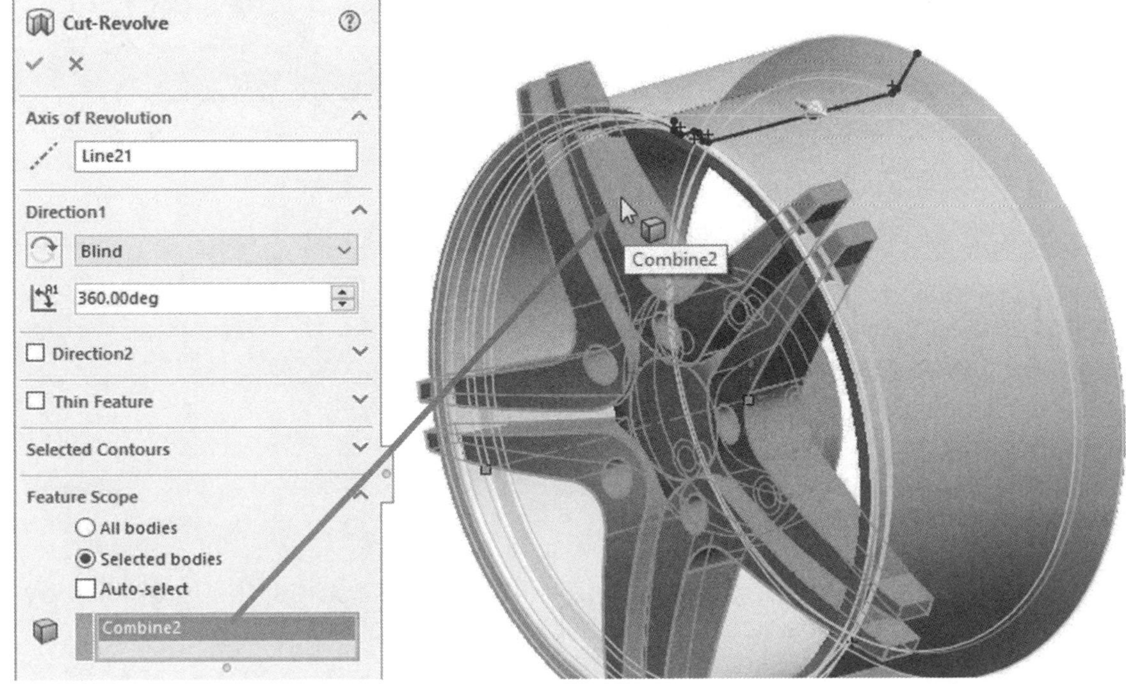

2.22. - After pressing OK to make the revolved cut, the sketch used generates more bodies by trimming the tips of the spokes, and now we are asked which bodies we want to keep. In the pop-up selection box use the "Selected Bodies" option, and pick the body at the center of the spokes to keep it. You will be able to see the preview before selecting it. Click OK to finish.

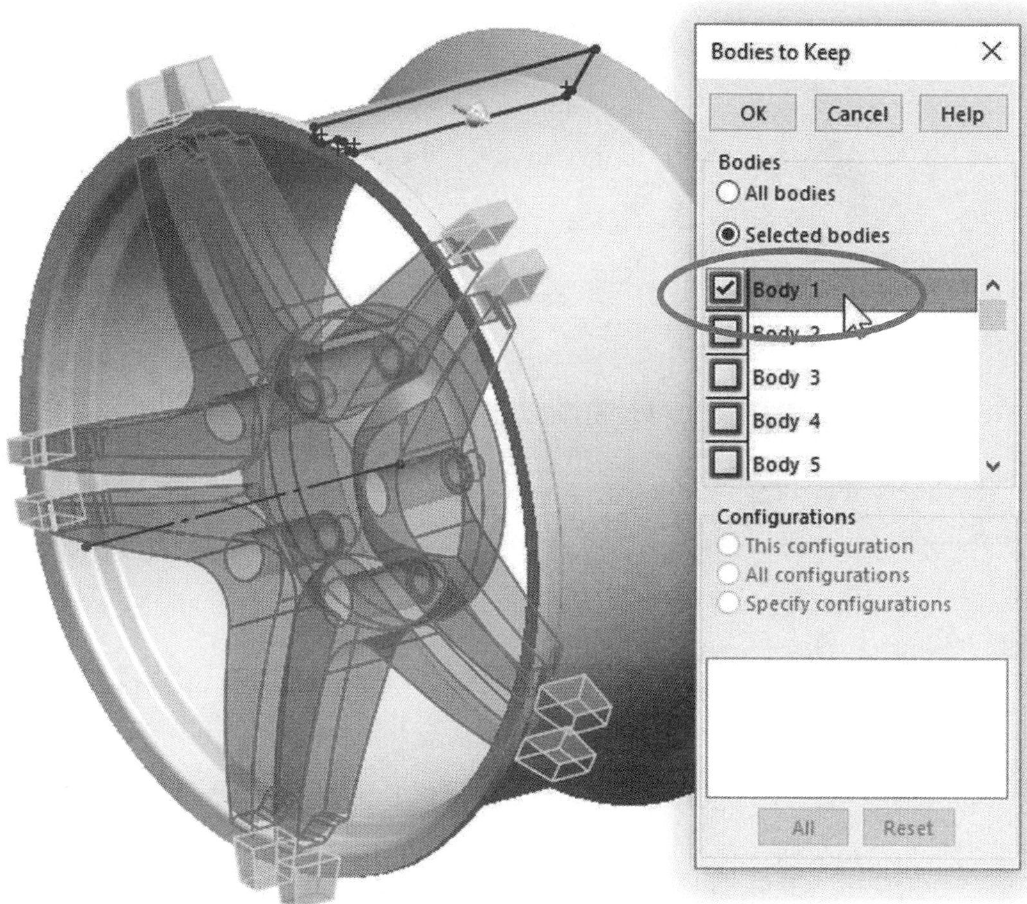

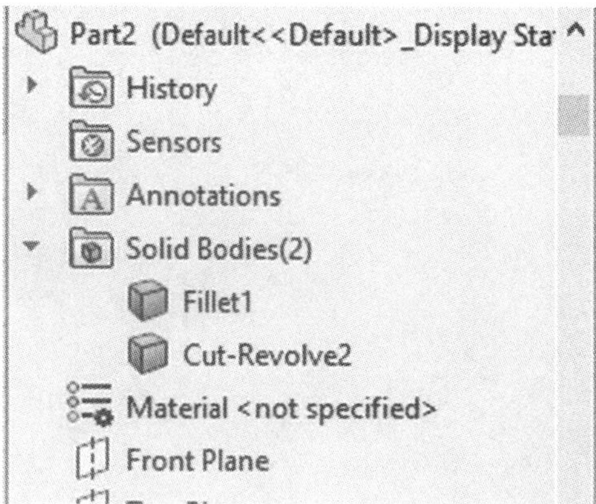

When a cut operation splits a part into multiple bodies, the largest body is usually the first one listed in the "Bodies to Keep" list.

If we keep all bodies, the unwanted bodies can be deleted by selecting them and from the right-mouse-button menu selecting “**Delete Bodies**.”

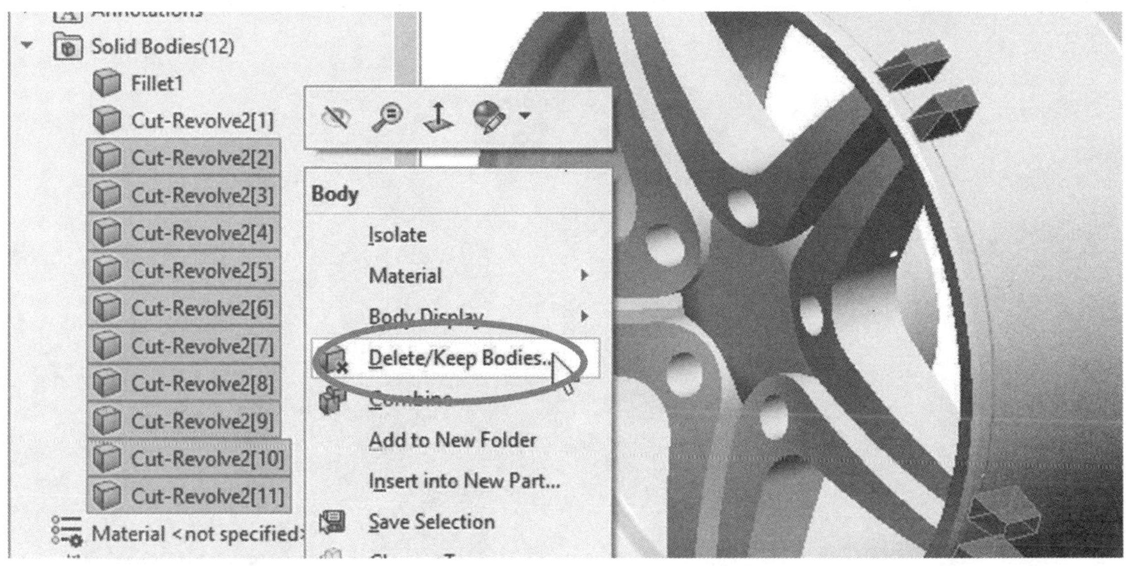

Then we are asked to confirm which bodies we want to delete and click OK to remove them. A new feature called *“Body-Delete”* is added to the part.

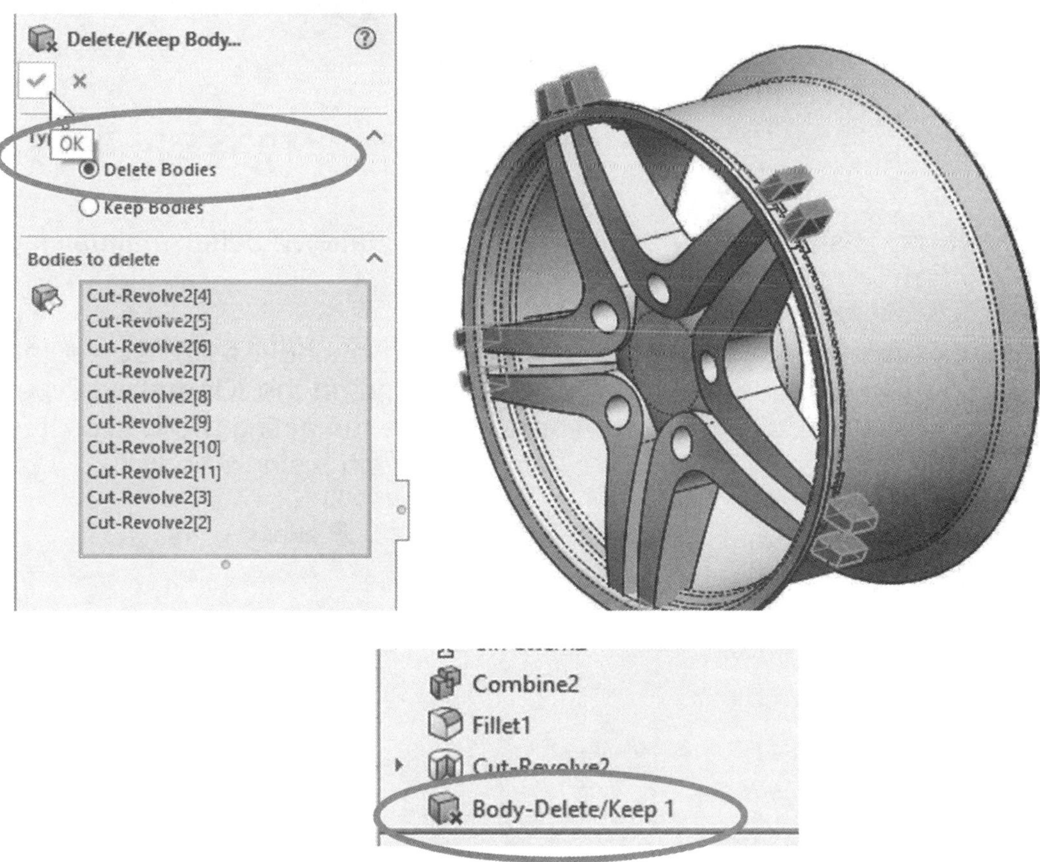

2.23. - Add a 5mm fillet to the spoke arms selecting the radial edges of the spokes and click OK to finish.

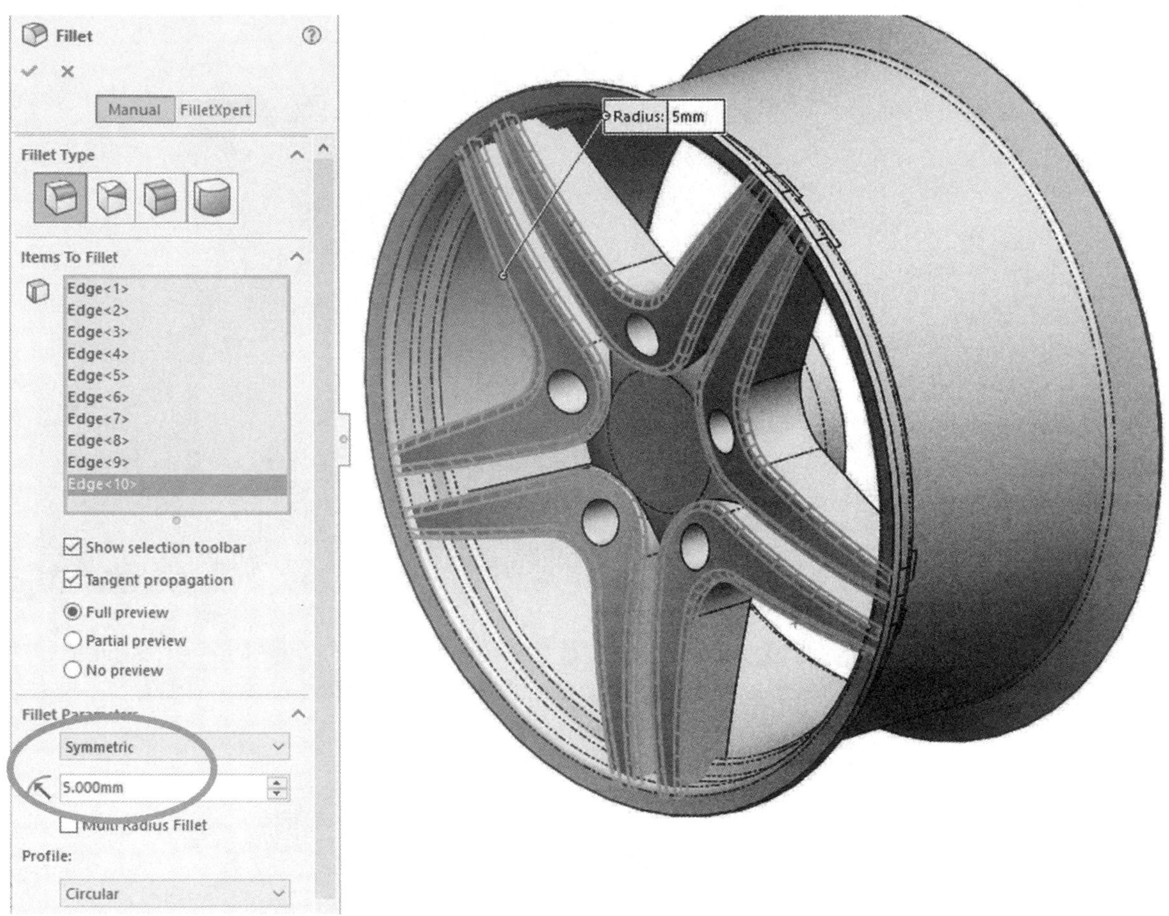

This fillet can also be done before the circular pattern, better maintaining design intent.

2.24. – The next step is to combine both bodies into a single solid. Select the remaining two bodies and use the "**Combine**" command. In the "Operation Type" select "Add" and click OK to finish. After combining the remaining two bodies, the *"Solid Bodies"* folder goes away since we have a part with a single body.

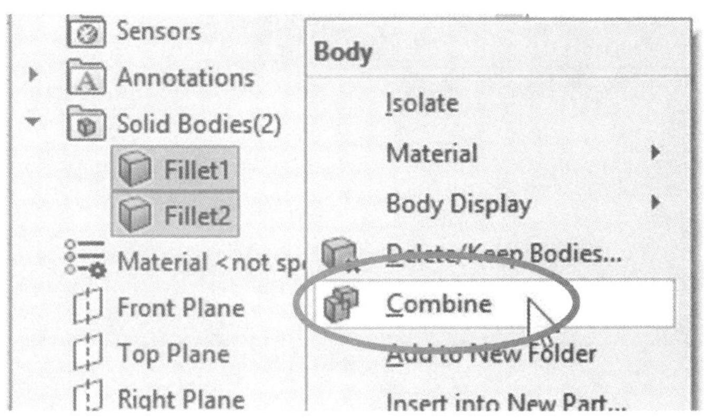

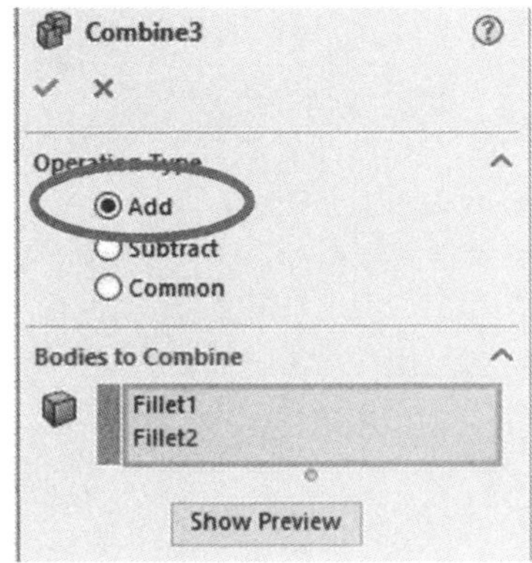

2.25. - Add a 5mm fillet to the edges in the center. After selecting the edge indicated, the expanded selection toolbar is displayed. Click the "Connected to start face" icon to automatically select all the edges needed. Click OK to finish the fillet.

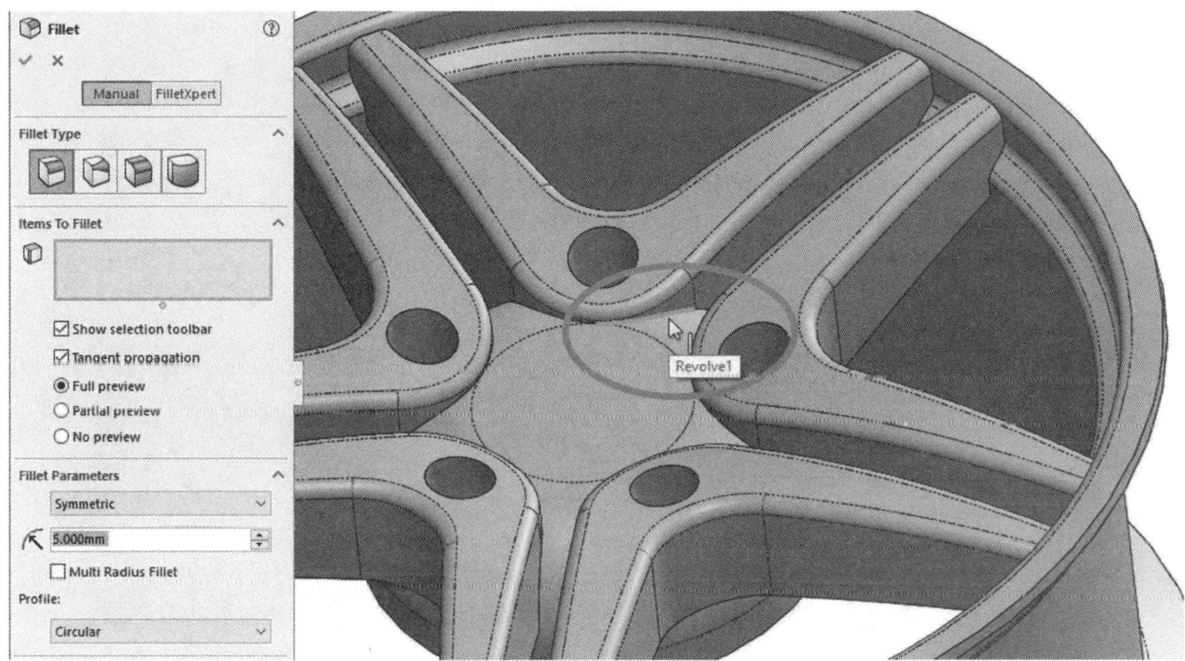

When we move the mouse over each of the icons in the selection toolbar we can see the edges that would be selected if we clicked on it. This is an easy way to make the selection of multiple edges easier.

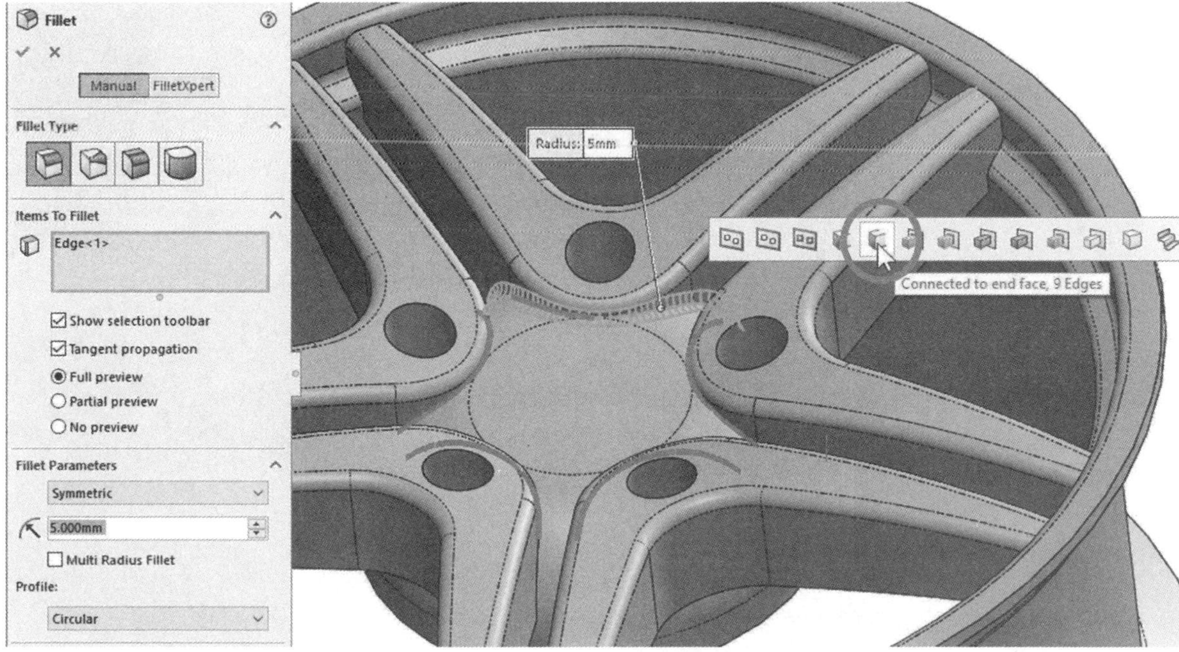

2.26. - Add a new 2.5mm fillet to the bolt holes and the connection between the spokes and the rim. Use the selection toolbar if needed.

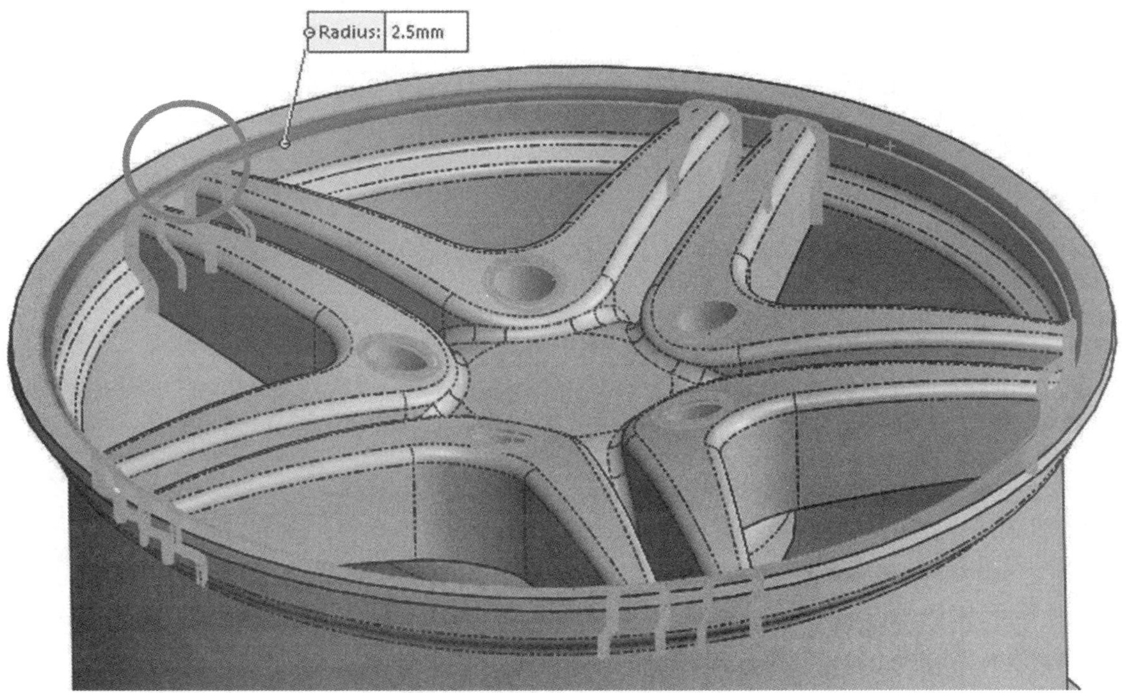

2.27. - To make the wheel lighter, add an 80 mm diameter hole in the rear of the hub. We want to make the cut deep enough to leave a thickness of 8mm at the other side. To accomplish this, we'll use the end condition called "Offset from Surface." Add the following sketch in the back face of the rim's hub.

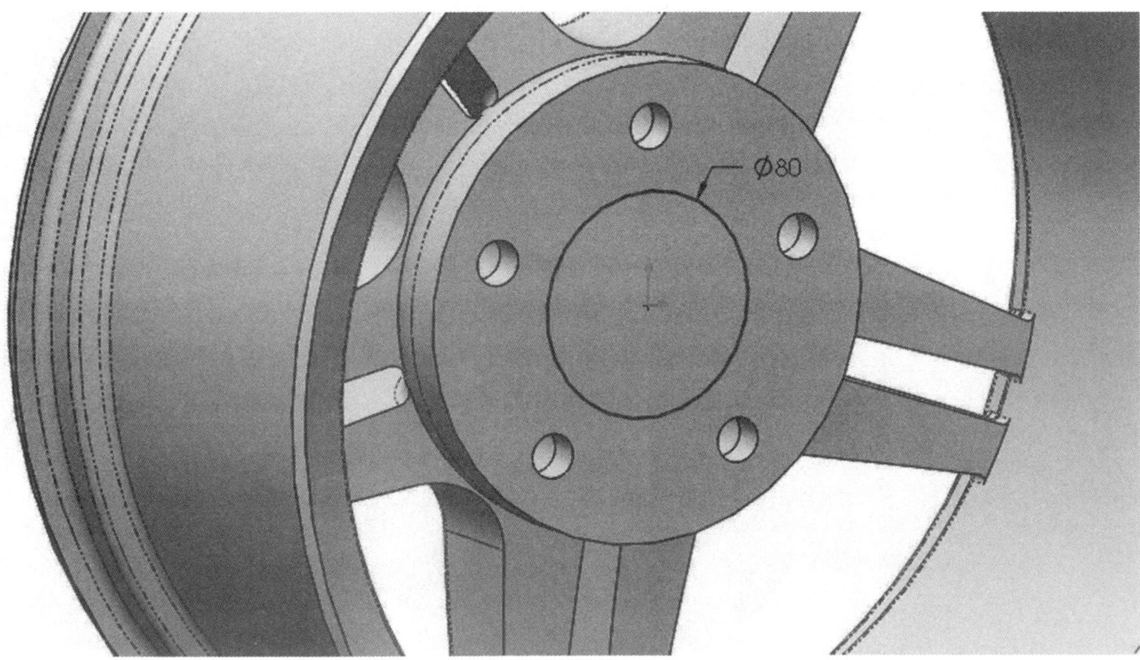

Using the “**Extruded Cut**” command select the “Offset from Surface” end condition and select the face on the opposite side of the sketch. Enter 8mm in the distance box. The cut made will be 8mm away from the selected face. The “Reverse offset” option allows us to measure the distance to one side of the selected face or the other. Rotate the view to select the face if needed and click OK to finish.

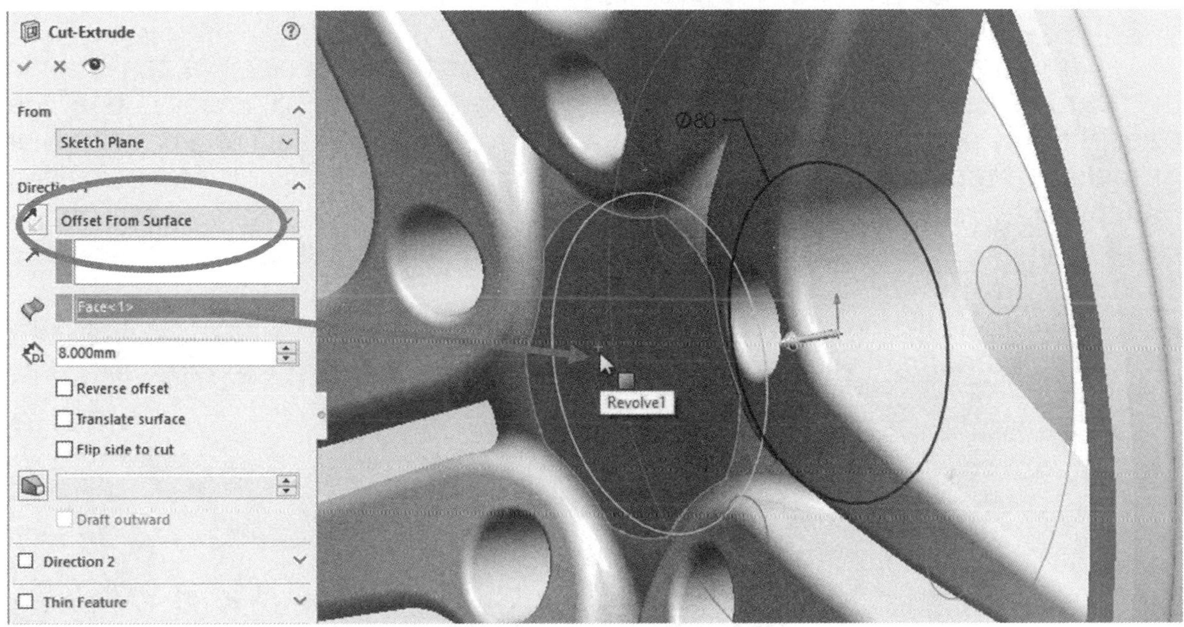

2.28. - As a finishing touch we’ll add a logo to our design. In SOLIDWORKS we can add a picture or image to our models for appearance or presentations. Add a new sketch in the front face of the wheel where the logo will be located, and go to the menu “**Tools, Sketch Tools, Sketch Picture**.”

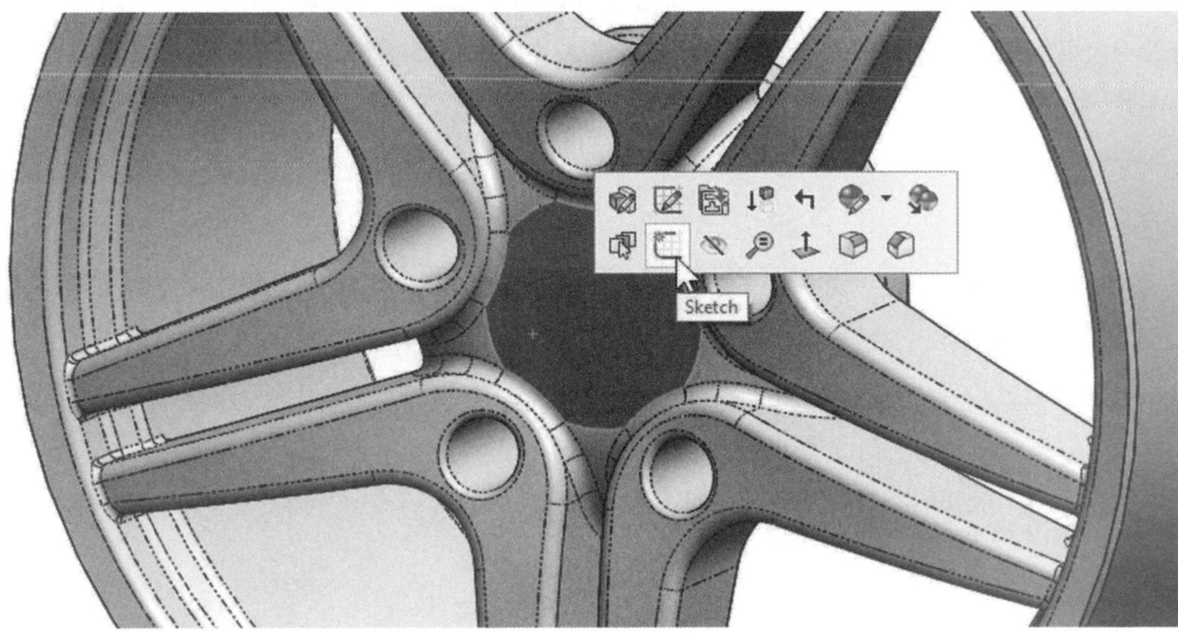

Browse to the included exercise files, locate the '*Needed Files*' folder and select '*logo.png*'. The image will be added to our sketch ready to adjust its properties. We can click and drag the image corners to resize and locate, or we can define its dimensions and location in the "Properties."

For this image, change its size to 70mm wide by 70mm height, and locate it 35mm to the left (-35mm) and 35mm down (-35mm).

Another property we can define is the image transparency. We can change the percentage of transparency for the entire image (Full image), use the transparency settings saved with the image (From file), or select an image's color to make it transparent (User defined). Since the image provided has a defined transparency, select the "Transparency" option "From file"; see how the image updates in the screen. Click OK to finish the image properties and exit the sketch.

File types that support transparency are GIF, PNG and TIFF. Transparency can be defined with many popular image editing software programs.

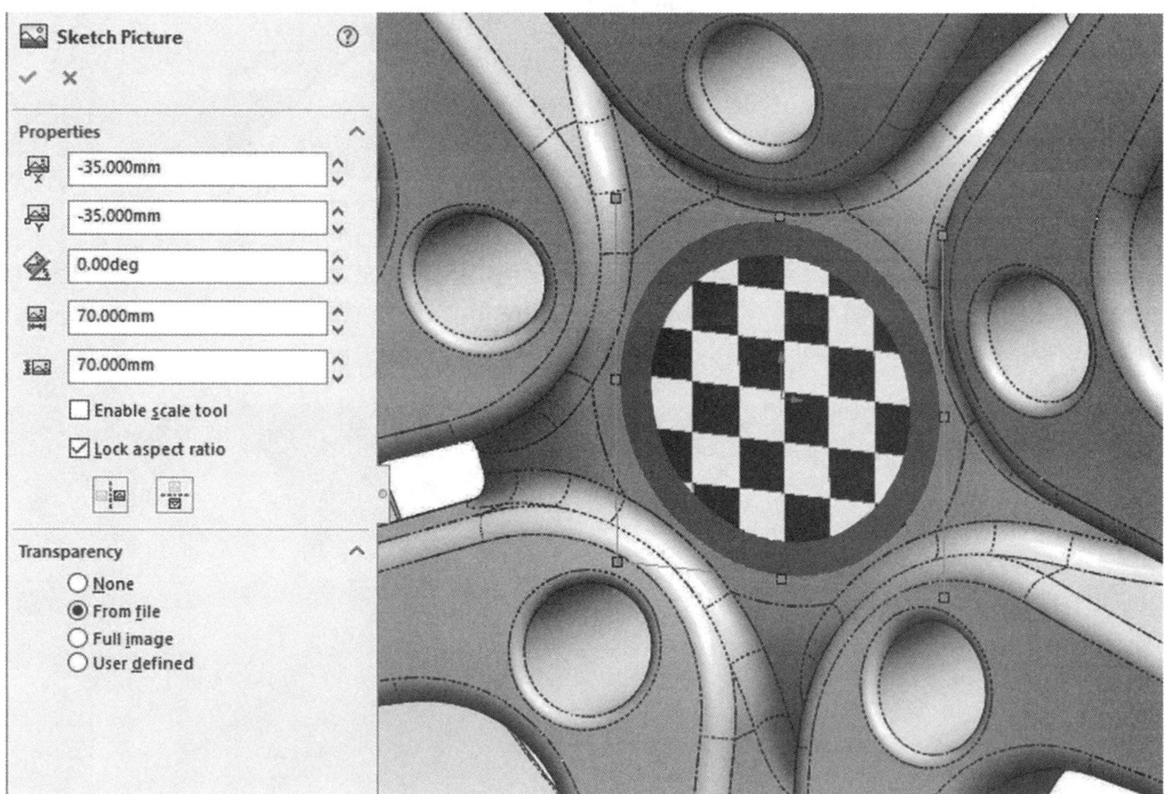

After the picture is added to the sketch it can be edited by double clicking in it in the FeatureManager.

Transparency: None

Transparency: From file

2.29. - In the FeatureManager the image is absorbed by the sketch, and, if needed, it can be hidden by hiding the sketch.

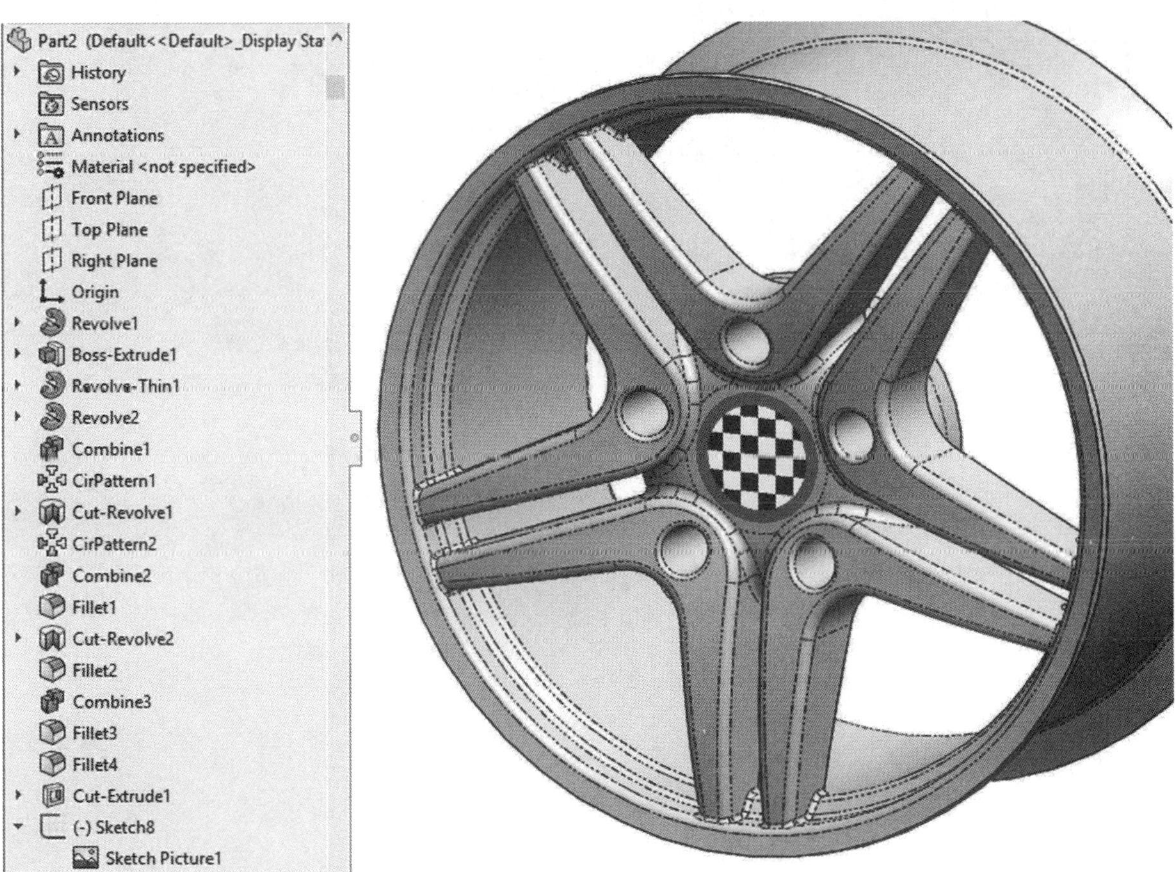

Feel free to add your favorite automotive logo. Due to copyright restrictions we cannot use registered logos in this book, but we can confirm that there are some that look VERY good!

2.30. – The finished part should look like this. Save the part as "*Multi Body Wheel*" and close it.

Combine Bodies

3.1. – Combining multiple bodies to calculate their common volume or subtracting one body from another are techniques frequently used when making odd shapes, or to calculate a container's volume. As we saw in the previous example, to obtain the common volume between bodies we need two intersecting bodies. In the first step, we'll make a "U" shaped link. Make a new part and draw the following sketch in the *"Right Plane."* Part dimensions are in inches.

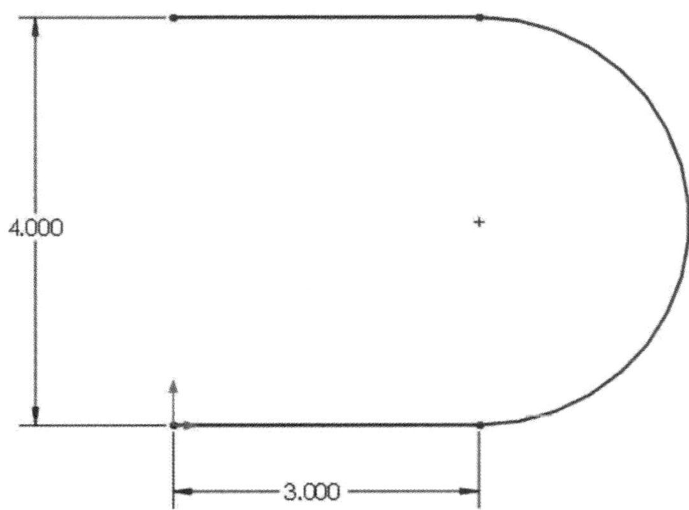

3.2. - Extrude the sketch as a Thin Feature. The extrusion's depth will be 3″ and the thickness of the part 0.5″ inside.

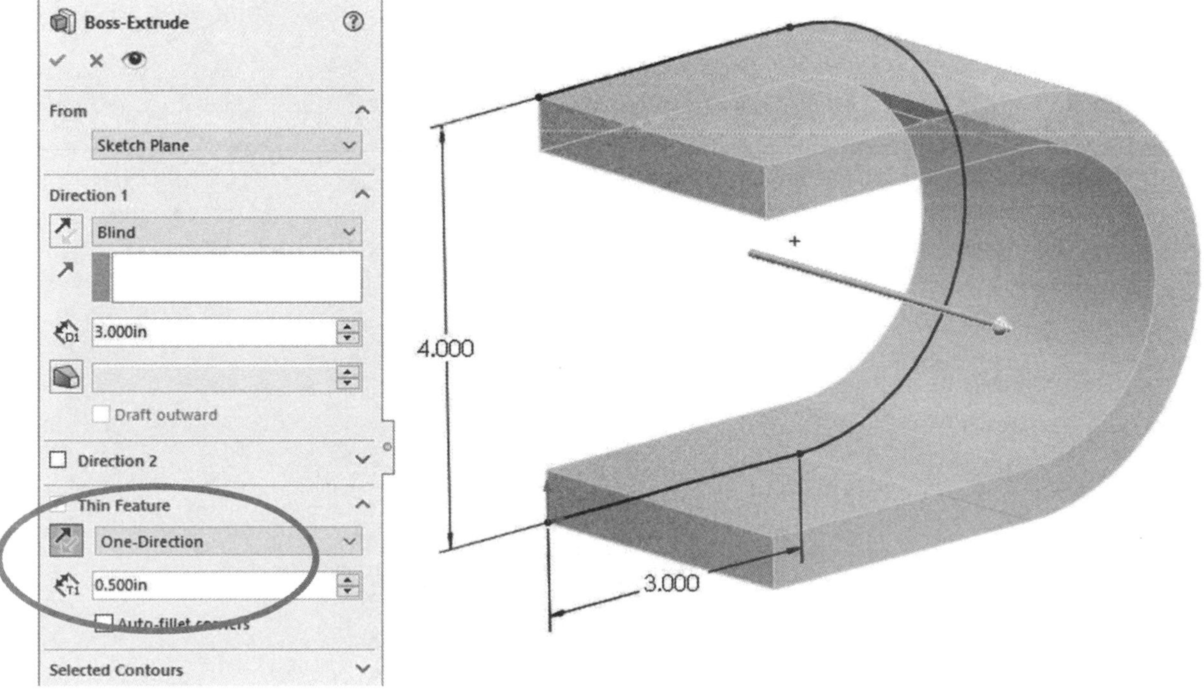

3.3. - For the next feature, add a sketch in the top face of the first feature. Looking at it from a top view should look like this:

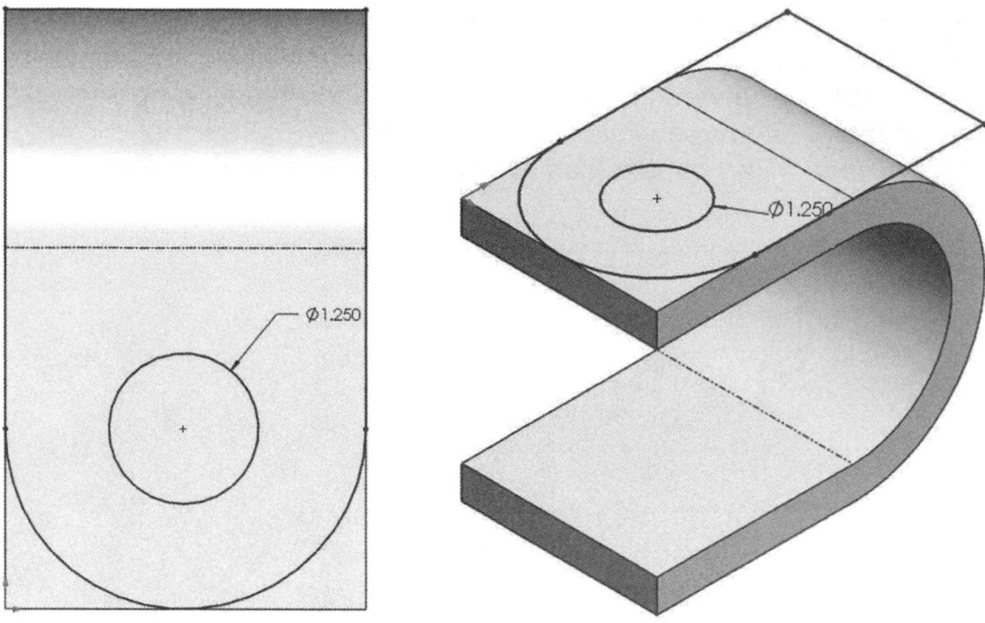

Notice the only dimension we need to add is the hole's diameter; everything else is defined with geometric relations (Coincident and Tangent relations).

3.4. - Extrude the second sketch downwards with the "**Through All**" end condition, and uncheck "**Merge result**." We'll have two overlapping bodies.

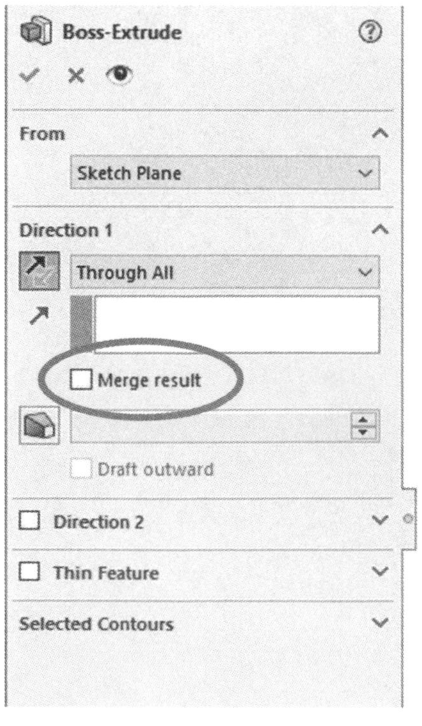

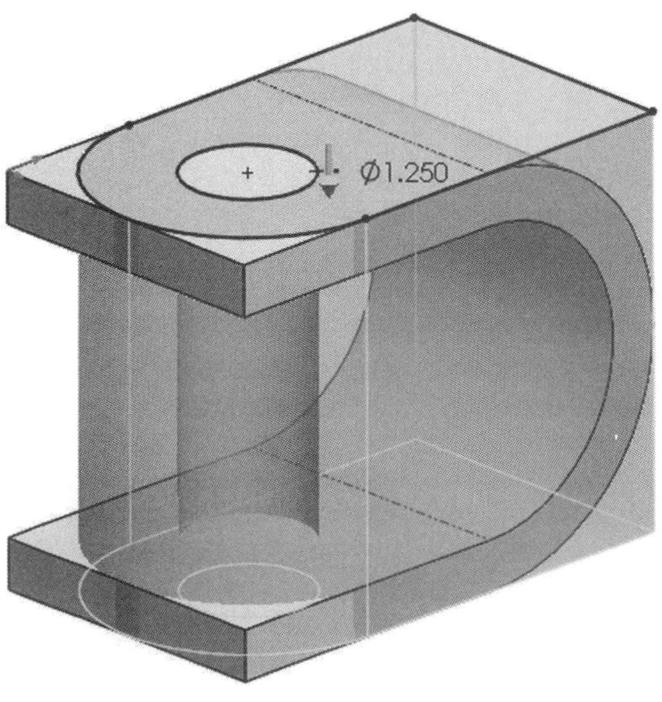

3.5. - Now that we have two separate bodies, we can combine them to get their common volume. Select the menu "**Insert, Features, Combine**" and select both bodies, or from the *"Solid Bodies"* folder in the FeatureManager pre-select both bodies, right-mouse-click and select "**Combine**" from the pop-up menu.

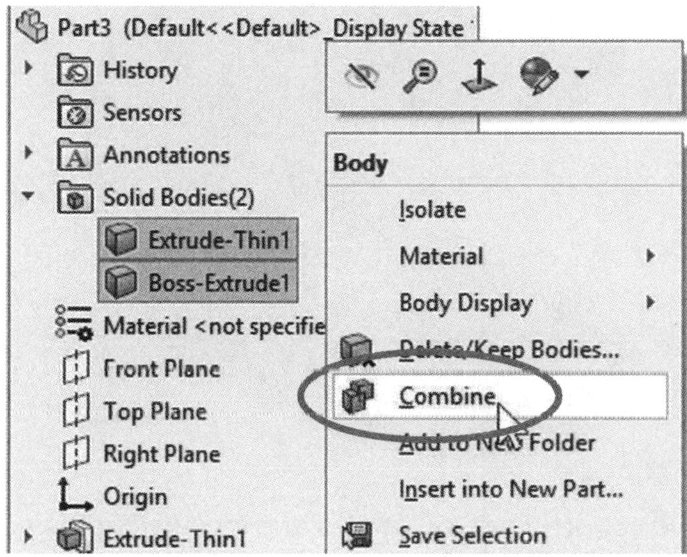

3.6. - In the Combine operation select "Common" under "Operation Type." Click in the "Show Preview" button to see what the resulting body will look like and click OK to finish.

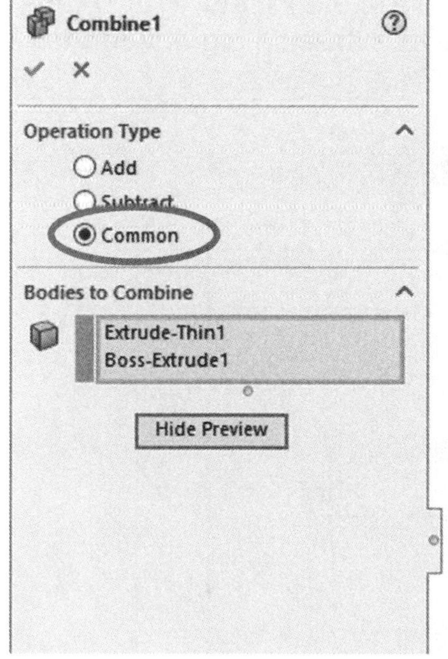

The finished part with the common volume of the two bodies will look like this.

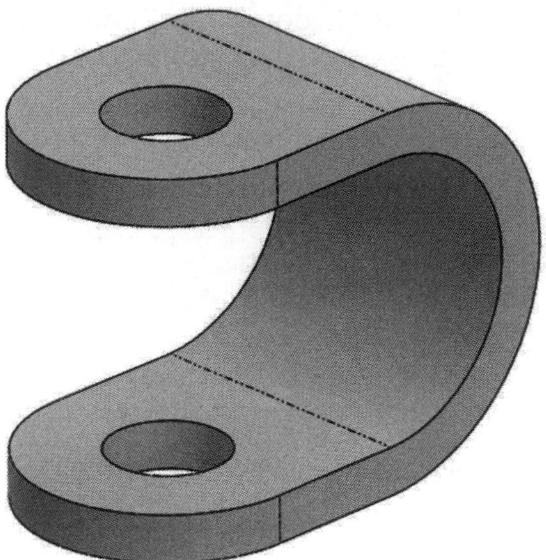

3.7. - So far we have covered adding bodies and calculating their common volume. The next multi body operation is to subtract one body from another. A common use is to obtain a mold's core and/or cavity (later in the book), but instead we'll learn how to obtain the volume capacity of an irregularly shaped container.

Locate the part *'Bottle.sldprt'* from the exercise files and open it. In order to obtain a body that represents the liquid inside the bottle, we need to make a new solid body bigger than the bottle up to the fill level. Add a sketch in the *"Front Plane"* and draw a rectangle as shown. Make the sketch and extrusion big enough to fit the bottle inside.

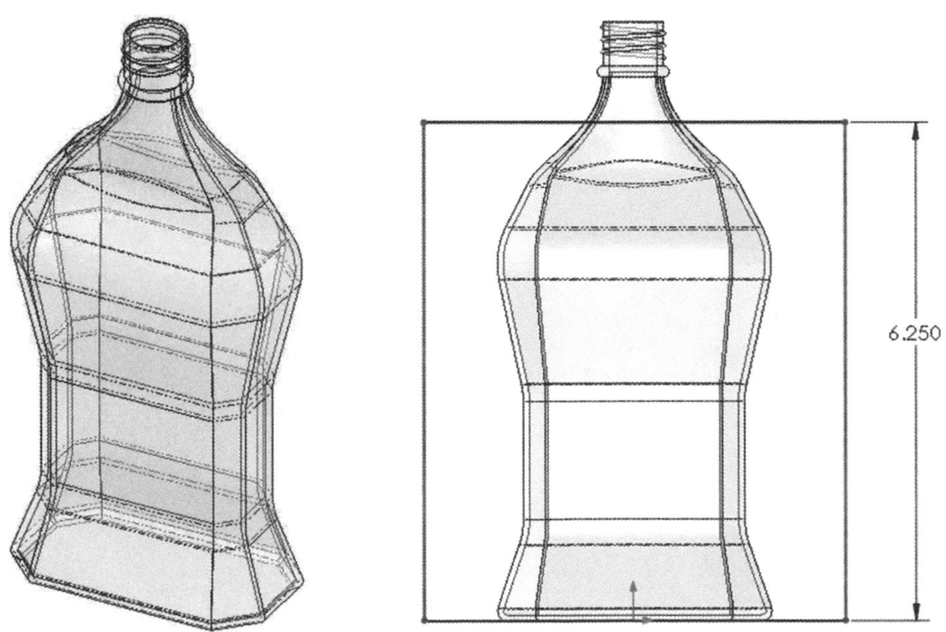

3.8. - Use the "Mid Plane" end condition and uncheck "Merge result" before extruding the new body.

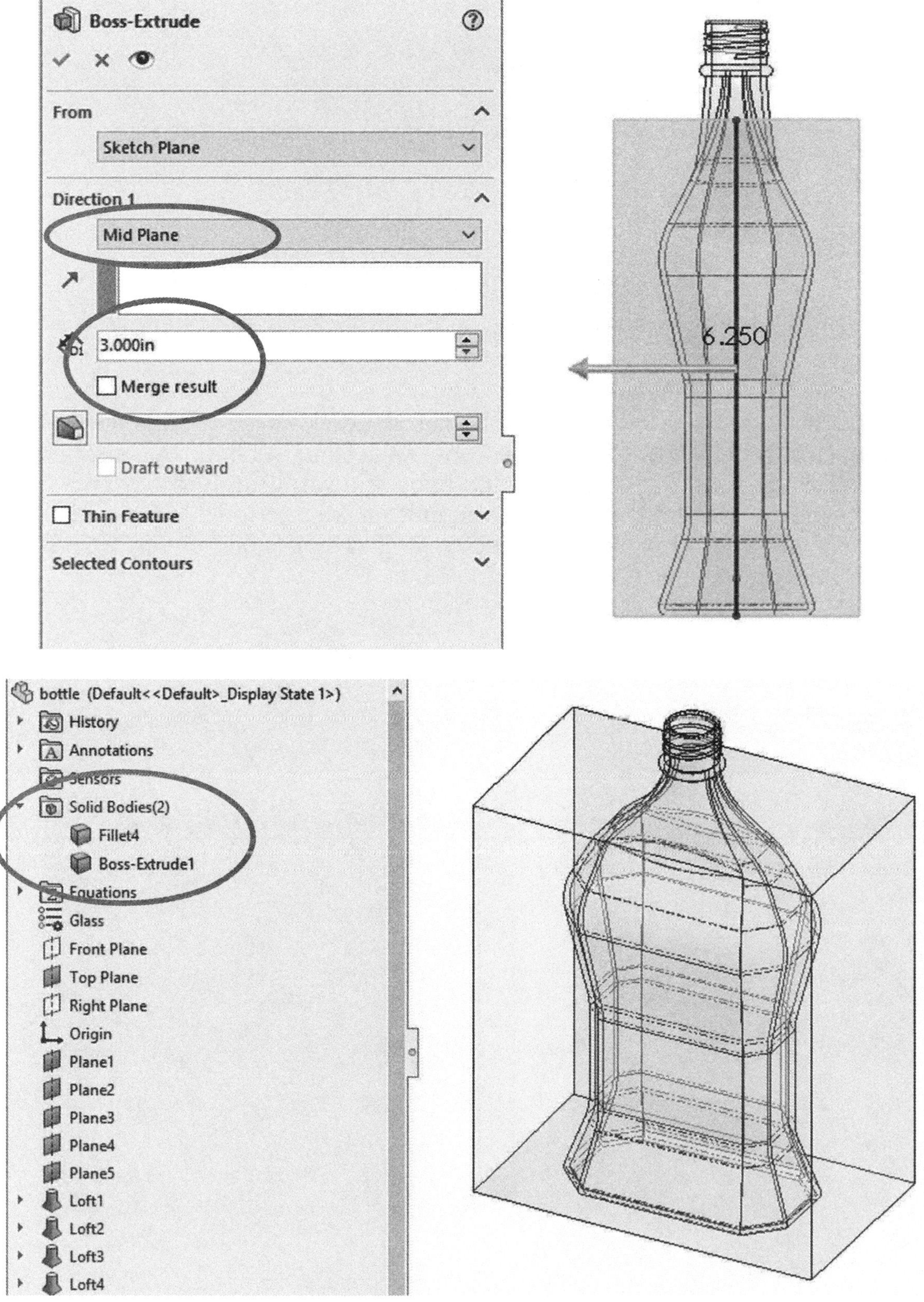

To change the part's transparency, use the "**Display Pane**." Activate the transparency at the part level (can also be done to a body individually).

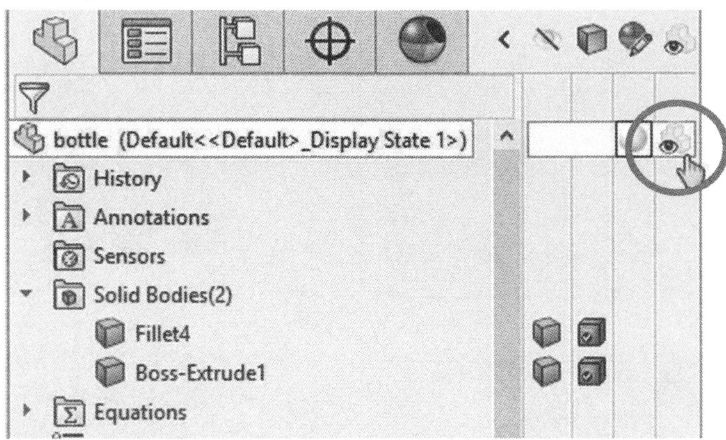

3.9. - Now that we have the two solid bodies needed, select the menu "**Insert, Features, Combine**." Under "Operation Type" select the "Subtract" option. In order to get a difference, we need to select the body that we want to remove material from (Main Body) and the body (or bodies) that we want to remove from it. In the "Main Body" selection box, select the body we just made, and under "Bodies to Subtract" select the bottle. Click OK to continue.

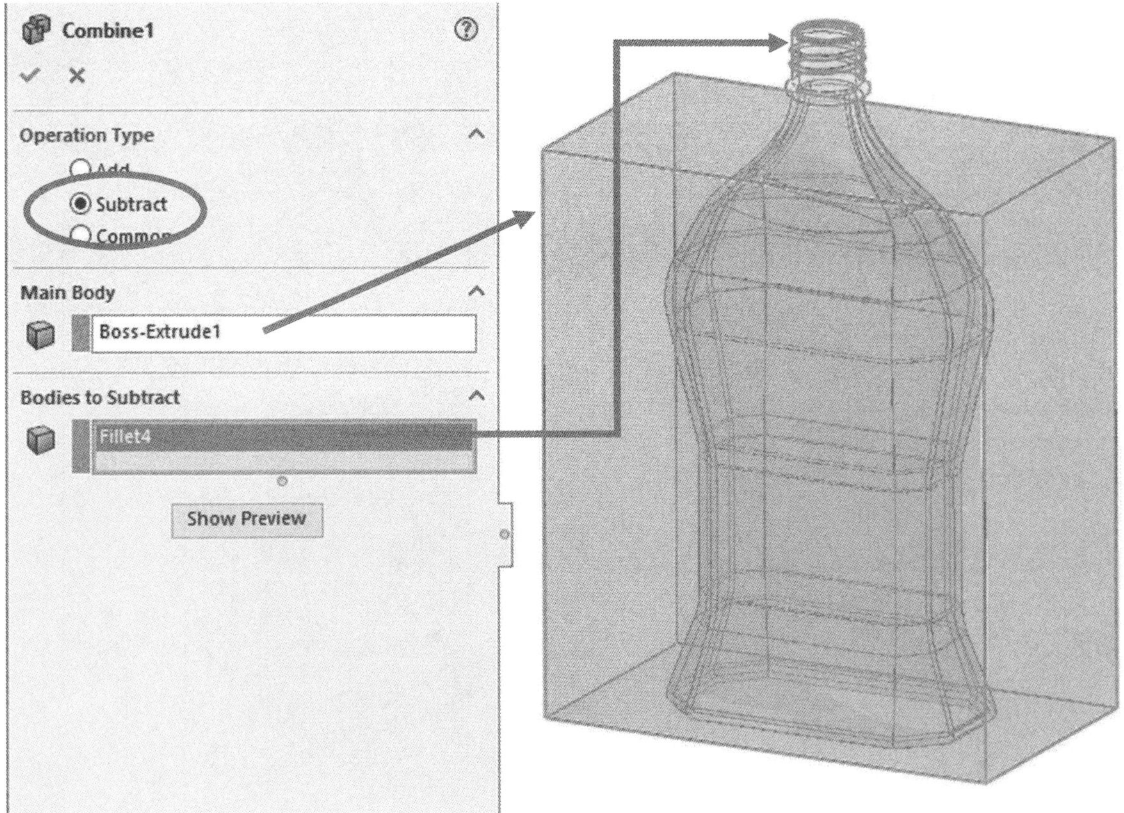

3.10. - After pressing OK we are immediately presented with the "Bodies to Keep" dialog, just like when we trimmed the ends of the spokes of the wheel. There are two bodies resulting from the operation, but we are interested only in the inside body. Select the inside body to keep it and click OK to finish.

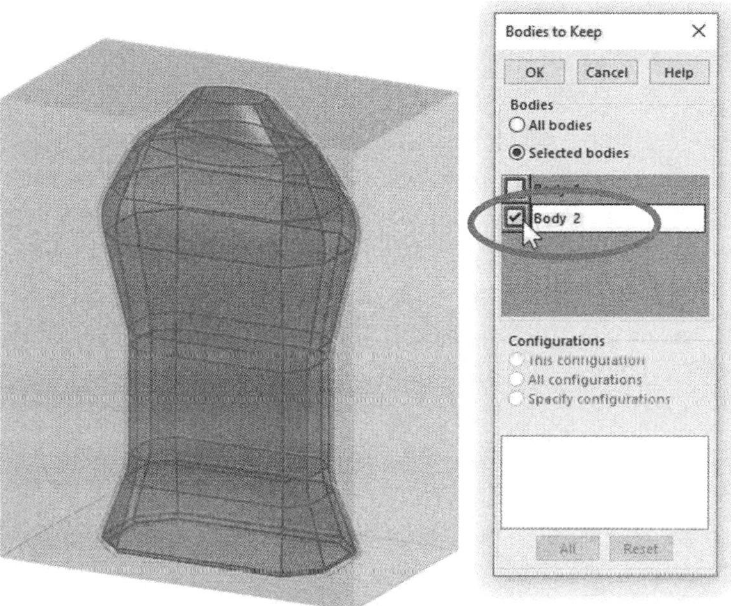

3.11. - Now we have the actual volume of the bottle up to the fill line. Notice that when we subtract one body from another, the original bodies are consumed and we are left only with the difference. After running a "**Mass Properties**" analysis, select the Options button, and select milliliters for volume; now we can see that the volume of liquid inside the bottle up to the fill line is 337.8 milliliters. Save and close the part.

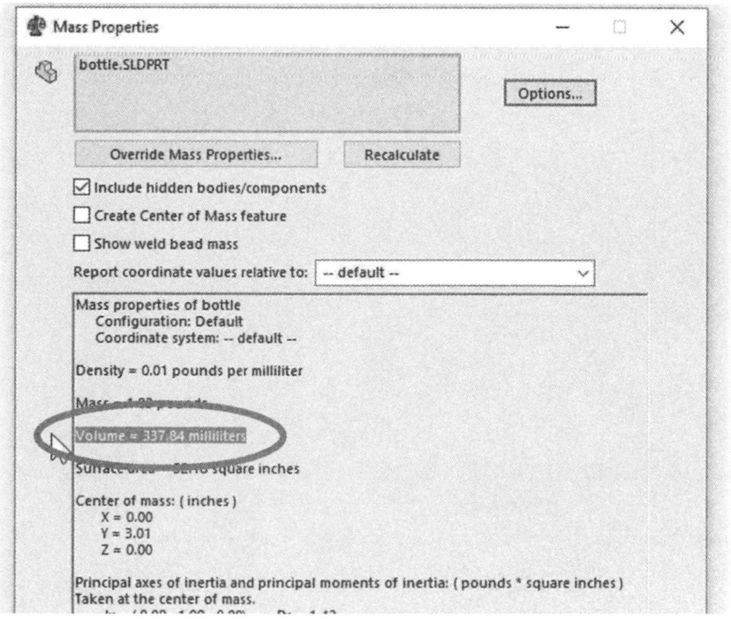

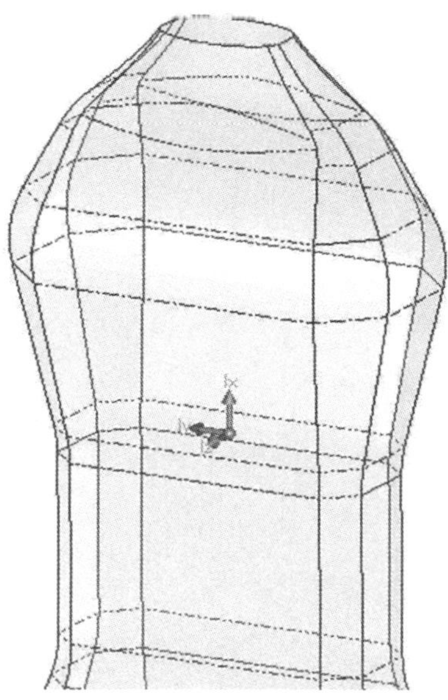

Notes:

Contour Selection

Contour Selection is a way to work with a sketch that has intersecting entities, endpoints shared by multiple entities and many of the common problems that would otherwise prevent us from using a sketch for a feature. We'll also learn how to reuse a sketch for multiple features and a new Extruded Boss/Cut option.

4.1. - Make a new part and add a sketch to the *"Front Plane"* as shown. In this example, we will not worry about dimensions to simplify the explanation and concentrate on how contour selection works.

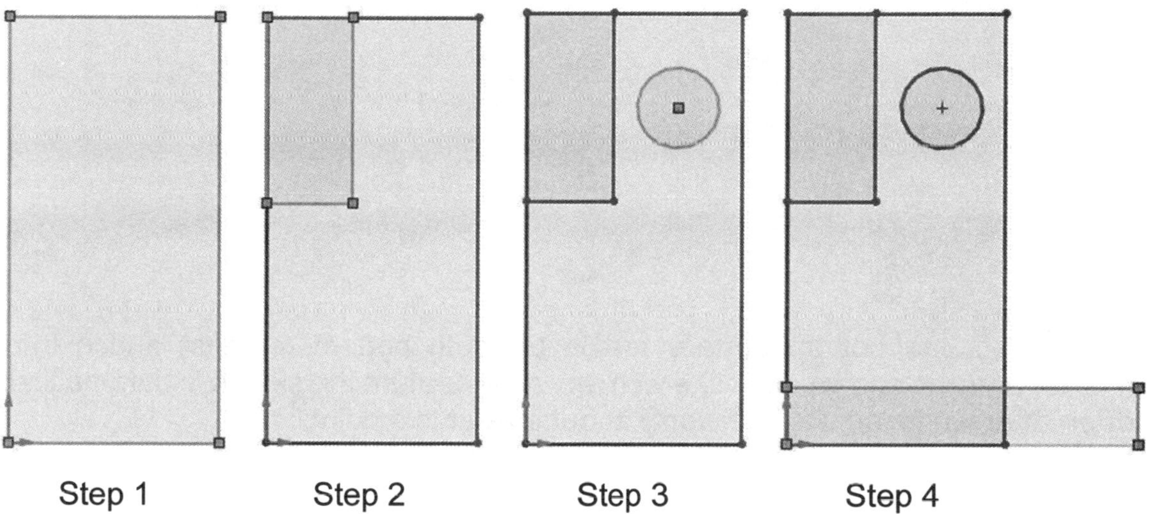

In this sketch we are drawing three rectangles and one circle. Note that the rectangles are overlapping and sharing endpoints.

4.2. - Select the "**Boss-Extrude**" command. Since our sketch has overlapping and shared endpoints, the Contour Selection tool is automatically activated and the "**Selected Contours**" selection box is opened; also, we won't get a preview until we select the region(s) or contour(s) that we want to use in the feature.

In the graphics area move the mouse pointer around the sketch to see the different regions available for selection. We can select single or multiple regions *and/or* complete closed contours by selecting their perimeter.

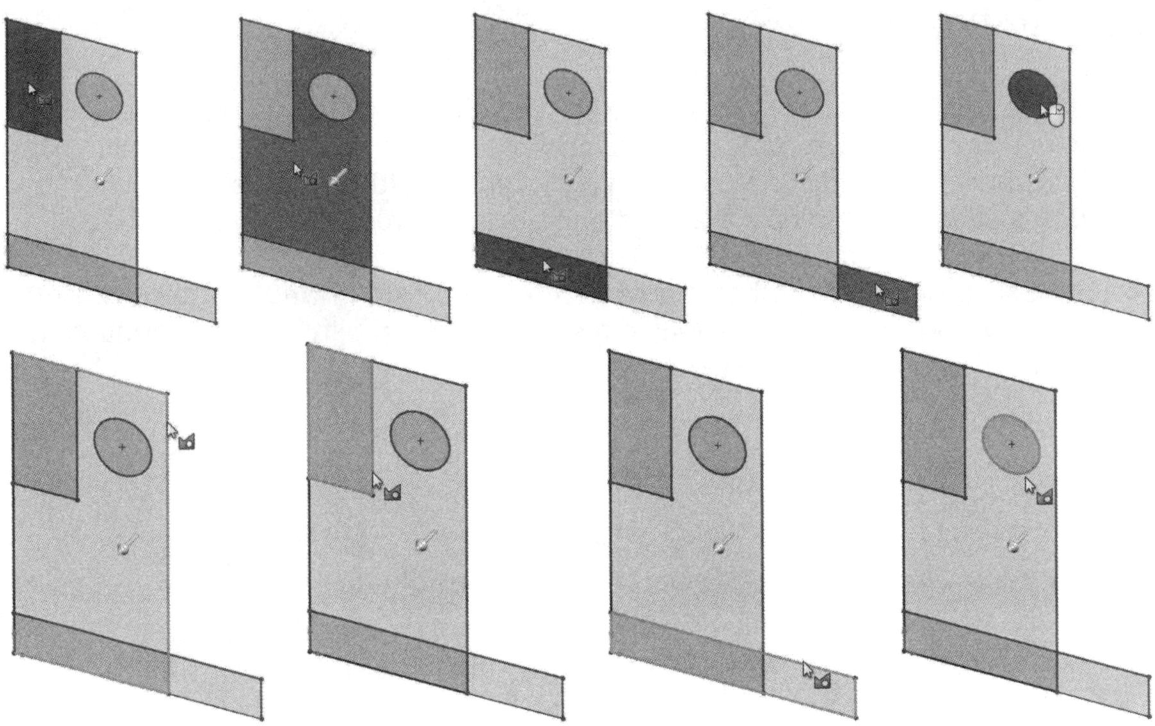

4.3. - Select the bottom square profile (or both bottom regions) and extrude *approximately* as shown. Just like with any other feature the sketch is automatically hidden after we finish. Do not worry about size at this point.

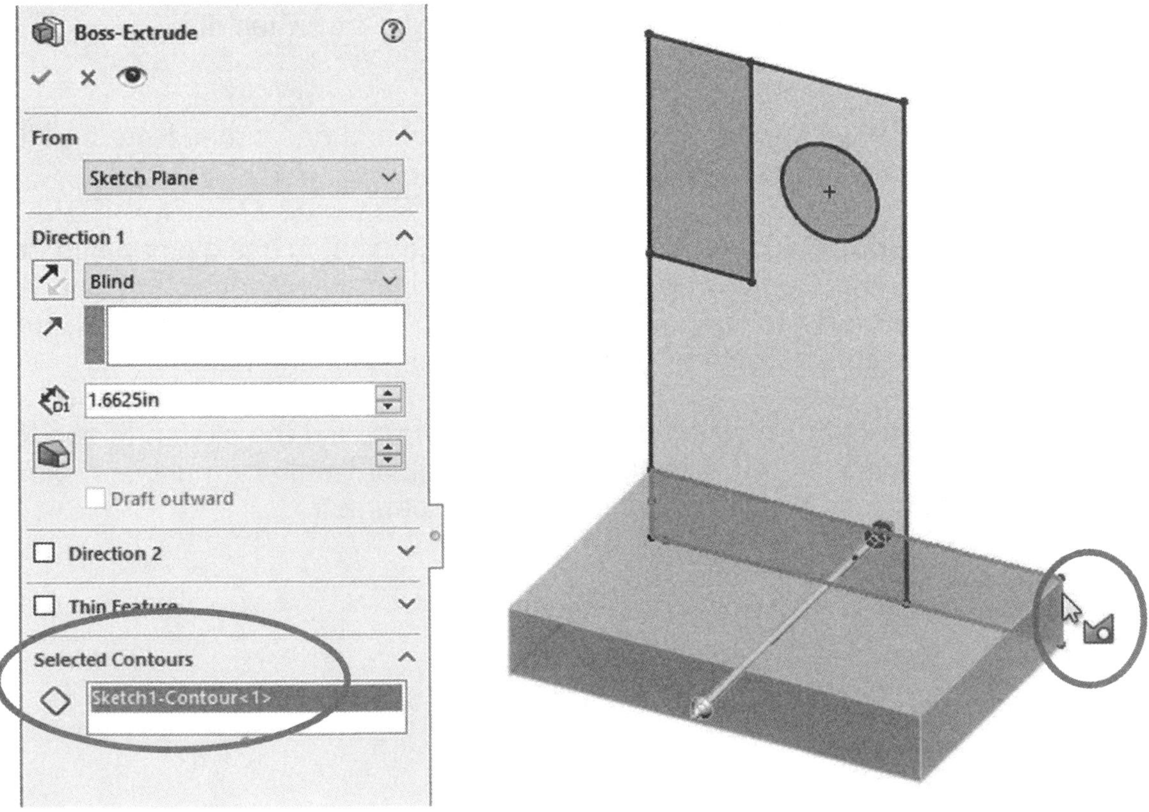

4.4. – After creating a feature using sketch contours, the sketch used remains visible and shows the "**Contours**" icon next to the sketch's name, letting us know the feature was made using contours. Optionally, we can hide the sketch when it is no longer needed.

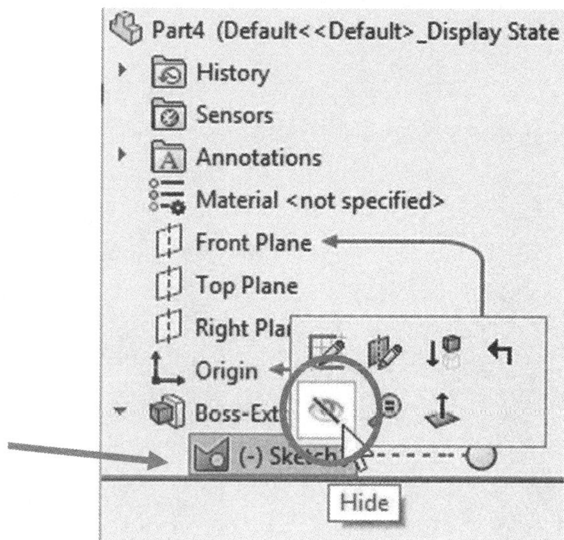

4.5. - We will now re-use the sketch from the first extrusion to create a new feature. To activate the "**Contour Selection Tool**" we must make a right-mouse-click in the graphics area (or the sketch itself). Be aware that by default the "**Contour Select Tool**" option is not available in the pop-up menu; you may have to expand the menu at the bottom to make it visible.

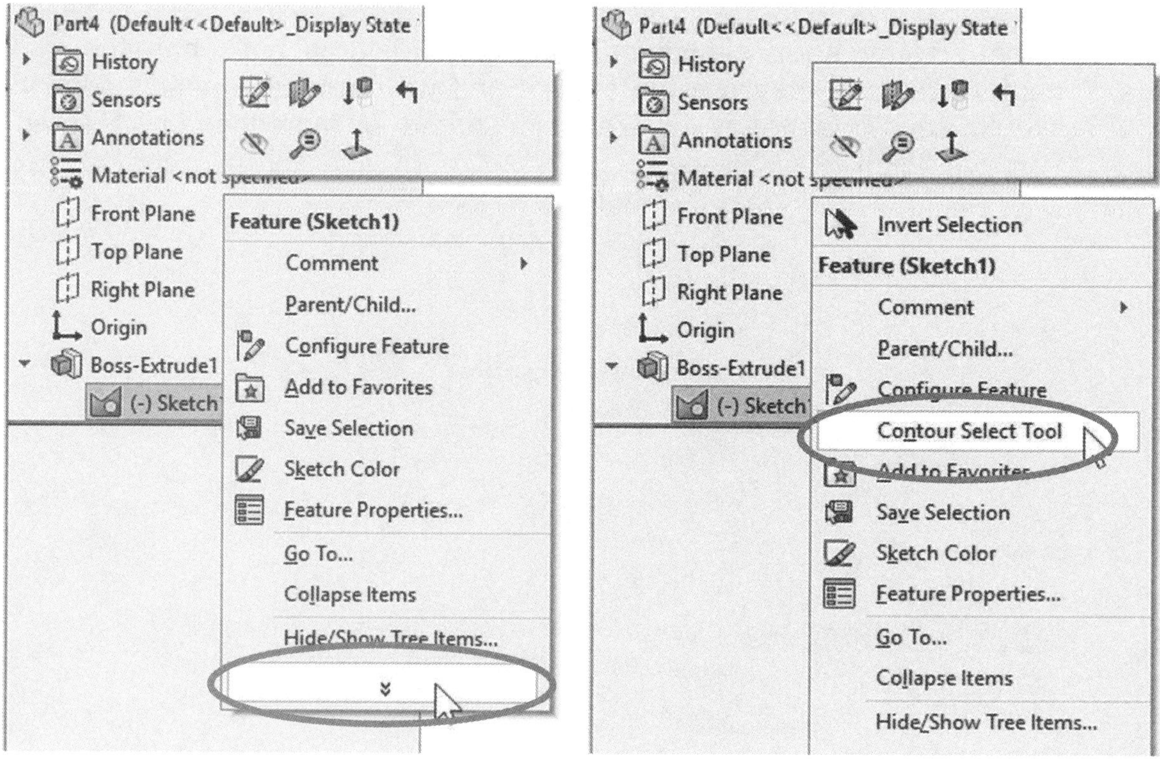

4.6. - After activating the "**Contour Select Tool**," select the regions indicated and extrude *approximately* as shown. You may have to click in the sketch to enable (activate) selection of regions, and either hold down the Ctrl key to pre-select all three regions and then extrude, OR select the Extruded Base command *and then* select the regions. Either way works the same.

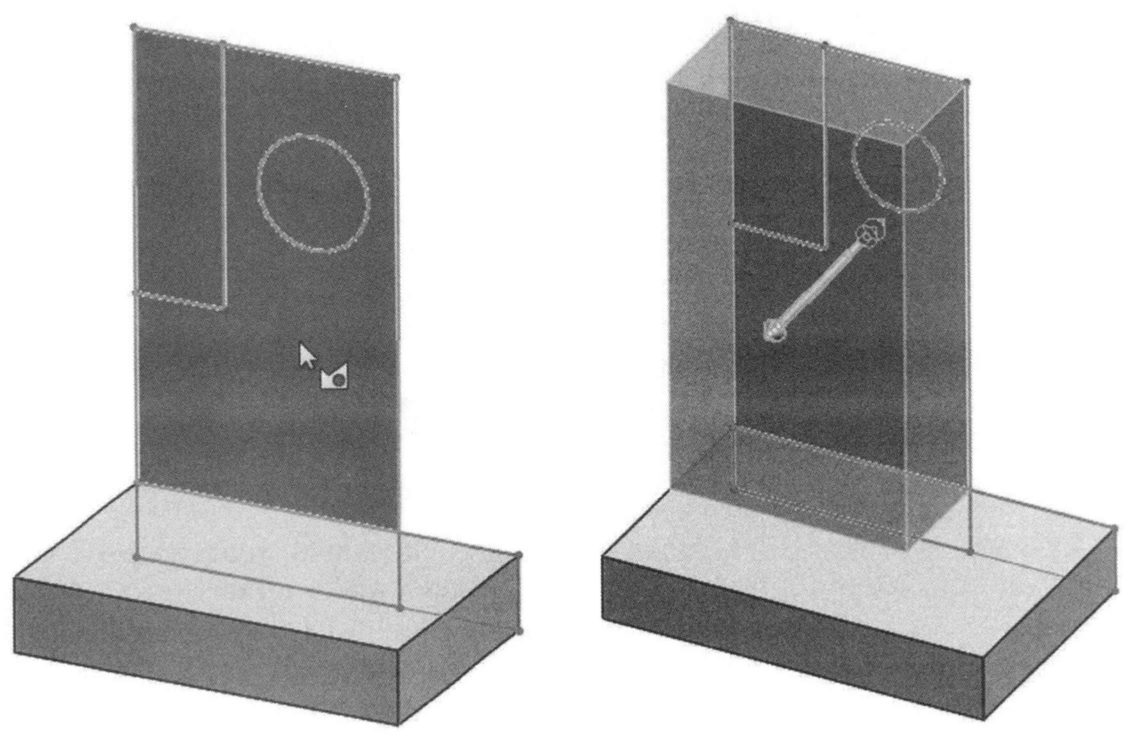

After using the same sketch for two or more features, both share the same name (because they are the same), and we can see a hand under the sketch icon. This means the sketch is "shared" by more than one feature.

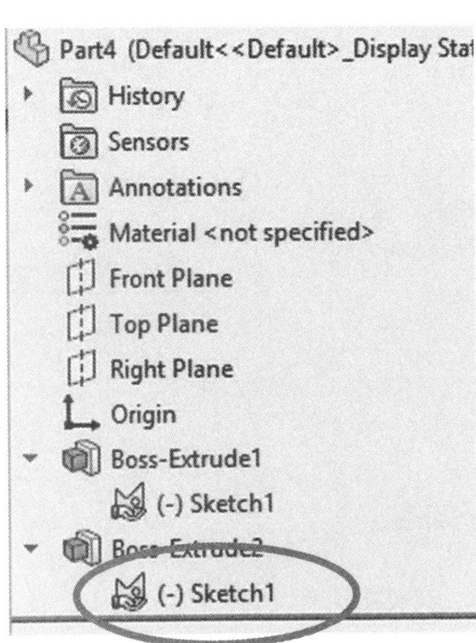

4.7. - Repeat the previous process to select the next contour and extrude as shown. Remember at this time we are only showing how it works and are not concerned about the dimensions.

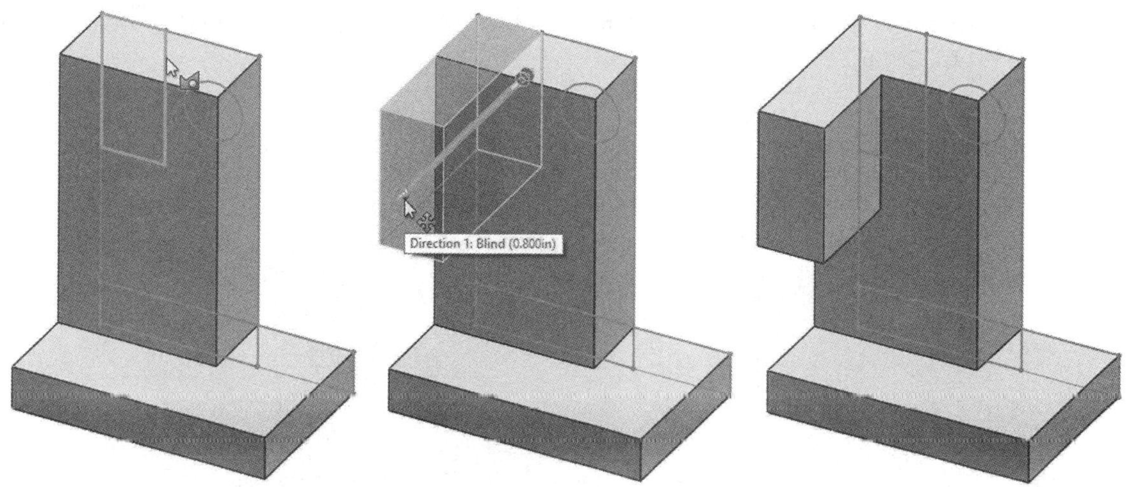

4.8. - Until now we have added features that start at the sketch plane. SOLIDWORKS has a powerful (yet sometimes under used) feature that allows us to add a feature starting somewhere *other* than the sketch plane.

For the last feature select the circle using the "**Contour Select Tool**," and use the "**Extruded Cut**" command. In the "From" start condition's drop-down menu select "Surface/Face/Plane." (We can also use a vertex or an offset distance from the sketch plane.)

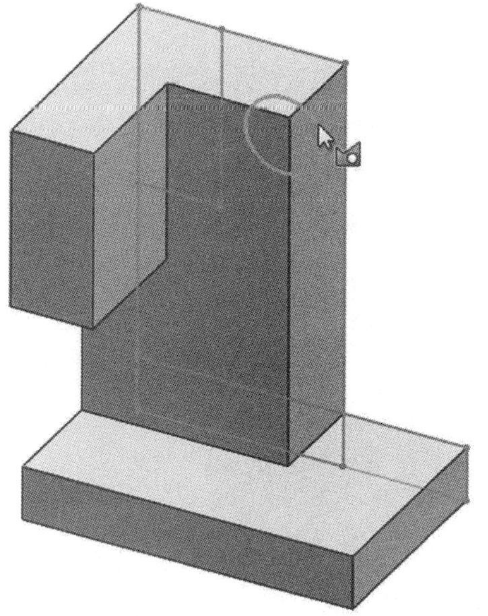

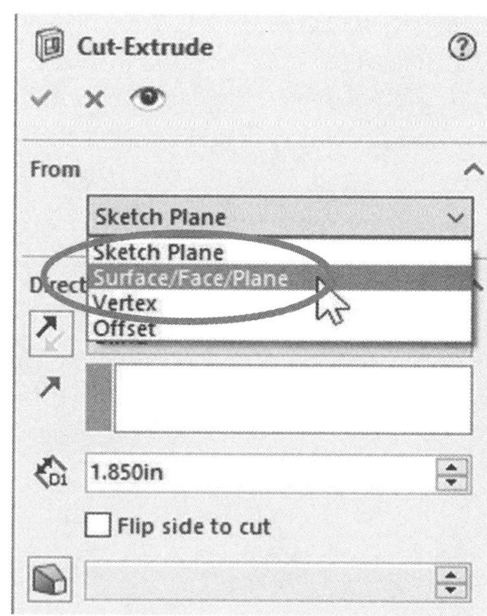

After activating "Surface/Face/Plane" a new selection box is revealed to select where we want the feature to start. Select the front face of the second extrusion as indicated to start the cut feature in this face. Make the feature's depth *about* half way deep, and click OK to finish. By using the "Start Condition" option for feature creation we can easily save time by not having to create auxiliary planes or additional geometry.

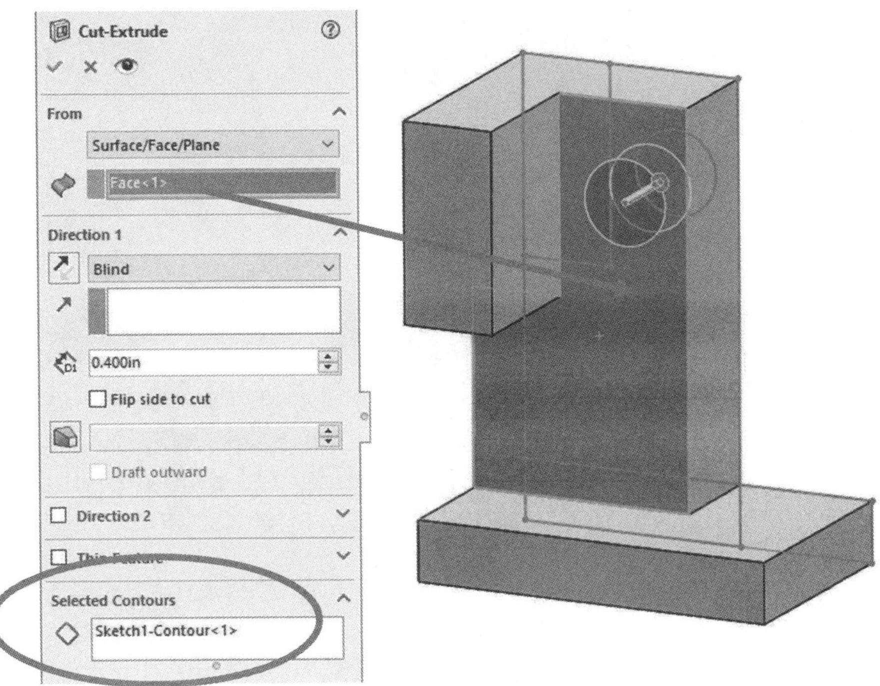

4.9. – After adding four features using the same sketch with multiple contours and self-intersecting geometry, select it in the FeatureManager and hide it.

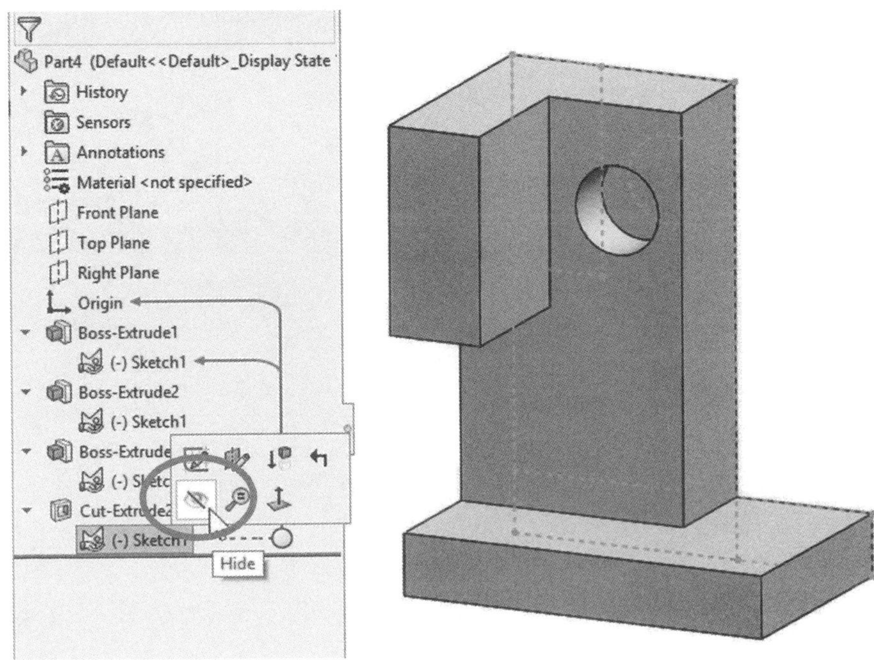

4.10. - When finished save the file and close it.

Exercises: Build the following parts using the knowledge acquired in this lesson. Try to use the most efficient method to complete the model. Open the required parts from the included exercise files.

Multi Body Exercise

Open the part *'Multibody Exercise.sldprt'* to build a mesh basket using a multi body part. Make a circular body pattern of the *"Extrude-Thin1"* feature (Blue body) using the centerline in the *"Circular Pattern Axis"* sketch with 7 instances spaced 30 degrees, and a linear body pattern using the *"Extrude-Thin2"* (Green body) with 9 instances spaced 1″.

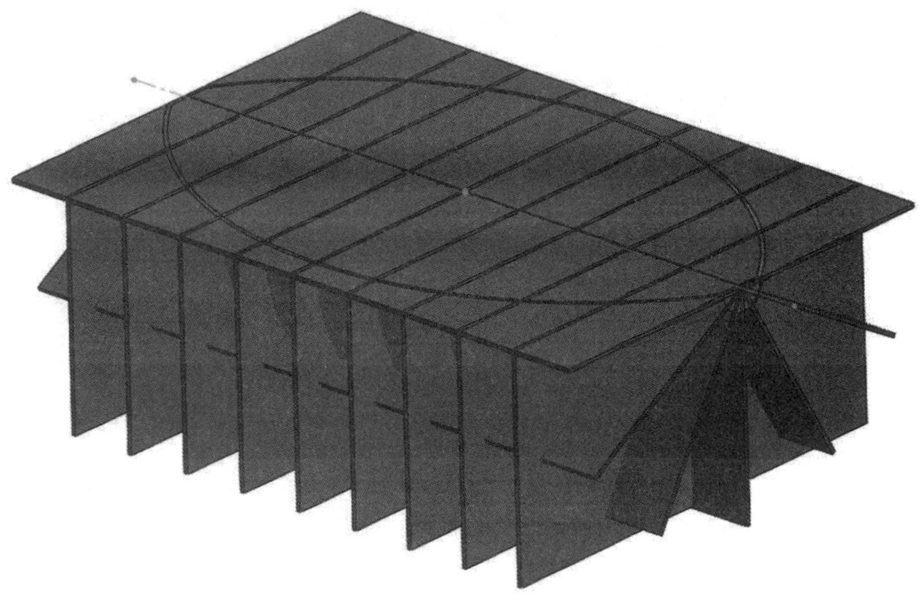

Combine (Add) all the bodies except the *"Revolve-Thin1"* (Red body) to make a single body with them.

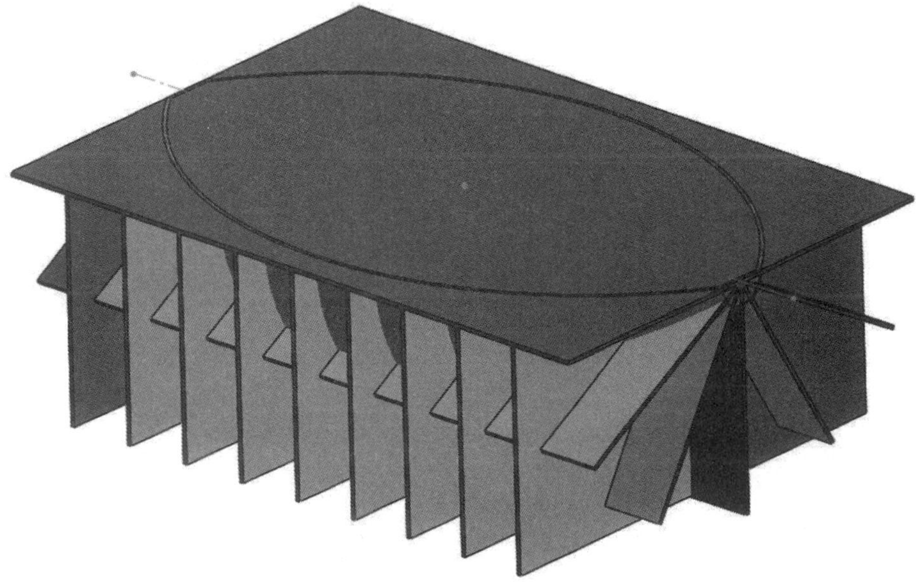

Finally combine the remaining two bodies to get their common volume.

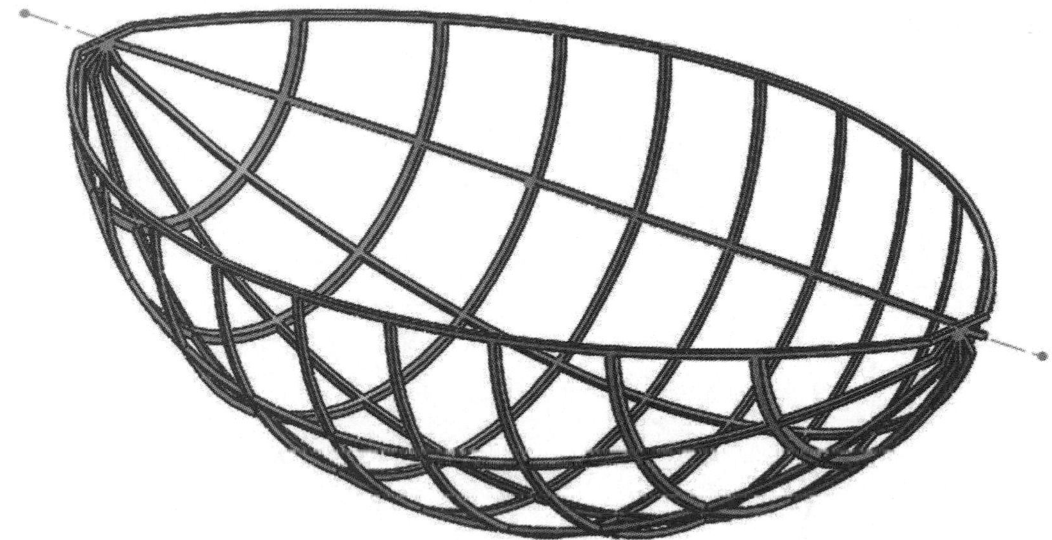

As an optional finishing touch, add a reinforcement in the ends of the basket and fillet the intersections with a 0.050″ radius. TIP: Use the expanded selection tool to automatically select all the small edges.

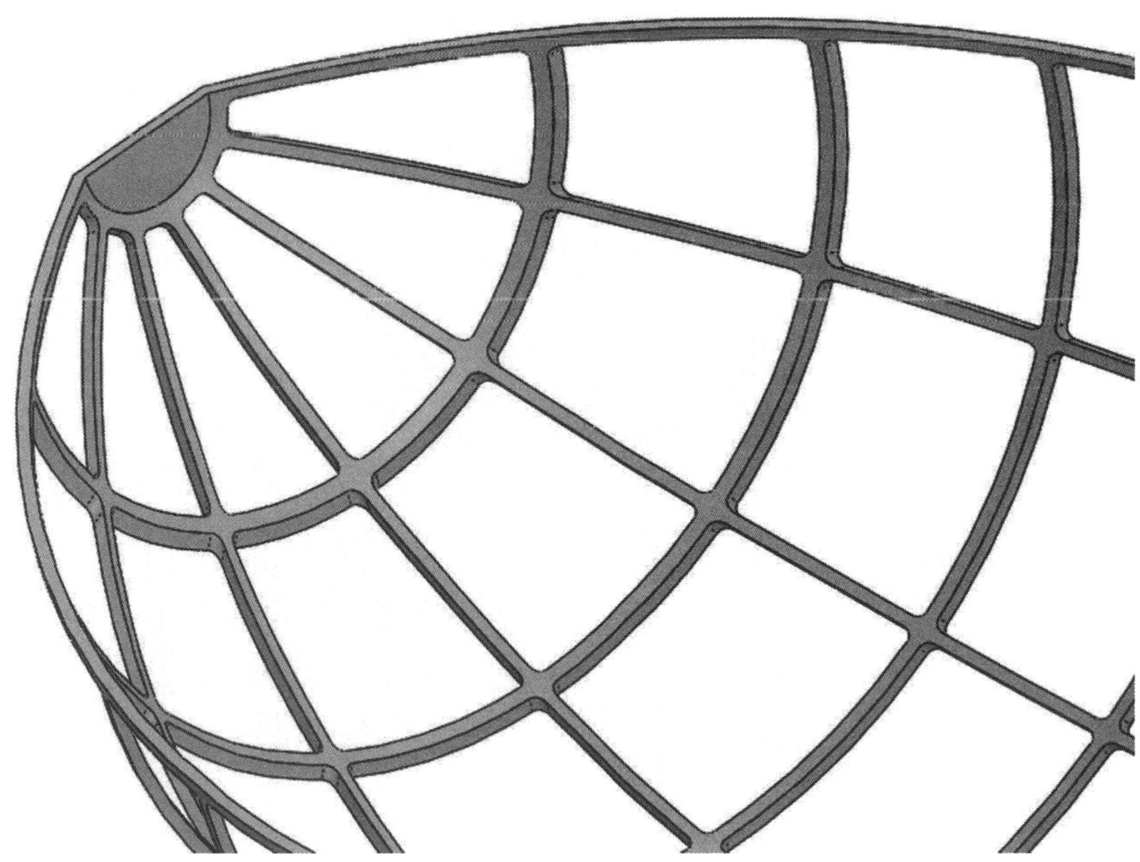

Contour Selection Exercise

Make the following sketch and make all four features off it using the "Contour Selection" tool.

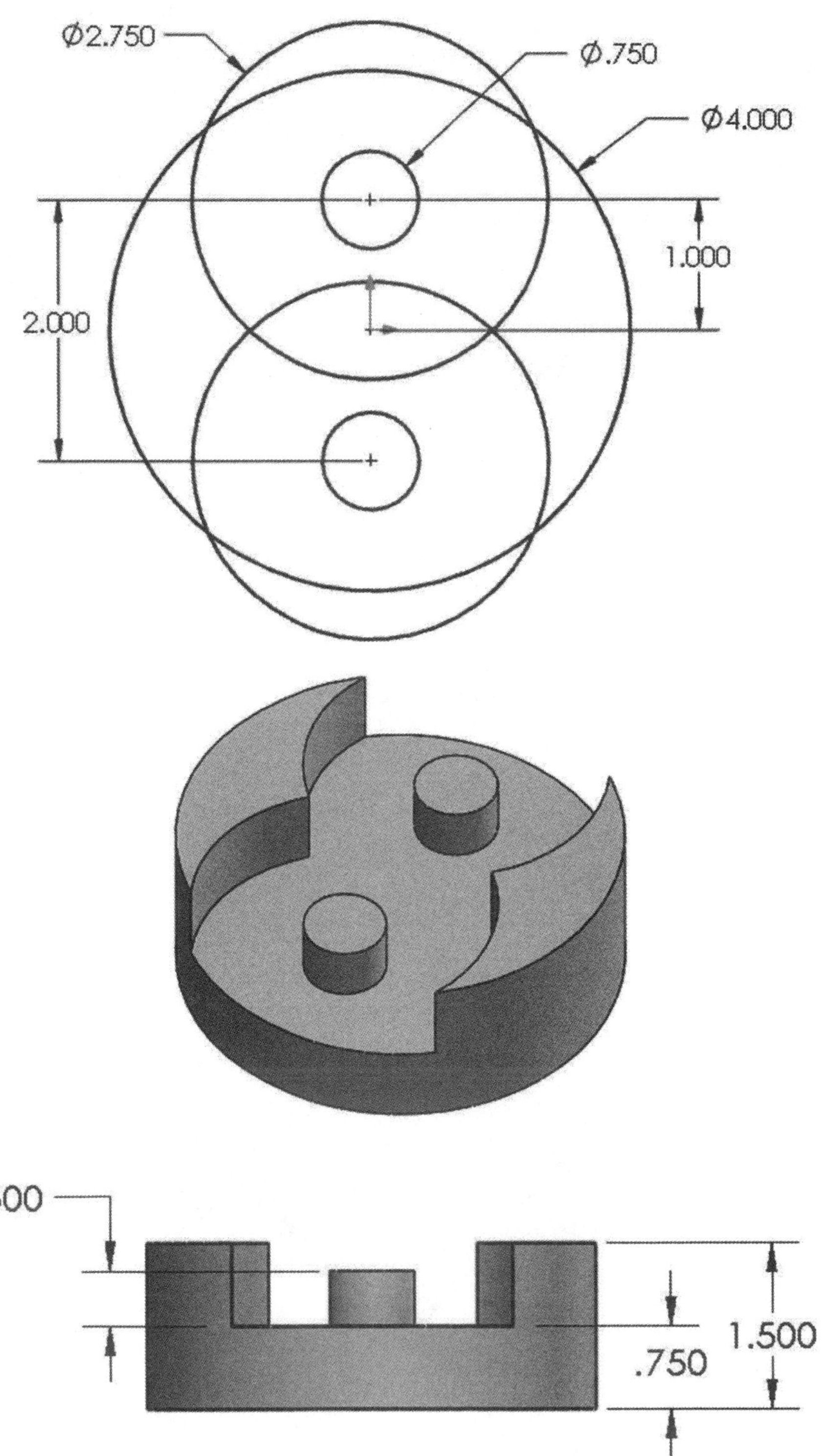

Part Editing

A very important skill to have when modeling in SOLIDWORKS (or any CAD package for that matter) is to be able to edit a model to change it and/or fix errors. Let's face it: the only constant in design is change, and when we make changes to our model, chances are we may cause errors down the road, that is, in the FeatureManager. For example, if we have a part with round edges (fillets) and we change a previous feature to eliminate an edge, the fillet will give us an error because it cannot find the edge. For the most part those are the types of errors we are talking about.

5.1. - To practice editing models and fixing errors, open the part *'Repair.sldprt'* from the exercise files. When we finish editing and fixing the part, it will look like this:

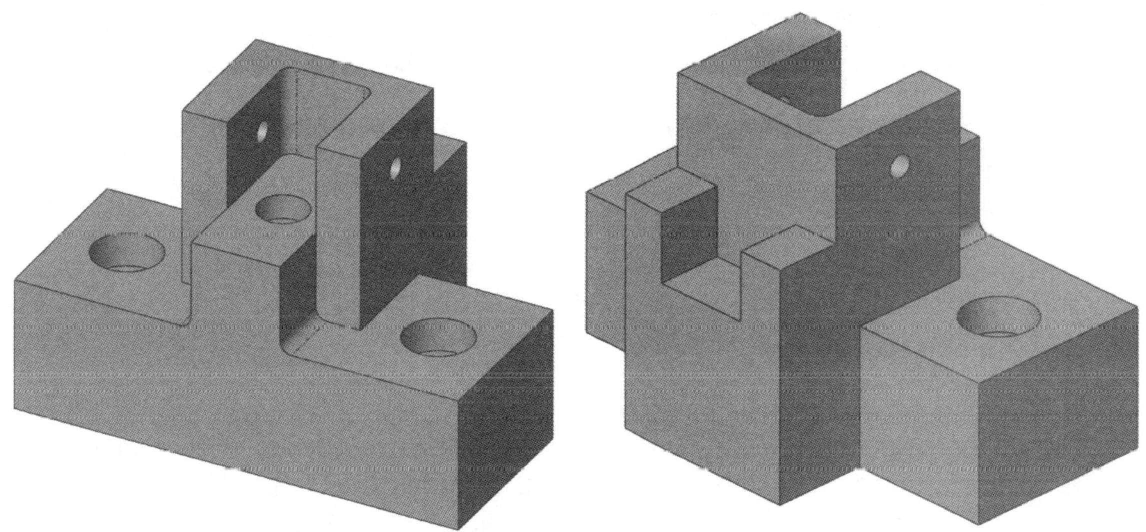

When we open the file we are asked if we want to rebuild it. Select "Rebuild" from the dialog to continue.

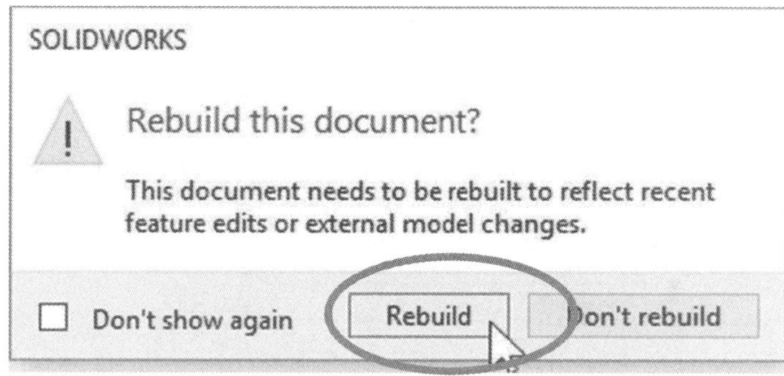

5.2. - When the model is rebuilt we see a list of errors and no geometry at all. The part has so many errors that no geometry can be generated, and the "**What's Wrong**" dialog box contains the full list of things that need to be fixed. Click the "Close" button for now. Since features are added chronologically starting at the top in the FeatureManager, the logical order to start fixing errors is from the top and work our way down.

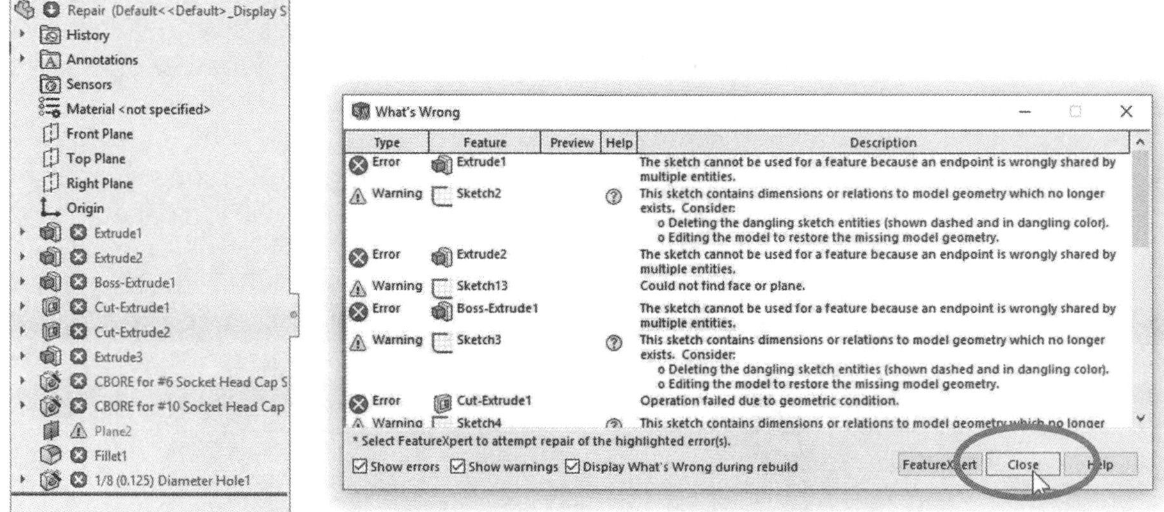

There are two types of errors: a red X means the feature failed to build; the yellow warning triangle means the feature has an error, but SOLIDWORKS was still able to build it.

The reasoning behind this order is that if a 'parent' feature has an error, it *may* cause problems in a 'child' feature; therefore, the dependency between features is referred to as "**Parent/Child**" relations. For example, if a sketch is added to a face of another feature, or a dimension references another feature's geometry, a "**Parent/Child**" relation is generated. To identify these relations, we can select a feature and look at the arrows indicating the relationships, or right-mouse-click a feature and select "**Parent/Child.**"

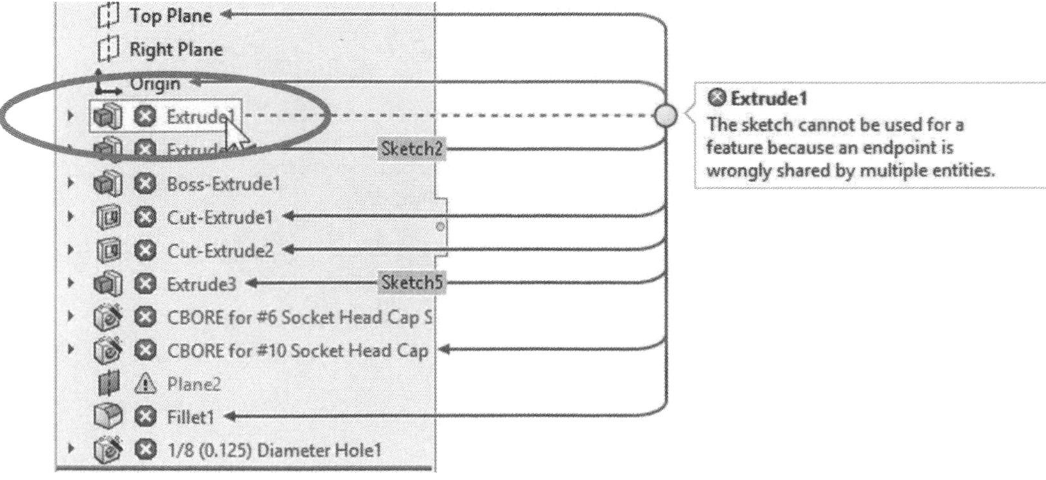

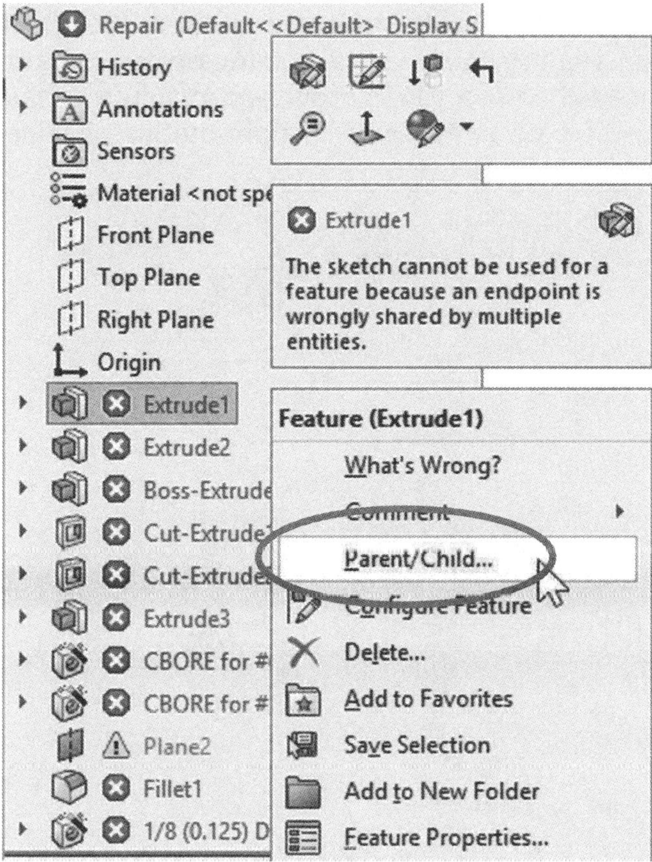

5.3. - Here we can see the features that *"Extrude1"* depends on (Parents), and which features depend on it (Children). The higher we go in the FeatureManager, the more Children a feature may have. Click "Close" to continue.

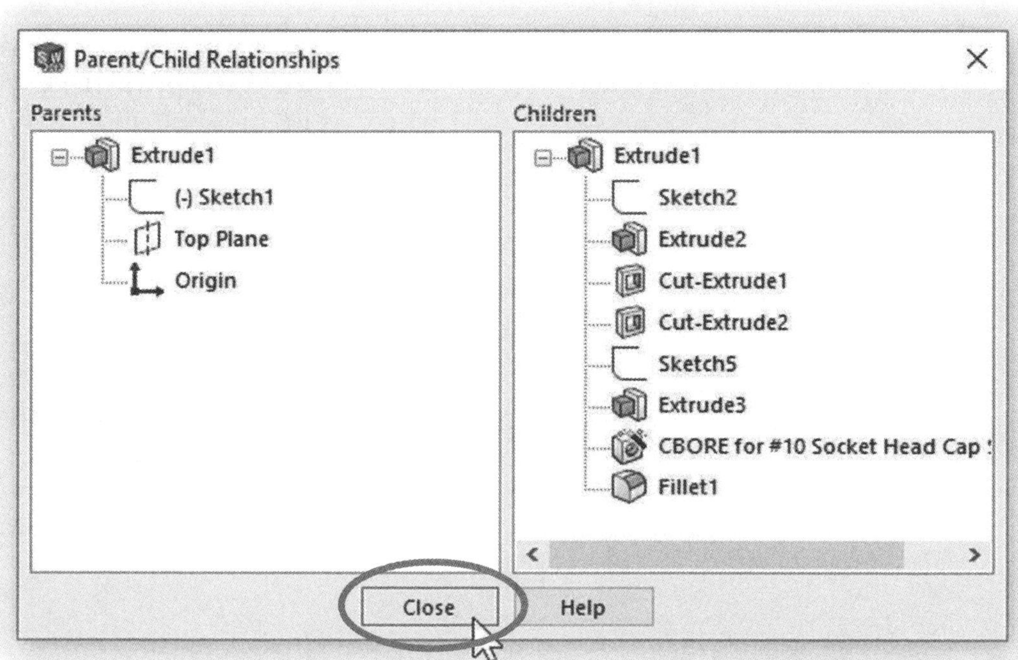

5.4. - A visual way to identify parent features is using the "**Dynamic Reference Visualization**" option resting the mouse pointer on top of a feature; its parent features will be indicated with a blue arrow and children features with a purple arrow. This option can be turned On/Off by right-mouse-clicking at the top of the FeatureManager.

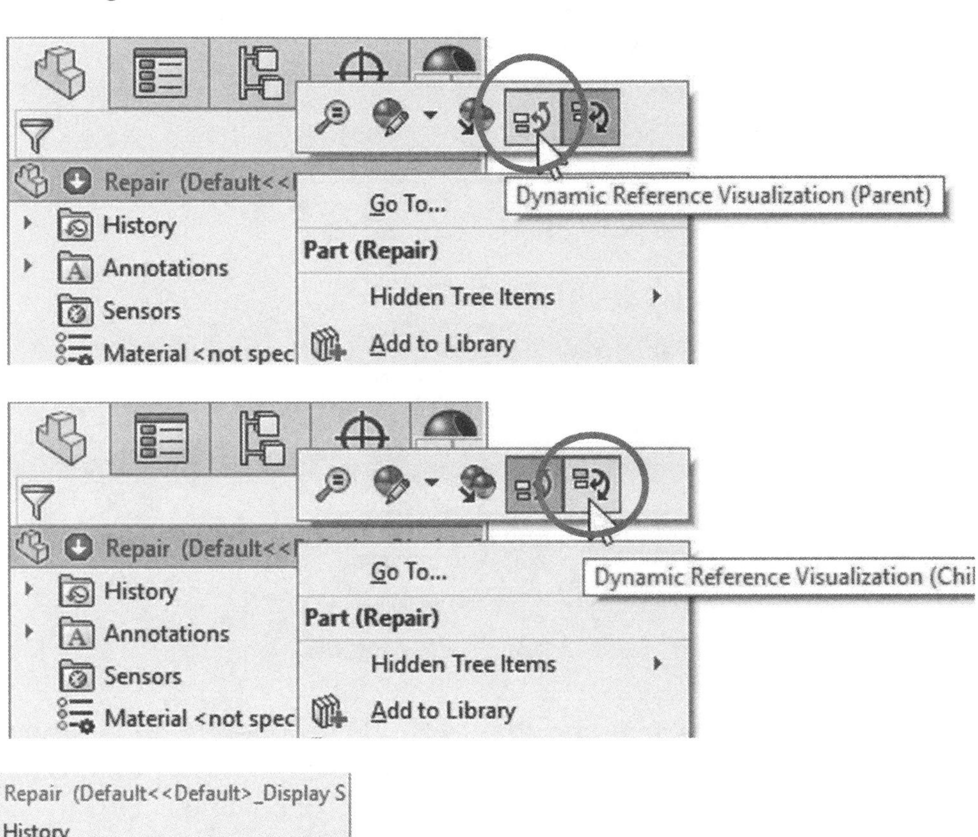

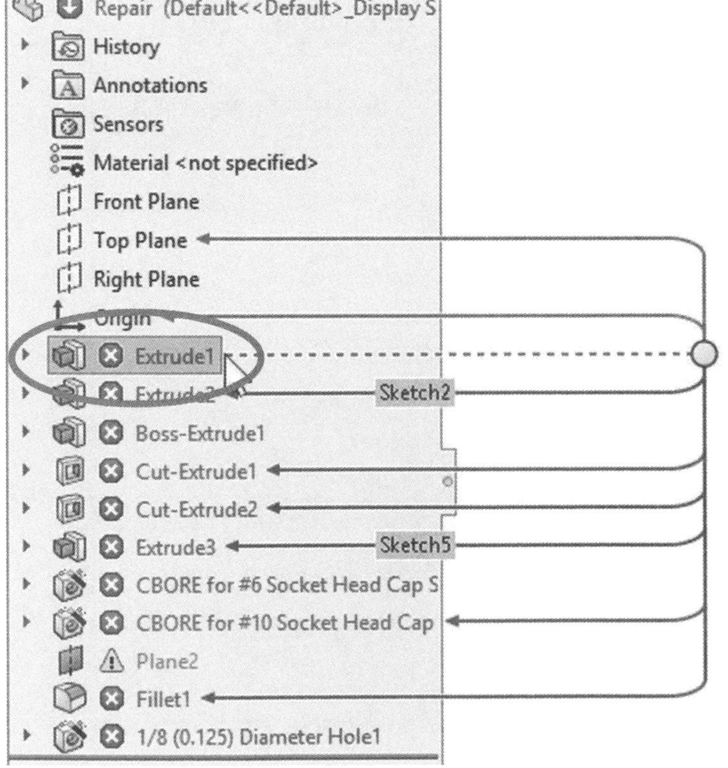

Parent features are indicated with blue arrows above the selected feature.

Child features are marked with purple arrows below the selected feature.

5.5. – By looking at the image above, we can see *"Boss-Extrude1"* has many child features; therefore, fixing this error will be our first step. By selecting the feature, we can see a preview of the error message, or we can right-mouse-click *"Boss-Extrude1"* and select **"What's Wrong?"** to see the message and in some instances, possible options to correct it.

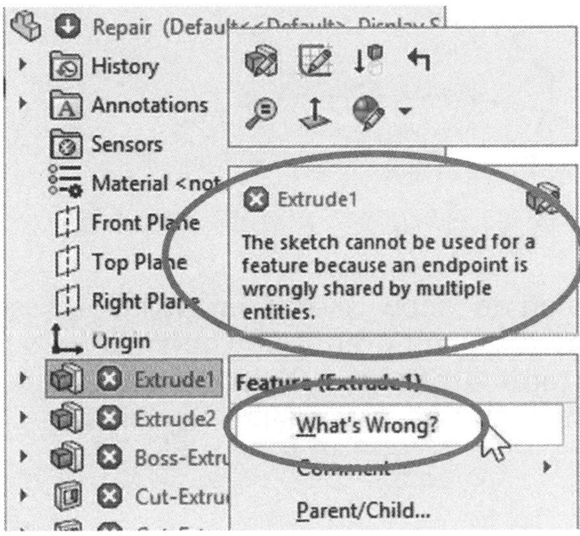

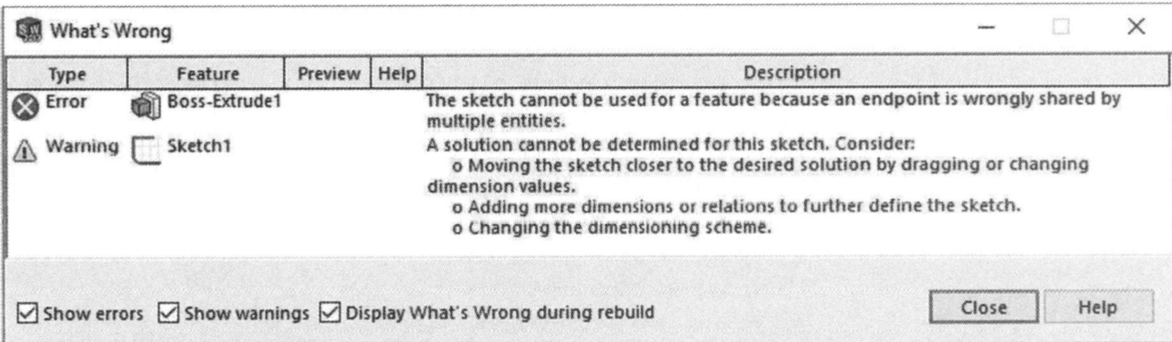

This error means that the **sketch** has intersecting lines and/or two or more lines are connected to the same endpoint. One common cause of this problem is when sketching we accidentally add overlapping lines. Close the "What's Wrong" dialog and edit the *"Boss-Extrude1"* sketch.

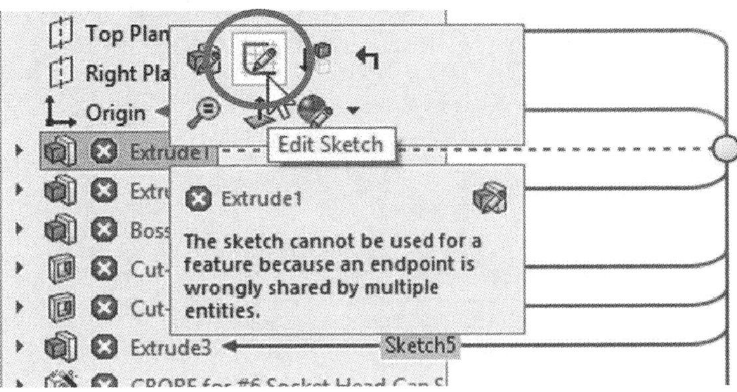

5.6. - Sometimes it's easy to see the geometric elements causing the problem in a sketch and we can correct it, but sometimes it's not that obvious. To help us identify the problem, select the menu "**Tools, Sketch Tools, Check Sketch for Feature**."

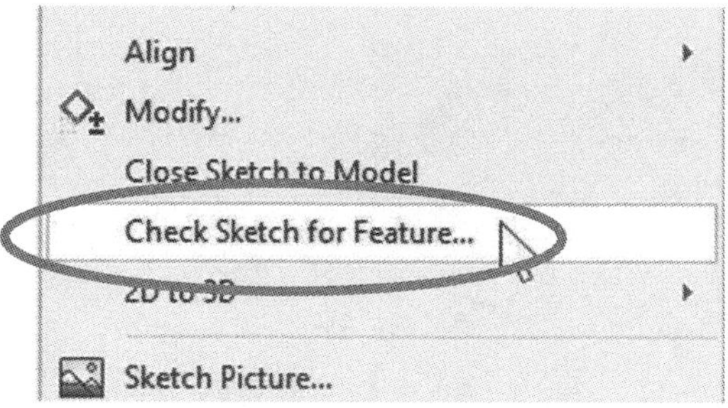

Setting the option to show sketch endpoints is usually a good idea, especially when looking for overlapping geometry. Go to the menu "**Tools, Options, System Options, Sketch, Display entity points in part/assembly sketches**" to identify overlapping geometry.

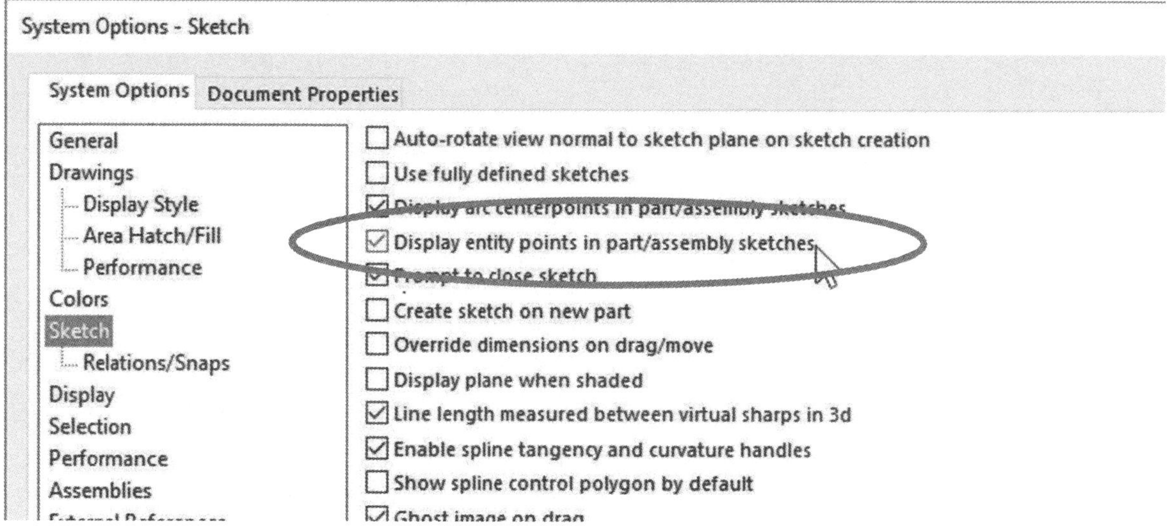

If the sketch has already been used for a feature, the feature type will be pre-selected in the drop-down menu. If the sketch has not been used for a feature, we have to select the type of feature that we intend to use it for. In our case "Base Extrude" is pre-selected. Click on "Check" to analyze the sketch.

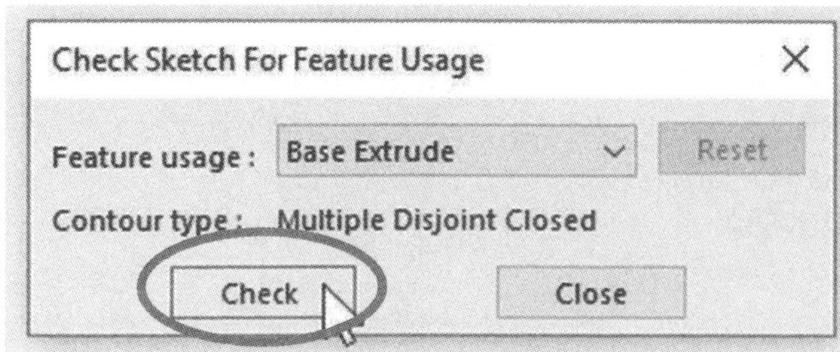

We are immediately presented with the same error message that we got using the "What's Wrong" command. Click OK to dismiss it and continue.

5.7. – After acknowledging the error message, SOLIDWORKS re-orients the sketch and places the magnifying glass on top of the geometric element suspected of causing the problem. We are given two areas of concern: the first problem is that we have overlapping entities, and the second is that multiple (2 or more) elements share the same endpoint.

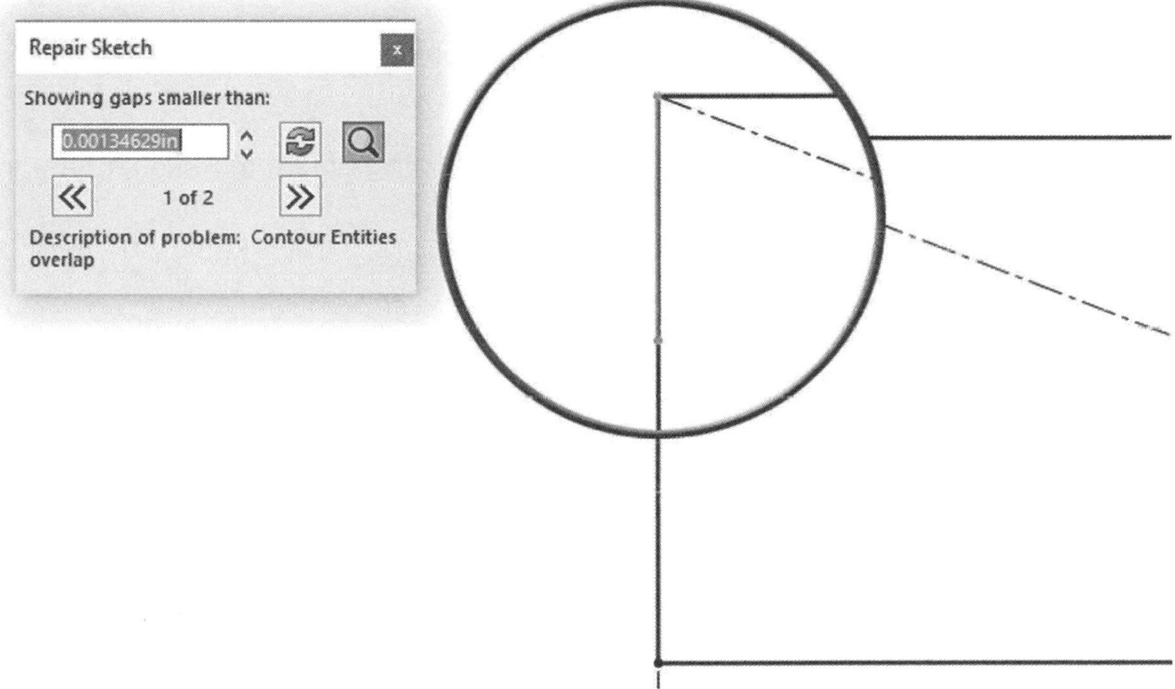

The advantage of using the "**Check Sketch for Feature**" command is that we can quickly identify where the problem is and correct it. If an identified line is not part of our design, usually it can be safely deleted. Window-select from left to right to select everything enclosed by the window; if an extra line is selected, delete it. Click "**Refresh**" to confirm that we don't have any more problems and close the "Repair Sketch" window.

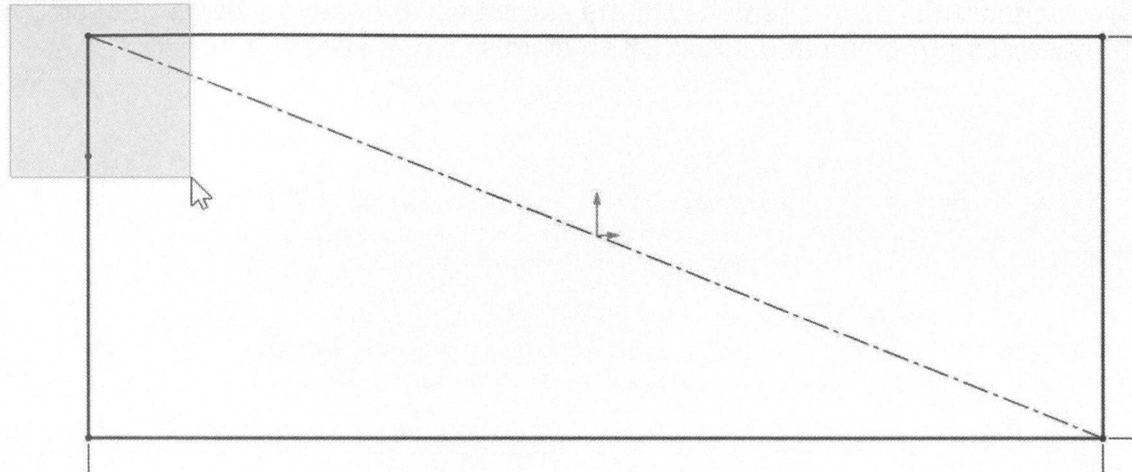

Window-selecting from left to right selects geometry completely enclosed by the window; window-selecting from right to left selects geometry crossed by the window.

If needed, turn off the magnifying lens to delete the extra geometry.

Run the "**Check Sketch for Feature**" again to confirm the sketch has no more problems and it can be used for a feature.

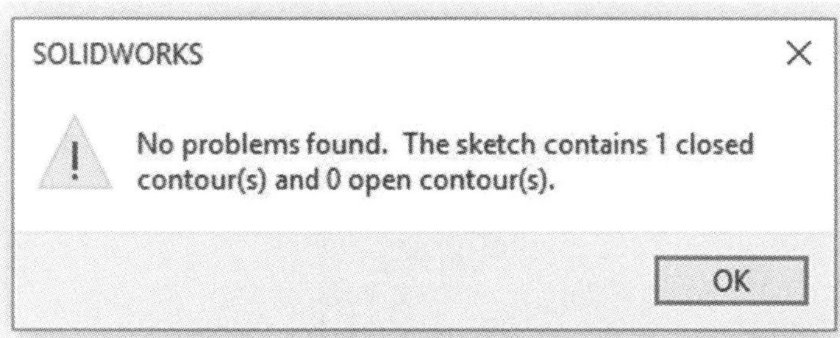

5.8. - Exit the sketch (or rebuild the model) to continue. We are warned another feature has an error and we are asked if we wish to repair it now or continue with the error. Click "Continue (Ignore Error)" and close the "**What's Wrong?**" message.

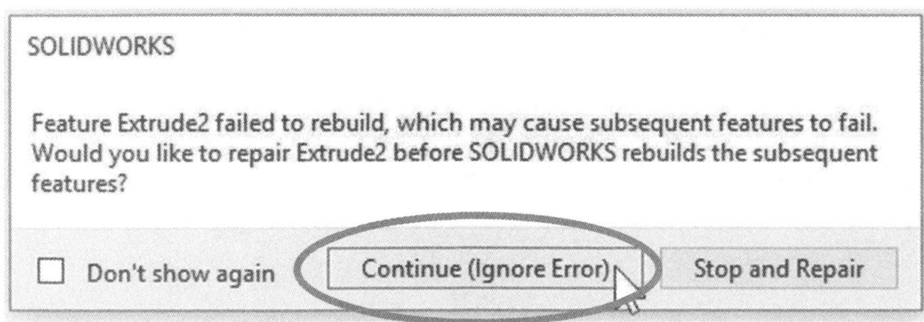

The part still has errors, but now we can see some features and (more importantly) the error from the first feature is fixed.

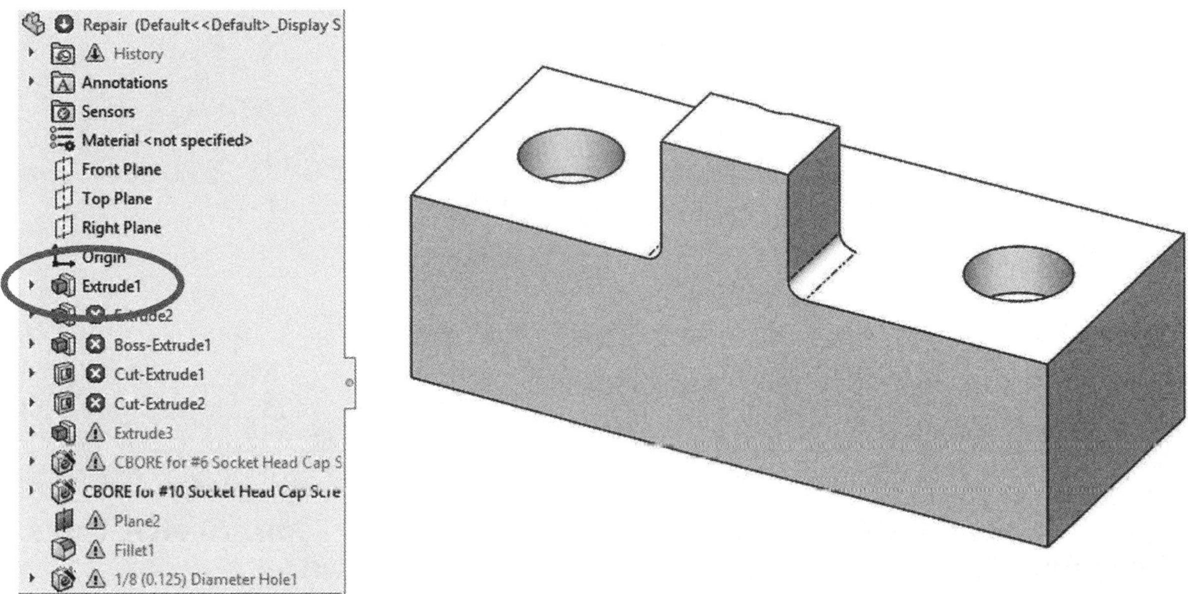

5.9. - Select the "Extrude2" feature, right-mouse-click and select "**What's Wrong?**" to view the error as before, *OR* select it to see the error in the pop-up bubble. Essentially we have the same problem as the first feature. Edit the "Extrude2" sketch to fix it.

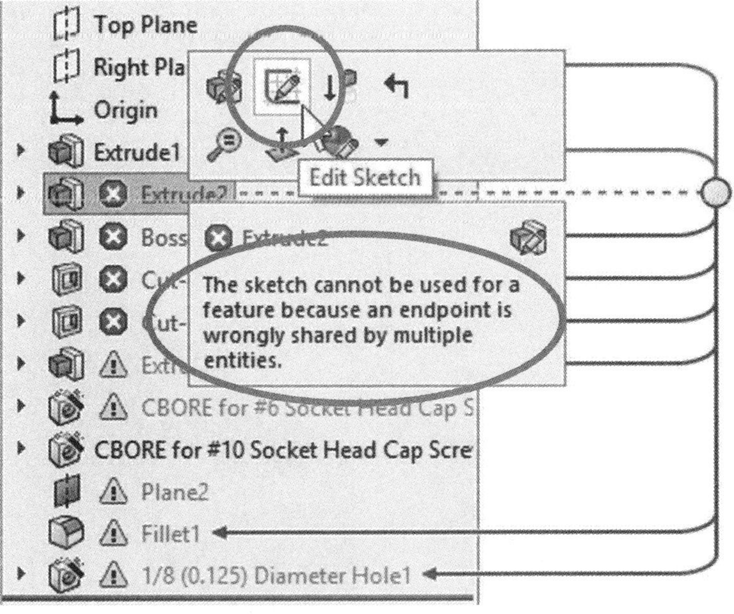

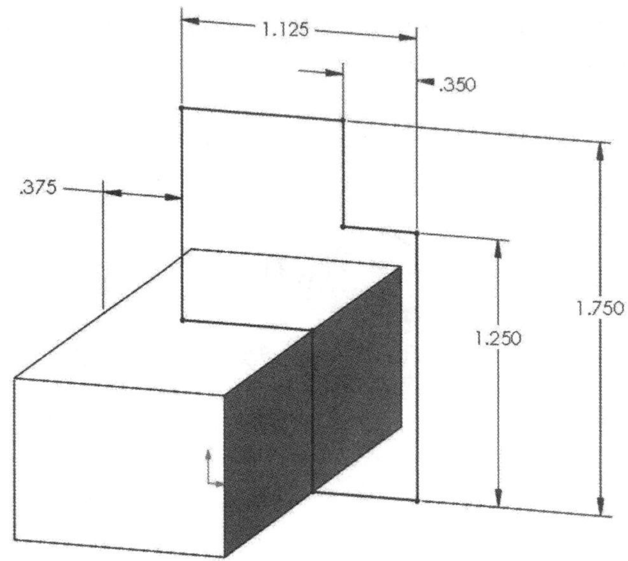

5.10. - Once we are editing the sketch, select the menu "**Tools, Sketch Tools, Check Sketch for Feature**" to find out what is the problem in the sketch. Dismiss the error message to continue.

In this case, looking at the three problems listed in the "Repair Sketch" dialog (one overlapping entity and two with more than two entities at an endpoint), the most likely problem is two identical lines overlapping. The entire line is selected, and an orange circle is indicating the extra overlapping line. Close the "Repair Sketch" dialog and delete one of the lines.

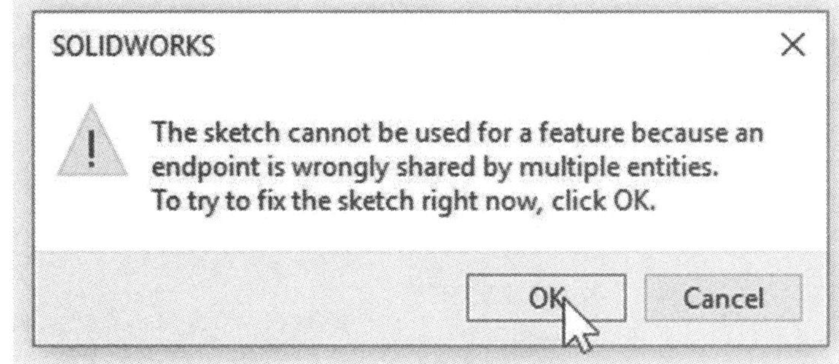

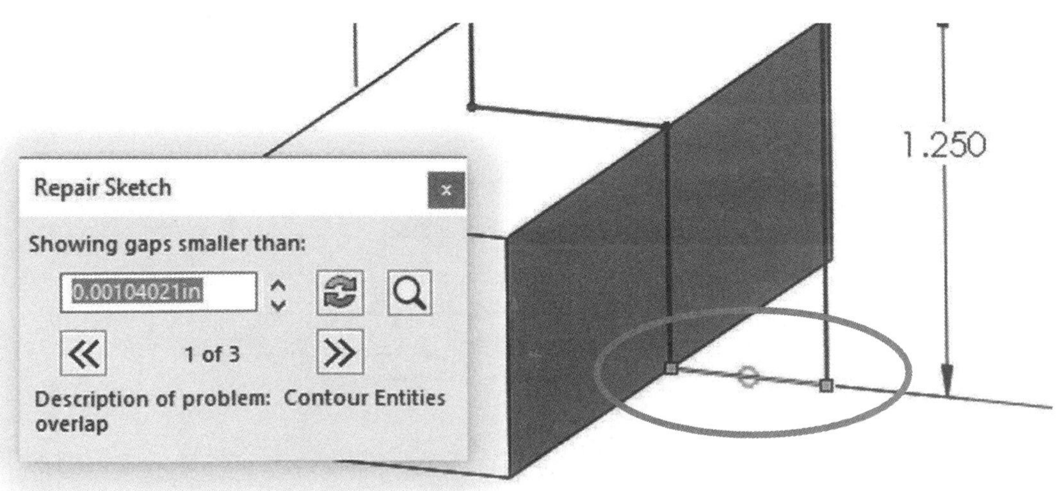

5.11. - If we use window-selection both lines will be selected, and we only need to select one. Click to select a line and then delete it. When a line with dimensions attached to it is deleted, these dimensions will also be deleted. After deleting it, we may be warned that other entities will also be deleted, most likely dimensions attached to the deleted line. If this is the case, click "Yes" to continue and delete the line and its dimensions.

If deleting the selected line also deletes a dimension, the top horizontal line will be blue (undefined) because it was referencing the deleted line. Add the missing dimension to fully define the sketch and rebuild the part. Select "Continue (Ignore Error)" after we exit the sketch and dismiss the "**What's Wrong?**" message.

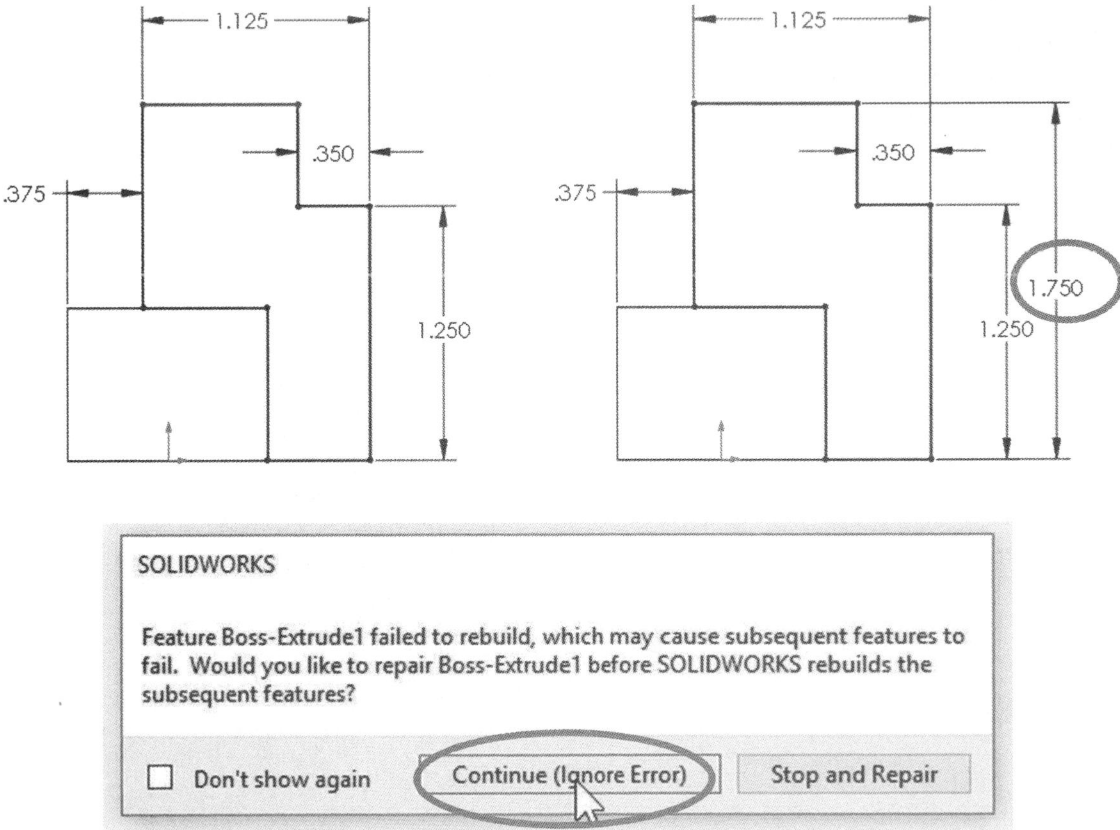

After fixing two problems, we have cleared multiple errors in our part, better illustrating the importance of understanding parent/child relations.

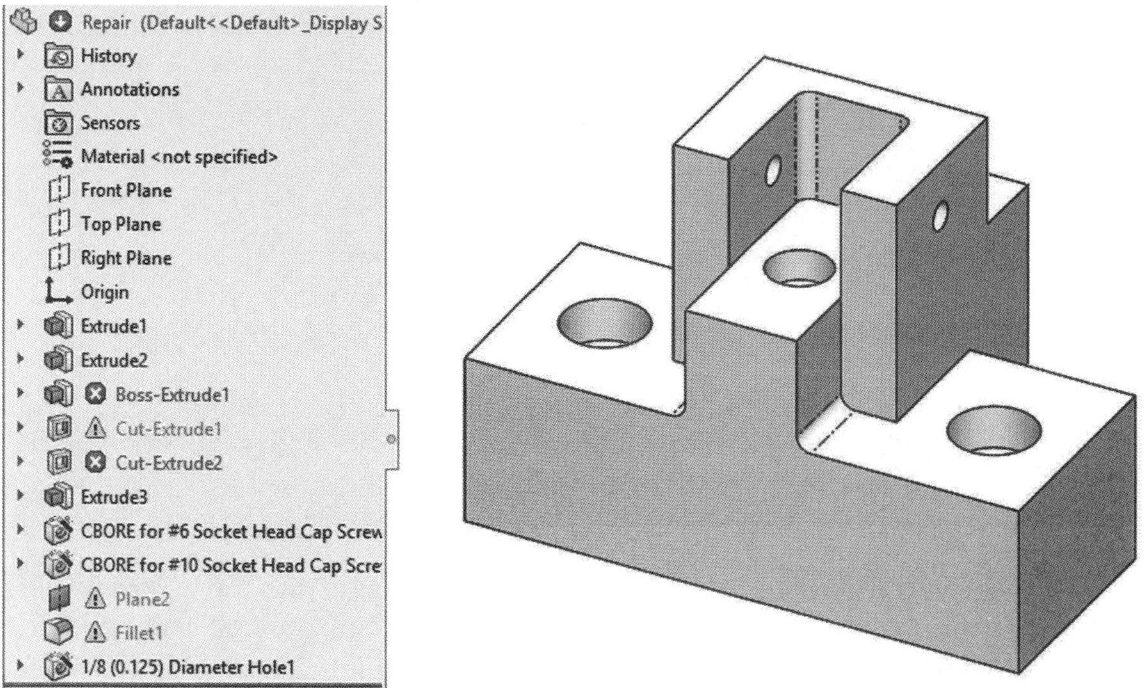

5.12. - Reviewing the error in the "*Boss-Extrude1*" feature we see the same error as the two previous features. After editing the sketch we can see a diagonal line. We can either delete it or convert it to **Construction Geometry**. The second option is usually safer, as we could lose dimensions and/or relations as in the previous step if we delete it; also, if needed, this change can be reversed. Switch to a Top view; edit the sketch, select the diagonal line and convert it to construction geometry from the pop-up toolbar.

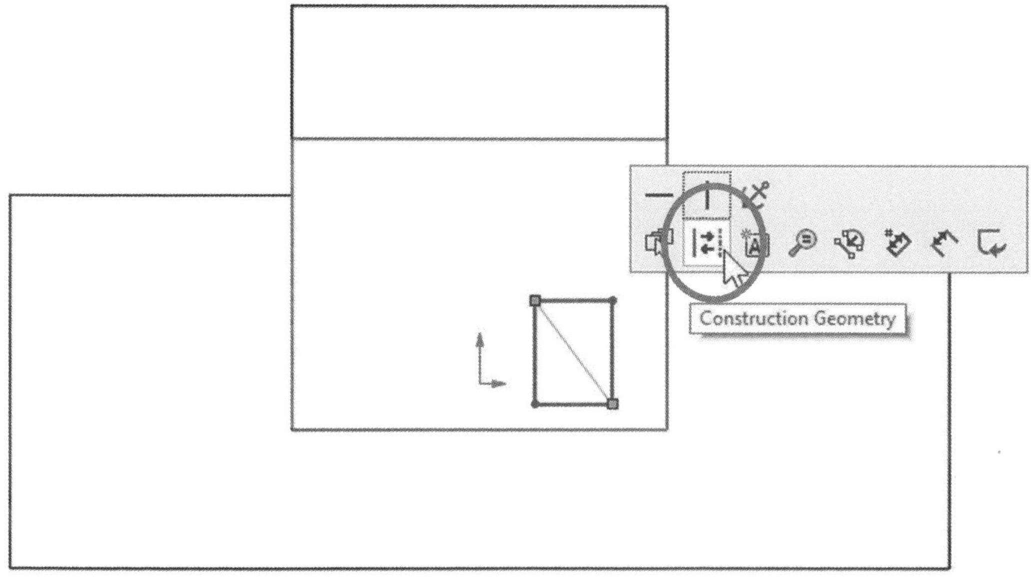

A different way to change a line (or any geometric element) to construction geometry is to turn On the “For construction” checkbox in the Property Manager after selecting it.

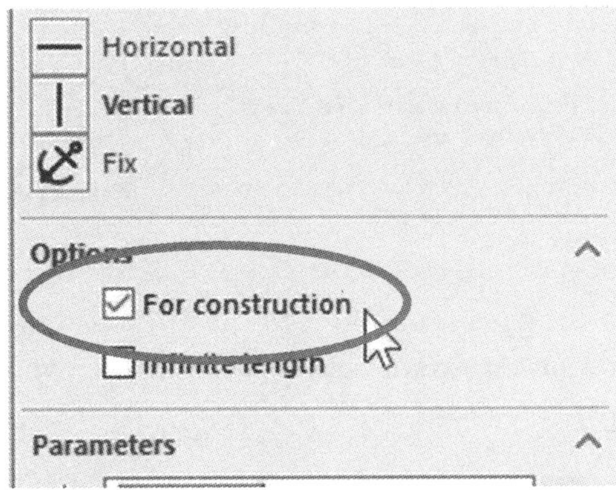

5.13. - Exit the sketch to rebuild the part and dismiss the error dialog to continue. Since the newly created feature does not seem to be part of the original design, the next logical step would be to delete it. Right-mouse-click in the feature, and after selecting “**Delete**” we get a “Confirm Delete” dialog:

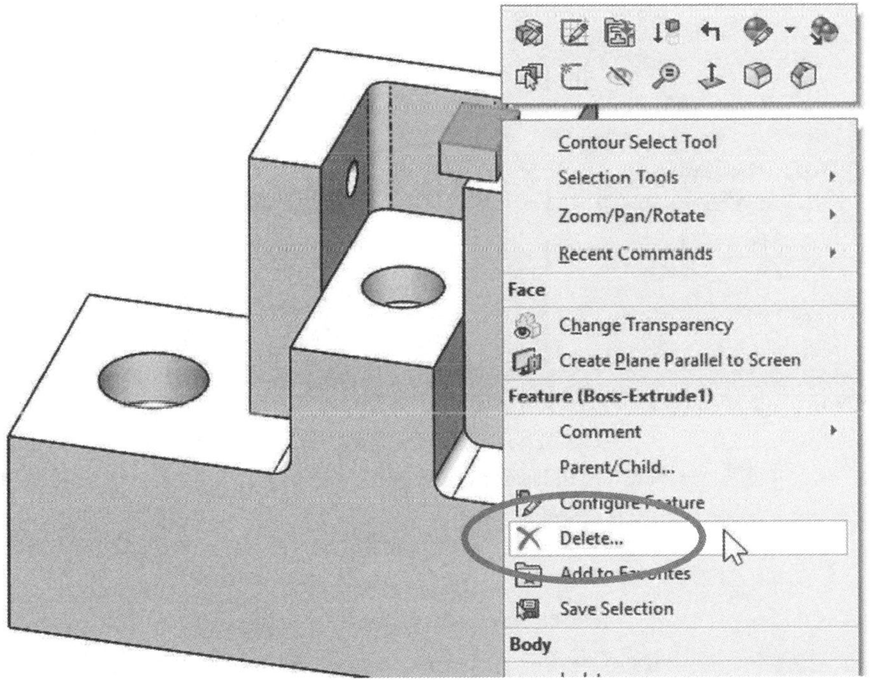

In the Delete confirmation dialog, we can see that deleting this feature would also delete its dependent features, all of which need to be in our part. The reason those features would also be deleted is because they are “children” features of *“Boss-Extrude1.”* Activating the “Delete child features” checkbox shows all the features that would be affected. Select “Cancel.” Since we don’t want to delete any of these features, we’ll edit those features to remove the dependencies *and then* delete this extra feature.

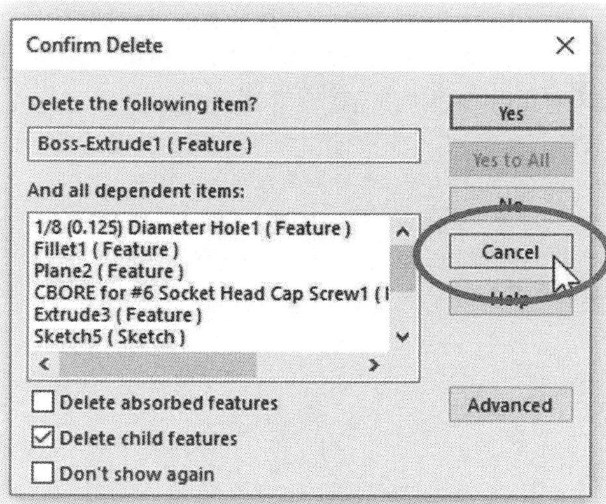

5.14. – Right-Mouse-Click *"Boss-Extrude1"* and select the "**Parent/Child**" command from the pop-up menu.

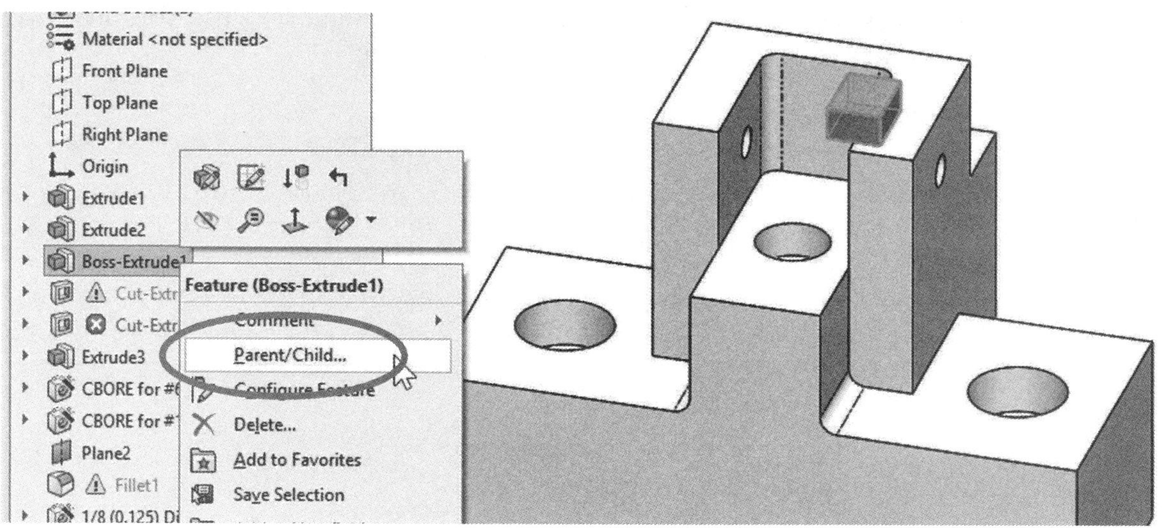

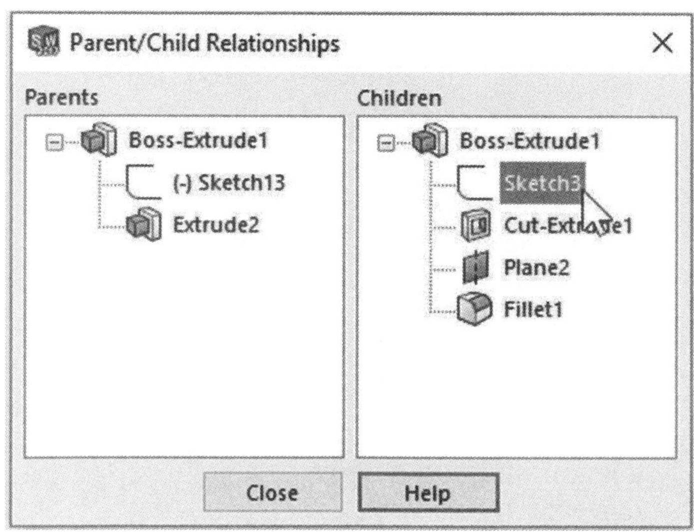

5.15. - The first dependent feature in the Children list is *"Sketch3"* (from the *"Cut-Extrude1"* feature). Close the "**Parent/Child Relationships**" window, select *"Sketch3"* in the FeatureManager, and edit it. The error in this sketch says that a sketch element(s) or dimension(s) are referencing geometry that no longer is there; therefore, we need to fix it.

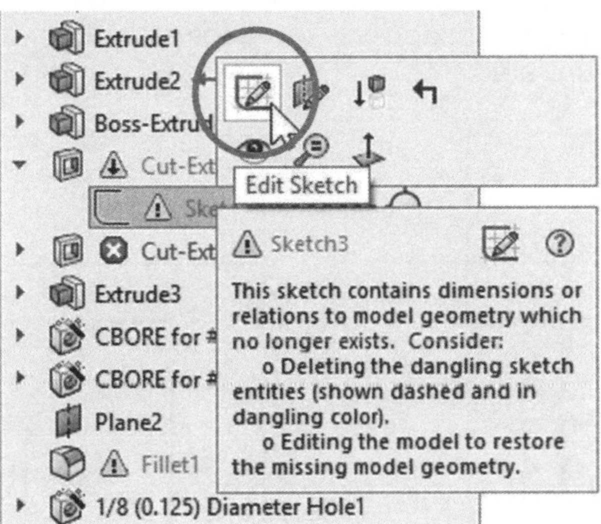

5.16. - Change to a Top view and "Hidden Lines Removed" mode for clarity. We can see a dimension colored in brown (0.603). The brown colored dimension (default color settings) tells us it is ***'dangling'***, which means it is referencing geometry that no longer exists; in other words, the dimension is attached to something that is no longer in the model. What we need to do is re-define what the dimension is referencing to fix the error.

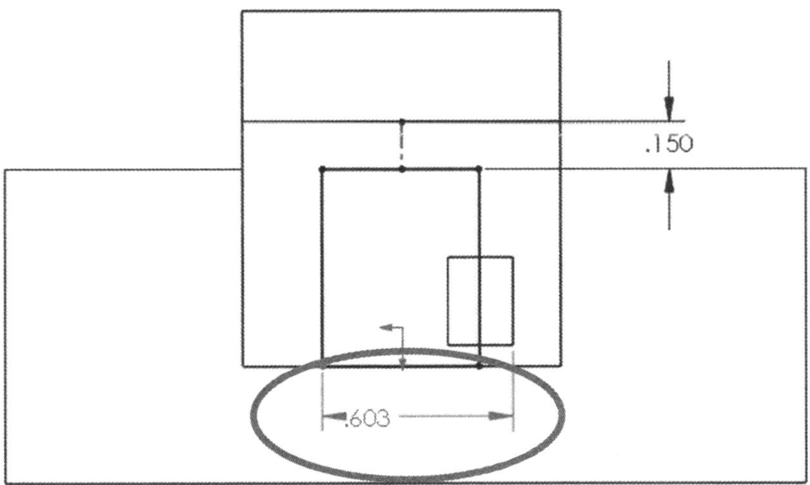

There are two ways to fix this error:

a) Delete the dimension and add it again to a valid reference, or
b) Re-attach the dimension to a valid reference.

Deleting the dimension is straightforward and often a good solution, so we'll talk about the second option. After selecting the dimension, we see a witness line ending with a red dot; this is the witness line missing a reference. It *may* look as if it was referencing the existing edge, but in fact it's referencing geometry that is no longer in the model. To re-attach it, **click & drag the red dot** onto a valid reference; it can be any model's edge or vertex. For this dimension, we'll use the right side edge. After re-attaching the dimension the geometry does not change, only the dimension's value is updated.

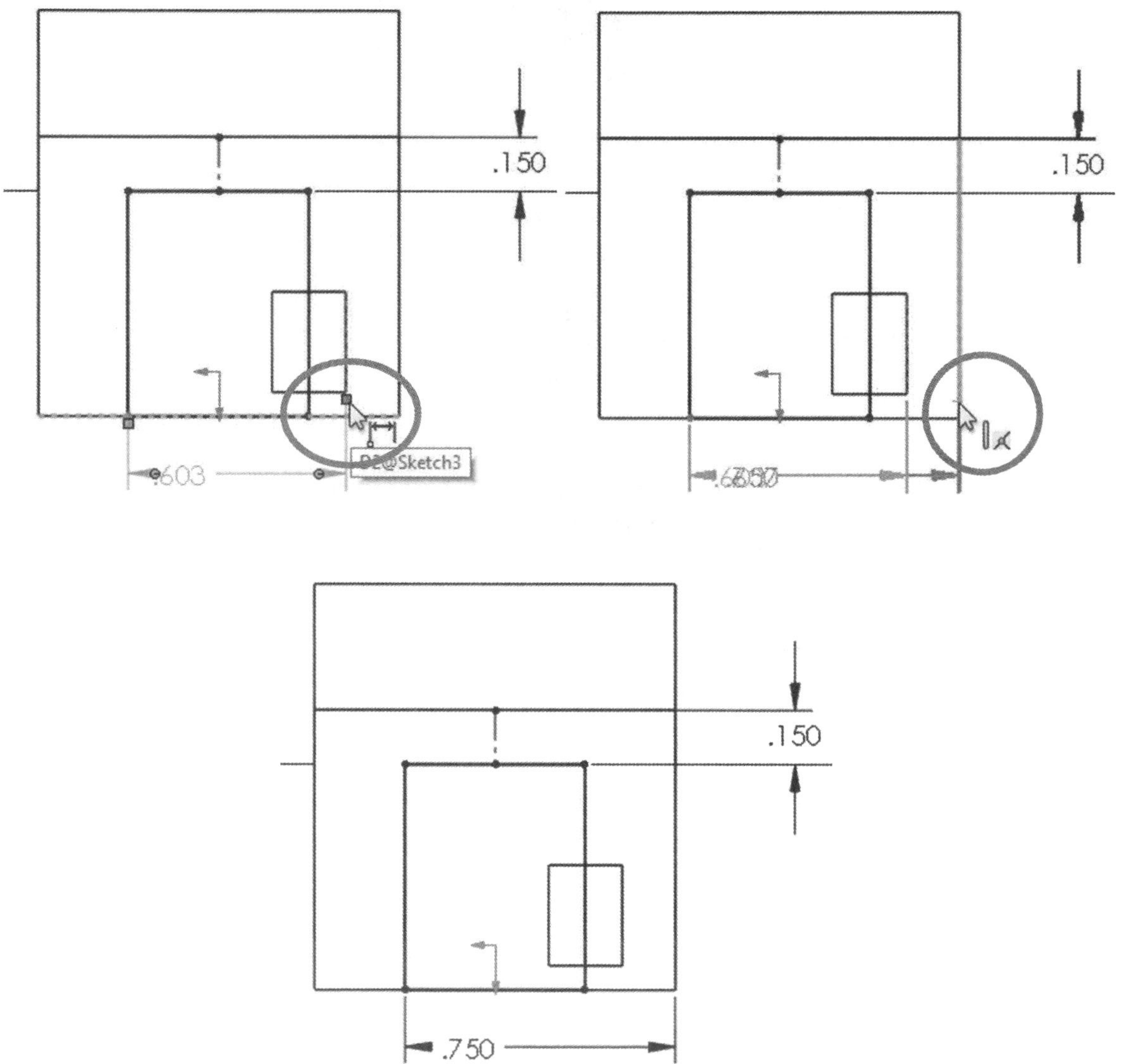

5.17. – Exit the sketch to rebuild the model and dismiss the "**What's Wrong**" dialog. The error in "*Sketch3*" is fixed, and after reviewing the "*Boss-Extrude1*" "**Parent/Child**" relationships *"Sketch3"* and *"Cut-Extrude1"* are no longer listed as children features of *"Boss-Extrude1"* because the dimension was changed to reference a different edge, effectively breaking the relation and fixing the error at the same time. From the relationships dialog, we can see that the only children left are *"Plane2"* and *"Fillet1."*

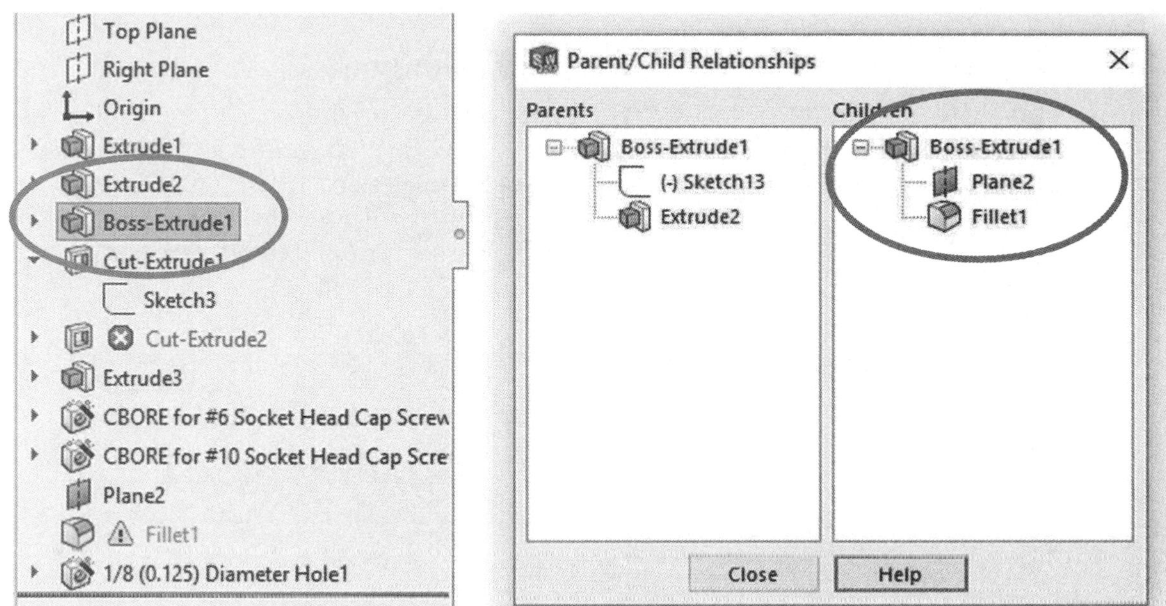

5.18. – Reviewing the "**Parent/Child**" relations of *"Plane2"* the only child feature is the *"1/8 (0.125) Diameter Hole1."* A **Hole Wizard** feature creates two sketches when added. The first sketch locates the hole's location(s) and is added in the plane/face where the hole is made; the second sketch is a profile for a revolved cut to create the hole, perpendicular to the first sketch.

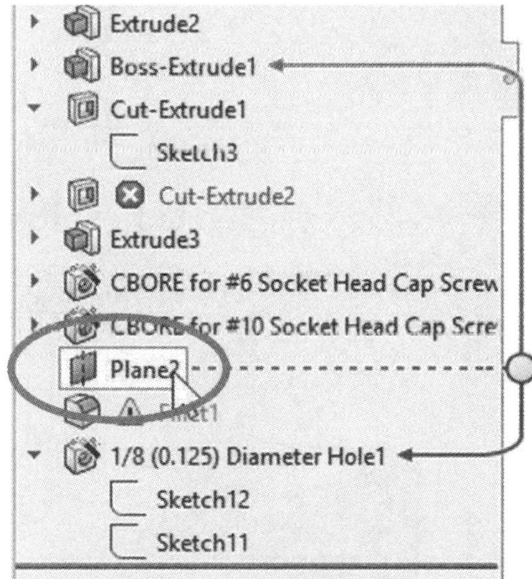

The most likely reason why the *"1/8 (0.125) Diameter Hole1"* feature is a child feature of *"Plane2"* is the hole wizard feature was located in *"Plane2."* We can find out which plane or face a sketch was added to by editing its sketch plane.

To find out, expand the Hole Wizard feature, select the first sketch (*"Sketch12"*) (the hole's location sketch), and from the pop-up menu select "**Edit Sketch Plane.**" In the "Sketch Plane" command we can see *"Plane2"* is selected, which confirms the sketch is located in "*Plane2.*"

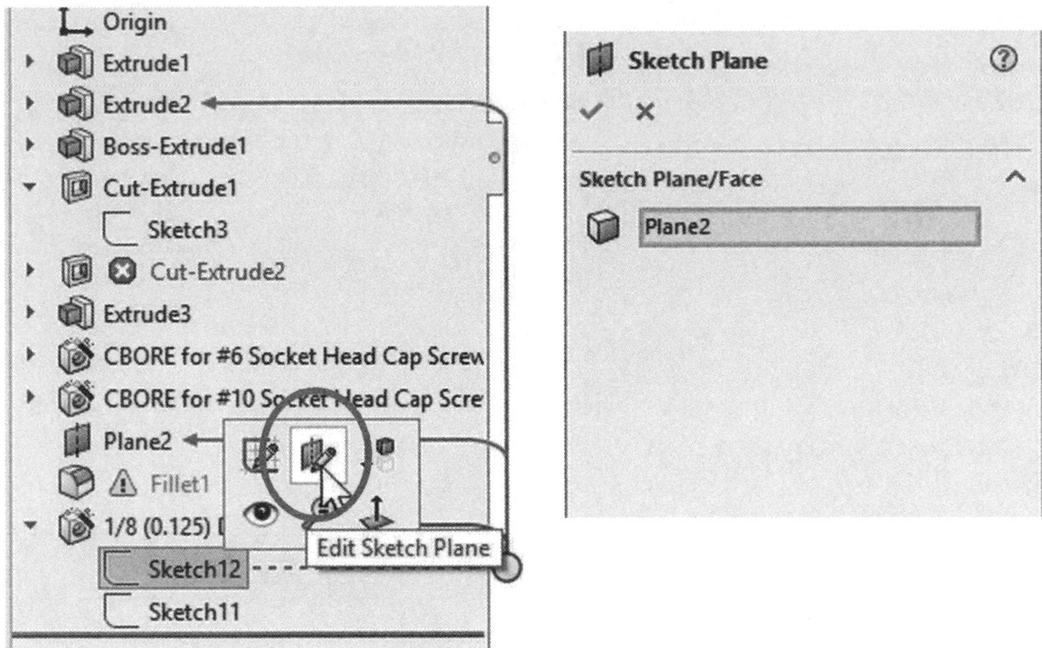

To change the sketch to a different plane (or face) and delete the relationship to *"Plane2"* (and by extension to*"Boss-Extrude1"*), select a different face (or plane) to locate the sketch. In "Sketch Plane" select the face indicated and click OK to finish. Close the "**What's Wrong?**" dialog to continue.

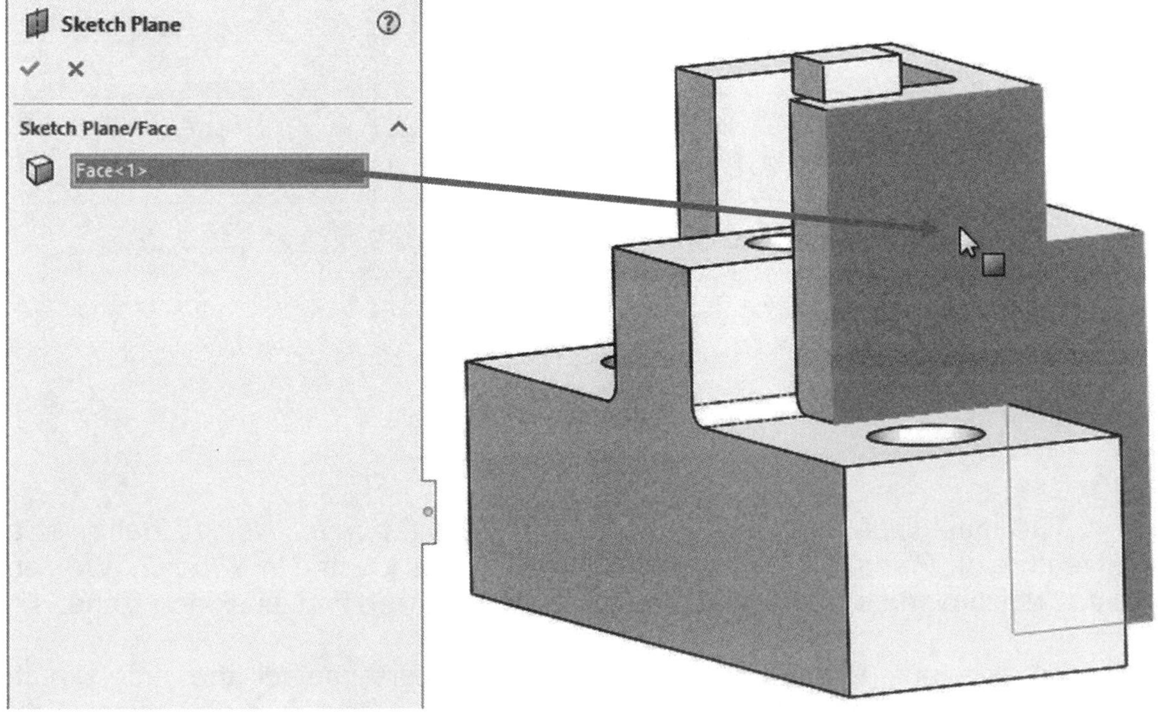

5.19. - Review the "**Parent/Child**" relationships again for *"Boss-Extrude1"* and *"Plane2."* Now the only children of *"Boss-Extrude1"* are *"Plane2"* and *"Fillet1,"* and *"Plane2"* has no children. Close the dialog to continue.

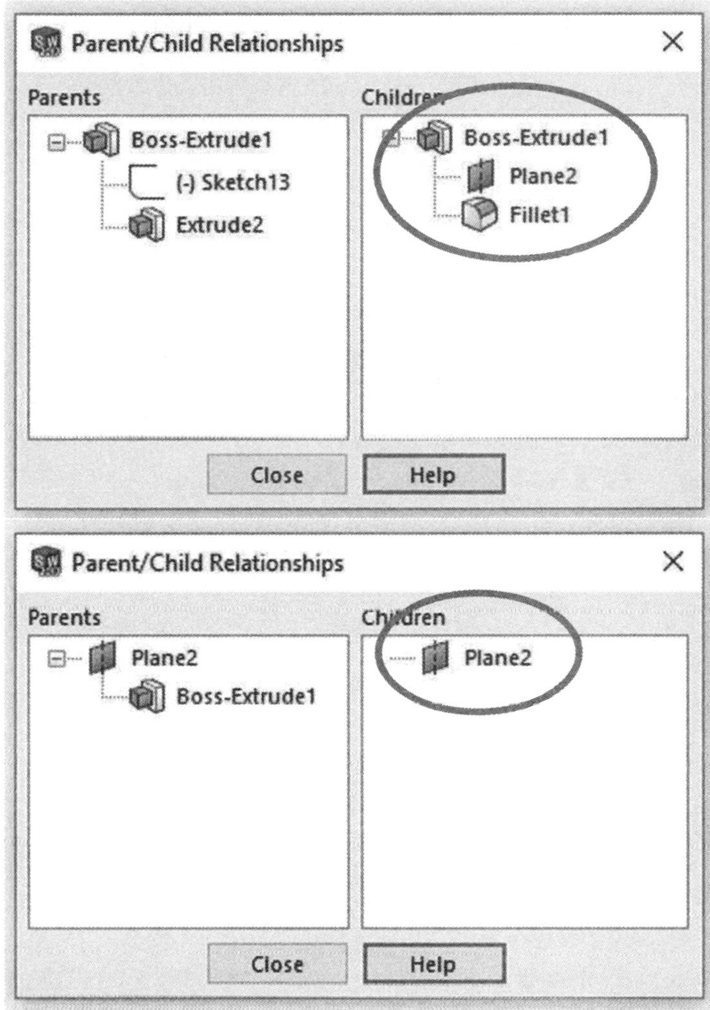

5.20. - Select *"Fillet1"* and check "**What's Wrong**?" Note that the icon is different; in this case the error is a warning. A warning means the feature was built but has an error; in this case the fillet was built but it's missing one or more edges.

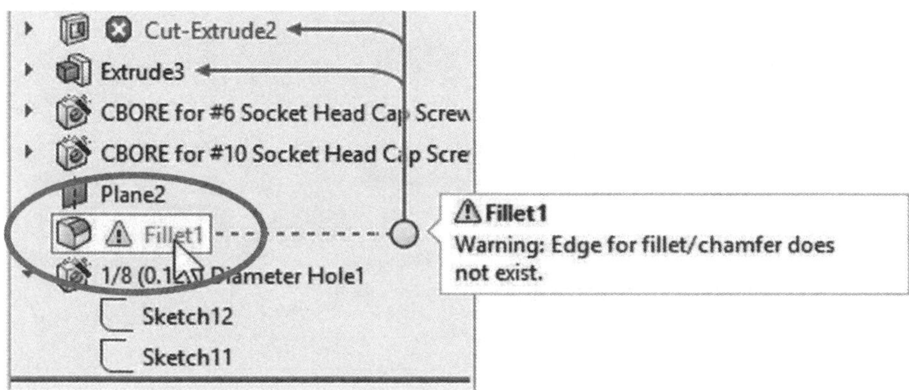

5.21. - It was mentioned that it is usually best to correct errors starting from top to bottom, in this case we can fix this error before the previous one. To fix the fillet, edit the "*Fillet1*" feature. When we edit the fillet, we get a message letting us know the fillet is missing items; this means that some edges that had been previously rounded are no longer in the model and the fillet could not find them.

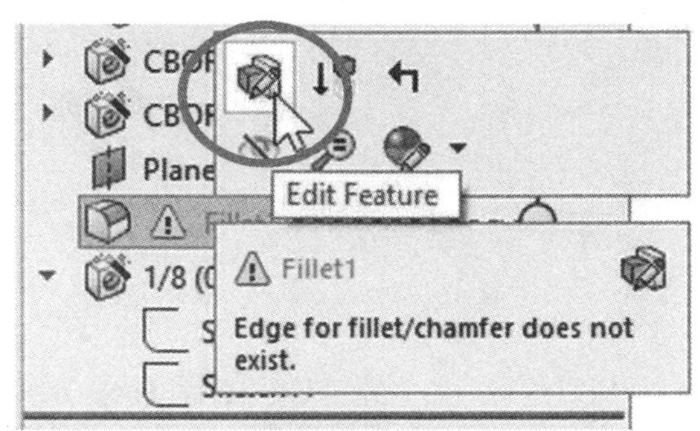

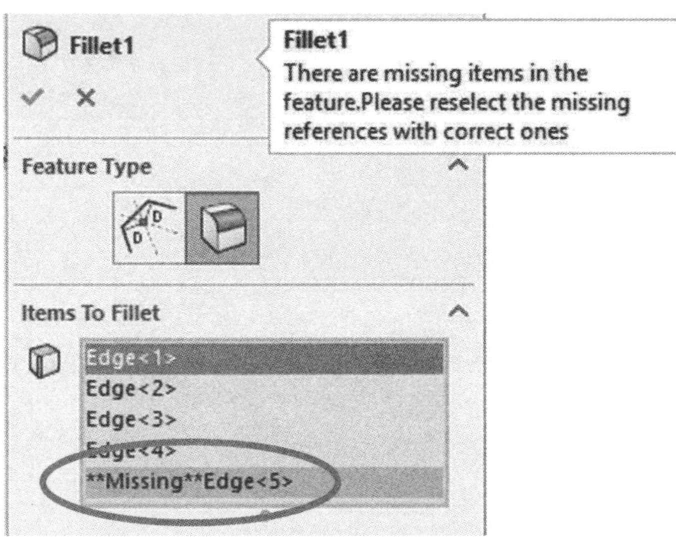

When we edit the fillet, the missing item(s) we were warned about are listed as "****Missing*****item***" in the "Items To Fillet" selection box. In this case we have a couple of options:

- Select the missing item in the list, delete it and click OK to continue.

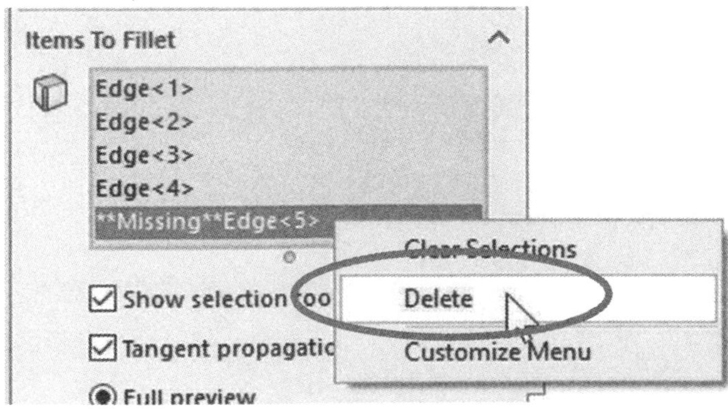

- Or simply select OK to continue, in this case we get a warning alerting us that the missing edges will be automatically removed. Click "Yes" to continue.

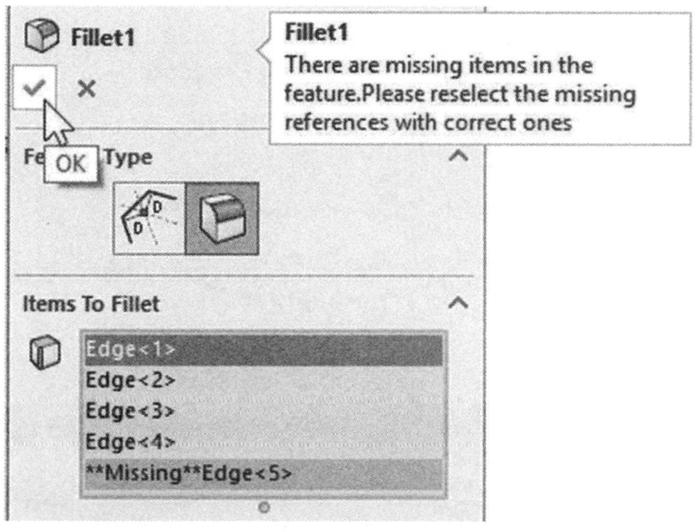

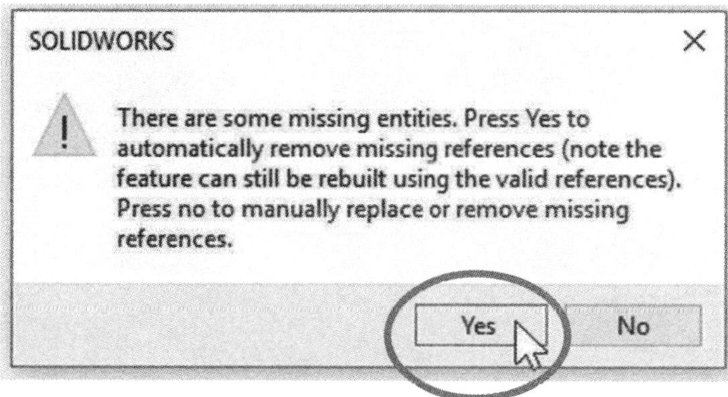

5.22. – After reviewing "**Parent/Child**" relations for *"Boss-Extrude1"* we can verify the only child feature is *"Plane2,"* and it is now safe to delete it.

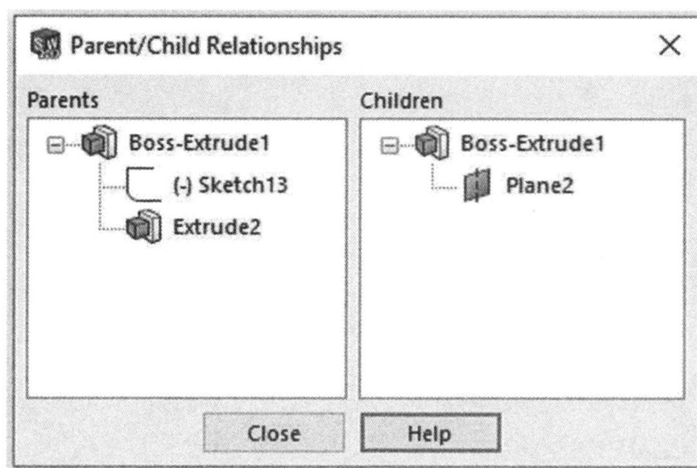

5.23. - Select *"Boss-Extrude1"* in the FeatureManager and delete it. This time the confirmation dialog only lists *"Plane2,"* which is not needed in the model and therefore, it's OK to delete. Be sure to activate the option "Also delete absorbed features" to delete its sketch, too.

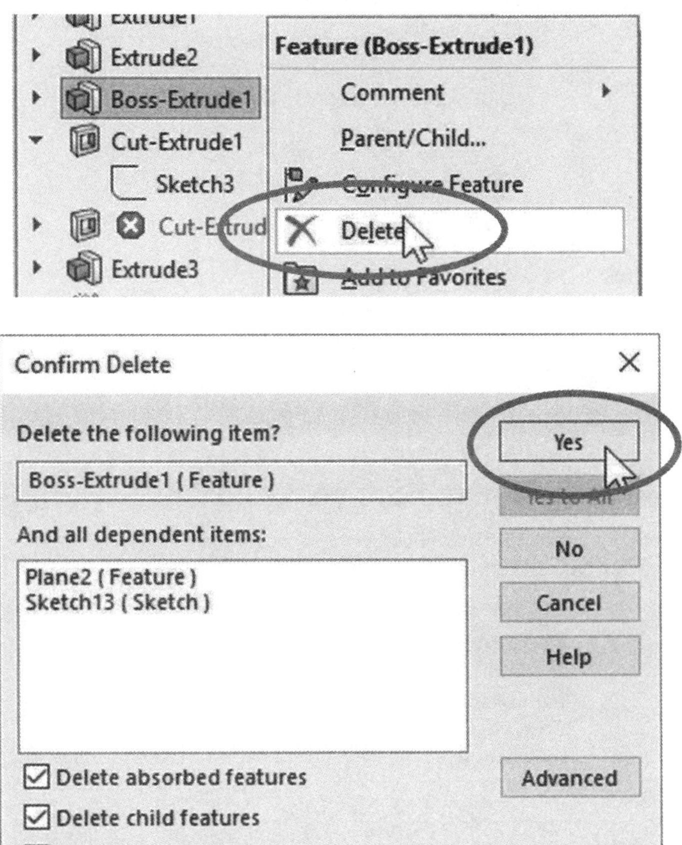

5.24. - The last error to fix is *"Cut-Extrude2."* The error we see using the "**What's Wrong?**" command means that while the feature is intended to remove material, it is not cutting the model. This is a feature error, not a sketch error. For our example, we'll assume the feature needs to cut 0.375″ into the lower step.

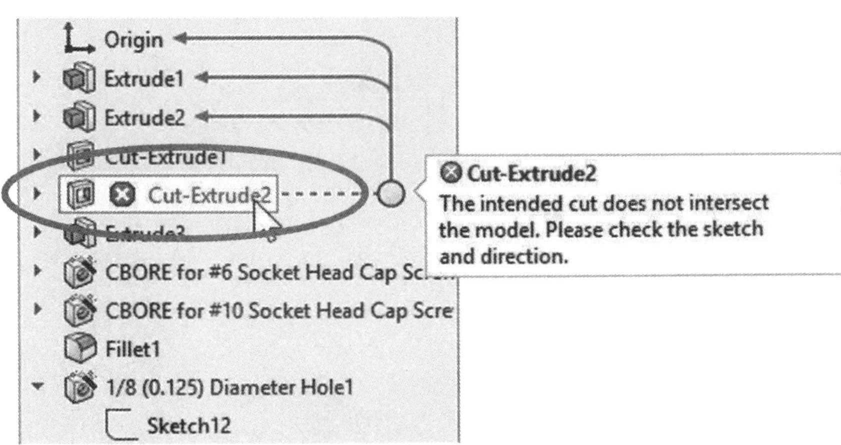

We have a couple of options to correct this error:

- Edit the sketch plane and change it to the lower step

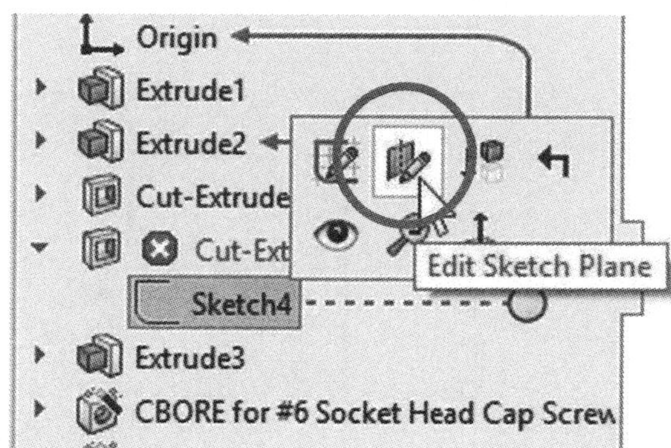

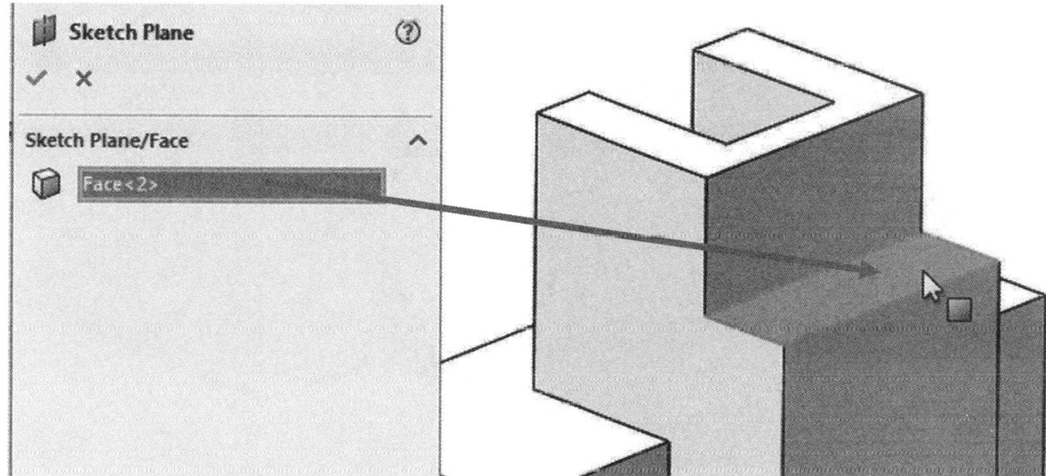

- We can edit *"Cut-Extrude2"* to make the feature start at a different surface than the sketch plane

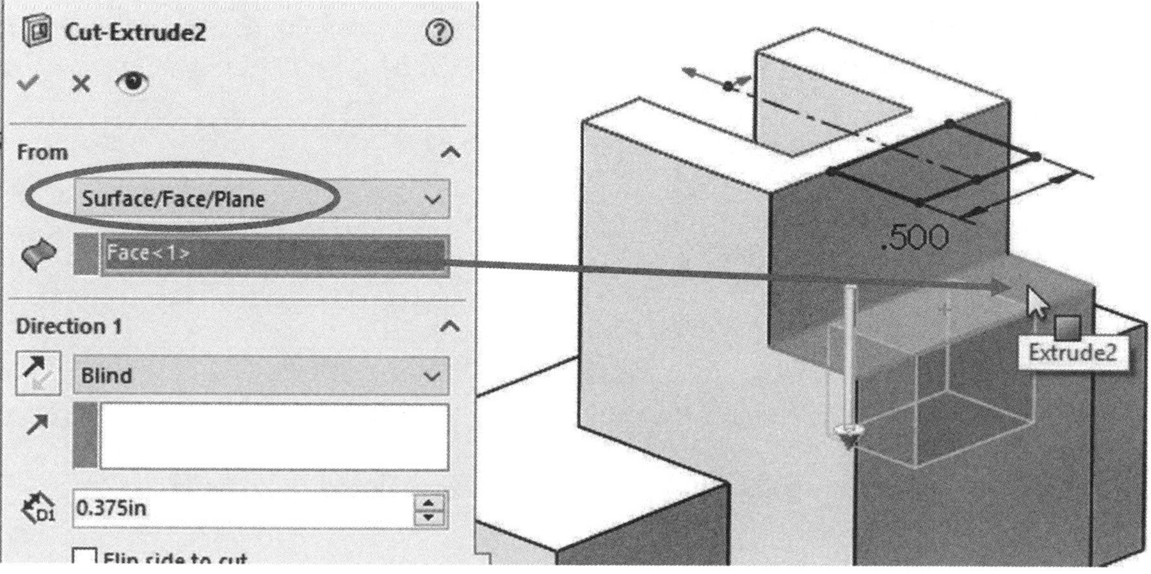

- Or we can change the feature's end condition to "**Offset from Surface**" and select the lower step's face. This way the cut will be the offset distance from either side of the selected face; if needed, turn on the "Reverse offset" checkbox to cut into the part as shown.

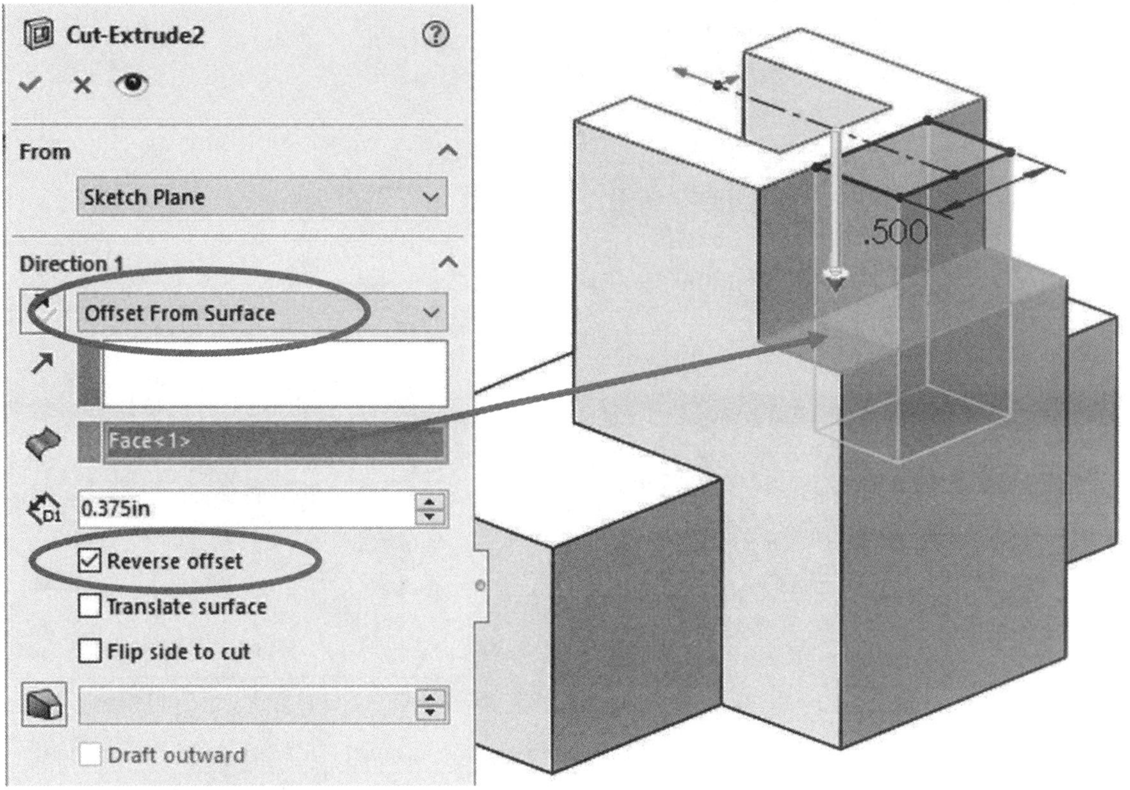

Select any option and click OK to finish repairing the errors in this part. The finished part fully repaired looks like this. Save and close the part.

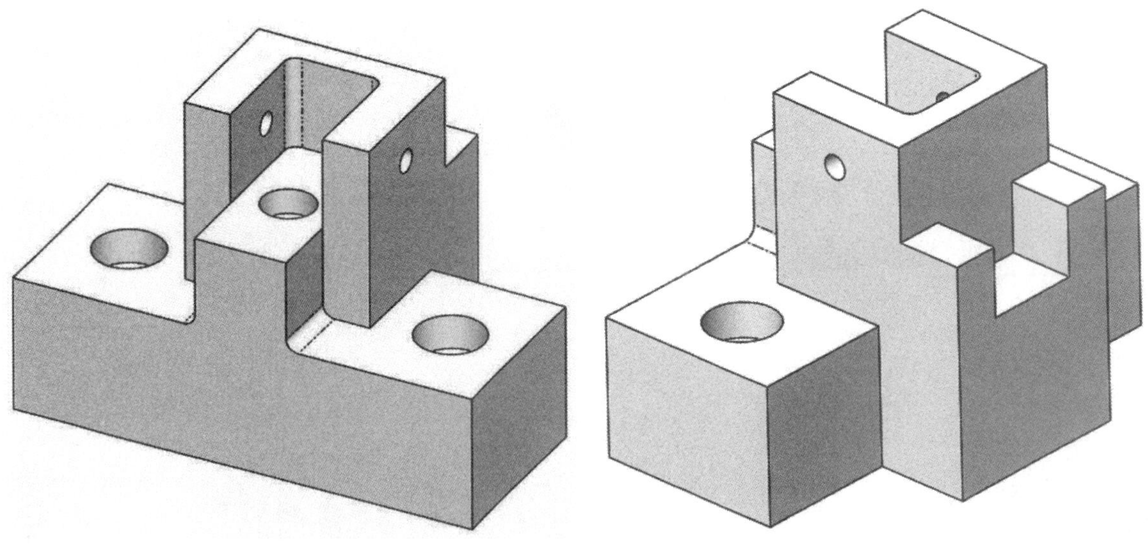

Sketch Editing

A sketch can have many different types of errors. In this lesson we'll learn how to use the sketch diagnostic tools to help us identify and correct these errors using a different part.

6.1. – Open the part *'Sketch Relations.sldprt'* from the exercise files. As before, this part has multiple errors, the difference is that they are all in the sketch. After opening the file, close the "**What's Wrong**" dialog and edit the "*Boss-Extrude1*" sketch.

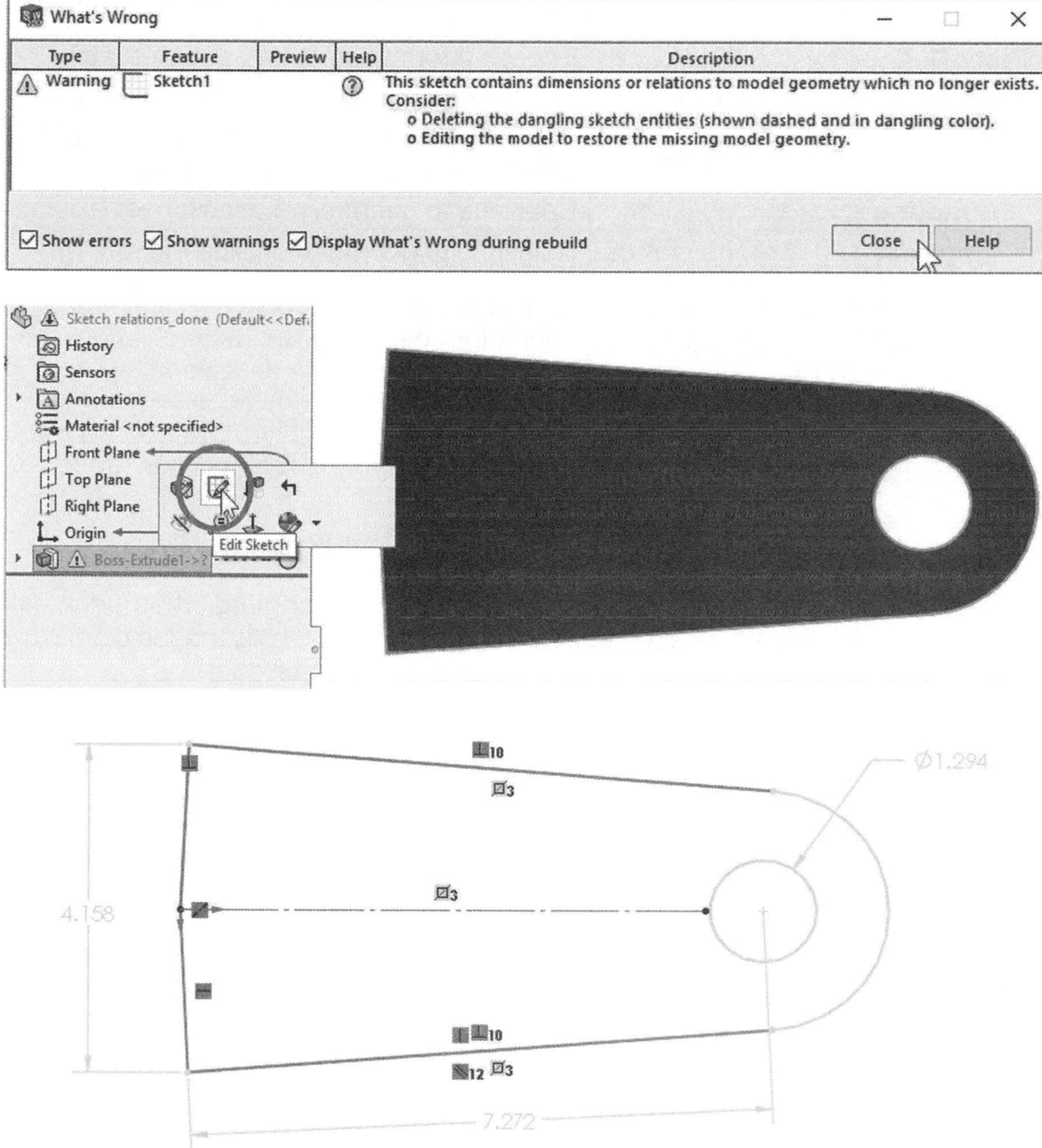

In a sketch, we can have many types of geometric relations. What we are going to focus on is when relations are not solved correctly and generate warnings and errors.

Under Defined or **Fully Defined** sketches can be used in a feature without a problem, the latter being the desired state. However, if we add conflicting geometric relations, for example, if we make a line both Horizontal *and* Vertical, the sketch will immediately give us an error warning. A sketch can be in any of the following states:

State	Sketch Color	Description
Under defined	**Blue**	Not enough relations to fully define sketch
Fully Defined	**Black**	Enough information to fully define sketch
Over defined	**Red**	Conflicting relations cannot be satisfied
Not Solved	**Yellow**	Relations cannot be solved; geometry cannot meet the required relations
Dangling	**Brown/Gold**	Relations to geometry that no longer exists
External	Can be under, fully or over defined, not solved or dangling	Relations to geometry outside the sketch
In Context *		Relations that reference other component's geometry; can only be added when the part/sketch is edited in an assembly
Locked *	Can be any color	*'Frozen' In context* relations
Broken *		*In context* relations that have been broken

* Will be covered more in depth in the **Top Down Design** section later.

A sketch can become **Over Defined**, **Not Solved** or **Dangling** when geometry referenced by the sketch is deleted or modified, or when the user adds conflicting relations and/or dimensions to the sketch that cannot be solved.

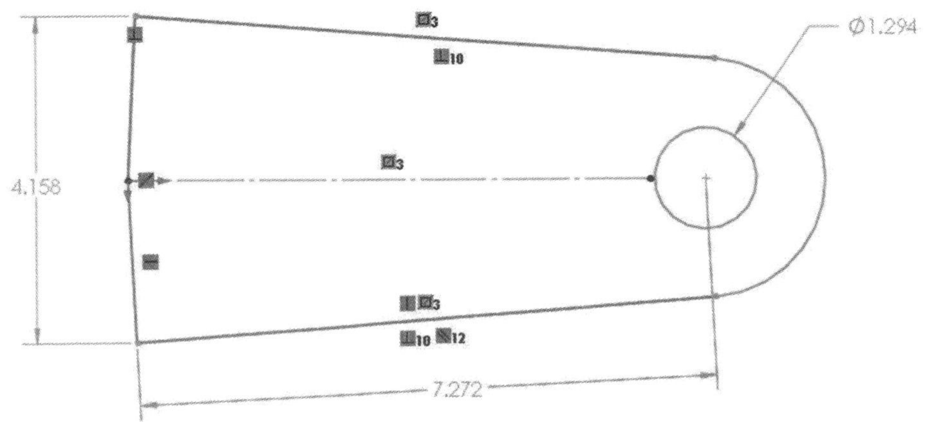

After we are done fixing the sketch it will look like the following image. In this case, even though it's easier to erase the sketch and recreate it, we'll take this opportunity to show how to identify errors and make changes to a sketch.

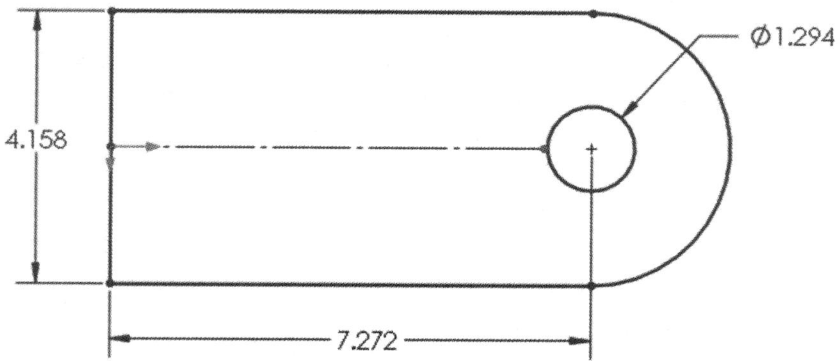

6.2. - After editing the sketch we see red and brown sketch elements. There are two ways we can approach this: we can manually sort through the geometric relations and delete the conflicting ones, or we can use the **SketchXpert** tool to fix them. We'll show the manual process first because this is a good way to understand what the **SketchXpert** does automatically. Select the "**Display/Delete Relations**" icon from the Sketch toolbar or the menu "**Tools, Relations, Display/Delete**."

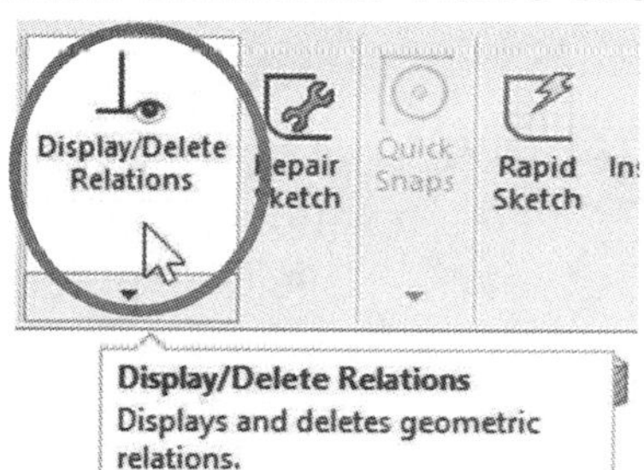

6.3. - In the PropertyManager we can see the existing sketch relations highlighted with different colors, and the selected relation's state at the bottom. The drop-down list at the top will help us filter the sketch relations and display them by state; alternatively, if we select a sketch element, its relations will be listed.

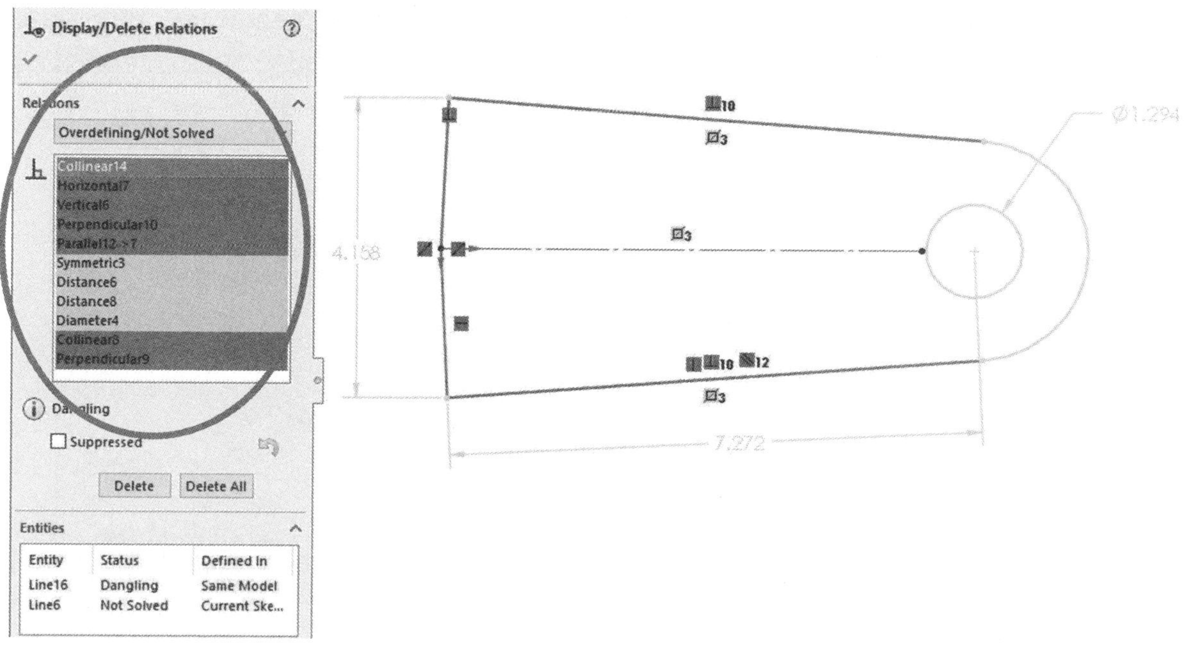

As we can see, having a large number of conflicting relations can be difficult to sort manually. This is a good diagnostic option when we have problems with a few relations. On the other hand, this is also a good tool to identify and/or delete relations, especially "External," "In context," "Locked" or "Broken," if needed.

6.4. - Using this tool we'll identify and remove the "**In Context**" relations, as we don't want our sketch to have relations to other parts' geometry. An "**In Context**" relation is added when we modify a sketch inside of an assembly and add references to other components. By deleting this relation, our part will no longer be referencing geometry outside the part. In Context relations will be covered extensively in the Top Down Design section.

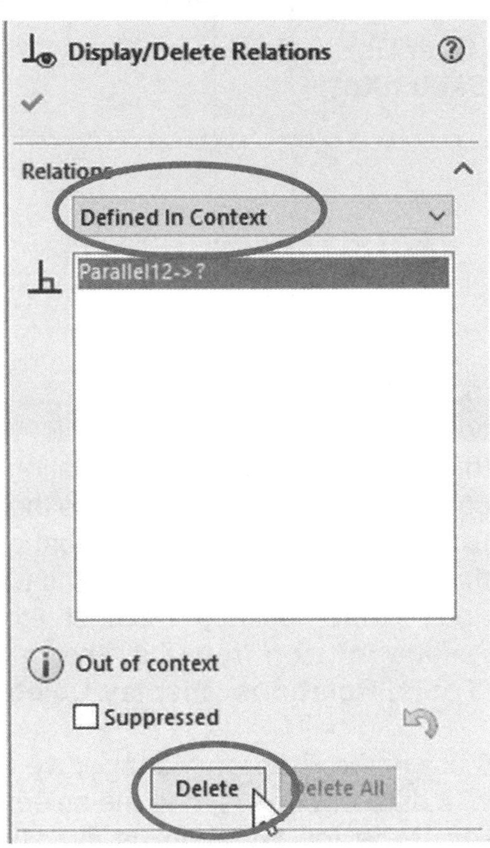

In the drop-down list select "**Defined In Context**." In our sketch, we only have one Parallel relation and its status is "Out of Context*." Click "Delete" at the bottom or select the relation and press "Delete" on the keyboard. Click OK to close the "**Display/Delete Relations**" dialog.

*More information will be provided later about relations in the context of an assembly in the "**Top Down Assembly**" lesson.

6.5. - After deleting this relation, our sketch is still in an undesired state, and we have both unsolvable and over defined geometry, as we can see in the status bar and by the red and brown sketch geometry and dimensions.

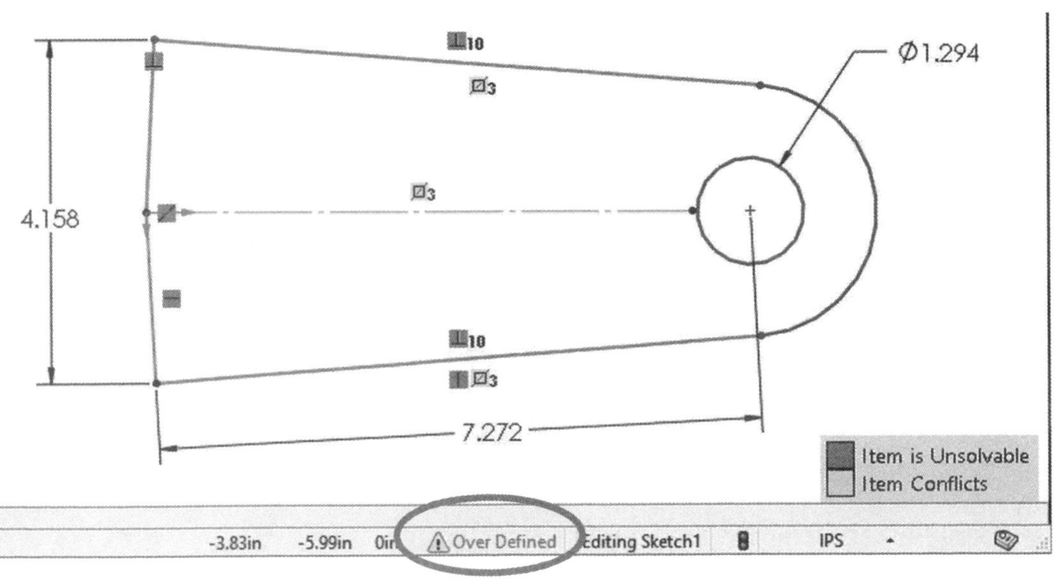

6.6. - The conflicting and over defined relations can be fixed using this tool, but it may take a long time to sort through all the possible options; this is a good tool when we only have a few relations that need to be fixed, or to find one or two relations that are causing problems. In this case, we'll use the **SketchXpert** to quickly evaluate multiple solutions to correct our sketch. To activate it click in the "**Over Defined**" message in the status bar, or use the menu "**Tools, Sketch Tools, SketchXpert**."

In the **SketchXpert** dialog select "Diagnose" to automatically analyze the sketch and evaluate possible solutions.

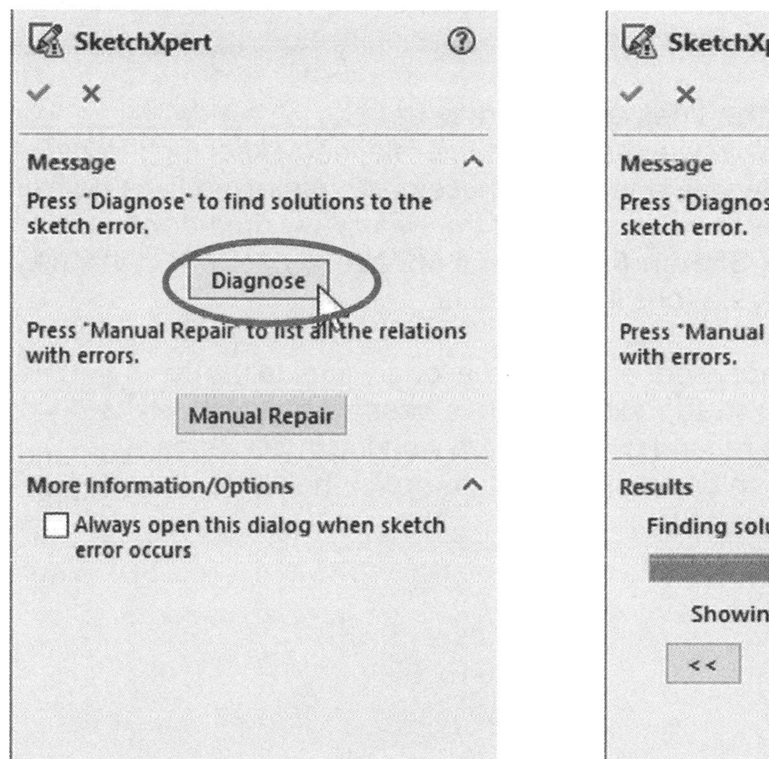

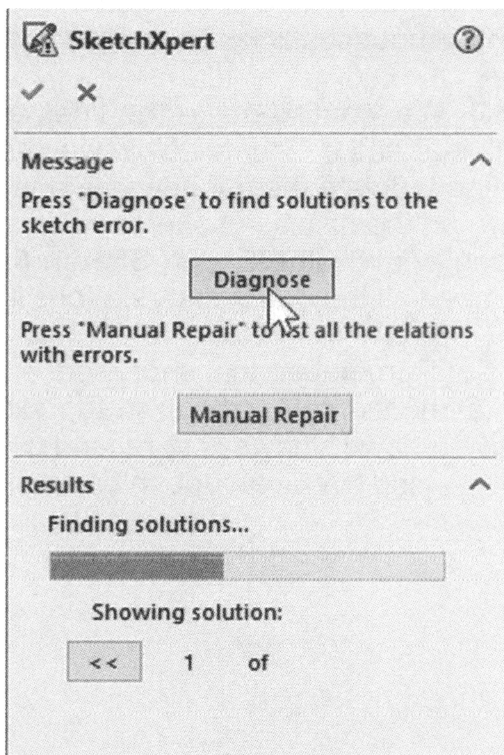

6.7. - SketchXpert quickly diagnoses the sketch and offers possible solutions which result in non-conflicting sketch geometry. We can view each option by advancing in the "Results" box. In the "More Information/Options" section we can see the relations and/or dimensions that would be deleted if we accept the displayed solution.

Click the arrow buttons to scroll the different options until you see the following solution (may be a different number) and click "Accept" when done. This is the solution that more closely resembles the desired result.

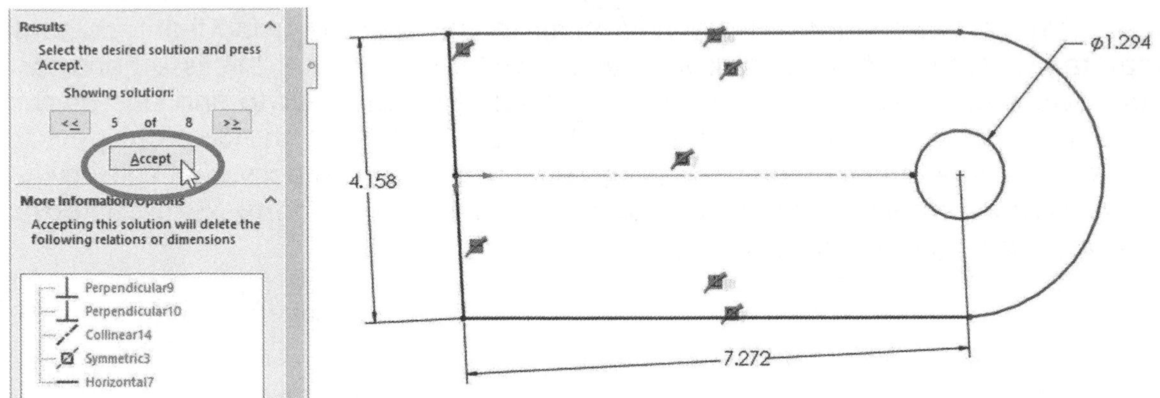

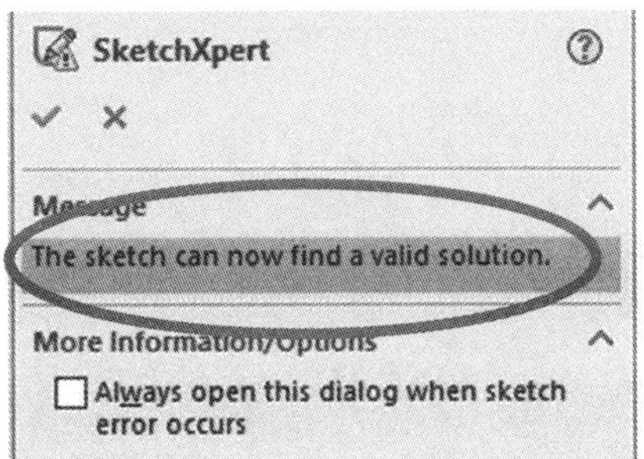

After accepting the solution we see the message "The sketch can now find a valid solution" with a green background to let us know that our sketch is free of errors. This doesn't mean the sketch fixed, it means the sketch's geometry can be built without conflicts. Click OK to continue.

6.8. - We want to make the lines on the left to be collinear and horizontal. Now we need to find the relation(s) keeping the lines fully defined, delete them if needed, and then, if necessary, make the lines horizontal. Turn on the display of sketch relations using the Hide/View drop-down icon or the menu "**View, Hide/Show, Sketch Relations**" if not already activated; this way we can see all the geometric relations in the sketch.

Remember the short red arrow at the origin indicates the horizontal direction in the sketch; in this case, the sketch is shown sideways to remind us that we are working in 3D space (and "Vertical" doesn't necessarily mean "*up*" in the screen).

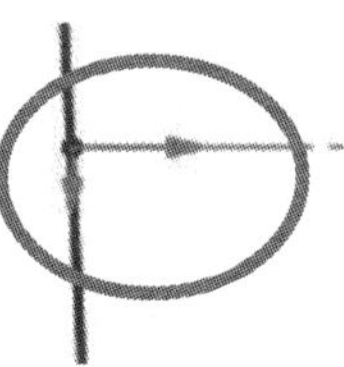

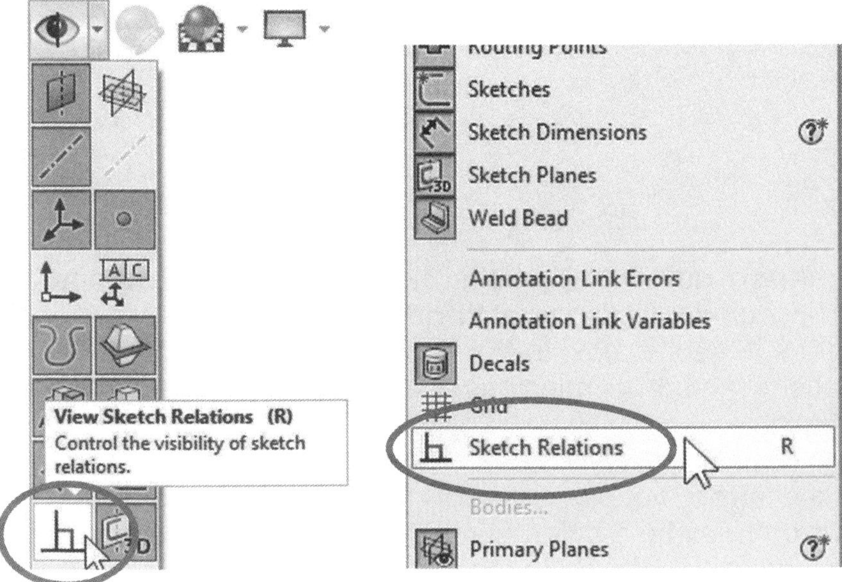

We can see that there are only three relations in the two lines on the left that we are interested in: "Collinear," "Coincident" and "**Fixed**." The "Fixed" relation completely constrains a sketch element. Think of it as putting a nail in a sketch element and hammering it in. Read: brute force. The "Fixed" relation should only be used as a temporary solution to prevent other elements in the sketch from moving, and removing it when it is no longer needed. Select the "Fixed" geometric relation in the screen and delete it.

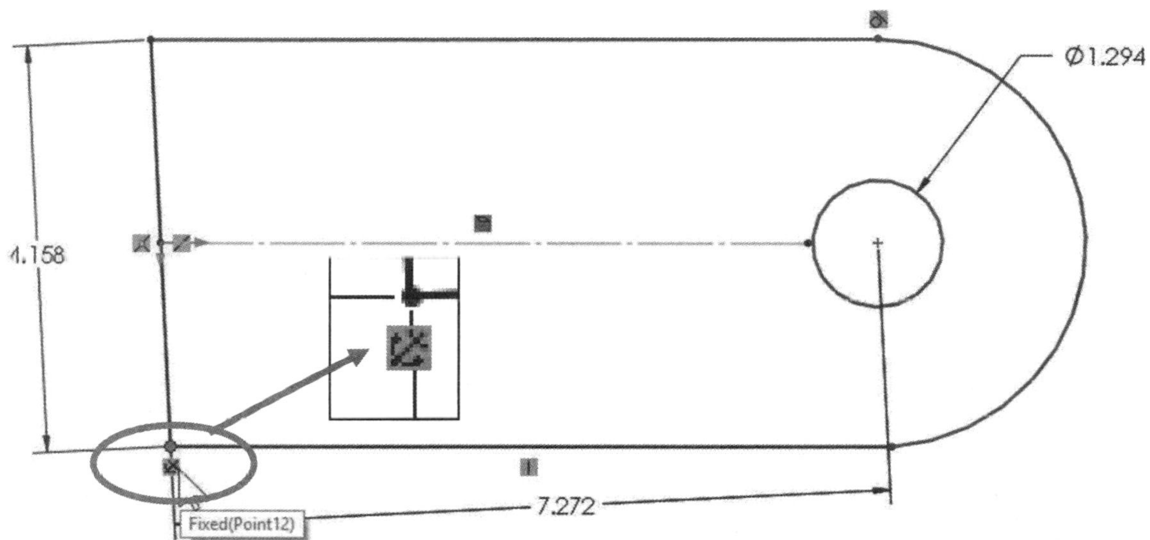

6.9. - The centerline has a relation colored brown/gold. This is an "**On Edge**" relation; this is the type of relation created when we use the "**Convert Entities**" command in a sketch to project model geometry. The brown/gold color means it is "dangling"; in other words, the edge or entity that was used to project this element from is no longer in the model, and the relation must be deleted. Select and delete the dangling "**On Edge**" relation to continue.

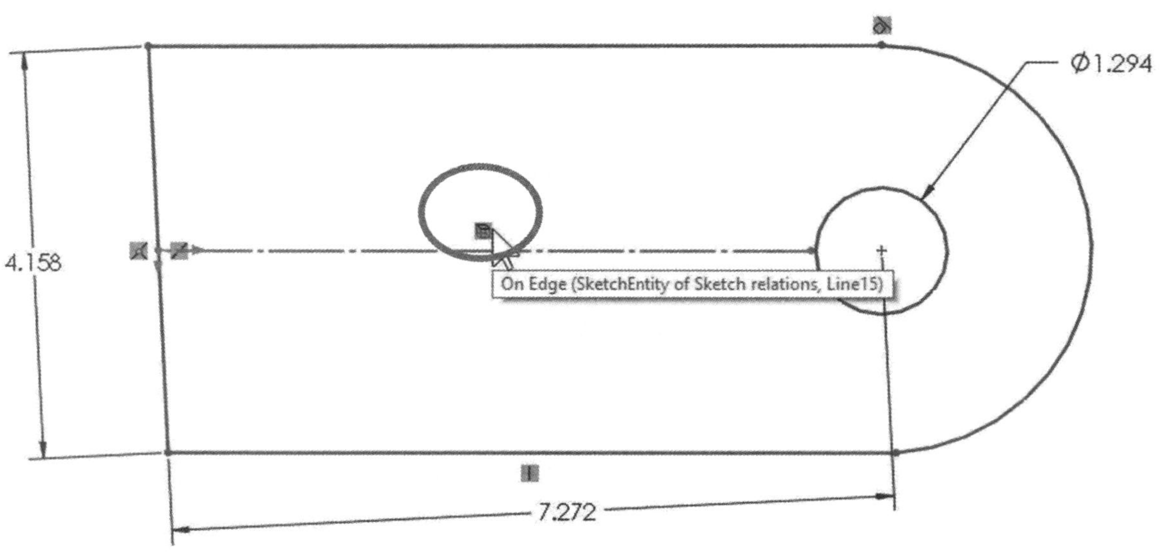

6.10. - Select one (or both) line(s) to the left of the sketch and add a "**Horizontal**" relation. Since both lines are already "Collinear," selecting one should be enough. (Remember the sketch is rotated 90° in the picture, and "*Vertical"* is sideways in this image.)

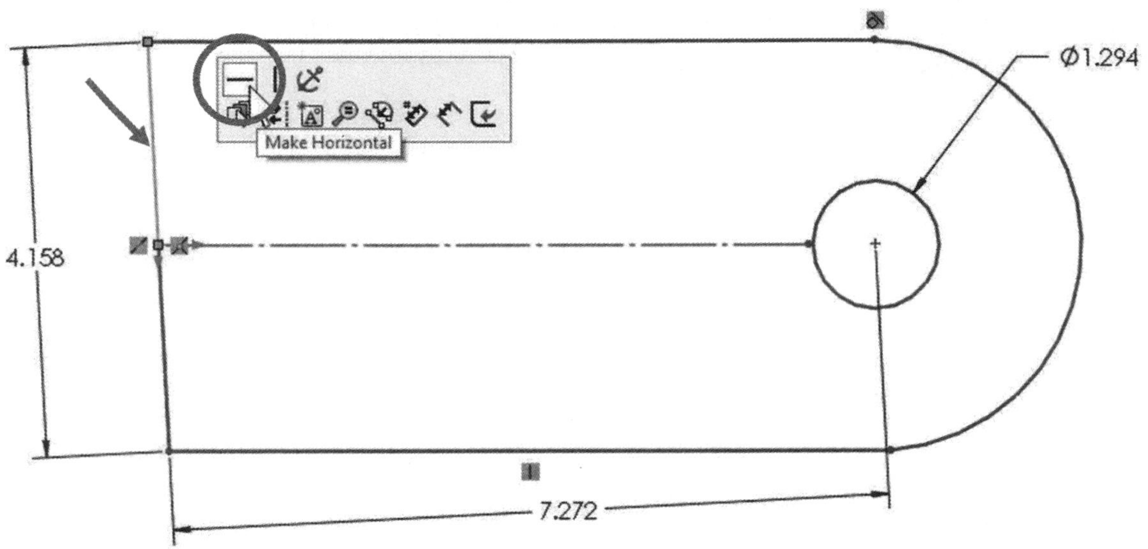

6.11. - Window select from right to left to select the centerline and the top line, and then add a "**Vertical**" relation.

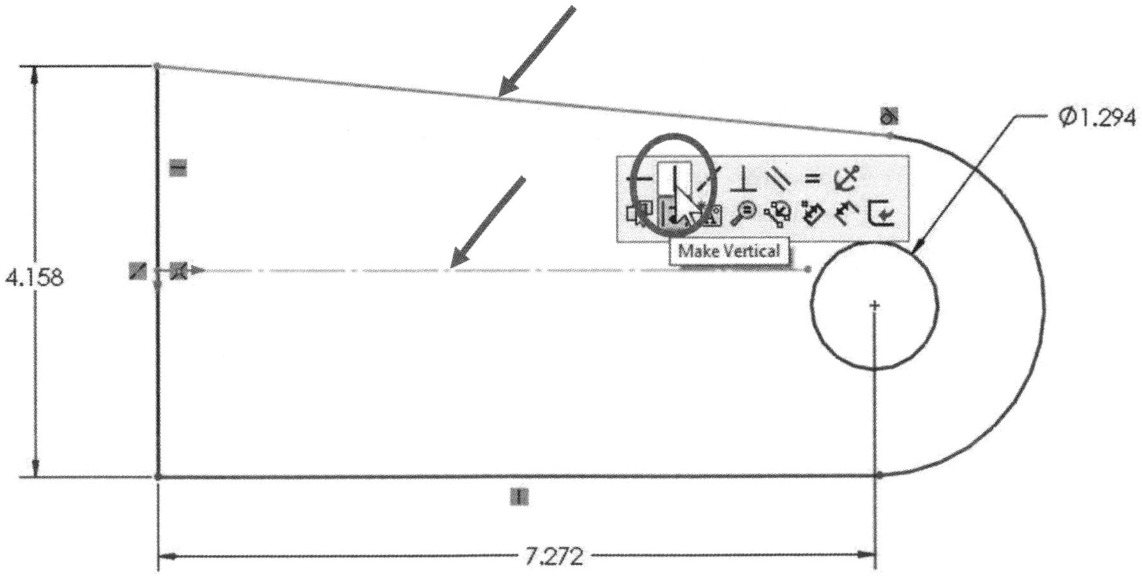

6.12. - After examining the sketch relations, we can see that the arc has a tangent relation to the top line, but not to the bottom one. To see the effect of not having this relation, click and drag the blue endpoint where the bottom line meets the arc. The next step is to add the missing "**Tangent**" relation. Press and hold the "Ctrl" key, select both the arc and bottom line, release the "Ctrl" key and click "Tangent" from the pop-up toolbar.

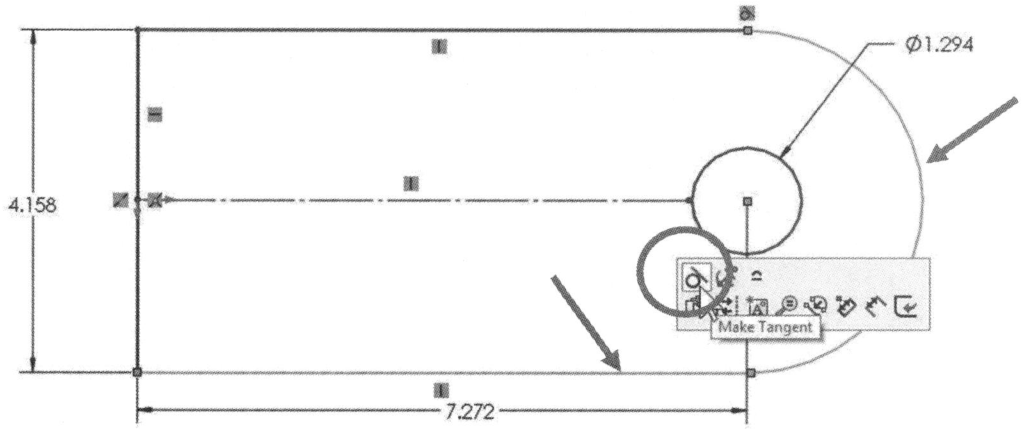

6.13. – After adding this relation the sketch looks very much like the geometry we are looking for, however, we still have blue geometric elements which means our sketch is under defined. To get an idea of what we may need to do to fully define it, click and drag any blue sketch element. Since the sketch moves up and down we must add another relation that will prevent this movement.

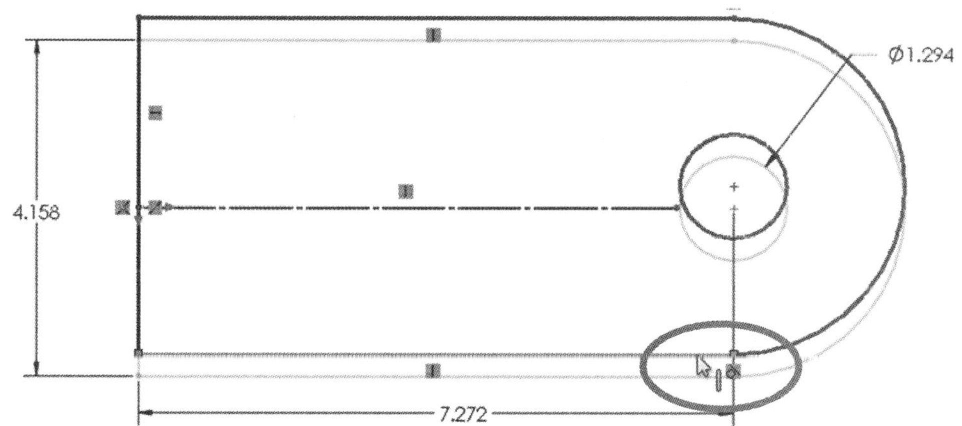

6.14. – To fully define our sketch we have a few options:

a) Select the center of the circle and make it coincident to the centerline.

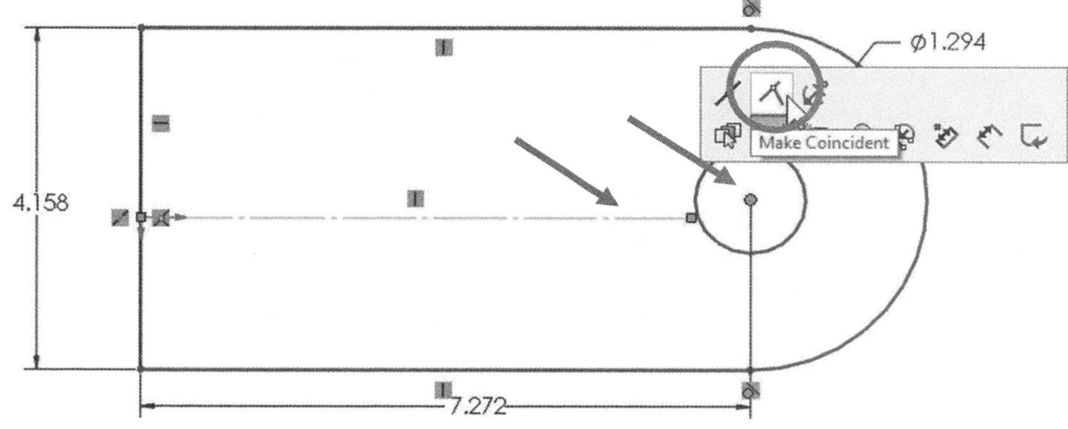

b) Add an "Equal" relation to both of the "horizontal" lines at the left; this way the sketch will be centered about the origin. (There are two line segments.)

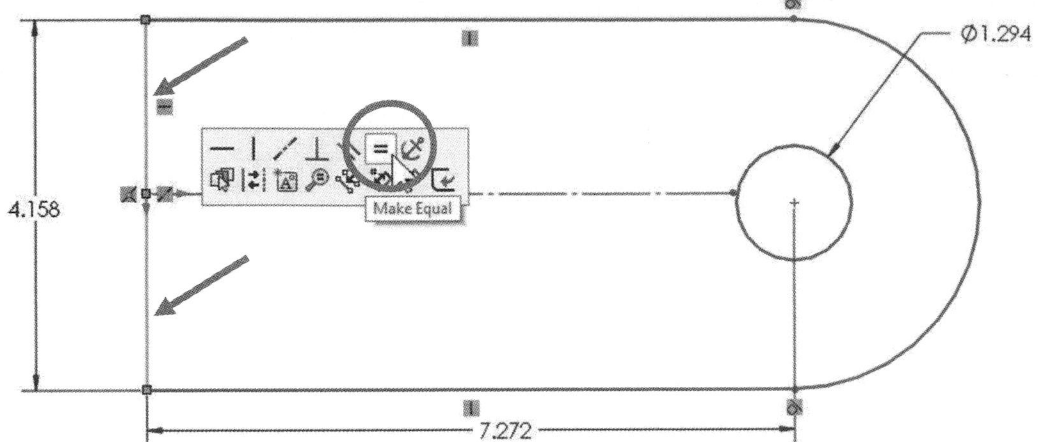

c) Select both "vertical" lines and the centerline, and add a "Symmetric" relation; this way the "vertical" lines will be a mirror image of each other about the centerline.

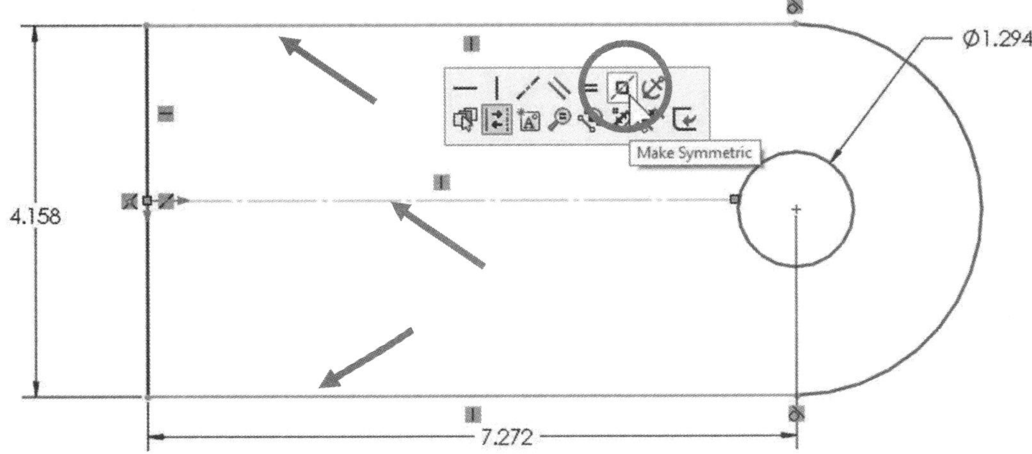

Adding any of these relations will fully define the sketch. Now that we have repaired the sketch, exit the sketch to finish. Save and close the file.

Exercise: Open the file *'Model Repair Exercise.sldprt'*. After opening and rebuilding it, a list of errors will be displayed; however, the part has so many problems that no geometry is built. Correct the errors in features and sketches using the knowledge acquired in this lesson. Use the following image of the finished part for reference. A high-resolution image is also included in the exercise files for your convenience.

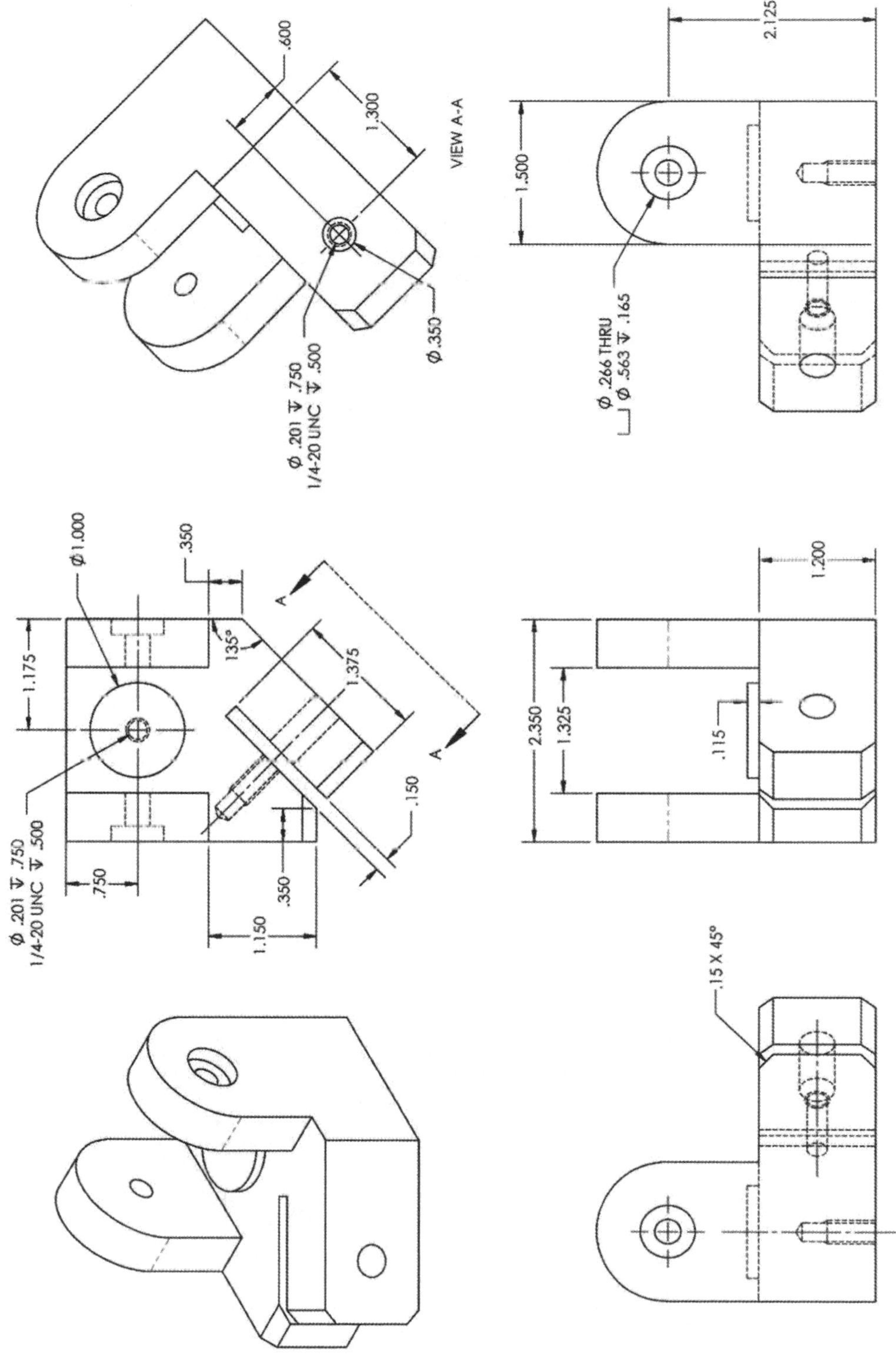

TIPS:

- **Boss-Extrude1**: Edit the sketch, convert two lines to construction geometry, and use SketchXpert to find a valid solution to the sketch.
- **Cut-Extrude1**: Use SketchXpert to find a valid solution to the sketch.
- **CBORE for 1/4 Binding Head Machine Screw1**: Edit Sketch6, delete the dangling concentric relation, add a new concentric relation to the round edge.
- **Boss-Extrude2**: Use SketchXpert to find a valid solution to the sketch, delete the dangling "On Edge" relation.

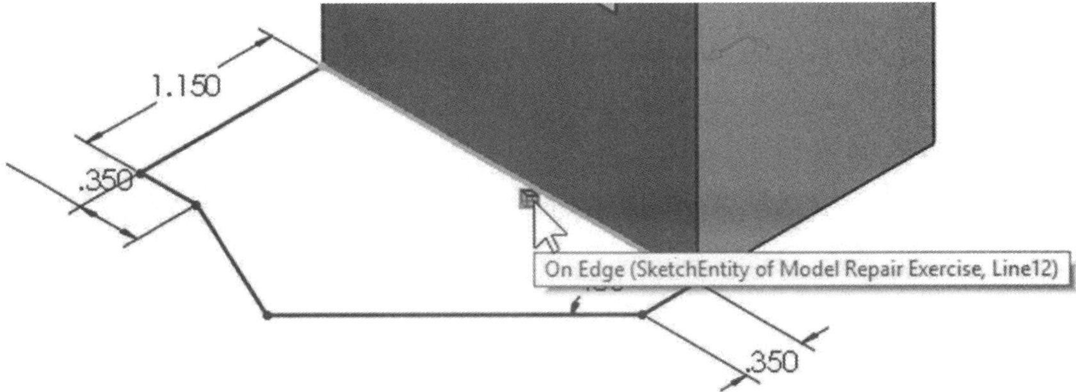

- Drag the line along the model's edge to separate it, and add the necessary relations to make the endpoints coincident to the model's endpoints.

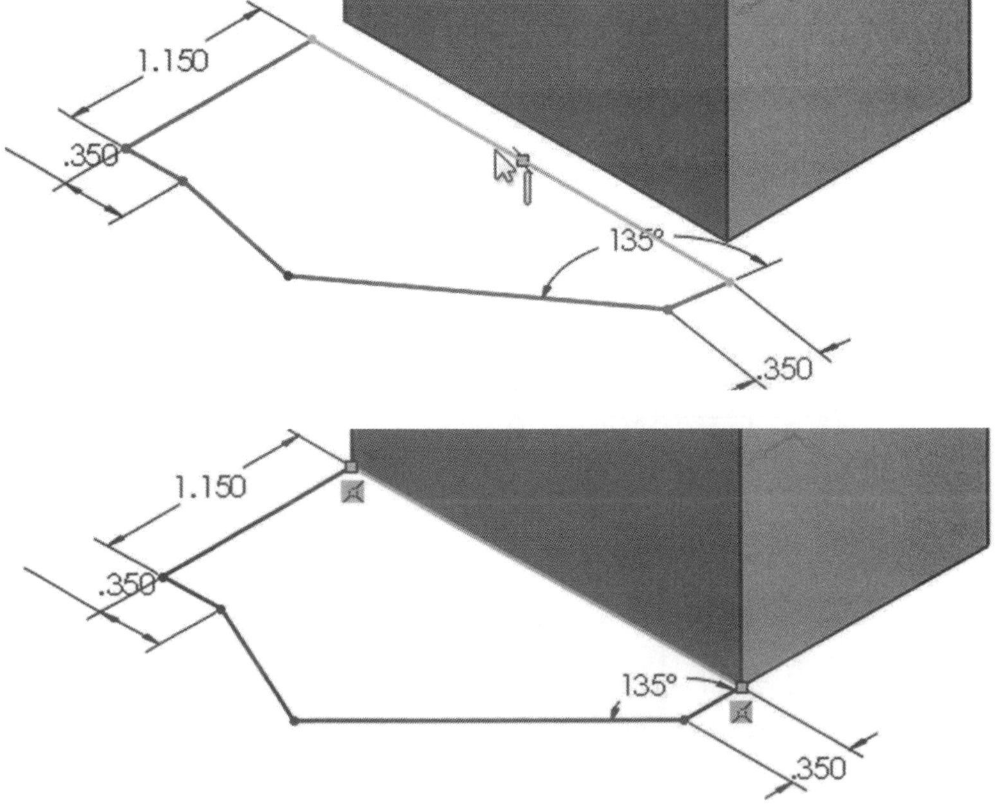

- Find and delete the "**Fixed**" relation, add a perpendicular and a vertical relation as needed.

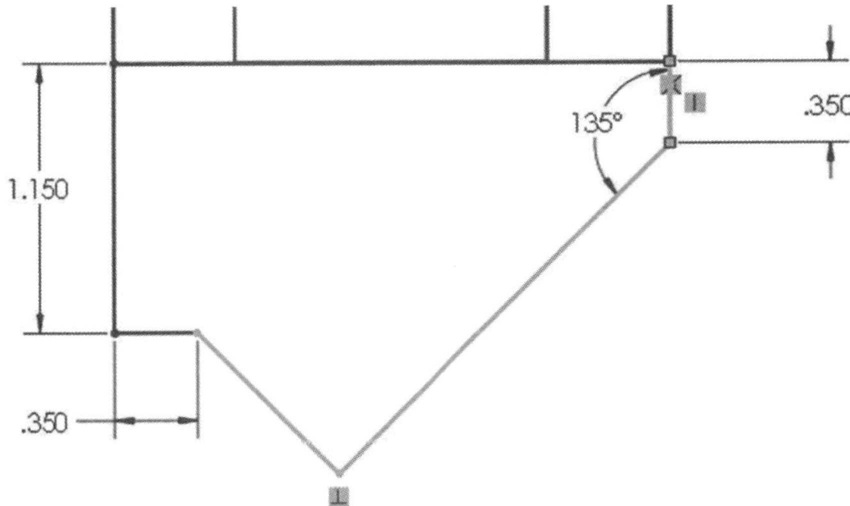

- **Boss-Extrude3**: Edit Sketch9, delete the overlapping arc, exit the sketch.
- **Boss-Extrude3**: Edit sketch plane for Sketch9, select the top of the part.

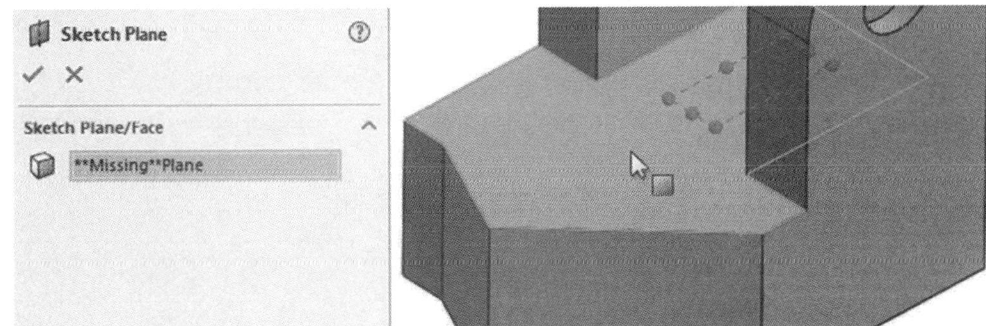

- **Cut-Extrude2**: Edit Sketch10, use "Check Sketch for Feature," delete extra line.
- **Cut-Extrude3**: Edit Sketch13, use SketchXpert to find a valid solution.
- **1/4-20 Tapped Hole2**: Edit Sketch15 (hole location sketch), delete the concentric relation, make sketch point concentric to Cut-Extrude2 edge.

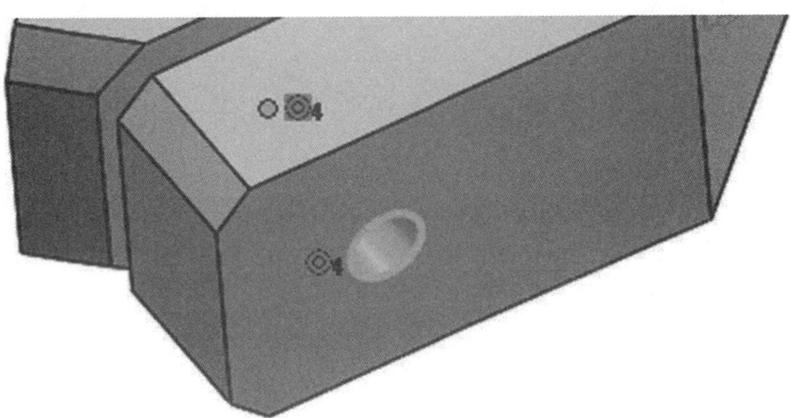

Notes:

Equations

One way to help us maintain design intent in our models is by using equations. Equations are commonly used to evenly space features, add or remove instances to a pattern's count, change dimensions when other features are modified, etc. All equations in SOLIDWORKS have the following format:

Variable = Expression

Where:

- ***Variable*** is the dimension/value to change (dependent value).
- ***Expression*** is the algebraic combination of other dimensions and values that will define the ***variable's*** value.

For example, if we have a part of length "**L**" where we want to evenly space a pattern of "**N**" number of holes spaced a dimension "**S**," the ***variable*** dimension will be "**S**" because that's the value we want to change when the number of holes "**N**" and/or the part's length "**L**" changes. The equation would be:

S = L / N

When working with equations, it's a good idea to rename dimensions and features; this way our equations would read:

"Spacing@Holes" = "Length@Base" / "Number@Holes"

Instead of:

"D1@LinearPattern1" = "D3@Sketch1" / "D2@LinearPattern1"

The former is more intuitive and easier to read. If we have one or two equations in a part it may not be difficult, but having multiple equations in a part it may become difficult to manage and modify them.

7.1. - To learn how to work with equations, we'll make a simple part, rename its features and dimensions, and finally add equations. Draw the following sketch (any plane) and extrude it 0.5″. Dimensions are in inches.

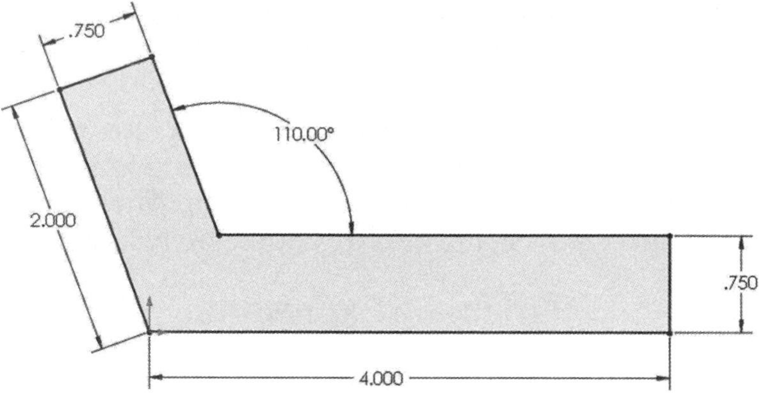

7.2. - Add a through hole in the corner...

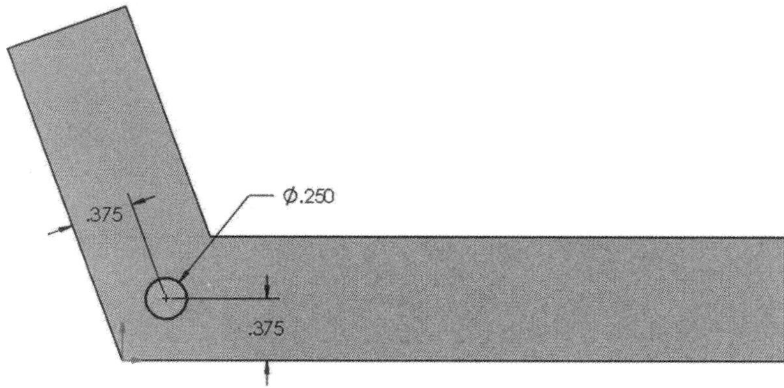

7.3. - Add a linear pattern in two directions. For "Direction 1" use the "Spacing and instances" option, select a diagonal edge and make three instances spaced 0.5″; for "Direction 2" select a horizontal edge and make six copies spaced 0.5″ (don't finish it yet).

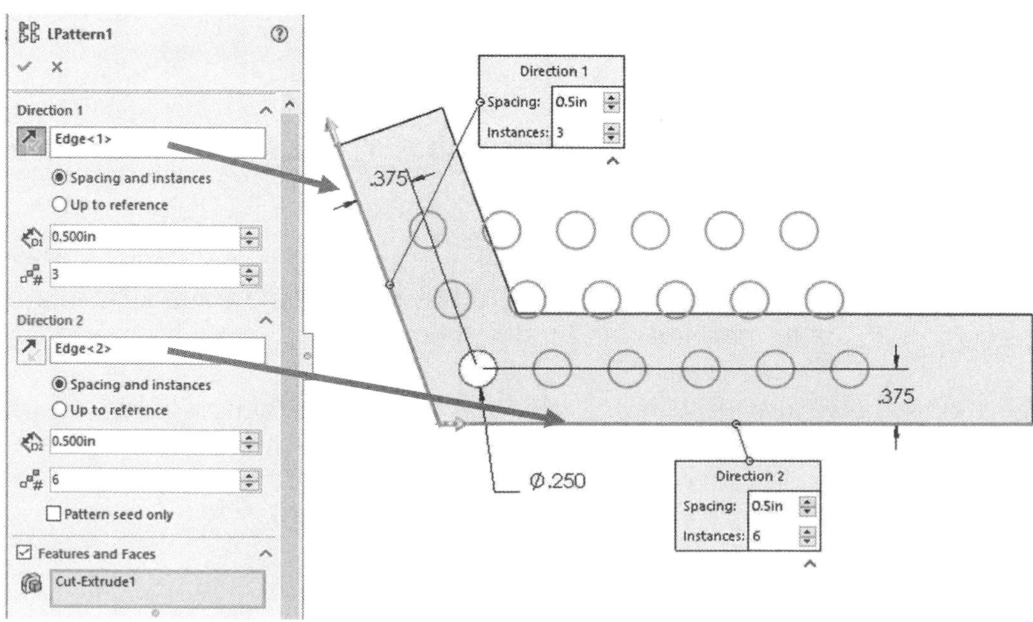

7.4. - Since we need a single row of holes in each direction, under the "Direction 2" options box check the option "**Pattern seed only**." This option allows us to only copy the original hole feature in both directions and not the "Direction 1" copies in "Direction 2" as seen in the preview. Click OK to finish the pattern.

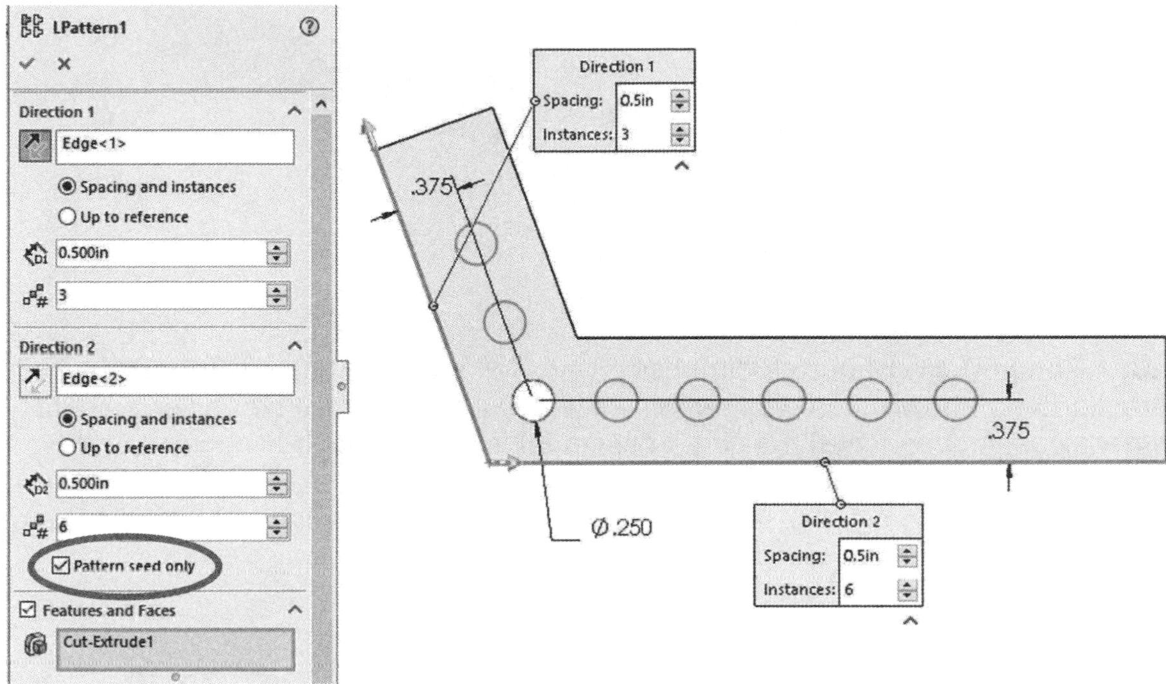

7.5. - Now we need to rename the dimensions, but first we must make feature dimensions visible. Right-mouse-click the "**Annotations**" folder and activate "**Show feature dimensions**," and to see dimension names, activate the option in the menu "**View, Hide/Show, Dimension Names**."

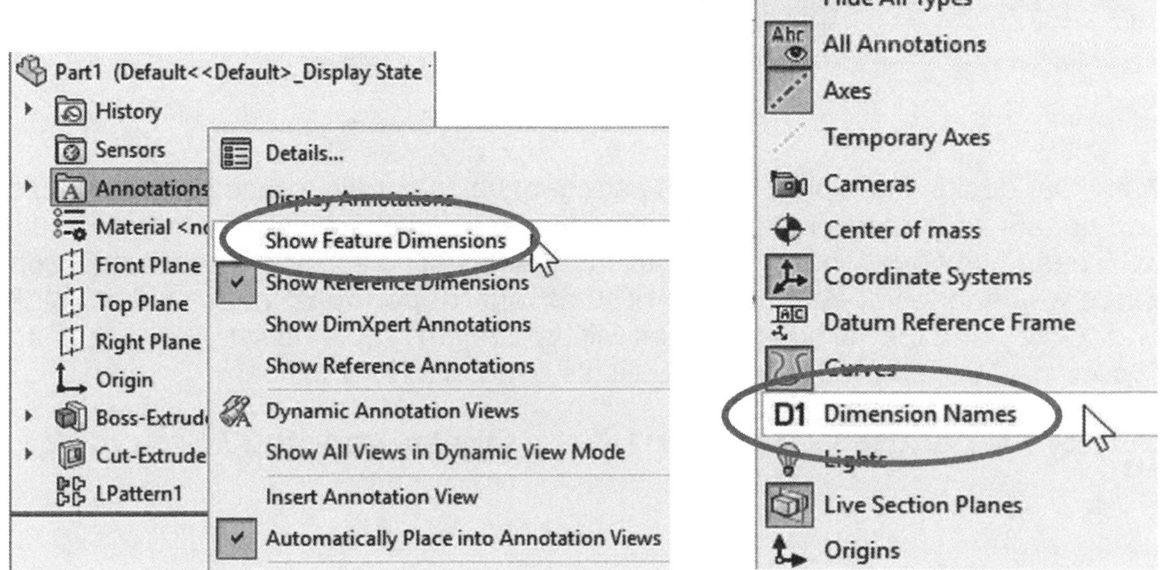

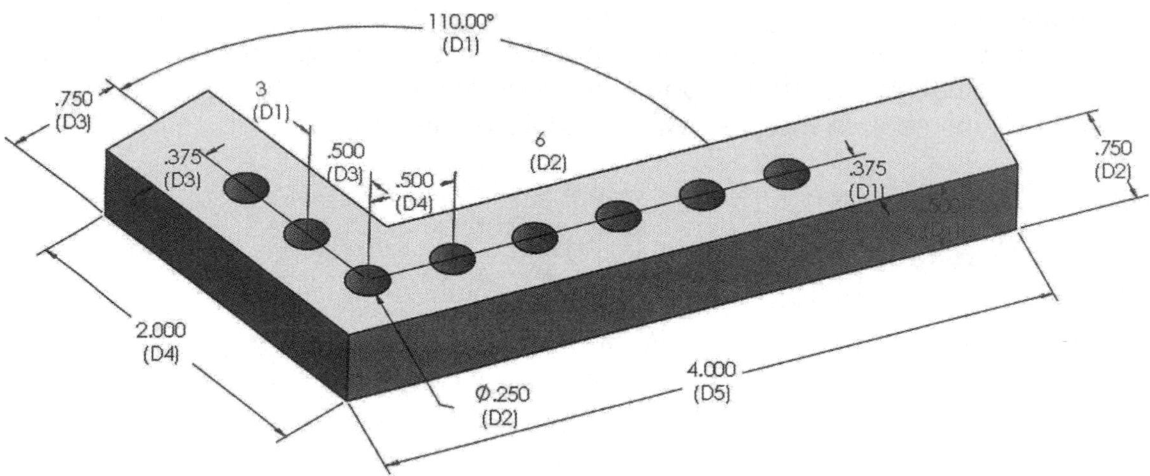

7.6. - Rename the pattern's dimensions as well as the length of the part in both directions as shown; this way, when we add equations it will be easier to identify dimensions, just as when making a design table. Select each dimension and type the name in its PropertyManager.

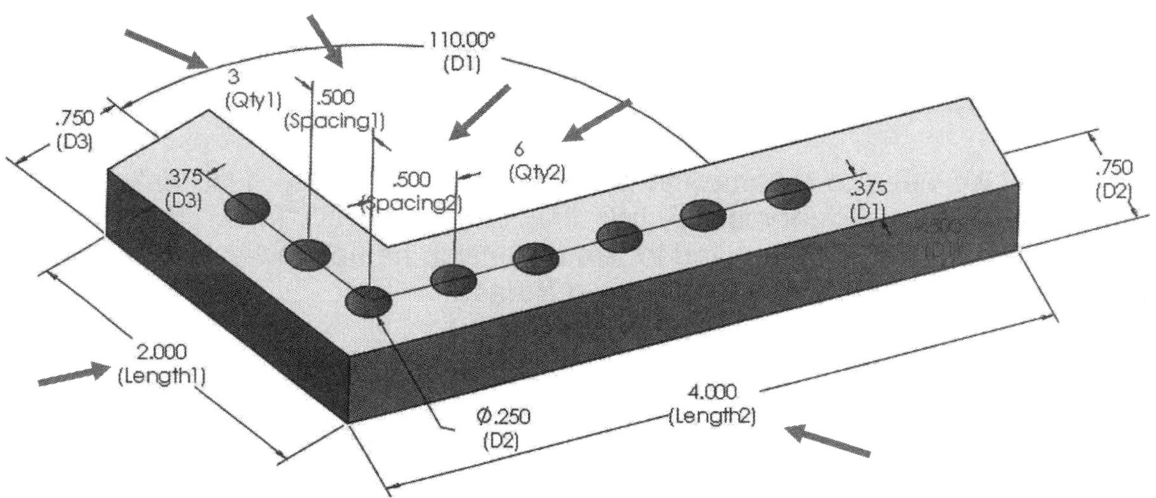

7.7. - Our design intent is to have a pattern of holes equally spaced to fill each side and update accordingly if either the length or spacing changes. To accomplish this, we'll need two equations to calculate the quantity of holes (one for each direction). Since we need to vary the number of holes, our "**dependent**" values will be *"Qty1"* and *"Qty2,"* and the "**driving**" values will be *"Length1," "Spacing1," "Length2,"* and "*Spacing2."* Based on this, our equation's general format will be:

Qty1 = Length1 / Spacing1 **Qty2 = Length2 / Spacing2**

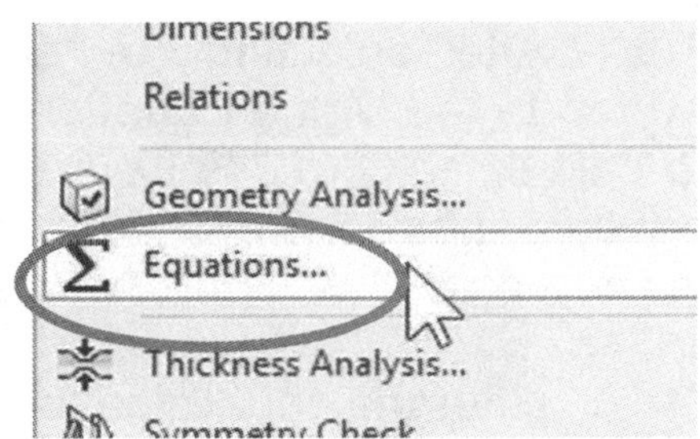

7.8. - There are two ways to add equations, and we'll add one equation with each method. For the first one, select the menu "**Tools, Equations**." We are immediately presented with the Equations dialog box. Here we can add equations, define Global Variables to be used as constants, and set features' suppression states.

To create the first equation, click inside the "Add equation" field; this is where the dependent variable will be listed. Here we can either type the dimension's name or select it in the screen. If the name is typed, it must be the exact full name of the dimension, therefore, selecting it is easier. Select *"Qty1"* in the screen to add its full name *"Qty1@LPattern1"* to the "Add Equation" field.

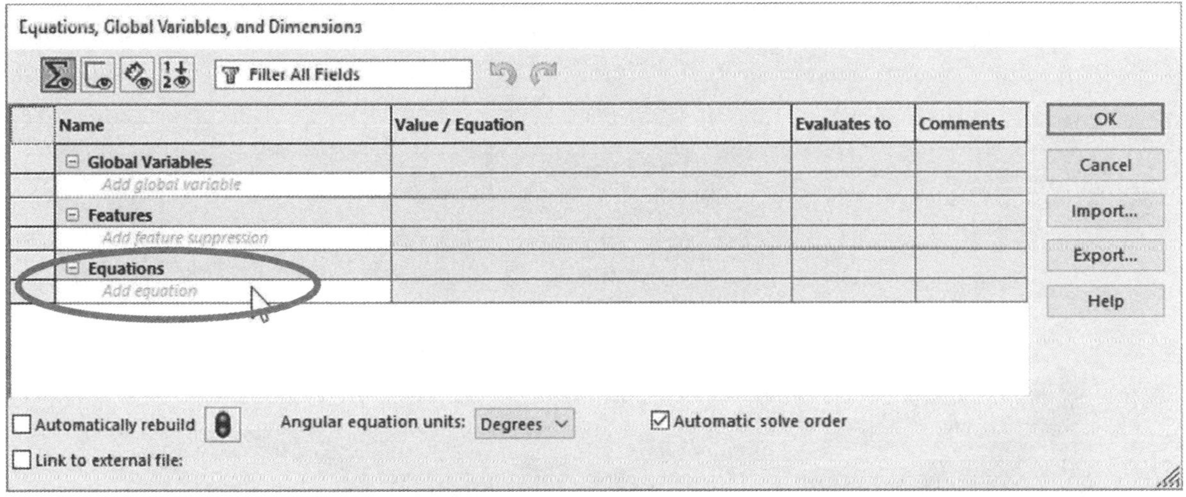

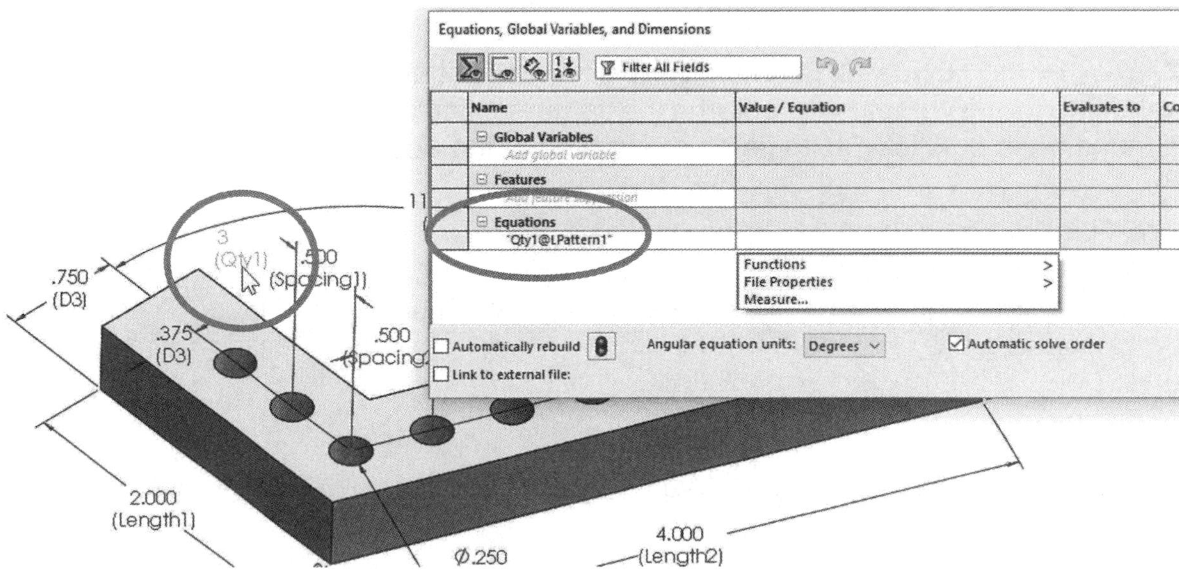

7.9. - After selecting the *"Qty1"* dimension, the equal sign (=) is automatically added to the Value/Equation field, waiting for us to enter the rest of the equation to calculate *"Qty1."* For our equation to calculate the correct number of holes, we have to subtract 0.375" from the length to compensate for the last hole's distance to the edge. If we don't, we may get an extra hole at the end.

After the equal sign, open a parenthesis, select the *"Length1"* dimension (its full name will be copied), type '**- 0.375**', close the parenthesis, type '/' to divide and finally select the *"Spacing1"* dimension. The green checkmark next to the equation will indicate this is a valid expression.

Our first equation looks like:

"Qty1@LPattern1" = ("Length1@Sketch1" - 0.375) / "Spacing1@LPattern1"

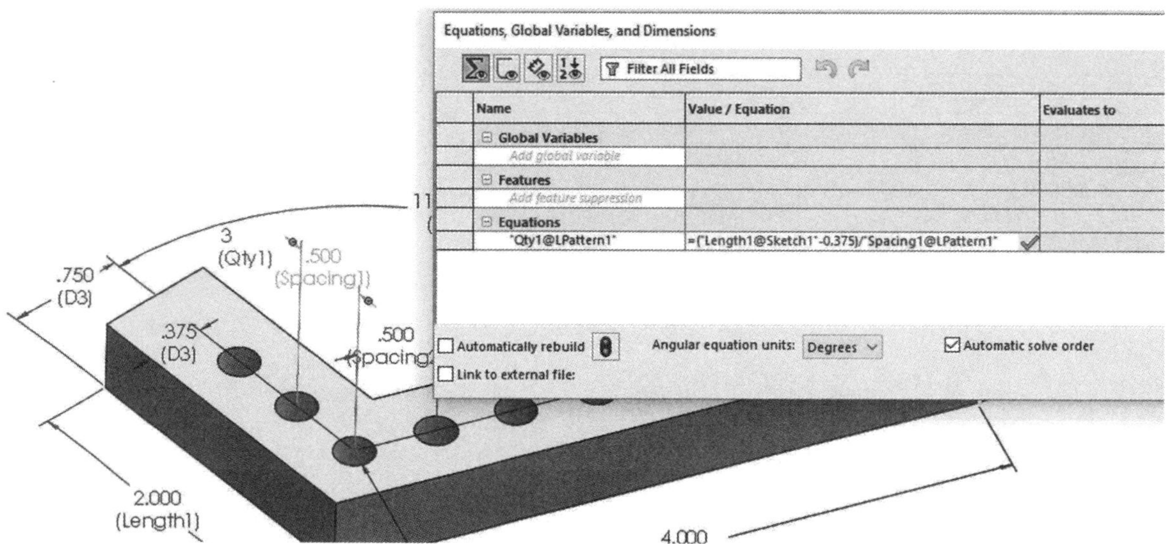

7.10. - In the "**Equations**" dialog we see the first equation is added, and after selecting the "Evaluates to" column the value is calculated. The value for the *"Qty1"* dimension is 3. Click OK to finish the first equation and continue.

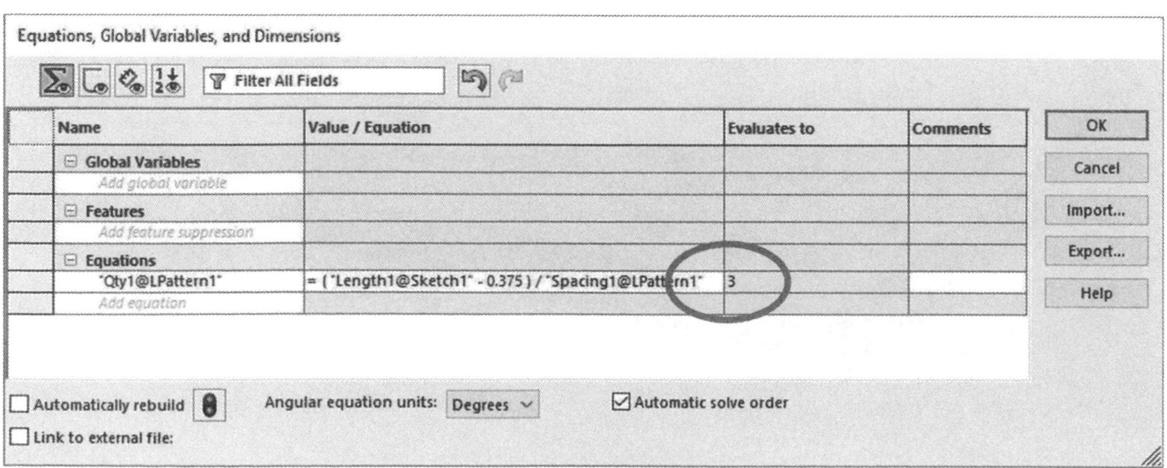

After adding the equation, we see a red ∑ symbol next to the *"Qty1"* dimension, letting us know that its value is driven by an equation and cannot be changed directly as other dimensions; it can only be changed when the driving dimensions ("*Length1"* and "*Spacing1"*) are changed.

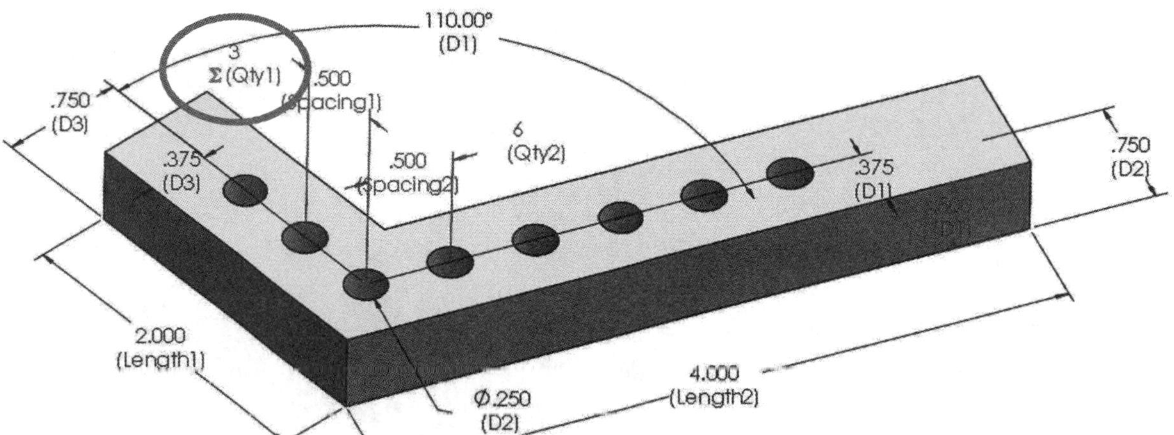

When a model has equations, a folder named *"Equations"* is automatically added to the FeatureManager.

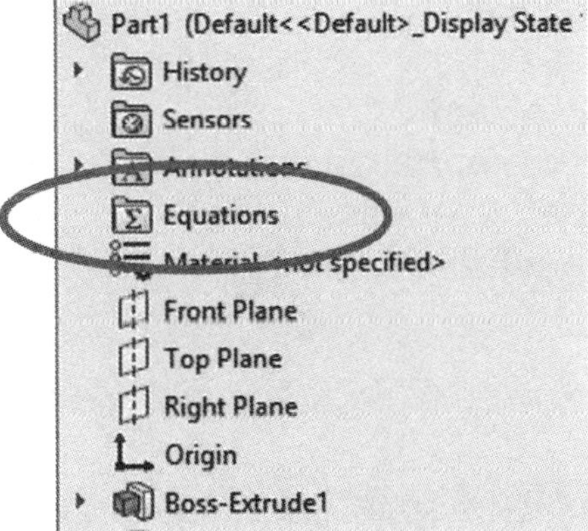

7.11. - To add the second equation we'll use a different approach. Double click in the *"Qty2"* dimension as if to change its value. In the second line, where we enter the dimension's value, type **'='** (equal sign); now we are ready to enter the equation just as we did in the previous step but using "*Length2*" and "*Spacing2.*" Remember we can select a dimension to copy its name into the equations editor.

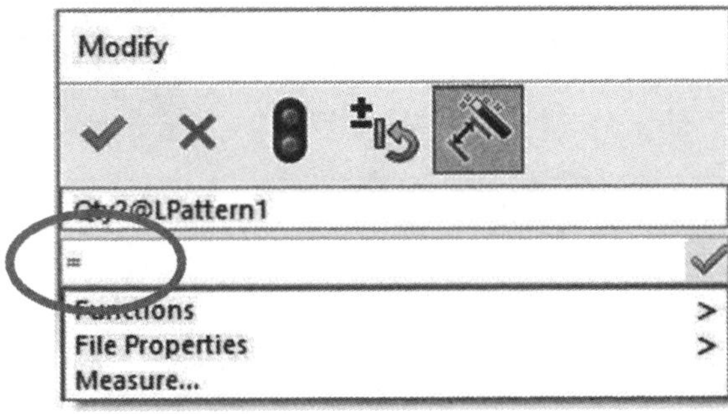

7.12. - Just as we did with the first equation, we'll open a parenthesis immediately after the equal sign, select the *"Length2"* dimension, enter '**- 0.375**', close the parenthesis, add '*/*' and select the *"Spacing2"* dimension. The completed second equation entered in the Modify dialog box looks like:

= ("Length2@Sketch1" - 0.375) / "Spacing2@LPattern1"

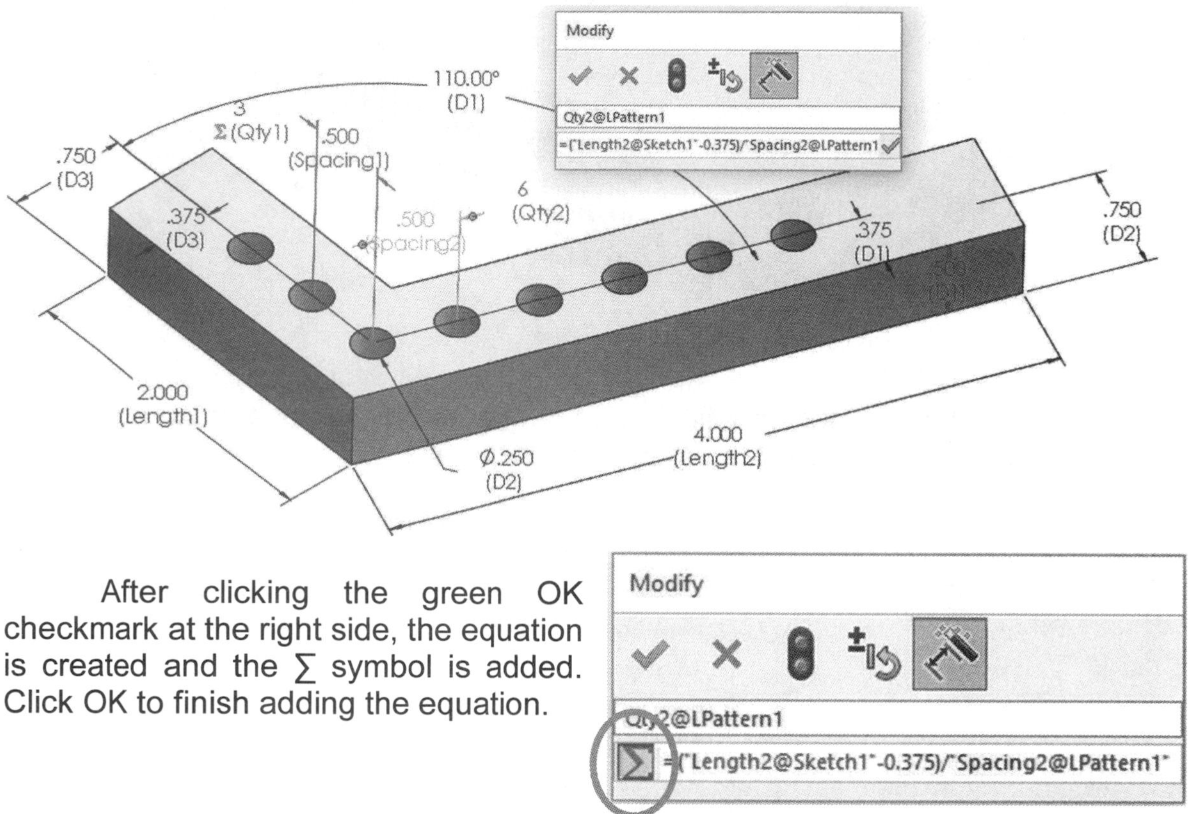

After clicking the green OK checkmark at the right side, the equation is created and the ∑ symbol is added. Click OK to finish adding the equation.

Now *"Qty2"* evaluates to 7. Both *"Qty1"* and *"Qty2"* have the red ∑ symbol letting us know these dimensions are driven by an equation.

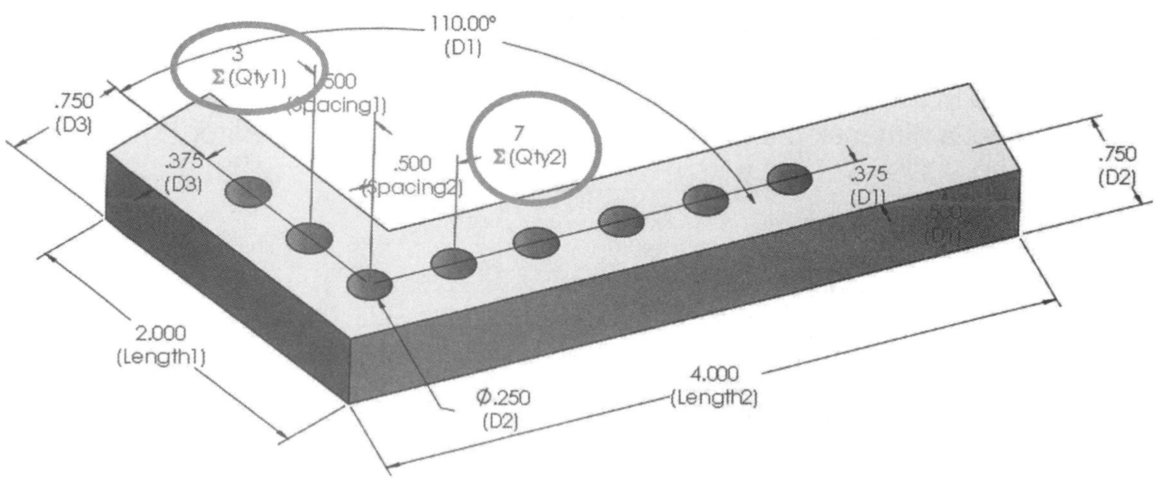

7.13. - To test the equations, change "*Length1"* to 2.75″ and *"Length2"* to 4.5″ and rebuild the model. Now we have 5 holes in the left side and 8 holes in the right side.

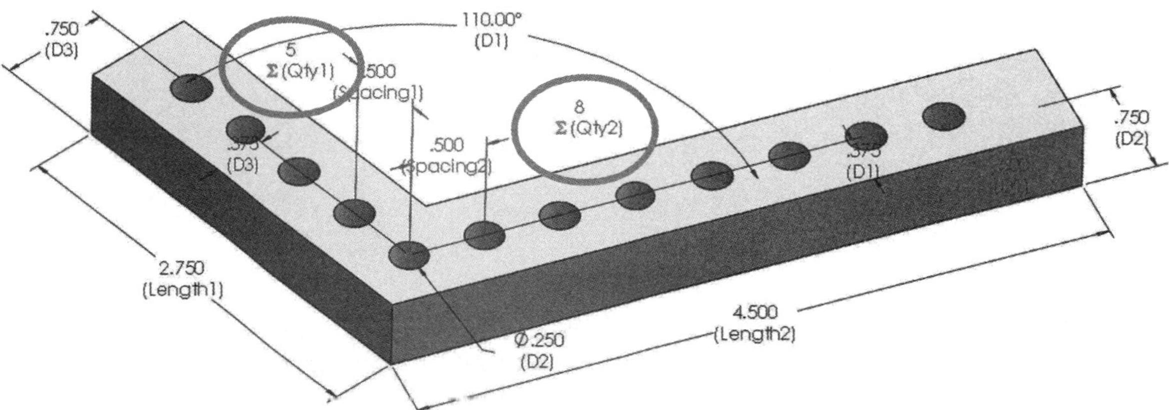

7.14. - Another way to maintain design intent is by using **Global Variables**. A Global Variable is essentially a constant with a name that can be used instead of a numeric value anywhere in a model; after creating a variable it can be used by itself in a value box and in equations. Double click in one of the 0.75″ dimensions and enter an '=' (equal sign) followed by the variable name *"Width."* After entering the name, a Global Variable icon is displayed. Clicking in the icon will create a new Global Variable and at the same time make the dimension's value equal to it. Click OK to continue.

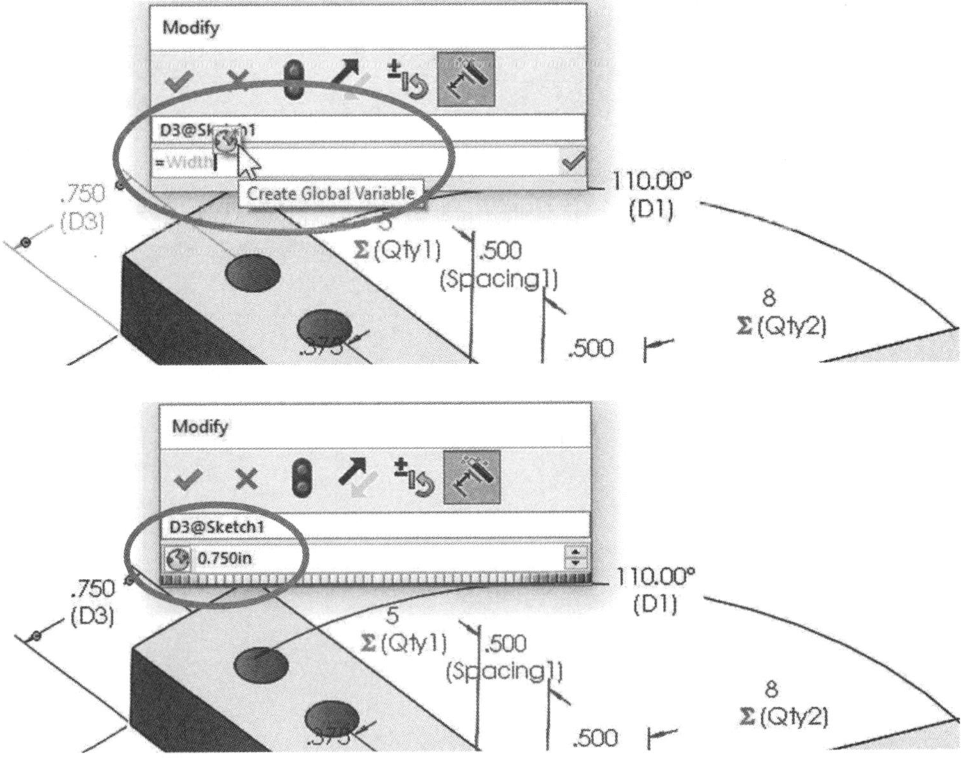

We can switch between the dimension's value and the Global Variable name using the world toggle button on the left side of the Modify dialog box.

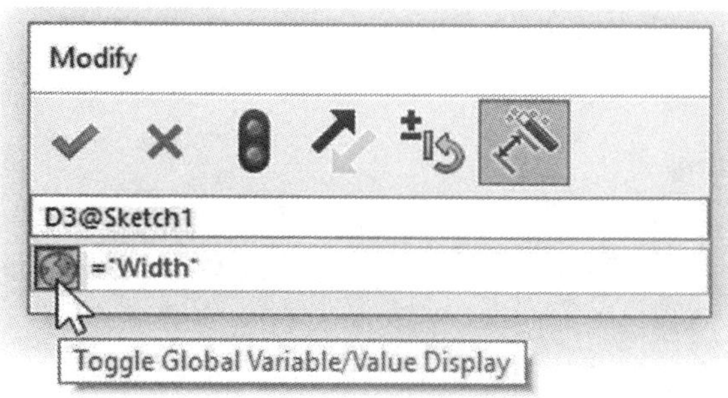

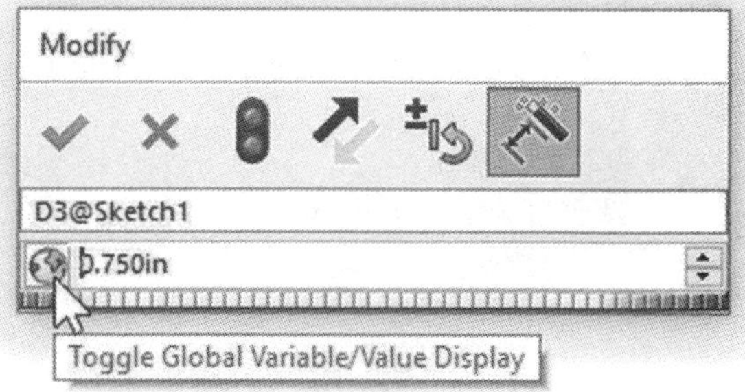

After creating the Global Variable, the value is displayed with the red ∑ symbol letting us know that the value is now driven by an equation. What happened was a Global Variable was created, and at the same time an equation was added making this dimension equal to the *"Width"* variable.

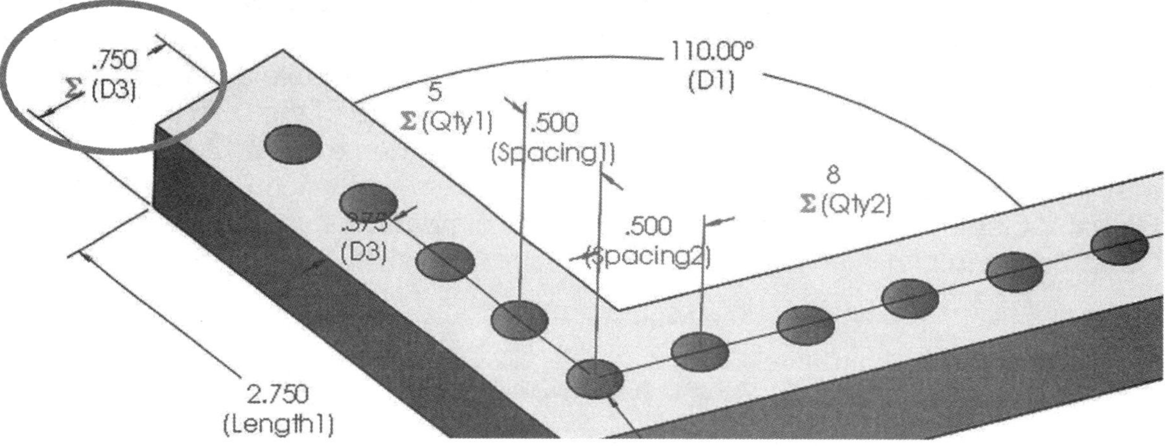

7.15. - To illustrate how to make a dimension equal to a Global Variable, double click in the other 0.75" dimension, enter an '**=**' (equal sign), and from the drop down menu select *"Global Variables, Width"* and click OK to finish.

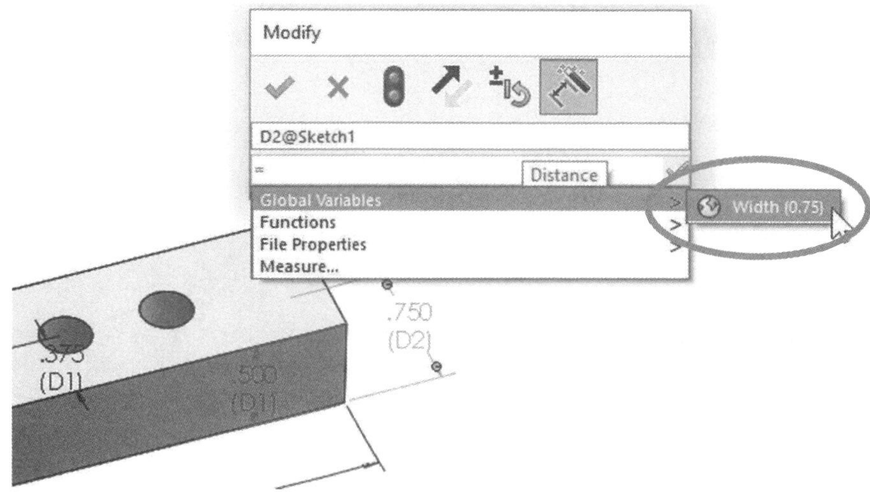

What we just did was to create a new equation making this dimension equal to the Global Variable called *"Width."* Global Variables are listed under the "**Equations**" folder in the FeatureManager and in the "**Equations**" dialog box. To change a Global Variable or an equation, right mouse click in the "**Equations**" folder and select "**Manage Equations**."

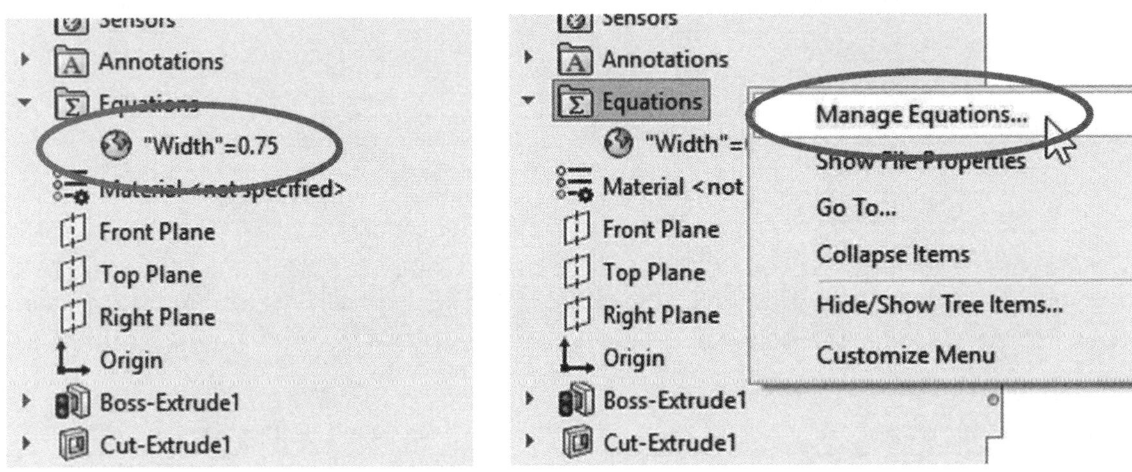

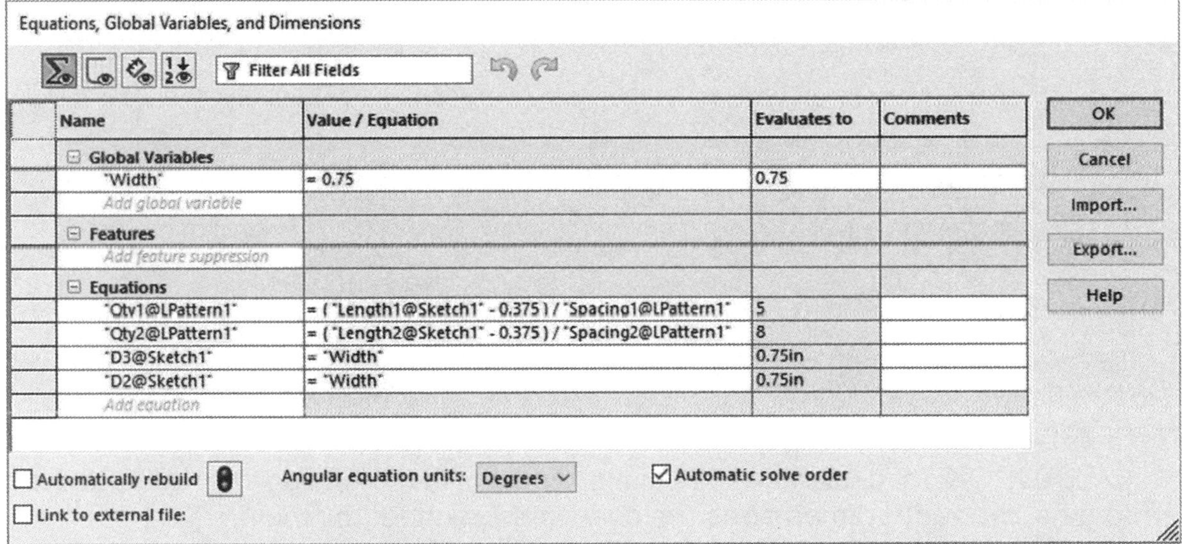

A dimension driven by an equation cannot be changed directly (*"Qty1"* and *"Qty2"*), only the referenced dimensions (*"Length1," "Length2," "Spacing1"* and *"Spacing2"*).

Using Global Variables in a model is an efficient way to update multiple dimensions with the same value at the same time.

7.16. - In the **"Equations"** dialog we can Add, Edit, or Delete equations. To delete an equation, right-mouse-click in the equation, and select the option from the menu. Delete the last two equations (the "Width" dimensions) to learn a different way to make multiple dimensions' value the same.

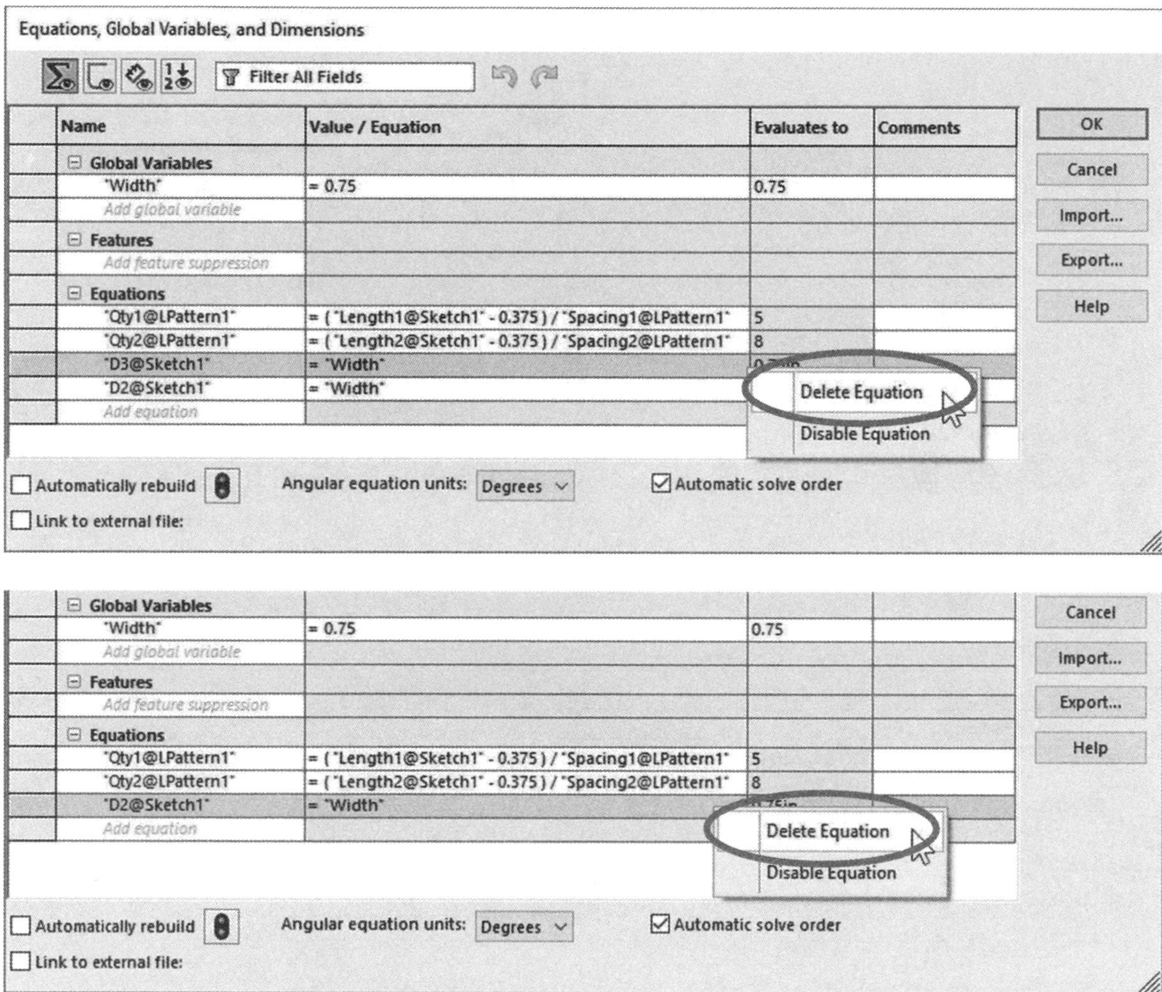

Click OK to close the equations dialog. After deleting the two previous equations, the width dimensions are no longer preceded by the red $\sum$.

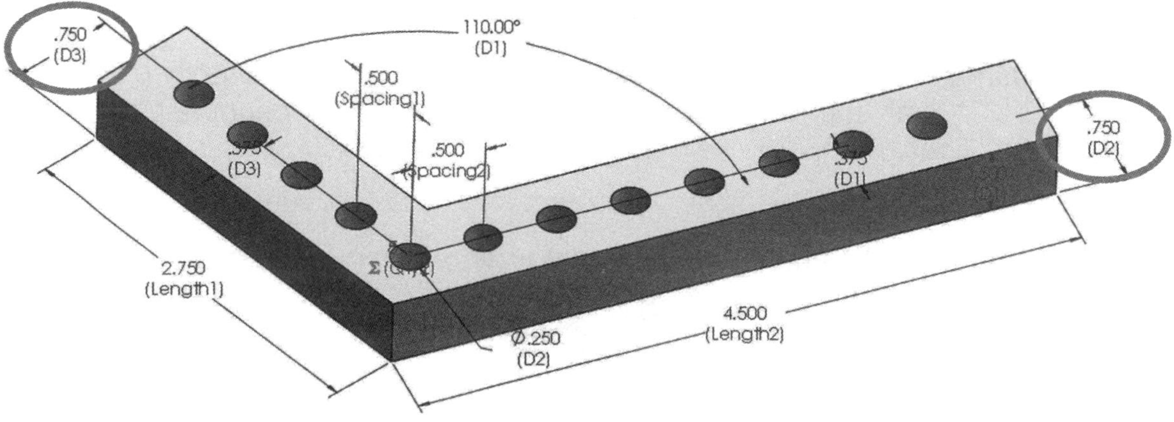

7.17. - A different way to make multiple dimensions equal to each other is by using the "**Link Values**" command. While similar to making dimensions equal to a global variable, the difference is that we don't have to create a variable, and unlike equations where the Global Variable must be modified in the equations editor, linked values can be changed directly in any linked dimension. Right-mouse-click in one of the width dimensions, and select "**Link Values**."

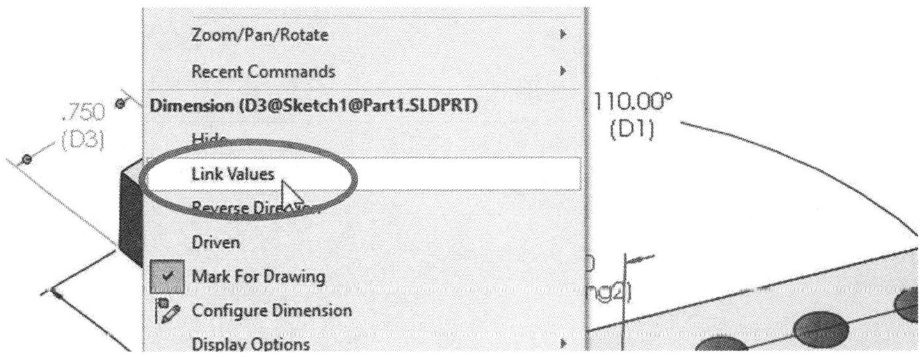

In the "Shared Values" window enter *'Link_Width* in the name field. (Do not use *'Width'*; the name was already used for a Global Variable.) Click OK to continue. Note that we cannot change the dimension's value here.

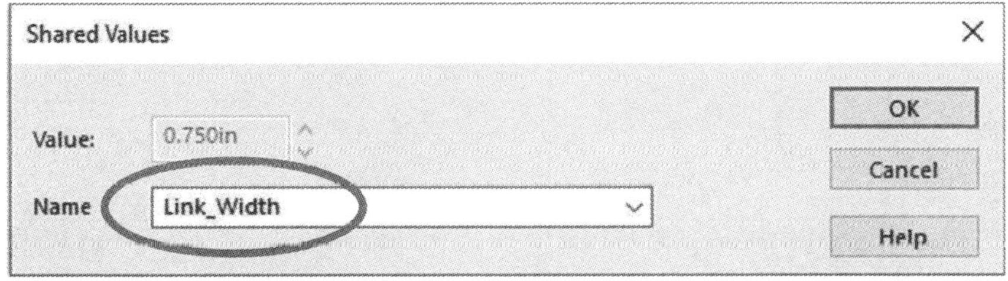

7.18. - After linking the dimension, its name is changed to *"Link_Width"* and a red chain-link icon is added letting us know it is linked to a shared value.

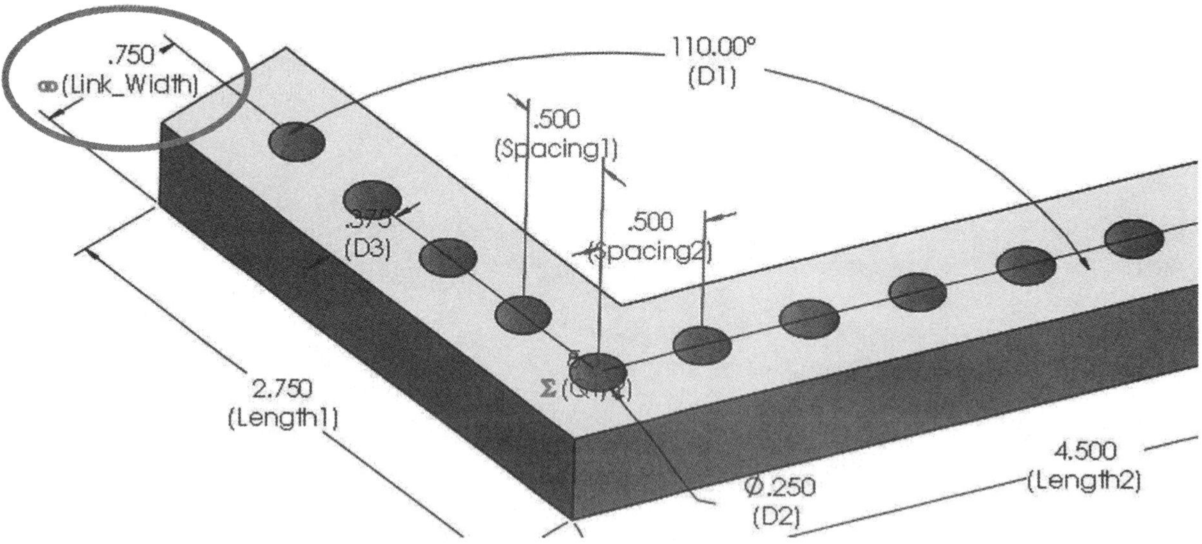

7.19. - To link the second dimension, right-mouse-click in it and select "**Link Values**" as we did for the first dimension, but instead of entering a name, select *"Link_Width"* from the drop-down list and click OK. Notice the previously made *"Width"* global variable is also listed as an option formatted as *"$VAR: var_name."*

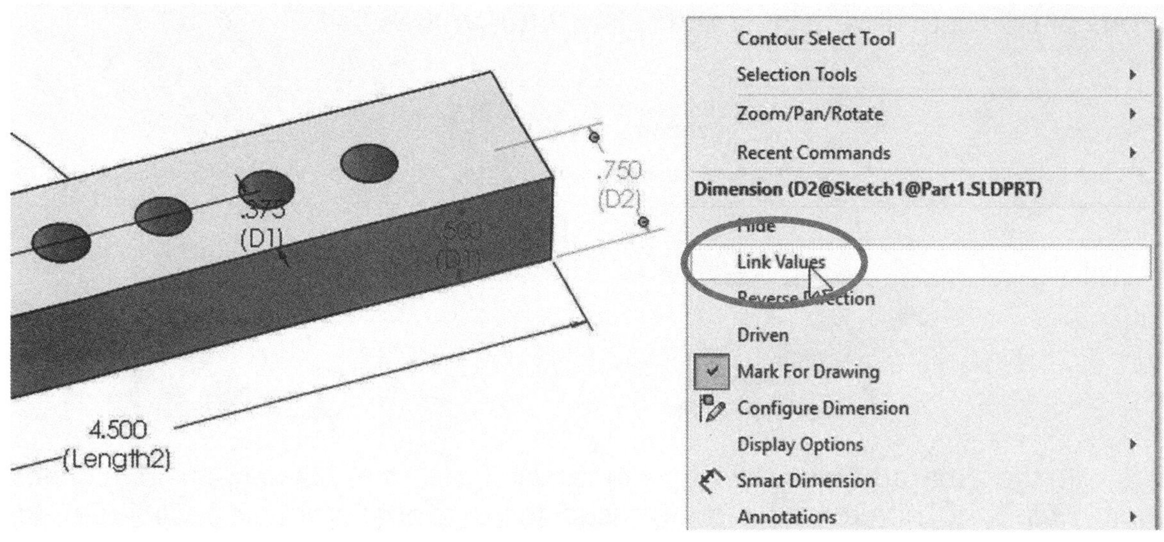

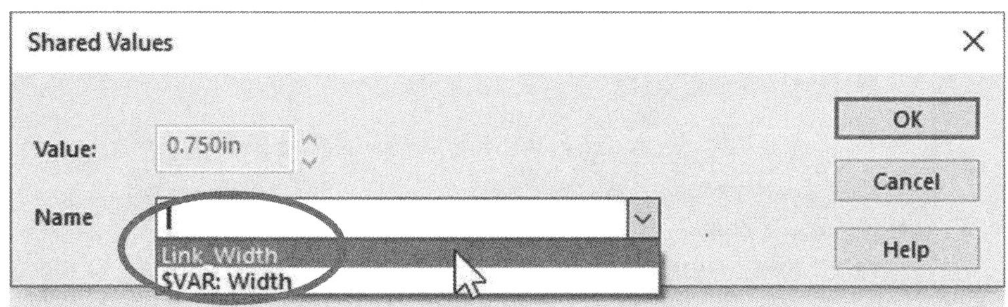

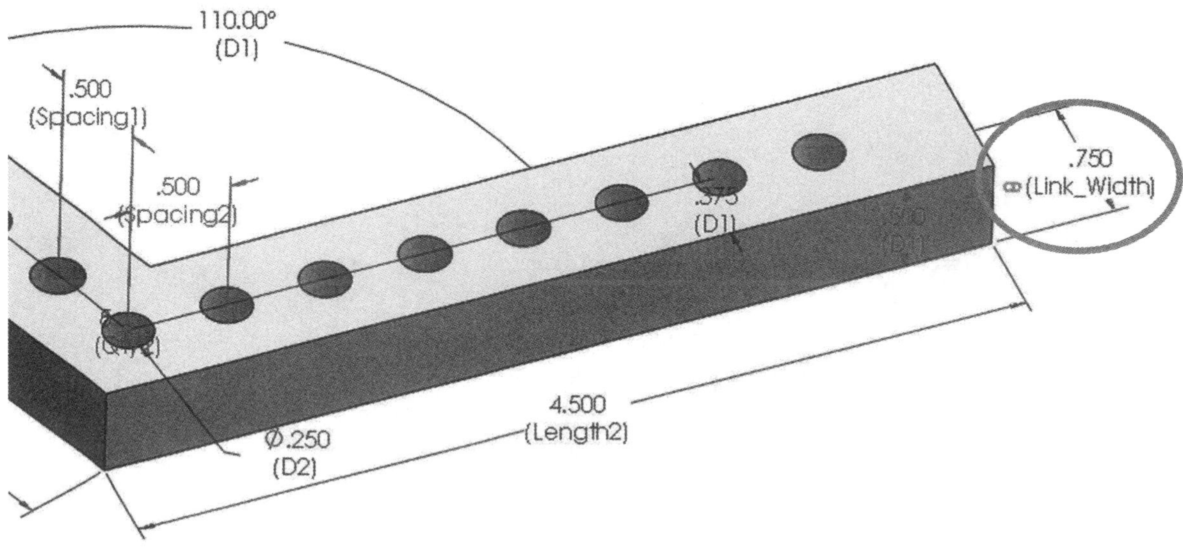

7.20. - Change the value of either linked dimension to 1″ and rebuild the part to see the other dimension update at the same time. Note the chain-link icon preceding the value alerting us that the dimension is linked to other dimensions.

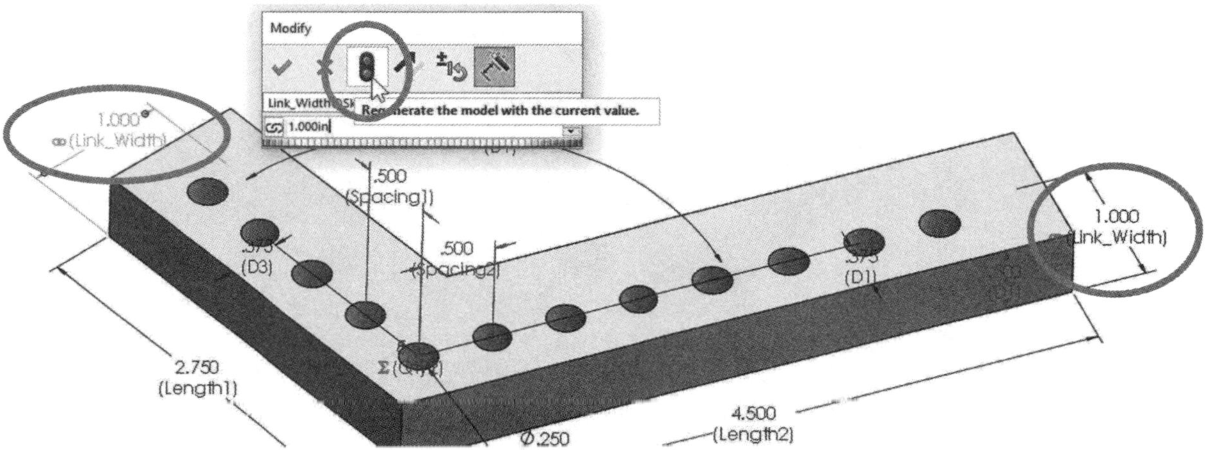

To un-link a value right-mouse-click and select "Unlink Value" from the menu.

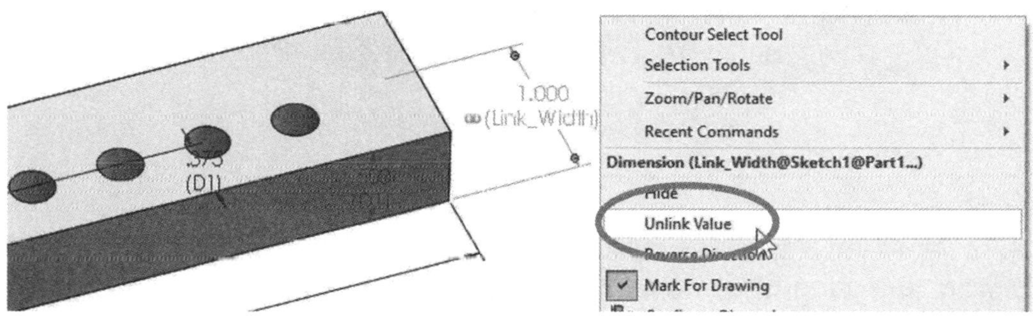

7.21. - Add two new equations to center the holes about the *"Link_Width"* dimension. Optionally rename the 0.375" dimensions. After rebuilding the part, the holes will be centered.

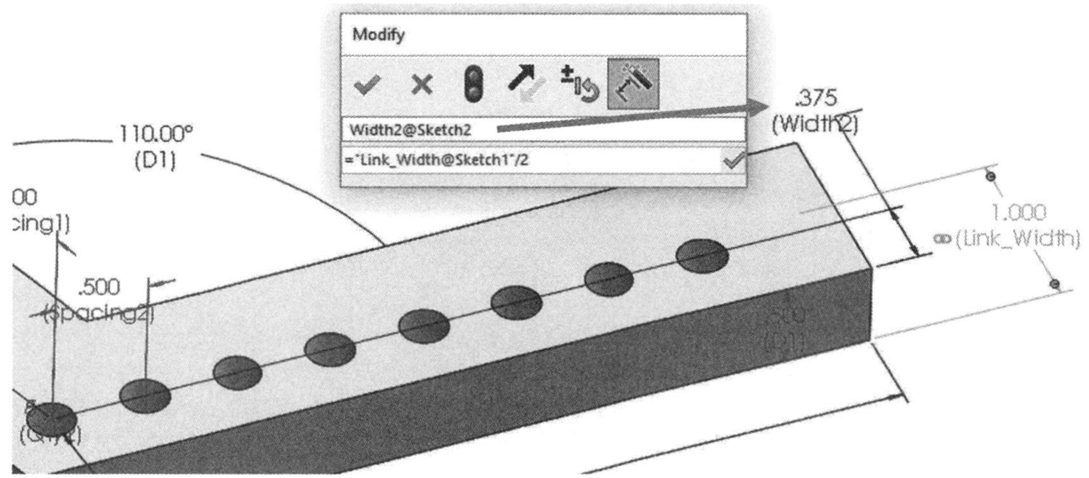

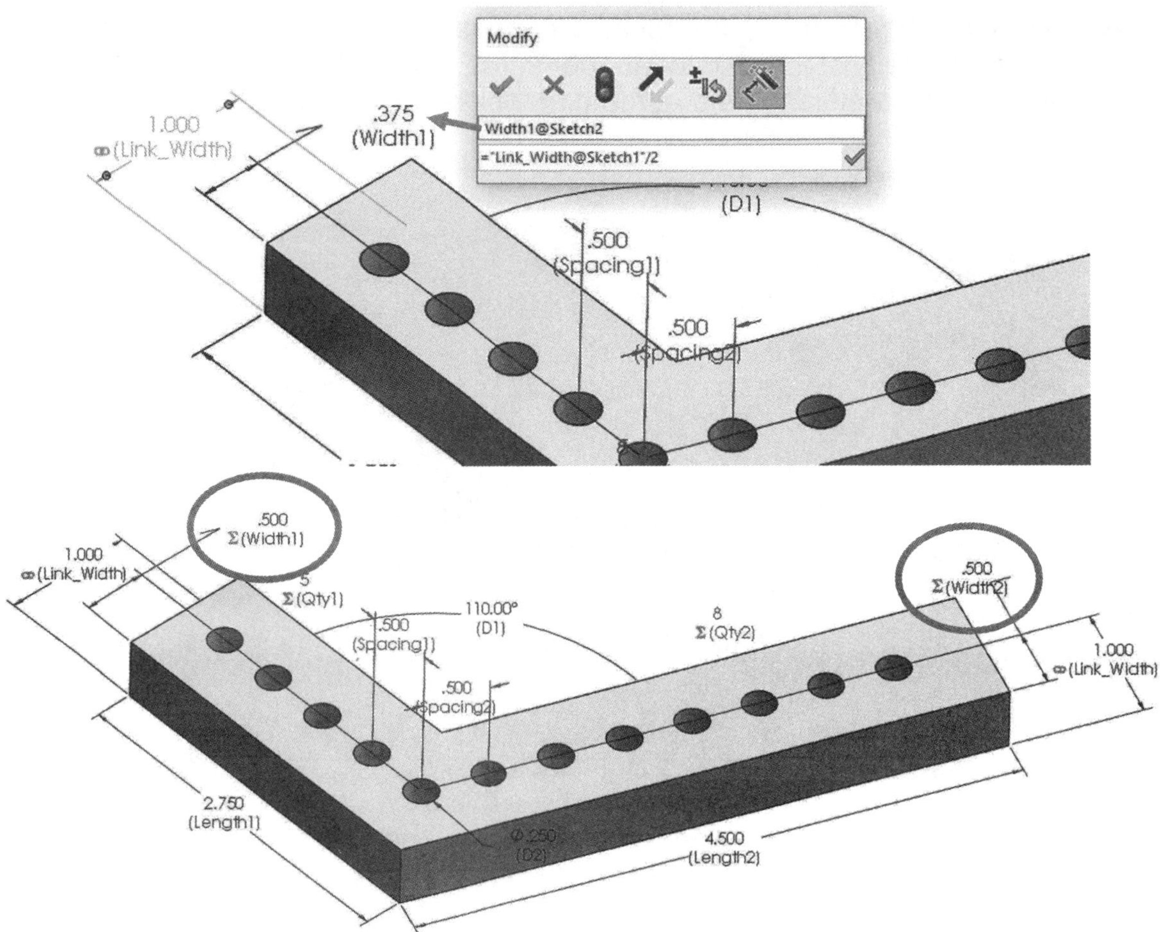

7.22. - To learn how to edit equations, and illustrate the effect of not subtracting 0.375″ from the length, remove the '**- 0.375**' portion from the two equations (it doesn't matter if we leave the parenthesis or not). Right-mouse-click in the "*Equations*" folder and select "Manage Equations."

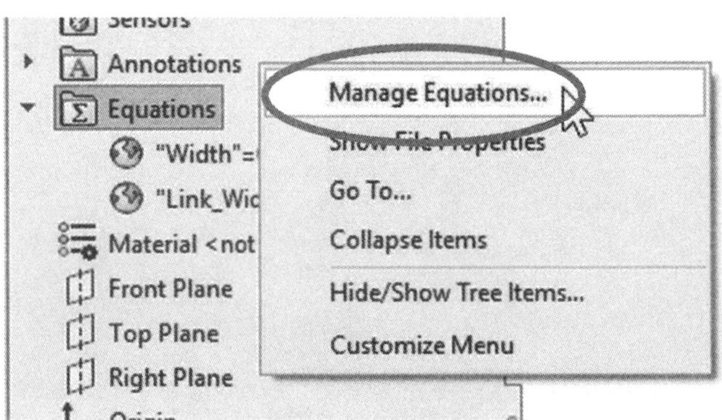

The equations should now read:

"Qty1@LPattern1" = ("Length1@Sketch1") / "Spacing1@LPattern1"
"Qty2@LPattern1" = ("Length2@Sketch1") / "Spacing2@LPattern1"

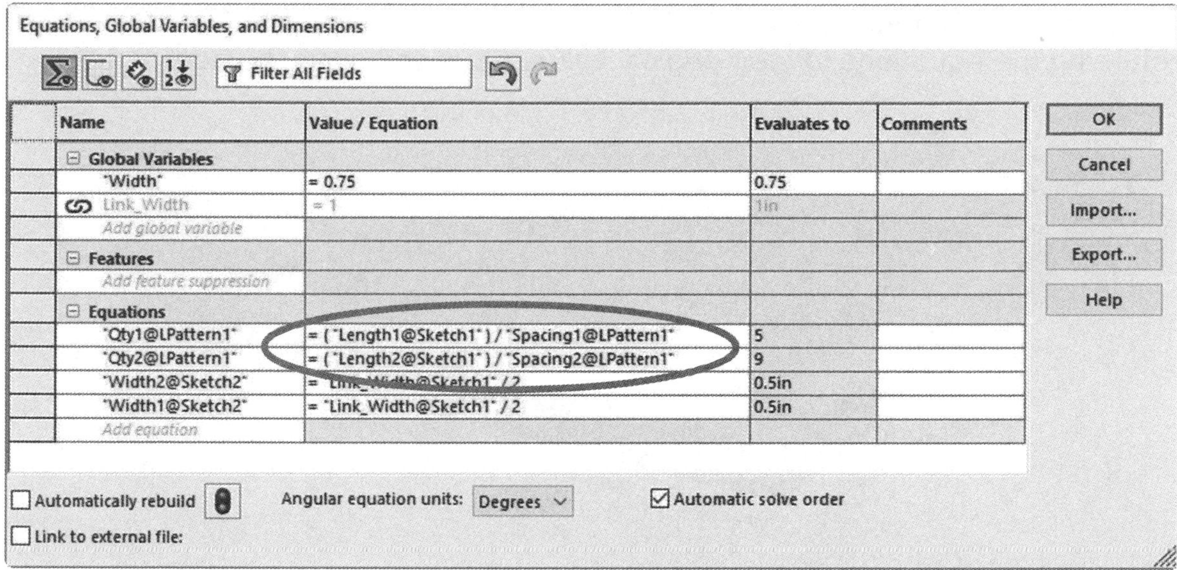

7.23. - Now the equations evaluate to 5 and 9 respectively. We can see the effects of the changes by rebuilding the model at the bottom of the dialog or having them automatically rebuild as we make changes to them. Click OK to finish and rebuild the model to see the effect of the change.

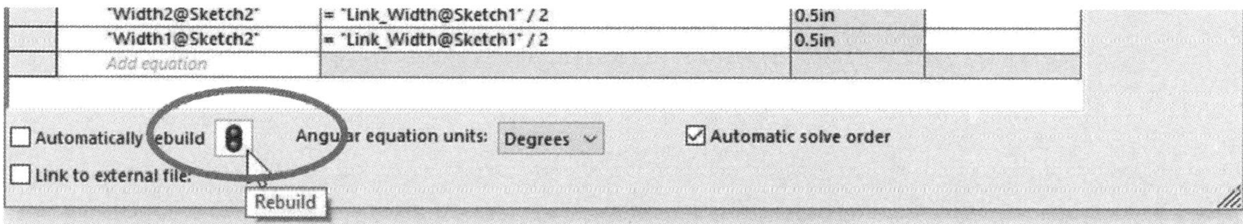

What happened with this change is that a sixth instance is added that breaks the edge of the part in the short side and gets very close to the edge in the long side; the equation resolves to 5.5 and is rounded up to 6. That is the reason why we subtracted 0.375″ from the length in the equations.

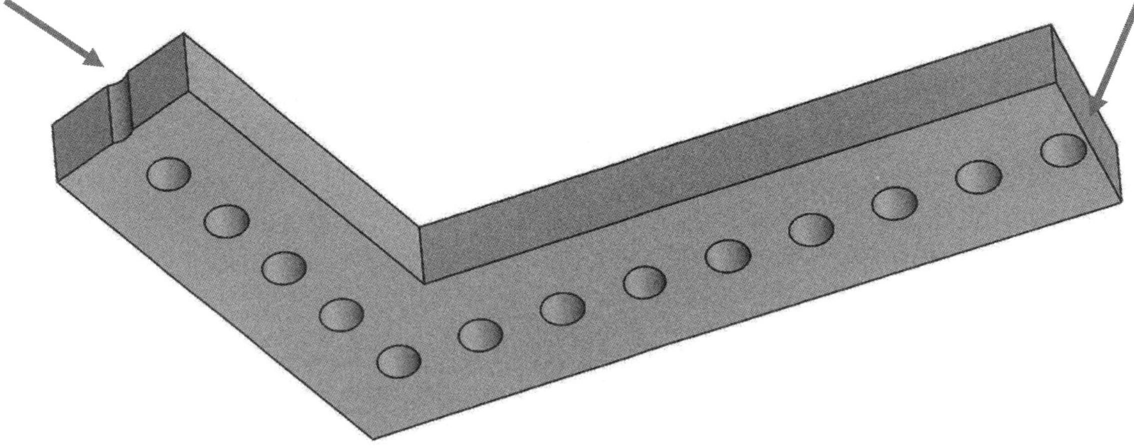

7.24. - Edit both equations again to subtract the 0.375″ distance from the length, returning the equations to their original value. Save and close the part.

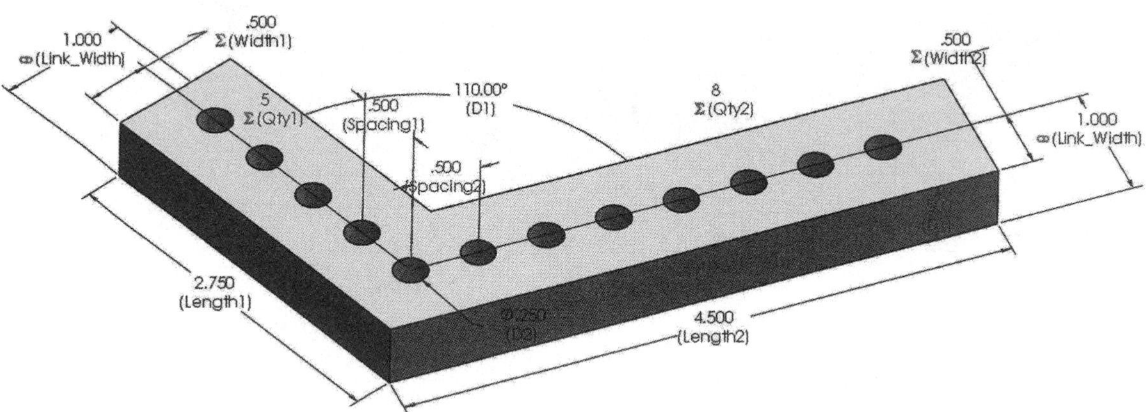

Exercise: Open the file *'Equations Exercise.sldprt'* from the exercise files and give all dimensions used in an equation a meaningful name. Add equations and use either Global Variables or Link Values to:

a) Make the bottom thickness equal to the wall thickness (10mm dimension).
b) Make the number of "Copies" equal to the outside diameter divided by 8.
c) Make the "*ShaftCut*" feature diameter ¼ of the "Body" inside diameter.

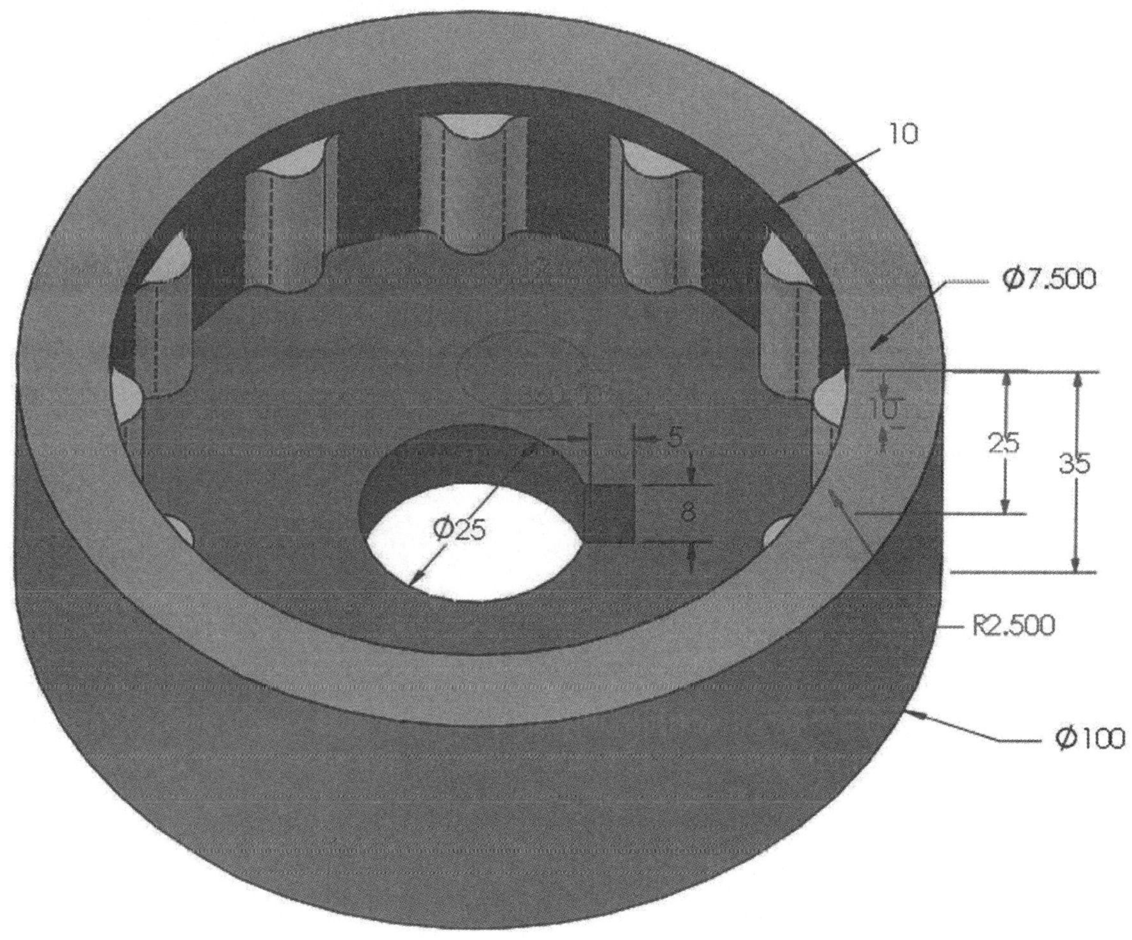

In the SOLIDWORKS equations, we can use almost any algebraic expression to evaluate values. As a general guide, you can write equations with the same algebraic format as you would in Excel formulas.

Notes:

Sheet Metal and Top Down Design

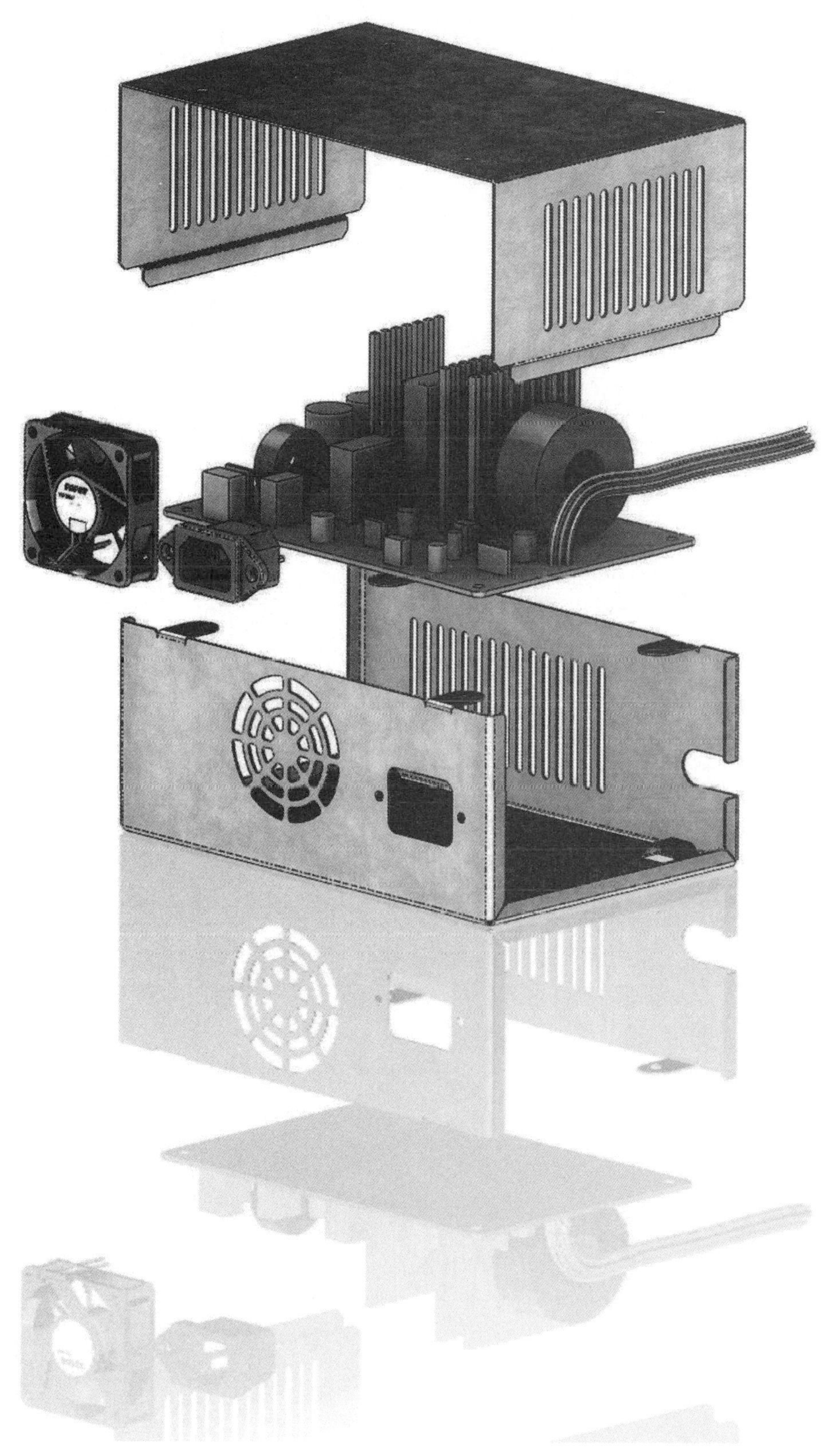

Notes:

Sheet Metal Design

When we talk about sheet metal components, we are talking about trimming and bending thin sheets of metal beyond their yield strength causing a permanent deformation. Sheet metal components are made from steel, aluminum, bronze or pretty much any malleable metal and can produce components with medium to high tolerances; this manufacturing process is well suited for high volume production where components need to or can have flexibility in an assembly process, and/or the final product's appearance does not negatively affect the cosmetic requirements of the final product, like the power supply or the locker we'll be making in this lesson.

Due to the inherent nature of the process of cutting and bending metal, this manufacturing technique is used mostly for internal components that are "out of sight" or don't have high aesthetical requirements, since the resulting components are not always as pretty or good-looking as a plastic injection molded part and have a more "industrial" look. A very good example is a computer's case. If you look at the back of your PC's case, most likely you'll see it is made from sheet metal, and the front (more visible parts) are probably made of plastic to give it a more attractive look as a consumer product.

Sheet metal tends to be a relatively cheap manufacturing process, and lends itself very well for mass produced components. One way of making large quantities of sheet metal components is using a progressive die. In this process, multiple dies are set up one after the other in a large press, and a long metal strip is fed in one end. Every time the press comes down, each die in the press performs a different cut or bend operation. When the press goes up, the metal strip advances one station, the press comes down again, a new bend or cut is added to the previous step and so on, until a finished part is completed with the last die, which usually cuts the part out of the strip. A progressive die is best suited for smaller component manufacturing.

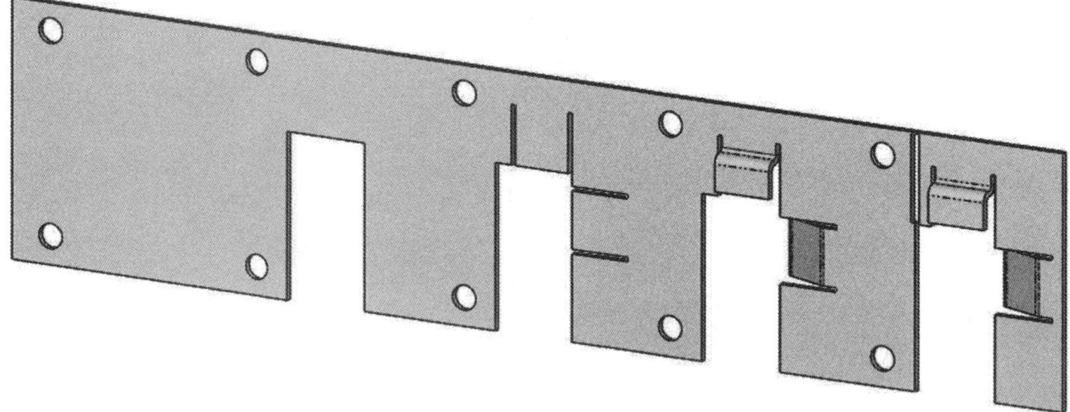

Example of a progressive die sheet metal part. The holes at the top and bottom are used to feed the metal strip in the press, and advance to a new station.

A different way to manufacture sheet metal components is to cut the sheet metal part's flat pattern from a larger sheet using a computer controlled torch, plasma, laser, water jet machines or even by hand, and then bend the part using different presses and bending tools; this process is often used with larger components where using a progressive die is not practical due to size.

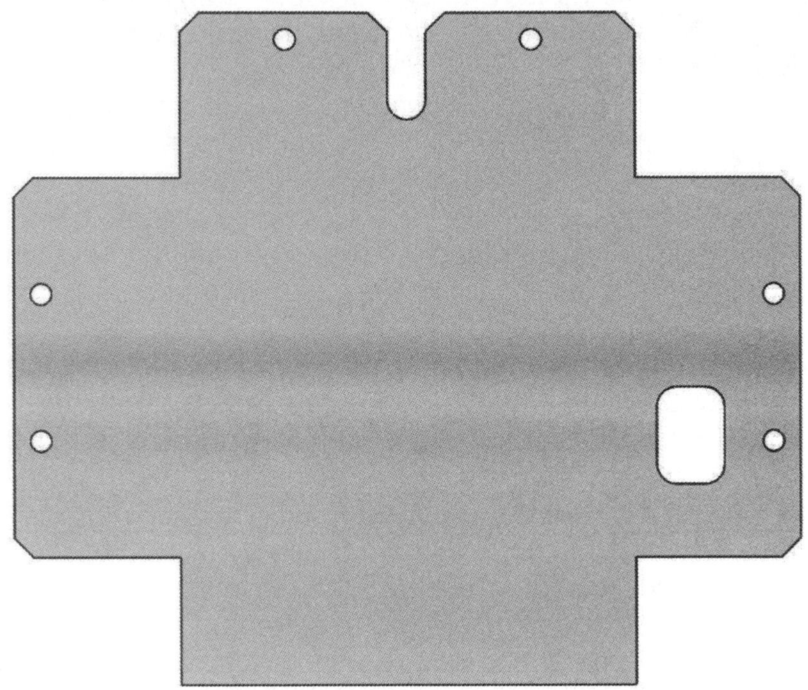

Flat pattern before bending

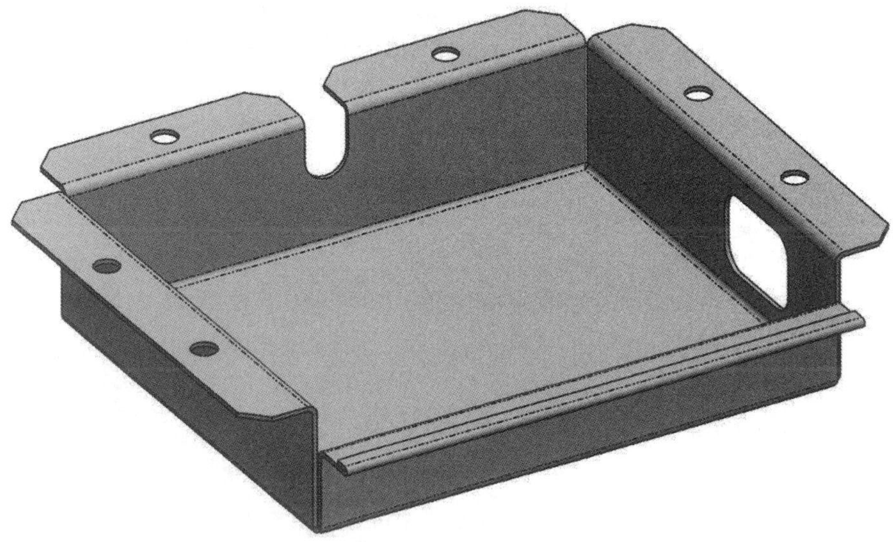

Finished part after bending the flat pattern

8.1. – In SOLIDWORKS, the Sheet Metal environment has its own set of commands and process specific tools. We'll learn how to use most of them in this lesson.

Create a new part, and if not already visible, turn on the Sheet Metal toolbar in the CommandManager and/or individual toolbar either by a right-mouse-click in the CommandManager's tabs and selecting "**Sheet Metal**" from the pop-up menu, or in the menu "**View, Toolbars, Sheet Metal**."

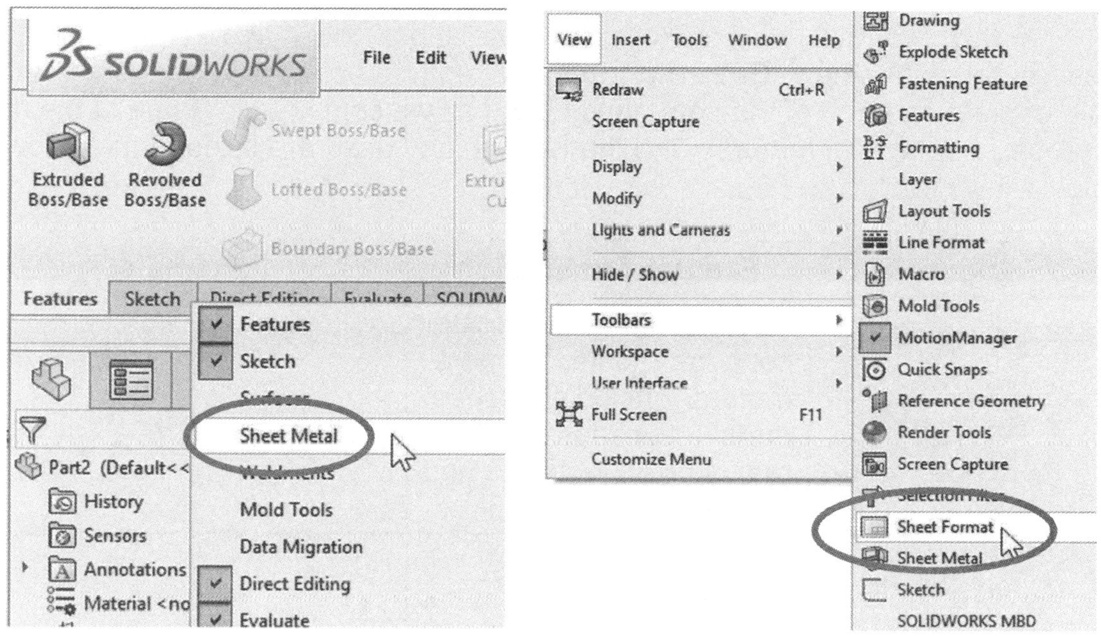

The Sheet Metal toolbar has most of the commands required to make a complete sheet metal part, and looks like this:

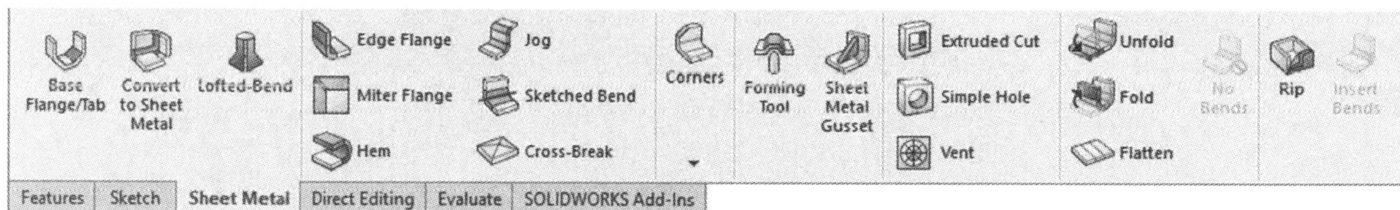

Command Manager

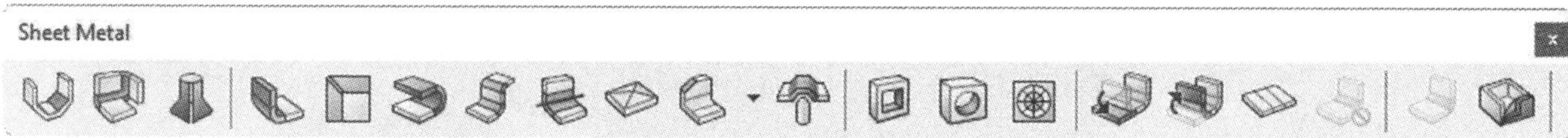

Floating Toolbar

8.2. – To design sheet metal components SOLIDWORKS offers a full set of process specific commands to complete just about any sheet metal feature needed. A sheet metal part can be done by:

a) Making a part in the bent shape and add other sheet metal features to it.

b) Start with an unbent sheet metal part (flat pattern), and add bends and features to complete the part.

c) Convert a solid model to sheet metal by adding bends and rips at the edges.

In this example, we'll make a sheet metal box cover. We'll start making our sheet metal part using a "**Base Flange**," and from this point we'll add more features while explaining each of the commands used as we complete it. Some of the most common sheet metal features and the terminology used are listed in the following table:

Icon	Feature Description	Example
Sheet-Metal	The Sheet-Metal folder contains default parameters including sheet metal thickness, bend radius, relief type and related options.	N/A
Base Flange/Tab	The Base Flange/Tab will be the first sheet metal feature (or additional tabs) in the part. If it's the first feature, it will automatically add the required Sheet Metal features to the Feature-Manager (Sheet-Metal, Base-Flange and Flat-Pattern) and define the sheet metal parameters (thickness, bend radius, etc.).	
Base Flange/Tab	Tab: Tabs are used to add more material to an existing sheet metal part using a sketch. Tabs can be bent later.	

Flat-Pattern Flat-Pattern	The Flat-Pattern folder includes the part's flat pattern (unbent). It is suppressed when the part is bent and un-suppressed when flat.	
Convert to Sheet Metal	Used to convert a solid body, surface or imported model to Sheet Metal by adding rips and bends to allow the part to be flattened. When converting a solid body, the faces not connected by a bend are automatically removed from the model.	
Edge Flange	Used to add flanges to model edges. Flange shapes can be modified and do not need to be the full length of the edge. The flange's angle can be defined too.	
Miter Flange	Used to automatically add flanges to a series of edges connected by bends. A miter flange does not have to be flat, but it has to run the full length of the edge.	

Hem	A hem feature is used to fold an edge onto itself; it's used to eliminate sharp edges and reinforce them at the same time.	
Jog	A jog will add two bends on a model using a single sketch line.	
Sketched Bend	A sketched bend will bend a flat area using a sketch line.	
Closed Corner	Used to close corners after relief cuts are made. (Relief cuts are cuts made in a sheet metal part to facilitate its fabrication; these cuts can be added manually or automatically when other sheet metal features are added.)	

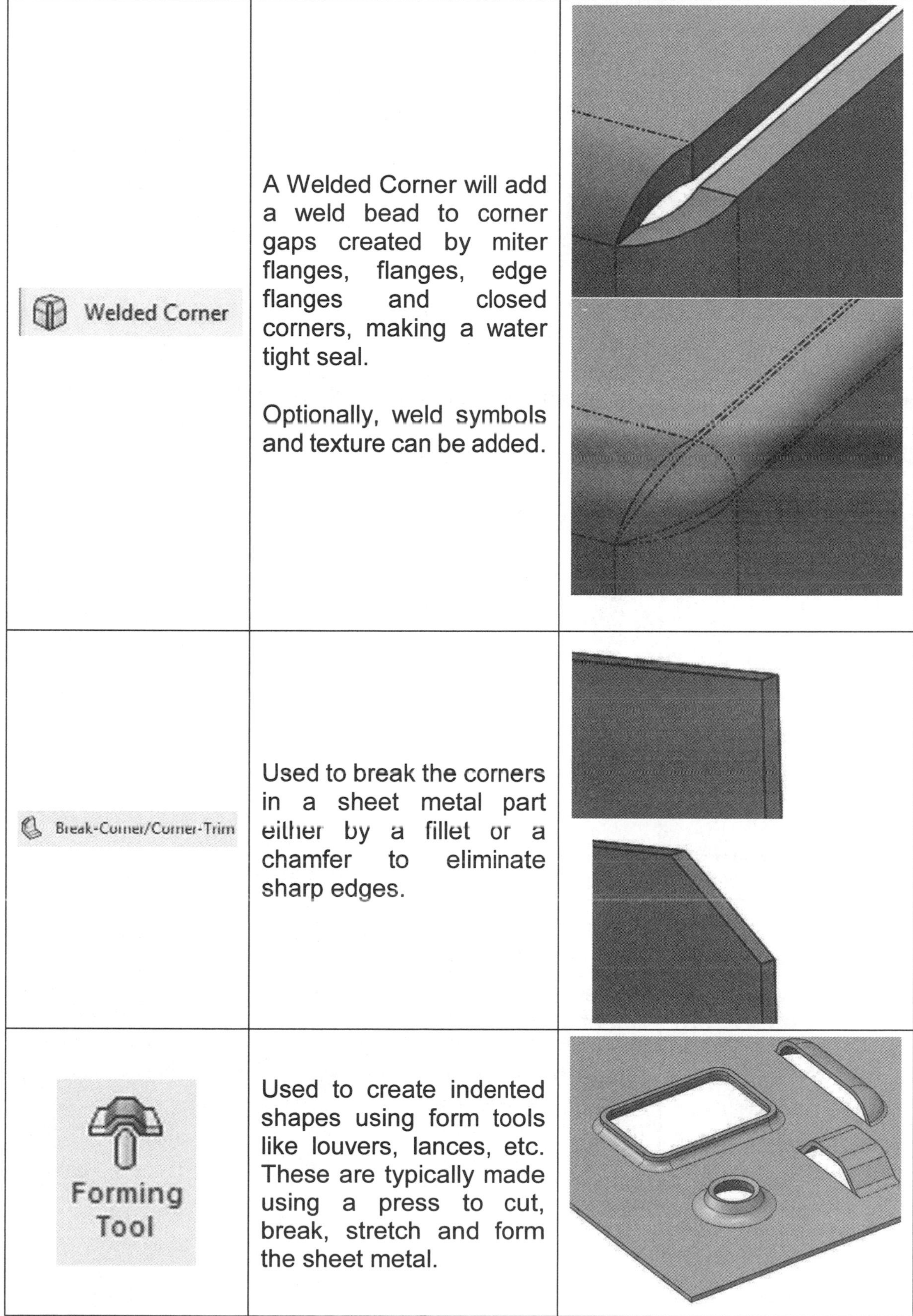

Welded Corner	A Welded Corner will add a weld bead to corner gaps created by miter flanges, flanges, edge flanges and closed corners, making a water tight seal. Optionally, weld symbols and texture can be added.	
Break-Corner/Corner-Trim	Used to break the corners in a sheet metal part either by a fillet or a chamfer to eliminate sharp edges.	
Forming Tool	Used to create indented shapes using form tools like louvers, lances, etc. These are typically made using a press to cut, break, stretch and form the sheet metal.	

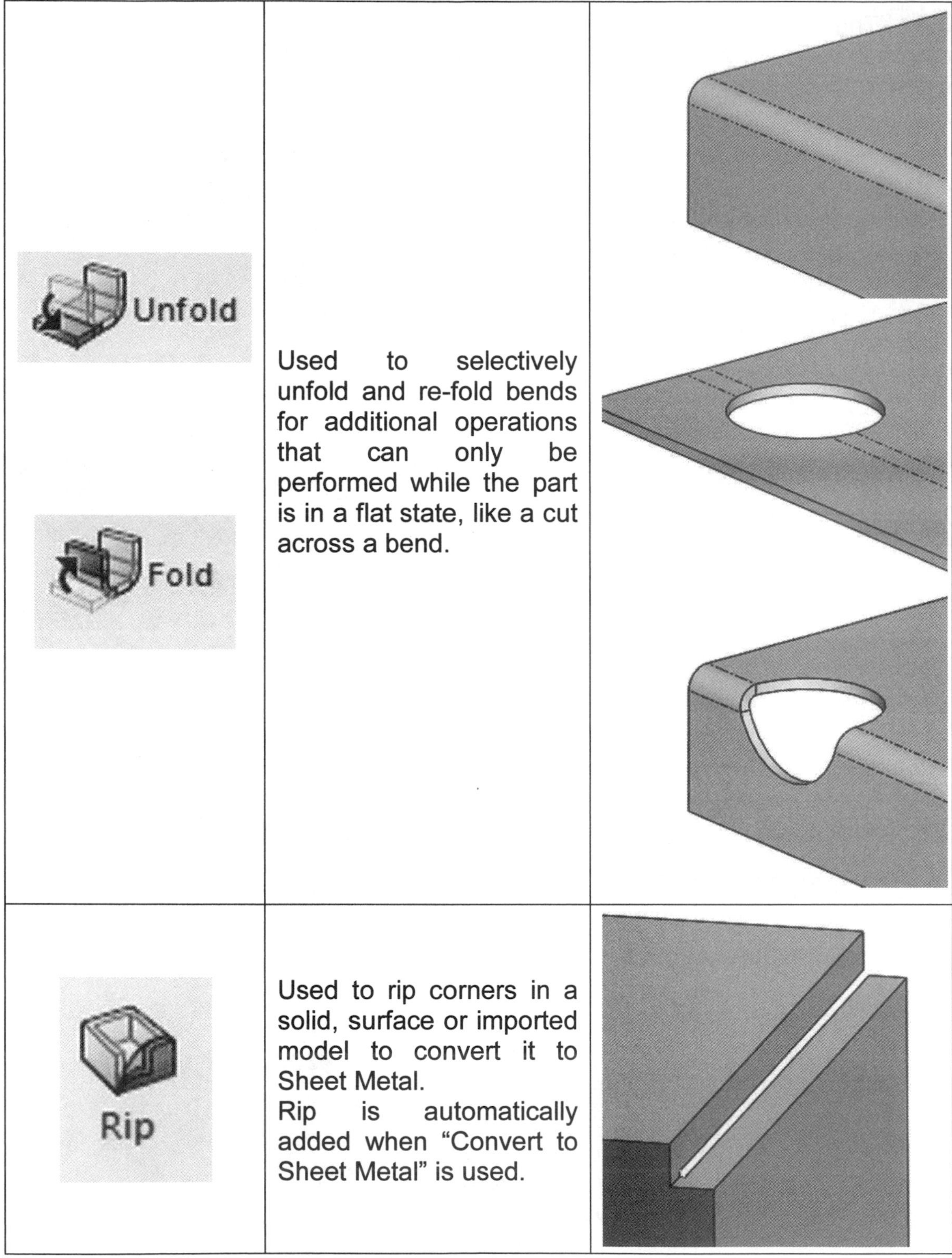

Unfold Fold	Used to selectively unfold and re-fold bends for additional operations that can only be performed while the part is in a flat state, like a cut across a bend.	
Rip	Used to rip corners in a solid, surface or imported model to convert it to Sheet Metal. Rip is automatically added when "Convert to Sheet Metal" is used.	

8.3. – The first feature of our sheet metal box will be a "Base-Flange." Add a new sketch in the Front plane and dimension as shown. Select the "**Base-Flange/Tab**" icon from the Sheet Metal toolbar in the CommandManager or the menu "**Insert, Sheet Metal, Base Flange**."

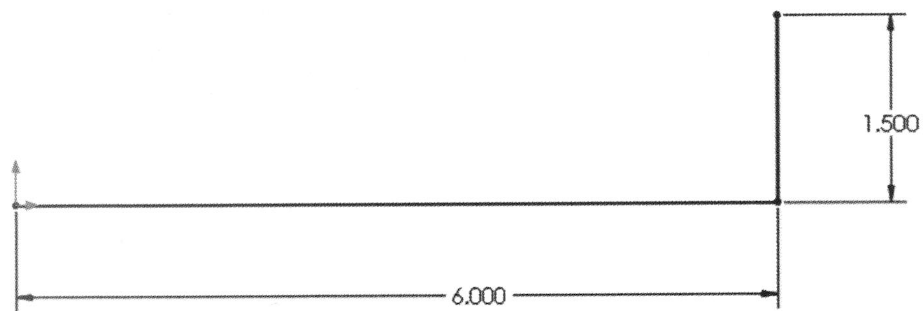

Since this is the first sheet metal feature, we are asked to define both the extrusion and the sheet metal parameters for the part. Enter the following values for the base flange, all values will be explained in the next section.

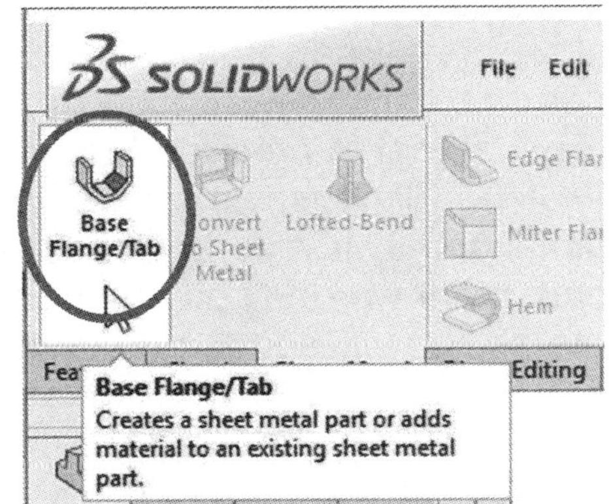

Make a 5" extrusion. In the "Sheet Metal Parameters" enter a **Thickness** of 0.03", a **Bend Radius** of 0.05" and set the "Reverse direction" checkbox ON to add the material's thickness inside the sketch. The sketch dimensions will be the part's outside dimensions. For **Bend Allowance** use a "**K-Factor**" of 0.5 (default value) and select the Rectangular "**Auto Relief**" type; check "Use relief ratio" with a valuc of 0.5.

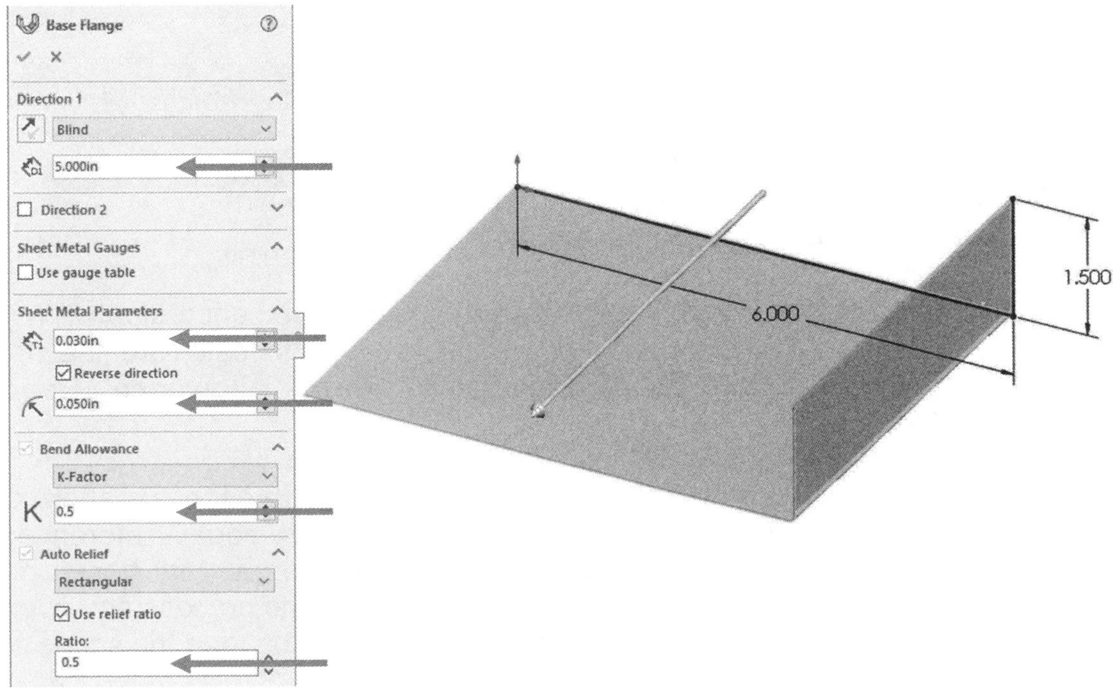

What the options in the “Sheet Metal Parameters” section mean:

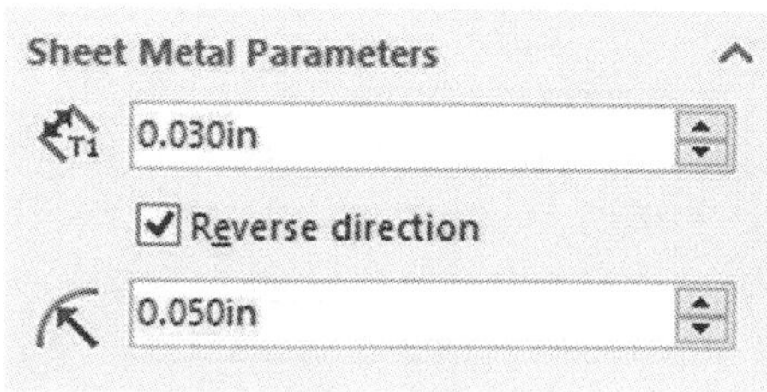

-Thickness of the material: This is the uniform thickness of the sheet metal part.
-Reverse direction defines the side of the sketch the material will be added to. In our case, our sketch defines the model’s outside dimensions.
-Bend Radius is the default inside radius for all bends in the part. Can be overridden for subsequent operations.

About the “**Reverse direction**” **checkbox:** A sheet metal part is a Thin Feature, therefore we need to define which side of the sketch we want the material to be added to; in other words, if the sketch dimensions are *inside* or *outside* dimensions. It may not seem to be important at first, but if we don't pay attention on which side of the sketch we add the material, the resulting part will be one material-thickness bigger or smaller, and that *may* ruin your design. Remember to pay attention to this important detail when designing sheet metal components, even if the sheet metal thickness is small and the part *may* allow it; after 2 or 3 bends are added and material thickness accumulates it will most definitively be a problem.

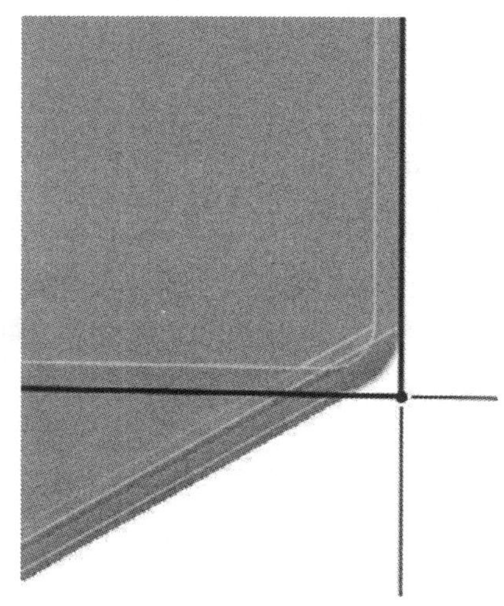

The “**Bend Allowance**” section refers to the way SOLIDWORKS calculates how much material will be consumed by a bend. When we bend a sheet metal component, the material on the inside face of the bend is compressed and the material outside is stretched. For example, if we take a 4″ long strip of metal, and bend it 90 degrees in the middle, the length of the flat portions plus the bent region will not necessarily add to 4″. SOLIDWORKS has multiple ways to calculate the flat pattern’s length: **Bend Table, Bend Calculation**, **Bend Allowance**, **Bend Deduction** and **K-Factor**.

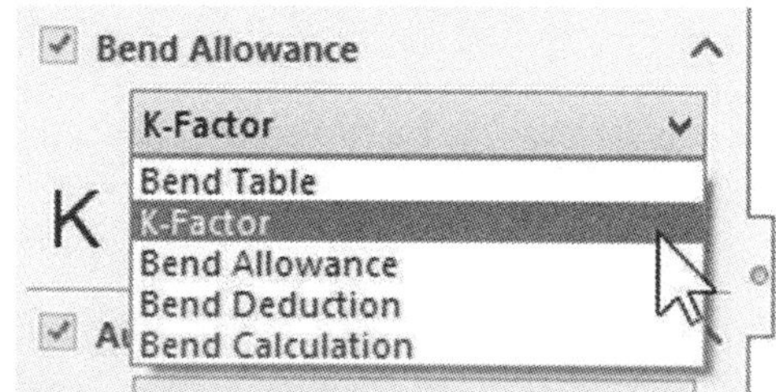

The method to calculate the bend allowance varies by preference, experience, availability of data, material, etc. There are numerous tables, books, references and guides to help us calculate it, and since these parameters vary based on material, thickness, bend angle, material’s grain direction, bend process and even temperature, we will not go into those details as they are well beyond the scope of

this book; to simplify our examples we'll direct the user to use a given bend allowance value for our examples.
Bend Allowance Methods:

- **Bend Table:** A Bend Table is an Excel or text file (usually one per material) with either a defined bend allowance or deduction; the table lists multiple combinations of material's thickness, bend radius and angle, from which SOLIDWORKS selects the correct value. Be aware that the Bend Tables included with SOLIDWORKS have a limited number of materials, and may not be accurate for everyone's needs, but a guide to correctly fill and format the table with your own data.

Unit:	Inches												
Type:	Bend Allowance												
Material:	**Soft Copper and Soft Brass**												
Comment:	Values specified are for 90-degree bends												
Radius	**Thickness**												
	1/64	**1/32**	**3/64**	**1/16**	**5/64**	**3/32**	**1/8**	**5/32**	**3/16**	**7/32**	**1/4**	**9/32**	**5/16**
1/32	0.058	0.066	0.075	0.083	0.092	0.101	0.118	0.135	0.152	0.169	0.187	0.204	0.221
3/64	0.083	0.091	0.1	0.108	0.117	0.126	0.143	0.16	0.177	0.194	0.212	0.229	0.246
1/16	0.107	0.115	0.124	0.132	0.141	0.15	0.167	0.184	0.201	0.218	0.236	0.253	0.27
3/32	0.156	0.164	0.173	0.181	0.19	0.199	0.216	0.233	0.25	0.267	0.285	0.302	0.319
1/8	0.205	0.213	0.222	0.23	0.239	0.248	0.265	0.282	0.299	0.316	0.334	0.351	0.368
5/32	0.254	0.262	0.271	0.279	0.288	0.297	0.314	0.331	0.348	0.365	0.383	0.4	0.417
3/16	0.303	0.311	0.32	0.328	0.337	0.346	0.363	0.38	0.397	0.414	0.432	0.449	0.466
7/32	0.353	0.361	0.37	0.378	0.387	0.396	0.413	0.43	0.447	0.464	0.482	0.499	0.516
1/4	0.401	0.409	0.418	0.426	0.435	0.444	0.461	0.478	0.495	0.512	0.53	0.547	0.564
9/32	0.45	0.458	0.467	0.475	0.484	0.493	0.51	0.527	0.544	0.561	0.579	0.596	0.613
5/16	0.499	0.507	0.516	0.524	0.533	0.542	0.559	0.576	0.593	0.61	0.628	0.645	0.662
Comment:	Extracted from Machinery Handbook - 26th Edition with permission from Industrial Press, Inc.												

- **Bend Allowance:** When we use the Bend Allowance method we specify a length of metal that will become the bend region, and it is calculated by adding a specified length to the flat regions.

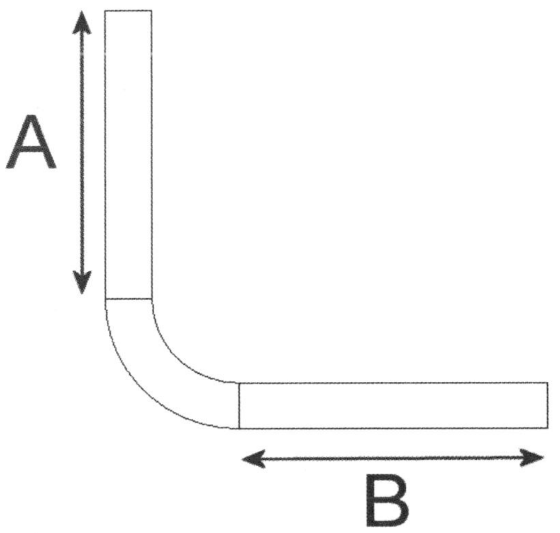

Flat Pattern Length = A + B + Bend Allowance (BA)

- **Bend Deduction:** If we use a Bend Deduction, we <u>subtract</u> a specified length to obtain the flat pattern length.

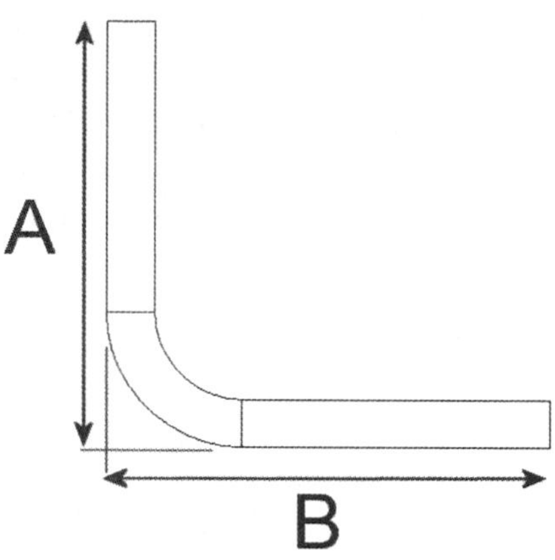

Flat Pattern Length = A + B – Bend Deduction (BD)

- **K-Factor:** As explained before, the material inside the bend is compressed and the material outside is stretched. The line between these regions is known as the **neutral axis**, which is the line that does not change length. The ratio between the distance of the neutral axis to the inside face and the sheet metal's thickness is known as the K-Factor.

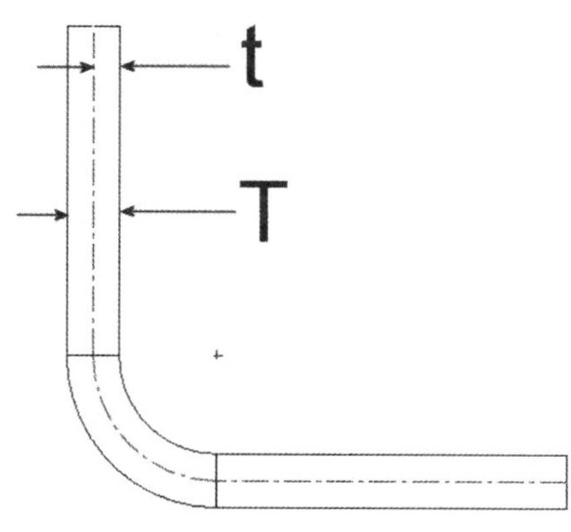

K-Factor = t / T

The bend allowance (BA) is then calculated by:

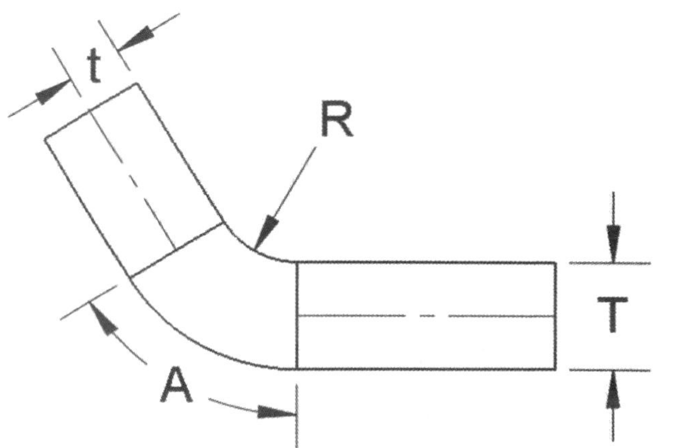

BA= π (R + KT) A/180

Where:

R= inside radius (Bend Radius)
A= Bend angle in Degrees
T= Material Thickness
K = K-Factor (t/T)

- **Bend Calculation:** The Bend Calculation method uses an Excel file which calculates the length of the sheet metal bend using equations, bend angle ranges, material, bend radius and K-Factor.

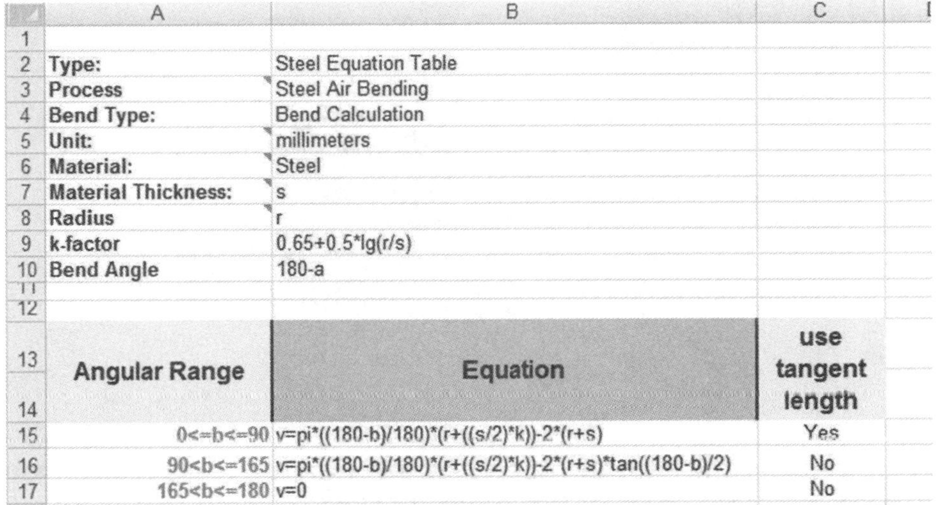

	A	B	C
1			
2	Type:	Steel Equation Table	
3	Process	Steel Air Bending	
4	Bend Type:	Bend Calculation	
5	Unit:	millimeters	
6	Material:	Steel	
7	Material Thickness:	s	
8	Radius	r	
9	k-factor	0.65+0.5*lg(r/s)	
10	Bend Angle	180-a	
11			
12			
13–14	Angular Range	Equation	use tangent length
15	0<=b<=90	v=pi*((180-b)/180)*(r+((s/2)*k))-2*(r+s)	Yes
16	90<b<=165	v=pi*((180-b)/180)*(r+((s/2)*k))-2*(r+s)*tan((180-b)/2)	No
17	165<b<=180	v=0	No

The "**Auto Relief**" section refers to the relief cuts generated by SOLIDWORKS for sheet metal fabrication processes. These small cuts (reliefs) are added at the end of bends to allow the metal to bend properly and not deform the material unexpectedly. The different types of Auto Relief available are:

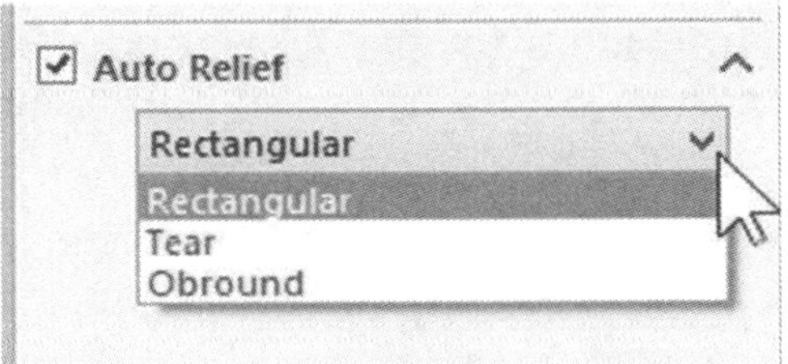

- **Rectangular:** In a rectangular relief, a cut is made using a thickness ratio or specifying relief dimensions. The ratio determines the cut's size (Dim) based on the sheet metal's thickness:

Dim = (Material Thickness) x (Relief Ratio)

- **Obround:** Is just like the Rectangular relief, but the relief's cut is rounded.

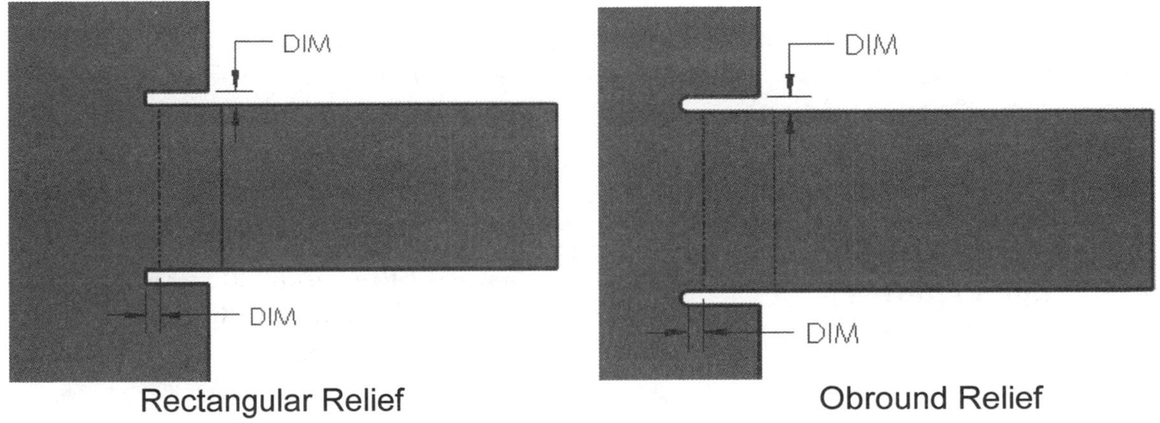

Rectangular Relief

Obround Relief

- **Tear:** In a tear the smallest possible relief cut is made, just enough to allow the metal to bend as intended. This is like ripping the metal at the time of bending during fabrication.

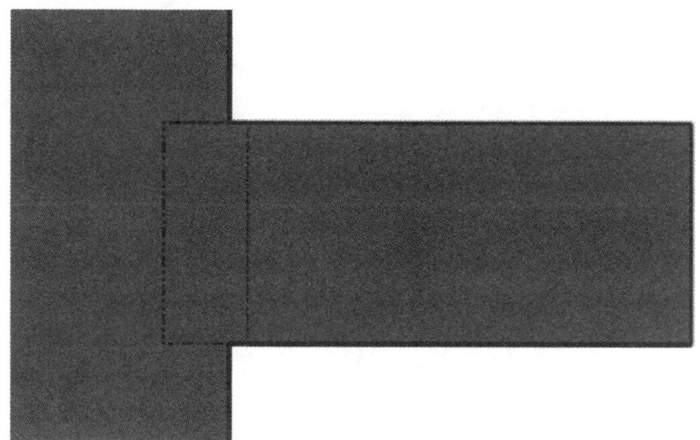

8.4. – The sheet metal features added to the FeatureManager include "*Sheet-Metal1,*" where we can change the part's thickness, default bend radius, bend allowance and auto relief type. The "*Base-Flange1*" is the extrusion's definition and parameters for the first bend, "*BaseBend1.*" The "*Flat-Pattern1*" feature contains options for the flattened part, which will be covered later in more detail.

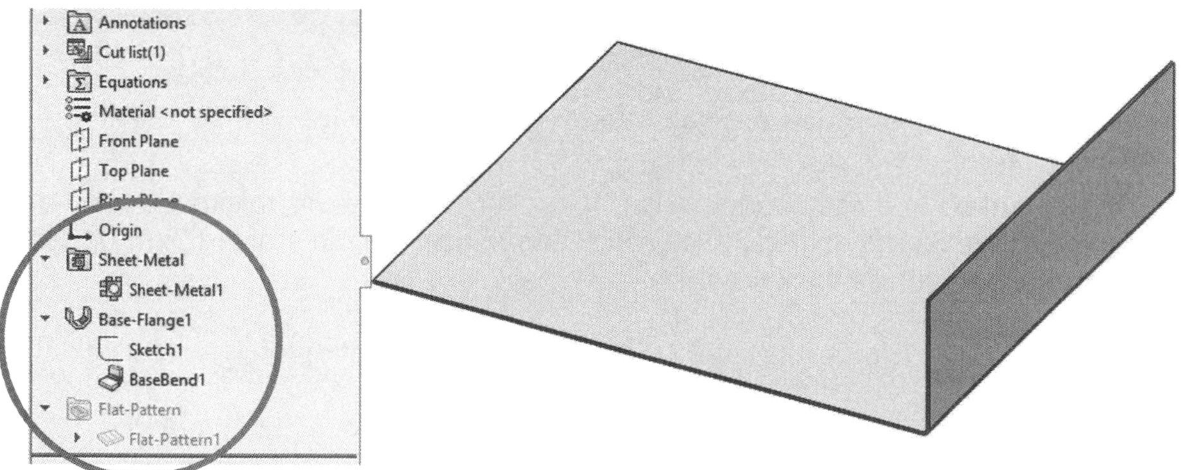

8.5. – At this point our component looks like a thin extrusion, except for having the sheet metal features added to the FeatureManager. In the next step, we'll add three vertical flanges. Select the "**Edge Flange**" command from the Sheet Metal tab in the CommandManager, or the menu "**Insert, Sheet Metal, Edge Flange**."

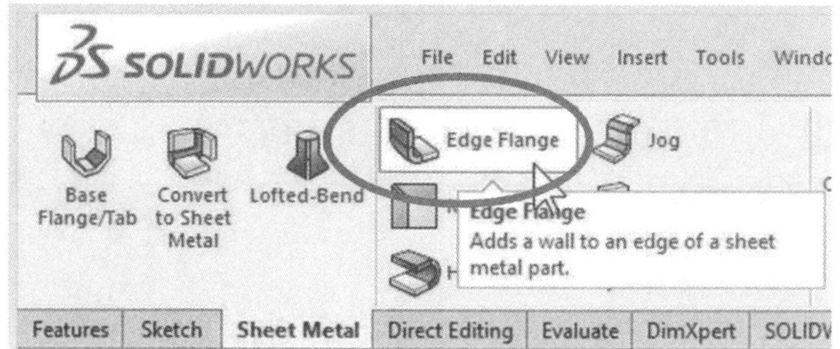

Select the three edges of the base feature to the “Flange Parameters” selection box. Immediately after selecting the first edge we can dynamically define the length and direction of the flange or enter a value in the “Flange Length” value. Click above the part to define the flange’s direction, and enter a value of 1.7”. Set the “Gap Distance” to 0.01”; this is the gap between the flanges at the corners.

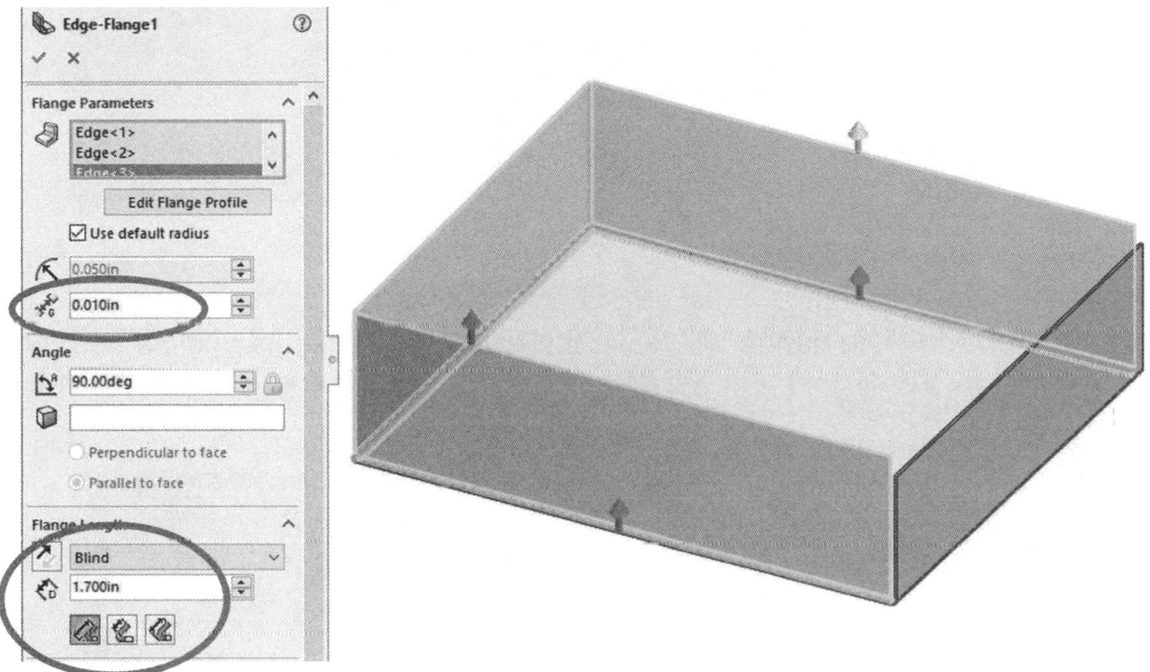

It doesn’t matter if we select the top or bottom edge, it works the same.

8.6. – We are interested in maintaining the original dimensions of the part, then, in the “Flange Position” section select the “Material Inside” option. Click OK to add the flanges and continue. For reference, the different options available are:

- **Material Inside**: The Flange added will not extend beyond the existing edge of the part (indicated by the arrow), maintaining the original part dimensions.

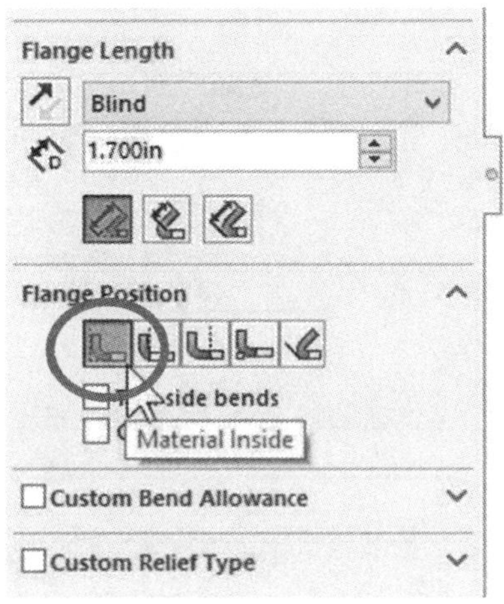

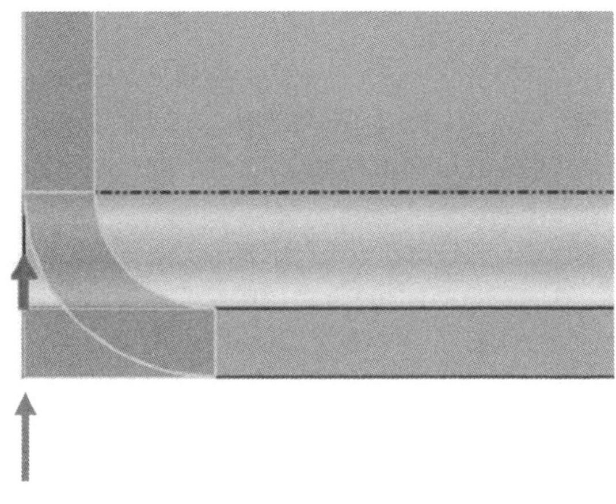

- **Material Outside**: The flange added will be one material thickness outside the existing edge (indicated by the arrow).

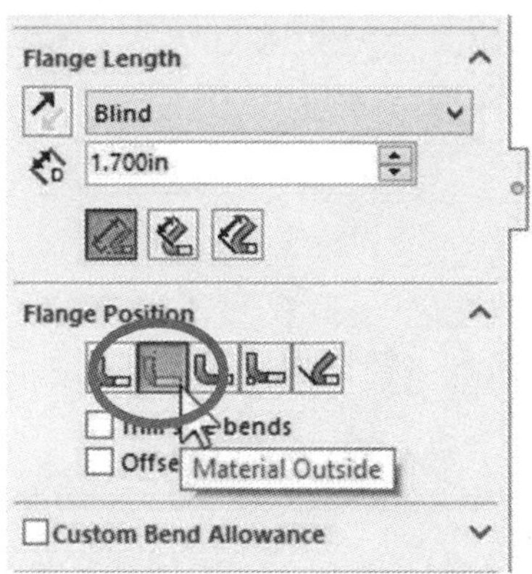

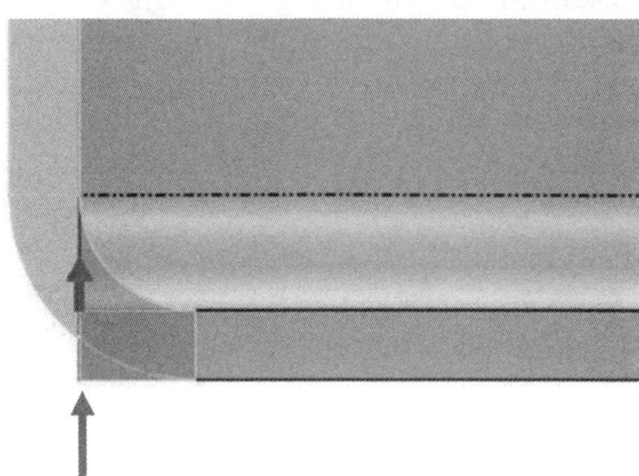

- **Bend Outside**: The flange will leave the existing edge in its original position (indicated by the arrow) and will add the material and the bent region outside. The part will increase in size by one thickness and one bend radius.

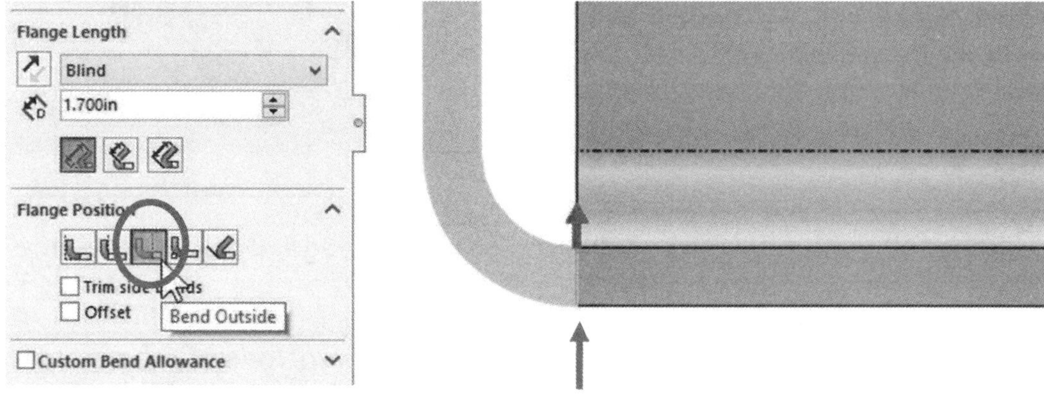

- **Bend from Virtual Sharp** and **Outer Virtual Sharp**: The virtual intersection of the outside face of the flange and the lower edge meet at the outside edge. This option's effect is visible when the angle of the flange is other than 90°. For a 90° flange the Material Inside option has the same effect.

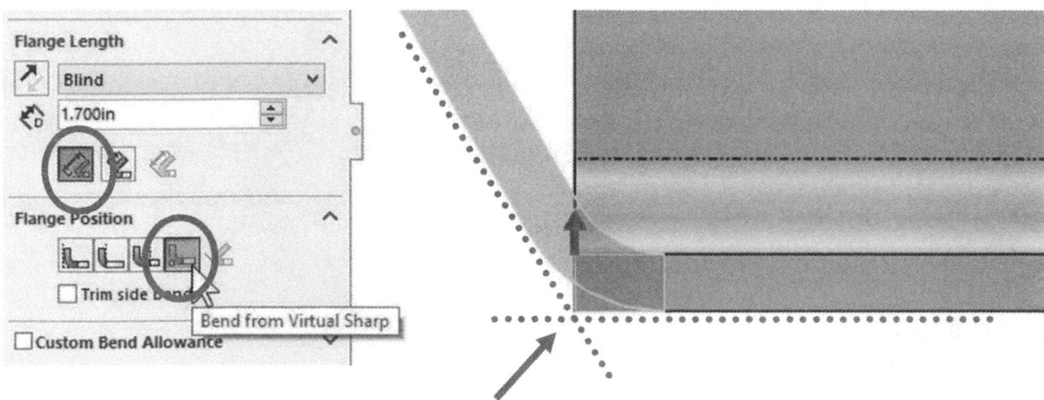

- **Bend from Virtual Sharp** and **Inner Virtual Sharp**: The virtual intersection of the inside face of the flange and the upper edge meet at the inside edge. Just as with Outer Virtual Sharp, this option makes a difference when the angle of the flange is different than 90°.

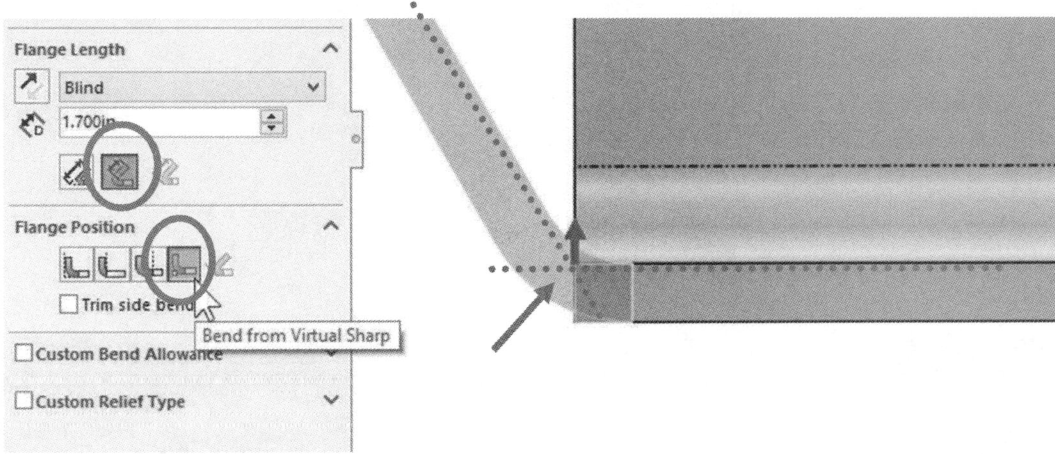

- **Tangent to Bend**: This option is only available when the flange forms an acute angle, making the bend tangent to the part's edge.

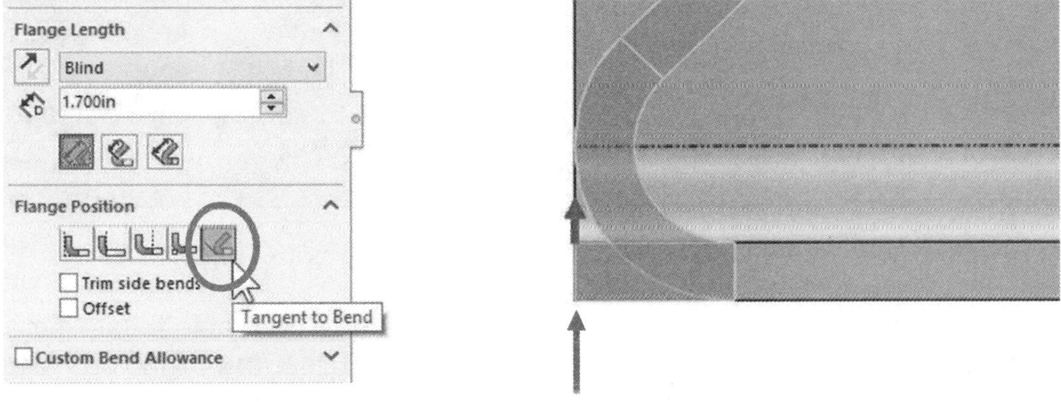

8.7. – After finishing the Edge Flange feature, we can see the corners formed by the new flanges with the "*Base-Flange1*" have an undesirable shape.

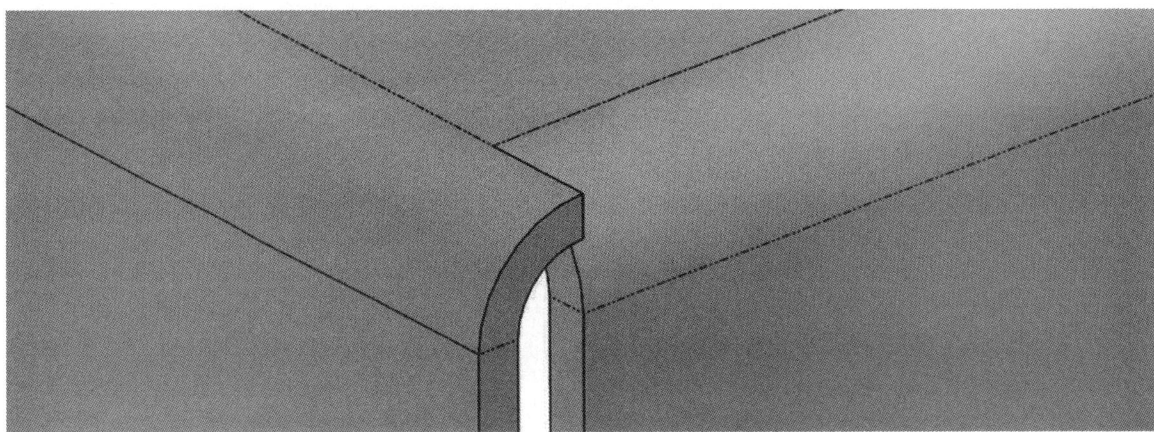

To improve the relief cuts, edit the previously made "Edge-Flange1" feature, and under the "Flange-Position" options turn "Trim side bends" ON. Click OK to finish.

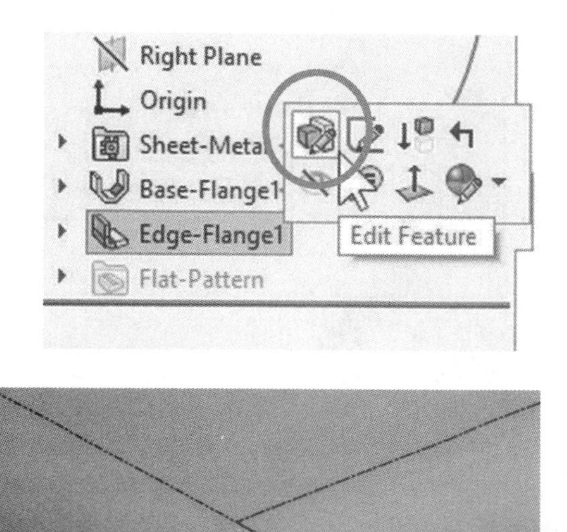

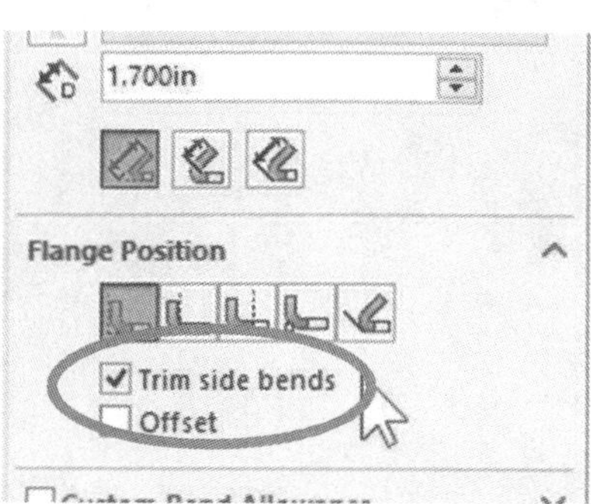

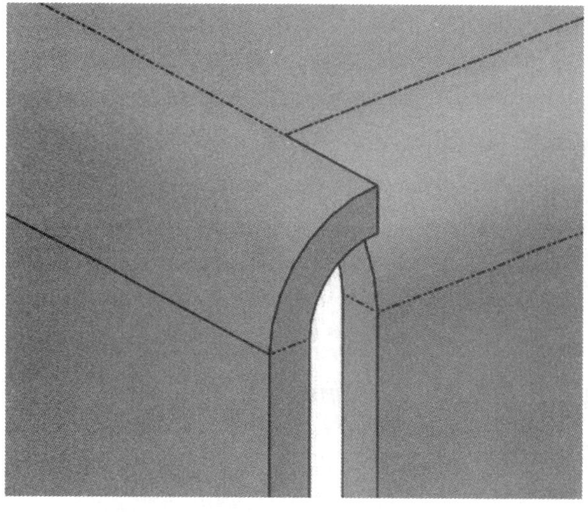

Without Trim Side Bends

With Trim Side Bends

8.8. – While the relief cut is better, we still don't have a desirable finish. To close the corner and reduce the gap in the edge we'll use the "**Closed Corner**" command. Select "**Closed Corner**" from the "Corners" drop-down command or the menu "**Insert, Sheet Metal, Closed Corner**."

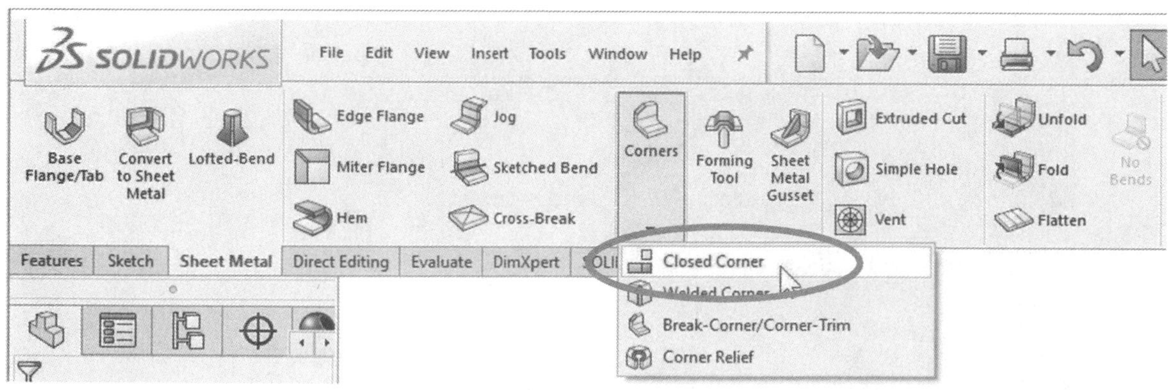

In the “Faces to Extend” selection box add a face of “*Edge-Flange1*” in the gap; the corresponding “Faces to match” is automatically selected. Use the “Underlap” corner type with a gap of 0.01”. Select the opposite edge’s face and its corresponding “face to match” to close both corners at the same time. Leave the “Open bend region” unchecked, otherwise the corner will remain open. If the corners are not closed correctly, you may have reversed the “Faces to Extend” and “Faces to match” selections.

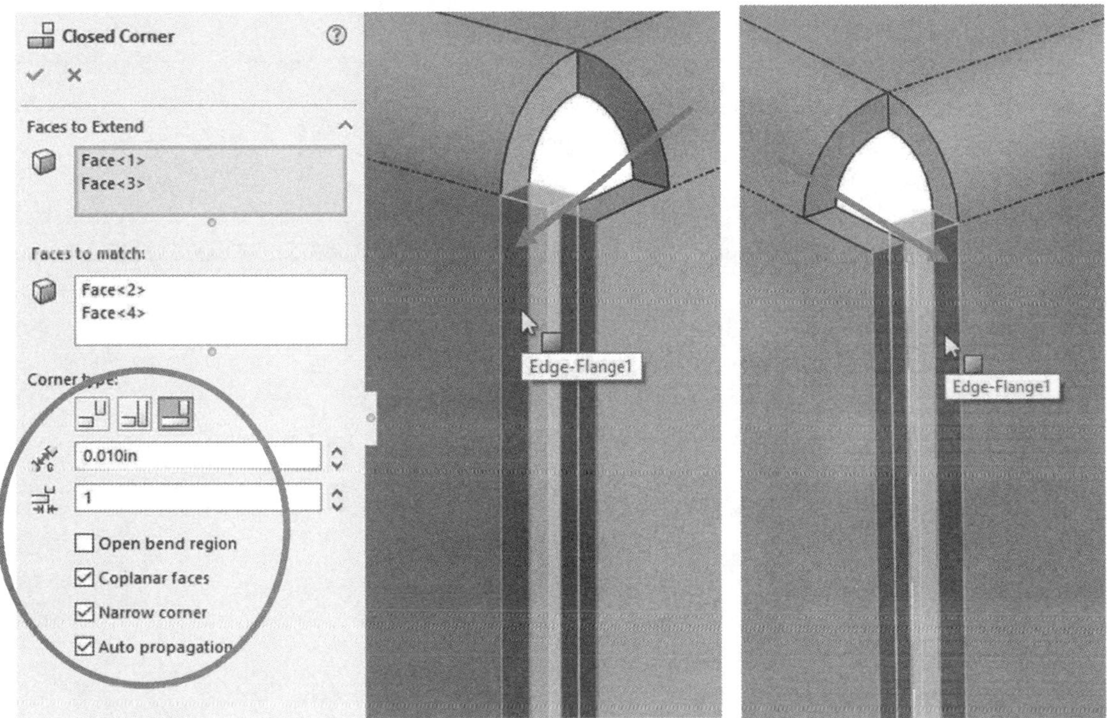

The finished closed corners will look like this.

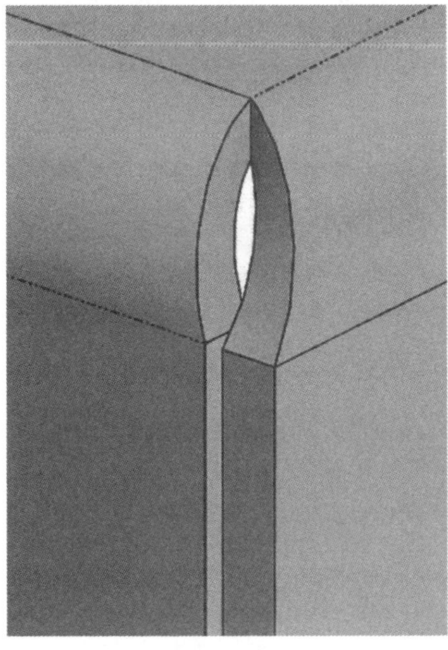

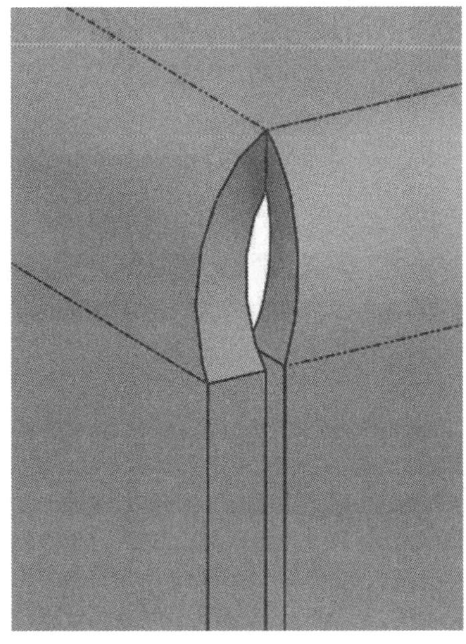

8.9. – For the next step we need to bend the first flange outward at a 45° angle. Switch to a Right view, add the next sketch in the short flange and dimension it. Select the "**Sketched Bend**" command from the Sheet Metal tab in the Command-Manager or in the menu "**Insert, Sheet Metal, Sketched Bend.**"

For a Sketched Bend feature, the sketch line doesn't need to cross the entire face. We did it to fully define the sketch.

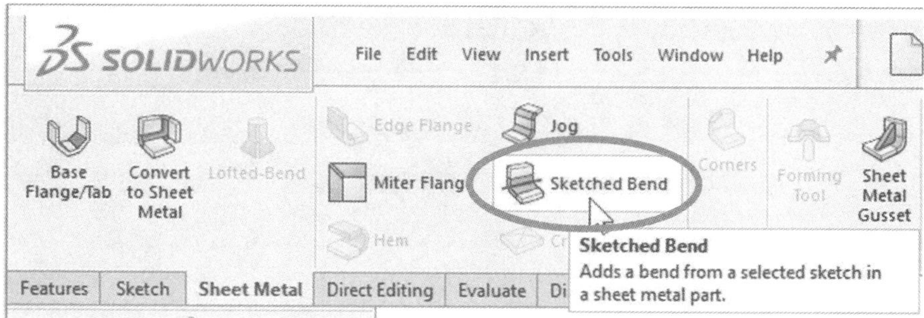

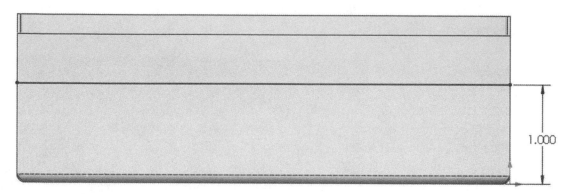

For a sketched bend, we must select the side of the face that will not move. Select the face, below the sketch line at the bottom half of the flange, enter the 45° angle for the bend, and if needed, reverse the direction. Use the Bend Centerline option and click OK to finish.

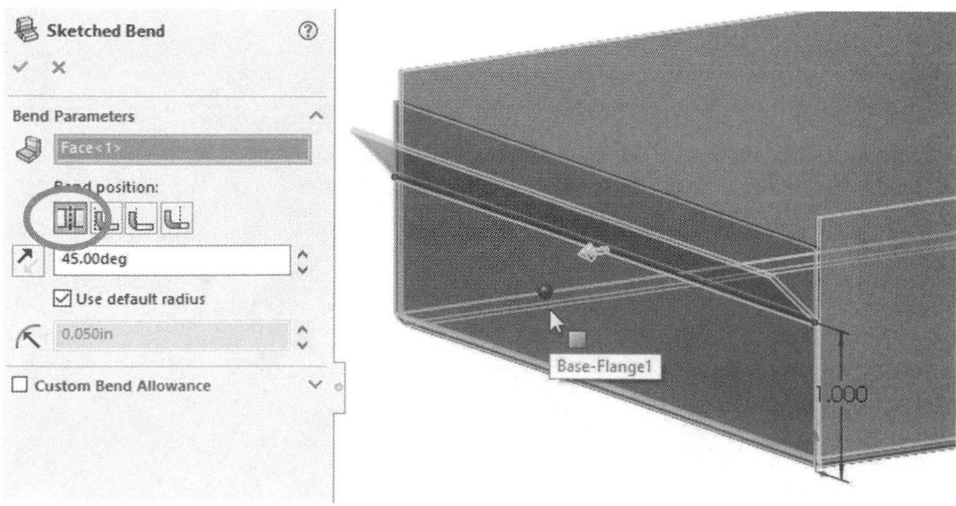

8.10. – To eliminate the sharp edge and reinforce the last bend we need to add a hem to it. Select the "**Hem**" command from the Sheet Metal tab or the menu "**Insert, Sheet Metal, Hem**."

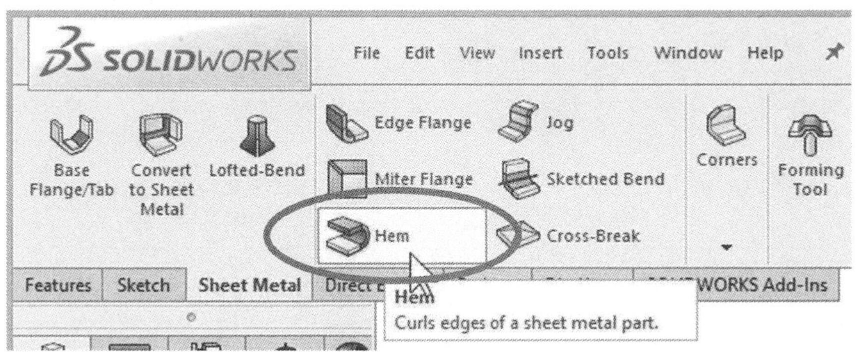

Select the bent flange's edge, and if necessary turn the "Reverse Direction" ON to add the hem inside. Use the "Material Inside" option, set the type to "Closed" and its length to 0.1". Click OK to finish.

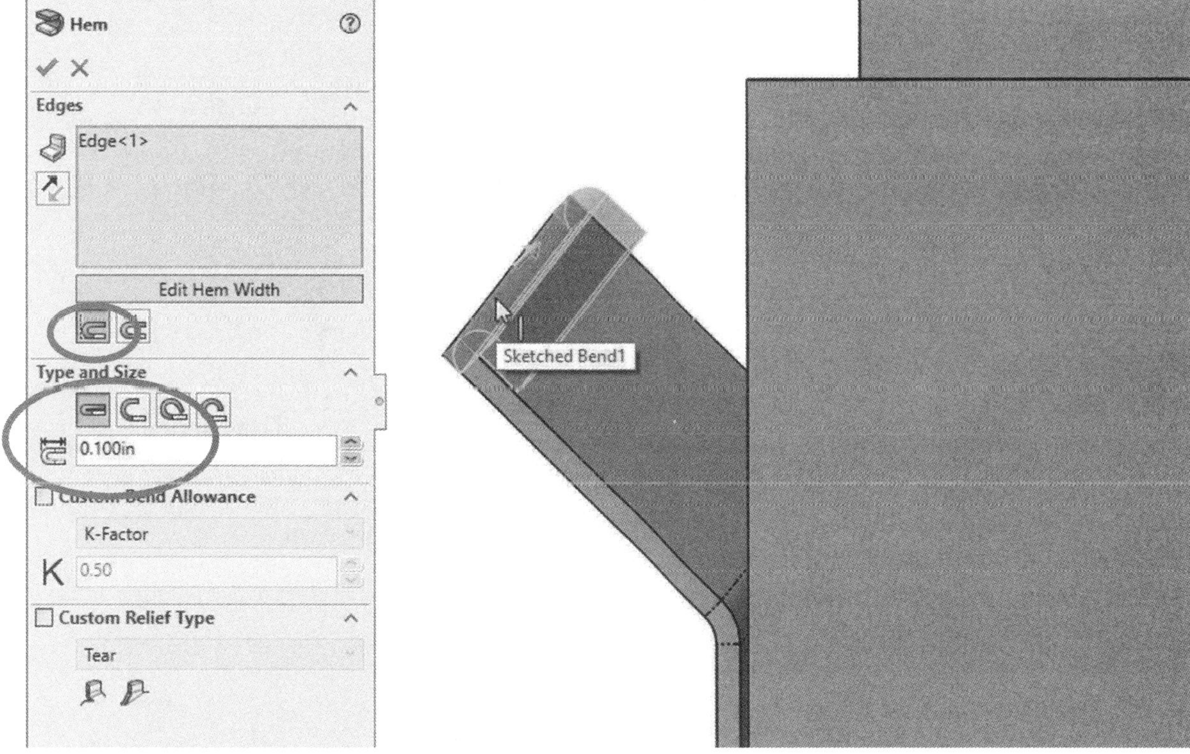

The "**Hem**" command has the following options available:

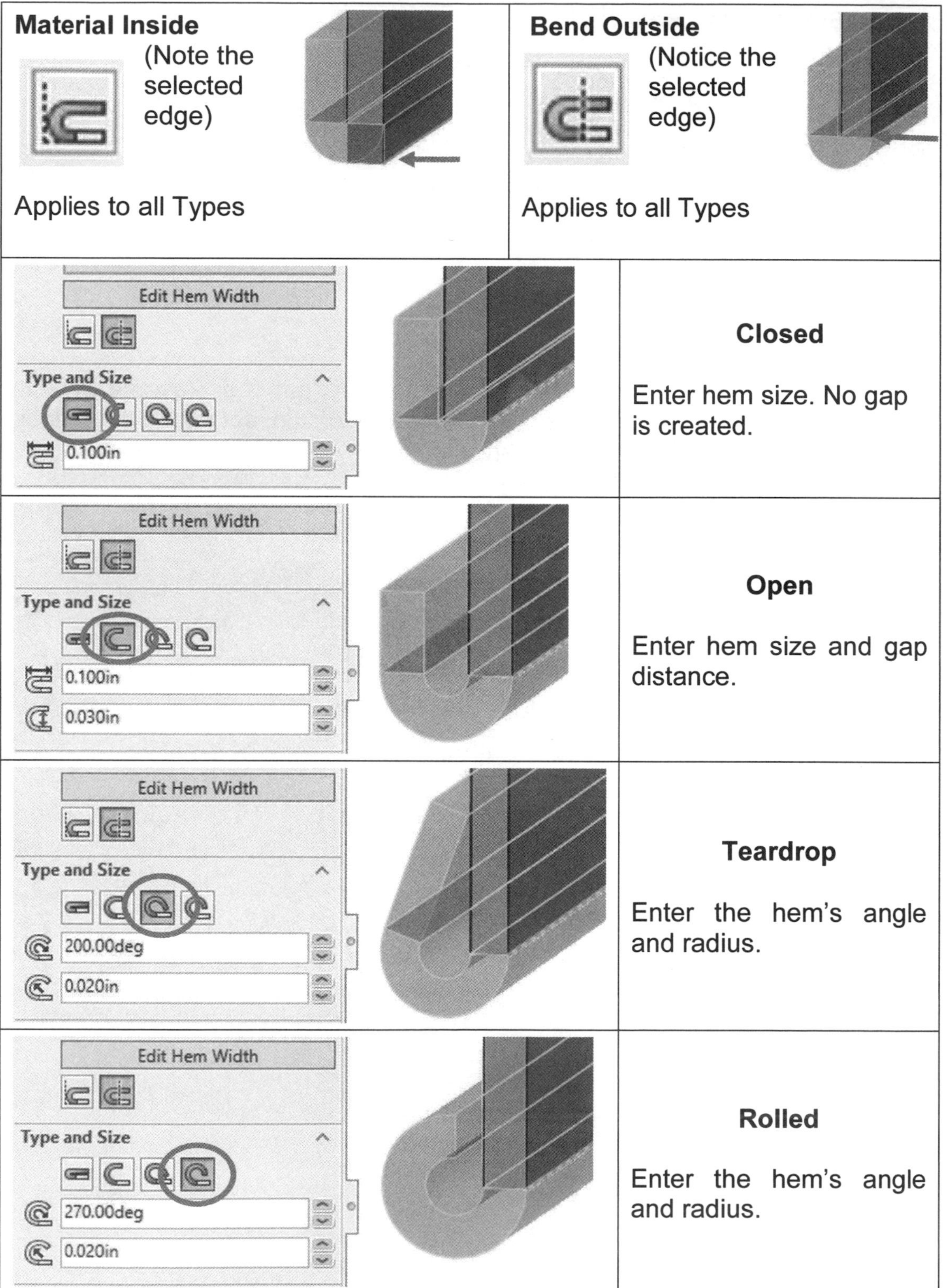

Material Inside (Note the selected edge) Applies to all Types	**Bend Outside** (Notice the selected edge) Applies to all Types
	Closed Enter hem size. No gap is created.
	Open Enter hem size and gap distance.
	Teardrop Enter the hem's angle and radius.
	Rolled Enter the hem's angle and radius.

8.11. – Now we'll add a flange going outward to the other 3 edges, and the flange opposite to the hem will be narrower than the edge. Use the "**Edge Flange**" command from the Sheet Metal tab and select the three edges indicated. Make the flanges 0.75" outward using the "Material Inside" option.

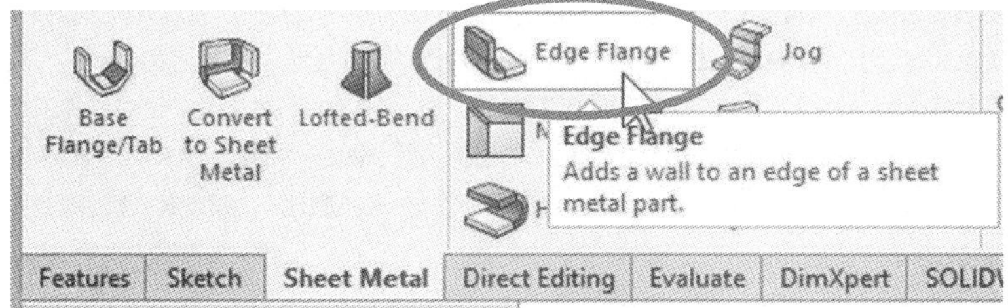

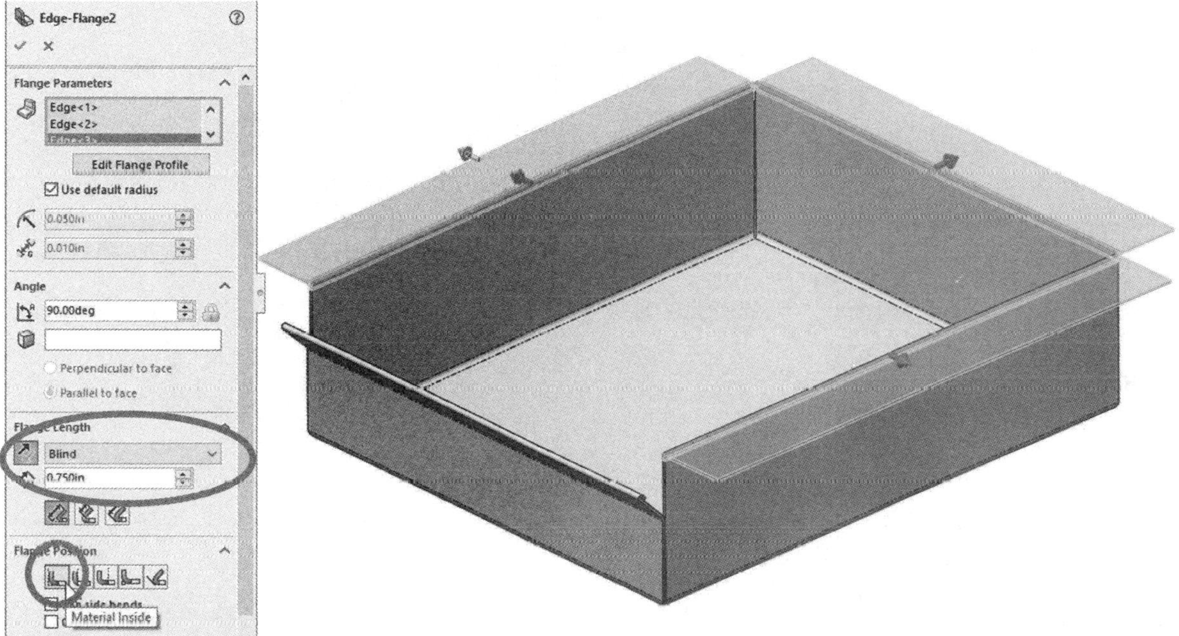

Before finishing the flanges, we need to change the width of the middle one. Select the middle edge in the "Flange Parameters" list and click "**Edit Flange Profile**." Depending on the order in which the edges were selected, the middle edge may or may not be "*Edge<2>*".

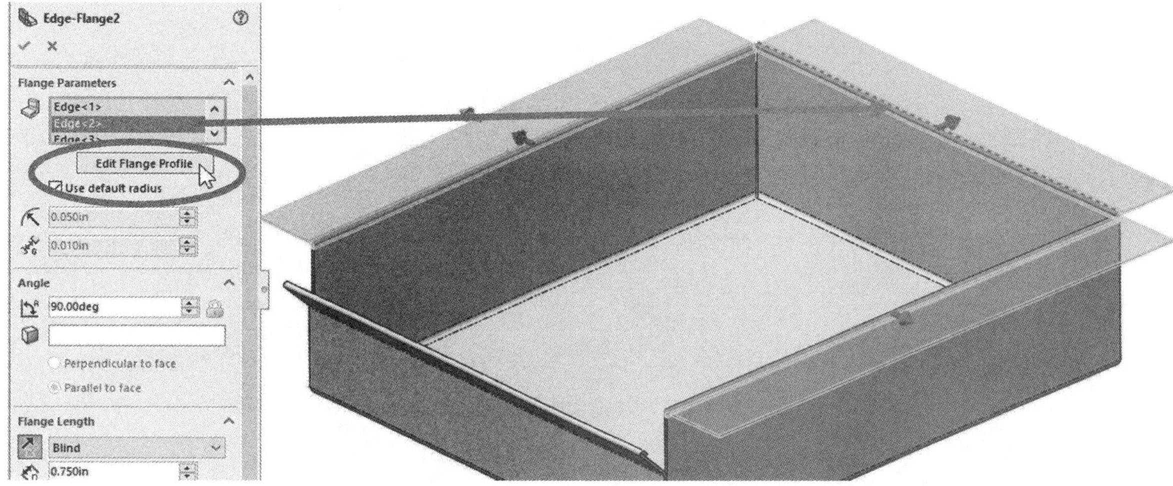

8.12. – When we edit the flange profile we are editing its sketch. The lines at the sides are colored black, which means they are fully defined; however, if we click and drag them *while editing the flange's profile* they will behave as if they were under defined. Click and drag the short side lines and move them towards the center and dimension them 1" from the edge. After modifying the sketch click "**Finish**" to add the flanges. If we still need to make changes to the flange parameters after completing the sketch, click "**Back**."

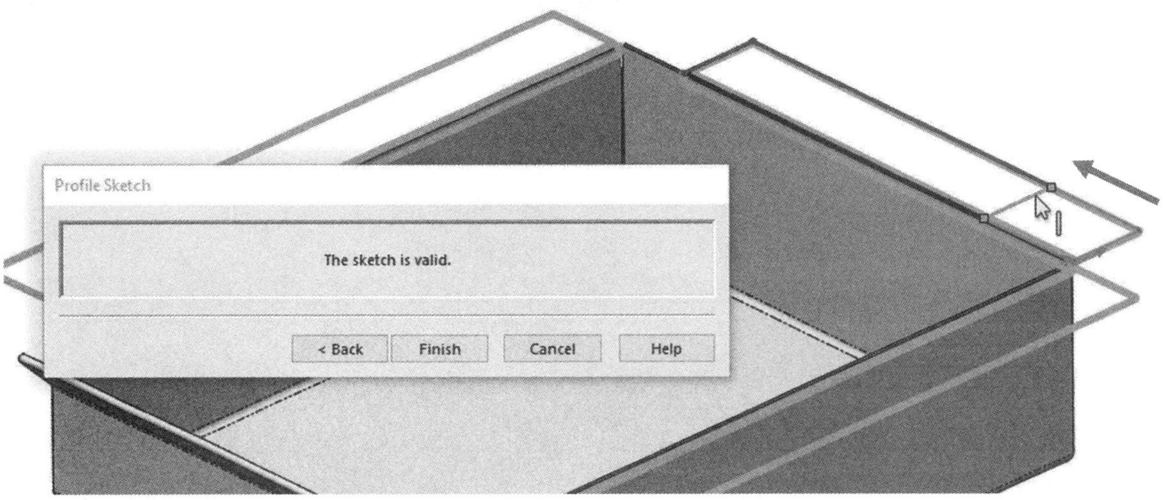

When we modify a flange's profile, we can change it as much as we want as long as we end with a closed profile connected to the edge; the message will alert you if the sketch is valid or not.

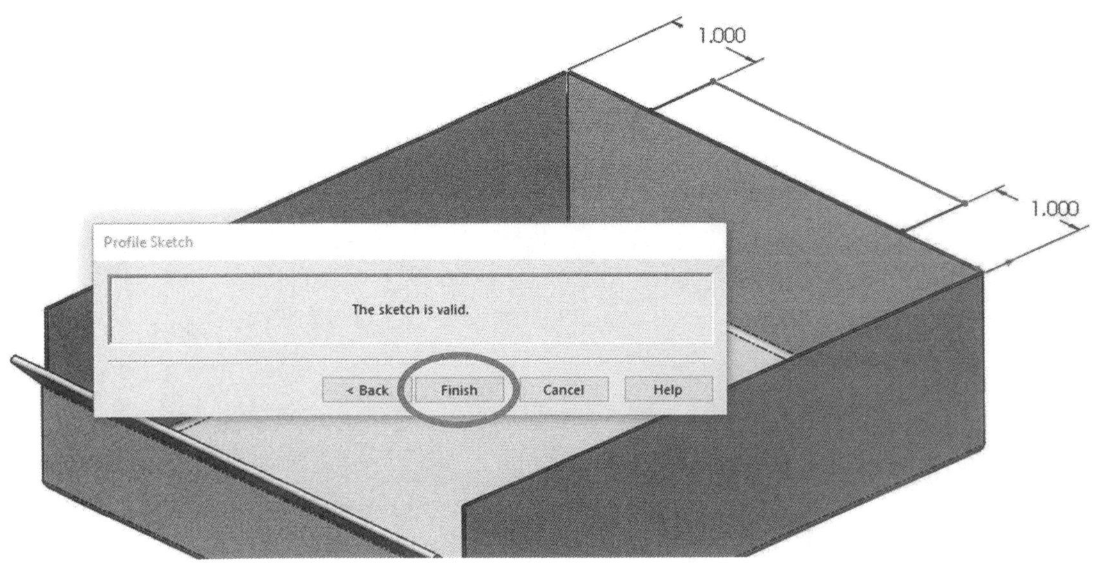

8.13. – The next step is to add an opening on one side. Adding a cut-extrude to a sheet metal part is like cutting any other part; one difference is in a sheet metal part we have a "Link to Thickness" option if we use the "Blind" end condition. Using this option makes the cut as deep as the sheet metal's thickness.

Switch to a Back view and add the following sketch to the indicated face. The dimensions are referencing the outer edges of the model. Make an "**Extruded Cut**" from the Features or Sheet Metal tab (it's the same) using the "Link to Thickness" option and click OK to finish.

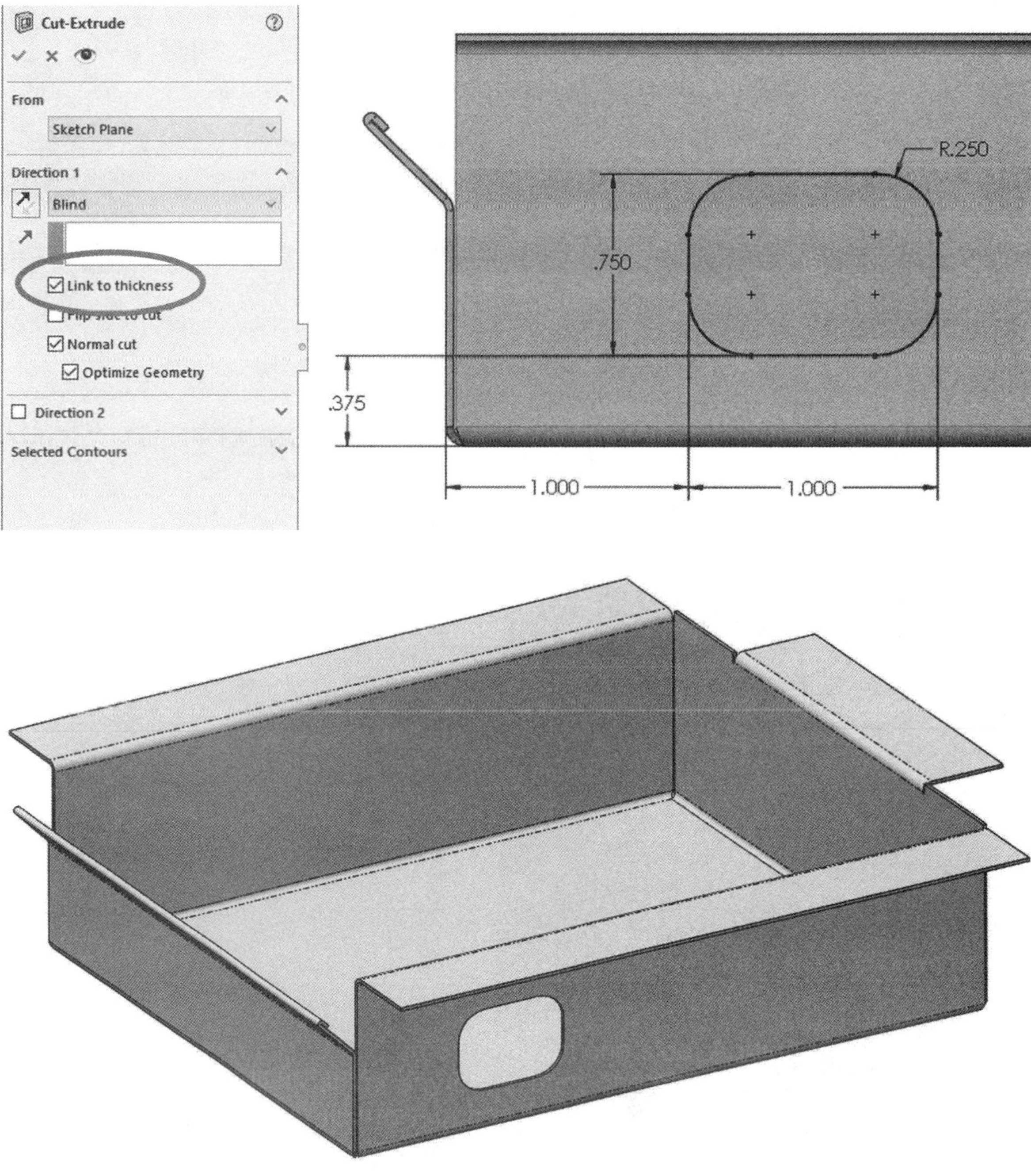

8.14. – When working with sheet metal, it is common to add forming features using pre-fabricated tools or dies. These features are added by pressing a die onto the metal using a press, which forms and cuts the part at the same time. Since the metal is ripped, deformed and stretched, formed features cannot be flattened in SOLIDWORKS. In this step, we'll learn how to use a built-in forming tool to add a feature to a sheet metal part. From the Design Library expand "**Forming Tools, louvers**," select the built in "Louver" and drag and drop it to the face indicated.

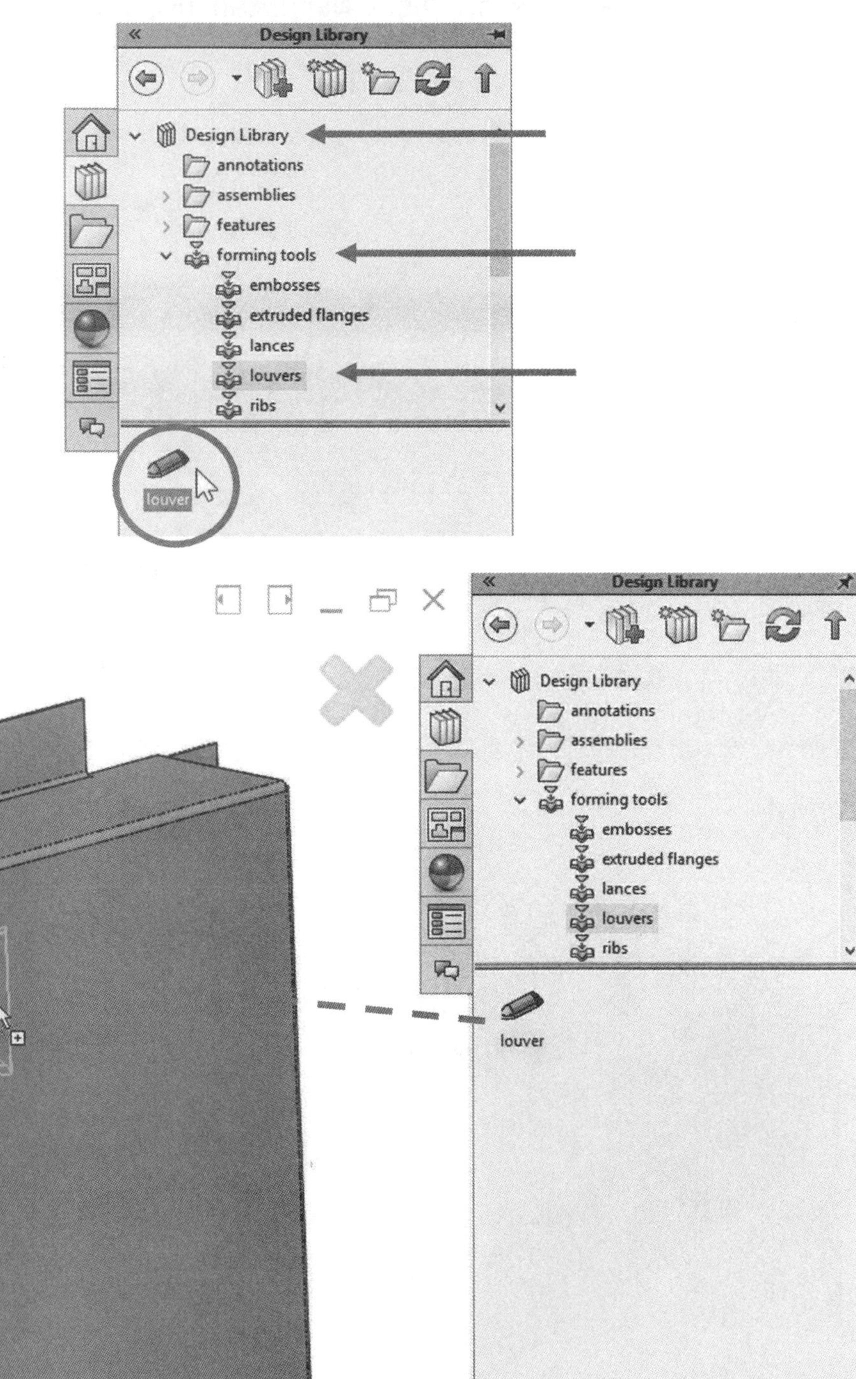

After dropping the louver feature we can change its direction, rotation and location. Click "Flip Tool" to make the louver go *out* of the part and if needed, change the rotation angle to match the next image. After orienting it correctly, select the "Position" tab at the top to dimension its location.

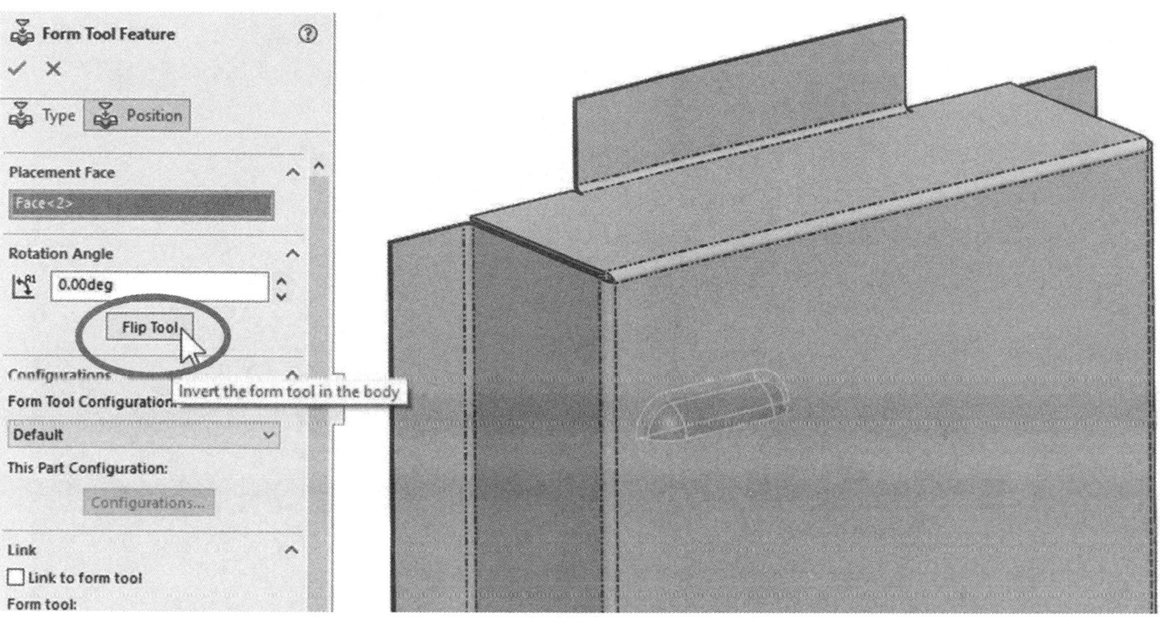

The "Position" tab reveals a sketch to locate the forming tool feature. Dimension it to the outside part edges as shown and click OK to finish.

When the "Position" tab is activated we get a behavior like the "Hole Wizard": At each location where a sketch Point is added, a new instance of the louver will be made.

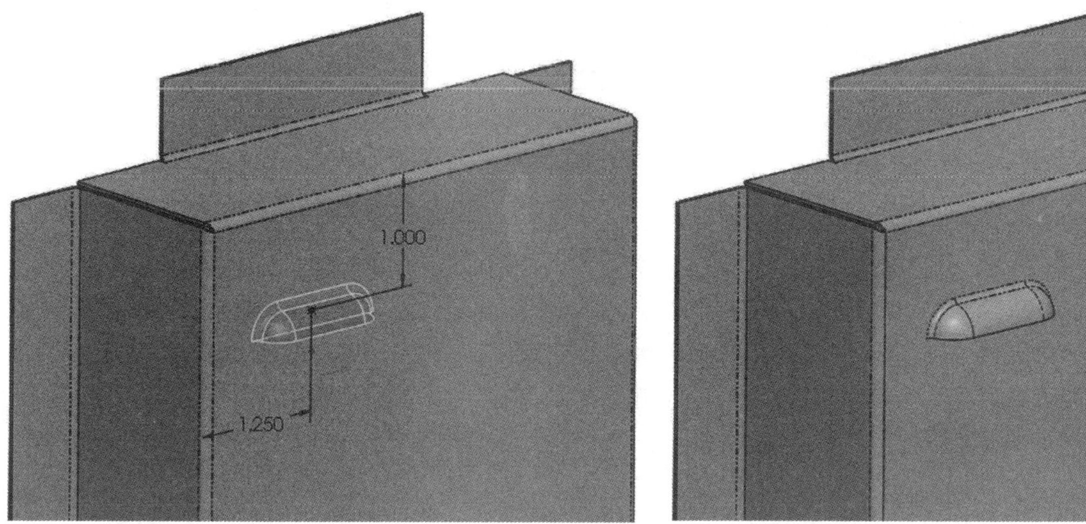

8.15. – Add a linear pattern in two directions with 2 instances across spaced 2.5" and 3 instances down spaced 0.75".

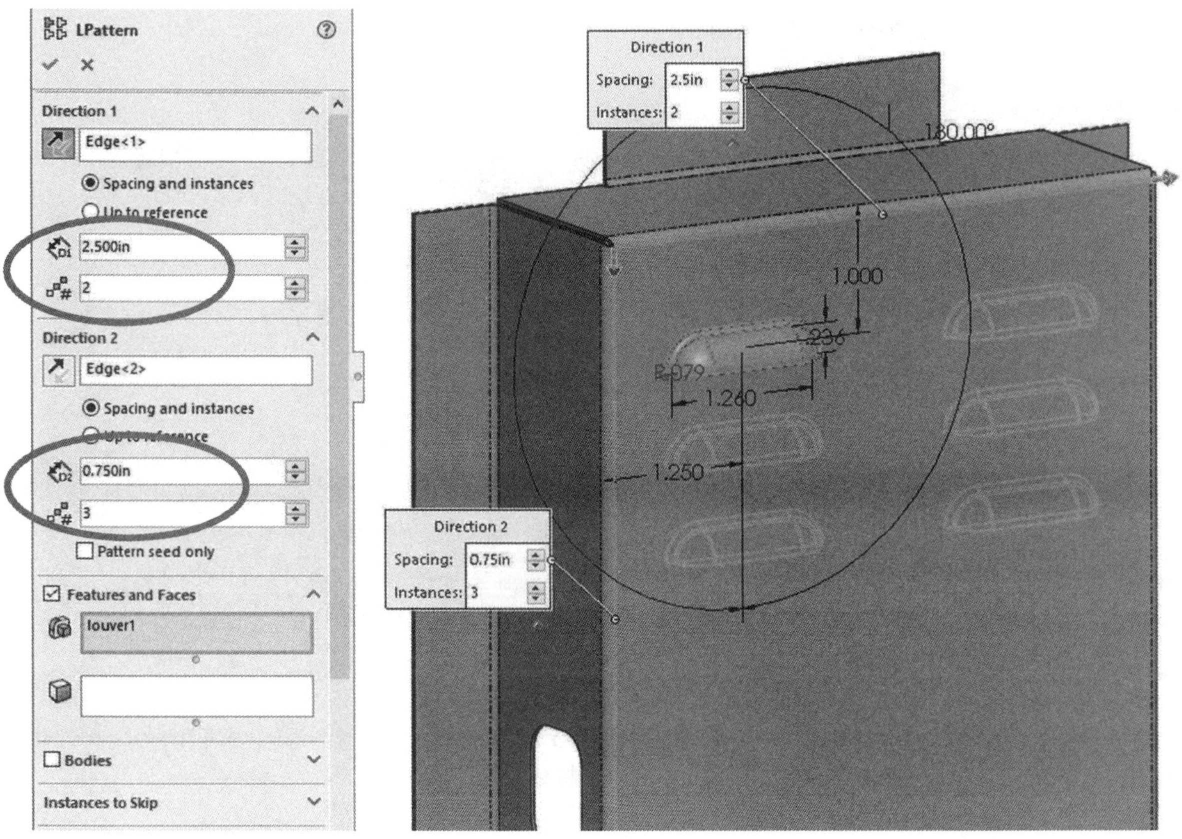

8.16. – The next feature will be a cut across a bend. This cut can be made while the part is bent; however, we will unfold one bend, add the cut, and finally re-fold it to show new functionality. Select the "**Unfold**" command form the Sheet Metal tab or the menu "**Insert, Sheet Metal, Unfold.**"

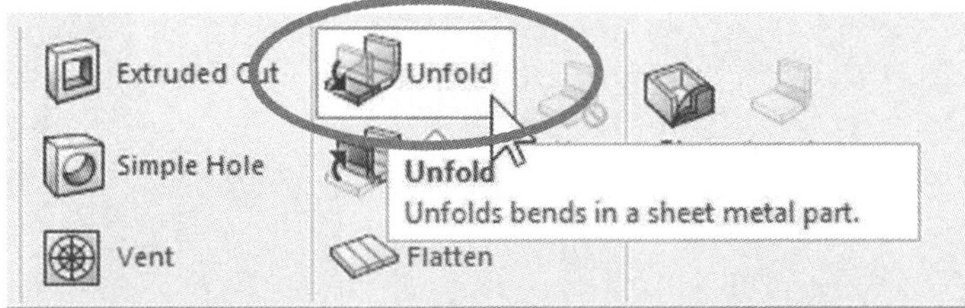

When unfolding a bend, we have to select a face to remain immovable, and the bend(s) that we wish to unfold. Select the fixed face indicated, and then the bend to unfold.

When we use the "**Fold**" or "**Unfold**" commands, SOLIDWORKS will only let us select bends.

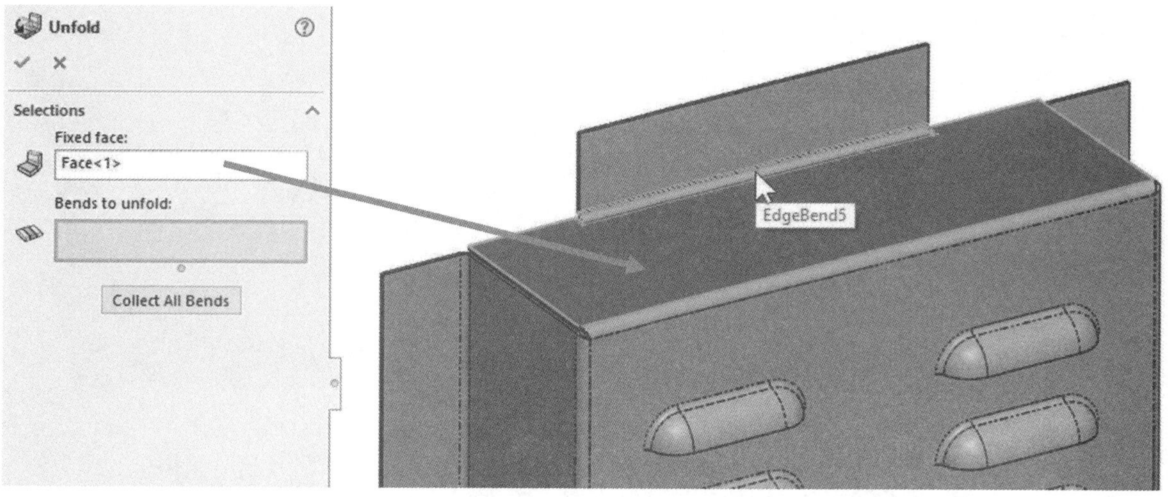

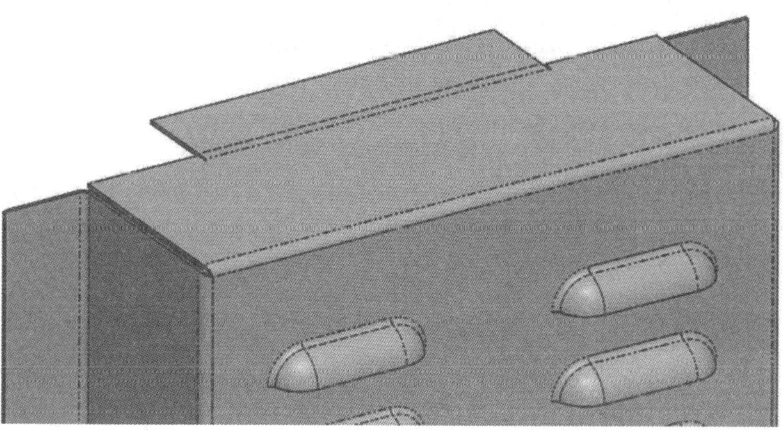

8.17. – Add the following sketch centered in the unbent flange and make an extruded cut using the "Link to Thickness" option.

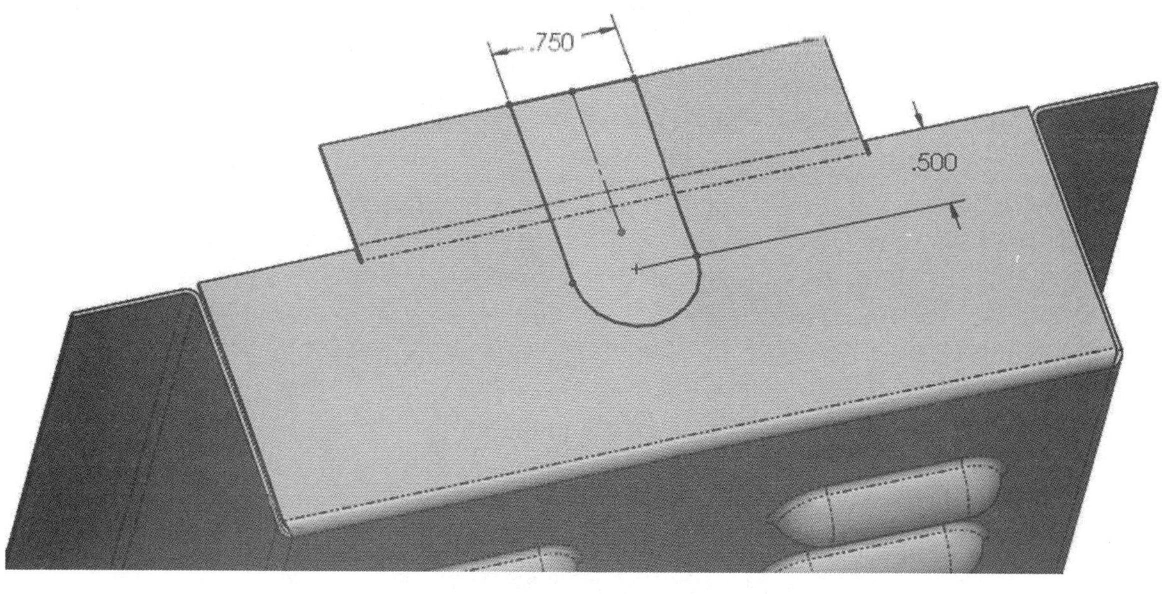

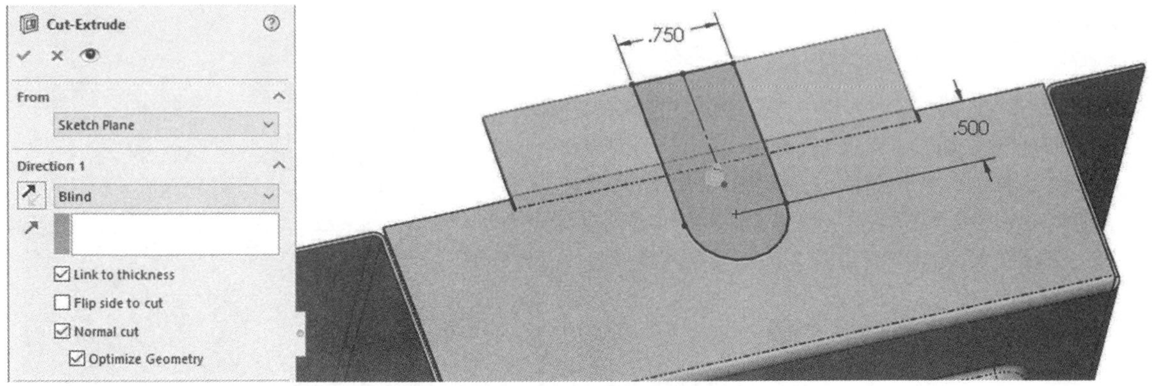

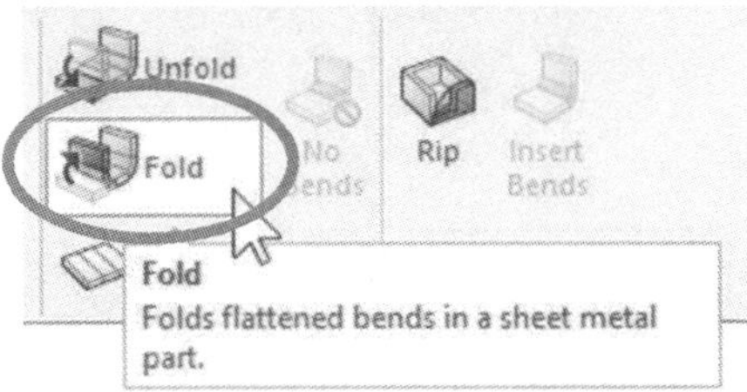

8.18. – After making the cut we need to re-fold the bend. Select the "**Fold**" command from the Sheet Metal tab or the menu "**Insert, Sheet Metal, Fold.**"

The "Fixed face" field is automatically selected based on the face selected when the flange was unfolded. In the "Bends to fold" select the flattened bend; to make selection easier, SOLIDWORKS activates a selection filter that only allows us to select bends. Optionally click "Collect All Bends" to add all flattened bends to the selection box. Click OK to re-fold the flange and continue.

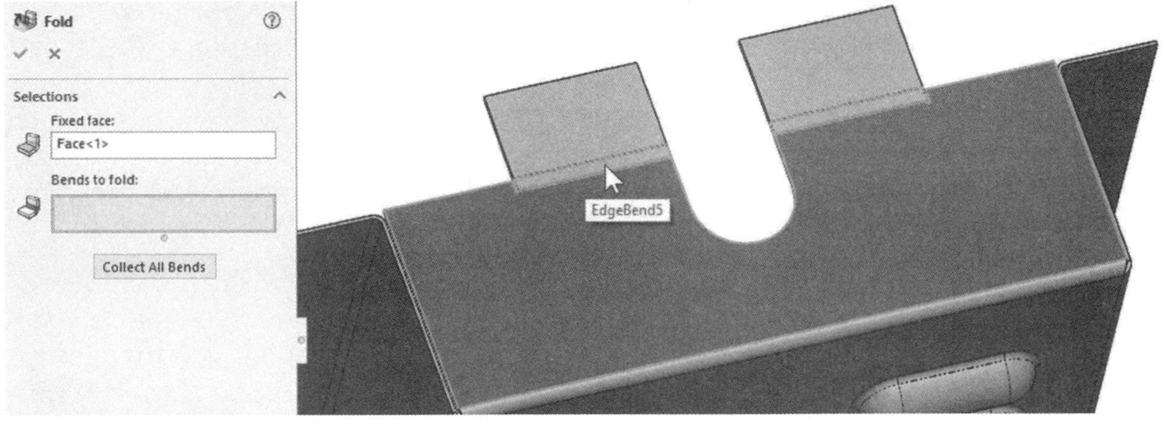

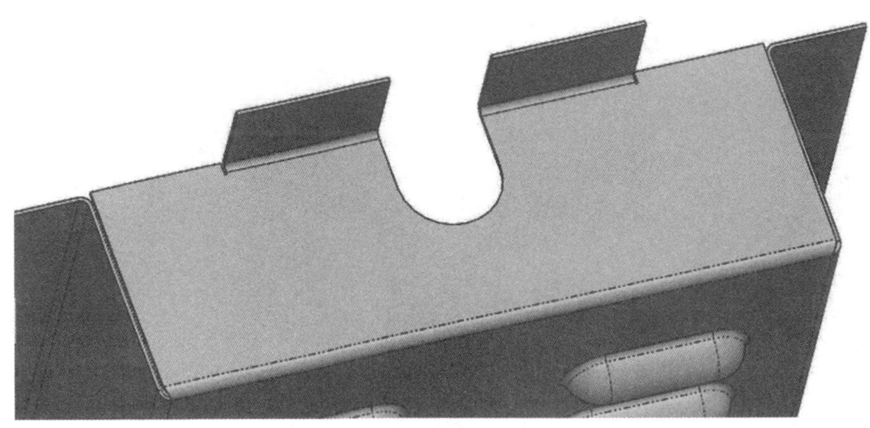

8.19. – Now we need to make holes to mount the part. Add the following sketch and make an extruded cut using "Link to Thickness." Make the construction lines centered in each face to save time. Add the first half of the sketch and use "**Mirror Entities**" to copy the other half about the middle centerline.

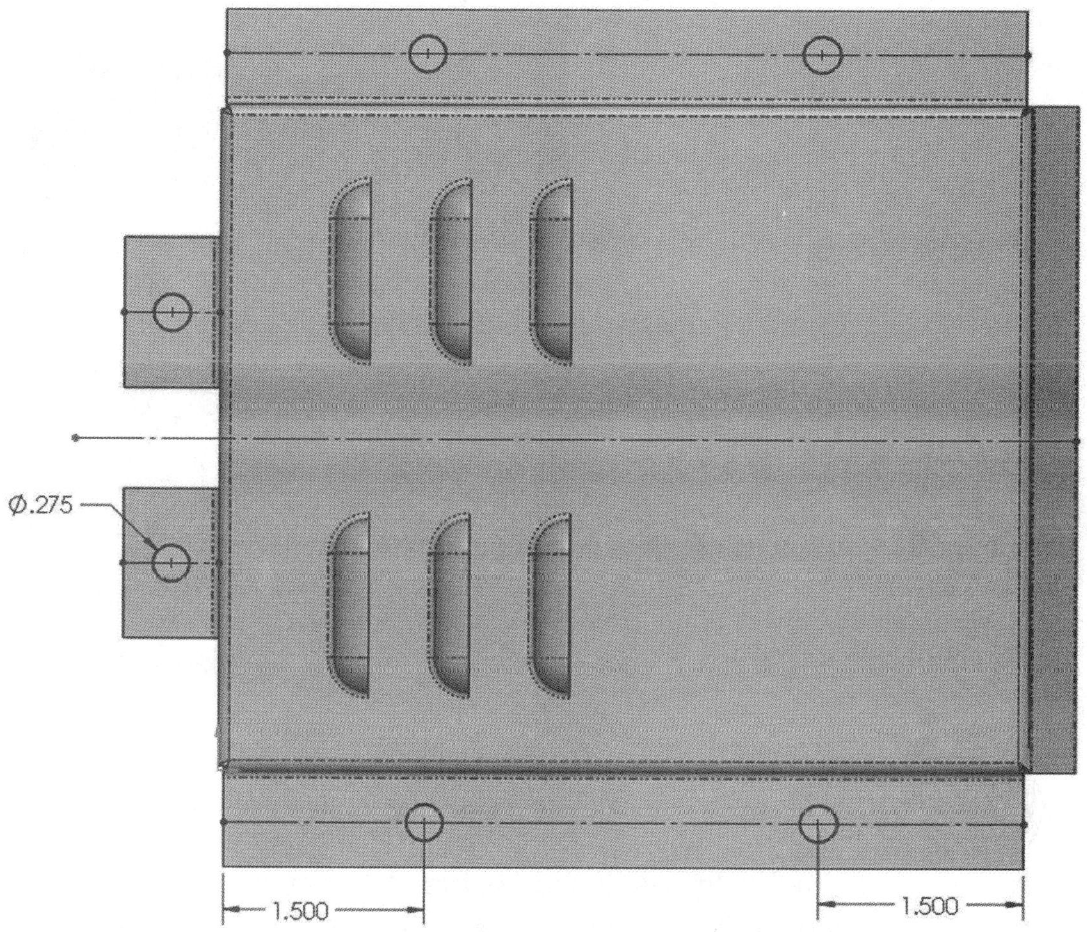

8.20. – Because of the nature of sheet metal manufacturing, when the metal is cut the resulting edges and corners usually end up having sharp edges. To eliminate them in our part, use the "**Corners, Break Corner**" command from the Sheet Metal tab or the menu "**Insert, Sheet Metal, Break Corner**."

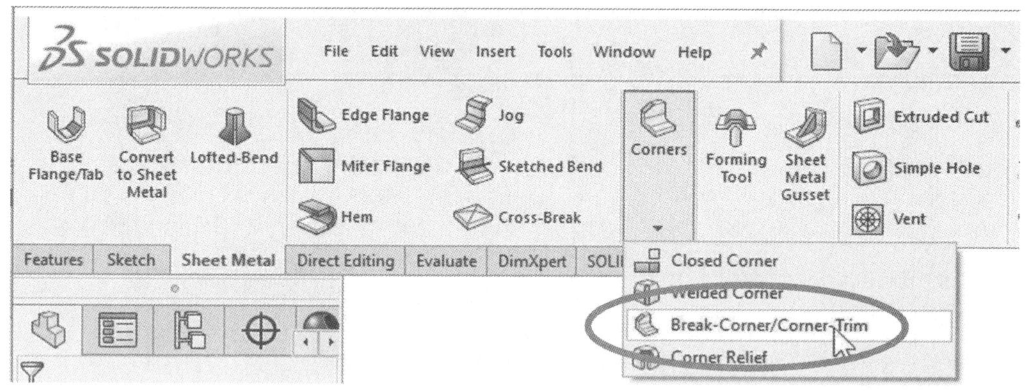

We have two options to remove the corners: a chamfer or a fillet. In our example we'll add a 0.2" chamfer to remove all corners. We can select each corner individually (SOLIDWORKS makes this easy by adding a small edge selection filter), or, when a face is selected all of its corners are automatically added. Select the indicated faces and click OK to finish.

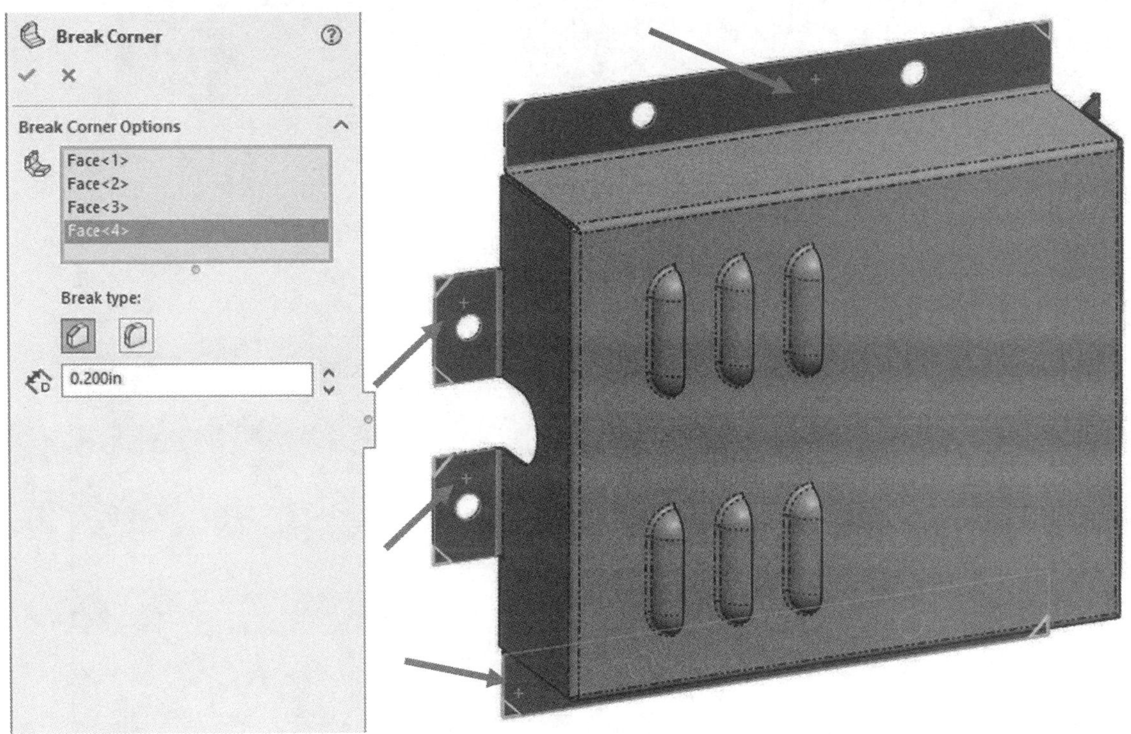

8.21. – After the part is finished, we need to generate a flat pattern for manufacturing. As soon as we make the sheet metal part the flat pattern is automatically added, the last folder in the FeatureManager is called "*Flat-Pattern*" and contains the "*Flat-Pattern1*" feature; it is suppressed when the part is bent. Press the "**Flatten**" command from the Sheet Metal tab to see the flat pattern to un-suppress "*Flat-Pattern1.*" Pressing "**Flatten**" again re-folds the part and automatically suppresses the "*Flat-Pattern1*" feature.

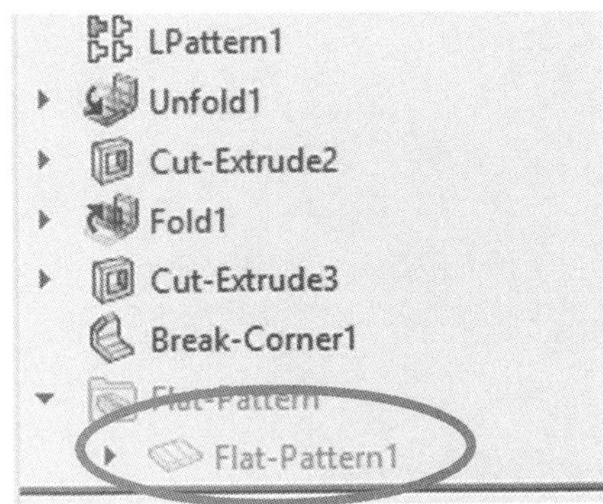

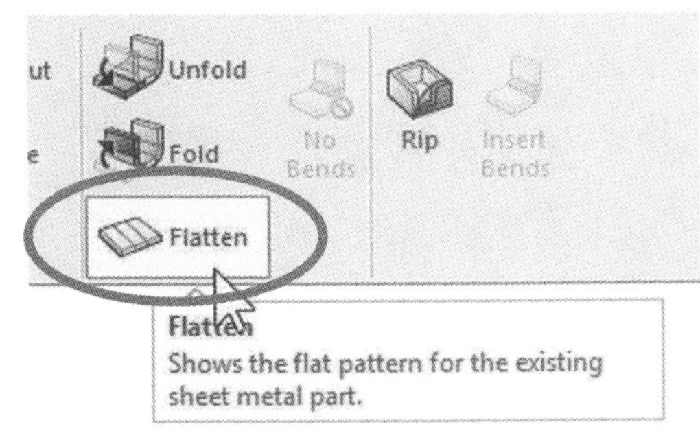

The "*Flat-Pattern1*" feature contains a sketch with bend lines, a sketch with the outline of the smallest sheet metal area from which we can make our part, and all the bend regions in our part.

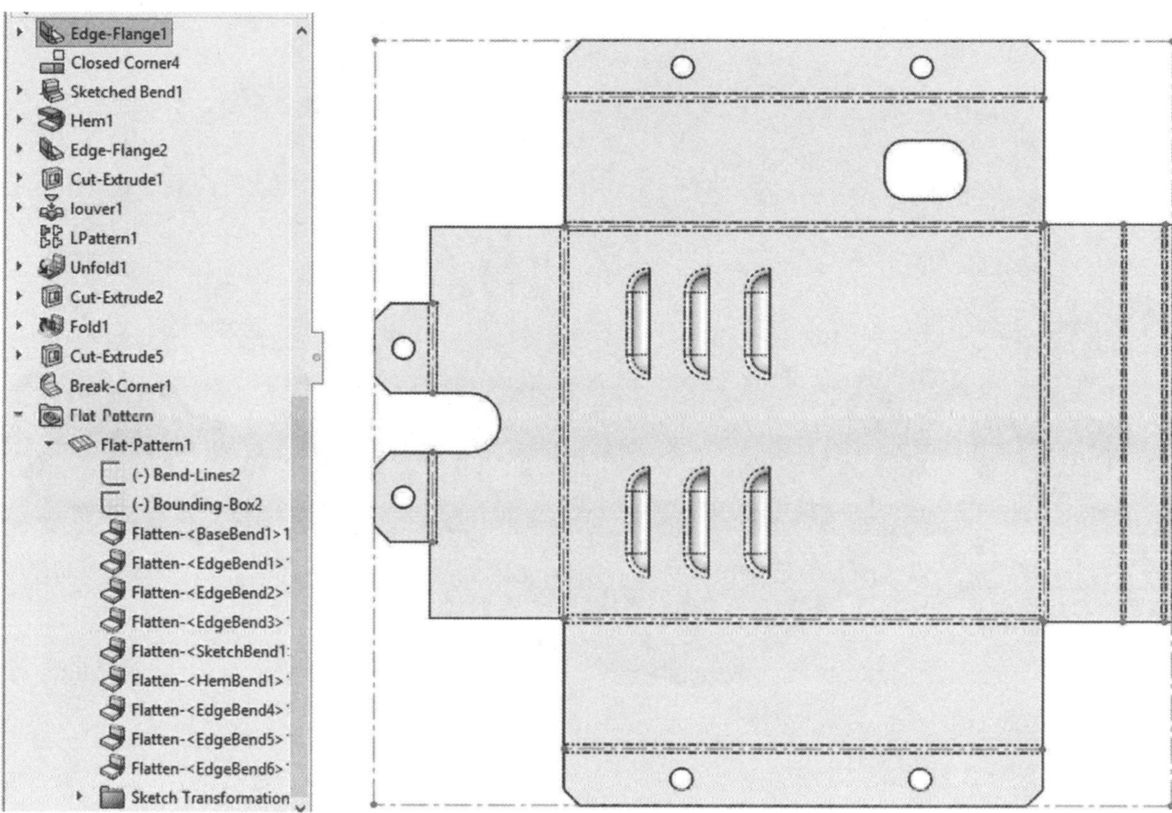

While the part is in the flattened state, the "*Flattened*" command is depressed and the "*Flat-Pattern1*" feature is unsuppressed.

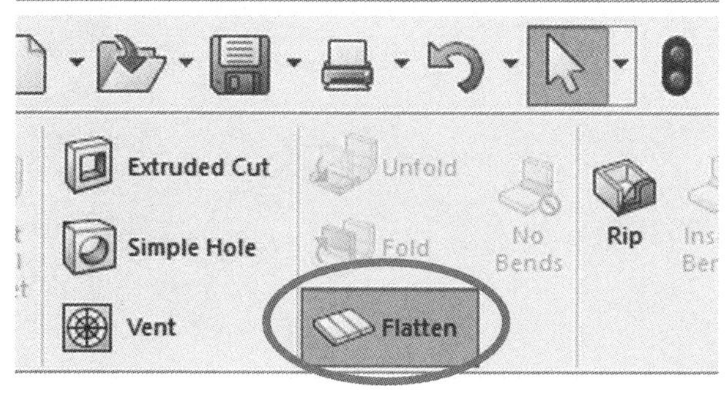

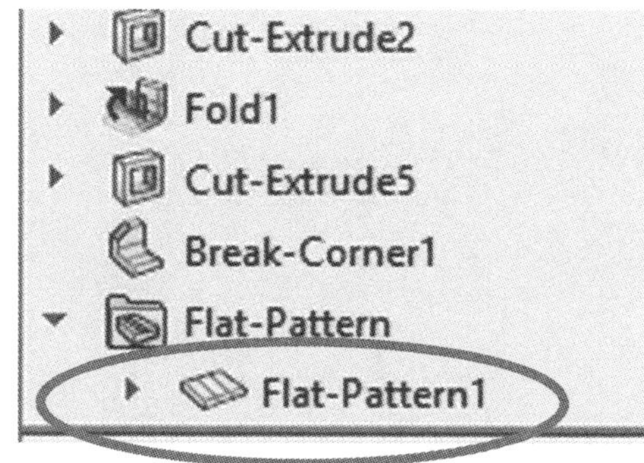

8.22. – Save the finished part as '*Sheet Metal Box*' and close.

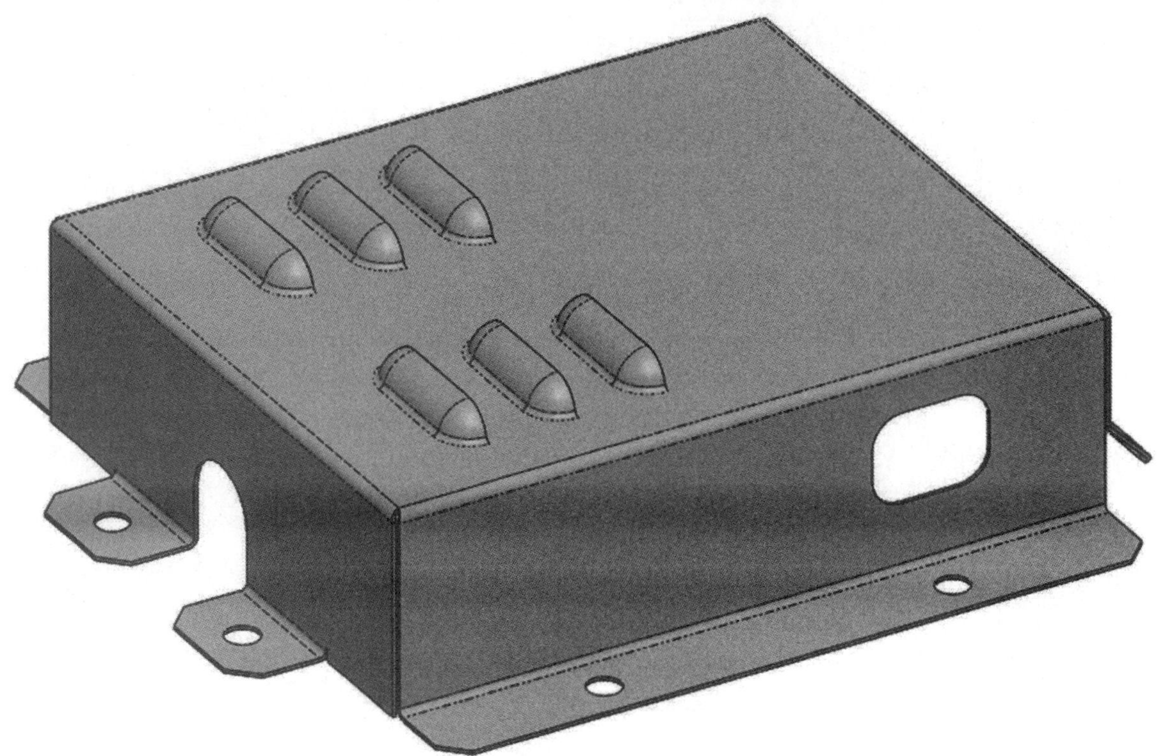

Understanding Top Down Design

In designing the power supply, we'll use different sheet metal and advanced assembly techniques including Top Down Design. The first thing we need to learn is the process of designing components in the context of an assembly.

Top Down Design is the process where we create parts and/or add features to a part using other components in the assembly as reference; in the Power Supply example, we'll design a sheet metal enclosure *around* the actual electronics of the power supply. In other words, we'll make an enclosure to fit the internal components instead of building an enclosure and *then* trying to fit the components inside. That's the main difference.

One of the biggest advantages of Top Down Design is that when a referenced (driving) component is modified, the components designed around it (driven) also change. Looking at this example, the base component is the DRIVING component, the circular hole's edge is the driving feature, and the *"DRIVEN"* part updates when the base part updates. In this example, the hole's diameter is changed from 1″ diameter to 1.625″ and the pin's size changes accordingly.

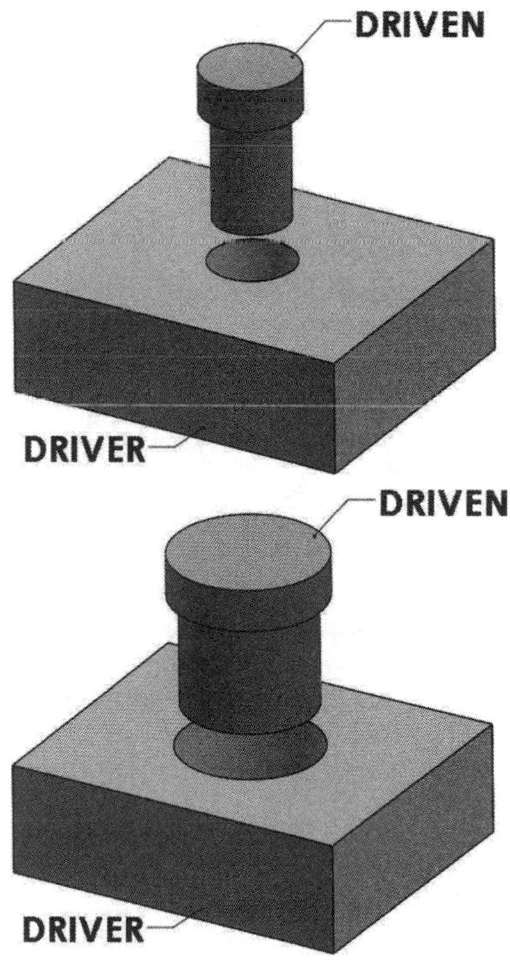

As a primer to **Top Down Design**, we'll make the components illustrated on the previous page as a simple example of how to work in the context of an assembly, or Top Down Design. In a **Bottom Up Design** approach (as we've done until now), we would make both parts and then assemble them together; the problem is that if a part is changed, the other must be manually modified to match the previous part.

In a **Top Down Design** approach, only one change is needed. For this example, we'll make the *'Base'* part, add it to a new assembly, create the *'Driven'* part while in the assembly and finally modify the size of the hole, which will propagate the changes to the pin (*'Driven'* part). The steps we'll follow for the Top Down Design example are the following:

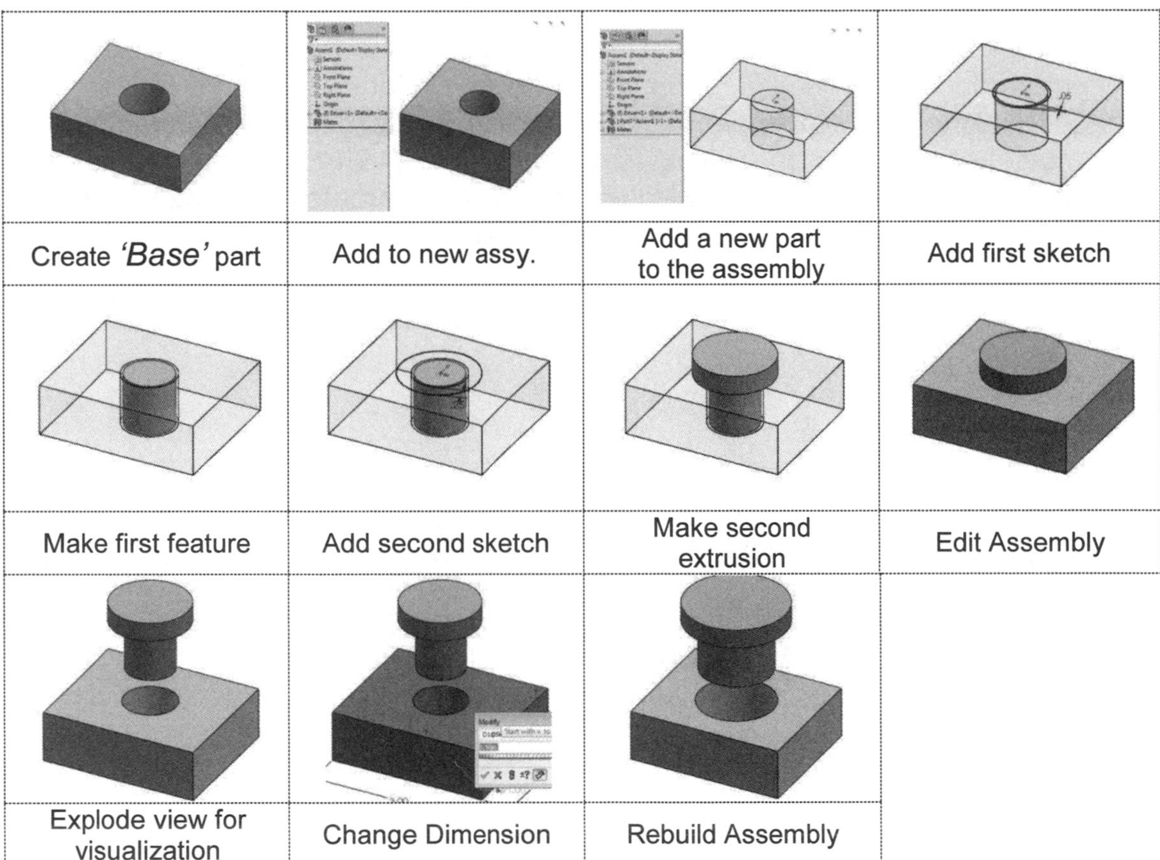

When designing components in the context of an assembly, you can reference other parts or assemblies including geometry, sketches, planes, axes, etc. for sketch relations, sketch planes and feature end conditions. The relations created this way are **External Relations** and are indicated in the Feature Manager with "**->**" both in the part's name and in each of the features that references external geometry. External relations are created every time a dimension, geometric relation, feature end condition or a sketch plane references another component (parts or assemblies) while being edited in the context of an assembly.

9.1. - Make the following part using the dimensions provided and save it as *'Top Down Base'* when done. Draw the base feature using the "Center Rectangle" tool to make the part centered about the origin, this way the hole will be centered in the part.

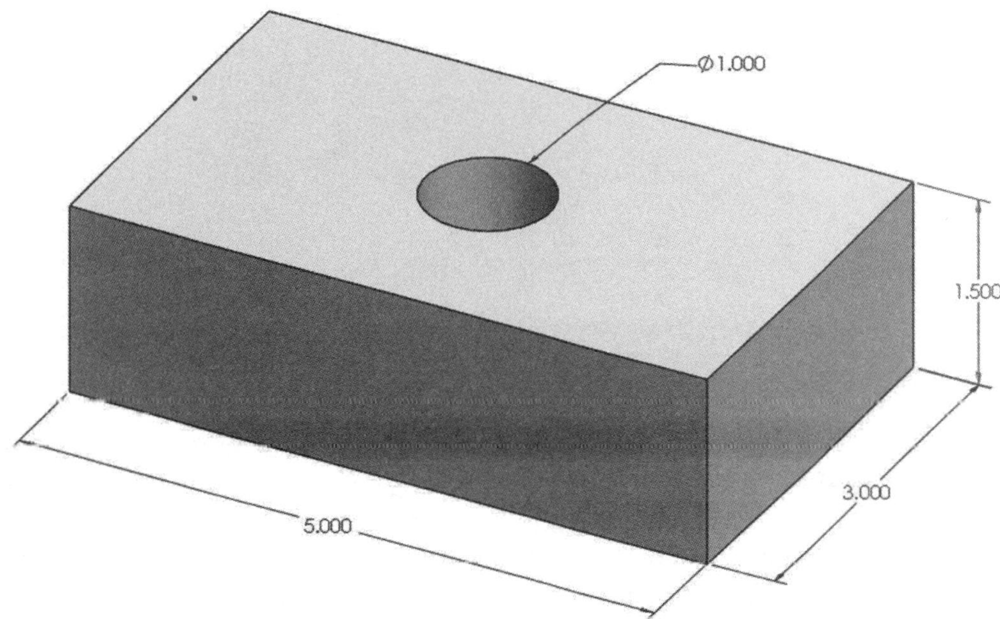

9.2. - Add the *'Top Down Base'* part to a new assembly. The location of this part in the assembly is not important for this example. Save the assembly with the name *'Top Down Assembly'.*

It is important to save the assembly before external references can be added. The reason for this is because SOLIDWORKS needs to know in which assembly the external references were created.

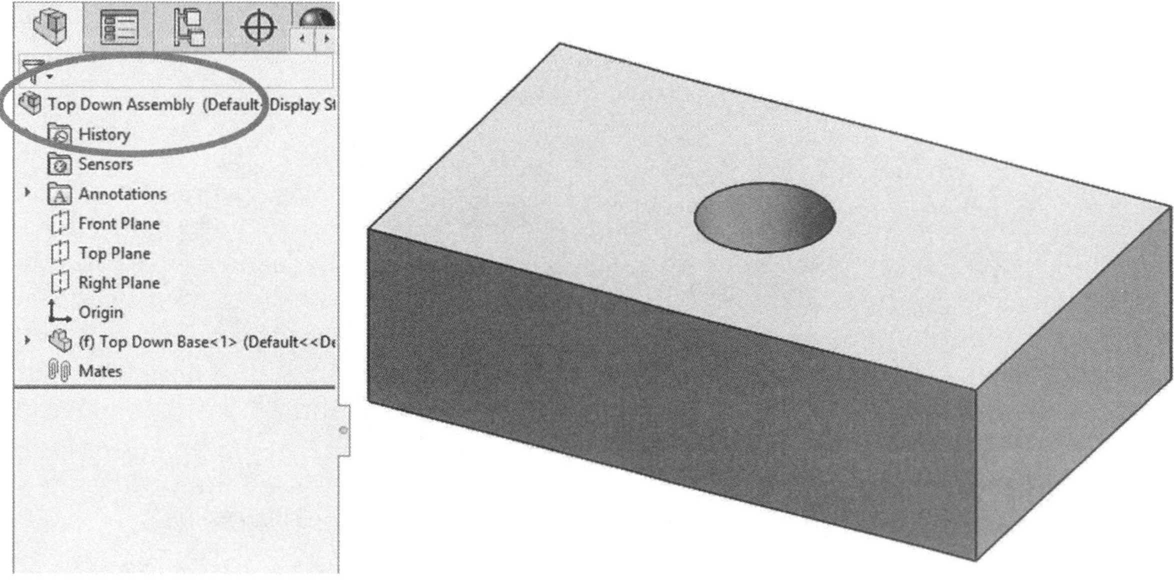

9.3. - The next step is to add a new component to the assembly. This is a new part that doesn't exist yet; it will be created *inside the assembly* or *in context*. From the "**Insert Components**" drop down menu in the Assembly toolbar select "**New Part**" or from the menu "**Insert, Component, New Part.**"

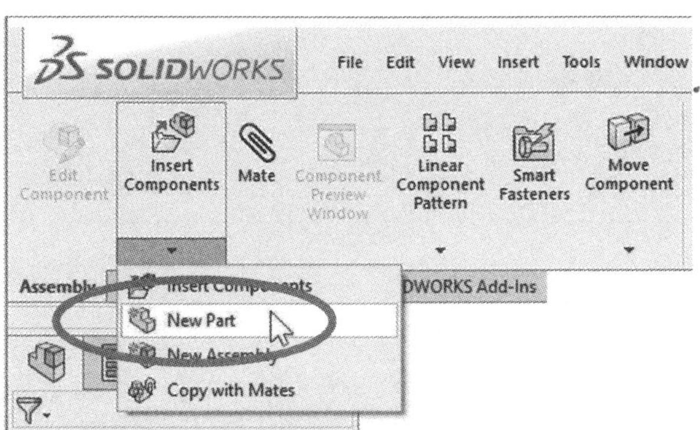

If after selecting the "**New Part**" command you are asked to select a template use the Part template. Immediately after SOLIDWORKS will ask us for a plane to locate the new part. Select the top face of *'Top Down Base'* in the assembly. By selecting this face, we are creating the first external reference in the new part; the *"Front Plane"* of the new part will have an "**InPlace**" mate with the face selected, which will fix the new part to that face. In other words, the new part will be immovable: all six degrees of freedom will be fixed to the face selected.

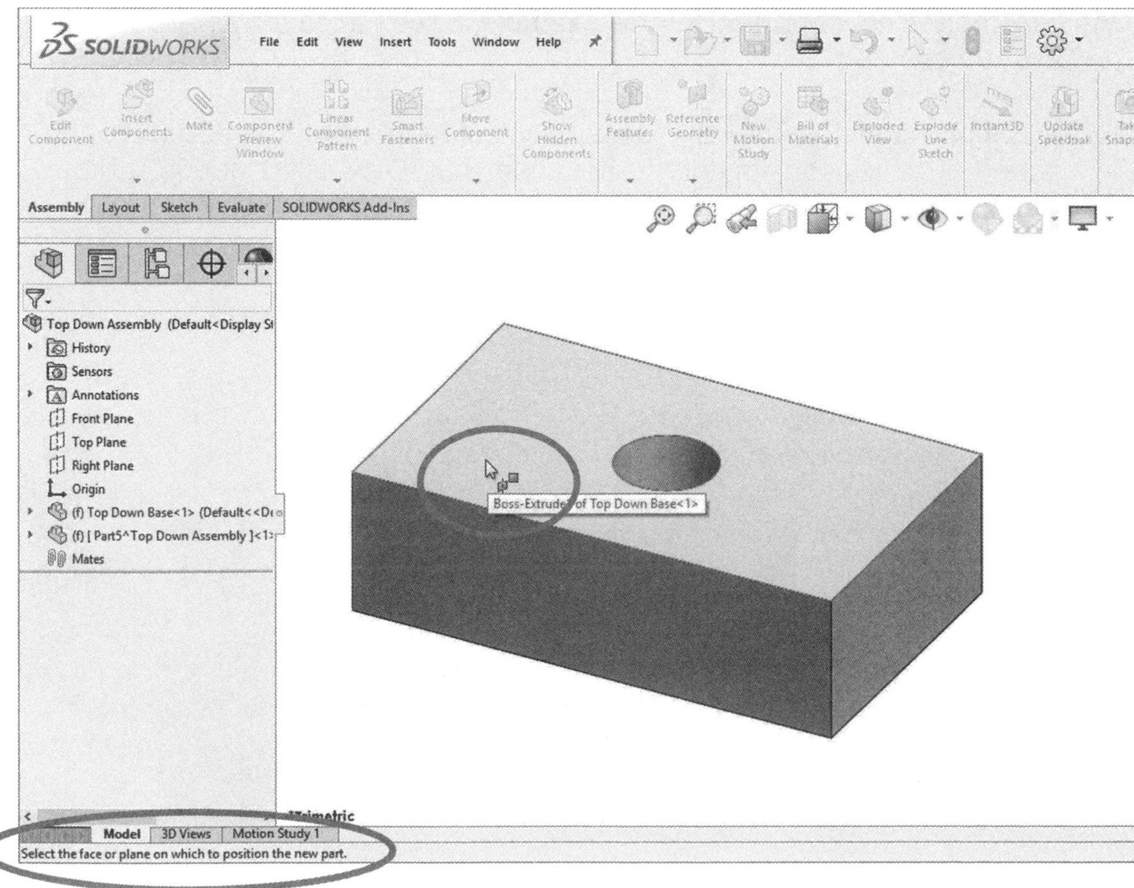

9.4. - After selecting the face in the assembly several things will happen:

A new part will be added to the FeatureManager tree and its name will be shown (by default) in blue. The name in blue indicates that we are *editing* this component *in* the assembly.

A new Sketch will be created in the *"Front Plane"* of the new part as indicated by the usual sketch editing indicators, plus an additional one: an "**Edit Component**" icon will be added to the CommandManager (or assembly toolbar) and will be activated. This indicates a component (part or subassembly) is being edited *in* the assembly. This icon will be visible in every CommandManager toolbar while a part or sub-assembly is being edited. The most obvious indicator to let us know we are editing a new part in the assembly is that existing assembly component(s) will become (by default) transparent.

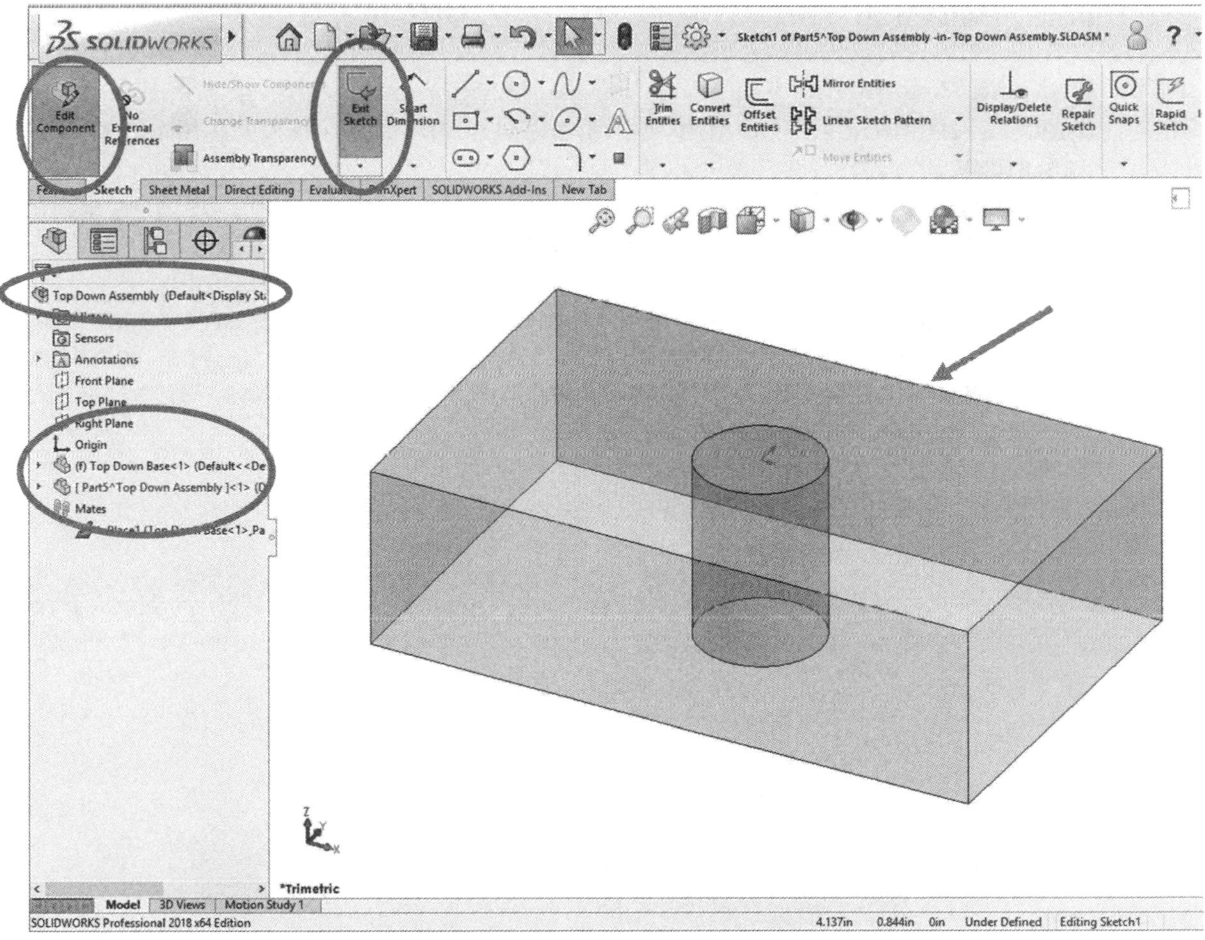

If, after selecting the face to insert a new part, you are asked to name the new part, the System Option "**Assemblies, Save new components to external files**" is set. The default setting is *off*.

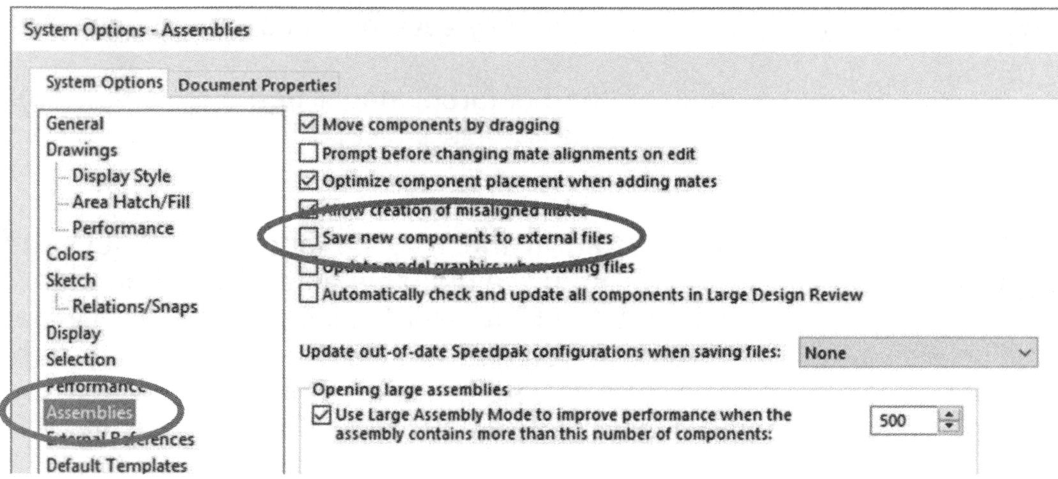

Making the other parts and/or sub-assemblies transparent is a System option in the "**Display/Selection**" section, and it has three different options:

- **Opaque Assembly** will force every component in the assembly to be opaque regardless of the component's transparency setting in the assembly.
- **Maintain Assembly Transparency** will leave components as they were; no changes to transparency will be made. Using this setting it will make it more difficult to know we are editing components in the assembly.
- **Force Assembly Transparency** (default setting) will make every assembly component transparent except the one being edited. The level of transparency can be set with the slider in the system options.

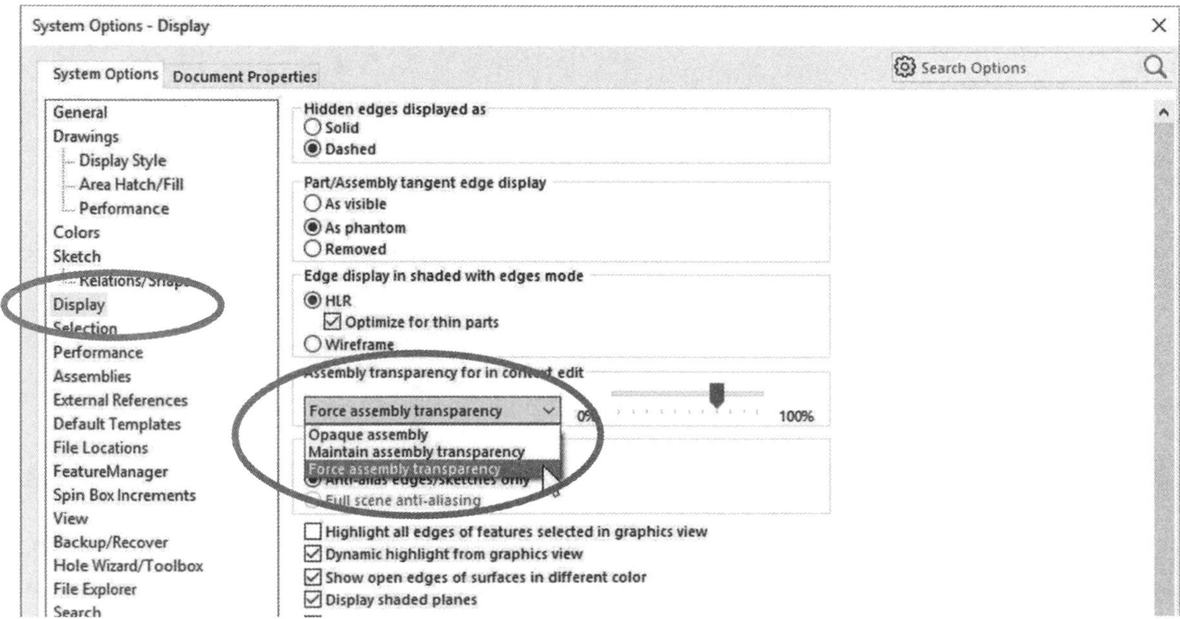

For this example we'll use the "Force assembly transparency" setting to help us better visualize if we are editing the assembly or a part.

9.5. - The next step is to design the *driven* part. What we want to do is to make the pin 0.05″ smaller than the hole. Since we are editing the first sketch of the new part, we'll use the **Offset Entities** command to offset the hole's edge. The only thing new in this step is we are making the offset selecting the edge of *another* component, other than the one we are working on. Select "Reverse" if needed to offset the edge inside of the hole; click OK to finish.

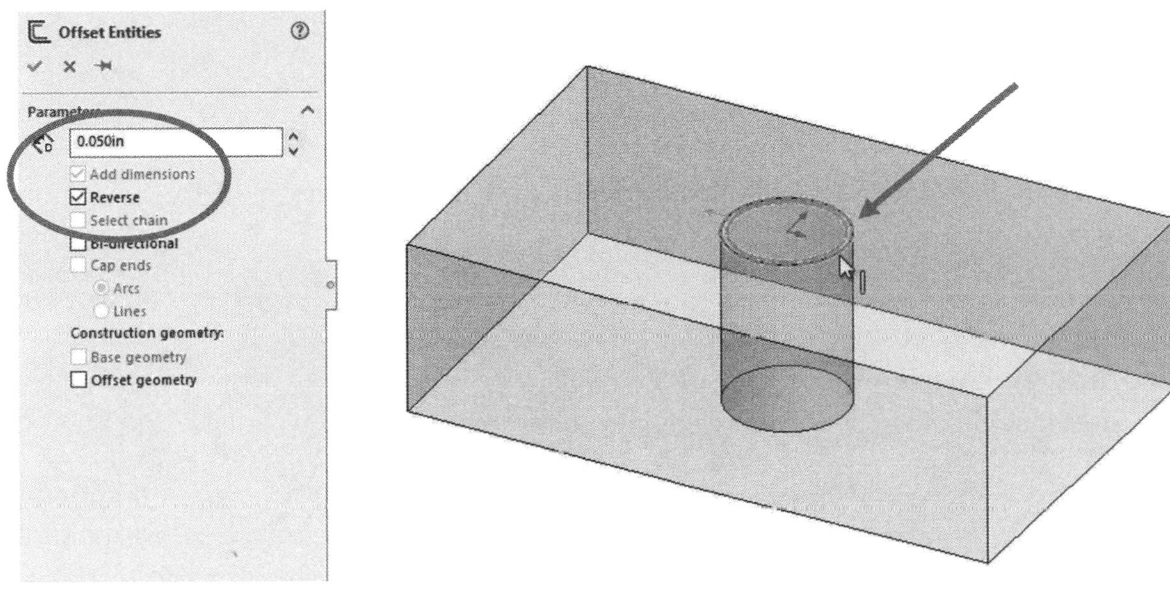

Your sketch will now look like this:

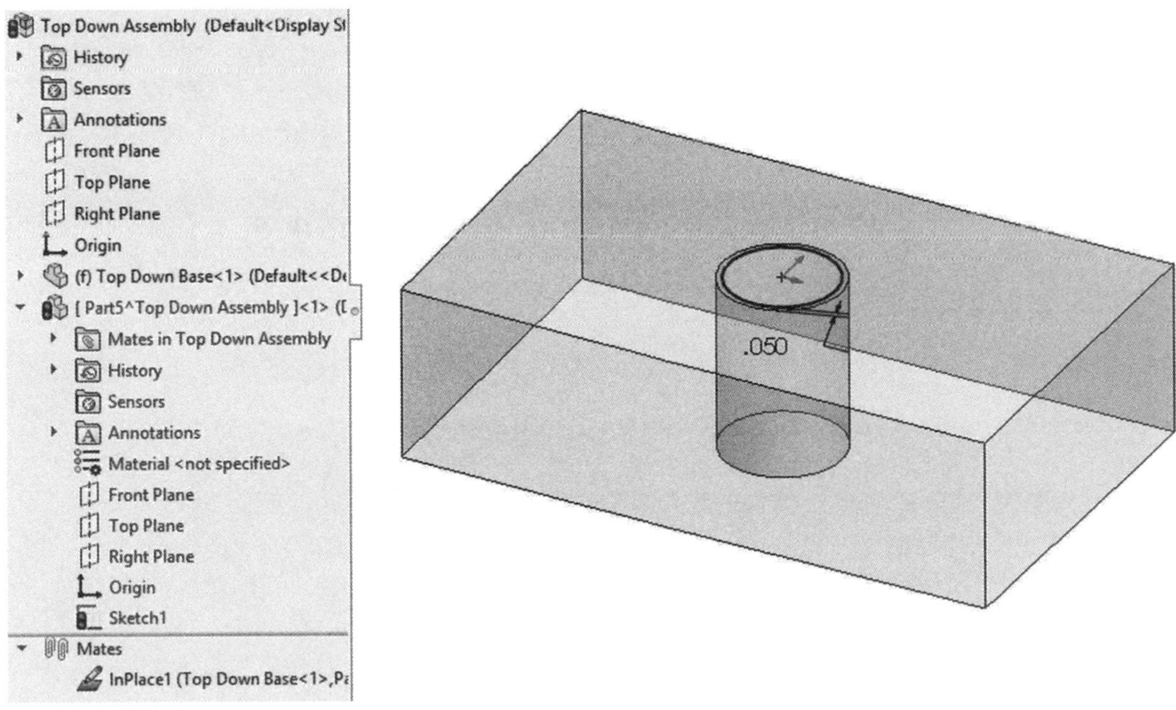

9.6. - Go to the Features tab, extrude the sketch using the "Up to Surface" end condition and select the bottom face of the *'Top Down Base'* part to make the new part's first extrusion the same thickness. Click OK to continue.

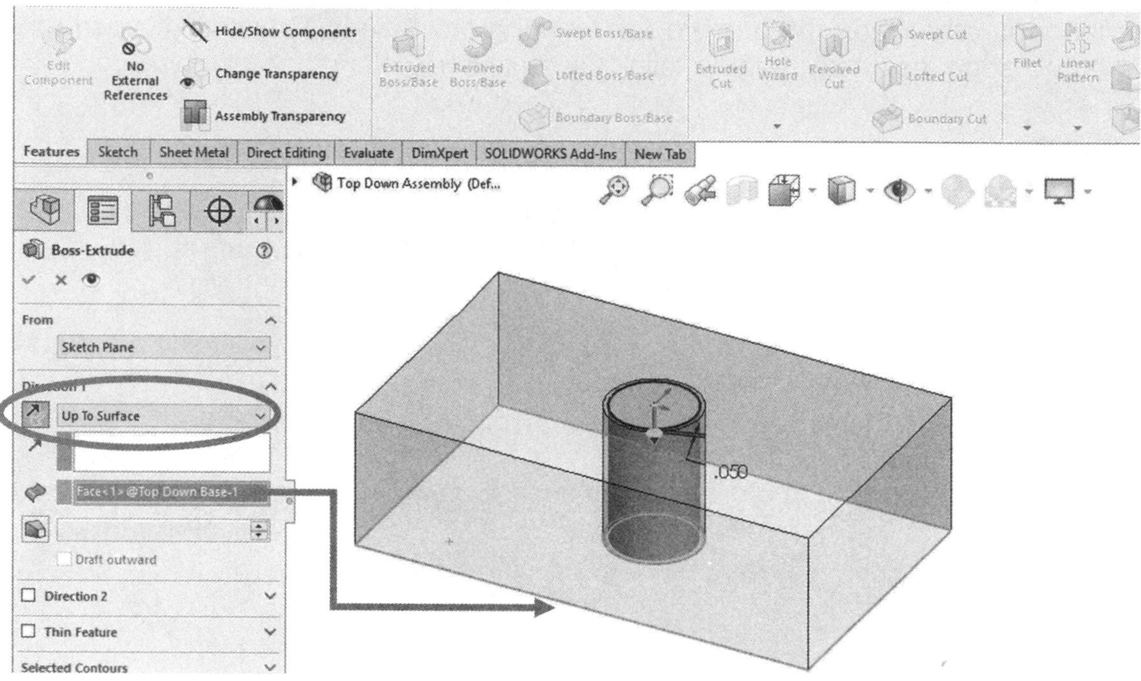

The new part and our assembly now look like this:

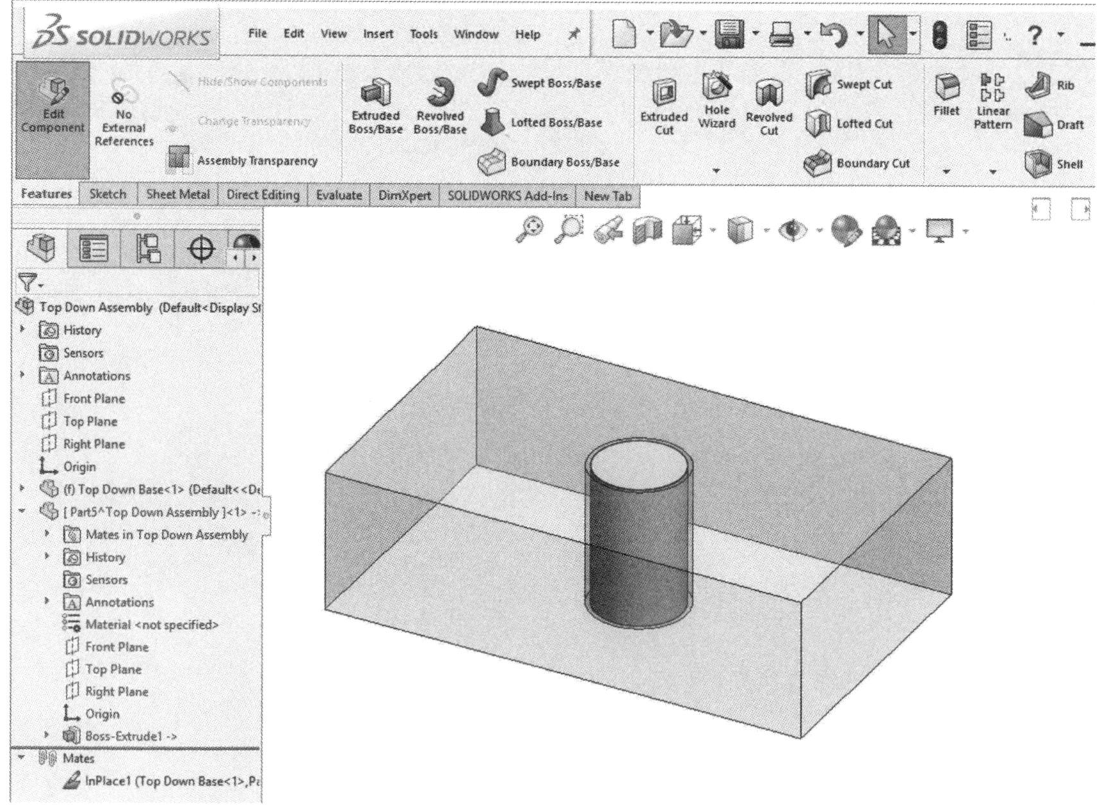

Let's look at the FeatureManager for a moment; there are a few things we need to pay attention to:

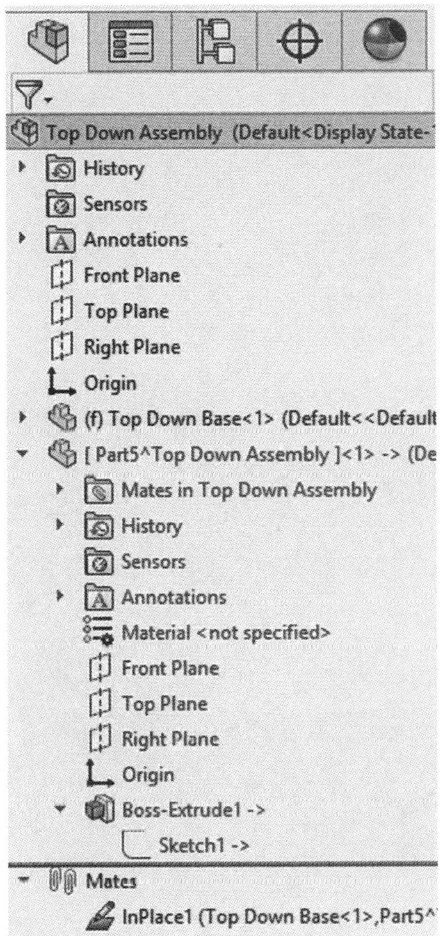

The part we added was automatically named, in this case **[Part5^Top Down Assembly]**. Notice the name is in rectangular brackets [ ]; this means that the part is *internal* to the assembly, meaning that while the part is a separate component, there is no external file for it (yet). In this case, we only have 2 actual files on the hard drive: the assembly (*'Top Down Assembly.sldasm'*) and the Base part (*'Top Down Base.sldprt'*).

The name of the new part and all of its features are shown in blue; this means that the part (or subassembly if that was the case) is being edited.

At the end of the new part's name, we can see *"Boss-Extrude1"* and *"Sketch1"* names are followed by **->**; this means there is *at least* one external reference. If you remember, *"Sketch1"* was made by offsetting the hole's edge and *"Boss-Extrude1"* was extruded up to the bottom of the *'Top Down Base'* part. Later in the lesson we'll go more in depth about external references, and internal and external components.

9.7. - The second feature of the new part will be the head of the pin. Remember that we are still editing the new part in the context of the assembly, and EVERY part modeling command is available to us. Add a new sketch on top of the extrusion we just made. Using the **Offset** command, make a sketch 0.5″ bigger than the hole in the *'Top Down Base'* part, just as we did for the first sketch.

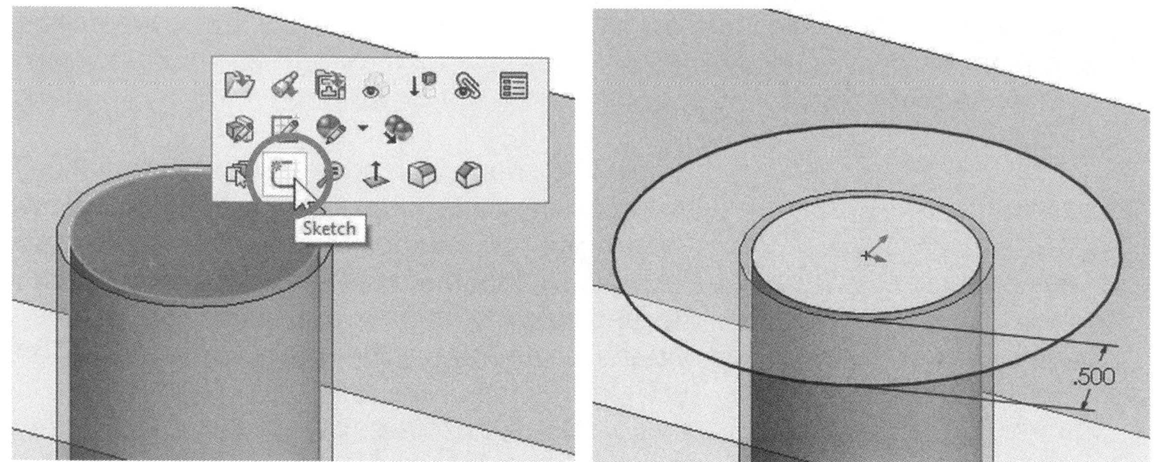

Extrude the sketch 0.5″ up to finish the part.

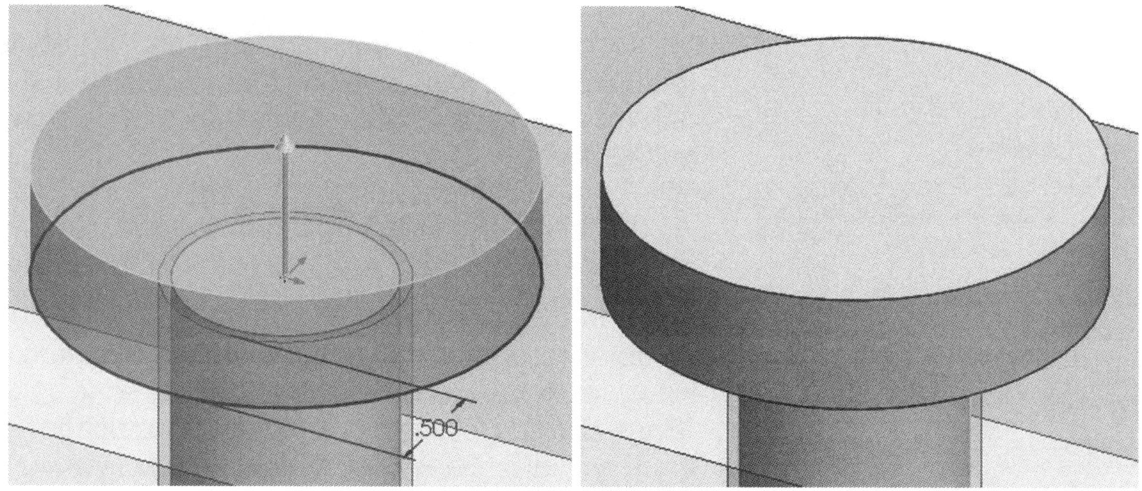

9.8. - We have finished the new part in the context of the assembly. Now we need to return to editing the assembly. As a reference, the different editing levels within SOLIDWORKS are:

TOP LEVEL EDITING	Icon	Editing Level
		Main Assembly
		Subassembly
	* As many Subassembly levels as needed.	
		Part in Assembly
		Part
		Part Feature
LOWER LEVEL EDITING		Part Sketch

Additionally, SOLIDWORKS has assembly sketch and assembly features in assemblies and subassemblies which will be covered later.

When we are at the part (or subassembly) editing level inside an assembly, we can right mouse click in the graphics area, and change the assembly transparency from the pop-up menu "**Assembly Transparency**" or select the "**Assembly Transparency**" icon in the CommandManager.

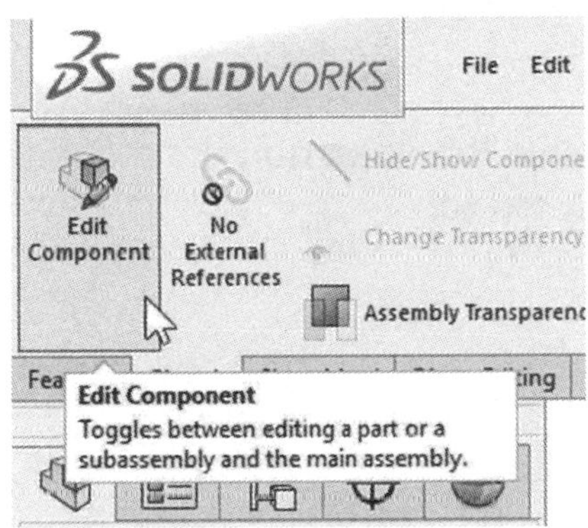

Currently we are editing the part *inside* the assembly. To return to editing the main assembly, select the "**Edit Component**" icon in the CommandManager (or Assembly toolbar if visible), or

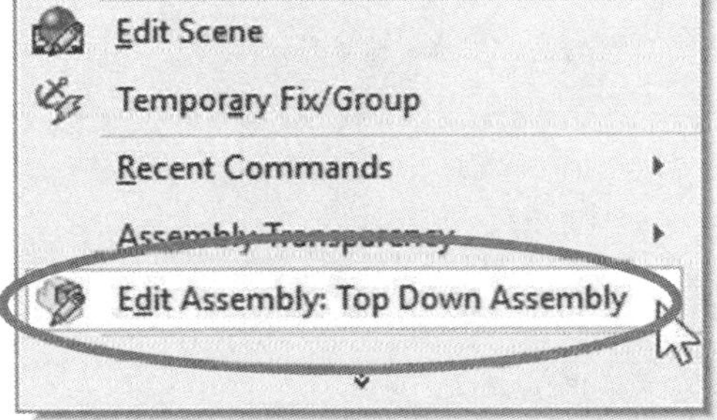

...right mouse click in the graphics area and select "**Edit Assembly: *name***," or

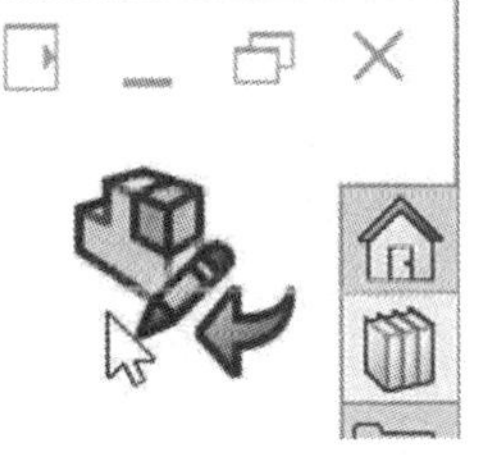

...select the **Edit Assembly** confirmation corner (just like when we close a Sketch).

While it may seem a little confusing at first, it's very simple: we are either editing the part or the assembly. The important thing is to pay attention to the indicators letting us know if we are editing a part or an assembly.

Now that we are editing the assembly, everything should look as we were used to, and the status bar at the bottom will read "**Editing Assembly**." Notice that when we are editing a part in the assembly, the assembly tools are disabled.

9.9. - Now that we are editing the assembly, add an exploded view and pull the pin out for visibility.

9.10. – To test the propagation of changes from the *Driver* part to the *Driven* part, change the hole's diameter from 1″ to 1.625″ and rebuild the assembly. The *Driven* part will update per the offset distance added in each sketch.

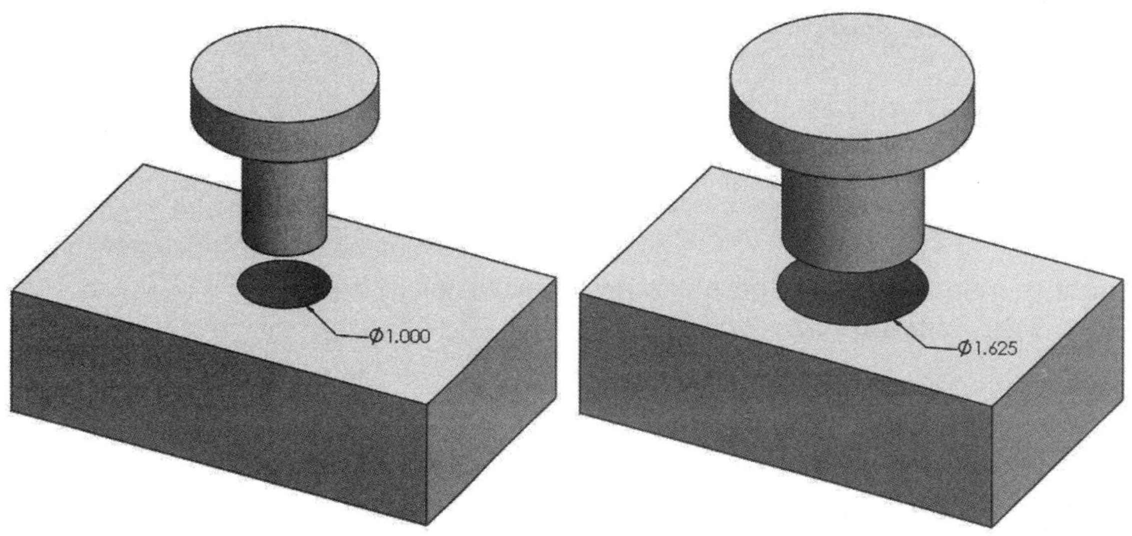

9.11. – When we save the assembly we are asked if we want to save all components, or select which components to save. Select "**Save All**."

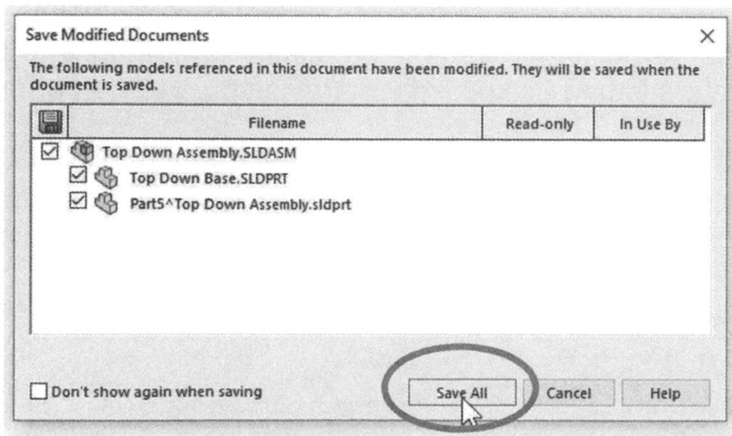

In the following message we are asked if we want to save the new part internally (inside the assembly) or externally to a new file. In this step select "**Save internally**" for now.

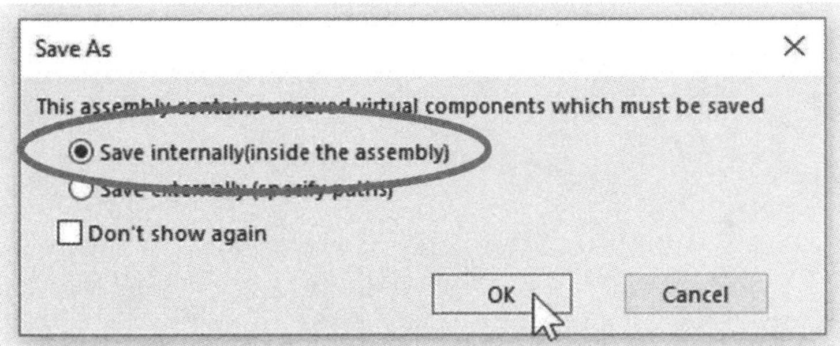

9.12. - Saving the part "internally" means that there is not an actual file we can reference, and therefore we cannot make a drawing of it while it remains an internal part. To *externalize* the part and make a file for it, we must (optionally) rename it like a regular feature with a slow double click or in its properties, and then save it. Rename the internal part as *'Top Down Driven'*. After renaming it right-mouse-click on it and select "**Save Part (in External File)**." Select the path to save the file and click OK.

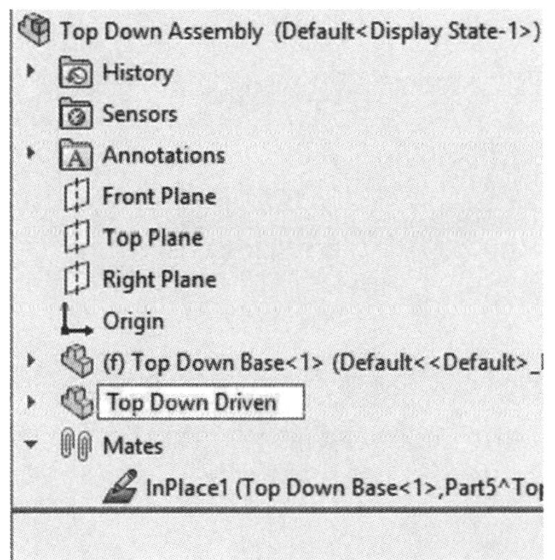

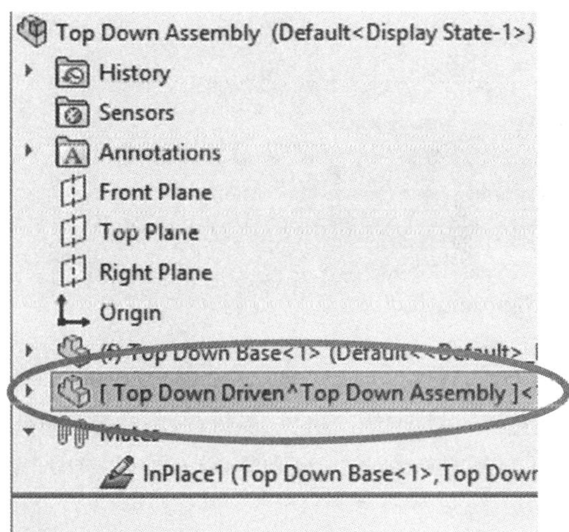

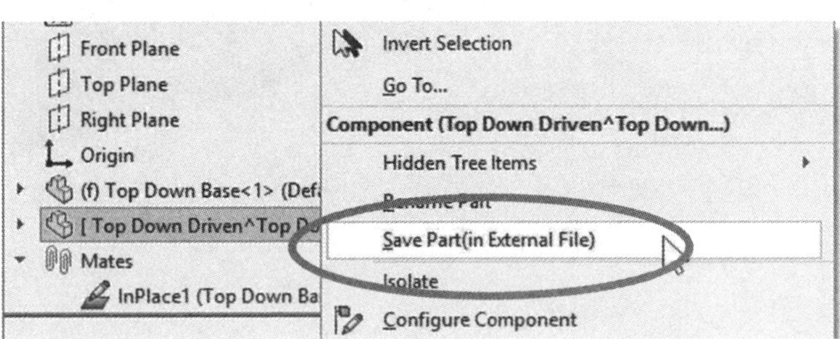

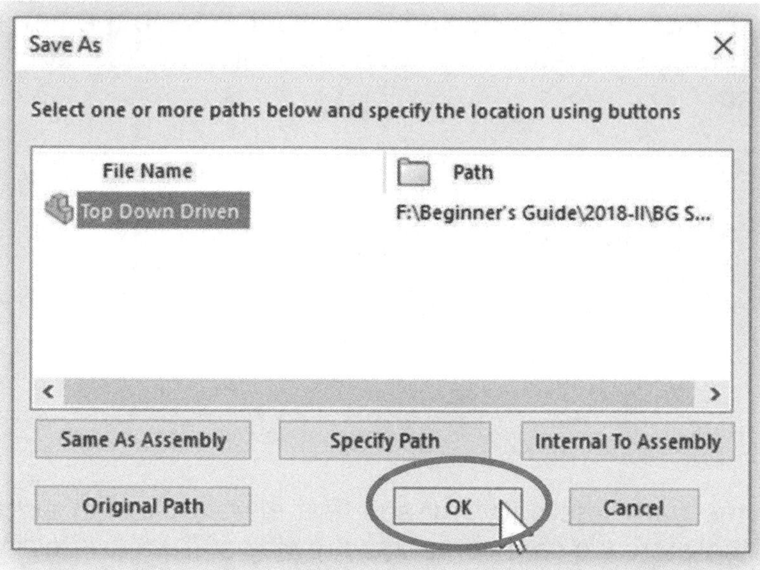

Now that the part has been "externalized" we have a file for this part and the FeatureManager shows the part's name as any other assembly component, except it shows that it also has external references (**->**). Save the assembly to finish.

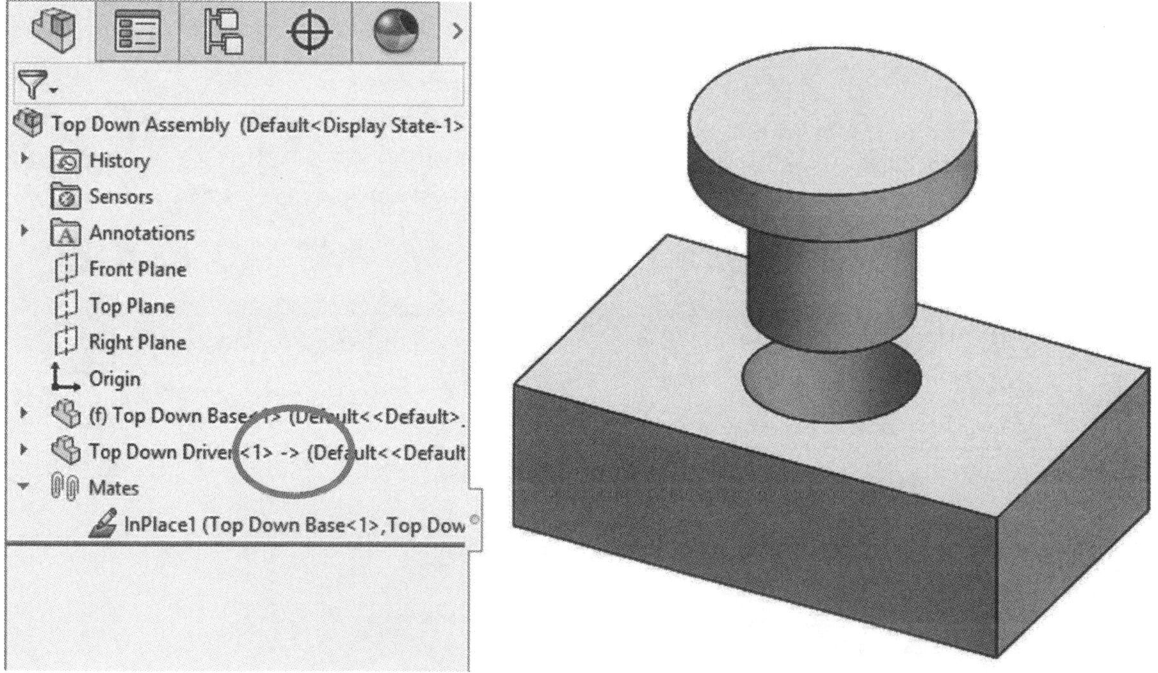

More About External References

External references can have four different states: In Context, Out of Context, Locked and Broken. Here is a description and example of each.

In Context:

When a part with external references and the assembly where those references were created are open in SOLIDWORKS (loaded in memory), if a referenced part (driving) is modified, the driven part is automatically updated, as in the previous example. **In context** external references show "**- >**" in the FeatureManager.

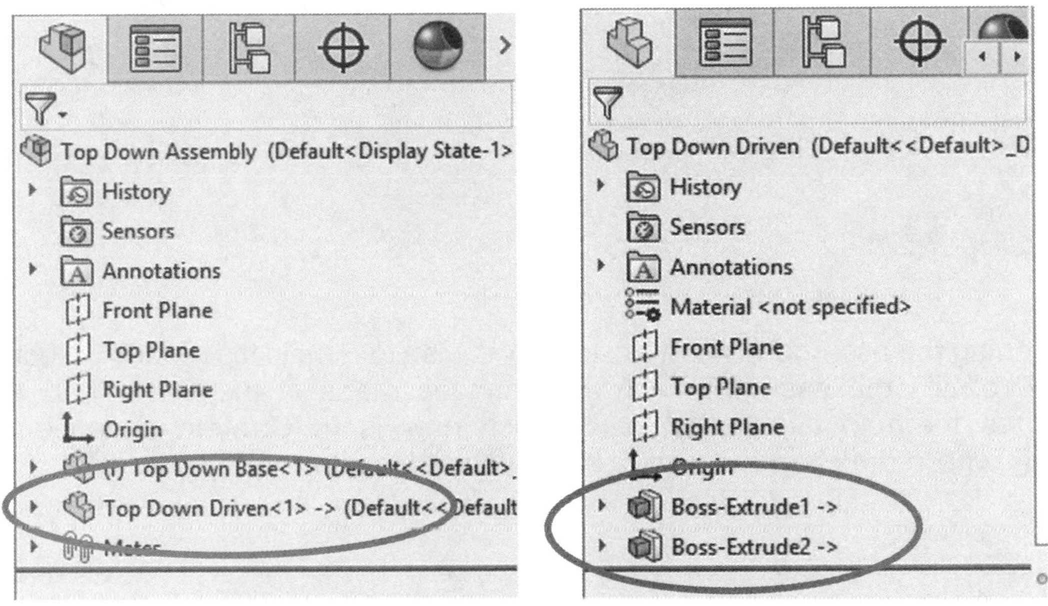

Assembly view Part view

Out of Context:

Out of Context is when a part with external references is open, but the assembly in which those references were created is not (loaded in memory). If a referenced part (driving) is modified, the driven part **WILL NOT** update until the assembly where the external references were added is opened (loaded in memory).

To better visualize it, close all parts and assemblies and open the *'Top Down Driven'* and *'Top Down Base'* parts only, **do not** open the assembly. Change the *'Top Down Base'* part's hole size and then go back to the *'Top Down Driven'* part; notice the changes do not propagate to the driven part. Finally, open the assembly to see the changes propagate. **Out of Context** references are displayed with "**- >?**" in the FeatureManager.

After changing the *Driver* part the *Driven* part is not updated, it's **Out of Context**.

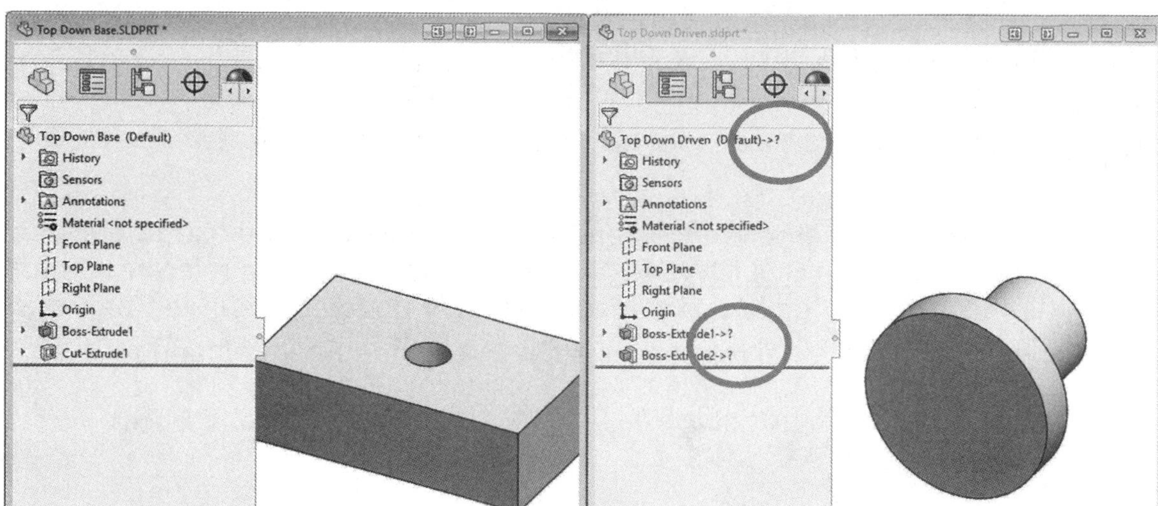

After opening the assembly, we are asked if we wish to rebuild the model. Selecting Yes will rebuild the assembly with the changes made in the *Driver* part and propagates them to the *Driven* part, which now is **In Context** because the assembly where the changes were made is loaded in memory.

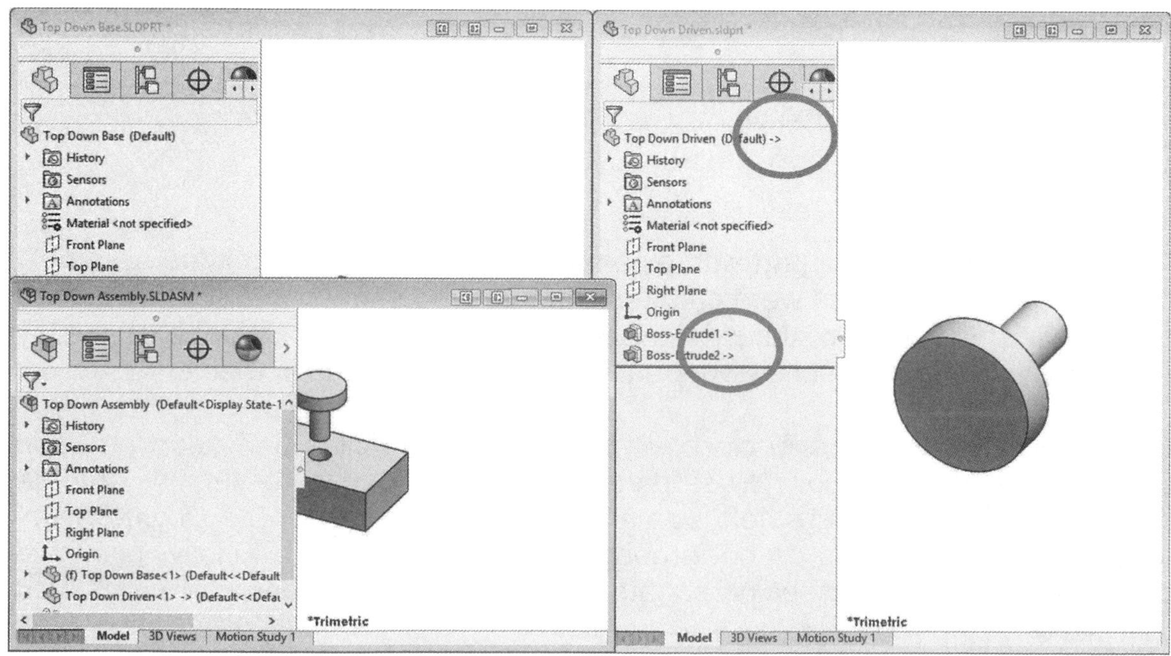

Locked:

When external references are locked, changes in the *Driver* part will not propagate to the *Driven* parts, even if the assembly is open (loaded in memory). To lock a part's external references, open the '*Top Down Driven*' part, right-mouse-click at the top of the part's FeatureManager and select **"List External Refs…"**

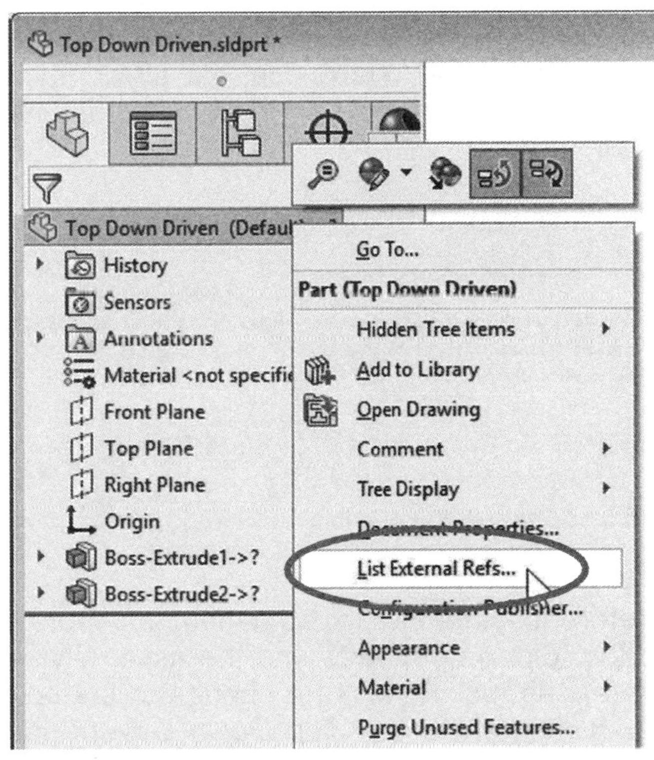

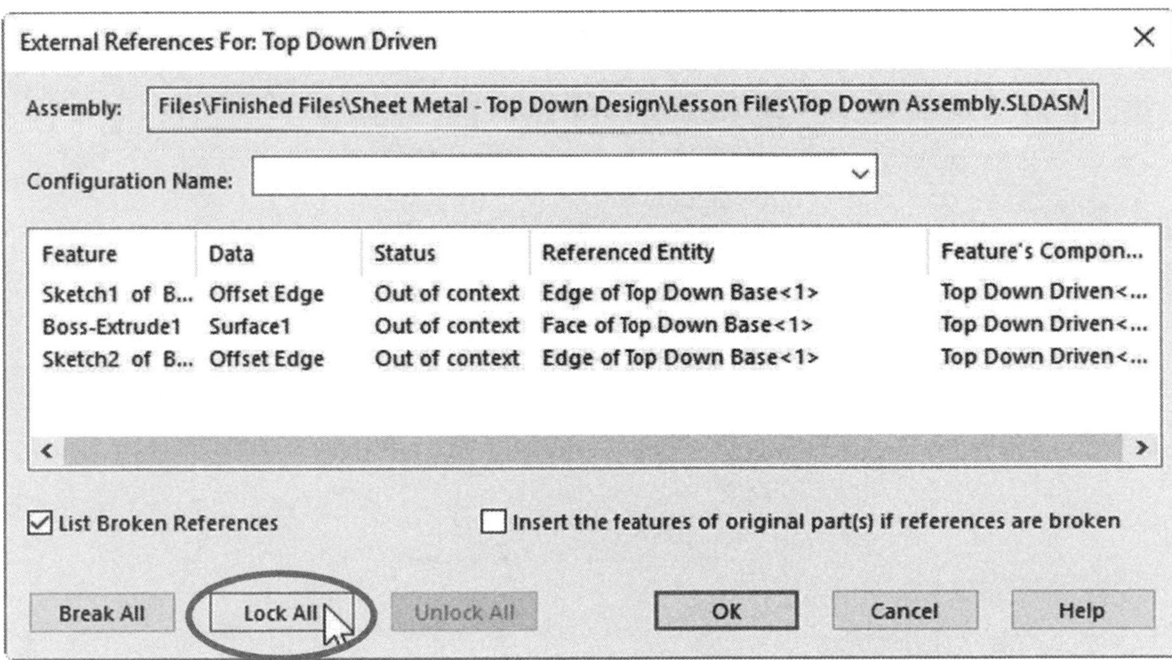

In the "**External References**" list we can see the features with external references, status (In context, Out of context, Locked or Broken) and which part and entity is being referenced. To lock references means that we can temporarily *freeze* them. This is useful when we are working in an assembly and we don't want to propagate changes immediately. After opening the *'Top Down Driven'* file, list the external references and select "**Lock All**."

When locking external references, ALL references are locked. There is no way to selectively lock individual references, it's all or nothing. When locking references, we'll see the following message; click OK to continue and close the external references window.

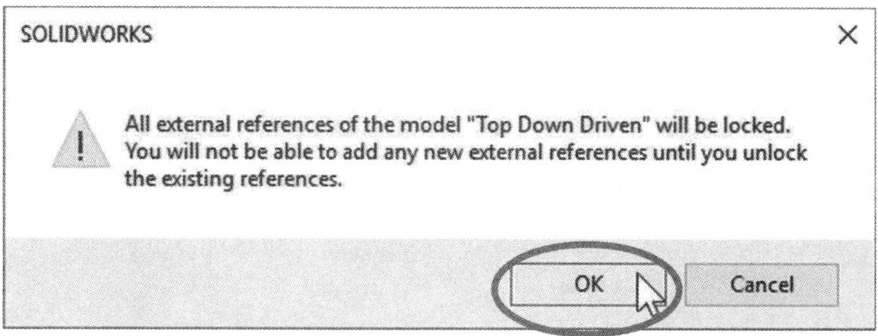

When the external references are locked, we cannot add more external references until they are unlocked. Open both parts and the assembly, with the *'Top Down Driven'* external references locked. Make a change to the hole's diameter in the *'Top Down Base'* part. Return to the *'Top Down Driven'* part and confirm that changes have not propagated, even when the assembly is open. Locked external references are listed in the Feature Manager followed by "**->***."

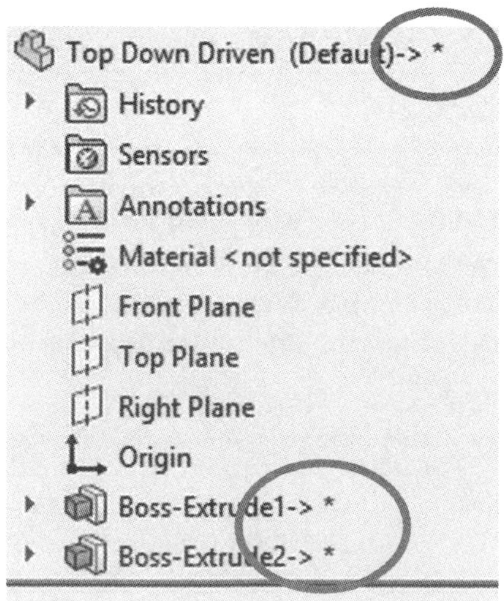

To unlock external references, list the part's external references again, and this time select "**Unlock All**." After unlocking, changes will propagate following the same rules when parts are "**In Context**" or "**Out of Context**."

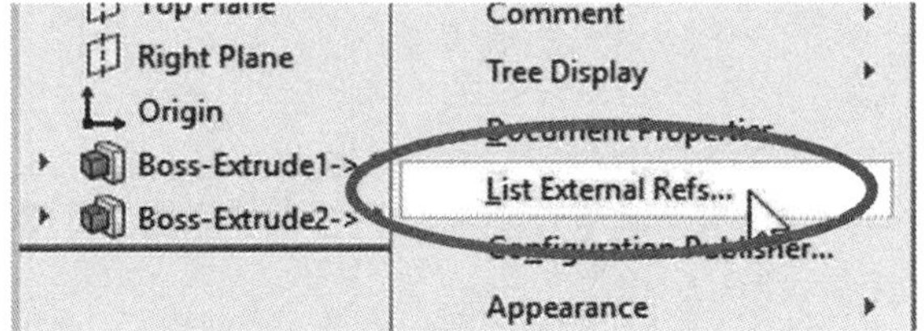

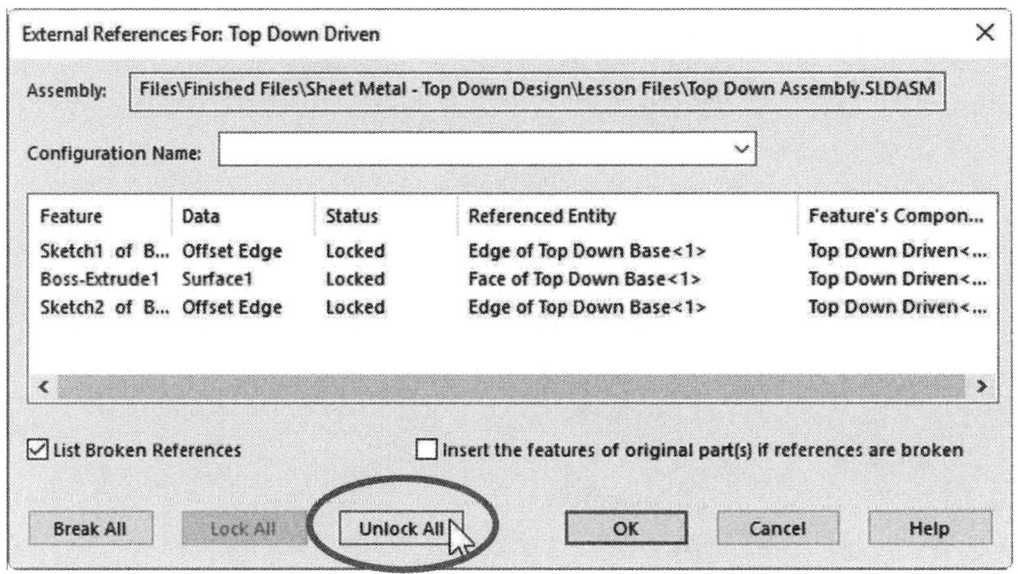

Broken:

Broken external references work the same way as "**Locked**"; the difference is that broken references **CANNOT** be re-established. Once a part's external references are broken, there is no way to recover them. To break external references, list the part's references as before, but select the "**Break All**" option. Broken external references are listed with "**- > x**" in the Feature Manager.

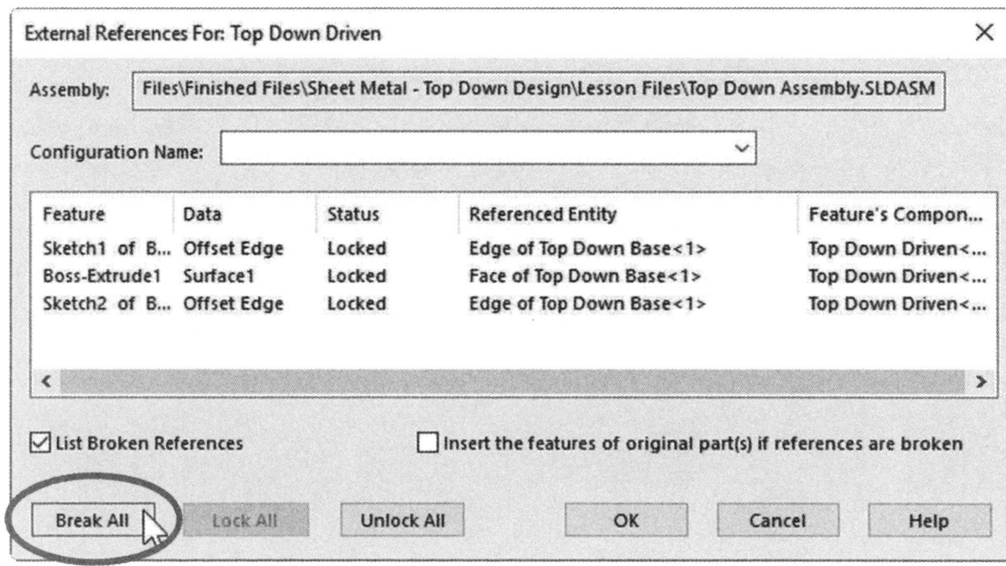

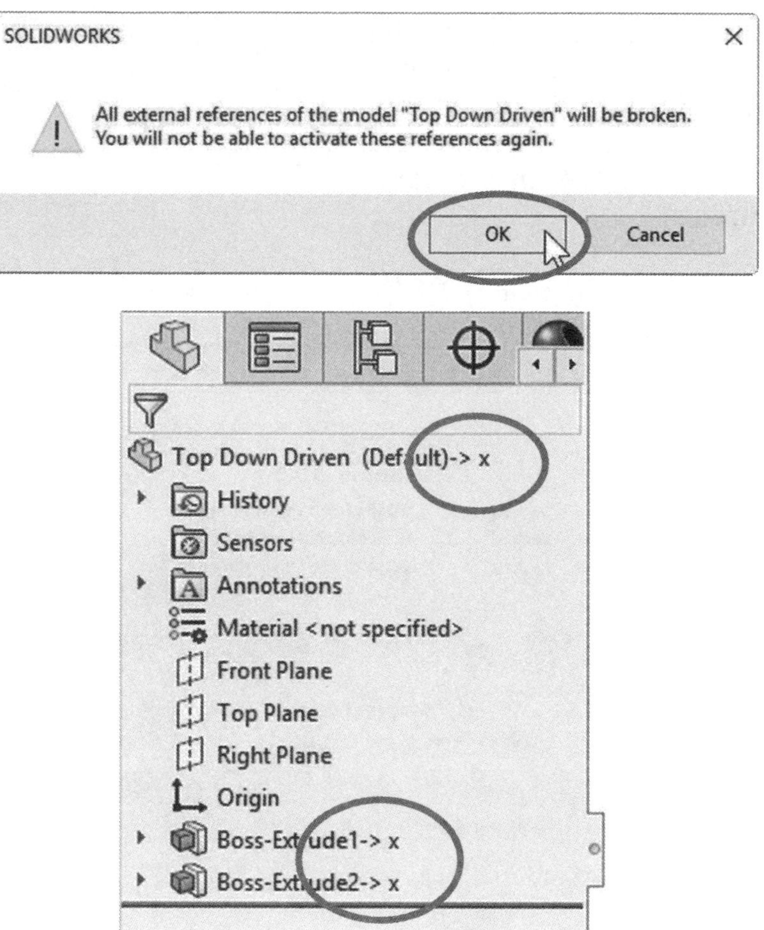

When listing External References for a part with broken references, the list will be empty unless we use the "**List Broken References**" checkbox.

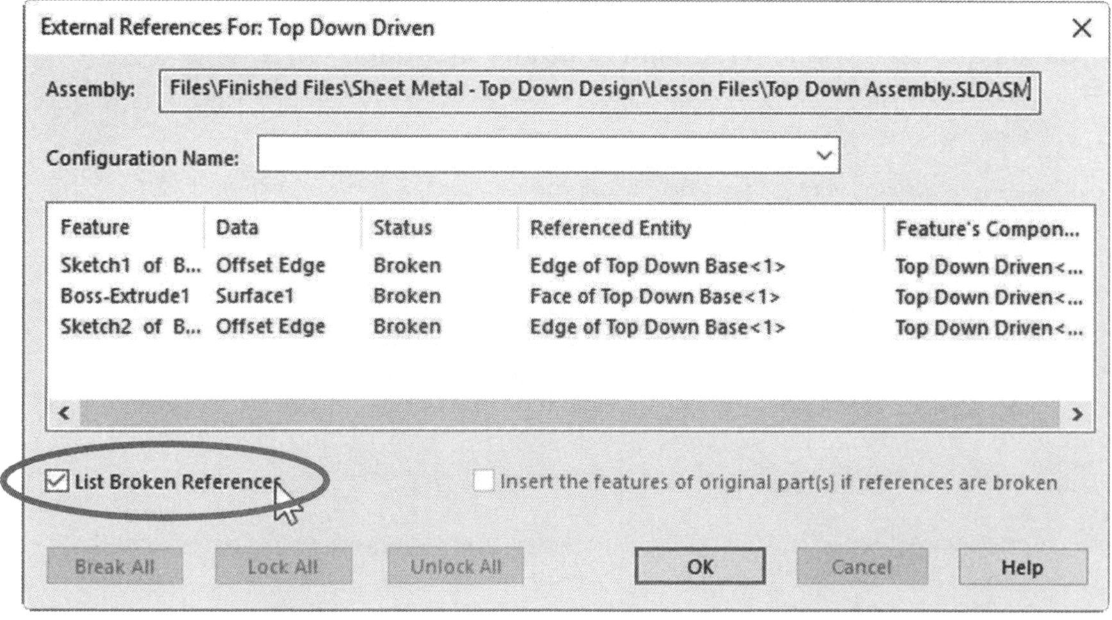

Sheet Metal and Top Down Design

10.1. - Now that we have a better understanding of how to make **Sheet Metal** components and **Top Down Design**, we'll design a computer's power supply enclosure using a Top Down approach and further practice how to make sheet metal parts. For our example, we'll start with the assumption that the electronic components are already designed and the enclosure will be designed *around* the power supply. From the files included with this book, locate and open the *'Electronics.sldprt'* part file. The electronics board looks like this:

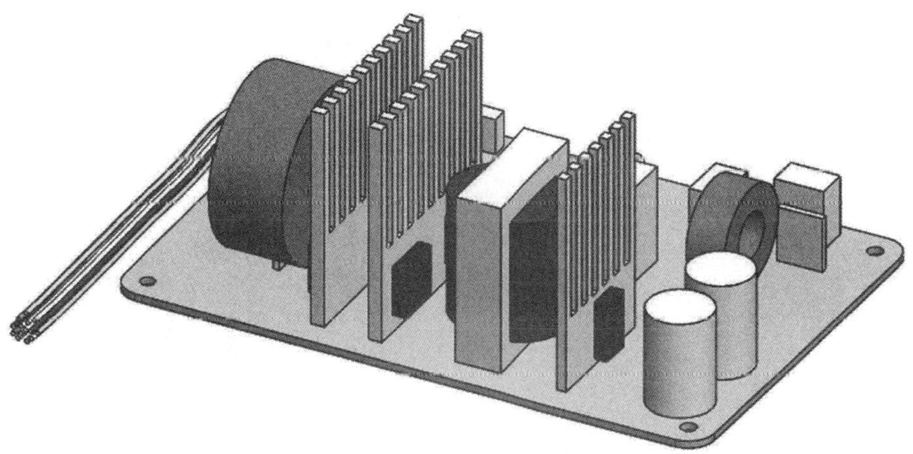

10.2. - First we need to make a new assembly, add the electronics board at the origin, and save the new assembly as *'Power Supply Assy'*. Remember that the assembly must be saved before we can add components and features in the context of the assembly.

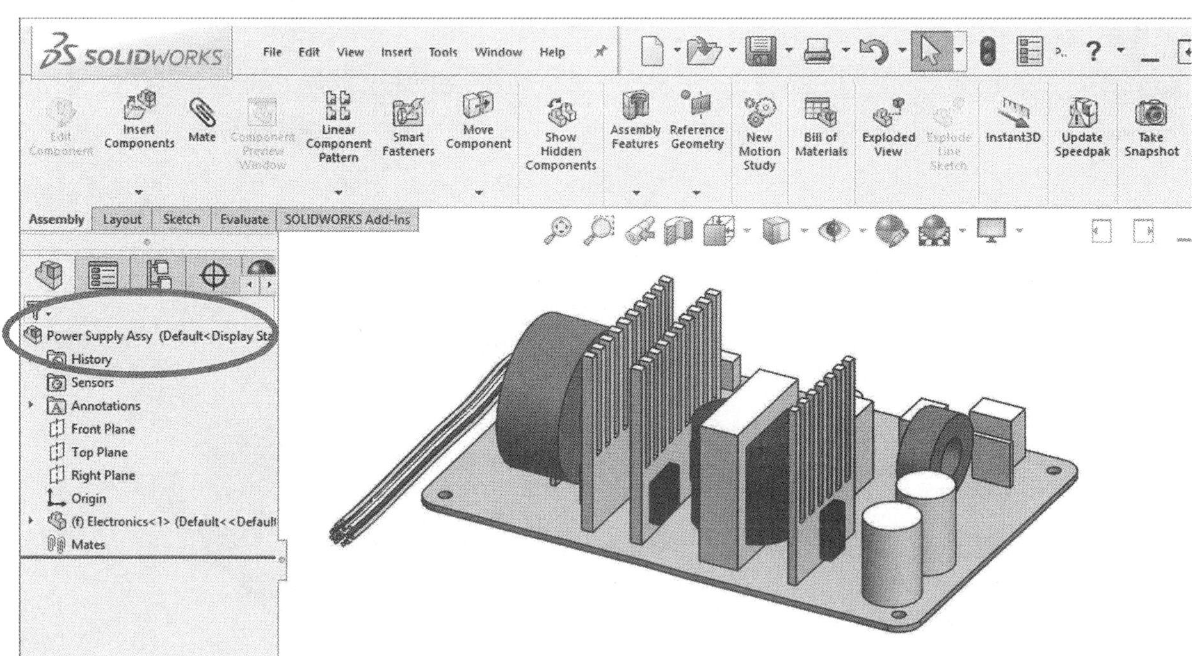

10.3. - The first component we are going to design in the context of the assembly is the bottom half of the enclosure. Select "**New Part**" from the drop-down menu in the "**Insert Components**" icon in the Assembly toolbar, or from the menu "**Insert, Component, New Part.**"

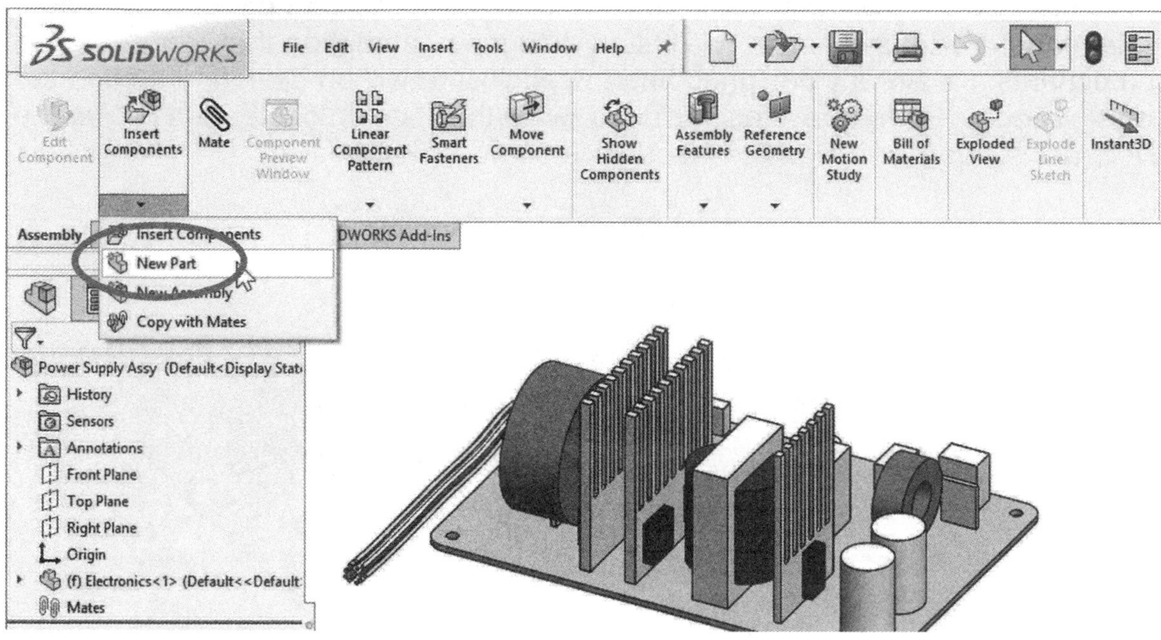

To locate the new part's *"Front Plane,"* select the assembly's *"Right Plane"* in the FeatureManager. Remember, when we locate the new part, at the same time we're adding an "**InPlace**" mate between the selected face/plane and the new part's *"Front Plane."*

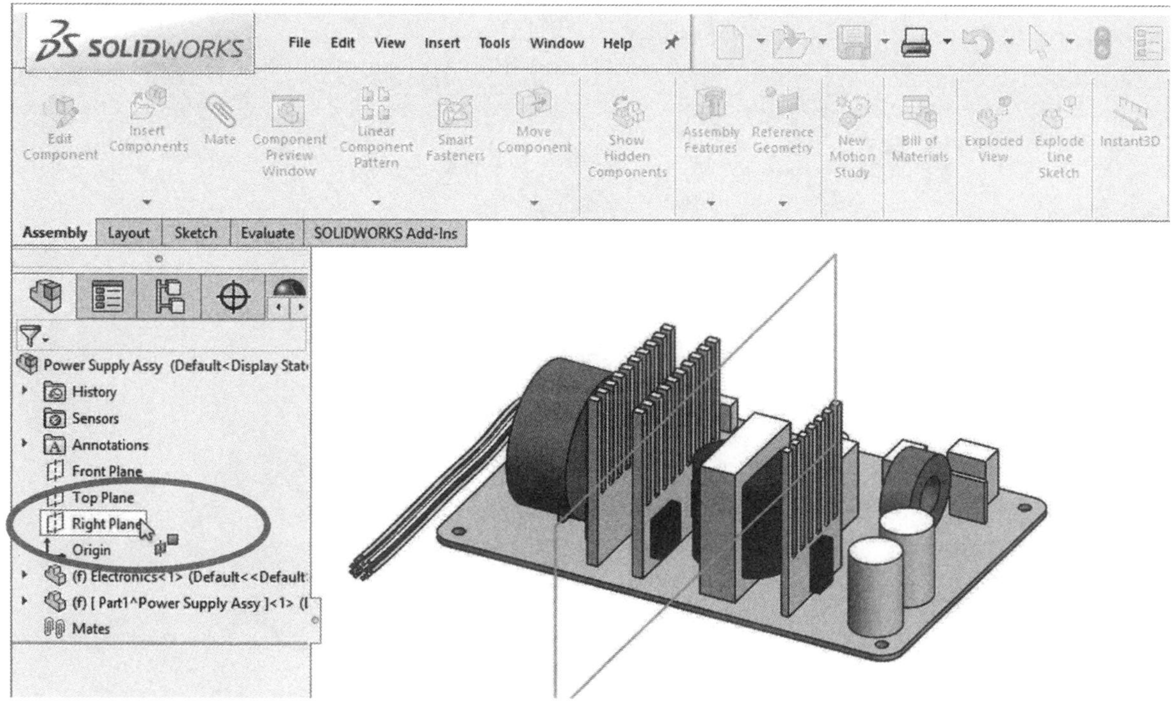

After selecting the assembly's *"Right Plane"* in the FeatureManager:

- The electronics board will become transparent (if the option is set in "**Tools, Options, Display/Selection, Force assembly Transparency**").
- The "**Edit Component"** icon in the CommandManager will be enabled.
- A name will be assigned to the new part in the FeatureManager tree if the option "**Tools, System Options: Assemblies, Save New Components to external files**" is NOT checked; if set, you will be asked for a new part name.
- The new part's FeatureManager will be blue (default setting).
- A new sketch will be created in the new part's *"Front Plane."*
- The status bar will read *"Under Defined"* since we are working in a new sketch.
- The title bar will read:

  "Sketch1 of Part#^Power Supply Assy -in- Power Supply Assy.SLDASM"

  Which means:

  Sketch1: We are editing the new part's first sketch (Sketch1)

  of Part#^Power Supply Assy: of **internal** *Part#* in the *'Power Supply Assy.'*

  -in- Power Supply Assy.SLDASM: in the *'Power Supply Assy.sldasm'* file.
- An "**InPlace**" mate will be automatically added, fixing the *"Front Plane"* of the new part to the selected plane or face, in our case, the assembly's *"Right Plane"* restricting the new part's six degrees of freedom.

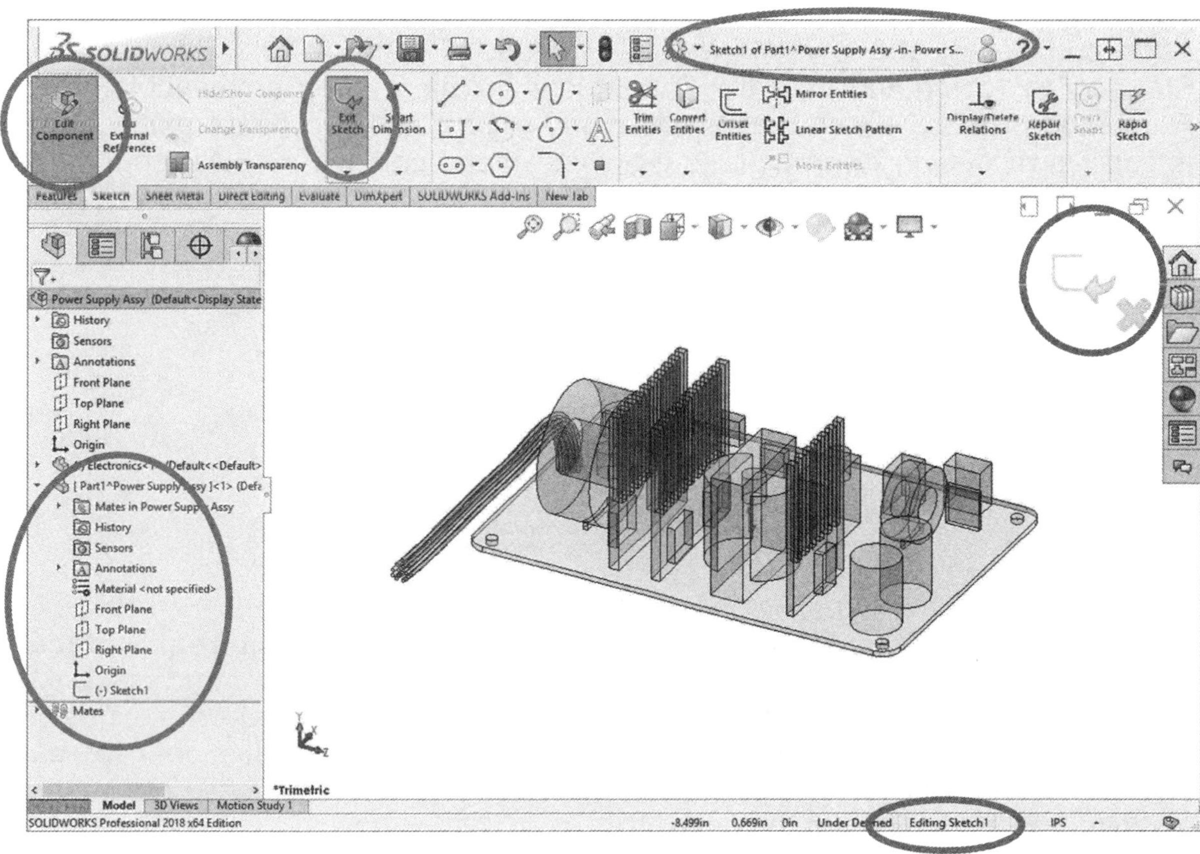

10.4. – Now we need to design the lower part of the sheet metal enclosure. Select a Left view orientation (even if we are editing a part in the assembly, view orientations are referenced to the main assembly, not the part), and draw the following sketch. Remember that we are already working in a new sketch in the new part inside the assembly. Note that this is an open sketch and it is dimensioned referencing the lower and outer edges of the electronic board. Both vertical lines are equal length. Dimension the top of the vertical line 0.3″ above the heat sink.

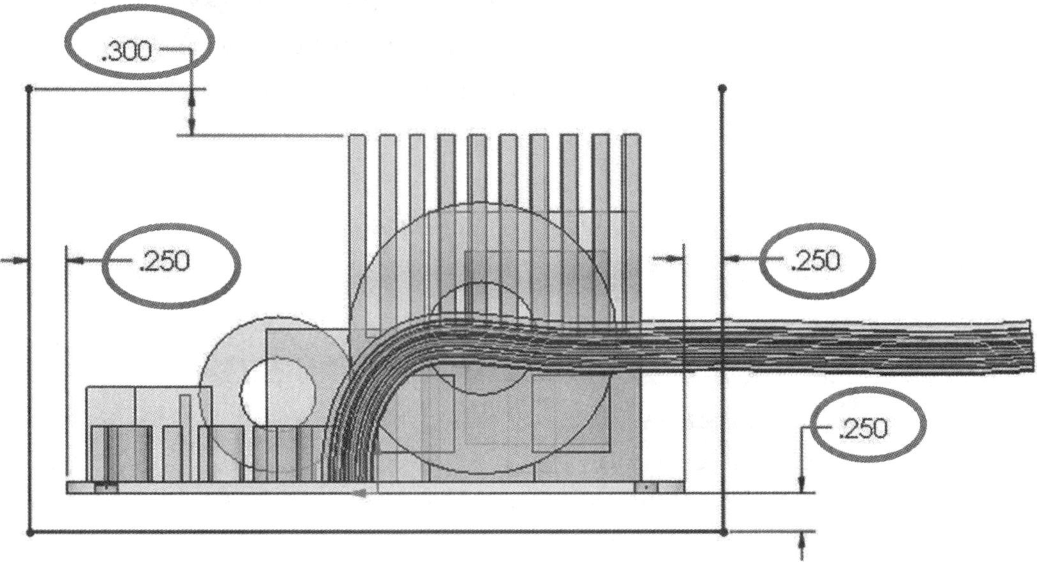

After completing the sketch, make the Base Flange selecting the "**Base-Flange/Tab**" command from the Sheet Metal toolbar in the CommandManager. Remember, we are designing the enclosure in the context of the assembly using a Top Down design approach, so the sheet metal part will be built around the electronics board.

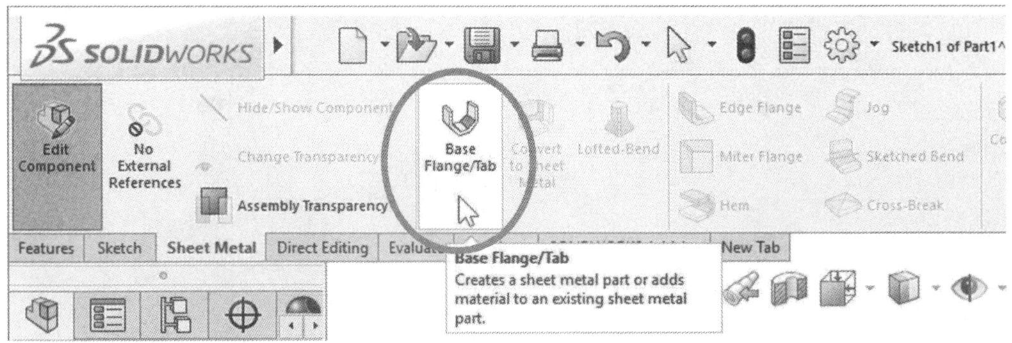

10.5. - Extrude the "**Base Flange**" in Direction 1 using the "**Offset from Surface**" end condition. Select the face at the end of the electronics board and make the offset 0.5", and check the "Reverse offset" option if needed to make the extrusion beyond the board. By adding this end condition, if the board's size changes our sheet metal enclosure will change, too; that is exactly the reason for creating the enclosure in the context of the assembly.

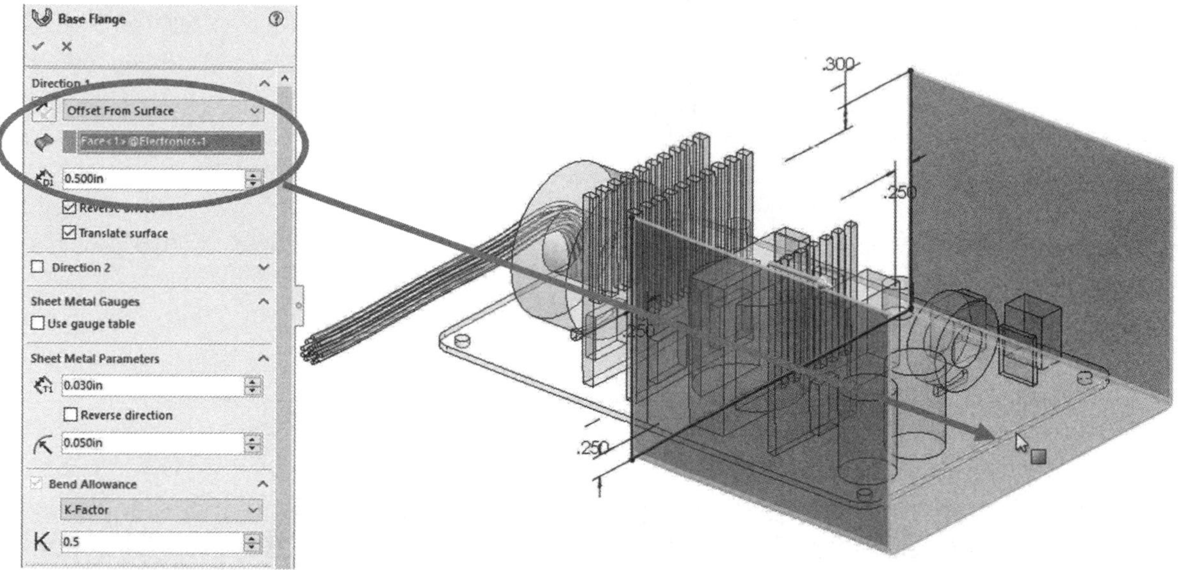

When the "Direction 2" option is activated, the preview is lost. For Direction 2 select the end condition "Offset From Surface" using the opposite face in the board; make the offset 0.5".

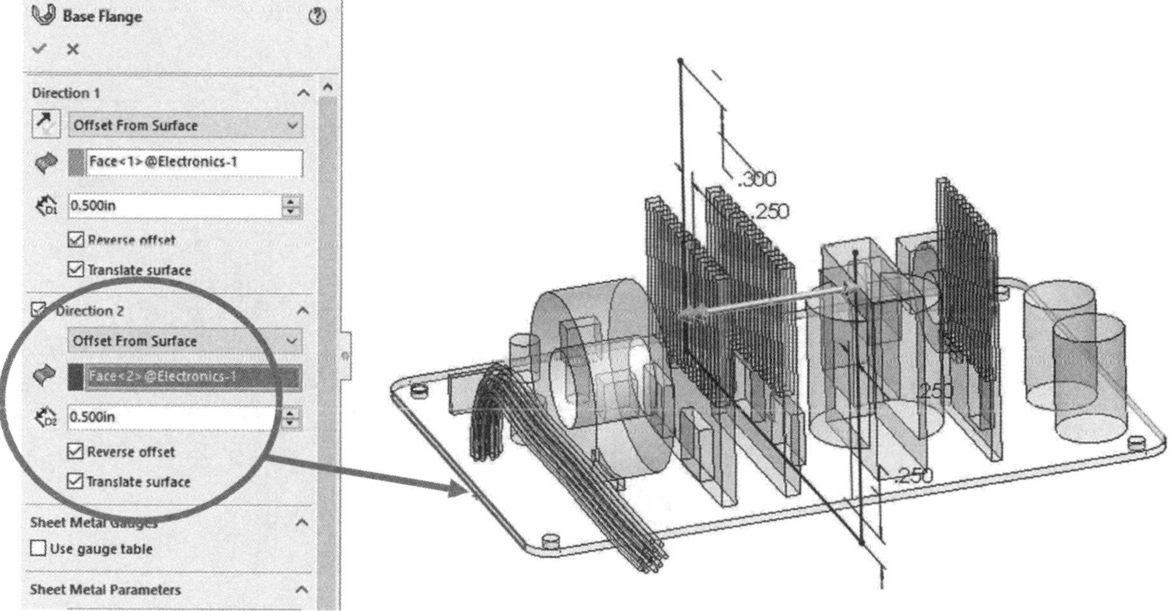

Now that we have defined the end conditions for the enclosure in both directions, we need to define the sheet metal options to control thickness, bend radius, bend allowance and auto relief type. For this example, use a **thickness** of 0.030″ with a **bend radius** of 0.05″ and a "Bend Allowance" **K-Factor** of 0.5. **Auto Relief** will be set to "Tear."

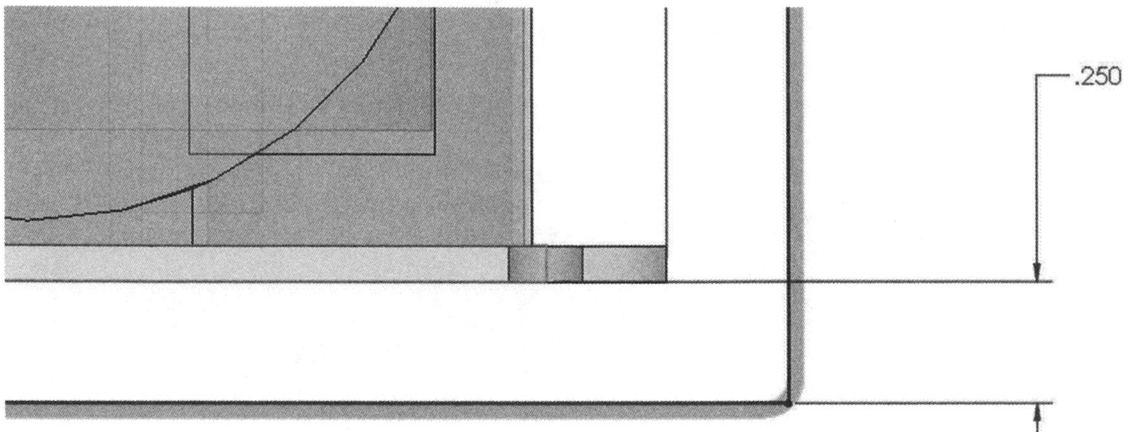

Since our design calls for a 0.25" space between the part and the electronics board, in the "Sheet Metal Parameters" section, activate the "Reverse Direction" checkbox to add the material outside the sketch and not inside, maintaining our intent. Click OK to add the Base Flange.

When the preview is lost as in this case, change to a "Blind" end condition temporarily to verify which side of the sketch the material is being added to, and then set the end condition back to "Offset From Surface."

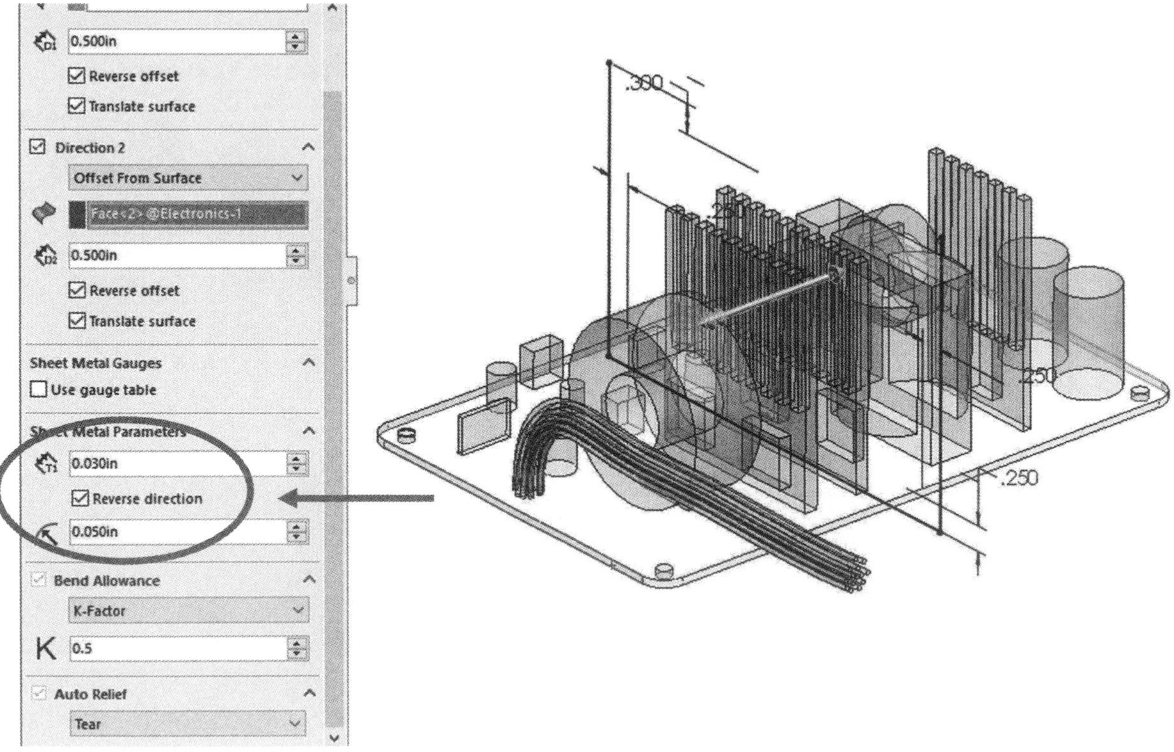

10.6. - After extruding the *"Base-Flange1"* with these parameters, our part looks like this. Keep in mind that we are still editing it in the context of the assembly, and that's the reason why the electronics board is still transparent.

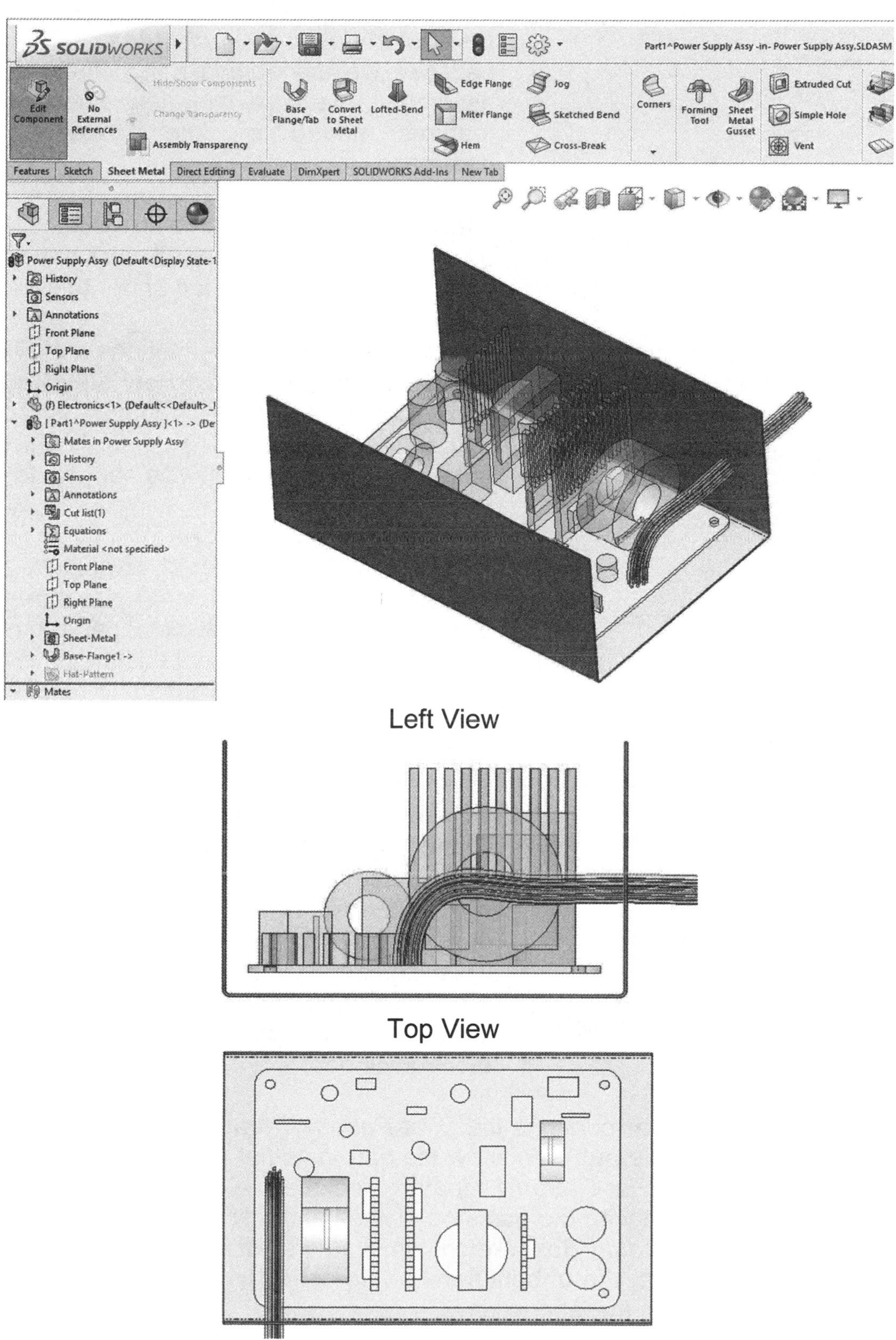

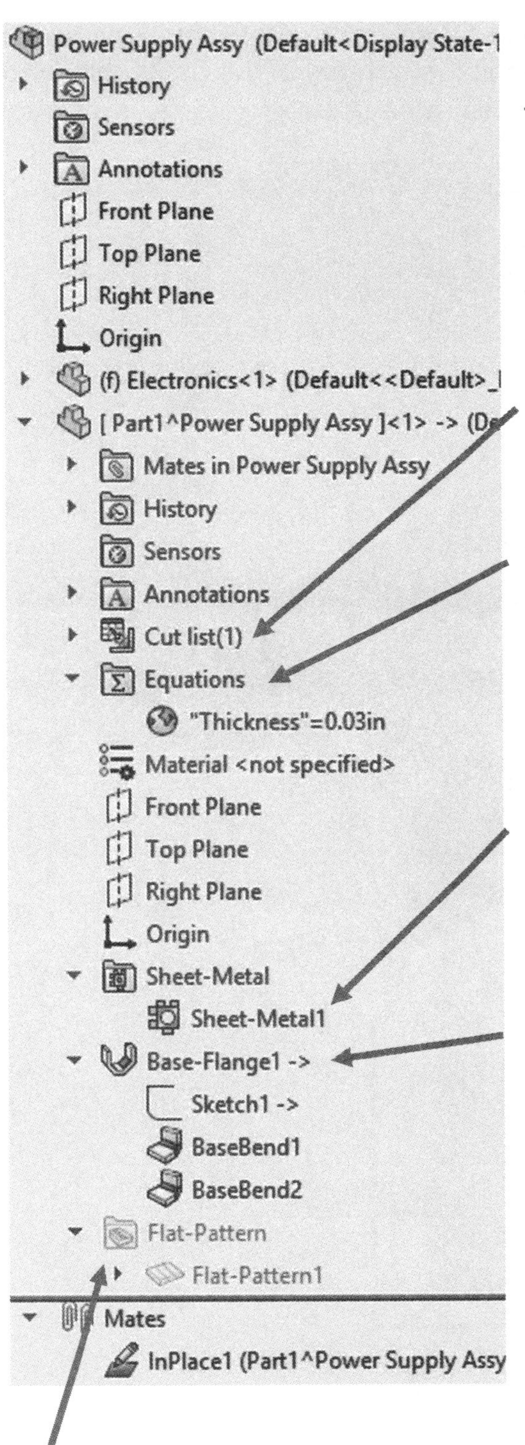

10.7. - The FeatureManager now shows the new part in the assembly and its Sheet Metal features.

In *"Part1^Power Supply Assy"* (in blue because we are editing it in context of the assembly), we see the new sheet metal features added by the "**Base-Flange**" command.

"**Cut list**": Shows the number of bodies in the part. If multiple sheet metal bodies are added to a part, they will be listed under this folder.

"**Equations**": Lists the equations and Global Variables in the part. For sheet metal components, a Global Variable is added automatically called *"Thickness"* which is used to make dimensions equal to the material's thickness.

Inside the *"Sheet Metal"* folder we see ***"Sheet-Metal1,"*** which is the feature that defines the default sheet metal parameters of the part including material thickness, default bend radius, bend allowance and relief type.

"Base-Flange1" is the first feature of the sheet metal part. Just like other sketch-based features, it includes the sketch used to make it and the bends created when the feature was made. Notice the "In Context" icon (**->**) next to the feature, letting us know that the part has external relations in the context of the assembly, in this case the "InPlace" mate, the dimensions added in the sketch and the extrusion's end conditions referencing the electronics board.

The ***"Flat-Pattern"*** folder contains the *"Flat-Pattern1"* feature and is suppressed when the part is bent (default). To show the part in its flat (unbent) state we must un-suppress it. The *"Flat-Pattern1"* feature includes a sketch (*"Bend-Lines"*) automatically generated with the part's bend lines and a list of all the bends in the part, as well as a *"Bounding-Box"* sketch, which is a sketch with the outline of the smallest square in which the unbent part can fit, useful to calculate the minimum size of material required for fabrication.

10.8. – The next step is to add mounting features to screw the electronics board to the box. These supports will be made using a "**Forming Tool**." As explained previously, a Forming Tool is essentially a tool (or die) that forms the sheet metal by bending, cutting and stretching as it is pressed into the part, hence the *"Forming"* part of the name. In this step, we'll use an existing forming tool. Later in this lesson we'll learn how to create one.

Rotate the model to view the assembly from below, approximately as the next image; in the "Task Pane" expand the Design Library, from the "**forming tools**" folder select the *"lances"* subfolder and look for the *"bridge lance"* forming tool. To apply it to our model, drag it onto the bottom face of the model.

Forming Tool folders are indicated by the icon

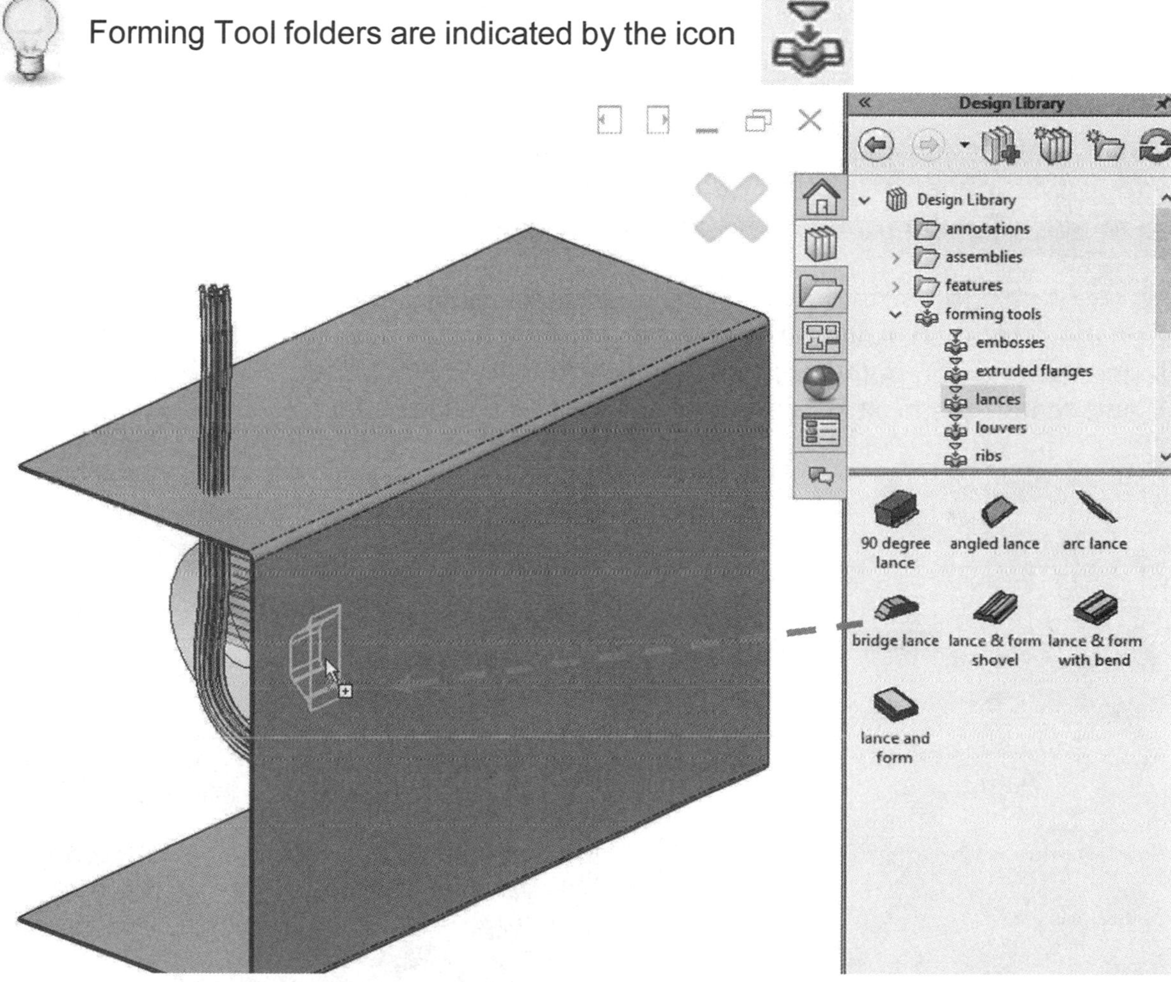

If the folder is not marked as a **Forming Tools** folder (indicated by the forming tool icon), after dropping a form tool in the sheet metal part SOLIDWORKS will ask *"Are you trying to make a derived part?"* The reason is the *"forming tools"* folder is not defined as such. To correct this behavior, right-mouse-click in the "**forming tools**" folder in the Design Library and turn on the "**Forming Tools Folder**" option. Now you can drag-and-drop forming tools into a sheet metal part.

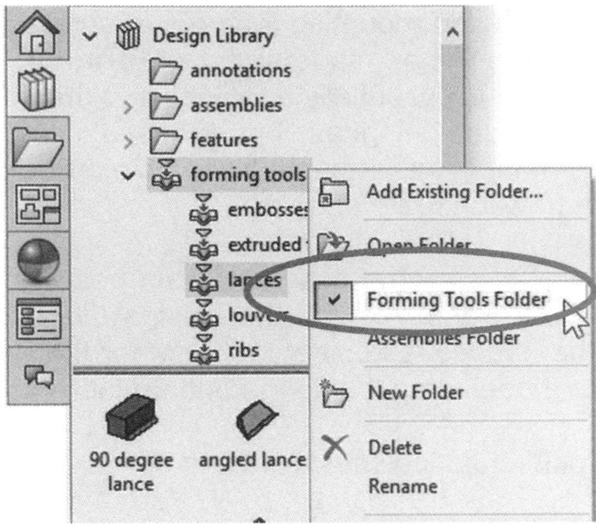

10.9. - After dropping the forming tool in the part, the "**Form Tool Feature**" PropertyManager is used to define the face where we want to locate it; if we want to rotate it, flip the form tool to go in or out of the part, etc.

If needed, rotate the form tool and change the direction the lance is going. For the "Rotation Angle" make sure the Rotation Angle is set to "90deg." If the lance is not going in the correct direction (into the part), press "Flip Tool" to change it and match the next image.

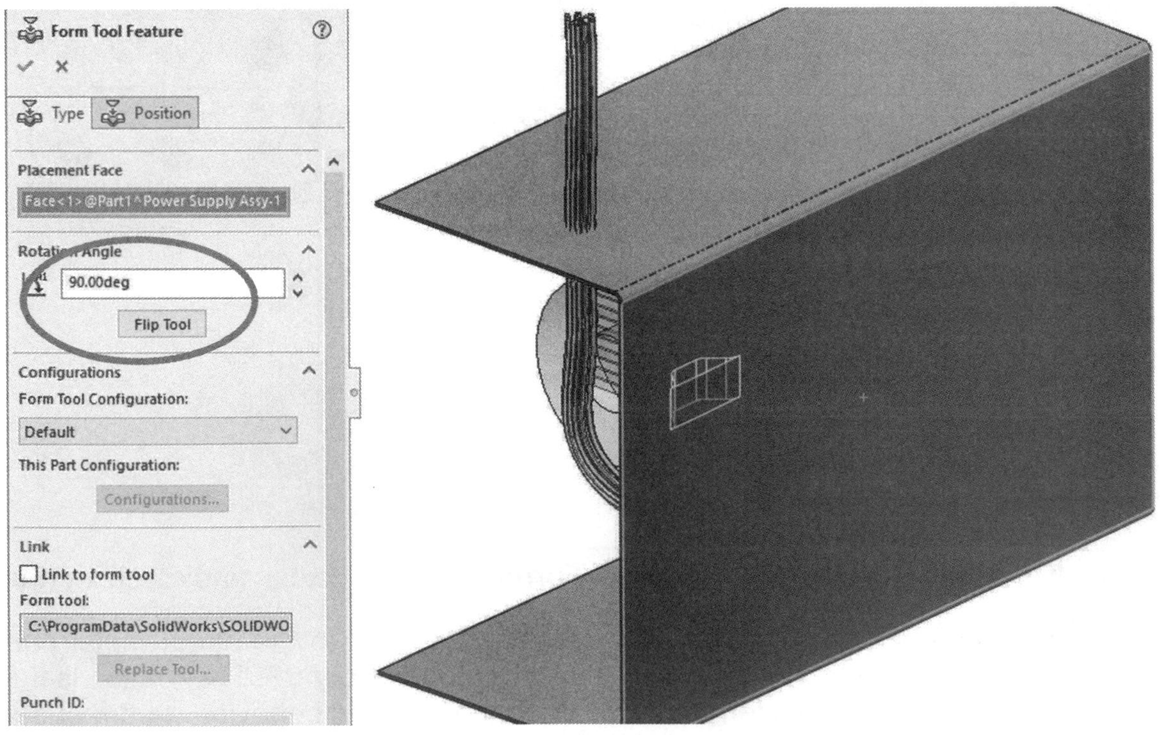

In the "Placement Face" selection box we can select a different face to add the forming tool, in case it was dropped in the wrong face.

10.10. - After making sure the form tool has the correct orientation and direction we need to dimension its location. Select the "Position" tab at the top of the properties and change to a bottom view with "Hidden Lines Visible" mode to easily locate it.

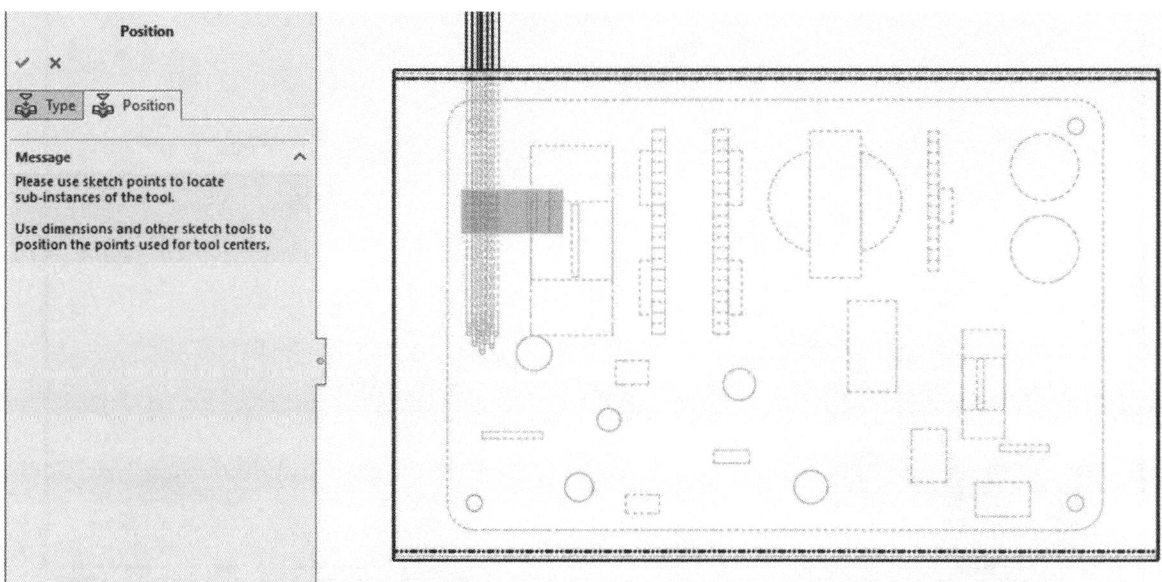

The process to locate the form tool feature is similar to the hole wizard. Notice the sketch point tool is active, then, wherever we add a **Sketch Point**, a new instance will be added. Since we want to locate the form tool features concentric with the mounting holes in the board, touch the edge of each of the three holes indicated in the electronics board to reveal their centers, and add a point concentric to each one.

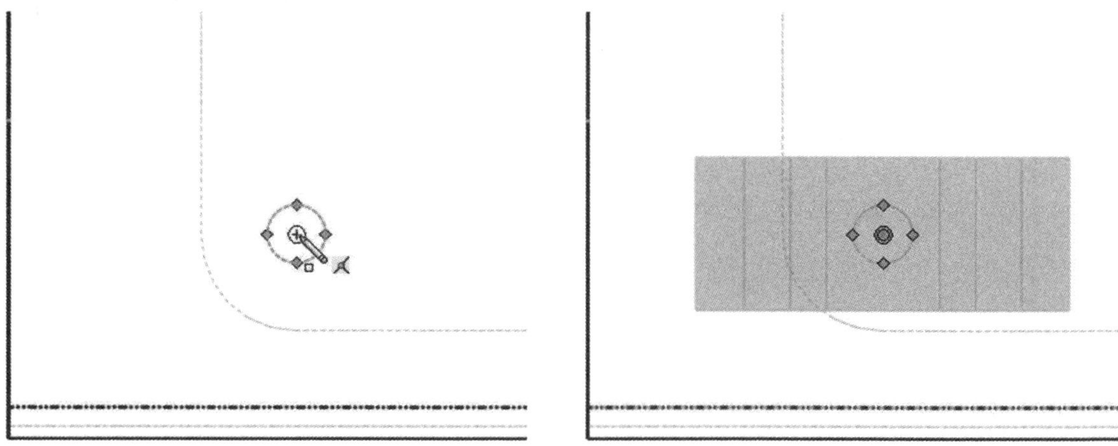

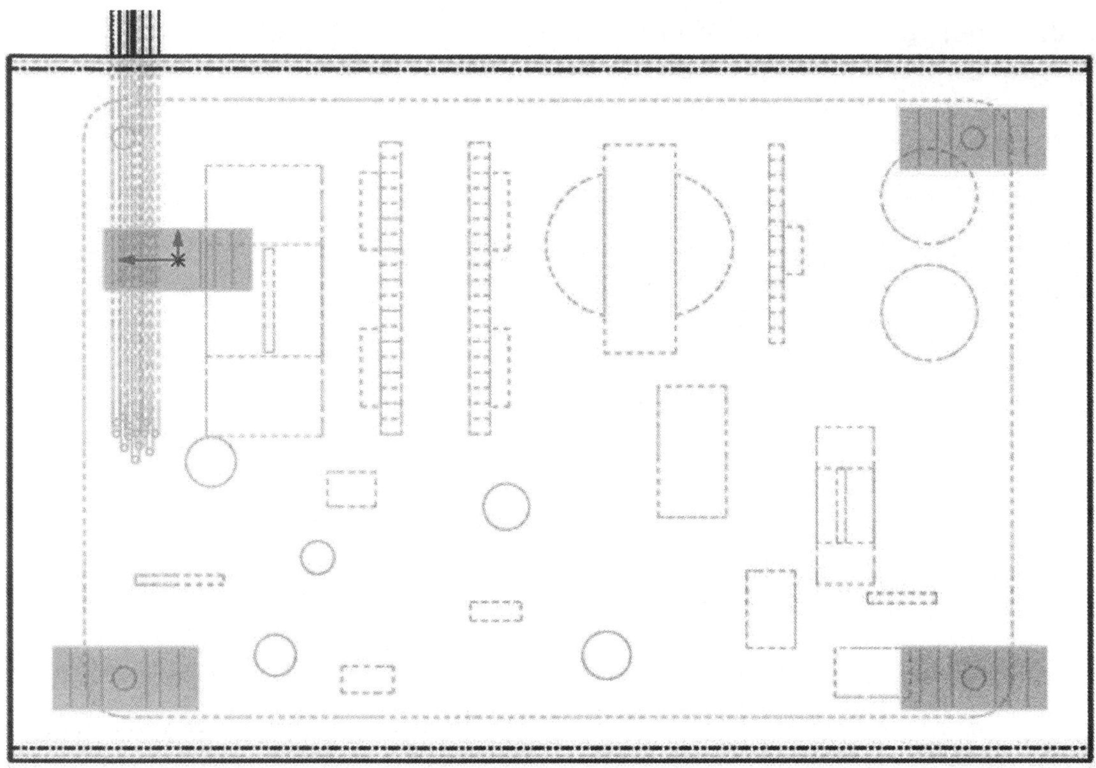

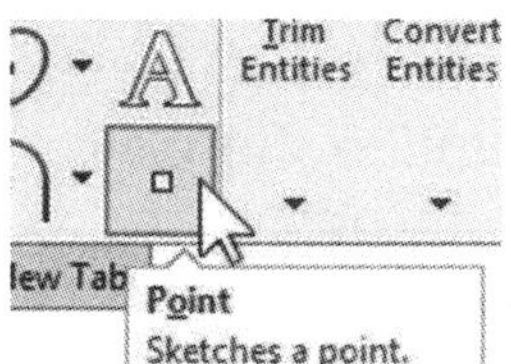

10.11. – To locate the point close to the corner where we dropped the form tool, turn off the sketch "**Point**" tool and add a concentric relation between the point and the edge of the last mounting hole.

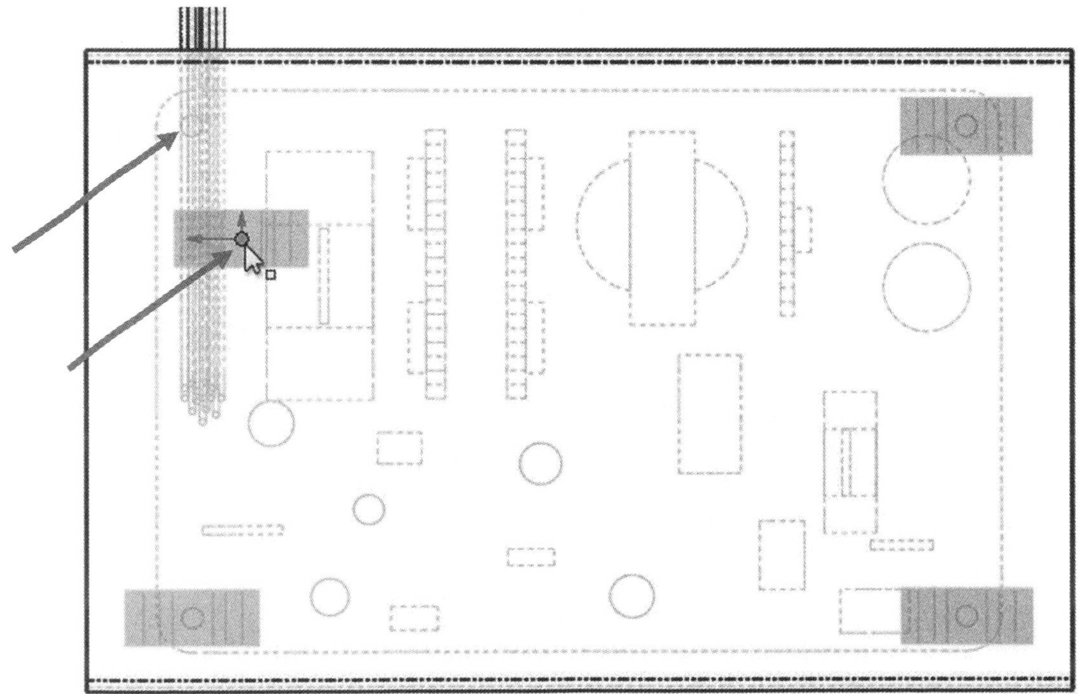

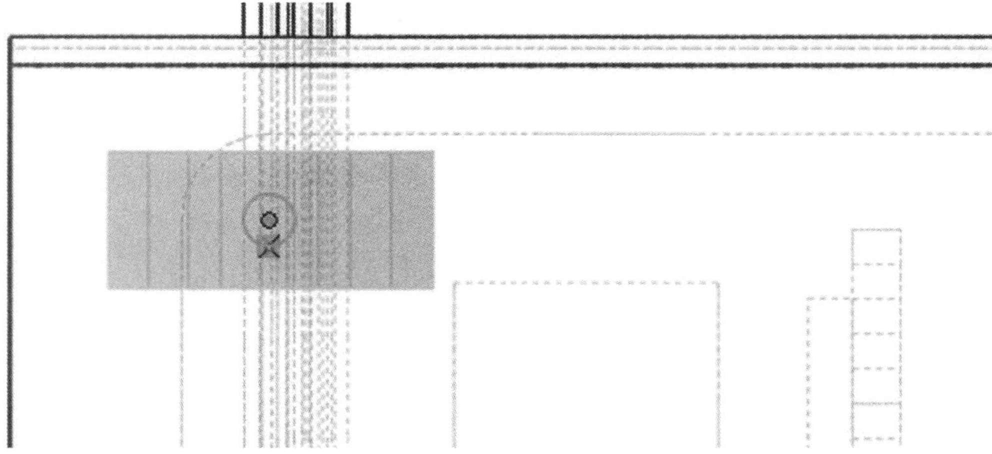

After locating the last form tool, click OK to add the lances to finish.

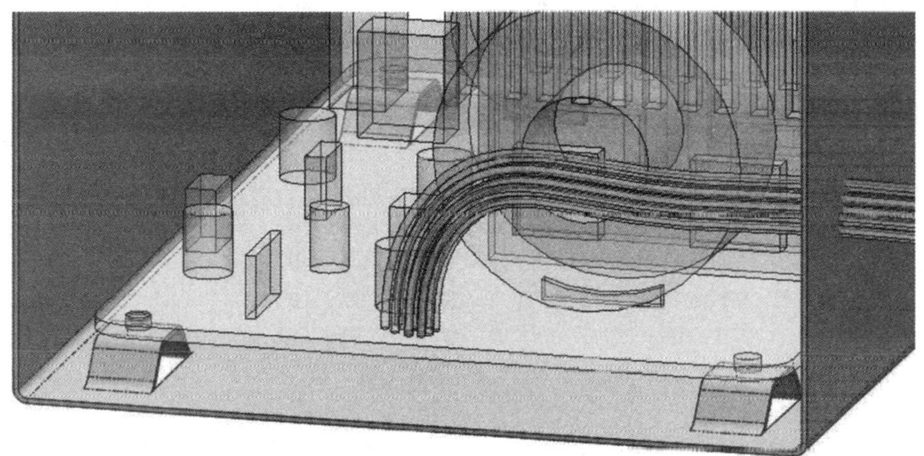

10.12. – Now we need to add a hole to each of the lances for the screws that will hold the board in place. Switch to a top view, select the top face of any of the lance features and add a new sketch. Draw four circles concentric to each of the board's mounting holes, make them all equal, and dimension one of them 0.1″ in diameter.

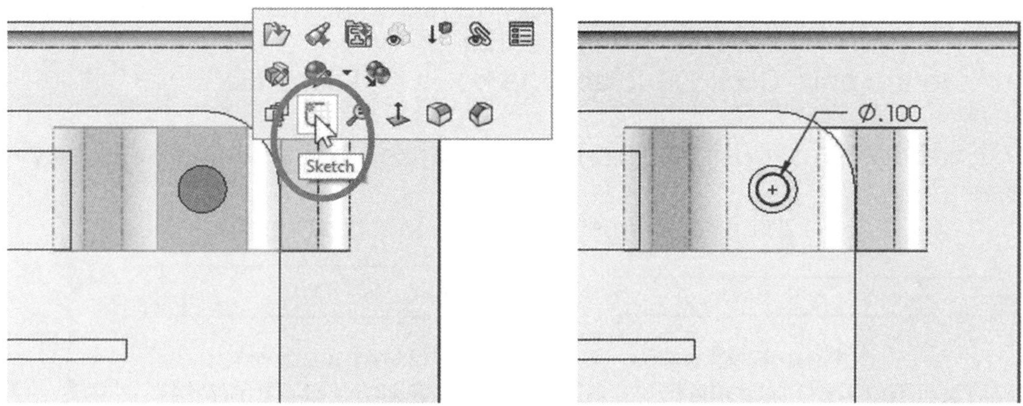

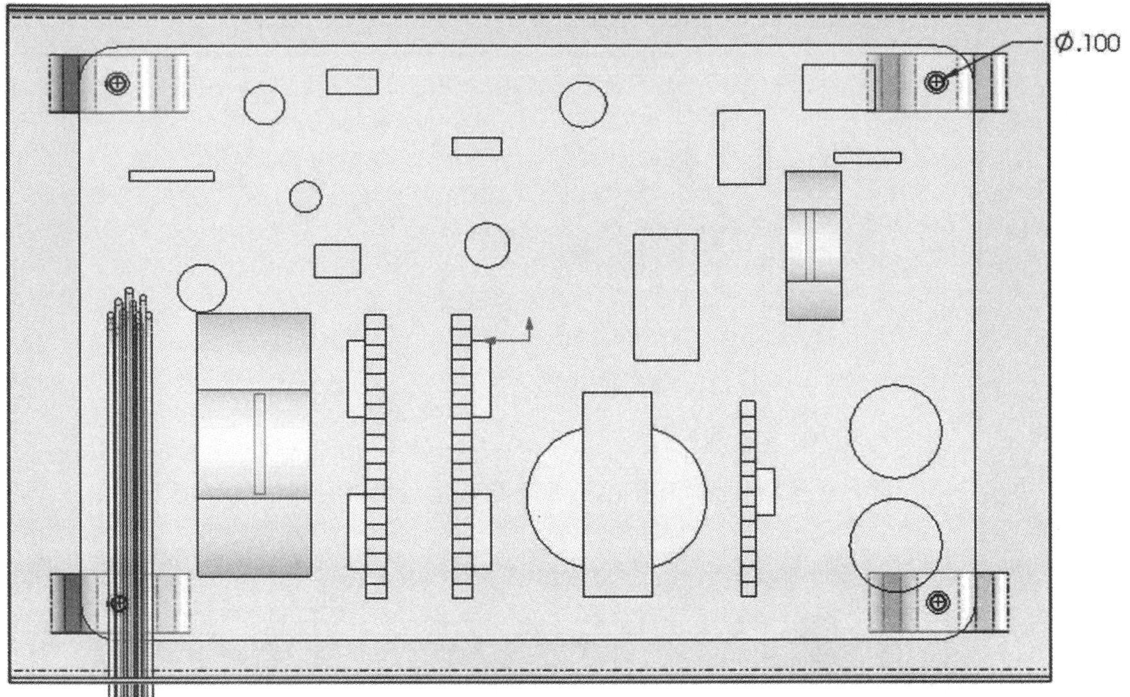

10.13. - After adding the four equal circles, select the "**Extruded Cut**" icon to make a cut. Use the "Link to Thickness" option to make the cut equal to the "Thickness" global variable created when we added the first sheet metal feature.

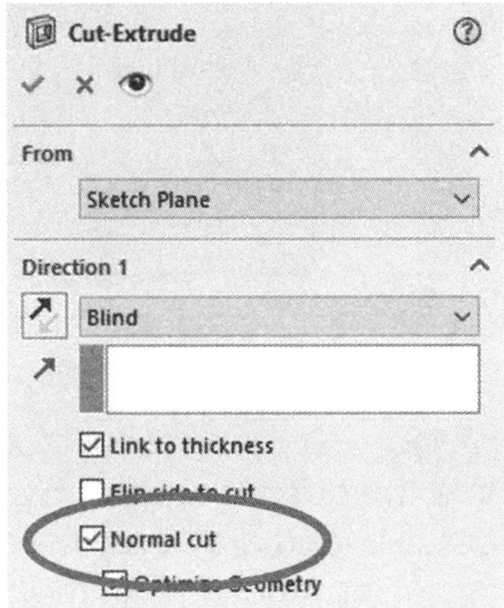

Another option specific to sheet metal is "**Normal Cut**." A normal cut means the cut will be made perpendicular to the sheet metal, even if the cut is being made at an angle. The reason for this is because in a sheet metal manufacturing process it is very difficult to make cuts at an angle in thin metal, it's not practical, and most likely not needed.

Therefore, this option exists by the very nature of the manufacturing process itself. In our example, it makes no difference because the cut's direction is perpendicular to the part. Click OK to add the holes and continue.

If an Extruded Cut is made oblique (at an angle) to the surface, this is what happens:

Oblique cut in sheet metal (section view)

Normal cut unchecked
(Difficult to fabricate)

Normal cut checked
(Easier to fabricate)

10.14. - After adding the mounting holes we need to add a fan to help us cool the power supply. In most cases, at some point we'll need to include purchased components in our designs, like electrical components, hardware, etc. Many suppliers provide 3D models of their products, saving designers the trouble of modeling purchased parts, reducing the risk of design errors and at the same time improving the chances of having their products selected for our designs, making it a win-win situation. There are multiple sources of 3D models, from manufacturer's web sites to portals that consolidate multiple suppliers and user submitted models like SOLIDWORKS' 3D Content Central. 3D Content Central is a free portal that is also integrated in SOLIDWORKS and can be accessed directly from within SOLIDWORKS or on the web at www.3dcontentcentral.com.

Remember that at this point we are editing the sheet metal part *in* the assembly, and we need to add a new component to the *'Power Supply Assy'* assembly. To stop editing the part and edit the assembly instead we need to select the "**Edit Component**" toggle button in the CommandManager, the confirmation corner, or right-mouse-click in the graphics area and select "**Edit Assembly: ....**" By doing this, we will stop editing the sheet metal part and return to the assembly editing level, where we can add more parts to the assembly.

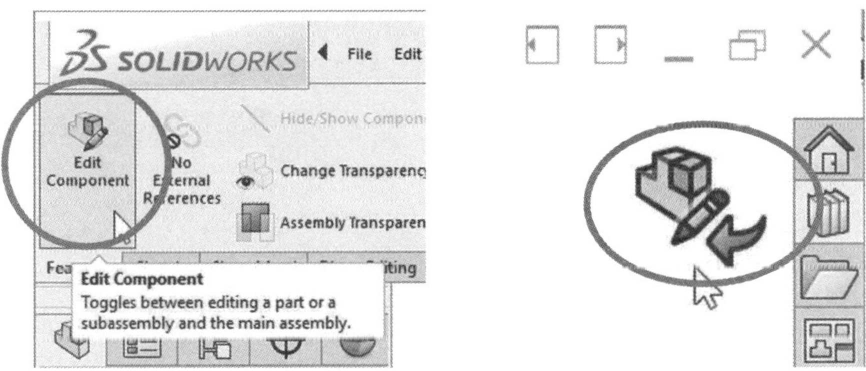

When we stop editing the part and return to editing the assembly:

- The sheet metal part's FeatureManager is no longer blue.
- The power supply electronics board will become opaque again (following the transparency rules set in the options).
- The status bar will read "Editing Assembly."
- The "Edit Component" button will be disabled (since no component is being edited).
- The confirmation corner will be turned Off.
- The title bar will read *'Power Supply Assy.sldasm'*.

In other words, the assembly looks like any other assembly and now we are ready to add more components to it.

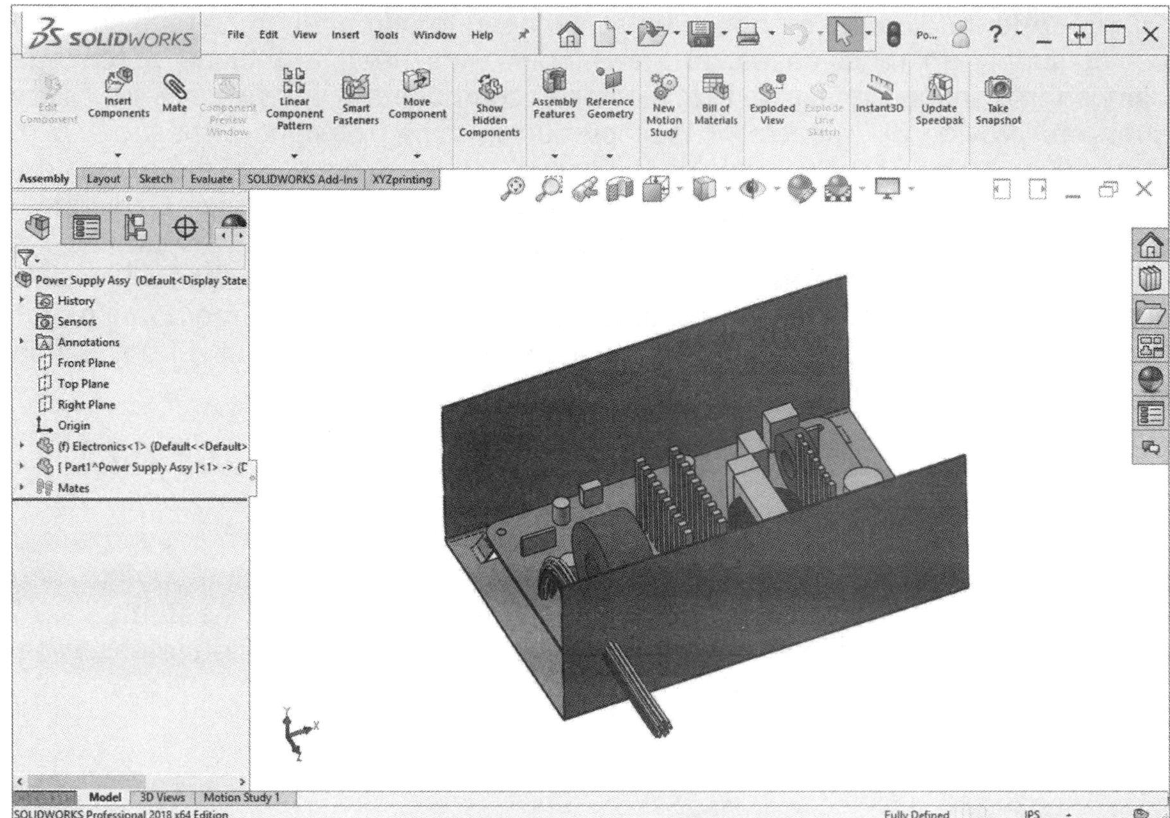

10.15. – Now that we are editing the assembly, open the "Task Pane" and select the Design Library tab, expand the "**3D ContentCentral**, **User Library, Home Page**" and click the icon in the bottom pane to display the **User Library** section.

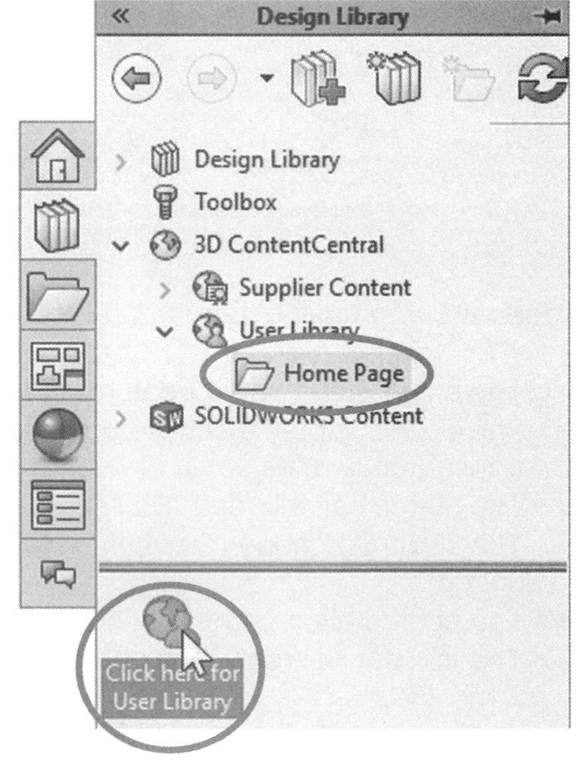

Immediately your web browser will open the 3dcontentcentral.com webpage, where we can search the online catalog for components. New users will need to register free of charge to download models.

Components are grouped by similarity, for example bearings, electrical parts, pneumatic or hydraulic components, power transmission, structural, etc.

For our power supply, we'll use a cooling fan model uploaded by a SOLIDWORKS user.

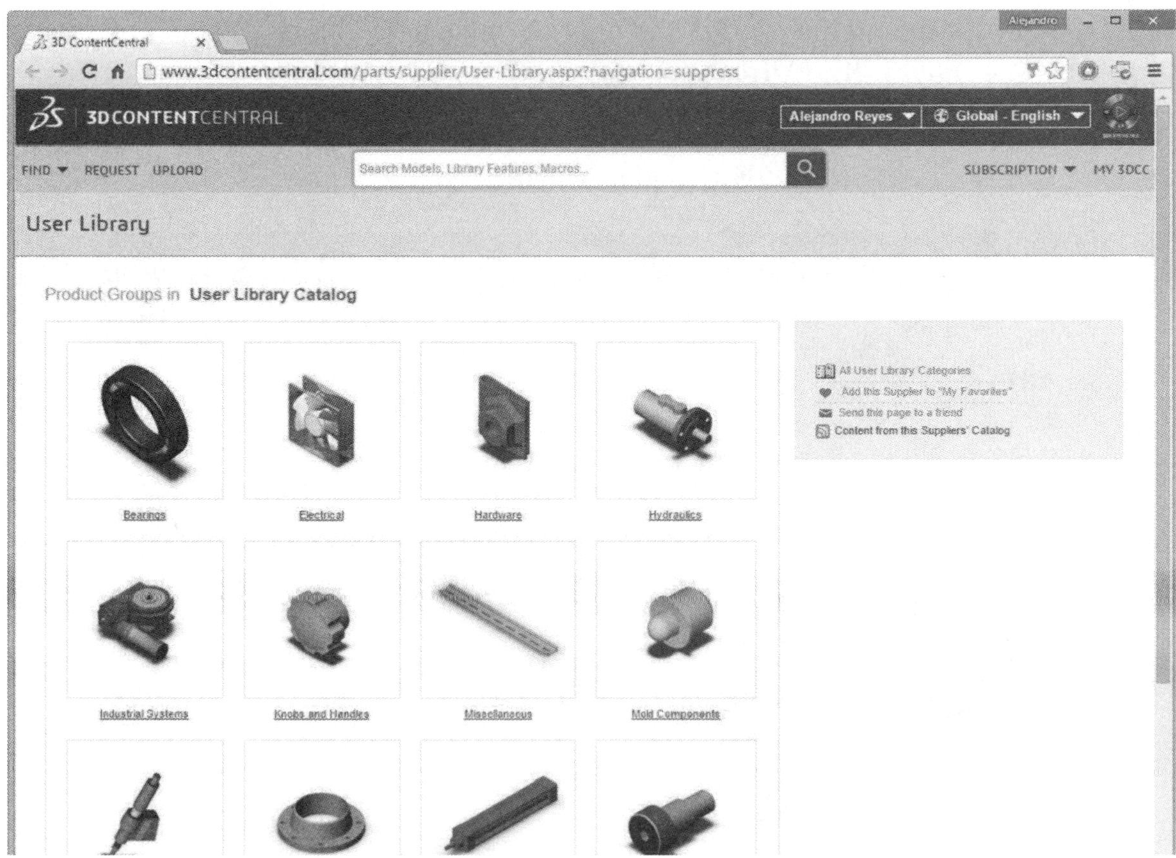

In the search box, type "*PABST 512F.*" This is the name of the model we need. If you know the exact model or part number you need, you may want to search the supplier content area, but if you are not sure or want to browse the models available, use a broad search term like "Fan" or "Cooling" and scroll through the available models in the library, until you find a suitable model for your design. The search results show the part we are interested in.

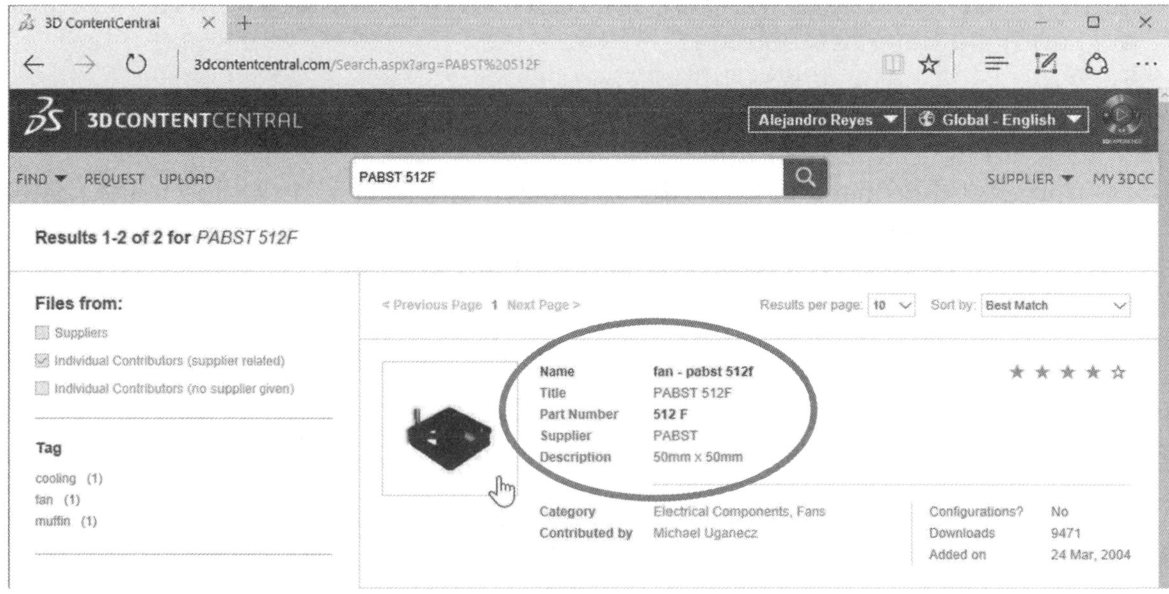

10.16. - After selecting the model, the next screen shows the details and a preview of the model; using the eDrawings plug-in for your web browser we can rotate, zoom, pan and animate the model. In the "Format" drop-down list select the target CAD system, SOLIDWORKS. From the "Version" list be sure to select the same version you are using (or earlier) and download the 3D model.

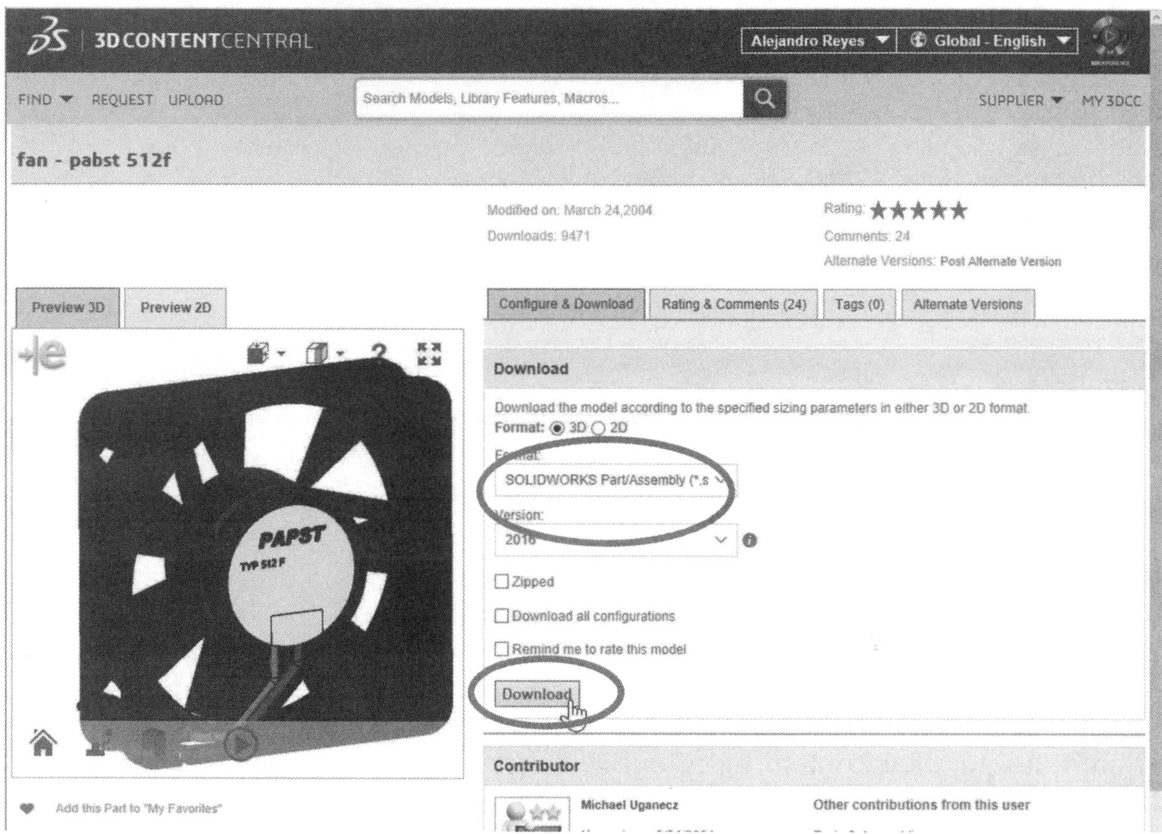

If you download a "Zipped" model you must unzip it before adding it to the assembly. Uncheck this option and download the model. After selecting "Download" we have two options:

- Click in the link to download the part to your PC and add it to the assembly as we have done before.
- Drag-and-drop the link into the assembly. The fan will be added to the power supply assembly, and we will be asked for a location and file name to save the downloaded part. Save it as "**Fan-PABST 512F**."

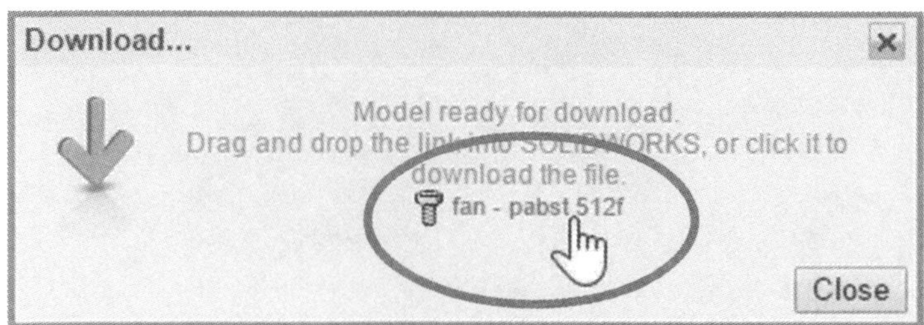

Using the click-and-drag method.

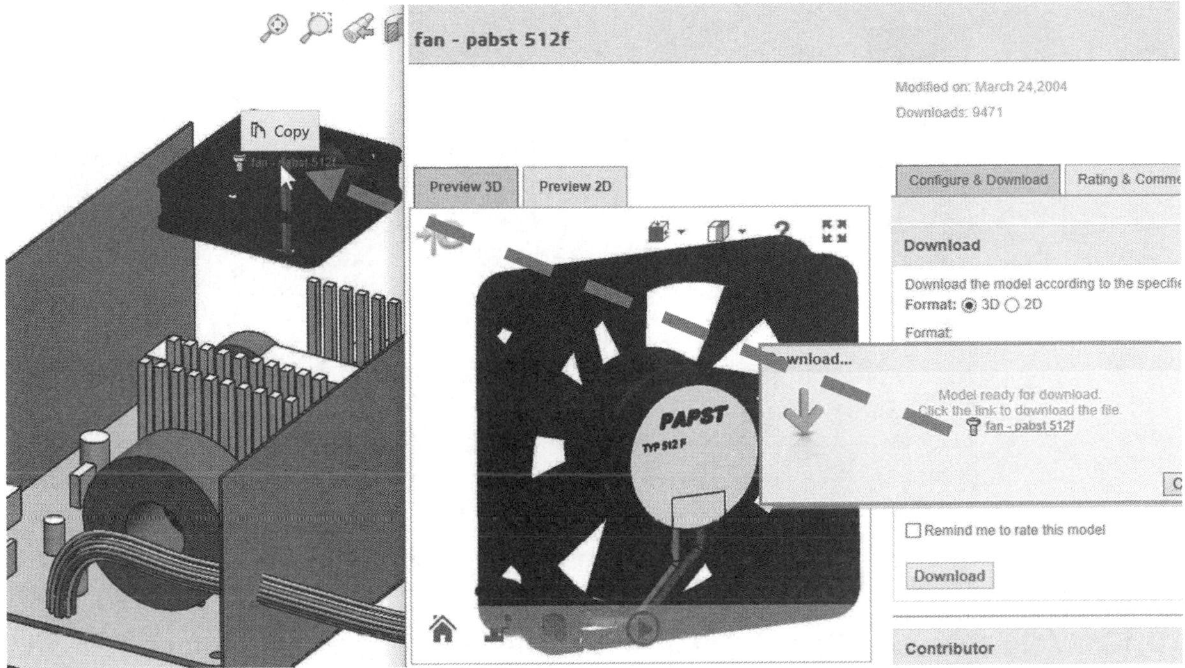

IMPORTANT: Drag and drop may only work using Internet Explorer; for other browsers, you may have to click in the link to save the file to disk and then add it to the assembly. If this is the case, select the folder to save the model, add it to the assembly and locate it *approximately* as shown.

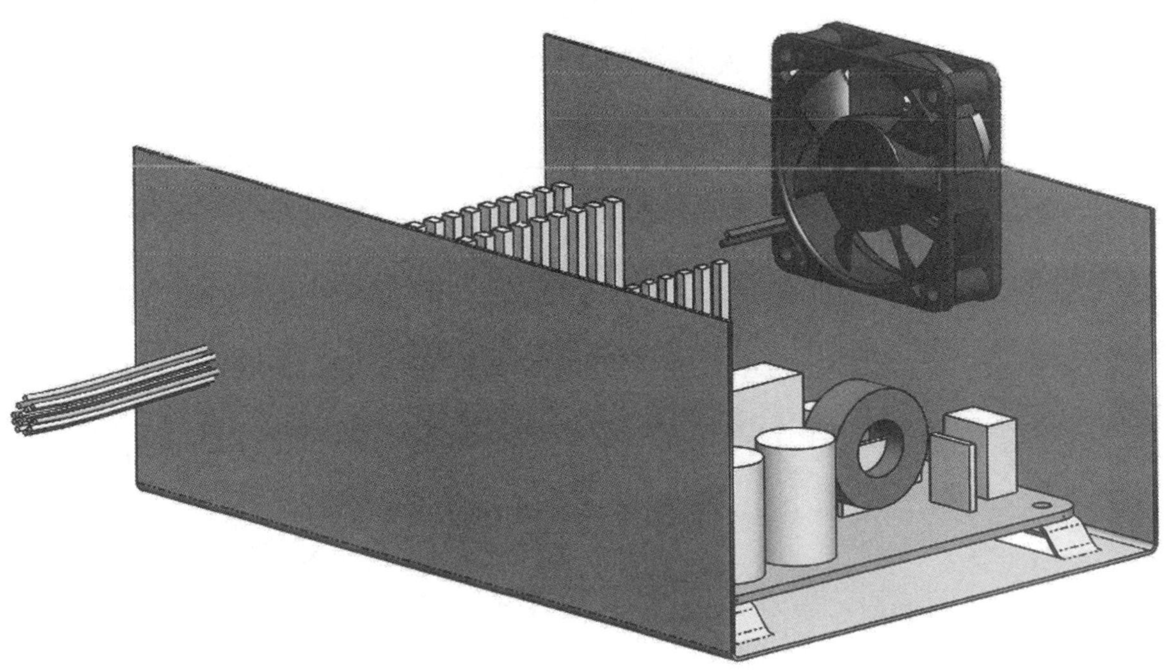

10.17. - Add a Coincident mate between the Fan and the inside wall of the enclosure.

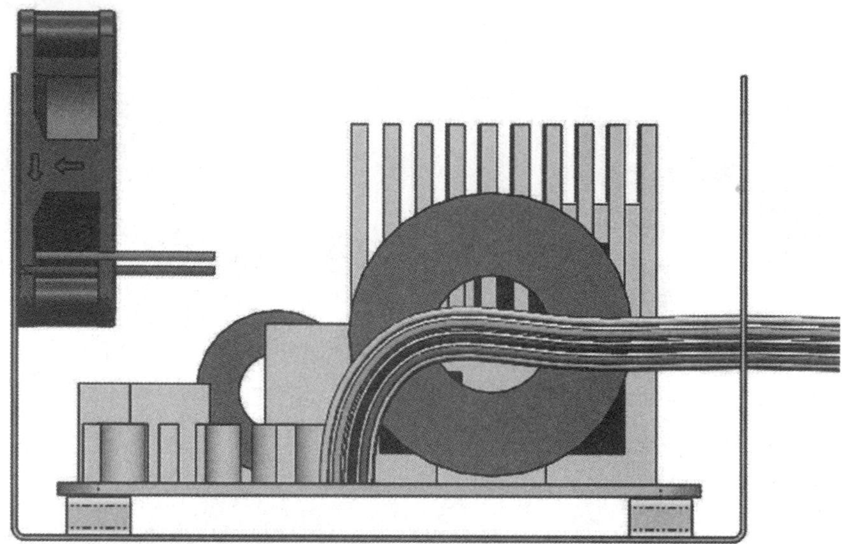

10.18. - Add a second Coincident mate between the *"Right Plane"* of the Fan and the face indicated in the electronics board, which places the fan about half way between the heat sinks.

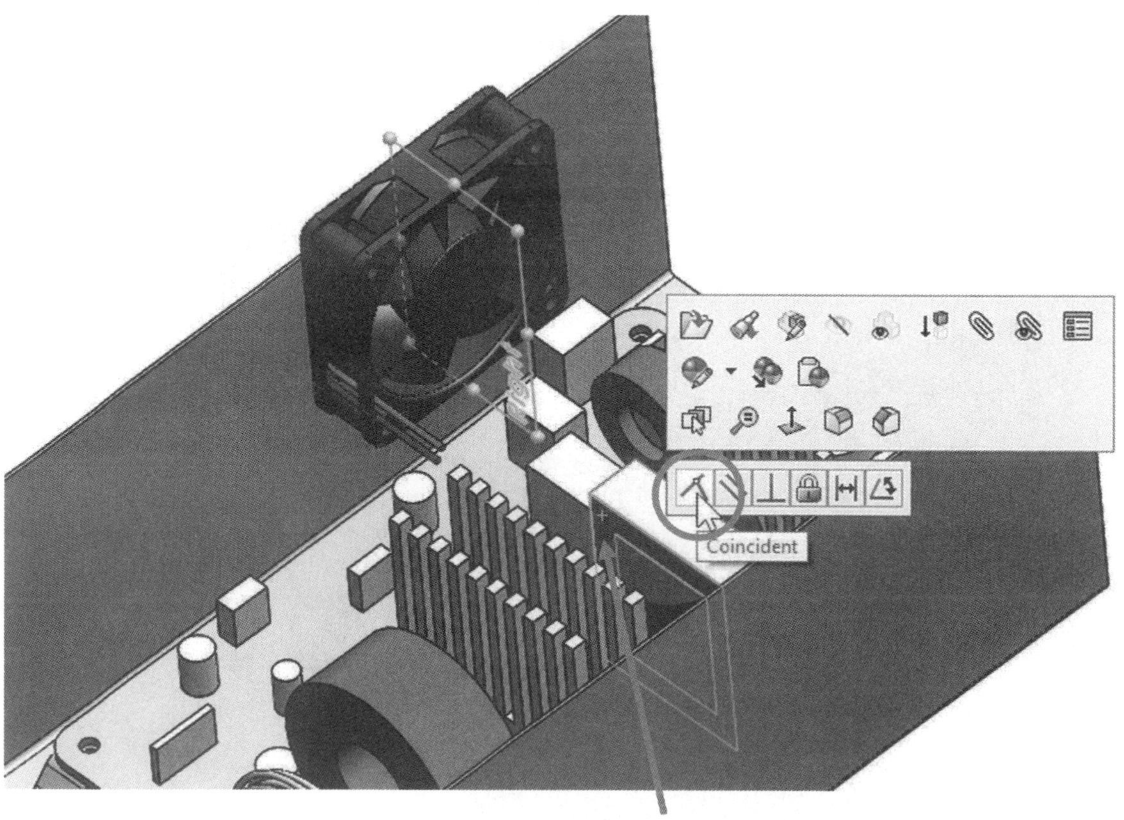

10.19. - To finish locating the Fan, add a 0.2″ Distance mate from the top face of the Fan to the top of the sheet metal part. Our Fan is now fully defined in the assembly.

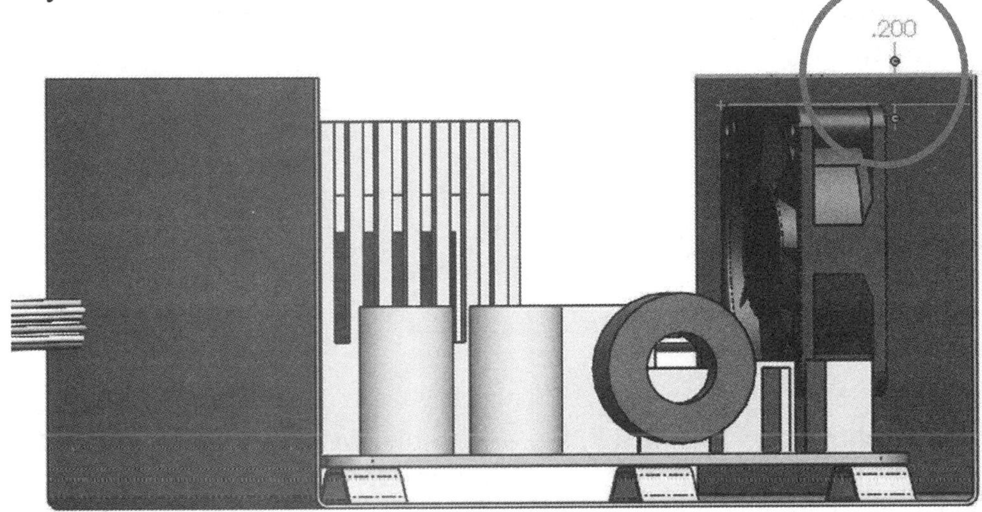

10.20. - With the fan fully defined in place, we need to edit the sheet metal part to make a ventilation cut. Select the sheet metal part in the FeatureManager or in the graphics area and select "**Edit Part**" from the pop-up menu, or use the "**EdIt Component**" command in the CommandManager.

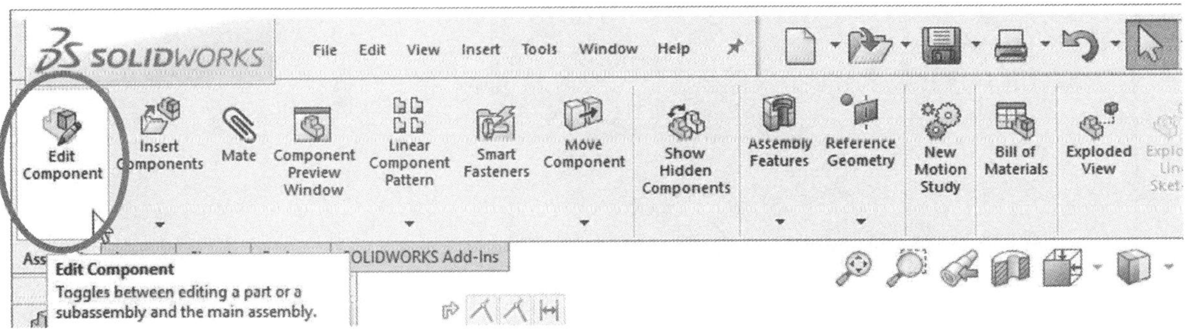

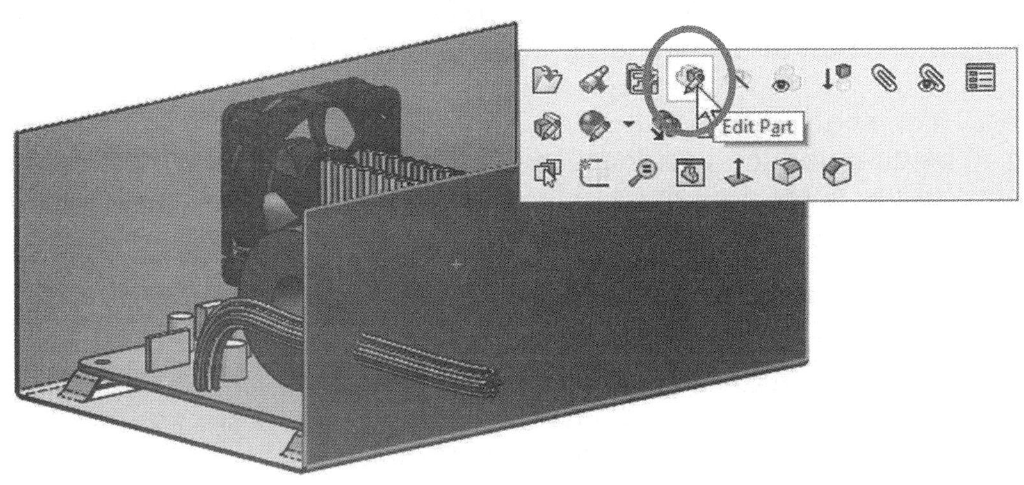

Now we are editing the sheet metal part in the assembly; the electronics board and the fan are transparent, and the box is opaque. To have better visibility and an uncluttered view in the next step, switch to a Back view, change the display mode to "Hidden Lines Visible," and hide the electronics board component to improve visibility.

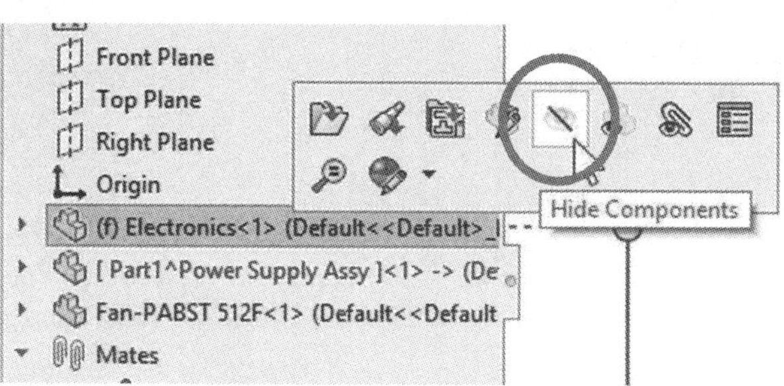

While editing a component in the assembly, we can hide, show and change the transparency of other components in the assembly at will.

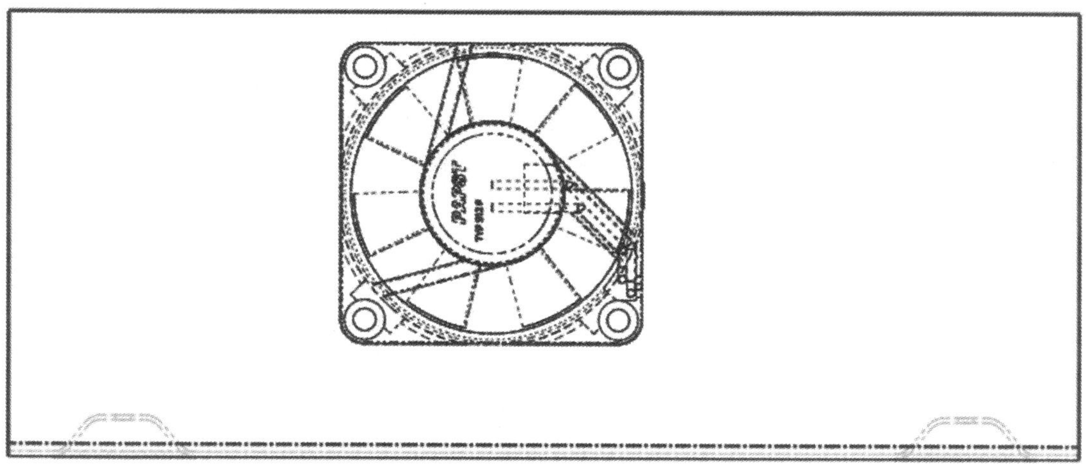

10.21. - To add ventilation to our power supply, we'll make a cut using a feature called "**Vent**" using the fan's location for reference. Select the face in the back of the sheet metal part and make the following sketch.

By defining the vent's location coincident to the fan, if the fan's position needs to be changed the vent location will change as well; that is the idea behind designing in the context of an assembly. Notice the sketch used for the vent has intersecting lines, and the only reference to the fan is a concentric relation to the fan's center.

To make this sketch, make a circle concentric to the fan, then either change to "Hidden Lines Removed" view mode and/or hide the Fan to finish the sketch and avoid adding unintended geometric relations to the Fan. We only need one concentric relation to the Fan for the entire sketch.

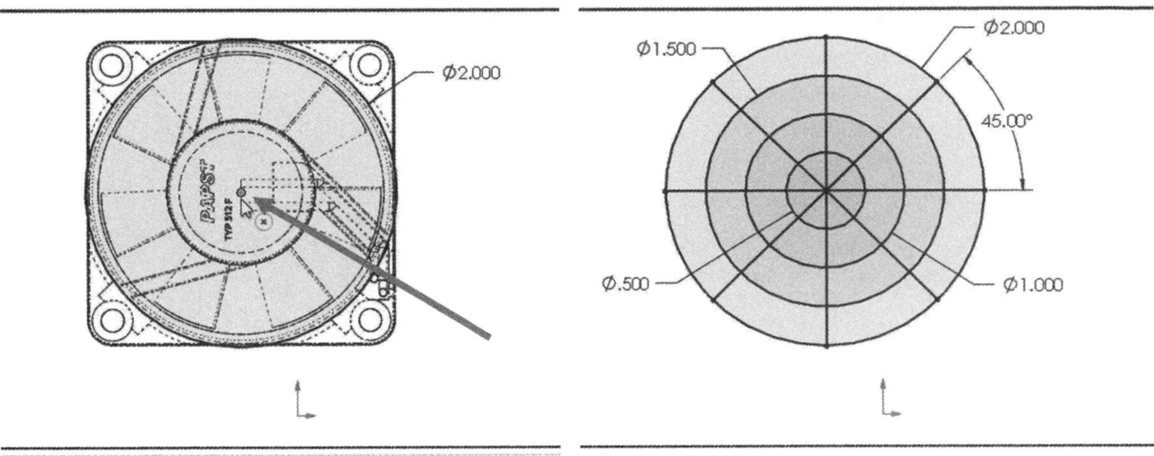

Hidden Lines Visible
(to add concentric relation to fan)

Hidden Lines Removed or hidden fan
(for clarity to finish the sketch)

10.22. - To add the vent select the menu "**Insert, Fastening Feature, Vent**" or from the Sheet Metal CommandManager tab.

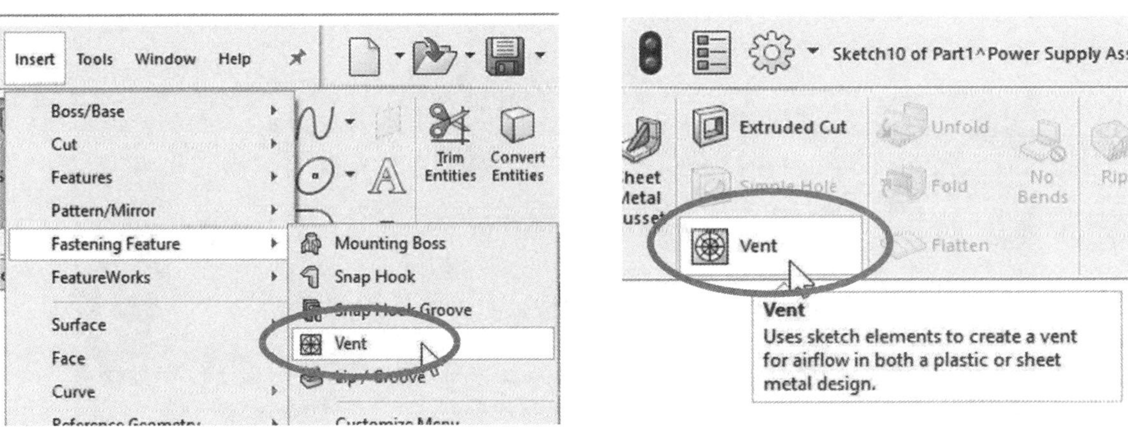

In this step the fan is hidden for clarity. In the "Boundary" selection box, use the outer (2″) circle and set the "Radius for the fillets" value equal to 0.05″ (this will be the fillet radius for all inside edges). In the "Ribs" selection box select all straight lines and make them 0.1″ wide. In the "Spars" selection box select the 1″ and 1.5″ diameter circles, and make them 0.1″ wide. Finally, in the "Fill-In Boundary" select the inner circle. The Fill-In Boundary is an area that will be covered.

Note the "Flow Area" section lists the total area and the open area percentage for reference. Click OK to finish the Vent feature.

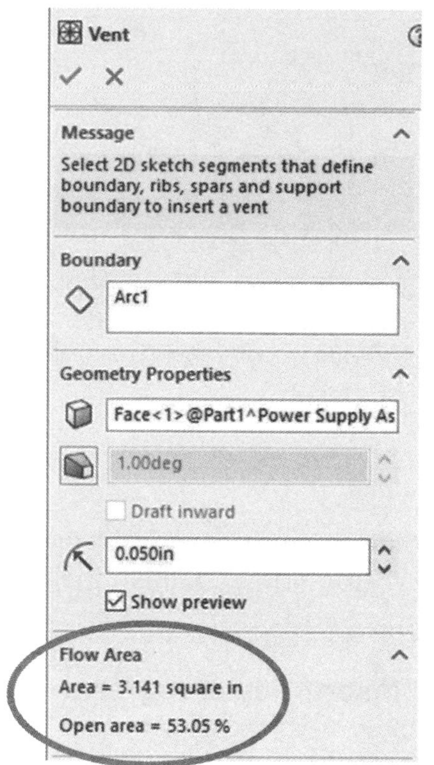

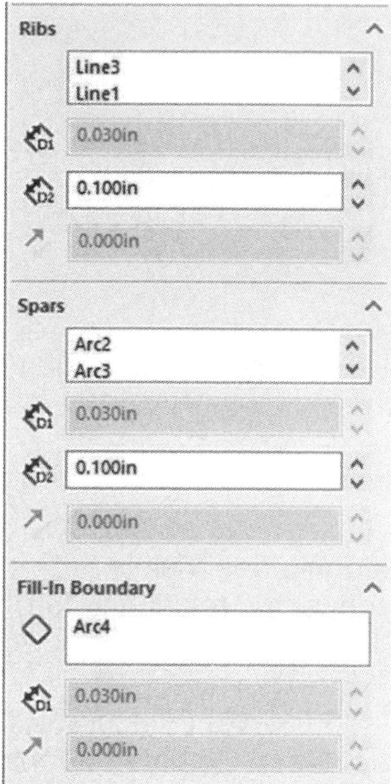

Fan and electronics board hidden for visibility

10.23. - So far the sheet metal component has been an *'internal'* component in the assembly (unless, if we were asked for a name when the component was added). If you remember, an internal part means that it only exists inside the assembly; there is no file on the hard disk that we can make a reference to. Since we need to have a *part file* to make a drawing, we need to *externalize* the part. (It's not possible to make a drawing of a part internal to an assembly.) Since we never gave the part a (file) name, it was automatically assigned by SOLIDWORKS. In this step, we'll make the part external. To externalize the part, select it in the FeatureManager with a right mouse click, then select "**Rename Part**" from the pop-up menu and change its name to *"Box Bottom."*

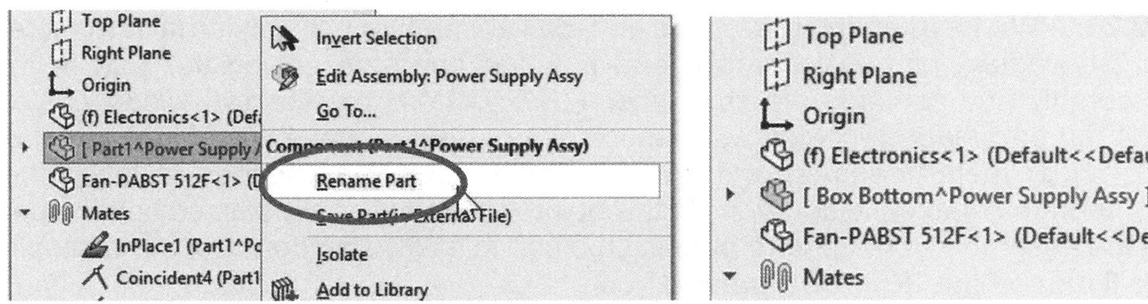

After renaming the part, select it again with a right mouse click and select "**Save Part (in External File)**" from the pop-up menu.

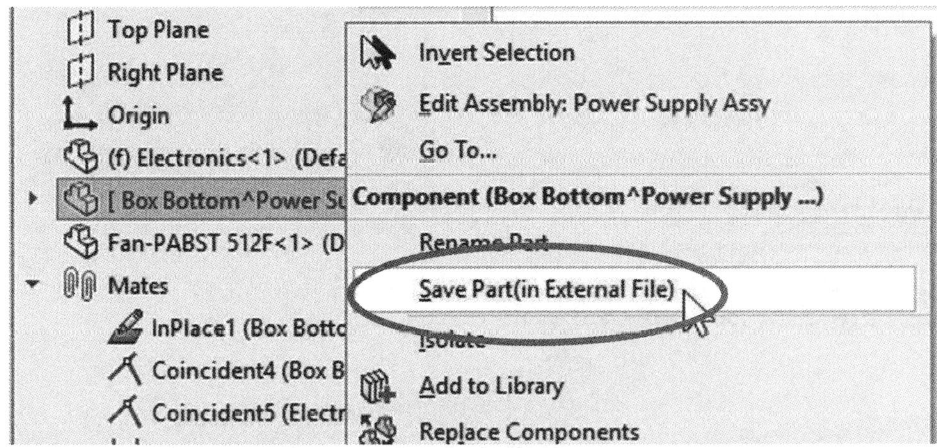

Make sure the path is correct and click OK. The *'Box Bottom'* has been saved as a part file and now looks the same as any other part in the assembly's FeatureManager. Note that we are still editing it inside the assembly.

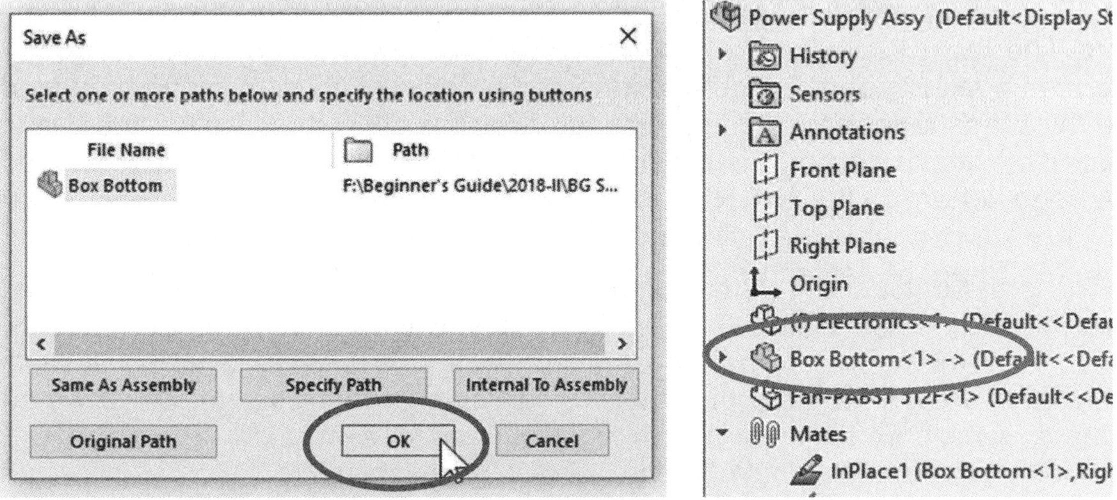

10.24. – We have added the lances and the vent feature using external references in the context of the assembly, and now we can stop editing the part in the assembly and continue the sheet metal '*Box Bottom*' part its own window. We'll edit the part in its own window as much as possible to make it easier to visualize the sheet metal features, and edit it in the assembly as needed to take other components into consideration. At this point we can optionally stop editing the part in the assembly or not before opening the part in its own window. For our example, we'll turn off the "**Edit Component**" icon…

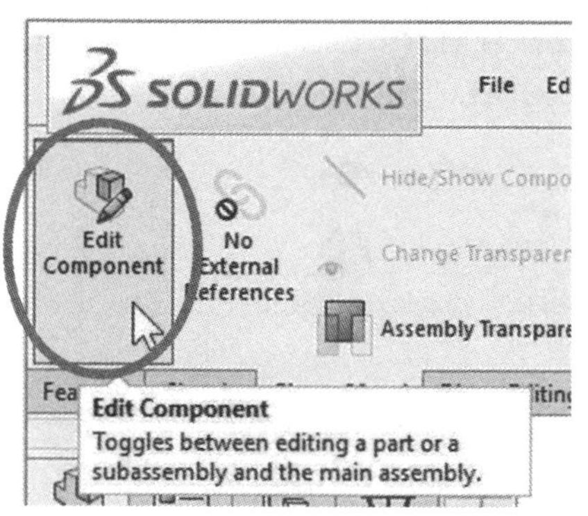

or

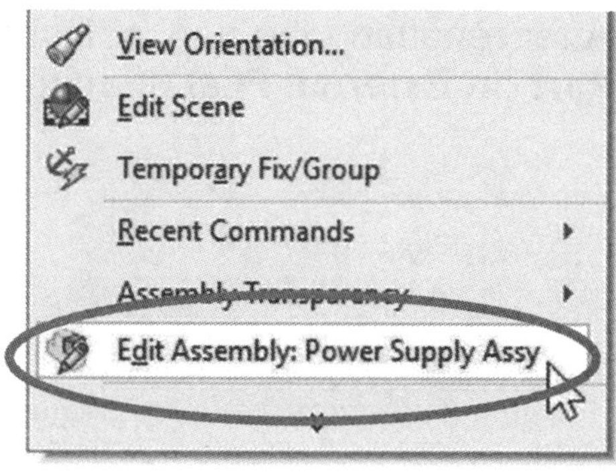

… and open the *'Box Bottom'* part in its own window.

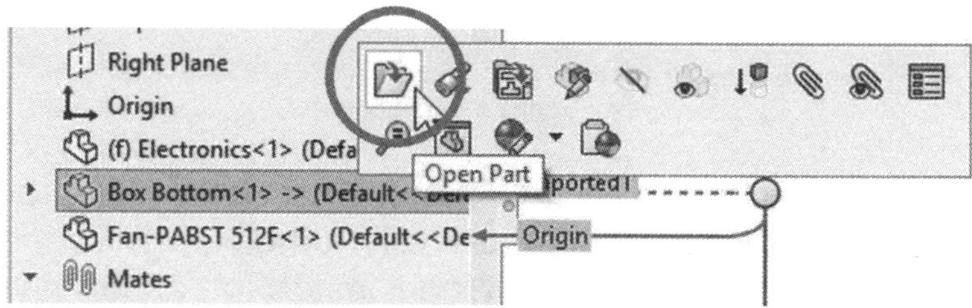

10.25. – In this step we need to add a miter flange to the sides of our sheet metal part. The miter flange is used to add flanges to one or more edges; this feature requires a sketch perpendicular to one of the edges. This sketch can be added to the thickness face at the end of the edge (which is difficult unless we zoom in very close), or, after selecting the **Miter Flange** command we can click near the end of the edge, where SOLIDWORKS will automatically create a plane perpendicular to it where we can draw the miter flange profile. Select the "**Miter Flange**" command from the Sheet Metal tab or the menu "**Insert, Sheet Metal, Miter Flange.**"

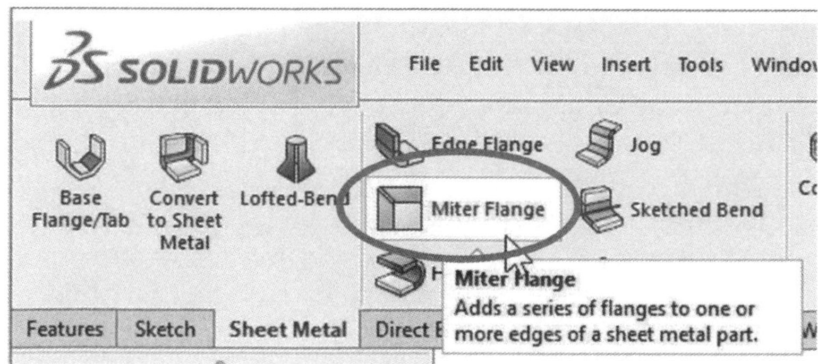

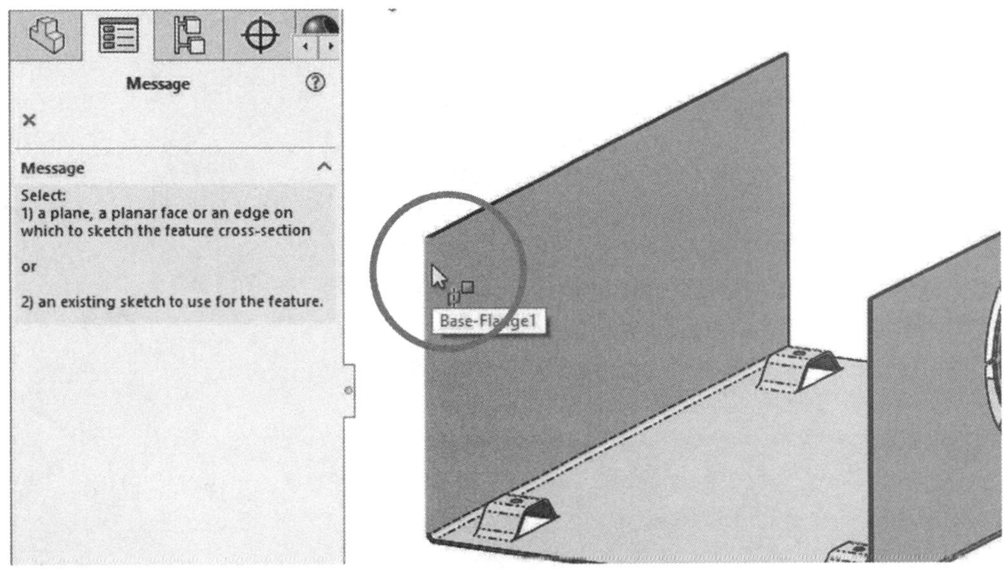

Select the edge indicated (a plane and a sketch will be created automatically) and draw a line perpendicular to the edge and dimension it as shown. This will be the profile for the miter flange. Notice the sketch was added in the newly created plane at the end of the edge.

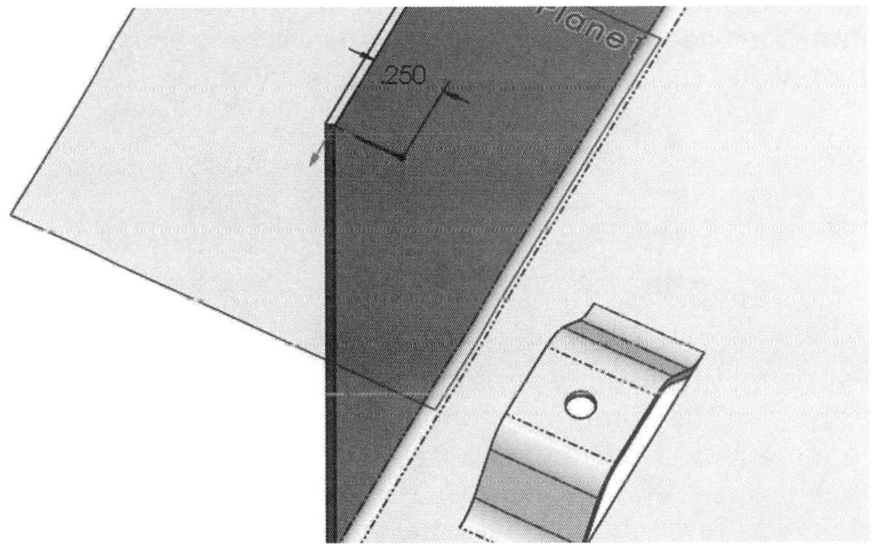

10.26. - Select the "**Miter Flange**" command icon (or Exit Sketch) to continue. The miter flange preview will visible in the first edge. To continue the flange along the bottom edge and opposite side, either select the "**Propagate**" icon to automatically select these edges, or manually select them in the screen. Leave the "**Use default radius**" checkbox selected to use the default sheet metal bend radius. "**Gap Distance**" is the spacing added at the corners of the miter between the flanges, use a value of 0.01." The "**Trim side bends**" option removes material when a new bend touches an existing bend (in this case there is no difference in the result). In "Flange Position" select the "Material inside" option to maintain the original part's dimension and click OK to complete the miter flange.

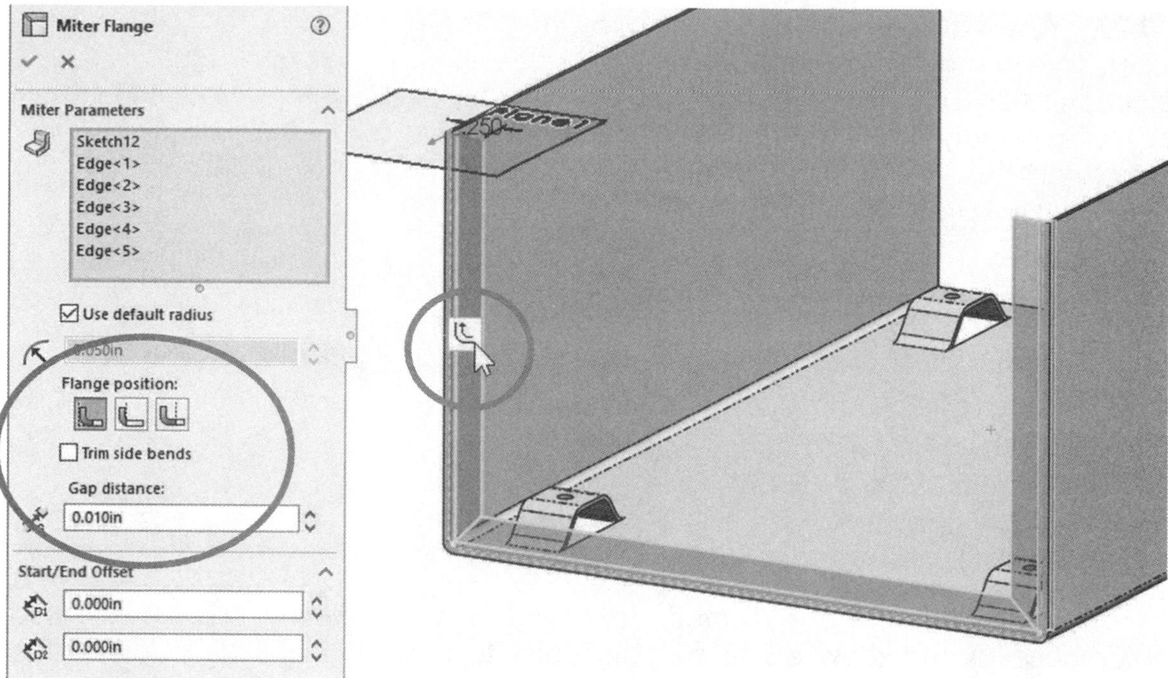

Repeat the previous steps to add a second miter flange in the other side using the same parameters as the first one. Keep in mind that if we select an edge after the "**Miter Flange**" command, a new plane will be created. To avoid this, we can use the previous (automatically added) plane if wanted.

10.27. - Now we need to add a cut on one side of the box to allow the power supply wires to exit the enclosure. Go back to the *'Power Supply Assy'* and make the electronics board visible. Select the *'Box Bottom'* part either in the FeatureManager or the assembly, and click in "**Edit Part**" to continue; this way we can add new features in the context of the assembly and reference other components. Hide the fan if wanted or needed.

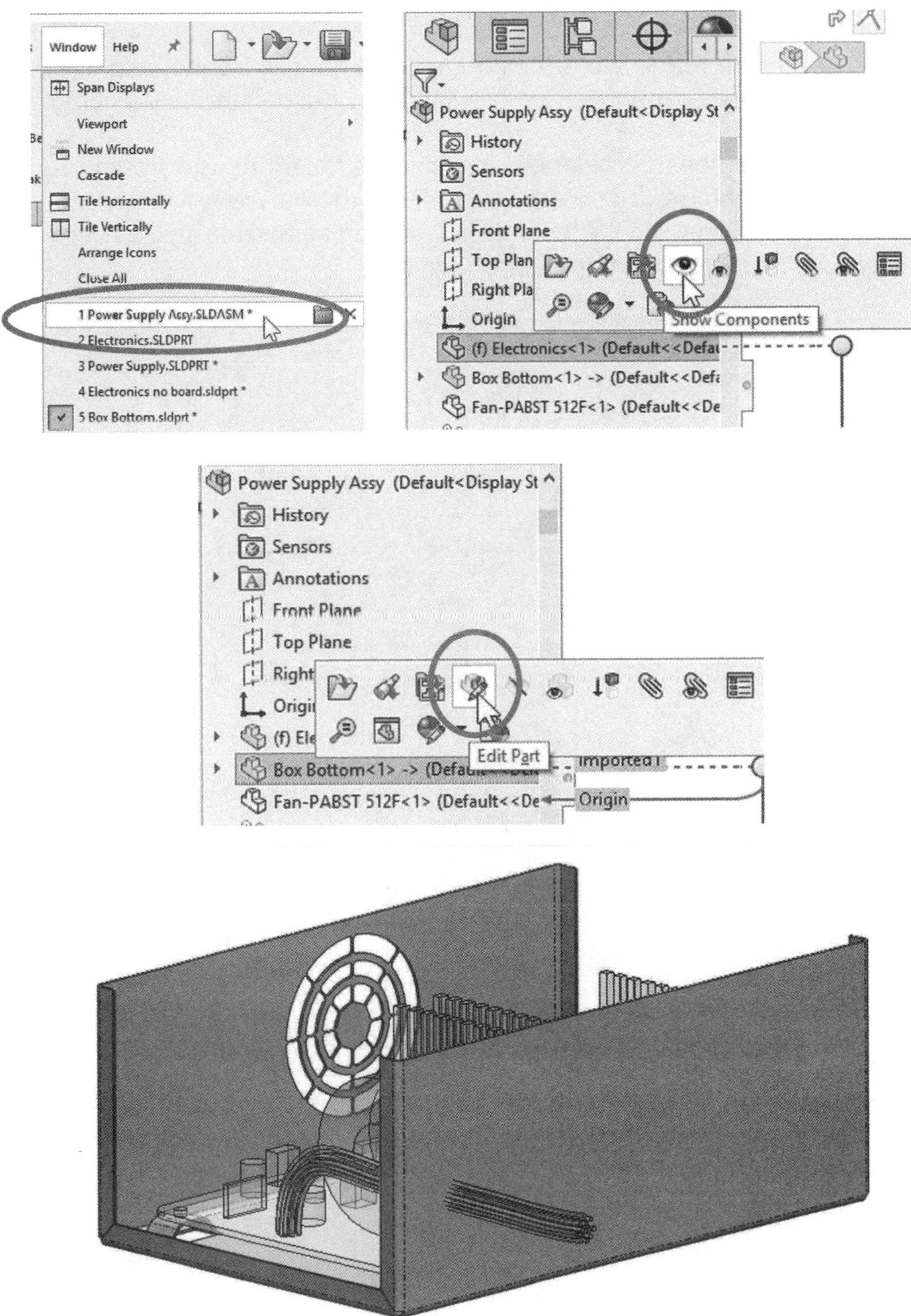

10.28. - To make the cut for the wires, first we need to unfold the miter edge closer to them. We'll use the "**Unfold**" command from the Sheet Metal tab or the menu "**Insert, Sheet Metal, Unfold**."

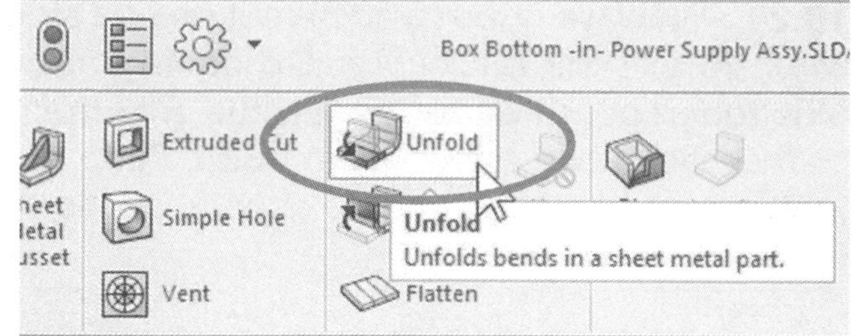

The cut can be made without unfolding the flange, but I chose to show the reader a different approach and review previously introduced commands.

In the "Fixed face" selection box, select the model's face that will remain fixed when we unfold the miter flange; by selecting this face, the flange will move out instead of the other way around. In the "Bends to unfold" select the bend region of the flange, click OK to unfold the flange and continue.

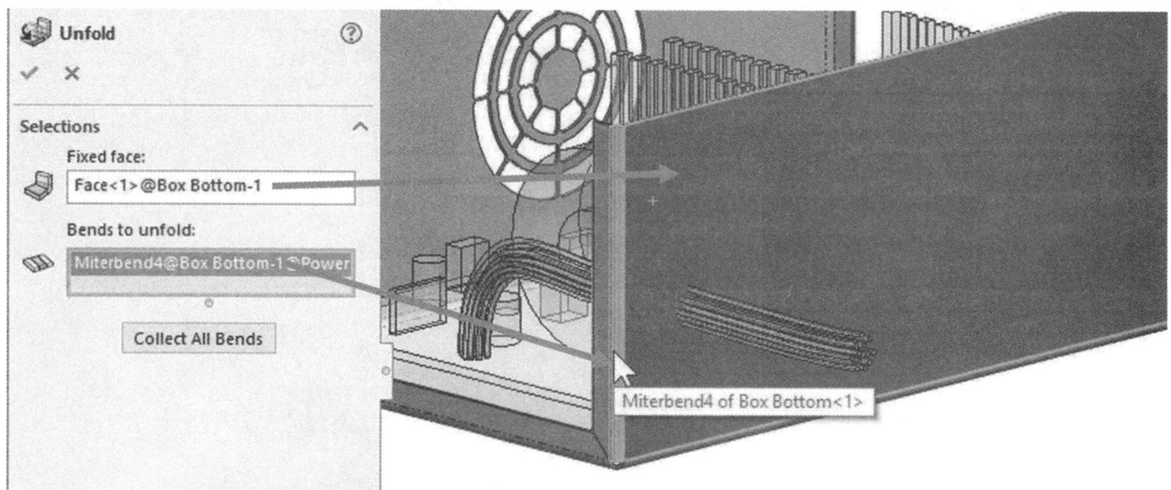

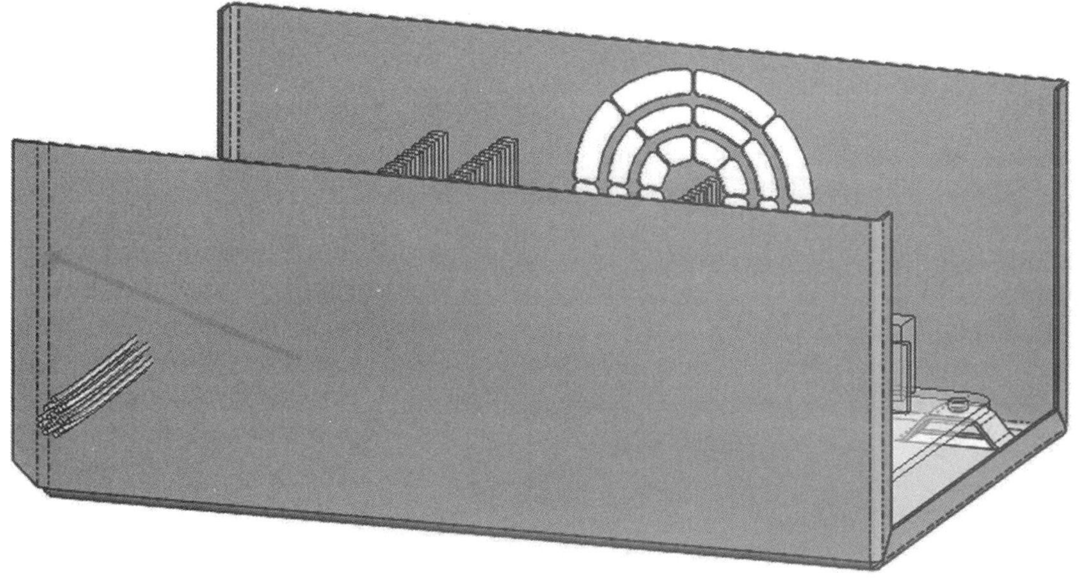

10.29. – With the flange unfolded we can add the cut for the wires and then re-fold the flange. Switch to a Front view and make the next sketch around the wires. Make the cut using the **"Link to thickness"** option. Remember we are editing the '*Box Bottom*' part in the assembly.

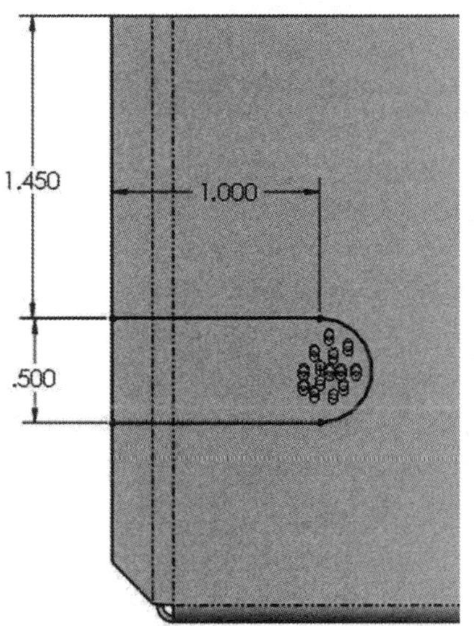

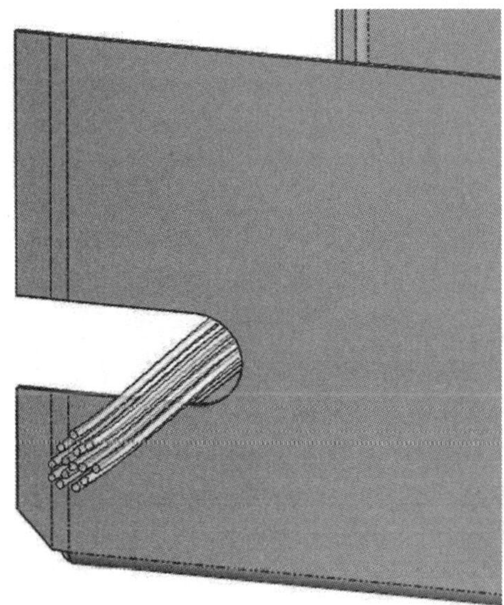

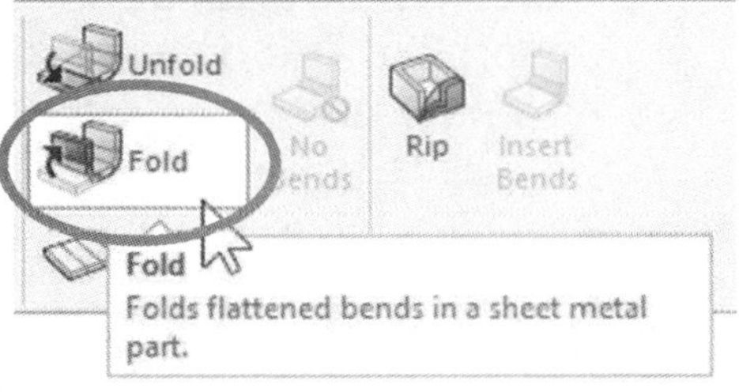

10.30. - To re-fold the flange after the cut, select the **"Fold"** command from the Sheet Metal toolbar or the menu **"Insert, Sheet Metal, Fold.**" The "Fixed face" is automatically selected, all we have to do is add the bend (SOLIDWORKS will only allow us to select bends in this step). Click in the "Collect All Bends" button to select all unfolded bends and click OK to re-fold the miter bend.

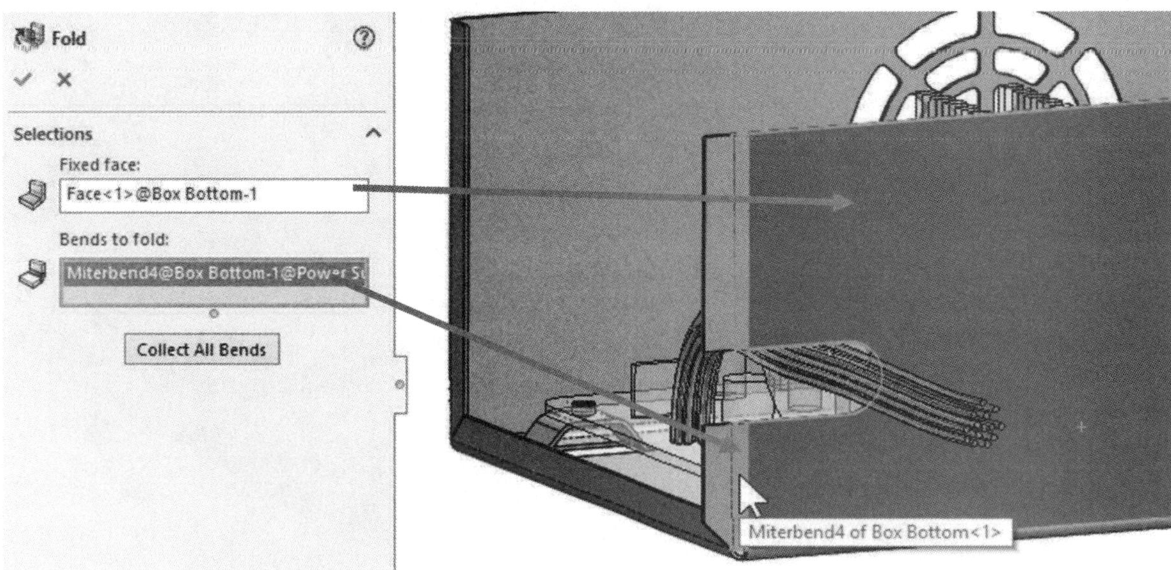

10.31. - The next step is to add flanges at the top of the box to screw the cover in place. In these flanges, we must modify their profile, as they will not run the full length of the edge and need to have a rounded edge. Select the "**Edge Flange**" command from the Sheet Metal toolbar, or the menu "**Insert, Sheet Metal, Edge Flange.**"

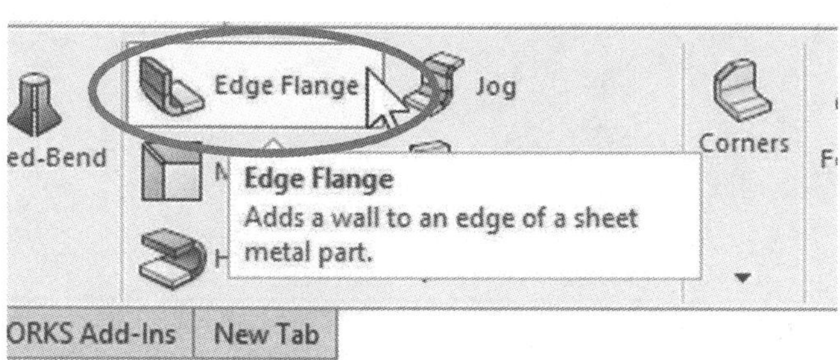

In the "Flange Parameters" selection box, select the top edge of the side with the vent. As soon as the mouse pointer is moved you can see a preview of the edge flange. Click towards the center of the box to add the flange inside; the size is not important since we are going to change its sketch. Select the "Edit Flange Profile" option to modify the shape and size of the flange.

If the intended flange runs the full length of the edge there is no need to edit the profile. The only reason to edit the flange's profile is if we need to change its length and/or profile.

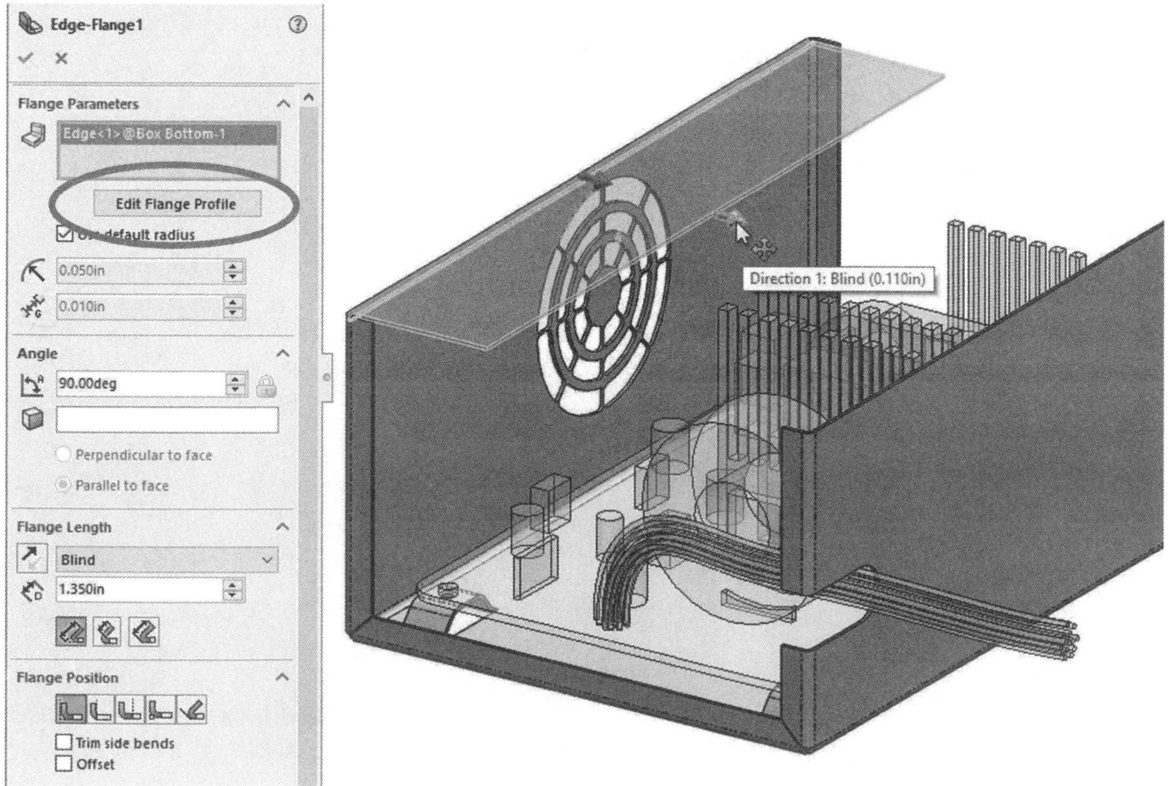

When we edit the flange profile we are editing its sketch, and we can change it as much as we want, as long as the sketch is a single closed profile. Drag the lines on the sides towards the middle, delete the short line in front and replace it with a tangent arc. When we edit the flange's profile the lines can be dragged even if they appear fully defined. Dimension as indicated to complete the sketch.

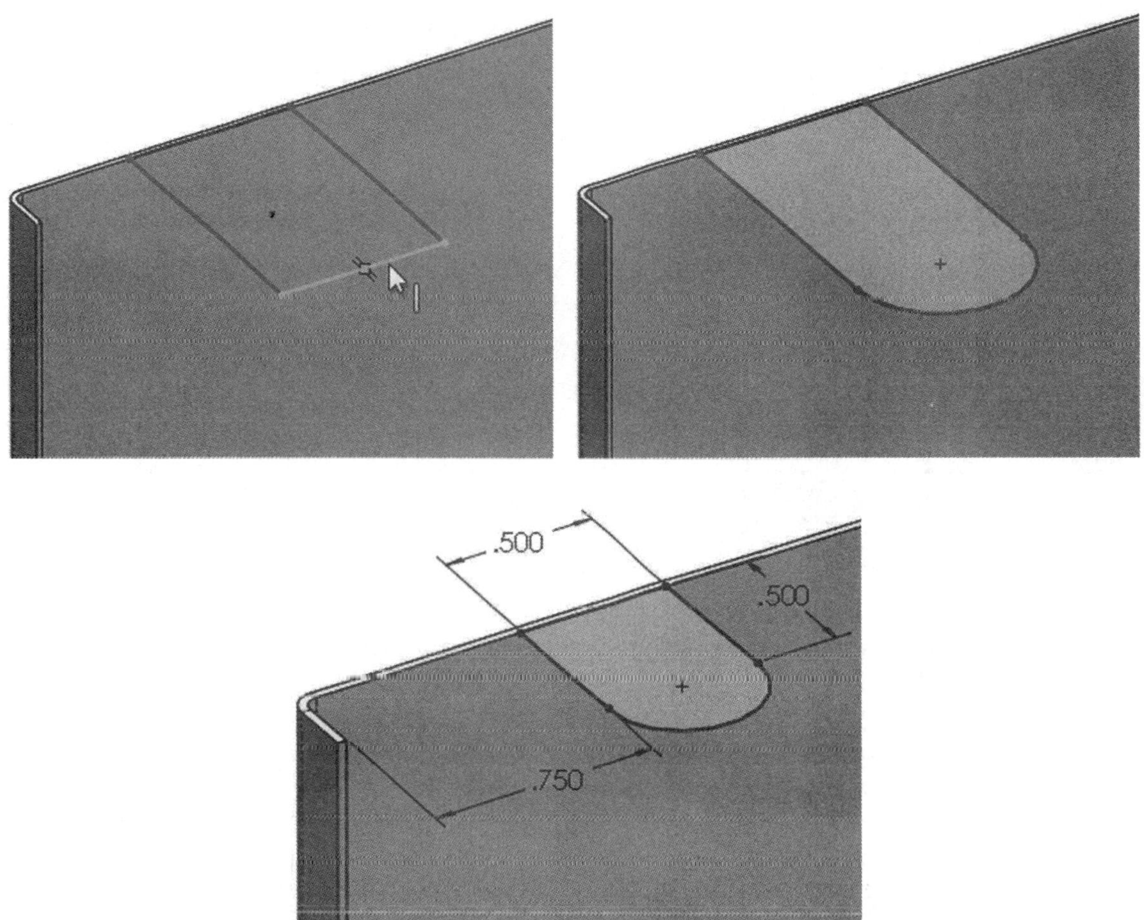

When the sketch is complete, click the "Back" button to return to the "Edge Flange" command to set the rest of the options.

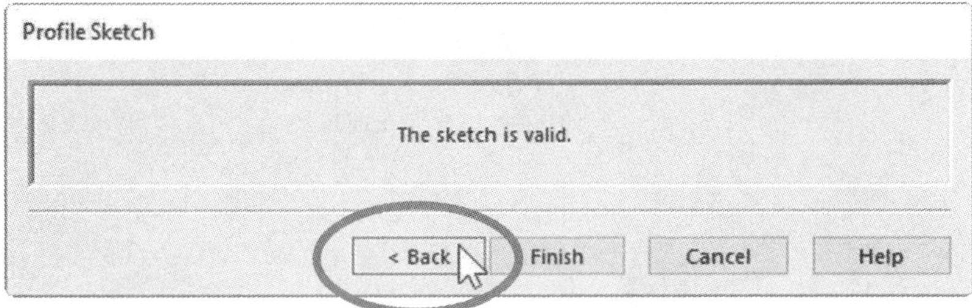

If we click "Finish" the flange will be completed with the parameters entered before editing the sketch. This message will also let us know if the sketch profile for the Edge Flange is valid or not.

10.32. - In the “Flange Parameters,” leave the angle to 90 degrees. In the “Flange Position” select the “Material Inside” option, and turn on the “Custom Relief Type” checkbox. Select an Obround relief for the flange with a 0.5 ratio to override the default settings. Click OK to add the edge flange.

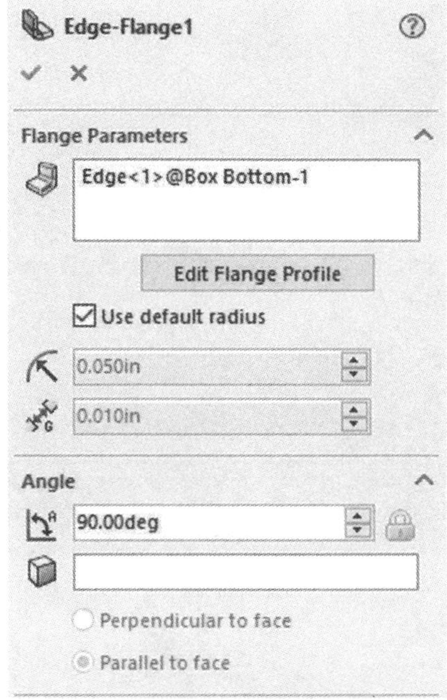

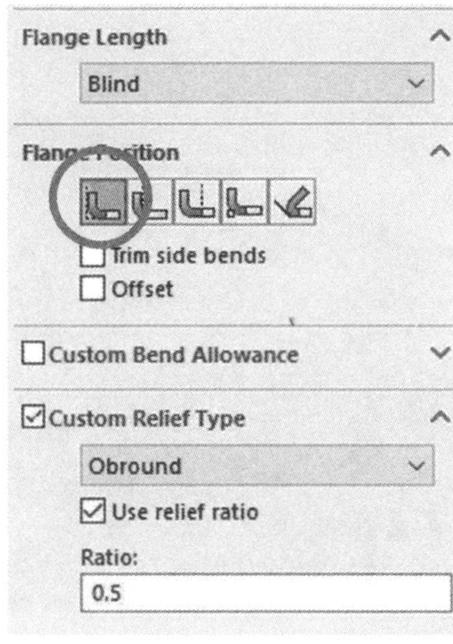

Rectangular or Obround relief types can also be defined by specifying the width and depth of the relief cuts unchecking the “Use relief ratio” checkbox.

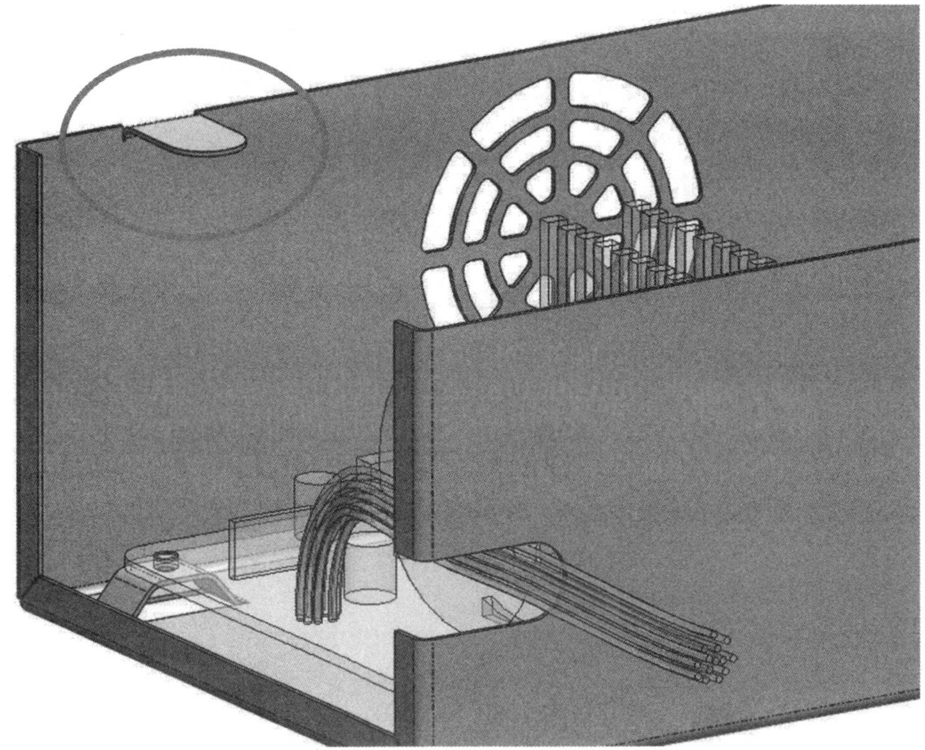

10.33. - Now we need to add a hole for a screw to the tab we just made. In this case, we'll use one of the built-in Form Tools to do it. Open the "Design Library" in the task pane, and scroll down to "forming tools, embosses." From the list of tools select "extruded hole." Drag-and-drop it in the top face of the tab just added. Make sure it is going *into* the part, and not out. Remember the form tool's direction can be reversed by pressing the Tab key before releasing the mouse button while we drag it in place, or using the "Flip tool" button after dropping it in place.

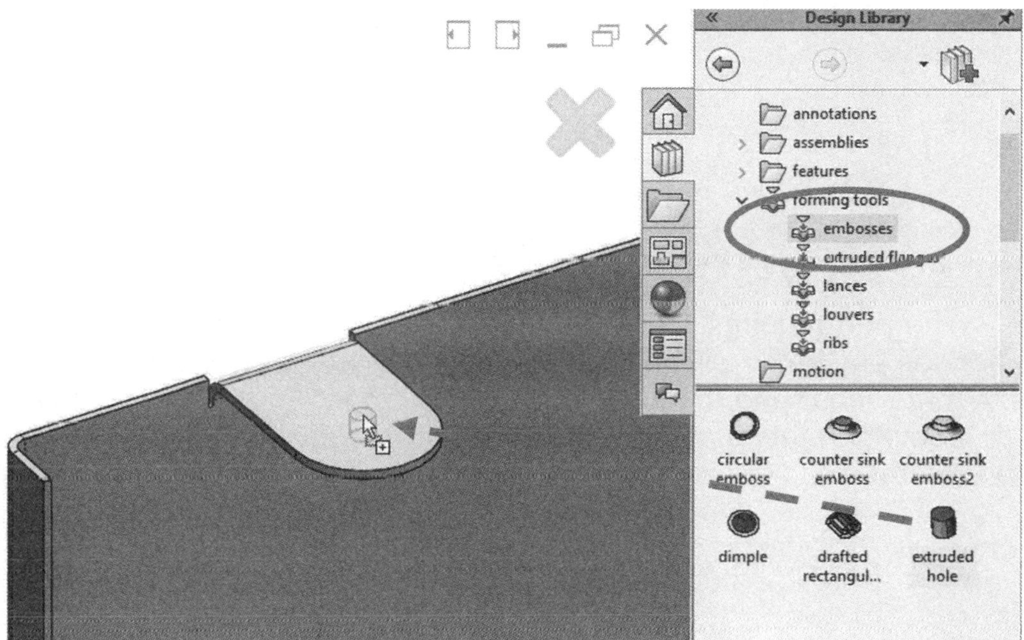

If the form tool is going into the part we don't need to change anything, if it is not, click "Flip tool" to correct it after dropping it in place. Now select the "Position" tab, add a concentric geometric relation between the locating point and the edge of the round tab, and click OK to finish.

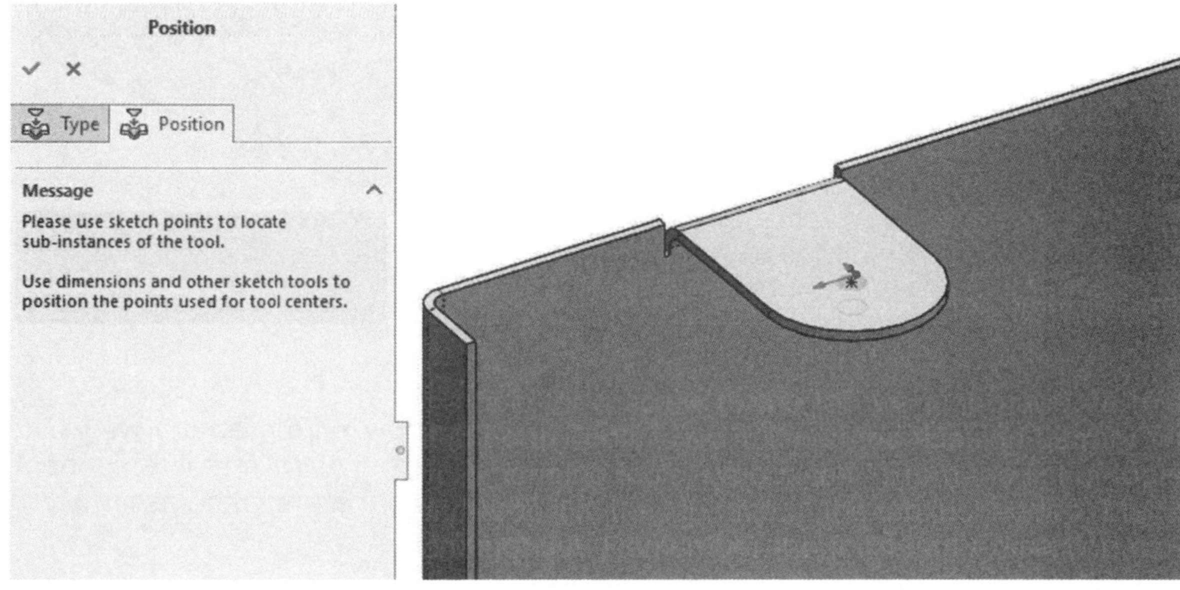

Our part now looks like this:

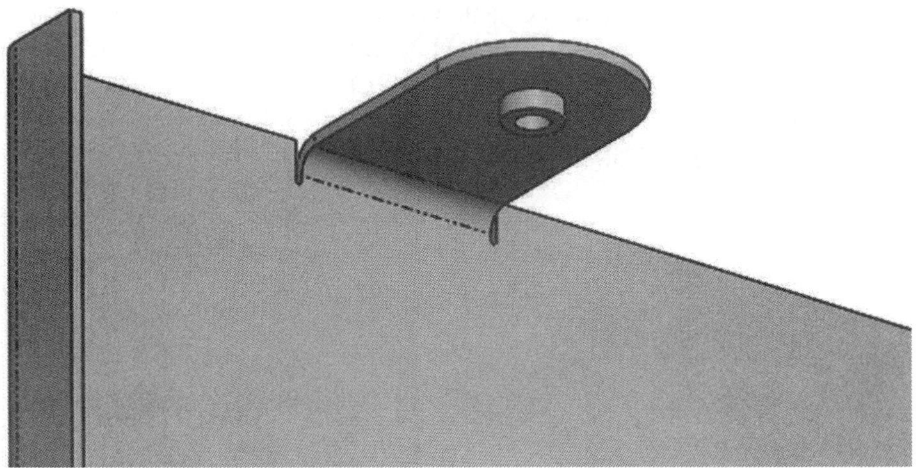

10.34. - The part requires four flanges with their corresponding holes. The other three flanges and holes will be added using two mirror features: the first about the part's *"Front Plane"* and the second about the *"Right Plane."* Remember that we are still working in the context of the assembly at this point, so be sure to select the part's planes and not the assembly planes from the fly out FeatureManager.

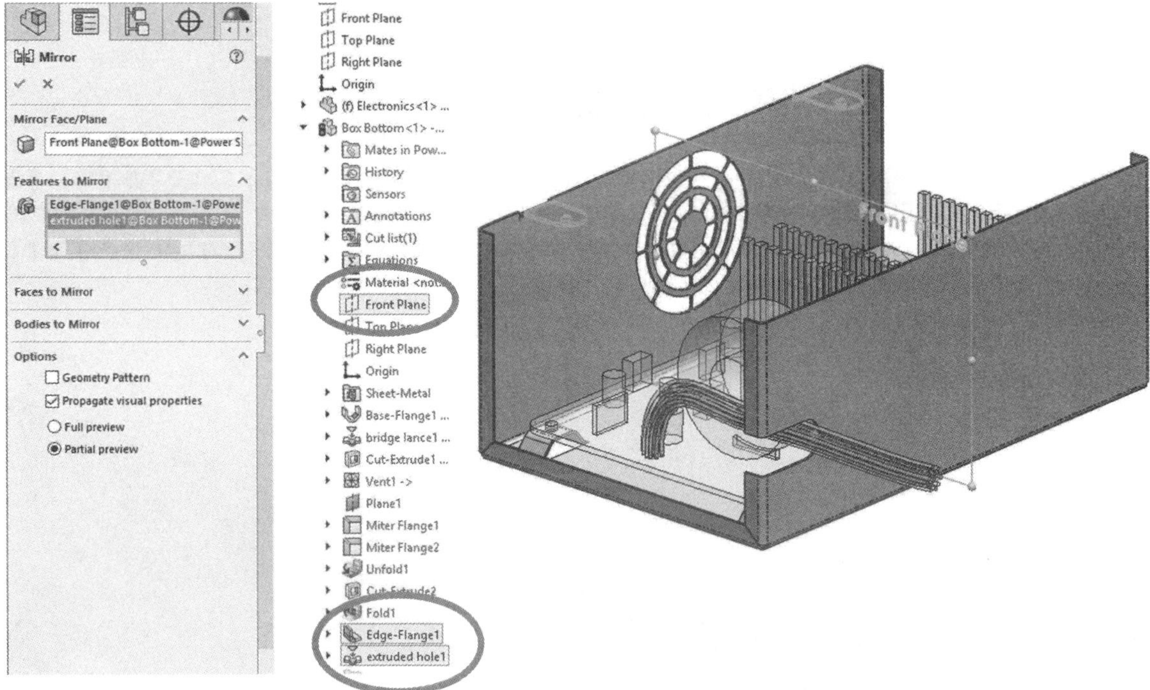

If we had selected the assembly's planes for the mirror feature we would have added an unnecessary external relation to our sheet metal component, since the mirror feature is not dependent on any feature in the assembly.

Add a second mirror feature now about the "*Right Plane*" to mirror both tabs.

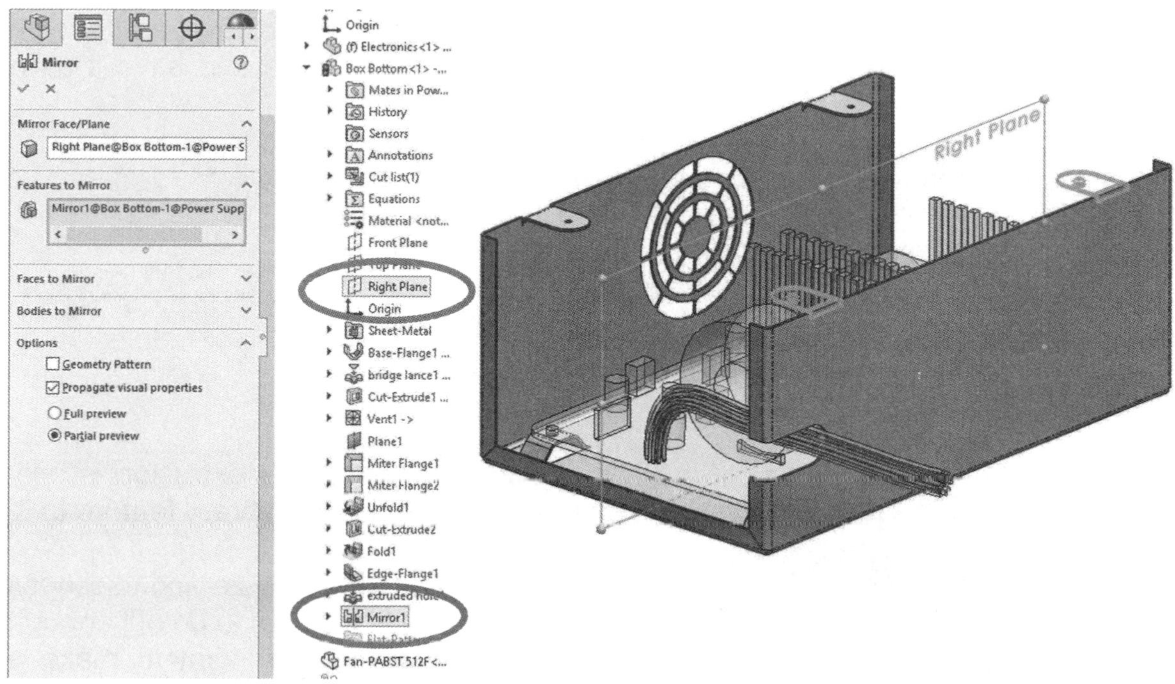

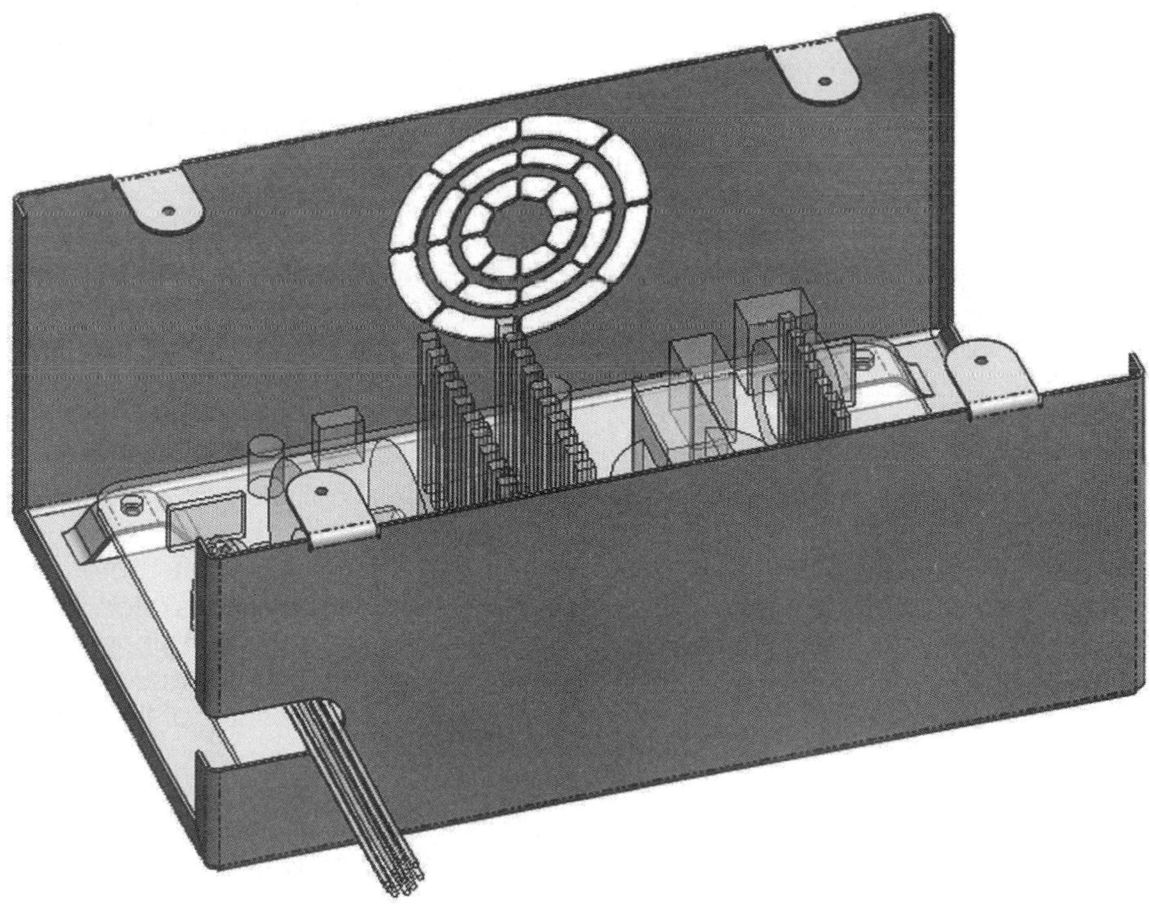

10.35. - Forming tools can only be used in a sheet metal part, but a **library feature** can be used in any part. In this step, we'll learn how to create a library feature that can be used in our designs. The process, in general, is to make a new part (outside the assembly), add the feature(s) we want the library to have and save it as a Library Feature.

In this step, we'll make a Library Feature of a cutout for a power supply connector, and add it as a feature to our sheet metal part next to the vent. To build a library feature, make a new part (*NOT* in the assembly), add a new sketch (any plane) with a 3″ wide by 2″ high square, locate the origin in a corner, and make a 0.125″ thick extrusion.

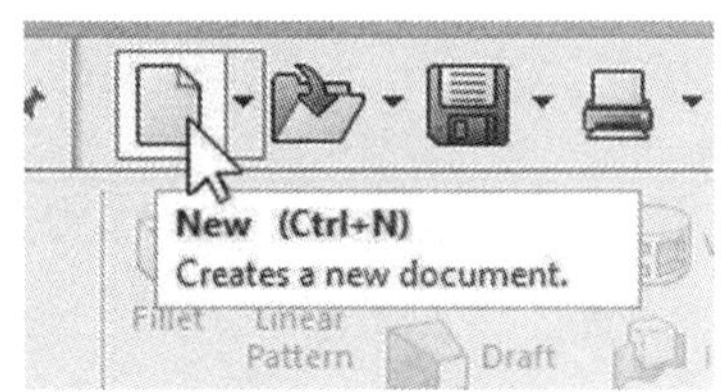

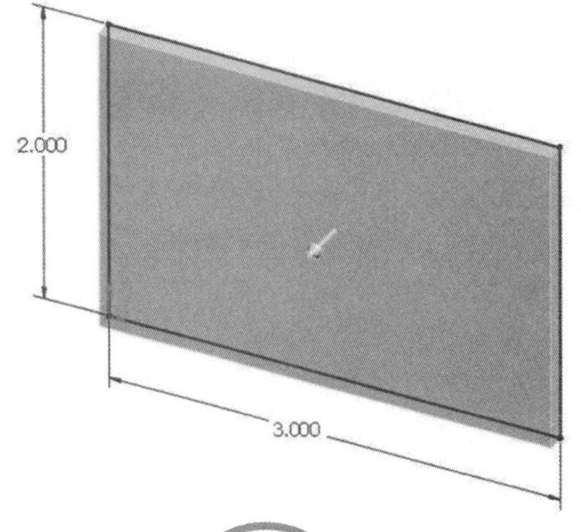

The purpose of this extrusion is only to have a base to add the library feature to it.

The base feature's size and thickness are not important, it just has to be big enough to hold the feature(s) we want to make a library with. Add the following sketch in the front face and make a cut using the "Up to next" end condition, the reason is to only cut up to the next face with it, and not a set depth or through the entire part when the library is used.

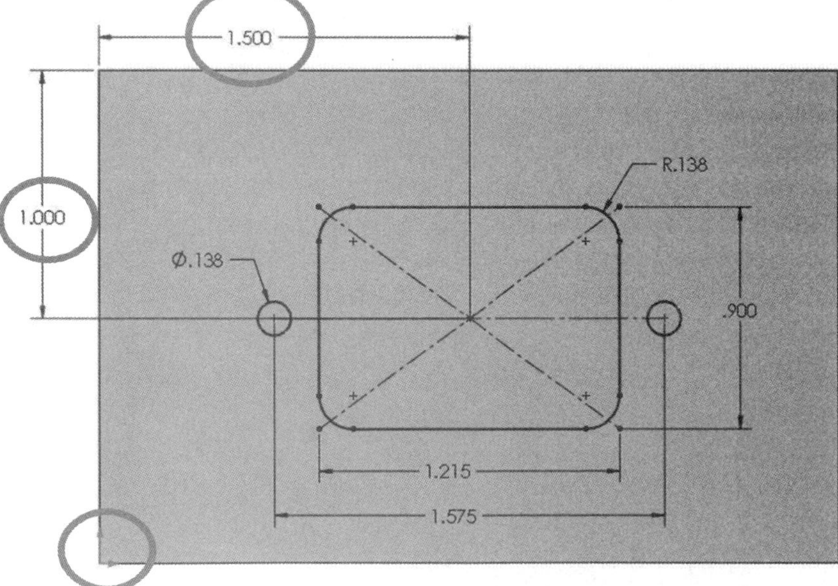

In this sketch, we are only using two dimensions to locate the sketch's position in reference to the edges of the first extrusion, creating only two external references in the sketch. Notice <u>the part's origin is in a corner</u> and not in the center of the part.

If the base feature rectangle's center is located at the origin, and the second sketch is also at the origin, adding either the 1″ or 1.5″ dimensions would over define the sketch.

10.36. - Once the cut is made using the "Up to next" end condition, select *"Cut-Extrude1"* in the Feature-Manager, select the menu "**File, Save as…**" and from the "Save as type" drop down selection box select "Lib Feat Part (*.sldlfp)." Name it *'Power Connector Cutout'*. Notice as soon as we select the library feature type, we are switched to the "Design Library" folder. Save the library in this folder.

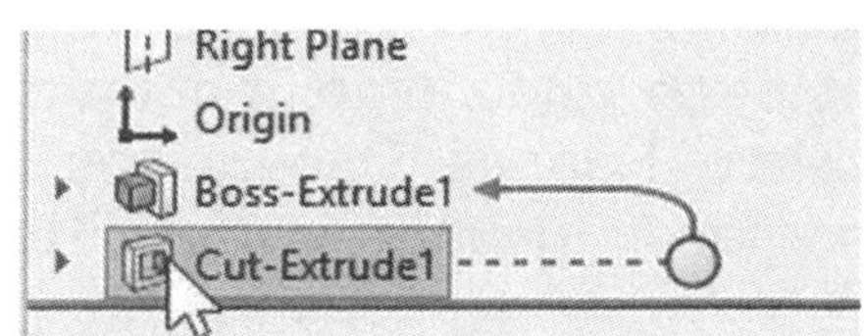

If the feature(s) to be saved as a library is (are) not pre-selected in the FeatureManager, saving a library feature will not work.

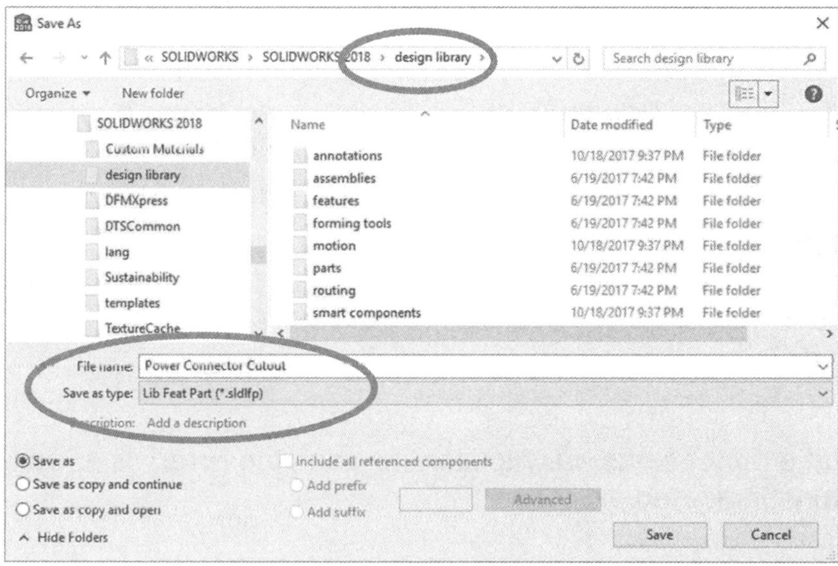

After saving the feature as a library, the FeatureManager changes to display it as such by superimposing a blue letter "L" in the feature(s) saved as a library, the part's icon changes to indicate this part is a library feature, and adds two folders called "*References*" and "*Dimensions*."

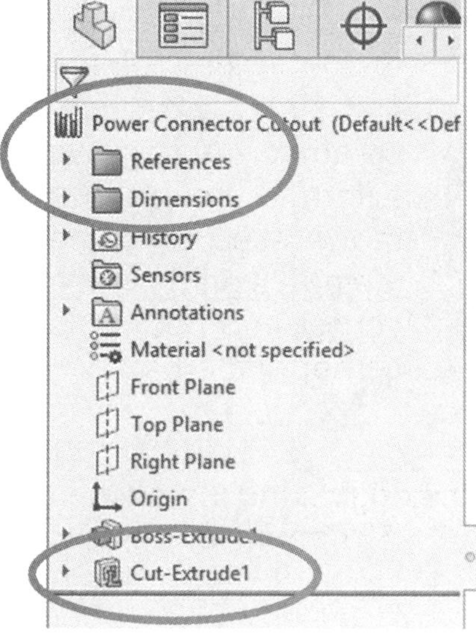

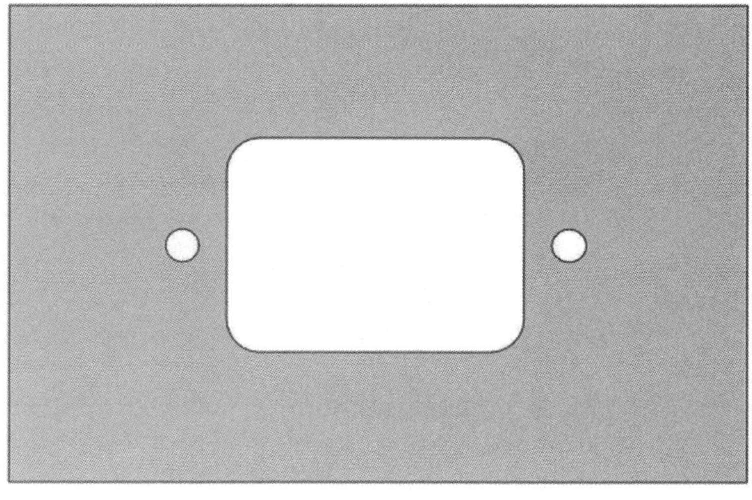

10.37. - The *"Dimensions"* folder is used to define which dimensions can be changed by the user, which dimensions will remain hidden, and which dimensions will be used to locate the library when it is added to a part. Before defining the role of each dimension, right-mouse-click the *"Annotations"* folder and turn on the "Show Feature Dimensions" option, as well as the option to display dimension names in the menu "**View, Hide/Show, Dimension Names**."

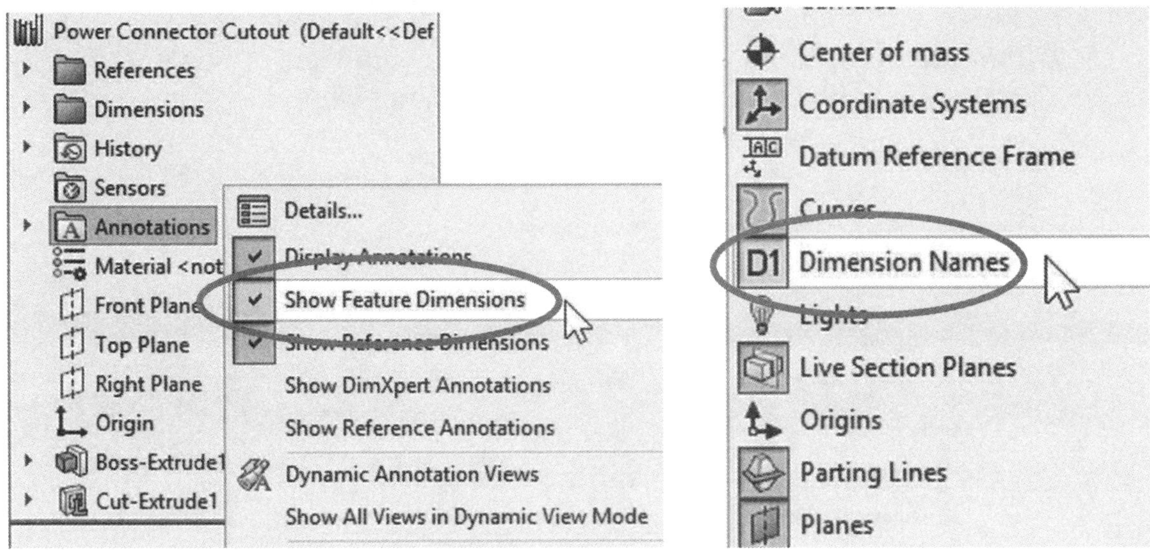

To make dimensions easy to identify when the library is added, rename the two dimensions indicated as shown.

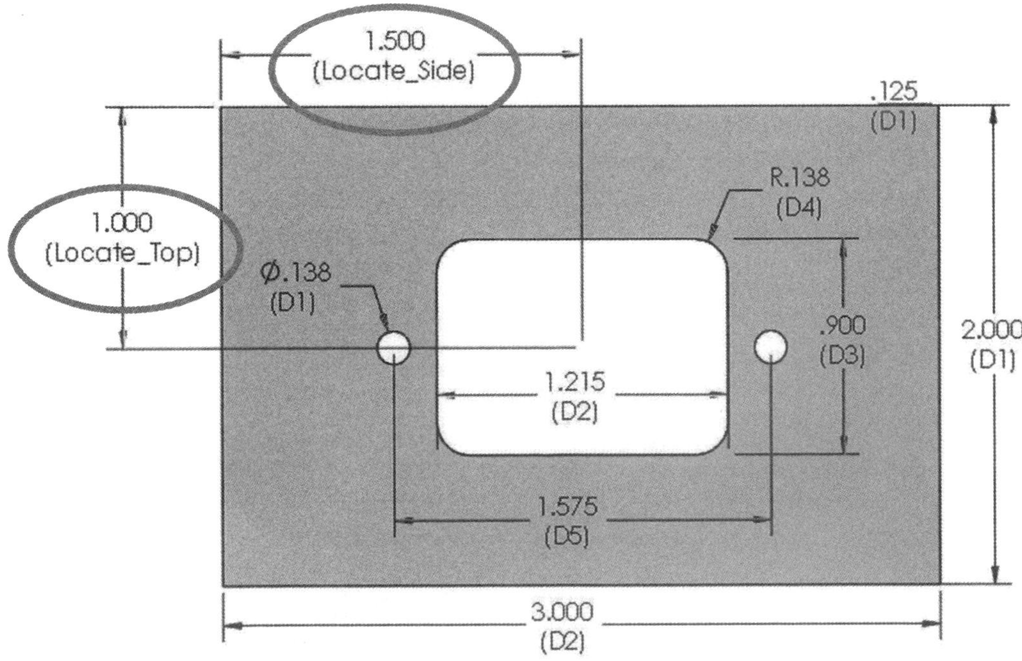

After renaming the dimensions, expand the *"Dimensions"* folder where we can see all the part's dimensions listed. Drag-and-drop the *"Locate_Top"* and *"Locate_Side"* dimensions to the *"Locating Dimensions"* folder and the rest to the *"Internal Dimensions"* folder.

- **Locating Dimensions** will be used to position the feature when we add the library to a part.
- **Internal Dimensions** will be hidden and cannot be changed by the user.
- Dimensions not added to either folder are visible and can be changed after the library is added to a part.
- The "References" folder contains the references that will be needed when adding the feature to a part. In our case a face to locate the sketch, and two edges to reference the locating dimensions.

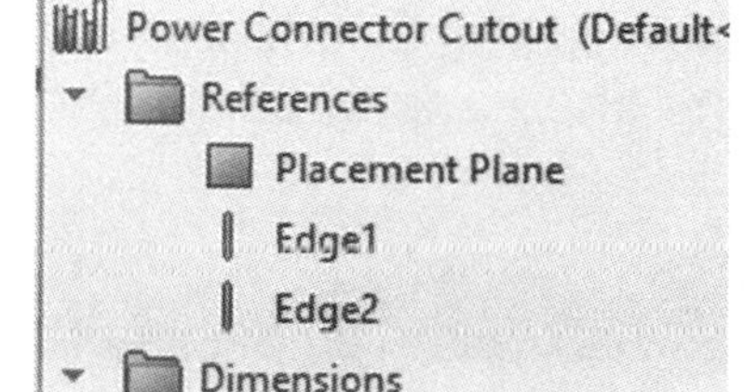

After moving the dimensions to the different folders, turn off dimension display, save the changes and close the library file. The library file we just made is now available under the main *"Design Library"* folder in the task pane (or the folder in which it was saved).

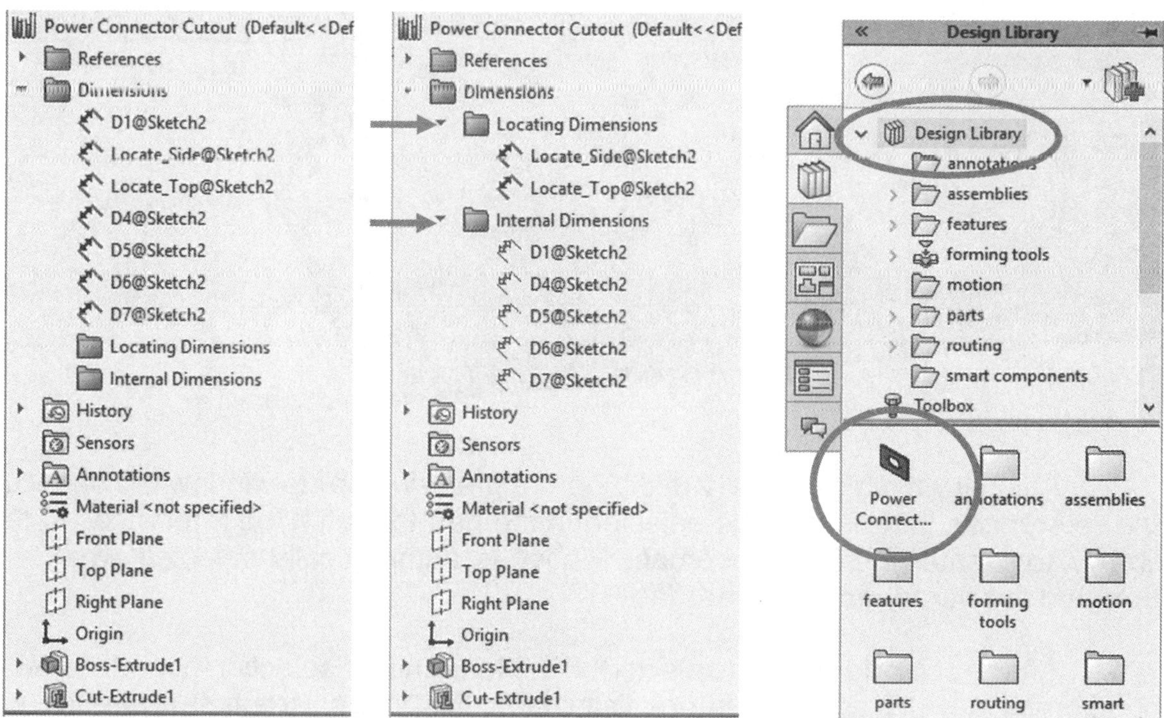

Library features can be organized by dragging-and-dropping them to a different folder in the Design Library, and more folders can be added as needed to keep libraries organized.

Library features can have multiple features; however, to make these features as easy to use as possible, be sure to add subsequent features referencing only the features included in the library, making the first feature a "parent" to all other features in the library. By doing this, the number of references needed when using the library feature will be kept to a minimum.

10.38. – After saving the changes and closing the library file, we are now ready to add the *'Power Connector Cutout'* library to the *'Box Bottom'* part, either in its own window or while editing it in the assembly (your choice).

Open the Task Pane, select the *"Design Library"* folder, and drag-and-drop the *'Power Connector Cutout'* library onto the part. In this image, we are editing the sheet metal part in the assembly.

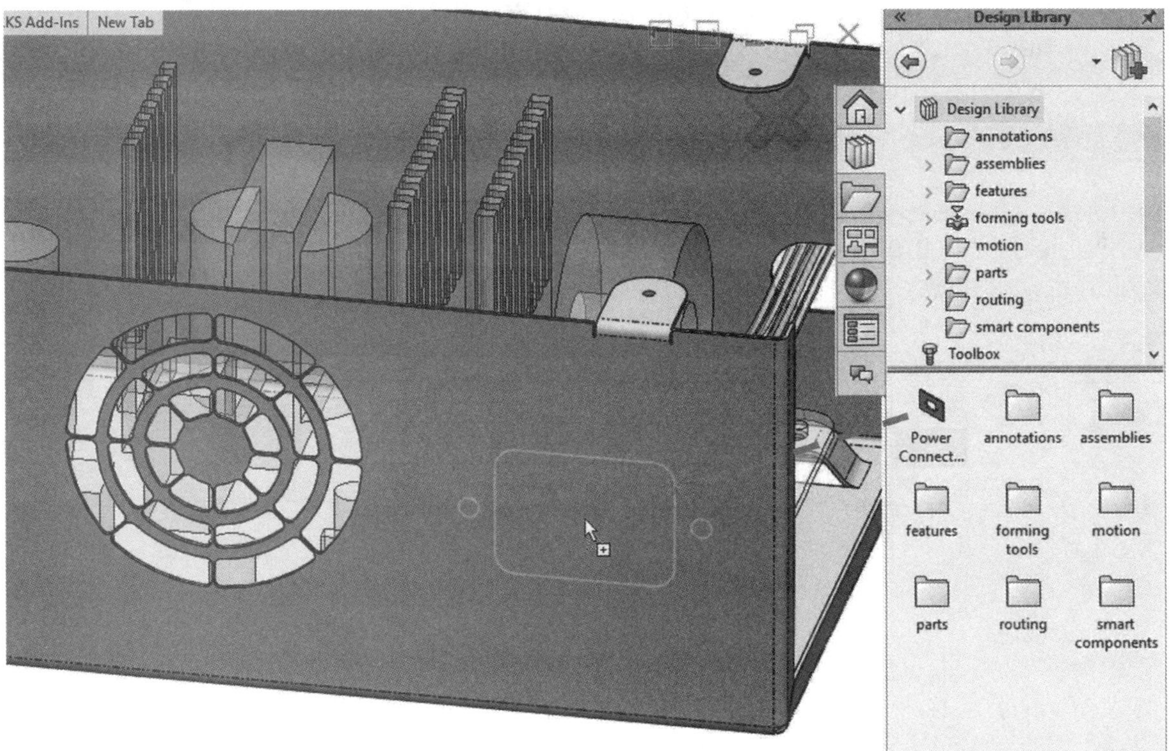

After dropping the library in the part, a preview pop-up window will show us the references that need to be selected in the part to match the references in the library to locate it. The "Placement Plane" is automatically selected when we dropped the library onto the rear face.

After each reference is selected, the "References" selection box will show a checkmark next to each reference listed. For this library, the first reference we need to select is *"Locate_Side"*; after selecting the indicated edge, the library's locating dimension will be referenced to it.

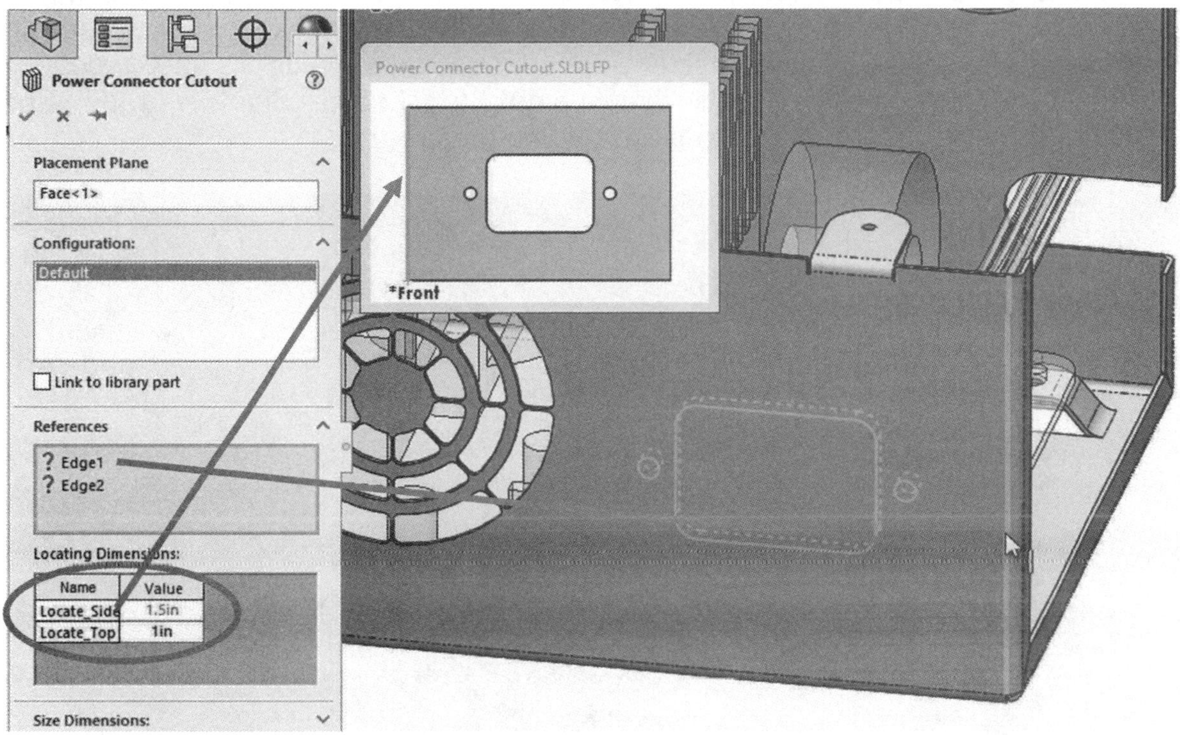

After selecting the first reference we are asked for the second reference in the preview window. Select the top edge of our part as reference to continue.

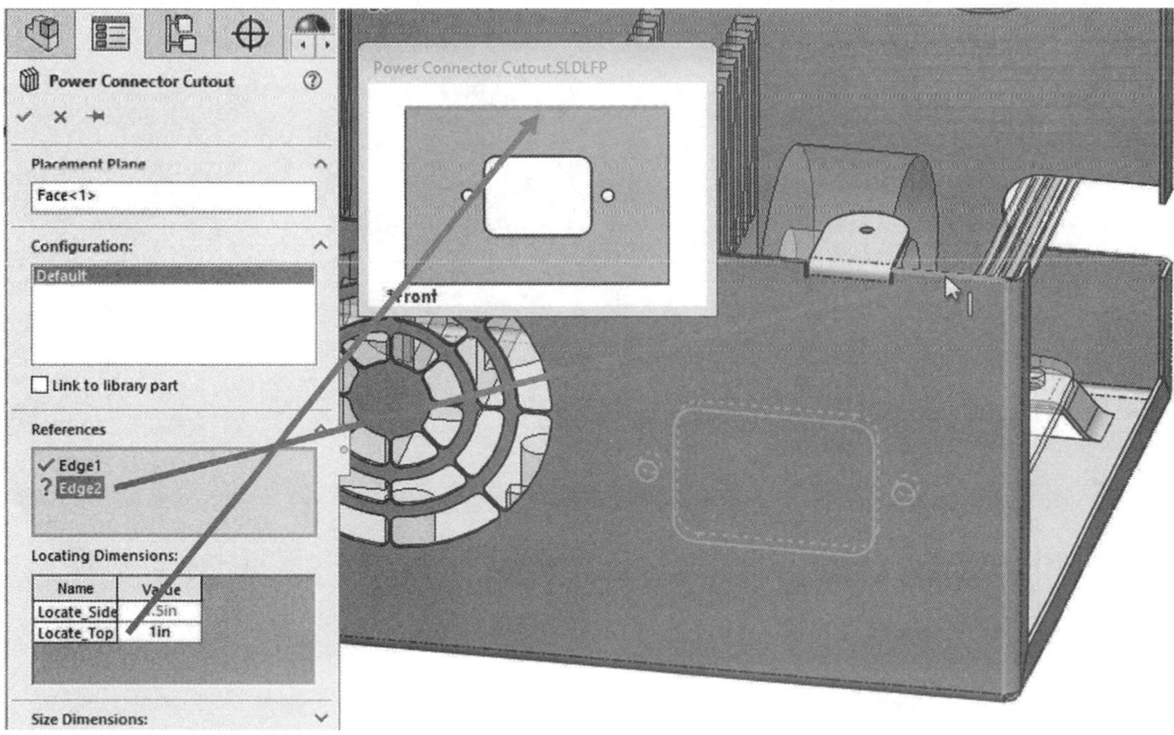

After the required locating references are selected, they are listed with a green checkmark, letting us know they are valid references. Now we need to change their values. Click inside each dimension's value in the "Locating Dimensions" box, and change their values to *"Locate_Side"* = 1.25″ and *"Locate_Top"* = 1.5″. Click OK to finish adding the library.

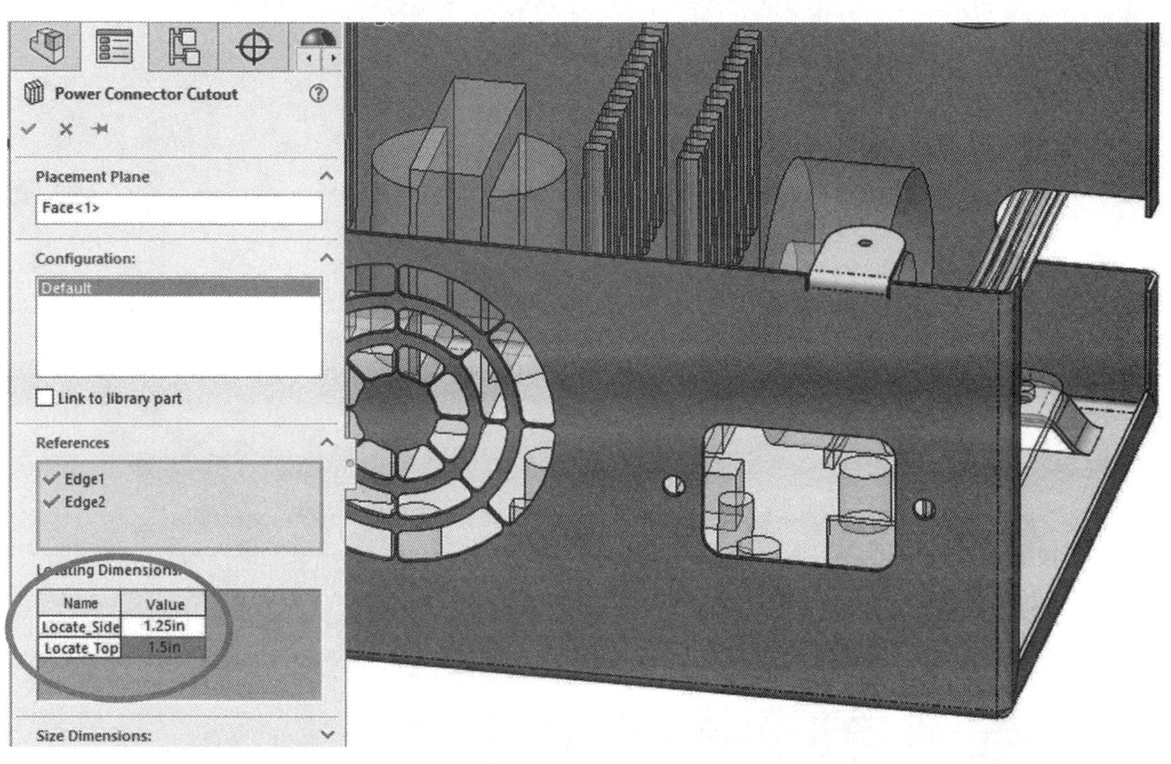

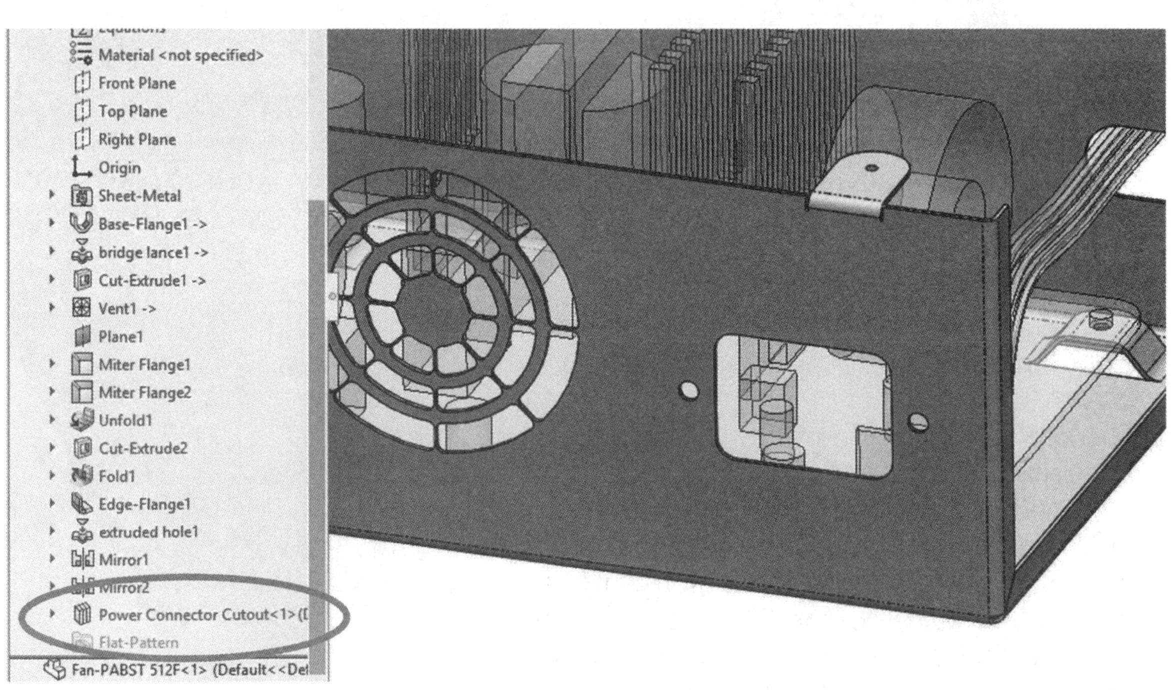

If a dimension is not designated as *"Internal"* or *"Locating"* in the library, its value can be changed in the "Size Dimensions" box after selecting "Override dimension values."

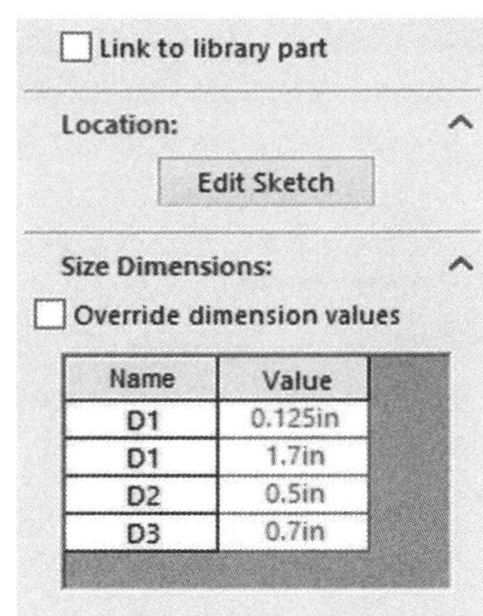

If the library has no locating dimensions, we'll see an "Edit Sketch" button to locate the library instead of the "References" selection box.

Library features can have multiple configurations, and every time the library is added, a different configuration can be used.

10.39. - Just as we can make libraries of individual features, we can also make libraries of complete parts and even assemblies. In the next step, we'll create a library part and use it in our assembly. The part we want to make a library of is the Power Connector for which we just made a cut out.

A library part can be any SOLIDWORKS part and can be used as such just by putting it in the *"Parts"* folder of the *"Design Library."* What makes library parts useful, more than keeping them organized and handy, is that we can add "**Mate References**" to them, and when they are dragged into an assembly, the mate references will look for a matching entity to mate with, saving us time. Open the part *'Power Connector.sldprt'* from the included files for this example.

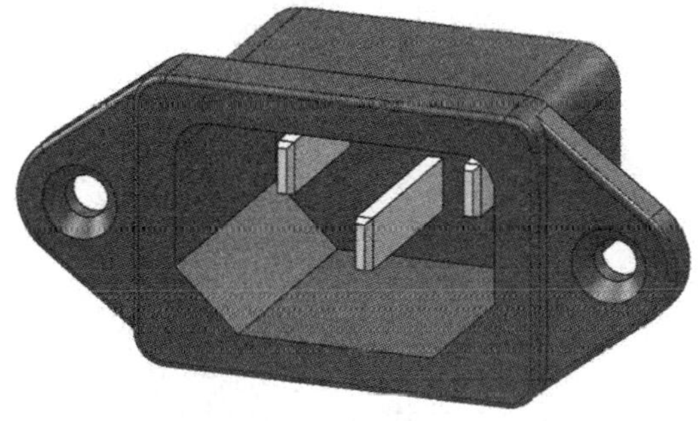

Once the connector is open, go to the menu "**Insert, Reference Geometry, Mate Reference**," or the drop-down menu in the "**Reference Geometry**" command in the Features tab of the Command-Manager.

We can add up to three mate references to a part; for our example, we'll only use a single reference.

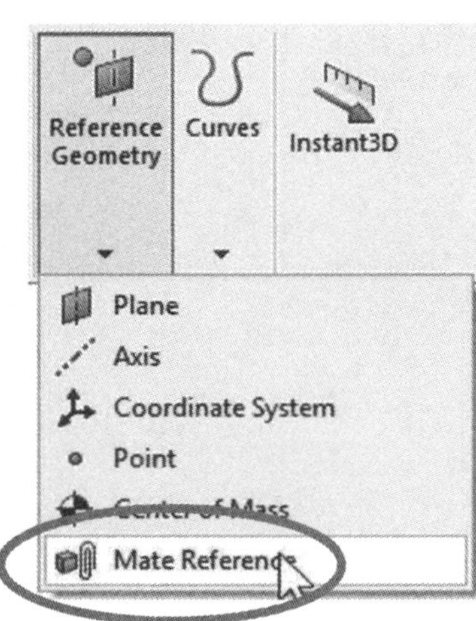

To add the mate reference, add the circular edge in the left mounting hole of the connector to the "Primary Reference Entity." The "Default" mate alignment is the only option available for circular edges. The alignment option refers to the direction of the mate; leave this option to "Any" to allow for different options when adding the part to the assembly. Optionally give a name to the mate reference and click OK to finish.

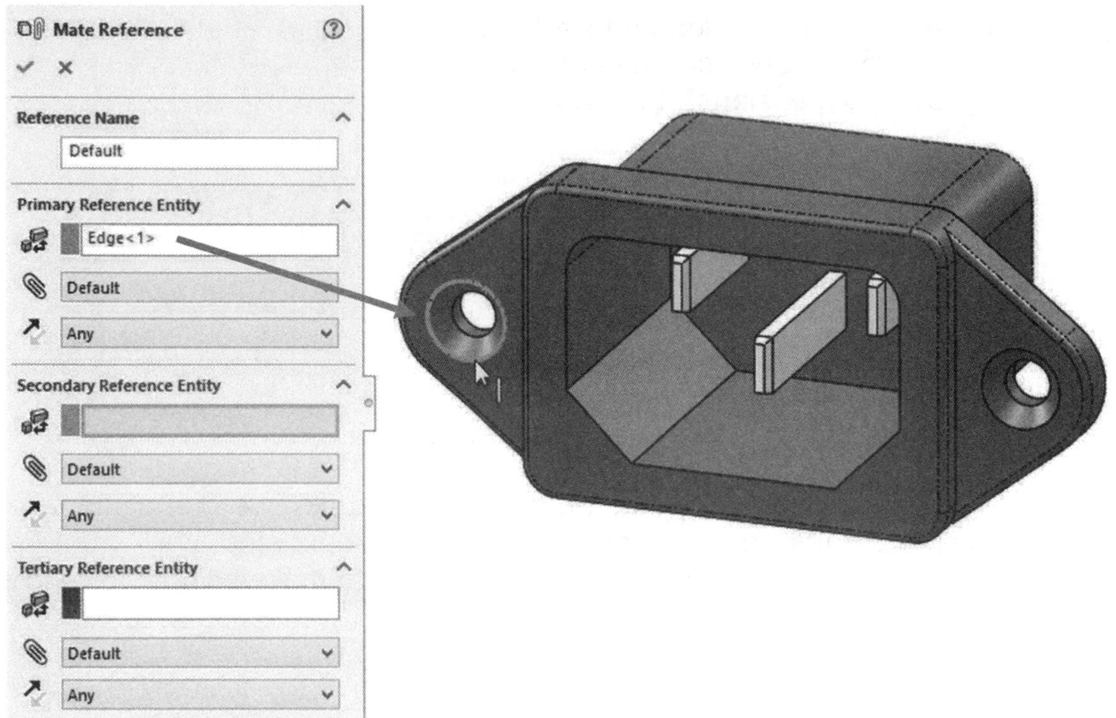

Mate References can be added to any entity that can be mated, including faces, edges, vertices, axes, sketch entities, points, planes, etc.

Mate Alignment refers to how the mated faces will be oriented. For flat faces imagine an arrow (Vector) going away from each face. "**Aligned**" means both arrows are pointing in the same direction; "**Anti-Aligned**" means they are pointing at each other. There is no analogy for cylindrical surfaces, just flip them as necessary. A good practice is to position parts as close to the desired orientation before adding mates, since SOLIDWORKS will not rotate a part more than 180 degrees to add a mate.

After adding the mate references, a new folder is added to the FeatureManager listing the mate reference added. Save the *'Power Connector'* part; now it's ready to be added to the *"Design Library."*

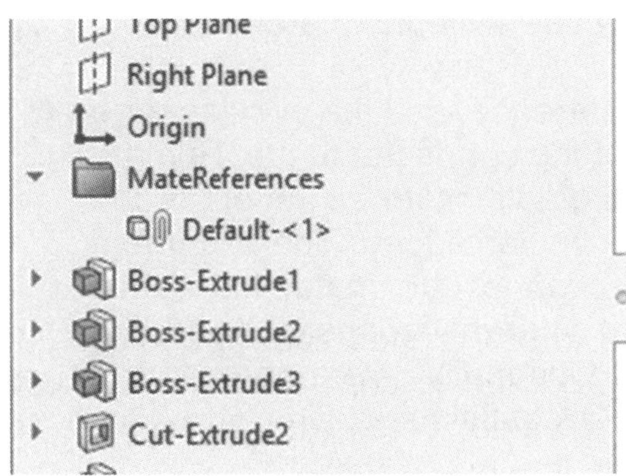

10.40. - To add the part to the library, right-mouse-click at the top of the FeatureManager in the part's name, and select "Add to Library" from the pop-up menu...

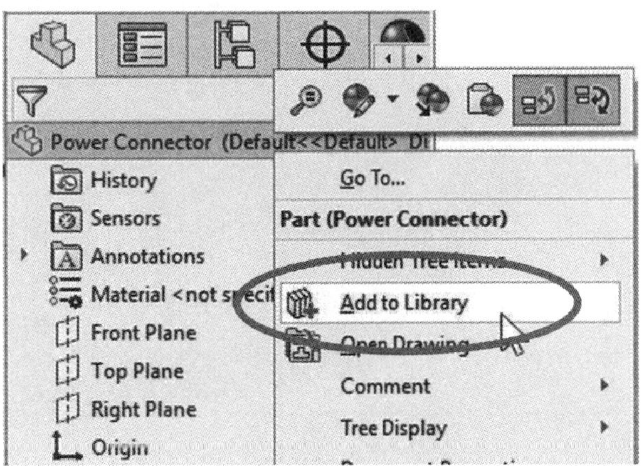

... or expand the *"Design Library"* and drag-and-drop the part from the top of the FeatureManager to the task pane in the desired folder; in this example we'll select the "parts" folder.

Before adding the part to the library using drag-and-drop, we need to keep the *"Design Library"* visible by clicking on the auto-show pin.

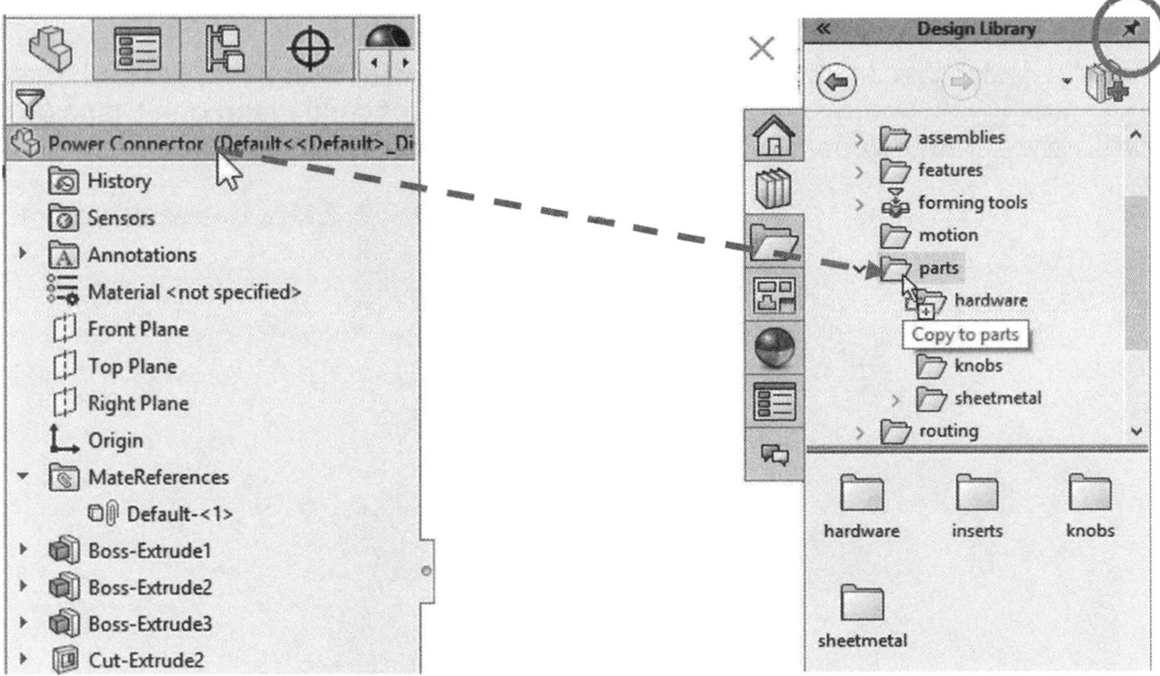

Now we are asked to select the folder to add the library to and optionally give it a different name. Click OK to copy the *'Power Connector'* to the selected library folder and close it. If we don't close the file, we'll be asked if we want to use the open part when adding it to the assembly.

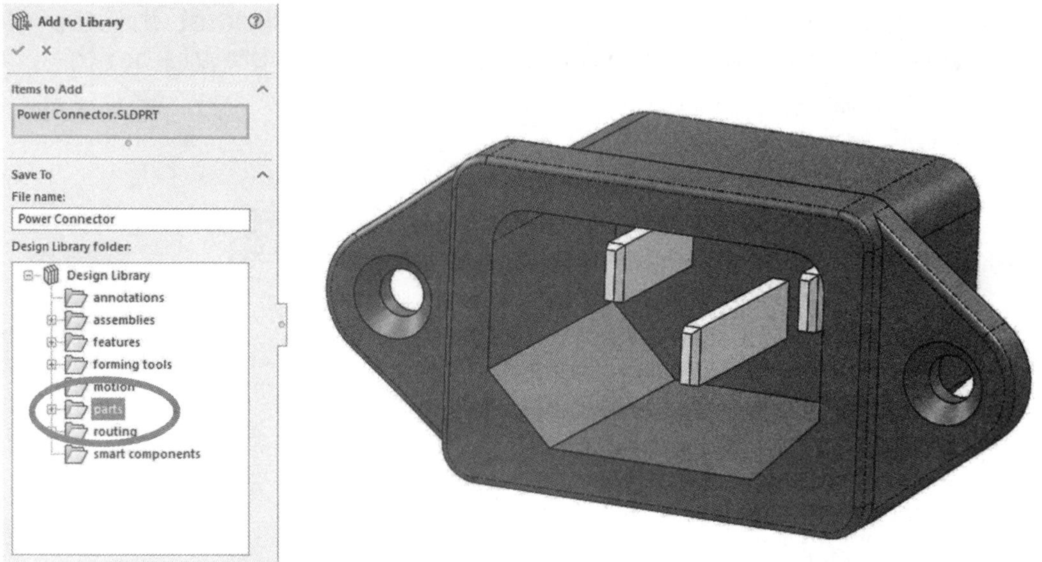

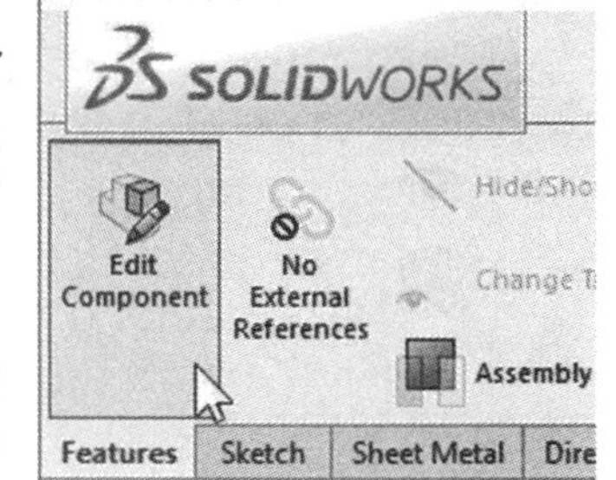

10.41. – After making the library and closing the '*Power Connector*', go back to the assembly. Be sure we are editing the assembly before adding the library part. If needed, turn Off the "**Edit Component**" command.

In the *"Design Library"* go to the "parts" folder and drag-and-drop the *'Power Connector'* in the hole of the cutout. Notice, as we move it close to a hole or a flat face, the part automatically snaps, trying to find a match using the defined mate reference. When we drop the part in the hole, one Concentric and one Coincident mates are added. Close the "Insert Components" message to continue.

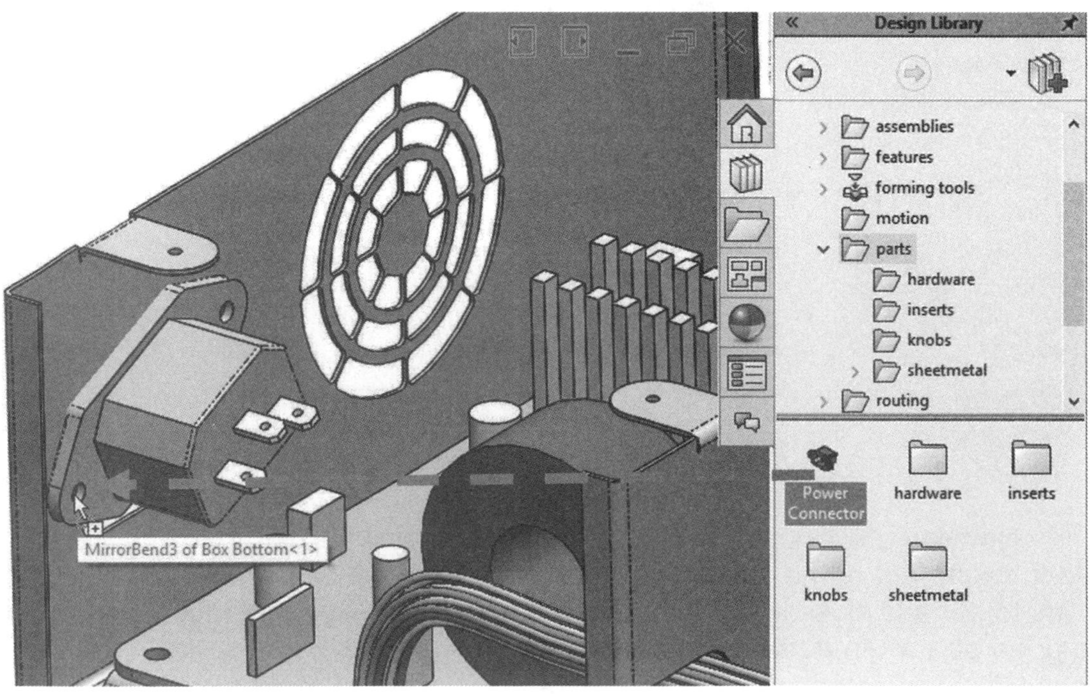

When adding library parts with mate references, if needed, we can press the Tab key once to flip the alignment and reverse the direction of the part before dropping it in place.

The last step is to add a Parallel mate to align the connector using the faces indicated.

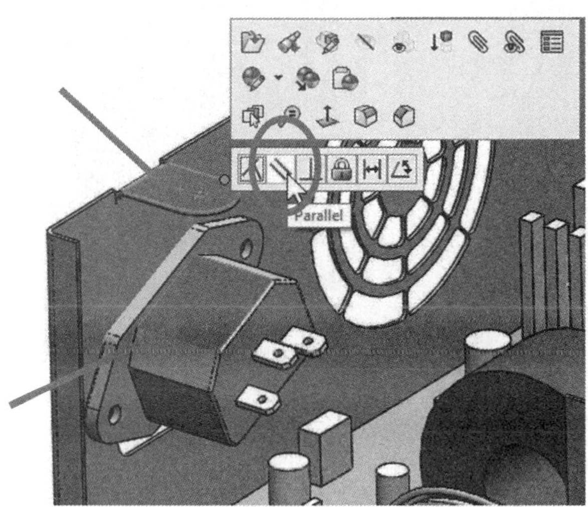

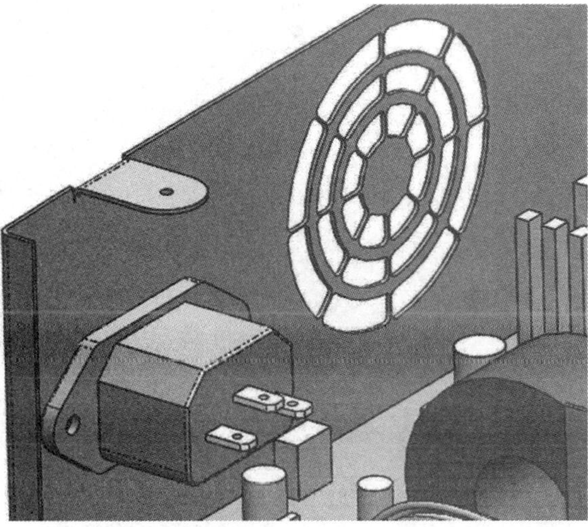

10.42. - For the fan to properly cool the power supply, we need to add a series of slots to let air into the enclosure. Open the *'Box Bottom'* part in its own window, add the following sketch using the "**Slot**" tool and make a cut using the "Link to thickness" option.

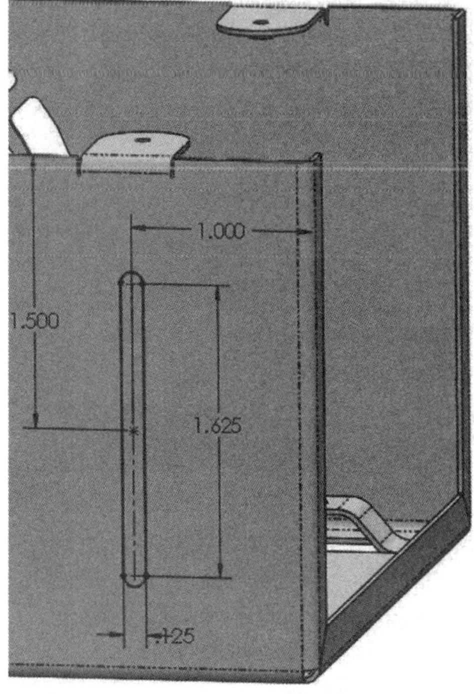

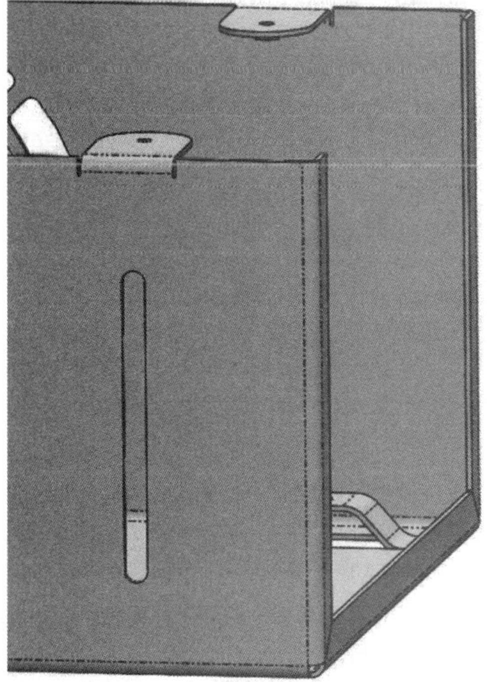

10.43. - Make a slot linear pattern with 15 copies spaced 0.25″.

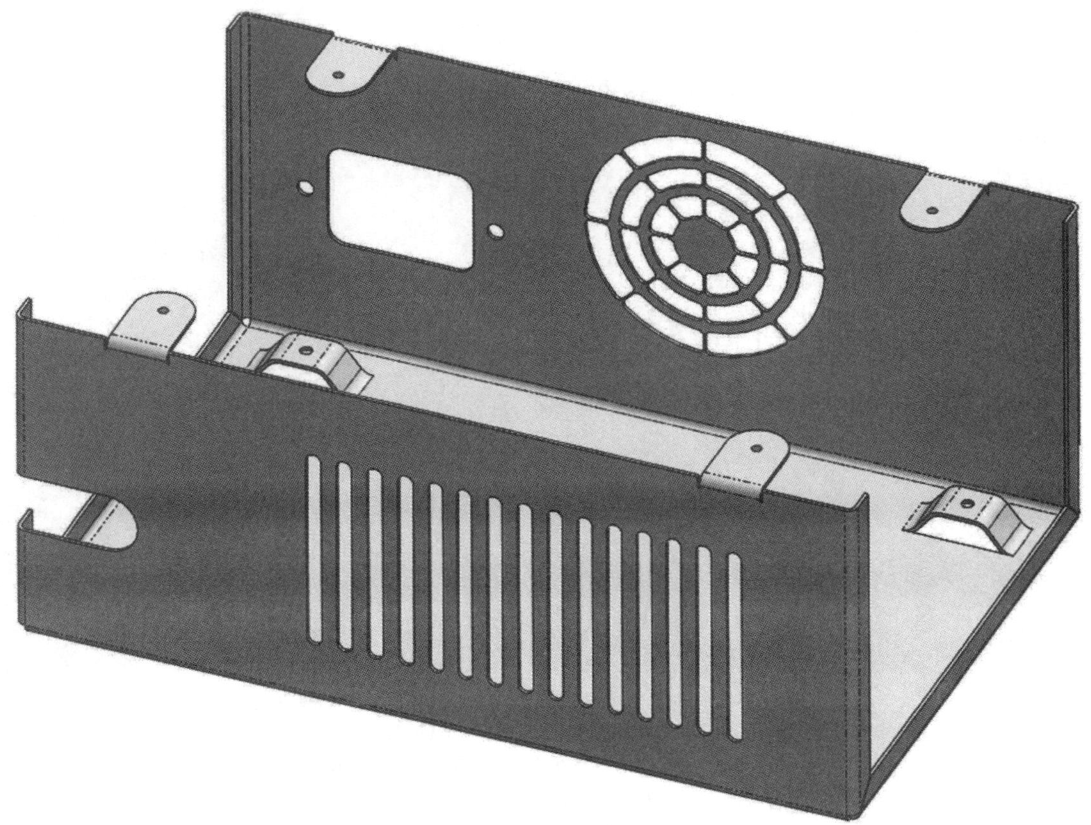

10.44. - The last operation to our sheet metal part is to eliminate all sharp edges. Select "**Break-Corner/Corner-Trim**" from the "**Corners**" drop-down command, or the menu "**Insert, Sheet Metal, Break Corner**." There are two options for the "**Break Corners**." For this part, we'll use a 0.050″ chamfer. By window-selecting the entire part, every small edge that could potentially have a sharp corner will be added to the selection list. Press OK to finish.

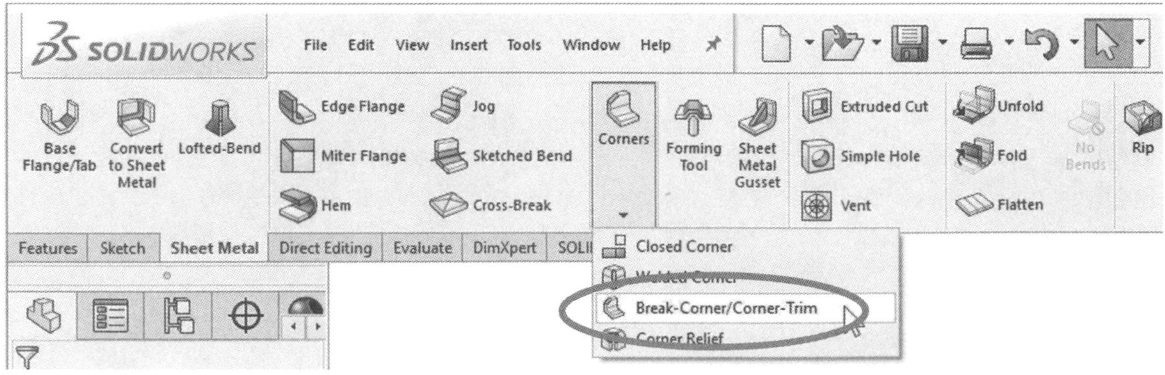

When using the "**Break Corner**" command a filter to select small edges is automatically enabled to make selection easier. Small edges can also be window-selected, or selecting a face will add all of its small edges, too.

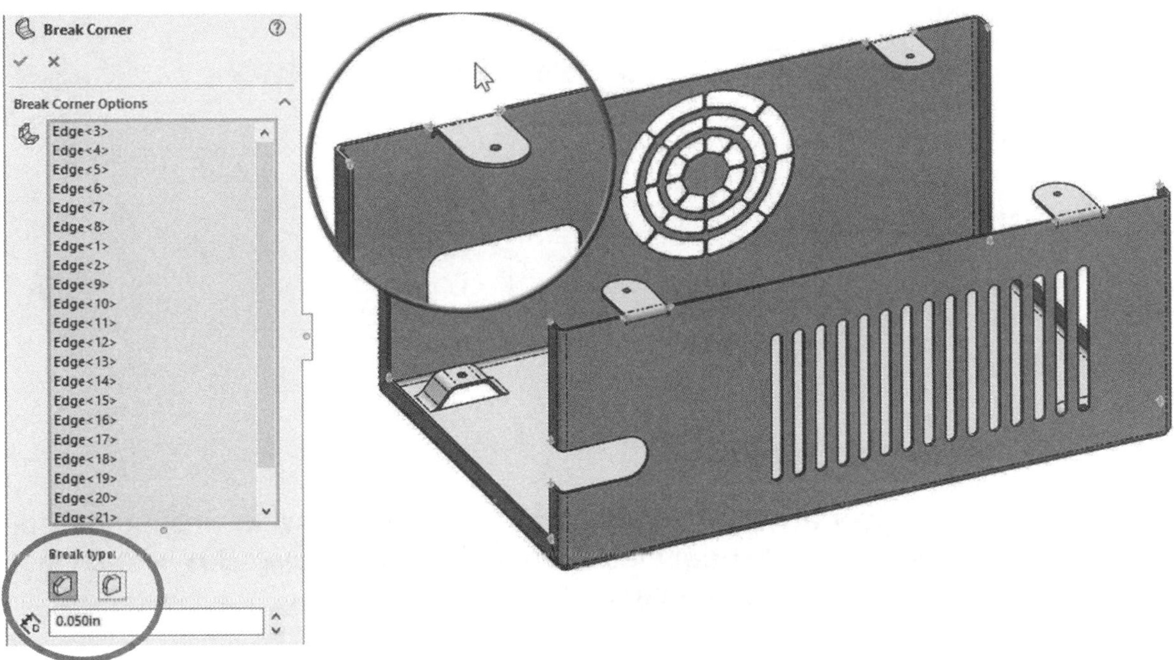

10.45. - Save the *'Box Bottom'* part. Our finished component now looks like this:

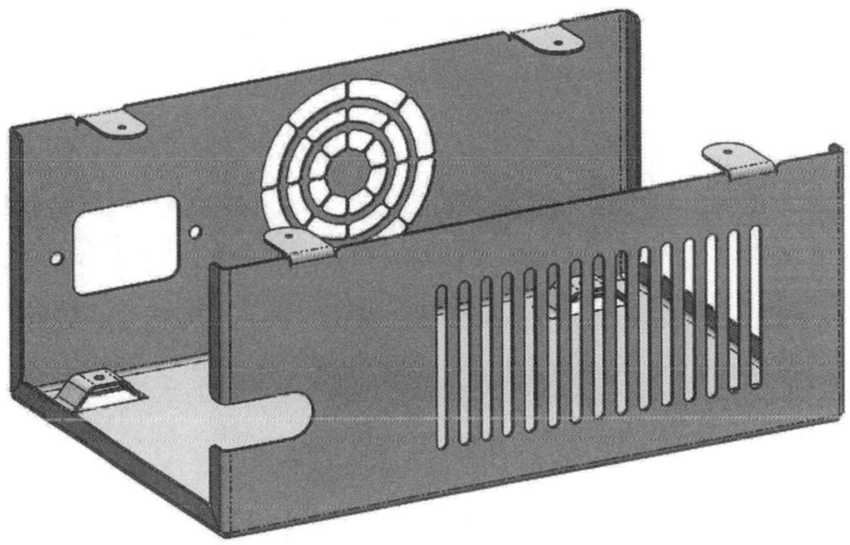

10.46. - Sheet metal components can be flattened at any time. The *"Flat-Pattern"* feature at the end of the FeatureManager is suppressed while the part is in the bent state, and when we select the "**Flatten**" command from the Sheet Metal toolbar, it is unsuppressed and the sheet metal part is unfolded.

A flat pattern is required in manufacturing to know the material size needed, how to cut and bend it. Using the correct bend allowance is extremely important; otherwise, when the calculated flat pattern is bent, the resulting part may not have the expected design dimensions. To unfold and see the flat pattern for the *'Box Bottom'* select the "**Flatten**" command from the Sheet Metal tab in the CommandManager.

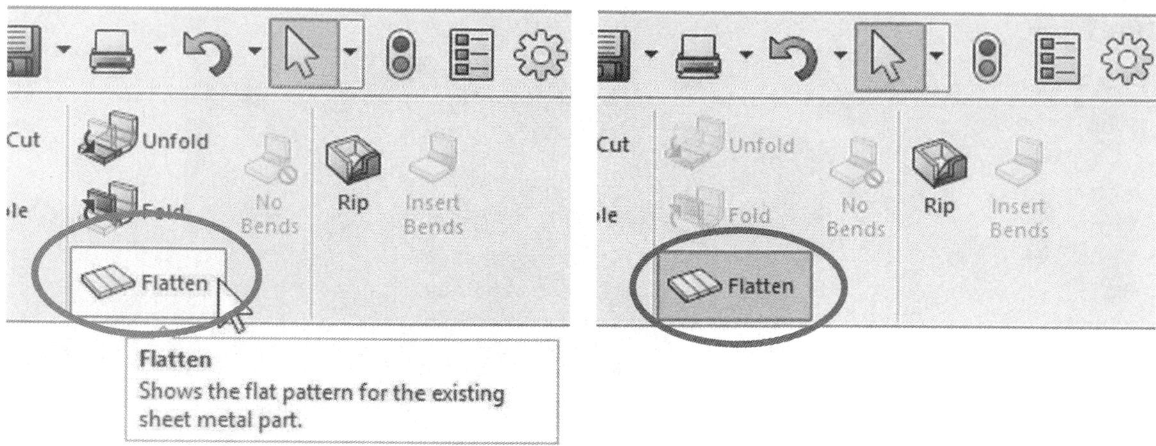

The *"Flat-Pattern"* folder and the *"Flat-Pattern(1)"* feature are now unsuppressed, and we can see the flattened sheet metal part and the *"Bounding-Box"* sketch that shows the minimum material size needed to make this part.

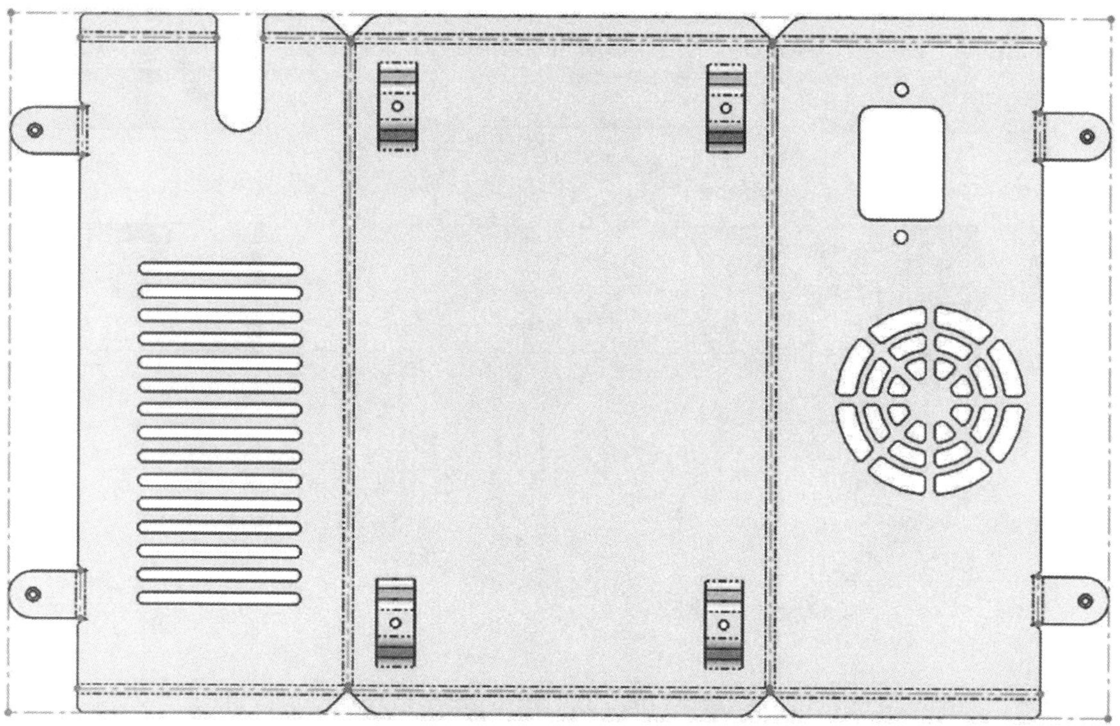

When working with sheet metal components, remember to turn off the "**Flatten**" command before returning to the assembly. If we don't, the part will be flattened in the assembly causing all kinds of errors because other parts are mated to the sheet metal, and in-context features were created in the assembly.

10.47. - Editing the *"Flat-Pattern1"* feature will show the following options:

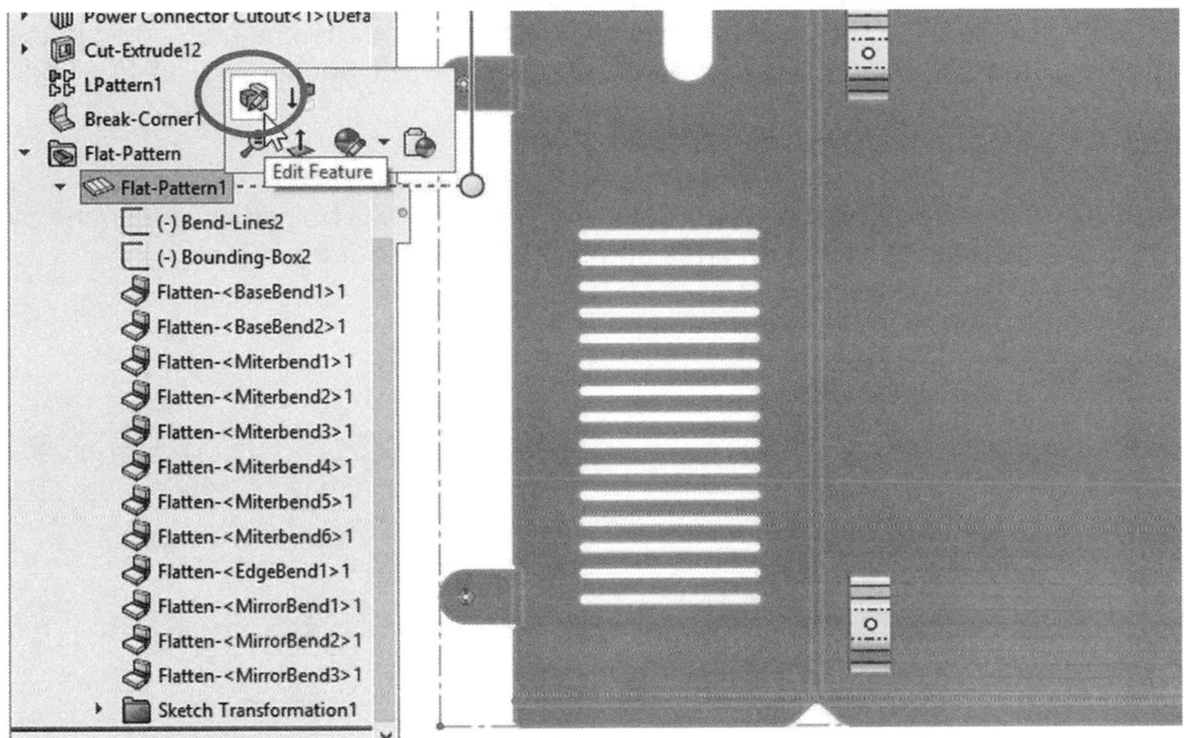

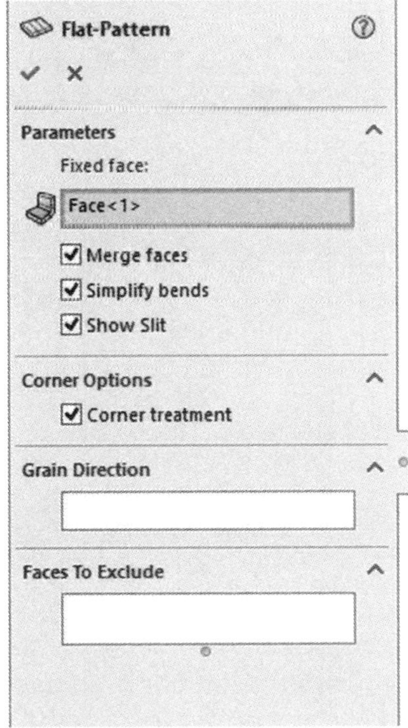

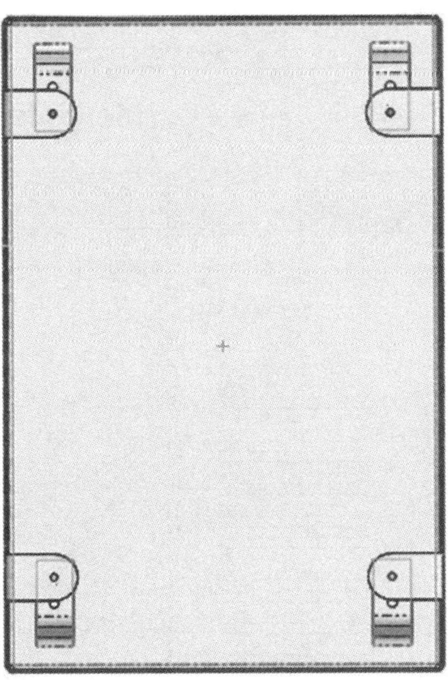

Flat Pattern Options Description

- **Merge faces**: When this option is checked, no bend regions are shown in the flat pattern, merging every flat face together. If left unchecked, all the bend regions will be displayed.

The "Merge faces" option is useful when the flat pattern is exported to a computer numerical control (CNC) software; this will help in programming by avoiding multiple broken lines and faces in every edge.

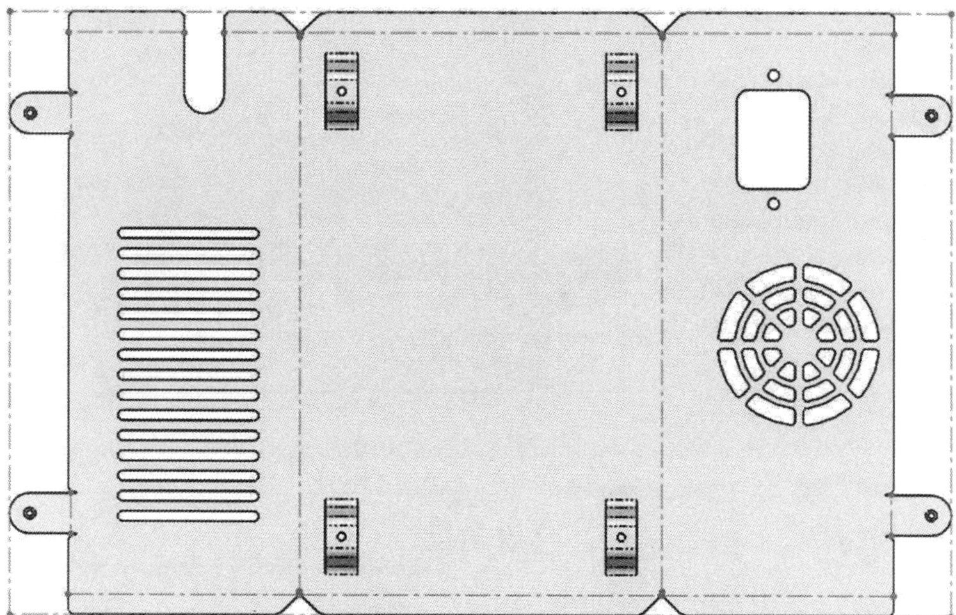

Merge faces checked

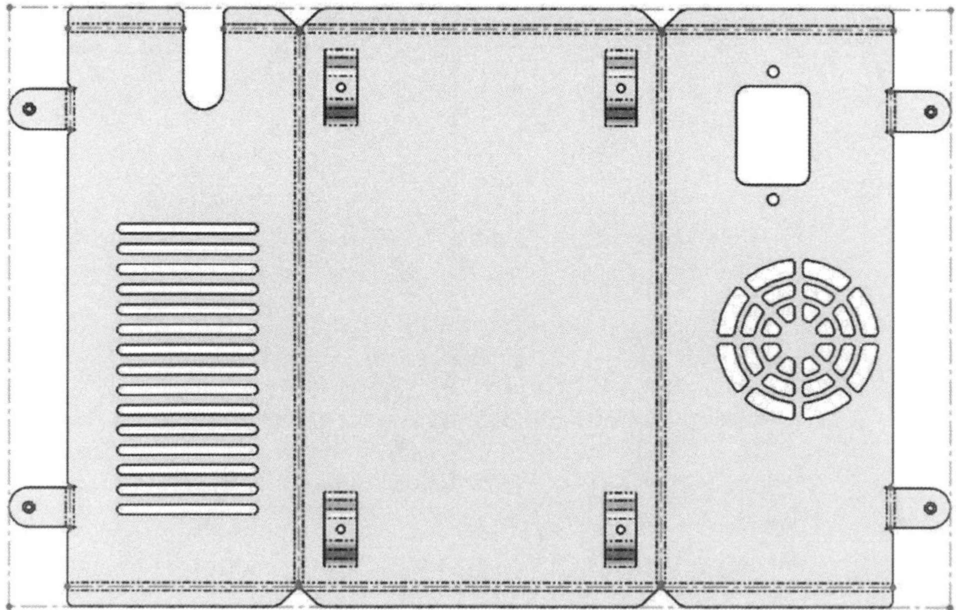

Merge faces unchecked

- **Simplify bends:** When flattening complex bends (like bends across a curved area), this option will make a straight edge in the bend region, making the manufacturing process easier.

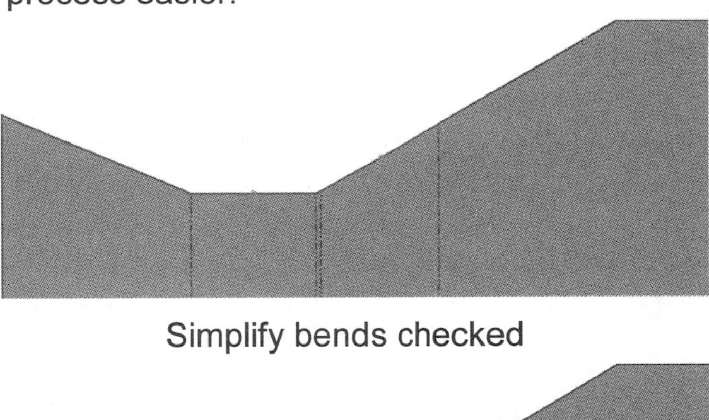

Simplify bends checked

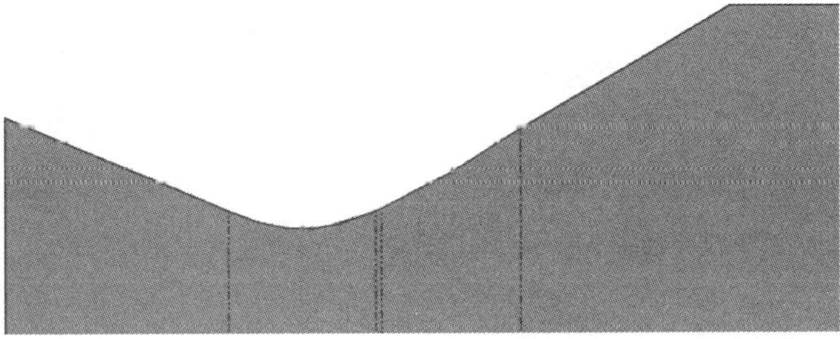

Simplify bends unchecked

- **Corner Treatment**: Will make a flat pattern easier for manufacturing to fabricate by eliminating complex cuts in bent corners.

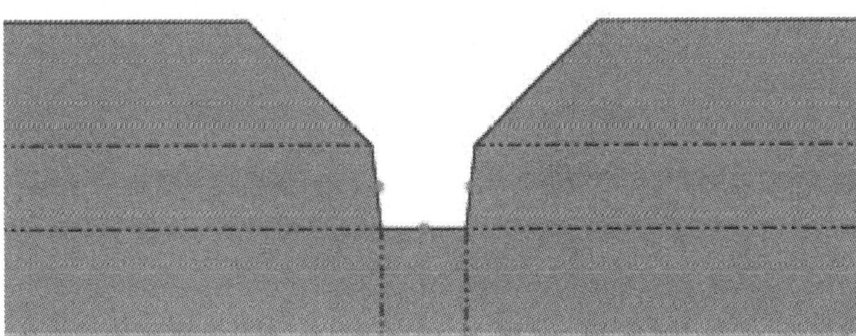

No corner treatment

With corner treatment

Notes:

Sheet Metal Box Cover

11.1. - After finishing the *'Box Bottom'* part, the second component to be made is the power supply's cover. Just as we did with the bottom part, we'll design the cover in the context of the assembly. Go to the assembly window if not already there, and make sure we are editing the assembly and not a part.

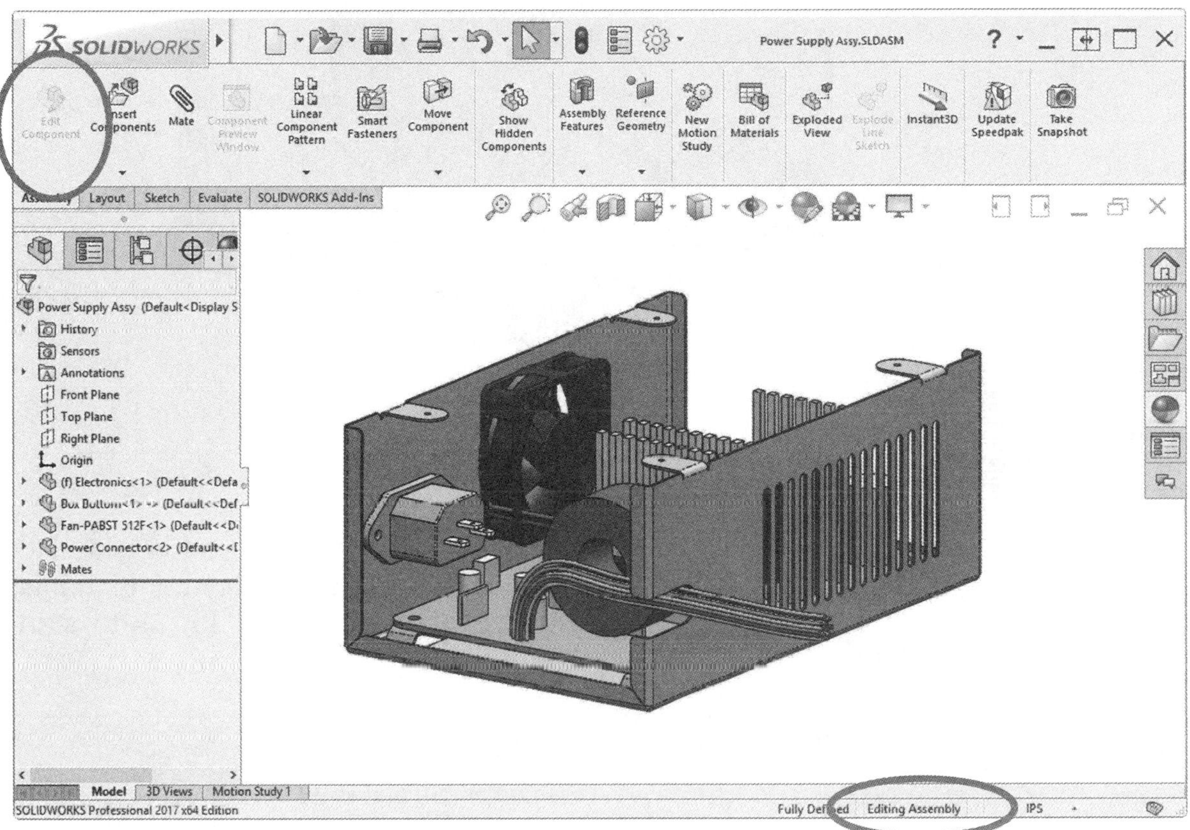

To add a new part to the assembly, select "**New Part**" from the "**Insert Components**" drop-down icon or from the menu "**Insert, Component, New Part**."

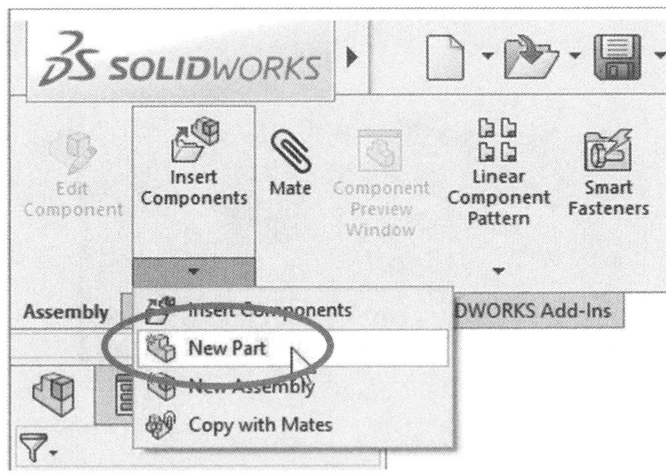

When asked to locate the new part, select the assembly's *"Front Plane"* in the FeatureManager. Remember, the *"Front Plane"* of the new part will be aligned with and constrained to the selected plane or face with an "**InPlace**" mate.

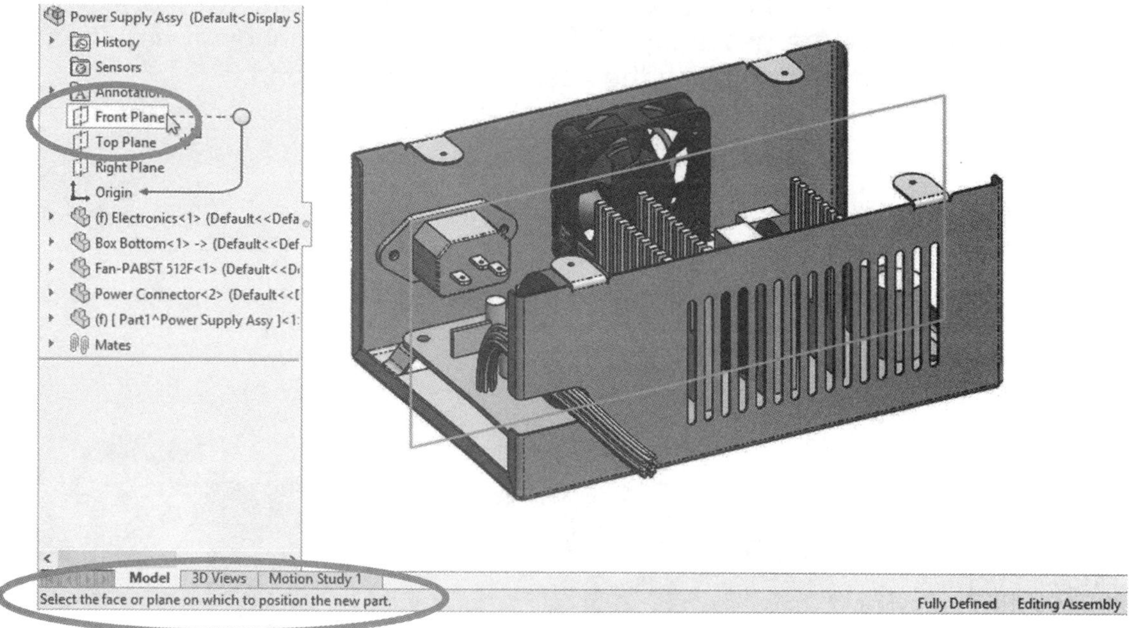

As soon as the assembly's *"Front Plane"* is selected, a new part is added to the assembly and its FeatureManager is blue. The other assembly components become transparent (if the option is set), and a sketch is ready for us to start working in the new part's *"Front Plane."*

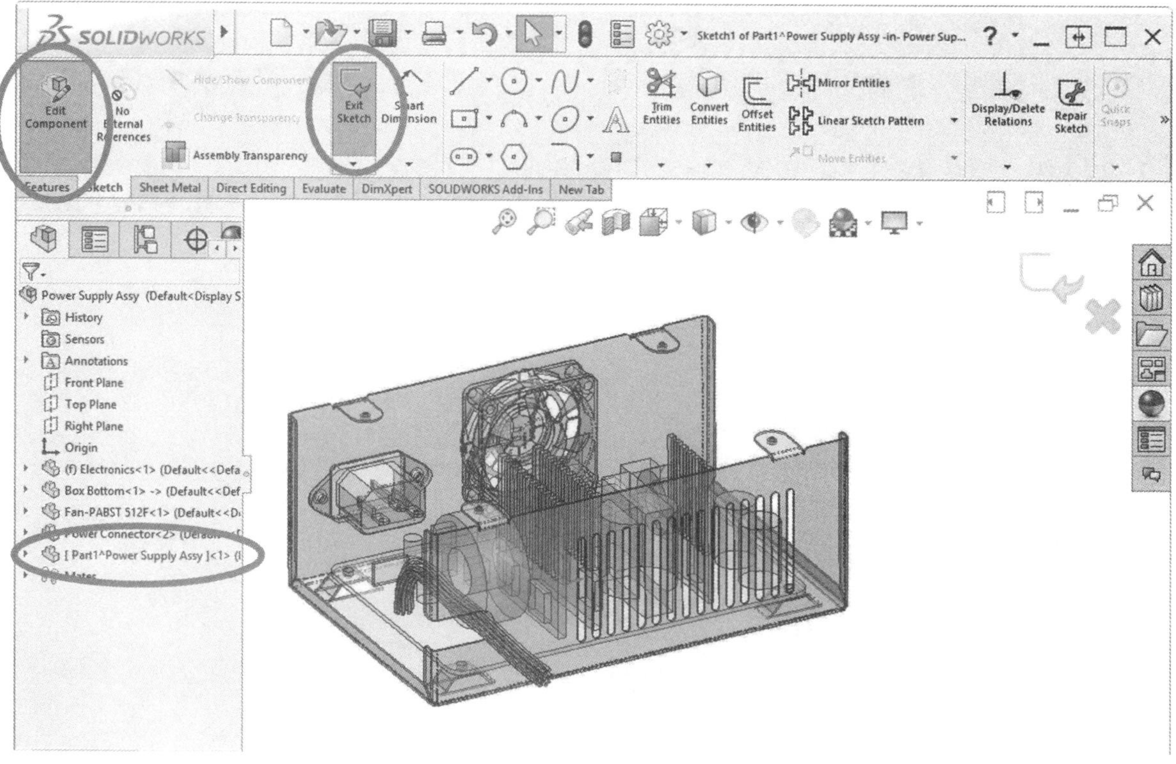

11.2. - Switch to a Front view, hide the electronics board, the fan and the power connector; they will not be needed to design the cover and keeping them visible makes the screen unnecessarily busy. Draw the next sketch using three lines, and add the necessary coincident relations to make the lines coincident to the top, bottom and side edges of the '*Box Bottom*'.

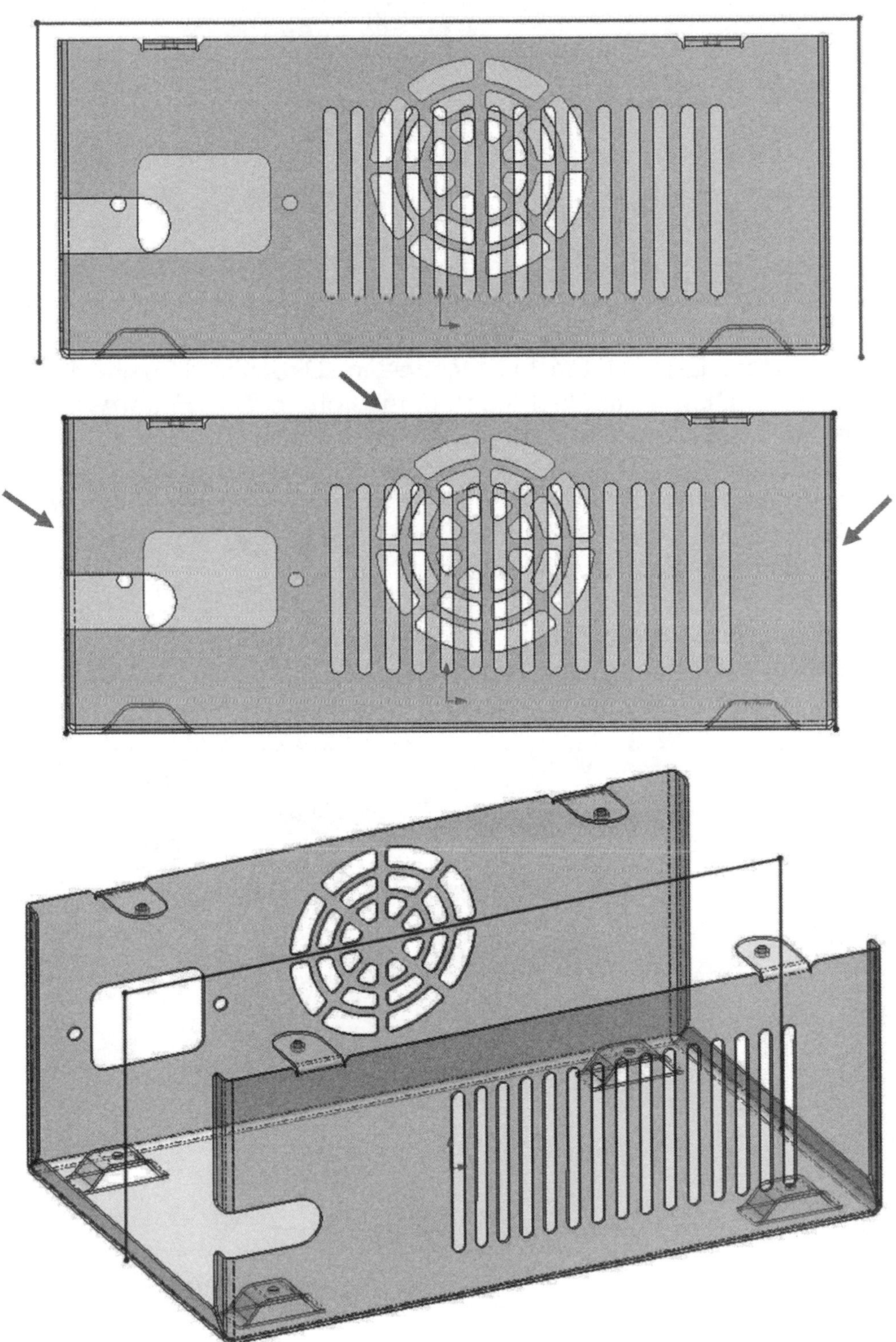

11.3. - After completing the sketch, select the "**Base-Flange**" icon from the Sheet Metal toolbar or the menu "**Insert, Sheet Metal, Base-Flange.**"

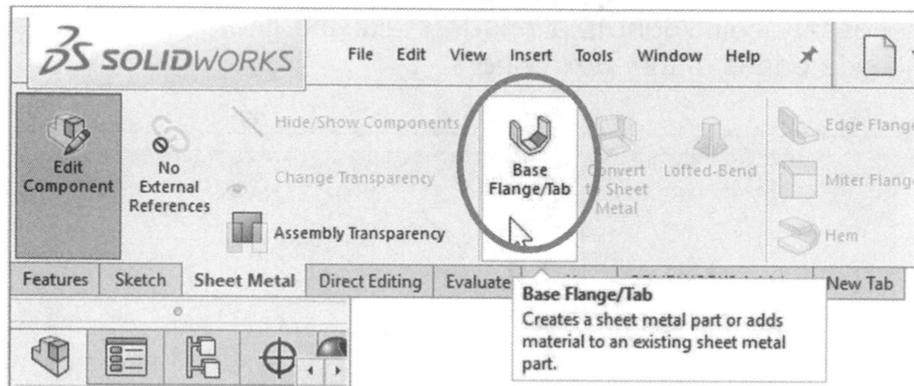

This sketch represents the inside dimensions of the sheet metal cover; therefore, we need to make the appropriate modifications to add the material *outside* the sketch. Extrude the base flange in "*Direction 1*" using the "**Up To Surface**" end condition and select the front face of the *'Box Bottom'*; in "*Direction 2*" also use "**Up To Surface**" and select the rear face of *'Box Bottom'*.

In the Sheet Metal options make the thickness equal to 0.030" with a default Bend Radius of 0.015". For "Bend Allowance" use a "*K-Factor*" of 0.5, and in the "Auto-Relief" section select Rectangular with a 0.5 ratio.

It is important to reverse the direction if needed, otherwise the sheet metal cover will overlap with the previous part.

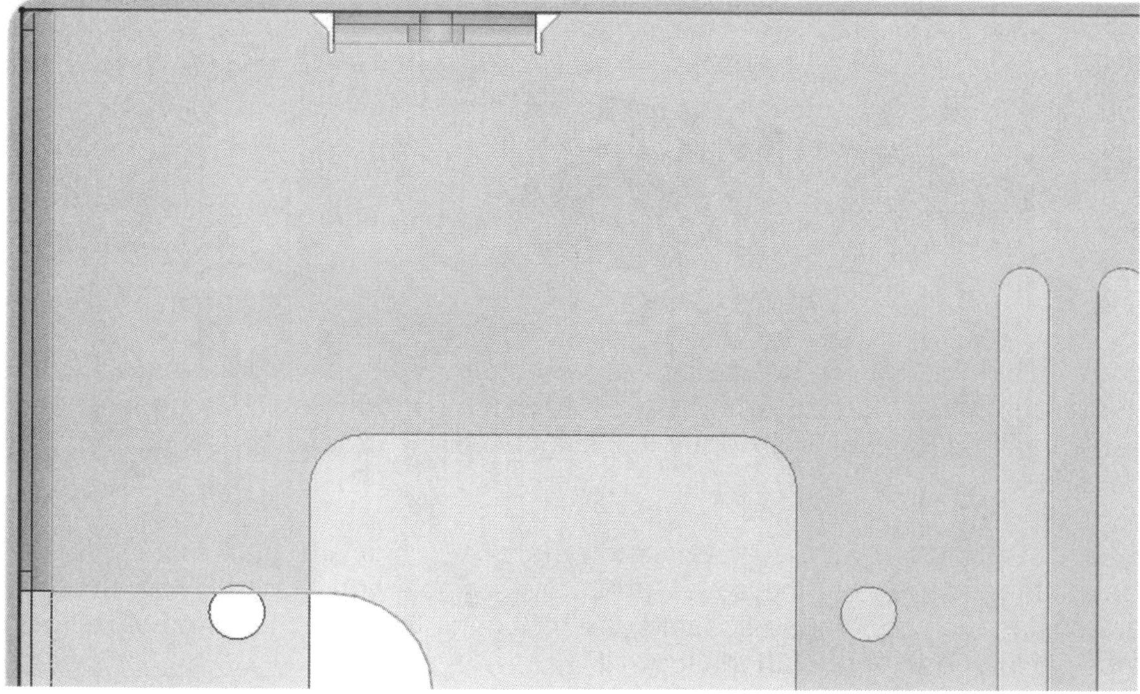

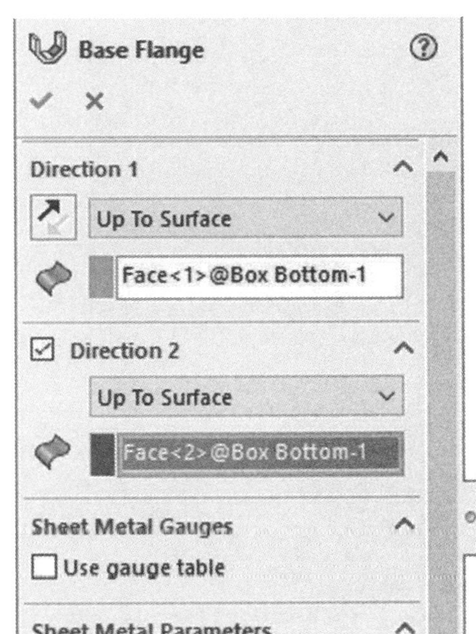

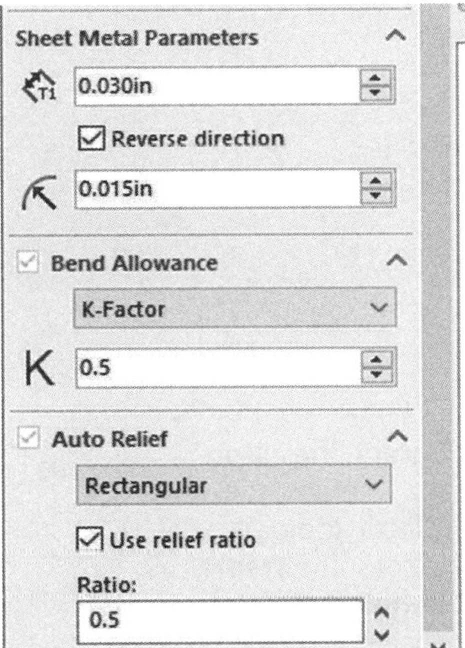

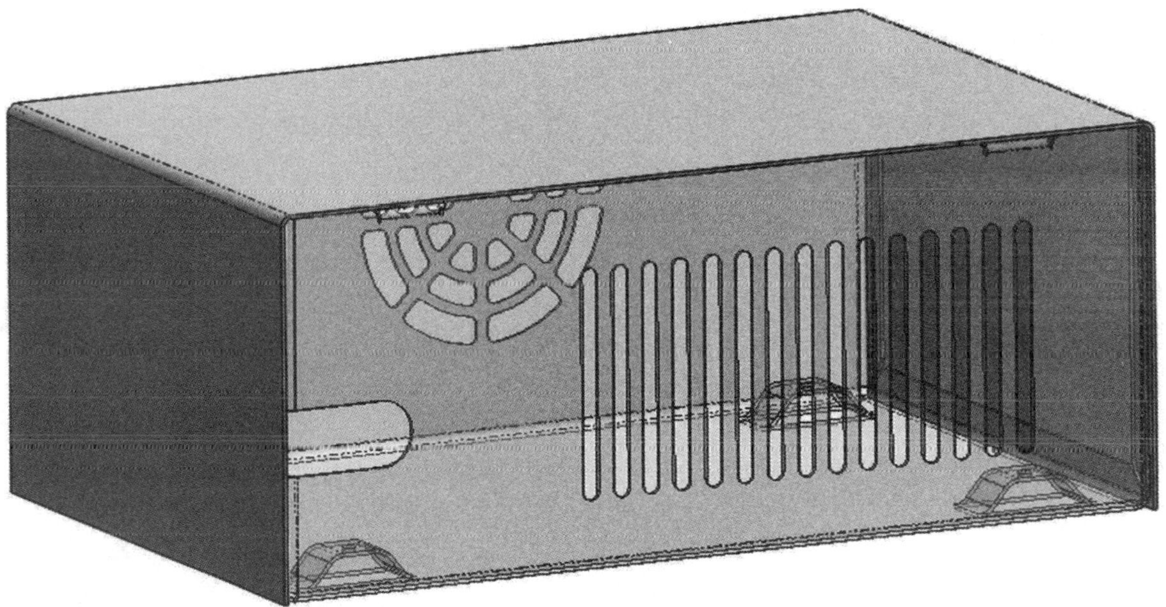

11.4. - The next step is to add jogs at the bottom of the cover to form a "Z" bend; these jogs will help us assemble the cover at the bottom. To add a jog, we need a sketch with a line; we can draw the sketch before or after selecting the "**Jog**" command.

The first jog will be done adding the sketch first and then making the jog. Select the side face of the cover, and draw a single line sketch. (Remember we are editing the cover in the assembly.) The sketch line does not need to cross the face we intend to bend, but the jog will.

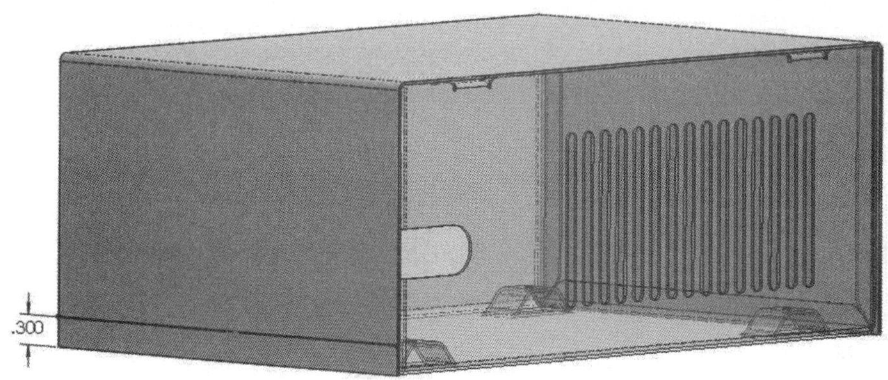

Select the "**Jog**" command from the Sheet Metal toolbar or from the menu "**Insert, Sheet Metal, Jog**."

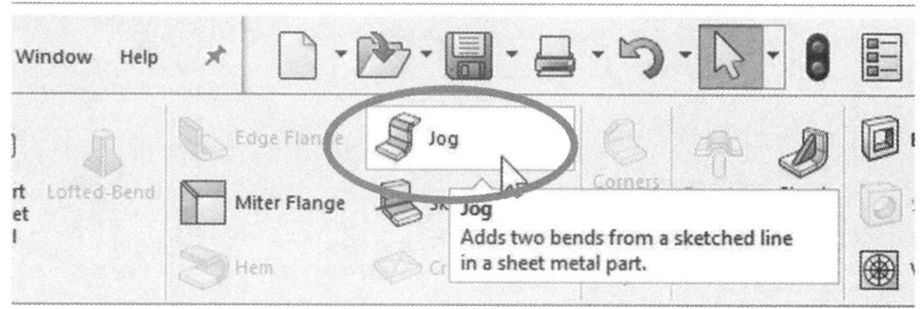

In the "**Jog**" command, for the "Fixed Face" selection box, click anywhere in the sketch face above the sketch line; this is the side of the face that will not move. Leave the option "Use default radius" checked to make the bends the same radius as the rest of the part.

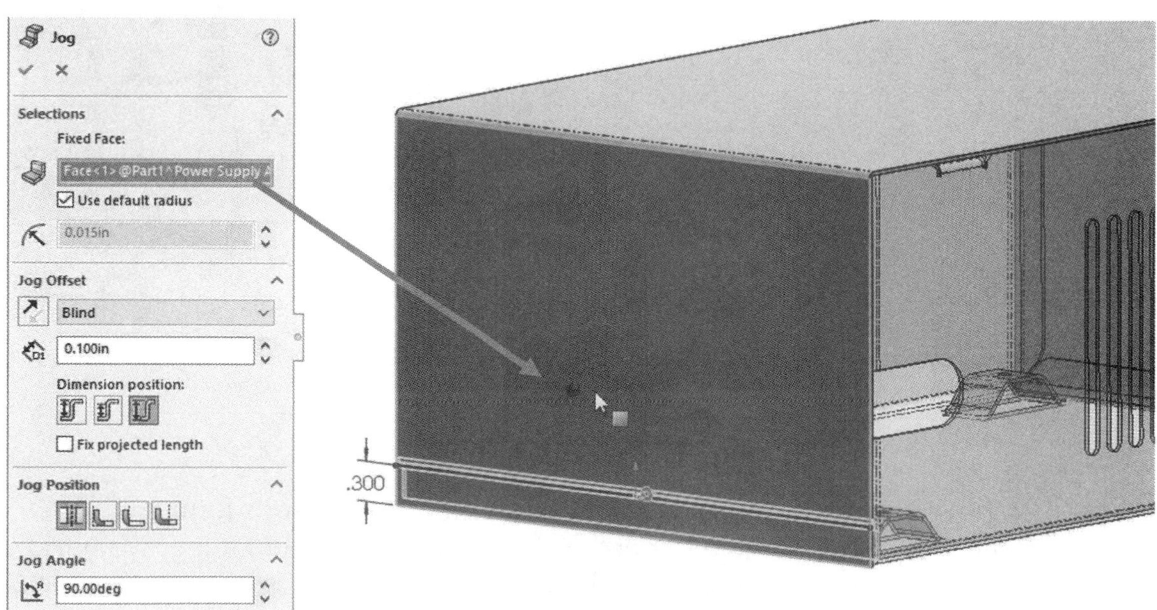

In the "Jog Offset" section we define the parameters for the jog bends. We can make the jog a fixed distance, up to a surface, vertex or offset. The purpose of the jog is to bend the cover inside the lower miter flange of the '*Base Bottom*' part to assemble them together.

When using the "Blind" end condition, there are three different ways to define how the jog distance is measured, and the icons for each type illustrate the difference between them:

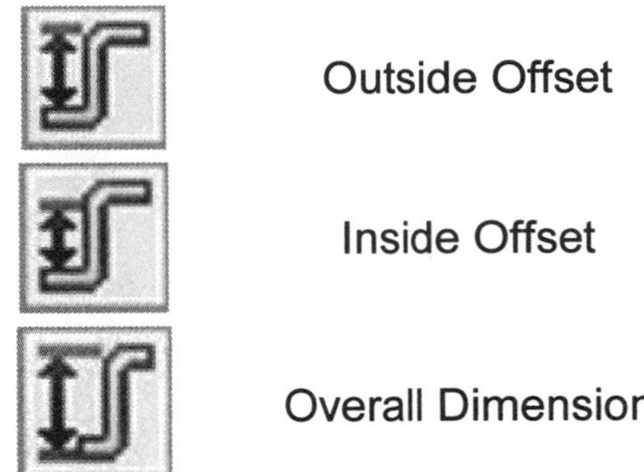

The "Fix projected length" option means that, if checked, material will be added to the flat pattern equaling the jog plus the jog bends, making the flat pattern bigger as seen below. If it's un-checked no material will be added, the flat pattern will remain the same, and the bends will be made with the existing material.

Fixed projected length option:

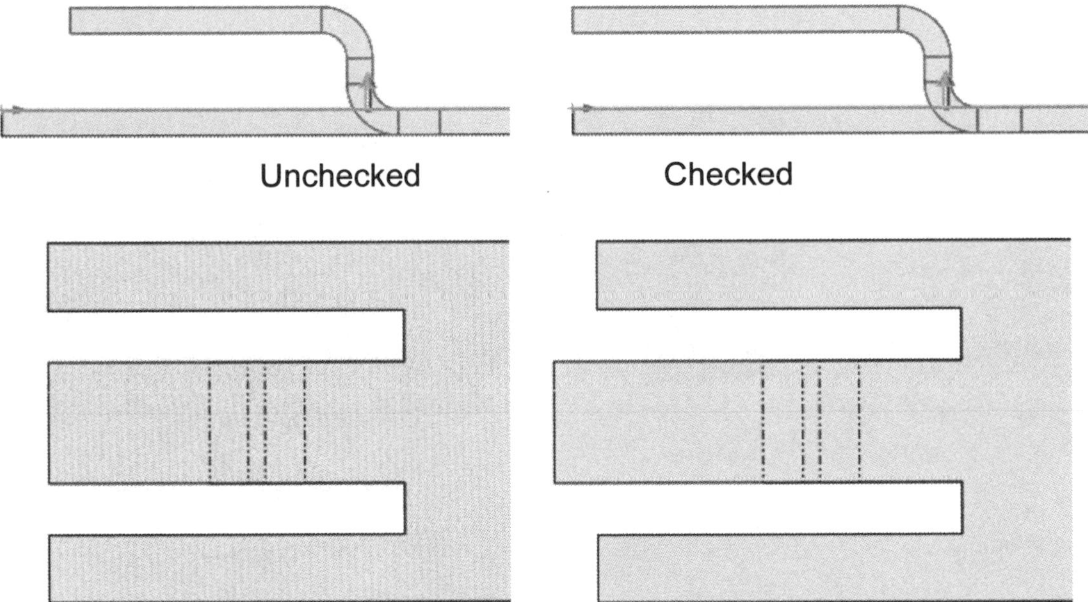

Flat pattern

The options for "Jog Position" are the same as a flange. Change to a Front view and zoom in the jog area for visibility. Select the "Blind" end condition and make it 0.1″ going into the enclosure (use reverse direction if needed); use the "Overall Dimension" and "Material Inside" options, and make the "Jog Angle" 90 degrees. Notice the option "Fix projected length" is not checked; if checked, the jog would extend and go through the bottom part. Click OK to add the jog and finish.

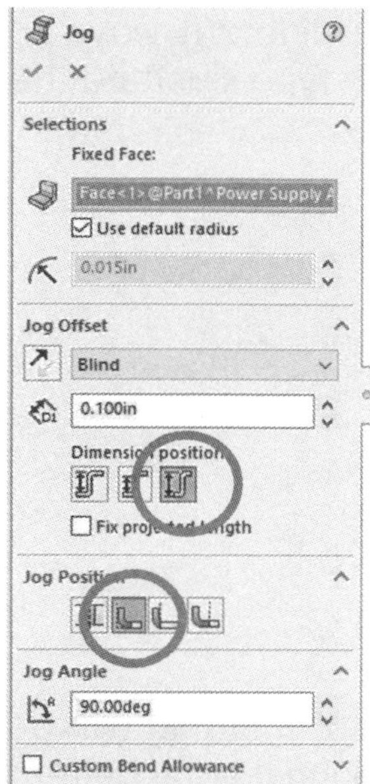

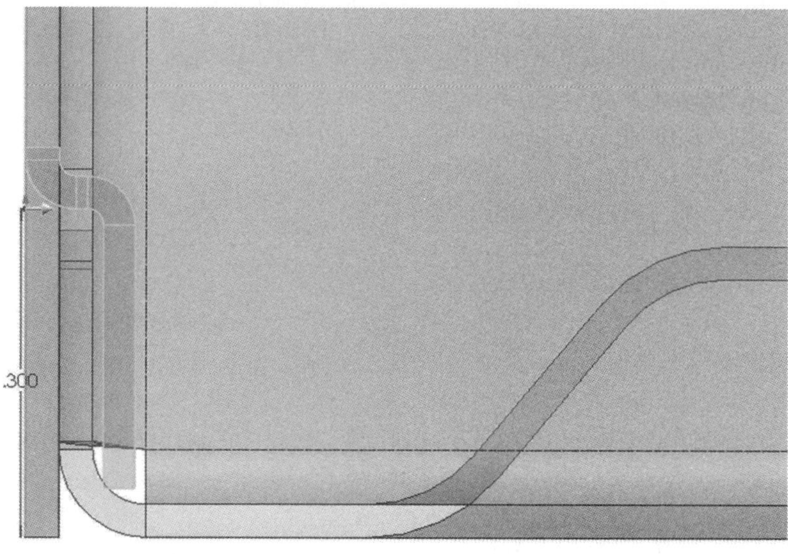

11.5. – To add the second jog in the other side, select the "**Jog**" command first, and when asked to select a face on which to draw the sketch, select the face in the opposite side of the cover. After adding the sketch line, click either "**Exit Sketch**" or "**Jog**" to continue. Use the same settings as before, also going inside.

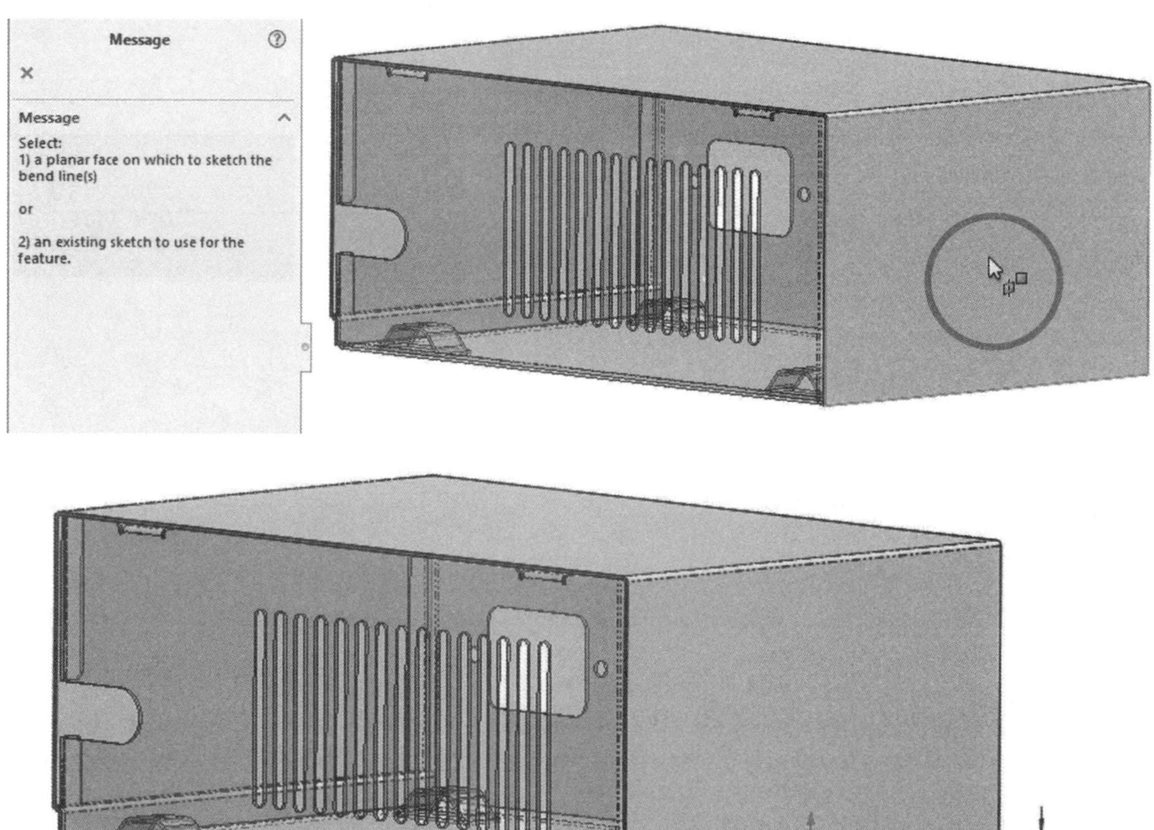

The second jog should now look like this when finished:

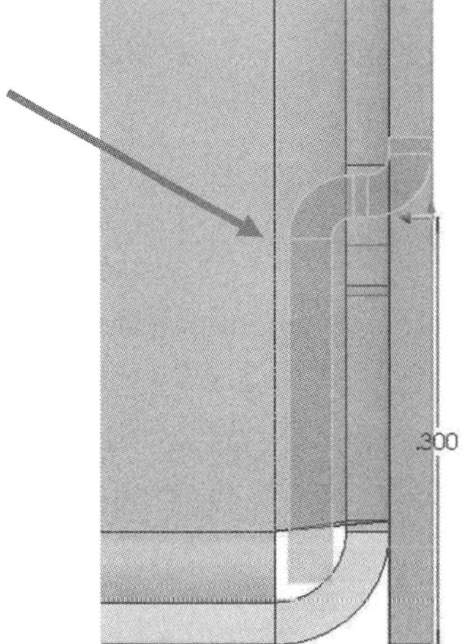

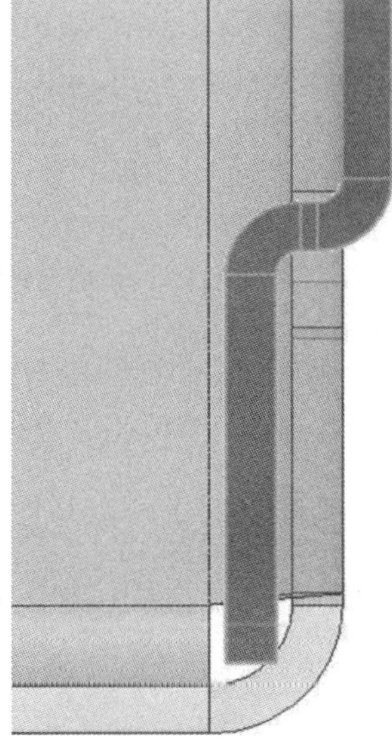

11.6. - With the jog we bent the entire length of the cover to fit behind the miter flange of the *'Box Bottom'* part, but now we have interferences at the corners. To see the overlapping areas better we need to make all components opaque. Click "**Assembly Transparency**" and select the "Opaque" option.

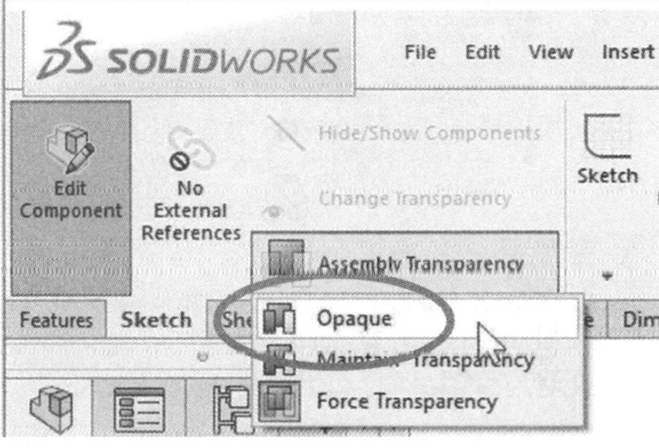

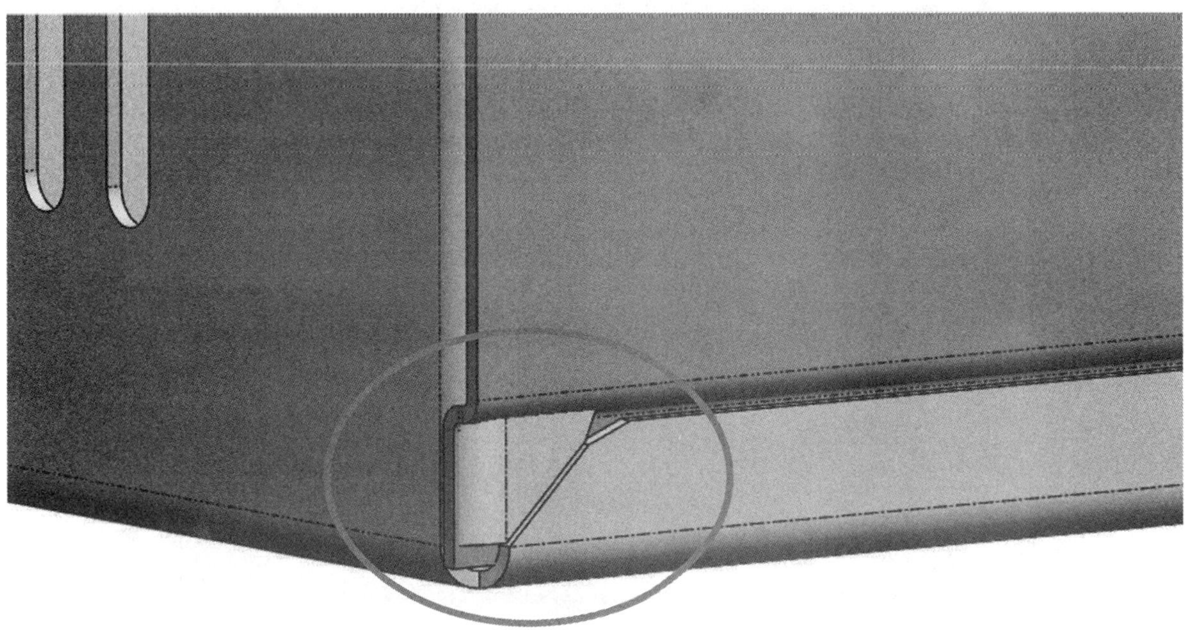

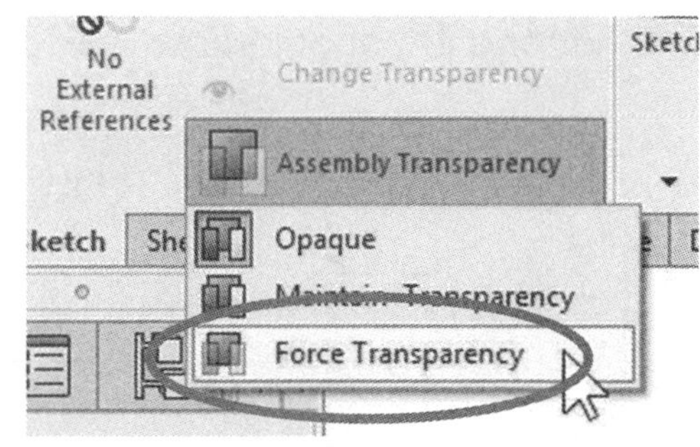

To correct this interference, we'll have to make a cut in the corners. Use the "**Assembly Transparency**" command and select "Force Transparency" to make the rest of the components transparent again. Add the following sketch on either side of the cover (left or right will work) making sure the rectangle is coincident with the jog's top bend line and the opposite corner coincident with the lower outside corner of the jog. To fully define the sketch, dimension the rectangle 0.050″ past the miter flange corner. (Note the miter corners were trimmed.)

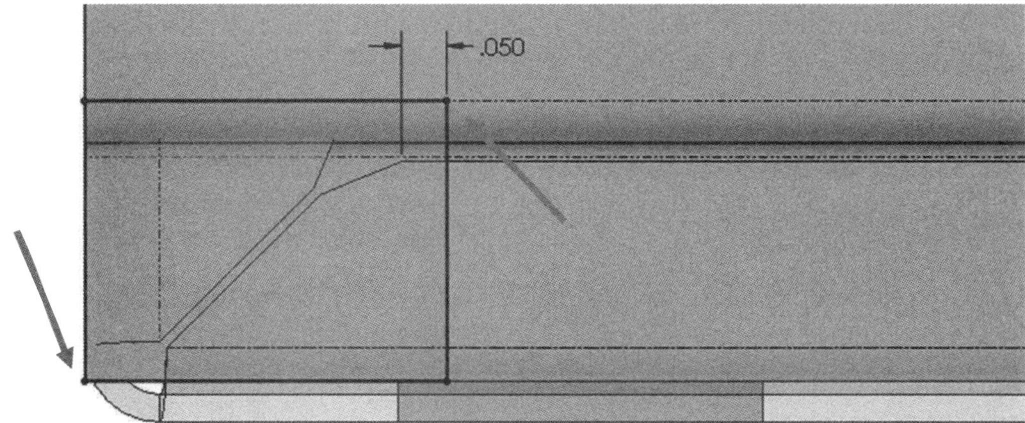

Mirror the sketch about the part's center and make a cut using the "Through All" end condition to cut the four corners at the same time (the next image is using "Shaded" mode for clarity, no edges).

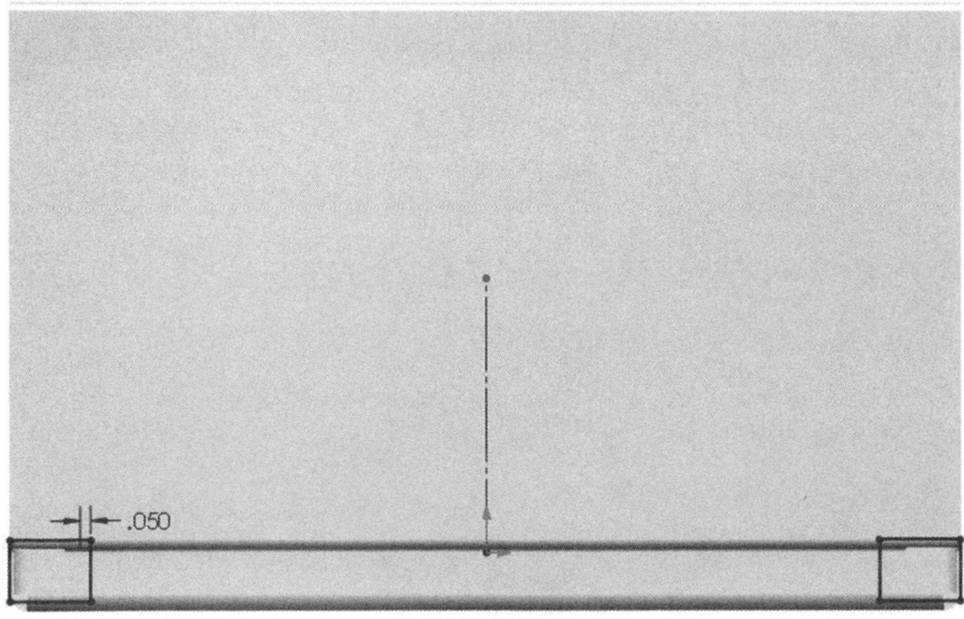

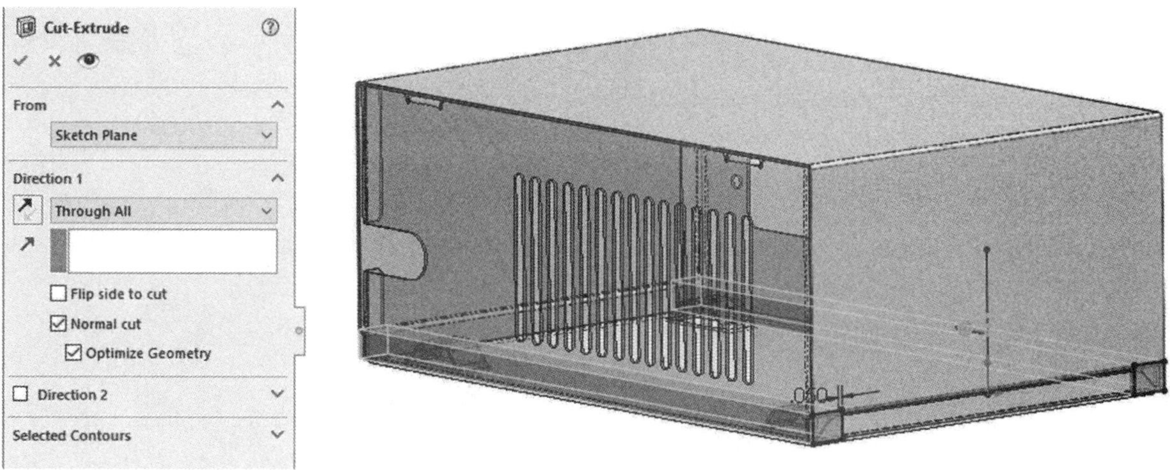

Cover part by itself to show the corner cuts

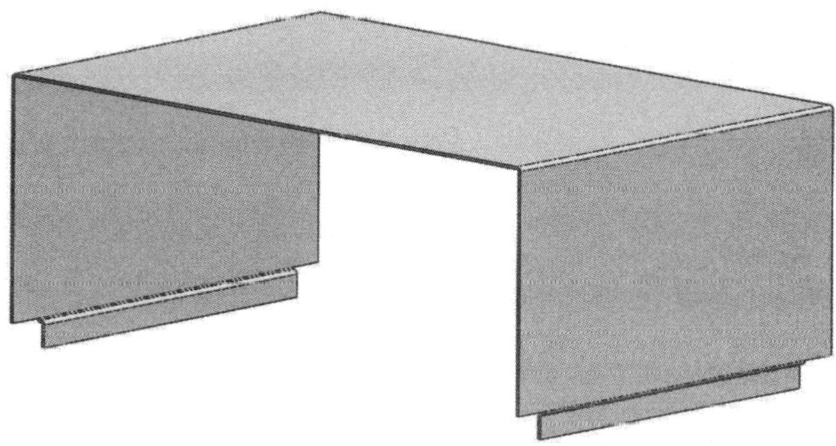

11.7. - The next step is to add holes to screw the cover to the *'Box Bottom'* part. In the assembly (remember we are still editing the cover in the assembly), switch to a top view and add a sketch on the top face of the cover. Change to "Hidden lines visible" mode and add four circles concentric to the holes in the tabs of the *'Box Bottom'*. Dimension one of the circles 0.150″ diameter and make all four circles equal. Make a cut using the "Link to thickness" option. (The "Shaded" view was added for clarity, as it's difficult to distinguish the sketch in hidden lines visible mode.)

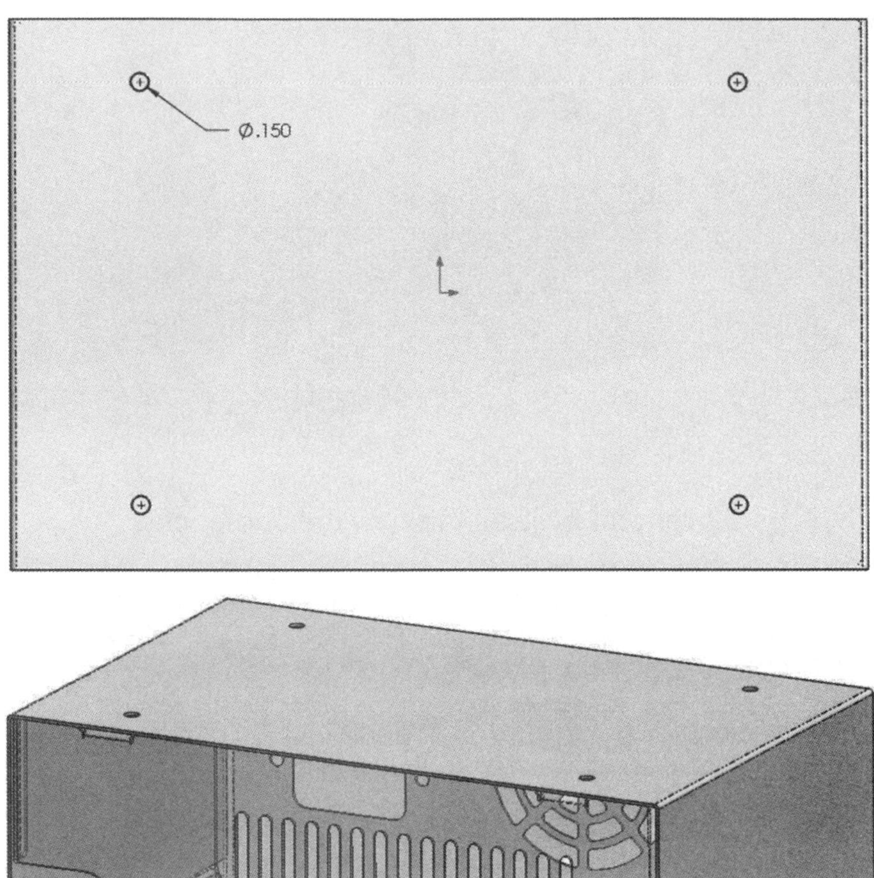

11.8. - Now we need to add ventilation slots to the cover. Add a new sketch in one side of the cover and dimension it as shown. Just as with the corner cuts, make a cut using the "Through All" end condition to cut both sides of the cover at the same time.

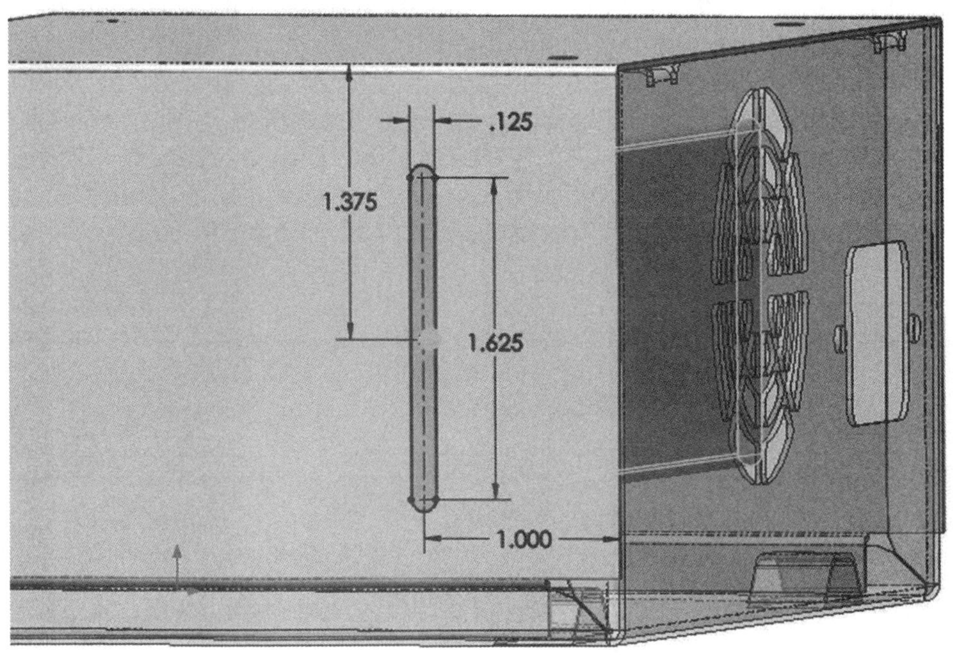

11.9. - Make a linear slot pattern with 12 instances spaced 0.250″.

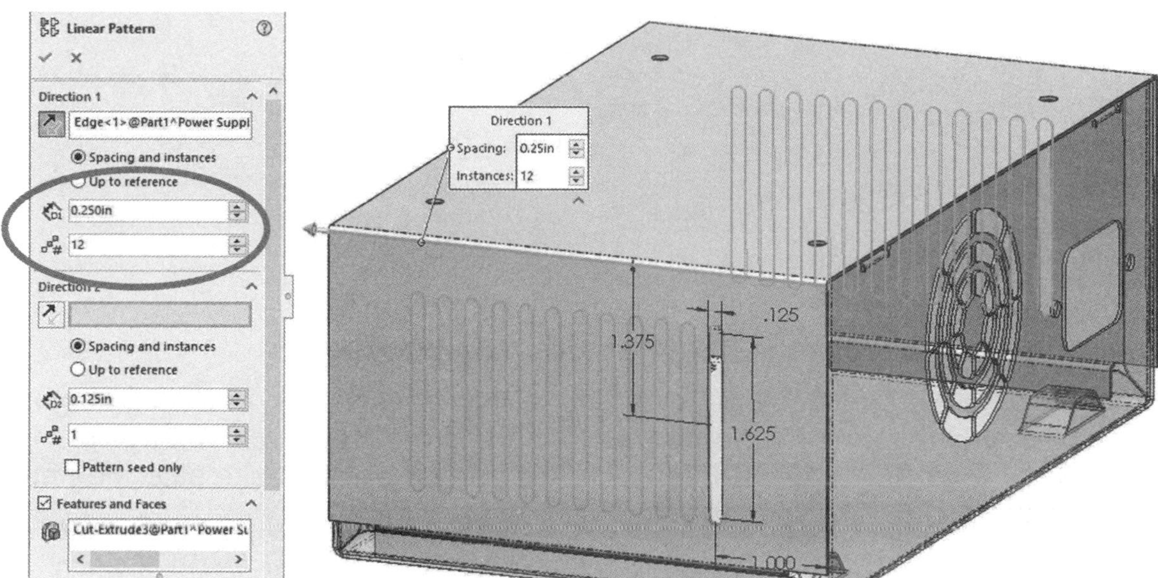

11.10. - As a final step, break all sharp corners using the "**Corners, Break-Corners**" command from the Sheet Metal toolbar. Add a 0.1″ chamfer break. Either select the sides and jog faces or window select the entire part to add all corners. Click OK to continue.

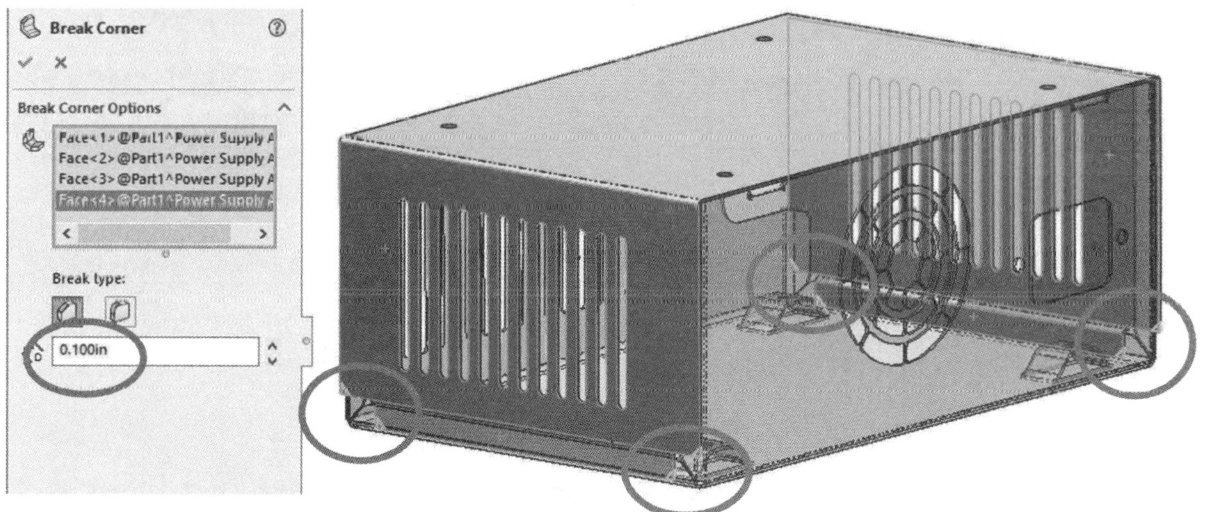

11.11. - Now the part is finished and we need to *externalize* it from the assembly. All this time the part has been internal, meaning it only exists in the assembly and there is no *'file'* we can reference for a drawing or other assemblies.

Stop editing the cover in context of the assembly and return to editing the assembly. Turn off the "**Edit Component**" button in the CommandManager, *or* right-mouse-click in the graphics area and select "**Edit assembly: Power Supply Assy**."

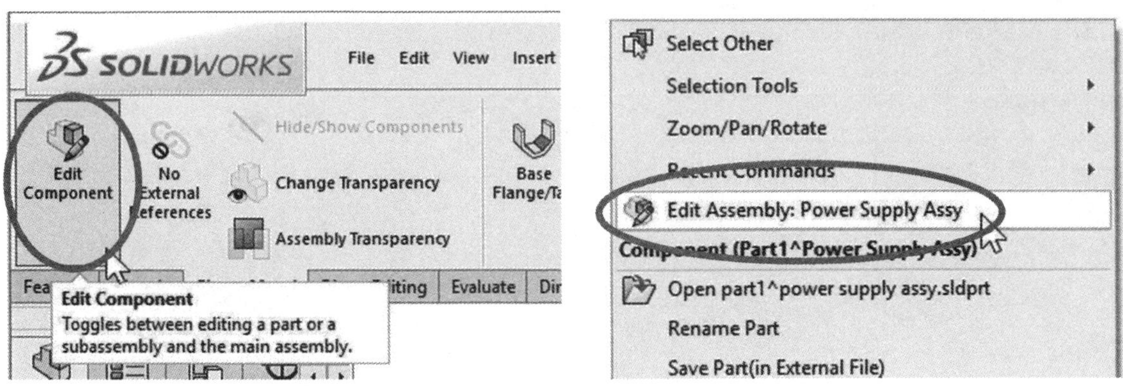

Rename the part as *'Box Cover'* with a slow double-click, or from the right-mouse-click menu select "**Rename Part**." Then right-mouse-click in it again and select "**Save Part (in External File)**" from the pop-up menu. Make sure the path is correct and click **OK** to save the file.

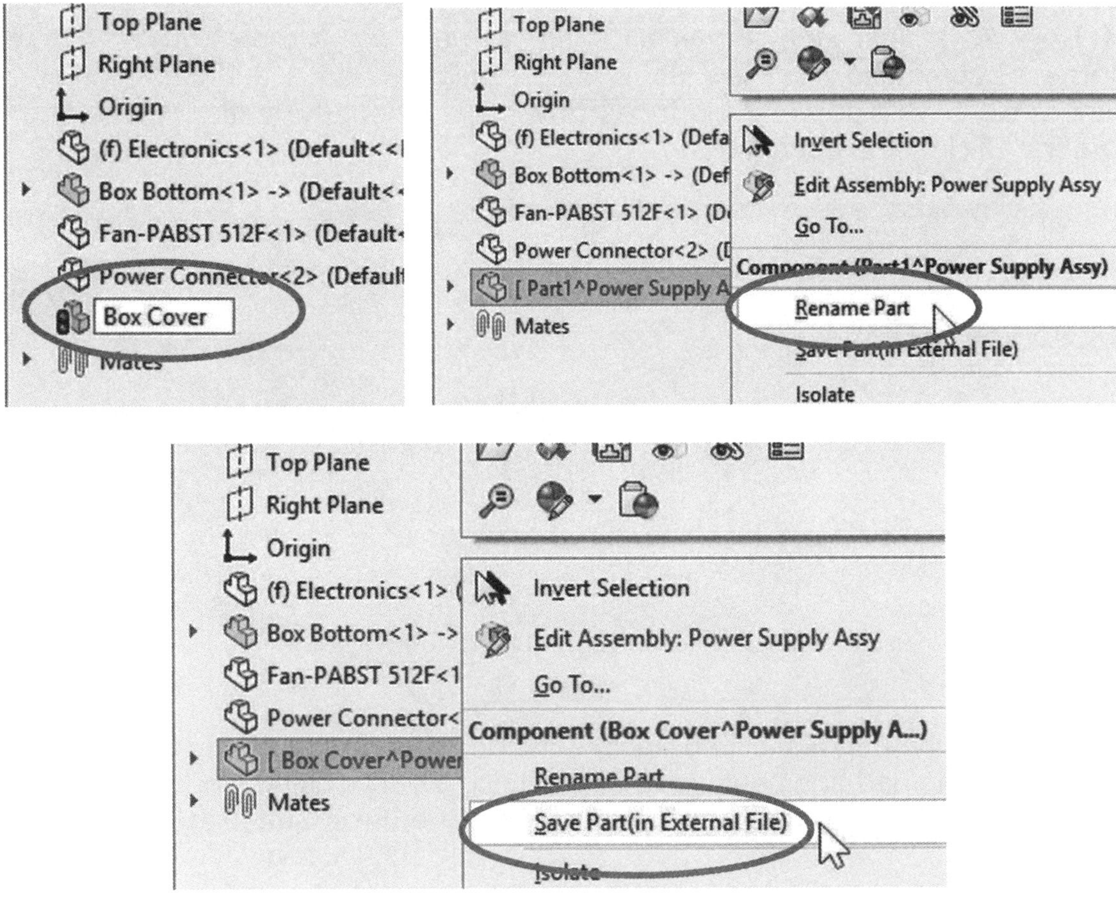

11.12. - Open the *'Box Cover'* part in its own window and view the flat pattern by selecting the "**Flatten**" command in the Sheet Metal toolbar. After reviewing the flat pattern, turn it off (re-fold the part) and return to the assembly.

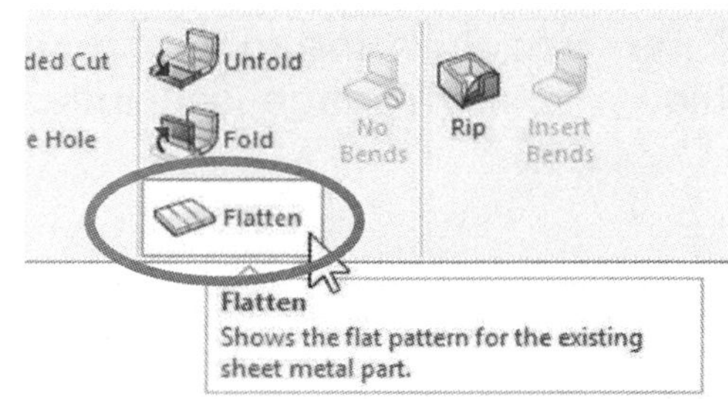

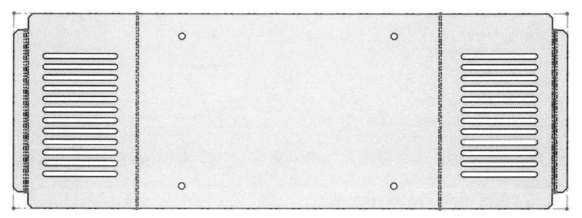

11.13. - Make all the components visible, save the assembly, and if wanted, add an exploded view to finish. We'll work on the sheet metal drawings next.

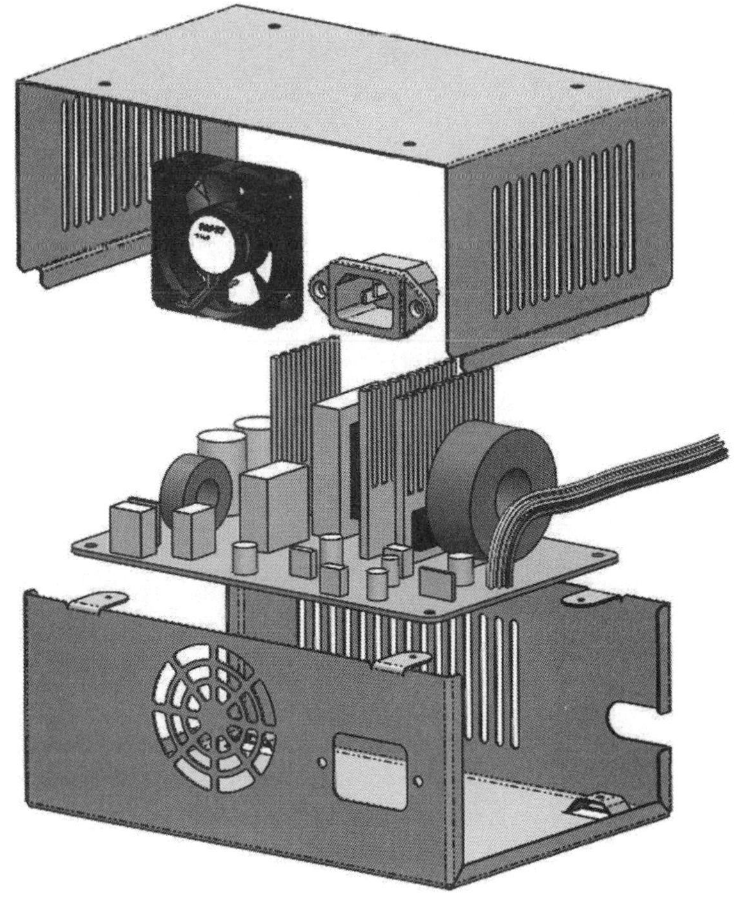

Exercise: Build the following sheet metal part using the information provided. A high-resolution image is included with the book files.

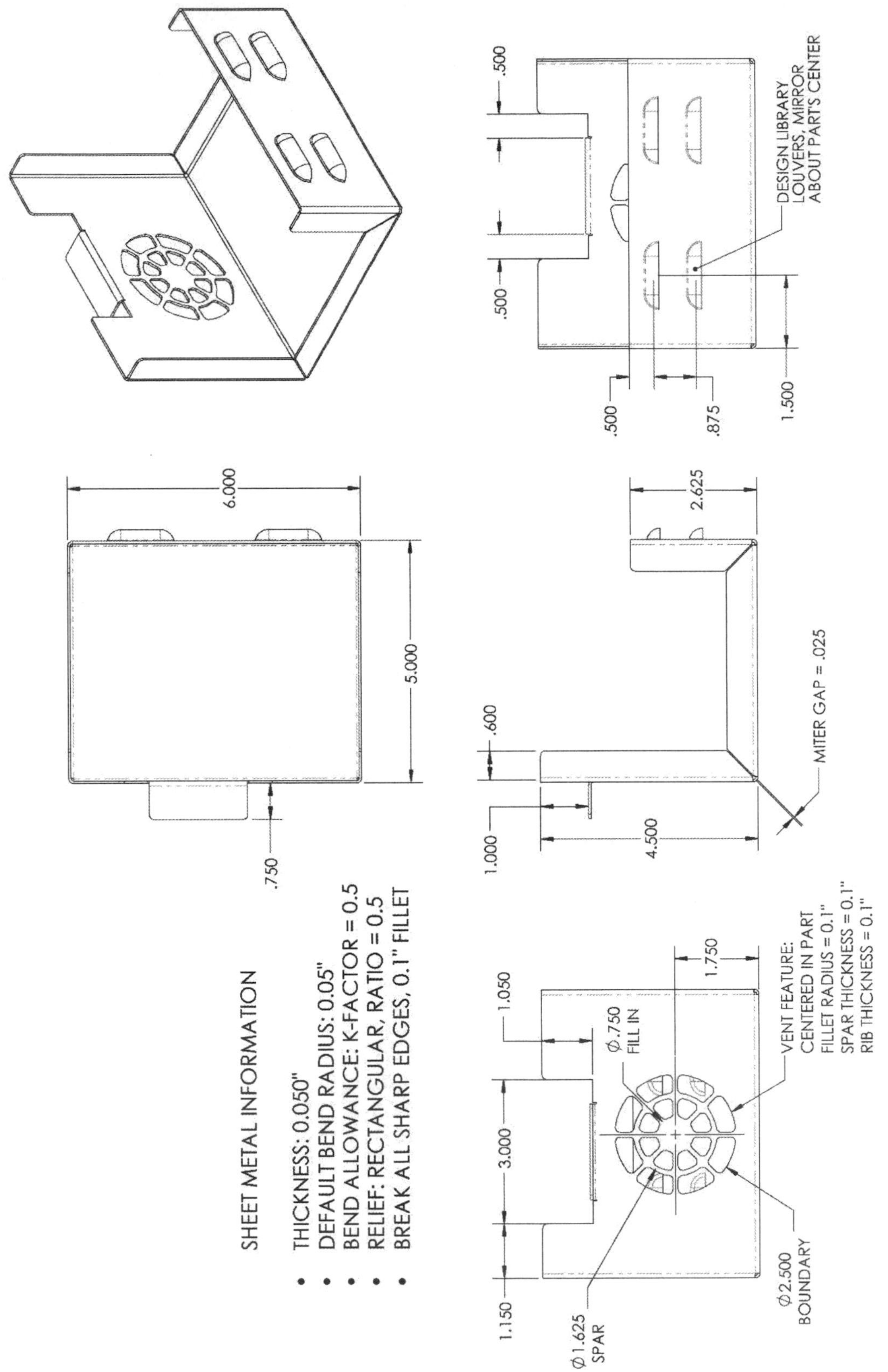

TIPS:

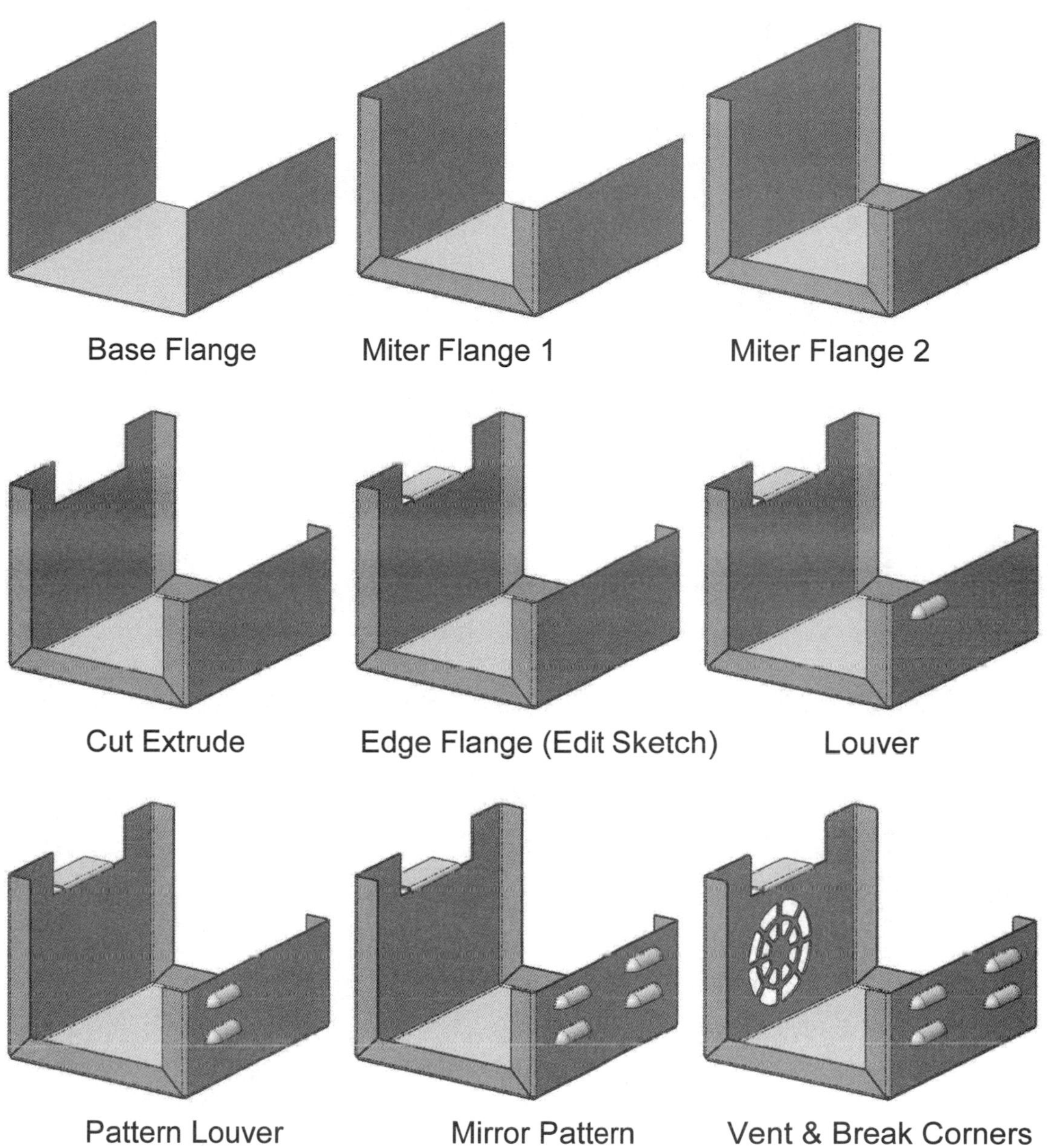
Base Flange
Miter Flange 1
Miter Flange 2
Cut Extrude
Edge Flange (Edit Sketch)
Louver
Pattern Louver
Mirror Pattern
Vent & Break Corners

Notes:

Sheet Metal Drawings

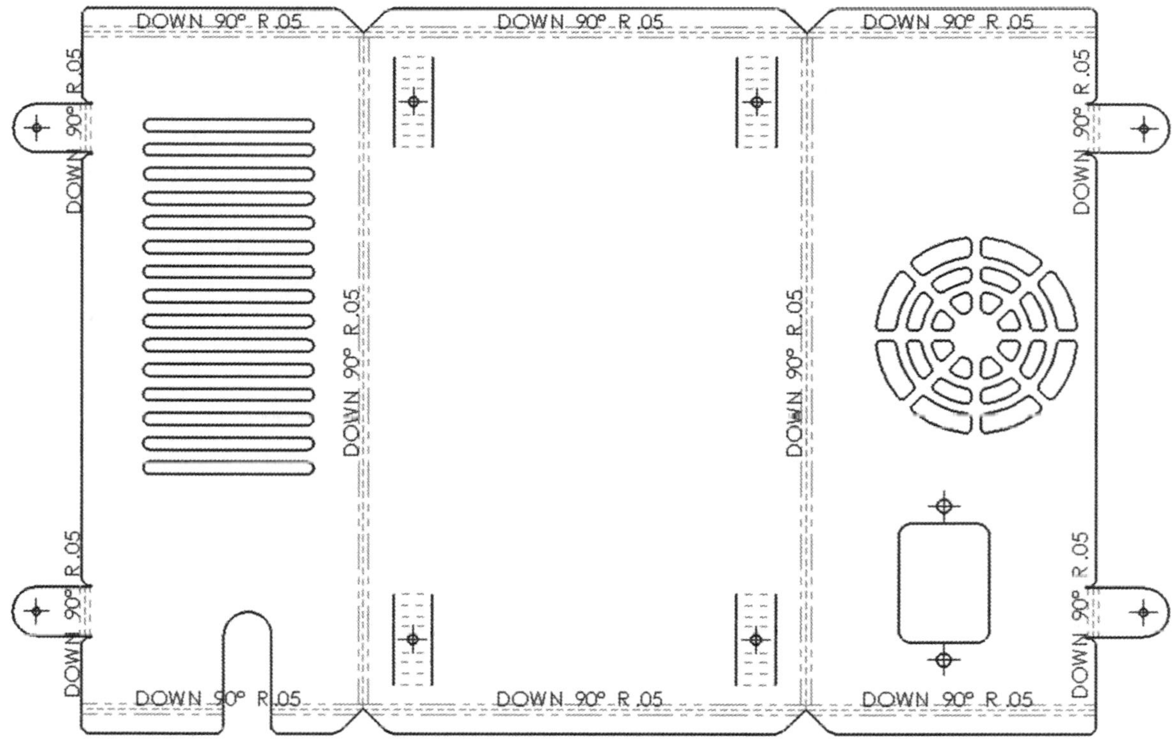

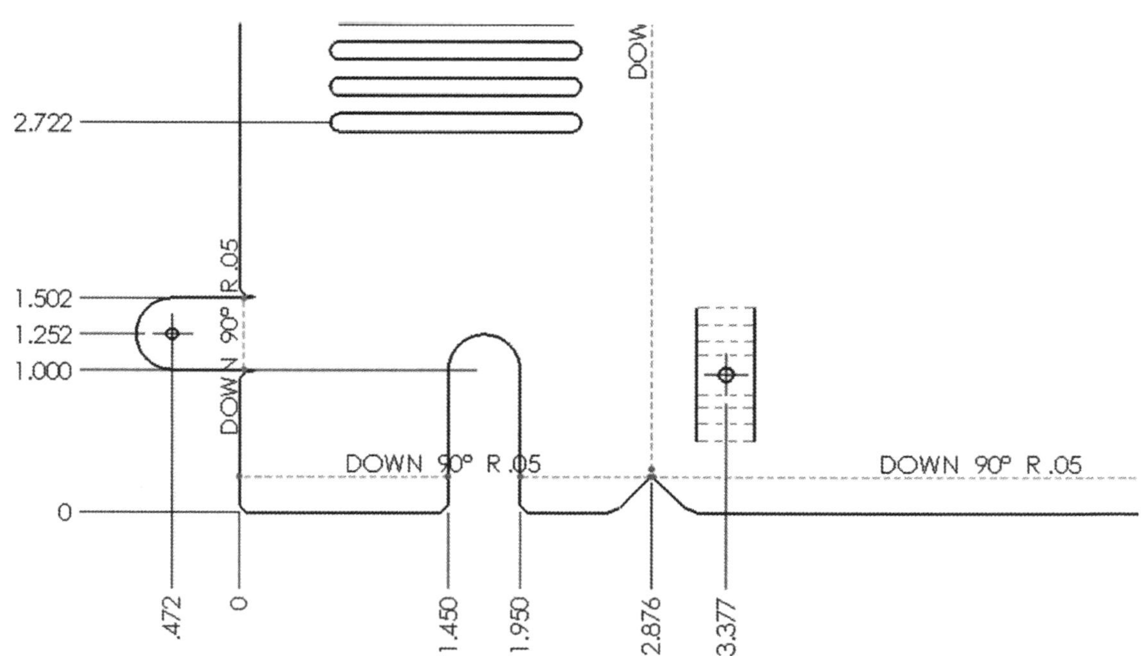

Notes:

12.1. - When making a sheet metal part's drawing, the first thing to notice is that additionally to the standard views, there is a new option to add a view of the part's "**Flat Pattern**".

Open the *'Box Bottom'* part and make a new drawing using a "B-Landscape" drawing template without a sheet format. From the "View Palette" drag the "Flat pattern" view into the drawing.

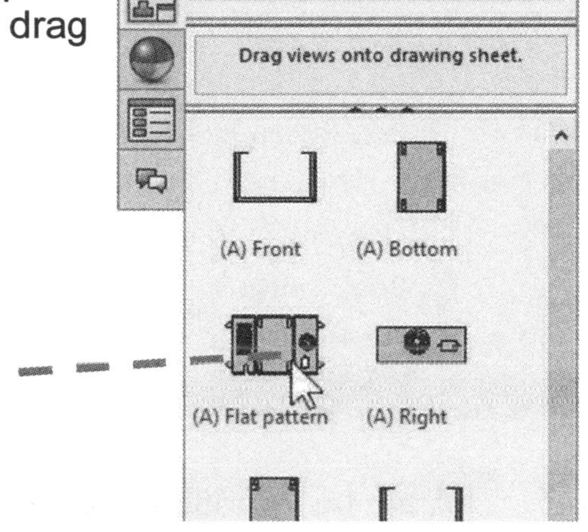

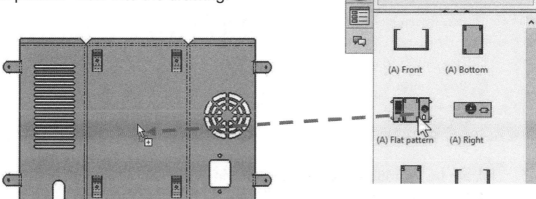

As soon as we drop the flat pattern in the drawing, a sketch with bend lines and bending notes is added automatically. These notes are a guide for manufacturing to bend the part correctly. Change the drawing's scale to 1:1 and adjust the bend note's font if needed.

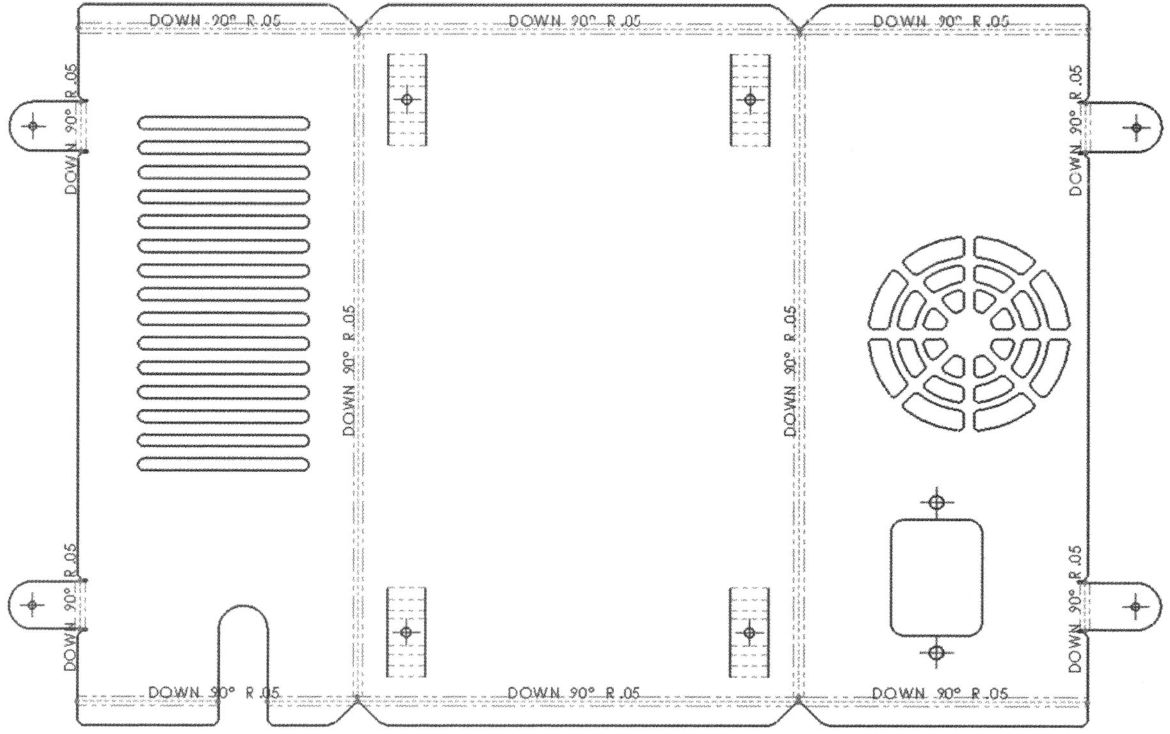

12.2. – When making a drawing of a sheet metal part, it is often common to also generate CNC files for manufacturing, and having a continuous line along the edges can be very helpful when programming a CNC machine to automatically cut the flat pattern. Open the *'Box Bottom'* part, expand the *"Flat-Pattern"* folder to edit the *"Flat-Pattern1"* feature and turn ON the "Merge faces" option. Click OK to accept the change and return to the drawing.

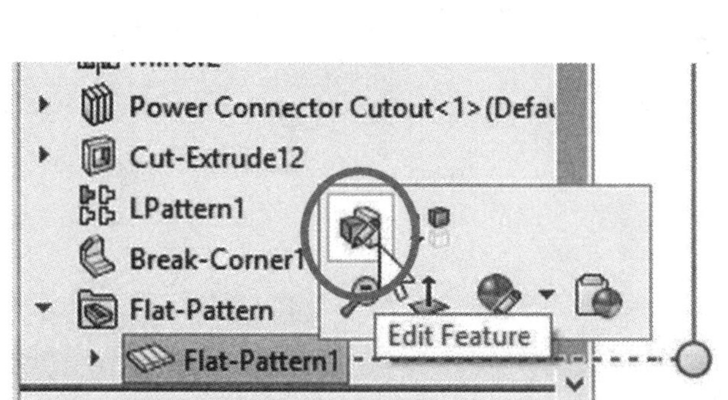

After this change we can see all the flat faces have been merged into a single face, and only the sketch with the bend lines remains visible.

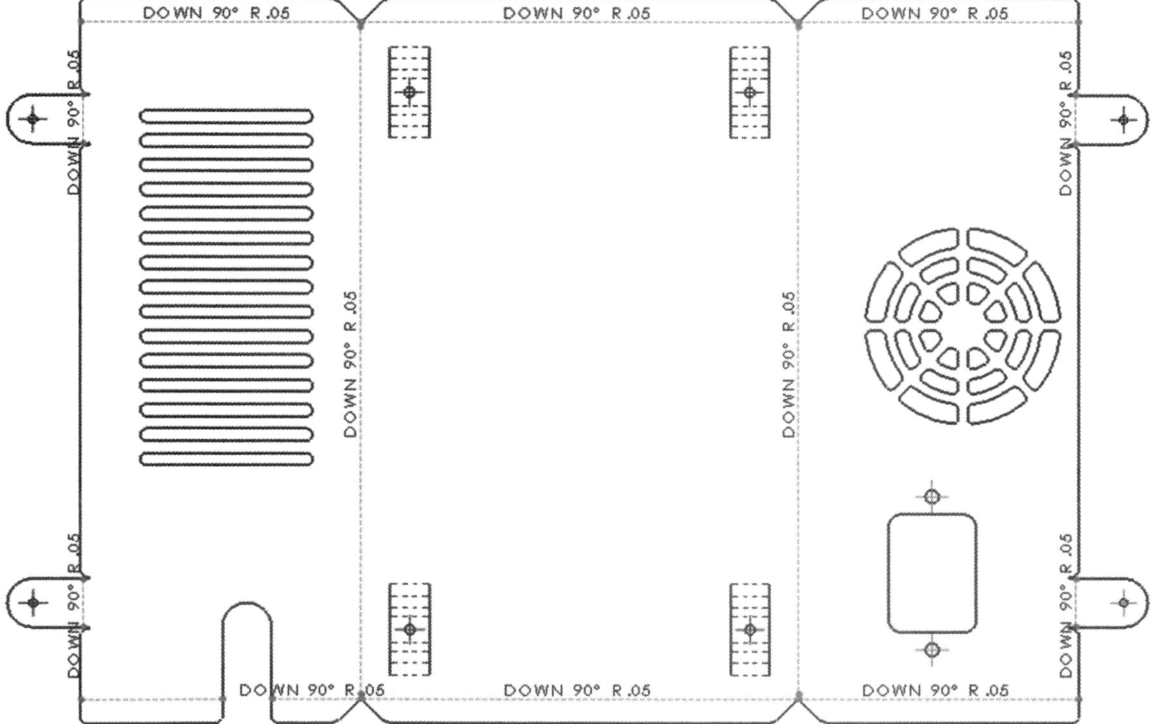

To enable or disable the option to add sheet metal notes in the flat pattern when making a drawing, go to the menu "**Tools, Options, Document Properties, Sheet Metal**, **Display sheet metal bend notes**."

12.3. – Computer Numerical Control (CNC) programming software can usually read a file with the sheet metal part's flat pattern profile for manufacturing; therefore, it is very common to export a 1:1 scale drawing in either DWG or DXF format using the "Merge faces" option. In this exported file, you may or may not want the bend lines and notes, since a CNC program only requires the contours to trace the part. To turn off all notes and bend lines in the drawing, you can:

- Right mouse click in the *"Annotations"* folder of the drawing and turn "Display Annotations" Off (Recommended).

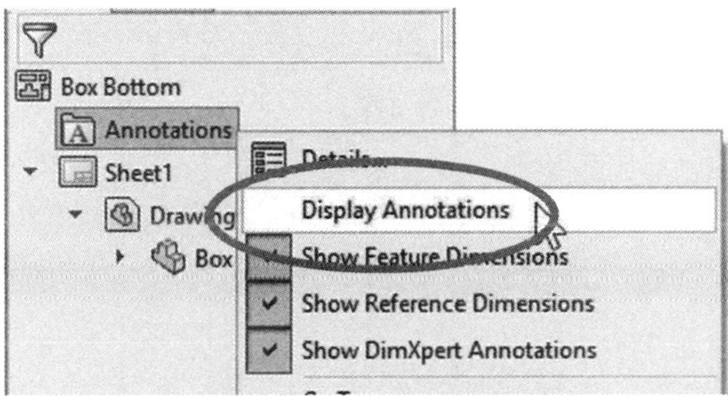

- Or expand *"Drawing View1"* and hide the sketch with bend lines in the part's *"Flat-pattern1"* feature (Not recommended).

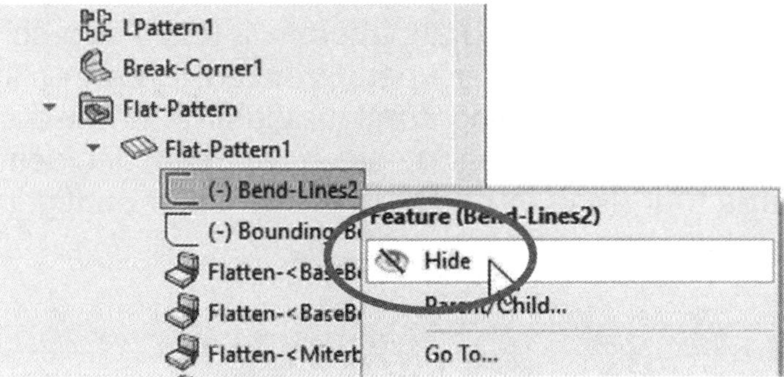

The advantage of the first option, turning "Display Annotations" off, is that centermarks, dimensions, and any other annotation in the drawing will also be hidden, and turning it back on will display everything back.

In the second option, hiding the "*Bend-Lines*" sketch will only hide the sketch and the bending annotations will be removed (deleted) from the drawing; centermarks and other annotations will remain visible. Even if we make the sketch visible again the bending annotations will not be displayed, only the sketch lines will be visible.

To properly document our design for manufacturing and/or inspection, the flat pattern can be manually dimensioned using either ordinate or baseline dimensions.

If the "Display Annotations" option is Off, the dimensions will also be hidden.

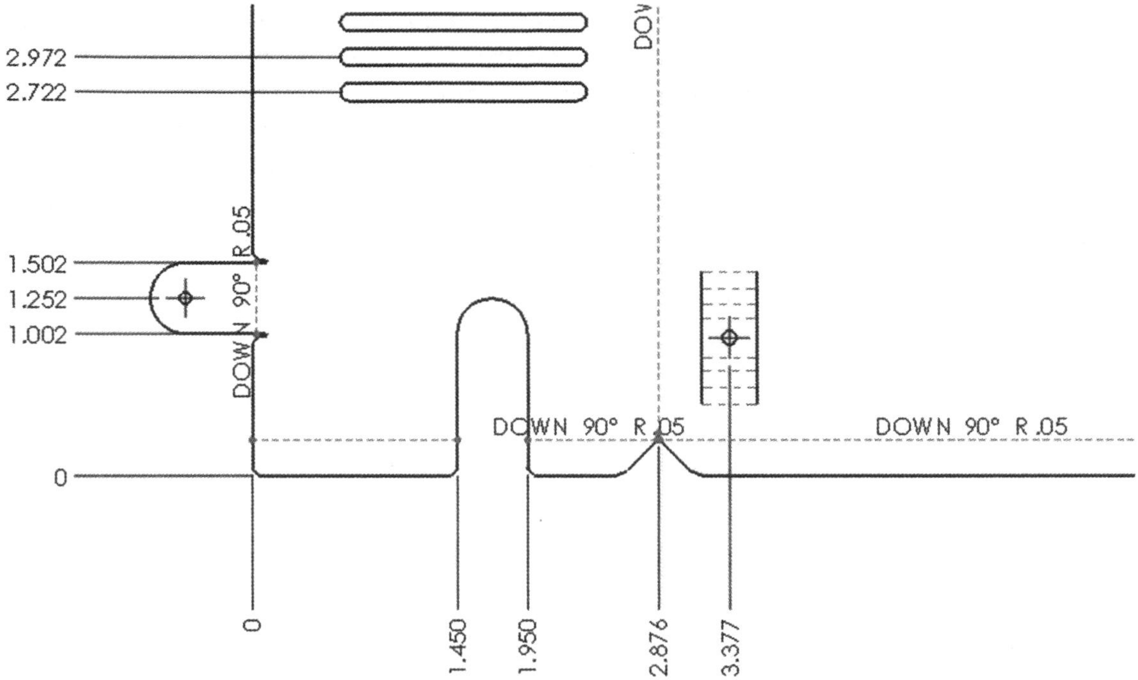

One more thing to know: the flat pattern is a part's sub-configuration, where the *"Flat-pattern1"* feature is unsuppressed, showing the part unfolded. Instead of manually changing configurations to show a sheet metal part's flat pattern, use the "**Flatten**" button, since we can accidentally suppress or unsuppress features that need to be On or Off in either configuration which could cause errors and confusion.

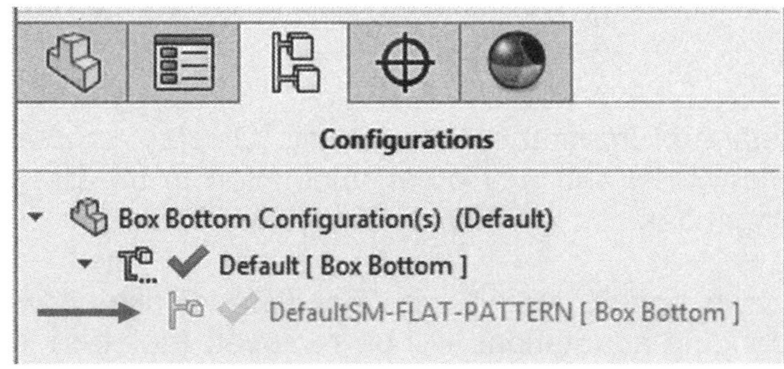

Exercises:

Make manufacturing drawings from the power supply sheet metal parts including a flat pattern in a separate sheet using a 1:1 scale. In a sheet metal drawing we usually find general dimensions in the folded part and detail dimensions in the flat pattern for manufacturing. The features driven by other components in the assembly do not have dimensions, but we can manually add and mark them with parentheses to know they are reference dimensions. The vent would probably be a pre-made tool, and we would just need to dimension its location and possibly orientation.

- Add a new sheet to the '*Box Bottom*' drawing, add the following views and dimension them accordingly.
- Use imported dimensions from the model and reference dimensions added manually.
- Make the Vent feature's sketch visible to dimension its location, and hide it afterwards if wanted/needed.

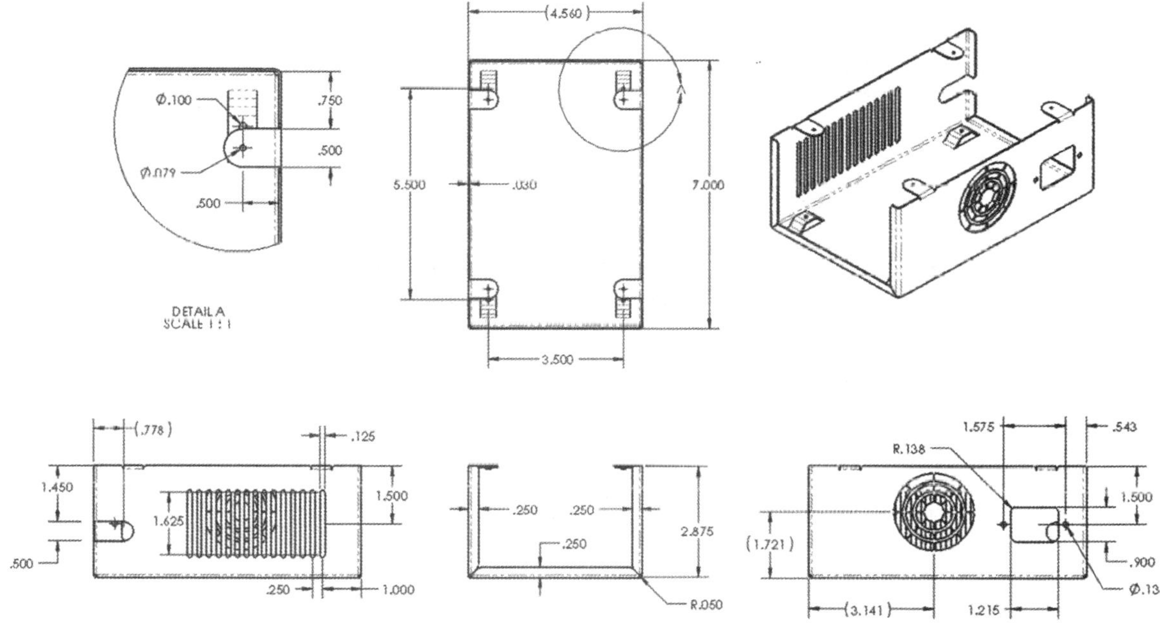

- Complete the flat pattern drawing sheet adding both horizontal and vertical ordinate dimensions.
- Add a second set of vertical ordinate dimensions to the right side to make the drawing easier to read.
- Right-mouse-click an ordinate dimension, and select the "**Break Alignment**" option to move the dimension to a different location.

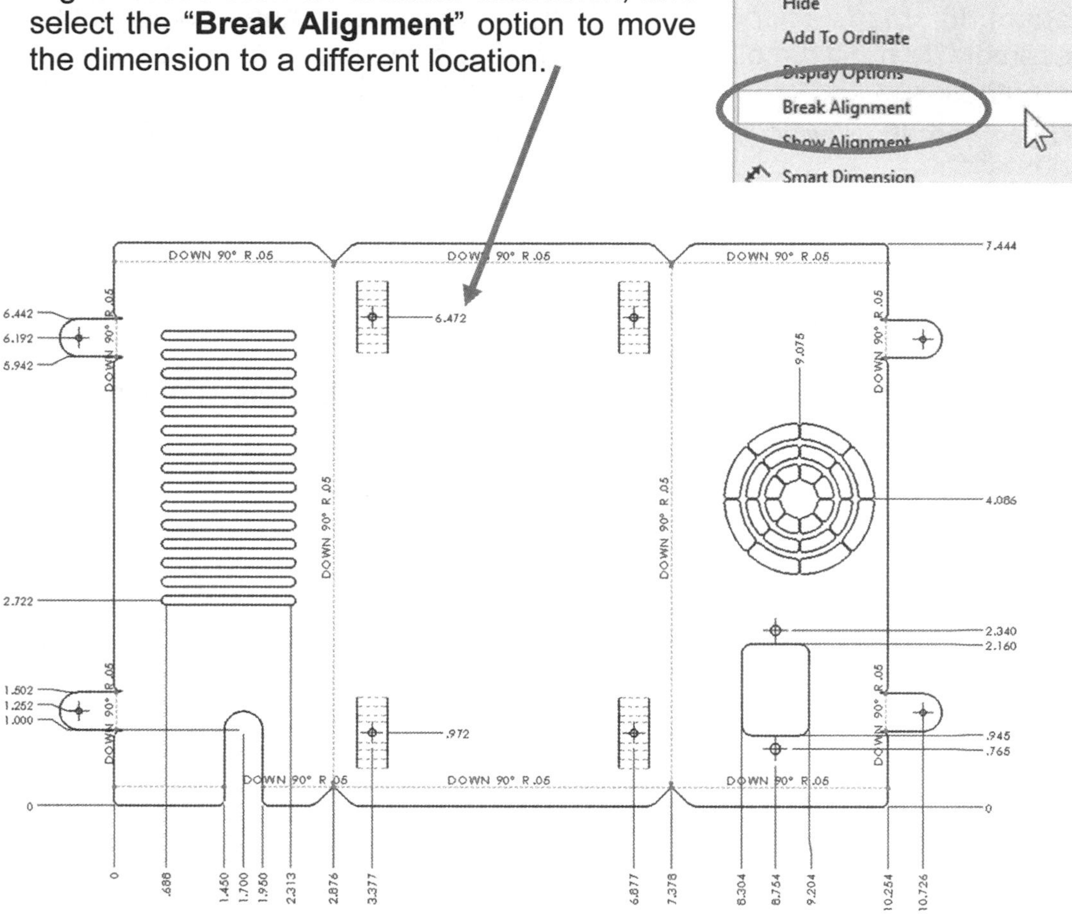

- To Jog a dimension right-mouse-click in it and select the option "**Display Options, Jog**," and then click and drag it.

- To add the dimensions to the vent in the flat pattern, select the inner arc of the filled in center to select its center.

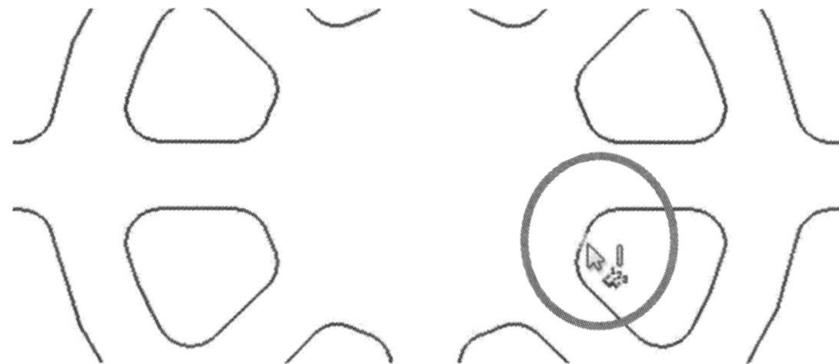

- In the '*Power Supply Assy*', change the '*Box Bottom*' to make it 0.4" higher than the tallest component in the electronics board.

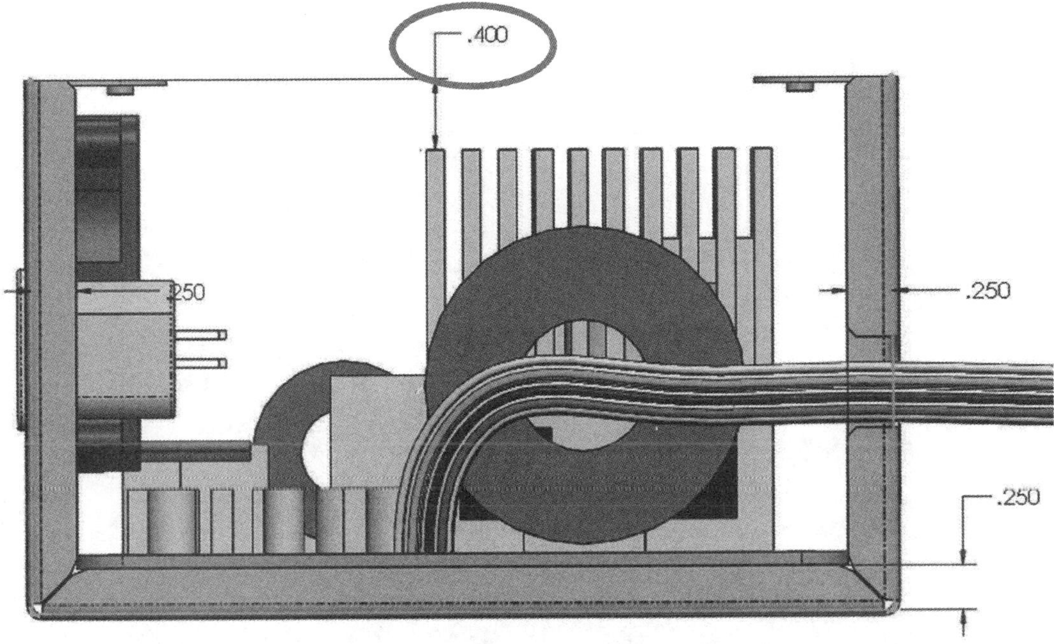

- Change the electronics board height to 4.25". After rebuilding the assembly there should be no errors, the '*Box Bottom*' and '*Box Cover*' must resize accordingly maintaining the design intent. The lances must remain concentric to the electronics board mounting holes.

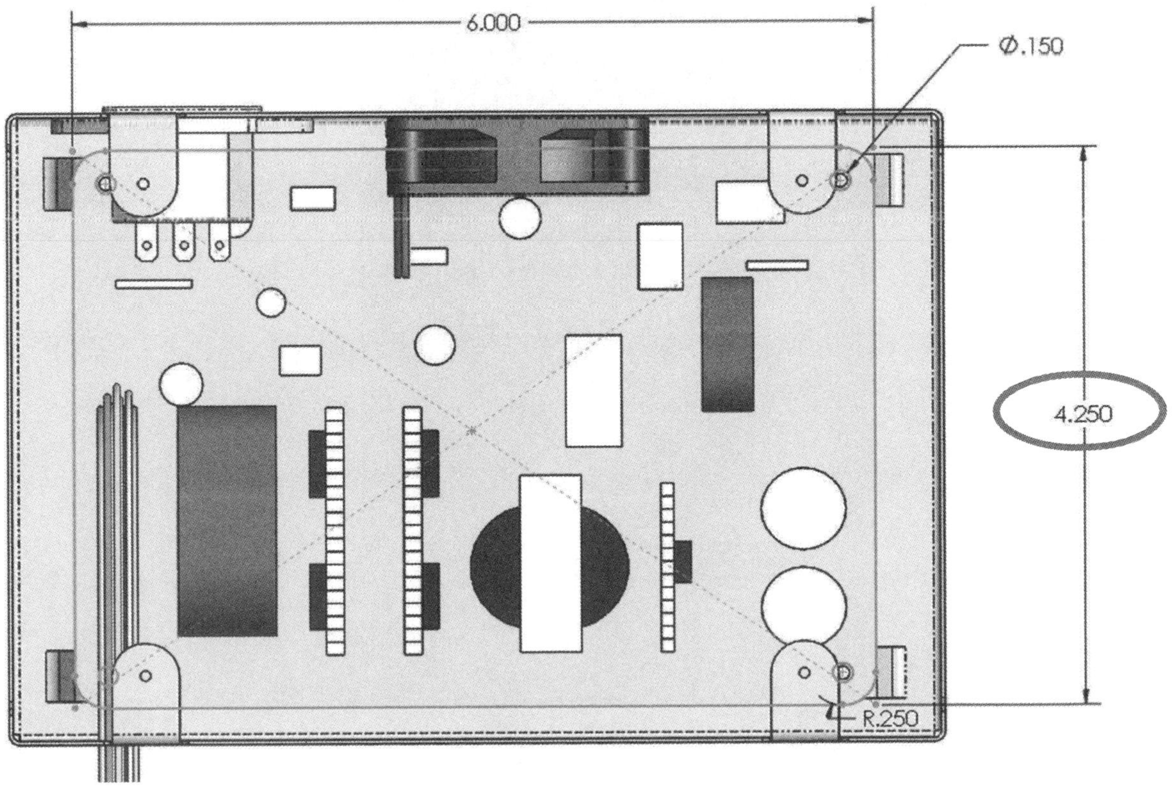

Sheet Metal Locker Top Down Design

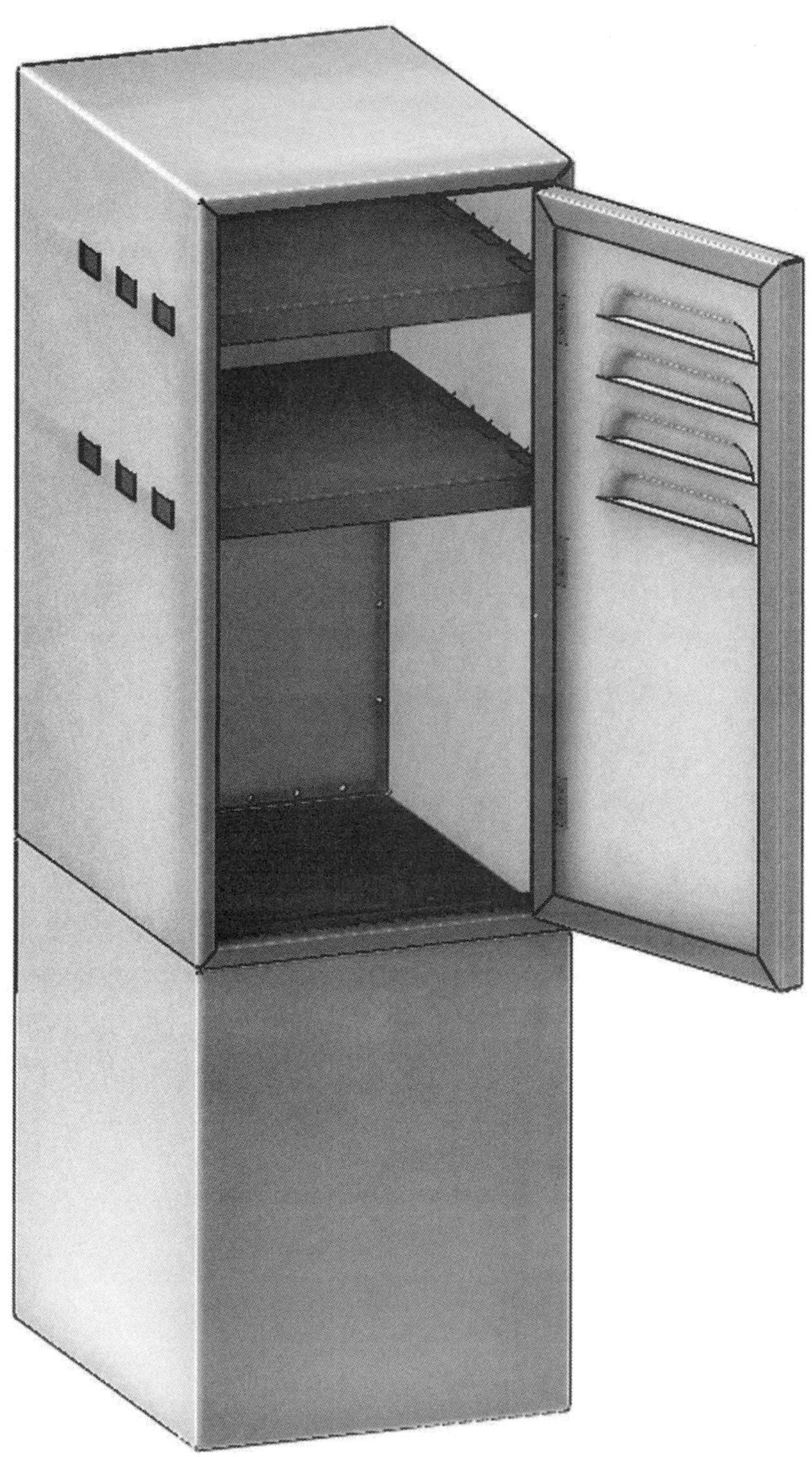

Notes:

For our next sheet metal project, we'll build a metal locker also using a top down design approach. The first part will be a base to raise the locker from the floor. In this example, we'll learn how to convert a solid body part to sheet metal. Solid and shelled parts can be converted directly. Note the bends in the corners:

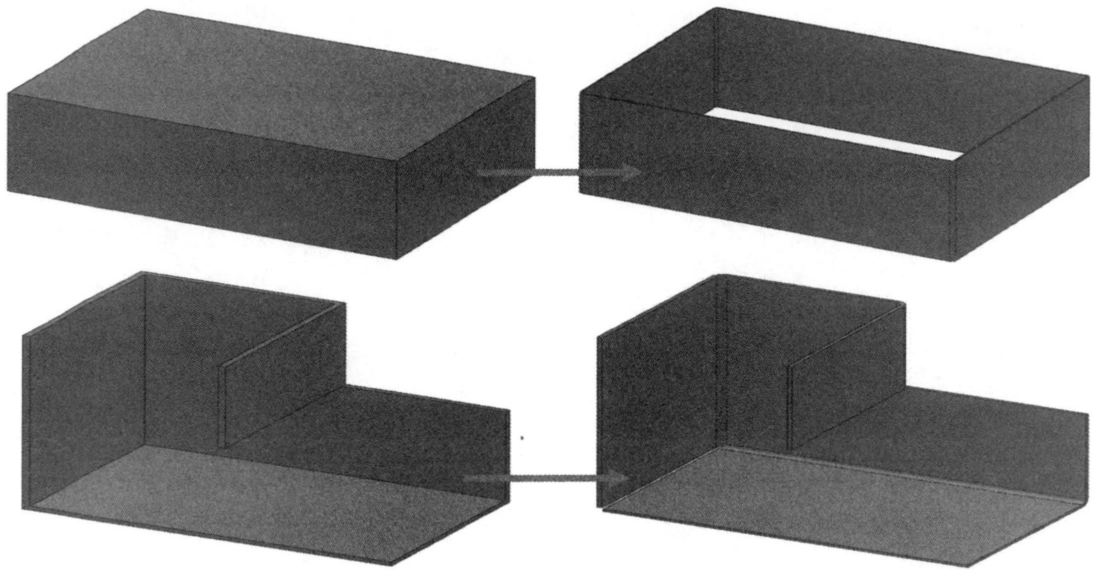

When a solid part is converted to sheet metal, we must select a face that will remain "fixed" or static, the faces connected to it by edges defined as "Bends" will be part of the sheet metal component, the rest of the faces will be deleted. If the part is made of surfaces or a shelled part before converting it to sheet metal, some of the faces not connected by "Bend" edges will also be removed. The size of our sheet metal locker is going to be designed smaller than a real one to better see the part's thickness and features.

13.1. - The first part of the assembly will be the locker's base. Make a new part, set the units to inches, draw the following sketch in the *"Top Plane,"* and extrude it 8″ up. The sketch must be centered with the origin to maintain symmetry.

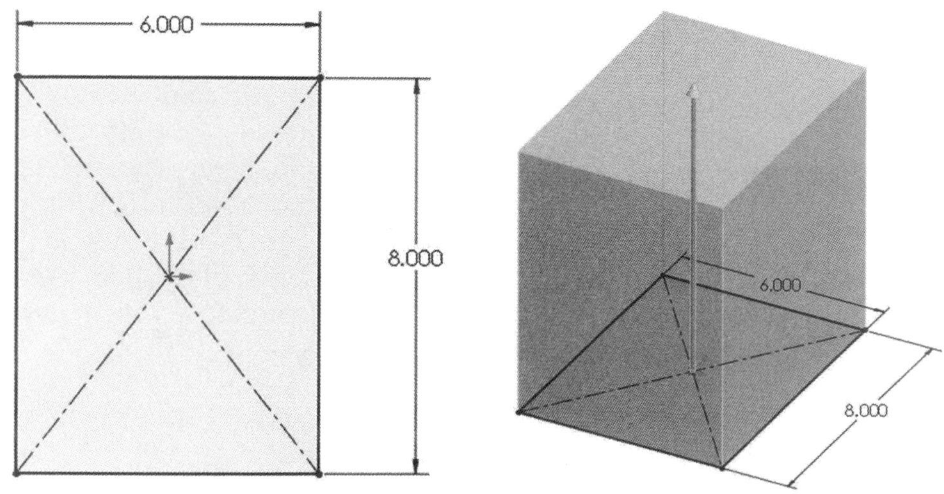

13.2. - From the Sheet Metal toolbar select "**Convert to Sheet Metal**." When we convert a part to sheet metal, at the same time we also need to define the sheet metal parameters. For the locker, we'll use an 18 gauge* steel sheet metal, but instead of entering the sheet metal values ourselves, we are going to use a "Sheet Metal Gauge" table. Gauge tables are Excel files that list the material's Gauge number, its thickness, and the available bend radii for each thickness.

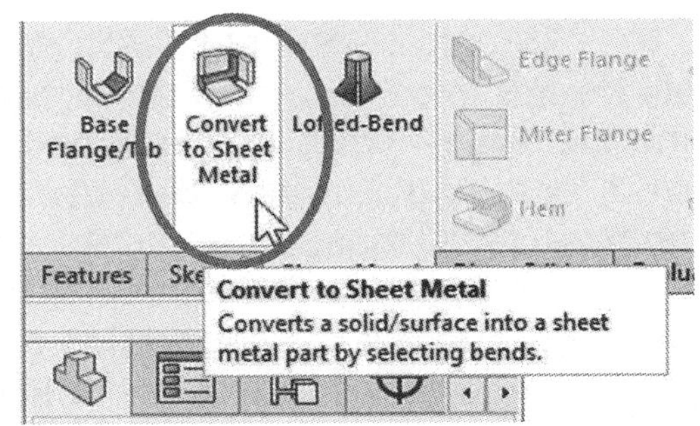

* *"Standard"* sheet metal gauge tables vary by country, material, standard, etc.

We can add, delete or modify existing gauge tables, which are located in the folder: ***"Install-folder"\SolidWorks\lang\english\Sheet Metal Gauge Tables*** The default gauge table location can be changed in the SOLIDWORKS options under "File Locations."

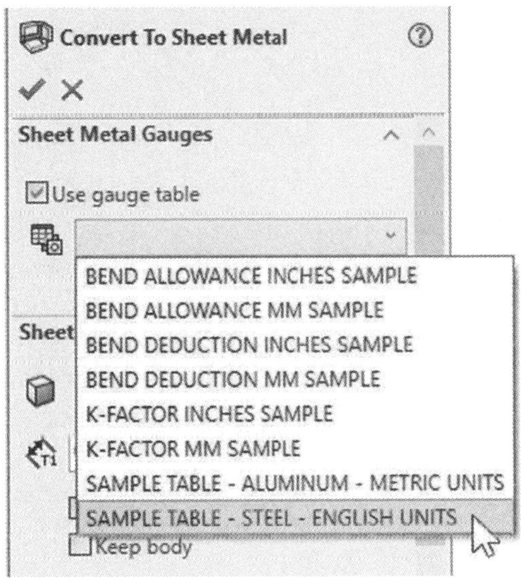

After selecting the "Convert to Sheet Metal" command, activate the "Use gauge table" option, and from the drop-down list select "SAMPLE TABLE–STEEL-ENGLISH UNITS."

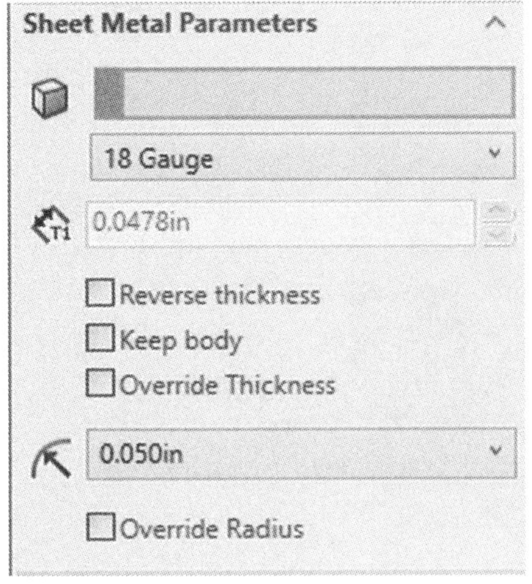

After loading the gauge table, select "18 Gauge" from the drop-down list. The thickness will be automatically filled from the gauge table. Then select 0.050" from the bend radius drop-down list to define the bend radius for our part. If needed, the material's thickness and bend radius can be overridden by selecting the corresponding checkboxes.

To convert the solid into a sheet metal part, we have to select a face that will be the fixed face, the edge(s) of the part that will be converted into bend(s), and the edge(s) that will be ripped (open). Select the left face of the part to be the fixed face (you can use "**Select other**" to select the hidden face). After selecting the fixed face, make sure the thickness of the sheet metal is added *'inside'* as we want to maintain the dimensions given as external dimensions. You can zoom into a corner to see if the material is added inside or out, and check (or uncheck) the "Reverse Thickness" option if needed.

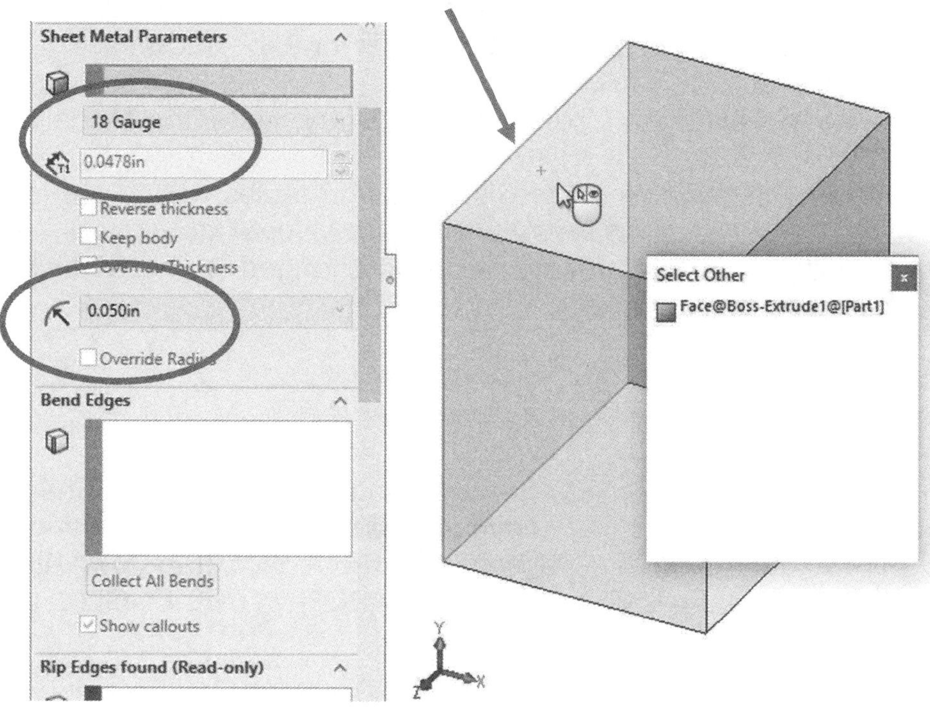

13.3. - Select the 3 vertical edges (labeled "Radius") that will become bends. The remaining vertical edge (labeled "Gap") is automatically selected as a "Rip Edge" which will be an open edge in the sheet metal part.

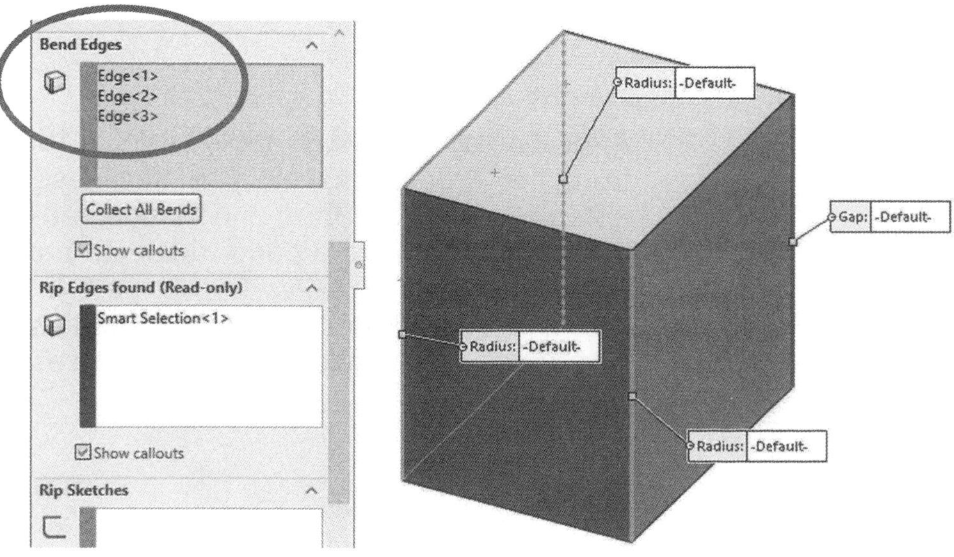

Set the default gap for rips to 0.01″, the "Auto Relief" type to Obround with a 0.5 ratio and for bend allowance use a K-Factor of 0.5. Click OK to convert the part to sheet metal with the given parameters.

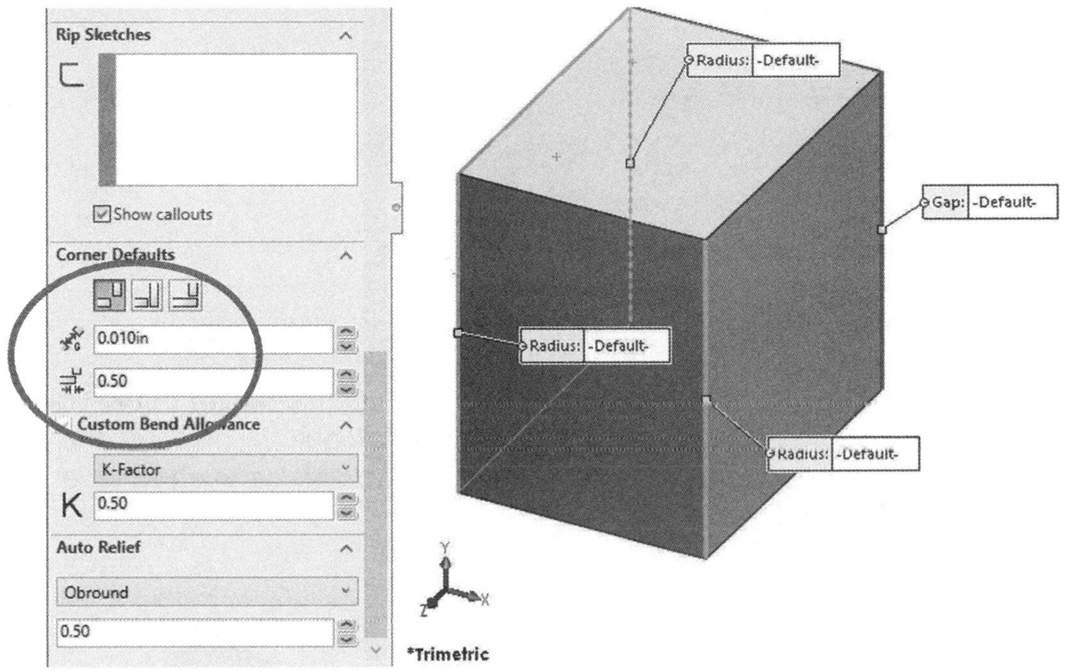

If needed, we can click the "Radius" and "Gap" labels to override the default bend radius and gap distance.

After entering the sheet metal parameters, click OK to convert the part to sheet metal; the top and bottom faces are automatically removed and the sheet metal features are added to the Feature Manager.

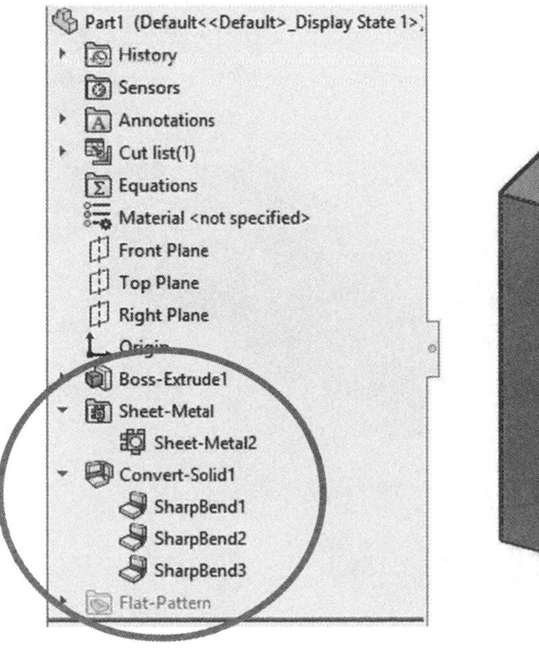

13.4. - Now we need to add miter flanges at the top and bottom, but instead of using the "**Miter Flange**," we'll use the "**Edge Flange**" command to show a different approach to accomplish the same result. Using the "**Edge Flange**" command, select all the edges indicated.

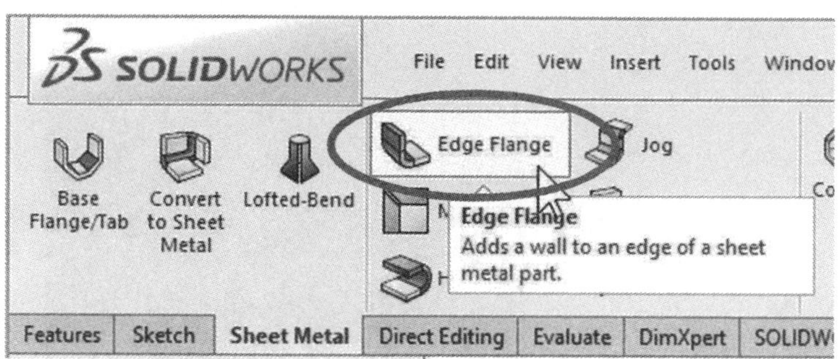

Using this technique, the miter cuts are automatically added to the corners. Make the flange 0.5″ wide, use the "Material Inside" option, and set the "Gap Distance" to 0.01″. If needed, the arrow in each flange can be used to reverse its direction. Click OK to add the flanges.

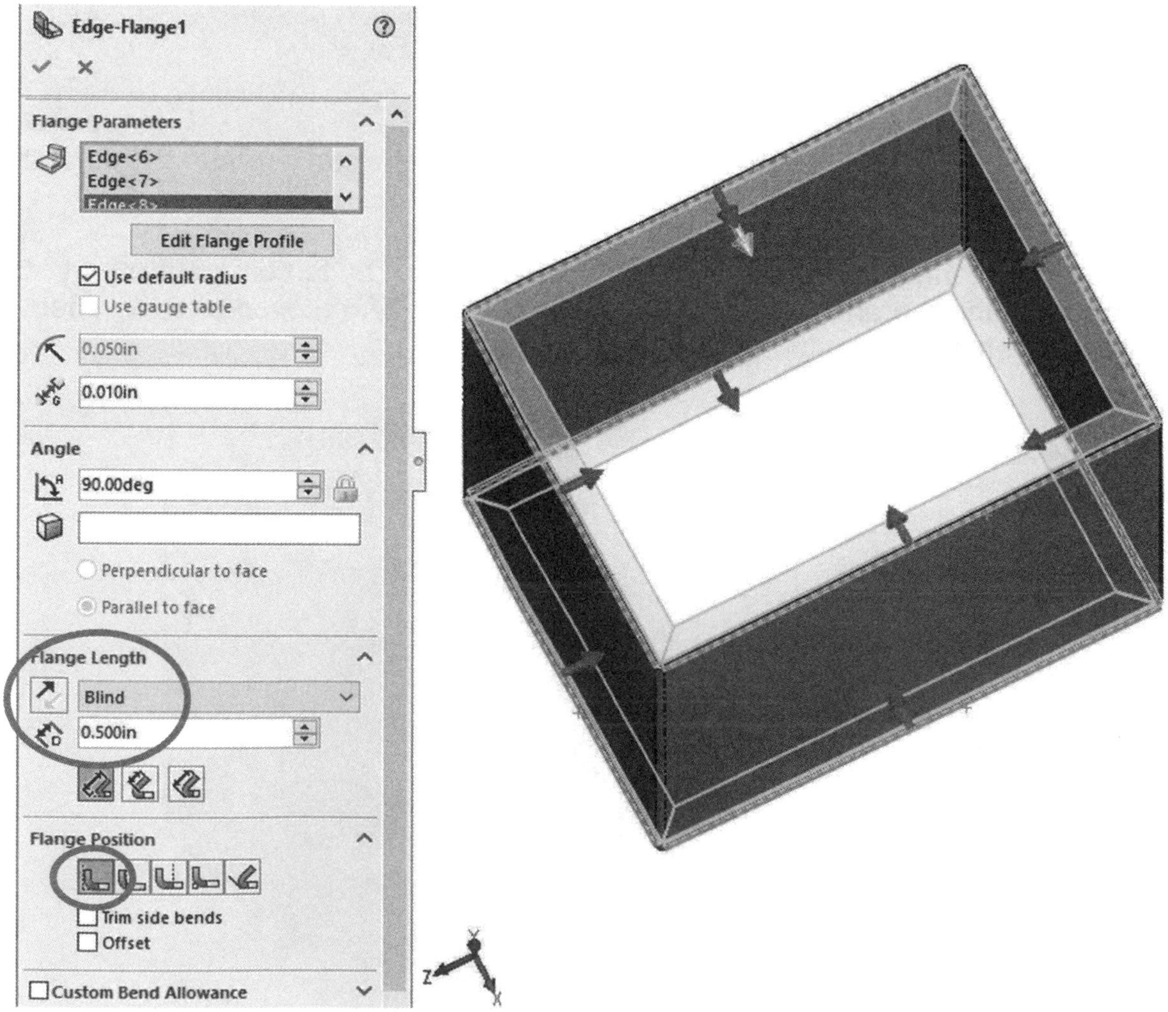

In this case there is no difference between using a "**Miter Flange**" or an "**Edge Flange**," but remember, the "Miter Flange" runs the length of the edge and can change the profile, and an "Edge Flange" can change the length, shape and angle of the flange but not the profile.

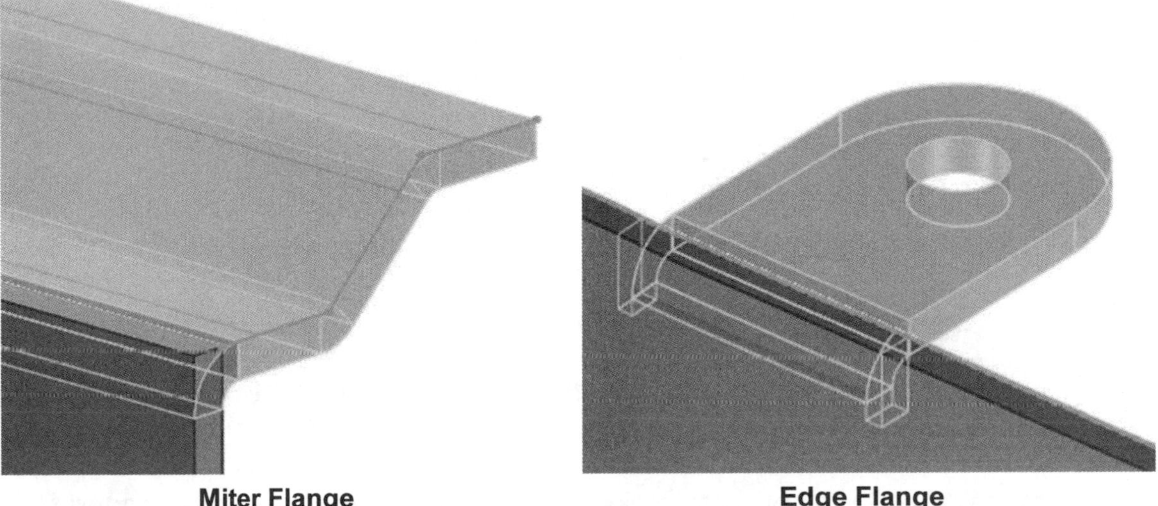

Miter Flange
Different profile, runs the length of the edge

Edge Flange
Different shape, not necessarily full length

13.5. - After adding the flanges, we need to close the gap in the corner that was ripped when we converted the solid to sheet metal.

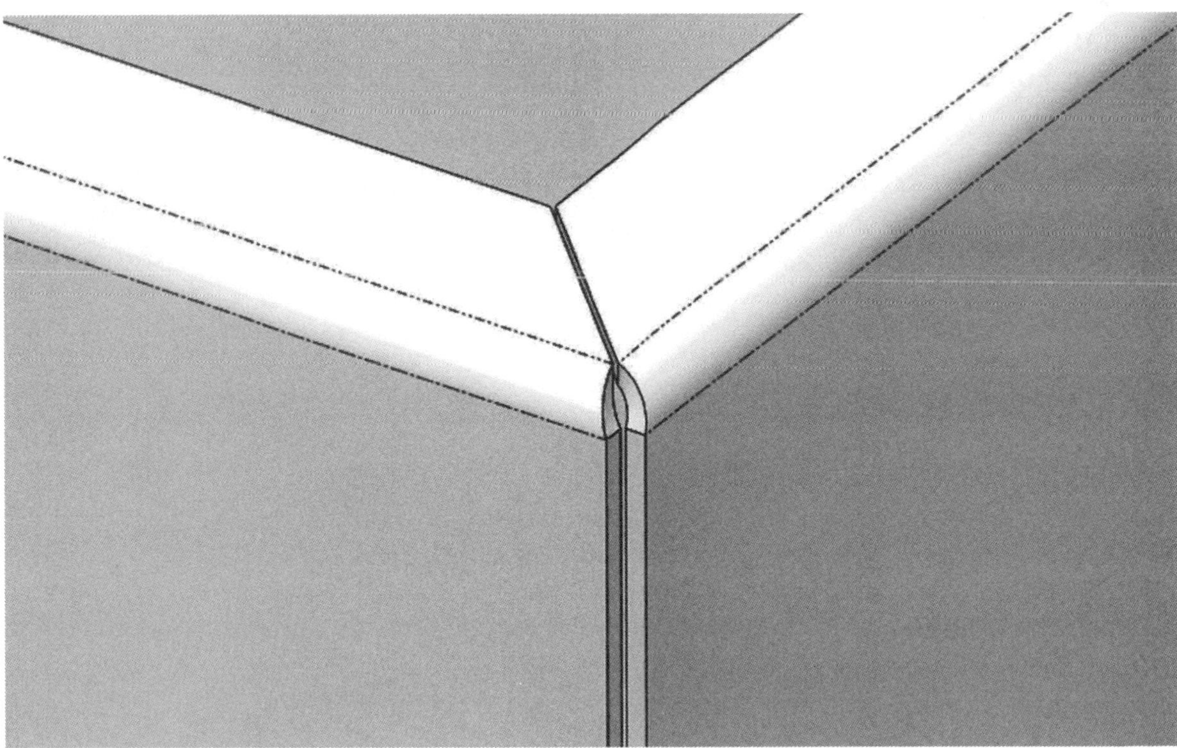

Select the "**Corners, Closed Corner**" drop down command. This feature will extend one side of the sheet metal to match the other using a butt, overlap, or underlap extension. In the "**Faces to Extend**" selection box, select the flat face on one side, and in the "**Faces to Match**" select the other flat face.

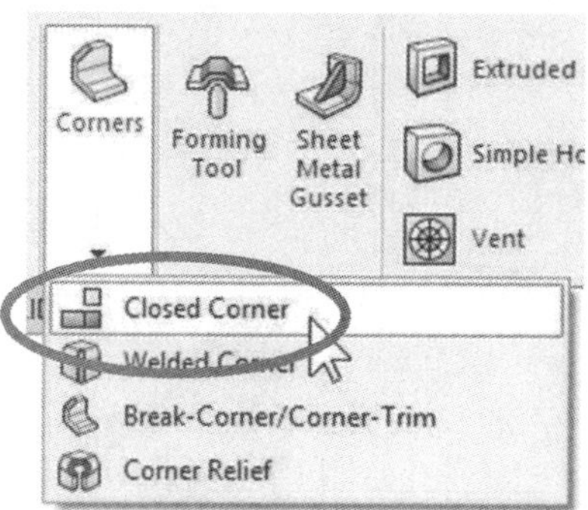

The corner can be closed when the solid is converted to Sheet Metal, but instead we decided to show how to use this command separately.

- Select the "Overlap" option to make the first face extend over the second face.
- The "Open bend region" option will close the bend region too if selected.
- "Coplanar faces" will automatically select faces that are coplanar to the one selected.
- The "Narrow Corner" option will attempt to close the gap when using a large bend radius.
- In the "Gap Distance" we can make the gap in the corner as small as needed. Use 0.005″ for our example.

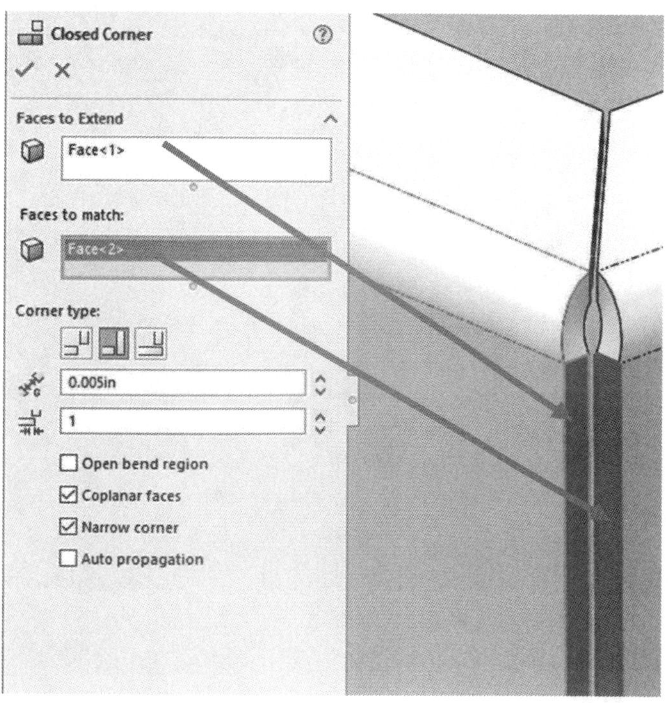

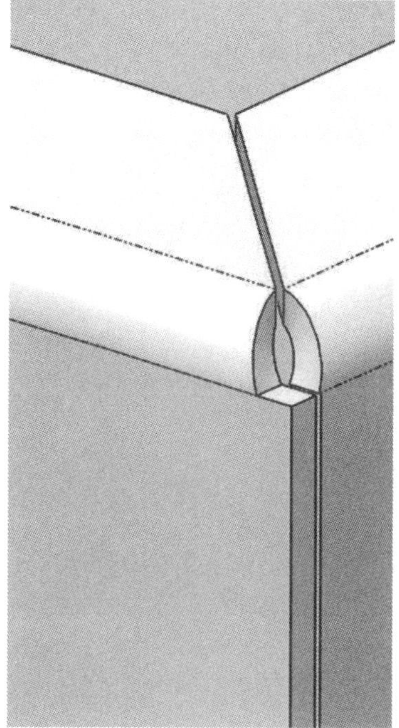

13.6. - After reviewing the other corners made when we added the edge flange we can see an extra piece of metal in the bend area. To eliminate it, select the *"Edge-Flange1"* feature, click "Edit Feature" in the pop-up menu and turn On the "Trim side bends" option under "Flange Position." Click OK to finish. Now all the corners made with this Edge Flange command are trimmed. This option was not turned on before because this way the difference of not activating it is better illustrated.

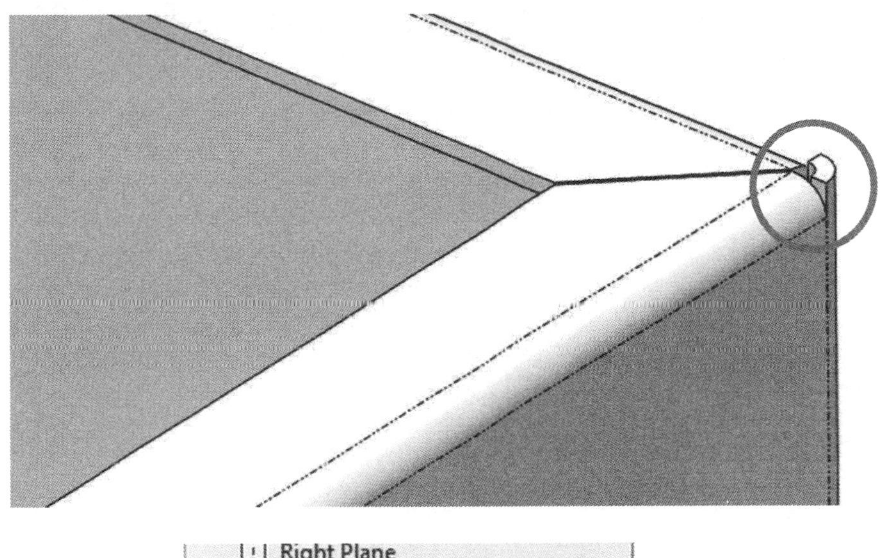

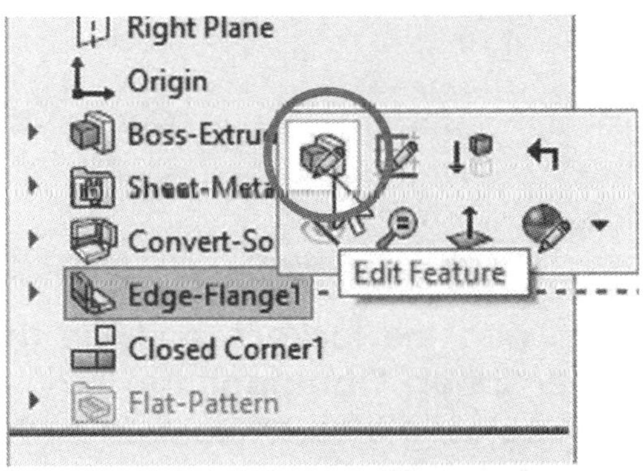

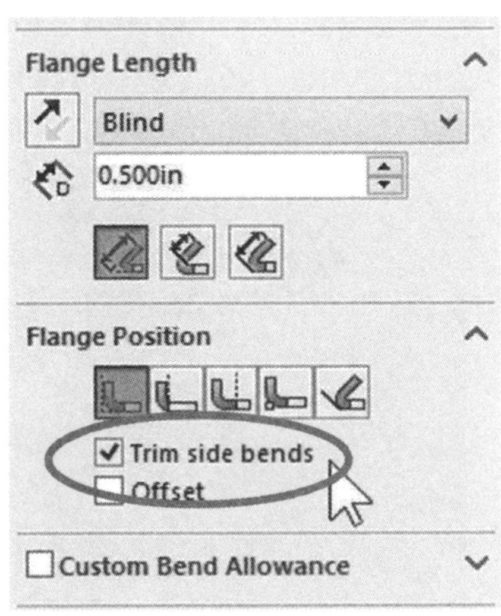

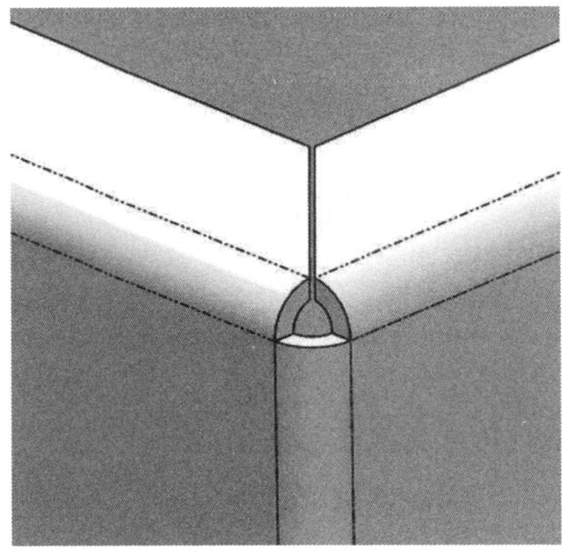

13.7. - Save the part as *'Locker Base'* and add it to a new assembly, and be sure to locate it at the origin, making it symmetric about the assembly's planes. The rest of the locker will be designed in the context of the assembly using the base as reference. Remember we must save the assembly before adding components in the context of the assembly. Save the new assembly as *'Locker Assembly'*.

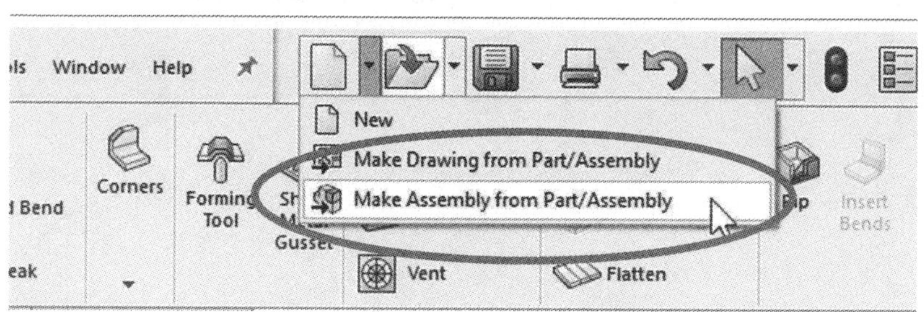

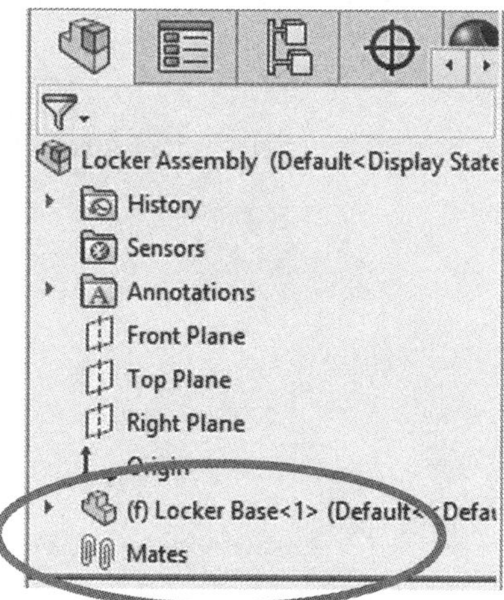

13.8. - Our next step is to add the locker's body to the assembly. From the Assembly toolbar select the "**Insert Components**" drop-down menu and select "**New Part**." When asked to select a face or plane to add the new part select the face indicated at the top of the *'Locker Base'*.

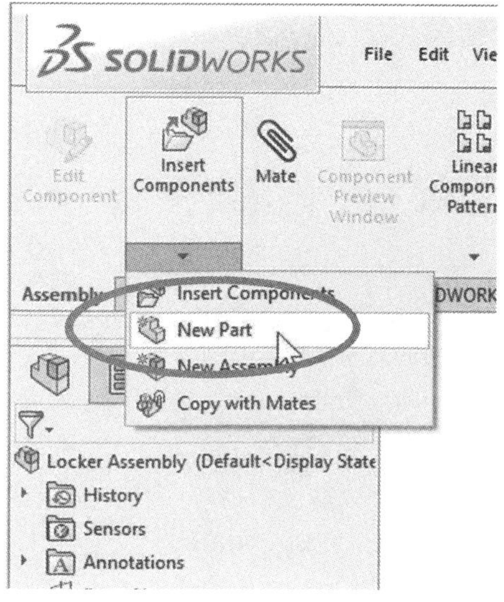

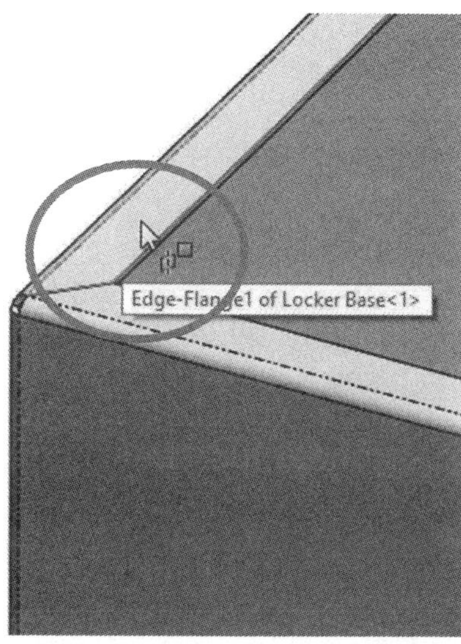

Immediately after selecting the Top face, a new part is added and, as we've seen before, it's listed in blue in the FeatureManager, the "**Edit Component**" icon is active, the *'Locker Base'* becomes transparent, and a new sketch is added.

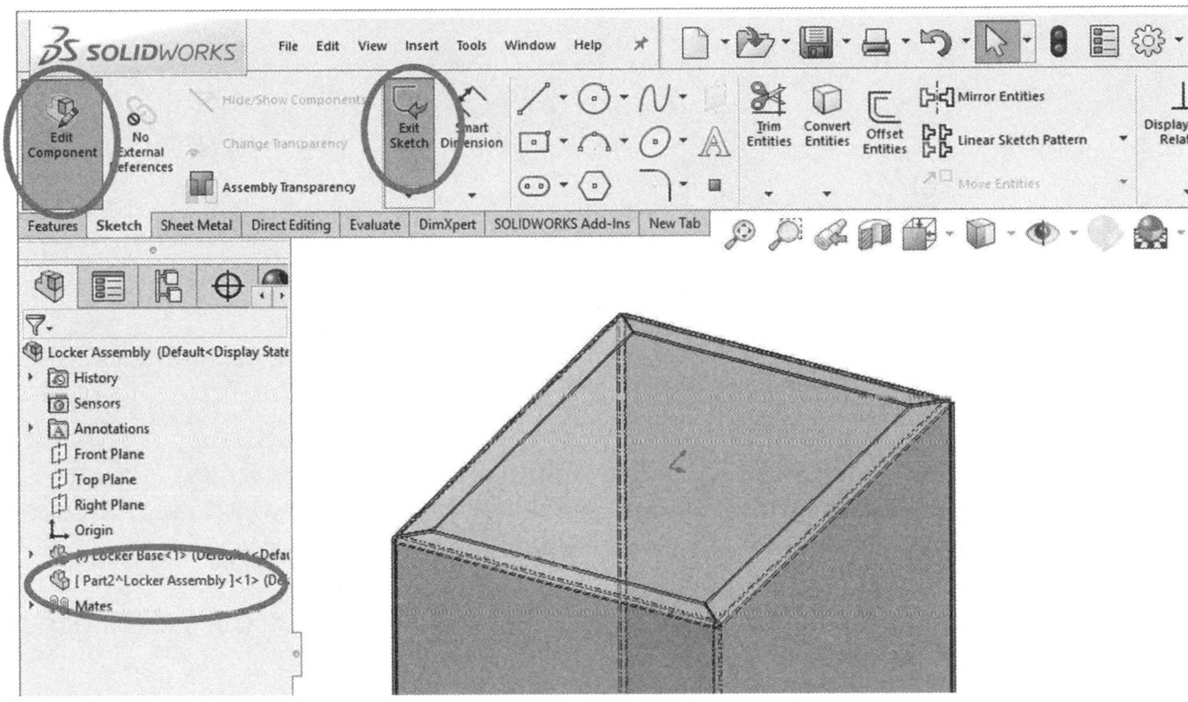

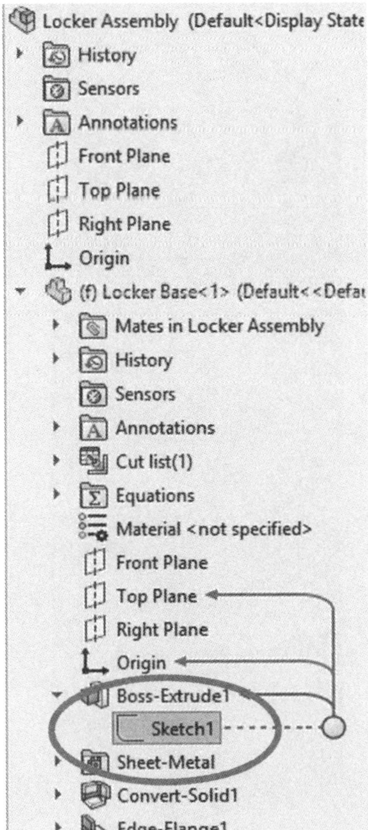

13.9. – Our intention is to make the locker's body the same size as the '*Locker Base*'; in this case we'll use the "**Convert Entities**" command to project the first sketch of the '*Locker Base*' to the new part. In the FeatureManager expand *'Locker Base'*, find the *"Boss-Extrude1"* feature and select *"Sketch1."* After selecting the sketch click "**Convert Entities**" from the Sketch toolbar to project the first sketch of the 'Locker Base' to the first sketch of the new part.

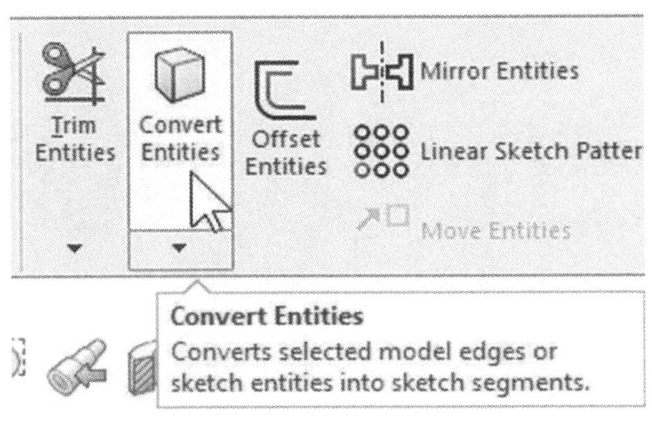

When working with sheet metal parts, using "Shaded" view mode helps to distinguish between sketch entities and model edges.

By selecting the sketch in the FeatureManager, the entire sketch is projected to the new sketch. To selectively convert sketch entities, we can make the sketch visible (Show) to select individual lines to convert.

13.10 - After converting the edges, extrude the sketch 16″. Switch back to "Shaded with Edges" mode for visibility.

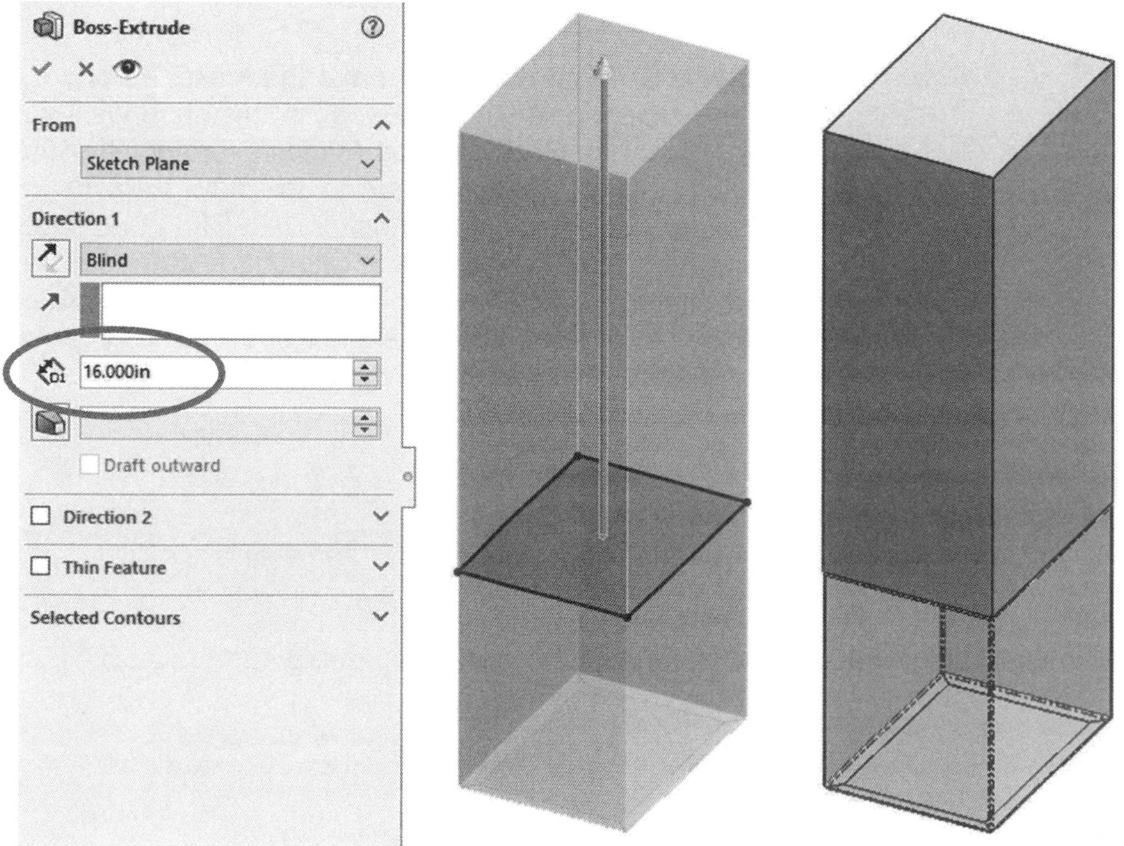

13.11. - When adding features that are not related to other components, we can either continue working in the assembly or open them in their own window. To some extent it's really a personal preference; however, when adding references to other components' geometry we must do it in the context of the assembly. Open the locker's body (the new part) in its own window to convert it into a sheet metal part; even though it's an internal component in the assembly now, it can be opened by itself.

After opening the part change to an isometric view and use the "**Shell**" command to hollow it; set the thickness to 0.05″ and remove the top and bottom faces, which will be the front and back of the locker's body. To convert a solid part into a sheet metal component using this technique, the shell's thickness must be equal to, or thicker than the intended sheet metal part's thickness.

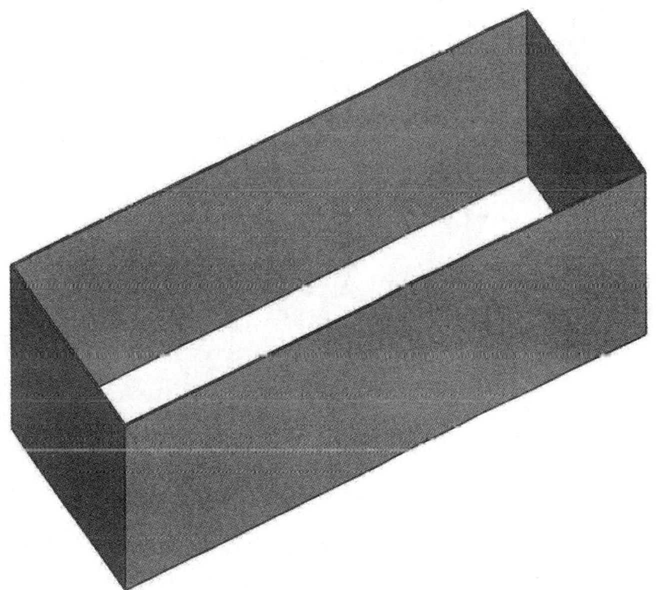

13.12. – We can convert this model to sheet metal without first making a shell; however, we are using this approach to show a different method. From the Sheet Metal toolbar select "**Convert to Sheet Metal**."

For this part, we'll use the same sheet metal parameters as the '*Locker Base*'. Select the <u>outside</u> face on the left side as the fixed face.

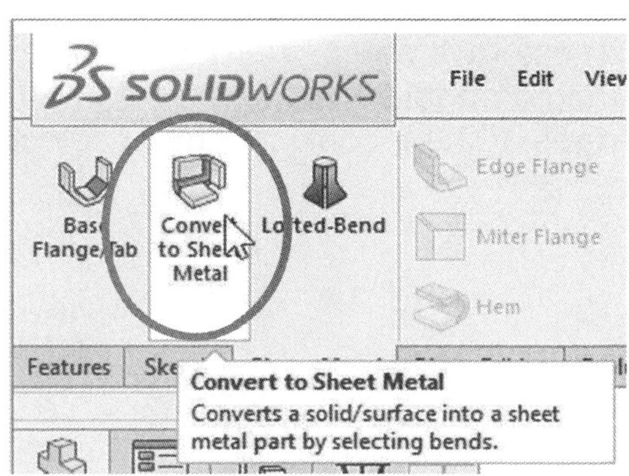

Activate the "Use gauge table" option, and from the sample steel table select "18 gauge" and a bend radius of 0.05". You *may* have to check the "**Reverse Thickness**" checkbox. In this example the material needs to be added inside; remember the part's dimensions are outside dimensions. If needed, zoom in a corner to verify material is added in the correct side. Essentially the sheet metal preview will overlap the shell's thickness. Now we must define which edges to bend or rip.

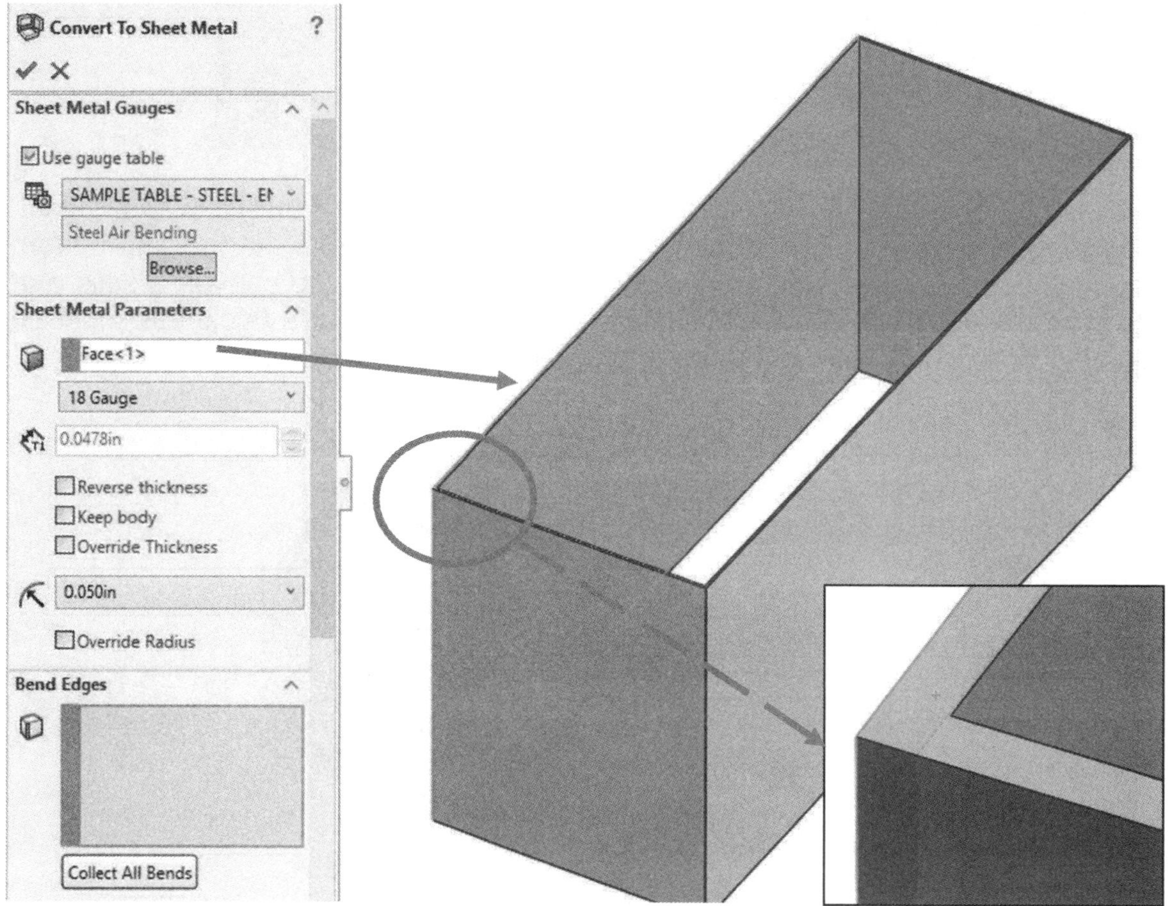

When a shelled part is converted to sheet metal, the edges selected for "**Bend Edges**" or "**Rip Edges**" must be connected to the "Fixed Face." For example, if we select the inside face of the shell, we must select the inside edges; if we select the outside face, we must select the outside edges. If the part is big compared to the shell's thickness, zoom into a corner to select the correct edges. If the incorrect edge is selected, you'll see the following message:

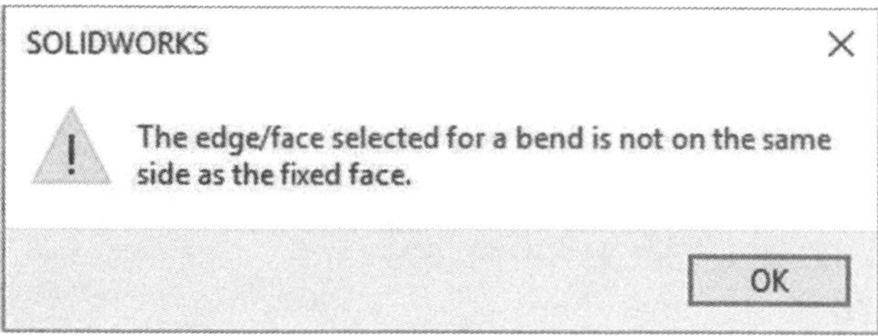

13.13. - After selecting three edges to convert to bends, the fourth edge will be automatically added to the "Rip Edges found" selection box. Set the default rips gap to 0.01", use a K-Factor ratio of 0.5, and set the default relief type to "Obround" with a 0.5 ratio. Click OK to complete and convert the shelled part to sheet metal.

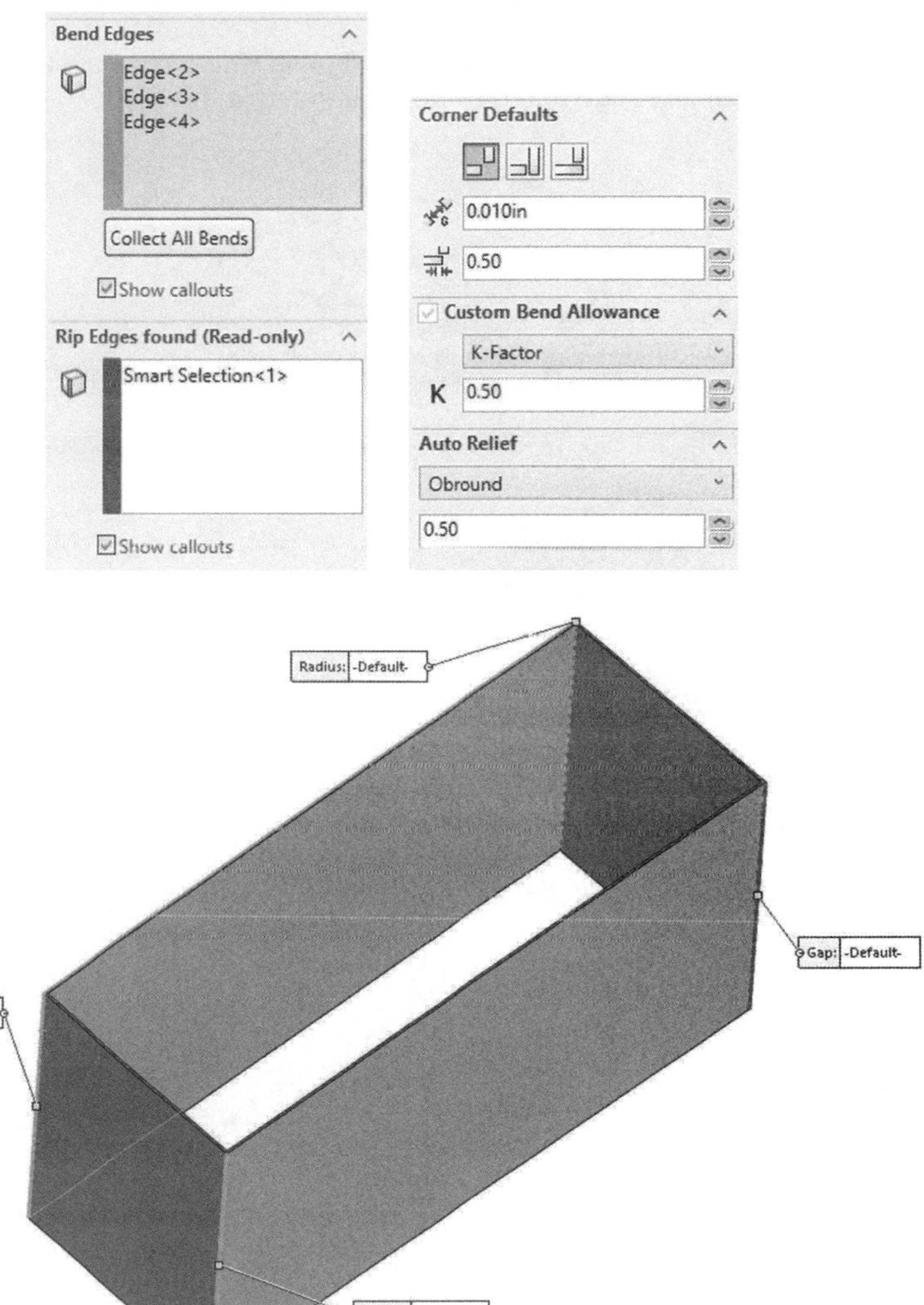

13.14. - The next step is to add a miter flange in the front and back of the locker. Immediately after selecting the "**Miter Flange**" icon from the Sheet Metal toolbar, SOLIDWORKS will ask us to select a face or plane to sketch the flange's profile, or an existing sketch with the profile. Rotate the model and use the magnifying glass if needed (Shortcut "G") to select the indicated face of the ripped edge.

Remember we can also select an edge, in which case a plane is added at the end of the edge and a sketch is created in this plane.

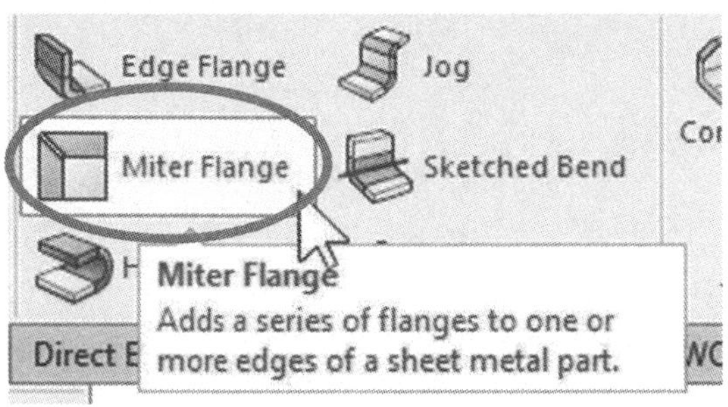

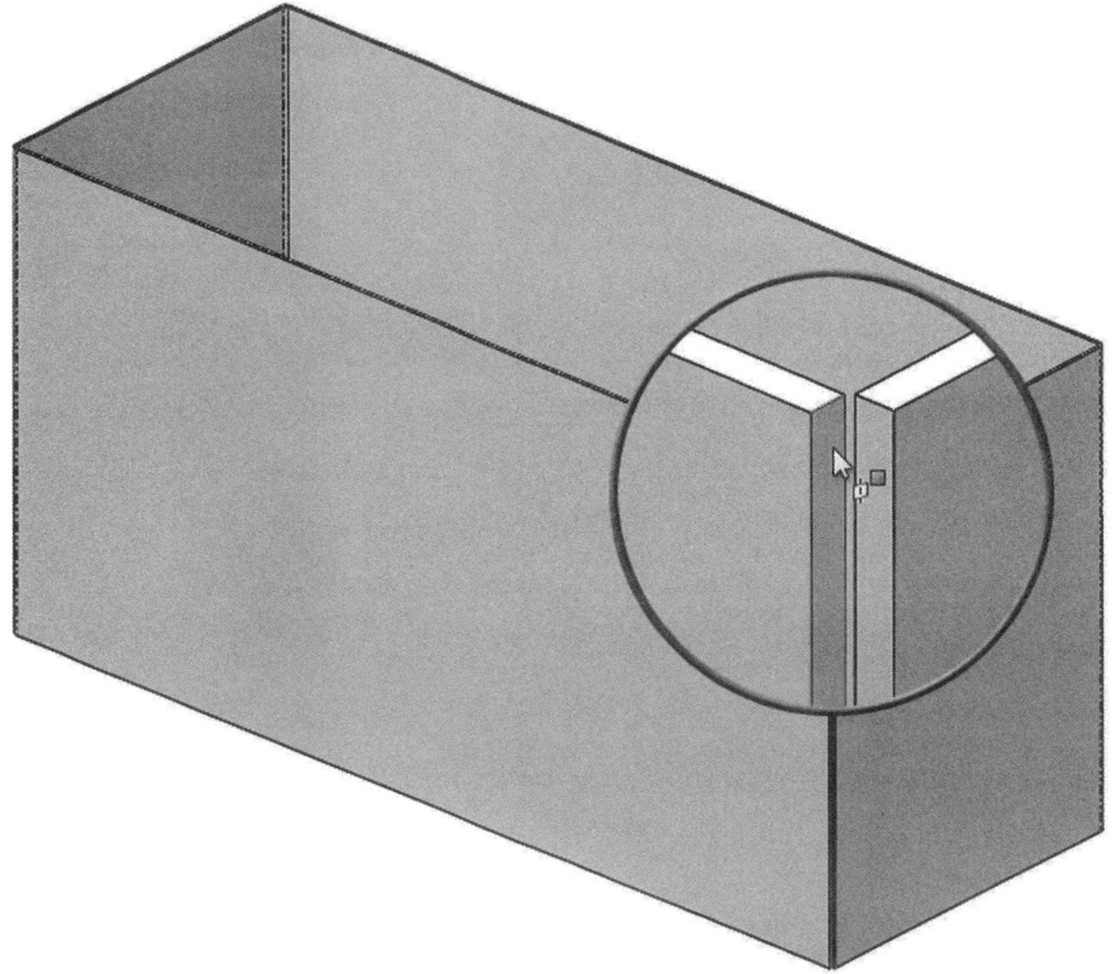

Select either the flat face to add the sketch, or the top edge (the inside or outside edge) to add a plane perpendicular to it and a new sketch in it.

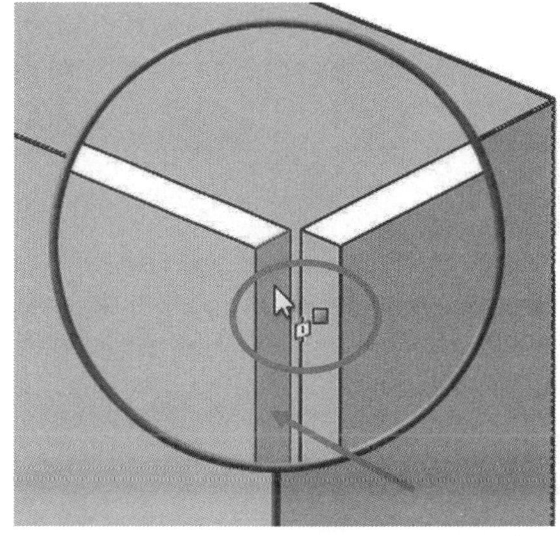

Face selected

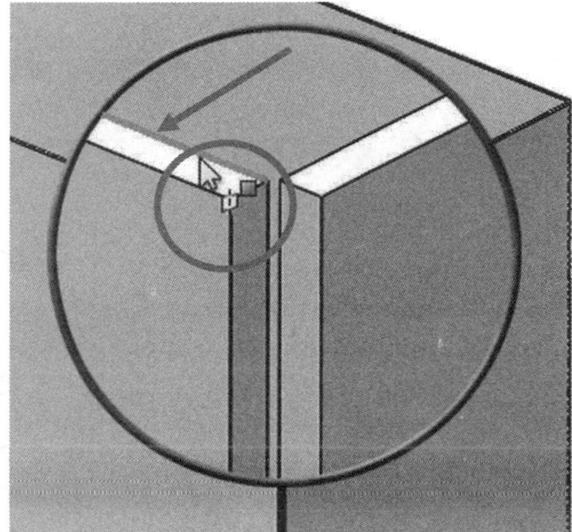

Edge selected

Selecting small faces is easier using a "**Selection Filter**." Selection filters allow us to select only the type of geometry or element we need. In this case press "**X**" on the keyboard (default filter shortcut for faces) to turn it on (press "**X**" again to turn it off when done). We know the selection filter is active when the mouse pointer has a small funnel added next to it. The Selection Filters toolbar can be enabled selecting the menu "**View, Toolbars, Selection Filter**," a right mouse click in any toolbar and selecting "**Selection Filter**," or using the default keyboard shortcut "**F5**." In this toolbar, we can turn on/off any combination of selection filters as needed.

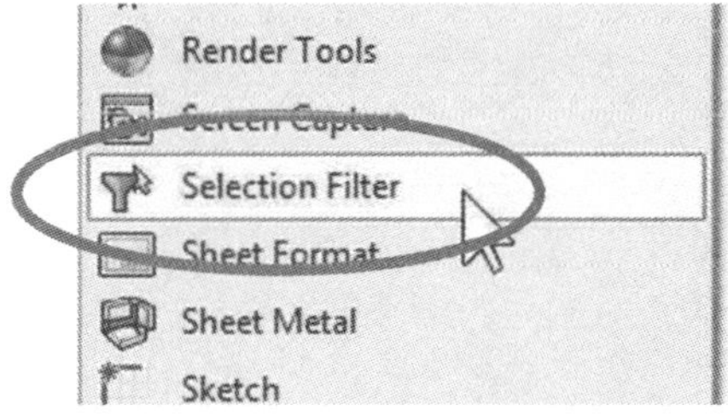

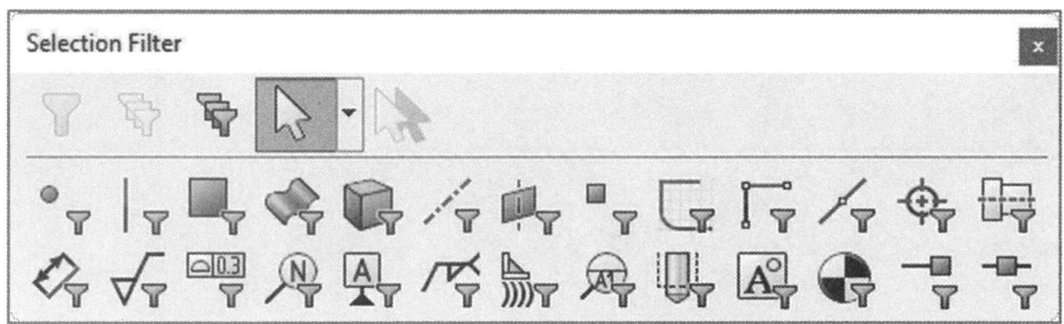

After selecting the face, draw the profile for the miter flange as shown. It's a single line 0.375″ long.

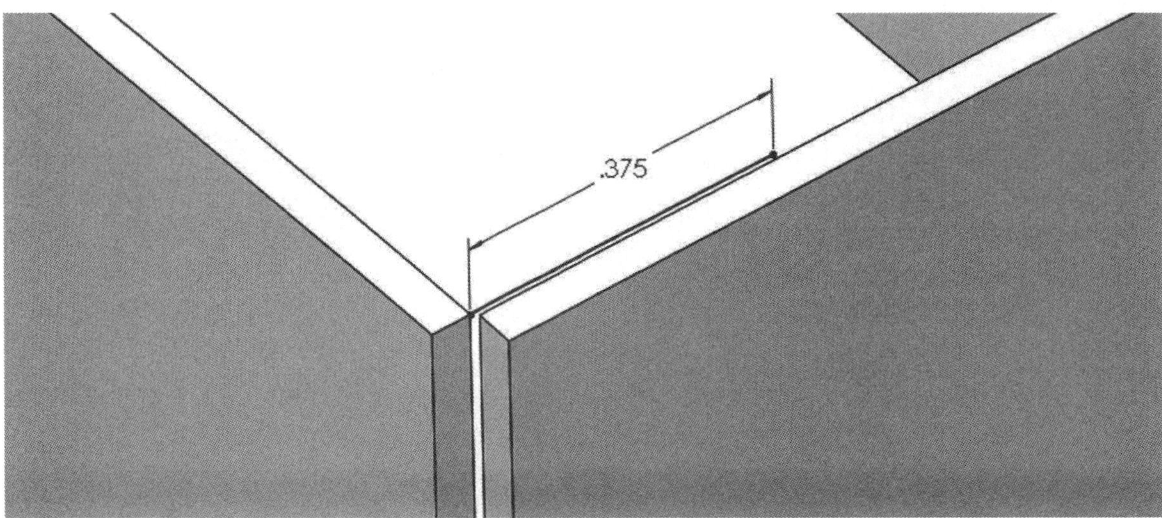

13.15. - After the sketch is finished, select "**Exit Sketch**" or the "**Miter Flange**" command to see the miter flange preview. Make the flange with the option "**Material Inside**" and a 0.01″ gap distance for the corner openings. Select the edges on the top face individually, or click the "**Propagate**" icon to automatically select the three tangent edges and click OK to finish.

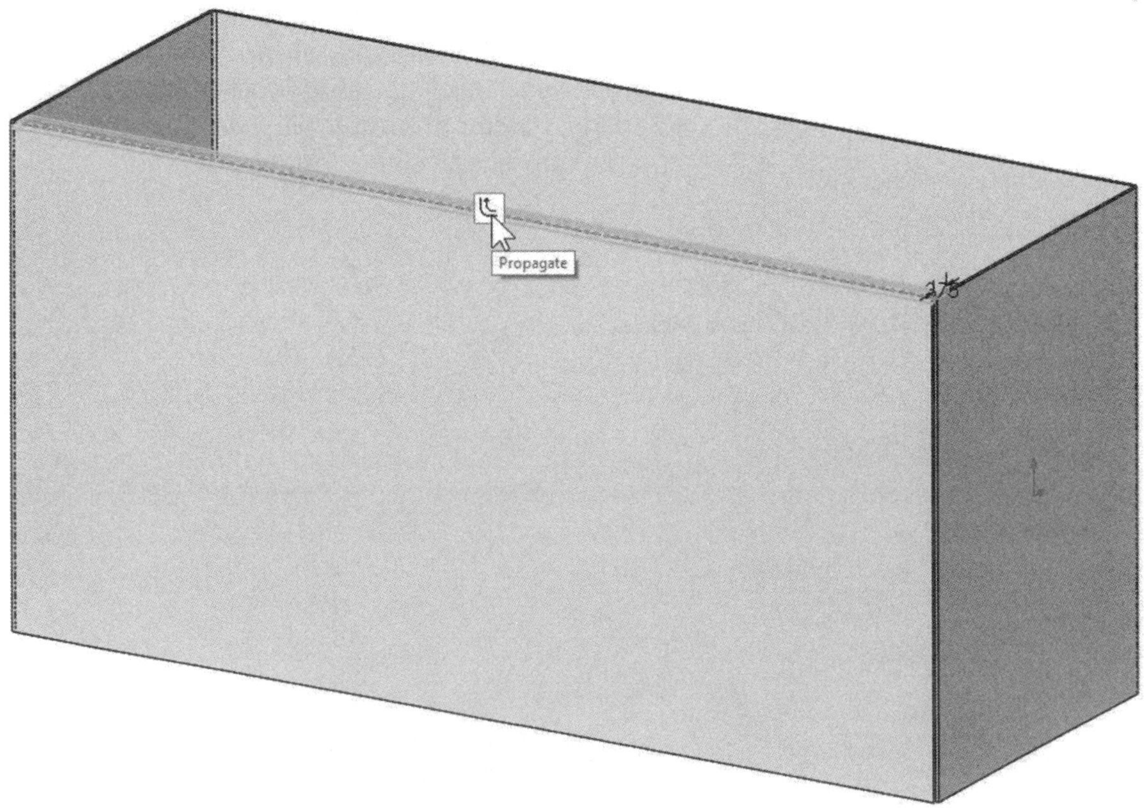

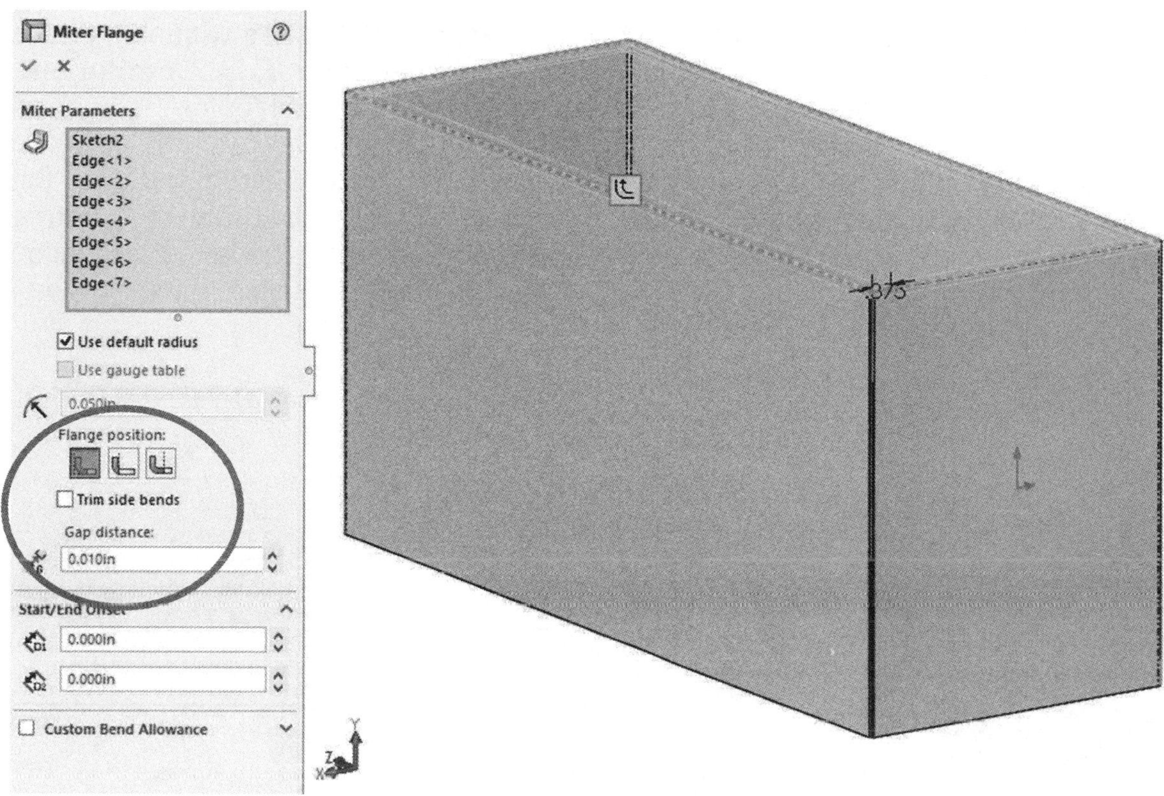

13.16. - Repeat the same "**Miter Flange**" command on the other side using the same options. Your part should now have miter flanges in the front and back and should look like this:

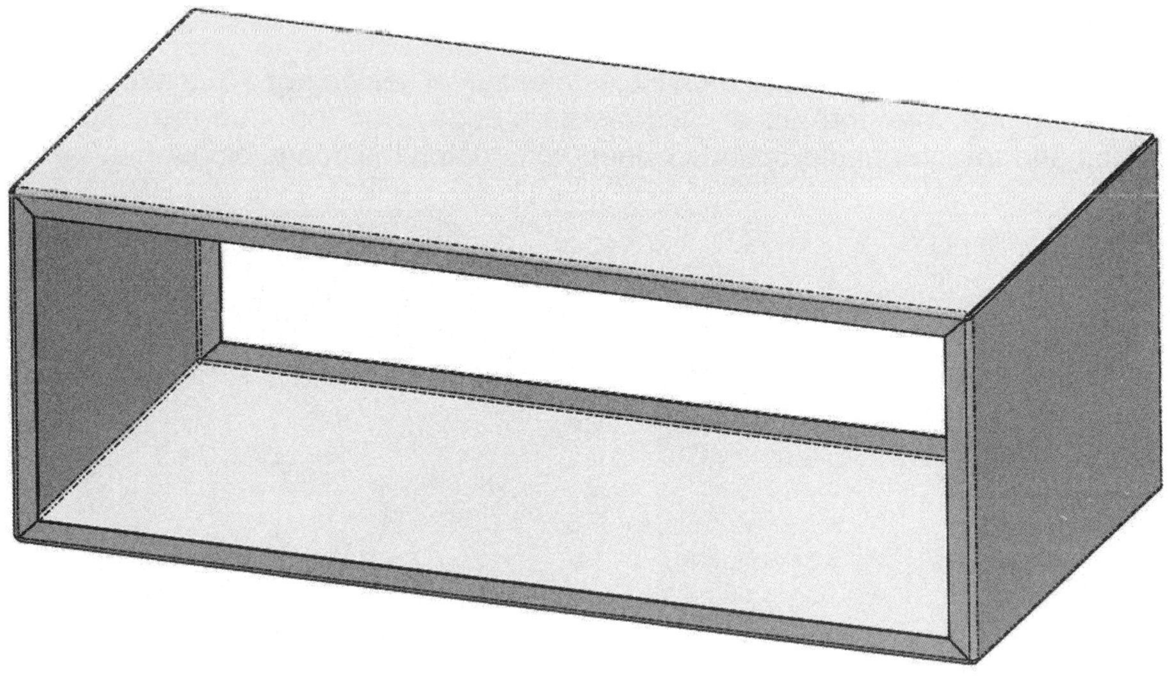

13.17. - Go back to the assembly window (if working in the part's window); notice we are still editing the part, since we never turned the "**Edit Component**" command off in the assembly. Now we need to add a form tool feature to support shelves inside the locker. Expand the "**Design Library**" and scroll down to "**forming tools, lances**" and find the "**90 degree lance**" tool. Drag-and-drop the lance in the right side of the locker. When adding form tool features, before releasing the mouse button we can press the "Tab" key to flip the lance's direction to go 'into' the part or 'out' of it. We can also flip the tool's direction in the Property Manager after dropping it.

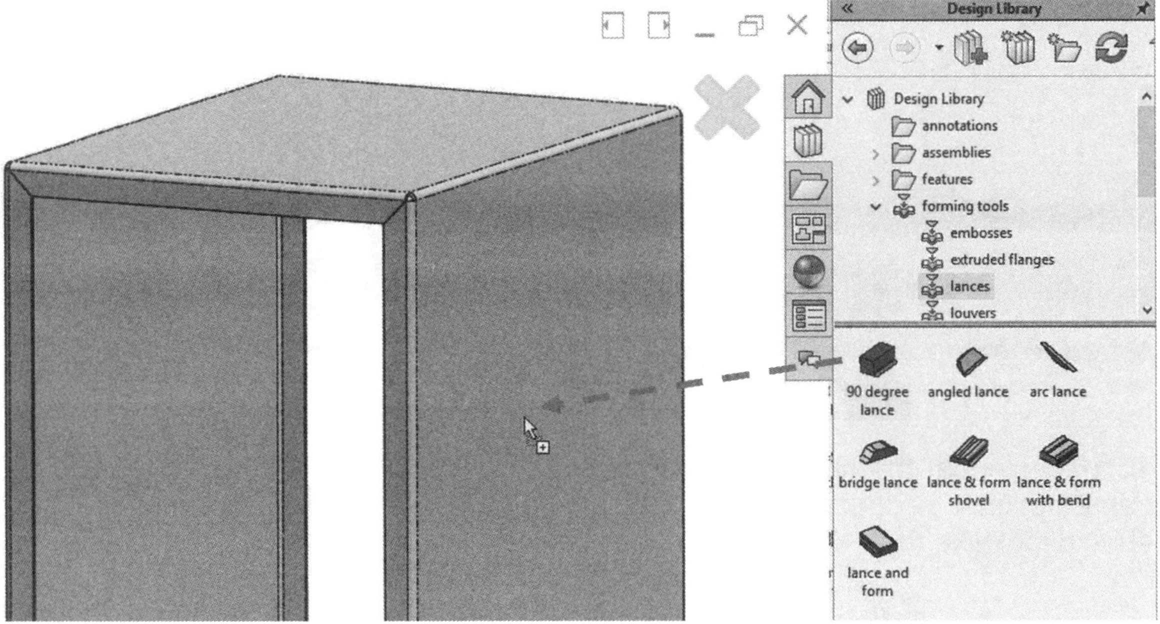

After dropping the lance on the side of the locker we need to make sure the lance is going 'into' the locker, and the bent edge is on top. Use the "Flip Tool" button and rotation angle options if needed to match the image below.

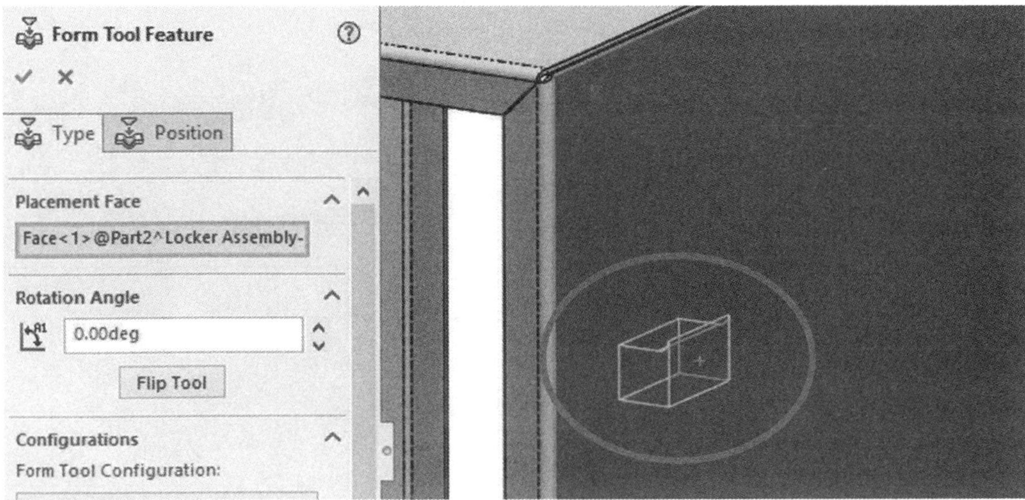

After adjusting direction and rotation, select the "Position" tab and dimension the feature's center 2″ from the front face and 9″ from the bottom of the locker's body. After dimensioning the location click OK to finish.

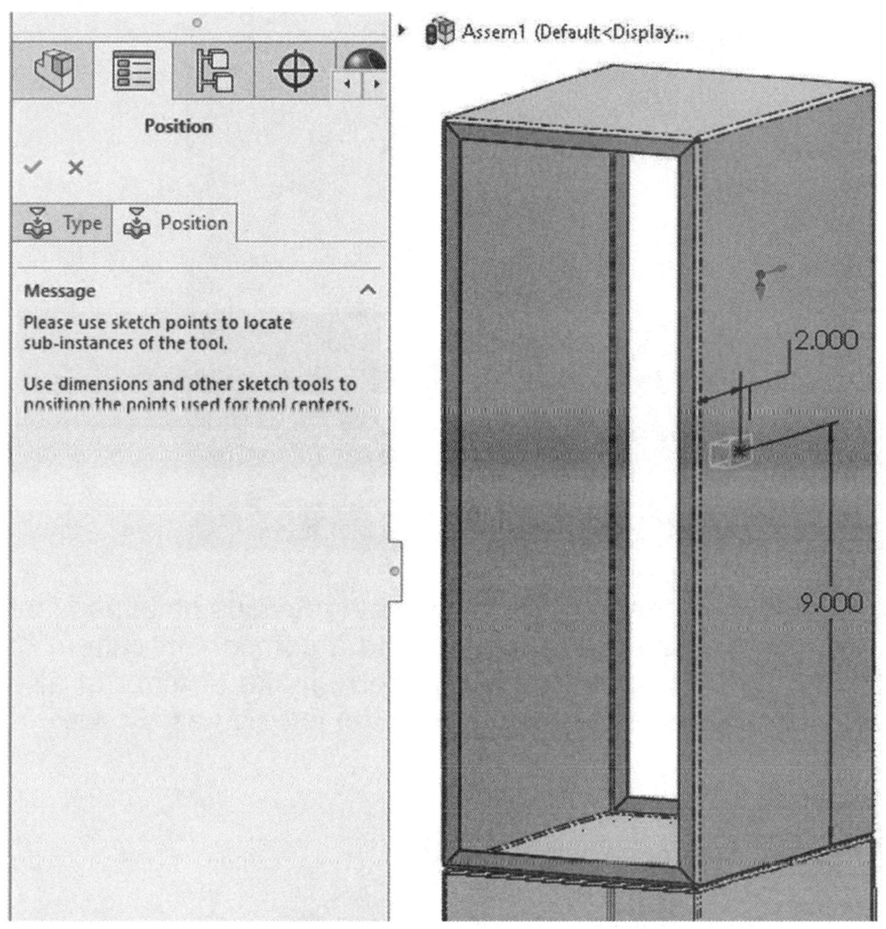

The lance should look like this from inside the locker when finished.

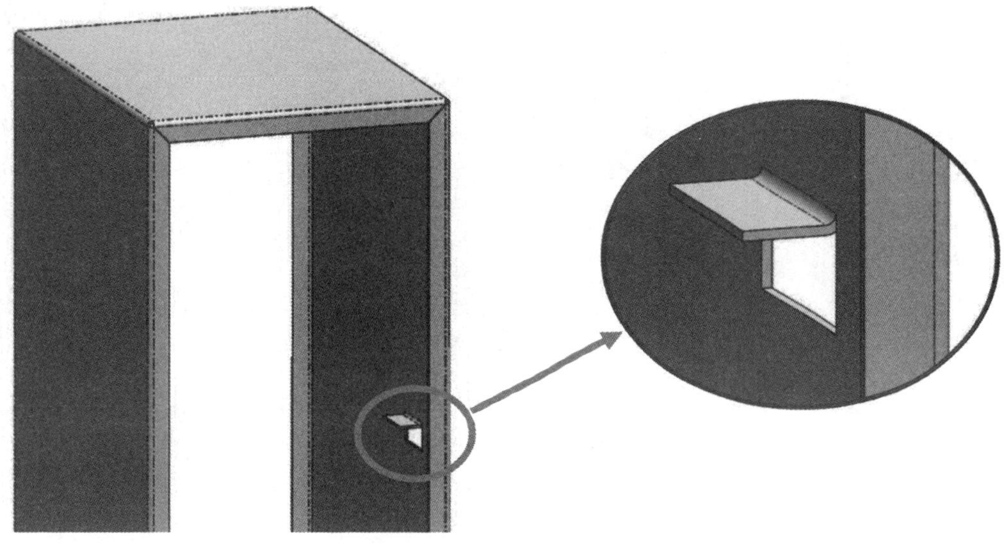

13.18. - Make a linear pattern with the lance going towards the back with 3 copies spaced 2″ apart.

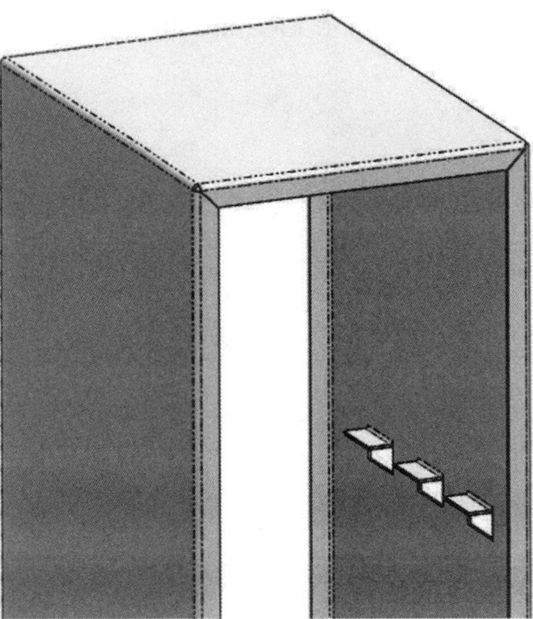

13.19. - And now add a linear pattern to copy the previous patterned lances 3″ up. The reason to add two patterns is to later add a *component pattern* driven by a *feature pattern*. After adding the vertical pattern, make a mirror of all lance cuts about the part's *"Right Plane"* to add them on the left side as shown.

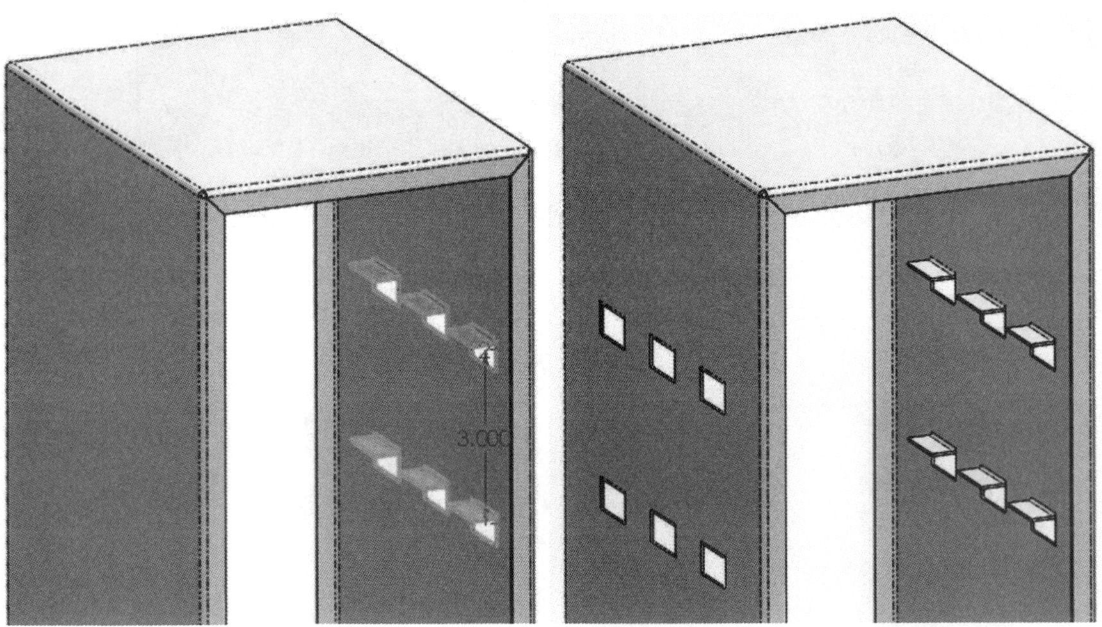

13.20. – After completing the lances, we must rename and externalize the part. Rename the internal part as *'Locker Body'* and save it to an external file.

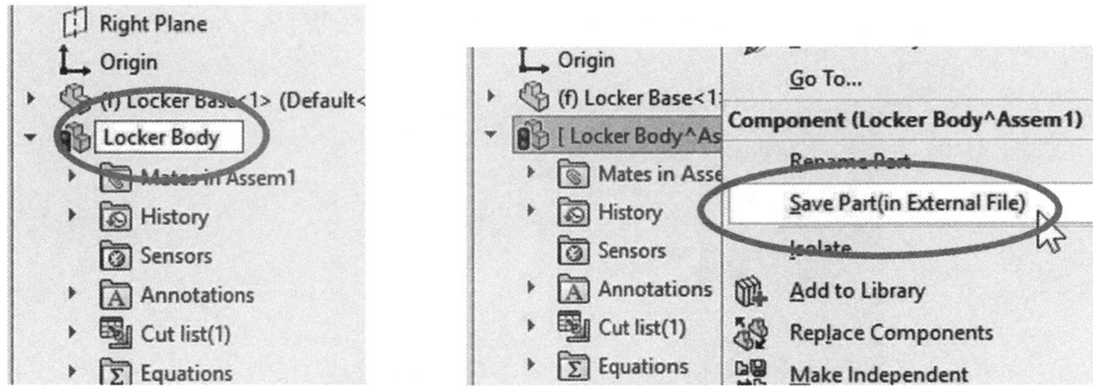

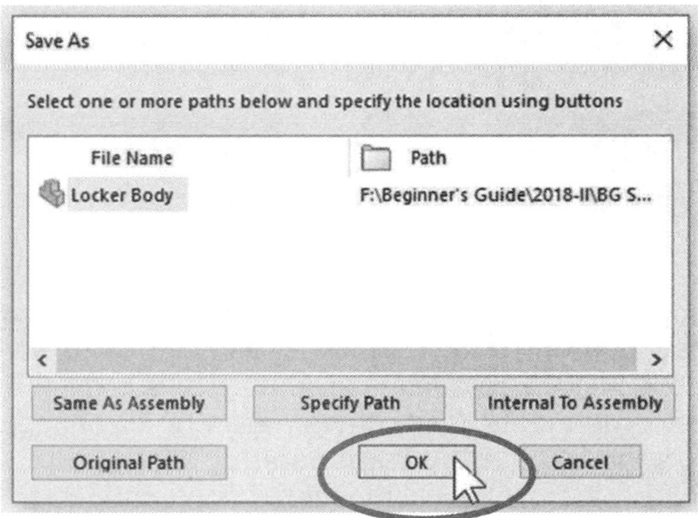

13.21. - Now that we are finished with the locker's body we can turn off "**Edit Component**" and go back to editing the assembly to add the next components, a cover in the back and a door in the front.

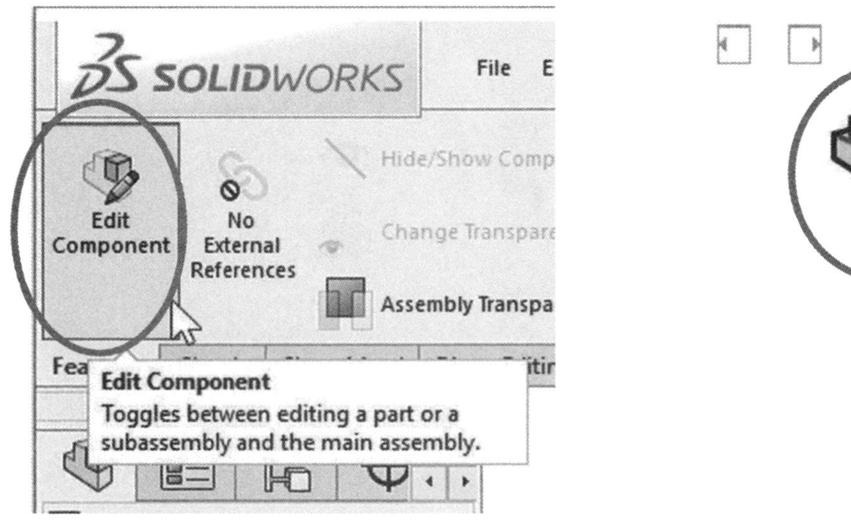

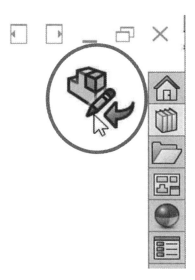

13.22. - The back is just a sheet metal cover that will be welded or riveted in place; it will be the same size as the locker's body, and will also be made in the context of the assembly. Select "**Insert Components, New Part**" drop down menu and select the inside face of the rear miter flange to add the back cover.

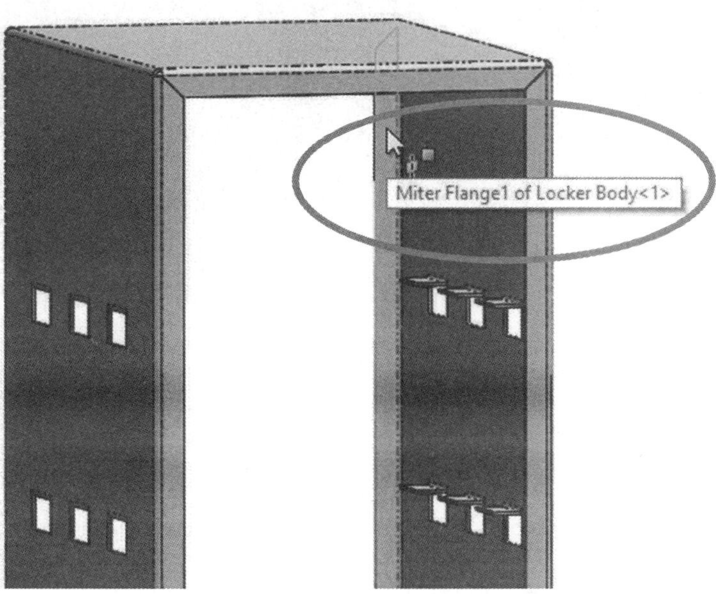

After adding the new part, the locker's base and body turn transparent; we are editing the new part and a new sketch is ready for us to work on it, but in this step, it is easier to work with an opaque assembly, as it is difficult and/or confusing to reference other components if they are transparent. To change the component's appearance, select the "**Assembly Transparency**" icon in the CommandManager and select "Opaque." If the other components don't change to opaque, we may have to rebuild the assembly to update the component's display. If we rebuild the assembly, we will exit the sketch. In this case, add a new sketch to the "*Front*" plane of the new part to continue.

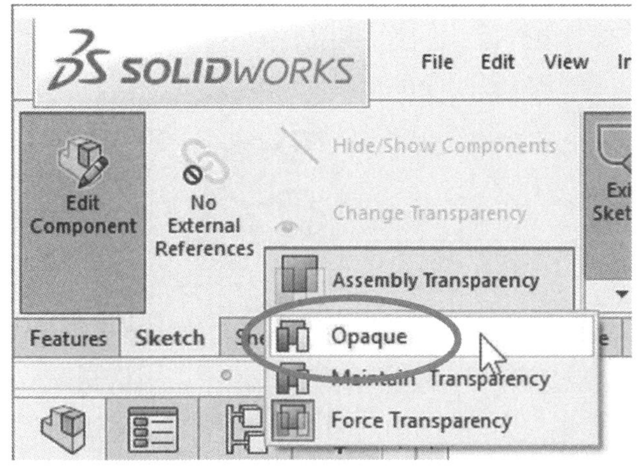

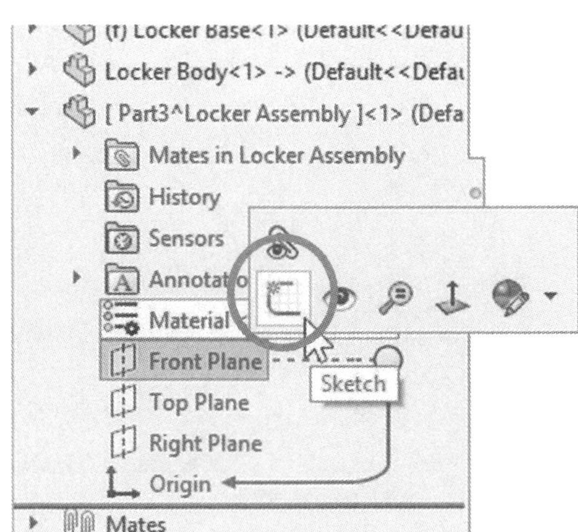

13.23. - Draw a rectangle starting in one of the lower corners (inside the locker) and finishing in the opposite upper corner. Be sure to capture a Coincident relation to a miter flange vertex. By adding these relations to the miter flange the sketch will be fully defined, and now we are ready to extrude the base flange.

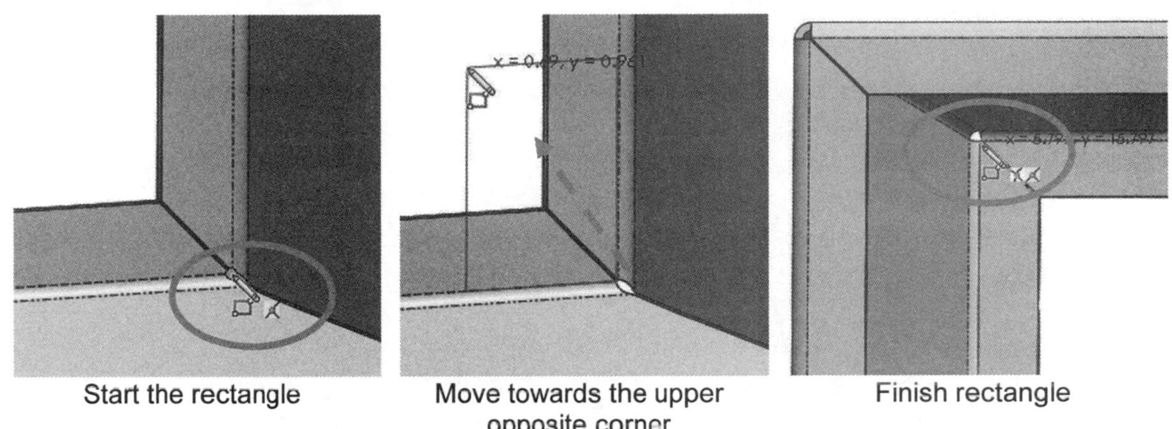

Start the rectangle | Move towards the upper opposite corner | Finish rectangle

Select "**Base Flange/Tab**" from the Sheet Metal toolbar or the menu "**Insert, Sheet Metal, Base Flange.**" Like the other parts, make the base flange from 18 gauge steel making sure the material is added inside the locker and does not overlap with the locker's body. This is a view from behind the locker, showing the new part is not overlapping with the '*Locker Body*'. Turn on the "Reverse Direction" option if needed.

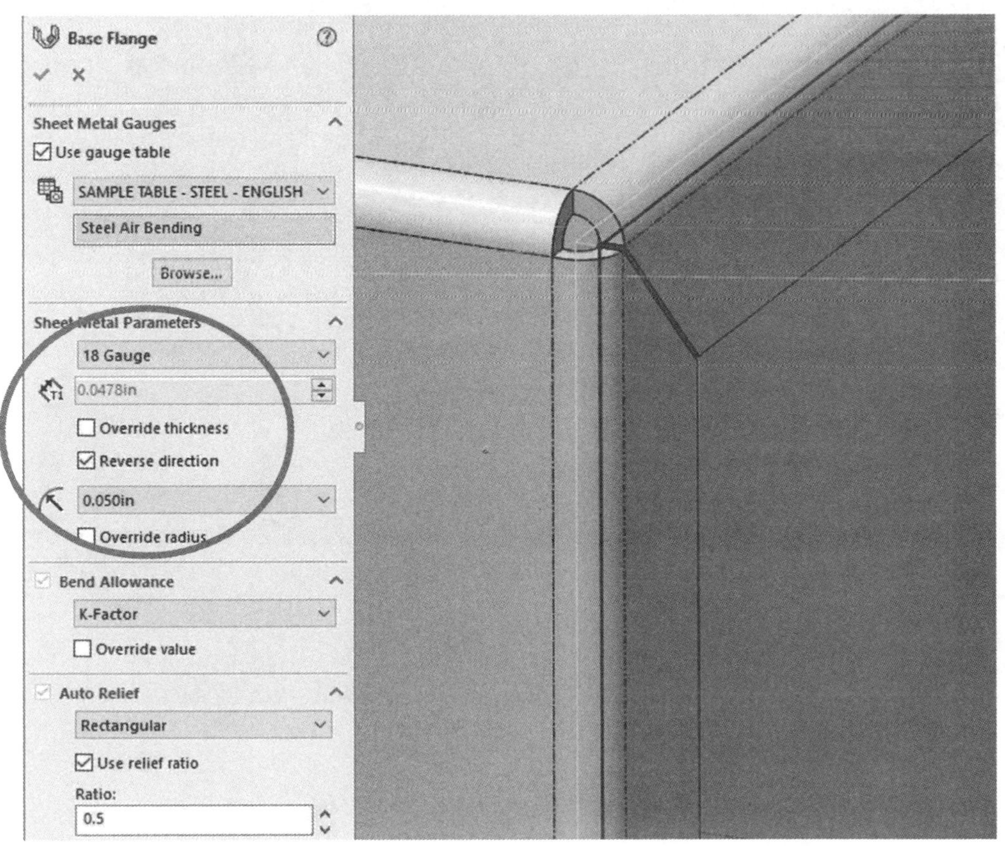

13.24. - Rename the part "**Locker Back**" in the FeatureManager, save it to an external file, and turn off "**Edit Component**" to continue editing the assembly.

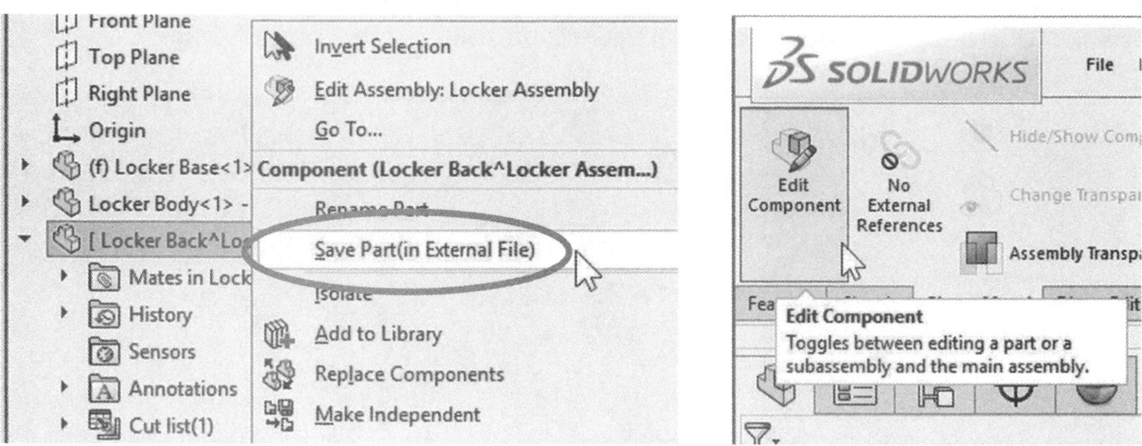

13.25. - The next component to be made will be a shelf. Select "**Insert Components, New part**" and select the top face of the lower lance to locate it. The magnifying lens can be used to zoom if needed to locate the new part.

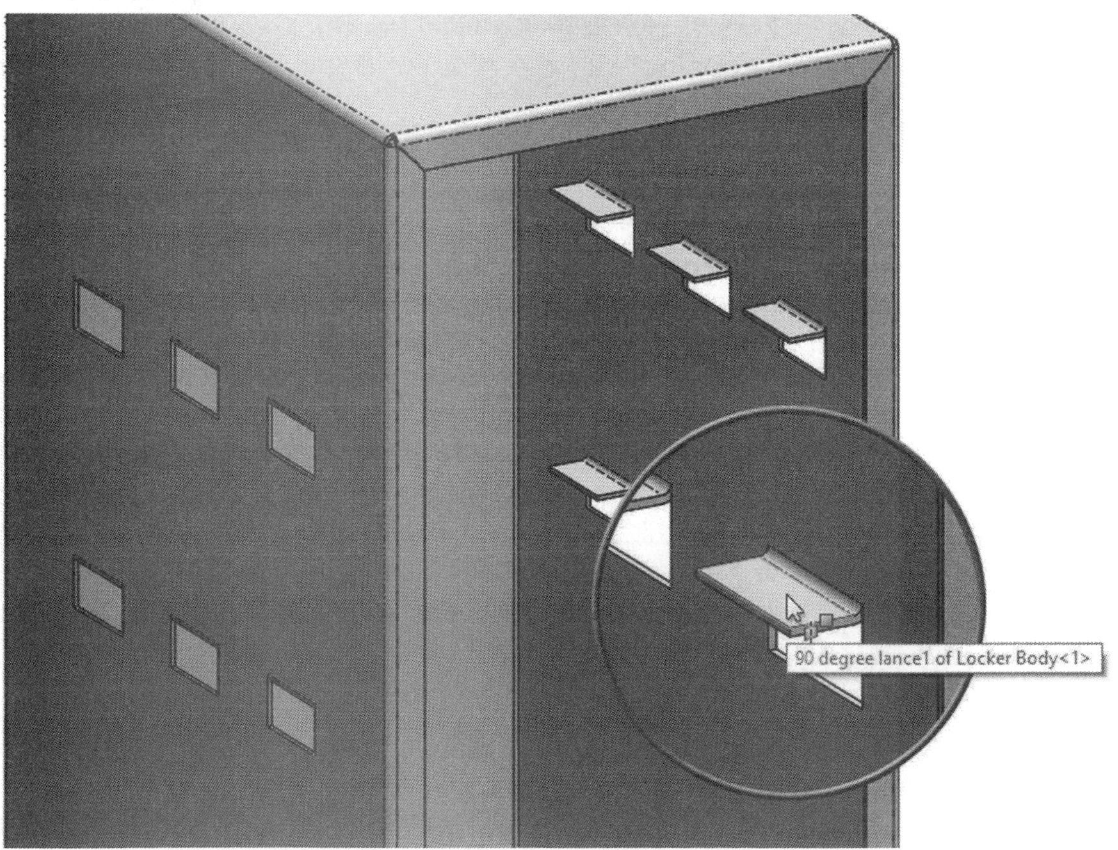

13.26. - Since we changed the assembly transparency setting to "Opaque" in the previous part, the other components will not become transparent when the new part is added. Draw a rectangle starting close to the lower front right lance and finishing close to the rear left lance as indicated; add a collinear relation to the bend lines and dimensions as indicated.

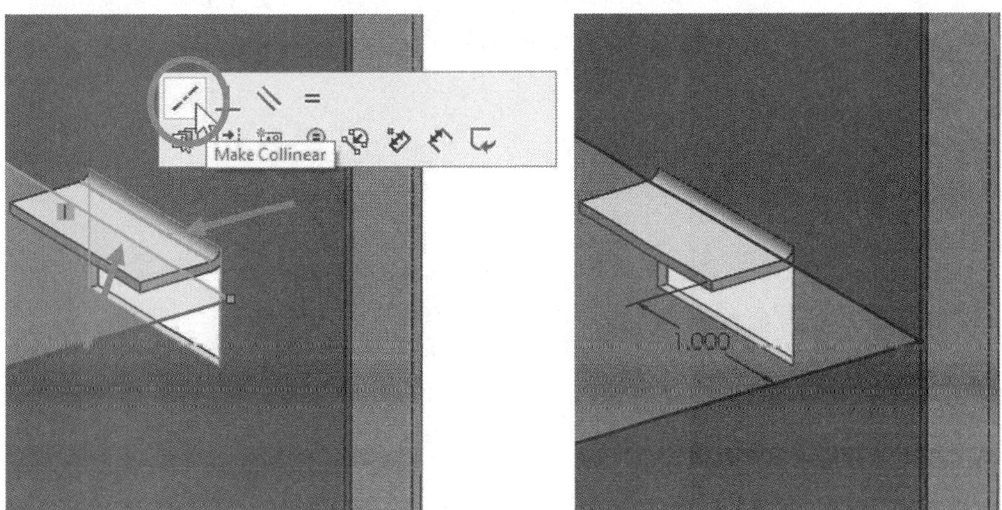

Lower front right lance

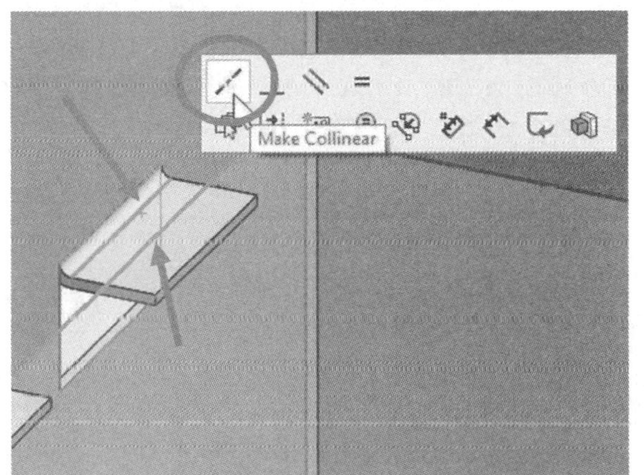

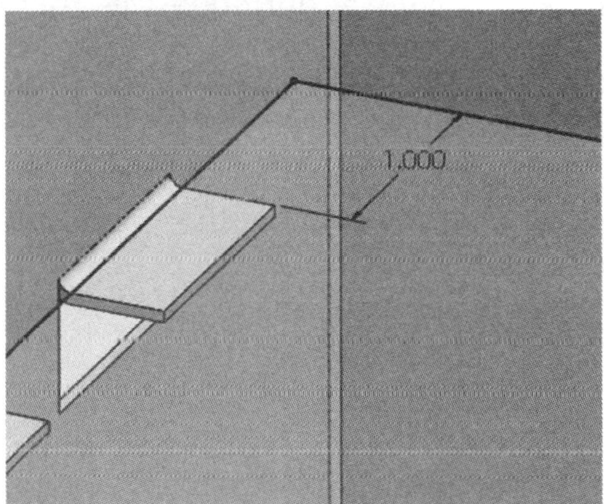

Lower rear left lance

13.27. - Click the "**Base Flange/Tab**" command from the Sheet Metal toolbar and make the shelf using 16-gauge steel. Since the shelf will be holding the locker's contents, it needs to be thicker. Be sure to add the material going up and not overlap with the lances. Set the bend radius to 0.075", and a Rectangular auto relief with a 0.5 ratio. Click OK to create the base flange and continue.

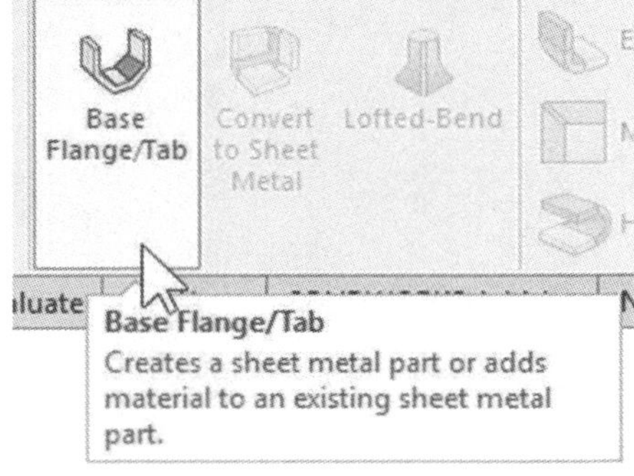

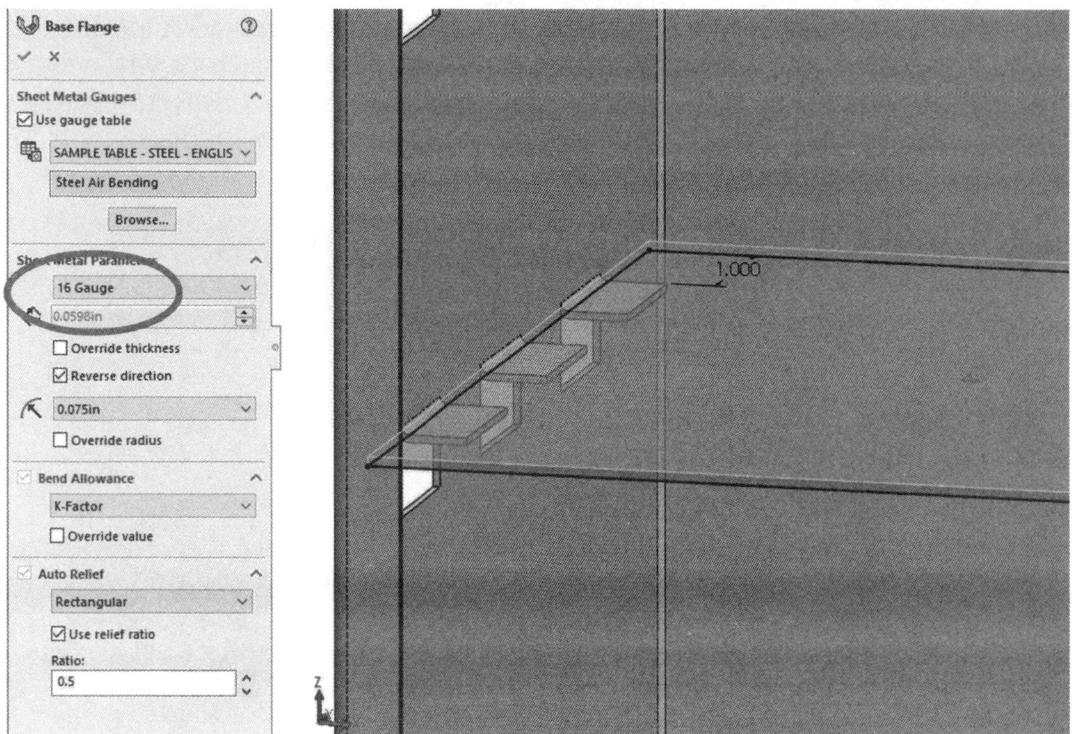

13.28. - Since a flat shelf like this will not be strong enough, we need to add a flange around for reinforcement. Click the "**Edge Flange**" command and add all four edges. Immediately after selecting the first edge a preview will be displayed, click towards the bottom to define the direction of the flange, then select the other three edges. Set the flange length to 0.625″; leave the flange gap to the default value of 0.01″ and set the flange position to "Material Inside."

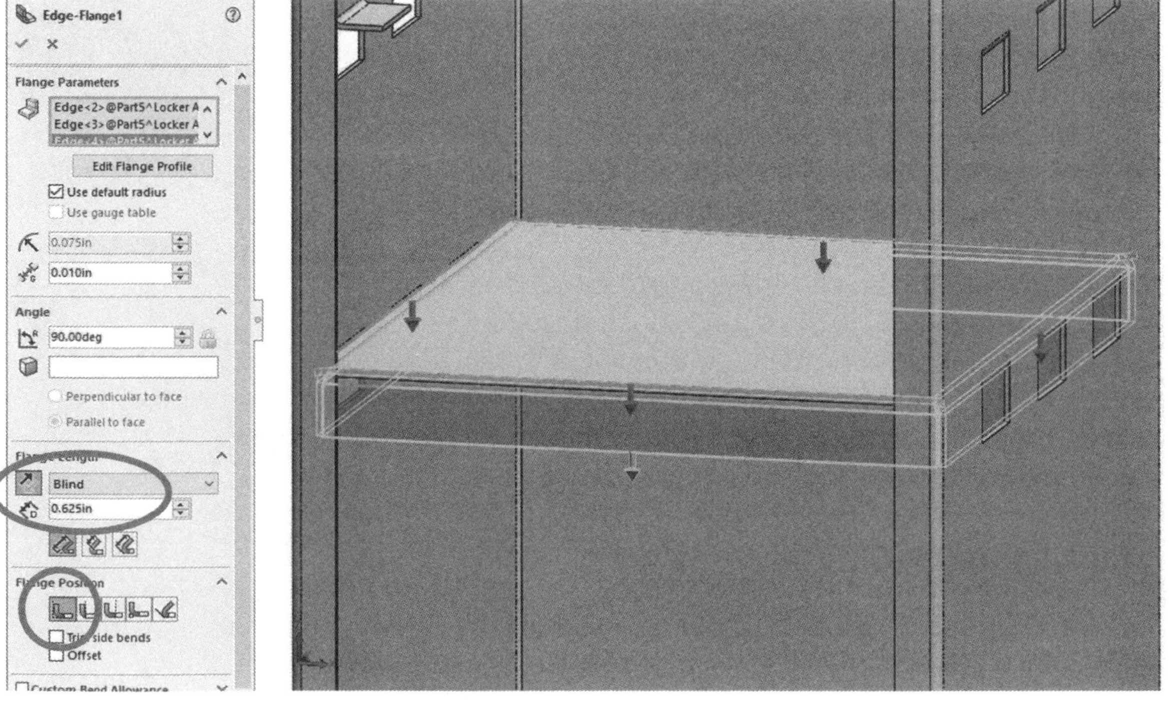

13.29. - The flanges just made interfere with the lances in the *'Locker Body'.* To eliminate the interferences, we'll unfold the side flanges, add cuts to clear the lances and then re-fold them. Select the "**Unfold**" command from the Sheet Metal toolbar and click in the top face of the shelf to select it as the fixed face.

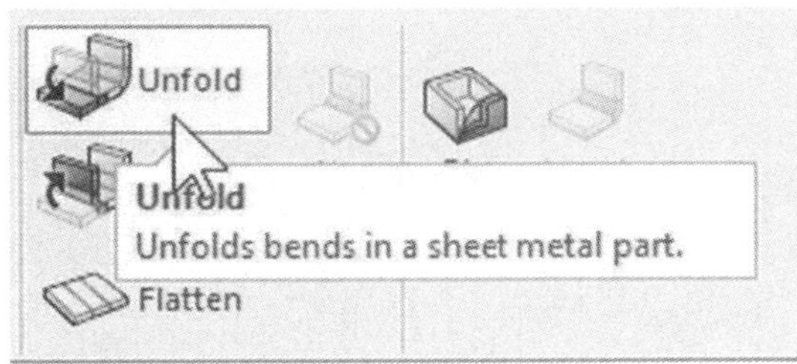

After selecting the fixed face, select the two side bends (the ones interfering with the lances) to unfold them. SOLIDWORKS will activate a selection filter to make it easier to select the bends. Click OK when done. After the bends are unfolded, the shelf will temporarily interfere with the *'Locker Body'*.

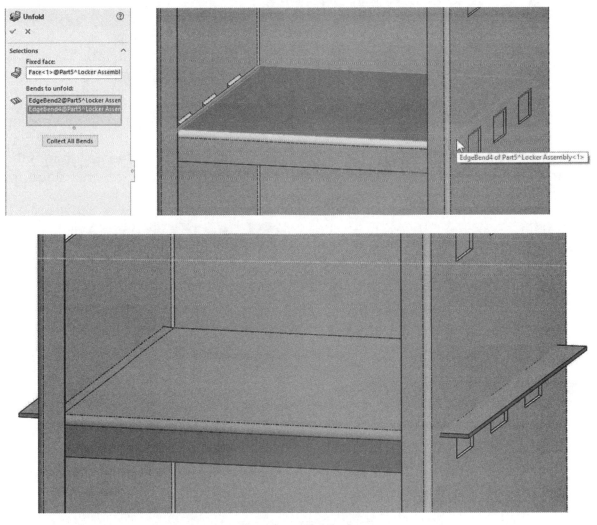

Bends unfolded

13.30. - To better visualize the next operation, we'll tilt the locker to view it from the bottom to make the cuts needed to clear the lances in the *'Locker Body'*. The reason is because from this angle we can see the lances we need to clear. Rotate your locker assembly until you can see the lances from below in one side, and change the shelf's color to differentiate it from the other components.

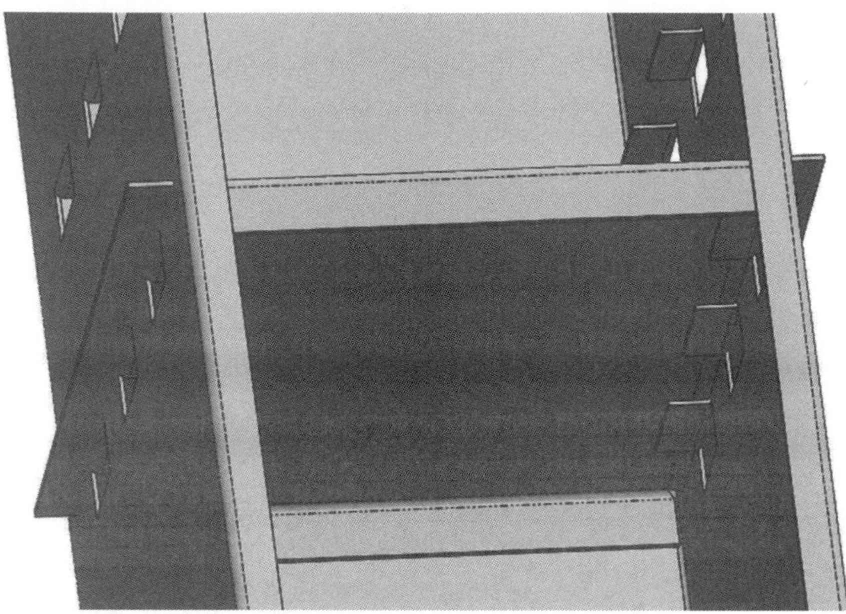

13.31. - Add a new sketch in the bottom face of the shelf. Draw a rectangle adding the necessary dimensions and relations to clear the lances as shown. Make one side of the rectangle coincident to the outside edge of the unfolded flange and the other side collinear to the bend line as indicated. Leave 0.1" on the sides for clearance.

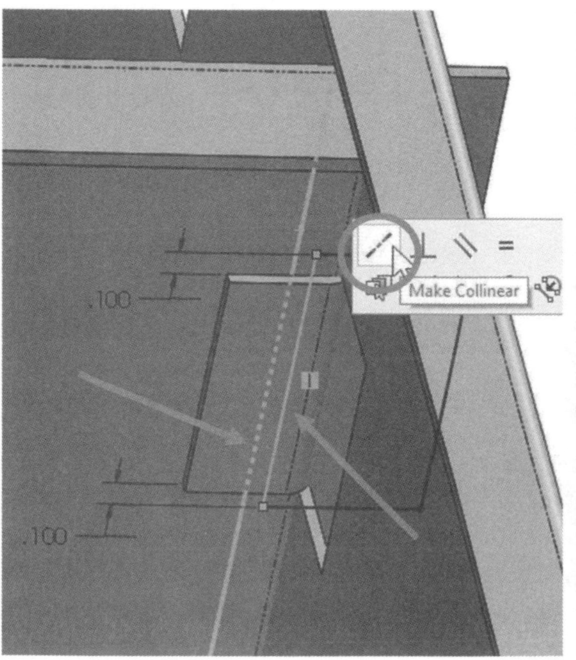

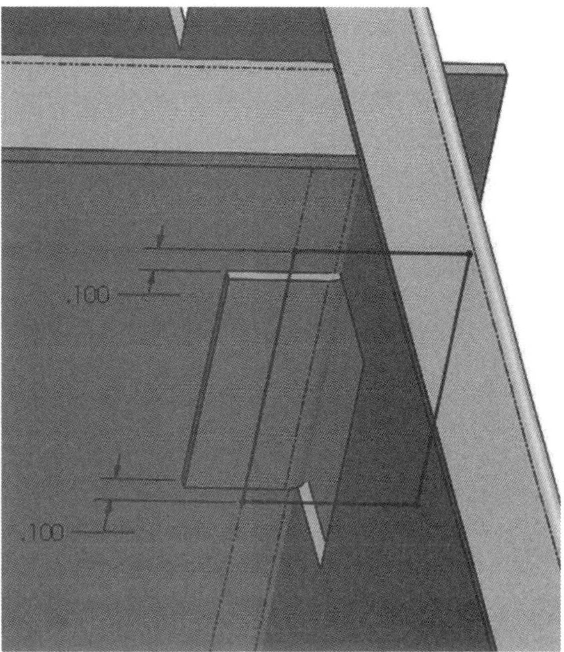

13.32. - With the sketch fully defined, our next step is to add a "**Sketch Pattern**." Not only can we make patterns of features, we can also make linear and circular patterns of sketch entities in the sketch. Select the menu "**Tools, Sketch Tools, Linear Pattern**" (the icon may not be in the Sketch toolbar by default).

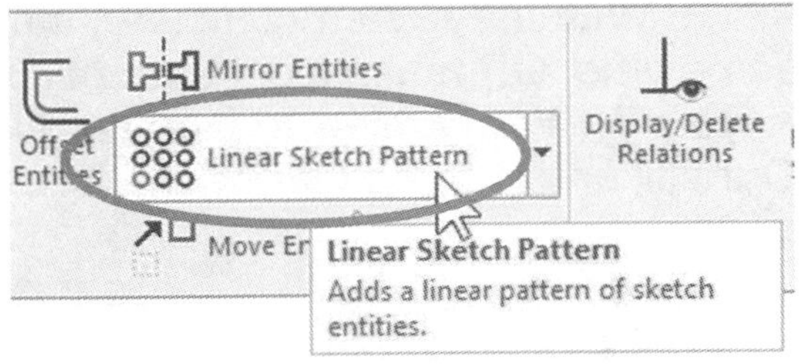

The "**Linear Sketch Pattern**" behaves like a feature pattern. By default, the "**X-axis**" direction is the sketch's horizontal direction (the short red origin arrow) and the "**Y-axis**" direction the vertical (long red origin arrow). For this pattern, we need to have 3 copies along the "Y-axis" spaced 2″ apart.

Turn on the option "**Dimension Y spacing**" to add a dimension between copies, otherwise the spacing between copies can only be changed by selecting a pattern instance and using the menu "**Tools, Sketch Tools, Edit Linear Pattern**."

Add the rectangle's lines to the "Entities to Pattern" selection box, enter the values for the "Y-axis" (*or* select an edge for the direction) and click OK to finish.

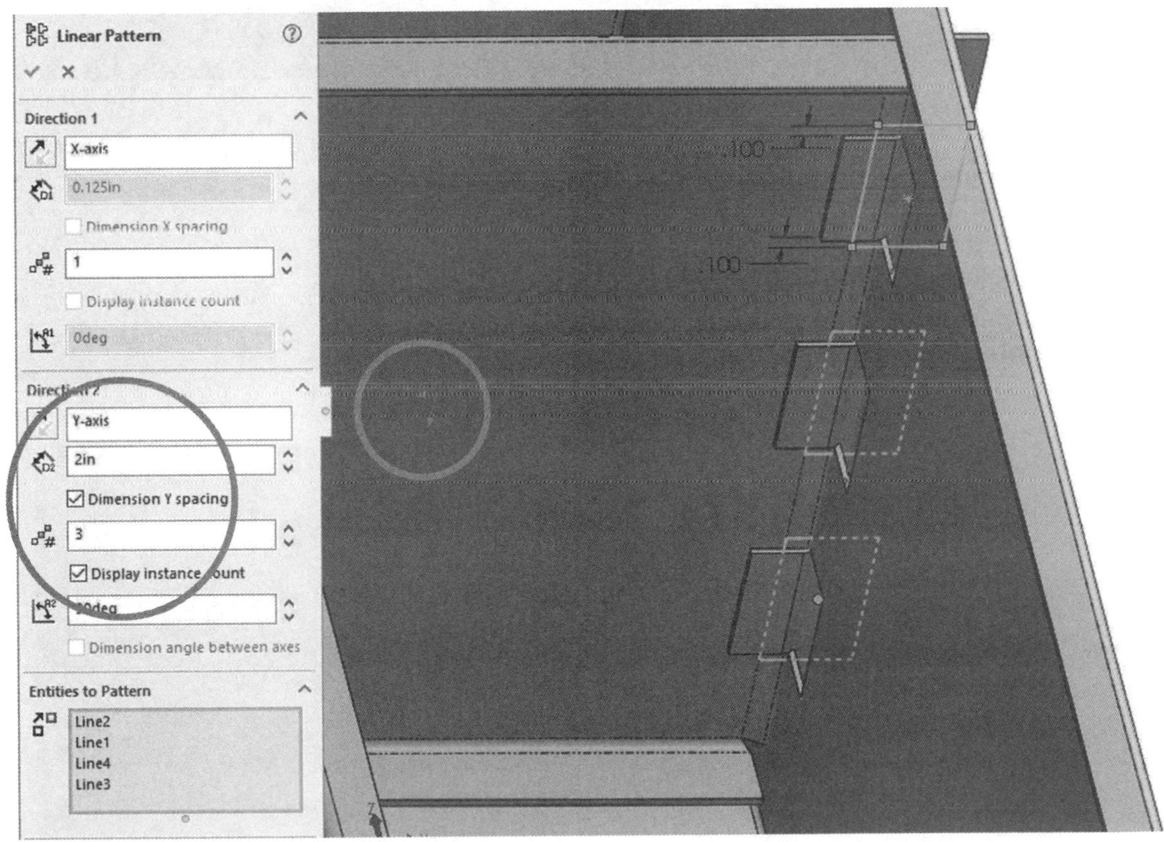

The direction of a "Sketch Pattern" can also be defined by selecting a model edge or a sketch line.

With the sketch complete, make a sketch mirror to make all six cuts at the same time, OR make a cut *and then* mirror the extruded cut. We'll leave the option to you. Just remember to use the "Link to thickness" option when making the cut. Our still unfolded part looks like this:

13.33. - To finish the shelf part, select the "**Fold**" command from the sheet metal toolbar. Either select both unfolded bends (the automatically enabled selection filter will only let us select bends) or click in the "**Collect All Bends**" button to add all unfolded bends to the selection list. After clicking OK the part is refolded. The next image is shown using the "**Force Transparency**" option.

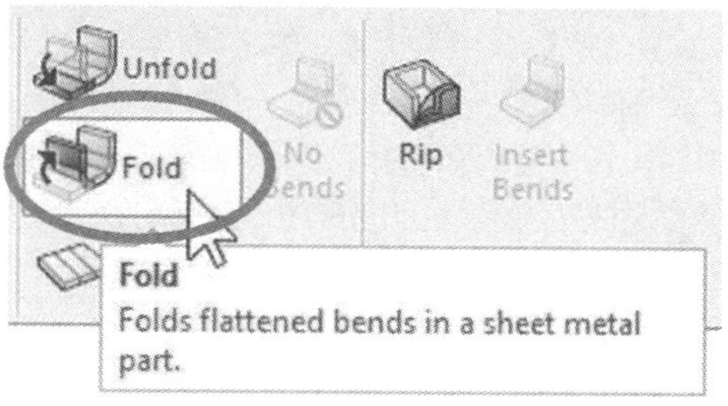

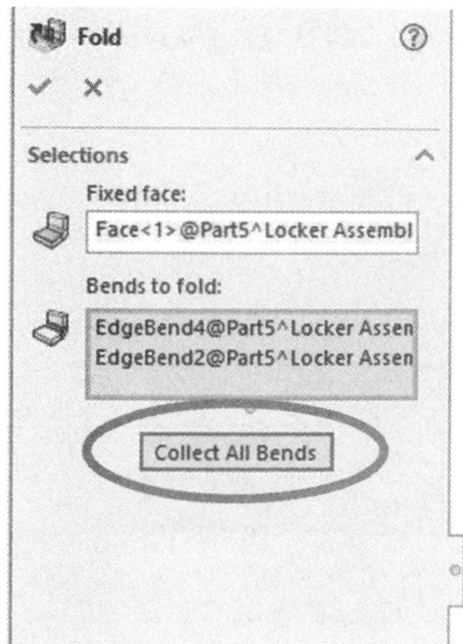

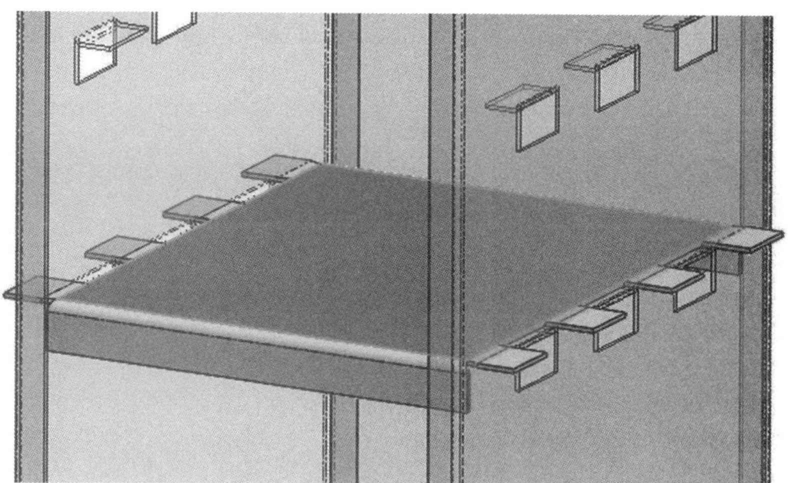

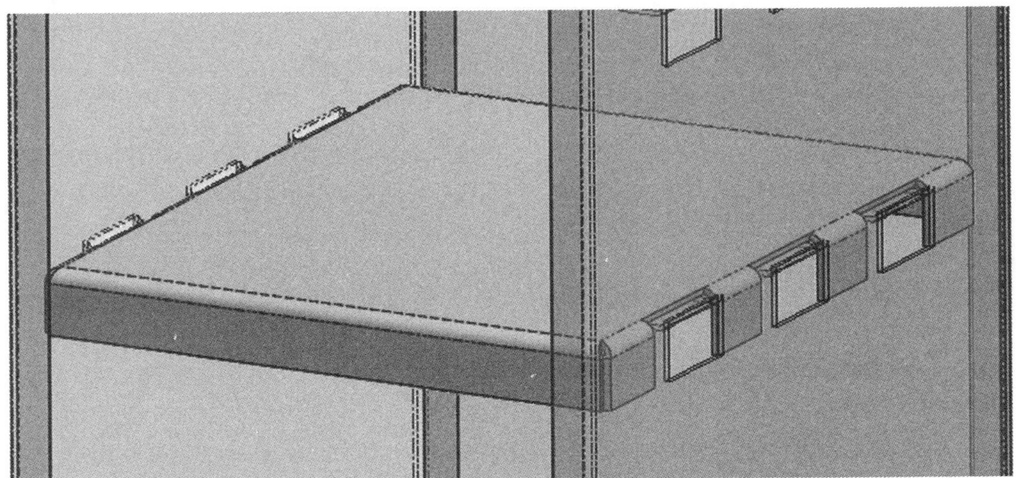

13.34. – At this point we want to externalize the shelf part. Stop editing the part in the assembly by turning off the "**Edit Component**" icon. Rename it *'Locker Shelf'* in the FeatureManager and save as an external part.

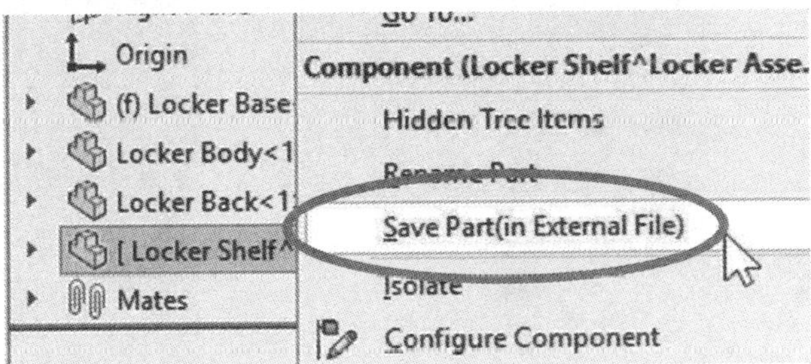

13.35. - The next step is to add a second *'Locker Shelf'* at the top. One way to do this is either by using the "**Insert Component**" command and mating it to the *'Locker Body'* or using a "**Component Pattern**." Parts and sub-assemblies can be patterned in an assembly using linear or circular patterns just like features in a part. In this example, a better option to maintain design intent is to use a "**Pattern Driven Component Pattern**." While editing the assembly, select "**Pattern Driven Component Pattern**" from the "**Linear Component Pattern**" drop down menu, or the menu "**Insert, Component Pattern, Pattern Driven.**"

In a "**Pattern Driven Component Pattern**" a copy of the component is added at each instance of a patterned feature. This is the reason why we patterned the lances vertically; for example, if we change the vertical pattern of lances to 3 copies, we would automatically get 3 shelves in the assembly.

In the "Components to Pattern" selection box select the *'Locker Shelf'* part, and for the "Driving Feature" select one of the lances in the top position. Turn on the "Propagate component level visual properties" checkbox to make the patterned copies look like the seed part. Click OK to finish the pattern.

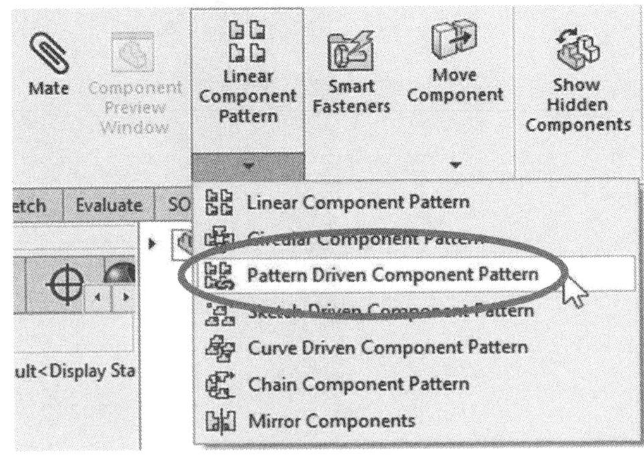

Be sure to select a lance in the top right side of the locker, and remember the lances in the left are a mirror copy, not a pattern copy.

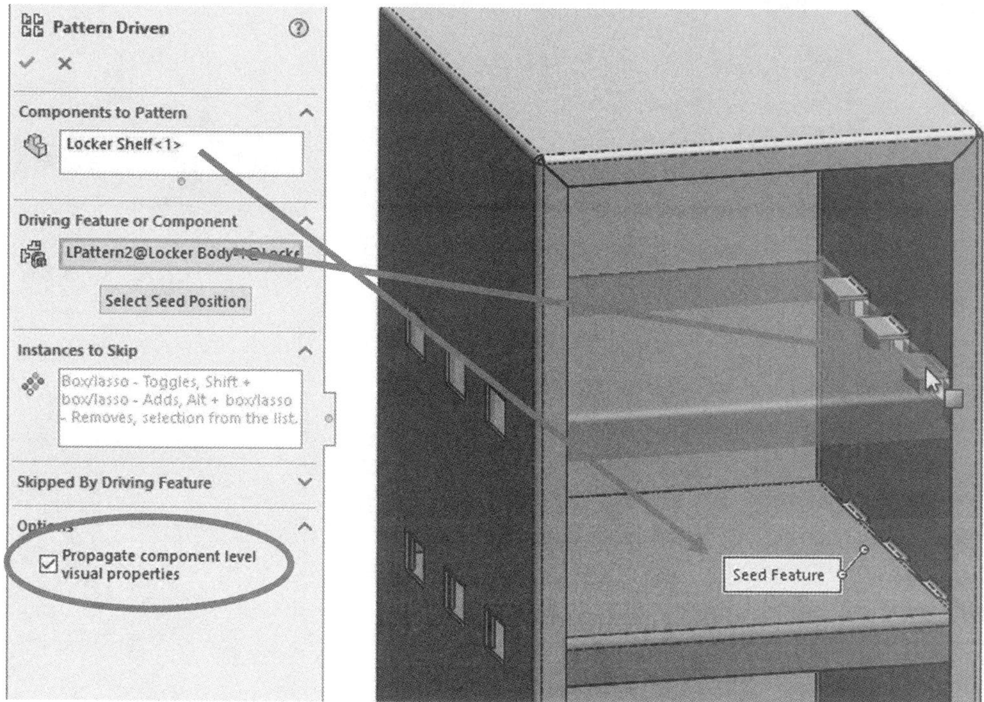

This pattern is very handy, especially when adding fasteners to an assembly; we can add one fastener to one hole and use the pattern of holes to add the rest of the fasteners, so that if the pattern of holes changes, the number of fasteners is also updated.

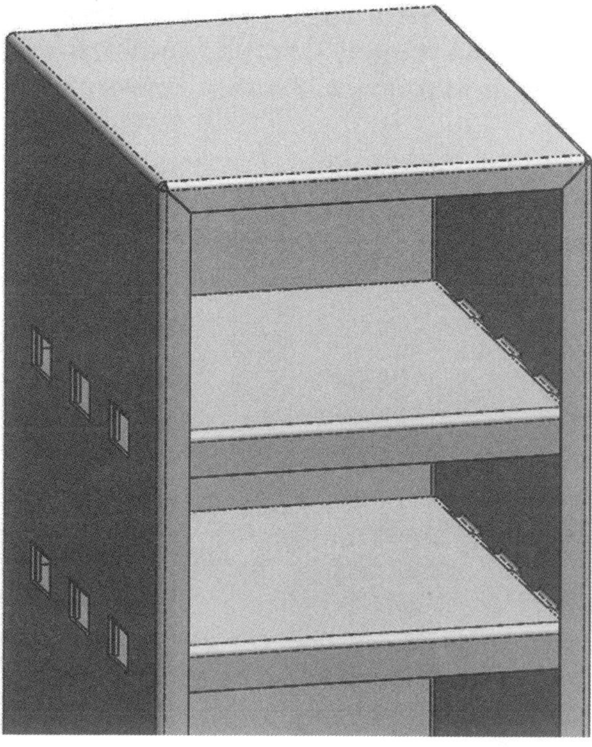

13.36. - After adding the top shelf, we realize the top space needs to be made smaller, and now we need to make a change to our design. Expand the FeatureManager of the *'Locker Body'* and find the *"LPattern2"* feature (the vertical pattern feature). Double click in it to display its dimensions and change the spacing from 3″ high to 4″. Rebuild the model after the change and notice both the lances pattern and the number of shelves is updated.

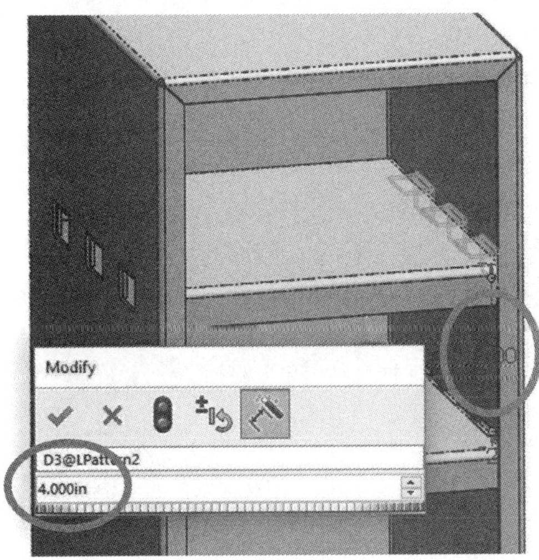

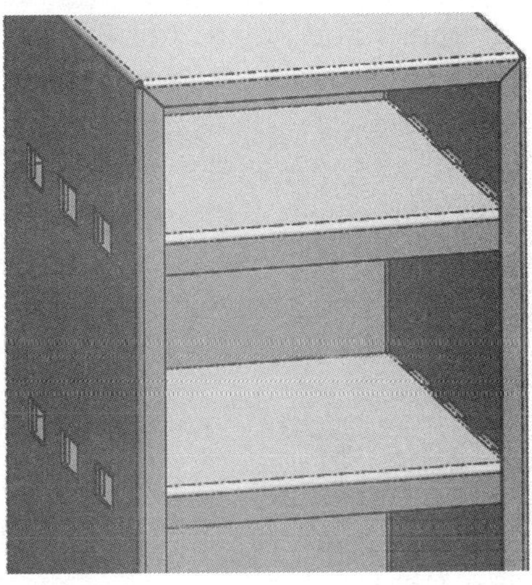

13.37. - The next step is to add a "**Hem**" to the front flange of the *'Locker Shelf'*. This hem will eliminate a sharp edge to reduce the risk to the end user, and at the same time reinforce it. The first thing we need to do is edit the *'Locker Shelf'* in the assembly. Select it in the graphics area or the FeatureManager and click the "**Edit Part**" command. If needed, set the Assembly Transparency option to "Force Transparency." We can edit either the top or bottom shelf, since it's the same part.

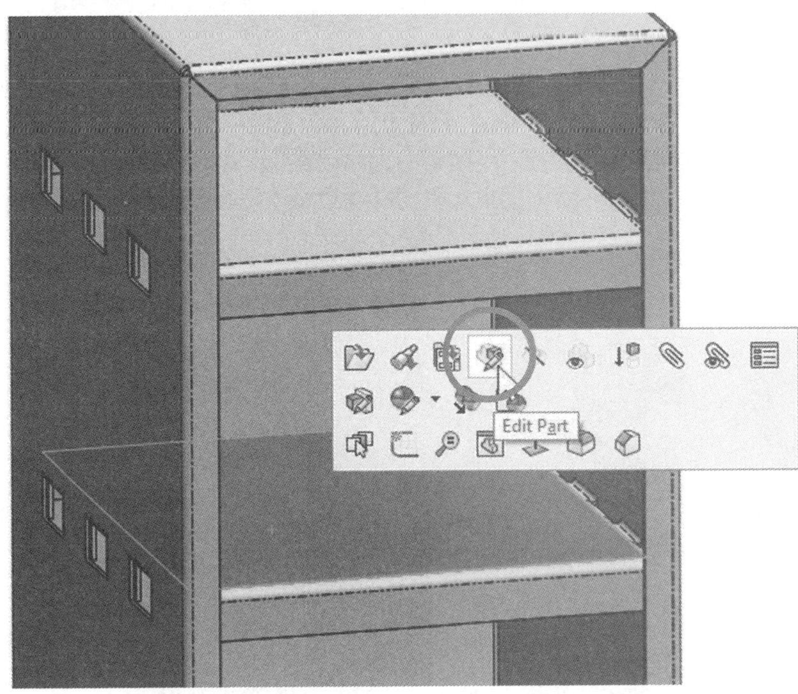

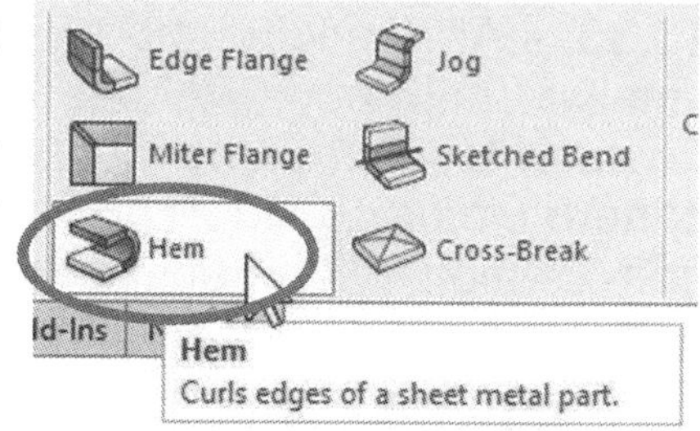

13.38. - Select the "**Hem**" command from the Sheet Metal toolbar. In the "Edges" selection box, pick the front edge of the *'Locker Shelf'*. A preview will show the hem's direction; make sure it's inside the locker and, if needed, use the "Reverse Direction" checkbox. For the *'Locker Shelf'* use a "Closed" hem 0.20″ long with the "Material inside" option. Click OK to add the hem and finish. After completing the hem notice that the top copy of the *'Locker Shelf'* also shows the hem just added.

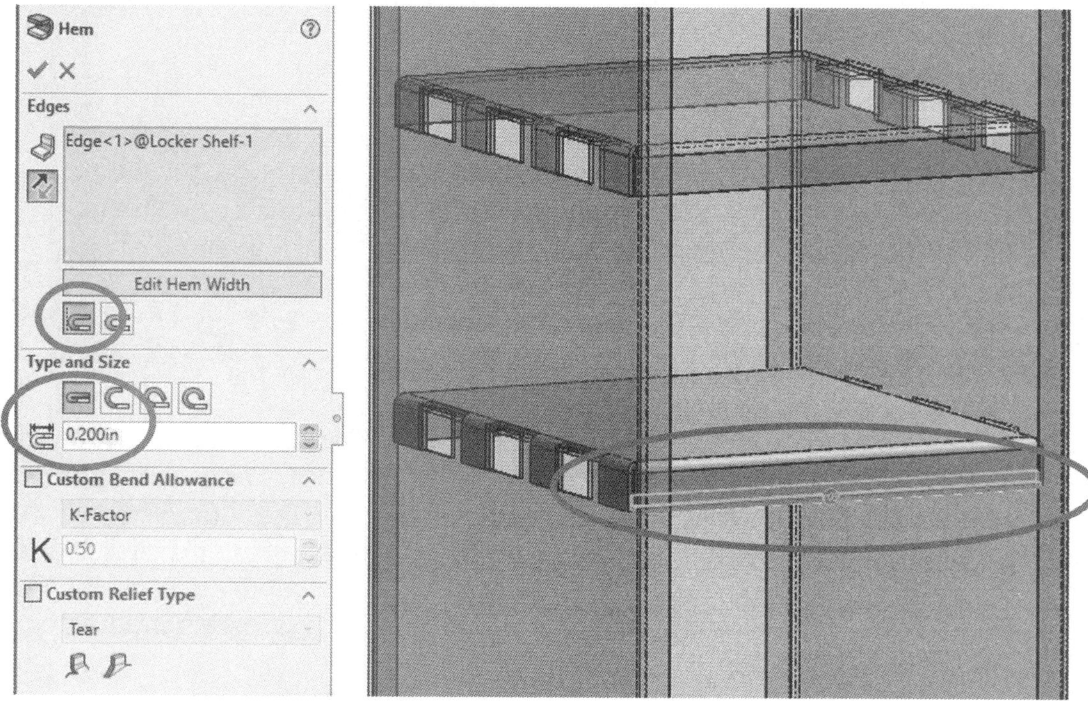

13.39. - Turn off "**Edit Component**" to stop editing the '*Locker Shelf*' and edit the assembly. The last component to add to our assembly will be the locker's door, which will also be designed in the context of the assembly. Select "**Insert Components, New Part"** and click in the front face of the *'Locker Body'* to locate the door.

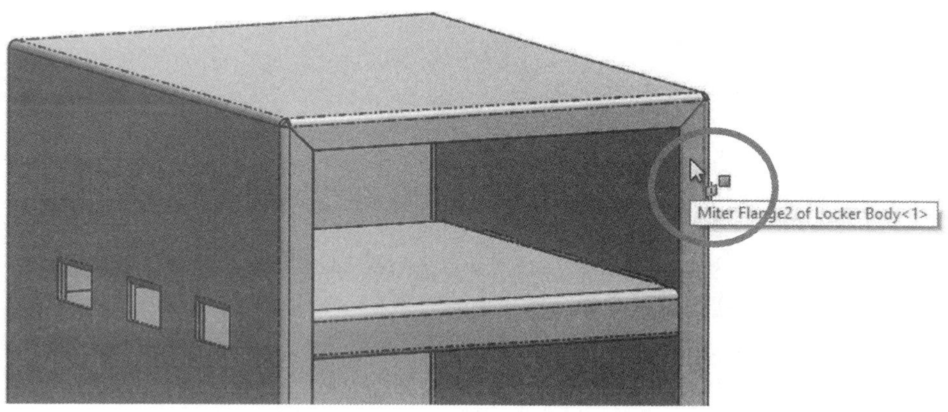

The purpose of adding relations in the context of the assembly is to propagate changes when other components change; however, in the case of the door, we have a slightly different situation. We want to be able to open the door and move it; however, if we capture relations to make the door the same size as the *'Locker Body'*, later when we open the door (moving the door in the assembly), it will be rebuilt and its size will be incorrect, fail to rebuild giving errors, and/or produce unexpected results when SOLIDWORKS tries to maintain the in-context relations. Notice how the door is smaller when it's opened after adding in-context relations to make it the same size of the *'Locker Body'*.

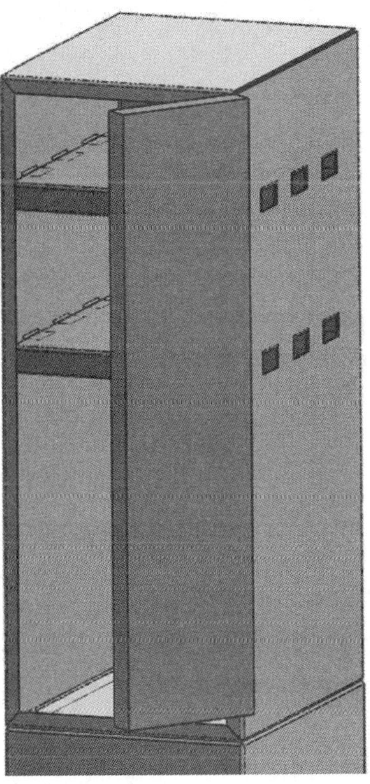

13.40. - One thing we can do to prevent capturing external relations is to activate the "**No External References**" option. While this command is active, no external references are captured to other components, automatically or manually. After the new part is added to the assembly, activate "**No External References**" and draw a rectangle for the door as indicated. Dimension it to match the locker's size and then add a Midpoint relation between the rectangle's lower horizontal line and the part's origin to center it about the origin (almost always a good idea). This relation will fully define the sketch. To be sure we are adding the relation to the part's origin, select it in the FeatureManager. Make the assembly opaque for clarity.

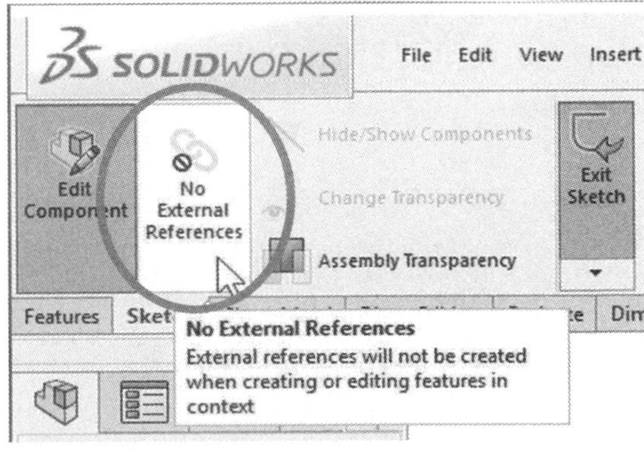

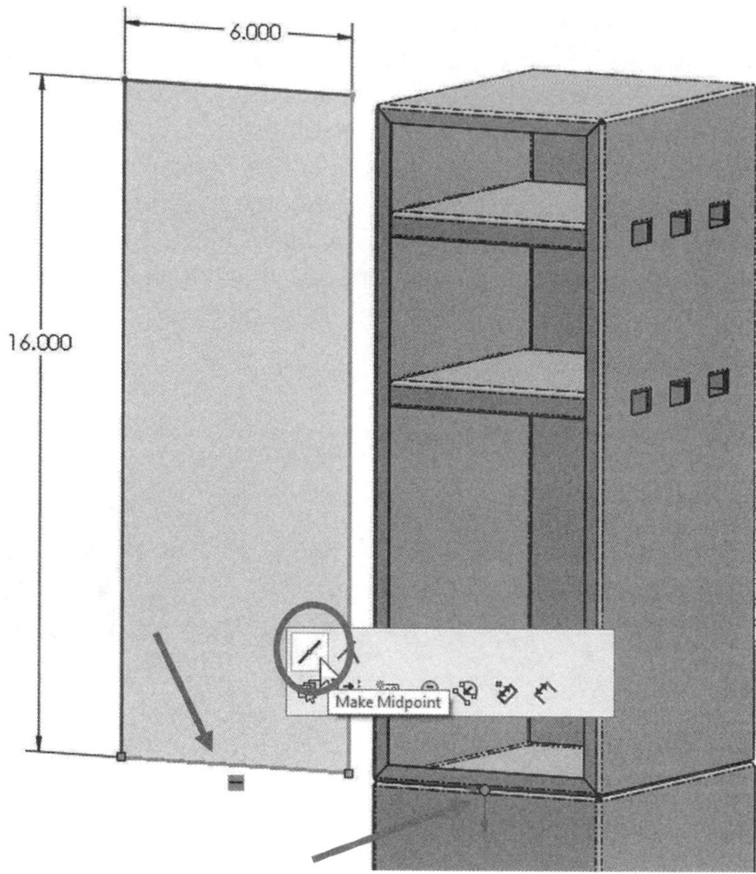

After adding the rectangle, extrude it 0.625″ outward. When finished, we can see in the FeatureManager that no external references have been added to the extrusion or its sketch.

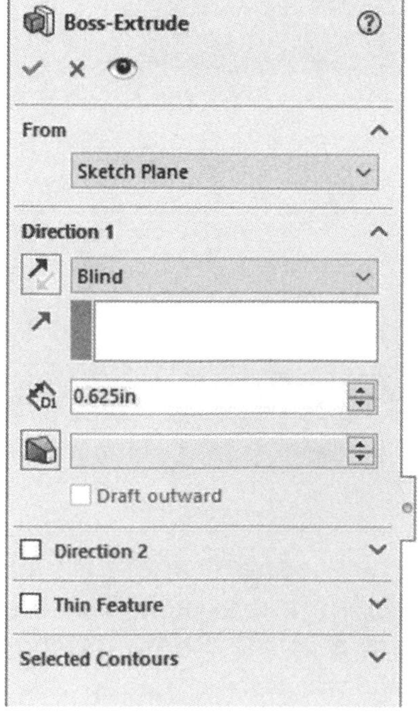

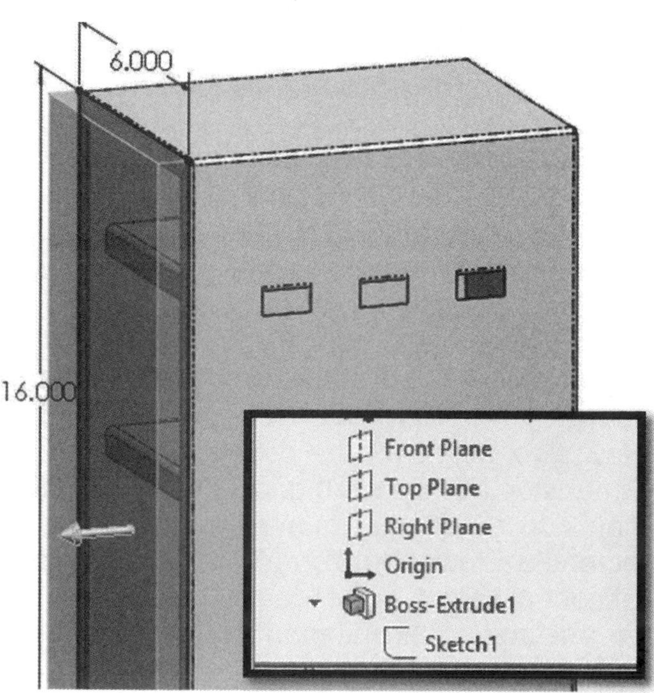

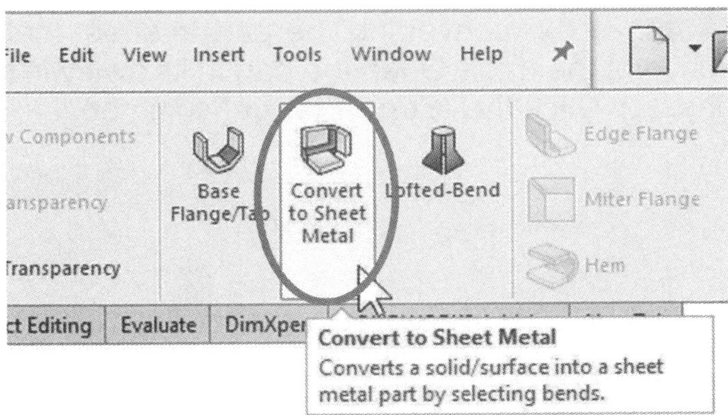

13.41. - The next step is to convert the solid door extrusion into a sheet metal part. From the Sheet Metal toolbar, select "**Convert to Sheet Metal**" to continue.

To convert the door to sheet metal, select the front face of the door as the fixed face and add the four edges around it to the "Bend Edges" selection box. Once the bend edges are selected, the small corner edges will be automatically added to the "Rip Edges found" selection box. Use the sample steel sheet metal gauge table as before, make the door using 18-gauge thickness, default bend radius of 0.05", default gap for rips of 0.02", "K-Factor" of 0.5, and set the "Auto Relief" option to "Tear."

Make sure the material thickness is added inside, making our part's dimensions external. Click OK to finish.

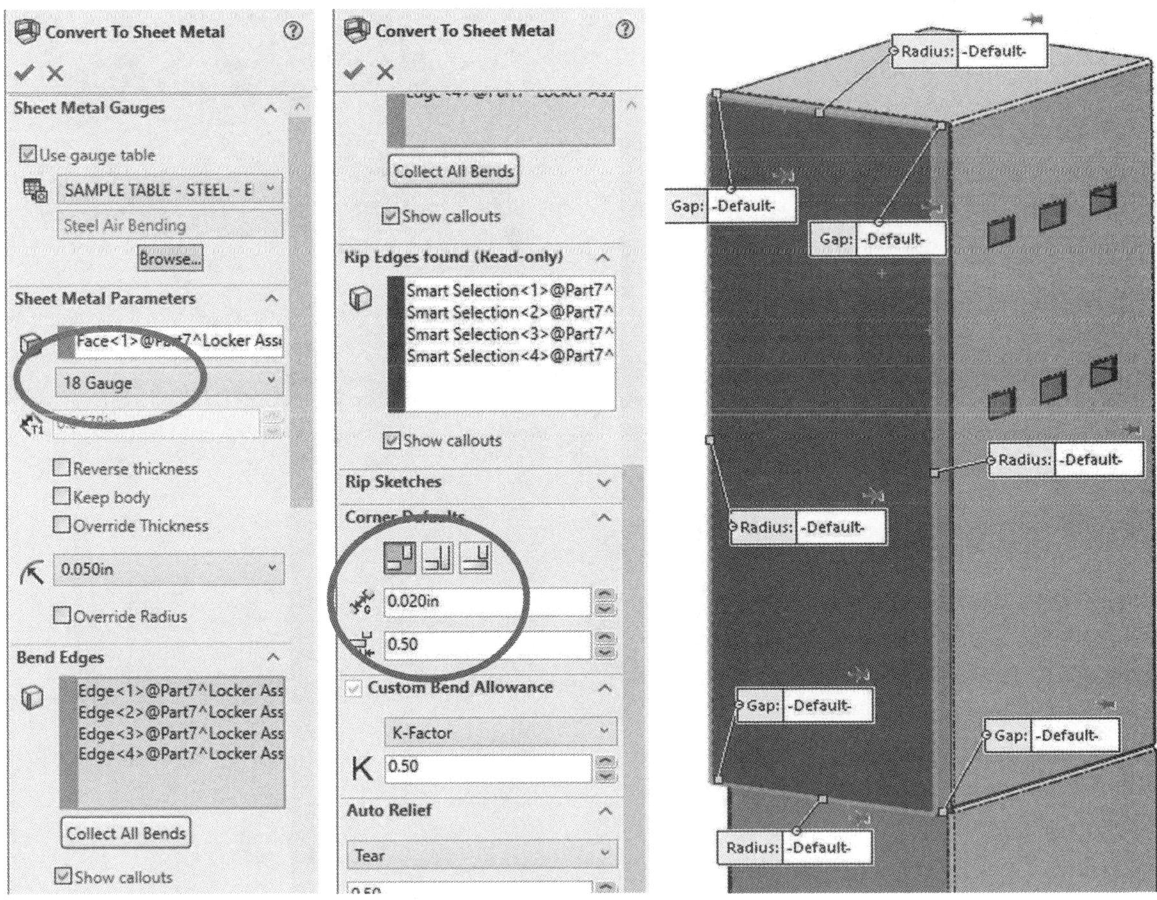

13.42. - After converting the part to sheet metal, we need to add flanges to the inside of the door. Open the part in its own window and turn it over to see the back (inside). Click the "**Edge Flange**" command.

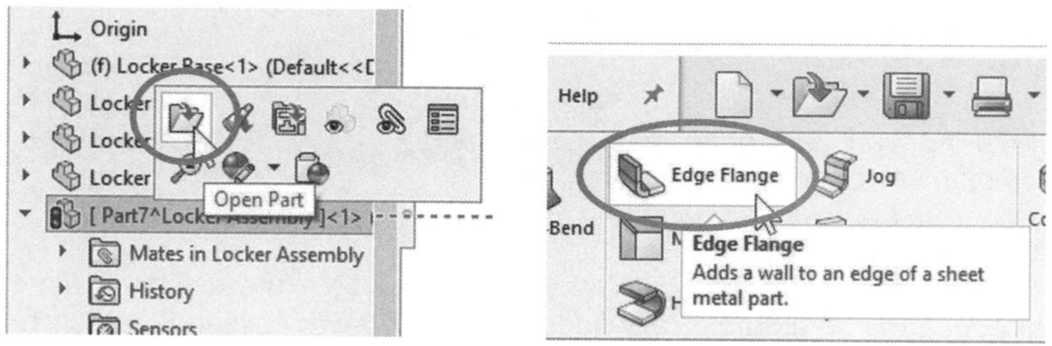

Add all four edges in the back to the "Edge" selection box. Make the flange 0.625″ long with a gap distance of 0.01″ using the "Material Inside" option.

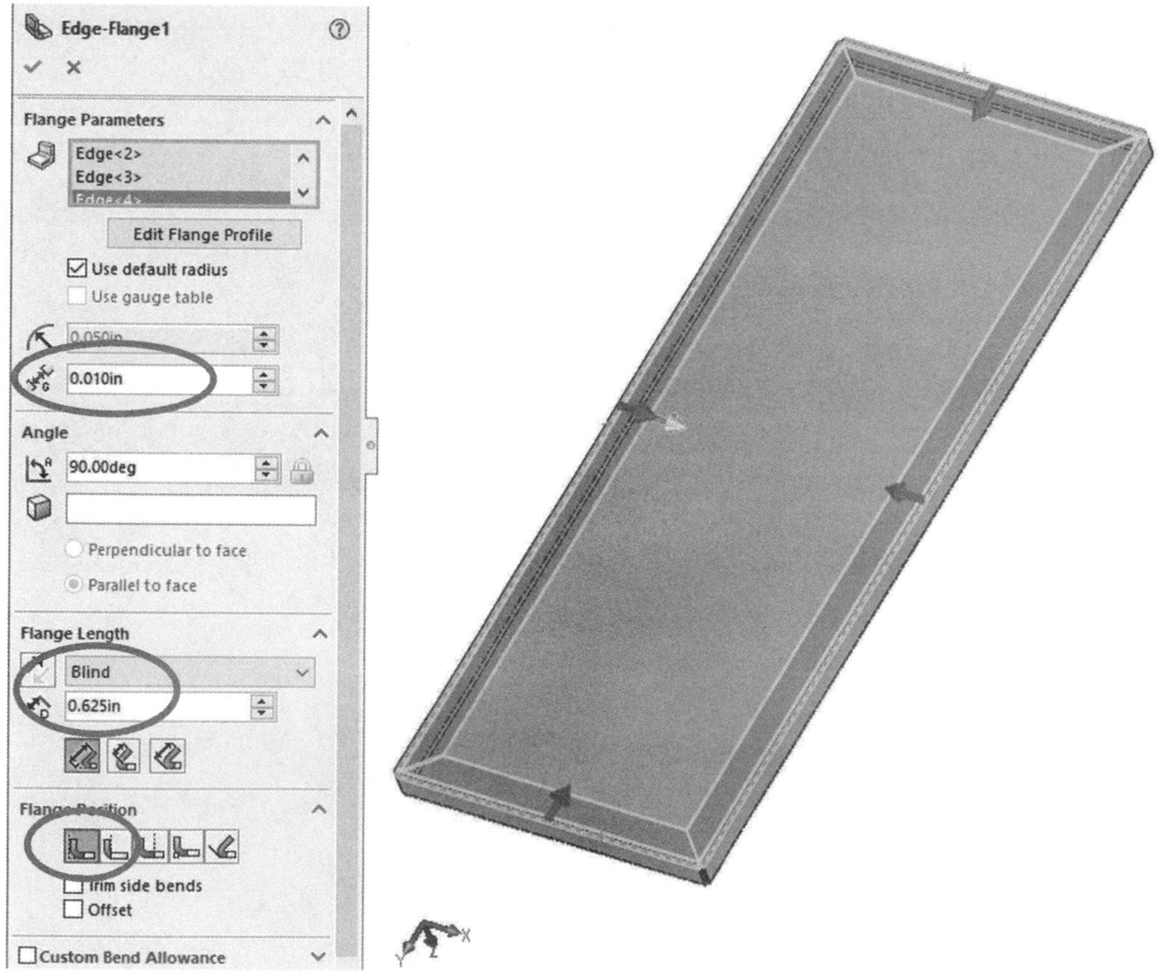

13.43. - Go back to the assembly, rename the new part as *'Locker Door'* and save it to an external file.

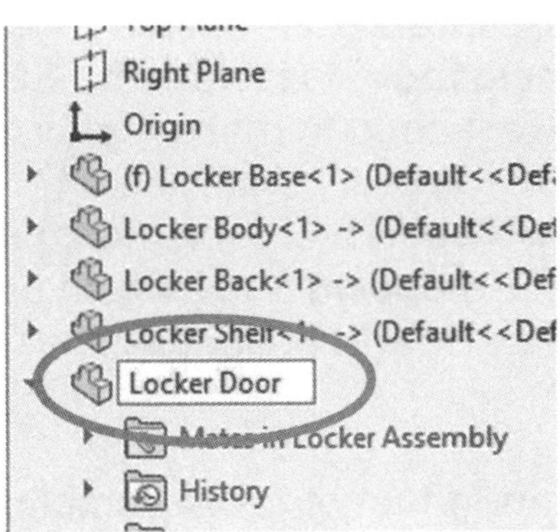

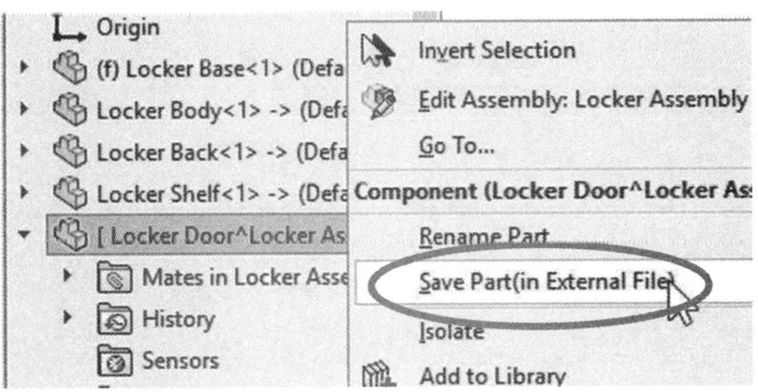

Create a Forming Tool

13.44. - Our next step is to add a set of louvers to the door to help ventilate the locker. Since the louver forming tool available in the SOLIDWORKS Design Library is too small for our design, we'll need to make a new one.

To create a "**Forming Tool**" library, the first thing we need to do is to build the shape of the tool, define a *'stopping face'*, which determines how deep the tool will go into the sheet metal part, and optionally the face(s) that will be removed (cut) as openings in the part.

Make a new part independent of the assembly, add the following sketch in the *"Top Plane"* and extrude it 0.375" up. This extrusion will serve as a base to build our library on top of it.

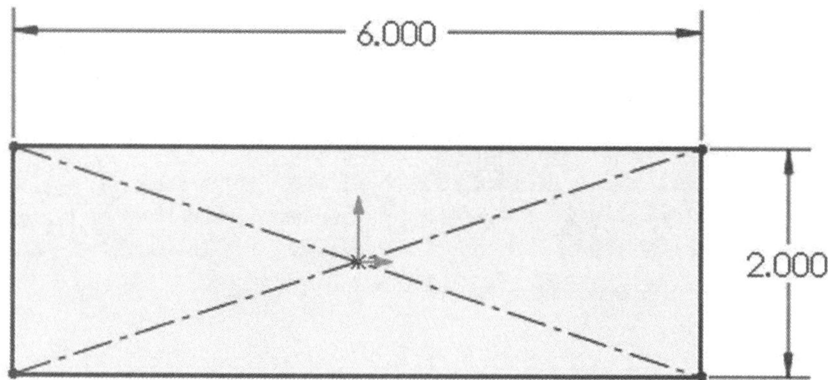

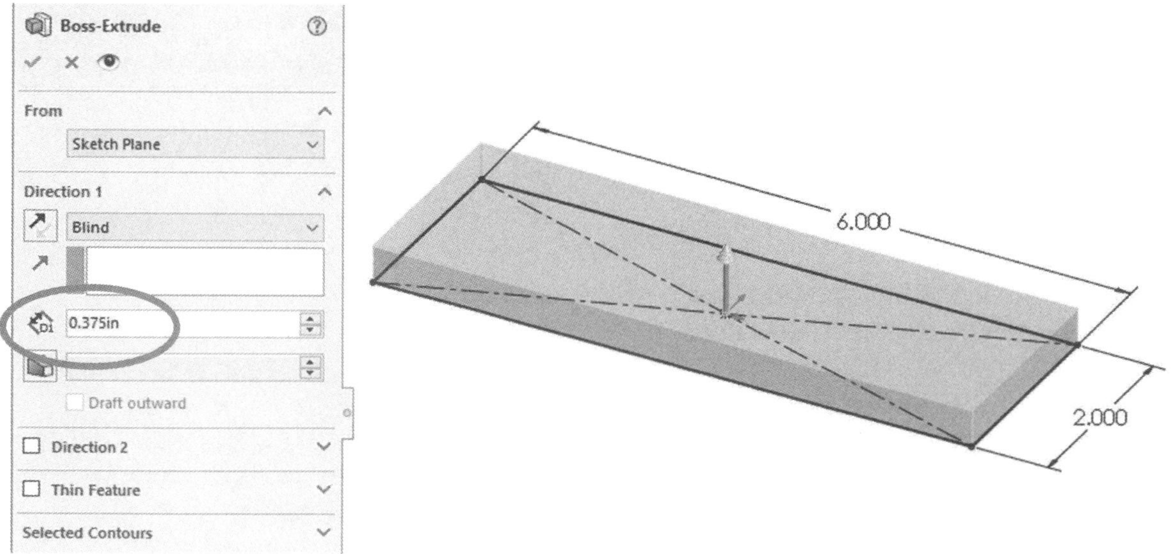

13.45. - Add a new sketch on the top face, extrude it 0.375″, and add a fillet to round the two indicated corners with a 0.625″ radius.

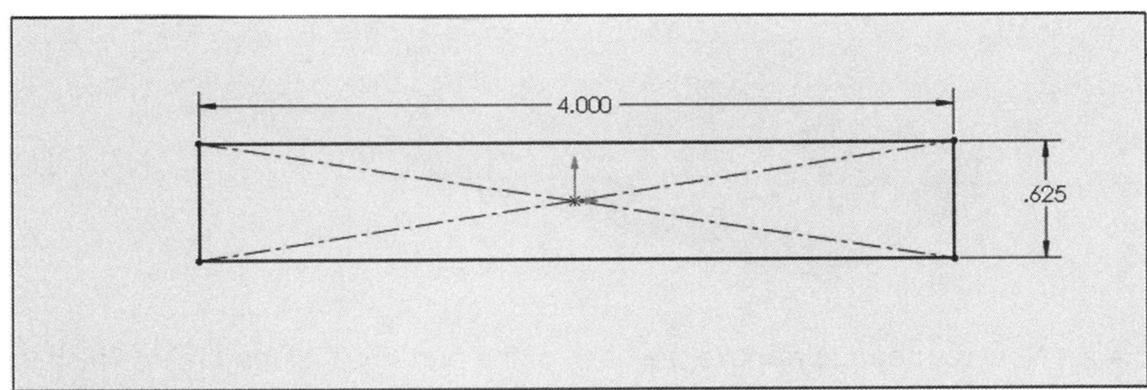

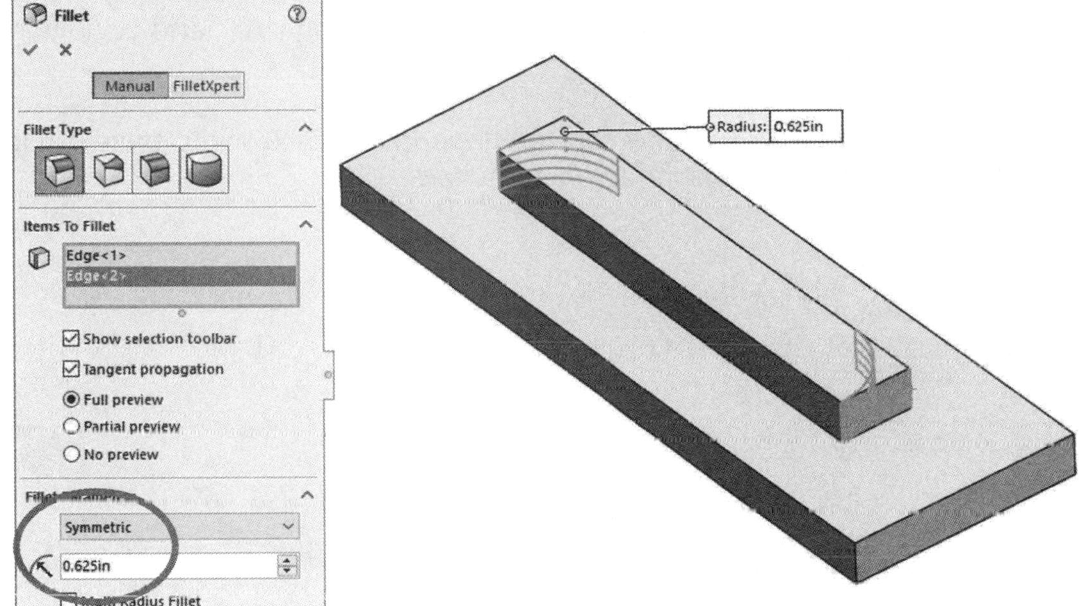

Add a second Fillet to the top edge with a 0.25″ radius.

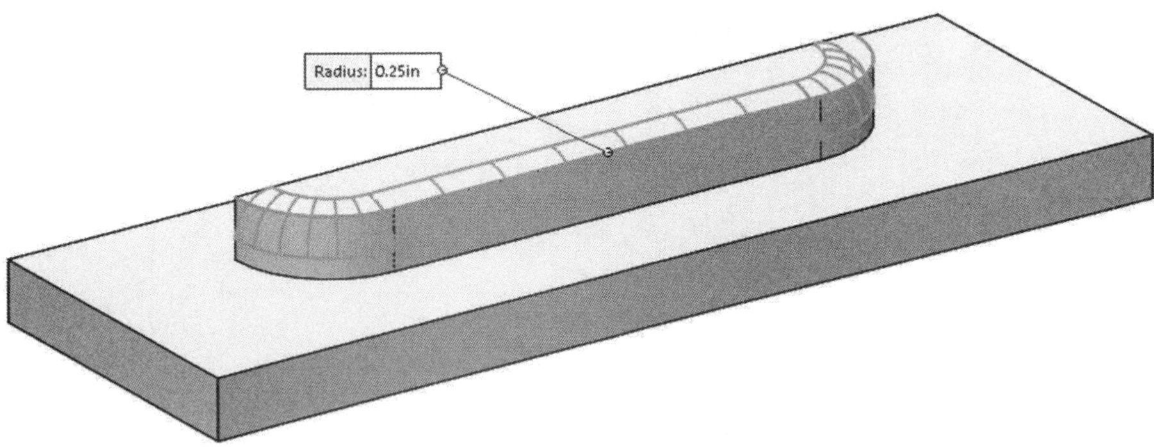

And finally add another fillet to the bottom edge with a 0.375″ radius; our part now looks like this:

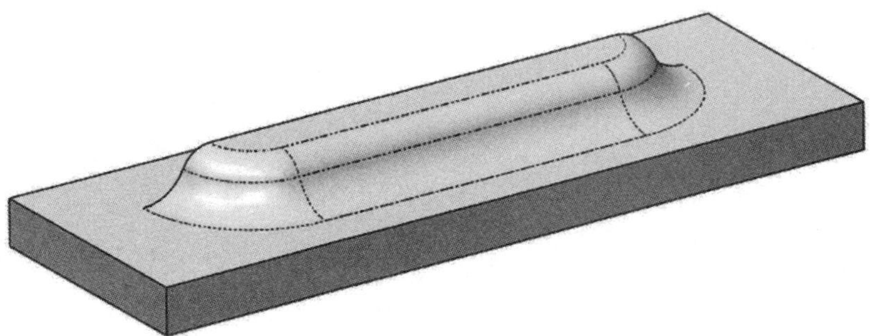

13.46. - Now we need to remove the base of the part (the first feature) to finish the forming tool's geometry. Add a sketch on the top face of the first feature, use "**Convert Entities**" to project the edges (or draw a square coincident to the base corners, your choice) and make a cut using the "Through All" end condition to remove the entire base.

This sketch and Cut-Extrude can be made in any face of the base, as long as the first extrusion is completely removed.

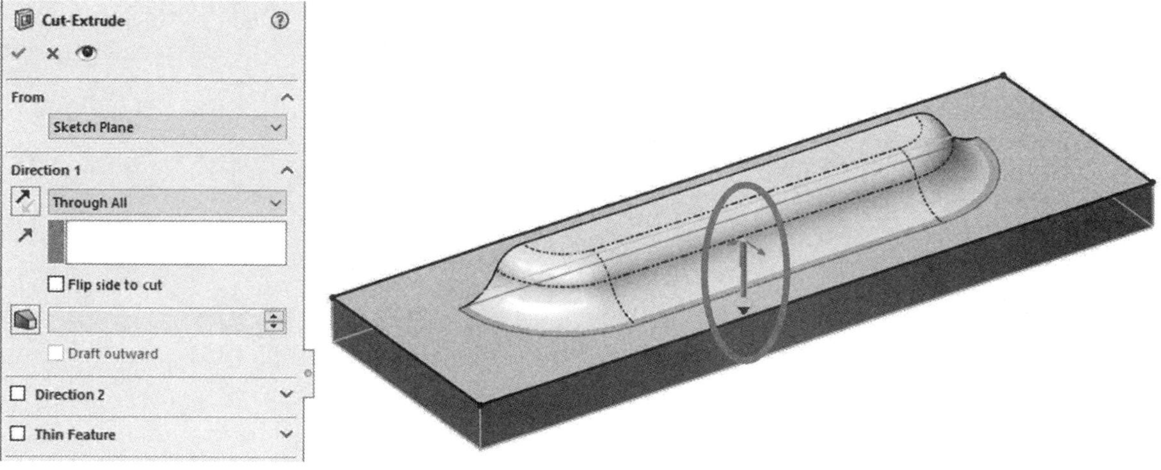

Our form tool now looks like this:

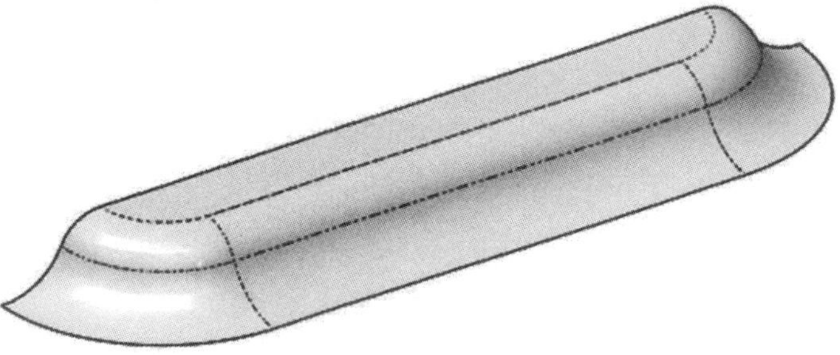

Cutting the base is not required to make a form tool; in our case, we had to do it to add the fillet at the bottom for our form tool.

13.47. - After the part is complete, select the "**Forming Tool**" icon in the Sheet Metal toolbar or the menu "**Insert, Sheet Metal, Forming Tool**."

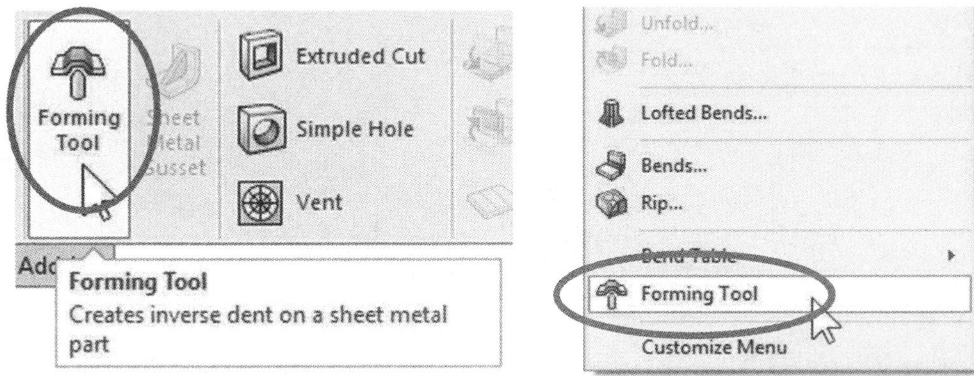

In the form tool's properties add the bottom face to the "Stopping Face" selection box as indicated. This is the face that will match the sheet metal part's surface and will define how deep the tool will go *into* the part.

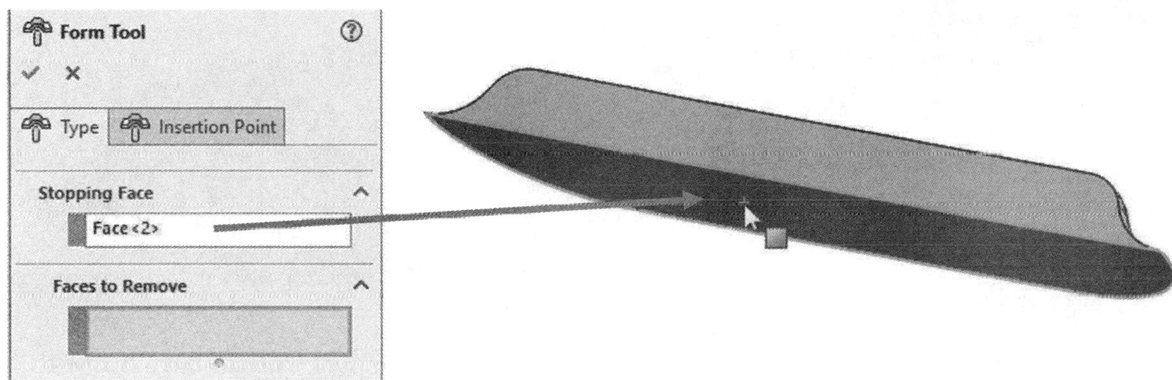

In the "Faces to Remove" selection box add the front face. Adding this face will create an opening in the sheet metal part when the forming tool is inserted. If no face is selected, the form tool will make a simple indentation in the sheet metal. Click OK when done.

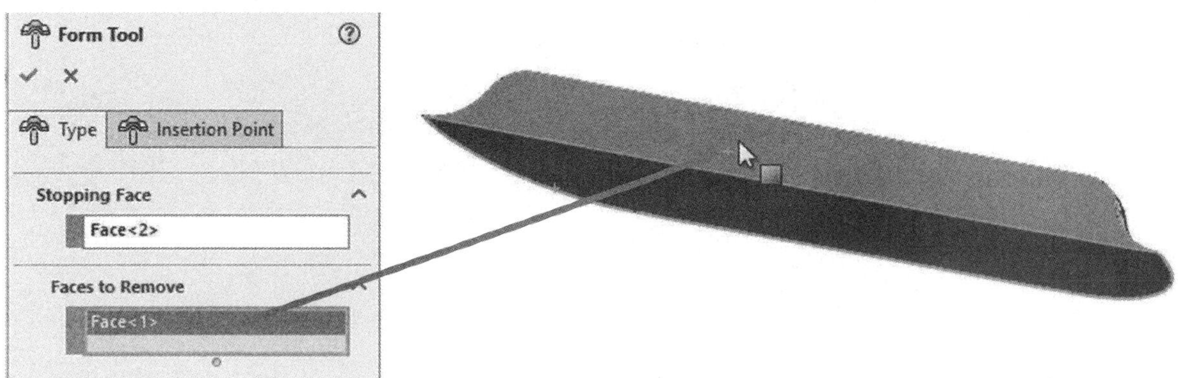

After finishing the form tool, SOLIDWORKS automatically changes the face's colors to indicate the stopping, forming, and open faces.

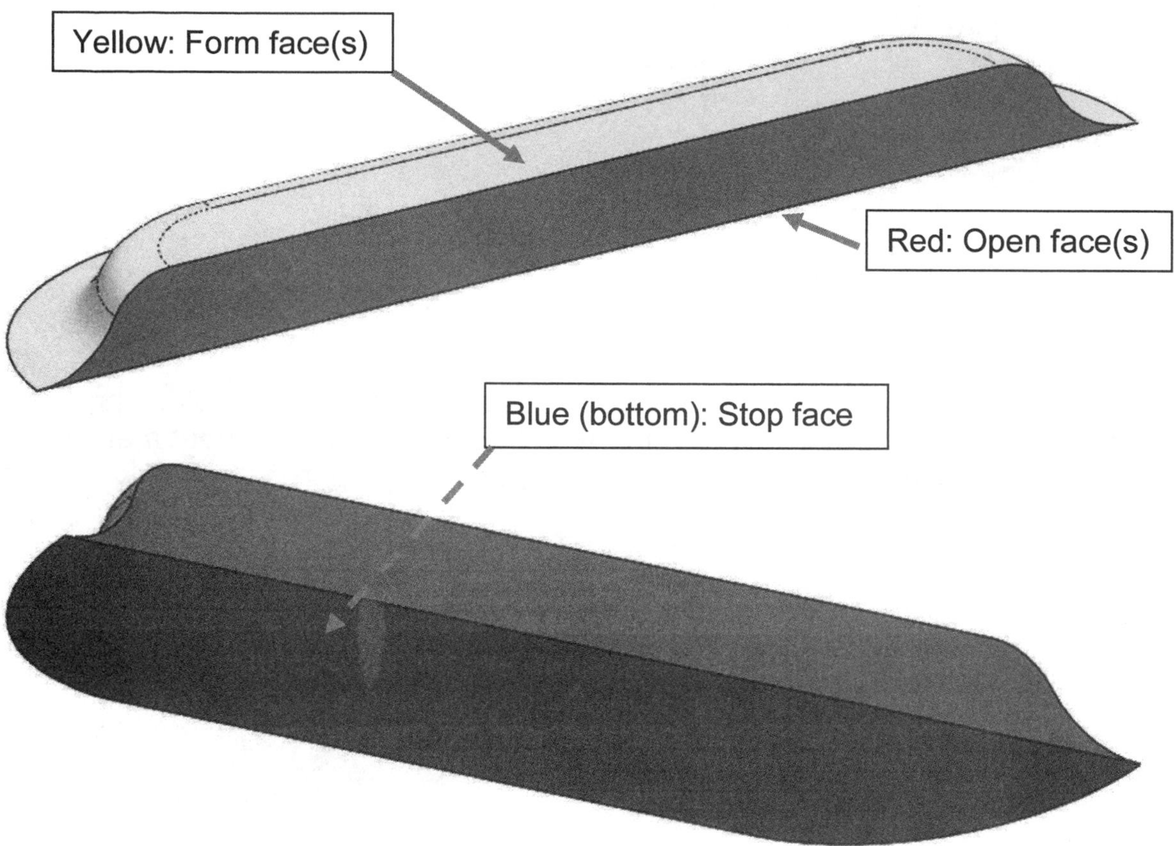

13.48. – After completing the form tool, we need to add it to our library. Save the file as '*4in Louver*' as we would a regular part. After saving it, right-mouse-click at the top of the FeatureManager and select "**Add to Library**."

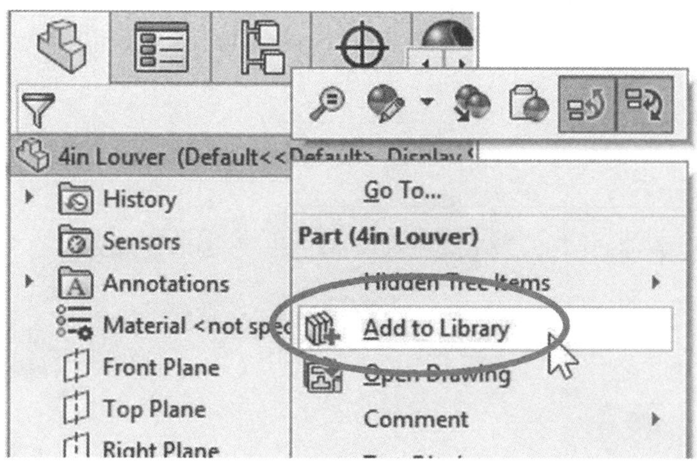

In the "File name:" field enter *'4in Louver'* and select the "louvers" folder under "forming tools." In the "File type:" drop down selection box, leave the option as "Part (*.sldprt)." Click OK to add the form feature to the library.

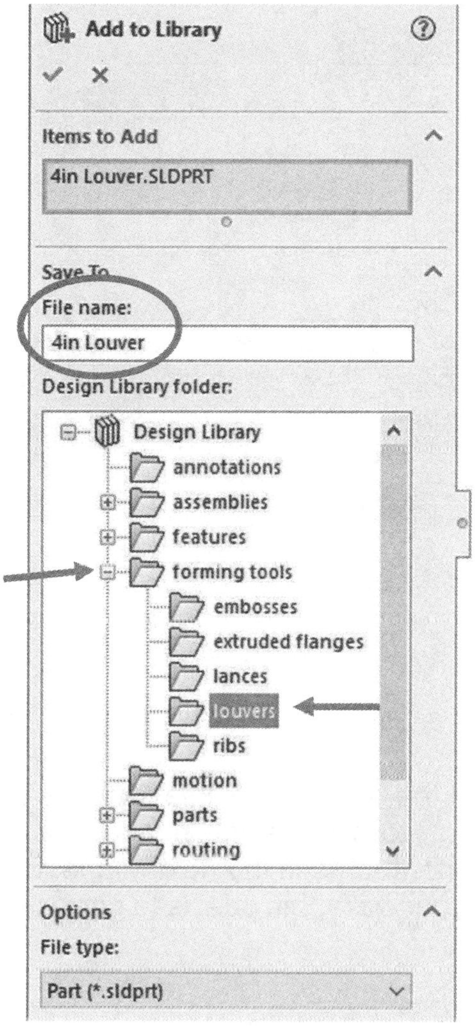

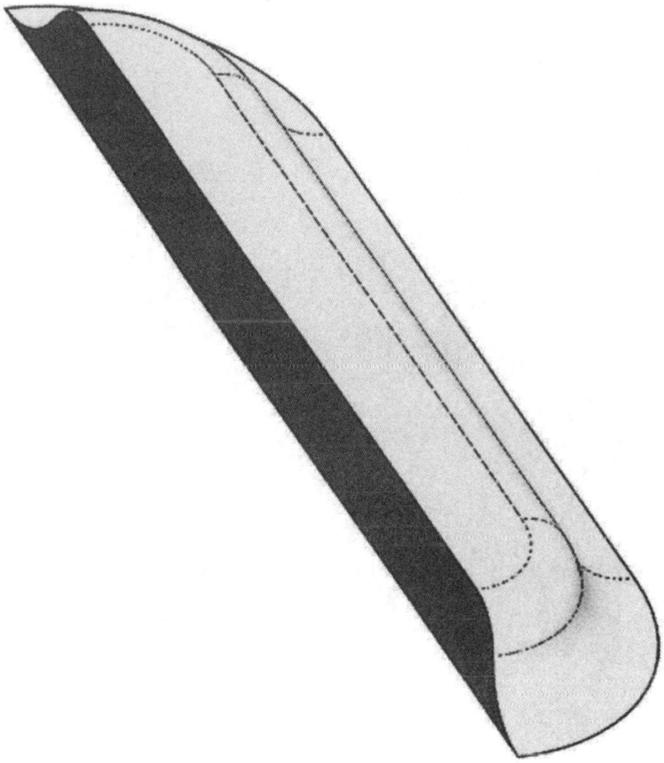

The form tool feature automatically adds a sketch called *"Orientation Sketch."* This is the sketch that will be used to locate the form tool when it is inserted into a sheet metal part.

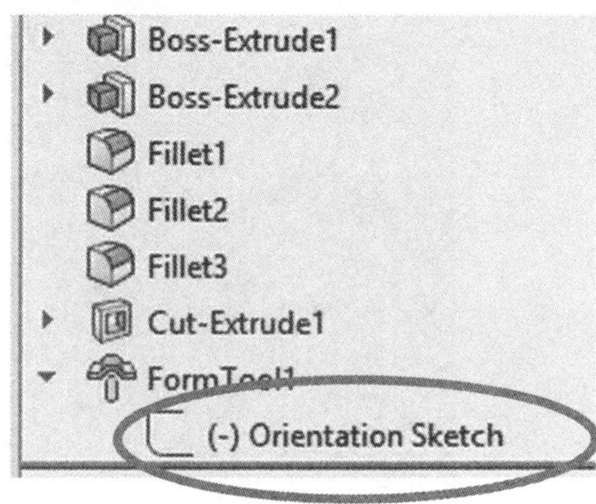

13.49. - Close the *'4in Louver'* library part and return to the assembly. Make sure we are still editing the *'Locker Door'* inside the assembly to add the library to it. After opening the Design Library we can see the new *'4in Louver'* form tool has been added to the available tools; click and drag it onto the locker's door to insert it.

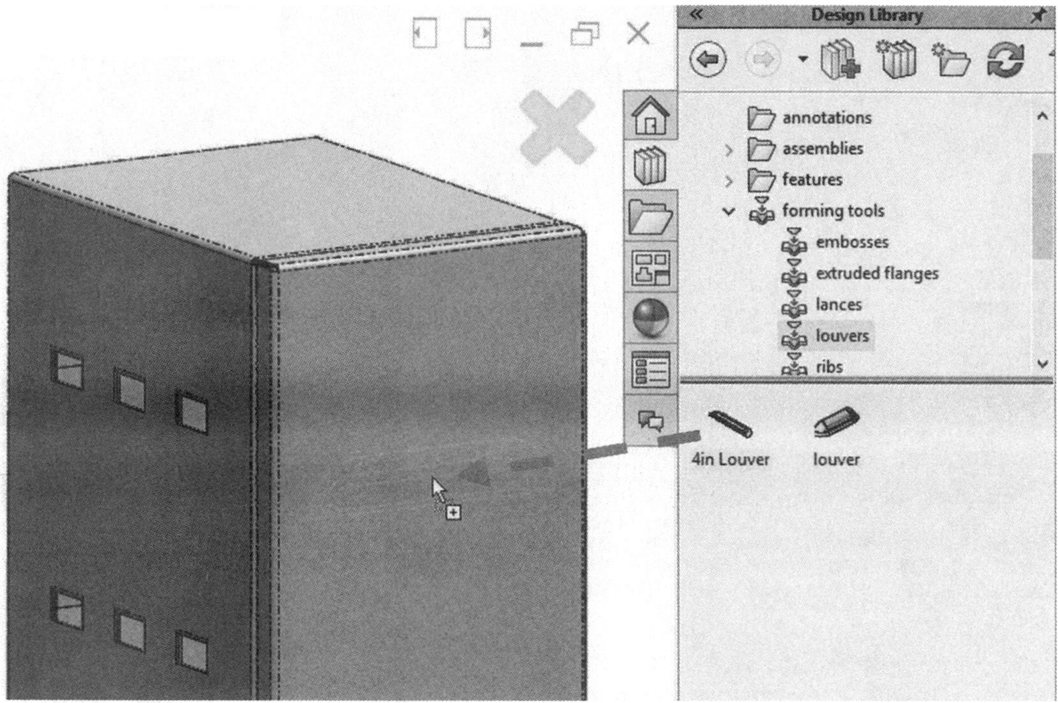

After dropping the form tool in the door use the "Rotation Angle" options to rotate the form tool, and if needed, use "Flip Tool" to reverse the direction and make it go outwards, getting the louver correctly oriented as in the next image. Be aware that the values entered *may* be different in your case; the goal is to correctly orient the louver to match the preview.

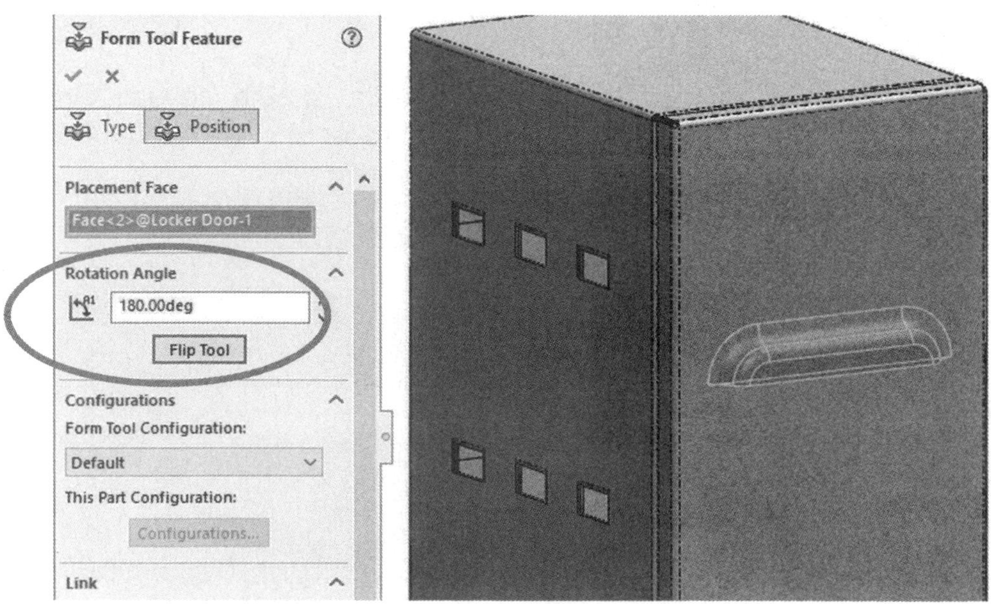

To locate the louver in the door, select the “Position” tab at the top of the Property Manager. Add a 2” dimension from the louver's center point to the top edge of the door; make the part’s origins visible and add a “Vertical” geometric relation between the louver’s center and the door’s origin to center it. Click OK to finish when done.

13.50. - When the louver is finished make a linear pattern of 4 louvers spaced 1.25″. Stop editing the part (turn off “**Edit Component**”) and save the assembly and all components at the same time.

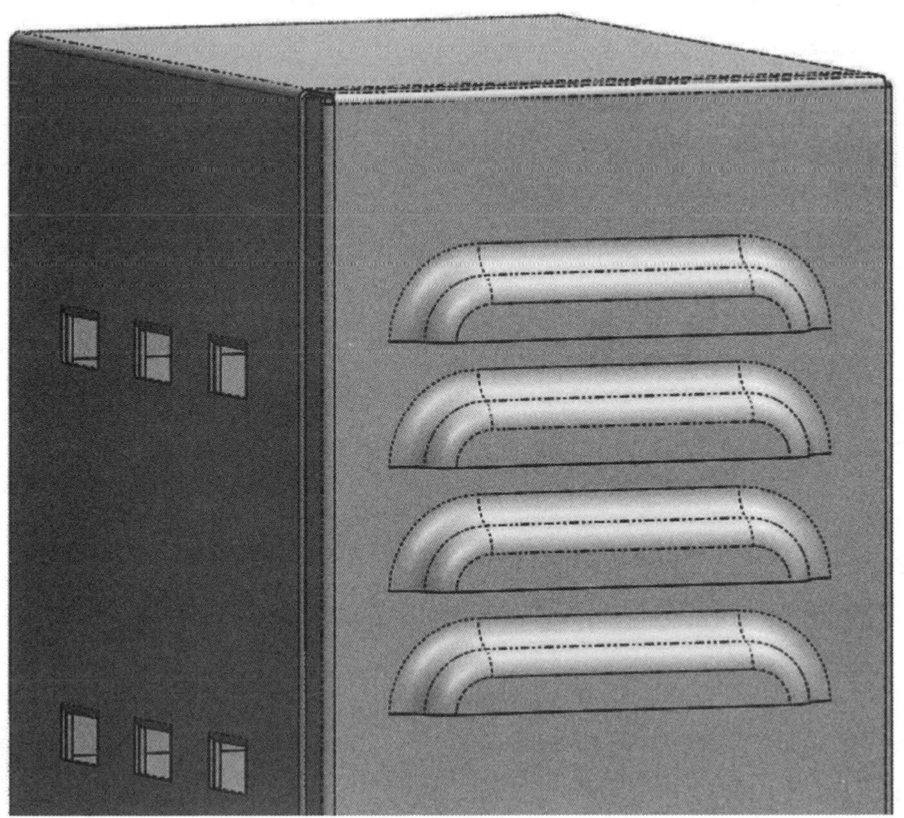

13.51. - After reviewing the assembly from a Left view, we notice the door protruding beyond the locker's base. Measuring the distance, we can see the *'Locker Body'* needs to be 0.625" smaller (the thickness of the door) to make the door flush with the base. To correct this, we need to edit the first sketch in the *'Locker Body'* to change its size.

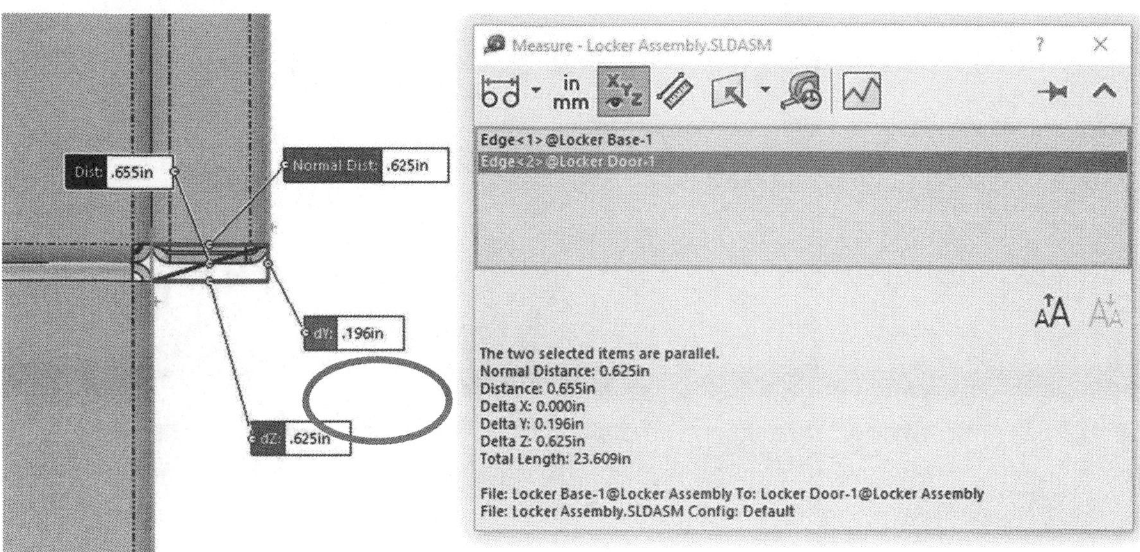

13.52. – Close the "Measure" tool, expand the *'Locker Body'* FeatureManager and edit the *"Boss-Extrude1"* sketch. When we edit a part's sketch while in the assembly, the "**Edit Component**" mode is automatically activated.

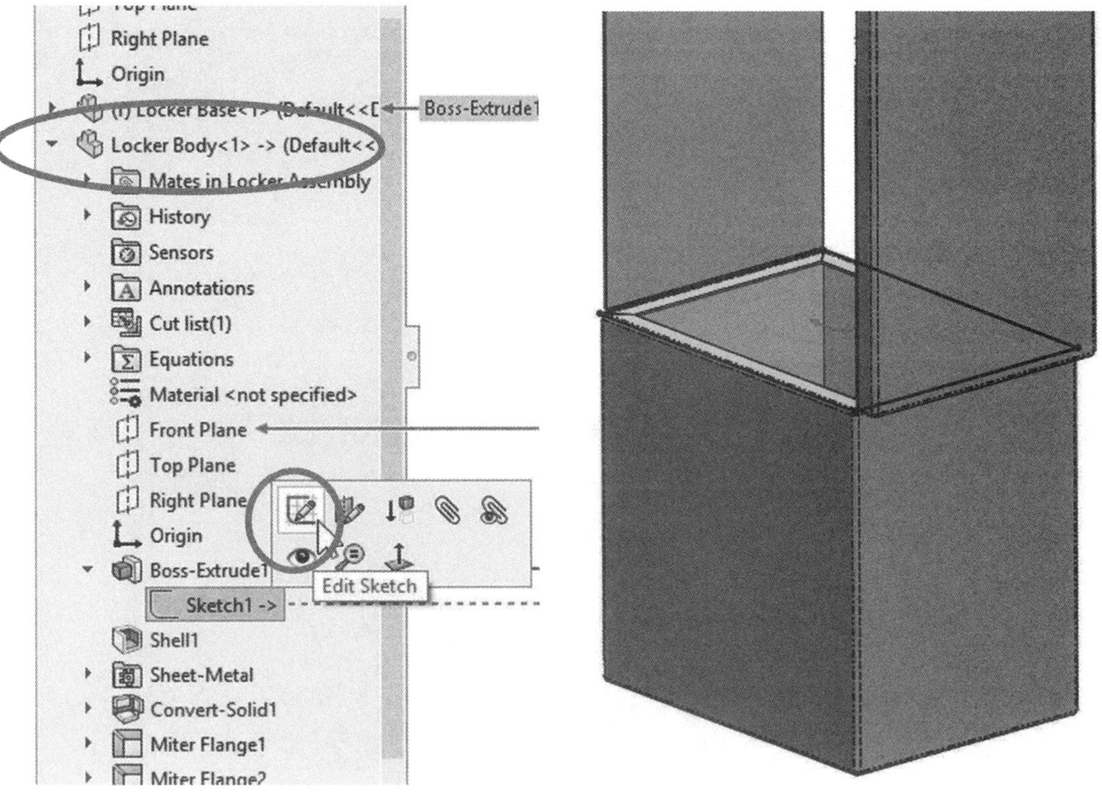

Select the front sketch line (close to the door) and delete the "On Edge" relation (created when we used the "**Convert Entities**" command).

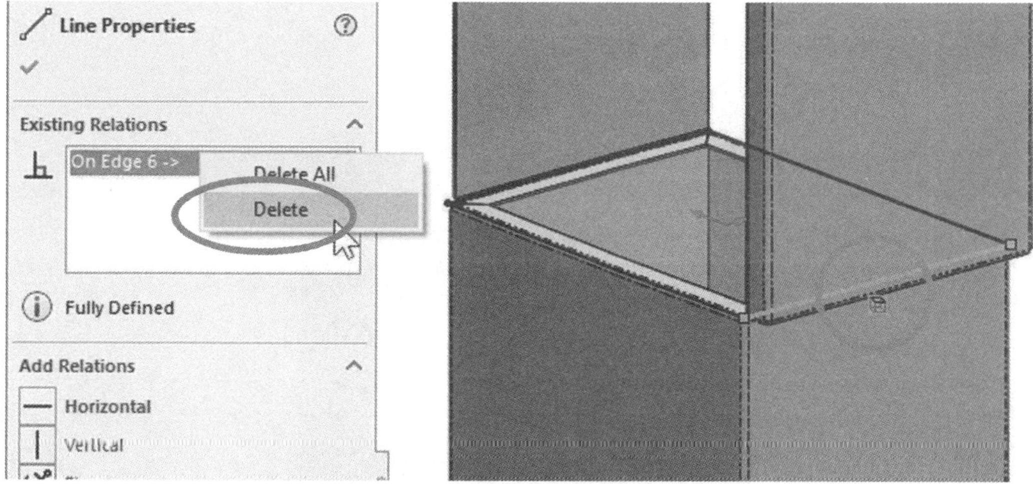

After deleting this external relation (we know it's external because it is followed by the "**->**" symbol), the line may still appear fully defined (in black), but now we can click-and-drag the front line towards the back to dimension it. When adding the dimension, be sure to select the edge of the base and not the door when adding the dimension.

If the "**No External References**" command is active you will NOT be able to add dimensions (or relations) to other components. Turn it off if it's active, otherwise you will not be able to add the dimension to the '*Locker Base'*.

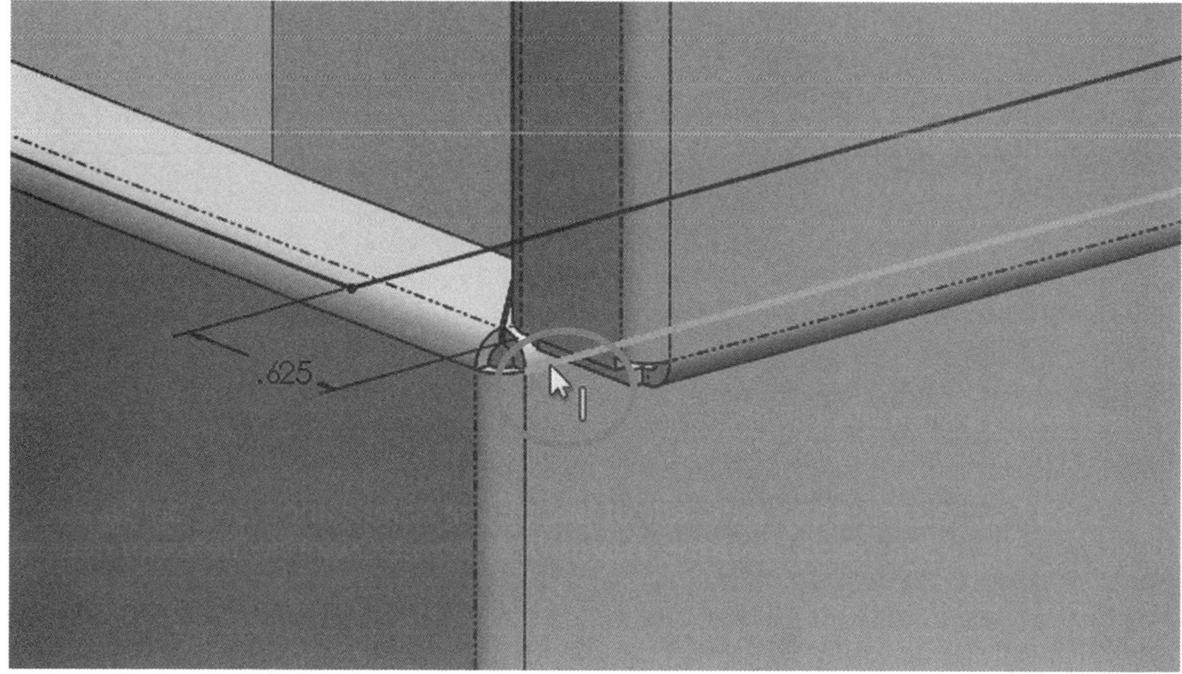

After the sketch changes are complete, exit the sketch or rebuild the model and turn off "**Edit Component**." When all components are rebuilt, your assembly should look like this:

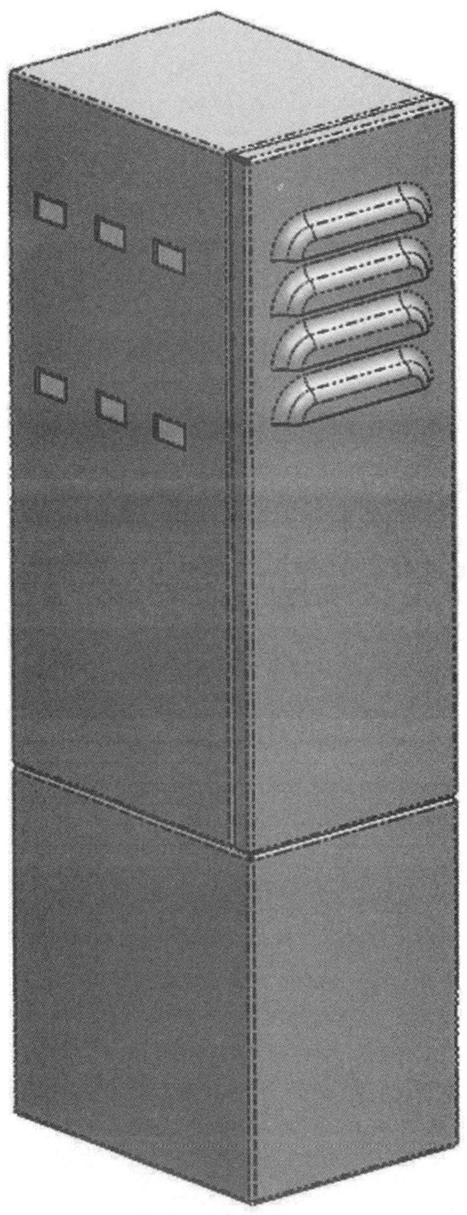

13.53. – After making this change, looking at the back of the assembly we can see the *'Locker Shelf'* is protruding through the back of the locker. The shelf was dimensioned to the lances in the body, and after the *'Locker Body'* was resized, the lances and the shelf moved backwards causing this problem. To fix it we need to edit the first *'Locker Shelf'* sketch and change the dimension to reference the back of the *'Locker Body'* instead of the lances. Expand the *'Locker Shelf'* FeatureManager, edit the *"Base-Flange1"* sketch and delete the dimension in the back. Hide the *'Locker Back'* part while editing the *'Locker Shelf'* for better visibility.

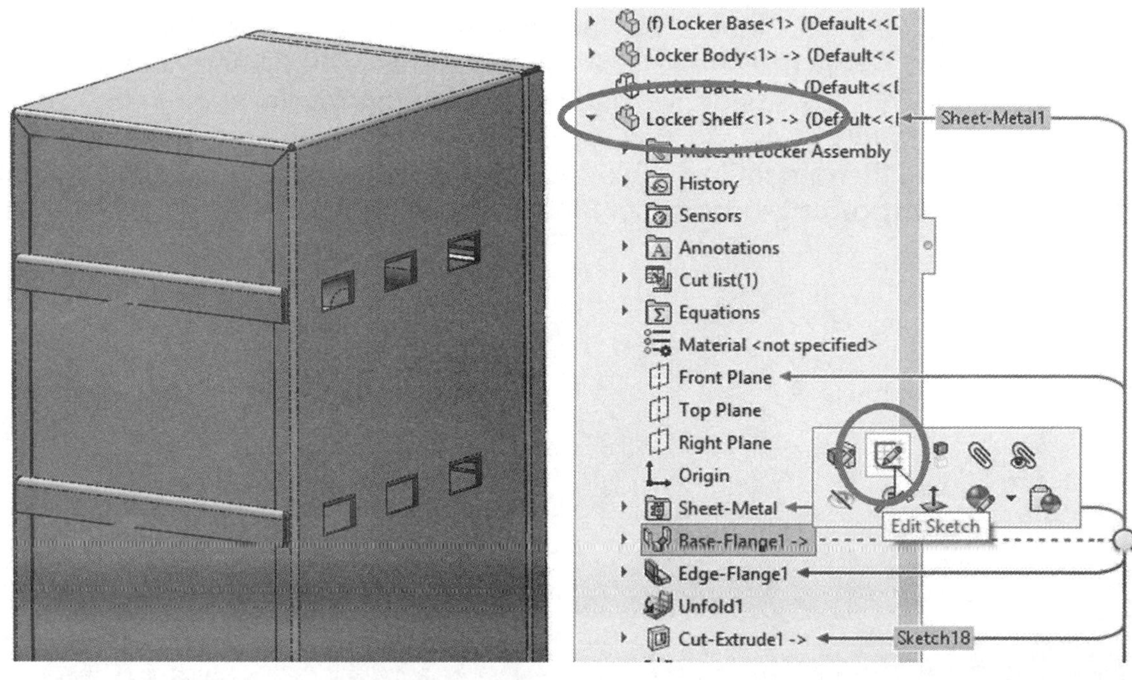
(f) Locker Base<1> (Default<<D
Locker Body<1> -> (Default<<
Locker Back<1> -> (Default<<I
Locker Shelf<1> -> (Default<<I
Sheet-Metal1
Mates in Locker Assembly
History
Sensors
Annotations
Cut list(1)
Equations
Material <not specified>
Front Plane
Top Plane
Right Plane
Origin
Sheet-Metal
Base-Flange1 ->
Edge-Flange1
Unfold1
Cut-Extrude1 ->
Sketch18
Edit Sketch

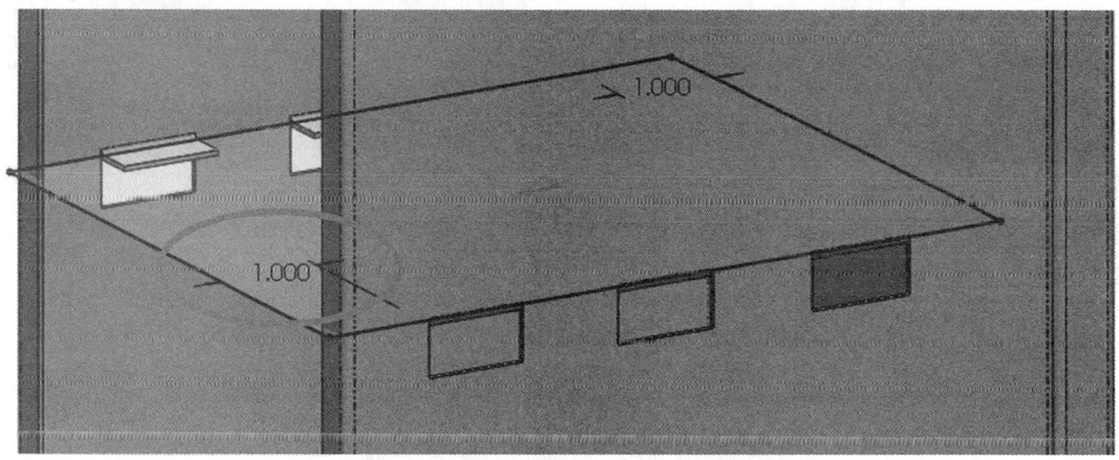
1.000
1.000

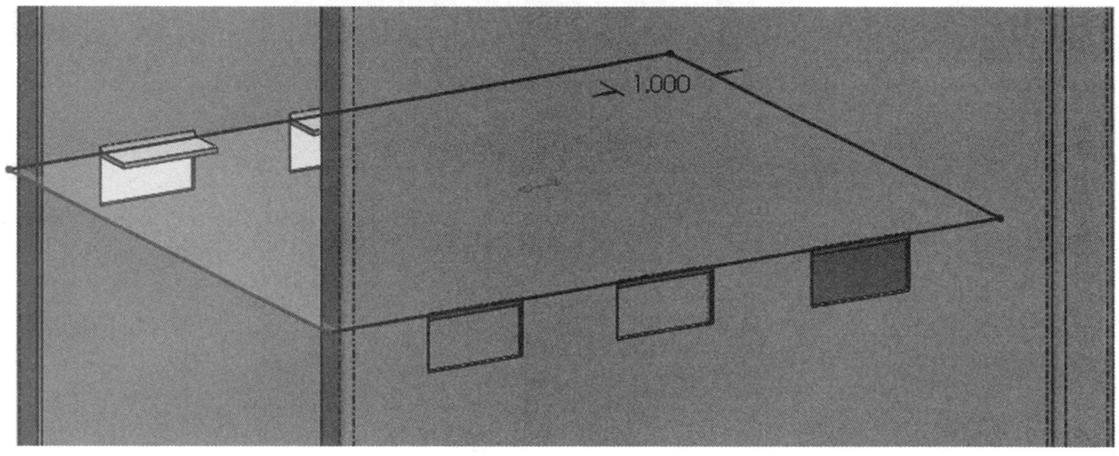
1.000

Drag the rear line inside the locker and dimension it to the edge of the *'Locker Body'*; this way the shelf's size will be referenced to it and not the lance's position. To add this dimension, we need to select a horizontal edge in the *'Locker Body'*; we can switch to a Top view to select the edge, or use the top or bottom miter edge. Make the dimension 0.375", exit the sketch to rebuild the shelf and turn off the "**Edit Component**" command. Finally make the '*Locker Back*' visible again to continue.

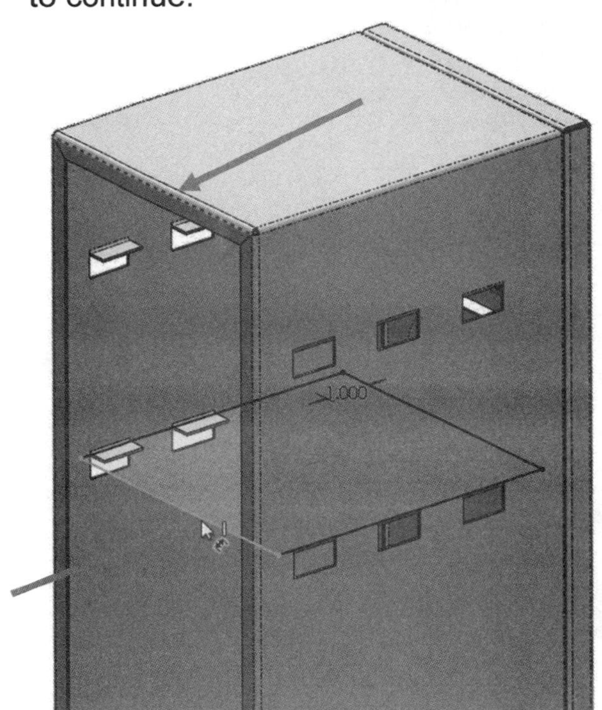

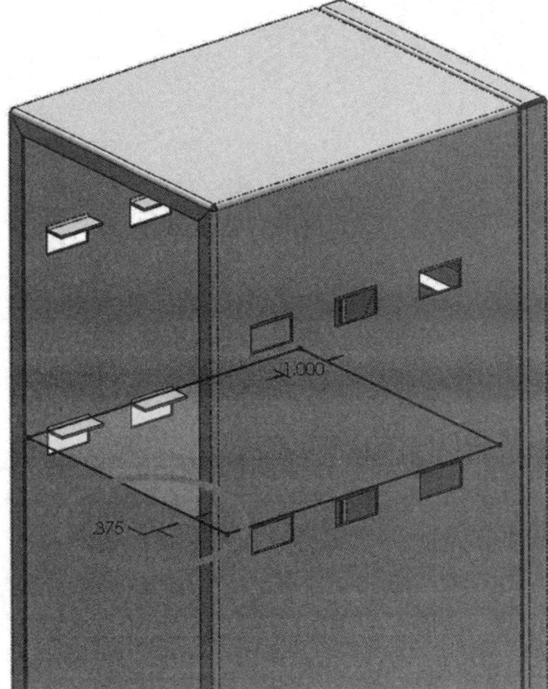

'*Locker Back*' part hidden for visibility

13.54. - After correcting the shelf the last step is to add hinges to the locker to open the door. While editing the assembly (not a part), if we try to click-and-drag the door to move it, we get the following message:

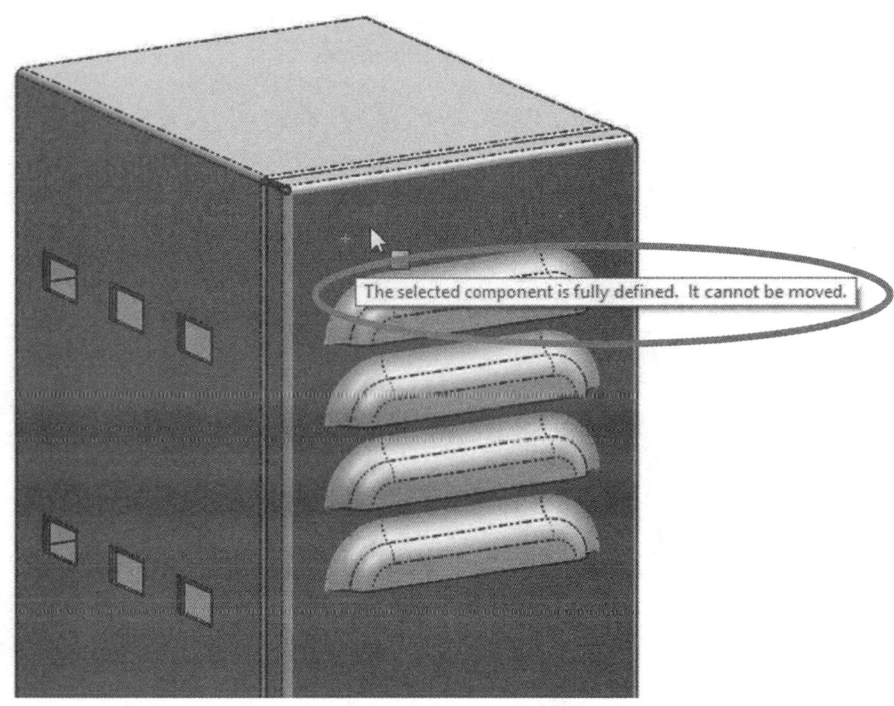

When we add components in the context of the assembly, an **InPlace** mate is automatically created when we select the face to locate the new part. The **InPlace** mate completely immobilizes a part, and to move it we must delete (or suppress) this mate.

Expand the *'Locker Door'* FeatureManager and locate the folder *"Mates in Locker Assembly."* This folder contains all the mates referencing the door in this assembly. If you remember, the door was made the same size as the body using dimensions, not external relations.

The reason we made it this way was to be able to move the door using assembly motion. If we had added external relations in the sketch, after moving the door its size would have changed and produced errors and/or undesired results.

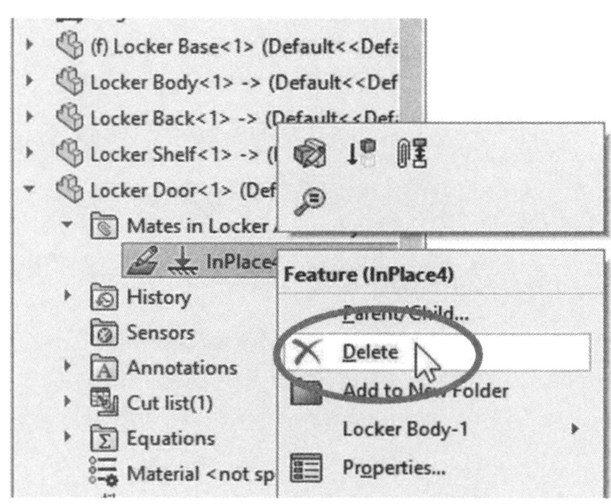

To move the door, we need to delete the **InPlace** mate, and then drag the door away from the locker.

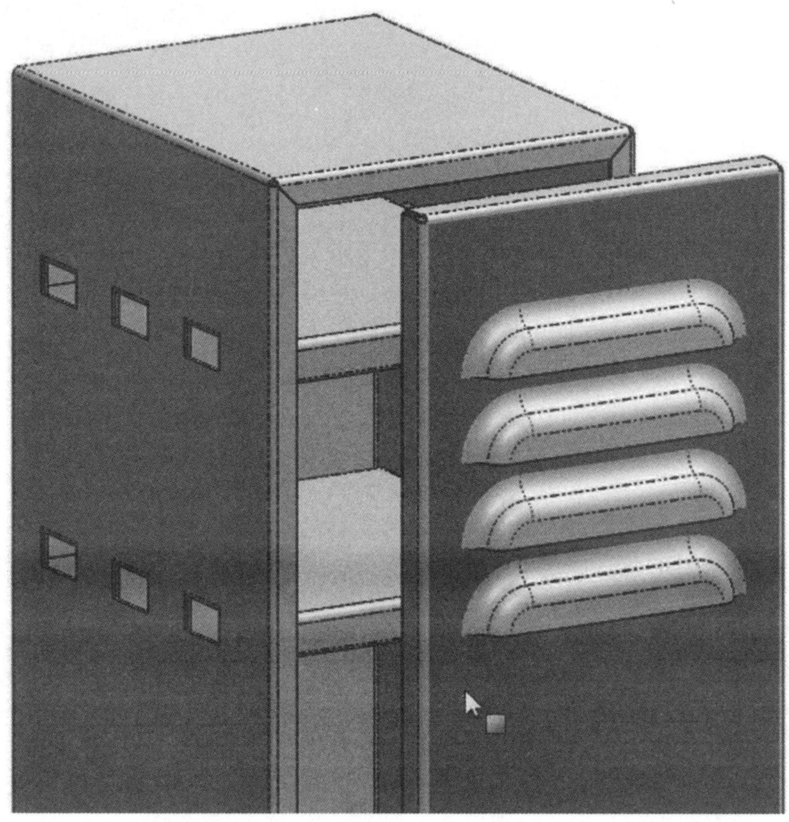

13.55. - The next step is to add the hinges to the assembly and mate them to the '*Locker Body*'. Using the "**Insert Components**" command, add the '*Hinge*' assembly from the accompanying files to the locker assembly.

13.56. - Move the door to the left and add a hinge near the right side of the locker. Mate the *'Hinge'* to the front flange of the *'Locker Body'* using a Coincident mate.

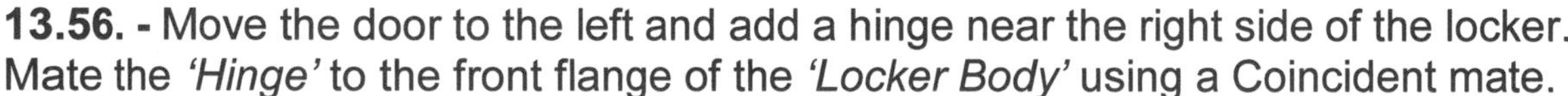

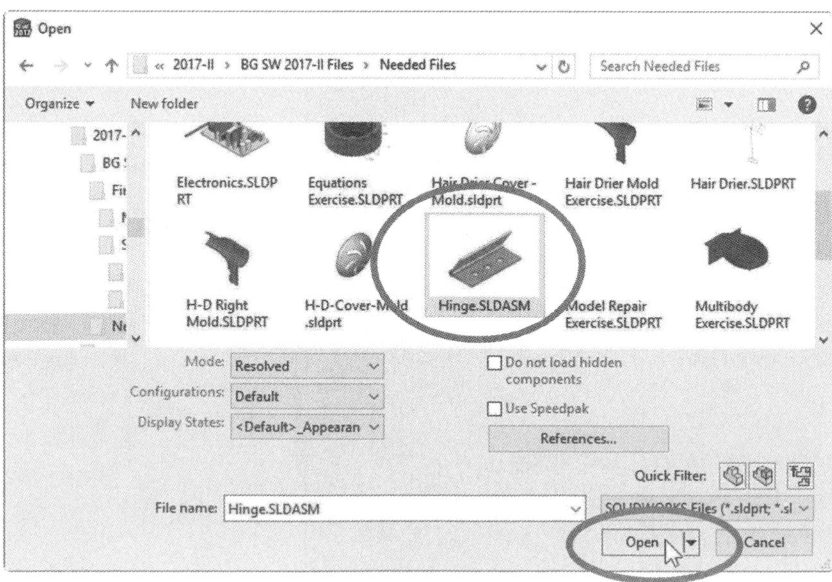

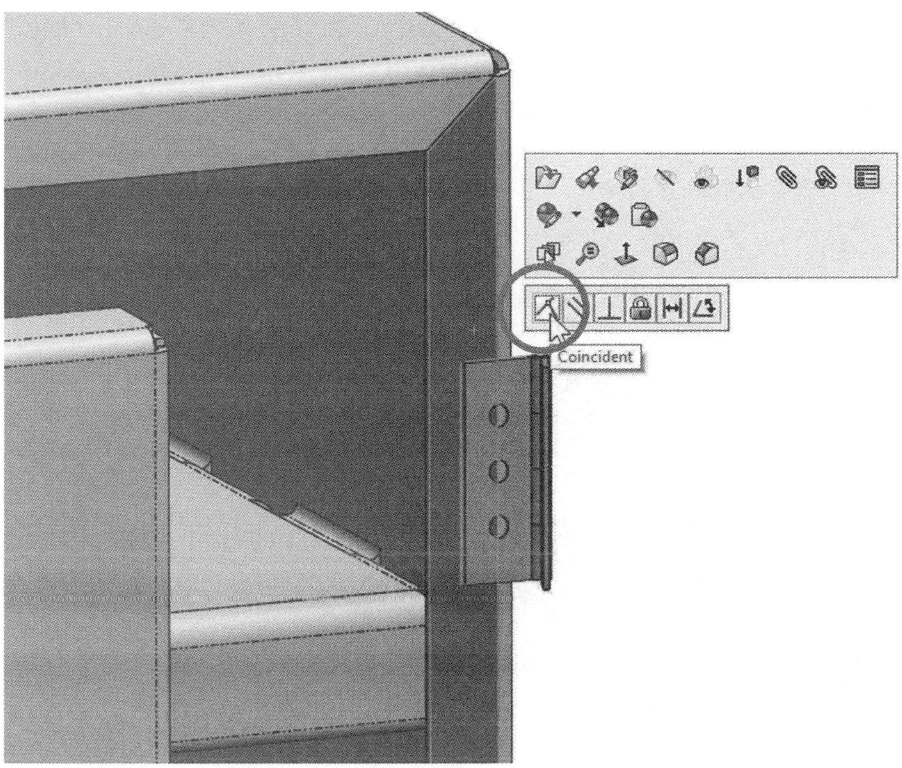

Add a second mate to make the top of the *'Hinge'* sub-assembly coincident to the top face of the upper shelf. Select an edge or a face of the hinge; in this case, either one gives the same result.

To select the smaller faces of the hinge we can activate the face Selection Filter. The default shortcut to activate the face filter is '**X**'. If you use a selection filter, don't forget to turn it off when you are finished using it, otherwise you may not be able to make other selections.

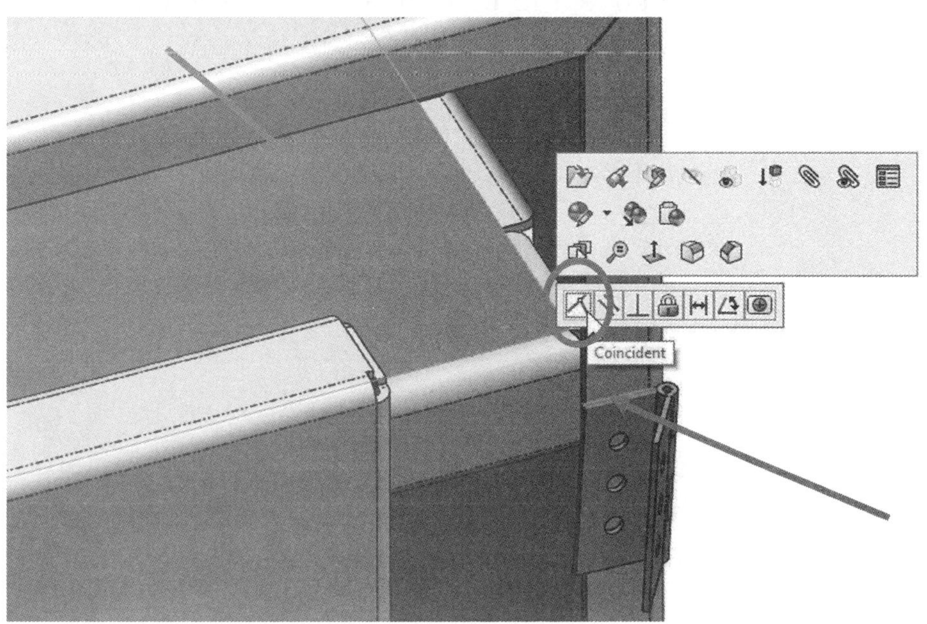

Finally add a coincident mate to make the hinge sub-assembly aligned to the inside edge of the miter flange as indicated. Select either faces or edges of the *'Hinge'* and the *'Locker Body'.* Use selection filters if needed (**X** for face filter).

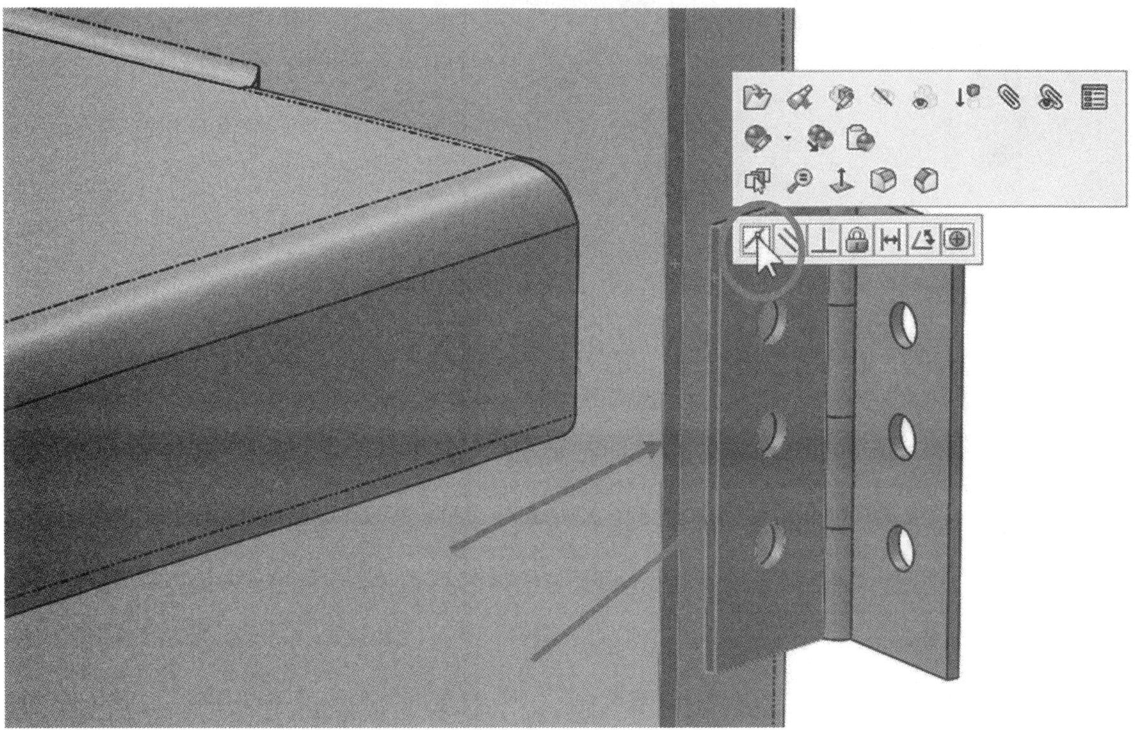

13.57. - With the hinge located in place, make a linear pattern going down to add two more *'Hinges'* spaced 5 inches apart.

Select the menu "**Insert, Component Pattern, Linear Pattern**" or the "**Linear Component Pattern**" command from the Assembly toolbar. Use the *'Hinge'* subassembly as the component to pattern and any vertical edge for the direction.

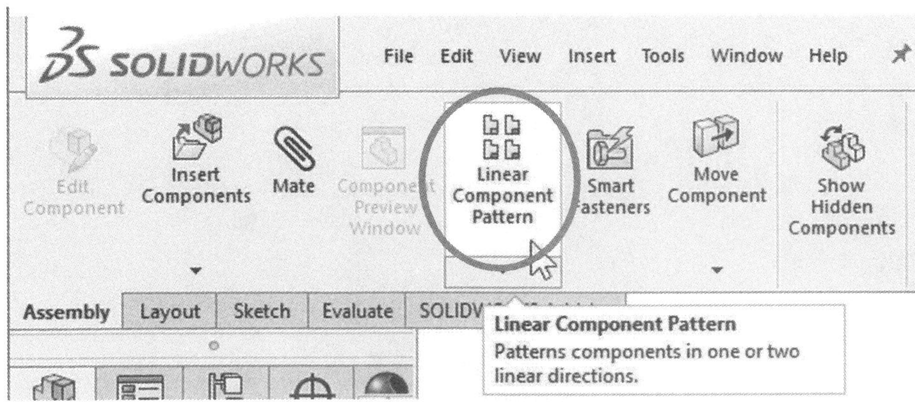

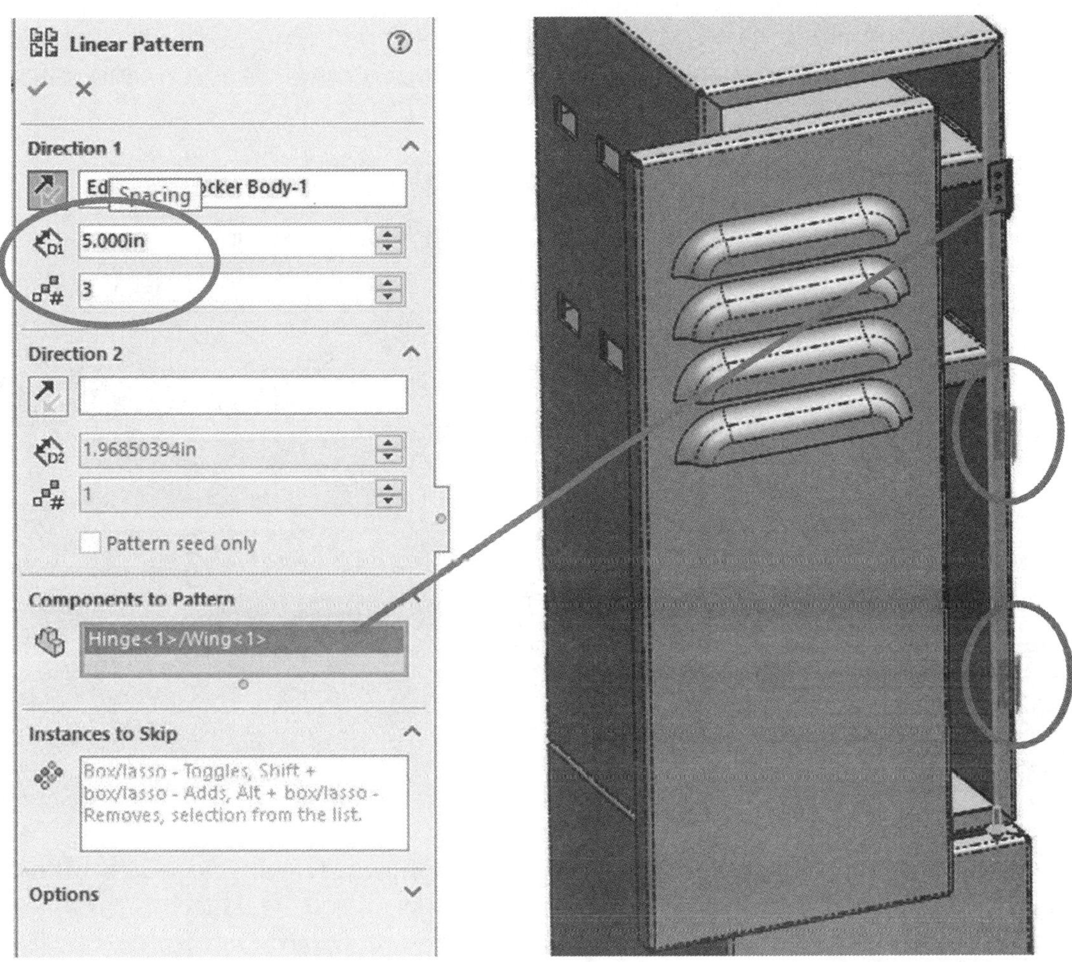

A component pattern feature is added to the assembly's FeatureManager with the two additional hinges.

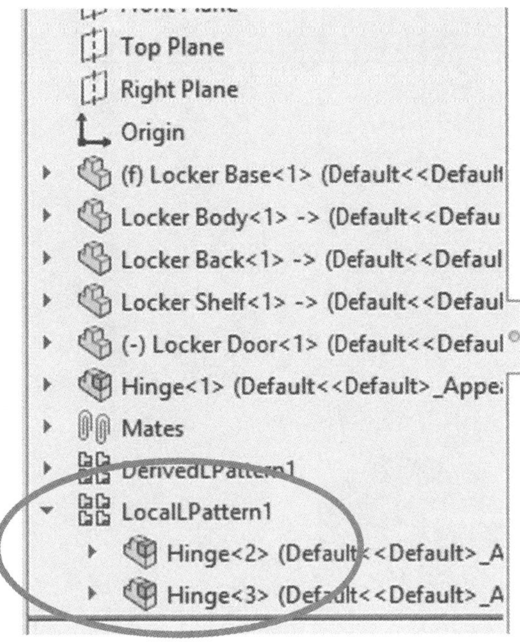

13.58. - Now we need to mate the door to the '*Hinge*'. Rotate the door to mate the outside face of the hinge and the face of the door's miter flange as indicated.

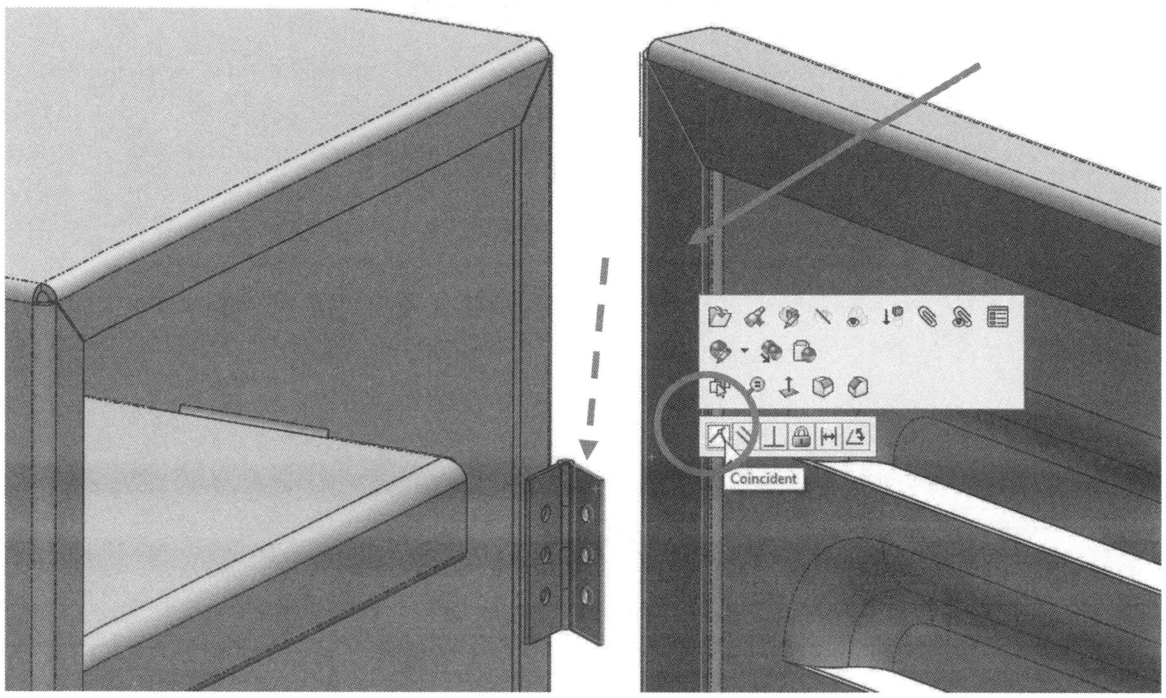

For the second mate, add a distance mate between the edge of the miter flange in the door and the '*Hinge*', make the distance 0.2″. Use the face selection filter if needed.

Finally, add a coincident mate between the top face of the door and the top face of the locker's body.

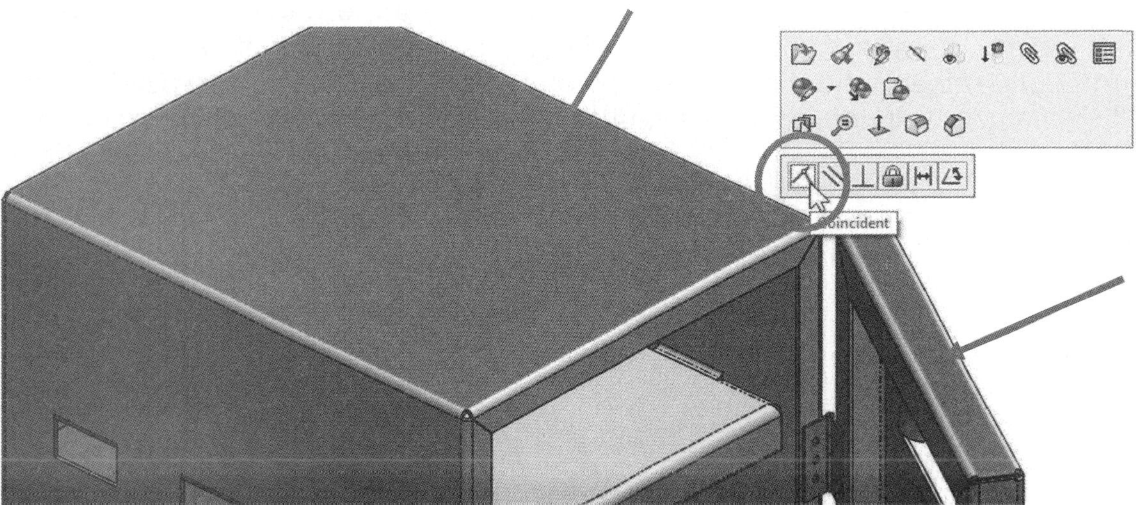

13.59. - After the door is mated to the hinge, if we click-and-drag to move it, we get the same message as before, alerting us that the component is fully defined and it cannot be moved. To illustrate how a subassembly behaves in SOLIDWORKS, open the *'Hinge'* assembly.

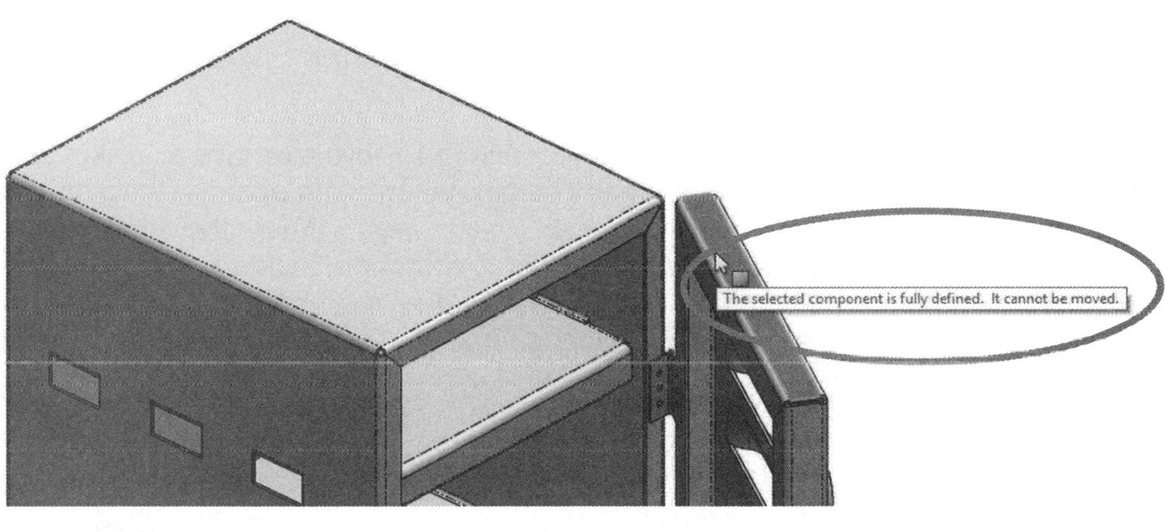

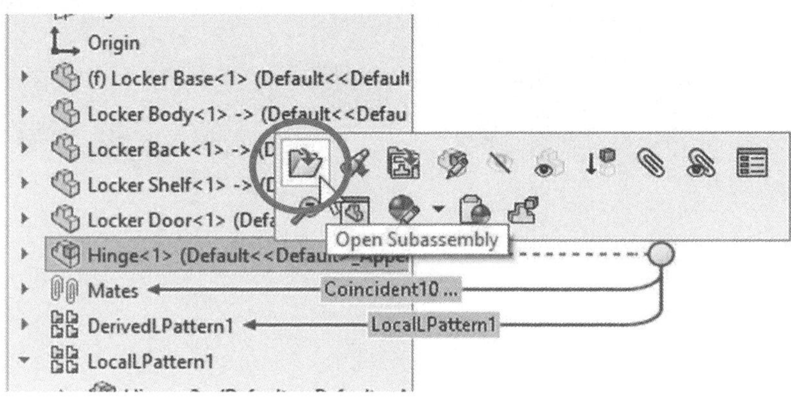

When mates are added to assemble components together, we are removing degrees of freedom (DOF). By default, a component has six degrees of freedom, three translations along the X, Y and Z axes, and three rotations about the X, Y and Z axes. When a component is under defined, it means that *at least* one DOF is not constrained, and it can move or rotate about one or more axes. If a part is fully defined, it will not move or rotate; this is the basis for assembly motion.

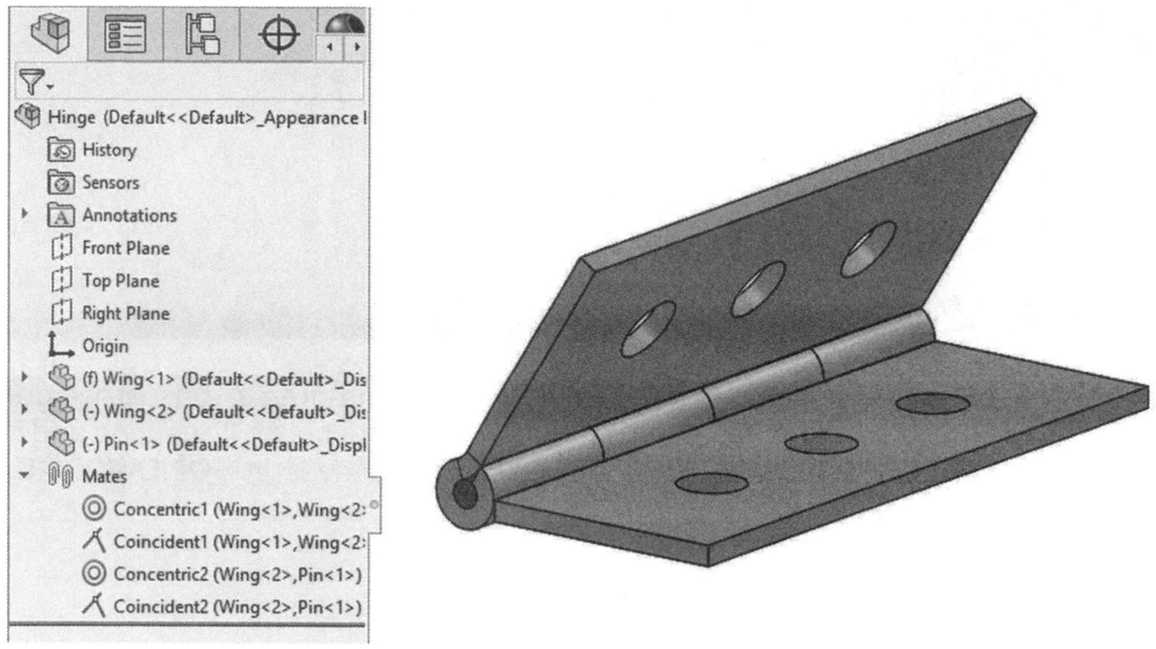

13.60. - In the *'Hinge'* assembly one *'Wing'* and the *'Pin'* parts have a (**-**) sign before their name. This means that these parts can move or rotate about *at least* one axis. In this case, both can rotate about the pin's axis. Since the first *'Wing'* is automatically fixed in place (because it was the first part added to the assembly) the other two can rotate about it. If we click-and-drag the under defined *'Wing'* part, it will move as a hinge as expected because it was constrained with one concentric and one coincident mate.

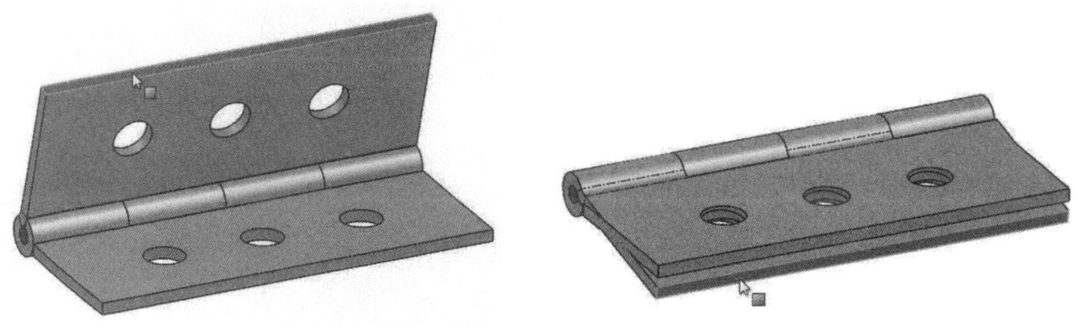

If we keep moving the *'Wing'*, it will move through the fixed *'Wing'*. This is the normal behavior. We can optionally activate "**Collision Detection**" to make parts stop when they touch each other.

Move the hinge to a position where the hinge's wings don't interfere with each other and click the "**Move Component**" icon in the Assembly toolbar; in the "Options" section select "Collision Detection." By default, "All components" is selected, and as we move a component, SOLIDWORKS will calculate collisions between all components in the assembly.

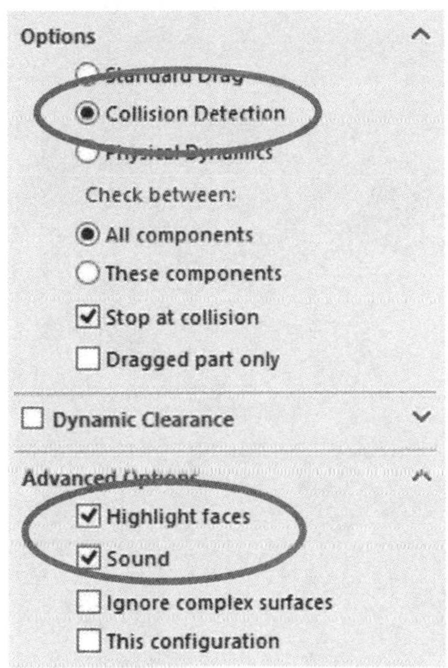

If the assembly has a small number of components, where *"small"* varies from a few to ten or more depending on the computer's capacity, using the "All components" option may be OK. However, if we have a large assembly with many components (and *"large"* also varies...) it is usually a good idea to limit the scope of collision detection using the option "These components" and selecting only the components we are interested in. Doing so will greatly enhance the speed and accuracy of calculation. With only three parts in this assembly using the "All components" option should be fast enough in just about any computer. Turn On the options "Stop at collision" to prevent the parts from going through each other, "Highlight faces" to visually identify the faces that are colliding, and "Sound" to hear when a collision is detected.

After setting the options drag the moving wing in the hinge towards the other wing part and see how it stops when the faces touch each other.

If we start "**Collision Detection**" with interference between the components being analyzed, the "Stop at Collision" option will be ignored, and the part's motion will continue. To re-enable collision detection the assembly must be set to a non-interfering position, or use the option "These components" and select non-interfering components in the selection box, then re-activate the option "Stop at Collision."

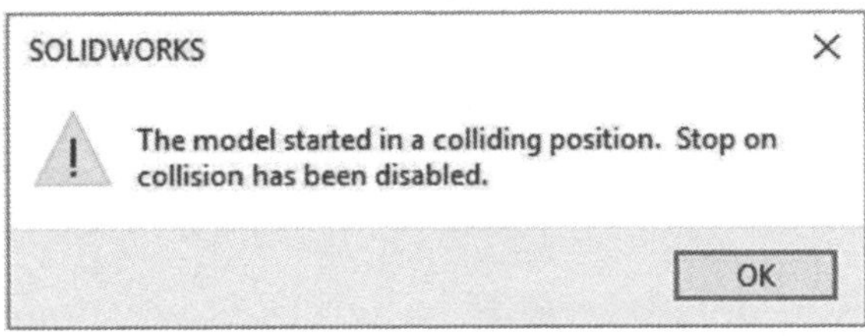

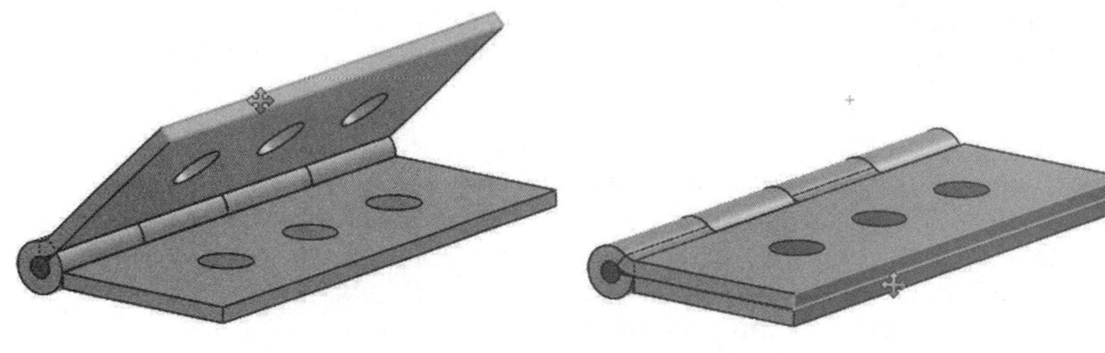

13.61. – After reviewing the Collision Detection, open the hinge approximately 90 degrees and go back to the locker's assembly window, where we'll be asked to rebuild it.

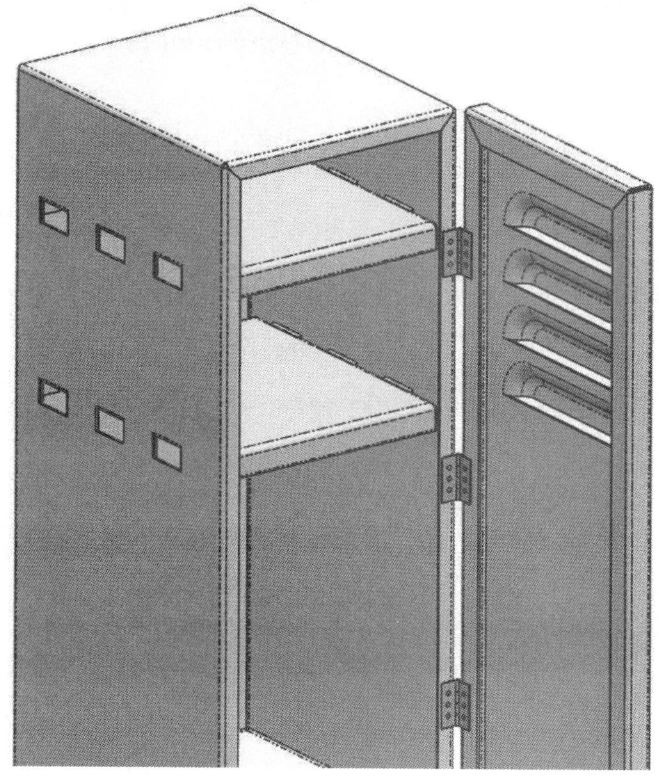

After rebuilding the assembly, the hinge will be in the same position it had in its own window. Moving the '*Hinge*' sub-assembly and going back to the main assembly is one way to move the door, but it's not really the easiest way, or the behavior we want.

What we need to do is to propagate the subassembly's motion to the main assembly. By default, subassemblies behave as a rigid body without motion to improve performance. Select the mated *'Hinge'* subassembly in the FeatureManager (not the patterned hinges), click in the "**Component Properties**" icon from the pop-up menu, set the option "Solve as" to "Flexible" and click OK to continue.

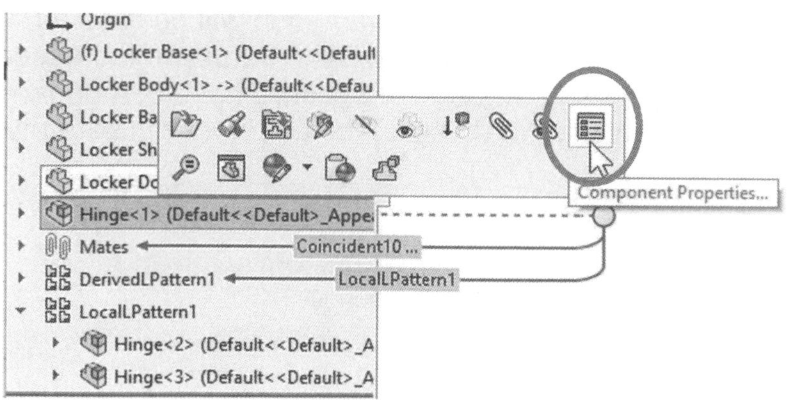

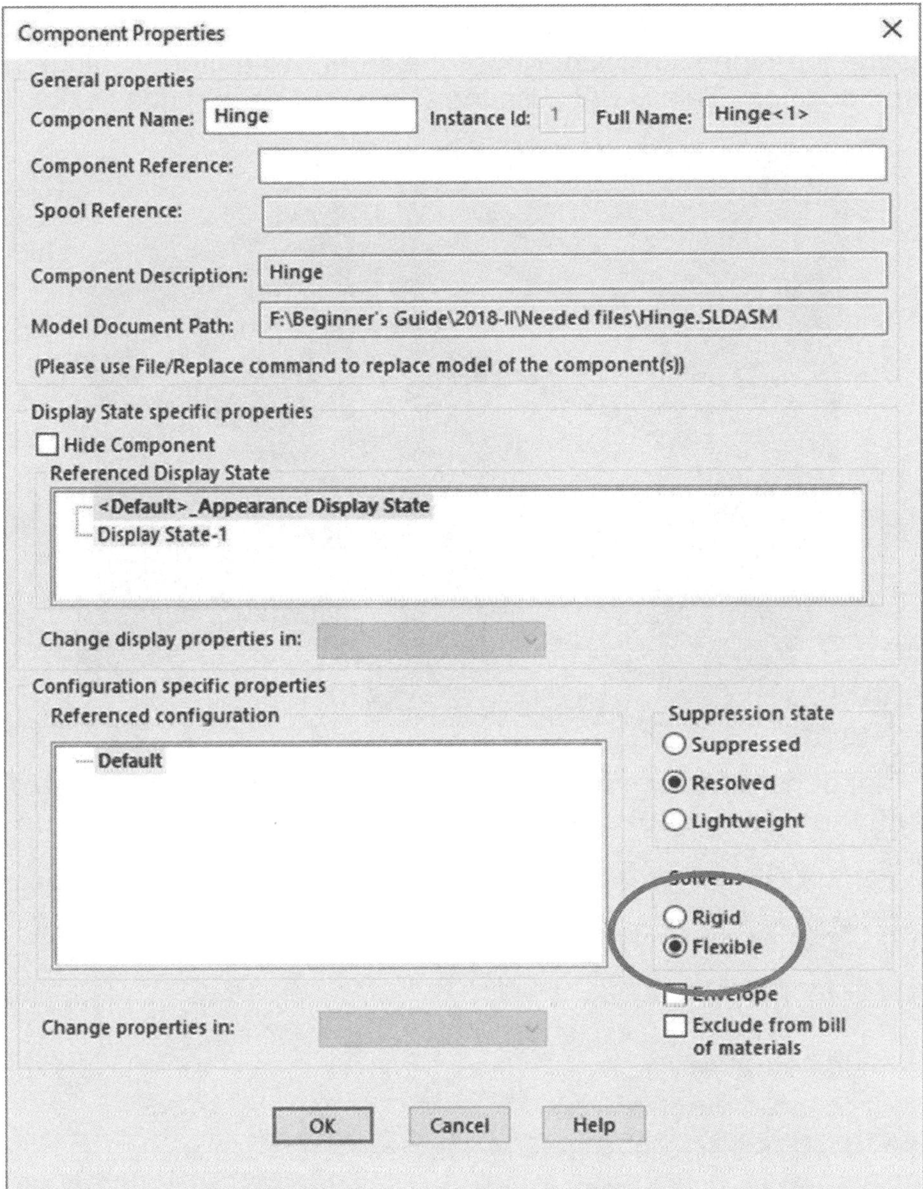

After the change, we notice the different icon, indicating a flexible sub-assembly; also, the '*Locker Door*' is now under defined, which means it has at least one un-constrained degree of freedom, which will allow us to move it.

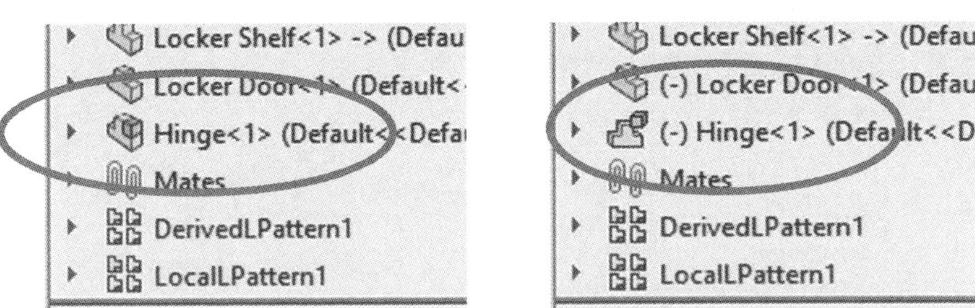

Rigid sub-assembly **Flexible sub-assembly**

13.62. – Once the *'Hinge'* subassembly has been made "Flexible" we can move the door using the hinge's motion. Notice the other two patterned hinges remain in their original position. This is because they have not been made flexible.

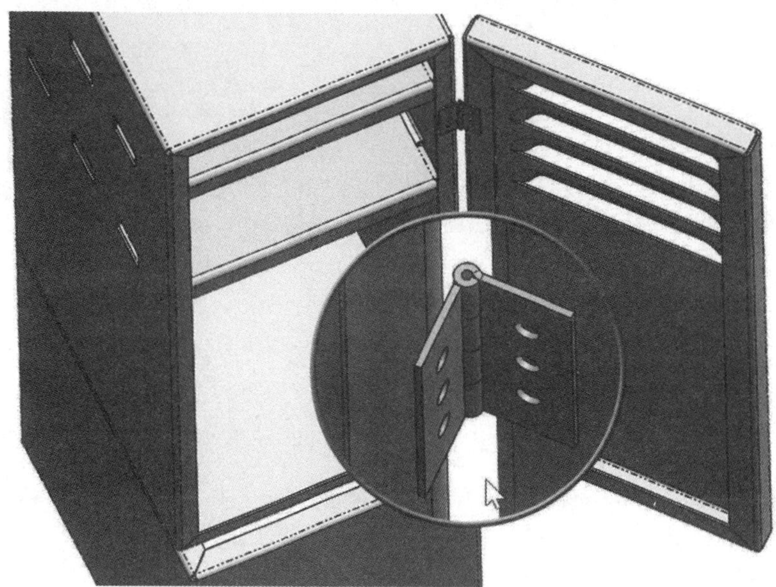

To make the other two hinges flexible, expand the "*Local Pattern1*" feature, and activate the "Make Subassembly Flexible" command in the pop-up menu for both hinges.

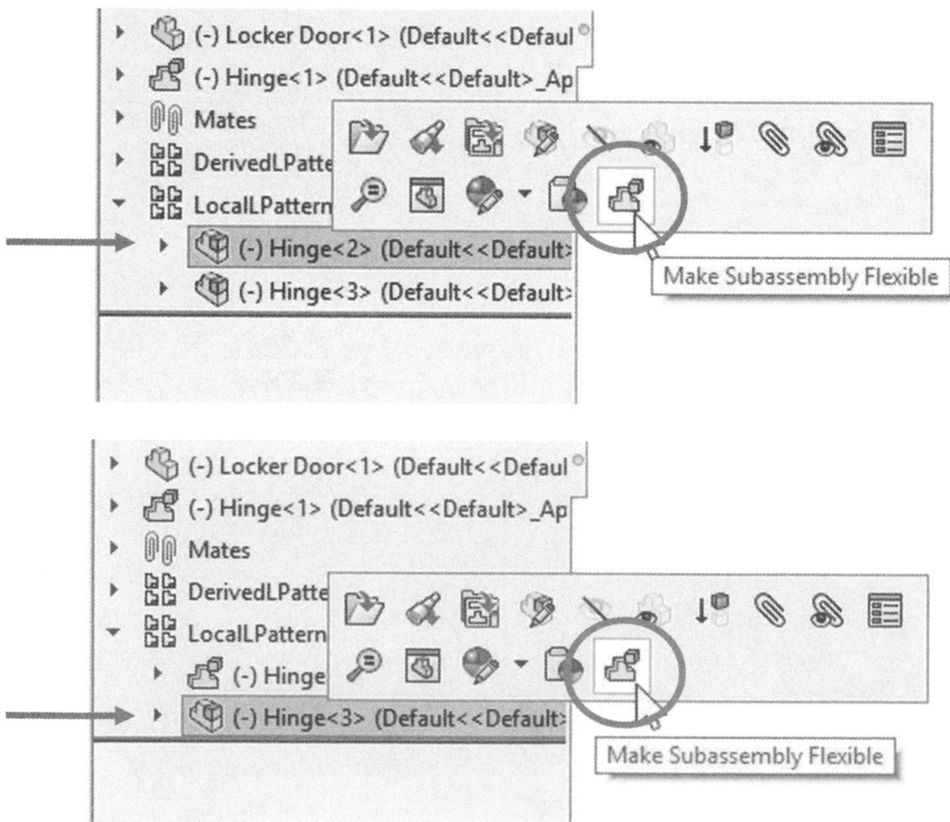

13.63. – After making the subassemblies flexible, mate the other wing of the hinge to the locker. To avoid over defining the assembly add a parallel mate. The assembly may have to be rebuilt to restore assembly motion after making the sub-assemblies flexible. If rebuilding the assembly does not restore motion, try a **force rebuild** using the shortcut Ctrl-Q.

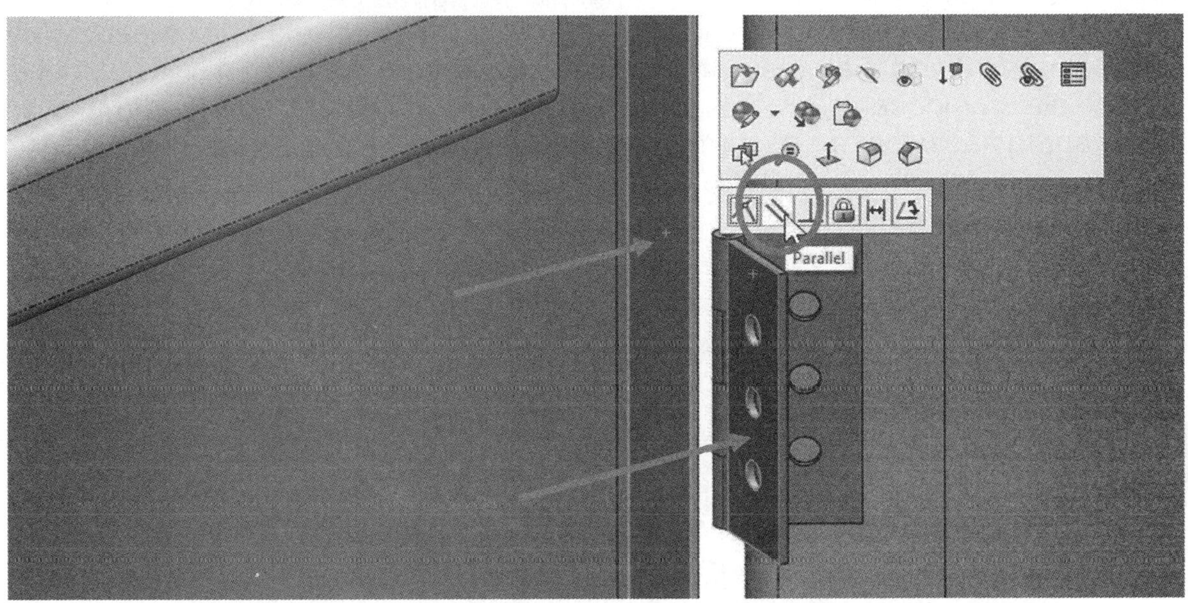

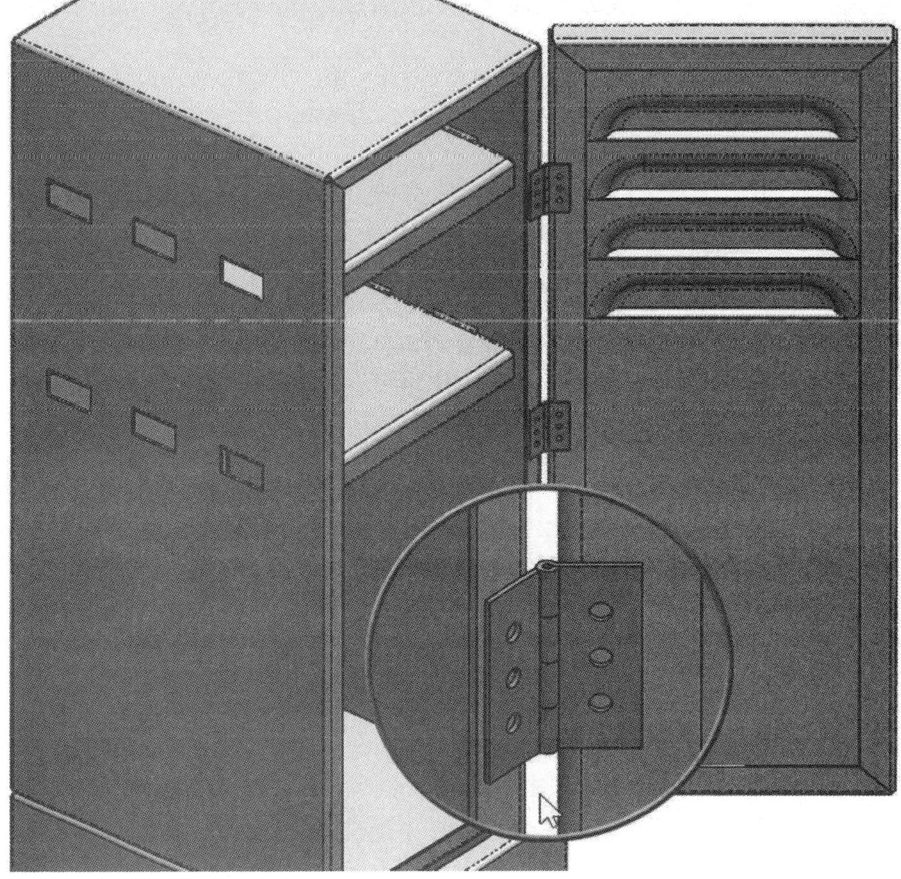

13.64. - The last thing we need to do is to add a series of holes to attach the *'Locker Back'* with rivets to the *'Locker Body'*. We can add the holes to one part and then make the holes in the other part using in-context relations, but we'll use a different approach. In some situations, like in sheet metal and welded components, it's common to perform certain operations after the components have been assembled, like holes. In this case the holes cannot be added to a part because their final position may not be known, or could vary until the final assembly is complete. We'll add the holes while editing the assembly, and they will be added to multiple components at the same time using an **Assembly Feature**. It is important to know that an assembly feature, by default, does not exist in the part, only in the assembly. Optionally, it can be changed and propagated to also exist in the part level.

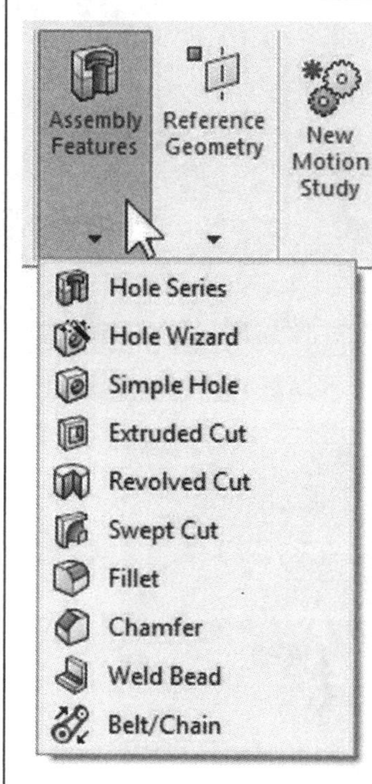

Most **Assembly Features** remove material, like a post-assembly operation, in our case is the equivalent of holding both parts together and drilling through them at the same time.

Assembly features include:

Hole Series to make holes using the Hole Wizard through several parts with the option to add different types of holes in the first, middle and last parts like counterbore, countersunk, clearance, threaded, etc.

Hole Wizard works the same way it works in a part but for multiple components.

Simple Hole makes single round holes in multiple components.

Extruded Cut, Revolved Cut and **Swept Cut** work as a cut feature in a part, but in the assembly cuts through multiple components.

Fillet and **Chamfer** work exactly as in a part, but in the assembly, they are used mostly in the weldments environment to prepare mated parts for welding.

Welded Bead is used to add welds in structural steel, including manufacturing annotations (covered later).

Belt/Chain is used to add Belt mates between pulleys or sprockets to maintain their relative motion and calculate belt/chain length, or to define the belt/chain length and locate components in an assembly.

We need to use ⅛″ rivets to secure the *'Locker Back'* to the *'Locker Body'*. Turn the locker view to see the upper corner from the back, and select the "**Hole Wizard**" option from the "**Assembly Features**" drop-down command. Using the "Hole" type, select a ⅛″ size. Set the hole's end condition to "Blind" with a depth of 0.25″, enough to go through both parts. The *'Locker Back'* color was changed for clarity.

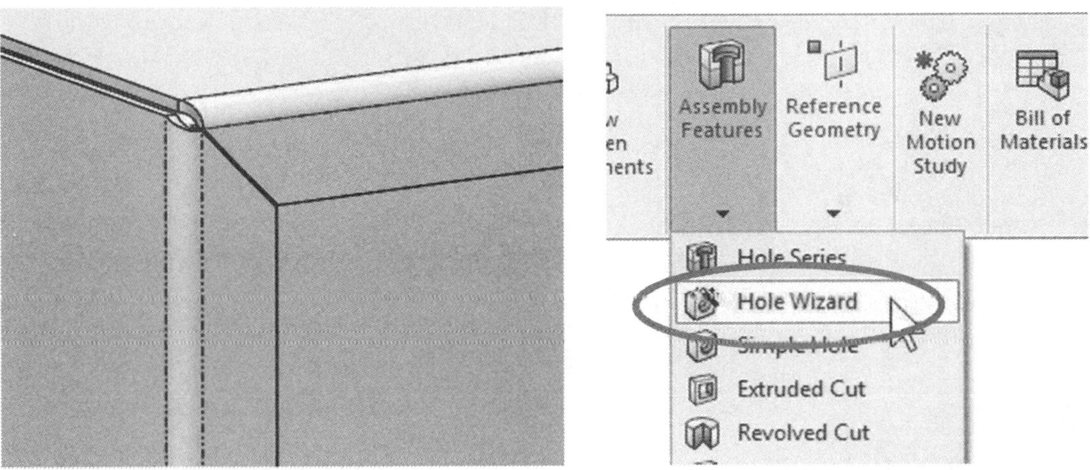

Notice the additional option at the bottom titled "Feature Scope." This is where we can define which parts will be affected by this feature and if we want the feature to be propagated to the individual parts. If we leave the "Propagate feature to parts" option unchecked, the holes will only exist in the assembly and not in the parts. This is useful when making post-assembly operations, like drilling. If the option is checked, the holes will be added at the part level, will have an external reference, and will exist in the assembly, like any other feature.

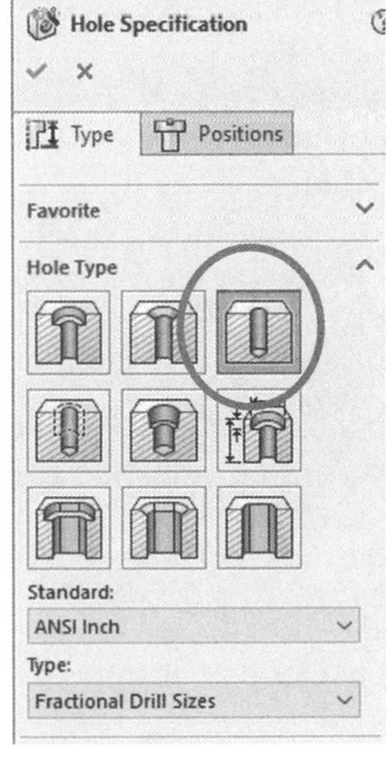

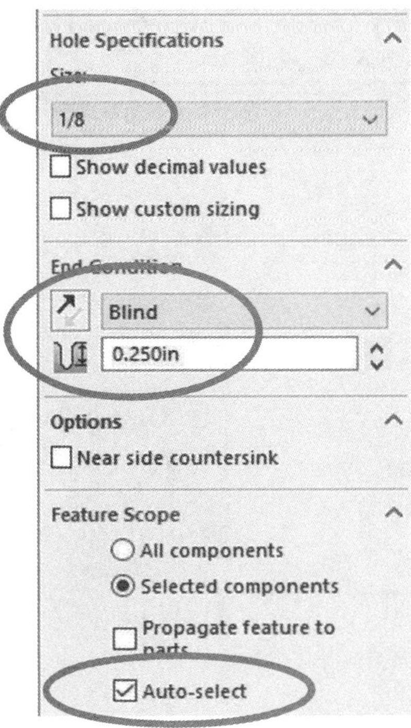

13.65. - To add the hole, select the "Positions" tab and then the back of the miter flange in the *'Locker Body'*. Immediately the "**Sketch Point**" tool is activated. Add a point and locate a single hole as indicated. We'll use a linear pattern to add the rest of the holes. Click OK when done to complete the hole.

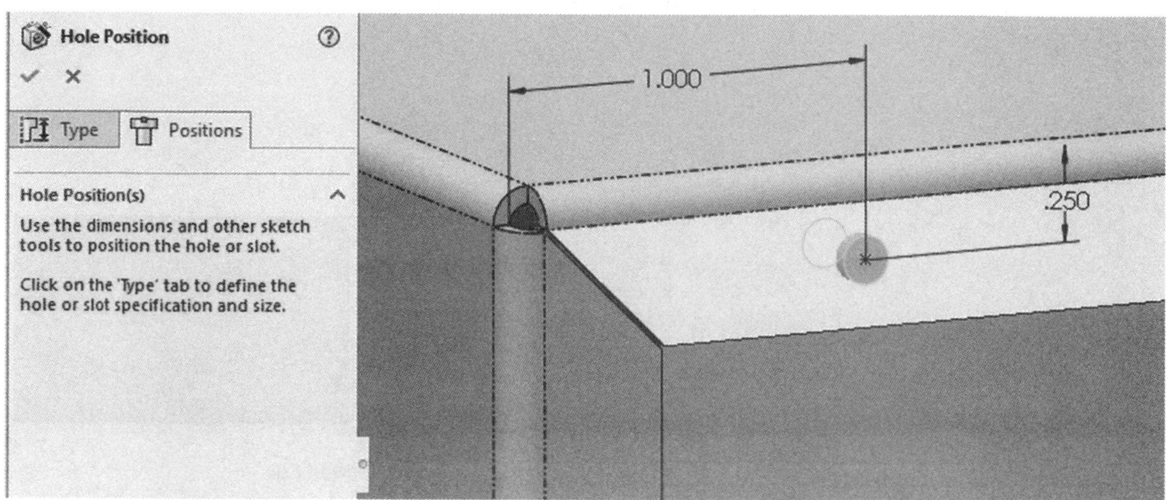

The hole is added to both parts in the assembly, and because the option "Propagate feature to part" was not checked in the Hole Wizard, when either part is opened by itself the hole will not exist.

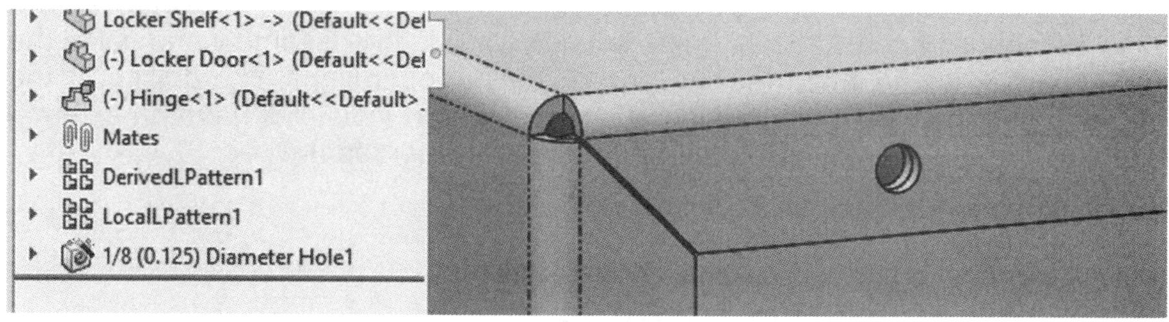

Assembly

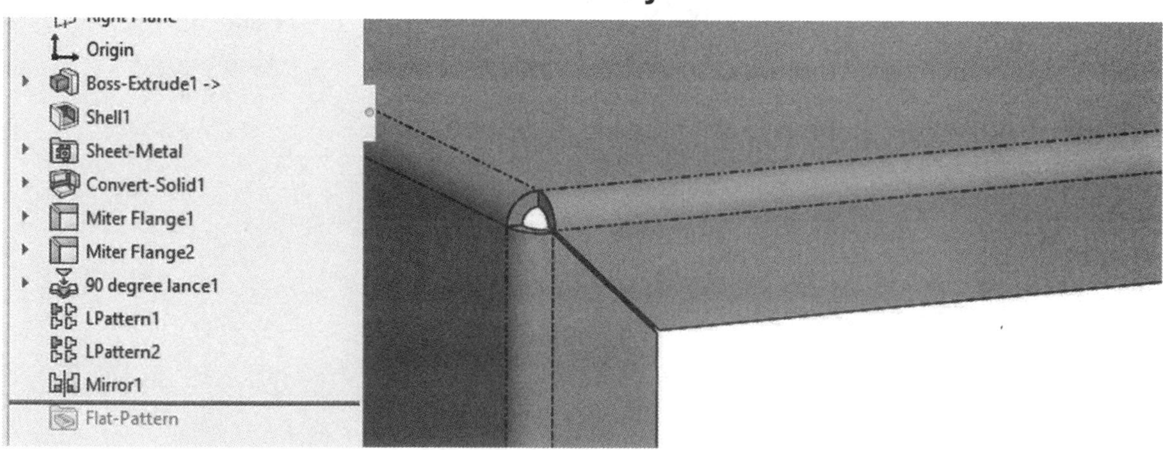

Part

When assembly features are added, the "**Assembly Features**" drop-down icon will be expanded to include linear, circular, table and sketch driven assembly patterns.

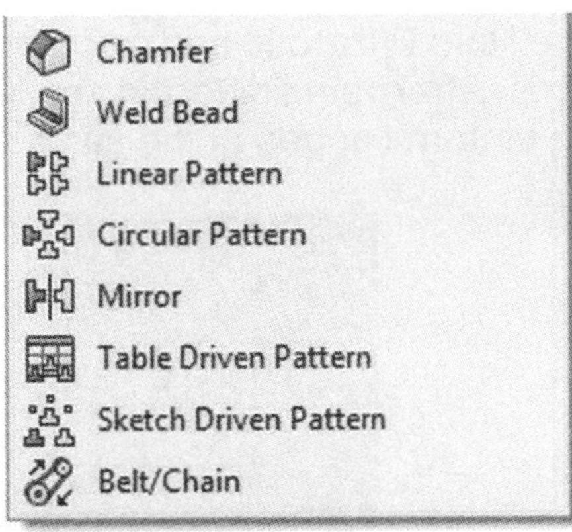

13.66. - Select the "**Assembly Features, Linear Pattern**" drop down command and add a horizontal pattern using the assembly feature hole with 5 instances spaced 1″ apart.

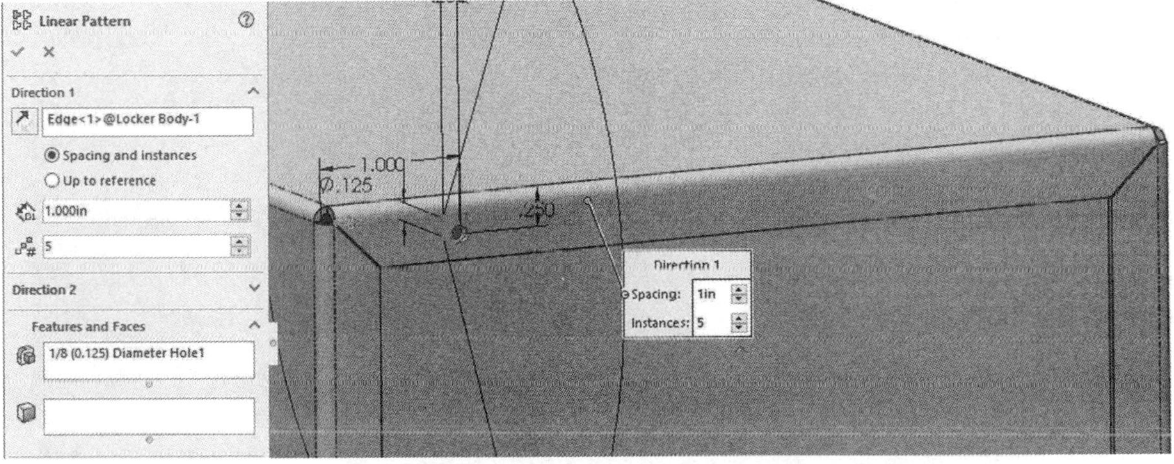

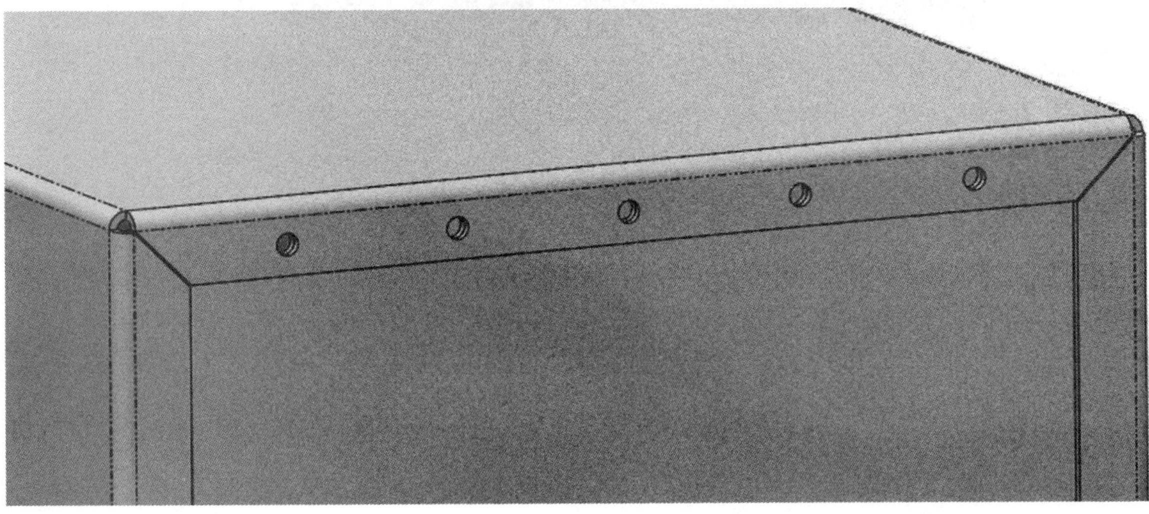

13.67. - Make similar assembly features and patterns in the other three flanges to complete the design. Add a vertical pattern in the side flanges with 8 instances spaced 2″ apart. Save the finished assembly when finished.

To save time, edit the Hole Wizard to add a second hole in the bottom flange using a sketch mirror. After rebuilding the model the pattern will add the holes at the top and bottom flanges at the same time.

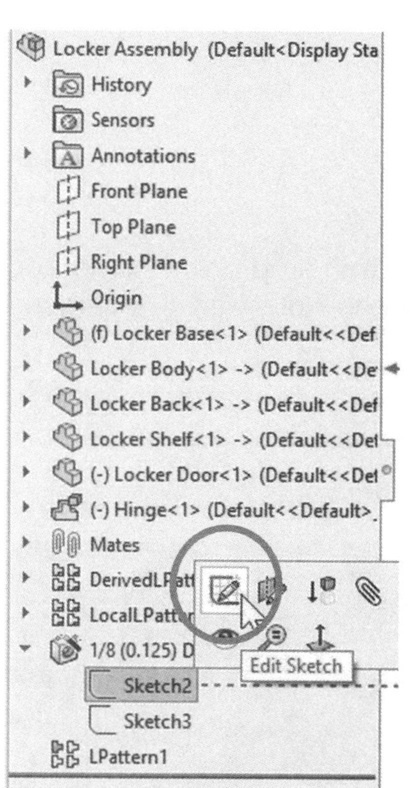

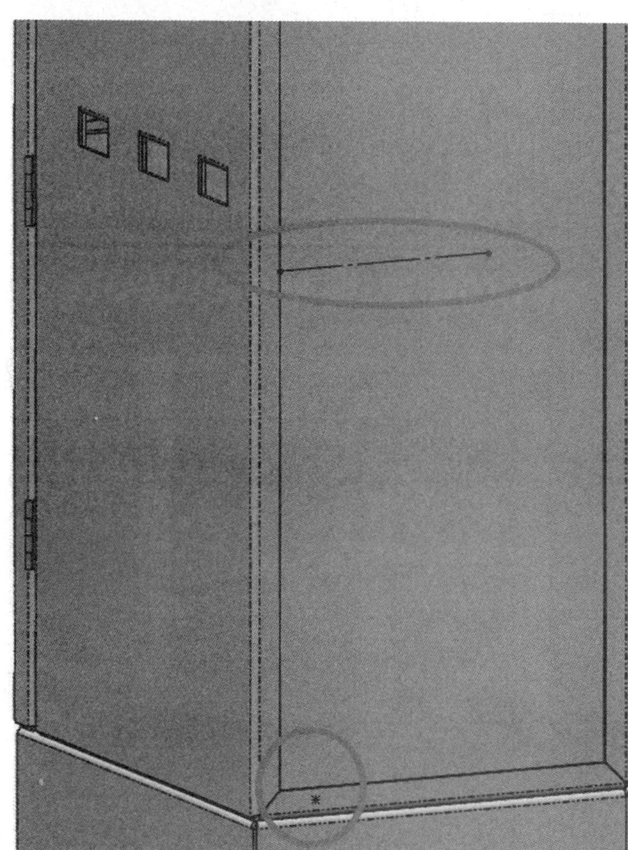

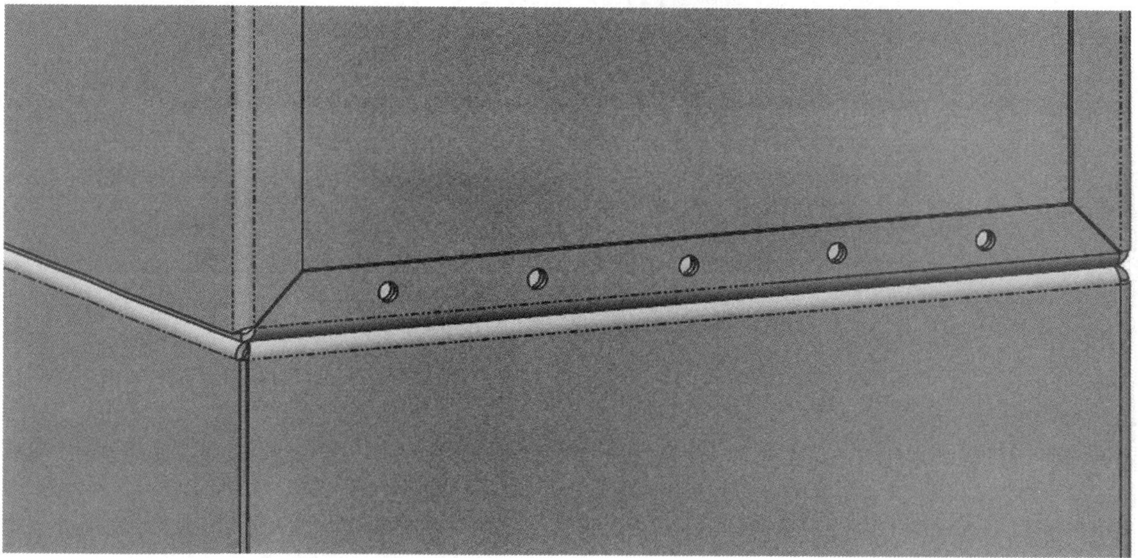

SOLIDWORKS Pack and Go

After finishing an assembly, sometimes we need to make a copy of the files to make a different version, send to a colleague, vendor, etc.

A common error when copying SOLIDWORKS files is to use the "Save as" option; the problem with "Save as" is that we either only make a copy of the assembly without the parts, drawings and sub-assemblies, or, if we are working in a part we only make a copy of the part without its drawing.

The real problem is if we have an assembly open, and then we open an assembly's part or sub-assembly and use "Save As." In this case, saving a component with a different name we can potentially change the references in the open assembly, and if we don't pay attention, we may be inadvertently changing references in assemblies, have duplicate files with different names, drawings that cannot find the original part or assembly, etc.

The best option when we need to make copies of entire assemblies complete with parts, drawings, sub-assemblies and even analysis results is using the built in "Pack and Go" command.

In this lesson, we are going to make a copy of the locker's assembly and its parts. The copied files will be used in this lesson's exercise.

14.1. – Open the '*Locker Assembly*' and go to the menu "**File, Pack and Go**."

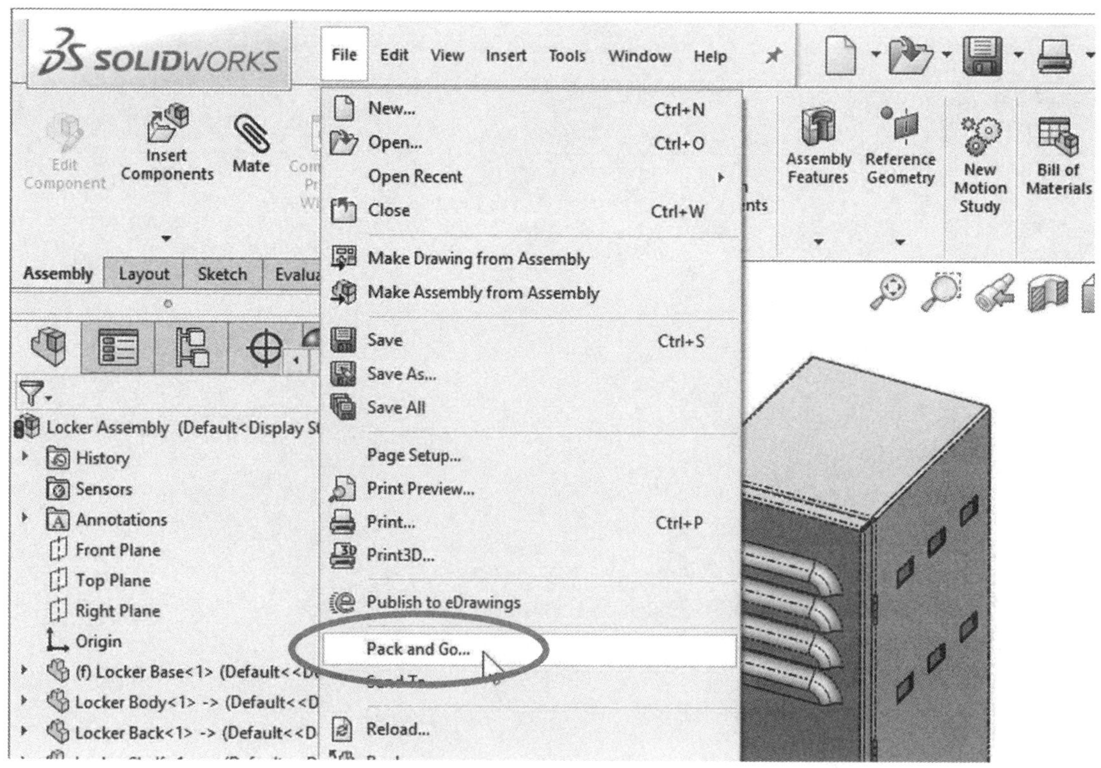

14.2. – In the “Pack and Go” window we can see the name of each file, the folder where each file is located, the new file name, the folder to save the copy to, file size, type and date modified.

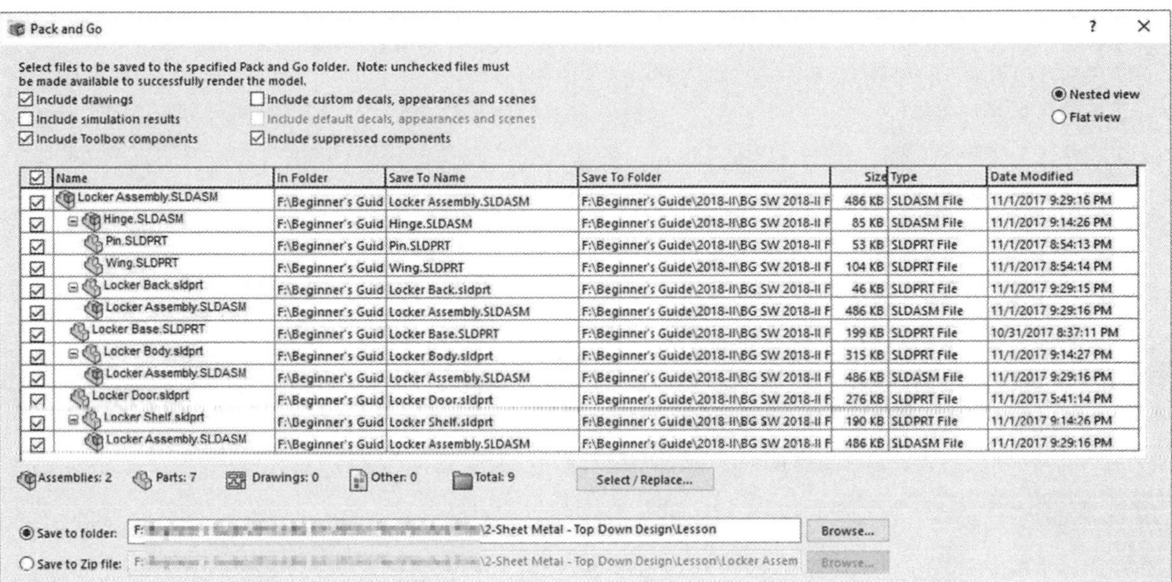

To differentiate the copied files, we will add a prefix to the new file names. Check the “Add prefix:” checkbox and type “EX-” in the field. After adding the prefix, the “Save To Name” field changes to reflect the new name. In the “Save to folder:” field, browse to the folder where you save your exercise files. Click OK to copy and rename the files to the destination folder to continue.

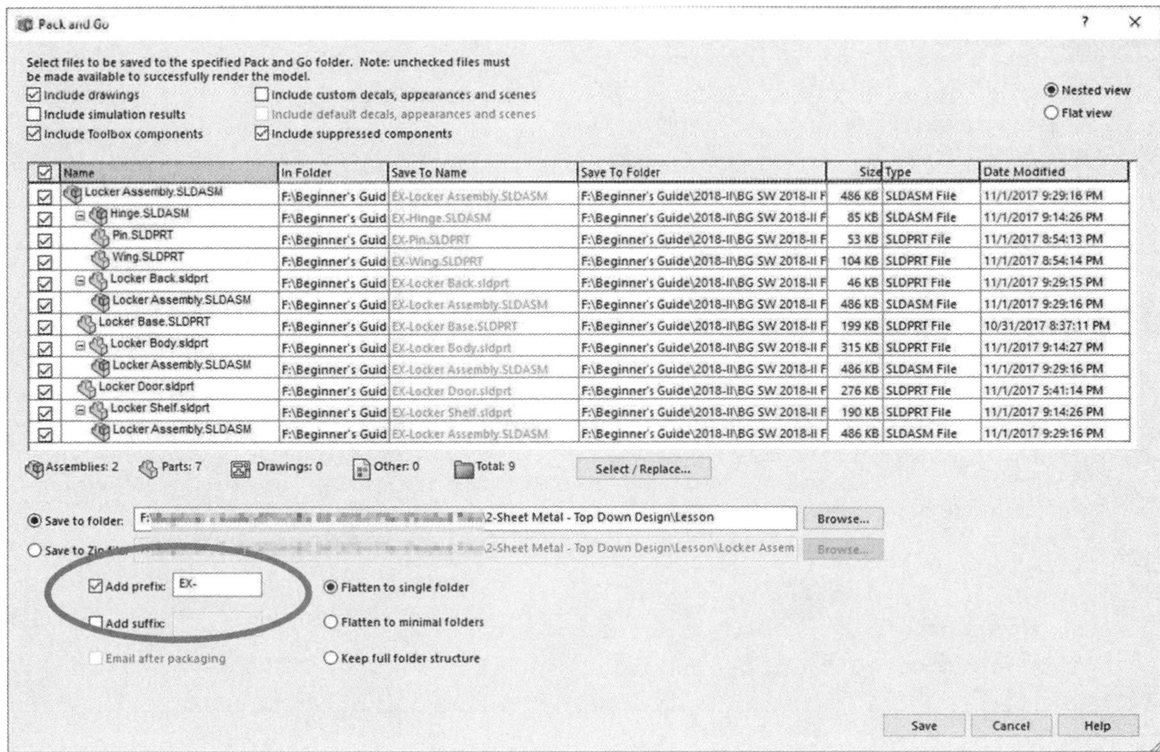

With "Pack and Go" we can also include drawings, simulation results, Toolbox components, custom decals, appearances and scenes at the same time.

Select files to be saved to the specified Pack and Go folder. Note: unchecked files must be made available to successfully render the model.

- [x] Include drawings
- [ ] Include simulation results
- [x] Include Toolbox components
- [ ] Include custom decals, appearances and scenes
- [ ] Include default decals, appearances and scenes
- [x] Include suppressed components

When copying files we can make a copy to a selected folder or create a Zip file, rename files with a prefix and/or a suffix, and optionally flatten the files to a single folder regardless of where the components are stored. This last option is helpful when we need to collect files scattered in multiple locations.

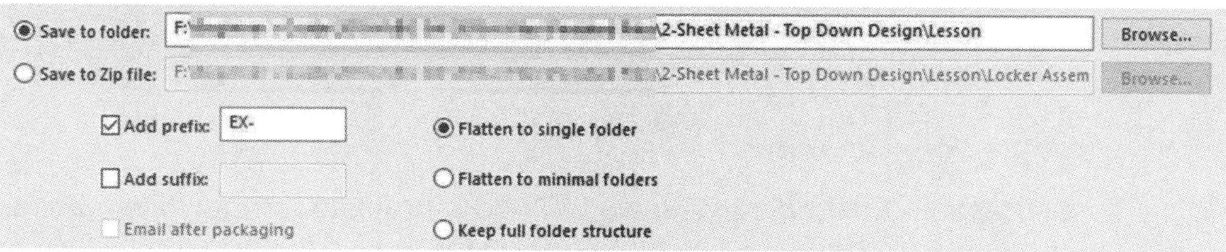

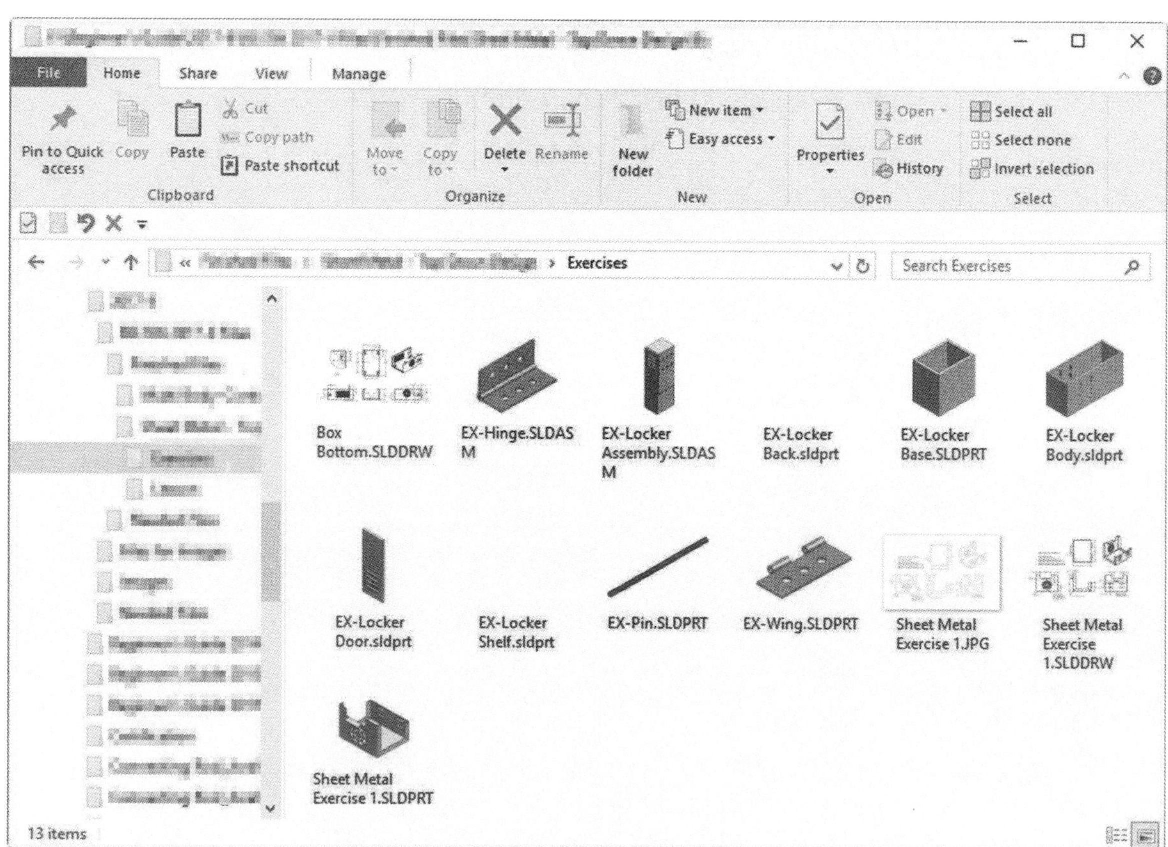

EXERCISES:

1: From the files copied using "Pack and Go" open the *'EX-Locker Assembly',* and change the *'EX-Locker Body'* to be the same size as the base adding a collinear relation to the base sketch of *'EX-Locker Base'* as originally designed.

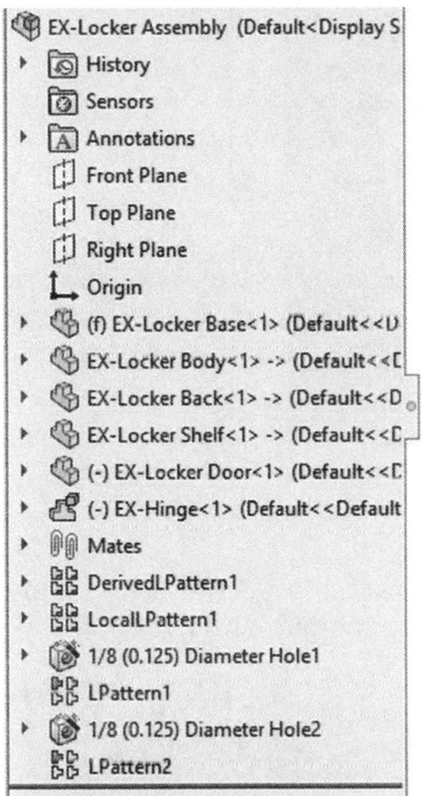

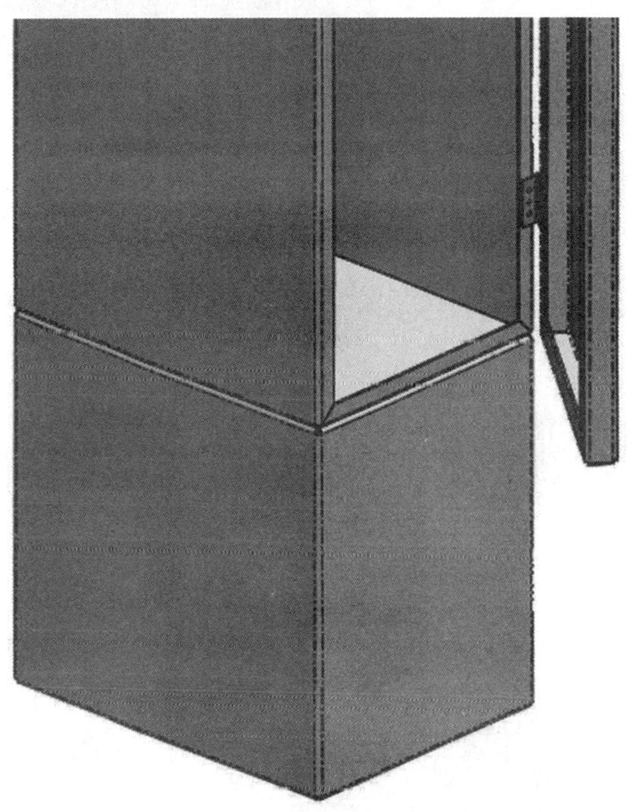

TIP: Edit the "Boss-Extrude1" sketch, delete the 0.625" dimension and add a collinear relation to the front edge of the base.

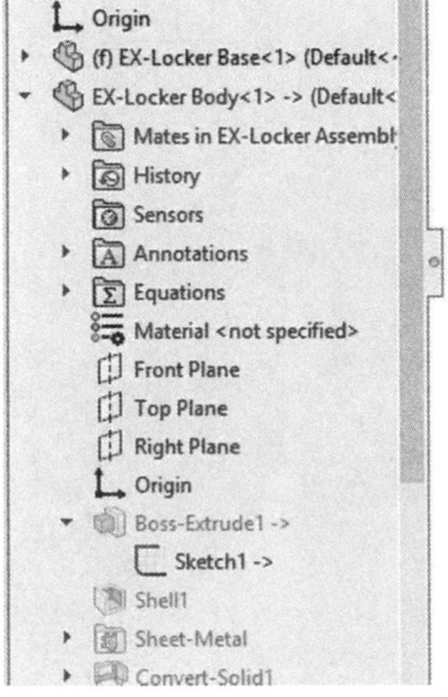

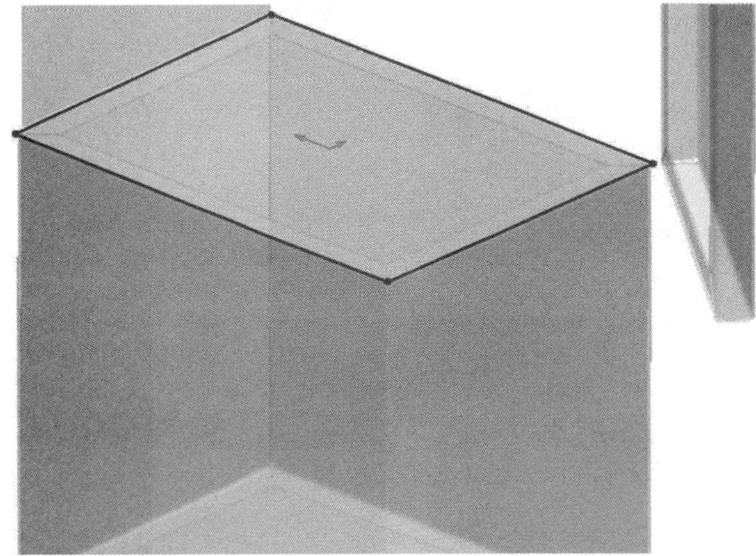

2: Change the *'EX-Locker Base'* dimension to 8″ wide x 10″ deep x 10″ high.

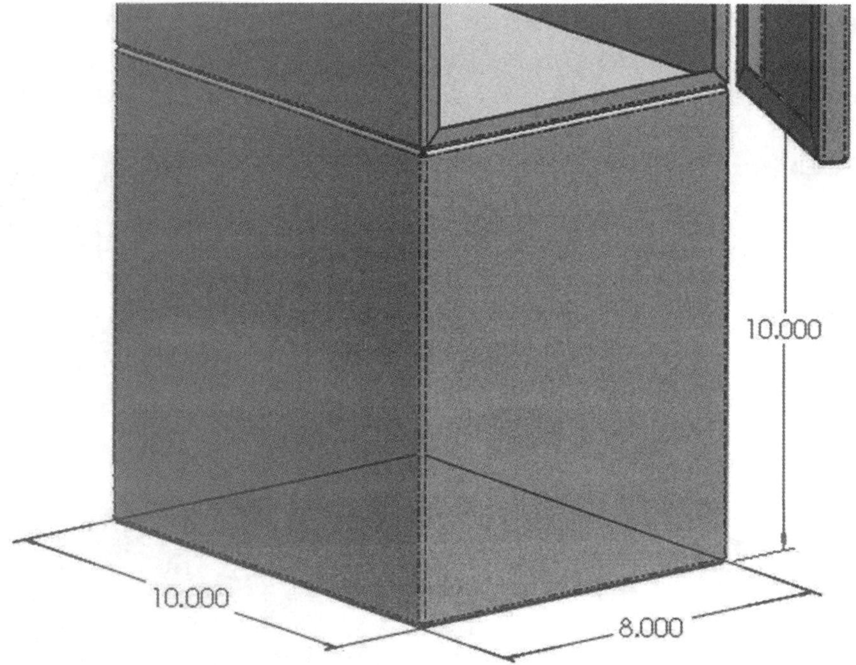

3: Add assembly features using the Hole Wizard to attach the *'Locker Body'* to the *'Locker Base'*. Choose the location, quantity and spacing for the rivet holes.

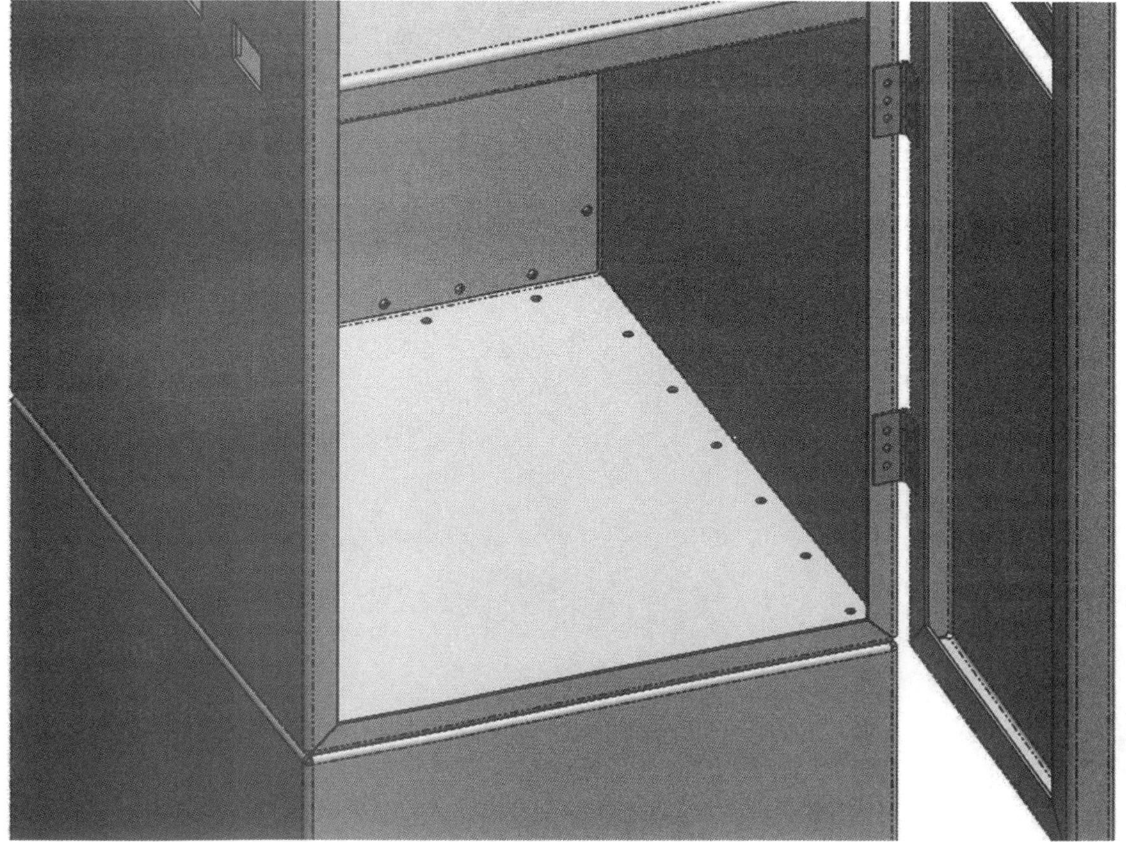

4: Add equations in the assembly to make the locker's door width 1″ narrower than the locker's base, and the locker's door height ⅞″ smaller than the locker's body. Equations in an assembly work just like in parts, except the dimension's name includes the component's name. Double click in the appropriate features to display their dimensions and rename them to make the equations easy to read. After rebuilding the assembly, the door's size updates.

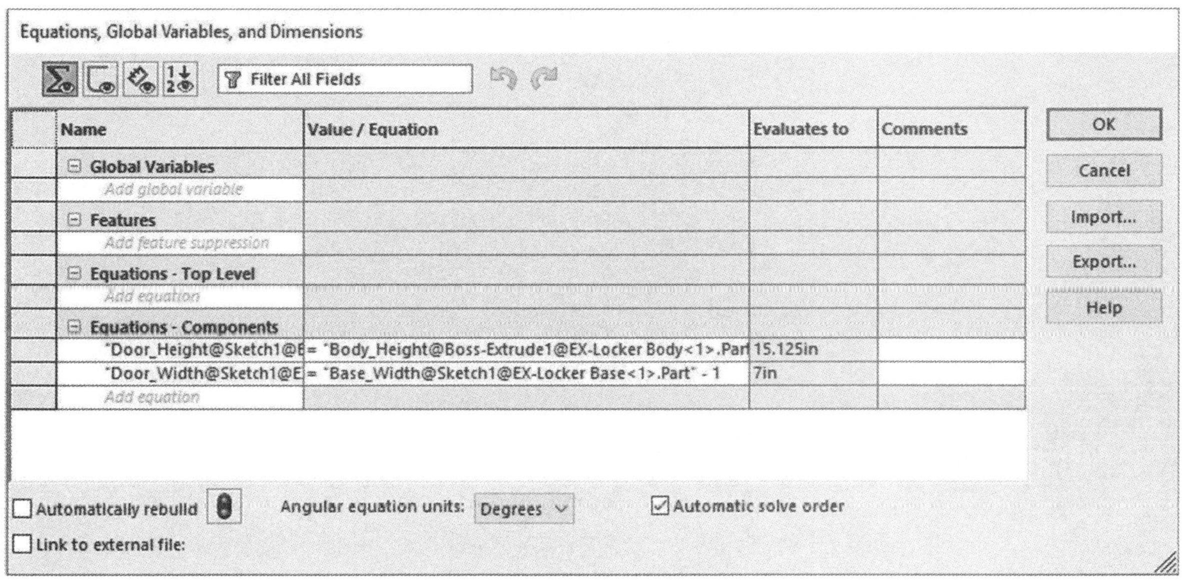

TIPS: Show the door's first extrusion dimensions and rename them "Door_Height" and "Door_Width."

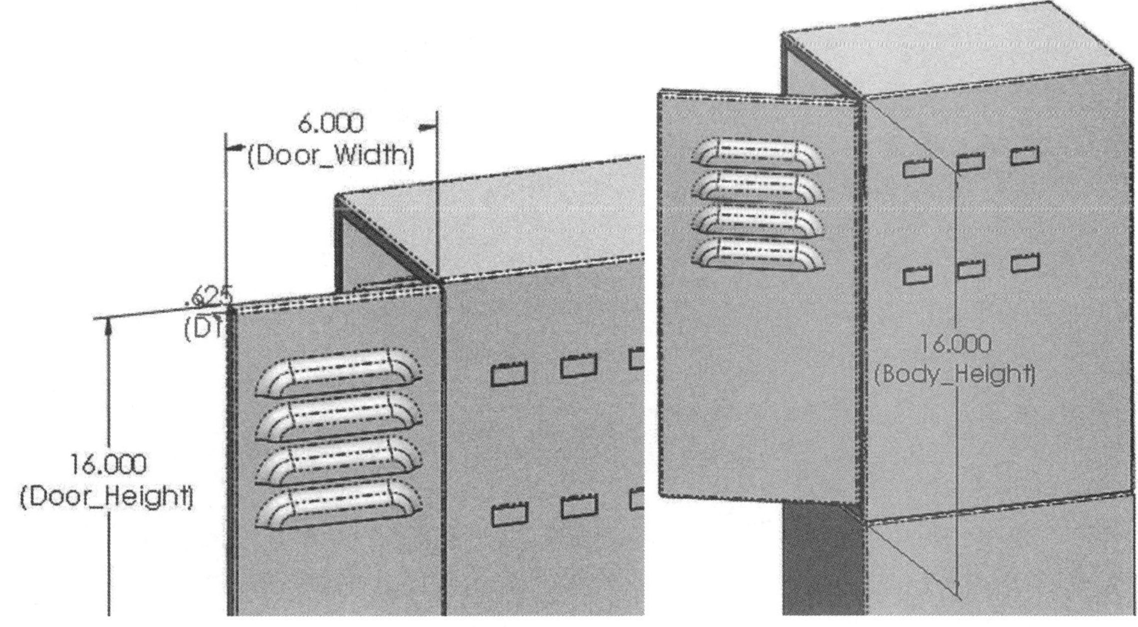

5: Modify the miter flange in the front of *'EX-Locker Body'* by editing its sketch as shown to accommodate the 'EX-*Locker Door*' and make it flush with the front of the assembly.

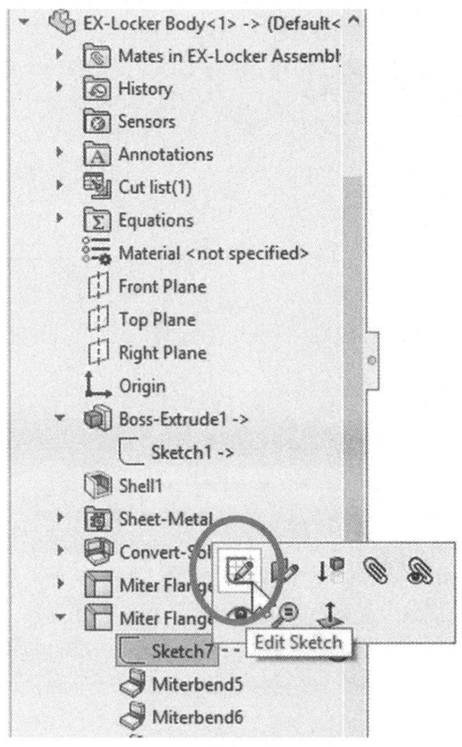

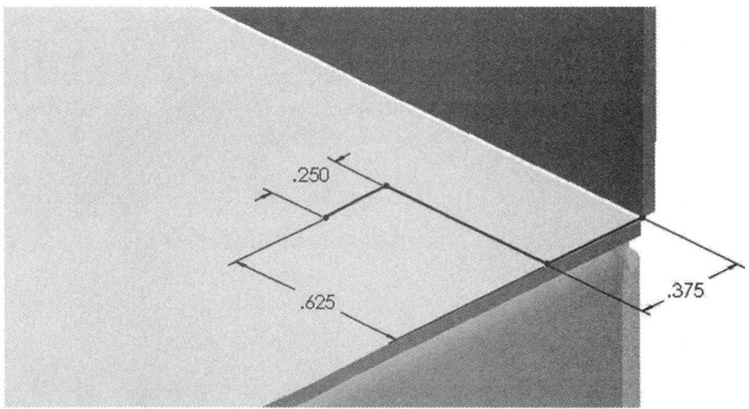

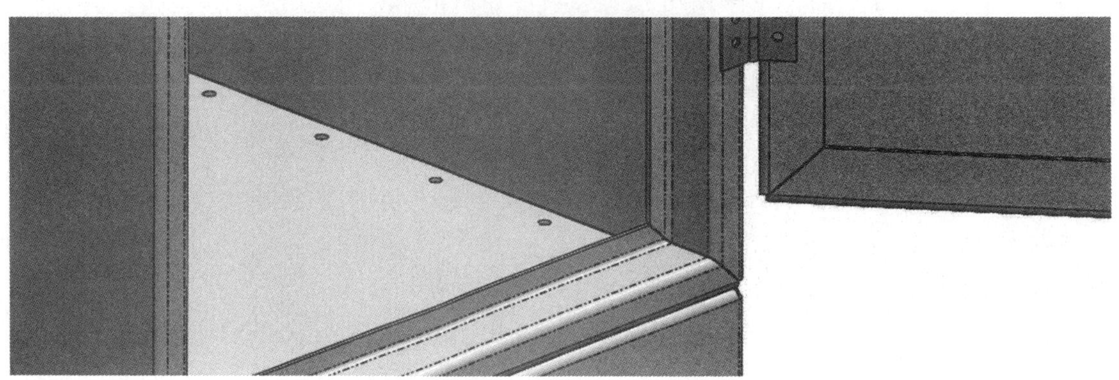

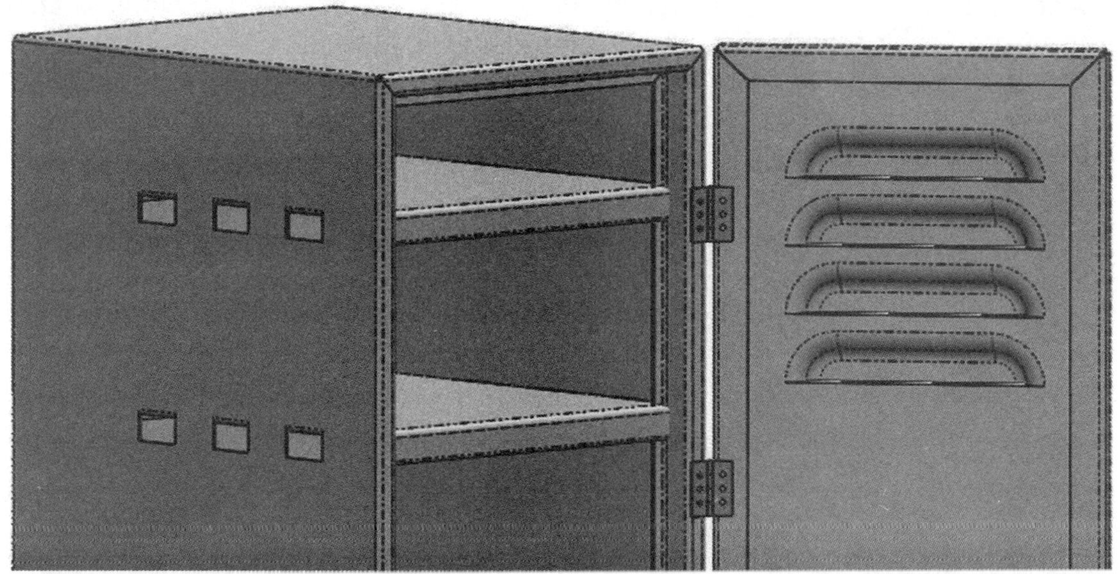

When the miter flange is edited, if an edge was used for a linear pattern's direction we may get an error at the assembly level. If this happens, edit the pattern and re-select the direction's edge to clear the error.

6: Modify the *'EX-Locker Shelf'* to fit correctly inside the *'EX-Locker Body'*. Edit the *"Base-Flange1"* sketch in the *'EX-Locker Shelf'* to be ¼" away from the inside face of the miter flange modified in the previous step.

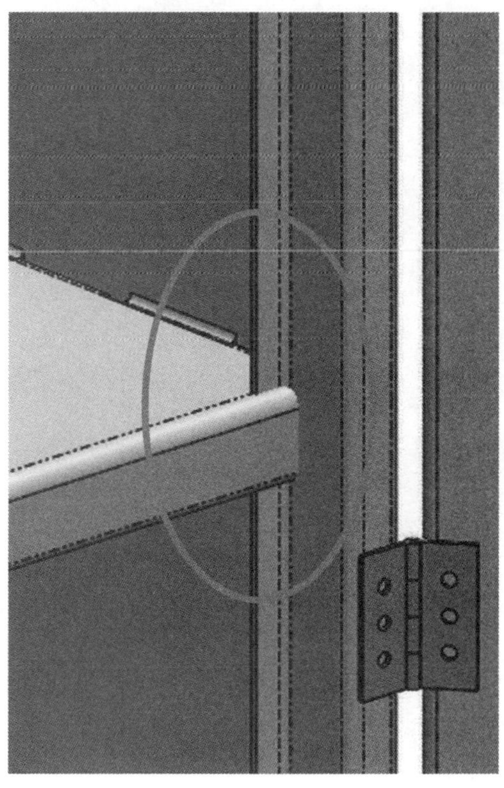

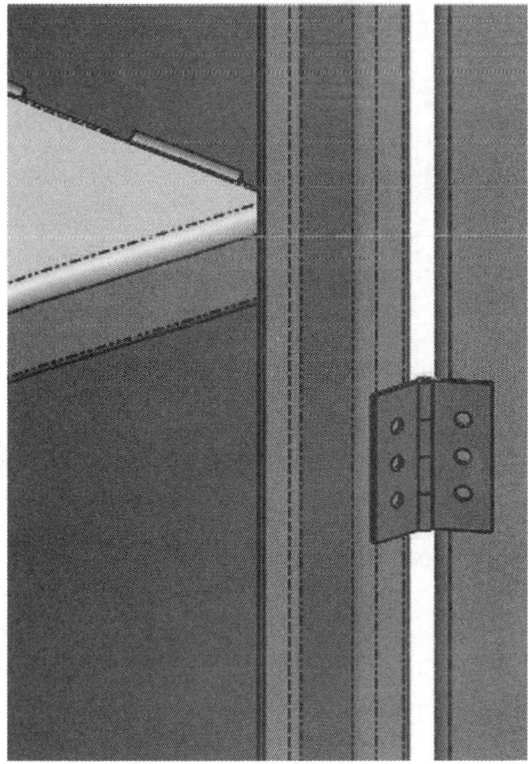

7: Add and delete the necessary mates to make the *'EX-Locker Door'* flush with the front face of the locker's body and mate the hinges in the front of the locker assembly. Use a Width mate to center the door vertically between the top and bottom Miter Flanges, and use "**Collision Detection**" to adjust the mates in the hinges and make sure the door opens and closes as intended without interferences.

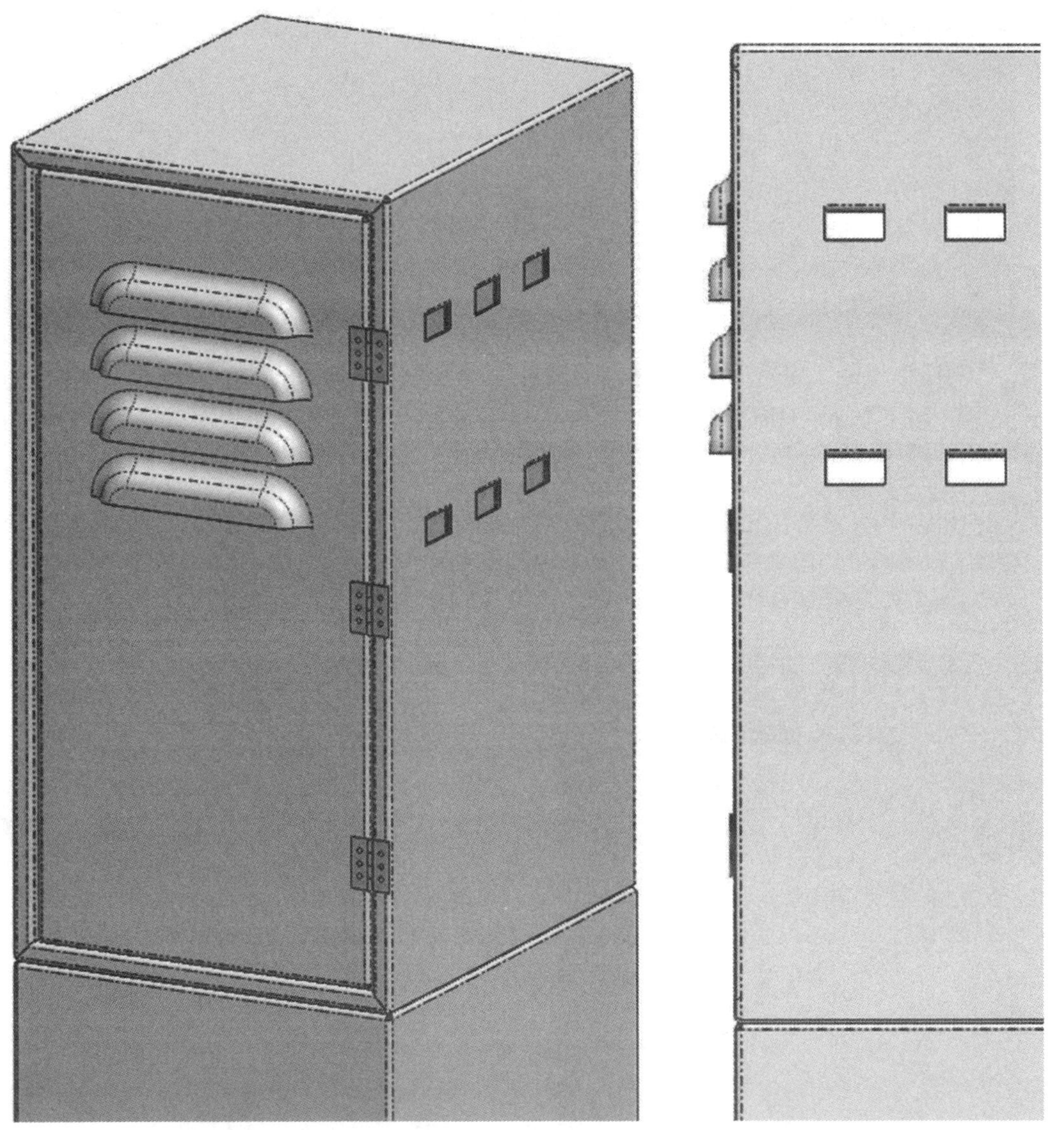

8: Make detail drawings of all parts including a flat pattern and an assembly drawing with an exploded view and bill of materials.

Extra Credit: Design a suitable lock for the door and incorporate it into the assembly.

Review and Questions:

a) Name the different options available in SOLIDWORKS to calculate sheet metal bend allowances.

b) How is the Relief Ratio calculated (Rectangular and Obround reliefs)?

c) What is and how is the K-factor calculated?

d) Which external reference state will propagate changes immediately in an assembly (In-Context, Out of Context, Broken, Locked)?

e) How can external references be temporarily "frozen"?

f) What does the option "Merge faces" do in a sheet metal component?

g) What is the "Link to thickness" option and what does it do?

h) What is the file extension used for library features?

i) What is the option "Flexible" in a subassembly used for?

j) Explain the difference between a Miter Flange and Edge Flange?

Answers:

a.- Bend allowance, Bend Deduction, K-Factor
b.- It's the relation between the sheet metal thickness and the width of the relief.
c.- It's the relation between the distance from the inside bend radius and the neutral plane to the thickness.
d.- In Context.
e.- Locking them using under "List External References."
f.- It eliminates all the bend lines and bend regions in the flat pattern making it a single surface.
g.- It's an option available when adding extrusions and cuts to sheet metal parts to make the extrusion (or cut) the same depth as the sheet metal's thickness.
h.- *.SLDLFP
i.- If a subassembly has mates that permit motion, the motion will be available at the top level assembly.
j.- In an Edge Flange, the width and shape of the flange can be changed but not its profile; in a Miter Flange, only the profile can be edited.

Notes:

3D Sketch and Weldments

A 3D sketch, as its name implies, exists in 3D space and is not limited to a 2D plane. 3D sketches are typically used for guide curves and paths for sweeps and lofts, as "centerlines" for structural profiles (weldments), as reference to create auxiliary geometry, etc.

A Weldment part in general refers to a multi body part that is made from extruded or swept profiles along a 2D or a 3D sketch. It is used for designing structures, welded components, tubular parts, etc. and includes special features for welding and manufacturing annotations like weld types, cut lists for each type and size of material used (very much like a Bill of Materials) and more trade specific tools to help us fully design and document welded structures.

Notes:

3D Sketch

To learn how to make a 3D sketch, we'll build a tubular frame for a chair. We'll need to make two sweeps, each using a 3D sketch as a path.

15.1. - Make a new part, add a 2D sketch and an auxiliary plane to help us make the 3D sketch. Add a new sketch in the *"Right Plane," add* a 14" construction line as shown, and exit the sketch.

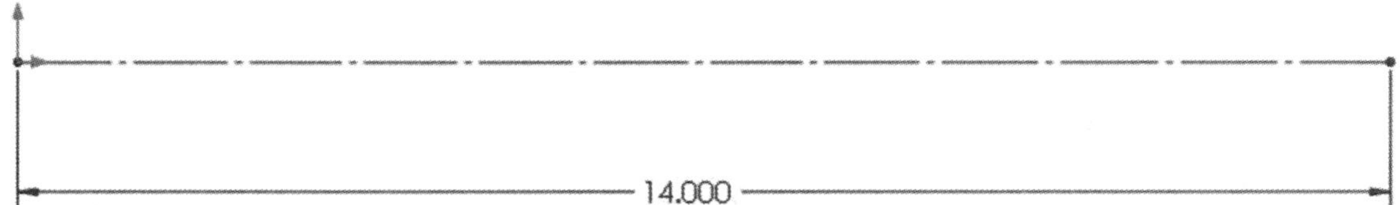

15.2. - Switch to an Isometric view and select the "**Reference Geometry, Plane**" command. Create an auxiliary plane 20 degrees from the *"Right Plane."* For the "First Reference" select the *"Right Plane"* and for the "Second Reference" use the sketch line we just drew. Make the plane 20 degrees to the left (you may have to "Flip" the direction) and click OK to create the plane. This plane will be used later for reference in the 3D Sketch. The references can be reversed.

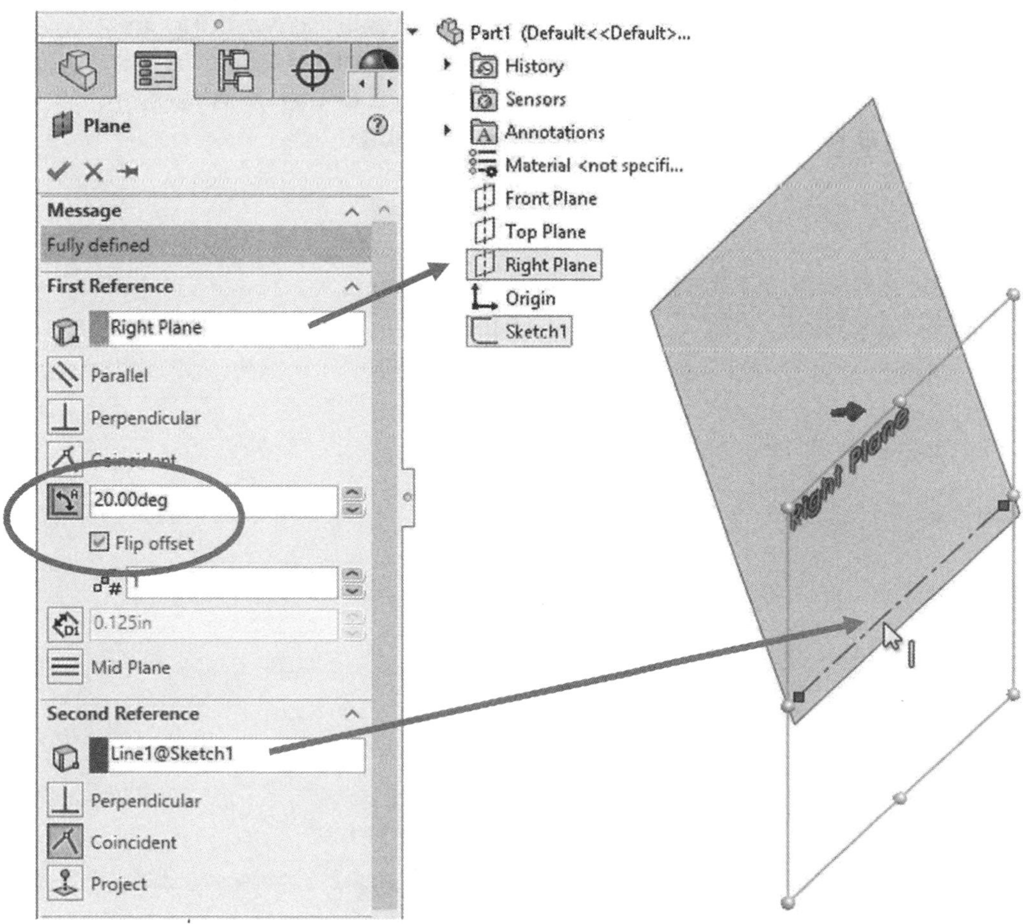

15.3. - The chair will be made from two bent tubular parts, each made with a sweep feature using a 3D sketch path. To make the first path for the tubular frame select the menu "**Insert, 3D Sketch**" or the "**3D Sketch**" command from the drop-down "**Sketch**" icon. When we start a 3D sketch, notice we are not asked to select a plane and the only difference from a regular sketch is that we get *"3DSketch1"* in the FeatureManager, status bar and window title.

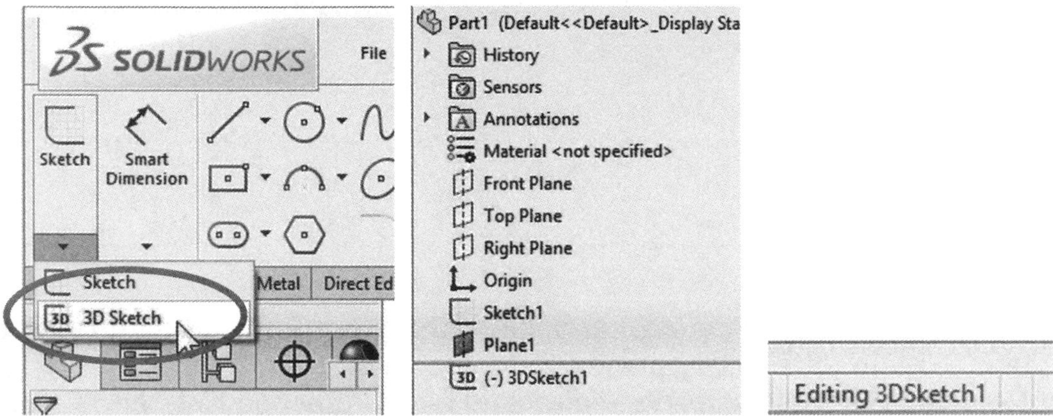

In a 3D sketch we have most of the tools available in a 2D sketch. After selecting the "**Line**" tool, we immediately see a bigger than usual sketch origin aligned parallel to the *"Front Plane"* and an "XY" label next to the mouse pointer. This is letting us know that when we draw a line, we'll be working in a 'plane' parallel to the *"Front Plane"* (XY-plane within the 3D Sketch). Click in the origin to start drawing a line, move towards the right and align it in the 'X' direction. The yellow inference icon is somewhat small, but visible.

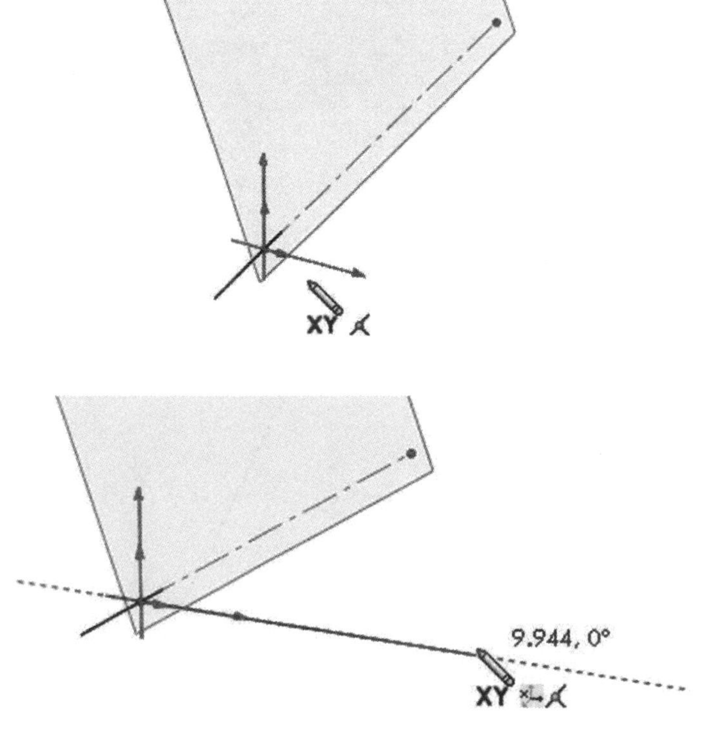

Click to complete the first horizontal line, go down to the right to add the second line; be aware we are still in the XY-plane. Notice the origin is now located at the start of the second line to indicate the orientation we are working on.

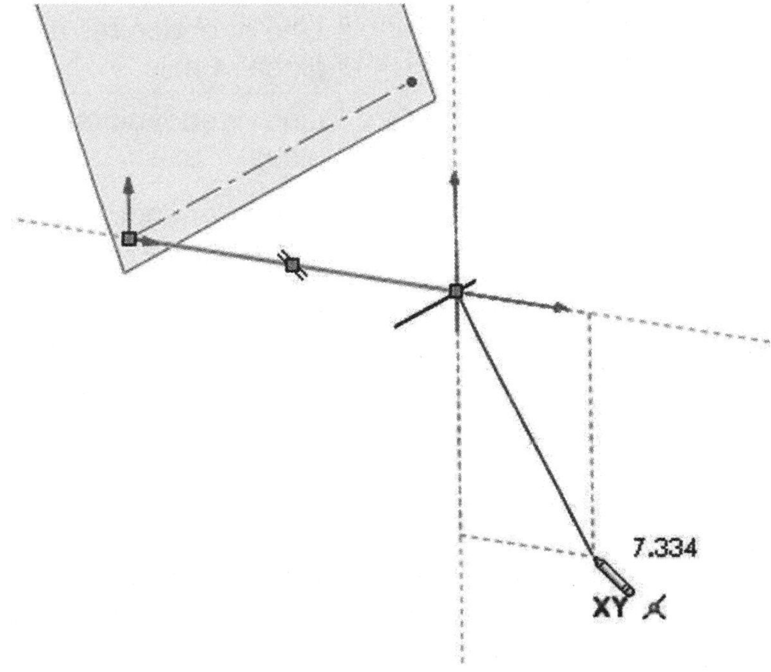

Repeat the same two lines starting at the other end of the construction sketch; the line tool will snap to the end of the line but will remain in a plane parallel to the XY-plane until we change it.

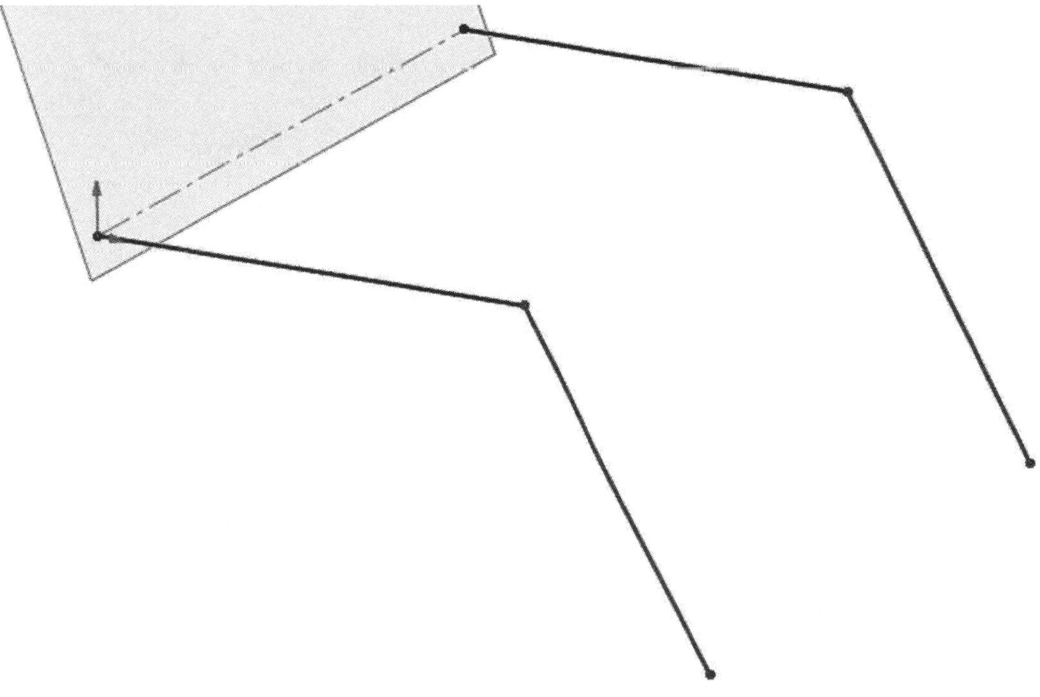

15.4. – Additionally to the 2D sketch geometric relations we know, we have a few more options to help us define geometry in a 3D sketch:

Relation:	Between entities:	Result:
Along X	One line or two points	Line or endpoints aligned along the global X axis.
Along Y		Line or endpoints aligned along the global Y axis.
Along Z		Line or endpoints aligned along the global Z axis.
Normal	Line and Plane or flat face	Makes the Line perpendicular to the plane or face.
On Plane	Plane or flat face and Line or point	Makes the line or point Coincident to the plane.
ParallelYZ	Plane or flat face and Line	Makes the selected line parallel to the YZ plane.
ParallelZX		Makes the selected line parallel to the ZX plane.

Now we need to add the relations required up to this point. Select both lines along the X-axis and add an **Equal** relation to make them the same length.

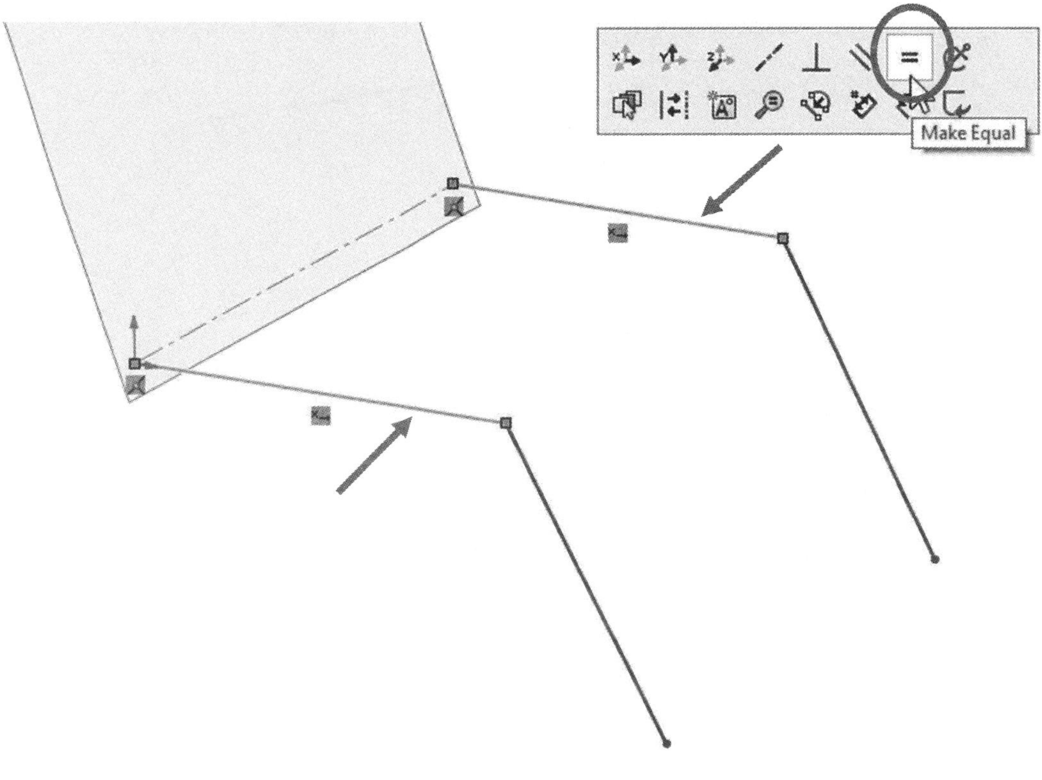

Now select both angled lines and add an **Equal** relation to them.

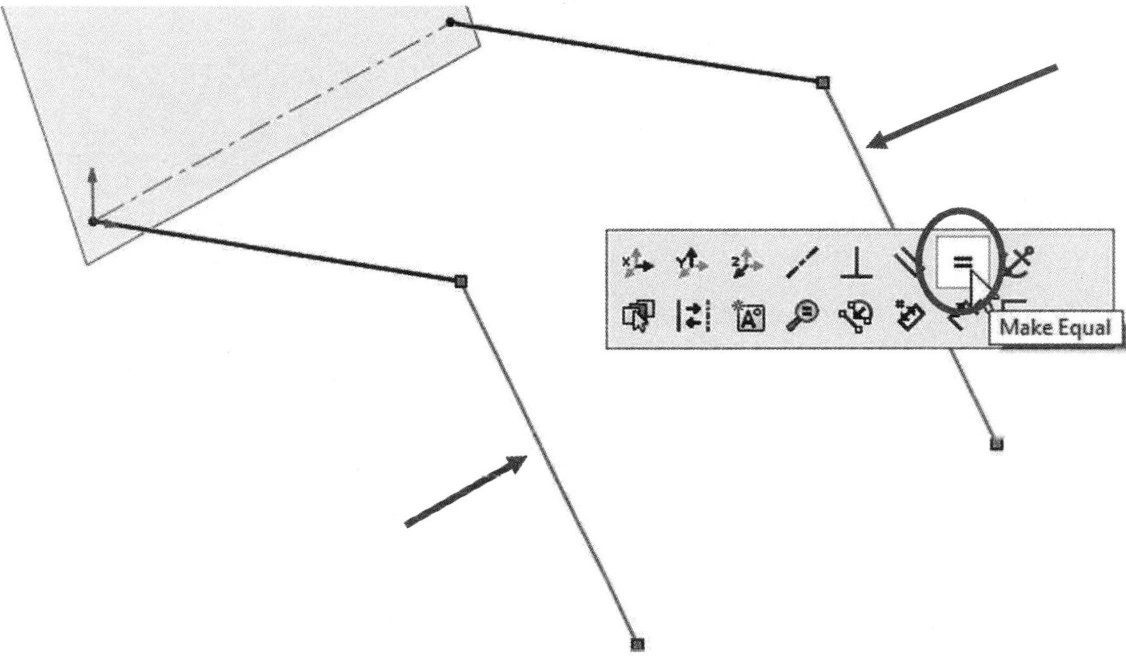

Add an "**On Plane**" relation between the first diagonal line and the *"Front Plane."* By adding this relation, the line will not 'move' along the Z-axis.

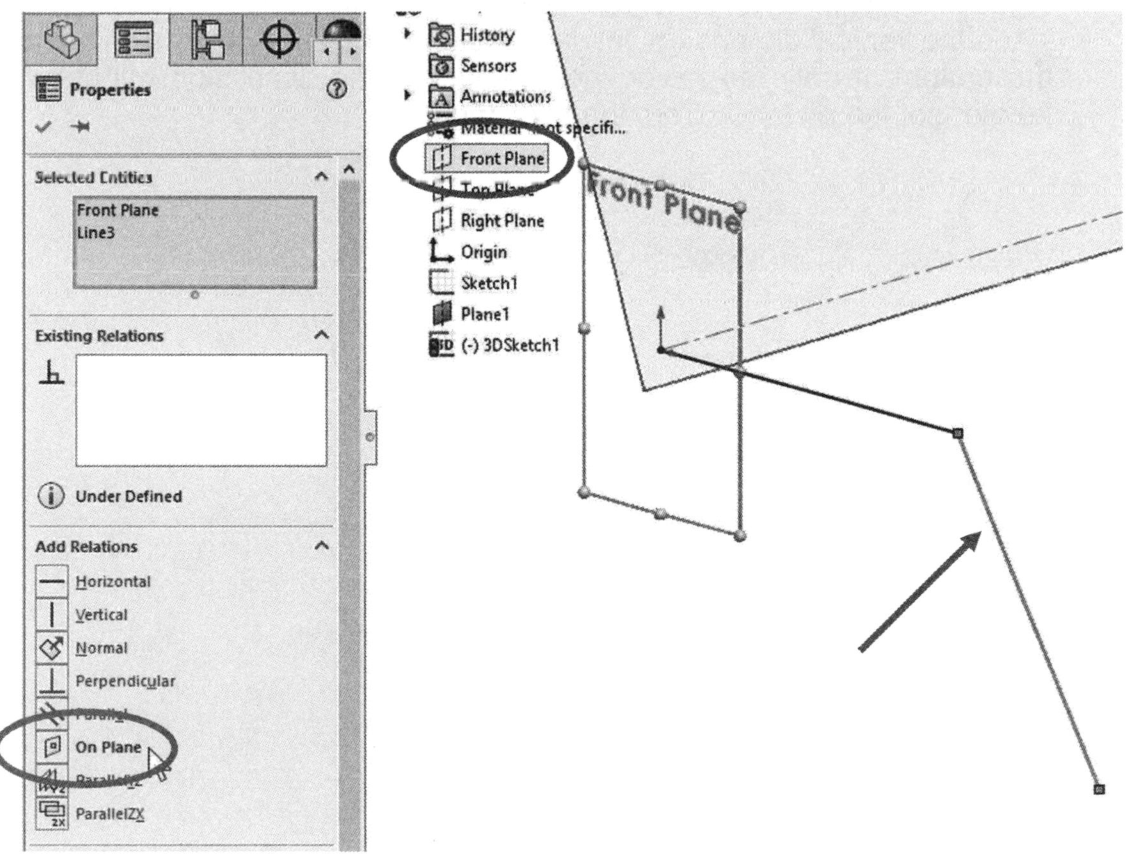

And finally add a Parallel relation between the two diagonal lines to keep them parallel to the XY plane.

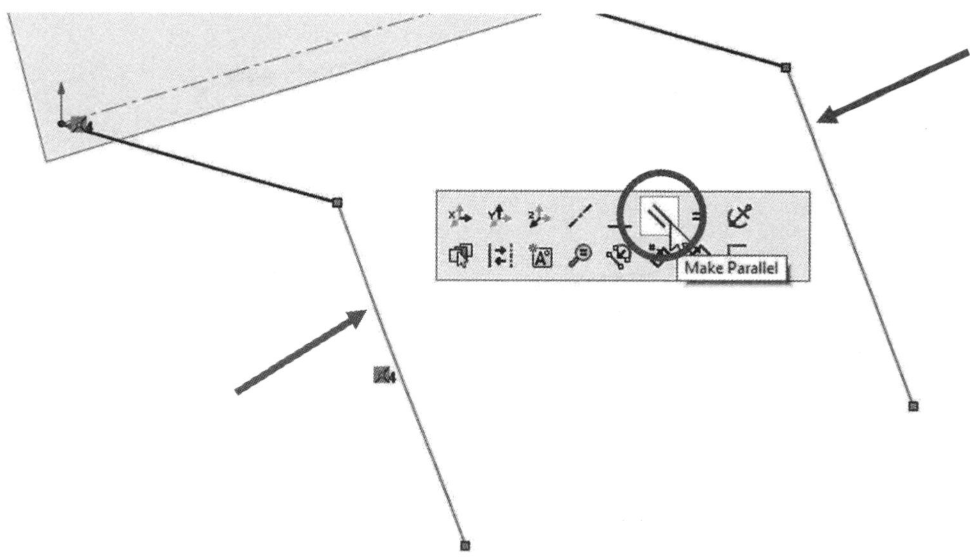

15.5. - Now we are going to add dimensions to the 3D Sketch. After adding the following dimensions, our sketch will be fully defined.

- Dimension one of the horizontal lines 12″ long.
- Add an angular 110° dimension between the two lines.
- Add the vertical dimension between the horizontal line and the endpoint at the bottom. If only the lower line is selected, the dimension will show its length, not the distance from the horizontal line.

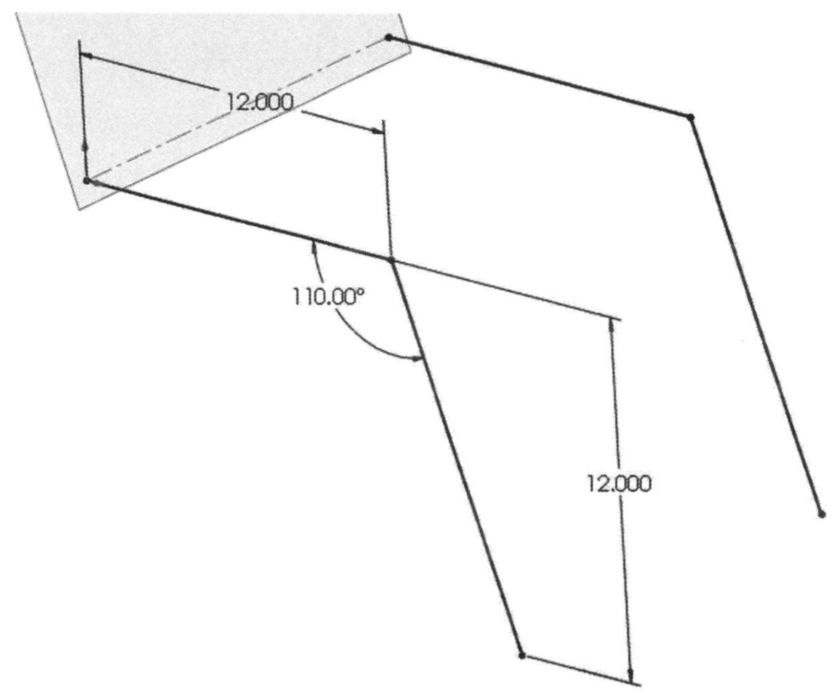

15.6. - These four lines will be part of the seat and front legs of the chair. The next step is to add the backrest of the chair. The plane we added at the beginning will help us align the 3D sketch's X, Y and Z axis directions about it. Pre-select *"Plane1"* and then turn on the "**Line**" tool. The sketch origin and the local directions will now be temporarily aligned to *"Plane1."*

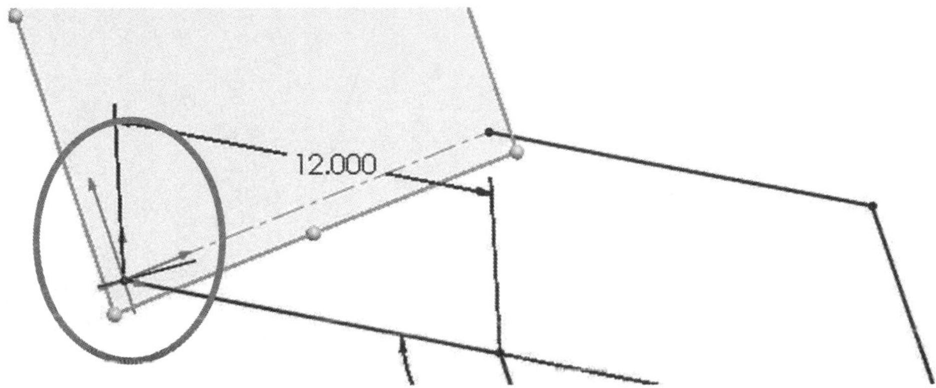

With the new plane alignment, draw a new line starting at the part's origin going 'up' along *"Plane1,"* then to the 'right' and 'down' ending at the start of the second set of lines to make them Coincident (don't worry if it's not parallel).

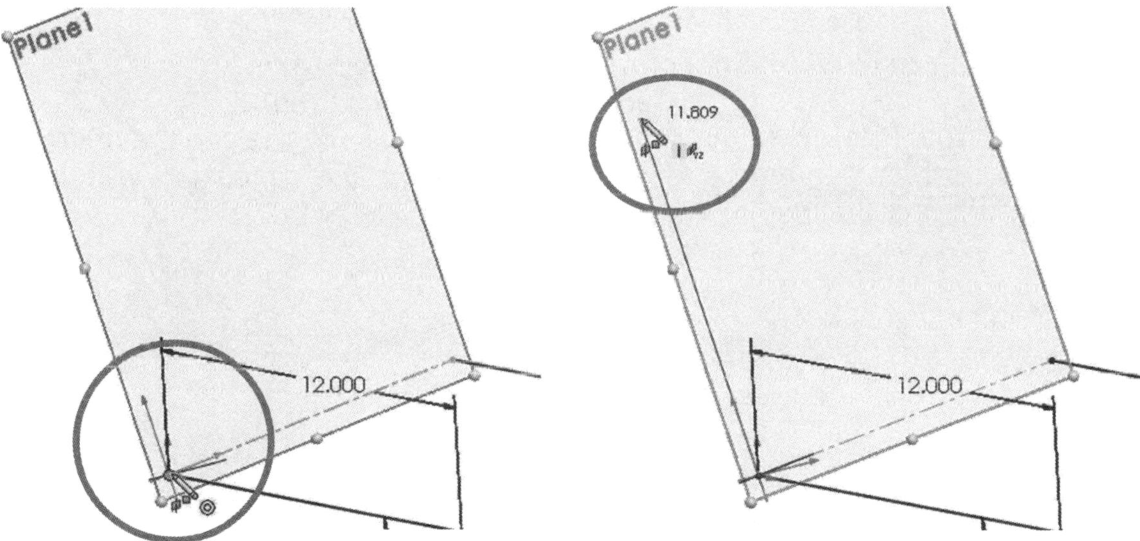

If after adding the first line the 3D Sketch origin is not aligned in the **XY Plane** direction, press the "Tab" key to change its orientation.

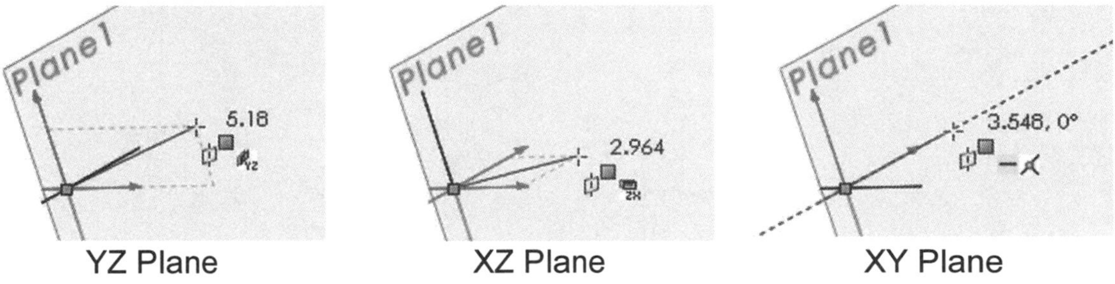

YZ Plane XZ Plane XY Plane

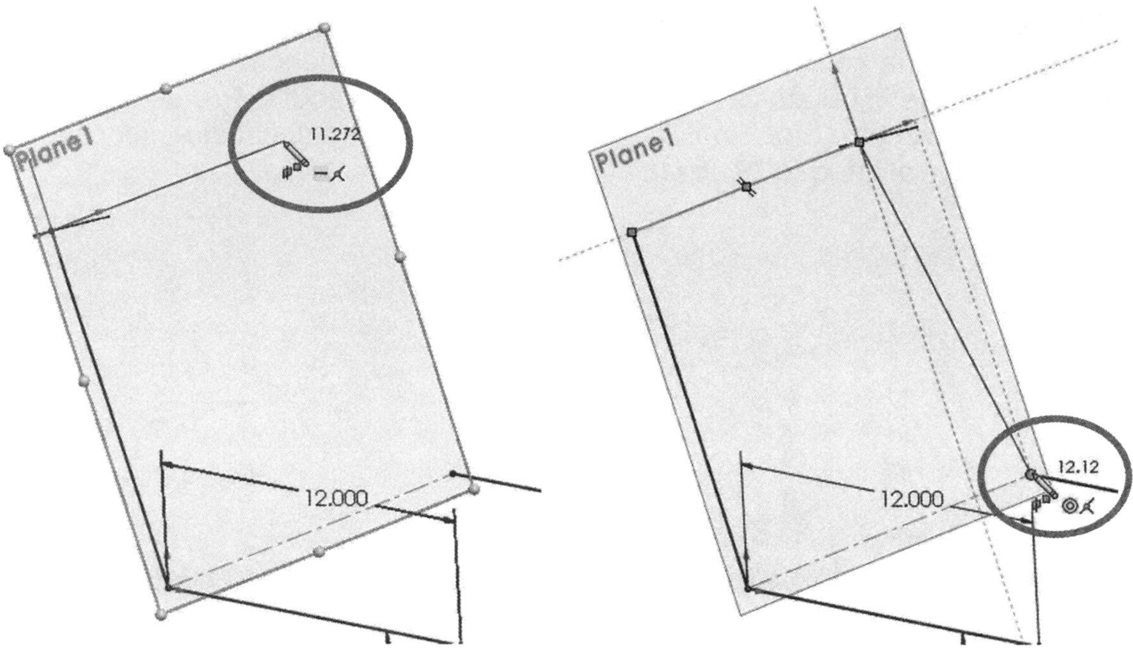

15.7. - Turn off the "**Line**" tool and add a Parallel relation between the two lines aligned with *"Plane1."*

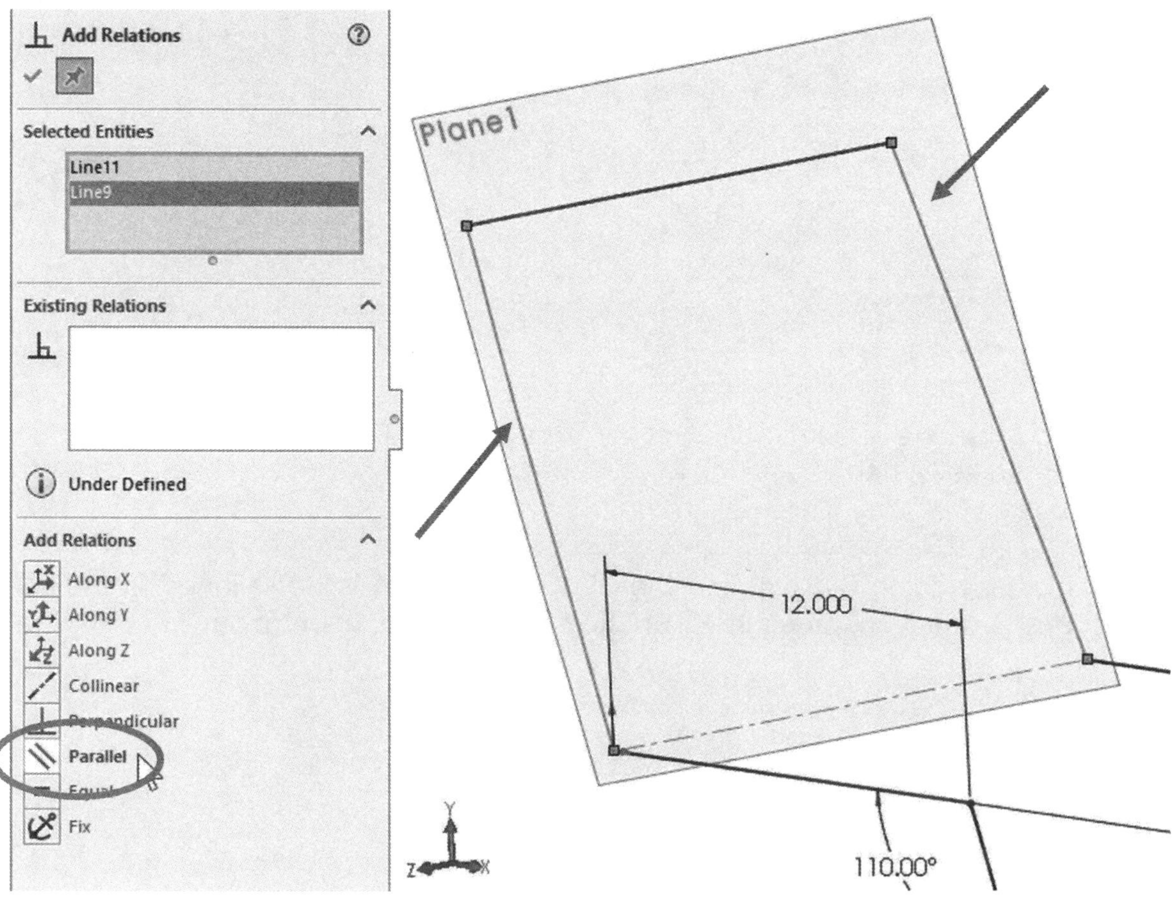

15.8. - Dimension the length of the backrest 18 inches.

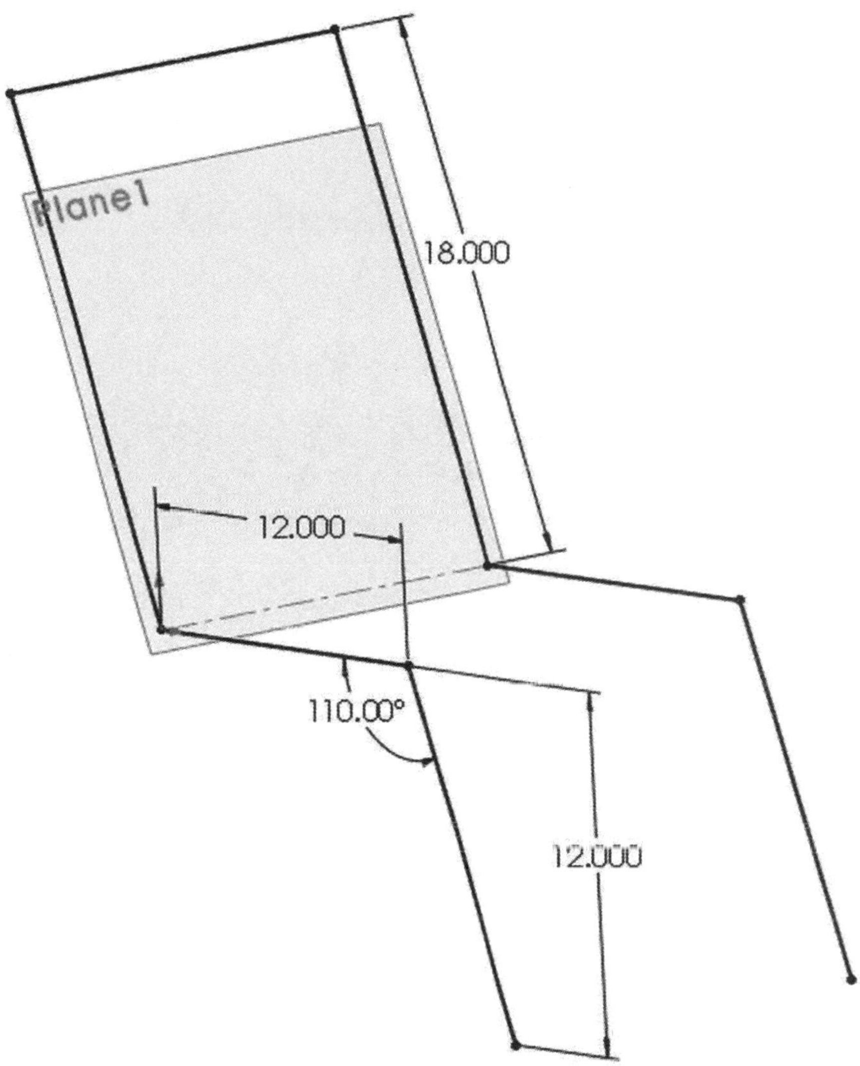

15.9. - To finish the 3D sketch, select the "**Sketch Fillet**" tool and round all corners with a 2″ radius. When adding the fillet, we are warned about segments with equal relations; keep adding fillets until done. Since we are rounding all segments, they will still be equal at the end. Click "Yes" to continue when asked.

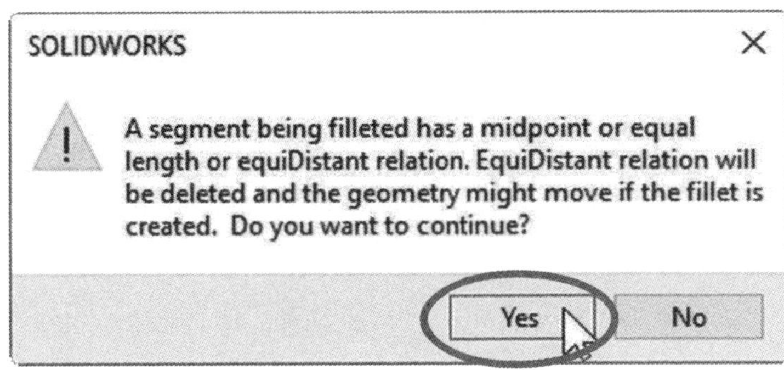

Click OK to finish the sketch fillets and exit the 3D sketch.

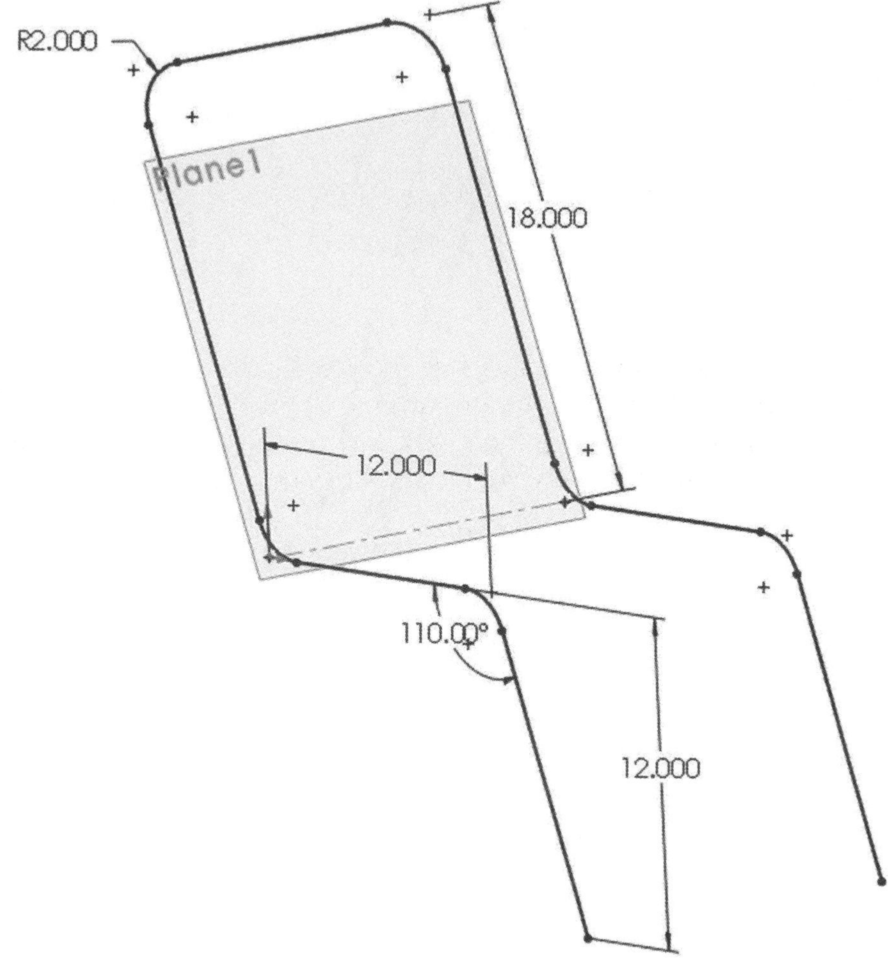

15.10. – Now we need to add the seat and rear legs of the chair. Start a new 3D Sketch (rotate the model as needed for clarity). Draw a line starting close to the origin coincident to the construction sketch and go along the X-axis for the first line.

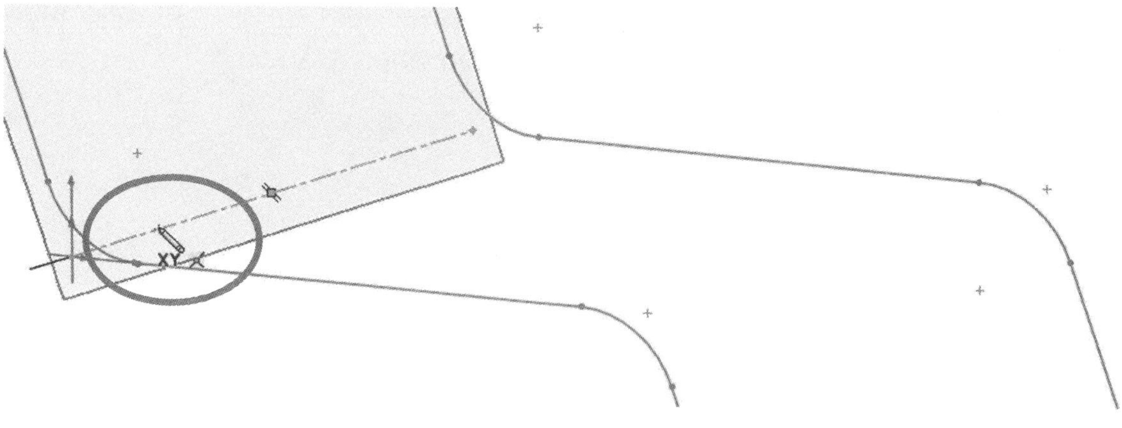

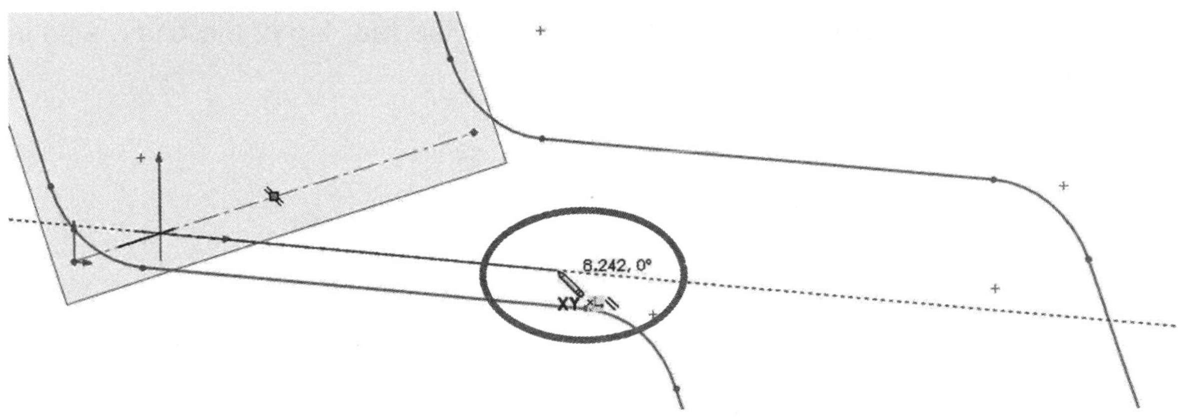

The second segment will be done along the Z-axis; to change the direction we need to change the working plane orientation to the YZ or XZ plane. To switch the working plane on-the-fly, press the "Tab" key once to rotate through the other work planes until we see either the YZ or XZ plane next to the mouse pointer and draw the line.

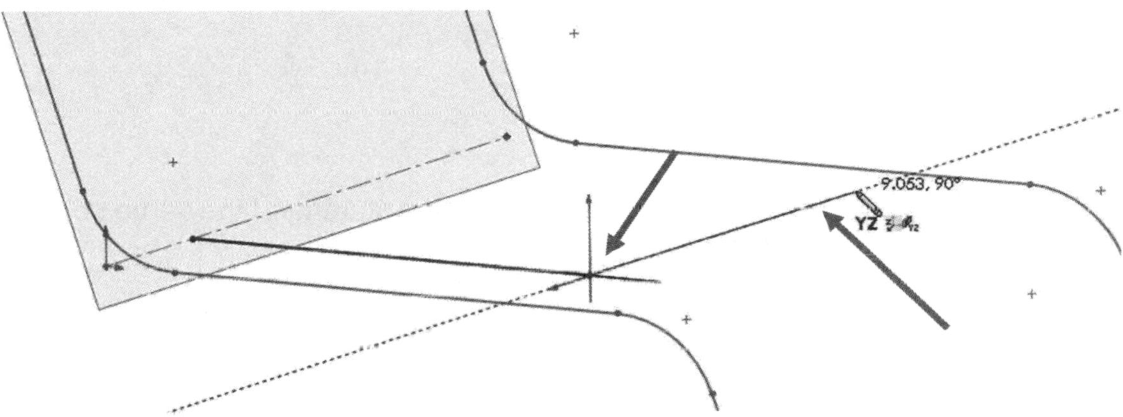

After adding the second line press the "Tab" key again to switch to the XY plane and draw the third line parallel to the first one making it coincident to the construction sketch line as shown.

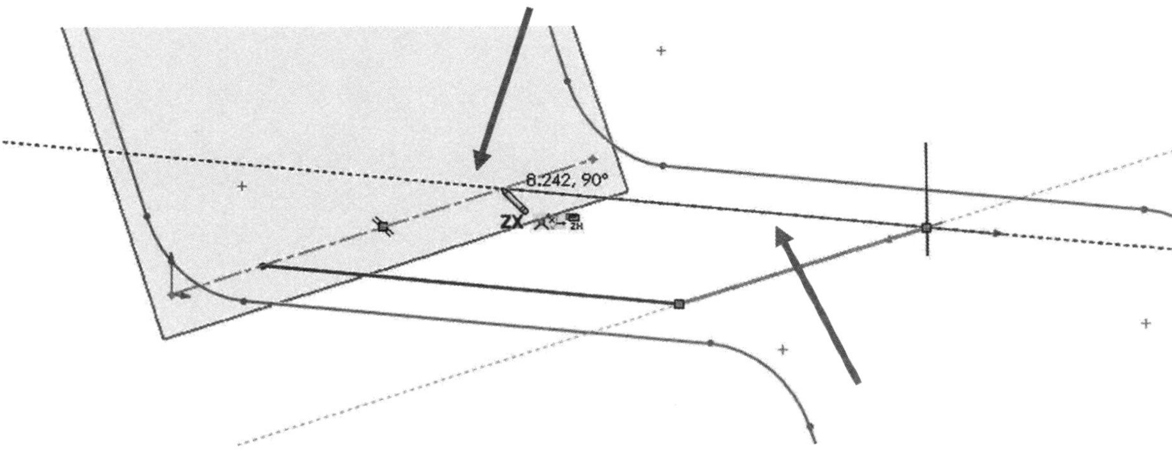

The next line will go 'down' at an angle for the rear leg of the chair, also in the XY plane.

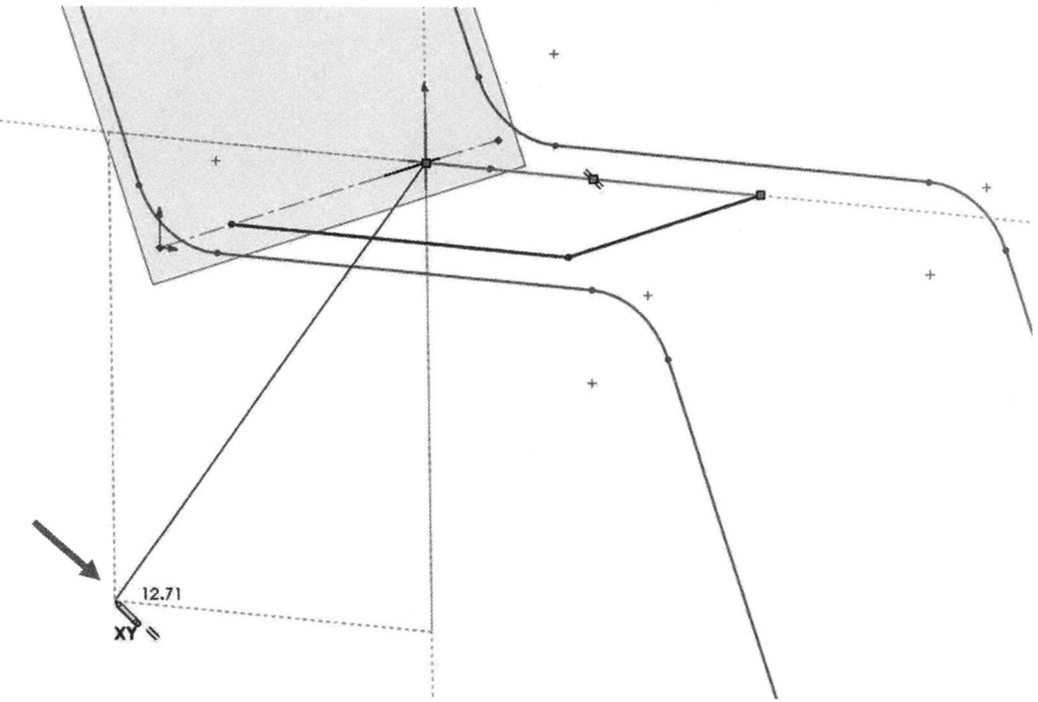

Add the final line similar to the previous one starting at the same point where the second 3D sketch started.

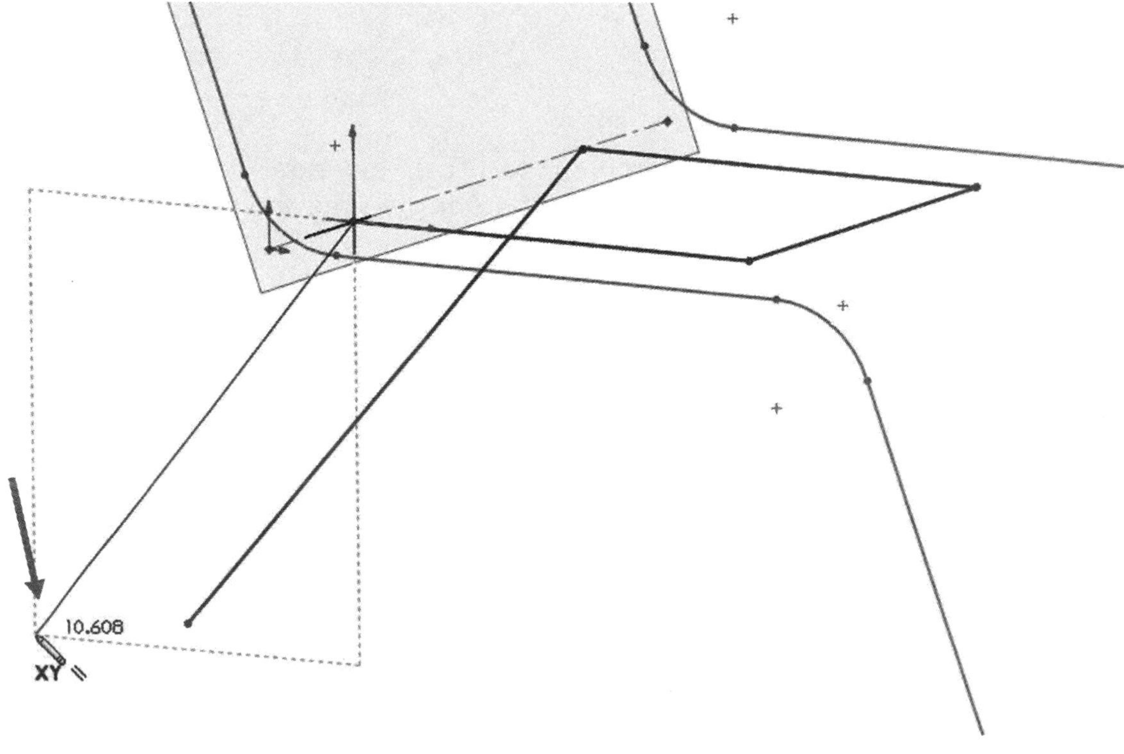

15.11. - Dimension the horizontal lines 0.75″ from the previous 3D Sketch; add the 110° angle between the two lines and the length of the horizontal as shown.

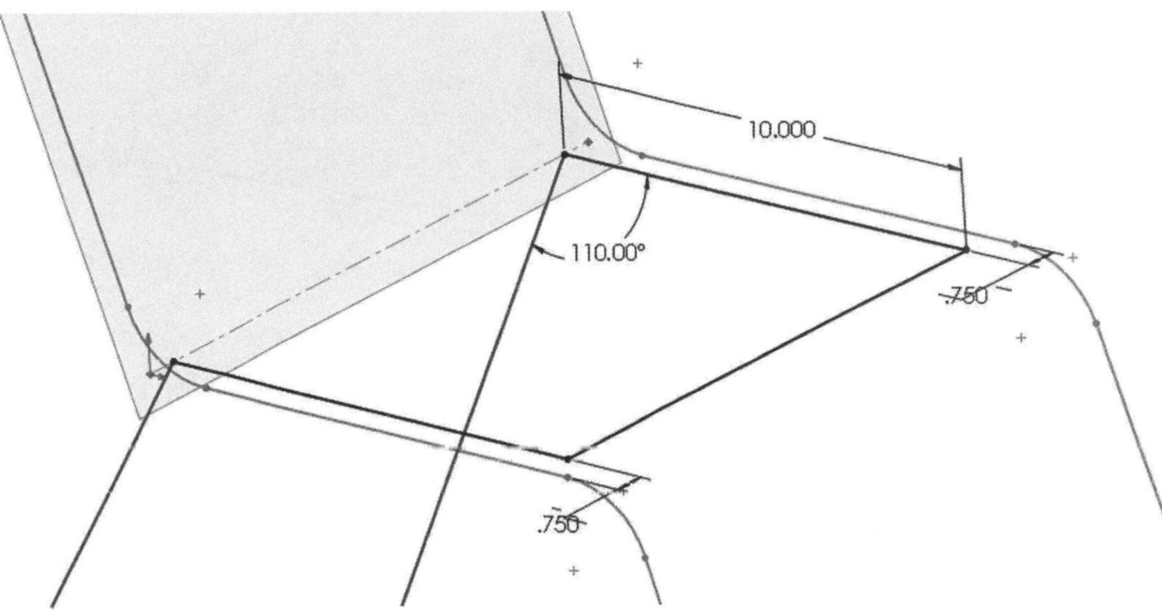

15.12. - Now we need to add some geometric relations. Select the endpoints at the bottom of the legs on one side of the chair and add an "**Along X**" relation; this way we'll make them lie on the same axis (X).

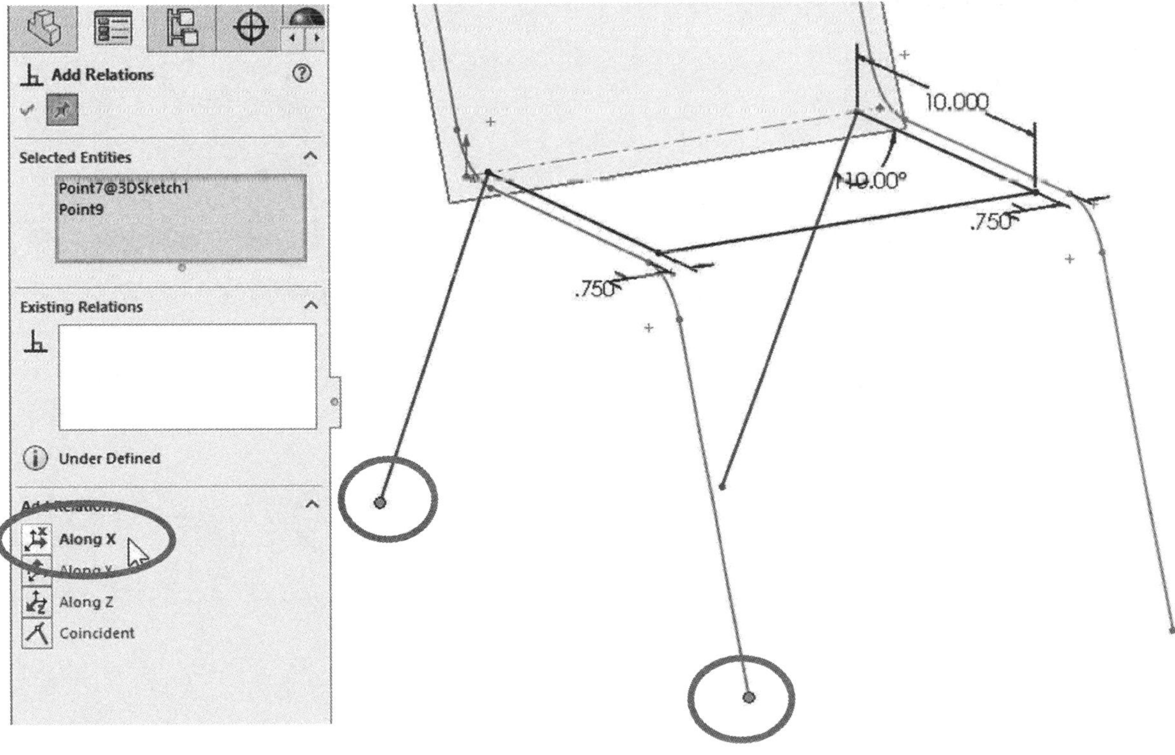

Repeat the same relation for the end points at the other side of the chair.

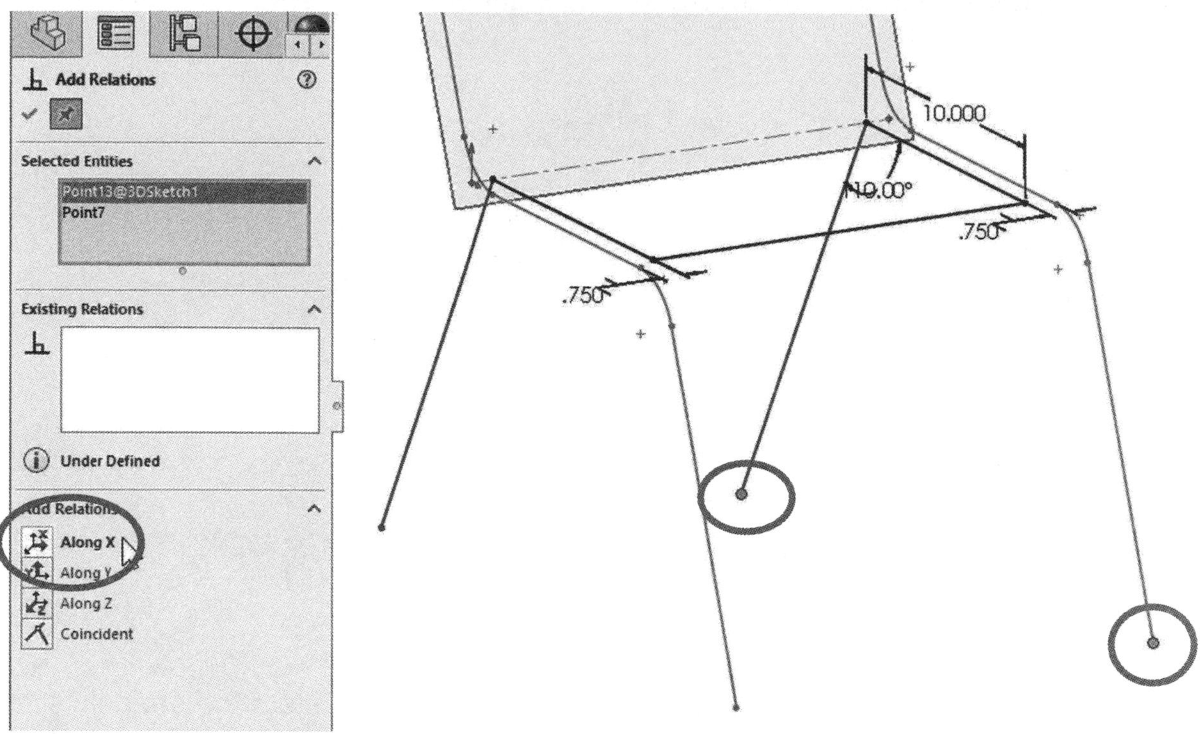

And finally, add an "**Along Z**" relation between the endpoints of the two lines in the back to fully define the sketch.

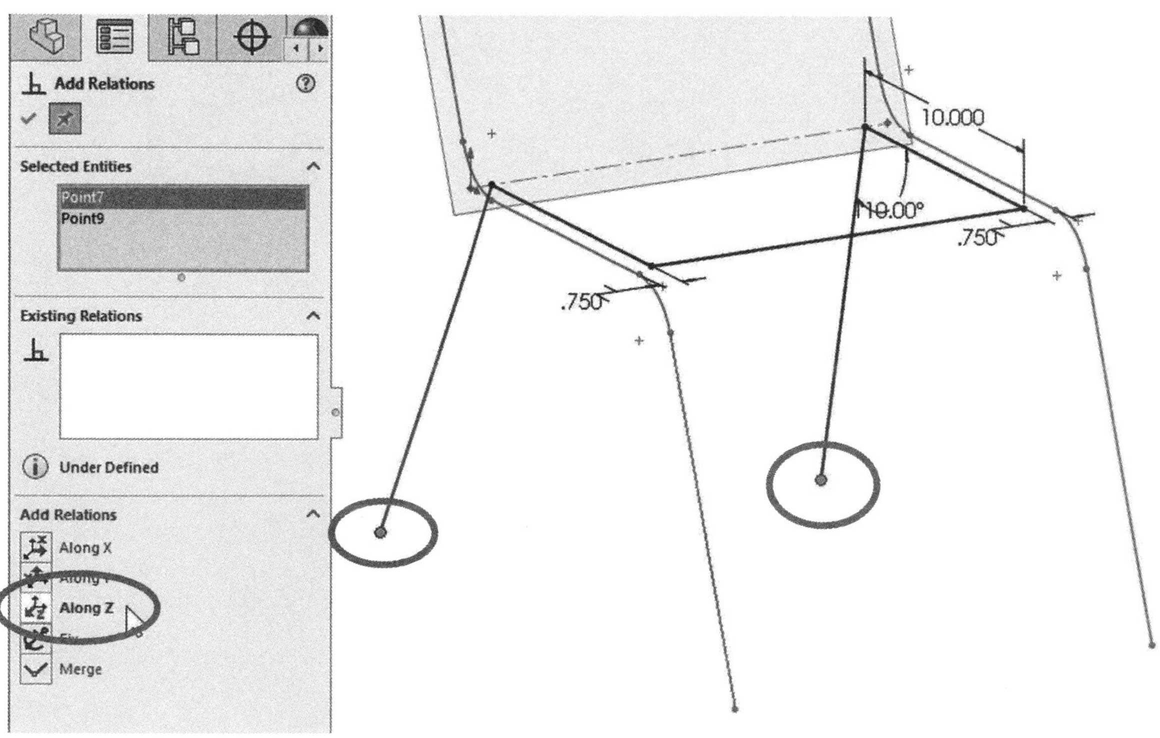

15.13. - To finish the second 3D Sketch, add a 2″ sketch fillet to all corners as we did in the first one, and Exit the Sketch to continue.

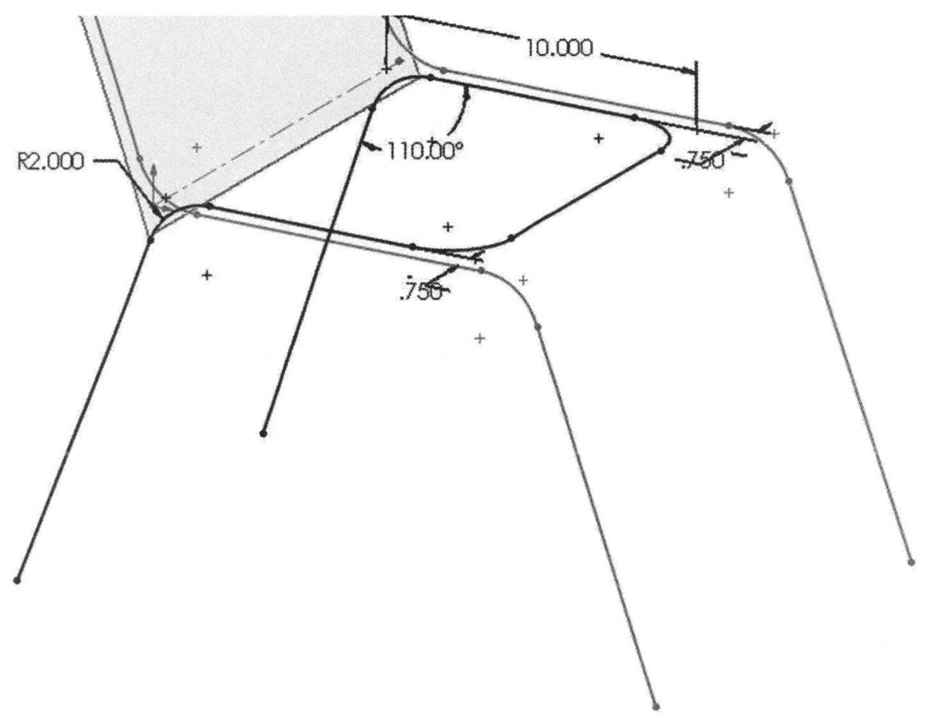

15.14. - To make the tubes of the chair, we need a plane at the endpoint of a leg to locate the sweep's profile. Add an auxiliary plane Parallel to the *"Top Plane"* ("First Reference") and Coincident to an endpoint of the chair's leg ("Second Reference").

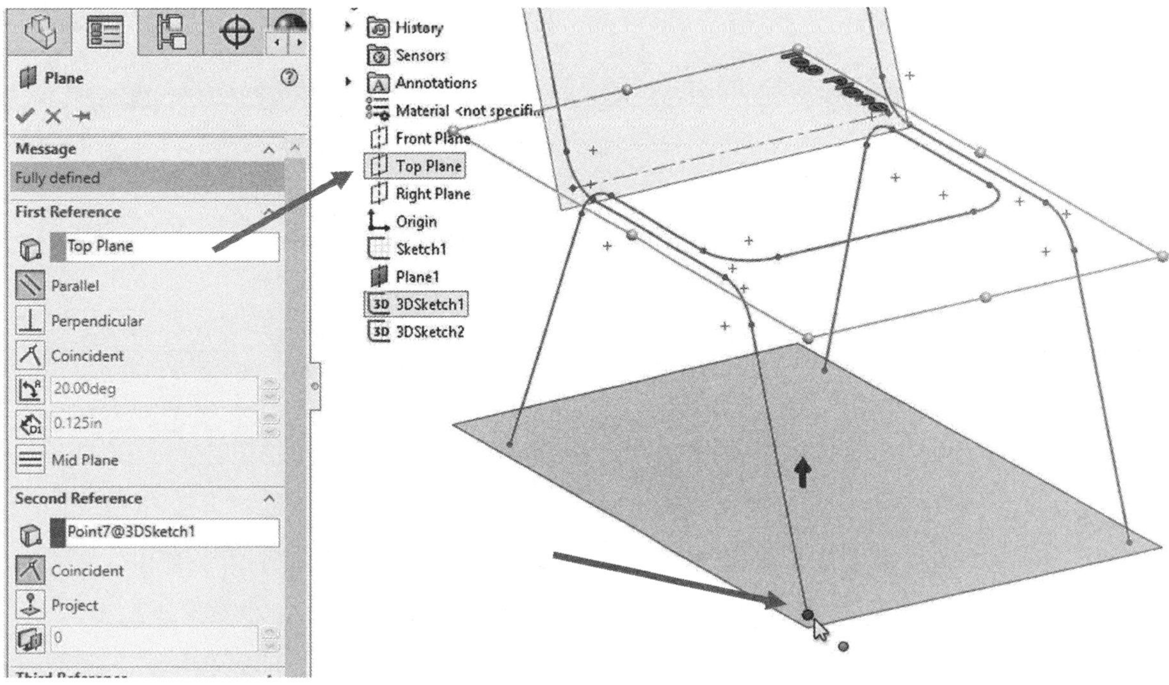

15.15. - When the new plane is finished, add a new 2D sketch in it. Be sure to make the circle coincident to the endpoint of the 3D sketch and exit the sketch.

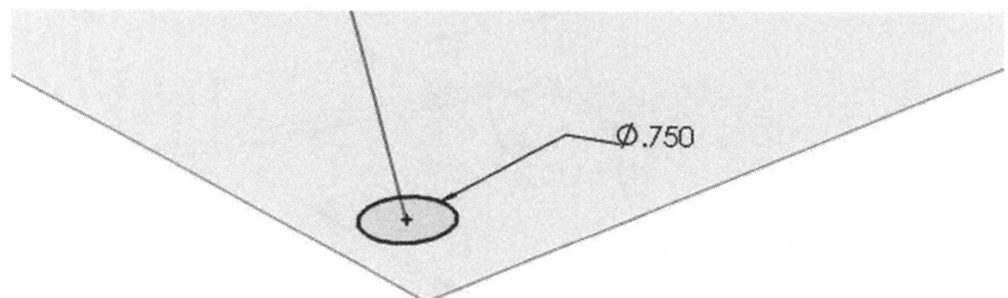

15.16. - Make a sweep using the last sketch as the "Profile" and the first 3D sketch as the "Path." Use the "Thin Feature" option to make the tube's wall 0.07″ thick with the material added inside. Click OK to finish the first tube.

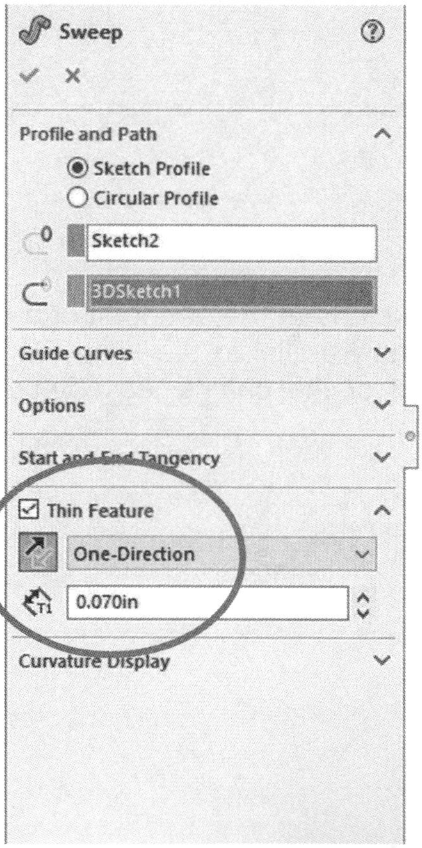

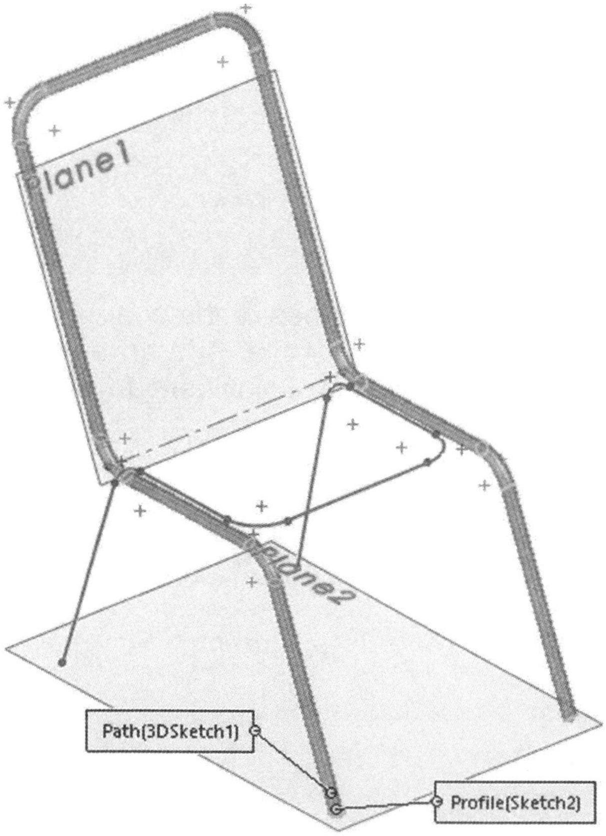

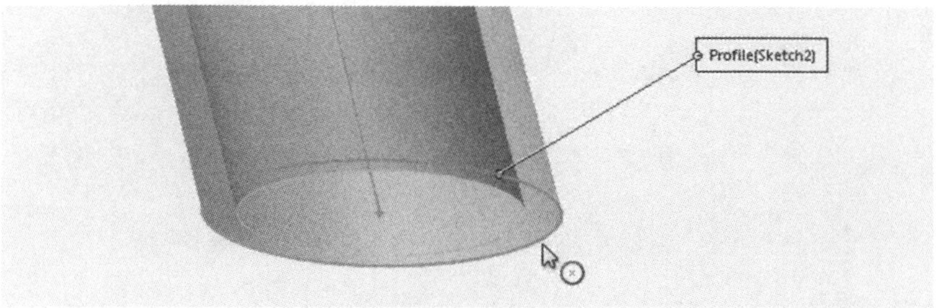

15.17. - To make the second sweep we need a new profile sketch. Since we want both profiles to be the same, we'll use a function called "**Derived Sketch**" to make the second profile equal to the first one. Using this approach, if the first sketch changes, the second one (Derived) changes too. However, this is a one-way direction: changes in the first sketch are reflected in the second sketch, and not the other way around. To add a derived sketch, we must pre-select the sketch we want to derive *and* the plane/face where we want to put it. Expand the *"Sweep-Thin1"* feature, select the profile sketch, hold down the "Ctrl" key and select *"Plane2"* in the FeatureManager or the graphics area. Now, select the menu "**Insert, Derived Sketch.**" (If we don't pre-select them, this command is disabled.)

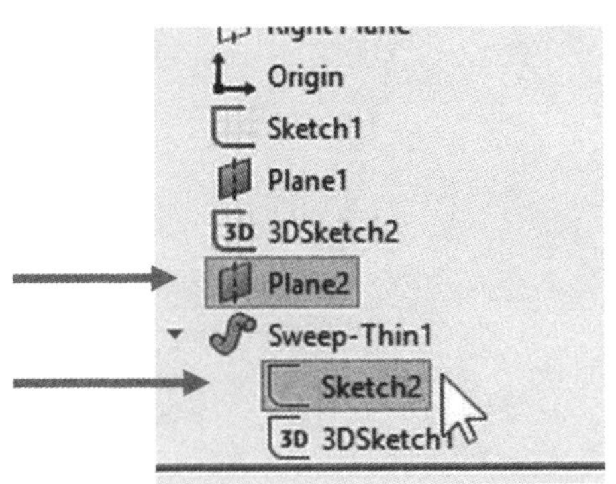

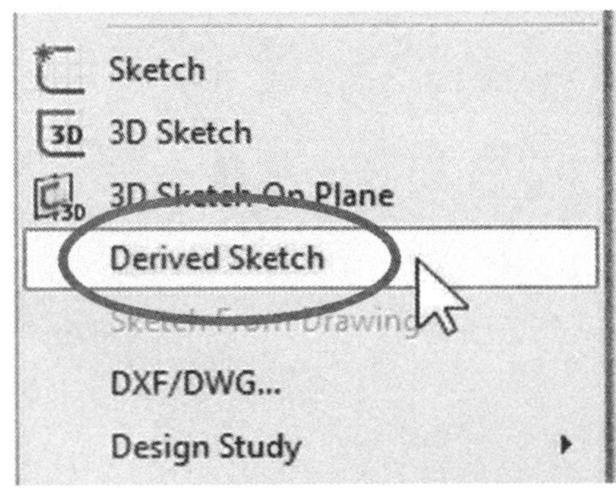

Immediately after selecting "**Derived Sketch**" we are editing the new sketch. Notice that most of the sketch tools are disabled except dimensioning and relations. In a derived sketch the geometry cannot be modified, it can only be located. The derived sketch is exactly on top of the sketch we derived it from; click and drag the circle away from the chair's leg for clarity.

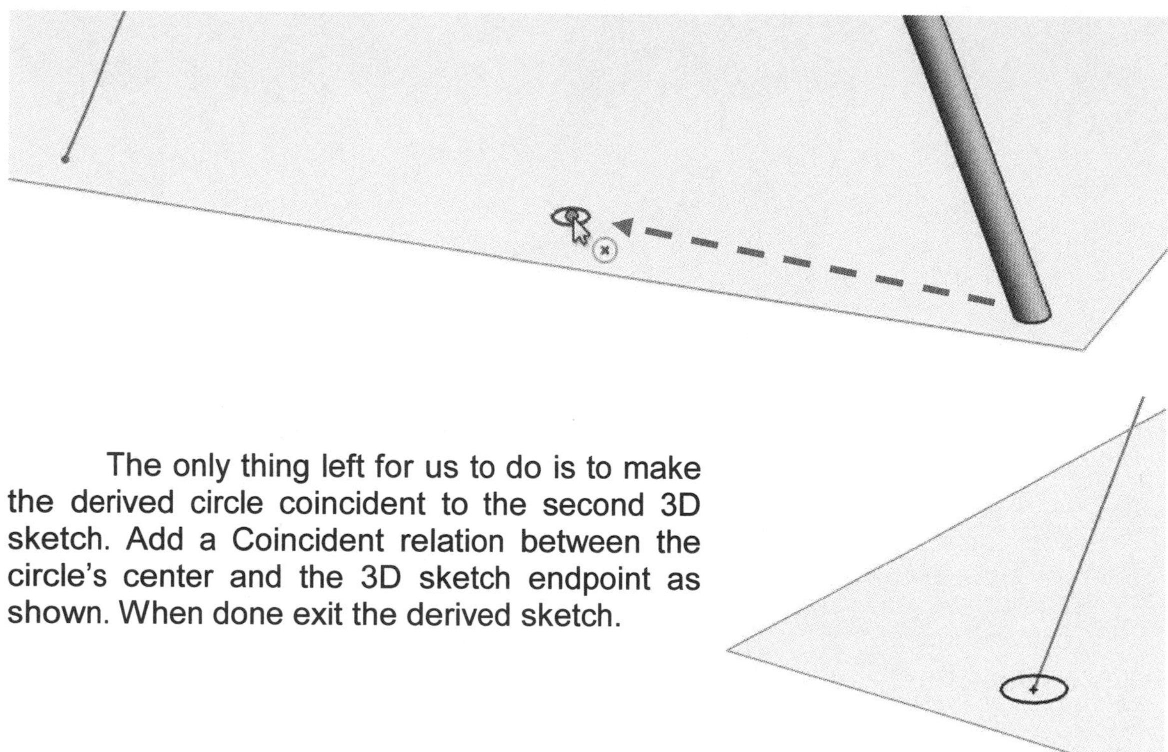

The only thing left for us to do is to make the derived circle coincident to the second 3D sketch. Add a Coincident relation between the circle's center and the 3D sketch endpoint as shown. When done exit the derived sketch.

15.18. - Make the second thin sweep with the same options as the first one; after finishing it hide all planes and sketches.

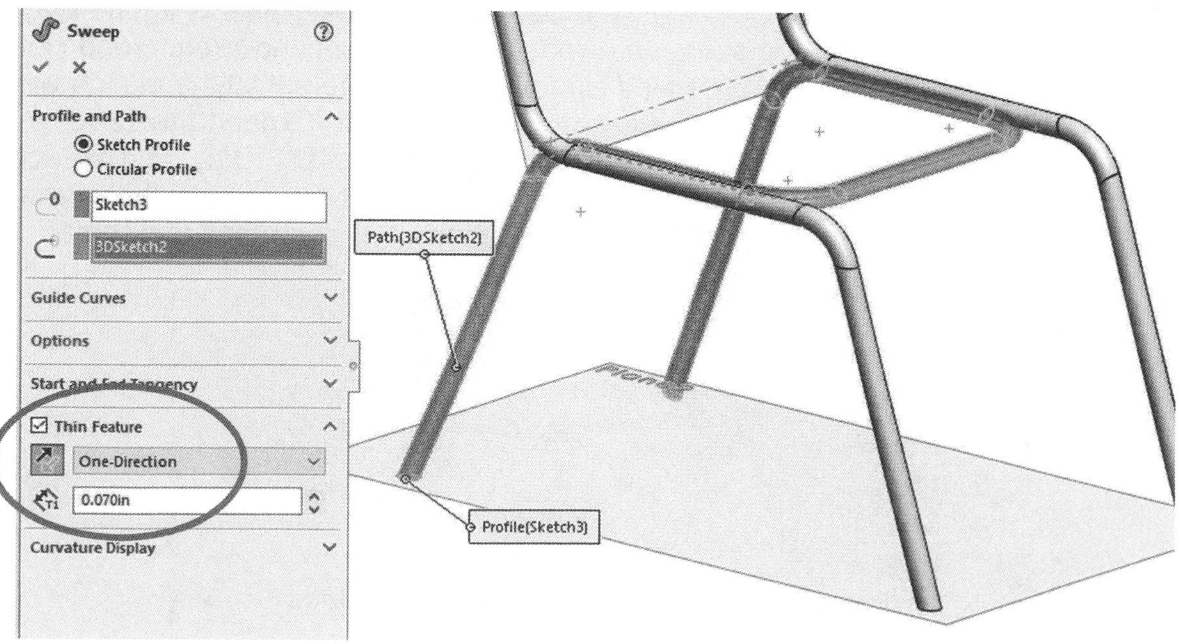

A derived sketch's link to its parent can be broken, or 'underived' from the original, by selecting the derived sketch with the right mouse button and selecting "**Underive**" from the pop-up menu, then it can be modified as needed. Un-deriving a sketch cannot be undone.

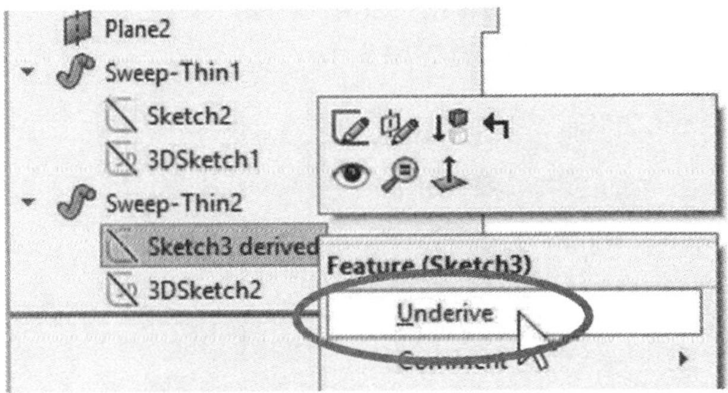

15.19. - After making the sweeps we need to finish their ends. Changing to a Front view and zooming into the leg ends we can see the chair legs are uneven at the end of the sweep features. To correct it we'll have to extend the legs just enough to go past the 'floor' (*"Plane2"*) and then trim them to size.

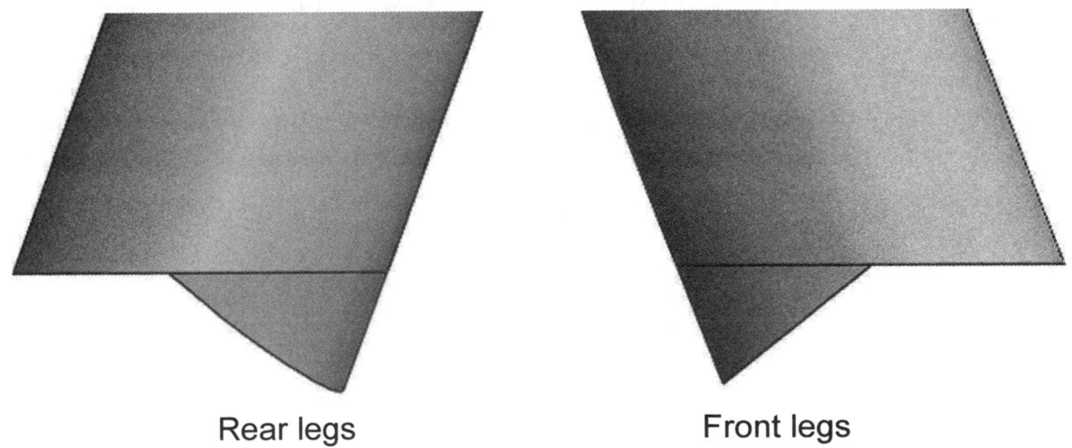

Rear legs Front legs

Select the flat bottom face at the end of the first sweep (the uneven leg), add a new sketch in it, and project both edges using "**Convert Entities**."

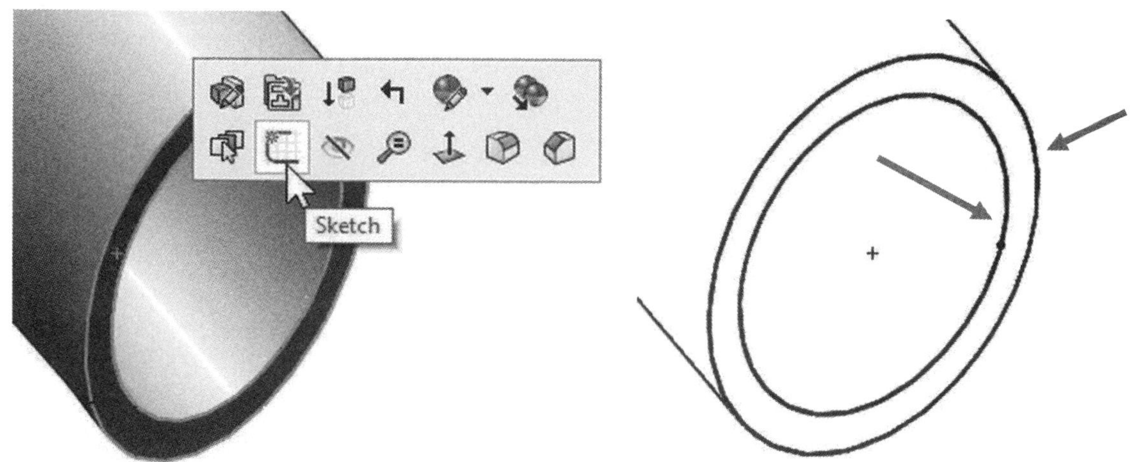

Expand *"Sweep-Thin1"* (and *"Sweep-Thin2"*), make *"3DSketch1"* (and *"3DSketch2"*) visible, and select the "**Boss Extrude**" command. If the option to globally show/hide sketches is set to hide, turn it back on (menu "**View, Hide/Show, Sketches**").

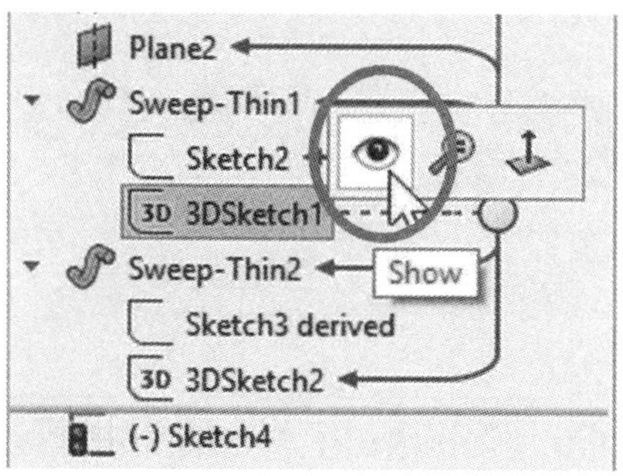

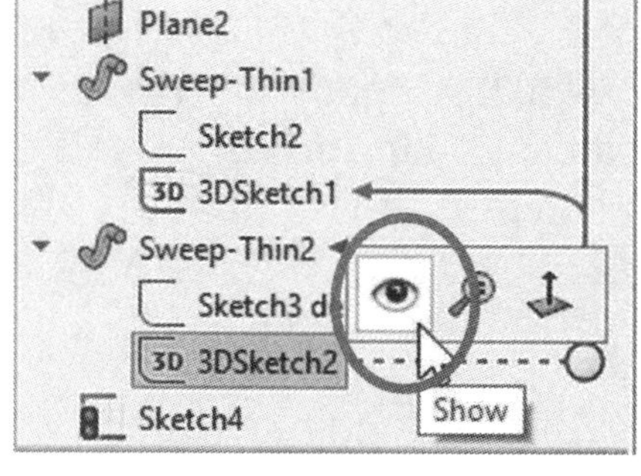

After selecting the "**Boss Extrude**" command we can see the extrusion's preview; when we add an extrusion to a part, by default, it's always perpendicular to the sketch plane; this setting, however, gives us an undesirable result. In our case, we need it to follow the same direction as the chair's leg. To get the correct result, select the "Direction of Extrusion" selection box, and select the now visible end of the 3DSketch, to make the extrusion follow its direction. Make the extrusion long enough to cross the floor level (*"Plane2"*) and click OK to finish. Repeat the same extrusion at the end of the other leg.

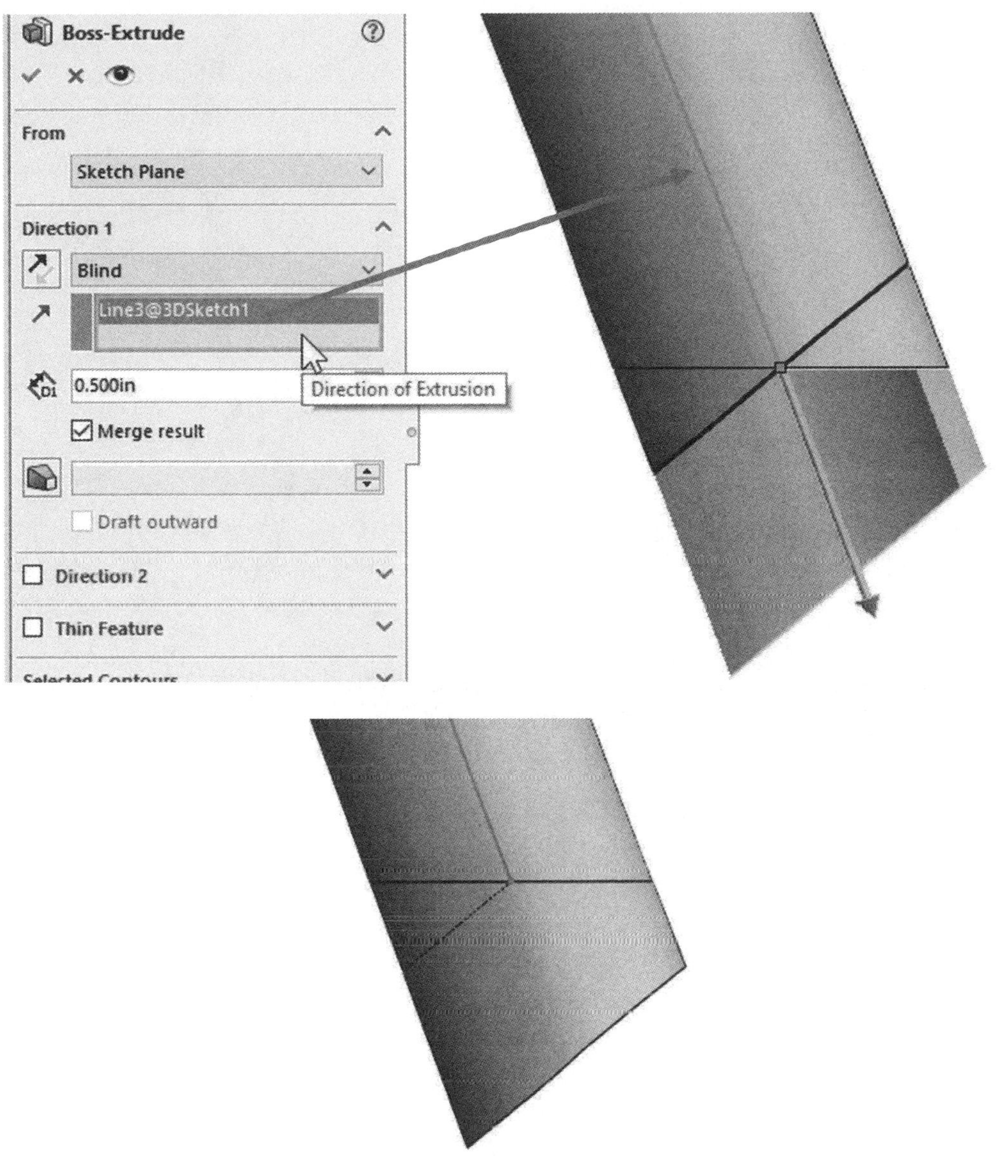

15.20. - After adding the extrusions, we need to trim the excess material added to level the legs. We can add a sketch and use an extruded cut, but we are going to use a different approached. Select the menu **"Insert, Cut, With Surface."** We can use any plane or surface as a cutting tool, and we can cut multiple bodies at the same time. Select *"Plane2"* as the Cut Surface and flip the direction if necessary. "**Cut With Surface**" will cut everything on the side of the plane/surface the arrow is pointing in. Click OK to trim the legs and finish.

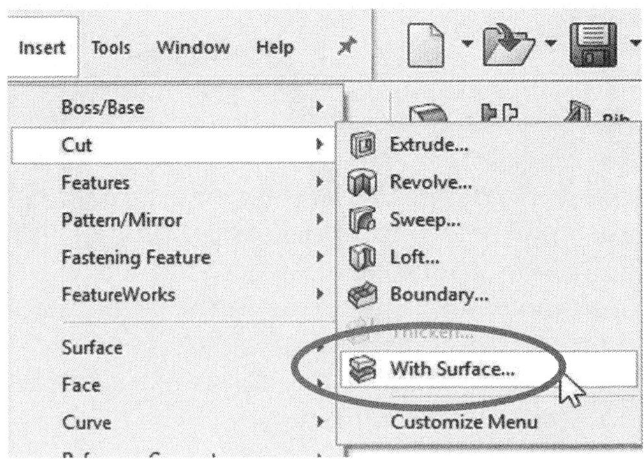

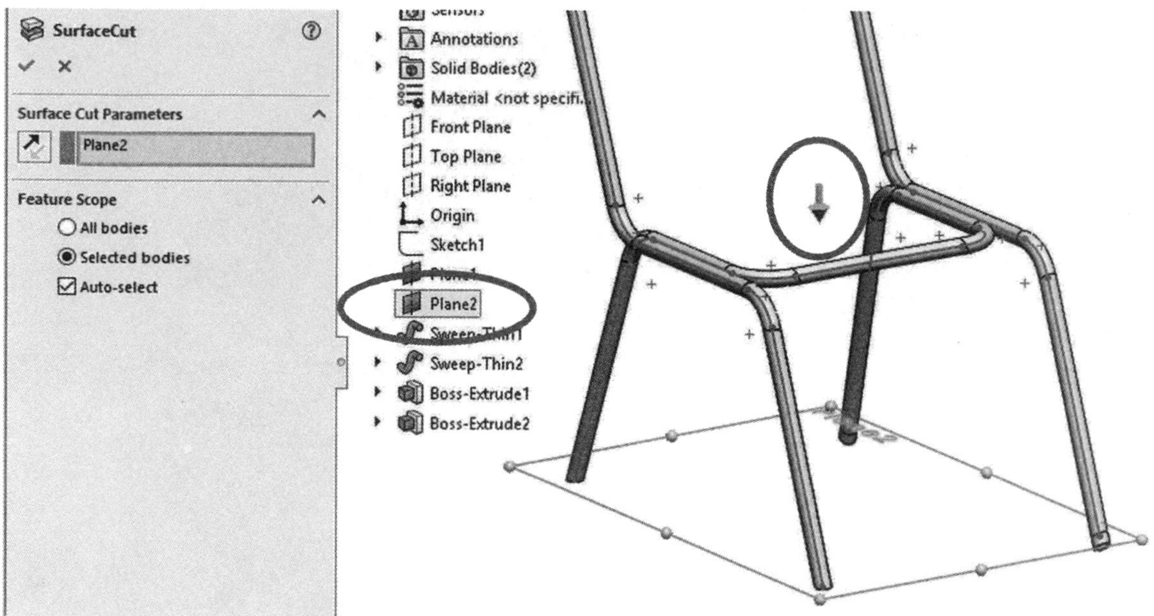

15.21. - Now our chair is correct and complete. Hide the sketches for a better look, save the part as *'Chair 3D Sketch'* and close it.

Projected Curve

16.1. - A different way to make the path to sweep the frame of the chair is by using a "**Projected Curve**." To build the projected curve we must draw the final curve's projections in the *"Front Plane"* and *"Top Plane,"* and then we can combine them into a 3D curve and use it as a sweep path. Make a new part and add the following sketch in the *"Front Plane."* The sketch's origin is at the bottom of the 18″ line. Exit the sketch when done.

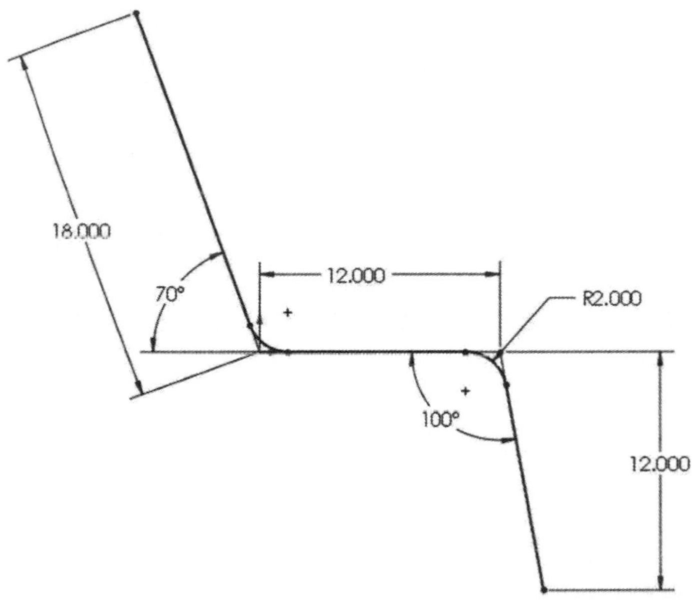

16.2. - Make a new sketch in the *"Top Plane"* as shown. Add geometric relations to the previous sketch endpoints to make it the same size. (You may have to rotate the model to select the endpoints.) Exit the sketch when done.

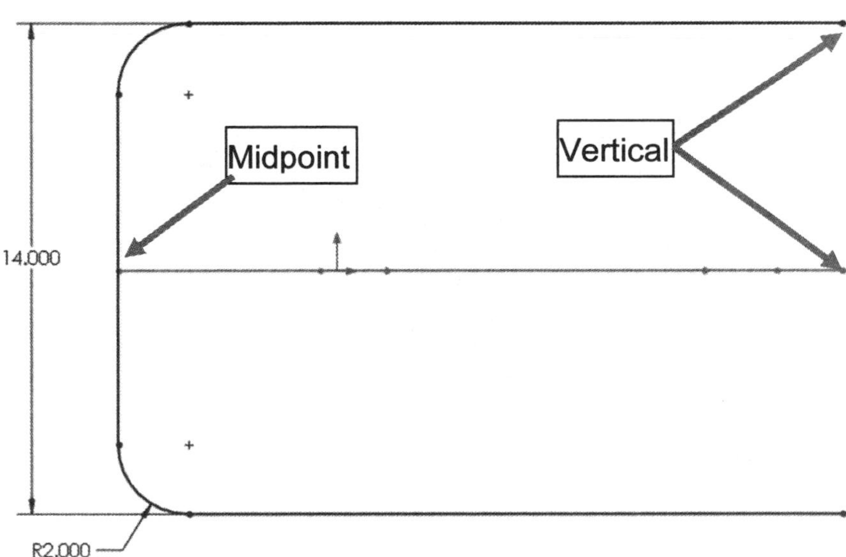

This is what the two sketches look like in an isometric view:

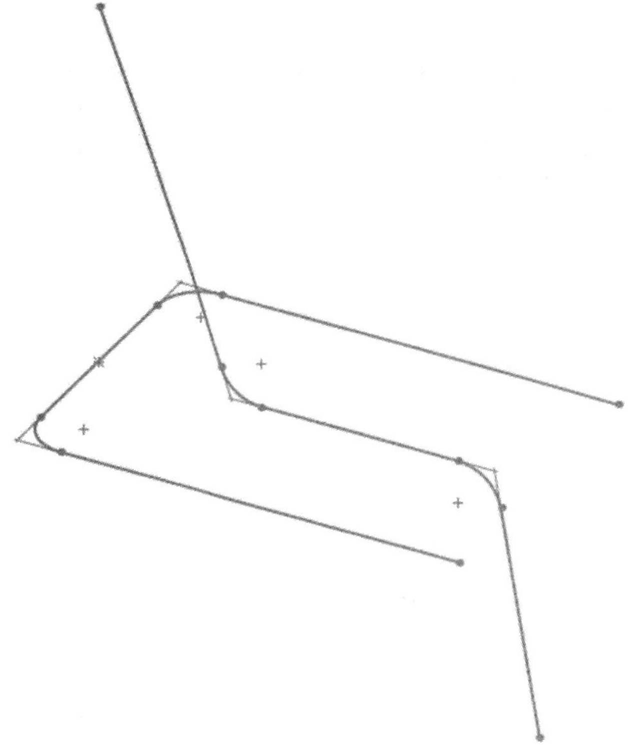

16.3. - Select the "**Projected Curve**" command from the "**Curves**" drop down menu in the Features toolbar or the menu "**Insert, Curve, Projected.**"

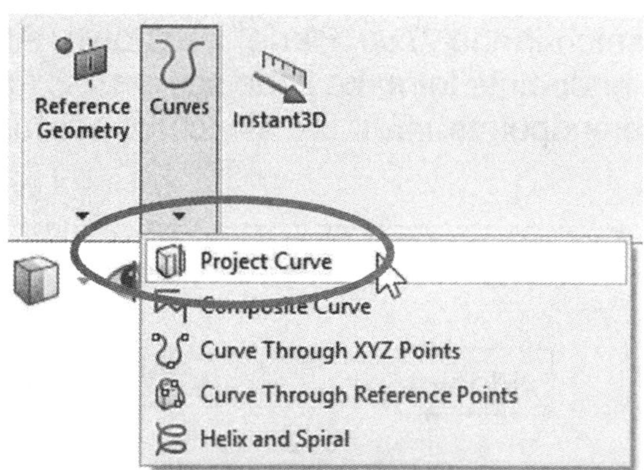

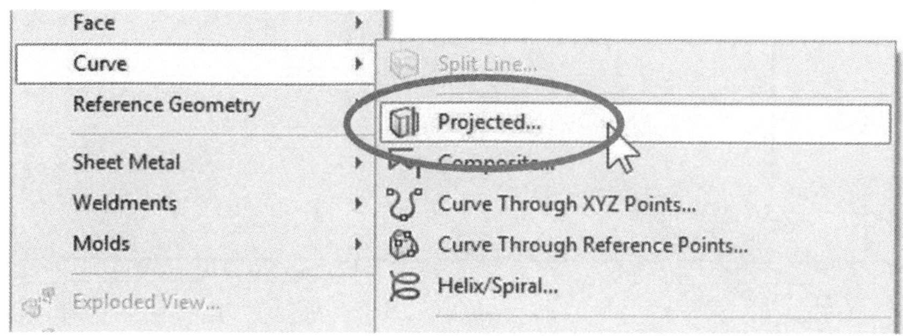

Select the "Sketch on Sketch" option and pick both sketches previously made; note the preview of the curve to be generated, and click OK to finish.

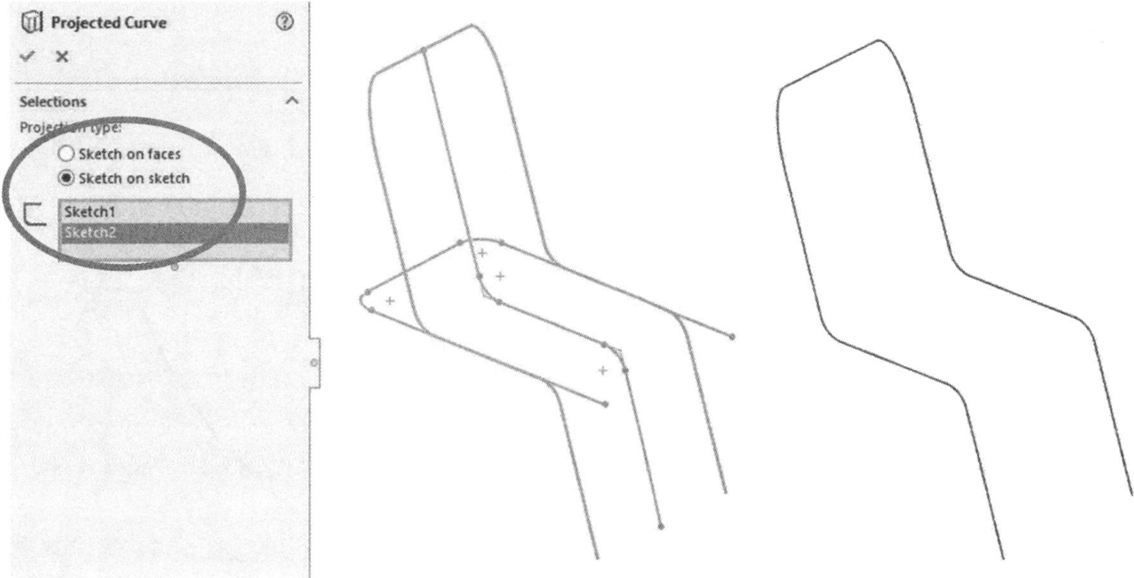

Now we can use the projected curve as a path for the sweep operation.

When projecting curves onto each other, depending on the angle, radii may not be correctly projected, as is the case with the curves at the top of the backrest. In this case, building the part with a 3D sketch produces the desired result, and a projected curve does not. This is a good example to learn how to create a projected curve, and also to show some of its limitations.

16.4. – The following part, curved along two planes, will be made with a "**Projected Curve**" using two sketches. Add the first sketch in the Front plane as indicated and Exit the sketch.

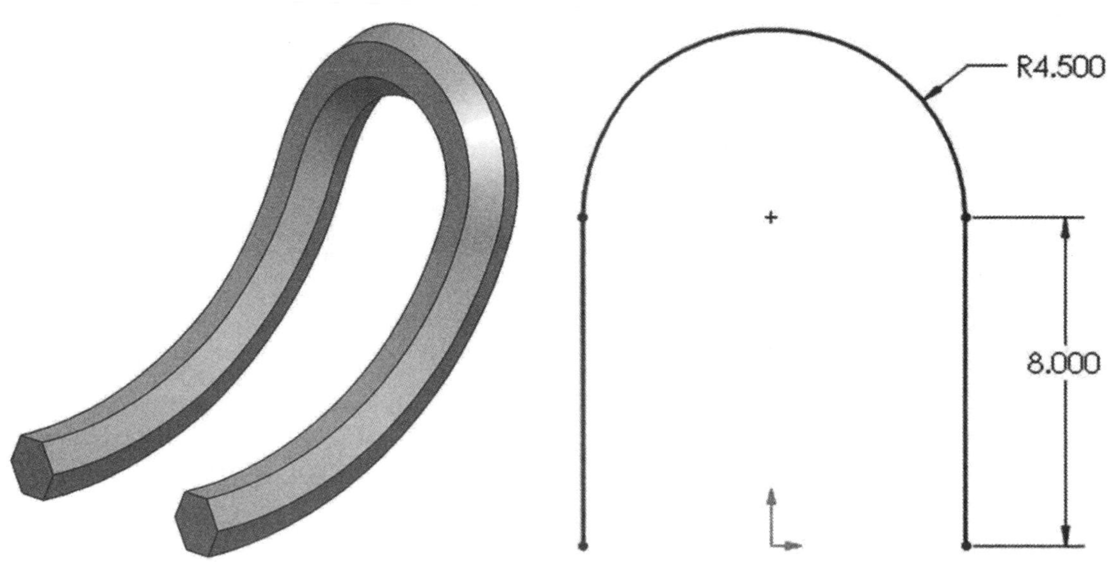

16.5. – Add a second sketch in the Right plane, add a pierce relation between the arc's center and the first sketch, add a horizontal relation between the arc's endpoint horizontal and the arc's center. This sketch is fully defined using only geometric relations. Exit the sketch when done.

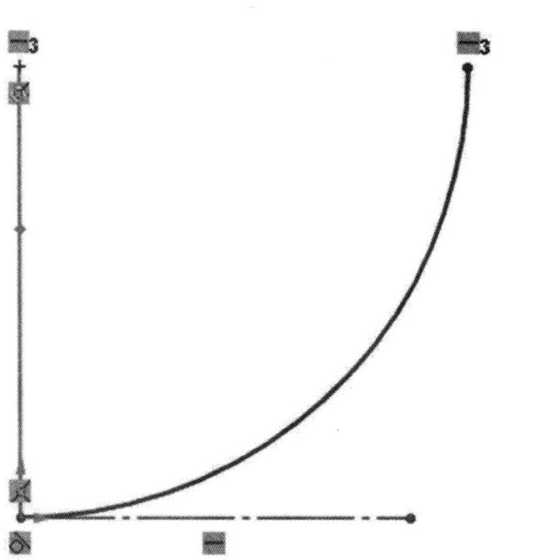

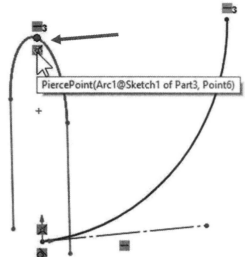

16.6. – Select the "**Projected Curve**" command from the "**Curves**" drop down menu in the Features toolbar or the menu "**Insert, Curve, Projected**." Using the Sketch on Sketch option, select both sketches; immediately the projected curve's preview is displayed. Click OK to finish the curve.

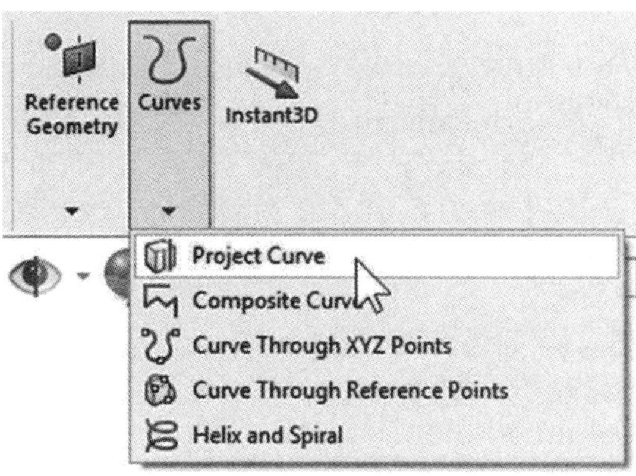

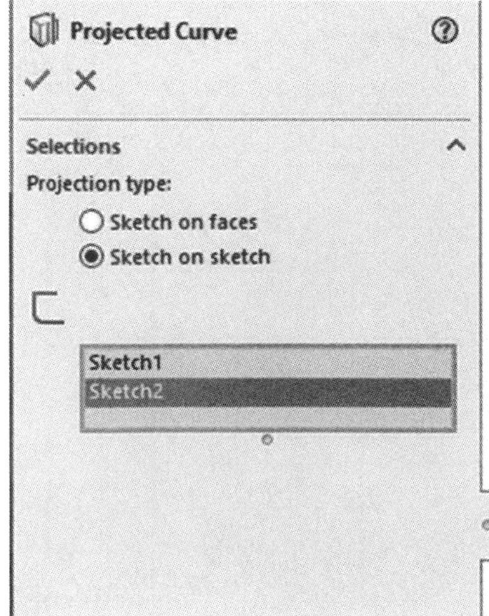

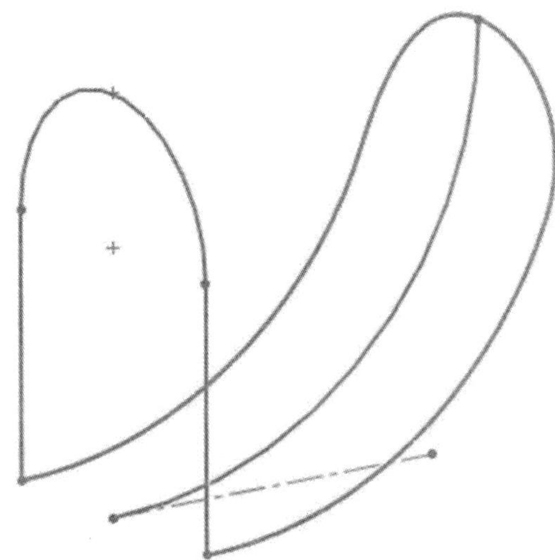

16.7. – For the sweep's profile, add a new sketch in the Front plane. Add a Pierce relation between the hexagon's center and one end of the curve. Exit the sketch when finished.

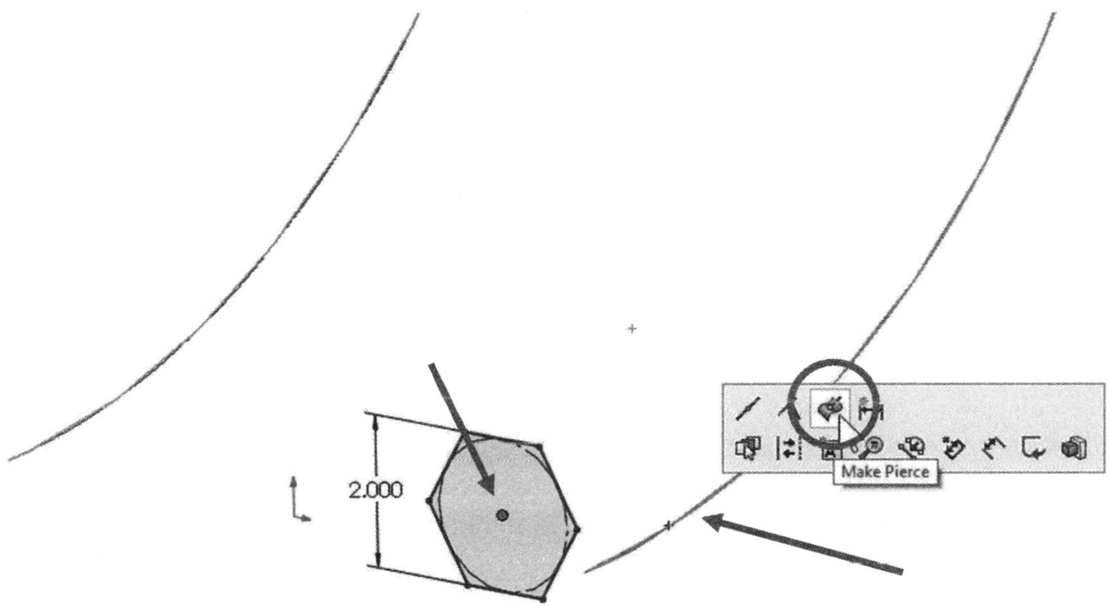

16.8. – Use the Sweep command with the profile and projected curves. Select the "Minimum Twist" and "Merge tangent faces" options. Click OK to finish and hide the projected curve. Save the finished part as "*Curve Sketch-on-Sketch*."

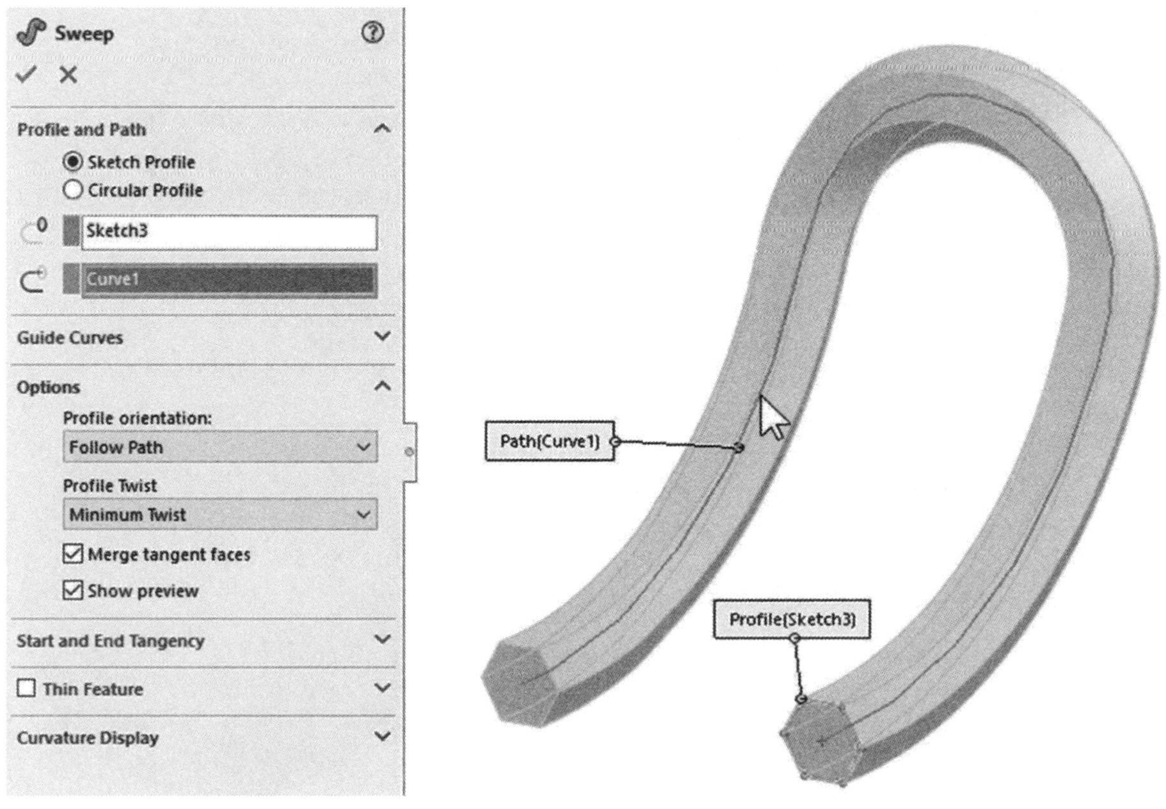

16.9. – To practice more "**Projected Curves**," and to learn how to make a "**Composite Curve**" by combining two or more curves and/or model edges, we'll make a baseball; the composite curve will be used as a path for a sweep feature along the spherical surface.

Start by creating a **Revolved Base** feature. Add the following sketch in the Front plane to make the ball's sphere first.

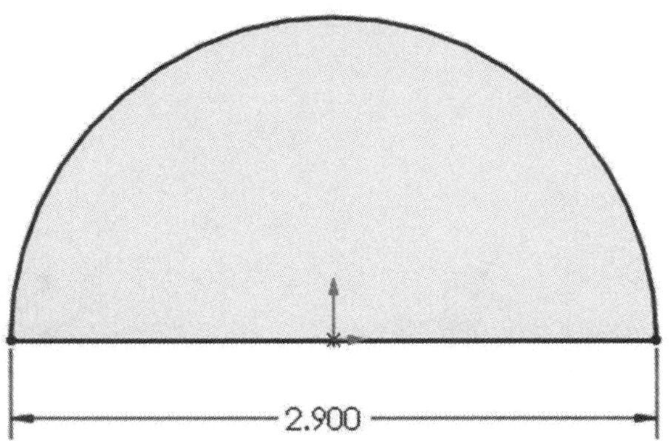

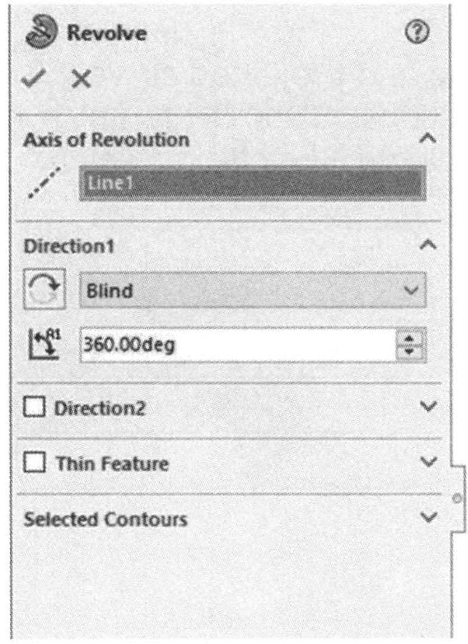

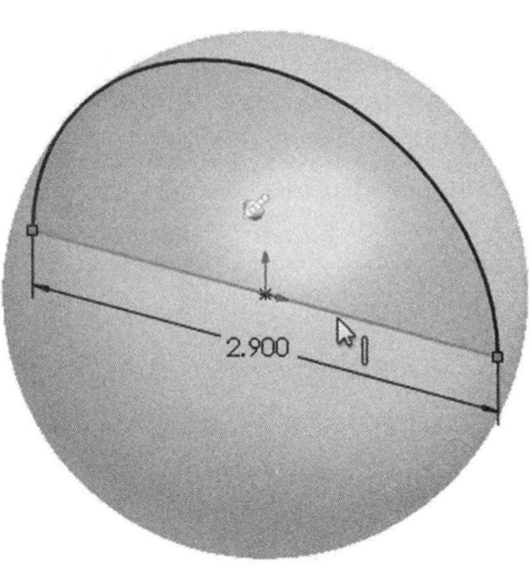

The sphere's sketch can be made in any plane, and the semi-circle be either horizontal or vertical; the resulting sphere is the same.

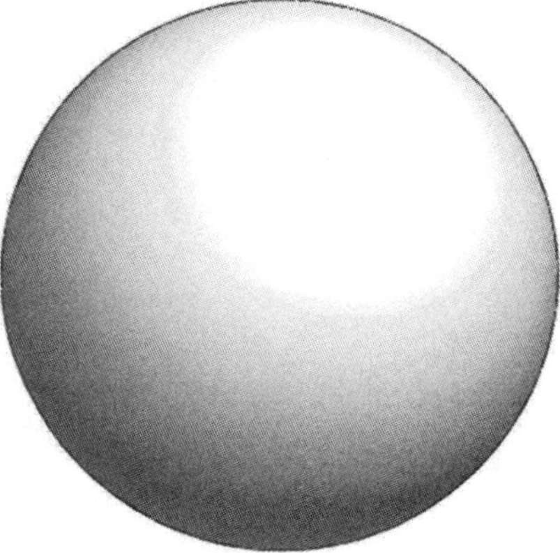

16.10. – To make the projected curve in the ball's surface, we need to add the following sketch in the Front plane. Exit the sketch when finished

- The arc's center is not at the origin.
- The arcs are tangent to each other.
- The arc's endpoints are vertical to each other.
- The vertical and horizontal construction lines are coincident to the quadrants of the sphere. Make them horizontal/vertical to the origin.
- The 0.400" and 0.350" dimensions are referenced to the arc.

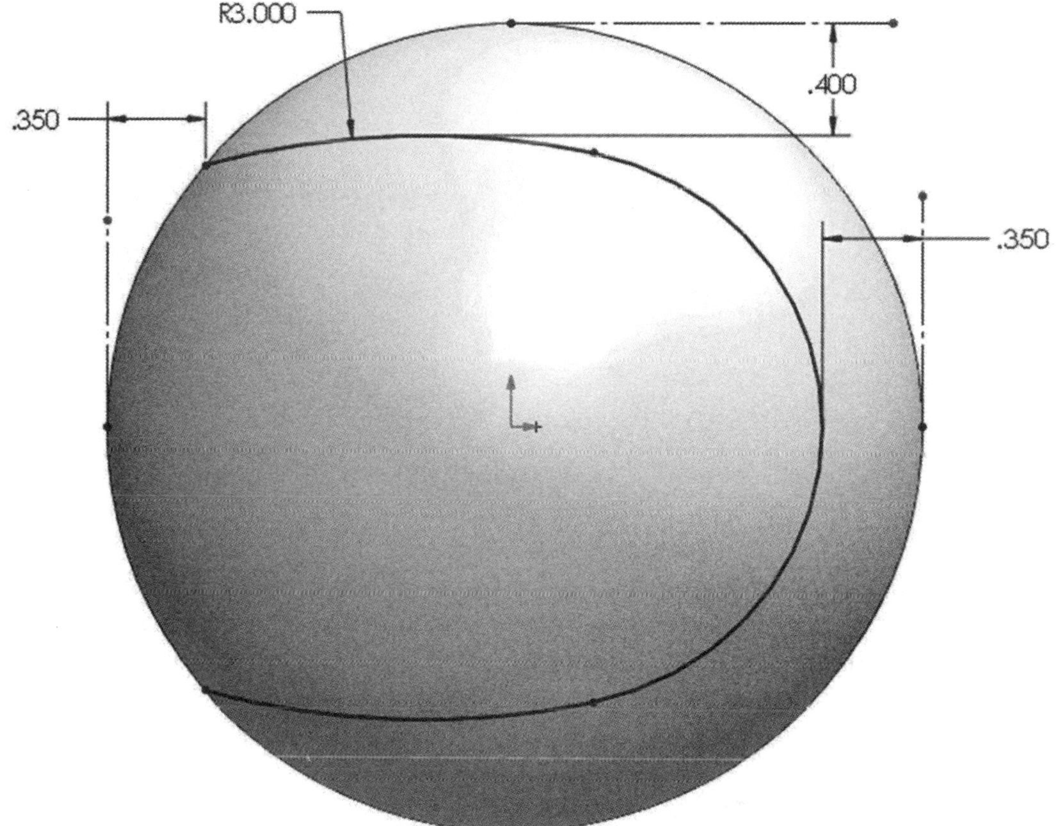

A different option to using the "**Projected Curve**" is to make a "**Split Line**" to divide the surface and get the necessary curve.

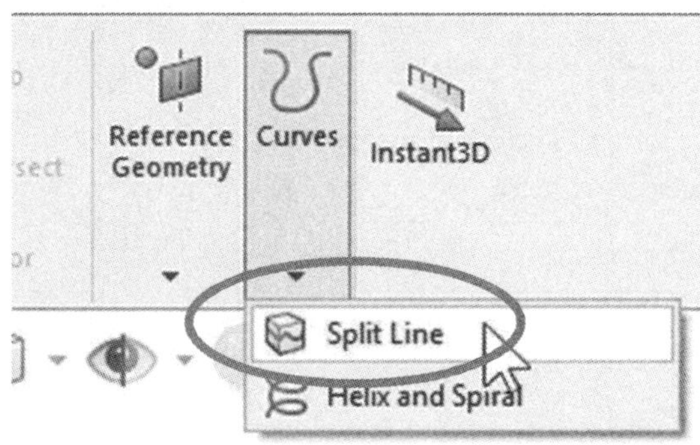

16.11. – From the "**Curves**" drop down menu, select the "**Projected Curve**" command or from the menu "**Insert, Curve, Projected.**" In the "Projection type" section select the "Sketch on faces" option, and select the previous sketch; in the Projection Faces selection box select the sphere's surface. The preview of the projected curve on the sphere's surface will be displayed. Click OK to finish the command.

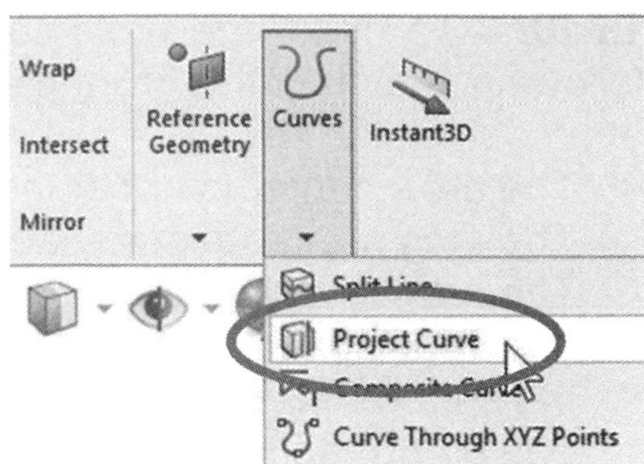

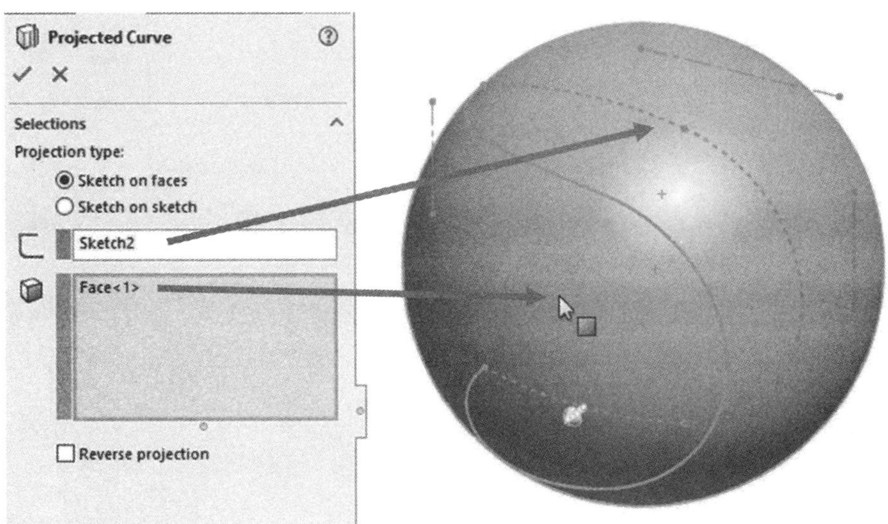

16.12. – The "**Projected Curve**" command can only project the curve in one direction—in our case we need the curve projected in both directions. Repeat the "**Projected Curve**" command using the same parameters, except this time we need to select the sketch from the Fly-out FeatureManager, and turn On the "Reverse Projection" check box to project the curve in the other side of the sphere. Click OK to finish the command.

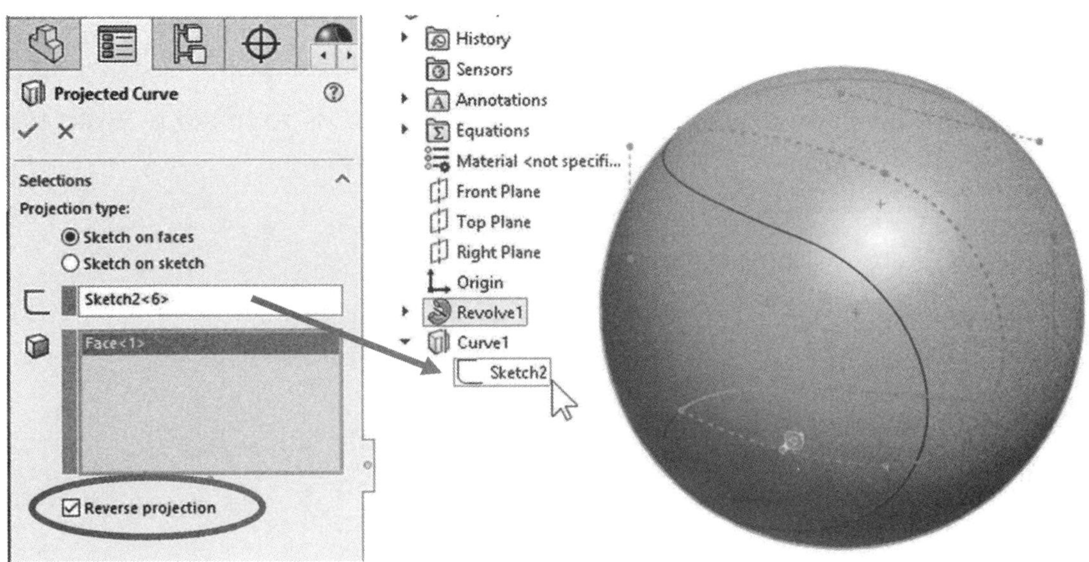

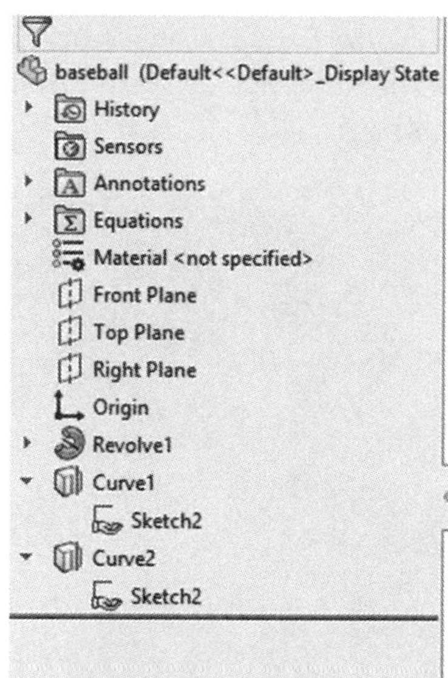

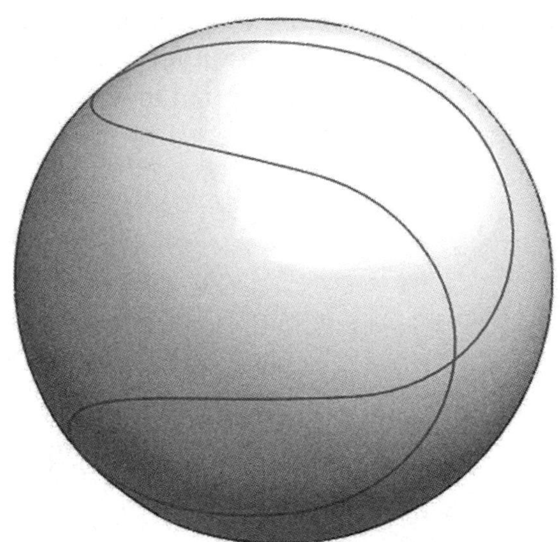

16.13. – After completing the two projected curves, from the "**Curves**" drop down menu, select the "**Composite Curve**" command or from the menu "**Insert, Curve, Composite**".

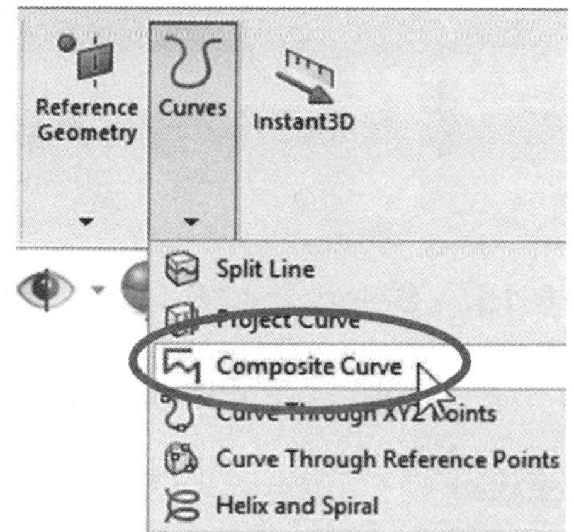

In the "Entities to Join" selection box, select both of the projected curves to combine them into a single curve. Click OK to finish the command.

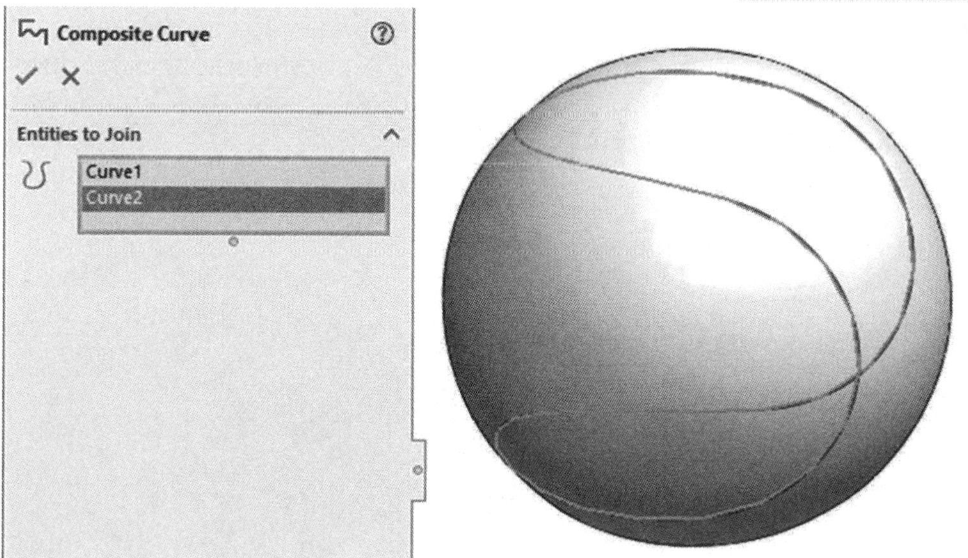

The reason to combine both curves in to a single one is to use it as a path for a Sweep command.

16.14. – Switch to a Front view, add a new sketch in the Front plane and draw the following sketch. Make the origin of the circle coincident to the sphere's silhouette and the composite curve. Exit the Sketch when finished.

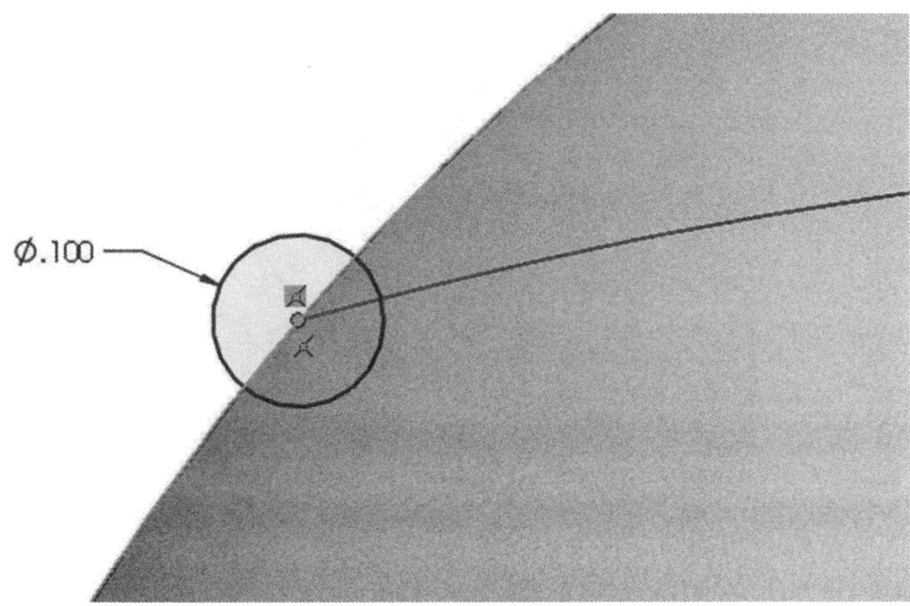

16.15. – Select the "Sweep" command, use the previous sketch as the profile, and the composite curve as the path. Turn On the "Merge tangent faces" option to make a single continuous surface. Click OK to finish and hide the composite curve.

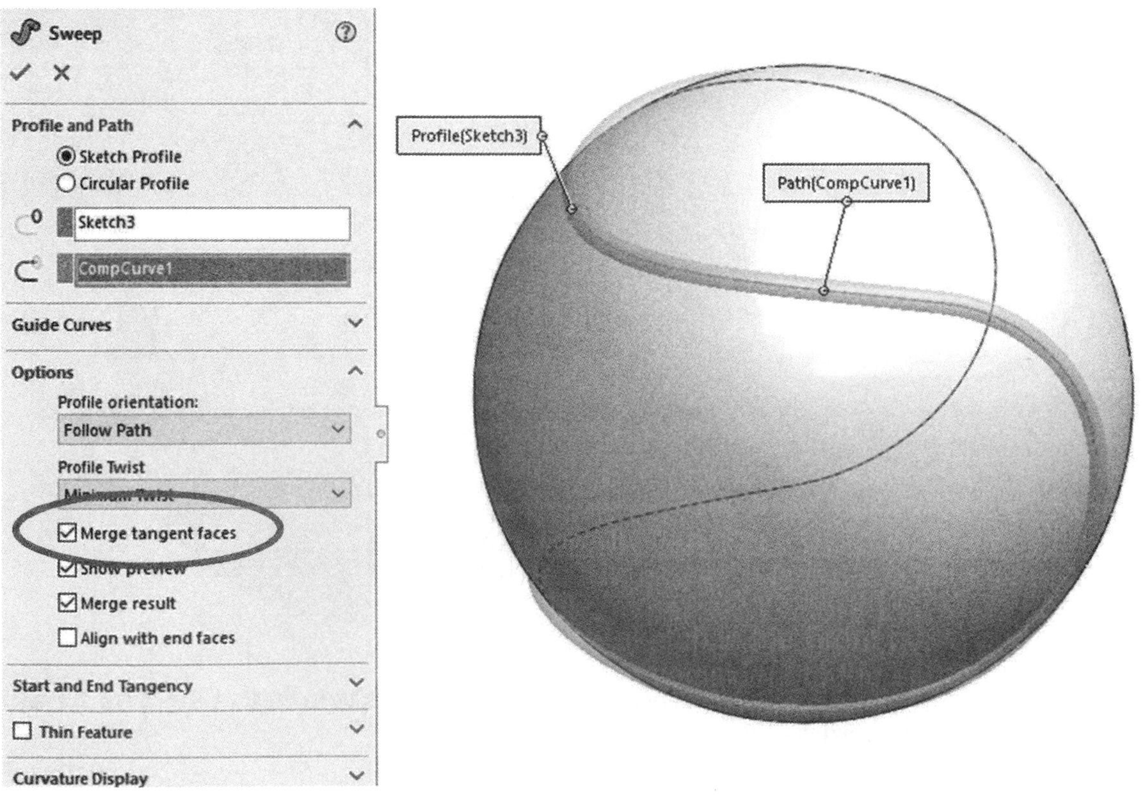

16.16. – Save the part as "Baseball" to finish.

EXERCISE: Create the second curve for the chair's frame using a projected curve and add the two sweep features.

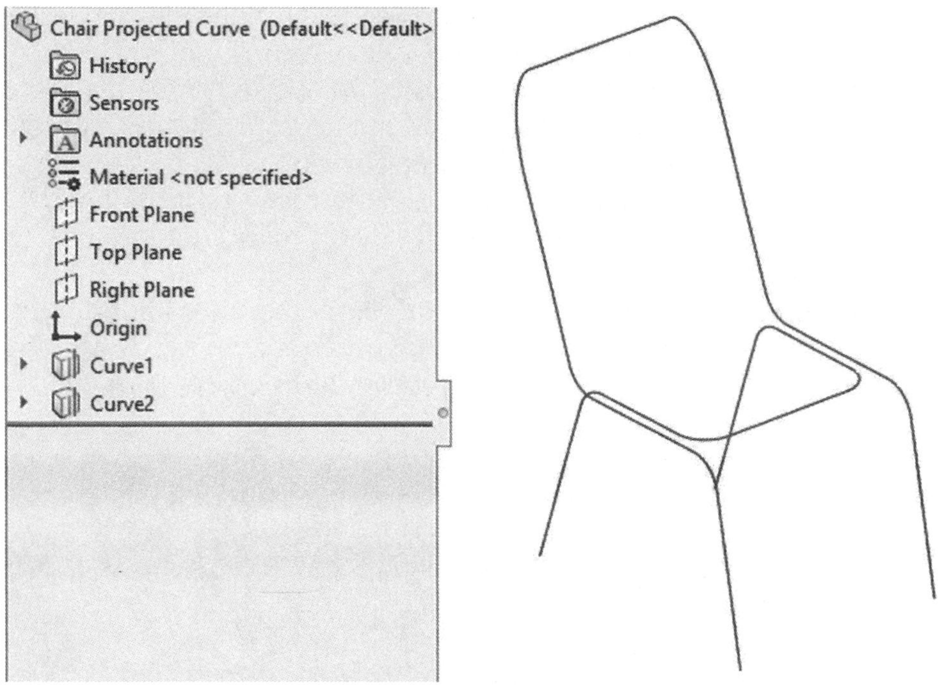

To make the extrusions at the end of the sweeps we must show the sketches used before projecting the curves to use them for the direction of extrusion.

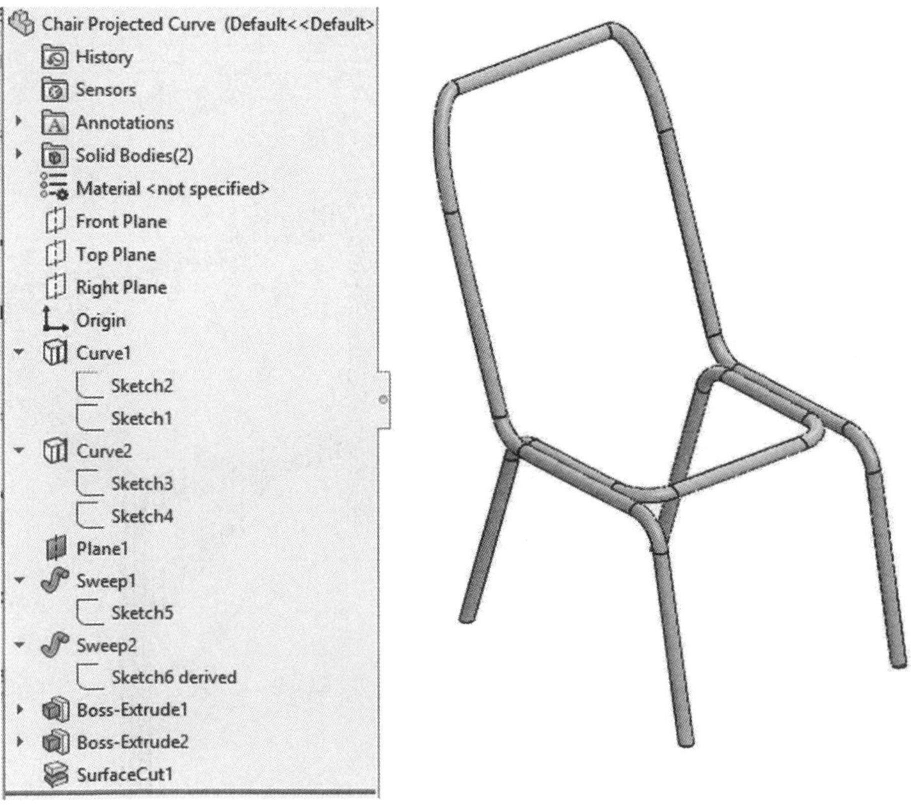

Weldments

In SOLIDWORKS, a welded structure (weldment) can be defined as a part whose elements are long and thin and are made with structural members, like extruded profiles. The weldments environment in SOLIDWORKS allows us to design welded, modular or similar structures made from constant cross section elements (i.e., round or square tubing), complete with weld beads, gussets and end caps using a multi body part. Detail drawings of weldments usually include material cut lists with length and cut angles ready for manufacturing.

SOLIDWORKS includes a library of structural members in both metric and imperial standards, and if needed, we can also add our own profiles to the built-in library or download hundreds of additional profiles from online libraries.

When we work with weldments, the first thing we need to do is to make one or more 2D and/or 3D sketches that will serve as the structure's 'skeleton', which will be used to define either the structure's centerline, inside, or outside dimensions, just like in sheet metal. In a weldments part, each sketch line is used to locate a structural element.

A weldment is one of the few instances where it's efficient to use a multi-body part file instead of an assembly to represent multiple *'parts'*.

To learn how to use the weldments environment, we'll make the following welded structure using a 2″ x 2″ structural steel profile.

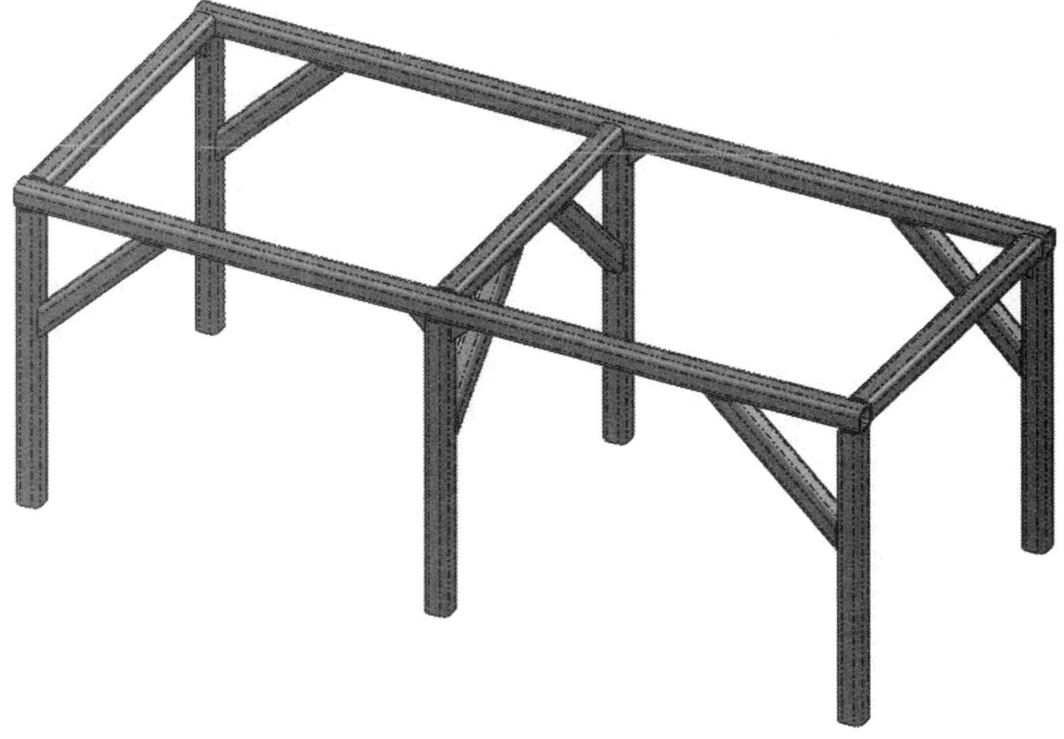

17.1. - The first thing we need to do is to make a 3D sketch (or a combination of multiple 2D sketches) that will form the *skeleton* for our welded structure. Just as with sheet metal, we need to decide if the sketch dimensions will represent inside, outside or centerline dimensions of the welded structure. In our case, the structure's leg dimensions will be at the centerline, and the height of the 3D sketch will be the top of the structure; in other words, the width and length of the conveyor are not as important, but the height is. Open a new part and start a 3D sketch (menu "**Insert, 3D Sketch**"). Select the "**Line**" tool and press the "Tab" key to switch to the "**YZ**" plane. By default, a 3D sketch always starts in the "**XY**" plane.

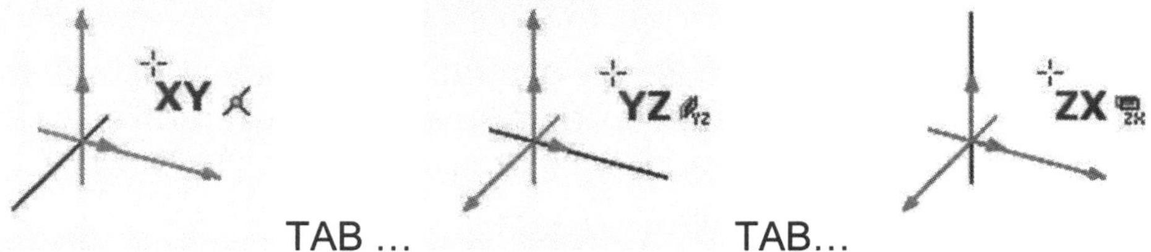

To make the 3D sketch instructions easier, the reader will be directed to go along the X, Y or Z axes in the positive or negative direction.

After selecting the **YZ** plane, start the 3D sketch line in the origin, first go up in the positive **Y** direction, then negative **Z,** finally negative **Y** and stop. Don't worry about the size; it will be dimensioned after adding some relations.

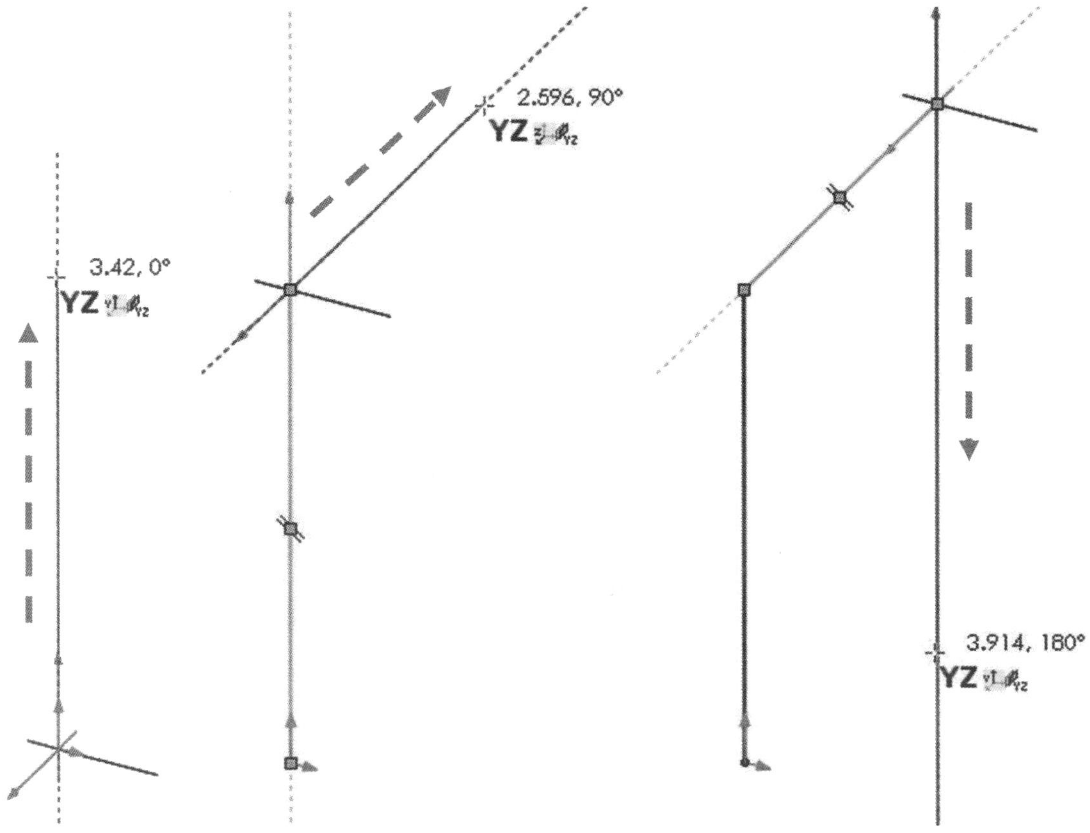

17.2. - Still with the "**Line**" tool active, start in the upper left and right corners and draw two lines going in the positive **X** direction. Press the "Tab" key if needed to switch to the **XY** or **XZ** plane; either one works in this case.

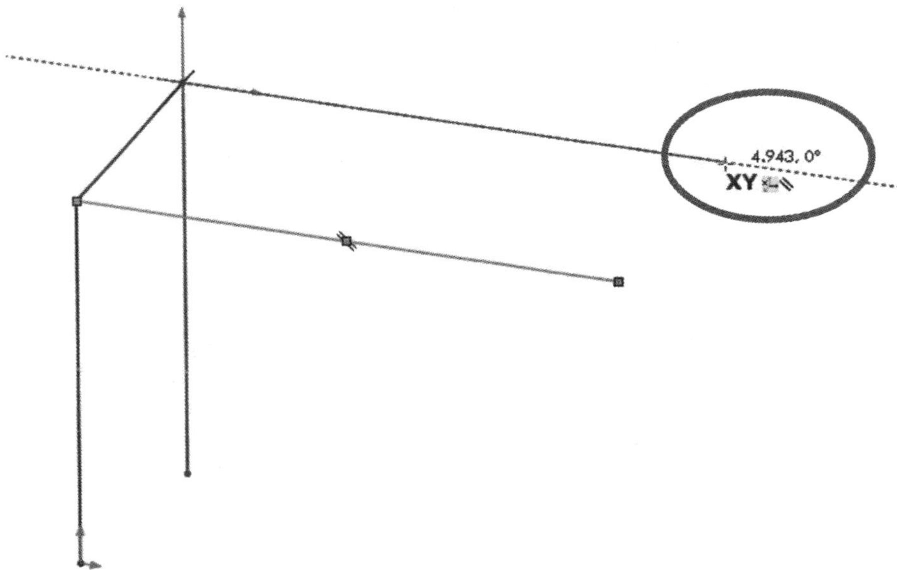

17.3. - When working in a 3D sketch we can also 'snap' to other geometric elements even if they are not in the same "plane" orientation. Draw the next four legs of the structure by capturing midpoint and coincident relations to the midpoints and endpoints of the two long horizontal lines as shown. Switch to a **XY** or **YZ** plane if needed by pressing the "Tab" key.

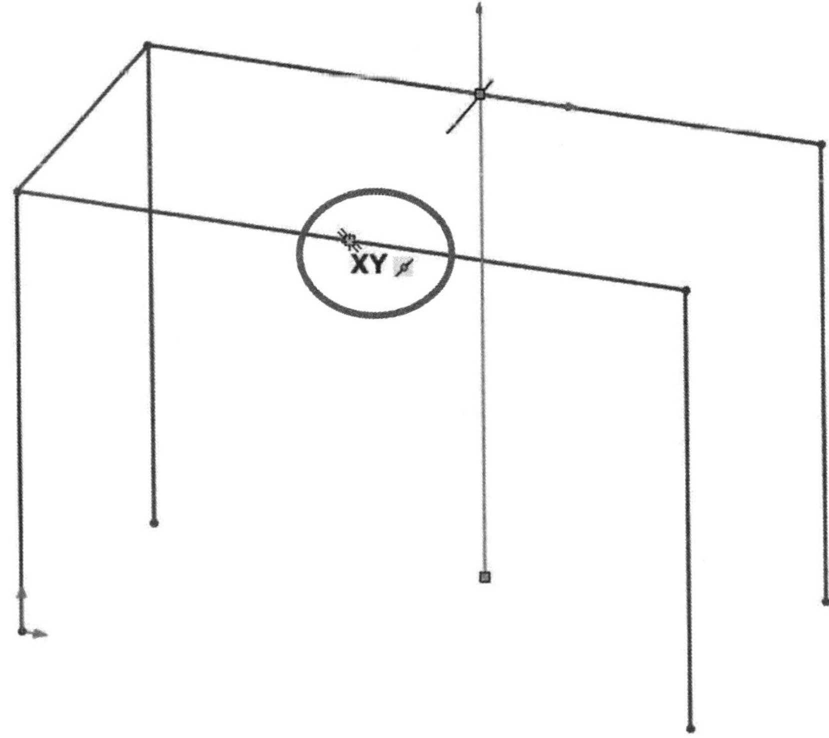

17.4. - Draw the next two lines along the **Z** direction to complete the geometry. Draw the lines by snapping to the endpoints, even if the lines are not exactly along the **Z** direction. We'll correct this in the next steps.

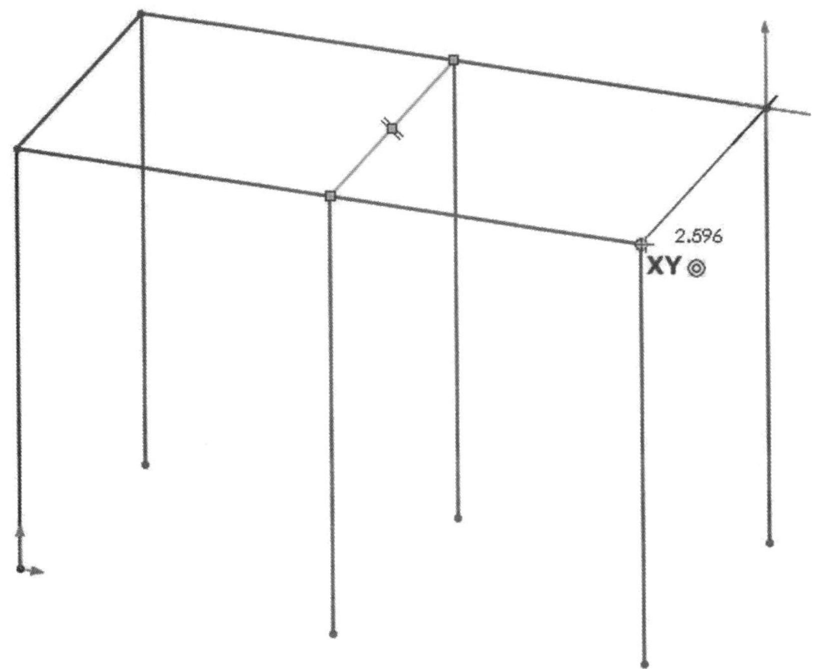

17.5. - Add **Along X** relations to the three endpoints of the legs on each side, one side at a time, *or* make all six legs Equal; the result is the same.

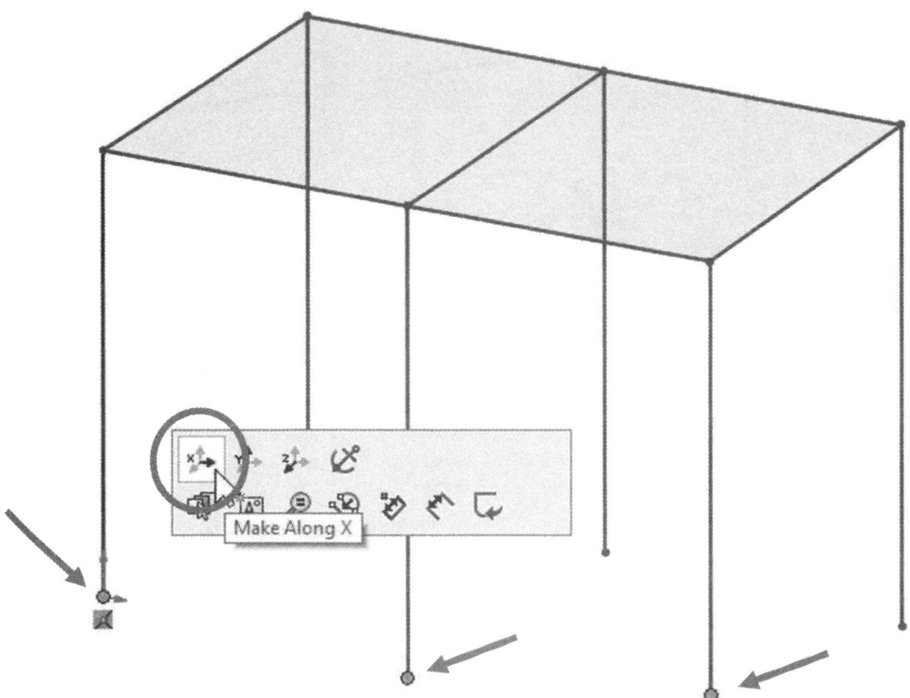

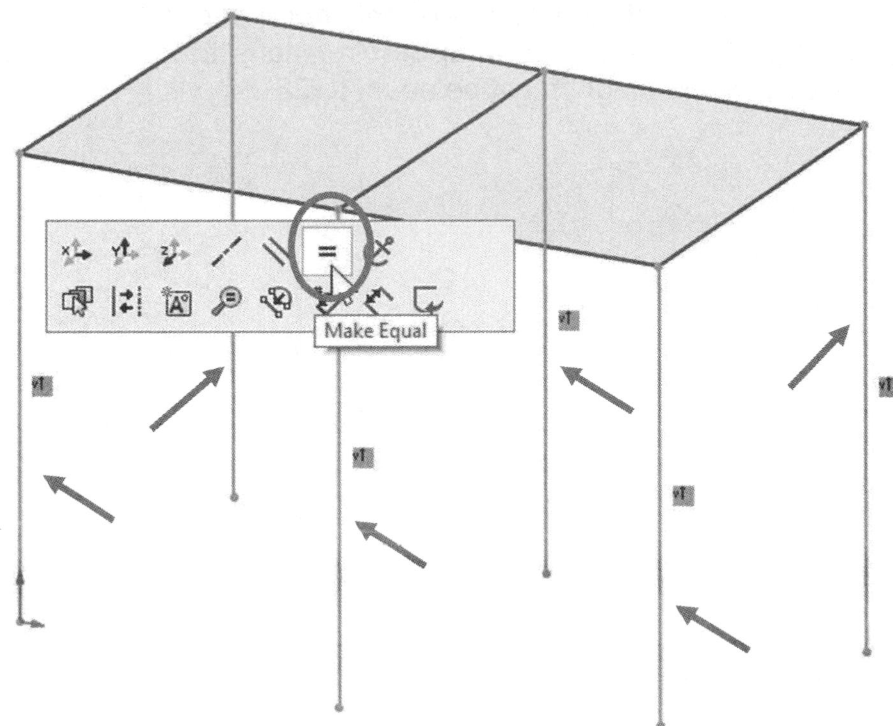

17.6. - Since we 'snapped' two of the transversal elements to existing endpoints and midpoints, they may not be exactly along the **Z** direction. We need to add relations to either align them **Along Z**, make them parallel to the first transversal line, or make the long horizontal elements equal.

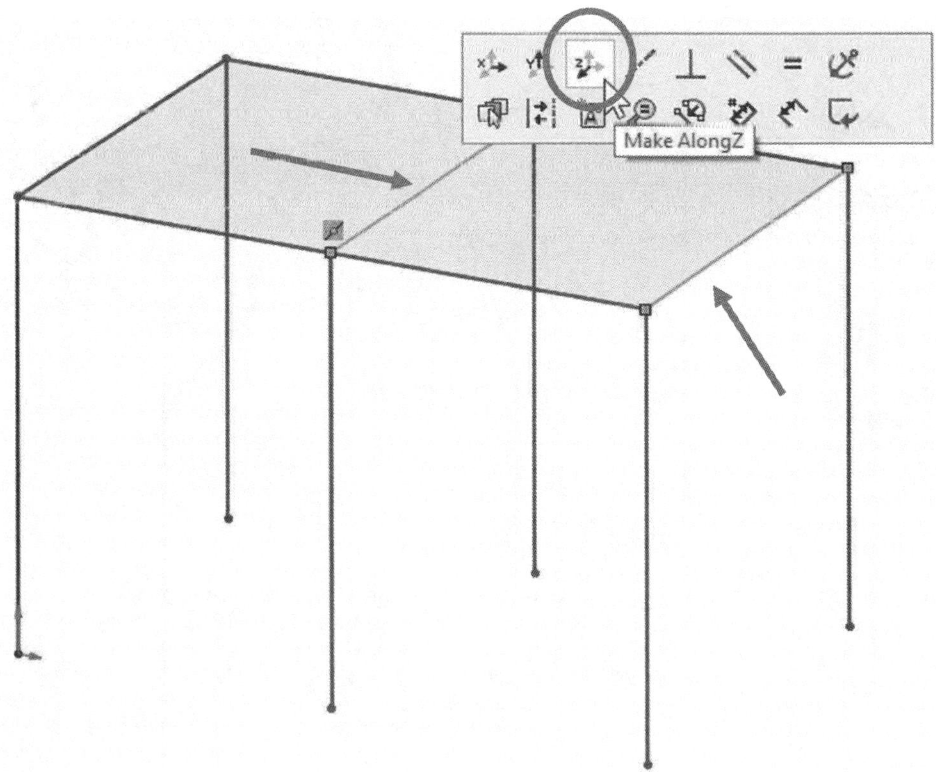

17.7. - Add the following dimensions. We only need one 24″ dimension between the 'leg' lines because we added **Midpoint** relations to the long horizontal line along **X**. The structure's height must be equal to 28″.

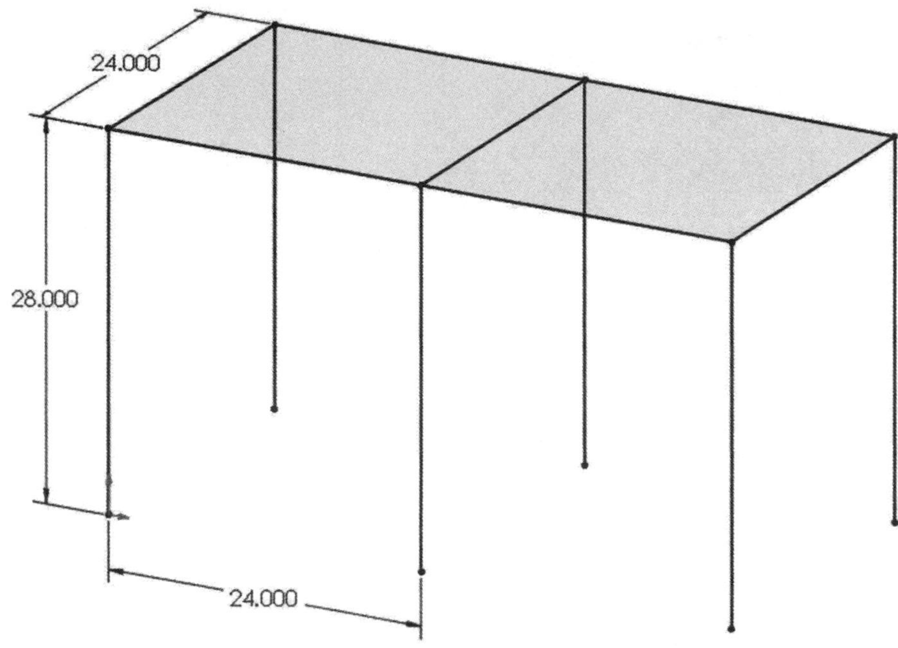

17.8. - Add the corner reinforcements, and dimension one corner as indicated. Draw the lines by 'snapping' to existing geometry (in this case it doesn't matter which plane we are working on). Make all the reinforcement lines **Equal**.

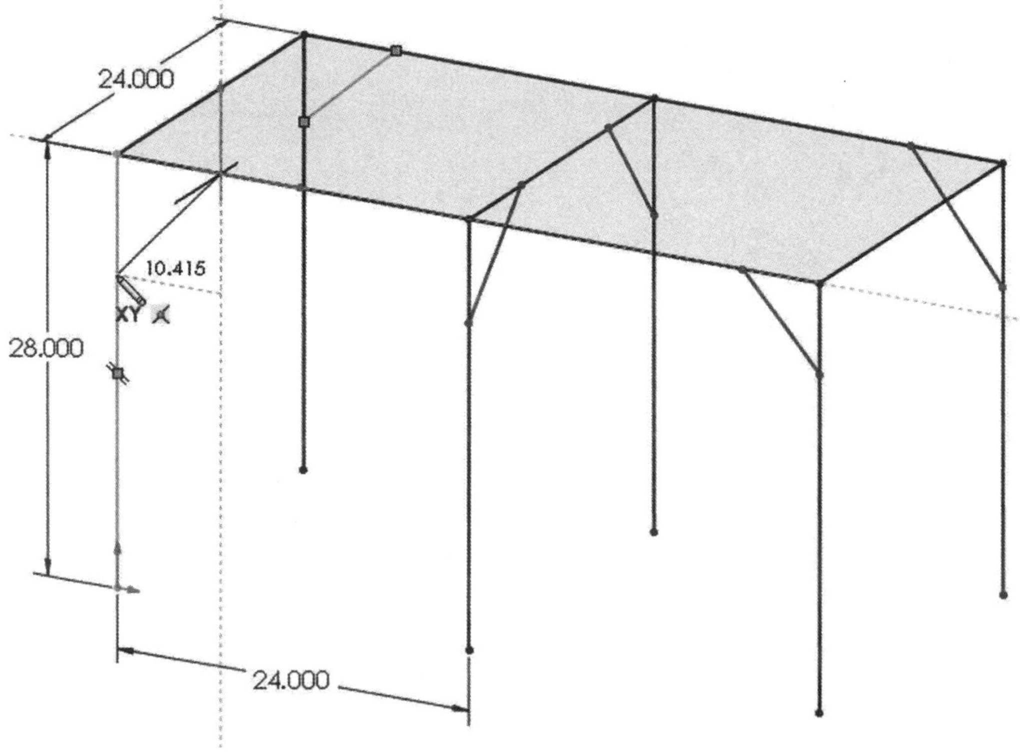

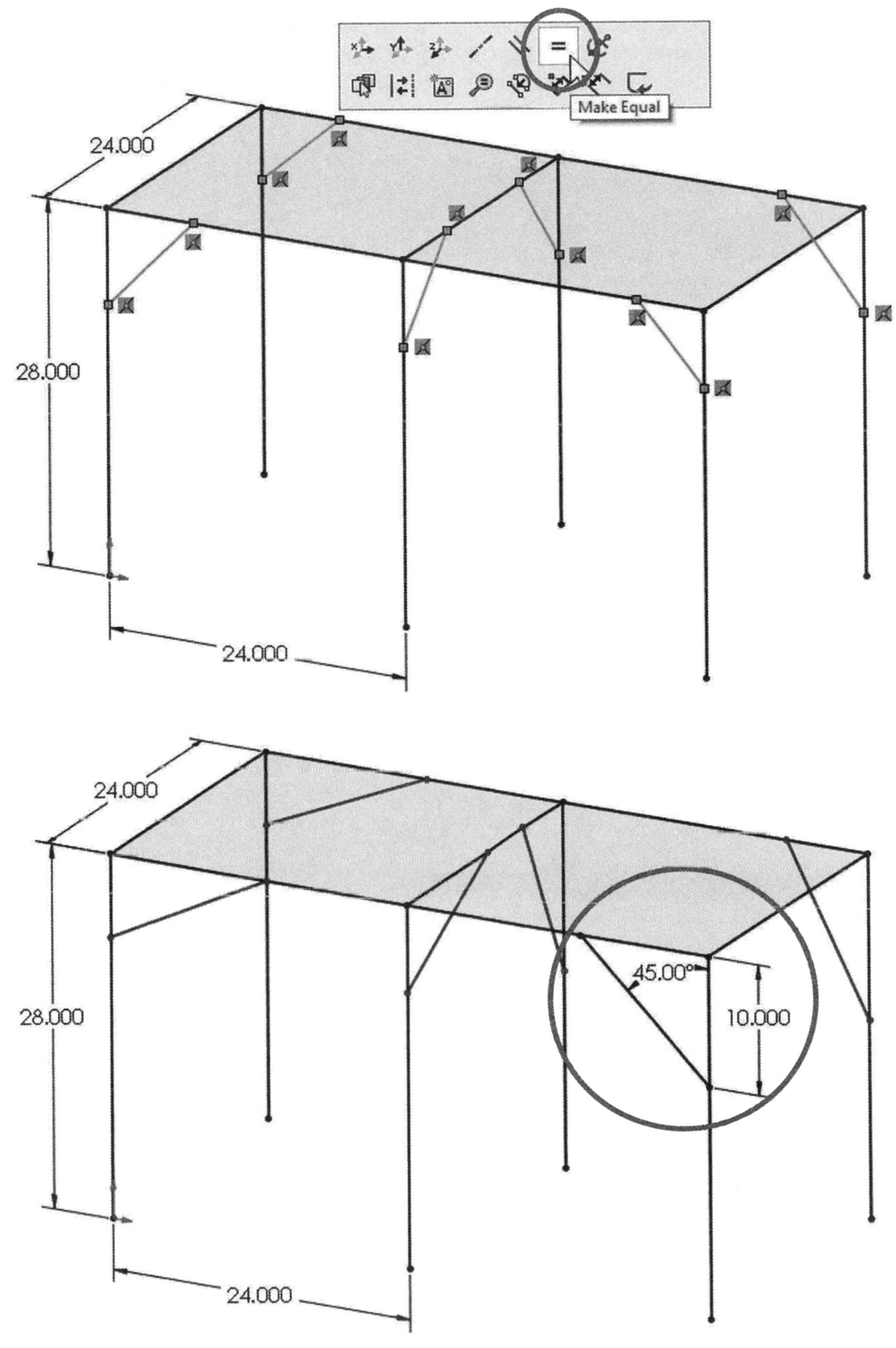
Make Equal
24.000
28.000
24.000
24.000
28.000
45.00°
10.000
24.000

17.9. - Make the lower three endpoints at the same level on one side by adding an **Along X** geometric relation, then the other side, and finally add an **Along Z** relation between the endpoint of the dimensioned line and the line to its 'right' to fully define the sketch.

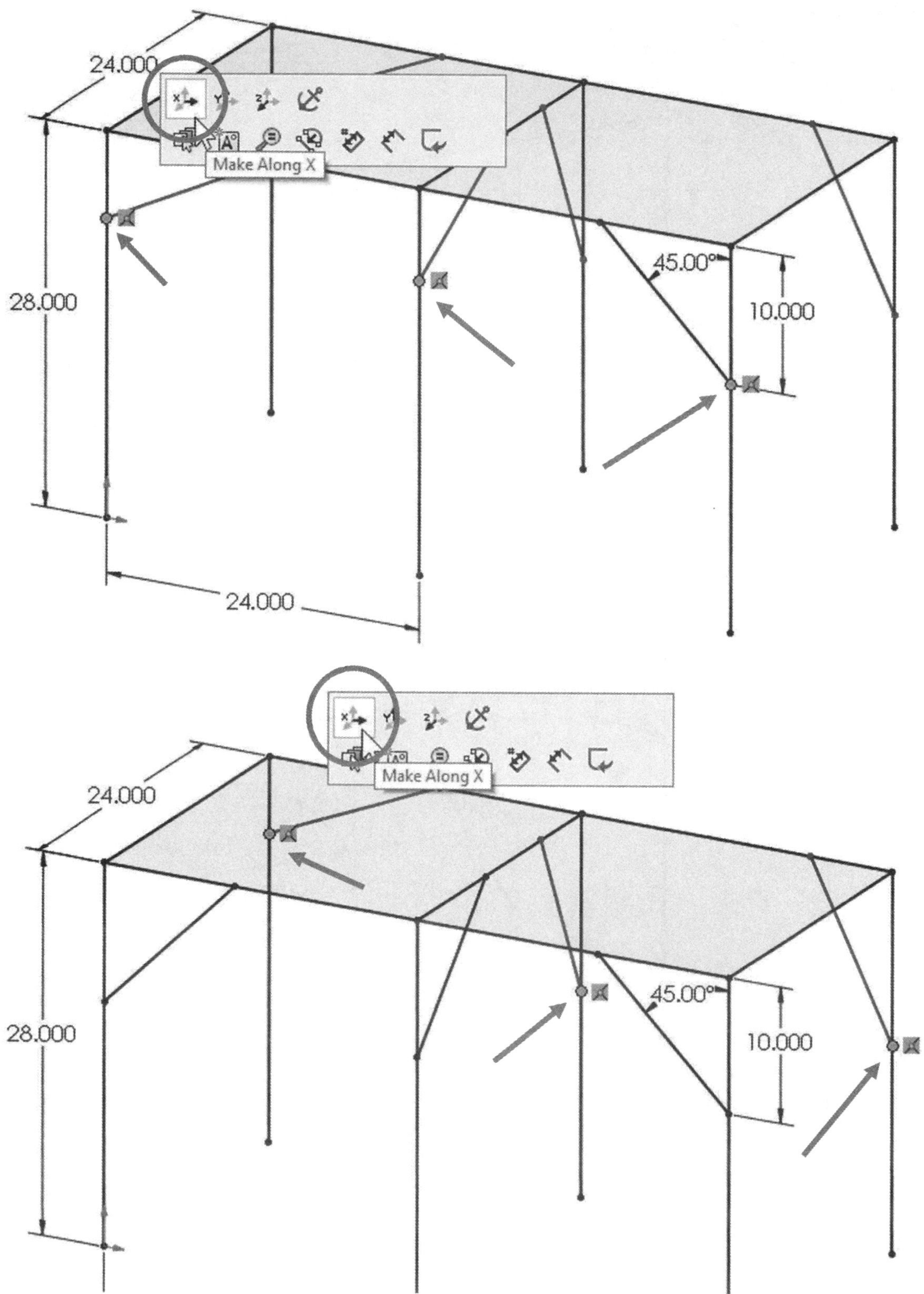

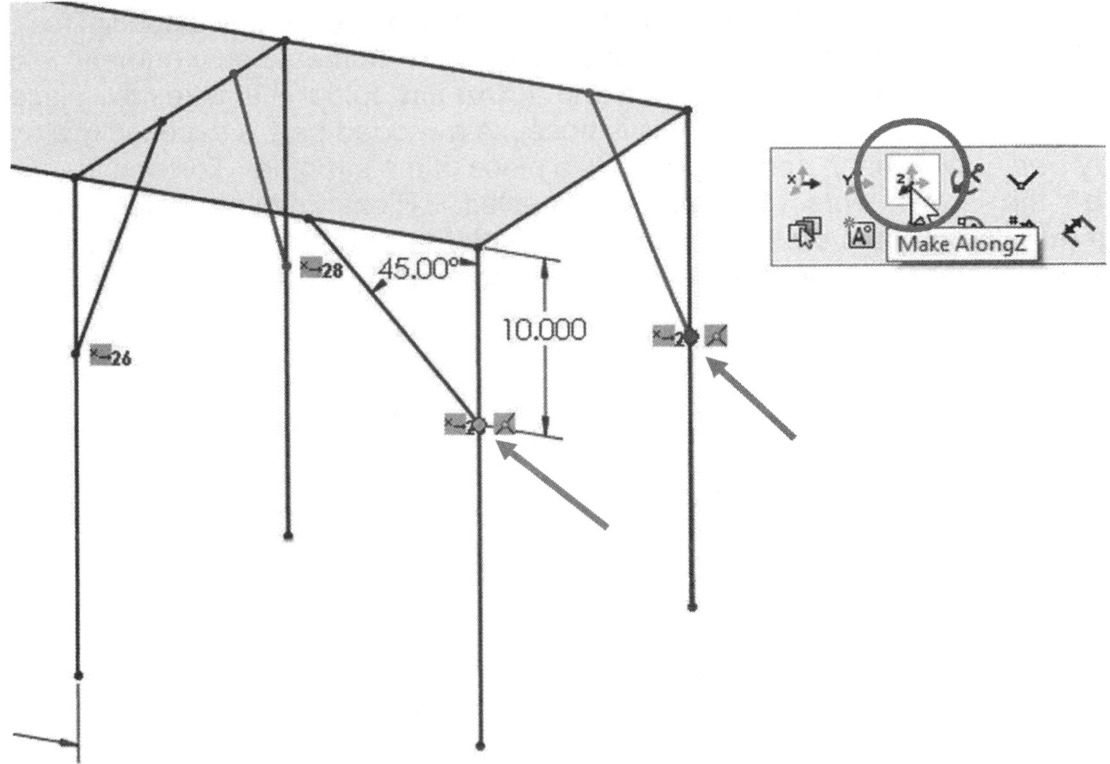

The fully defined 3D sketch now looks like this. Exit the 3D sketch to continue.

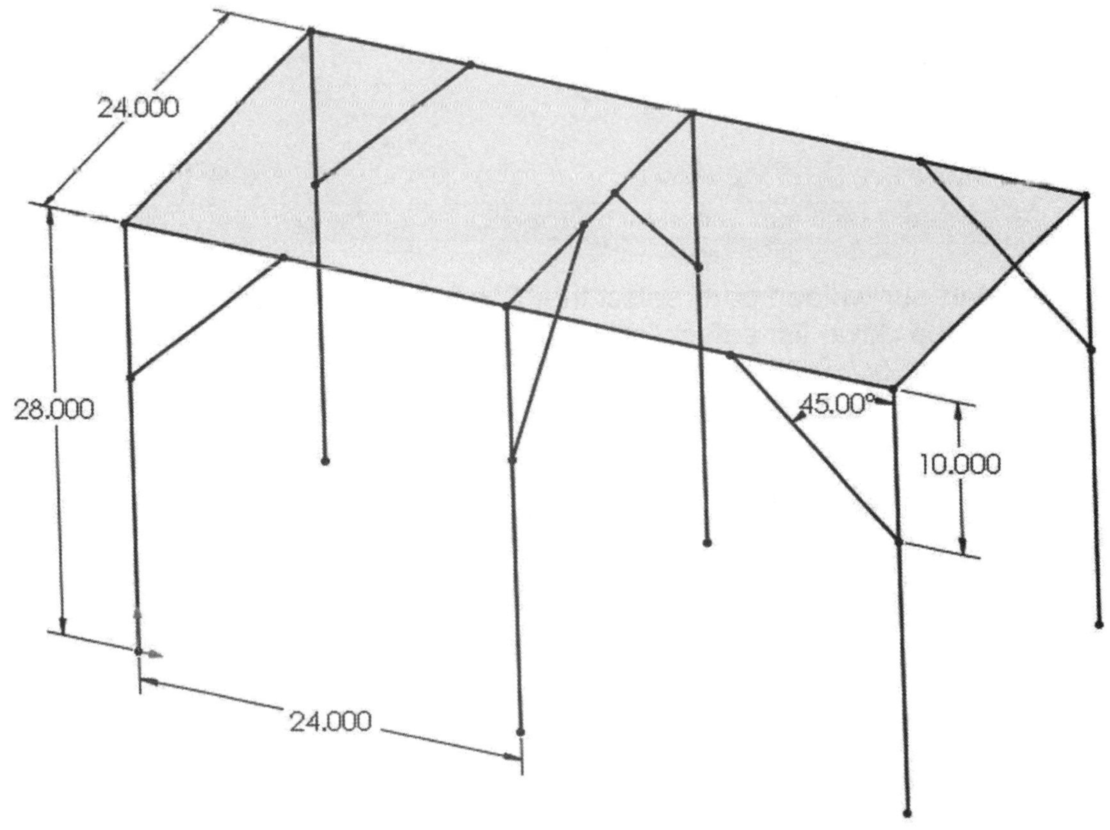

17.10. - Weldments work similarly to sheet metal in the sense that a special feature is added to the FeatureManager that enables the weldments environment, which in this case is a "**Weldment**" feature and a "**Cut list**" folder that takes the place of "**Solid Bodies**"; the reason for this is because a welded part is essentially a multi body part, where each body represents a piece of the structure. To continue make sure the weldments toolbar is enabled. Right mouse click in the CommandManager's tabs and activate the "**Weldments**" tab, and/or right mouse click in a toolbar and turn on the Weldments toolbar.

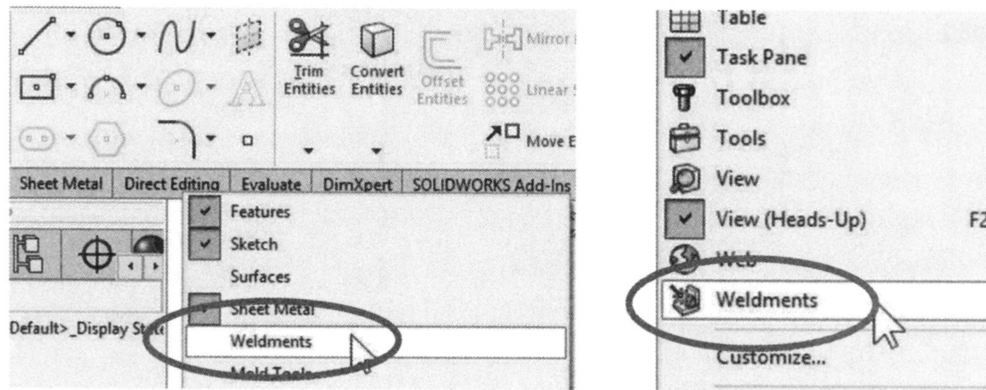

In the Weldments toolbar is a "**Weldment**" command; selecting it enables the weldments environment in the part but, just as with sheet metal, it is also automatically added when we make the first "**Structural Member**" feature.

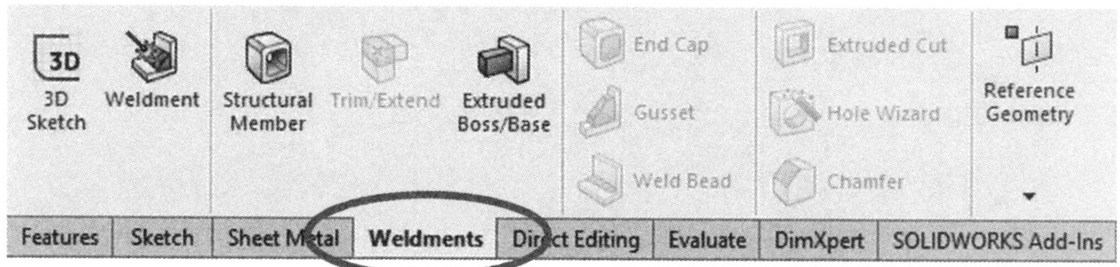

To start our welded part, select the "**Structural Member**" command. In the "Standard:" drop down list select "ansi inch," for "Type" select "square tube," and for "Size" select "2x2x0.25".

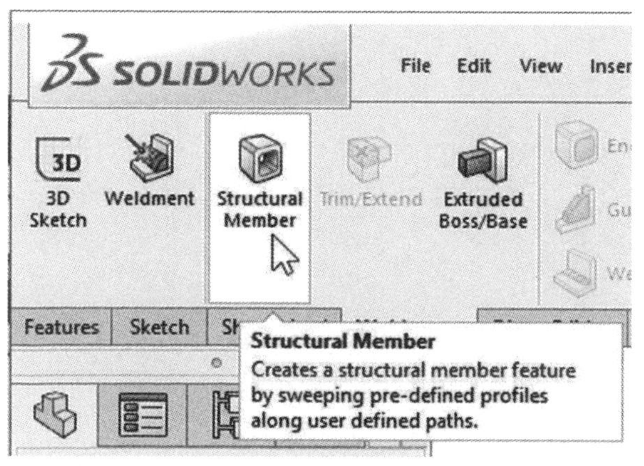

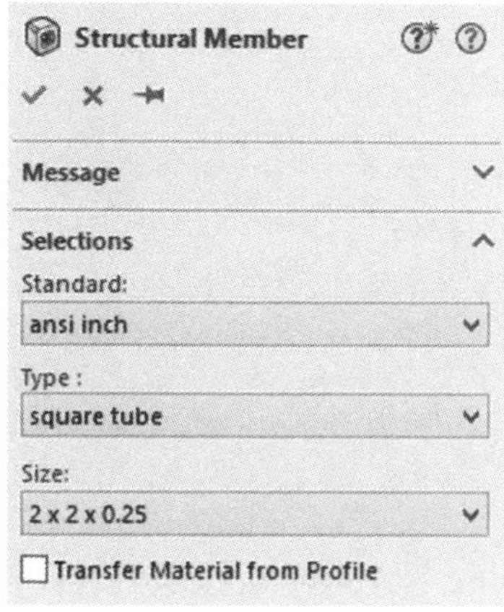

17.11. - When we add structural members they are added in groups oriented in the same direction or laying in the same plane.

There are two different ways to add the structural members; in the first case we'll add all the elements and then trim them to size. In the second one we'll show an optimized way of adding elements to minimize or even eliminate trimming. Both options will be shown to learn different functionality.

Using the first technique, we'll add the structure's '*legs*' first. As we select sketch segments, we get a preview of each one with the selected profile centered in each line. Click OK when finished selecting the legs to create the first group.

Since we'll be adding more structural members, and the legs are just the first group, activate the "Keep Visible" option, then click OK to create them; this way we can continue adding more groups without having to re-open the command.

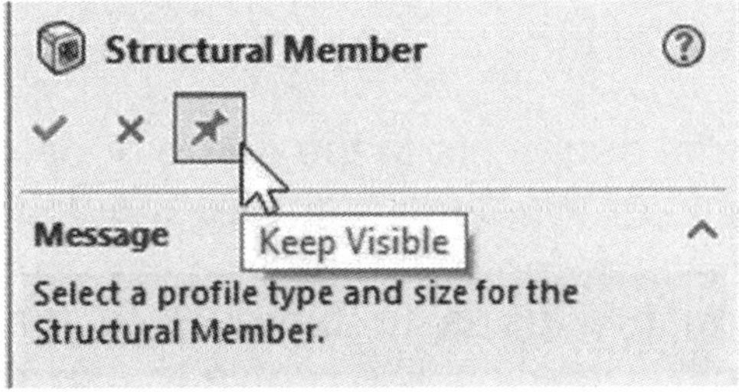

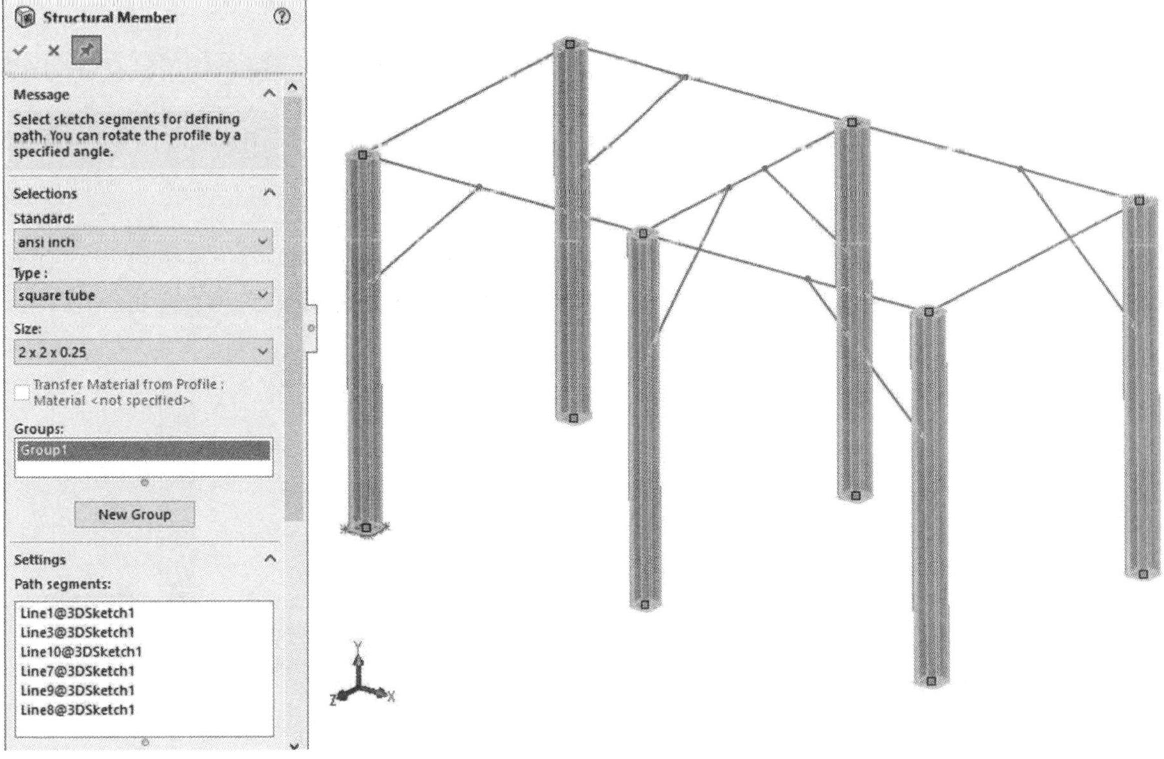

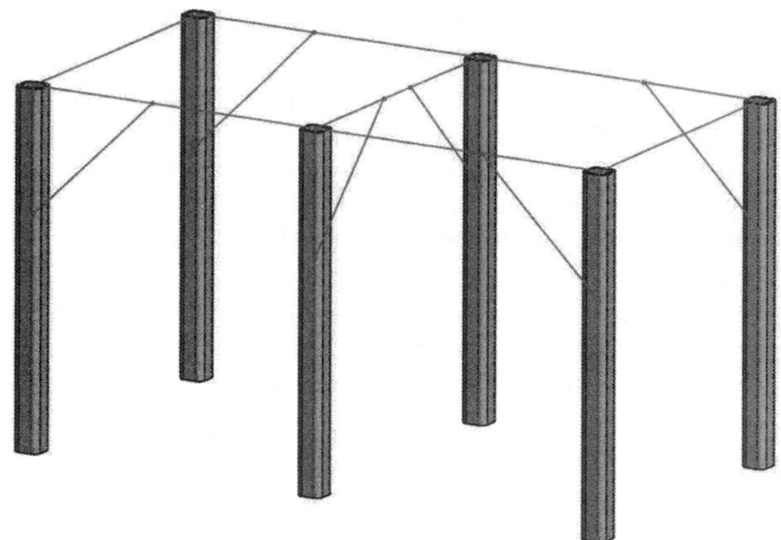

When selecting sketch lines to add Structural Members to a group, remember that the lines in the same group can be:

a) Discontinuous parallel segments in the same or different planes, or
b) Continuous segments in the same plane connected by endpoints.

The idea of having groups is to be able to modify multiple elements' properties in a single operation, like profile to use, orientation, location, etc. in a single step.

17.12. - The second group will be the four lines that make the outside perimeter at the top. We won't be able to select the cross member in the middle because it's not connected to an endpoint. Select the four segments, but do not click OK yet.

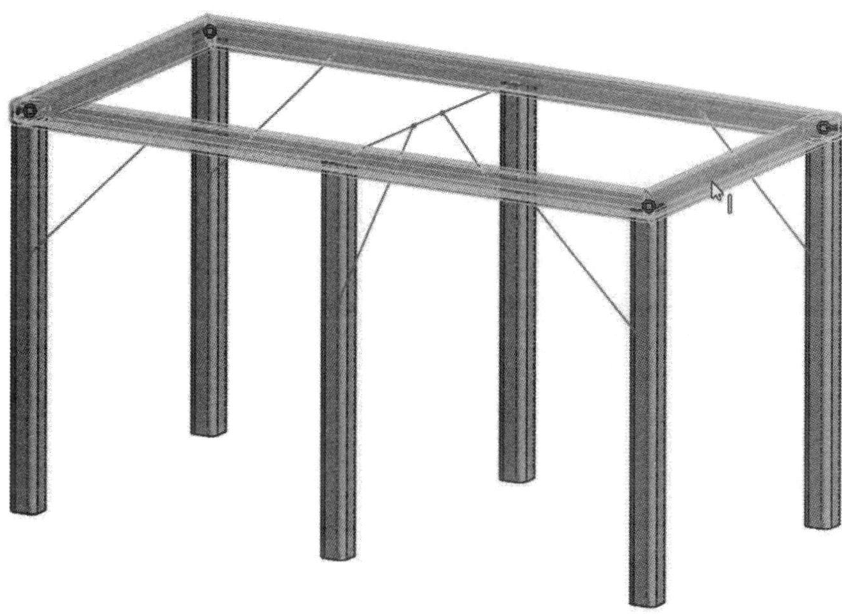

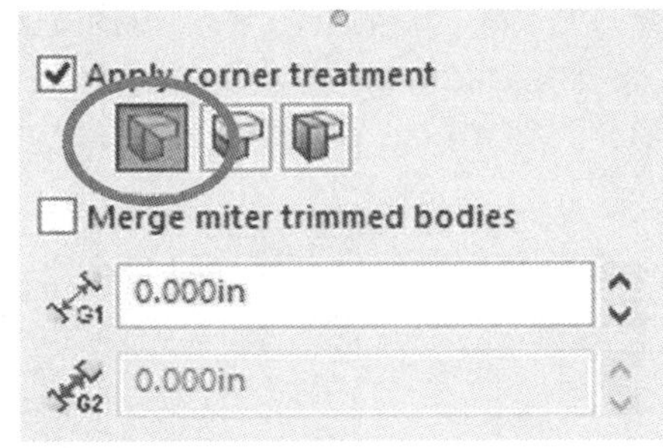

Before selecting OK to create the structural members, we need to make a few modifications to trim the corners and locate this group. Zoom in a corner to preview the corner treatment and select "End Miter." Using this option all the connecting elements <u>in this group</u> will get this treatment.

Be sure to turn off the option "Merge miter trimmed bodies," otherwise the four elements will be merged into a single body.

The first distance box after corner treatments is to specify a gap between the connected segments of the same group, and the second is to add a gap between this group and the previous one. This is usually done to add a gap to weld structural members together.

The different options for corner treatments available are:

End Miter End Butt 1 End Butt 2

If we use either of the "End Butt" corner treatments, we can select a simple cut or a coped cut end condition to match the other piece.

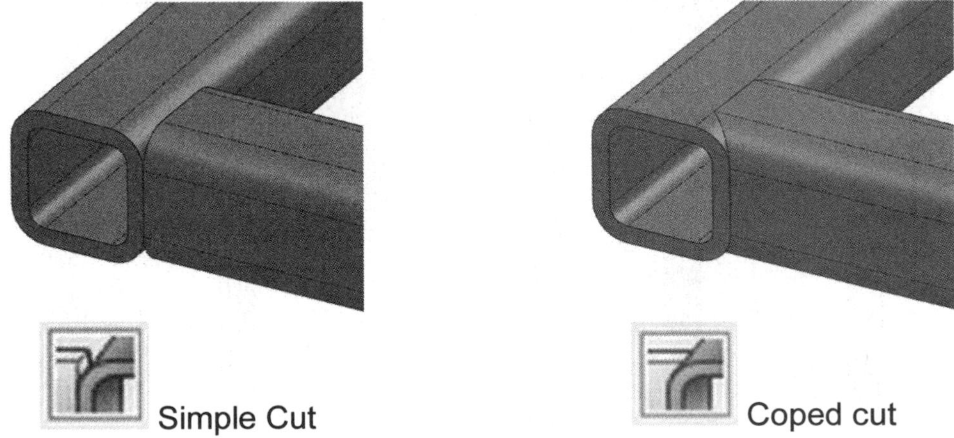

Simple Cut Coped cut

If needed, we can change the type of corner treatment individually by clicking in the pink point at the corner of the preview and selecting a different option of corner treatment, merge bodies or add weld gaps.

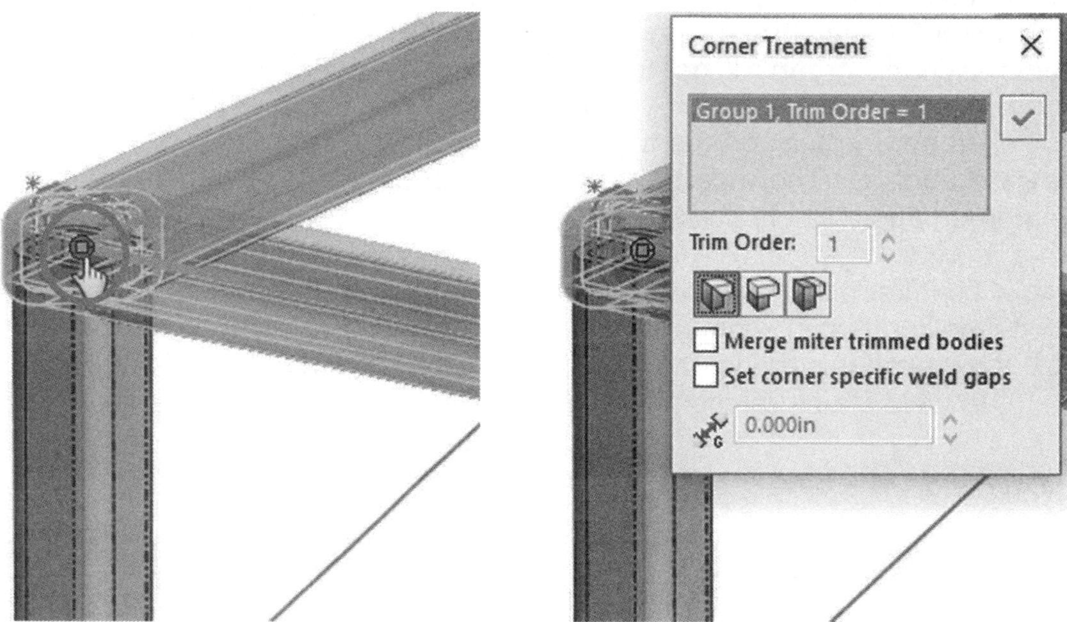

17.13. – By default, when adding new structural members, the sketch line pierces the profile's origin automatically, in this case the center. If we remember, the structure must have a height of 28″, which is the height of the 3D sketch; if we don't change the profile's location in reference to the sketch, the resulting height will be taller than needed one half of the structural member profile's height.

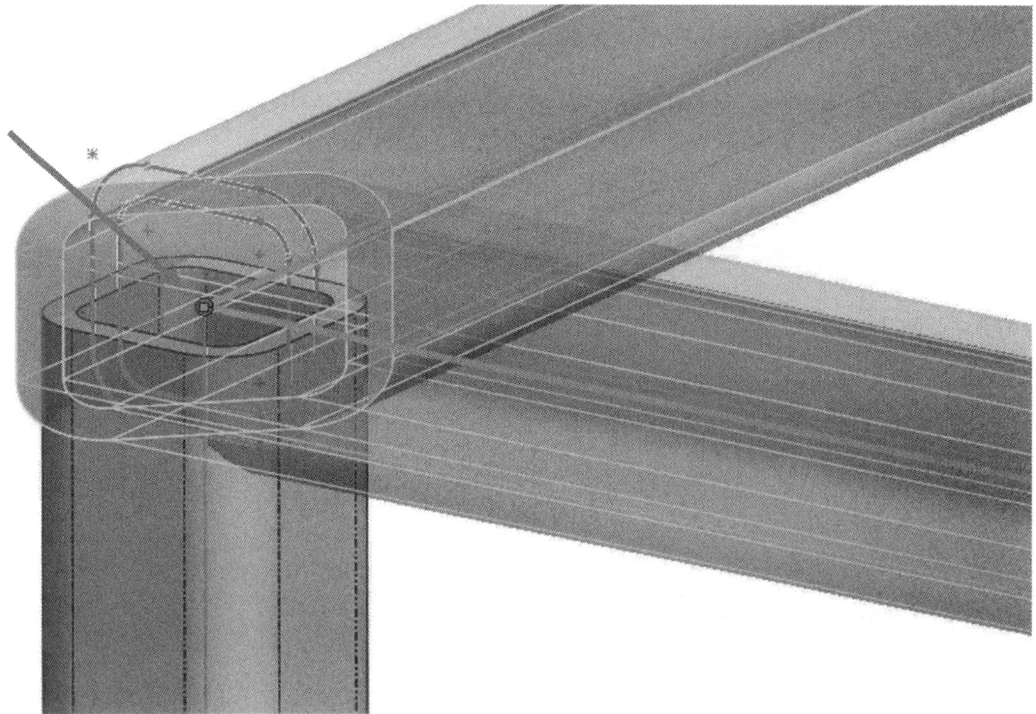

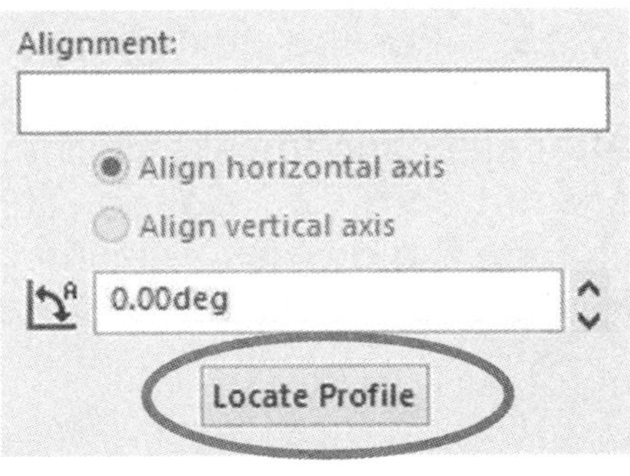

To change the profile's location we have to make the element's sketch line pierce the top of the profile instead of its center. At the bottom of the "Structural Members" options, click in the "Locate Profile" button. The part will zoom into the profile's sketch where we can select the point in the profile's sketch to be coincident to the sketch line. Any sketch endpoint in the profile can be selected as a pierce point.

Select the top middle point as indicated; the preview will update the profile's location to the desired position. Click OK to add the group in this location and continue. Do not worry about the members intersecting each other; we'll trim them as soon as we finish adding the last structural member group.

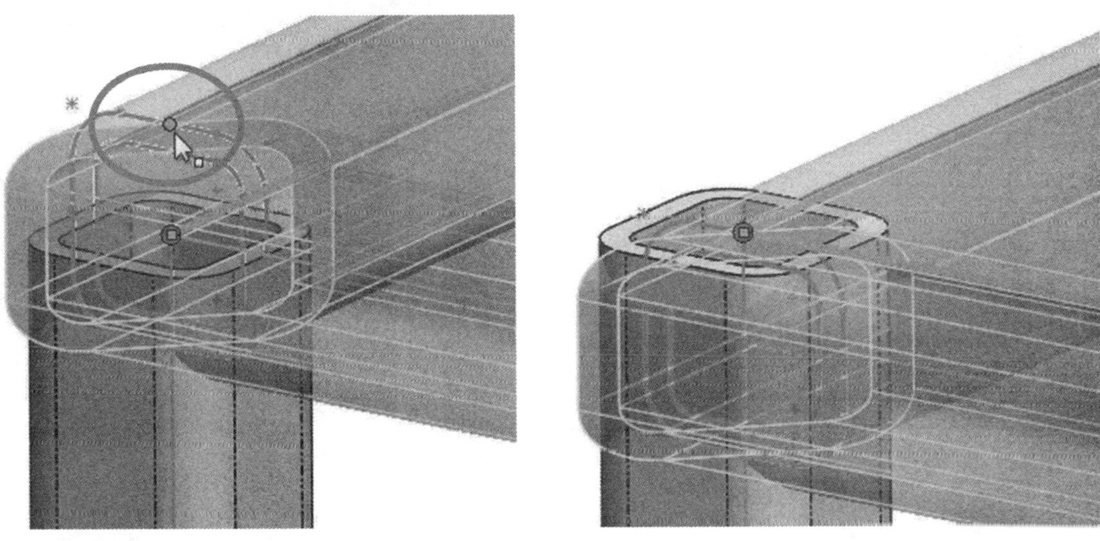

17.14. - For the next group, select the top middle line, and change the profile's location as we did in the previous step. Click OK to continue to the next group.

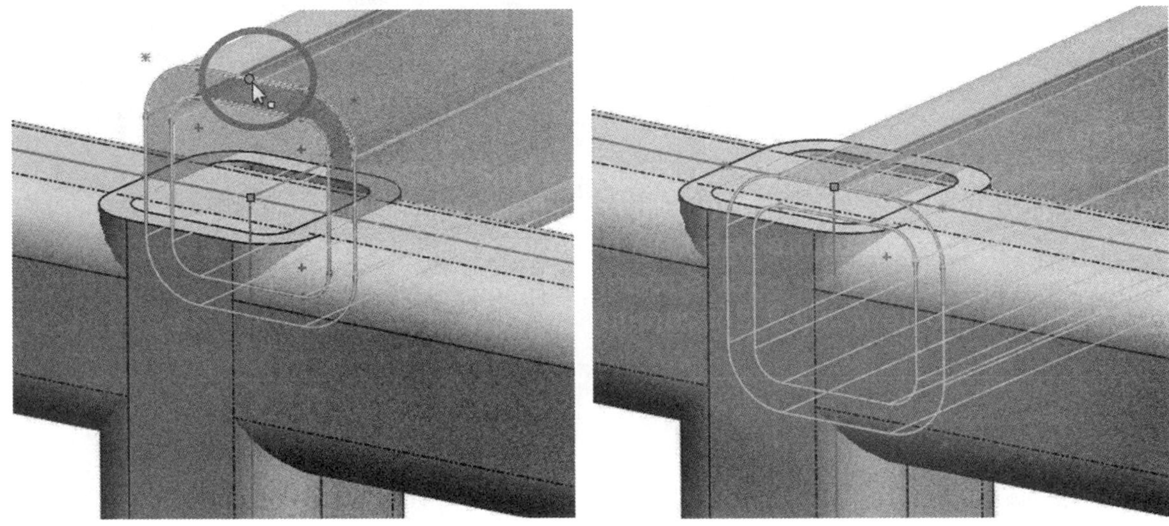

17.15. - For the next group, select two diagonal elements in one side. (We can only select two at a time since they are parallel and oriented in the same direction.) After selecting the sketch lines to add the structural members we notice the profiles are not correctly aligned. They are located in the centerline but are rotated. To change the profile's rotation and correctly align them to the structure, scroll down to "Rotation Angle," and enter "45" degrees.

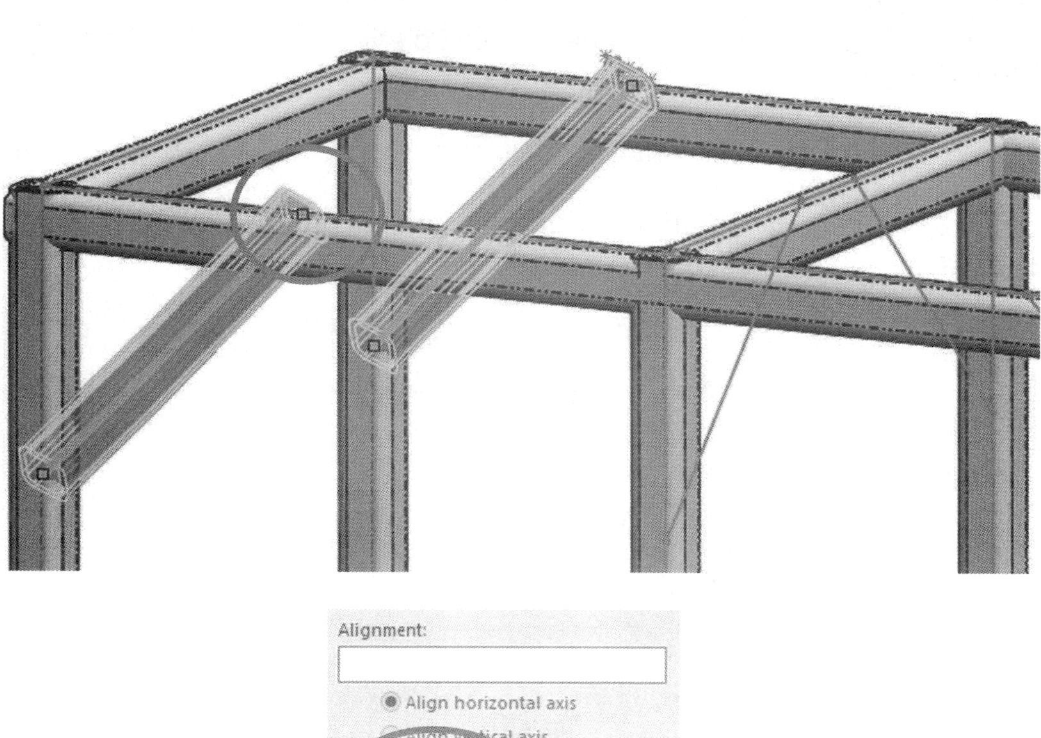

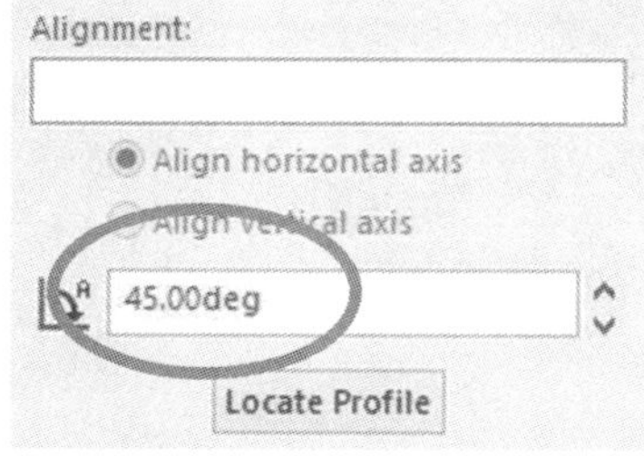

Other options to orient the profile include selecting a line or edge in the "Alignment" selection box and aligning the horizontal or vertical axis. We can also mirror the profile horizontally or vertically if needed.

17.16. - Repeat the same process with the reinforcements on the other side (another group of two) and finally make a group for each of the reinforcements in the center; since they are not parallel to each other or any other member, we can only add one at a time. Close the "Structural Member" command when done. Our structure now looks like this:

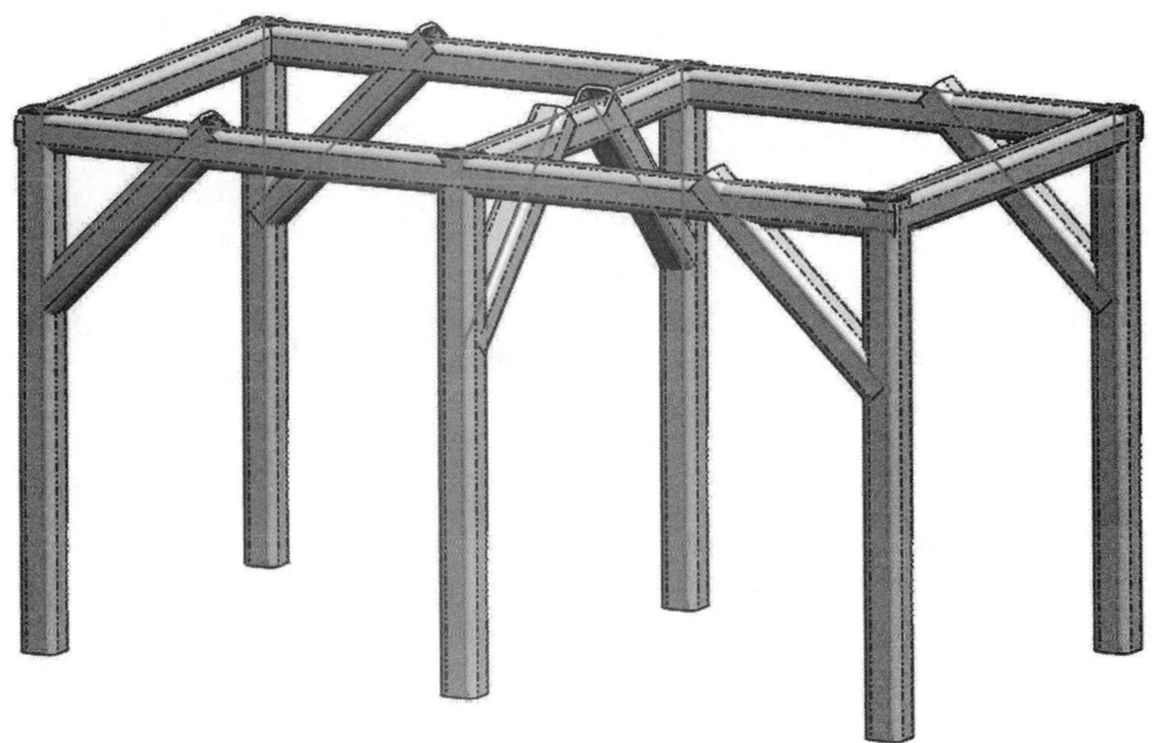

Notice the new "*Weldment*" feature automatically added to the Feature Manager after adding the Structural Members.

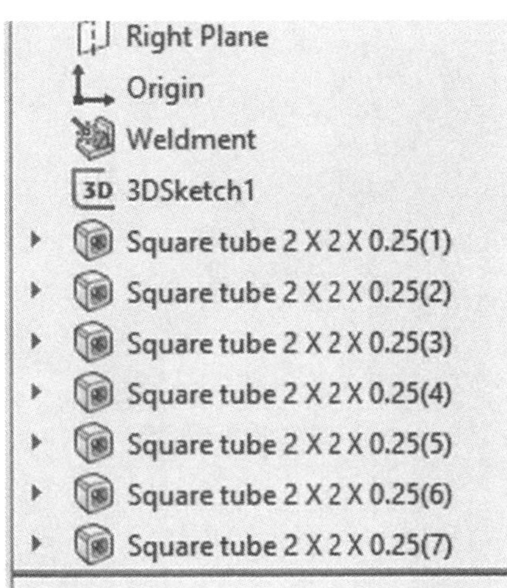

17.17. - Since the structural members are intersecting each other, the next step is to trim the excess material. From the "Weldments" toolbar, select the "**Trim/Extend**" command. We have the option to trim structural elements with one or more faces/planes or other bodies. In our example, we'll trim the structural members using a model face. Select the "End Trim" option to cut the bodies using a selected surface. In the "Bodies to be Trimmed" selection box, select two or four elements to trim at a time. (We can select all bodies to trim at the same time, but the screen will be cluttered with "keep" or "discard" labels.)

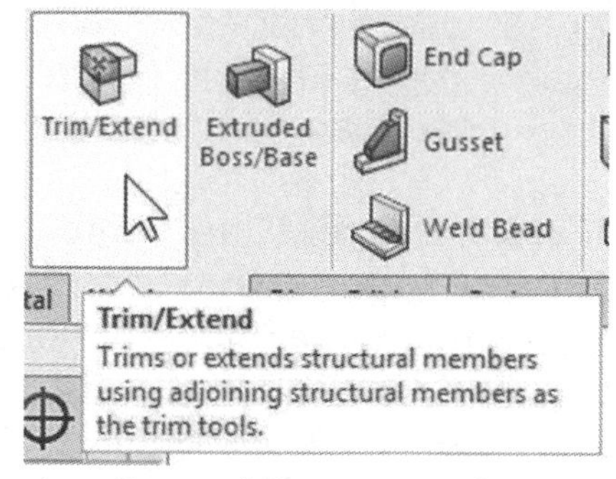

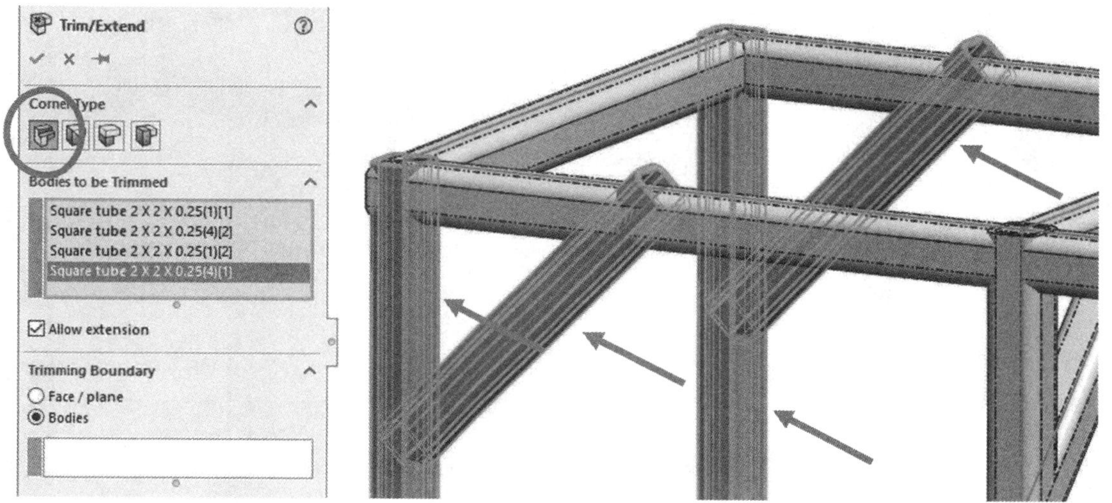

In the "Trimming Boundary" selection box, select the "Face/Plane" option and click the face under any of the top horizontal members to use it as the trim boundary. Rotate the model to select the correct face from below if needed.

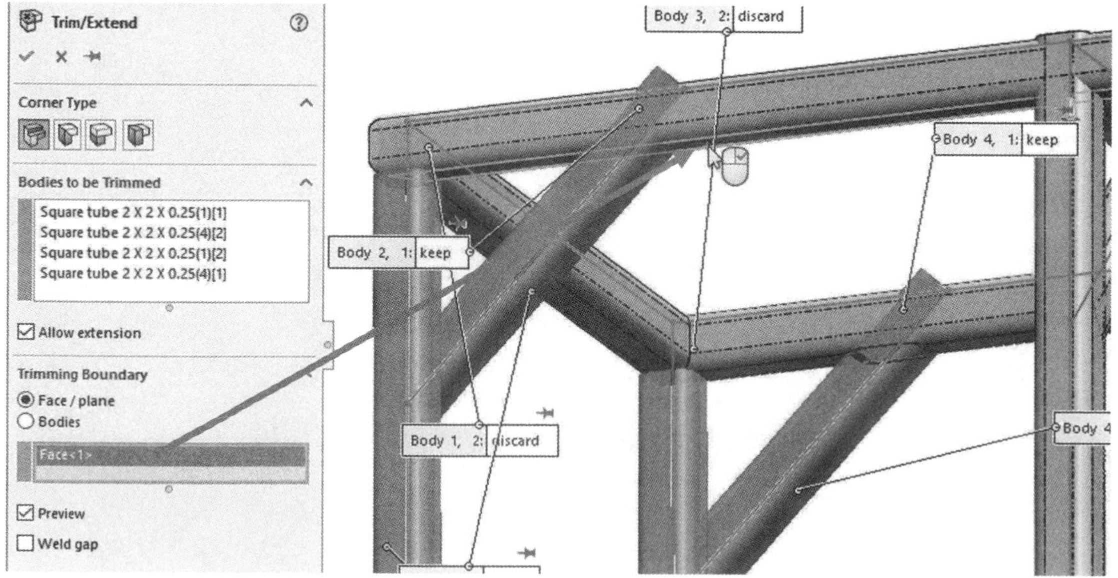

17.18. - After selecting the trim face, the bodies are 'cut' in the preview and we have to review which bodies we want to keep and which bodies to discard. Check each member being cut. Review the pop up labels of each segment to toggle the "keep" or "discard" message by clicking in it. The labels can be moved around for clarity. By default, smaller segments are *usually* marked correctly to be discarded. The discarded element's preview is transparent making it easier to identify which elements to delete. After correctly marking the elements to be discarded, press the "Keep visible" push pin in the "Trim/Extend" command and click OK to continue trimming the rest of the structure. Trim the top part of the other elements using the same trimming face.

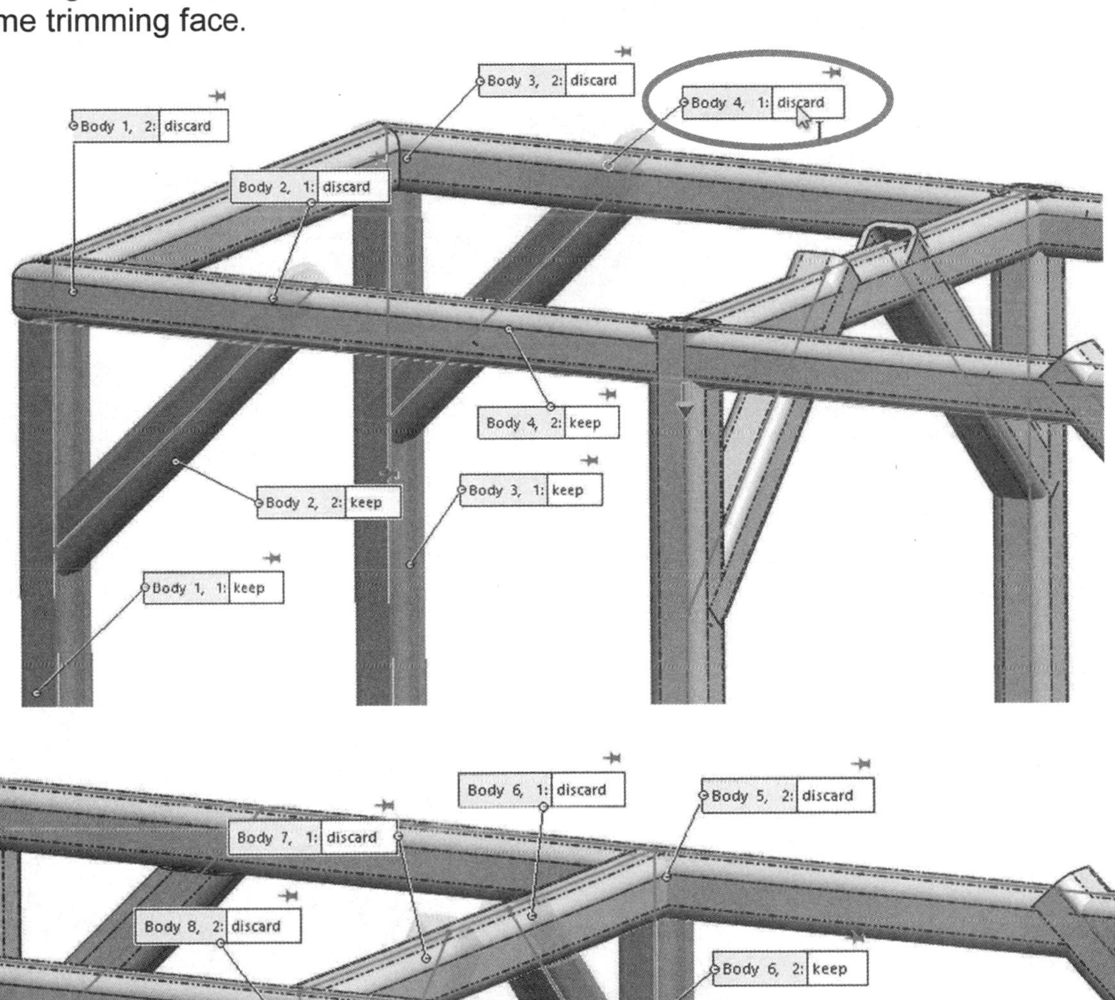

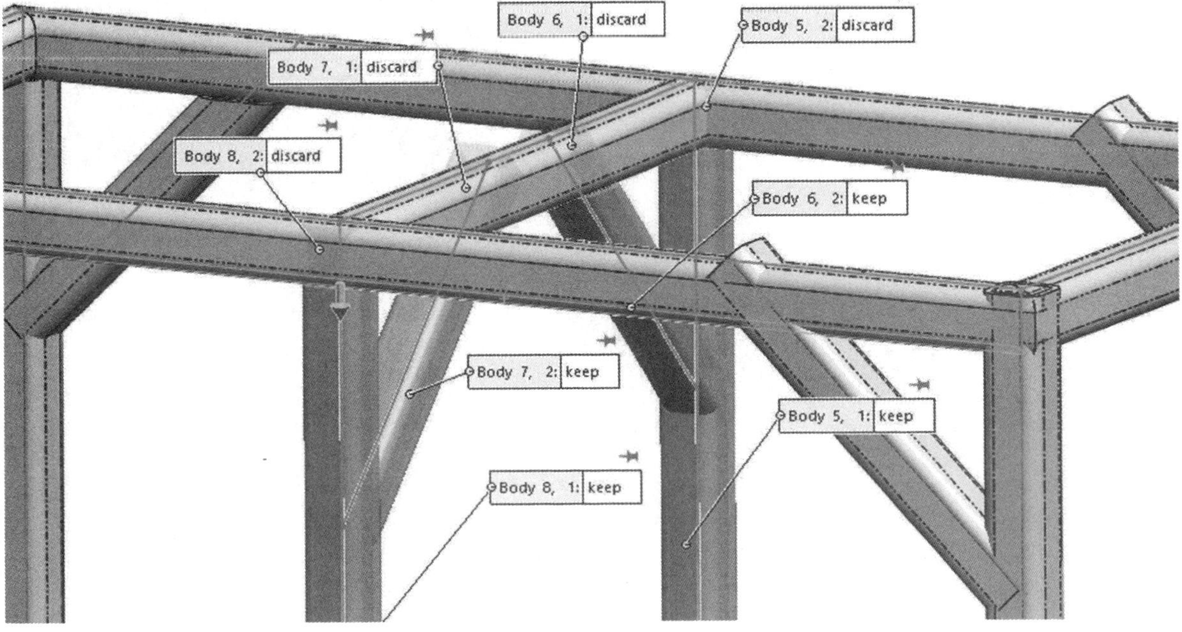

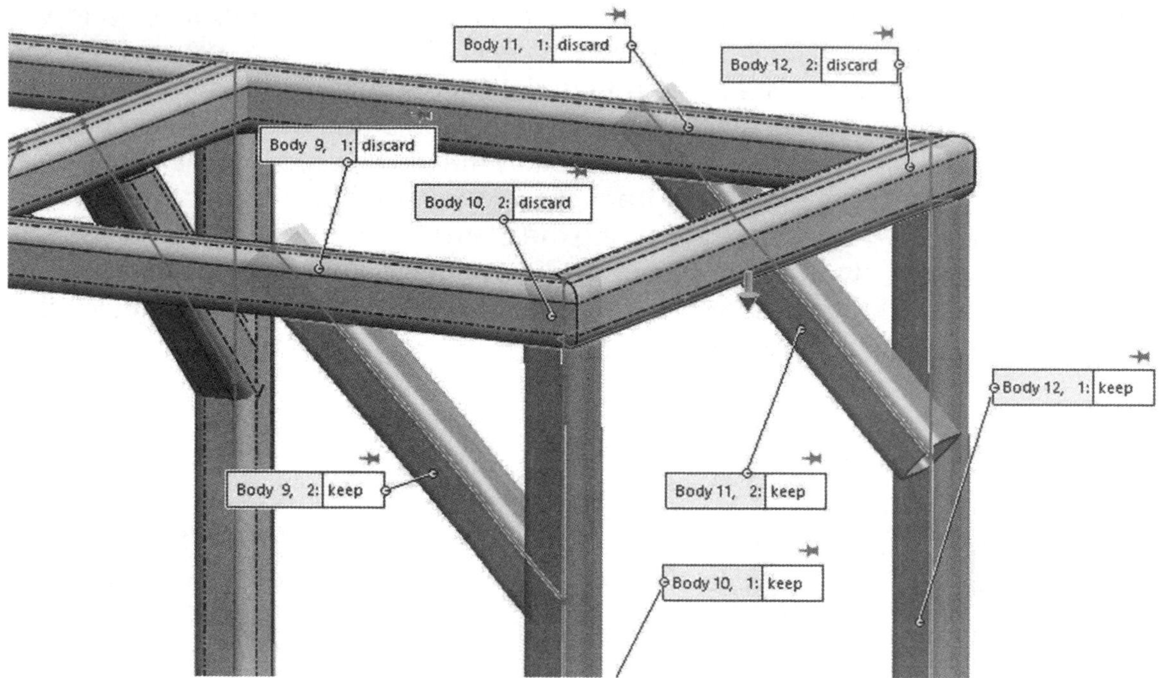

We could have chosen to trim all the legs at the same time, but the screen would have been too busy with all of the "keep/discard" labels; that's why it was shown with just a few elements at a time.

17.19. - Trim the lower part of the diagonal reinforcing elements using the faces indicated.

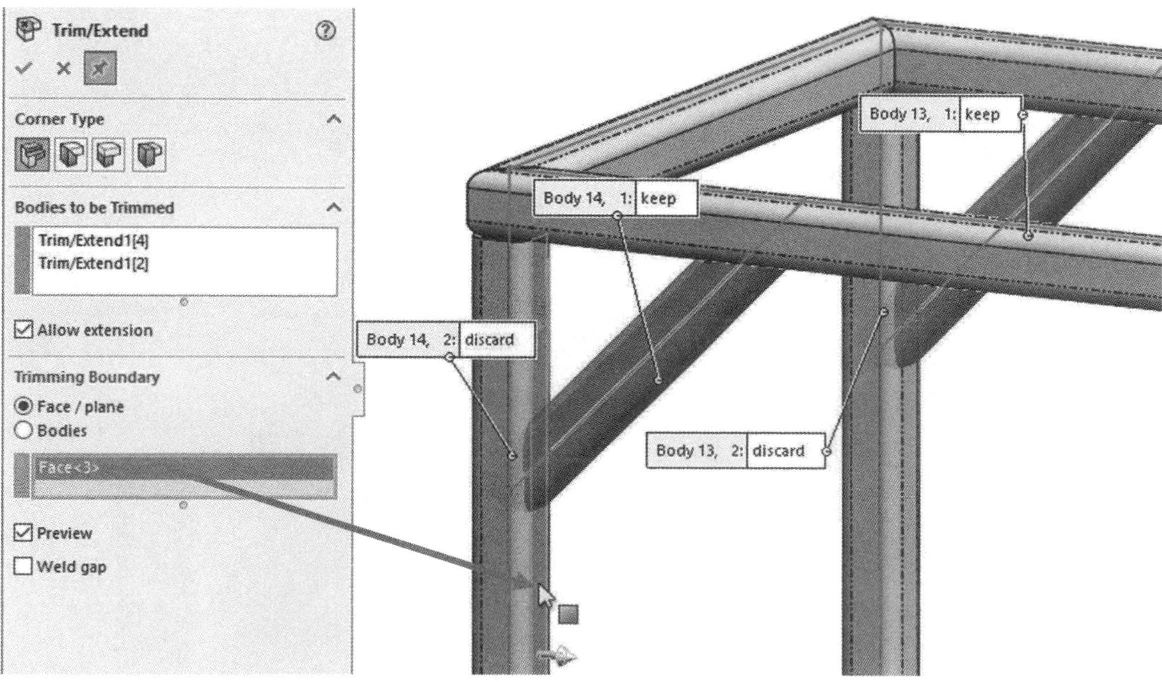

Trim the other side reinforcements.

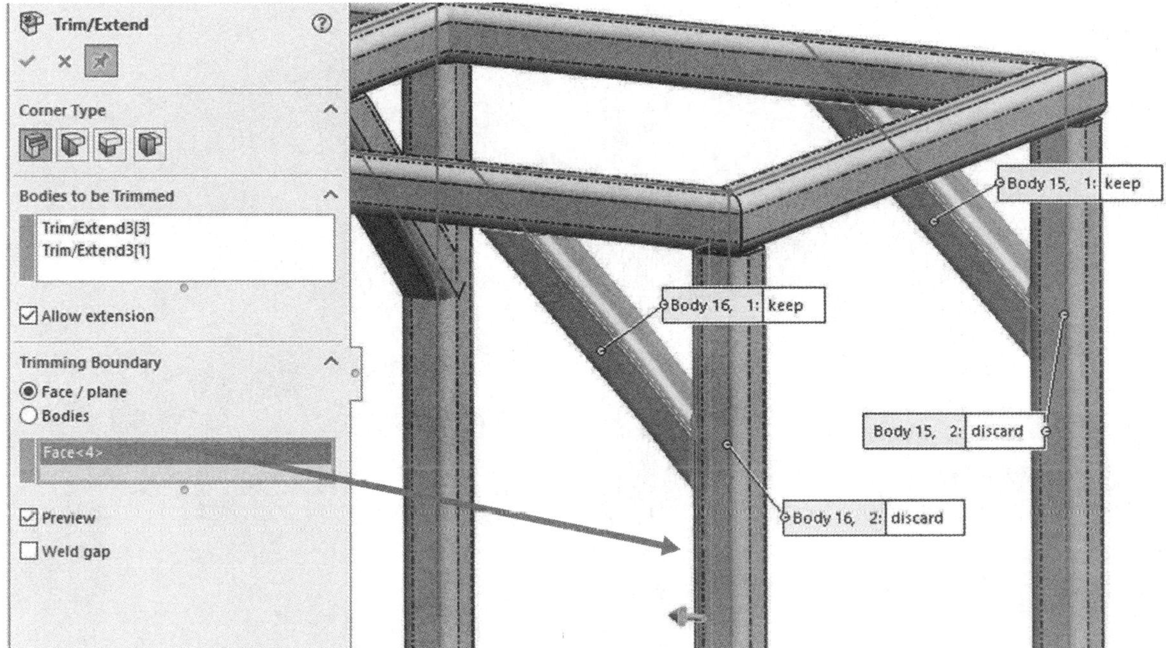

Trim the center bodies on one side,

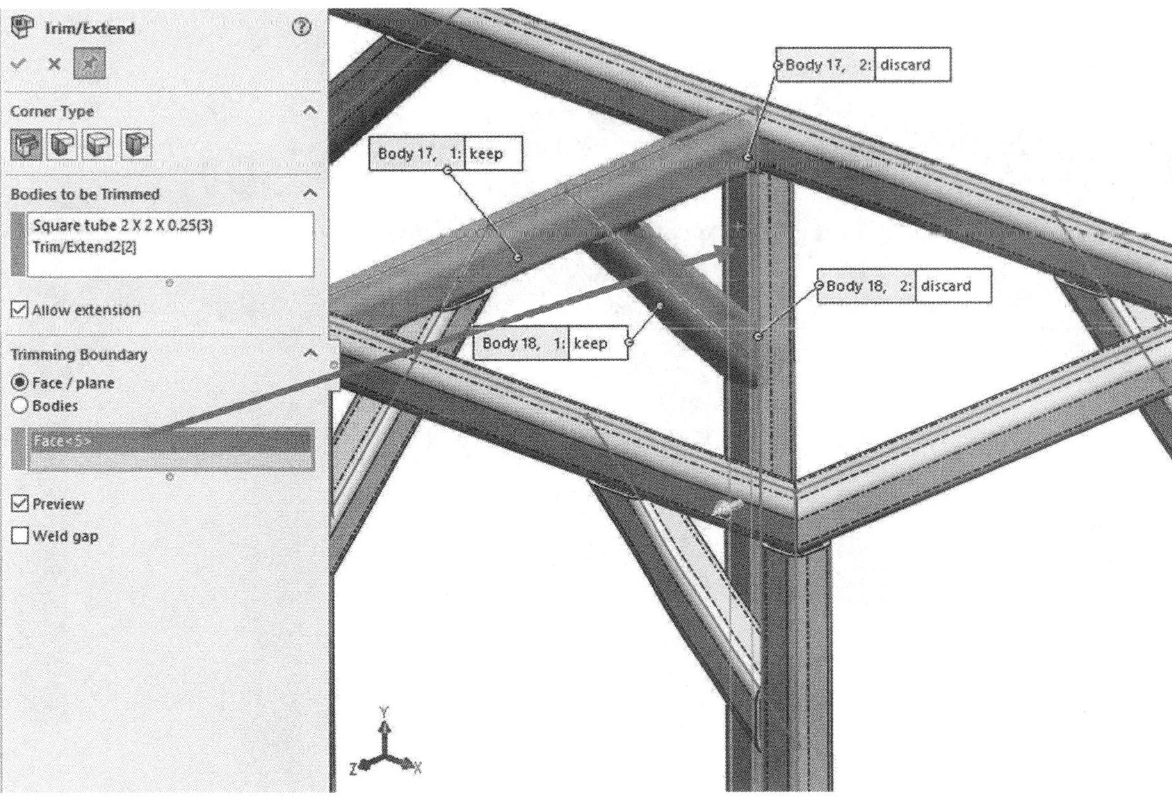

And then trim the other side of the center bodies. Close the "**Trim/Extend**" command when done to finish.

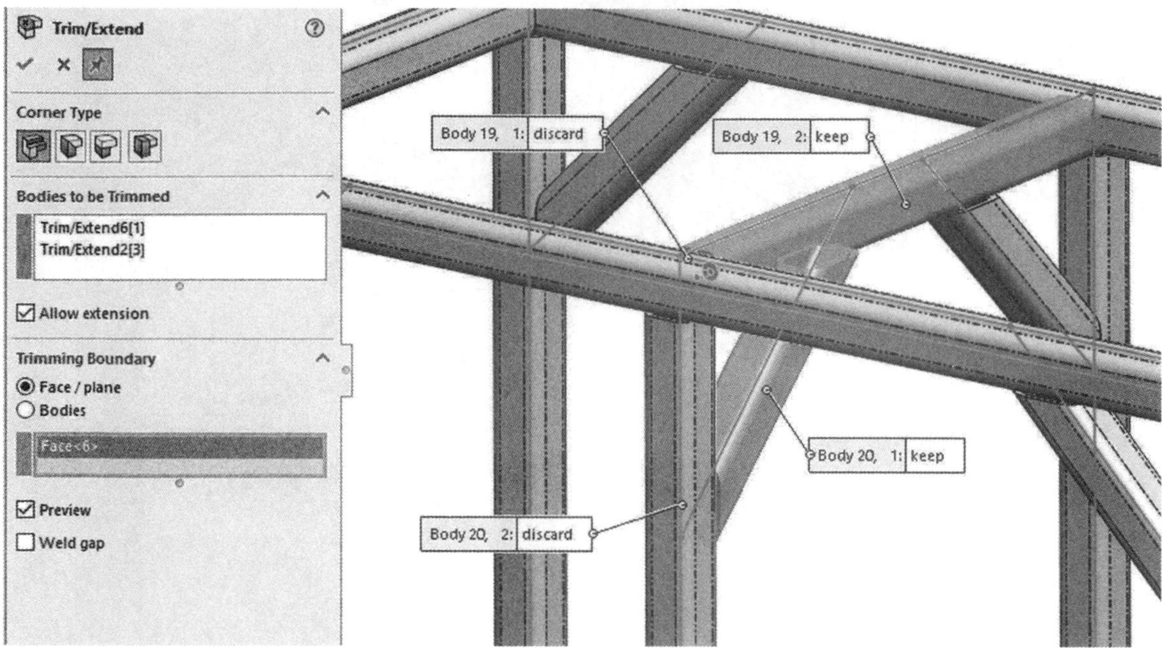

Our structure is now completely trimmed and no bodies intersect each other.

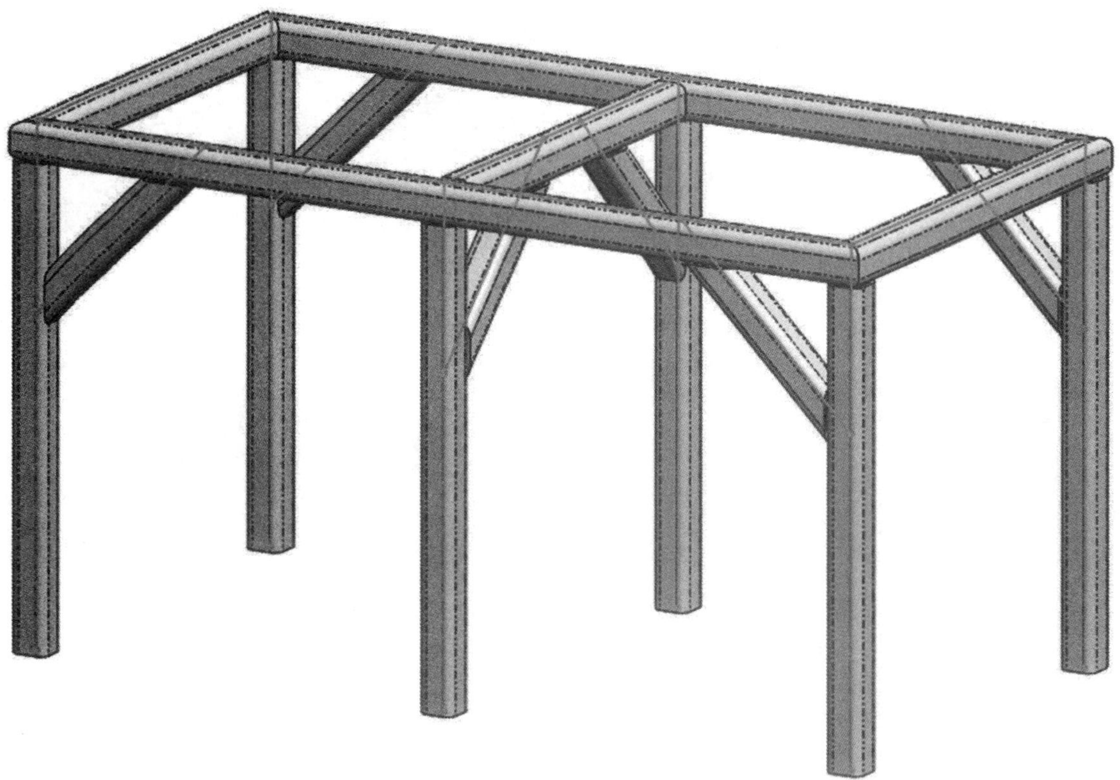

17.20. – After learning how to add structural elements and trim them, we'll cover the second technique to reduce, or even eliminate trimming altogether. Delete all structural members and trim operations, and only keep the original 3D sketch.

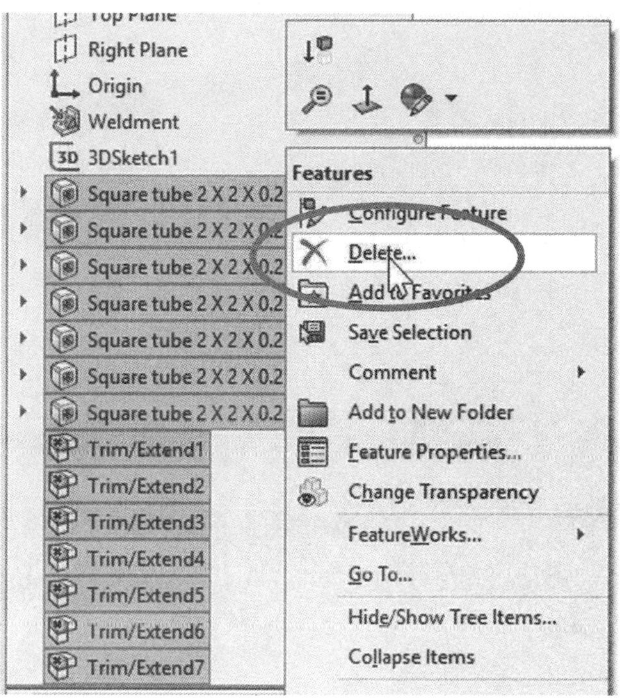

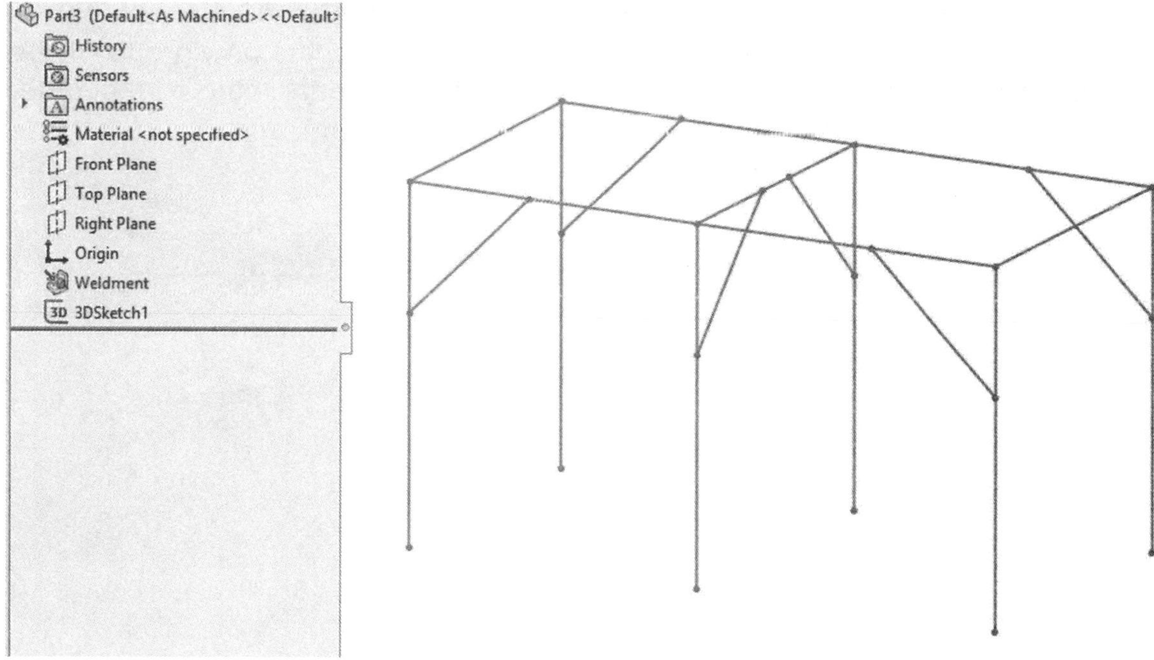

17.21. – Select the "**Structural Member**" command, and using the same settings as before select the top four elements, locating the profile to match the top of the structure to make the table 28" high, using the end miter corner treatment.

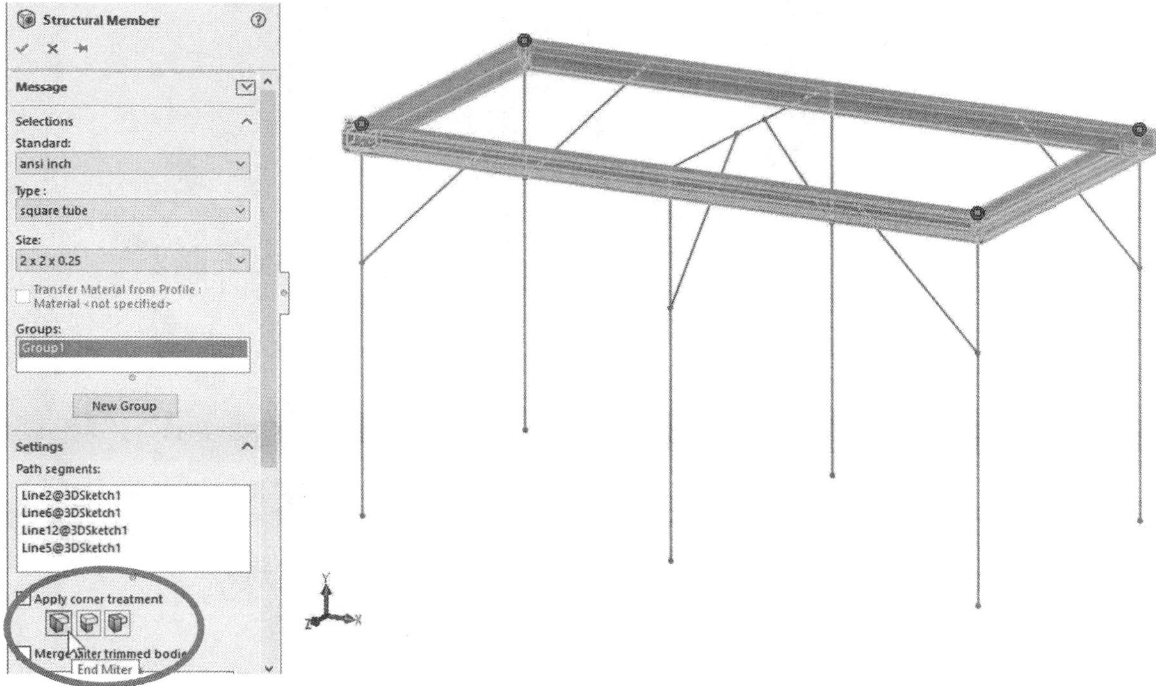

Now, instead of finishing the command to add the next elements group, select the "New Group" button. By adding groups this way, the new group will be automatically trimmed to existing groups. Adding the elements in the correct order can potentially eliminate trimming.

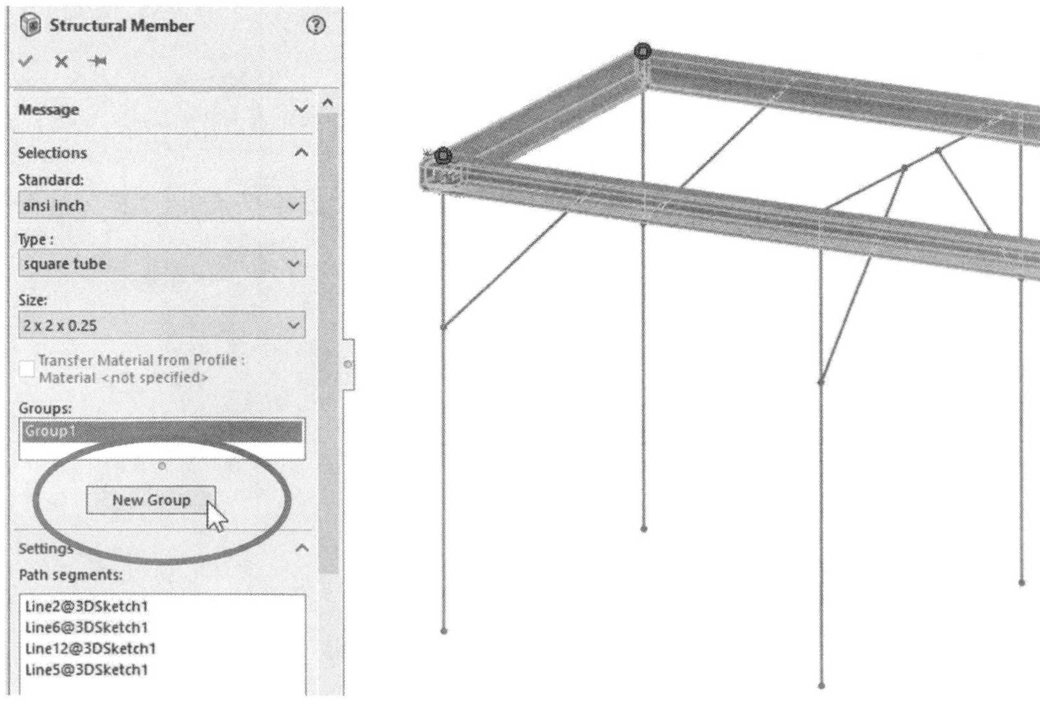

After adding the first group, its elements become translucent and we can start working on the second group. For the second group select the six vertical elements for the structure's legs. After selecting each sketch segment, the newly added elements are automatically trimmed to not overlap any previous group.

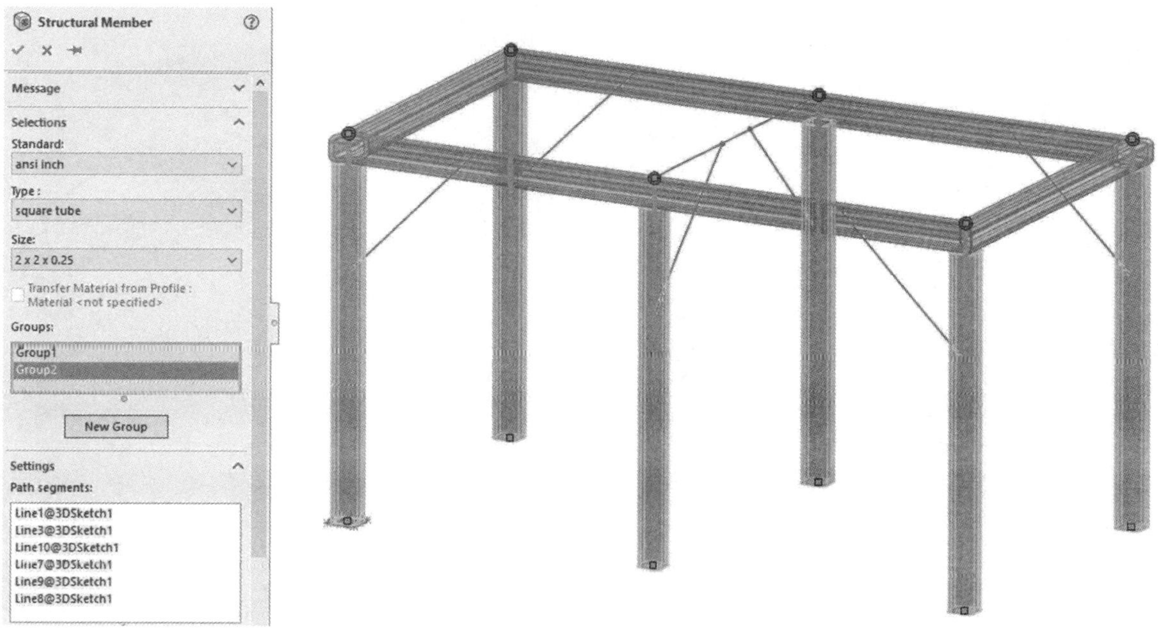

17.22. – Select the "New Group" button again, and this time select the horizontal element at the top of the structure. Remember to locate the profile to match the top of the structure. As with the legs, it is also trimmed to size to other elements.

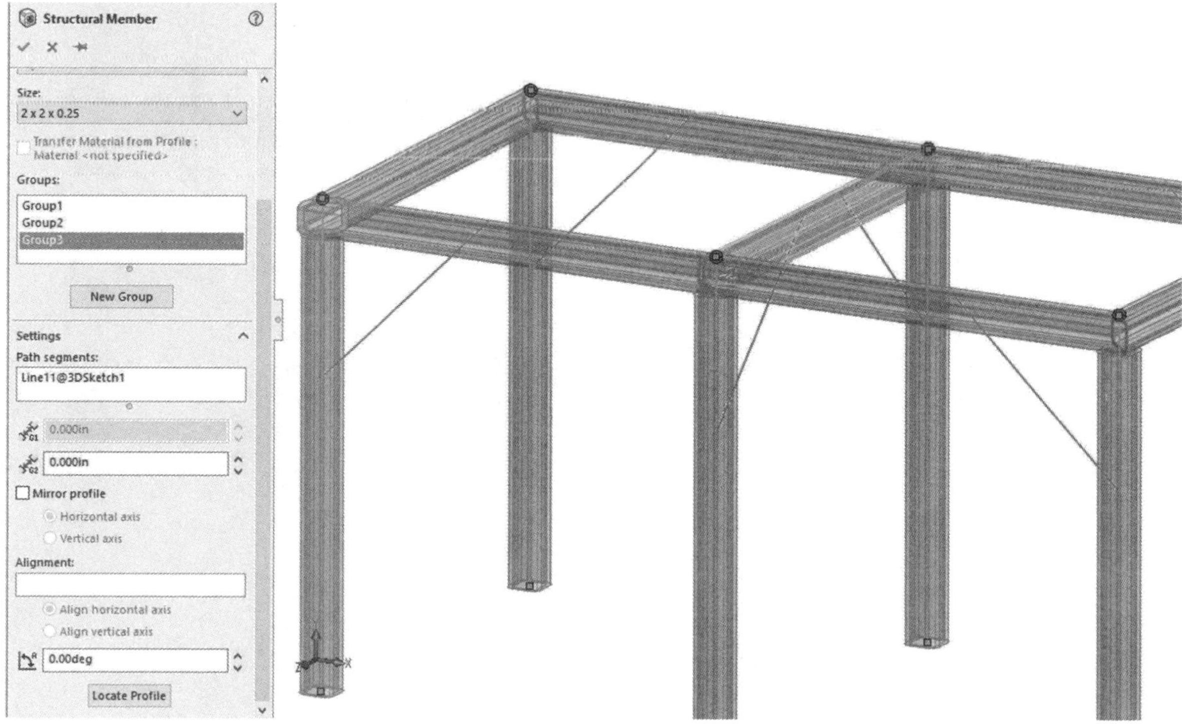

17.23. – Select the "New Group" button and continue adding the rest of the elements. Remember to rotate the elements in the corner 45°, and add the middle reinforcement elements one at a time.

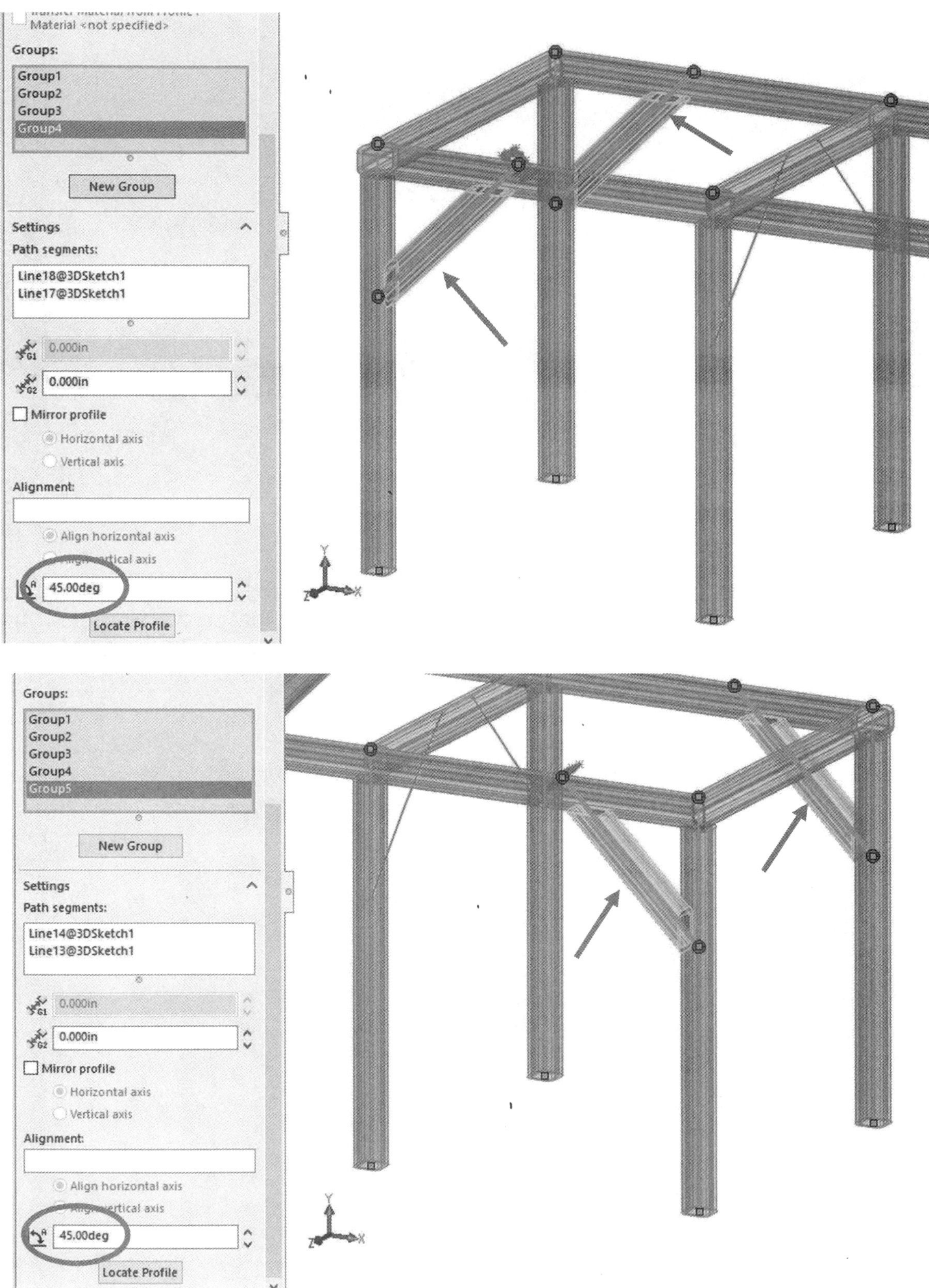

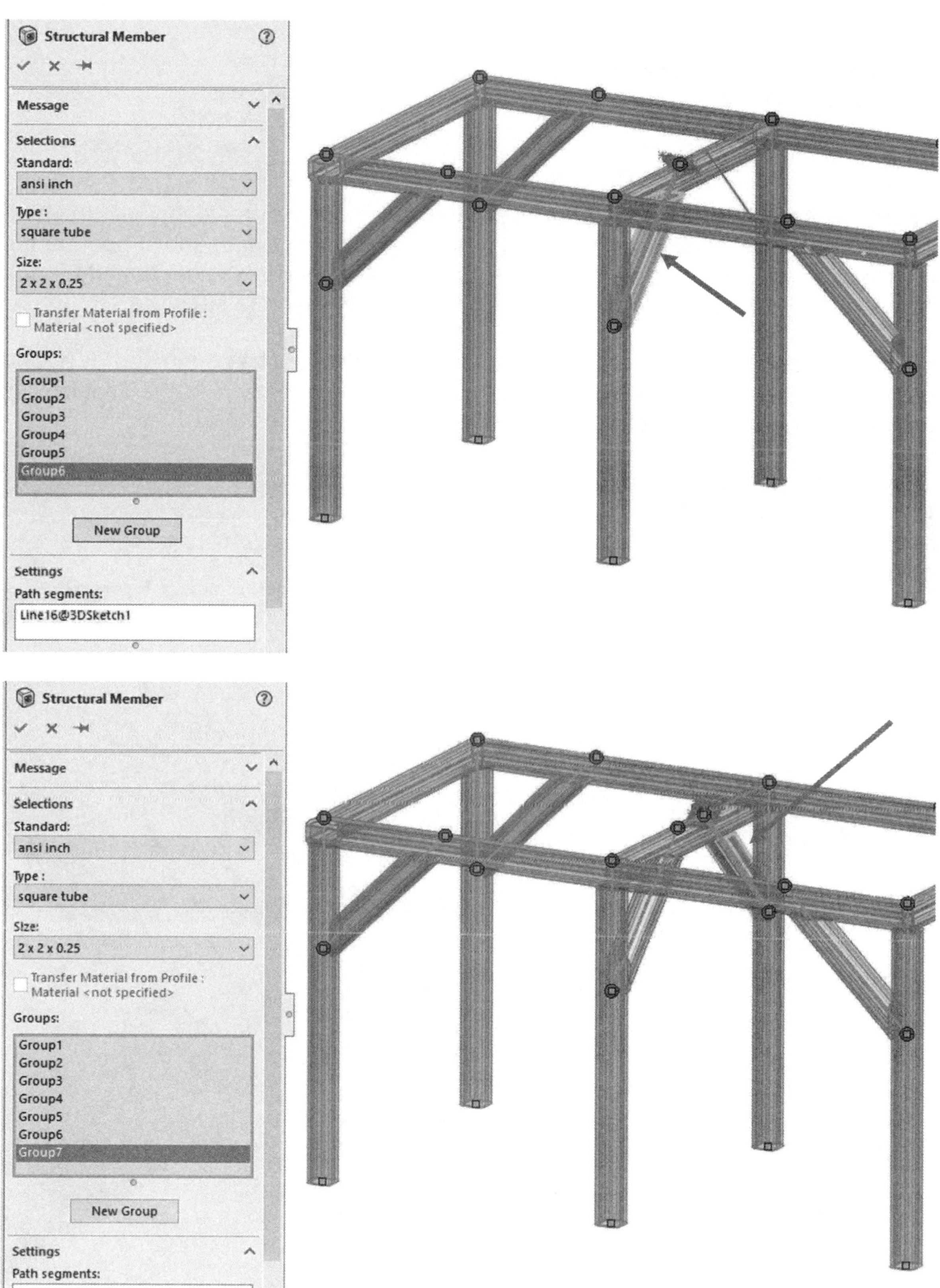
Structural Member
Message
Selections
Standard:
ansi inch
Type :
square tube
Size:
2 x 2 x 0.25
Transfer Material from Profile :
Material <not specified>
Groups:
Group1
Group2
Group3
Group4
Group5
Group6
New Group
Settings
Path segments:
Line16@3DSketch1
Structural Member
Message
Selections
Standard:
ansi inch
Type :
square tube
Size:
2 x 2 x 0.25
Transfer Material from Profile :
Material <not specified>
Groups:
Group1
Group2
Group3
Group4
Group5
Group6
Group7
New Group
Settings
Path segments:
Line15@3DSketch1

After adding all elements in this order using this method there is no excess material to trim and only one feature is created using the same settings.

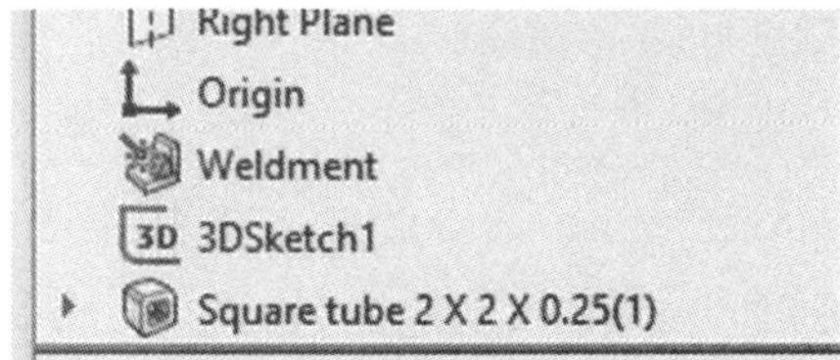

17.24. – After completing the structure to this point using either technique, the next step is to add gussets. Gussets are plates used to add strength and support at the intersections of structural components. For our example, we'll add gussets to the corners, and after that, weld beads will be added to finish the structure.

To add a gusset, we must select the faces that will be connected by it, and define the gusset's dimensions. Click the "**Gusset**" command in the Weldments toolbar, add the faces indicated below in the "Supporting Faces" selection box, activate the "Chamfer" option and enter the dimensions, thickness and location shown. Zoom in a corner from below to make selection easier.

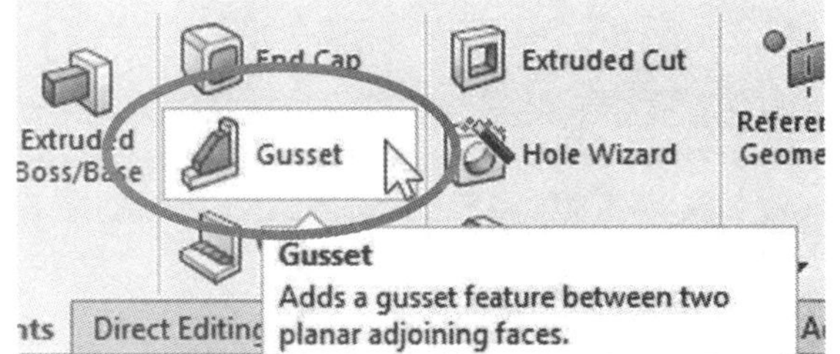

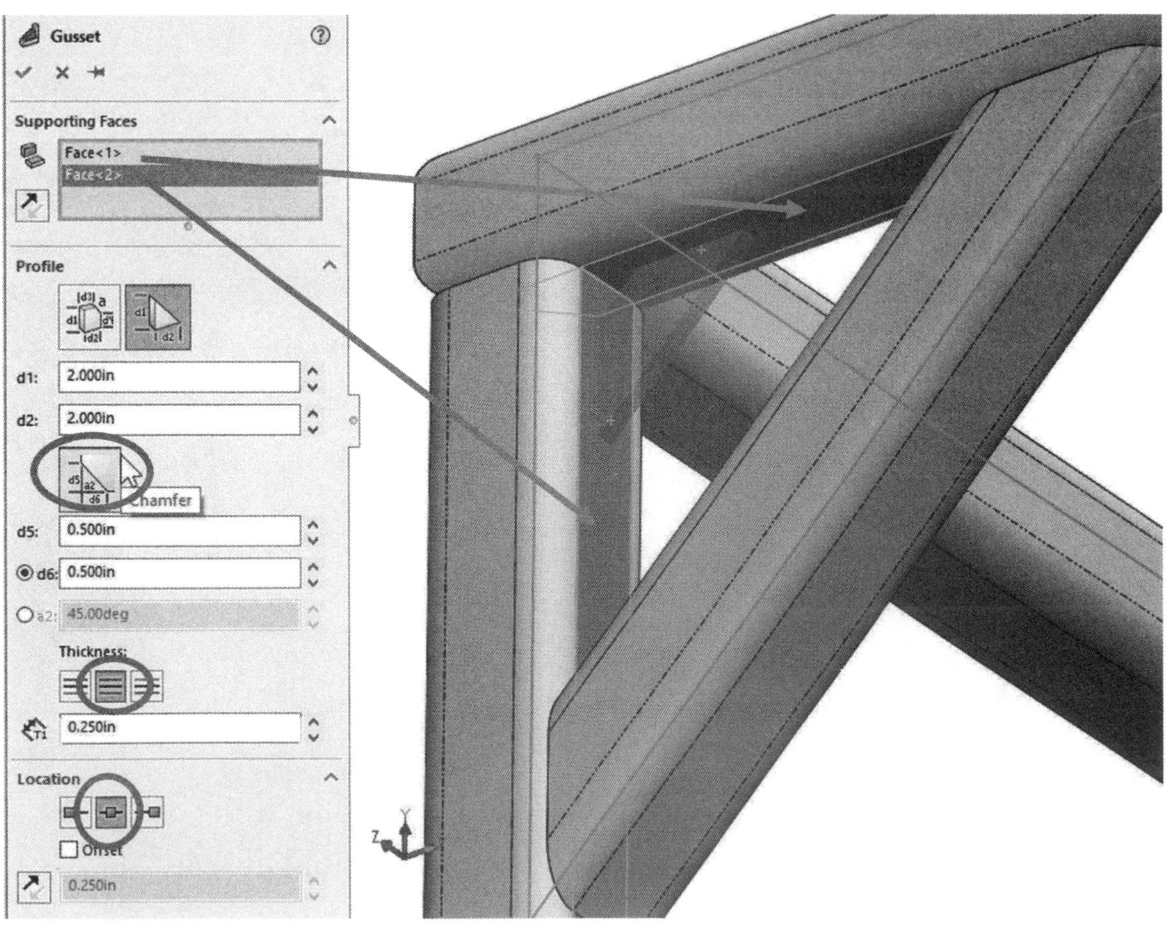

We can make polygonal or triangular gussets. In our example, we'll use the triangular option and dimension it 2″ x 2″ (measured from the corner).

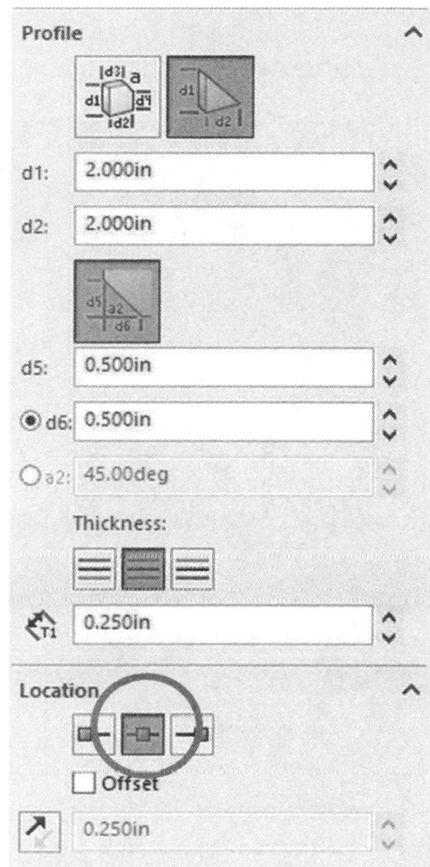

A chamfer is usually added to allow space for a weld bead in the corner and a better fit. Activate the "Chamfer" button and set the chamfer dimensions (*"d5"* and *"d6"*) to 0.5″ x 0.5″. (Chamfers can also be defined by a dimension and an angle.)

The gusset's thickness can be added on either side or both sides. We'll make our gusset 0.25″ thick using the midplane option ("Both Sides").

Finally, the "Location" option for the gusset can be set on one side of the face, the center, the other side, or at an offset from any of them. We'll make ours in the center. Press the "Keep Visible" push pin to keep adding gussets at other corners. Press OK to add the first gusset and continue.

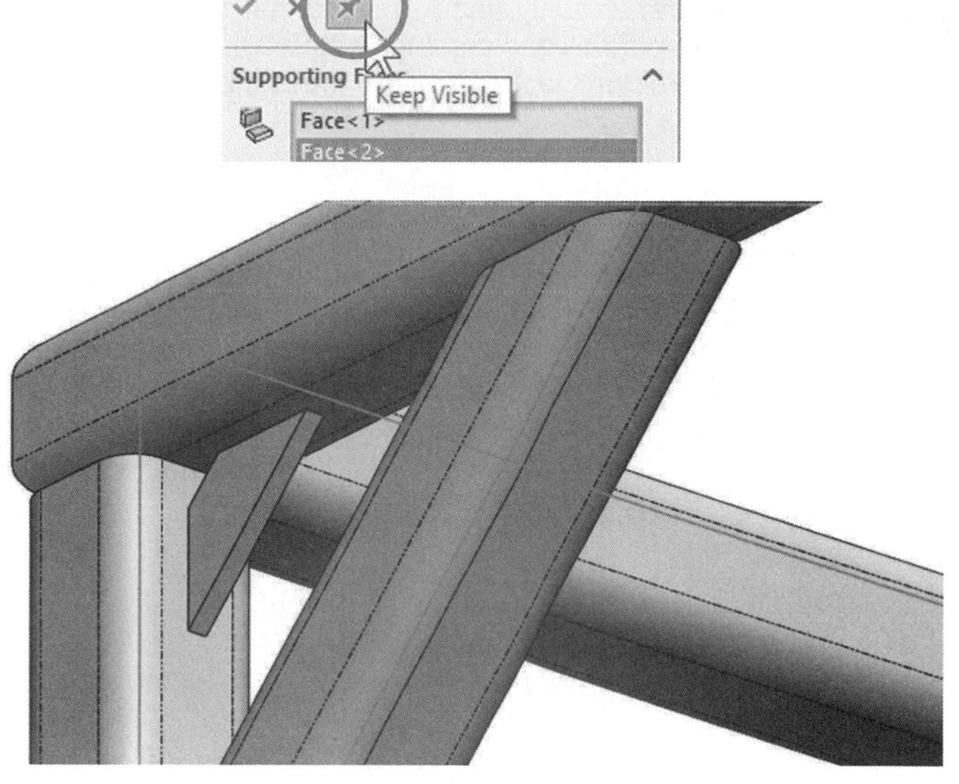

Gussets can be added using all possible combinations of thickness, location and offset options.

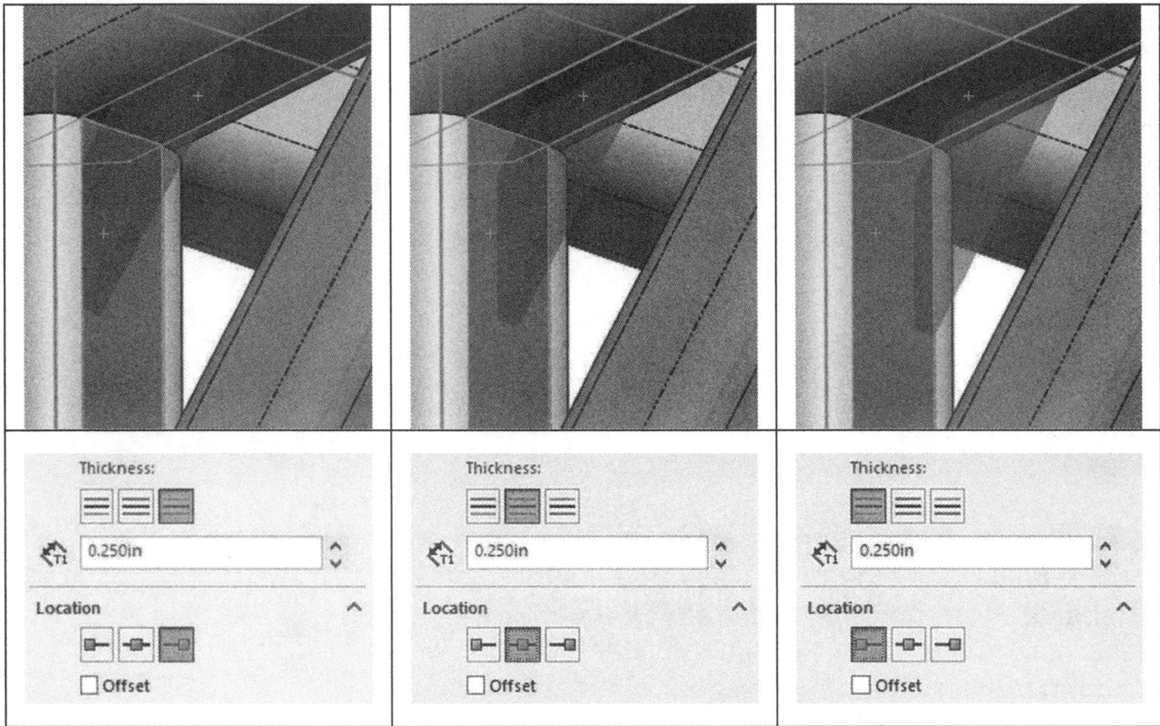

17.25. - Add the remaining gussets to the other inside corners using the same settings as the first one. After adding a gusset SOLIDWORKS remembers the previous settings; therefore, we only need to select the next set of faces for the new gusset and press OK to add it and continue until we are done. After all gussets have been added, the structure looks like this (gussets highlighted, 3D Sketch hidden):

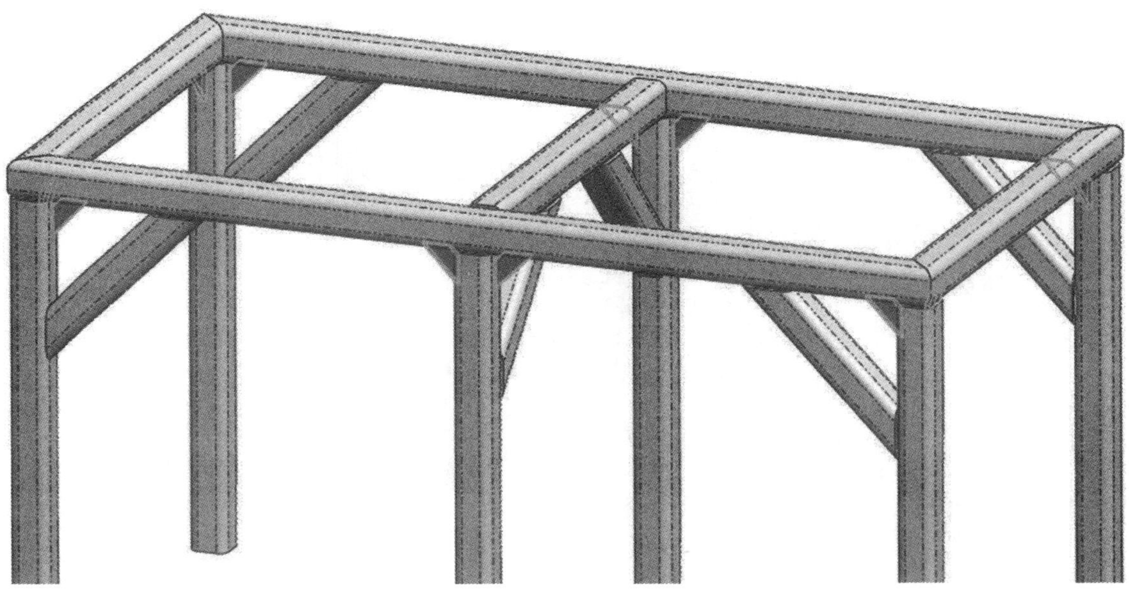

17.26. - Another feature unique to weldments is the "**End Cap**." An end cap will add a cover to close the open end of a structural member. Rotate the part to look at it from below and zoom at the end of one leg. Click the "**End Cap**" command in the Weldments tab and select the end face of a leg. For our example the end cap's thickness will be set to 0.125″. The **inward** option in "Thickness Direction" reduces the structural element's length by the end cap's thickness, essentially maintaining the original element's length including the end cap.

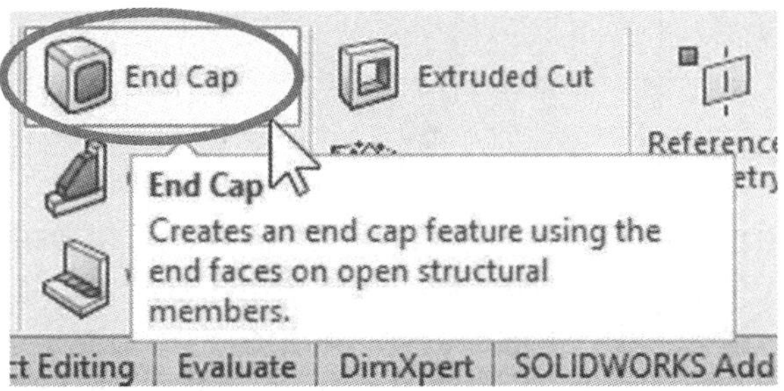

Set the offset ratio to **0.5** (half the structural element's thickness), activate the option "Corner Treatment" and select the "Chamfer" option with a value of 0.125″. Select the six legs' end faces to add all end caps with one command and click OK to finish. (A selection filter for faces is automatically activated while using the command, making it easier to pick the end face of the legs.)

The "Offset" option refers to the distance from the edge of the end cap to the edge of the structural member, essentially making it smaller than the structural member's profile, and is typically used to make room for a weld bead. This offset can be defined using a ratio of the structural member's thickness or a specified distance by turning on the "Offset Value" option.

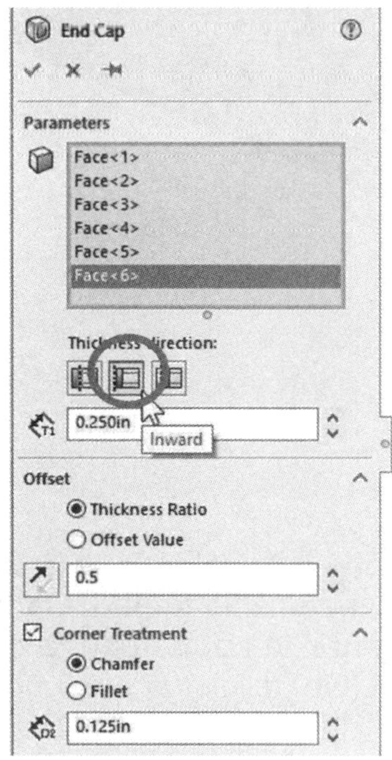

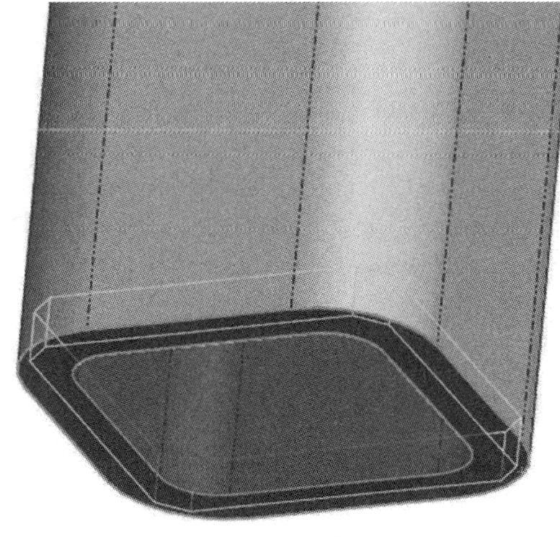

A note about "Thickness direction": If we make the end cap going "Outward," the cap's thickness is added to the length of the structural member. If we make it "Inward," the structural member will be shortened by the end cap's thickness, as in this case, to maintain the original height.

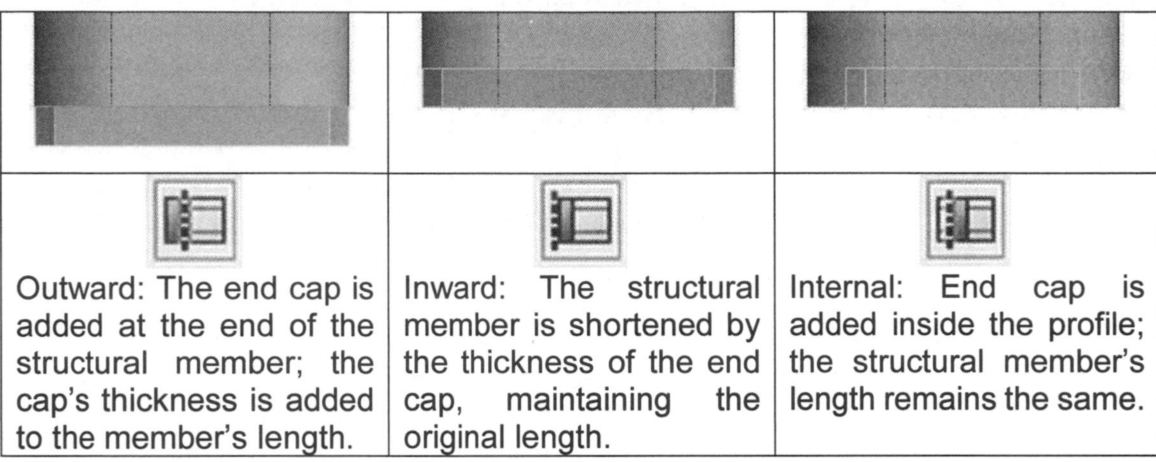

Outward	Inward	Internal
Outward: The end cap is added at the end of the structural member; the cap's thickness is added to the member's length.	Inward: The structural member is shortened by the thickness of the end cap, maintaining the original length.	Internal: End cap is added inside the profile; the structural member's length remains the same.

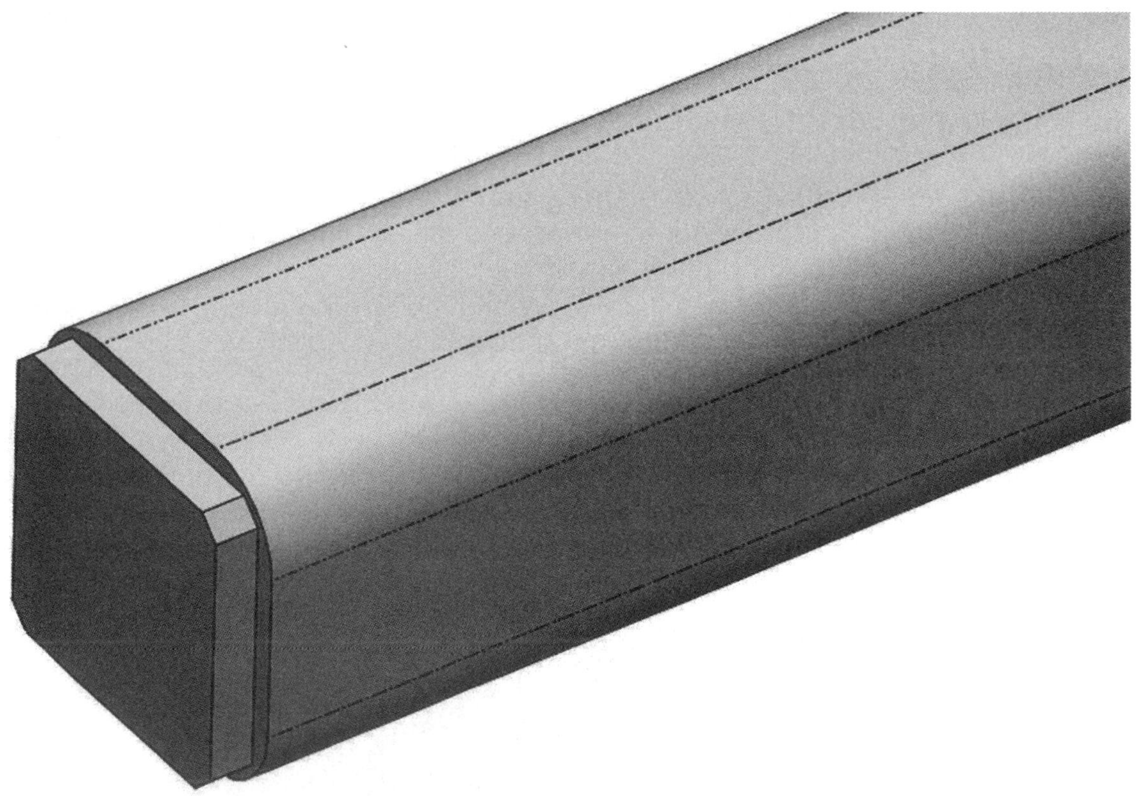

An important detail to keep in mind when we work with weldments is that the "Merge result" option in all features that add material is turned 'Off' by default. If we add an extrusion (any type) and want it to be merged to one or more bodies, we must turn the "Merge result" option 'On' following multi body part rules.

17.27. - To complete the structure, we'll add weld beads and the corresponding welding annotations. Some advantages of adding weld beads to our model include automatic creation of welding annotations for drawings, welding symbols for manufacturing, etc. Select the "**Weld Bead**" command from the Weldments tab or the menu "**Insert, Weldments, Weld Bead.**"

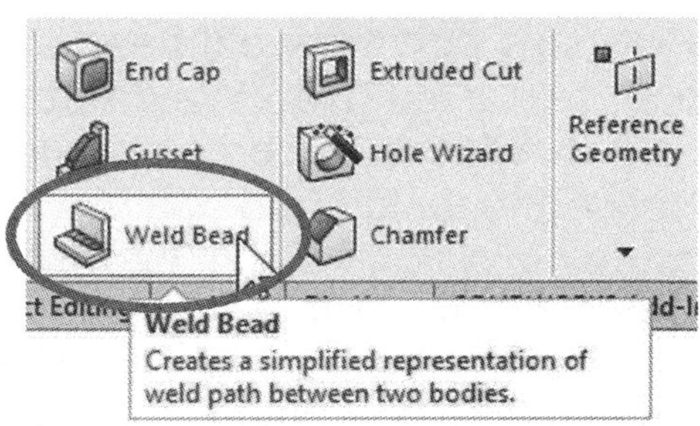

Zoom in a corner looking at it from the bottom. There are two ways to add a weld bead: using the "Weld Geometry" or "Weld Path" option. The main differences between them are:

- Using "**Weld Geometry**" we can select intersecting faces of different bodies, or a face and an edge of the same or different bodies to locate the weld bead. If we choose to select faces we can use the "Smart Weld Selection Tool."
- Using "**Weld Path**" we select edges that the weld bead will follow, either from the same or different bodies. The "Smart Weld Selection Tool" is not available in this option.

Using the "Weld Geometry" option, select the two faces indicated to add a weld bead at their intersection; notice the weld bead preview at the intersection of the faces to be welded. Set the weld bead size to 0.25″ and turn on the "Tangent Propagation" option. In this case we want the weld to go around the sides; it only goes half way because the weld bead can only be added between two bodies, and in this corner we have three bodies. Notice the preview goes around the tangency where both bodies touch. If we had merged the mitered bodies when the structural member was done, the weld bead would go all the way around.

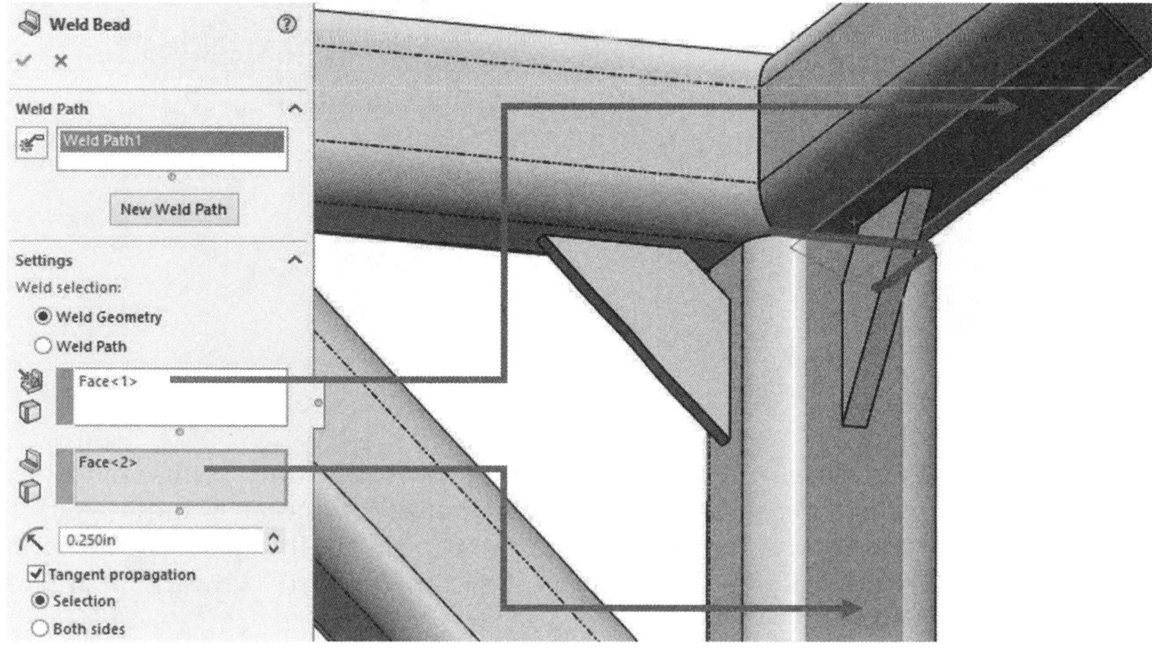

To add the weld bead and proceed to the next one, select the “New Weld Path” button and select the next two faces to complete the weld around the leg. Notice the weld bead settings remain the same. If we continue adding weld beads that share the same settings, they will be grouped together. Press the “New Weld Path” again to add the next two welds using the same settings, one at a time.

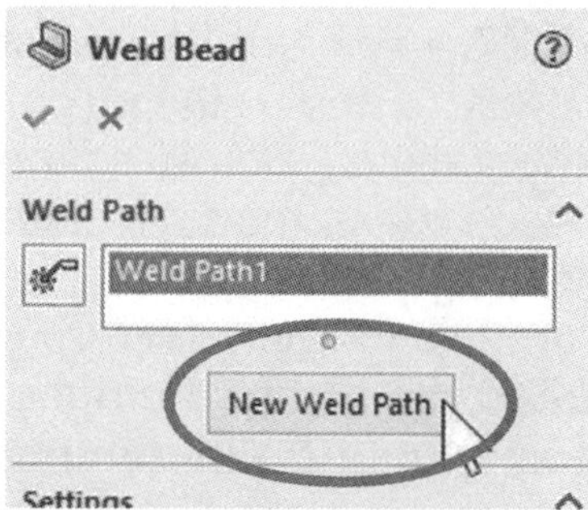

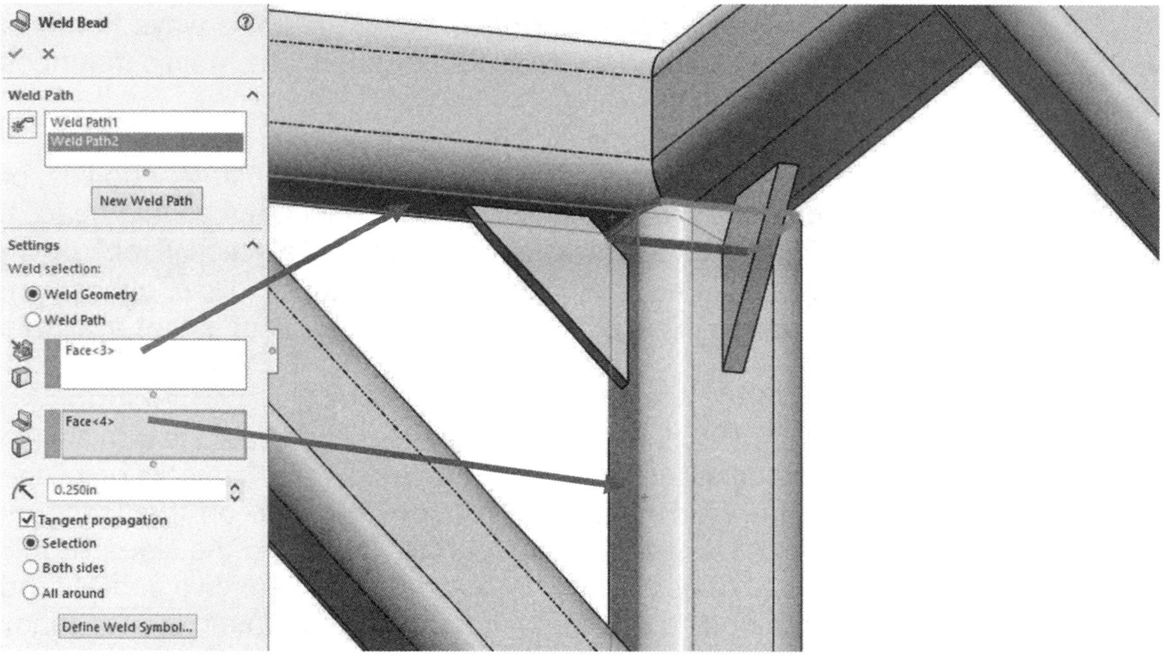

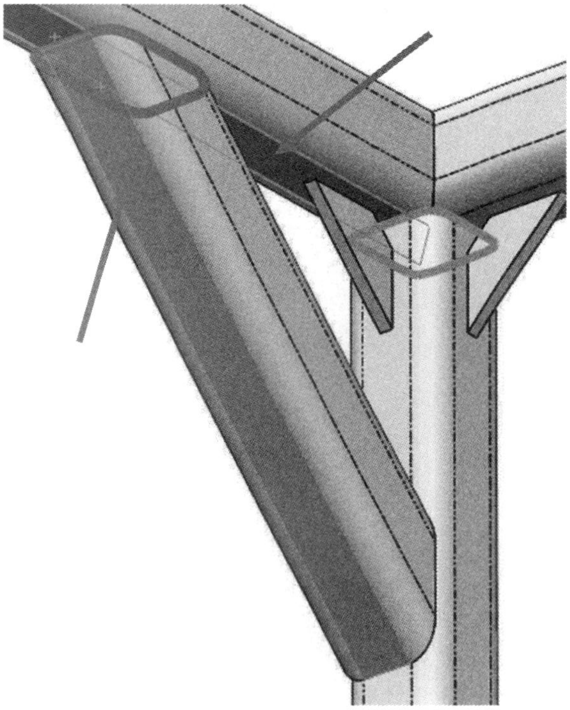

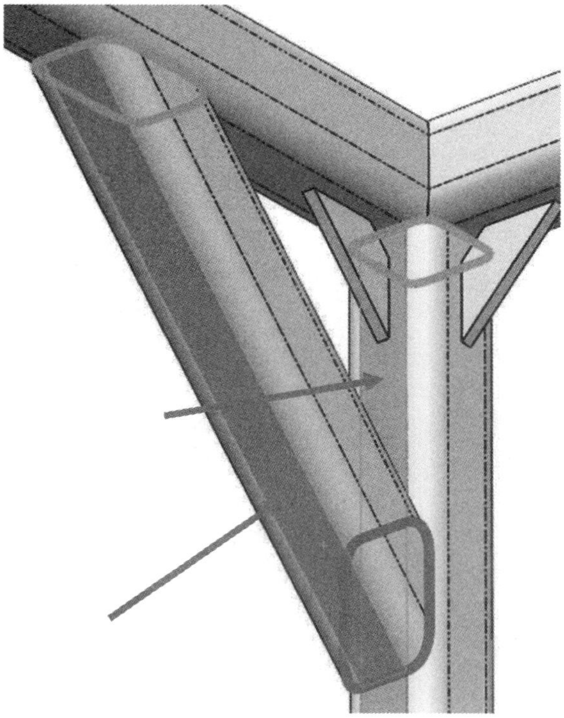

17.28. - We can keep adding the remaining weld beads selecting the faces, but now we are going to learn how to use the "Smart Weld Selection Tool." Click in the "New Weld Path" to add the next weld bead, and zoom in the second leg of the structure. Instead of selecting faces individually, we'll use the **Smart Weld Selection Tool**.

After activating the "Smart Weld Selection Tool" the mouse pointer changes to a pencil icon.

To use this tool, we need to click-and-drag across the intersection we wish to add a weld bead to, going from one face to the other. Notice that as we make a new selection, a new 'Weld Path' is automatically added to the list.

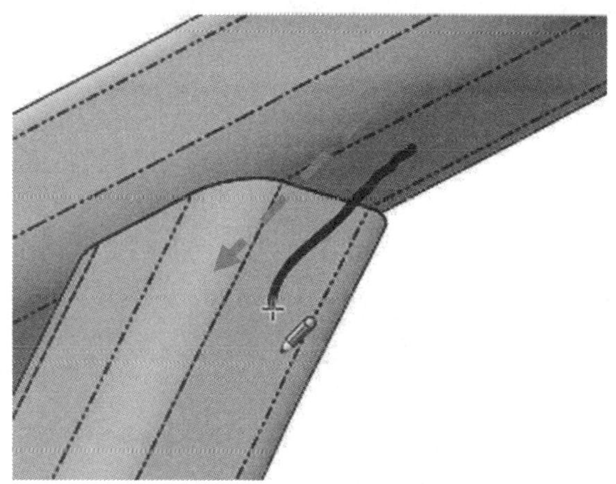

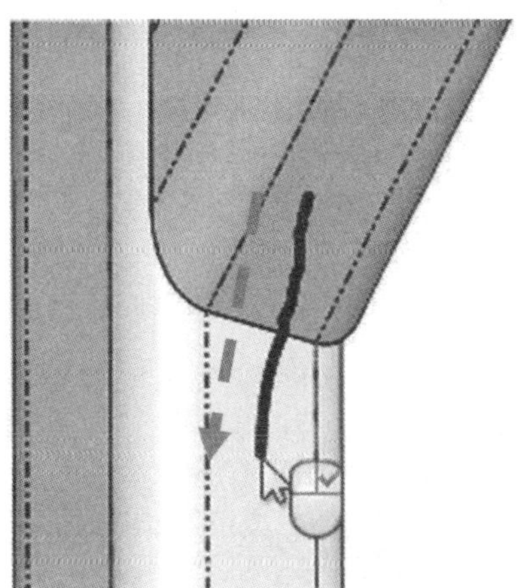

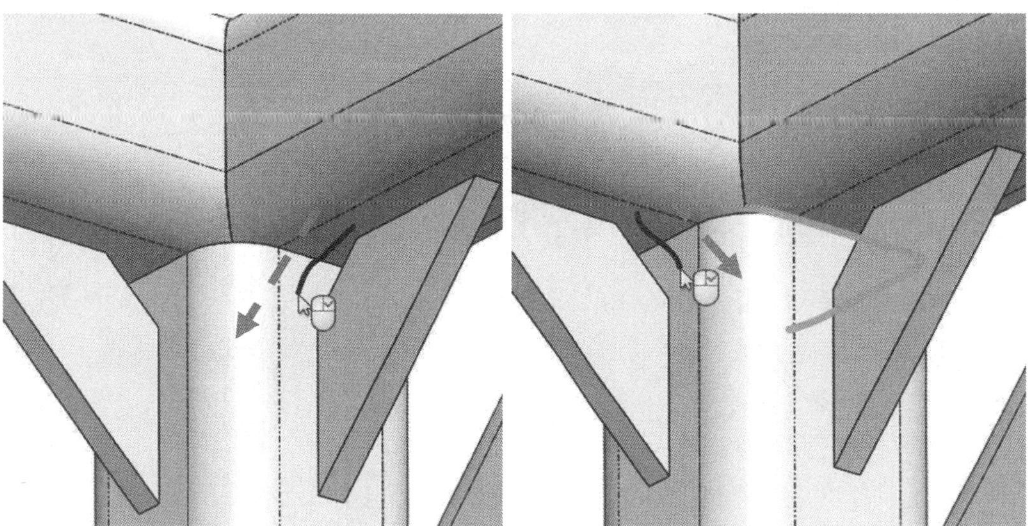

When adding weld beads using the "Smart Weld Selection Tool," click and drag starting in an unselected face; otherwise it will get unselected and you'll have to re-select again to add a weld bead.

Add the rest of the weld beads between all structural members including the corners and click OK to complete these weld beads. We'll add weld beads to the gussets in the following step.

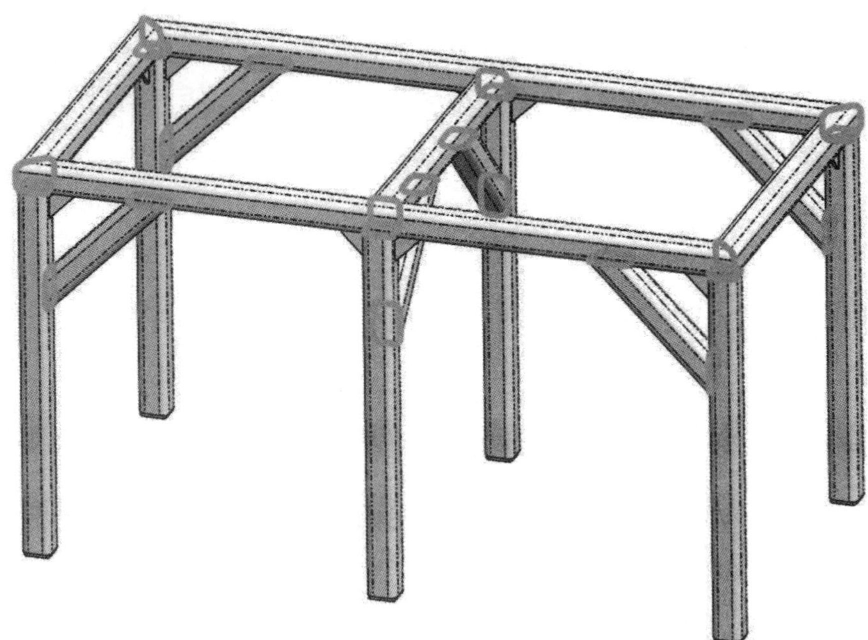

17.29. - After completing the weld beads the corresponding annotations are added to each weld. Right-mouse-click in the "*Weld Folder,*" and turn on the option "Show Cosmetic Welds."

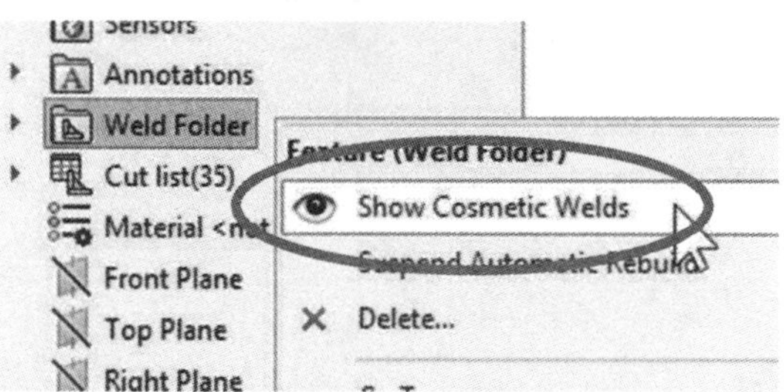

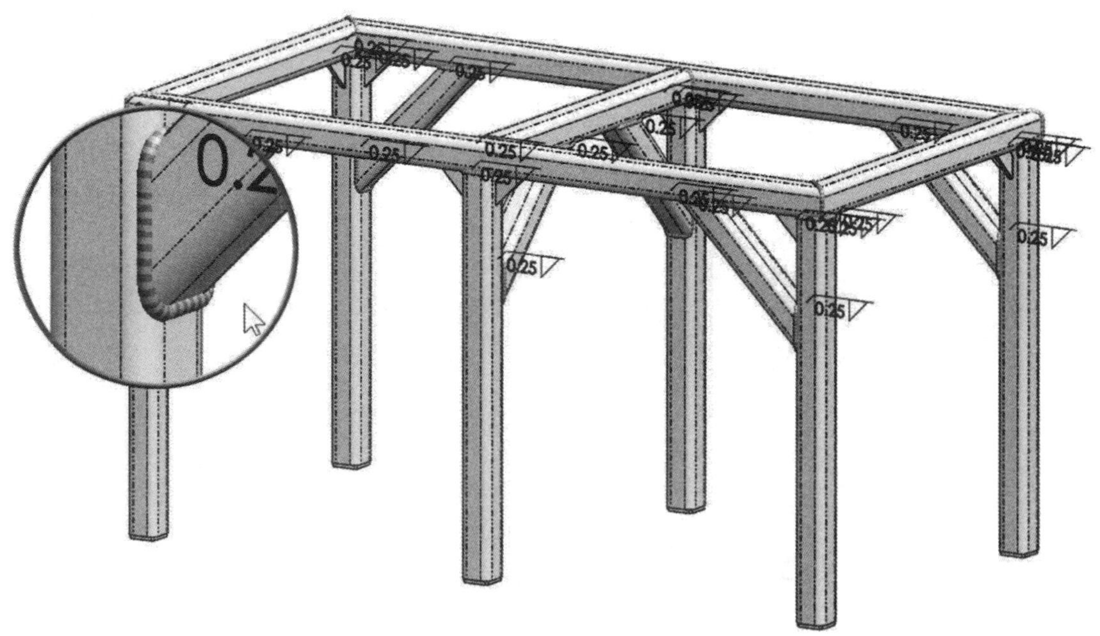

Under the *"Weld Folder"* we see the new *"0.25in Fillet Weld"* including all the weld beads added listing the length of each one; this is useful for cost estimating and manufacturing purposes. Depending on the selections made when the weld beads were added, you may have more or fewer welds.

To turn off the cosmetic weld beads right mouse click in the *"Weld Folder"* and select "Hide Cosmetic Welds." To hide the welding annotations, right mouse click in the *"Annotations"* folder and turn "Display Annotations" Off.

17.30. - To add the weld beads to the gussets, we'll use different options. Select the "**Weld Bead**" command from the Weldments toolbar and zoom in on one of the gussets (annotations have been turned off for clarity).

Select the two faces indicated to weld the gusset to the structural element using the "Smart Weld Selection Tool"; after selecting the faces make the bead size 0.125″ and select the now available option "Both sides" under "Settings." Notice a new weld bead is added to the opposite side of the gusset saving us time (similarly, if we select "All around," the entire loop is selected).

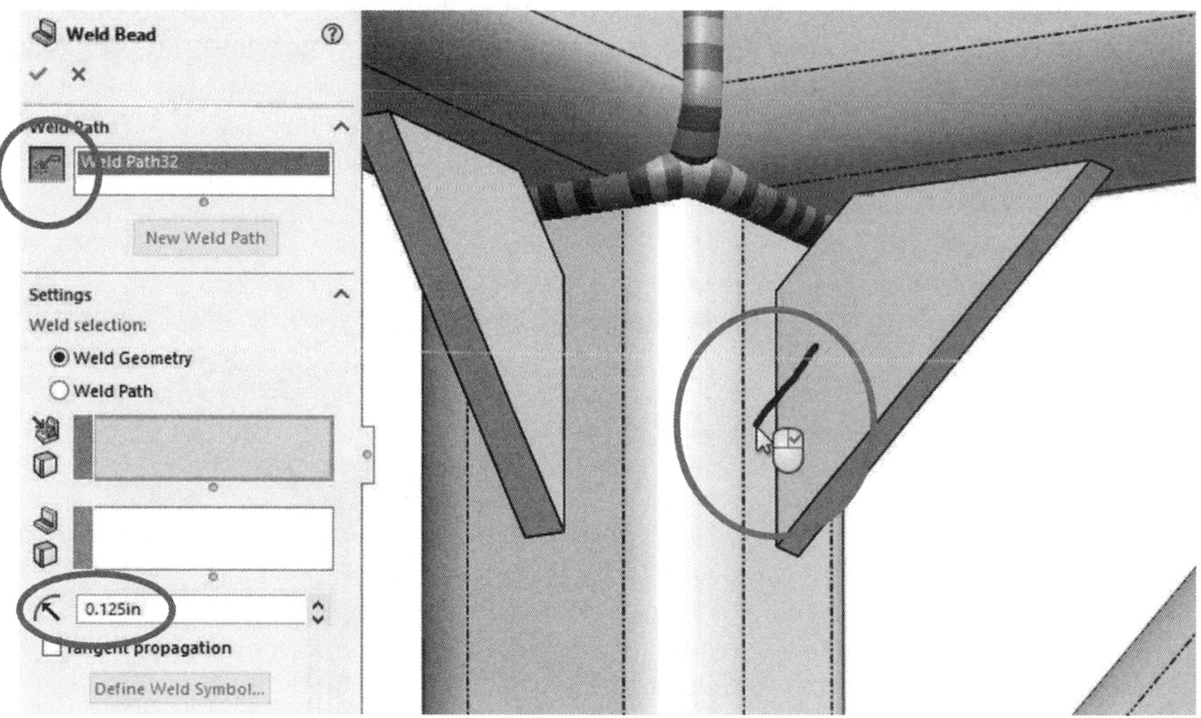

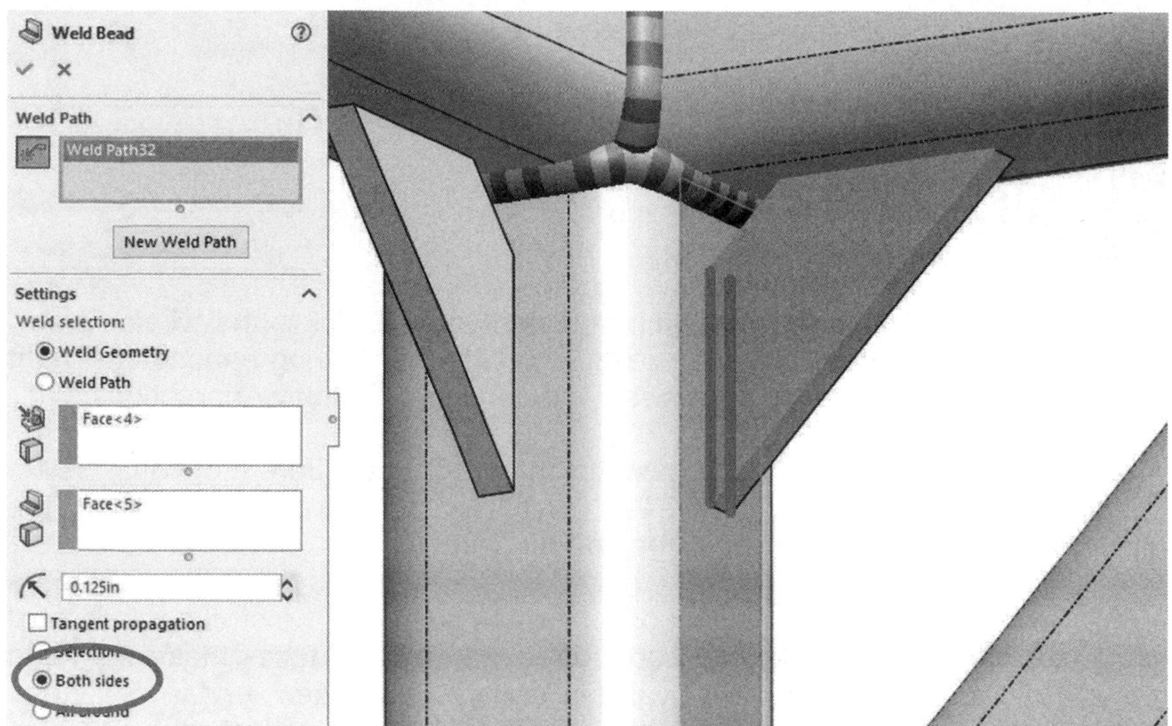

In these images, it's easier to see the reason to add the chamfer in the gusset is to let the weld bead pass through the corner. Weld Bead options include defining a starting offset, bead length and an intermittent (optionally staggered) weld bead.

Add weld beads at both ends of all gussets using the same settings and click OK to finish.

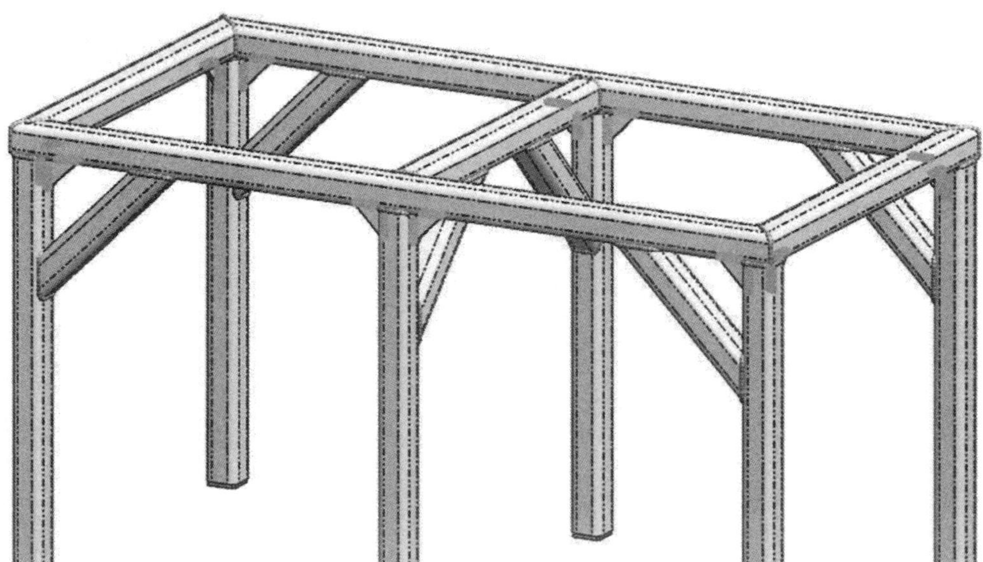

Every time a weld bead is added, the part's annotations are displayed again. Turn them off to avoid a busy screen.

17.31. - Now that we have completed the welded structure, SOLIDWORKS automatically creates a 'cut list' by grouping structural members of the same type and size together. If we expand the "**Cut list**" folder in the FeatureManager, we'll see the solid bodies already grouped by exact size.

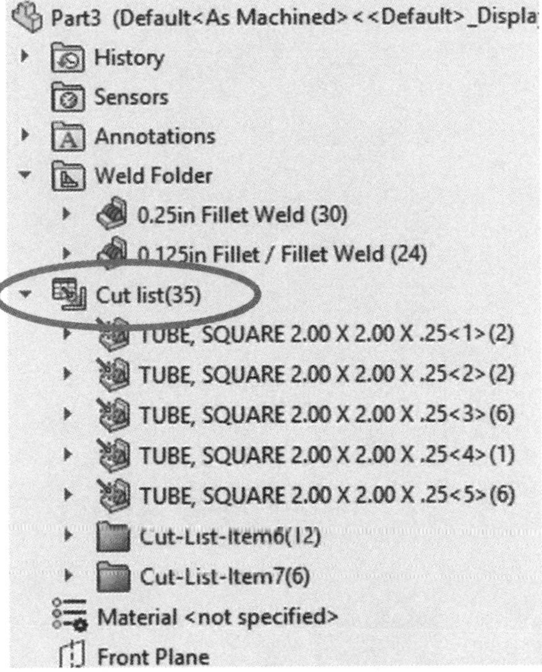

If desired or needed, the "Cut-List-Item" folders can be reordered; the reason to do this is because this is the order in which they will be listed when we import the cut list into the drawing. A good idea is to reorder the elements by type and size, and leave the gussets and end caps at the end. To reorder the items, simply drag-and-drop them up or down as needed. Save the part as *'Weldments Table'* to finish.

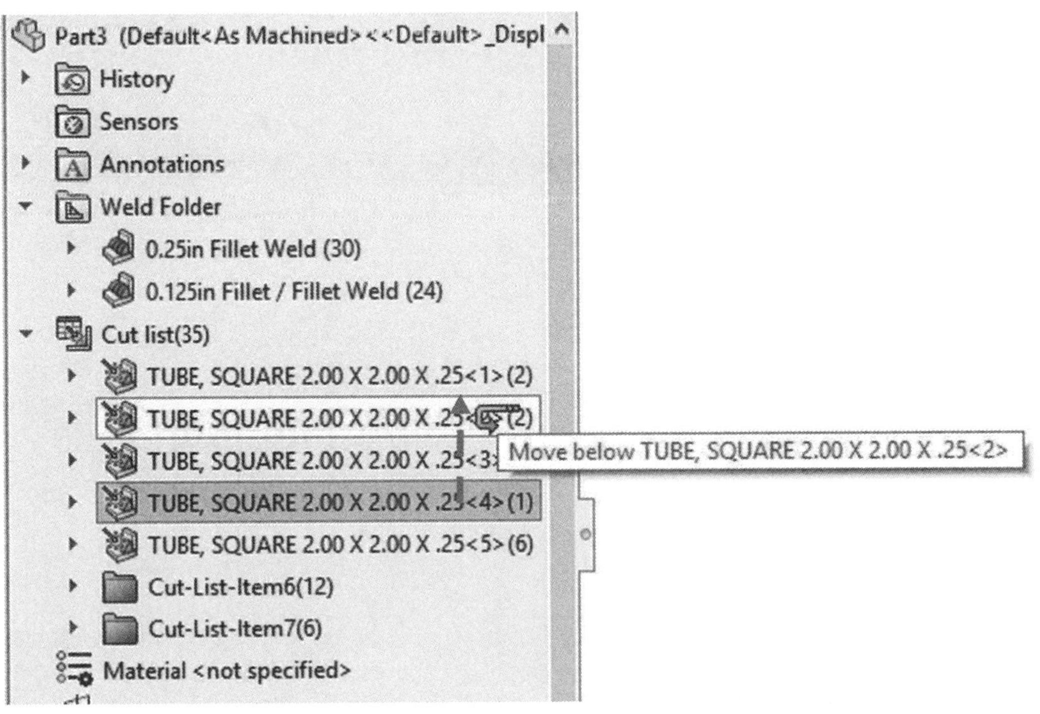

Notes:

Weldment Drawings

18.1. – Using the previously made welded structure make a new drawing using the B-Landscape sheet format. Add a trimetric view using shaded with edges mode and change it to tangent edges removed for a clear view. Import the model's dimensions to this view and arrange the title block.

When working with welded structures using imperial units, it is common to use fractional dimensions, since the accuracy requirements are not as high as those for machined components due to the nature of structural metal working. Change the drawing's units to fractions. Enter "16" in the "Fractions" field to define the lowest significant fraction, and use the option "Round to nearest fraction."

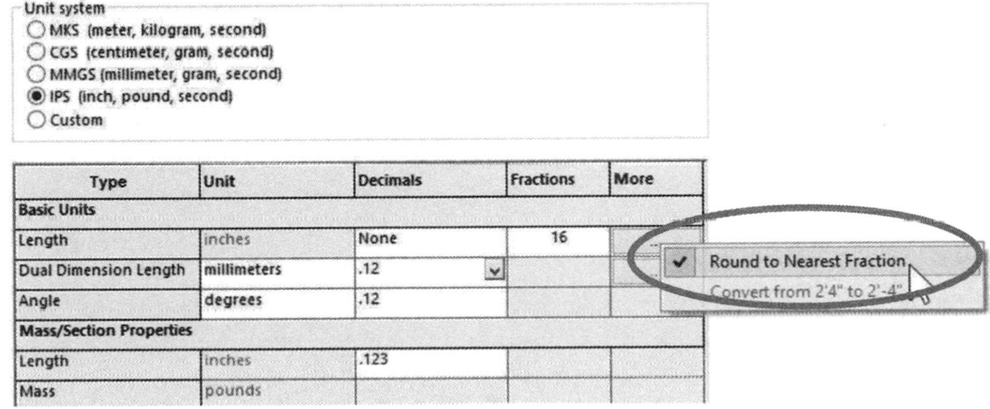

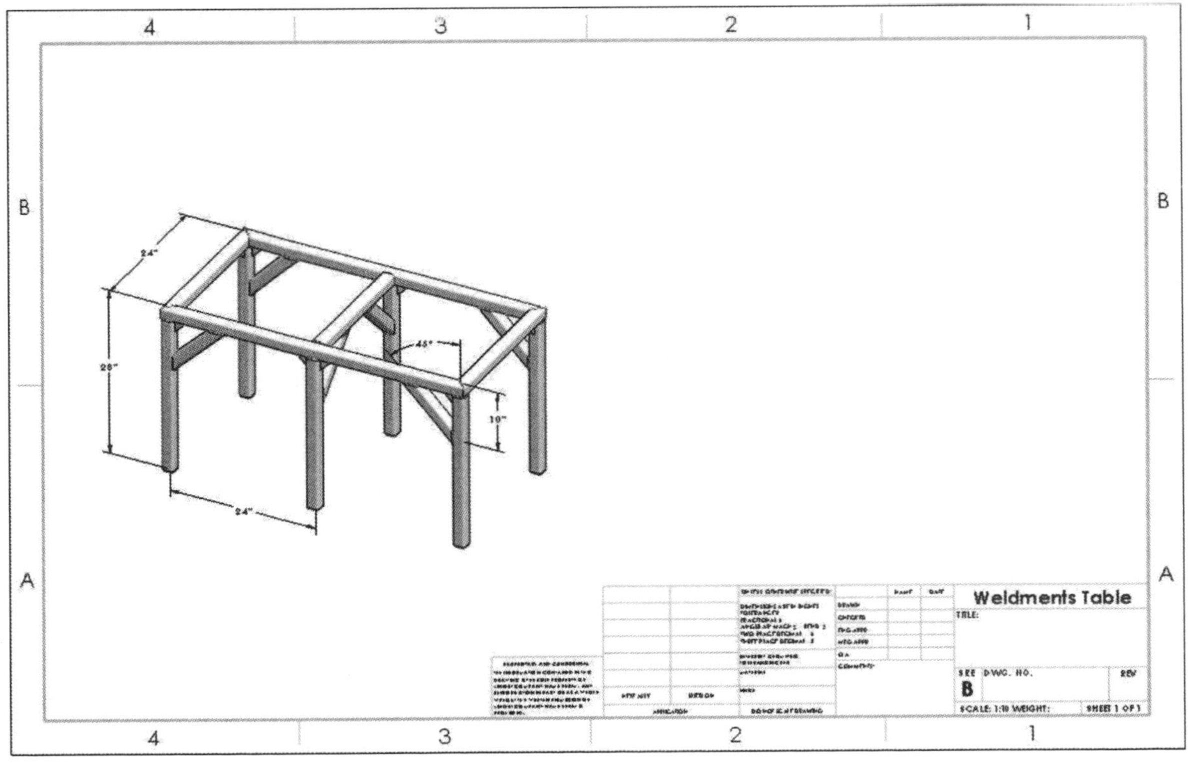

18.2. - Select the view and from the right-mouse-click menu, select "**Tables, Weldment Cut List**" or from the menu "**Insert, Tables, Weldment Cut List.**" Accept the defaults for the table, click OK and locate it in the upper right corner. The table template to be used must be "cut list." If needed, adjust the view's scale and the table's row height and width.

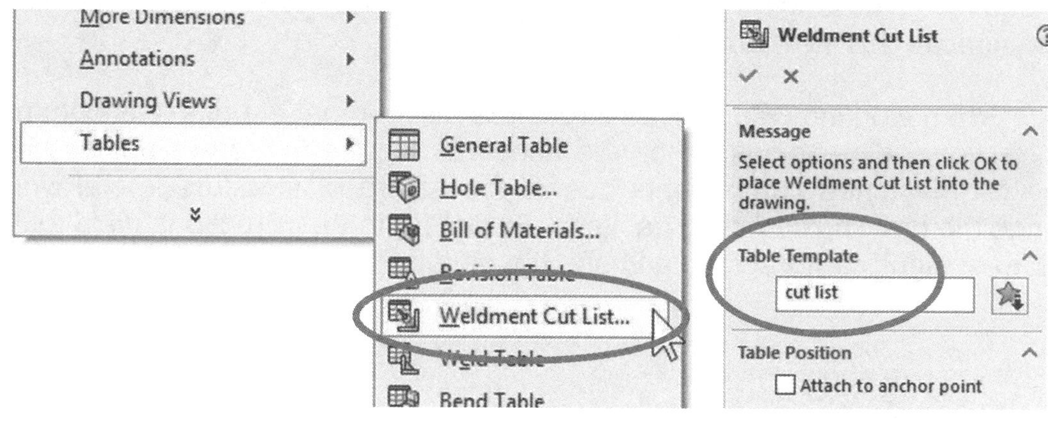

ITEM NO.	QTY.	DESCRIPTION	LENGTH
1	2	TUBE, SQUARE 2" X 2" X 1/4"	50"
2	2	TUBE, SQUARE 2" X 2" X 1/4"	26"
3	6	TUBE, SQUARE 2" X 2" X 1/4"	25 3/4"
4	1	TUBE, SQUARE 2" X 2" X 1/4"	22"
5	6	TUBE, SQUARE 2" X 2" X 1/4"	11 7/8"
6	12		
7	6		

18.3. - The table shows two rows without a description; these correspond to the gussets and end caps. For SOLIDWORKS to add this information, we need to go back to the welded part and fill the necessary fields. Switch back to the '*Weldments Table*' part, expand the *"Cut list,"* right-mouse-click in the gussets folder and select "Properties."

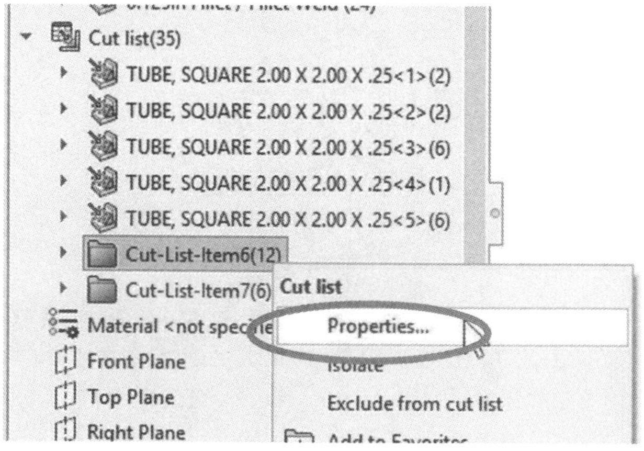

In the "Cut List Summary" tab we see a list of properties and the corresponding value for each group. Add a "Description" and "Length" properties for the Gusset and End Cap groups, and fill in the missing information for each one. When finished click OK to continue and go back to the weldment drawing to see the updated table.

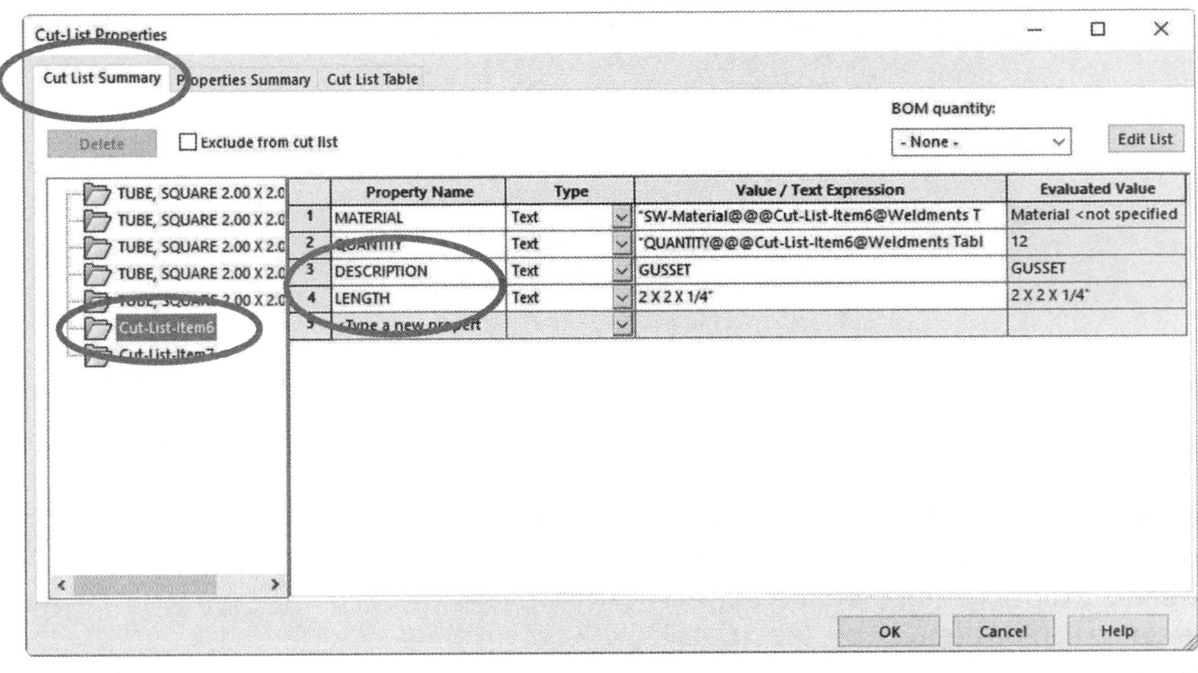

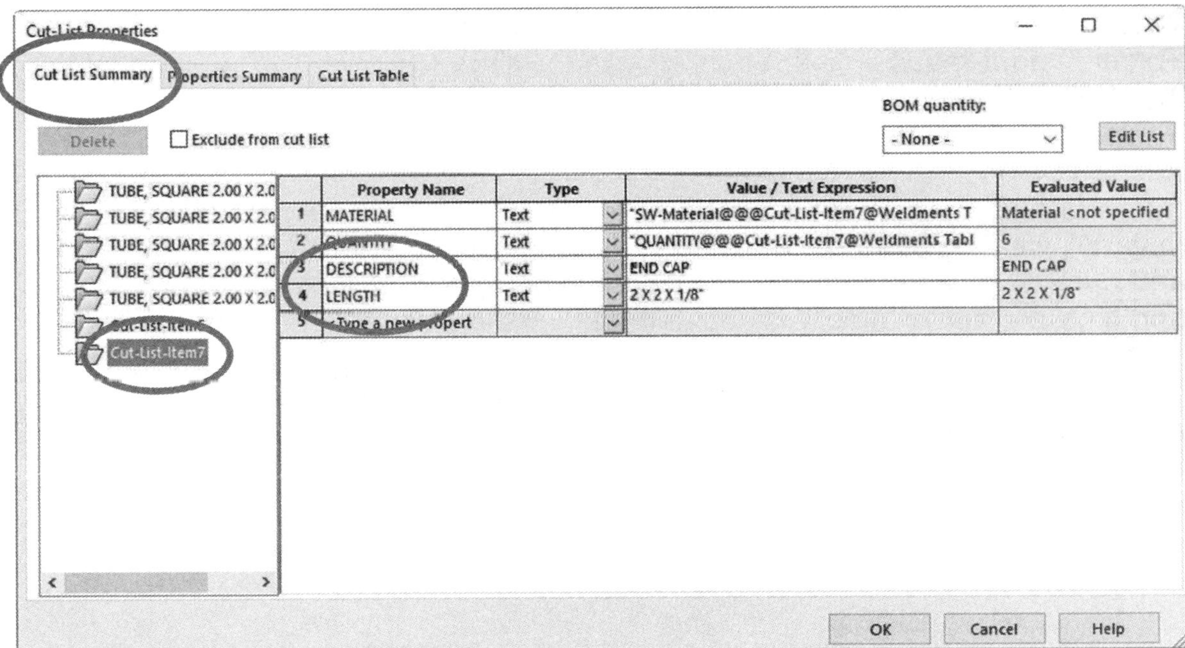

When a group's field is selected in the "Properties Summary," the items in the group are highlighted in the graphics area.

	Property Name	Type	Value / Text Expression	Evaluated Value
1	TUBE, SQUARE 2.00 X	Text	*LENGTH@@@TUBE, SQUARE 2.00 X 2.00 X .25<9	50
2	TUBE, SQUARE 2.00 X	Text	"LENGTH@@@TUBE, SQUARE 2.00 X 2.00 X .25<1	26
3	TUBE, SQUARE 2.00 X	Text	"LENGTH@@@TUBE, SQUARE 2.00 X 2.00 X .25<1	25.75
4	TUBE, SQUARE 2.00 X	Text	"LENGTH@@@TUBE, SQUARE 2.00 X 2.00 X .25<1	22
5	TUBE, SQUARE 2.00 X	Text	"LENGTH@@@TUBE, SQUARE 2.00 X 2.00 X .25<1	11.899
6	Cut-List-Item15	Text	2 X 2 X 1/4"	2 X 2 X 1/4"
7	Cut-List-Item16	Text	2 X 2 X 1/8"	2 X 2 X 1/8"

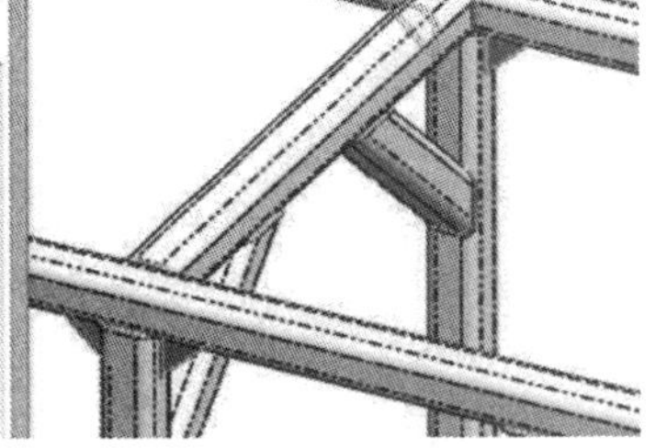

Updated weldments table in the drawing:

ITEM NO.	QTY.	DESCRIPTION	LENGTH
1	2	TUBE, SQUARE 2" X 2" X 1/4"	50"
2	2	TUBE, SQUARE 2" X 2" X 1/4"	26"
3	6	TUBE, SQUARE 2" X 2" X 1/4"	25 3/4"
4	1	TUBE, SQUARE 2" X 2" X 1/4"	22"
5	6	TUBE, SQUARE 2" X 2" X 1/4"	11 7/8"
6	12	GUSSET	2 X 2 X 1/4"
7	6	END CAP	2 X 2 X 1/8"

18.4. – After verifying the cut list is complete we can work on the weld bead's properties to have the correct welding information in the detail drawing. Go back to the *'Weldments Table'* part, expand the *"Weld Folder,"* right-mouse-click each of the *"Fillet Weld"* items and select "Properties" from the menu.

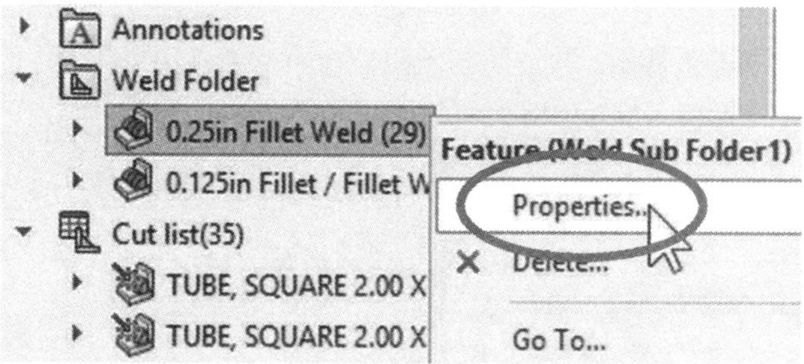

In the "Weld Bead Properties" we can fill in additional properties like weld material, welding process, weight per unit length, cost, welding time per unit length and number of weld passes. The total number of welds, weld length, mass, cost and time are automatically calculated. Adding this information is not required, but having a weld table in our drawing is useful for material and cost calculations especially because welding is a significant cost in a welded structure. The following are sample values for our example.

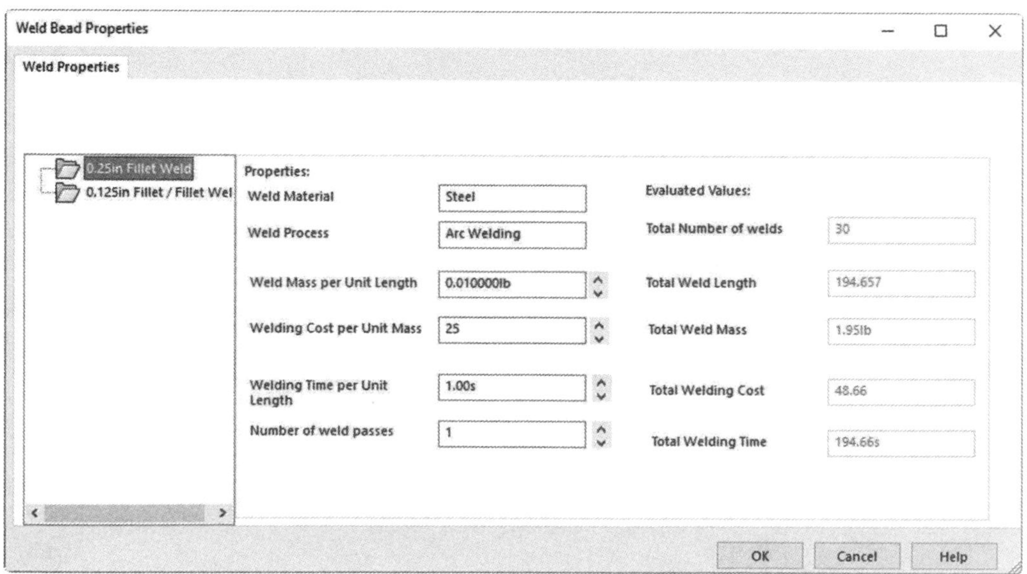

After adding the weld information, go back to the drawing to add a "**Weld Table**." Right-mouse-click in the structure's view, and from the pop-up menu, select "**Tables, Weld Table**." In the "Weld Table" options, make sure the template "weldtable-standard" is selected, turn the option "Combine same weld type" On and click OK to locate the table in the drawing. Change the table's font, size and format as needed.

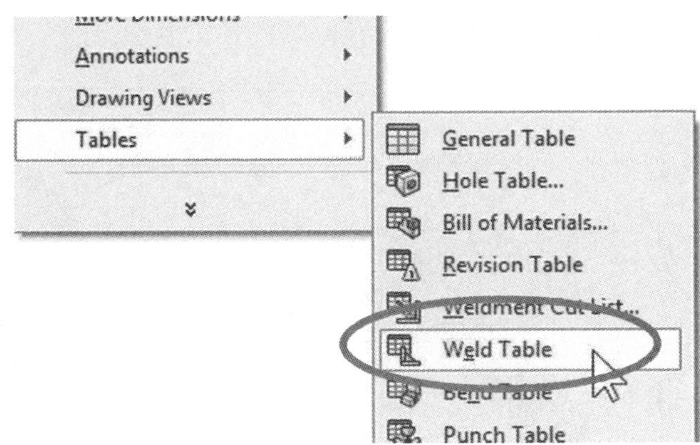

ITEM NO.	WELD SIZE	SYMBOL	WELD LENGTH	WELD MATERIAL	QTY.
1	0.25	◺	183"	Steel	1
2	0.125	◺	144"	Steel	1

18.5. – The next step is to add identification balloons to match each structural element to the cut list. Select the "**Auto Balloon**" command from the Annotation toolbar or the menu "**Insert, Annotations, Auto Balloons**." In the "Balloon Layout" options select "Faces" to attach balloon leaders to structural members' faces. Under "Balloon Settings" select the "Circular Split Line" style, and pick "Quantity" in the "Lower text" selection list. Arrange the balloons as needed to continue.

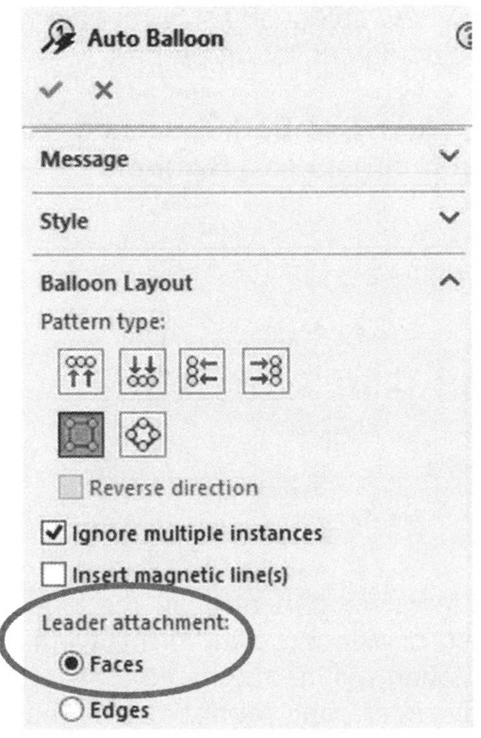

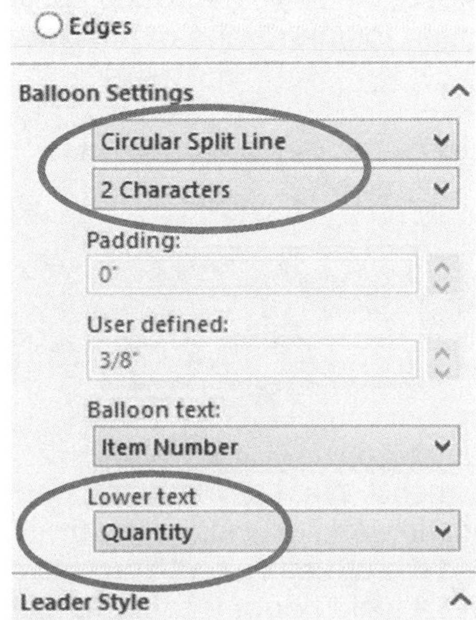

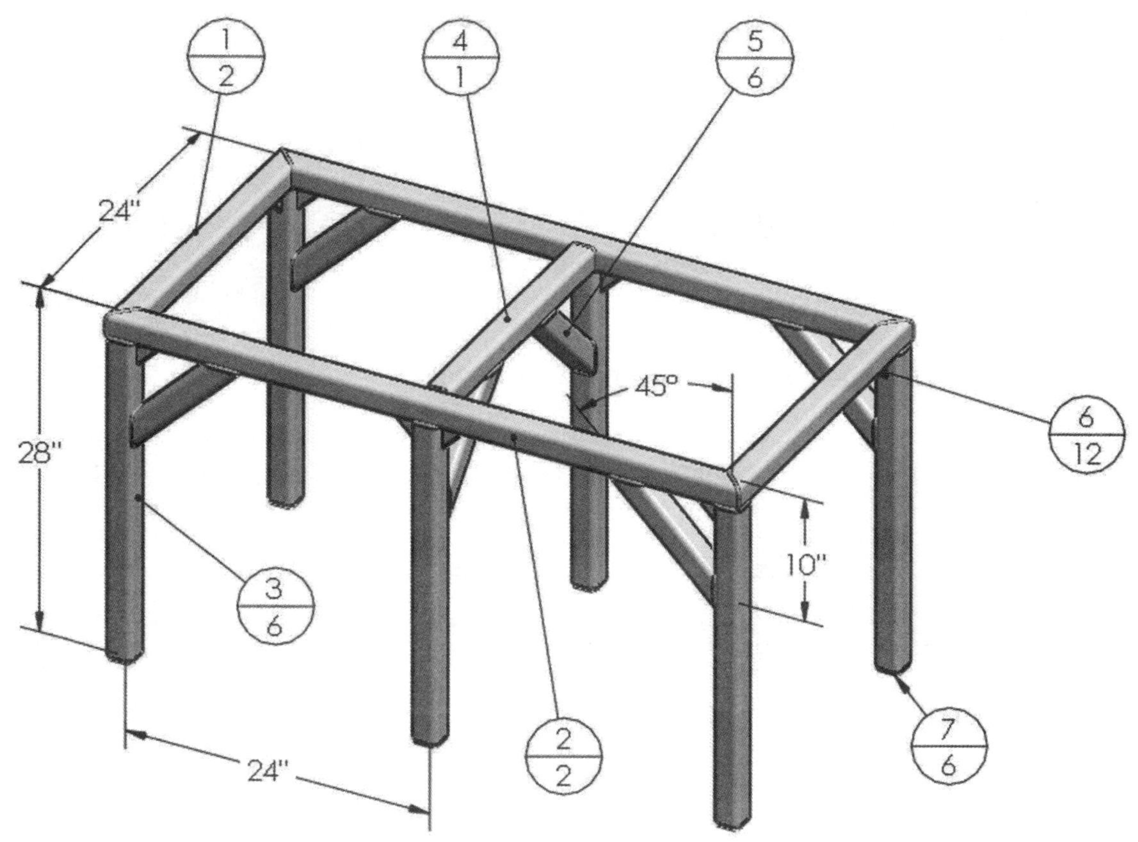

In the “Balloon text” or “Lower text” selection lists we have the option to select “Custom Properties.” If we do, a new selection list is revealed where we can select any custom property including cut list properties.

18.6. - To identify the welds, select the “**Weld Symbol**” command from the Annotation toolbar or the menu “**Insert, Annotations, Weld Symbol**.”

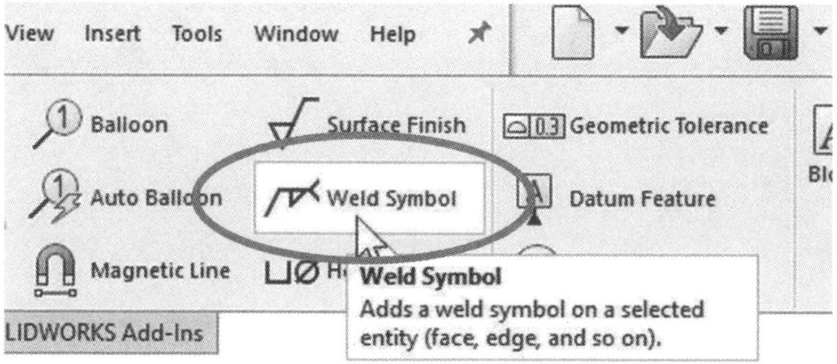

In the Weld Symbol properties window, we can add all the necessary specifications for a welding instruction sheet, or we can click in the weld beads added to the structure and the welding information will be automatically filled. Click to select a weld bead, locate the symbol in the sheet and repeat as needed.

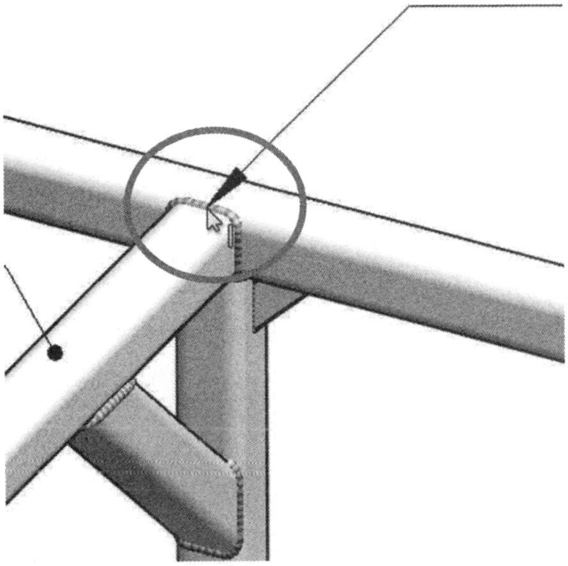

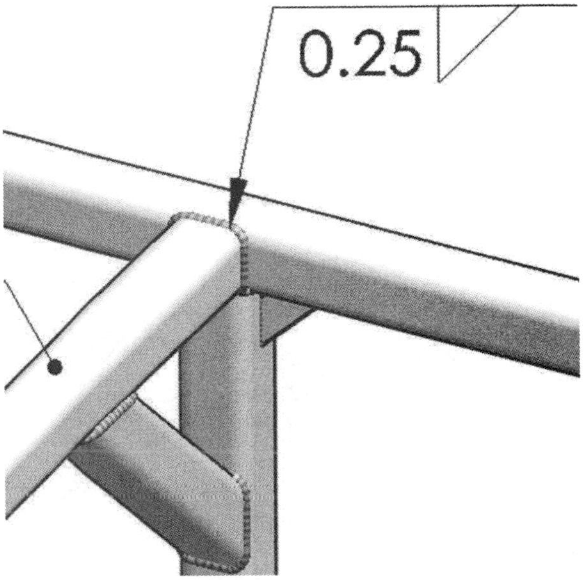

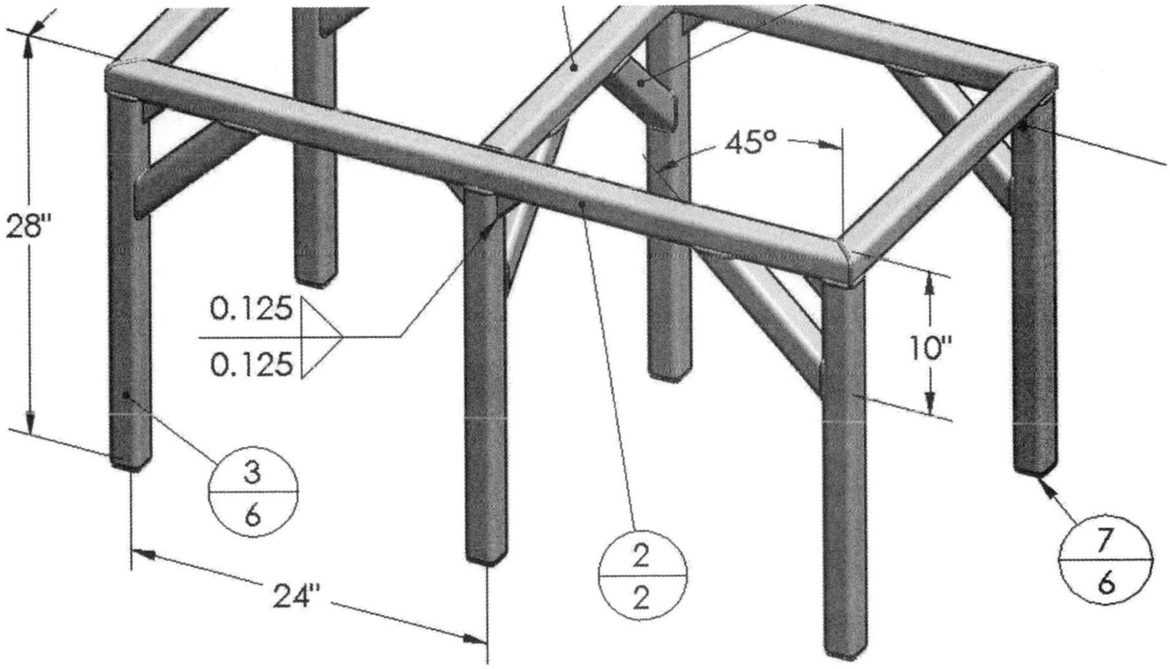

When adding more Weld Symbols, the last type used is remembered. When a different weld bead is selected, the symbol is updated to reflect the currently selected weld bead. Close the Weld Symbol properties window when done adding welding annotations.

18.7. - Save the finished drawing and close.

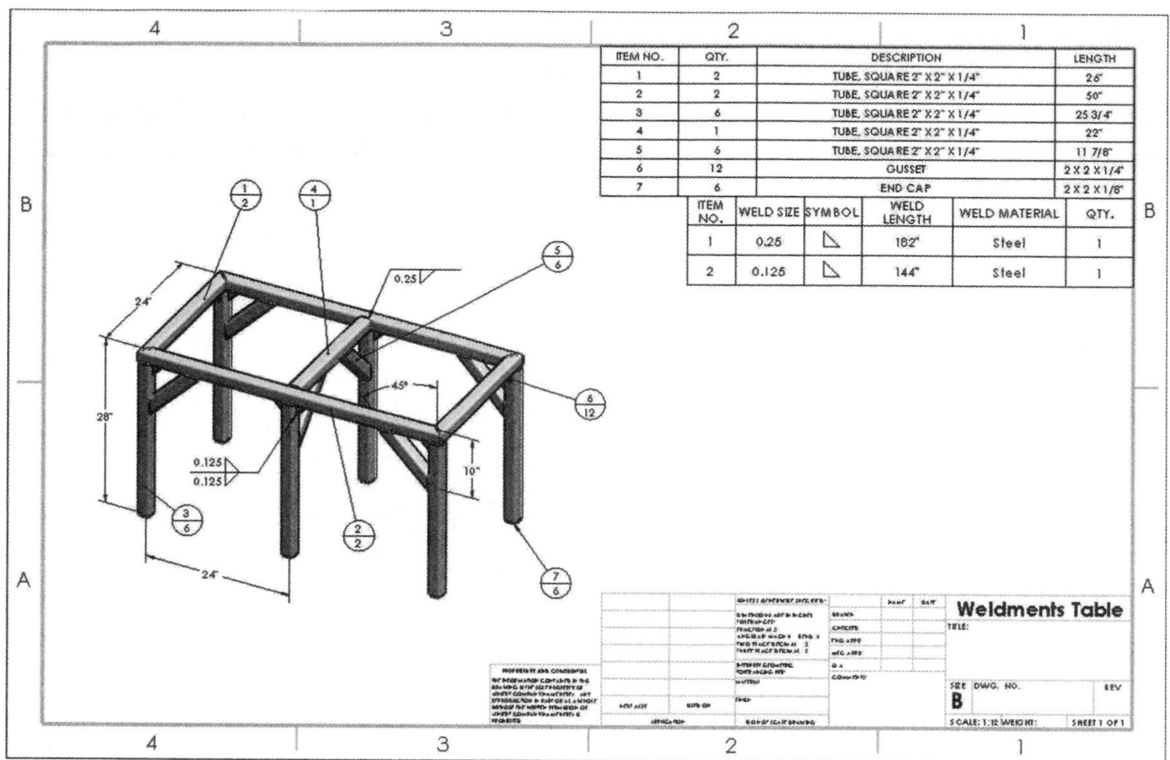

ITEM NO.	QTY.	DESCRIPTION	LENGTH
1	2	TUBE, SQUARE 2" X 2" X 1/4"	26"
2	2	TUBE, SQUARE 2" X 2" X 1/4"	50"
3	6	TUBE, SQUARE 2" X 2" X 1/4"	25 3/4"
4	1	TUBE, SQUARE 2" X 2" X 1/4"	22"
5	6	TUBE, SQUARE 2" X 2" X 1/4"	11 7/8"
6	12	GUSSET	2 X 2 X 1/4"
7	6	END CAP	2 X 2 X 1/8"

ITEM NO.	WELD SIZE	SYMBOL	WELD LENGTH	WELD MATERIAL	QTY.
1	0.25	◺	182"	Steel	1
2	0.125	◺	144"	Steel	1

0.25
45°
0.125
0.125
24"
28"
10"
24"
Weldments Table
TITLE:
SIZE
B
DWG. NO.
REV
SCALE: 1:12 WEIGHT:
SHEET 1 OF 1

Curved Elements

19.1. - To learn more weldment options, create a new part and add the following (2D) sketch in the *"Top Plane."* The arc's center is coincident to the origin; this will help us by not having to add an auxiliary plane for the next step. When working with structural members we can only use circular arcs and straight lines. Exit the sketch when finished.

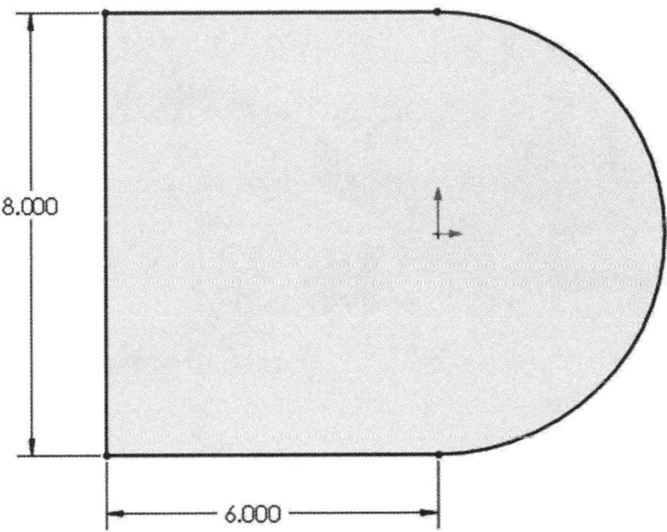

19.2. - Switch to an isometric view; add the following sketch in the *"Right Plane,"* and exit the sketch.

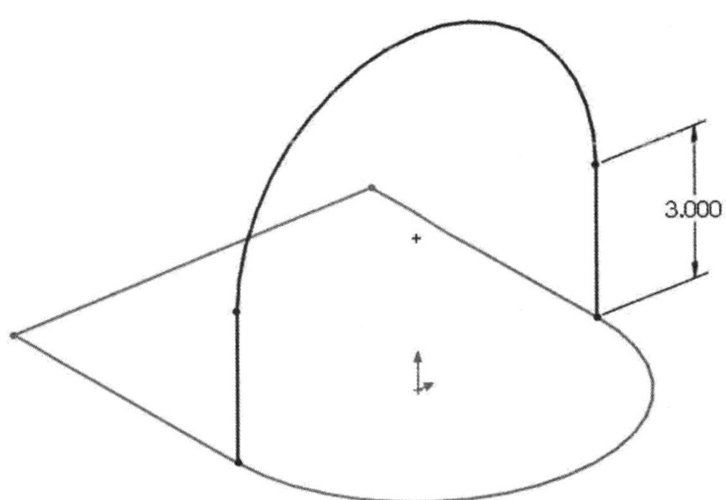

Alternatively, (or as a challenge), we can make a 3D sketch including the lines from the two sketches. When we work with structural members, it makes no difference if we use a 2D or a 3D sketch.

19.3. - Select the "**Structural Member**" command from the Weldments toolbar. Select the "ansi inch" standard; use the "pipe" type and the "0.5 sch 40" size. Select the lines indicated for the first group.

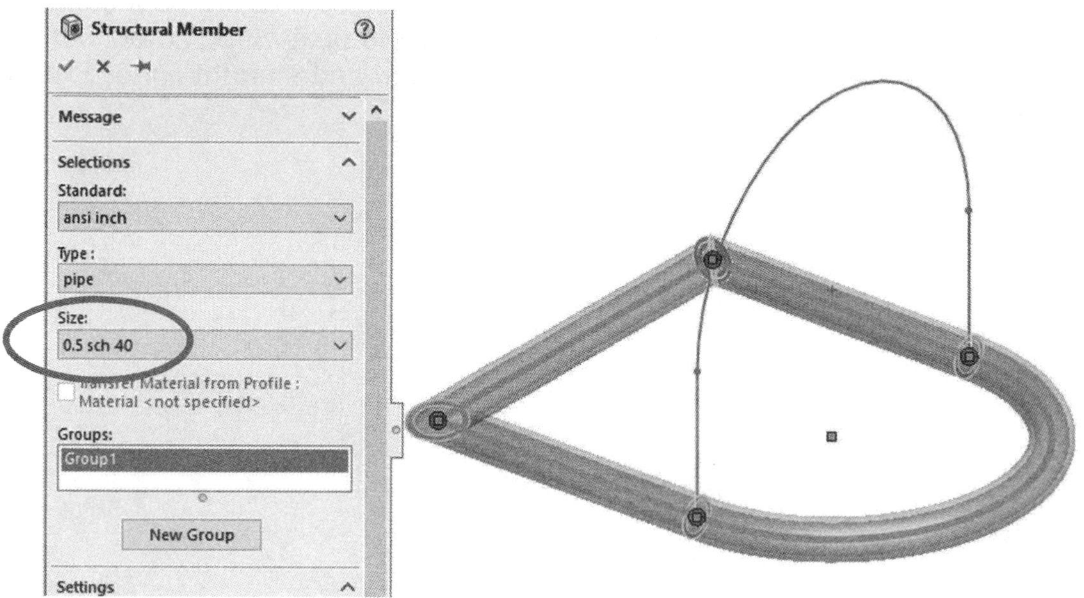

For "Corner treatment," instead of applying a global option, we'll manually select the treatment for each connection. Be sure the option "Merge arc segment bodies" and "Merge miter trimmed bodies" are turned Off in this exercise. Click the point in the corner and select the following corner treatments:

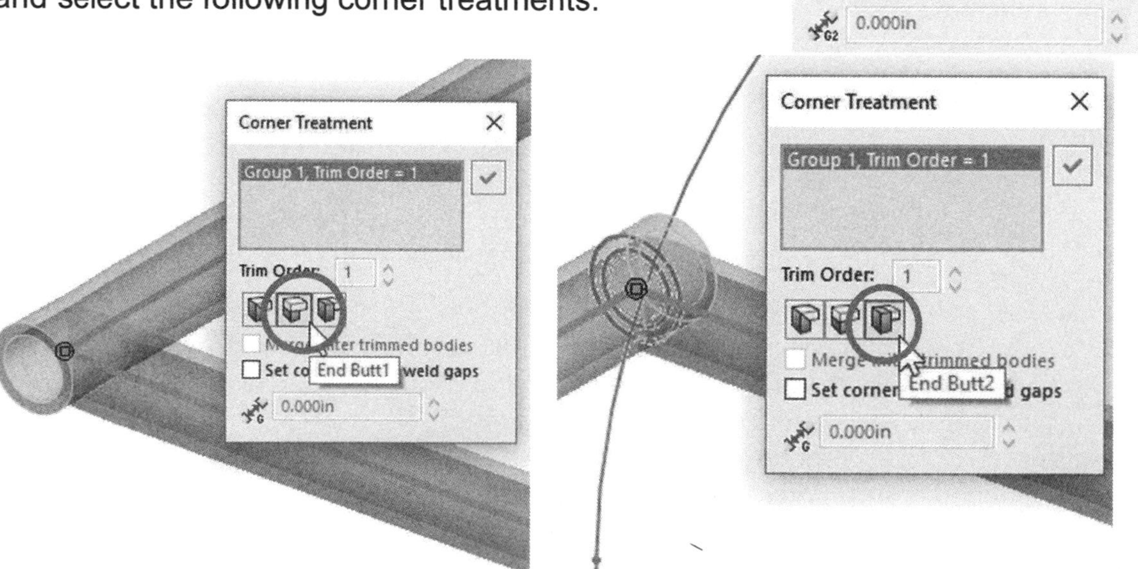

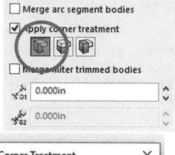

The option "Set corner specific weld gaps" adds the specified gap between the connecting elements to allow space for weld beads, useful when welding thick metal components.

In the case of the joint between the straight line and arc, changing the corner treatment gives us the option to merge the segments into a single element. In our example we'll leave this option unchecked.

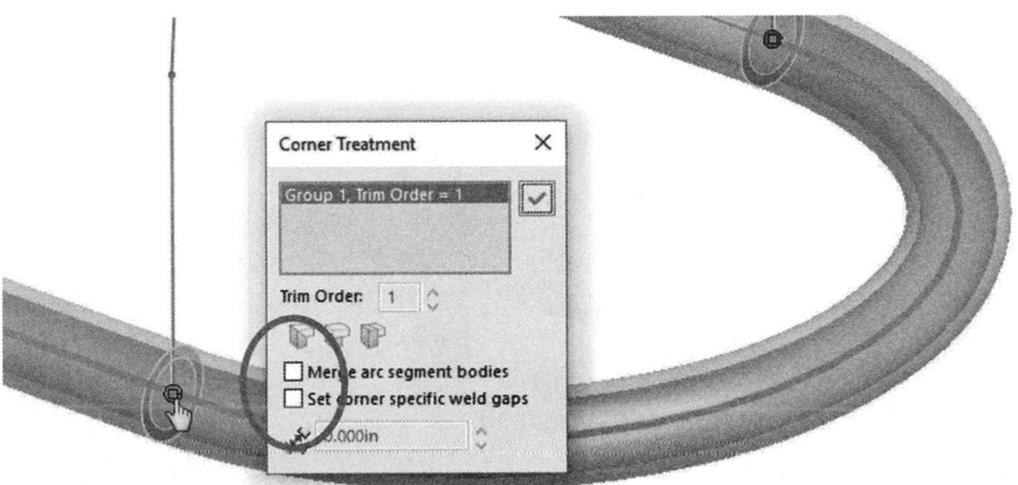

19.4. – Click "New Group" to add the second group. Use the same structural member size and settings for the vertical elements. Click OK to complete it and save the part as *'Welded Frame'*.

19.5. - Now select the "**Trim/Extend**" command to trim the overlapping elements. In the "Corner Type" settings select the "End Trim" option. Select the vertical elements in the "Bodies to be Trimmed" selection box. In "Trimming Boundary" use the option "Bodies" and select both of the bottom straight members and the curved one. Make sure the trimmed segments are correctly marked "keep" or "discard." Be sure to use the option "Simple cut between members" and click OK to finish.

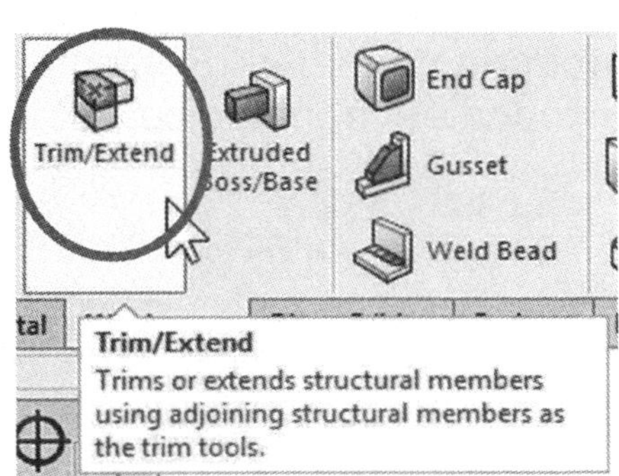

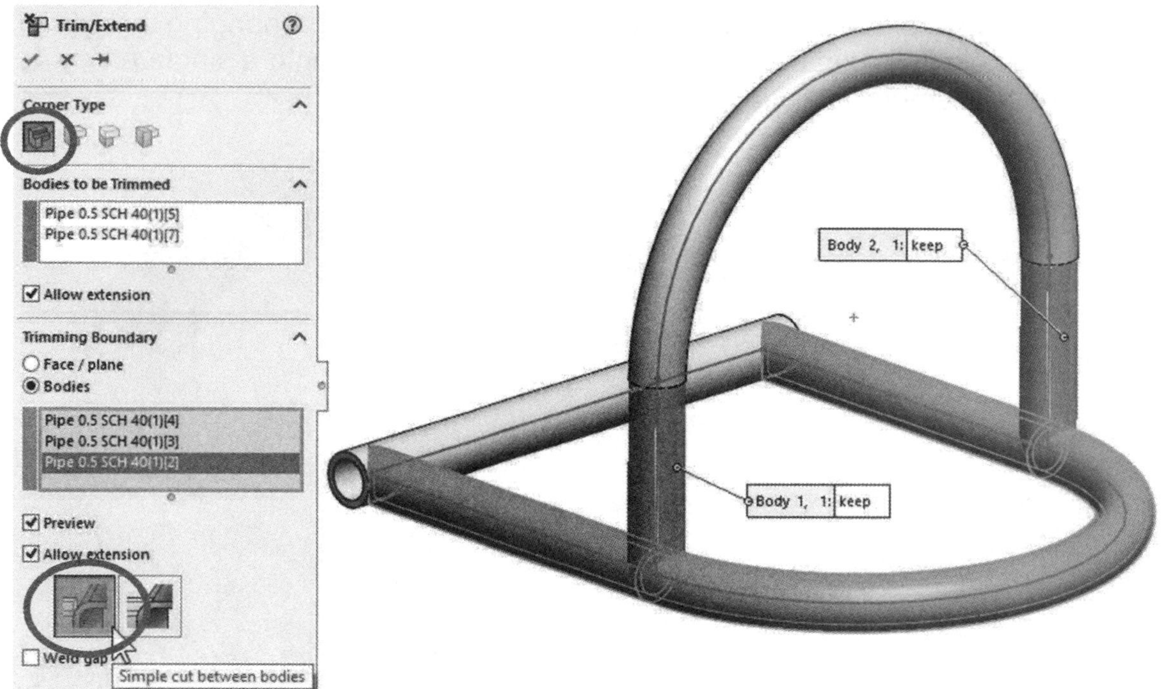

After completing the structural members hide the sketch(s) used.

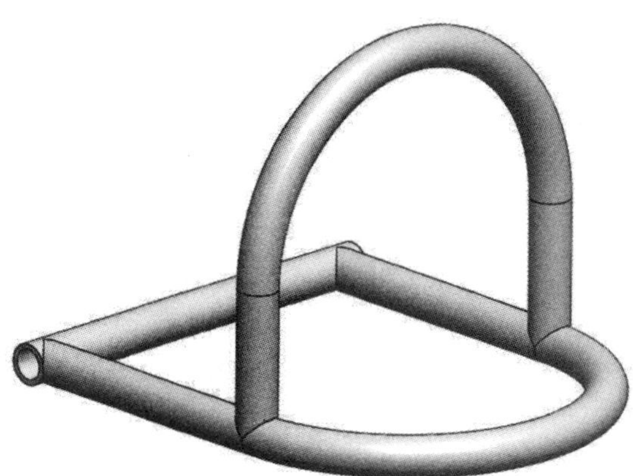

19.6. - Expand the *"Cut list"* folder and make sure the elements are correctly grouped together. Remember that we can preview the cut list by selecting "Properties" from the cut list item's right-mouse-button menu.

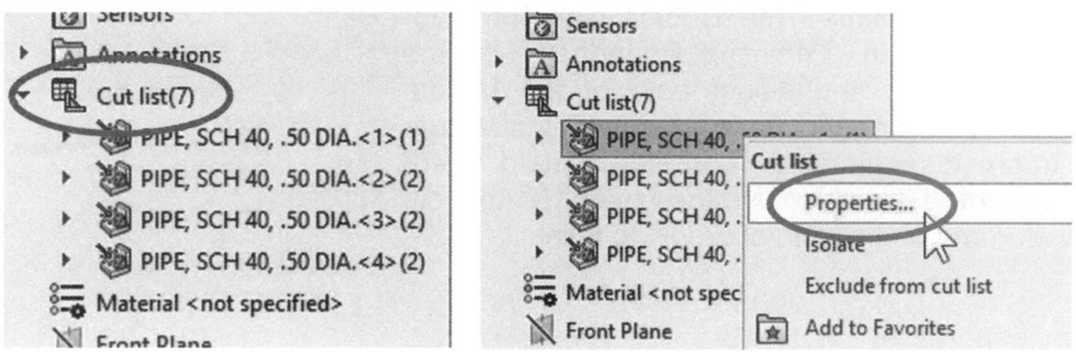

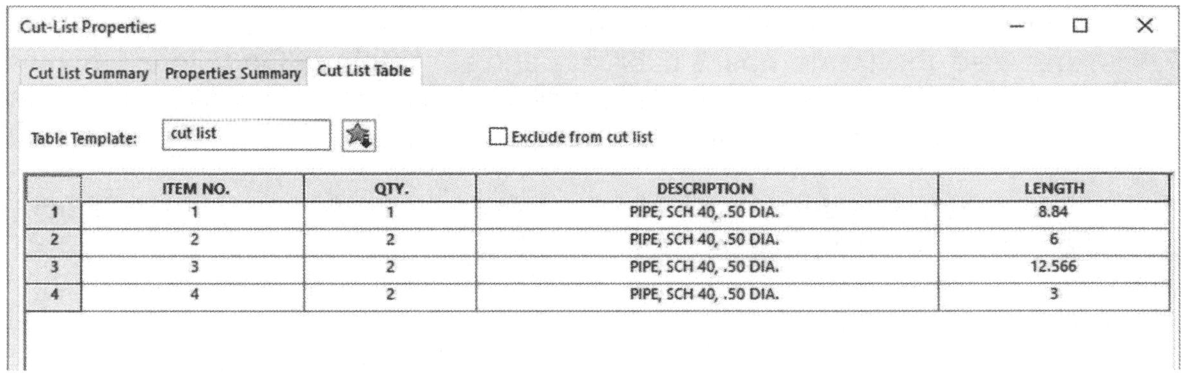

	ITEM NO.	QTY.	DESCRIPTION	LENGTH
1	1	1	PIPE, SCH 40, .50 DIA.	8.84
2	2	2	PIPE, SCH 40, .50 DIA.	6
3	3	2	PIPE, SCH 40, .50 DIA.	12.566
4	4	2	PIPE, SCH 40, .50 DIA.	3

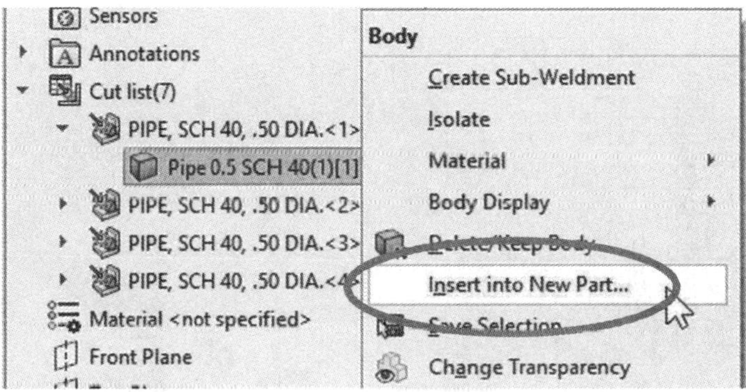

19.7. - When working with multi body parts (like weldments), we can save each body to a new part file and/or make a new assembly with those parts at the same time. To save a single body to a new part we can select it from the "Cut-List" folder or the graphics area with the right-mouse-button and select "Insert into New Part…" from the pop-up menu. In our case, what we want to do is to save all bodies each to a part file, and at the same time make an assembly with those parts. Save the part as *'Welded Frame'*, then right-mouse-click in the *"Cut list"* folder and select the "**Save Bodies**" option.

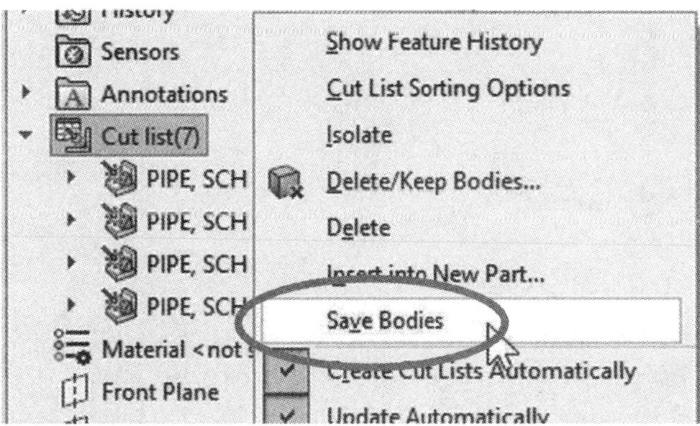

19.8. - In the "**Save Bodies**" command we can select one or more bodies to be saved to a part. If asked, select a template for the parts and the assembly. Select the "Browse…" button in the "Create Assembly" option box, locate and name it *'Welded Frame Assembly'*.

After naming the assembly, enable the option "Derive resulting parts from similar bodies or cut list." If this option is turned Off, each body in the cut list will be saved as an individual part, even identical bodies. By turning this option On, the number of part files created is reduced by only saving unique bodies, and reusing them in the assembly where the same body is located.

Once the option is selected, the number of bodies to save is reduced from 7 to 4 because the frame has 3 duplicate bodies. Rename the different bodies 'Body1' to 'Body4' by clicking in the label attached to each body or double-clicking in the row for each body in the list.

The option "Consume cut bodies" removes the bodies from the multi body part after saving them to an external file. We'll leave this option unchecked.

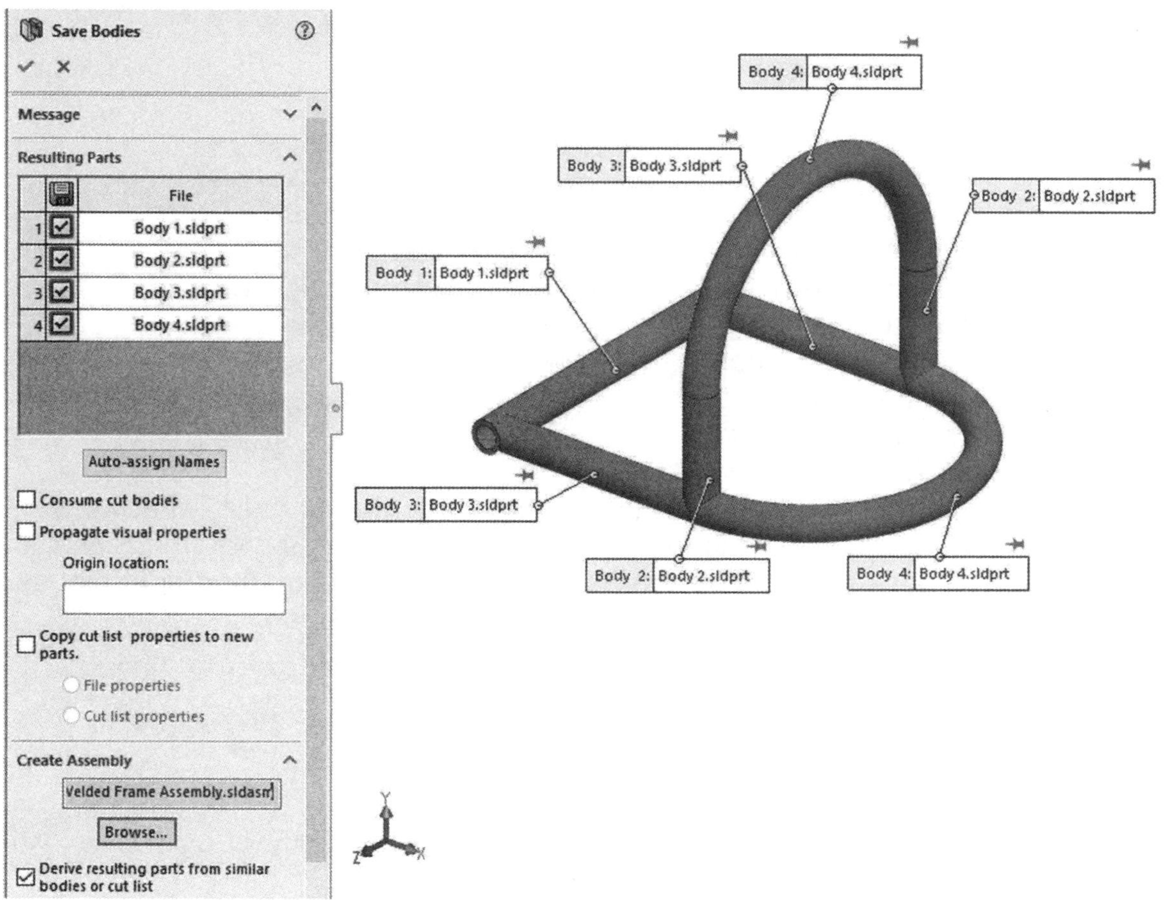

19.9. - After creating the assembly, a feature called *"Save Bodies1"* is added to the FeatureManager; a new part file is generated for each body (or group) and is saved as well as the new assembly which is already open. Switch to the new assembly window that was automatically made. A drawback of making an assembly using this technique is that all the parts will be automatically "fixed" in the assembly.

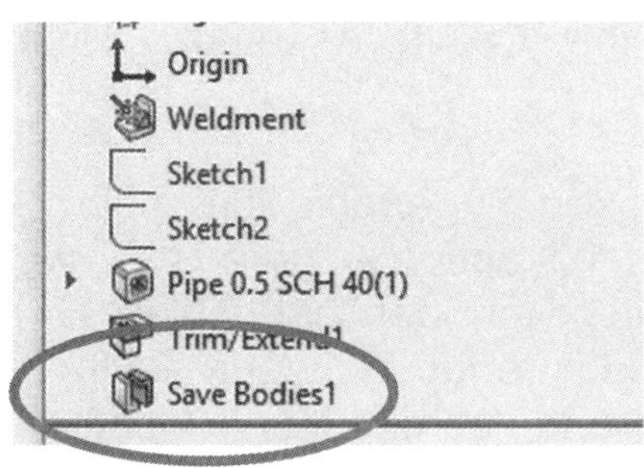

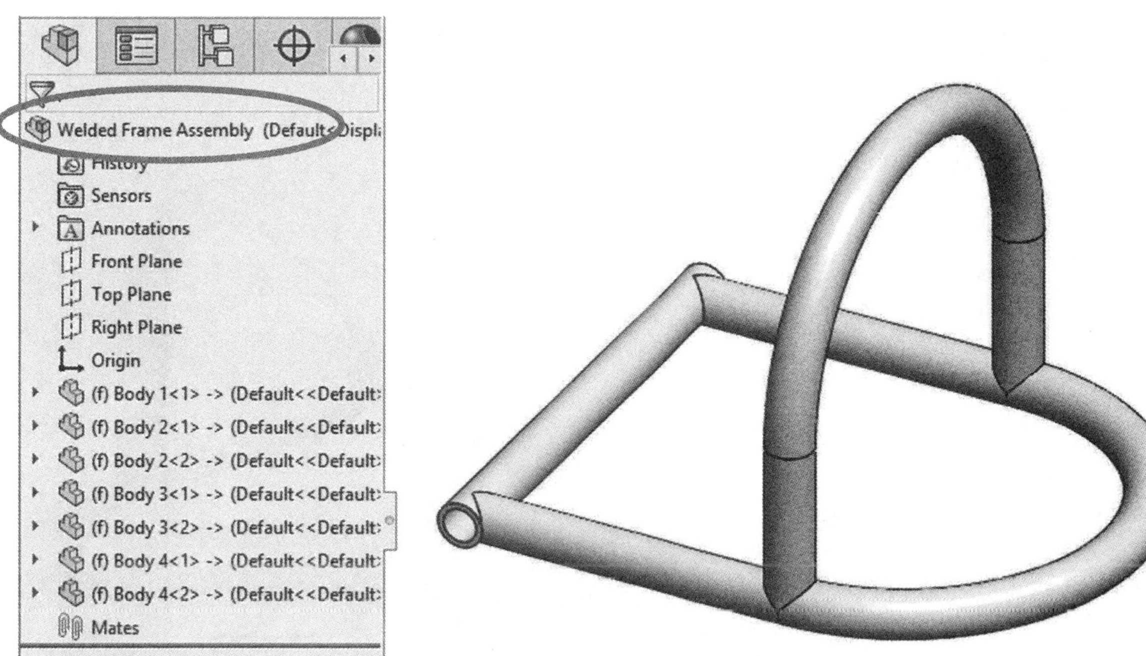

The bodies saved to a part have an external reference to the original welded part. In this case, the reference status is "In context" because the *'Welded Frame'* part is open and loaded in memory.

19.10. - Opening one of the assembly parts shows the only feature is a saved body from a multi body part, listed as *"Stock-part-name."* From this point on, we can add more features to a part if needed, and these changes will be reflected in the assembly but not in the original multi body part.

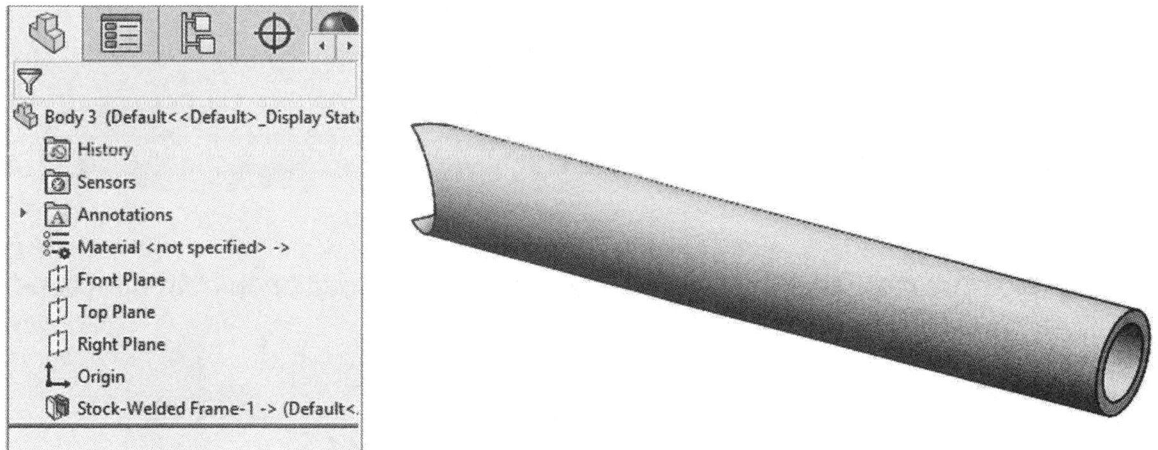

If we add a feature to the welded <u>multi body part</u>, like a cut across multiple bodies, the changes will be propagated to the externally saved parts **only** if the feature is added BEFORE the *"Save Bodies1"* feature. To add a feature before the *"Save Bodies1"* feature we must move the Rollback bar above it and then add the new feature.

19.11. – In the '*Welded Frame*' part, move the roll back bar above the "*Save Bodies1*" feature.

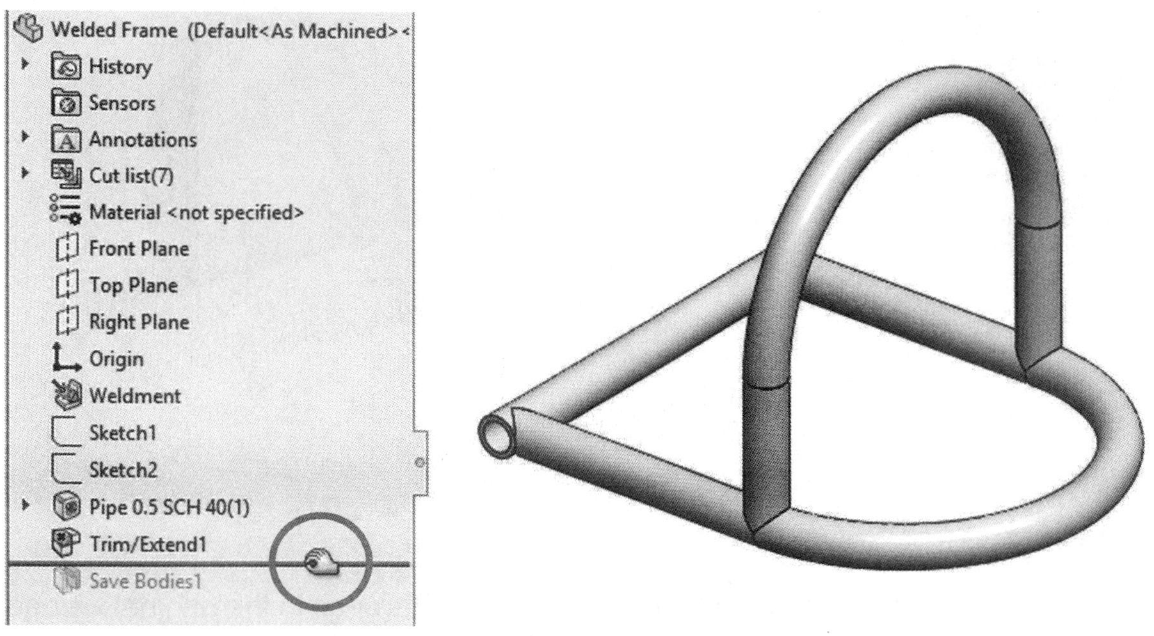

19.12. - Add the following sketch in the "Front Plane."

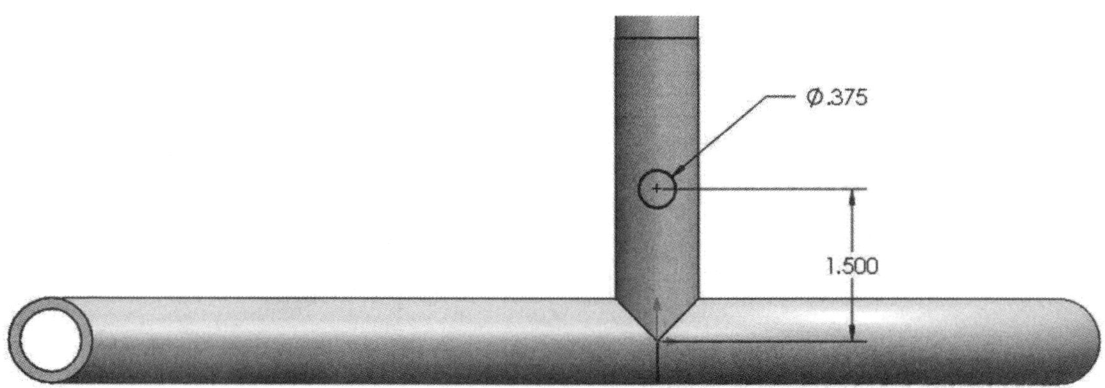

19.13. - Add an Extruded Cut using the though all end condition in both directions.

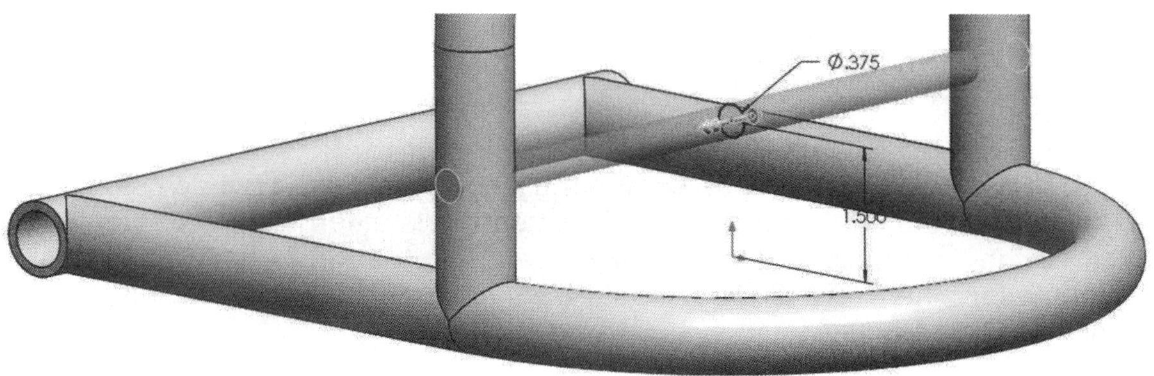

19.14. - Move roll back bar below the "*Save Bodies1*" feature.

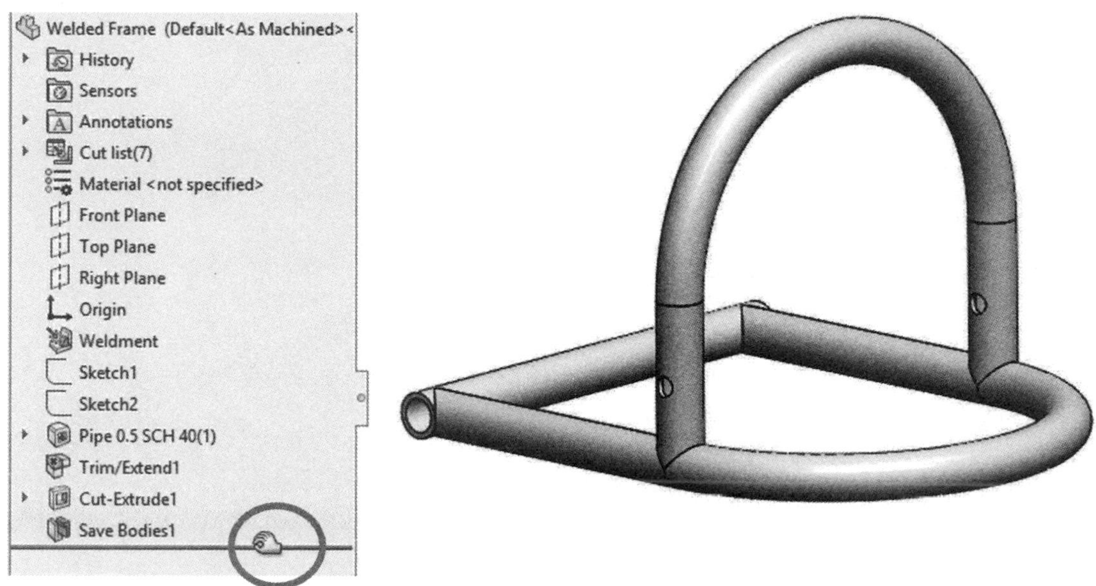

19.15. - After opening the assembly and the part, we can see the hole has propagated from the original multi body part to the saved body. Save and close all parts and the assembly.

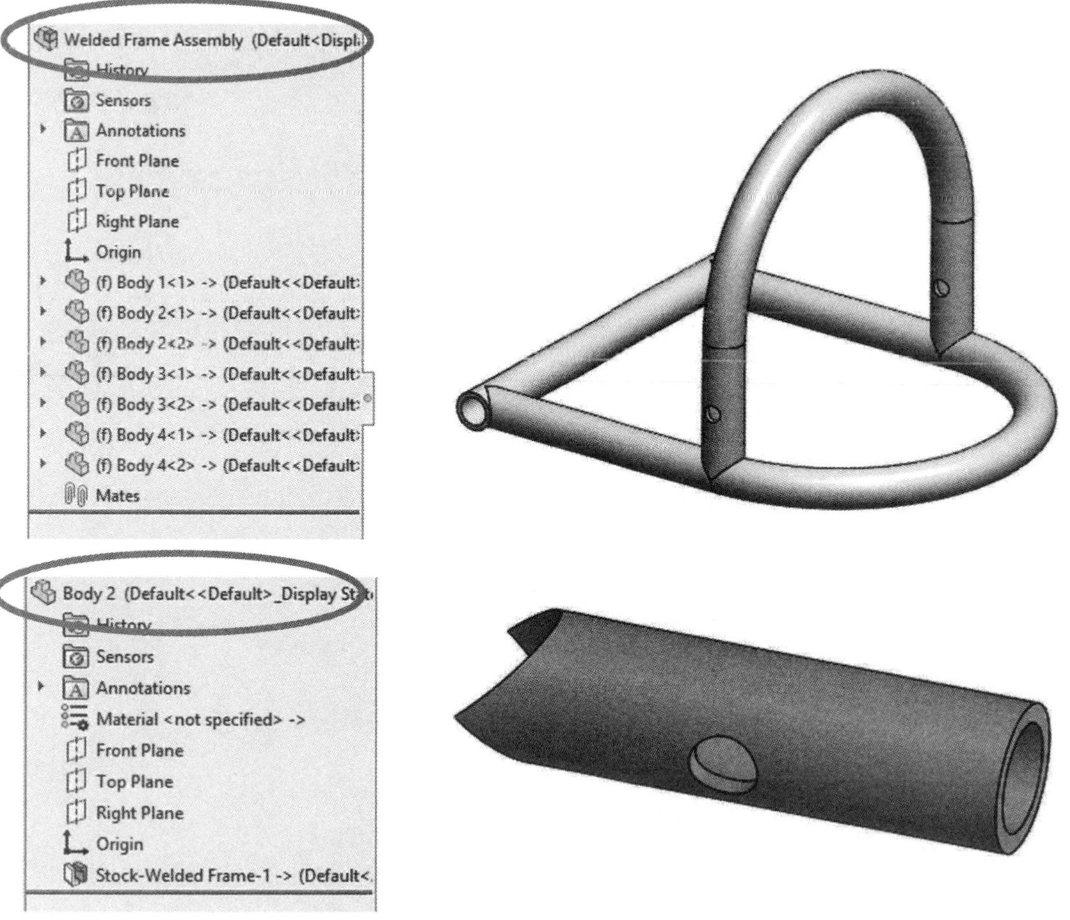

Notes:

Structural Member Libraries

SOLIDWORKS includes a small number of profiles pre-loaded in the "**Structural Member**" library. In this lesson we'll learn how to create a custom library, and later how to download and add more standard profiles. To add custom profiles, we must make a new part, draw the desired sketch profile, and save the sketch as a library in the designated folder for structural members.

20.1. – To make a custom structural profile, add the following sketch in a new part; we can use any plane. Set the part's units to millimeters (mm). Don't forget the sketch points in the middle of each side. The reason to add these points is to be able to use them as location points to position the profile in reference to the weldment sketch. Exit the sketch when done.

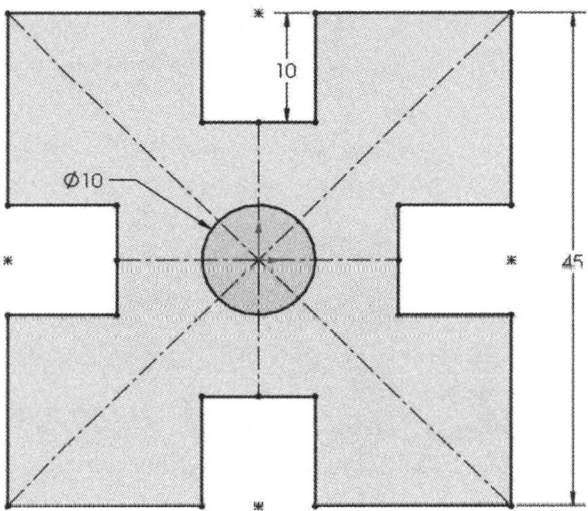

20.2. – In order to have a description in the cut list when using the library, we need to add a custom property (menu "**File, Properties**"). The Property's name has to be *"Description"* and enter *"Square Profile, 45mm x 45mm"* for the value. Click OK to add the description and continue.

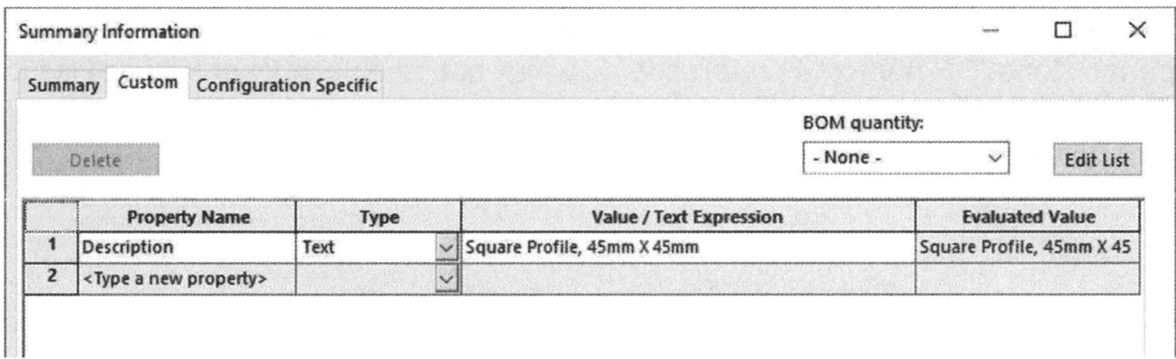

20.3. - To save the new weldment profile, we need to decide if we want to add it to the existing library folders, or if we want to create a new library folder. In the first case we need to locate the folder where the library files are located. To find it go to the menu "**Tools, Options, System Options, File Locations**," from the drop-down menu select "Weldment Profiles" and save it to this location. The default folder location is:

"**<*SW_Install_Folder*>\lang*language*\weldment profiles**"

The sub-folders in this location reflect the different standards and types of libraries available. To add the new profile, pre-select the sketch in the FeatureManager and save the file as a Library Feature Part (Lib Feat Part *.sldlfp). For our exercise we'll save the library as "**45 x 45**" in the following folder:

***<SW_Install_Folder>*\lang\english\weldment profiles \iso\rectangular tube**

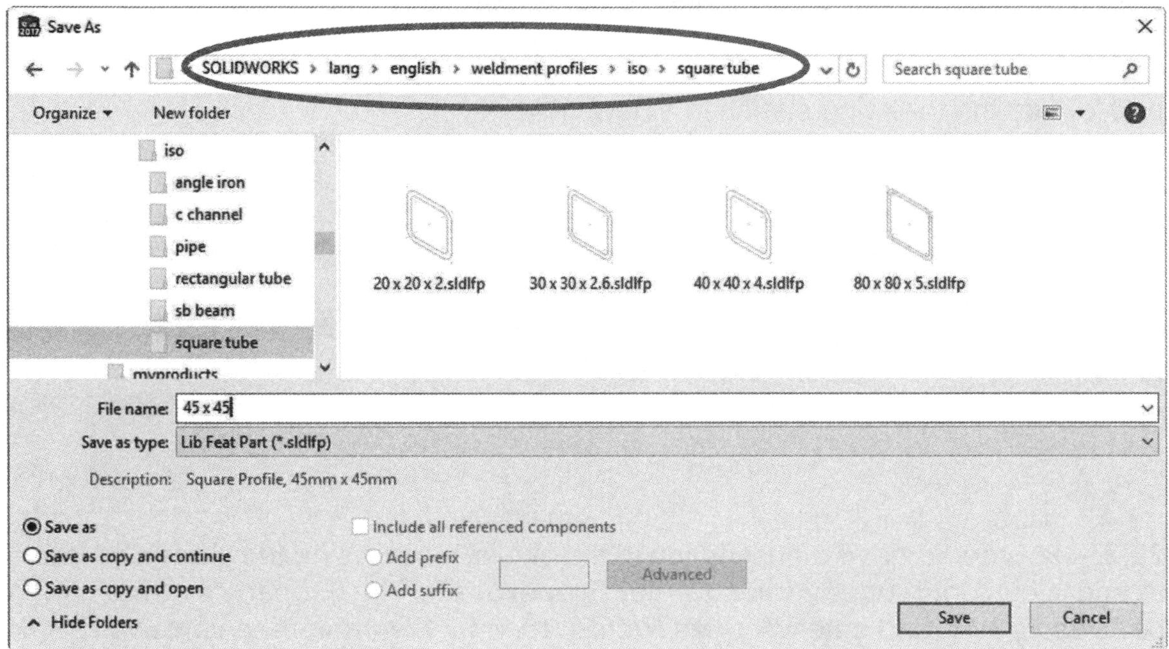

If the library can be saved to this folder, the profile is already available for structural members, but depending on your operating system and user permissions, OR personal preference, you may not be able or want to save files to the *'Program Files'* folder. If you get this warning the library must be saved to a new location.

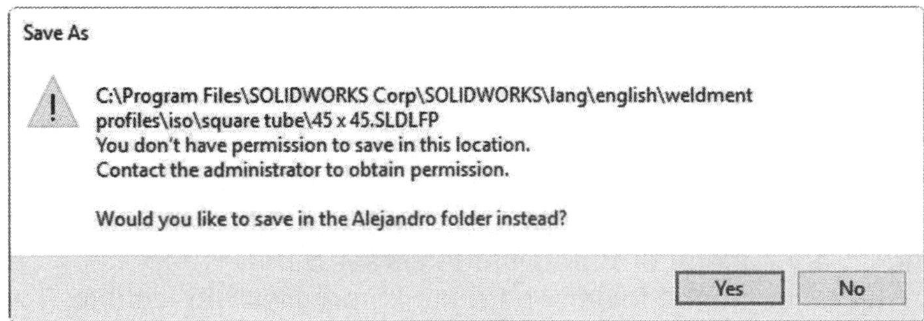

If this is the case, the new profile must be saved to a folder structure similar to the original weldment profiles location. We have a home folder with the "ansi inch" and "iso" folders which correspond to the "Standard." The sub-folders in each of these correspond to the "Type" field and the libraries in each folder are added to the "Size" field.

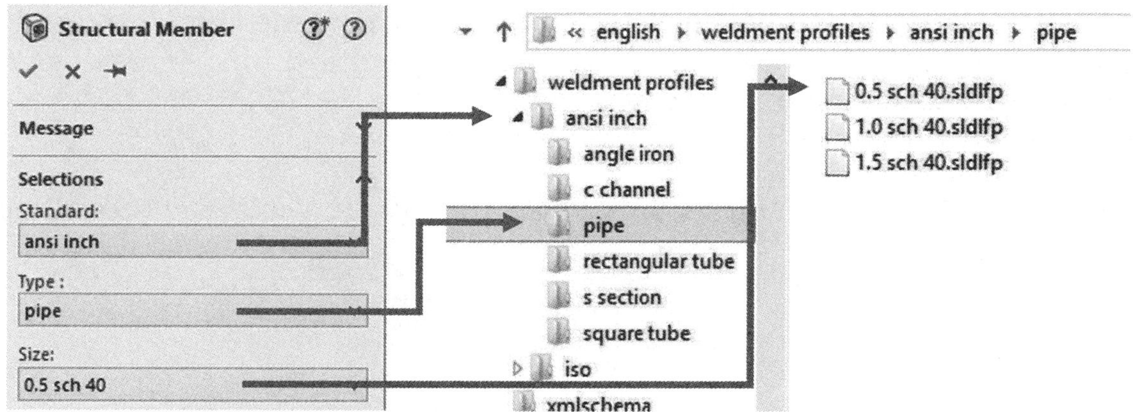

20.4. - In order to add a new location for weldment profiles, create a folder structure similar to the default location. The folder names can be different, but they must have a *'home'* folder (*MyProfiles*), one or more *'standard'* folders (*Metric*) and one or more *'type'* folders (*Square*). Using Windows Explorer, add the folder structure at a location of your choosing and save the library as "**45 x 45**" in the *'type'* folder.

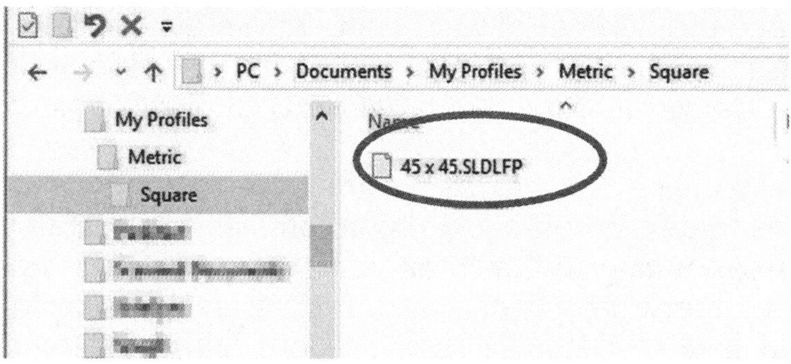

20.5. - To add the new folder as a weldment profiles library, go to the menu "**Tools, Options, System Options, File Locations**," select "Weldment Profiles" from the drop-down list; click the "Add" button, browse to the *'home'* folder (in this example "*MyProfiles*") and select OK to finish.

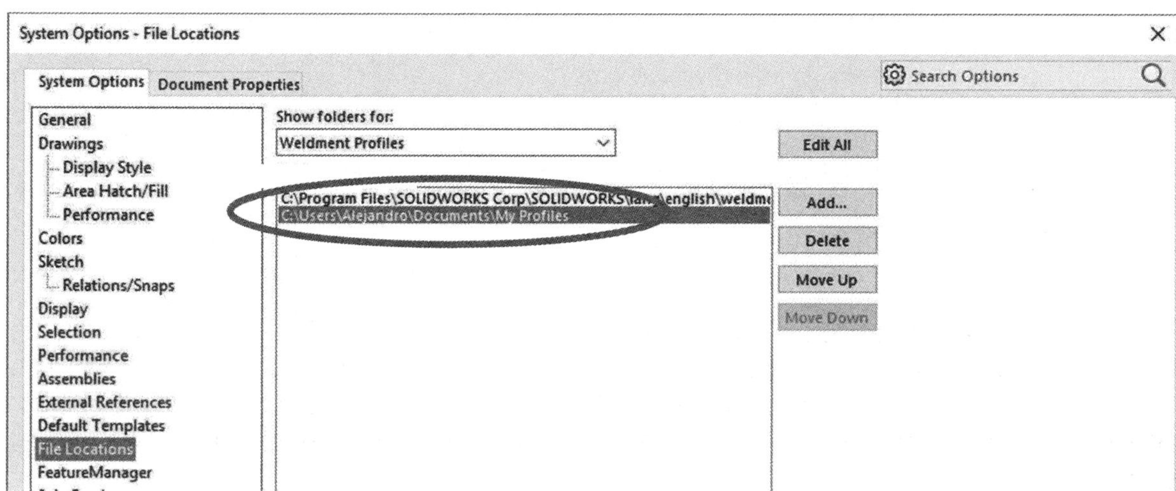

And now the new library is ready for use in the "**Structural Member**" command.

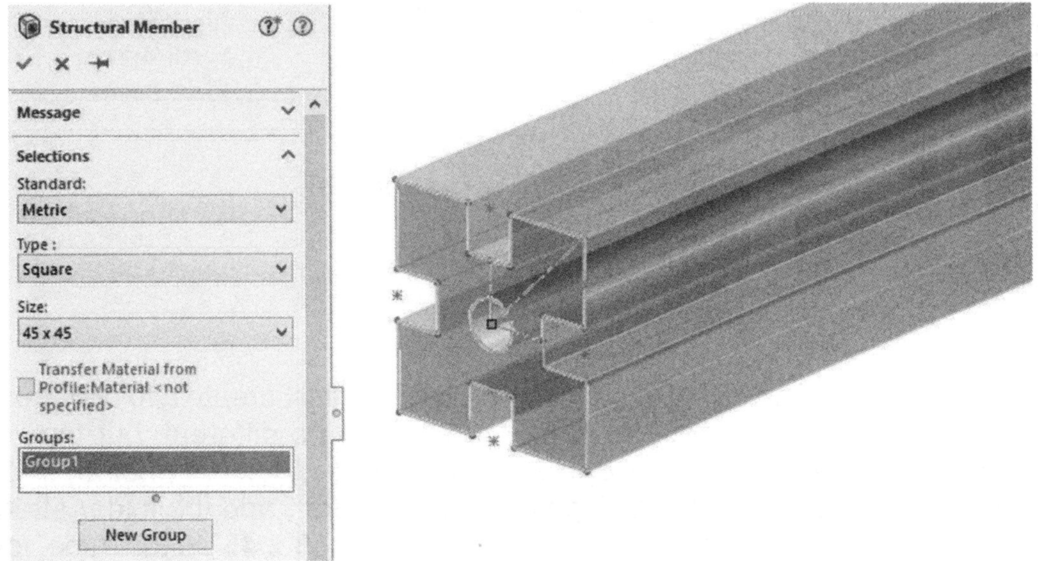

20.6. - To increase the number of weldment profiles available for our designs, we can download a zip file with a more complete list of weldment profiles from the SOLIDWORKS' "Design Library, SOLIDWORKS Content, Weldments" section.

After downloading the zip file, extract its contents to the newly added weldment profiles folder. When adding more profiles to our library, remember the folder name corresponding to the "Standard" cannot be the same as an existing "Standard" folder, otherwise only one will be listed; in this example the downloaded files were saved to the custom library location, and the "Standard" folder was renamed "ANSI Profiles."

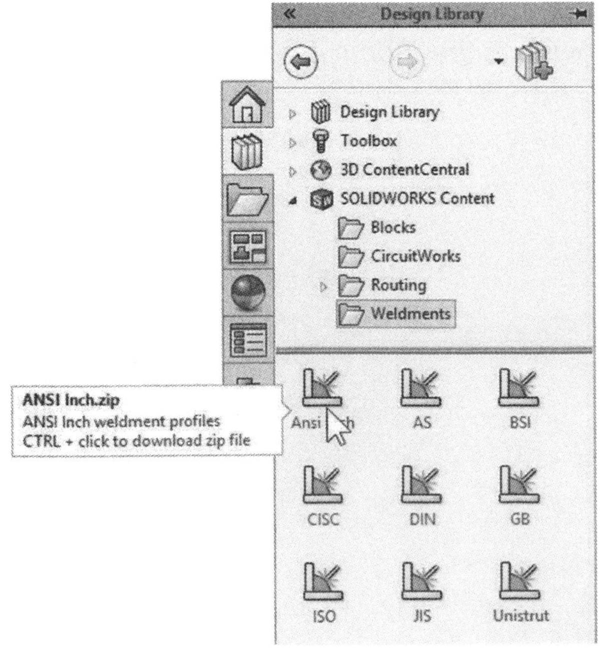

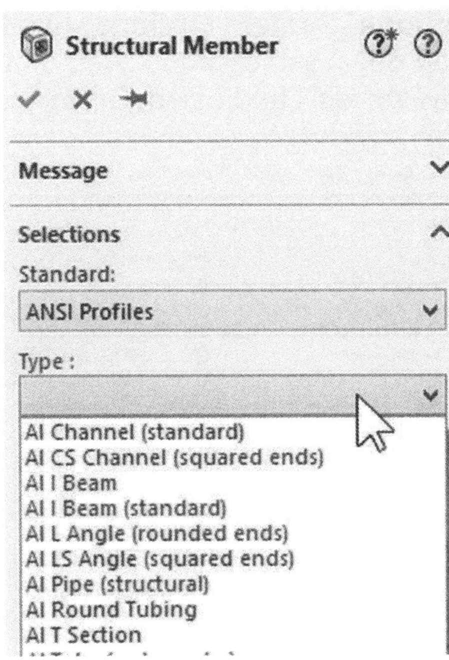

Exercise: Using the Weldment Profile library created, build the next structure. Note the dimensions given are in millimeters and are all external dimensions; this means that the finished structure cannot be bigger than these dimensions in any direction. After completing the structure, make a detail drawing including a cut list. This structure is not welded as it is usually assembled using special hardware. Make the horizontal elements continuous and the vertical elements intermittent.

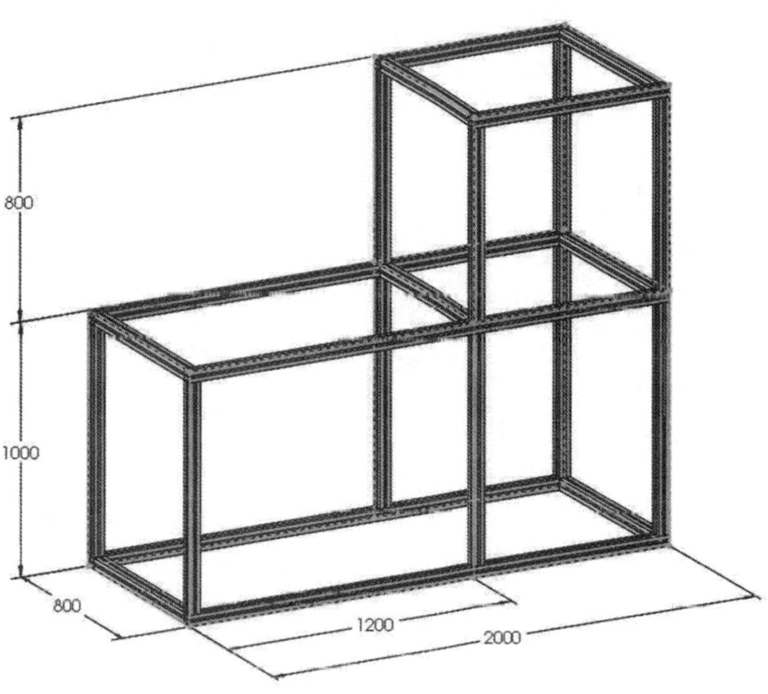

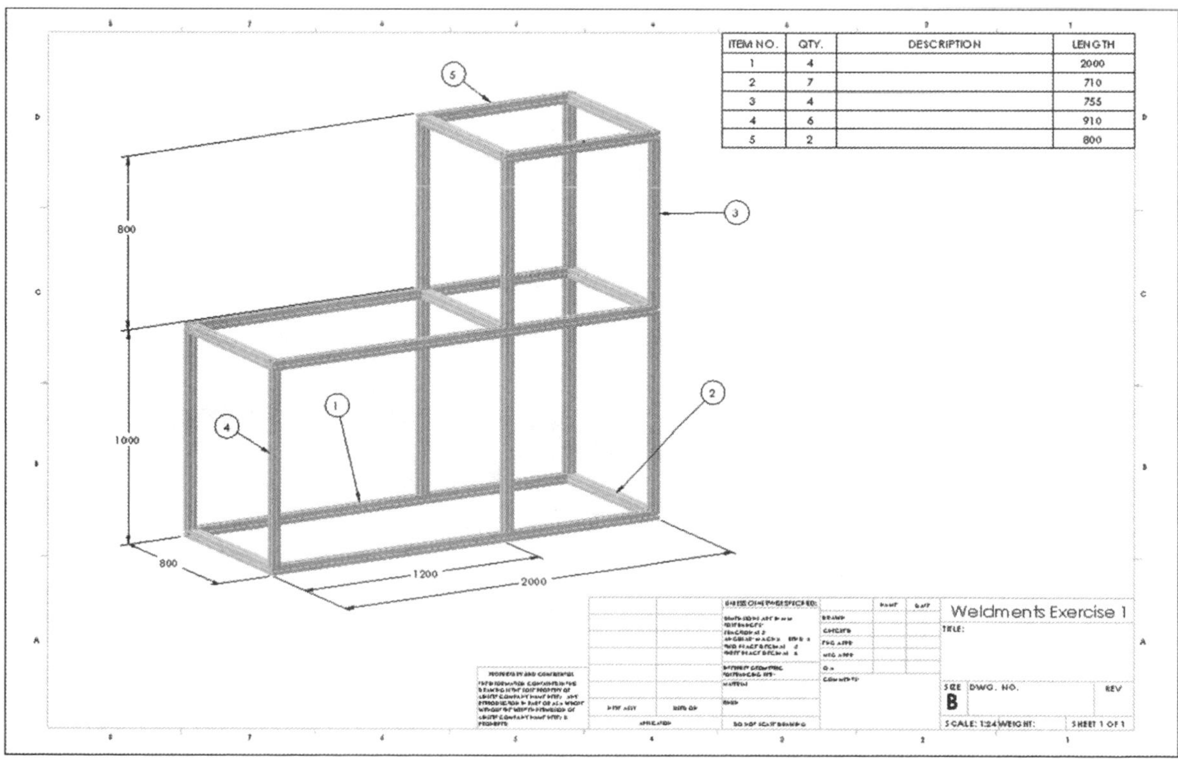

ITEM NO.	QTY.	DESCRIPTION	LENGTH
1	4		2000
2	7		710
3	4		755
4	6		910
5	2		800

Exercise: Build the following dune buggy frame using a 1″ sch-40 pipe profile; trim all ends and add 0.125" weld beads to all joints. Open the *'Dune Buggy'* frame sketch from the accompanying files.

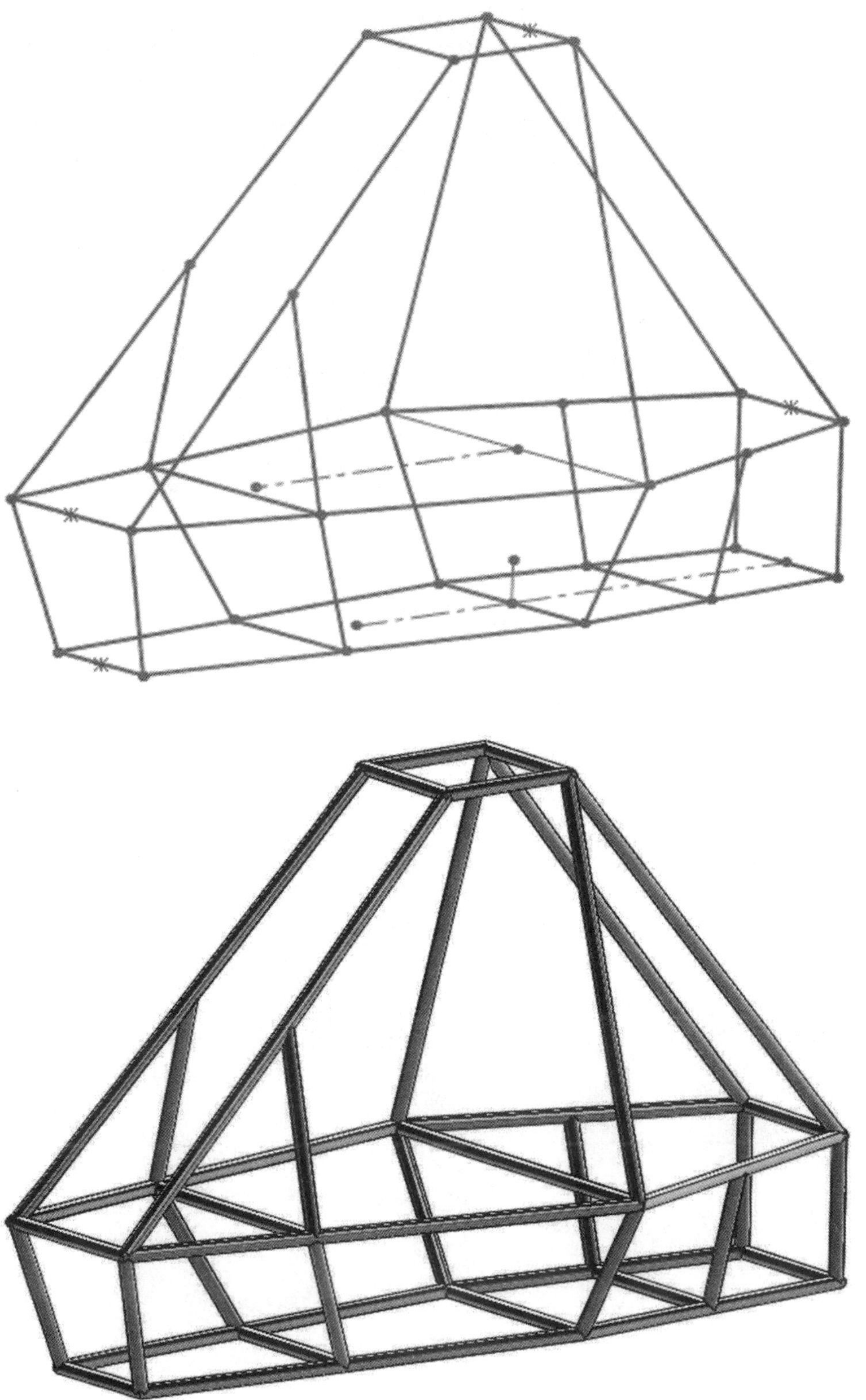

TIP: To correctly trim the unions where three or more elements meet, trim two at a time. You may have to turn off the "Allow extension" in the "Trim/Extend" command. If this option is turned off, the element is only trimmed and will not extend to the trimming boundary.

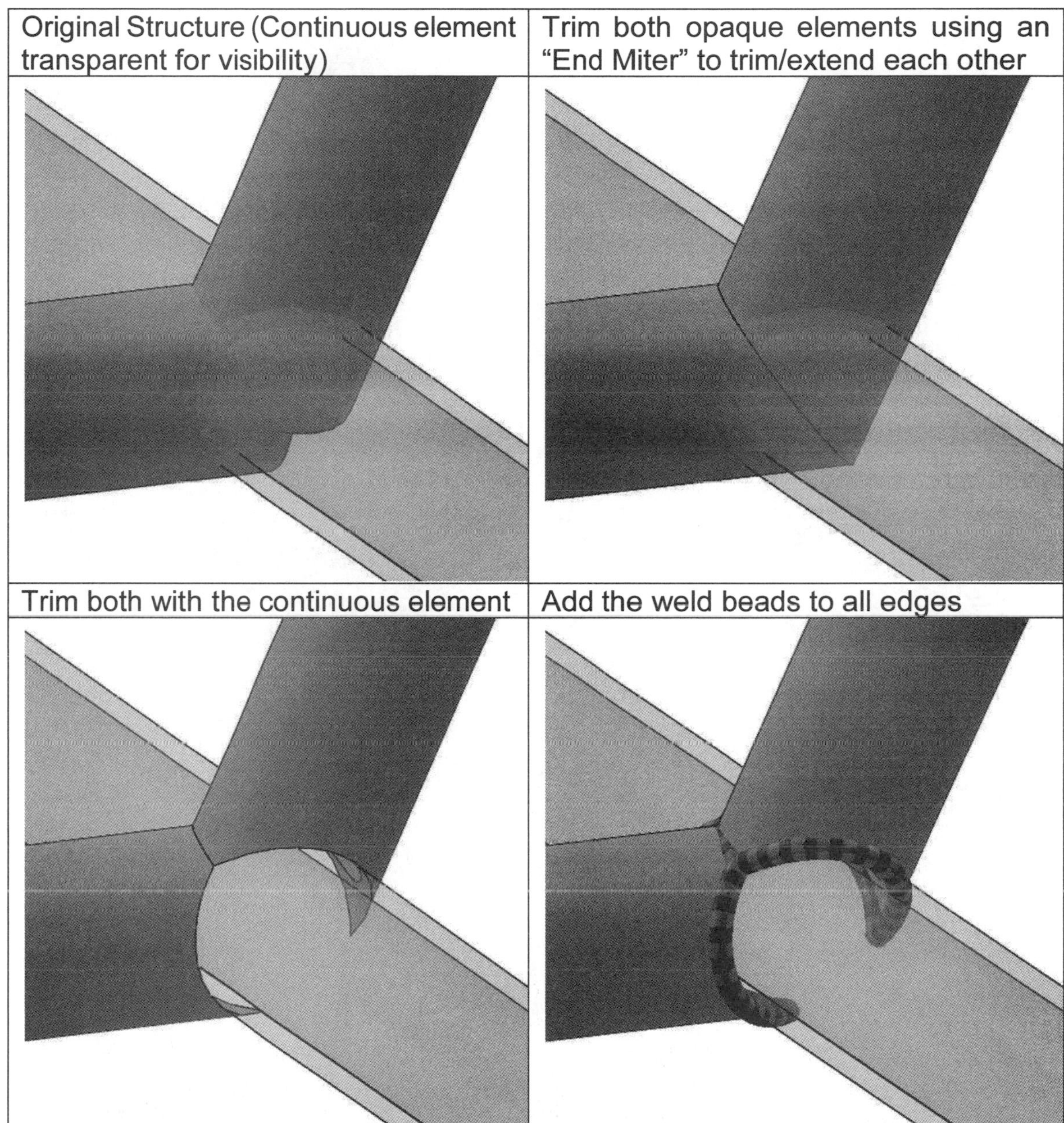

Original Structure (Continuous element transparent for visibility)	Trim both opaque elements using an "End Miter" to trim/extend each other
Trim both with the continuous element	Add the weld beads to all edges

Exercise: From the accompanying files open the *'Stage Steps'* frame sketch, build the following stage steps using a 1.5″ sch-40 pipe profile and change the material to AISI-1020 steel. Merge the arc segment bodies for a continuous bent tube. Trim the structure as shown in the following details, and add 0.125" weld beads to all joints.

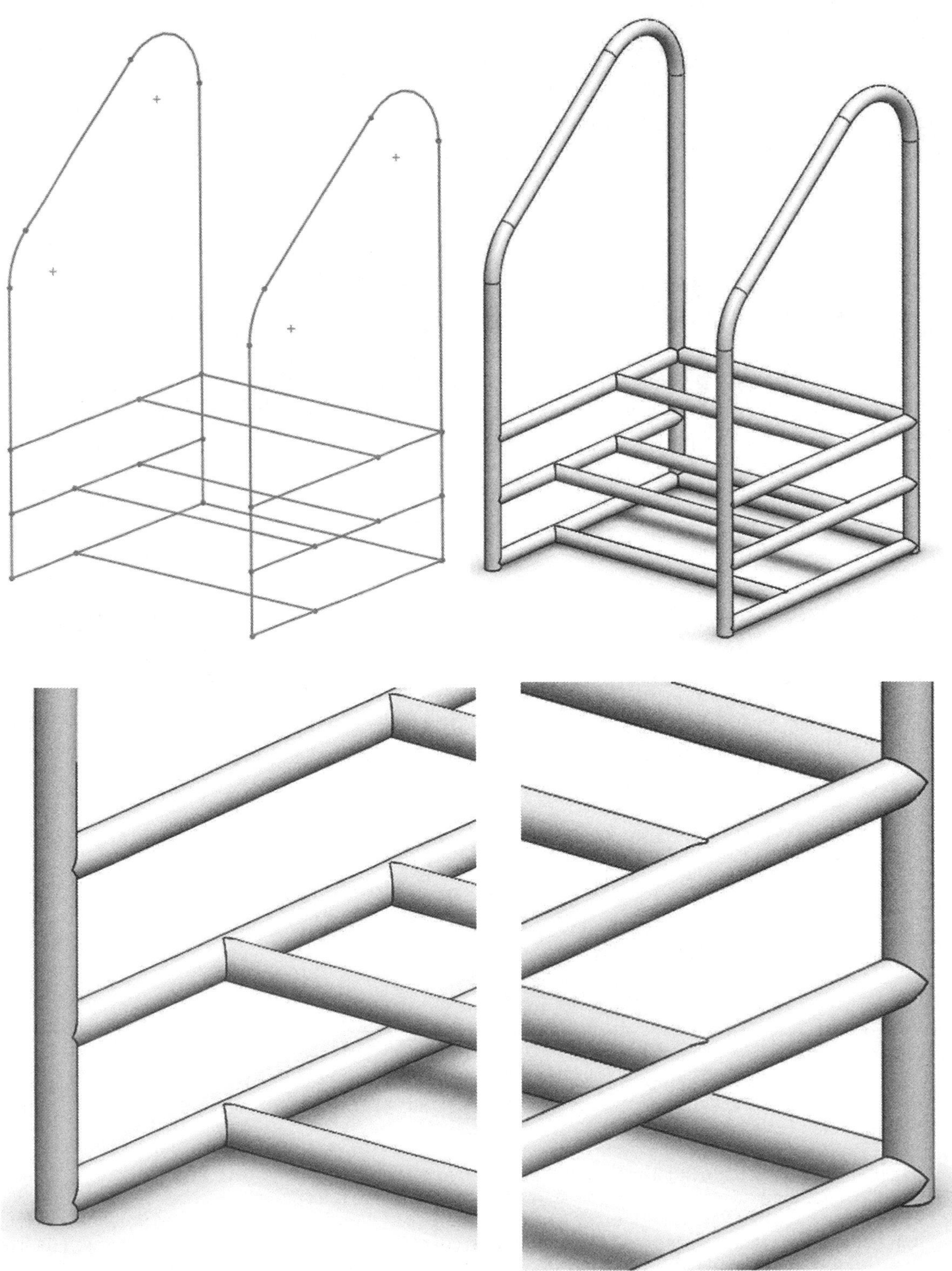

Review and Questions:

a) Name three geometric relations available in a 3D sketch that are not available in a 2D sketch.

b) Describe how to create a Derived sketch.

c) What is the difference between using a 2D or a 3D sketch for weldments?

d) Name the conditions to be able to select multiple sketch segments at the same time for a structural member group.

e) Can multiple gussets and end caps be added at the same time?

f) How can we add a description to gussets and end caps in a drawing's cut list?

g) If welded part bodies are saved to an assembly, are identical bodies saved as a unique file or different files?

Answers:

a - Along X, Y, Z, On-Plane, Normal, ParallelYZ, ParallelZX.
b - Pre-select the sketch to derive and the plane/face to place the derived sketch and click in the menu "Insert, Derived sketch."
c - None. They both work the same.
d - They must be parallel or continuous connected by an endpoint.
e - Gussets have to be added one at a time, end caps can be added multiple at a time.
f - Right mouse click in a "Cut list" group and select "Properties," then fill in the missing information.
g - Each body, even identical ones, are saved to a different file unless the option "Derive resulting parts from similar bodies or cut list" is turned on; in this case identical bodies are saved to a single part file.

Notes:

Surfacing and Mold Tools

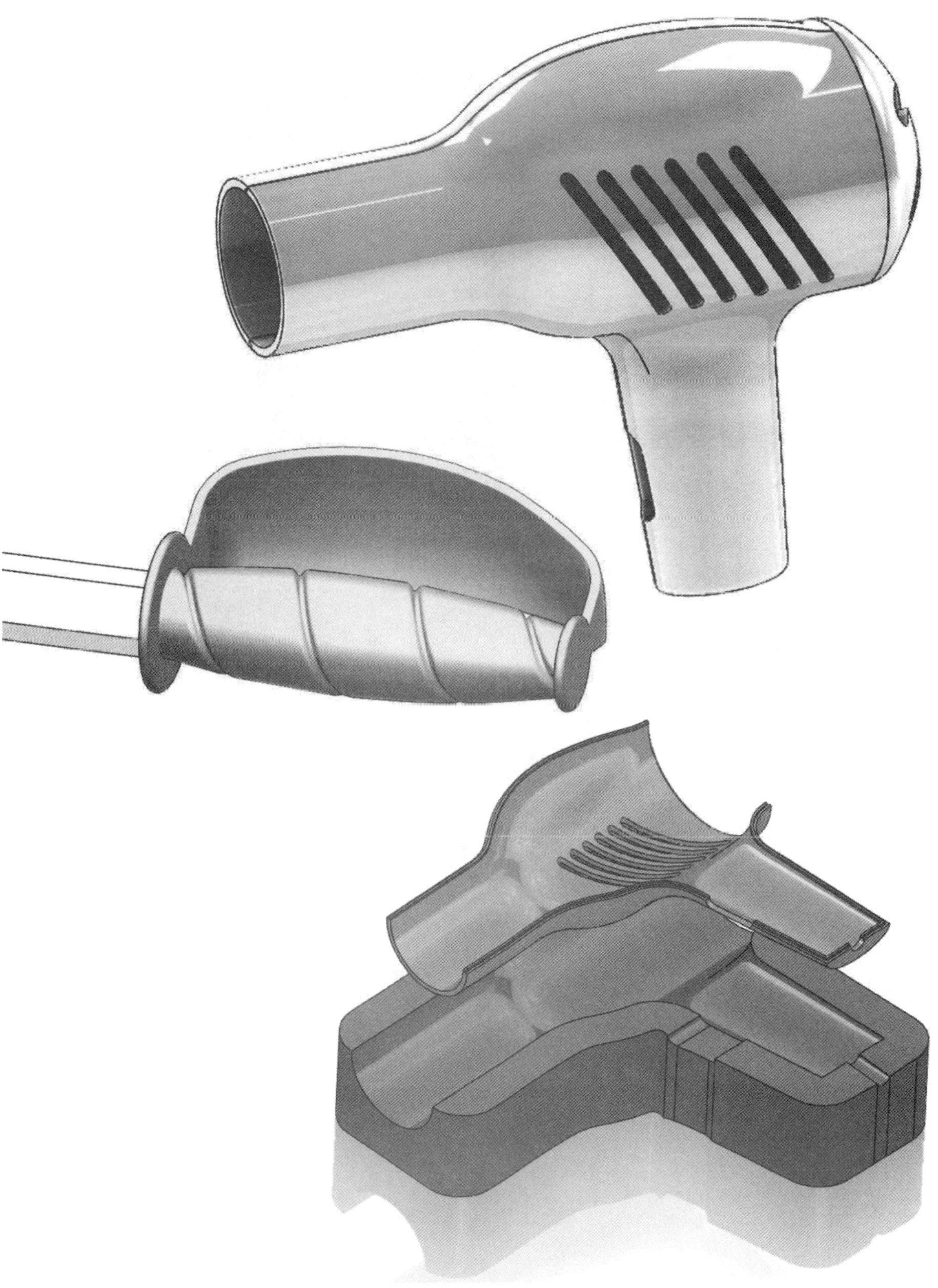

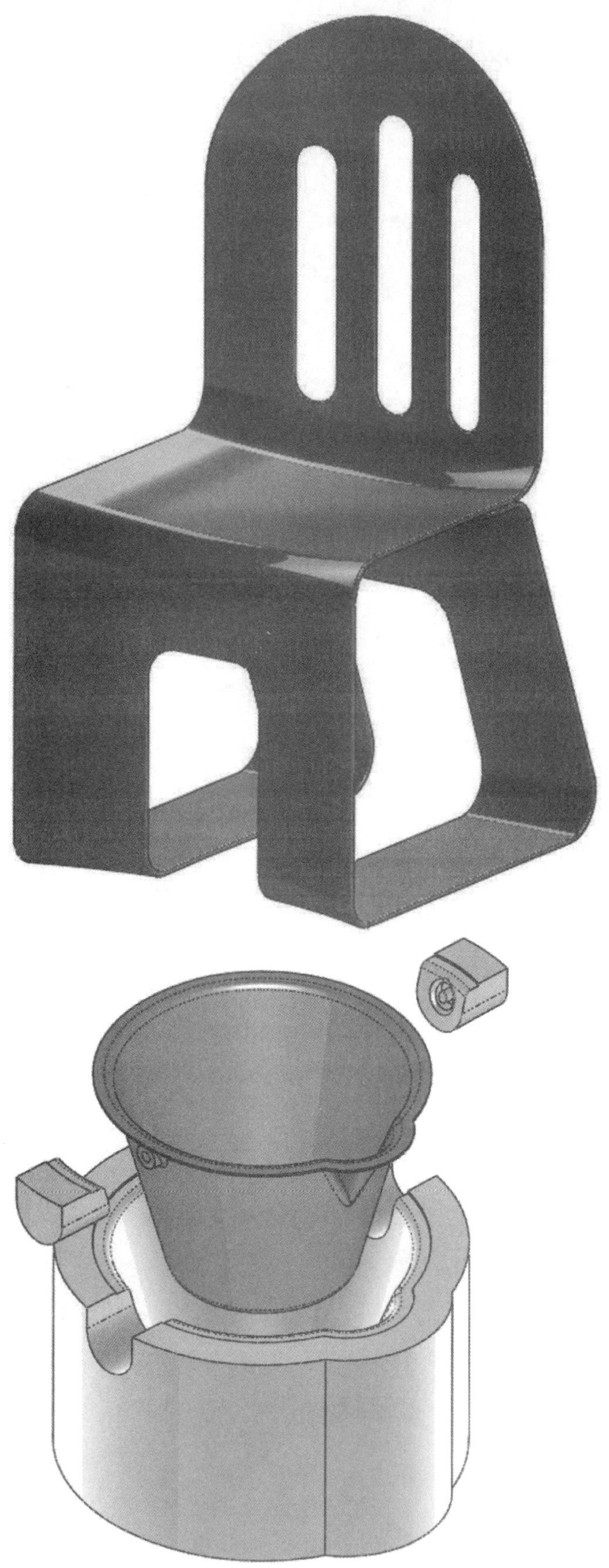

Surface Modeling

So far, we have been working with single and multi-body solid parts, and now we'll start working also with surfaces. By including surfaces in our models, we can create increasingly complex designs that would otherwise be difficult to model only with solid bodies. Working with both solids and surfaces is called 'hybrid modeling', as we incorporate both solid and surface bodies in the same model. When we work with a hybrid model, the surfaces are usually used as support geometry, as stopping or trimming boundaries for solids and/or other surfaces.

Background and a little history: The first computer models used for design (early 1970s) were wireframe models capable of defining the edges of a part and had severe limitations, especially when it came to defining a three dimensional surface, as it could only be approximated using a mesh of lines with little detail. Think of a wireframe model as a 'stick figure'. A wireframe model is basically what we are able to do now with a 3D sketch – just lines.

Some years later, surface models were developed that allowed more definition of a component, including curved faces, and were a huge improvement over the wireframe models but still had limitations when it came to calculating a part's volume, weight, etc. When we talk about a surface model, think of a parade's floating balloon. It has the surfaces (fabric) and the wireframe (stitches) and is essentially hollow. Similarly, it's safe to say that a surface model *may* have imperfections, like faces not matching correctly creating gaps (think of the balloon with an opening at a corner) or overlapping faces. This would create a problem if we wanted to calculate volume, weight, etc.

Later, in the 1980s (give or take a few years…), solid models became available, but (just like surface modelers) were difficult to use, required special training and were available in high end (read 'expensive') workstations. A design station's cost, complete with hardware and software, would be in the tens of thousands of dollars, affordable only by big corporations or government agencies. Design tools were greatly improved, producing better results faster, making it easier to integrate with computer numerical controlled (CNC) manufacturing.

Some years later (circa 1990s), solid modelers became available to the Windows operating system, making them considerably more affordable to mainstream designers, and being native Windows applications made them easier to learn and use. These solid modelers incorporated many of the advantages surface and solid models had to offer, including better integration with manufacturing, analysis, animation and many other downstream applications.

At present, with the advances in technology making computers faster than ever before (and continuing to get faster every day), we can take advantage of it and create bigger, ever more complex designs in record time, helping us bring better and safer products to market faster than ever before.

A word about hybrid modeling: It is not uncommon to see that dedicated surface modelers, usually available in high end CAD applications, can generate superior and better looking surfaces for automotive, aerospace and consumer product design than many solid modelers (like SOLIDWORKS), and they can also make changes easier and faster maintaining the intended shape, making it very attractive to create highly complex surfaces easier with those tools. With that said, it is common in those industries to generate complex surfaces in one software package and then import them into a mainstream modeler like SOLIDWORKS to complete the rest of the design, including tooling for manufacturing.

When we talk about surface modeling, we are referring to creating models using mainly surfaces, or the 'skins' of parts, and afterwards converting them into a solid or using them to build solids. The tools available to build surfaces include almost all the same features available to solid models like extruded, revolved, swept and lofted surface plus a few more tools specific to surface modeling, most of which will be covered in this lesson.

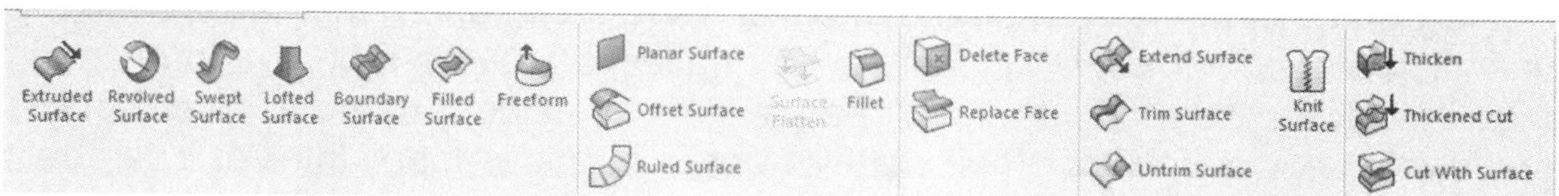

To start modeling with surfaces, we'll make a hair dryer using a few simple surfaces, and then we'll convert them into a solid model. After finishing, we'll split the model to make a file for each of the different parts to practice and learn more about multi body parts.

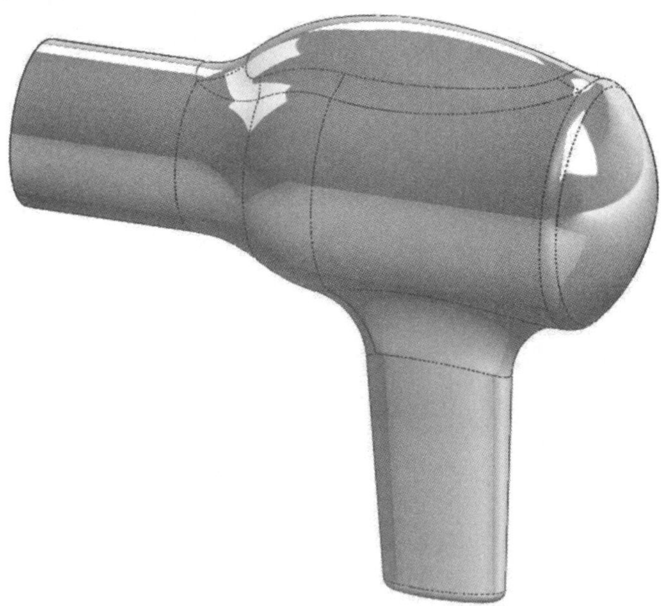

If the Surfaces toolbar is not visible, activate it by right mouse clicking in a CommandManager's tab and select "**Surfaces**," and/or the menu "**View, Toolbars, Surfaces**."

21.1. - To save time and focus on the surfacing functionality, open the file *'Hair Drier.sldprt'* from the included files. It has several pre-made sketches and planes for us to start working.

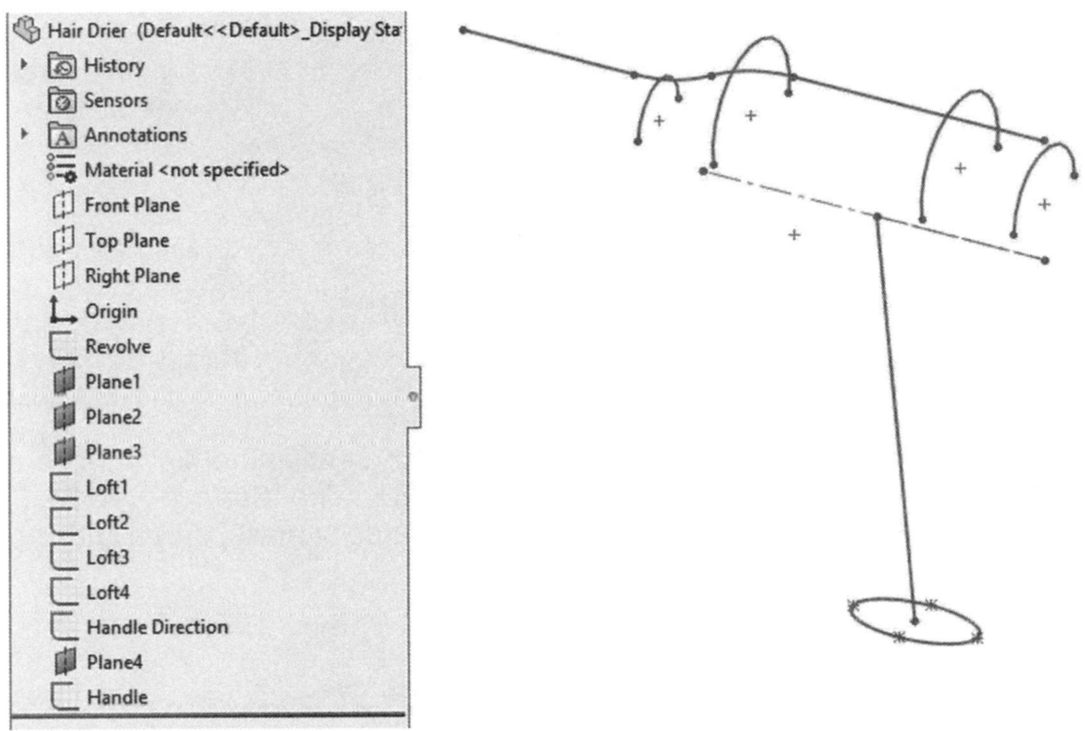

21.2. - The first feature will be a revolved surface. Select the sketch named *"Revolve"* in the FeatureManager and click the "**Revolved Surface**" command from the Surfaces tab in the CommandManager, or the menu "**Insert, Surface, Revolve.**" The sketch's centerline is automatically selected. Revolved surfaces follow the same rules as a revolved feature.

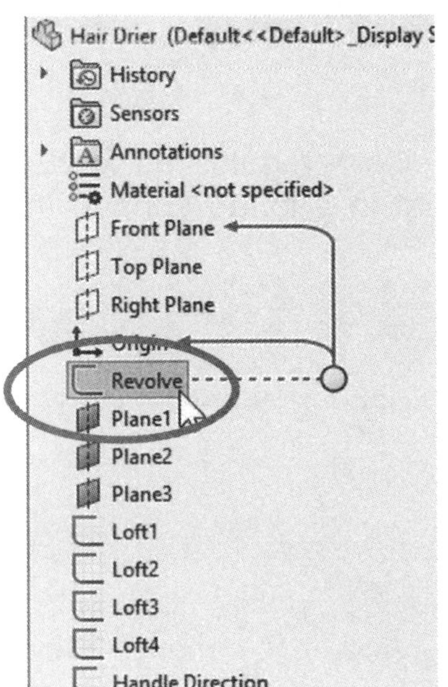

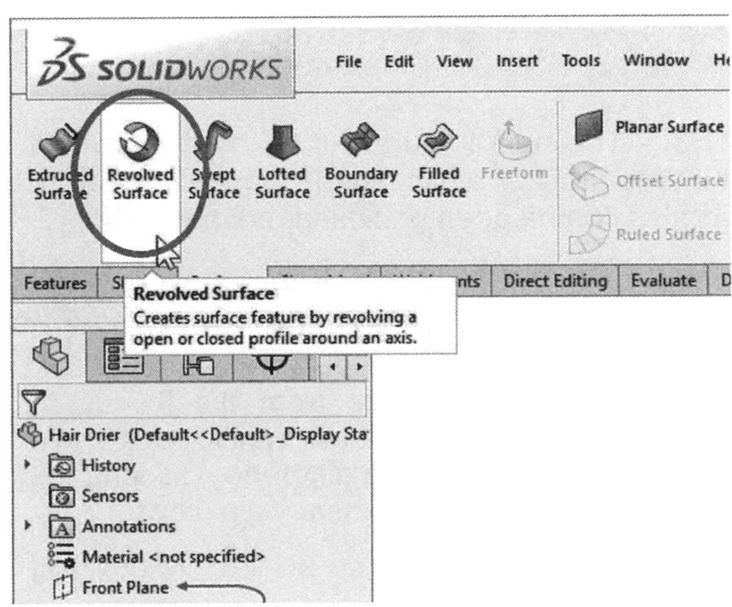

Make the revolved surface 360 degrees. Click OK to complete it.

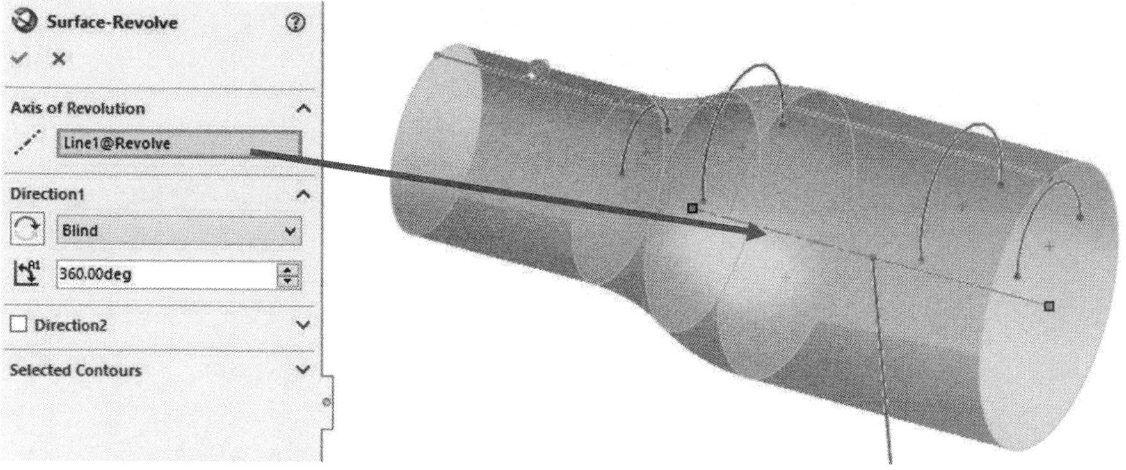

When we work with surfaces, immediately after creating the first surface a "**Surface Bodies**" folder is added to the FeatureManager. The *"Surface-Revolve1"* we just made is only a surface. It has no thickness, volume or mass, only a surface.

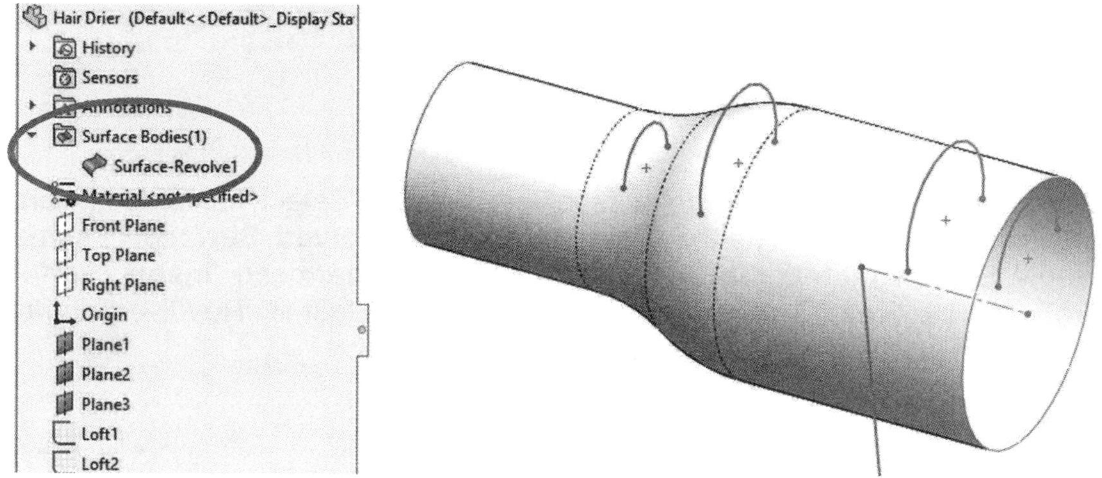

21.3. - The next feature will be a lofted surface. Just like a regular loft feature multiple profiles are needed, the main difference is that in a lofted surface the profiles can be open or closed, but not both.

Select the "**Lofted Surface**" icon or the menu "**Insert, Surface, Loft**," and select the four sketches named *"Loft1"* through *"Loft4"* on the screen or from the fly-out FeatureManager. Select them front to back or back to front, it makes no difference, as long as they are selected in consecutive order.

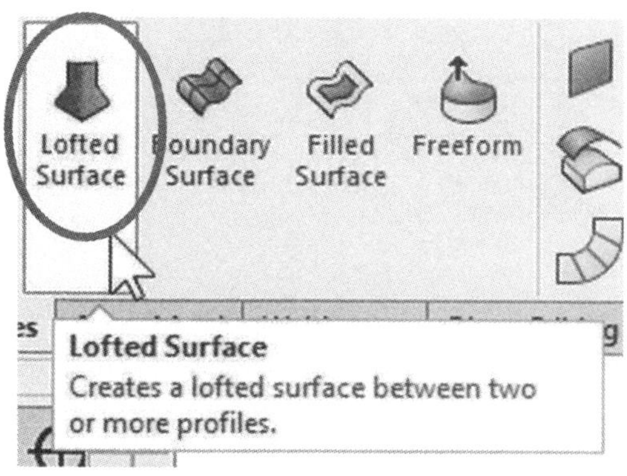

Just like solid body lofts, we must select the sketches near the vertices that will be connected by the loft to prevent twisting and avoid undesired results. A lofted surface can include guide curves, centerline, start and end constraints; our example will be a simple loft. Leave the rest of the loft options to their default value for this example and click OK to complete it.

The main difference between a lofted surface and a solid body loft is that in a lofted surface we can use open *or* closed sketches, curves, edges, etc. to create irregular, complex surfaces.

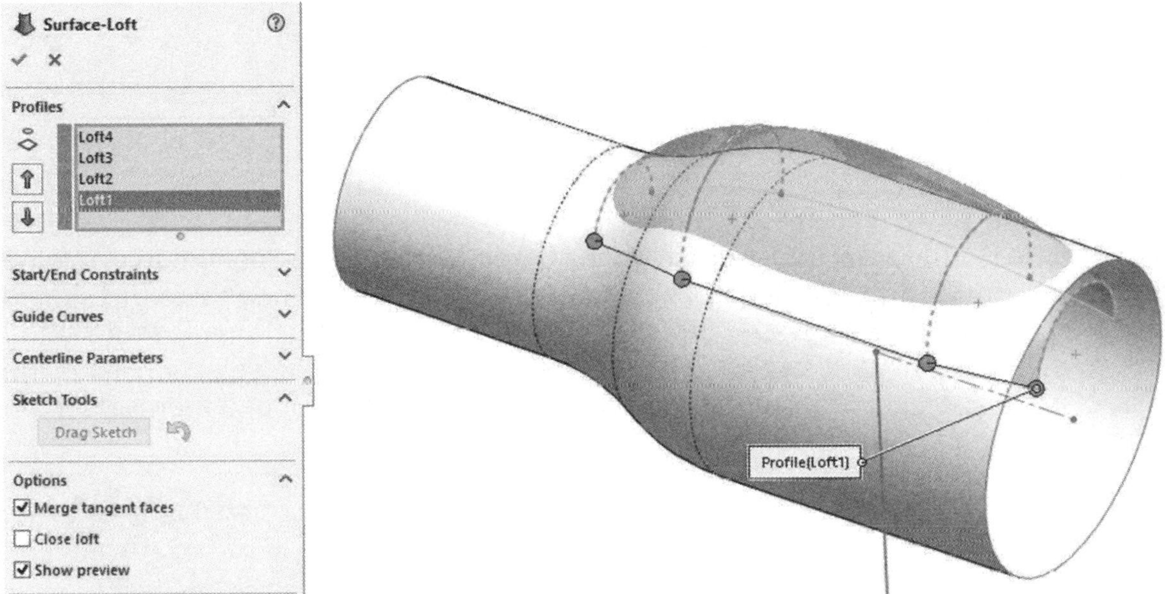

After the loft is created we have two surface bodies. Note that surfaces do not 'merge' automatically as solids do; they simply intersect each other and remain as separate bodies.

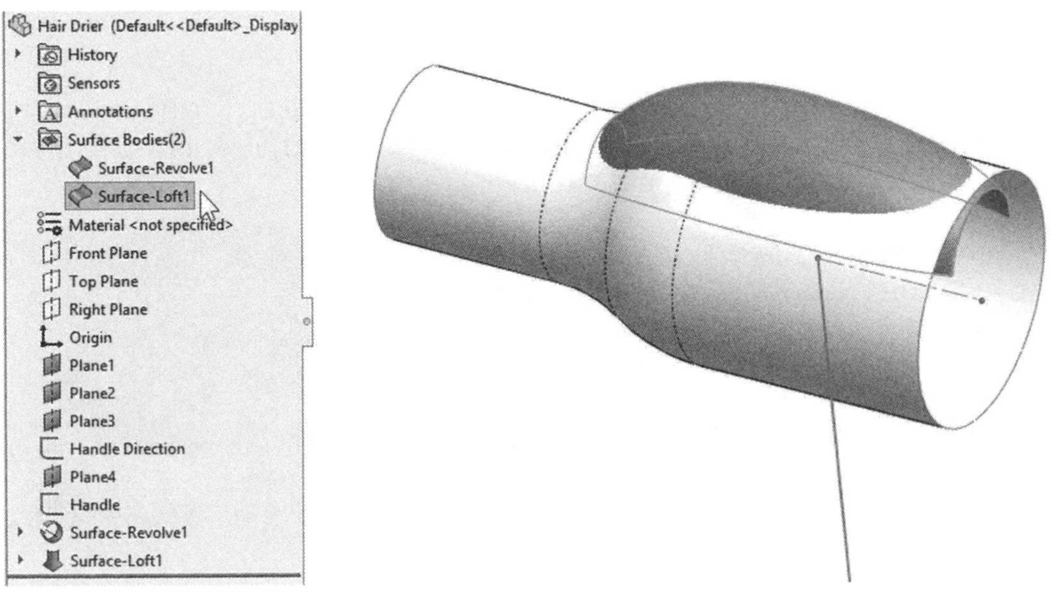

21.4. - For the next feature, we'll make the hair drier's handle using an extruded surface with a draft, along a defined direction. Select the sketch named *"Handle"* in the FeatureManager and click the "**Extruded Surface**" icon, or the menu "**Insert, Surface, Extrude**."

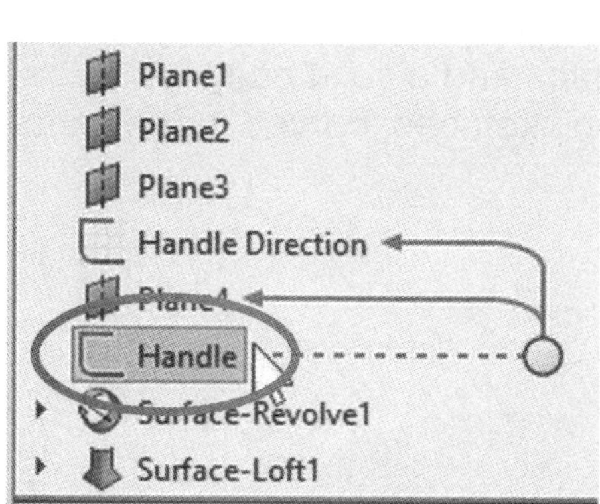

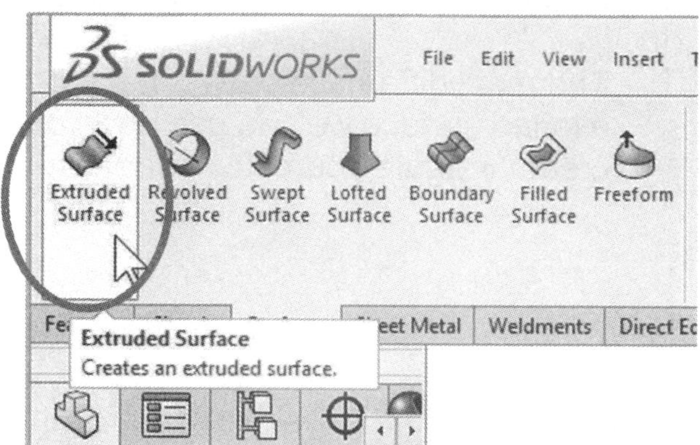

Extrude the new surface using the "**Up to Surface**" end condition up to the revolved surface. By default, extrusions (solid or surface) are normal to the sketch plane; in this case we'll make the extrusion along the direction indicated by the sketch called *"Handle Direction."* Right under the end condition selection list is the "Direction of Extrusion" selection box. Click inside to activate it and then select the *"Handle Direction"* sketch as shown.

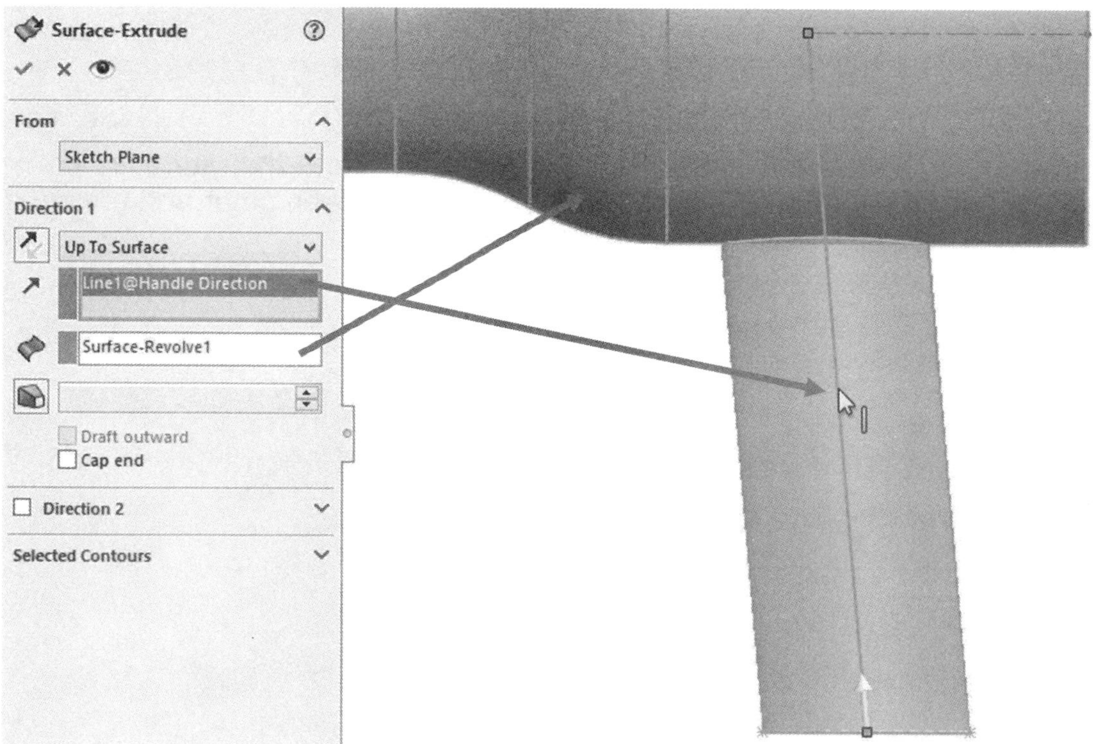

The "Direction of Extrusion" option is also available for extruded boss and cuts in solid bodies.

Another option we'll use in the extrusion is the option to add a draft as we extrude. Adding a draft makes the body (surface or solid) grow or shrink with a defined angle. To turn the draft option on, click in the "**Draft On/Off**" icon; the angle box will be enabled as well as the "**Draft outward**" checkbox. For our example, enter a 3 degrees angle and check the "Draft outward" checkbox to make the surface wider as it is extruded. Click OK to finish the surface.

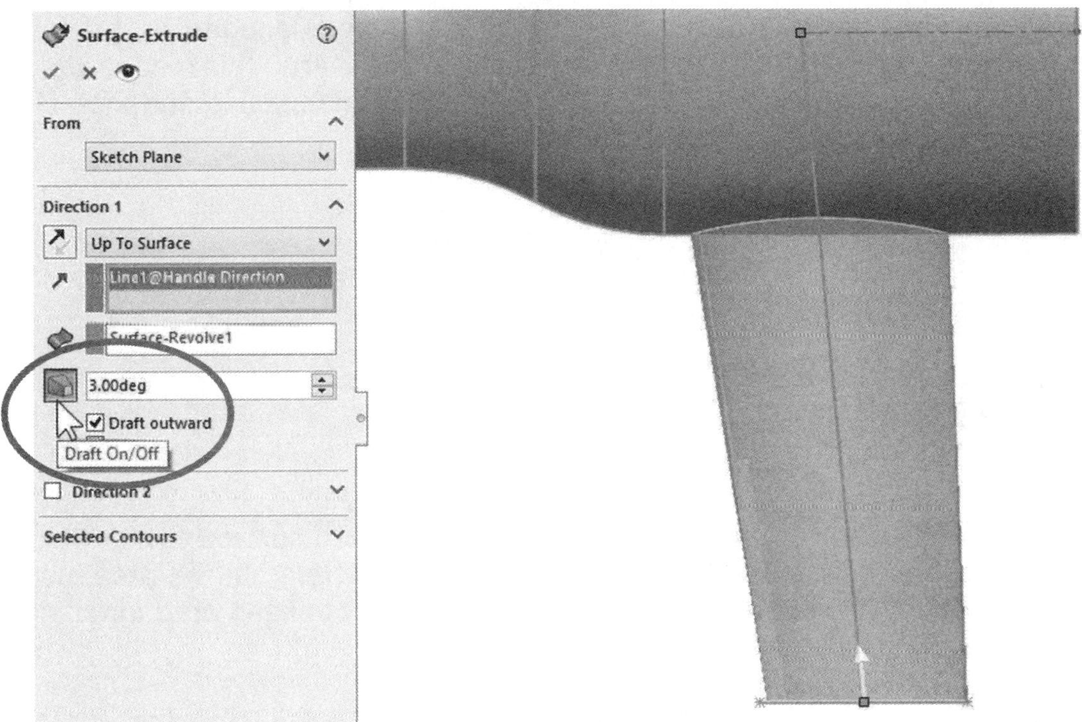

Now we have three surface bodies. Hide the *"Handle Direction"* sketch. It will not be needed anymore.

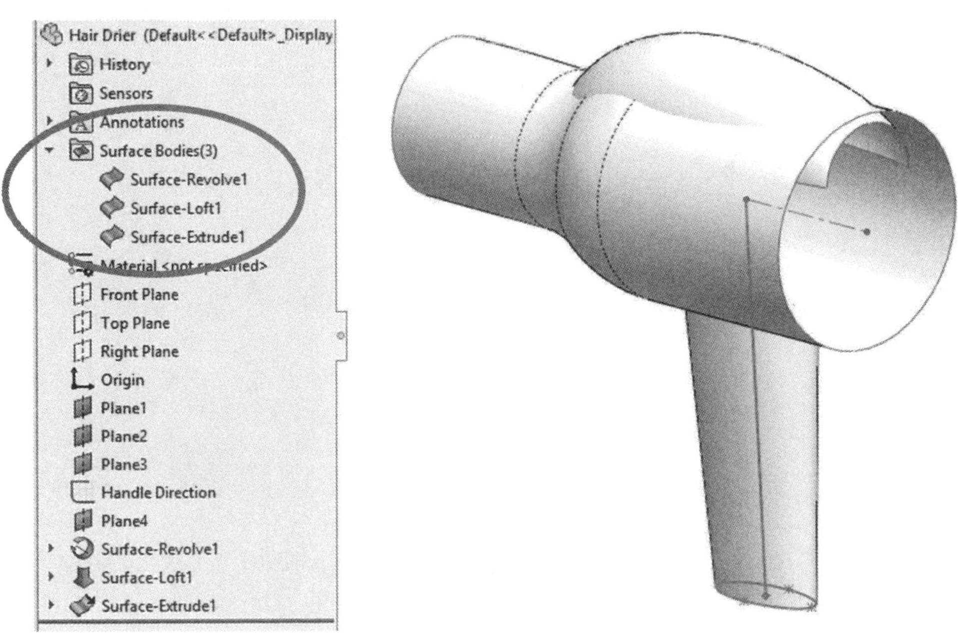

21.5. - The next step is to trim the surfaces to make them match exactly, and form the outside of the hair dryer. By trimming the surfaces, we are going to make all trimmed surfaces touch at the edges, with the goal to form a single, closed, continuous surface that will be converted into a solid body. There are several ways to trim surfaces, and they can be trimmed using other surfaces, planes, or a sketch. The first step will be to trim the revolved surface with the handle surface.

Select the "**Trim Surface**" icon from the CommandManager, or the menu "**Insert, Surface, Trim**." Under "**Trim Type**" select "**Standard**"; this means we will use a trim tool (Sketch, Plane or Surface) to cut other intersecting surfaces. Add the handle surface in the "**Trim tool**" selection box.

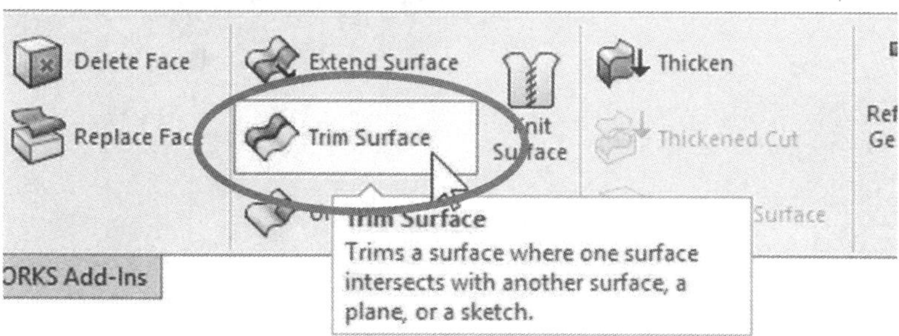

Using the "Keep selections" option, add the revolved surface to the "pieces to keep" selection box, select the revolved surface anywhere but the area inside the handle surface; this is the part of the surface we want to keep after trimming.

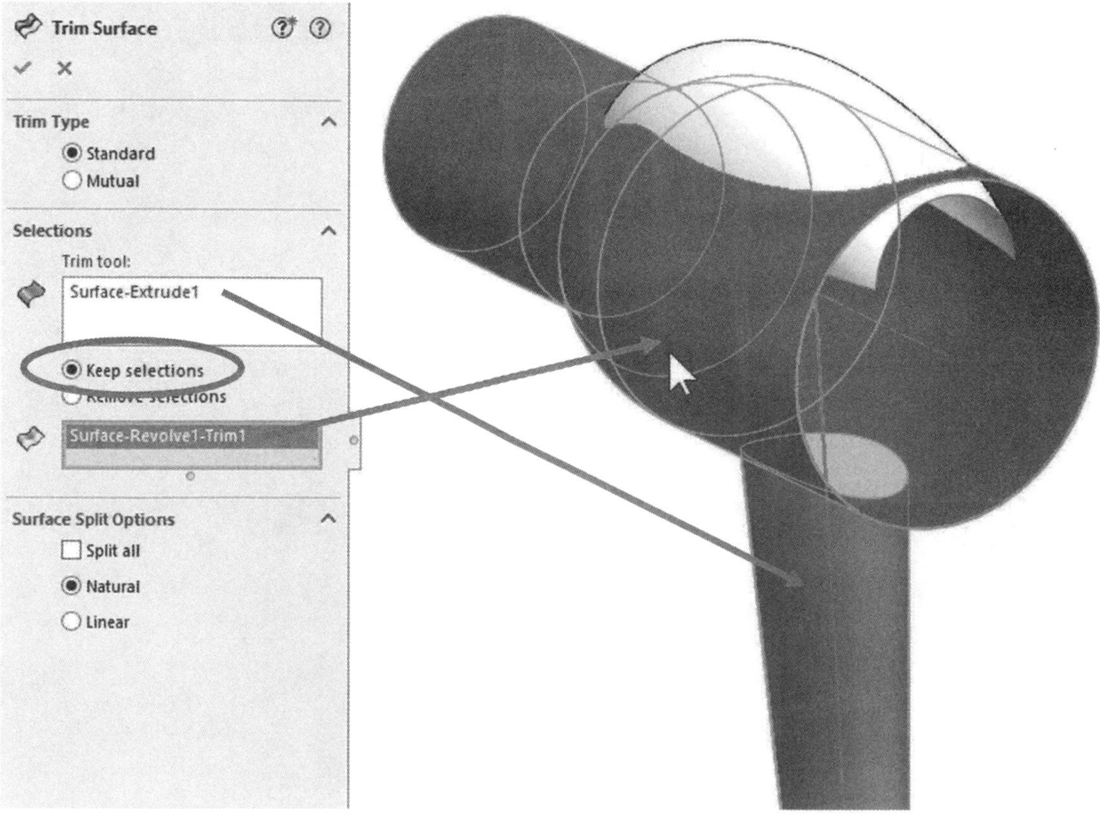

Optionally, if we use the “Remove selections” option, we must select the elliptical surface inside to remove it, leaving the rest of the surface. Use either approach, and click OK to complete. Trimming surfaces using this approach only trims one surface. In the next step, we'll see how to mutually trim two surfaces.

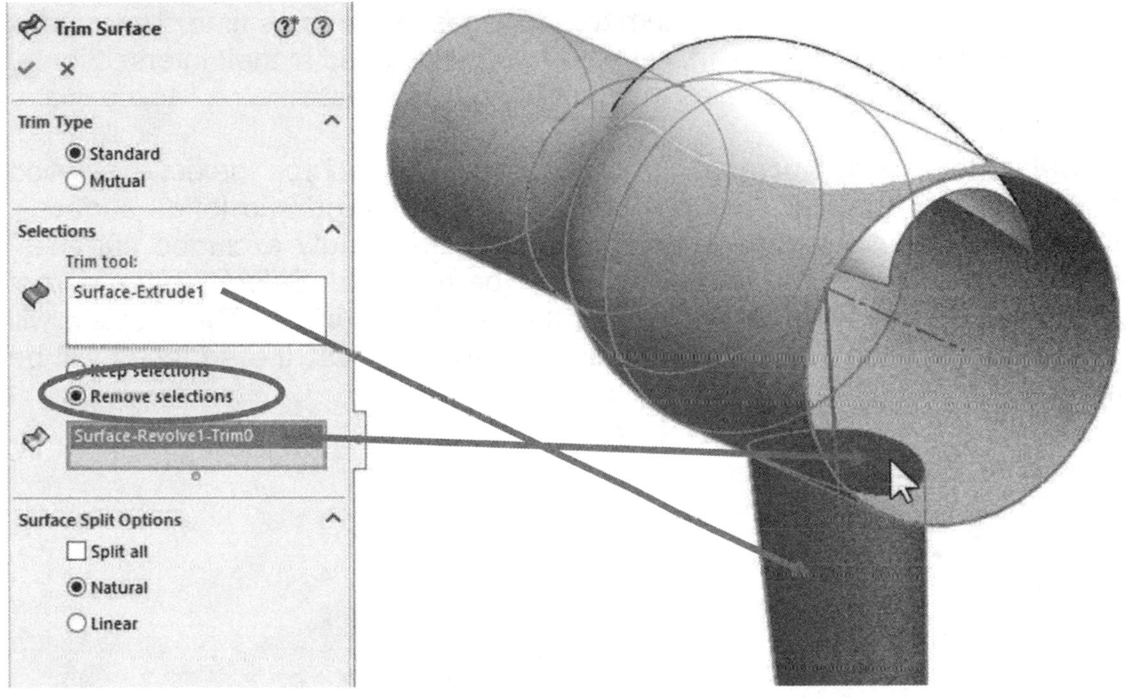

This is the result of trimming the first surface; we still have three surface bodies.

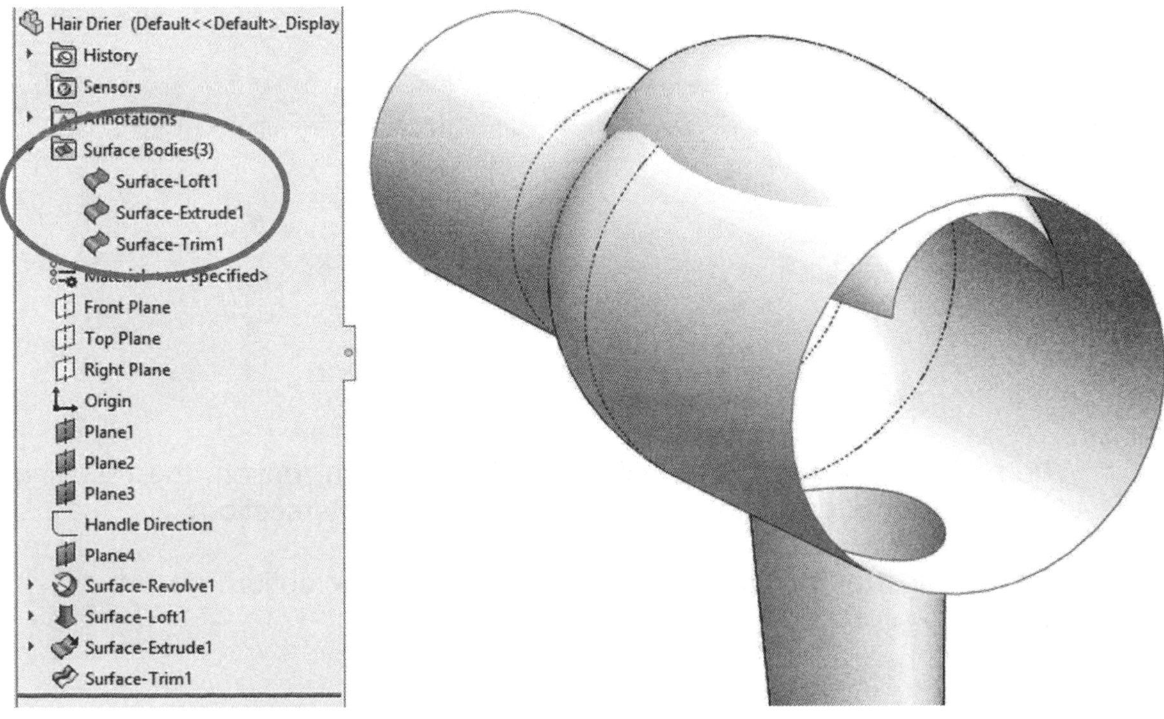

When we use a sketch as a trim tool, the surface is trimmed by projecting the sketch normal to its plane onto the surface. To understand it better, think of the sketch being extruded as a surface with the "Through All" end condition in both directions first, and then used as a trimming surface.

21.6. - For the next feature, we will trim two surfaces at the same time. This means that we'll select two (or more) surfaces, they will be trimmed at their intersection(s) and then we can select the bodies we want to keep or delete.

Select the "**Trim Surface**" command; under "Trim Type" select the option "Mutual." In the "Surfaces" selection box, pick the revolved and lofted surfaces. Selecting the "Keep selections" option turns on the "Show excluded surfaces" preview option, showing the surfaces that will be removed. Select the top part of the lofted surface and the larger body of the revolved surface. The preview will show the surfaces that will be removed. After we click OK the surfaces will be trimmed.

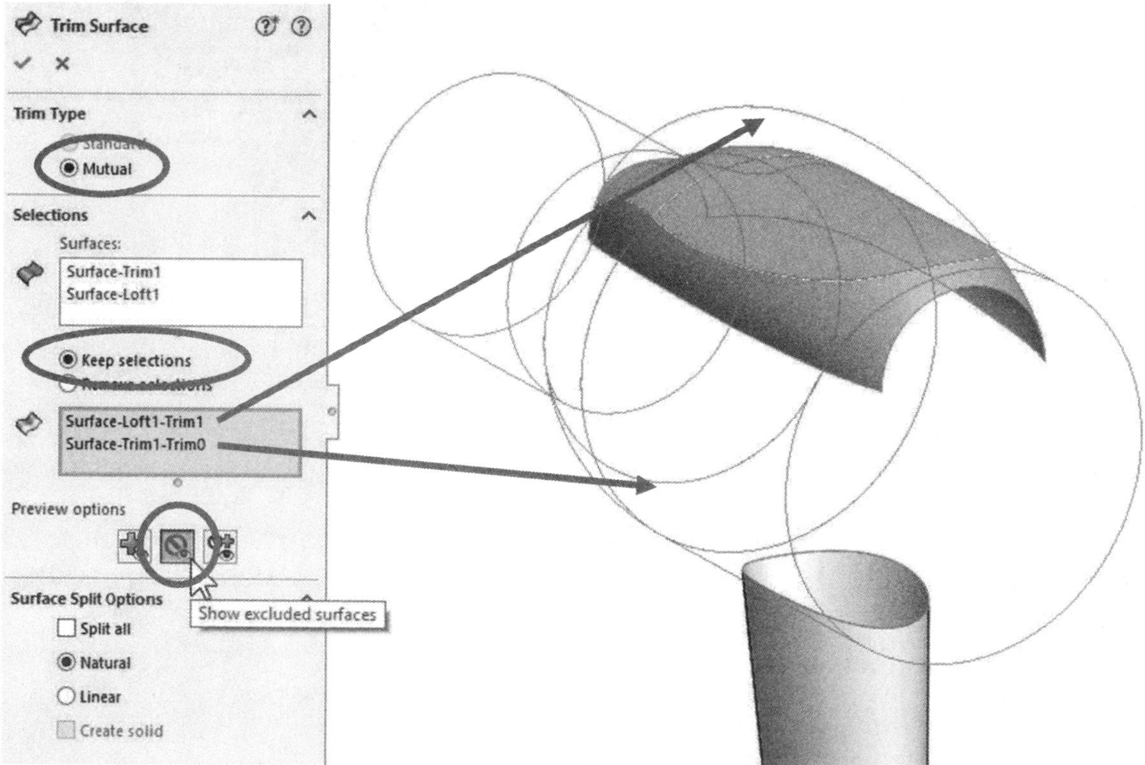

When we trim two surfaces using the "Mutual Trim" option, the resulting surfaces are merged into a single surface at their intersection.

If we select the "Show included surfaces" preview option the selection is reversed.

Our resulting model now looks like this:

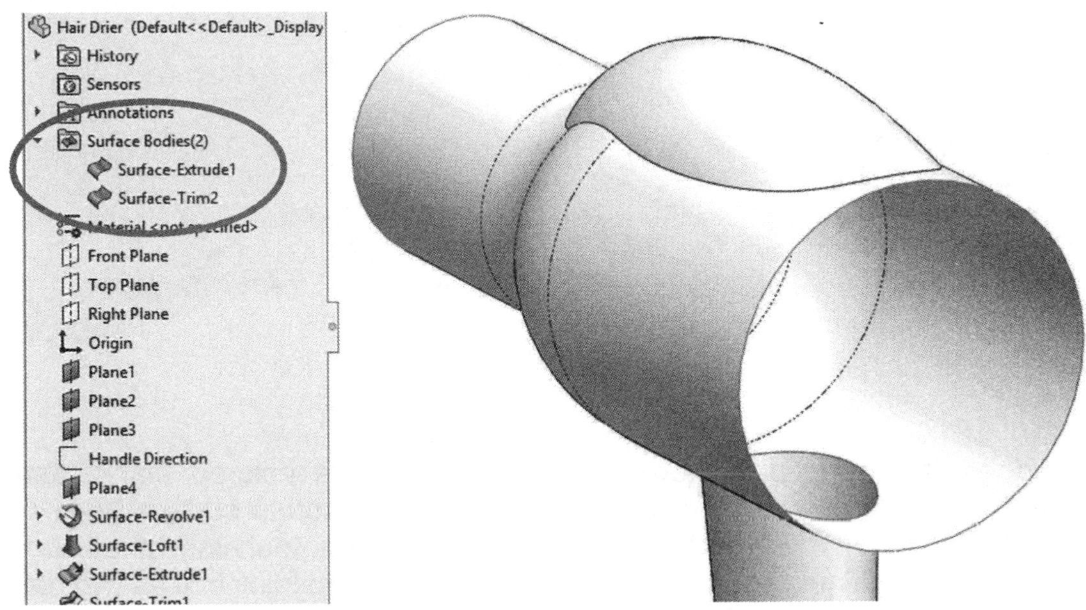

If we had used the "Remove Selections" option, we would have selected the inside faces instead.

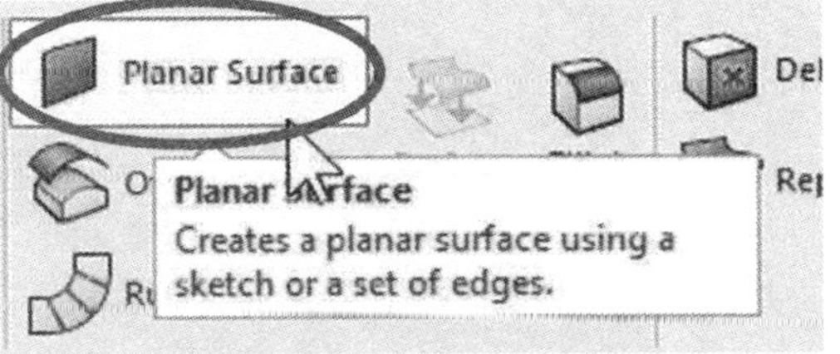

21.7. - Since we need to make a closed volume before we can convert it into a solid, we'll need to close the openings with surfaces, and we have a couple of options to close them. For the opening at the bottom of the handle, we'll make a flat surface. Select the "**Planar Surface**" command from the Command Manager or the menu "**Insert, Surface, Planar**."

To make a planar surface, we can select a set of closed edges (in the same 2D plane) or a closed 2D sketch. Turn the model over, select the edge at the bottom of the handle and click OK to add the new surface.

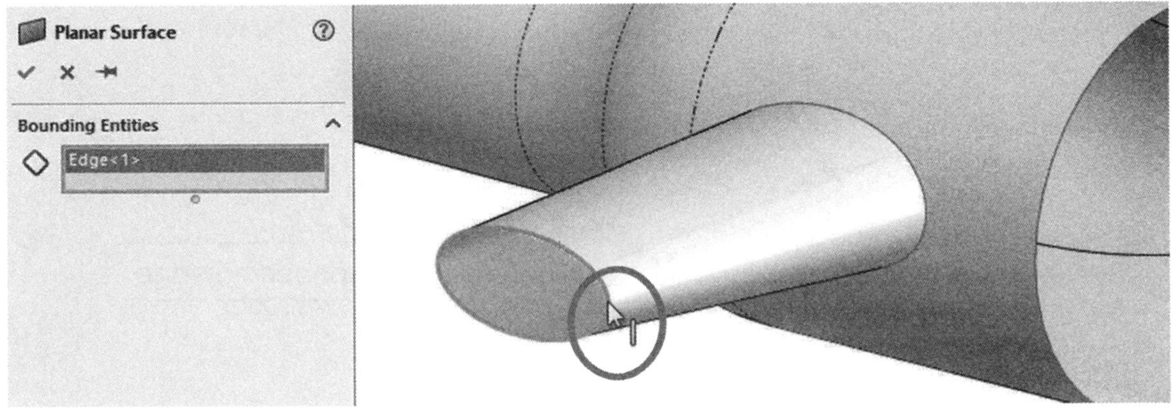

After adding the planar surface, we have three surface bodies again.

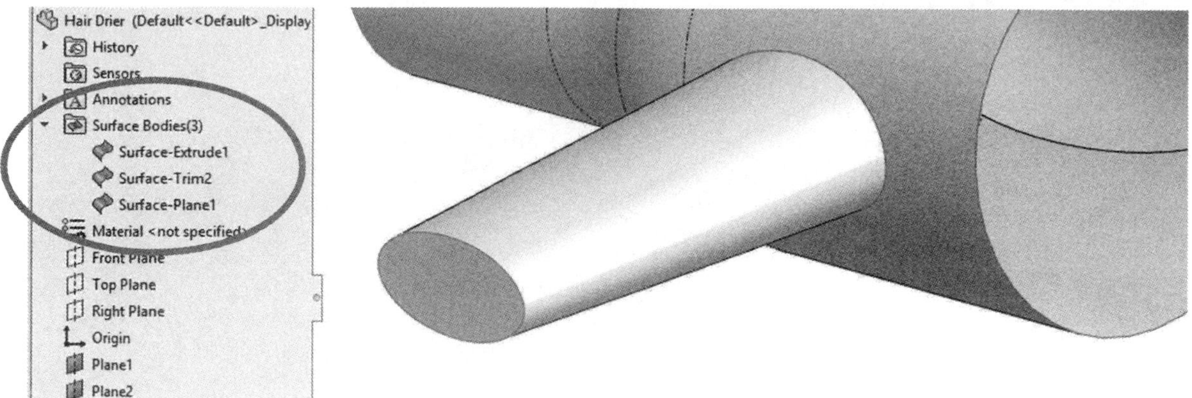

Edges that are connected to other surfaces are colored black (default setting) and open surface edges not connected to other surface bodies are a different color. Surface bodies may have edges touching other surface edges, but it doesn't necessarily imply they are connected; think of it as two pieces of fabric aligned but not stitched together.

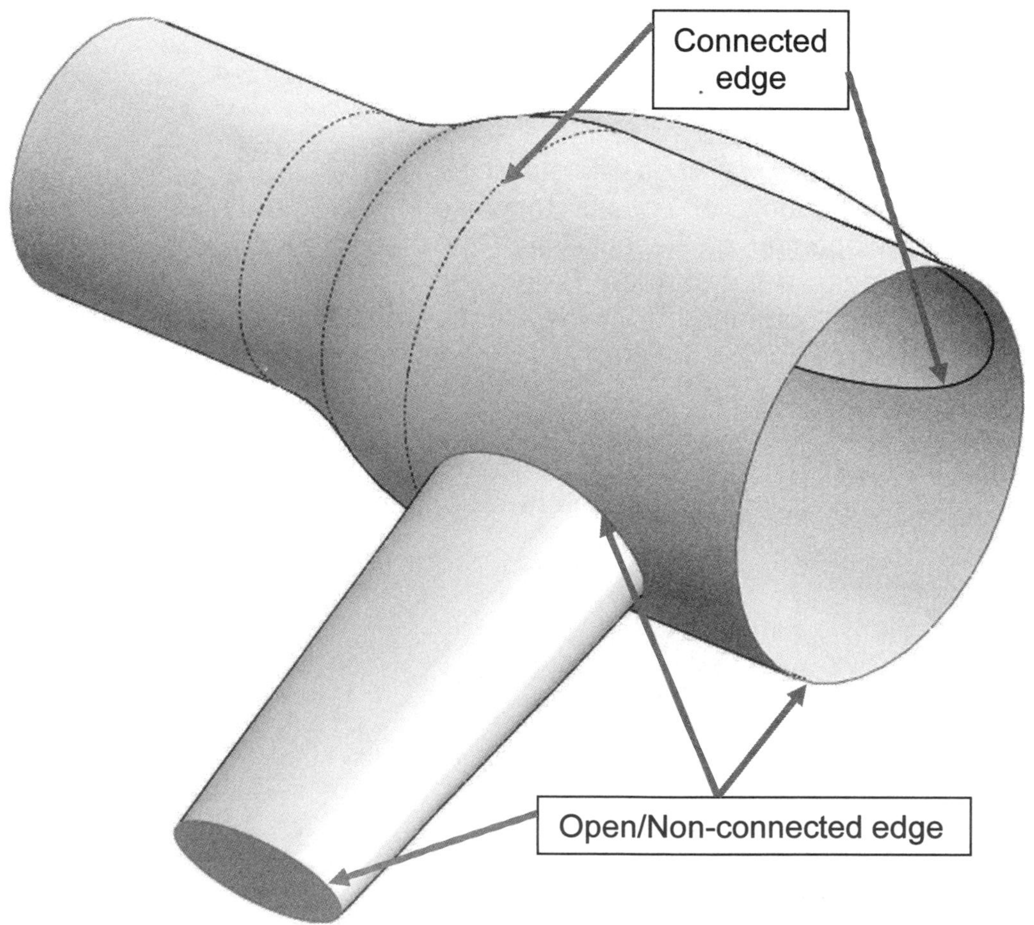

21.8. - For the front of the hair drier we'll use a different command. Select the "**Filled Surface**" command from the Command-Manager or the menu "**Insert, Surface, Fill.**" This command is more flexible than the planar surface; it allows us to use planar or non-planar sketches, edges or curves, make the new surface tangent to adjacent faces, or constrain the surface to one or more curves giving us more control over the resulting surface. For the first part of this command we'll add the open edge in front of the hair drier to the "Patch Boundary" selection box. In the "Curvature Control" selection list pick "Contact." Using this option, the new surface is made to only touch the selected edges, and since the edge is in a single plane, the resulting surface is flat. Click OK to complete this surface.

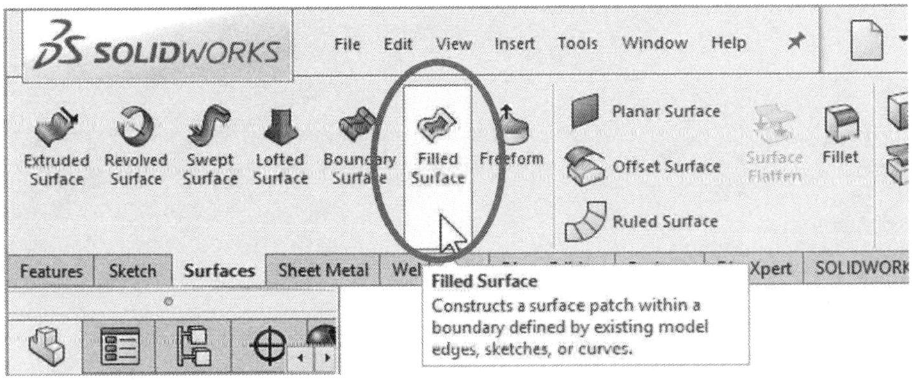

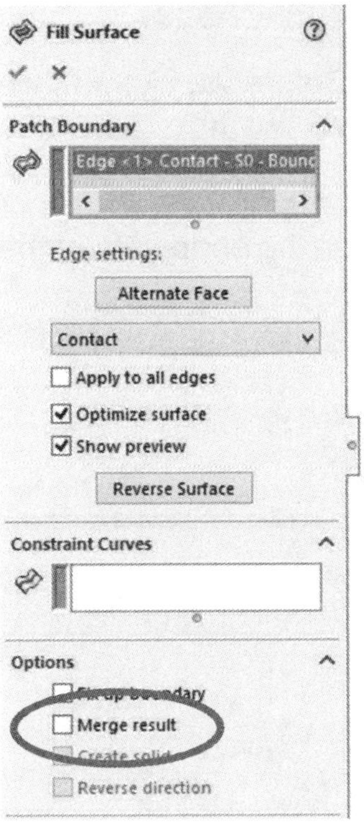

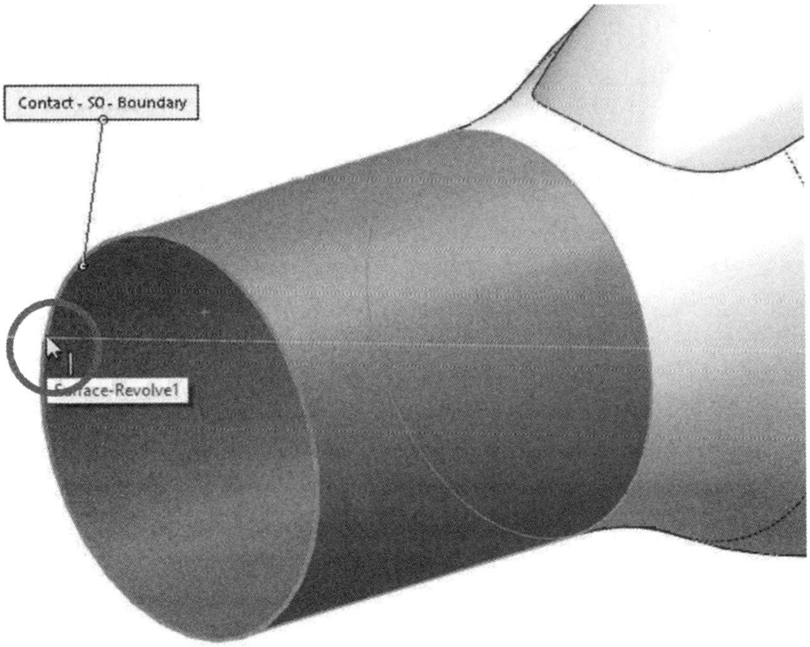

The Fill Surface command has an option to merge the resulting surface to other surfaces; in this step, we'll leave this setting unchecked.

The filled surface is the fourth surface body in the model.

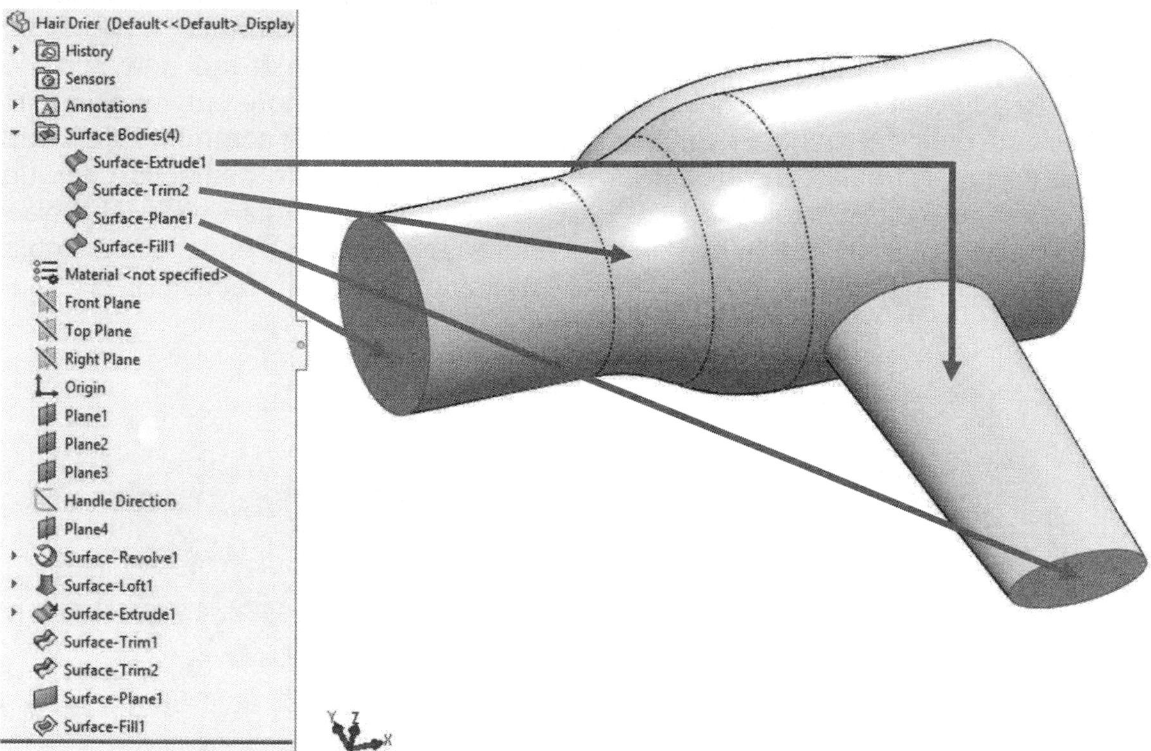

21.9. - The last place to close is the back of the hair drier. Before we close it, we need to add two sketches that will be used to control a "**Filled Surface**."

Select the *"Front Plane"* and add the following sketch. Use a **3 Point Arc** or a **Centerpoint Arc**. Be sure to add "Pierce" relations between the endpoints of the arc and the surface's edge.

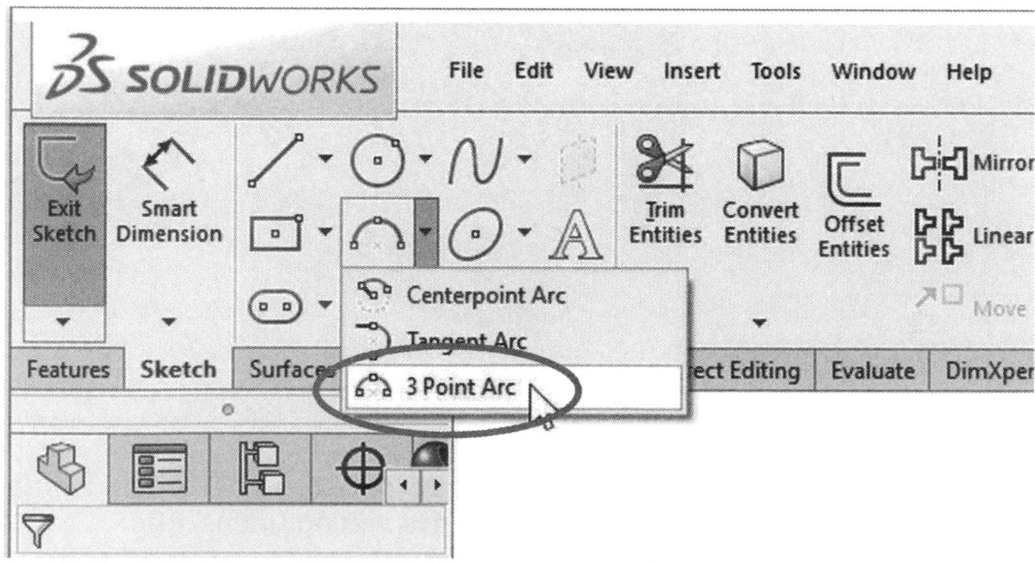

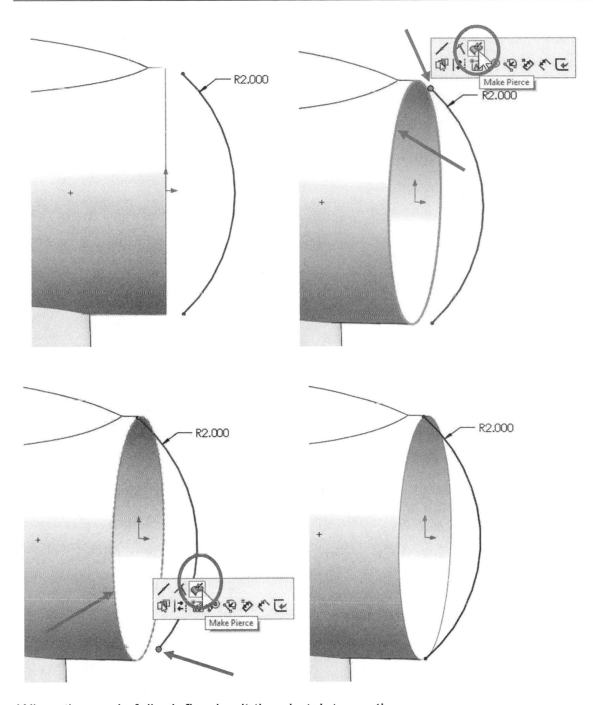

When the arc is fully defined exit the sketch to continue.

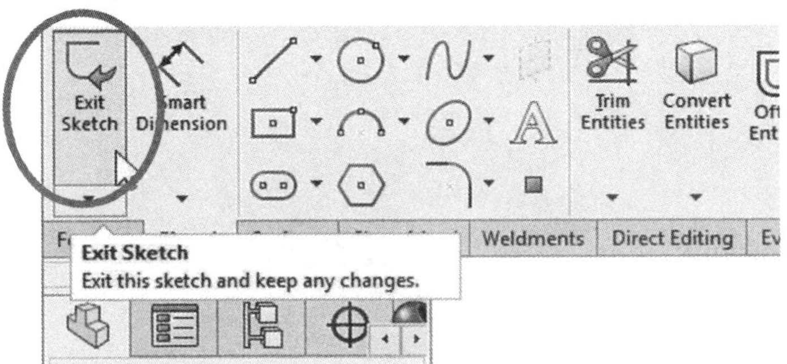

21.10. - Add a new sketch in the *"Top Plane"* with a similar arc and add "Pierce" relations to the revolved surface's edge on both sides just as in the previous step.

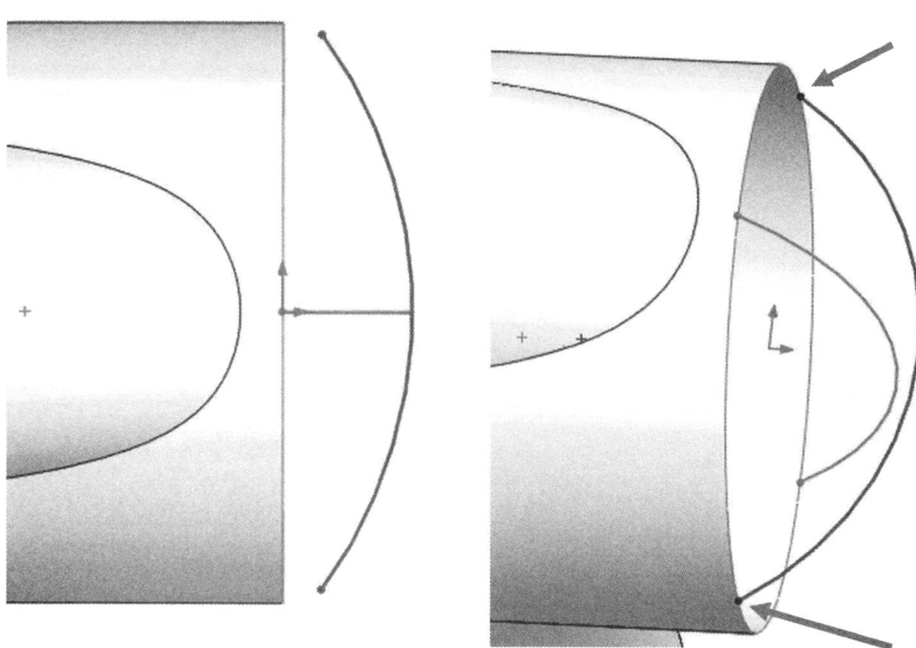

To make the arc coincident to the previous sketch, add a "**Point**" to its midpoint, and then a "Pierce" relation between the sketch point and the previous sketch. Exit the sketch when done.

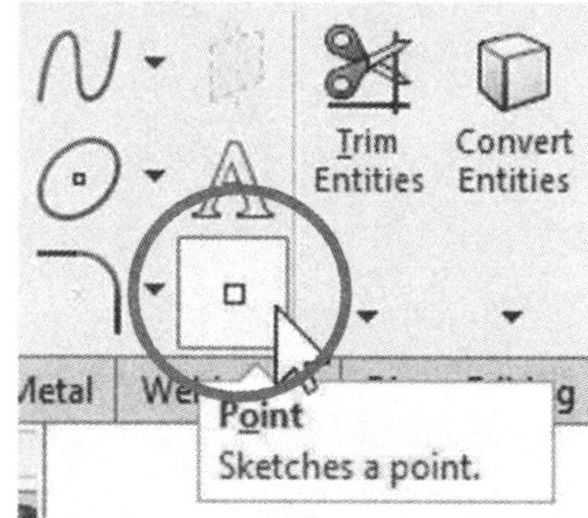

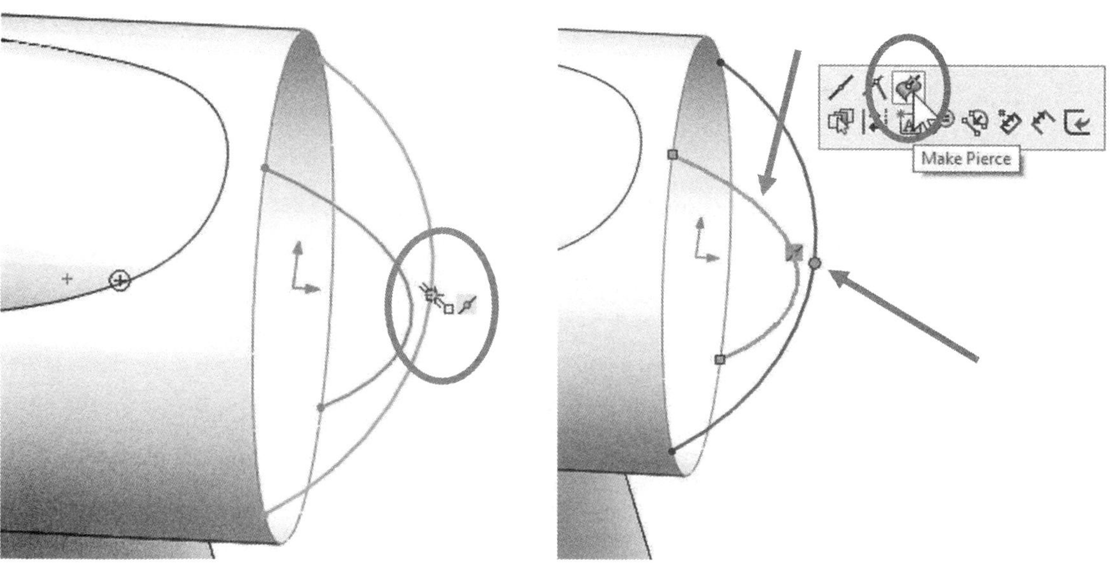

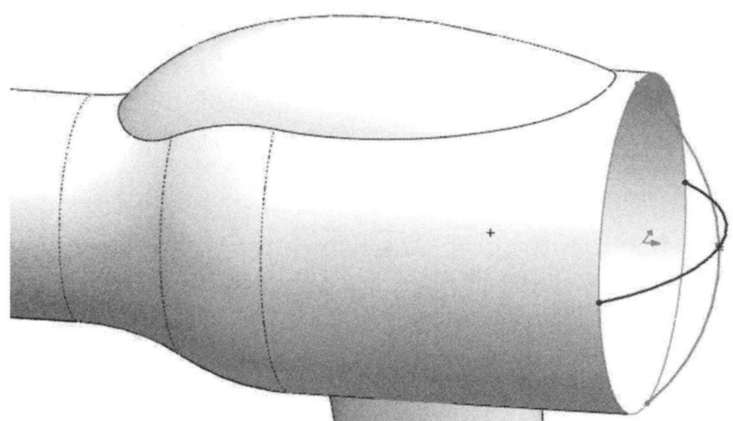

21.11. - Select the "**Filled Surface**" command from the Surfaces toolbar. Add the open edge to the "Patch Boundary" selection box and both sketches under "Constraint Curves." The surface will conform to the curves and match exactly with the edge. To get a better preview of the surface, turn on "Mesh Preview" in the "Curvature Display" options. By having constraint curves in the surface using the curvature control or not will make no difference in the resulting surface. Check the option "**Merge result**" to merge the new surface with the revolved surface and click OK to finish.

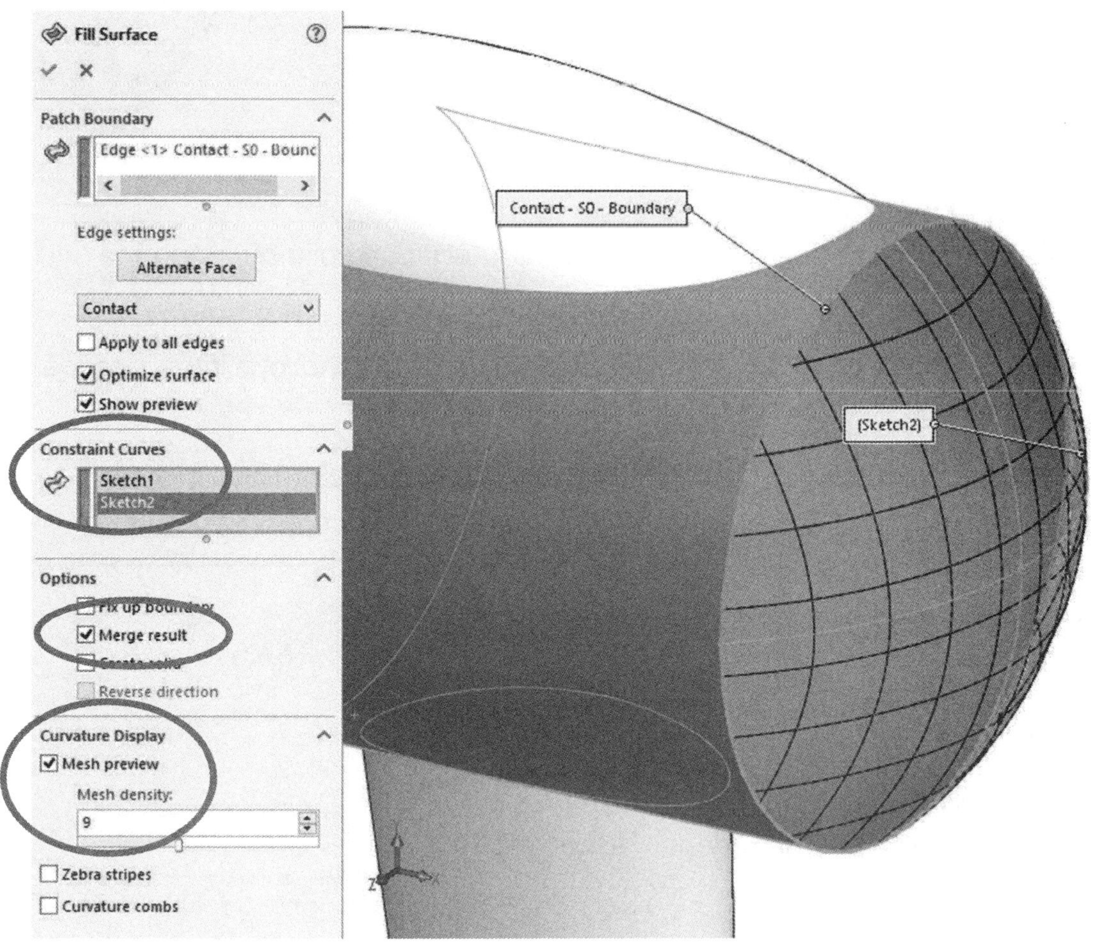

21.12. – At this point it looks like we have a closed surface to convert into a solid, but we still need to merge all four surface bodies into a single surface. To merge all bodies we'll use the "**Knit Surface**" command. Using the different edges' color we can identify which edges are 'open' (light blue), and which are connected (black).

The previous filled surface is not listed separately because we chose to merge it with the revolved surface by checking the "Merge result" option.

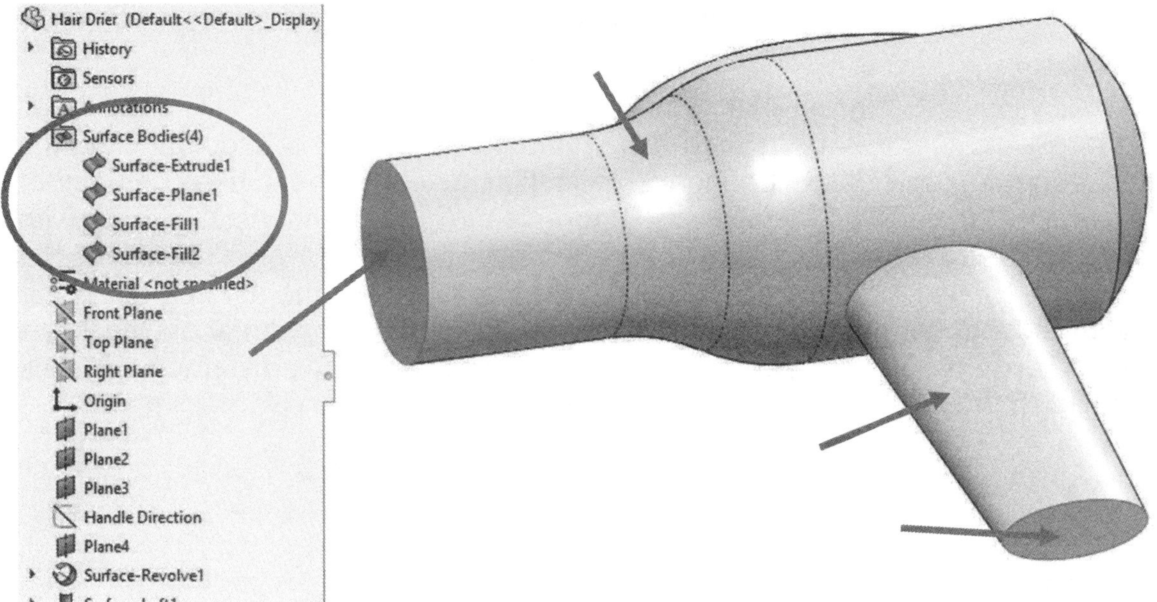

There are two ways to convert a surface into a solid body:

a) We can convert an open or closed surface into a solid by giving the surface a specified thickness.

b) We can convert a closed surface into a solid (non-hollow) if the surface is a single body defining a closed volume.

When we refer to a *closed surface*, we are talking about a *water tight* surface. In other words, it must be a completely enclosed volume, like a balloon.

21.13. - We'll show both methods to form a solid. The first option will be to convert a surface into a solid by making it thicker. The first thing we need to do is to merge the surfaces into a single surface body; select the "**Knit Surface**" command in the Surfaces toolbar, or the menu "**Insert, Surfaces, Knit.**"

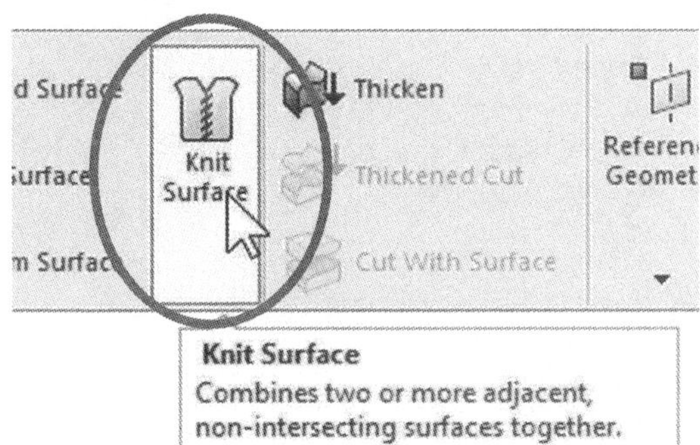

The hair drier will ultimately have the front open, so we'll knit (merge) the other three surfaces together. In the graphics area, select the handle, the flat surface at the bottom of the handle, and the main body surface. The "Gap Control" option allow us to change the tolerance to consider two edges coincident and merge them into a single edge; this option is useful when merging surfaces that may have a small gap between them. Click OK to finish.

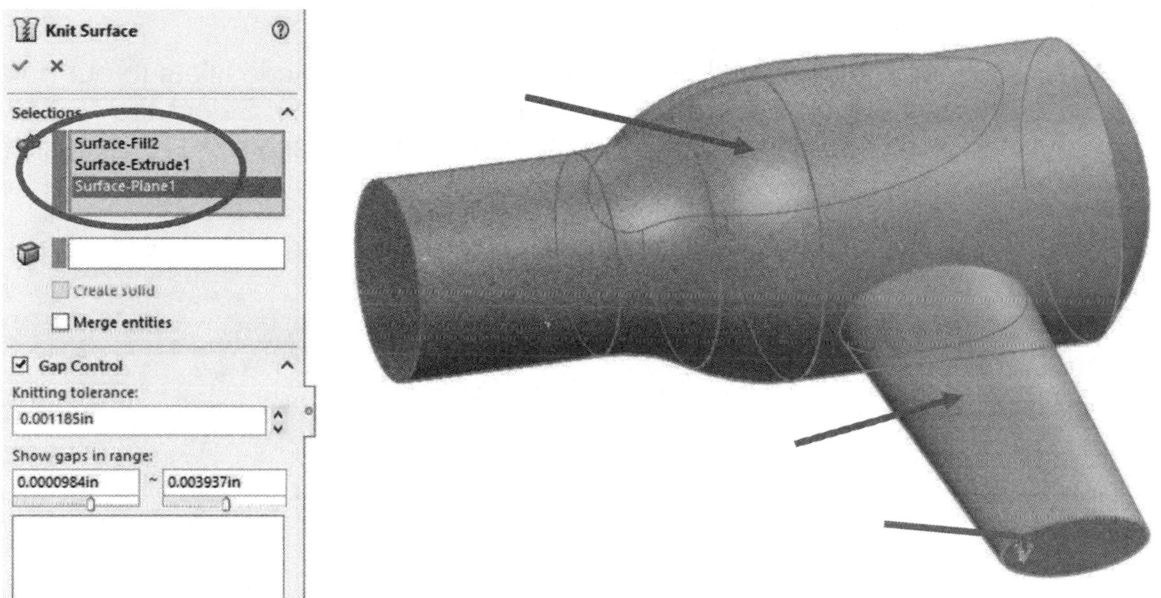

After knitting these surfaces our model has two surface bodies.

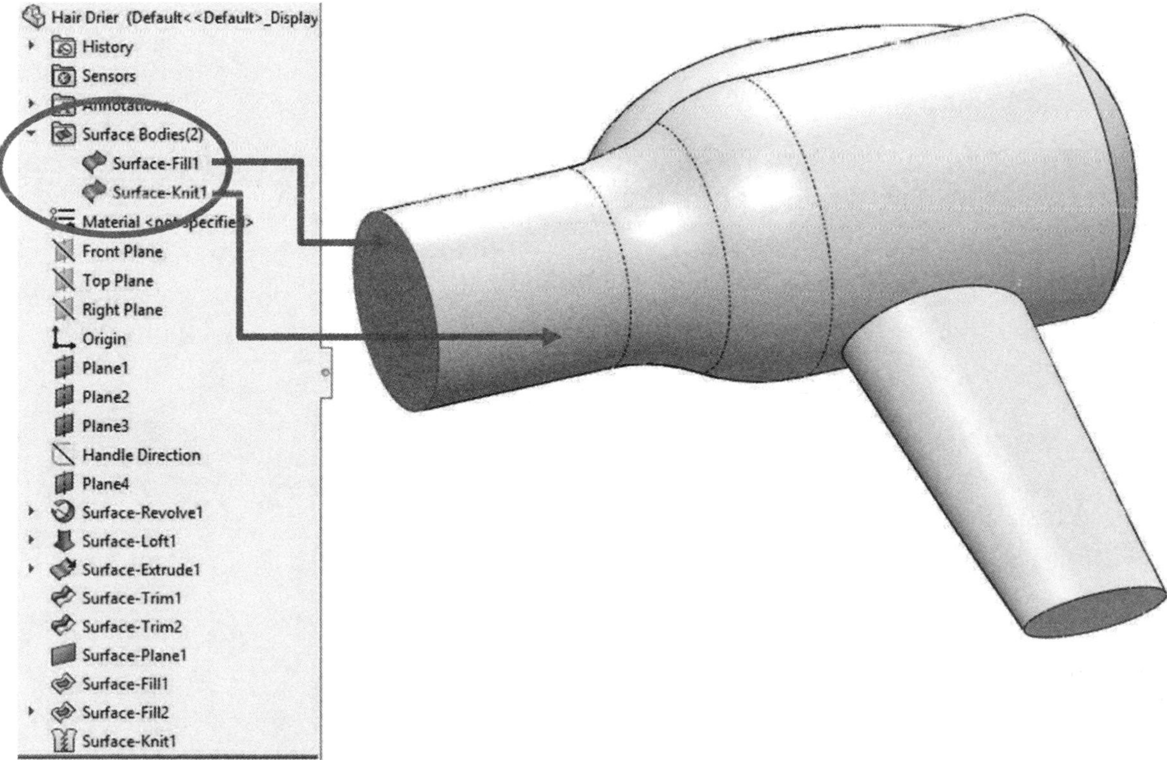

Things to keep in mind when knitting surfaces:

- Surfaces must be touching at the edges.
- Surfaces must not overlap.
- Surfaces knitted will be absorbed into a new surface body.
- If the selected surfaces form a closed volume, they can be turned into a solid body which will absorb the surface bodies.

21.14. - Before we make a solid with this surface we want to add fillets to our hair drier. We can add fillets to a surface just like to a solid body. Select the "**Fillet**" command and add a 0.25 inch fillet to the back of the drier and a 0.125 inch fillet at the bottom of the handle.

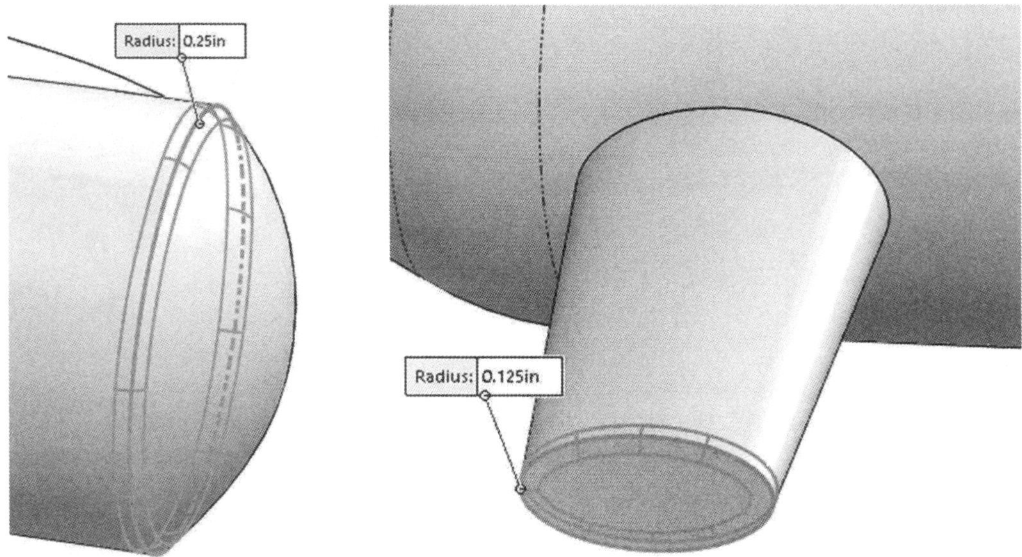

To make the surface more aesthetic, add a 1″ fillet at the top of the hair drier. Make sure the "Tangent propagation" option is turned on.

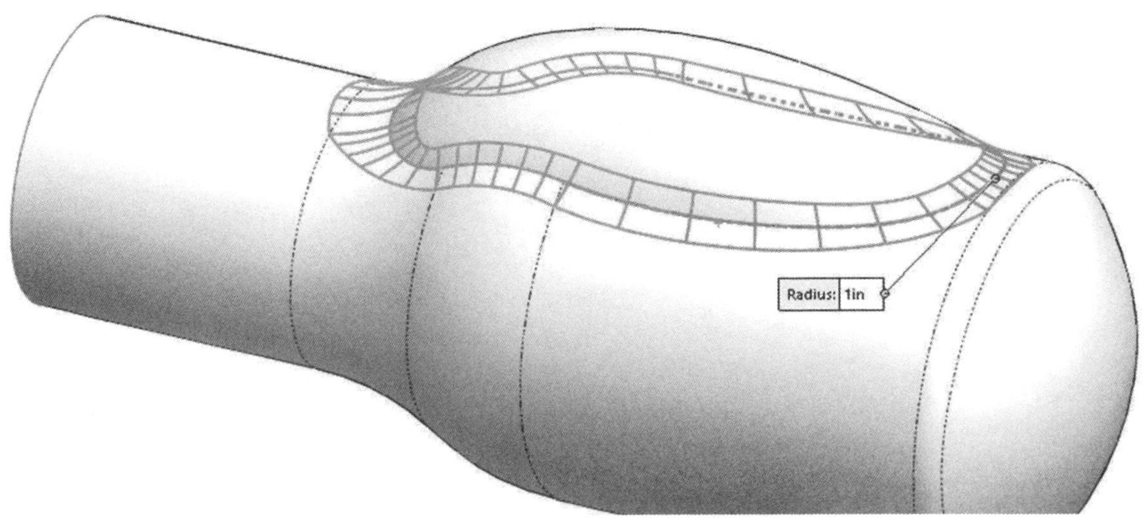

21.15. - For the last fillet, if we use a regular fillet with a constant radius, we get a fillet that covers a large surface and in general doesn't look good. Select the "**Fillet**" command. In this case use "**Face Fillet**" for the "Fillet Type" option and pick the two faces indicated, one in each selection box. If needed, click on the "Reverse Face Normal" option to make the arrows in the faces point to each other in order for the Fillet command to work. Make the fillet 1″ radius.

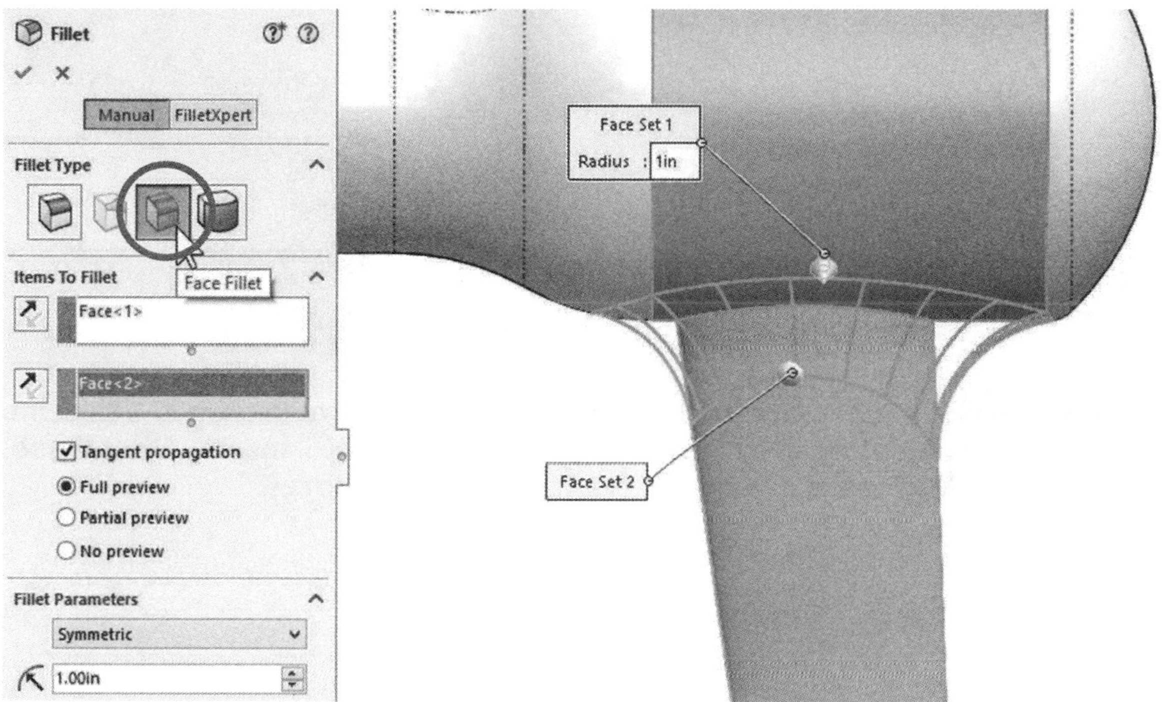

In the "Fillet Parameters" select the "**Chord Width**" option. Notice the difference in the fillet's preview; instead of defining a fillet's radius, we are defining the length of the chord created by the fillet, making it a constant width all around the edge. Under the "Fillet Options" make sure the "Trim and attach" option is selected to ensure the surfaces are trimmed and merged with the fillet's surface. Click OK to finish.

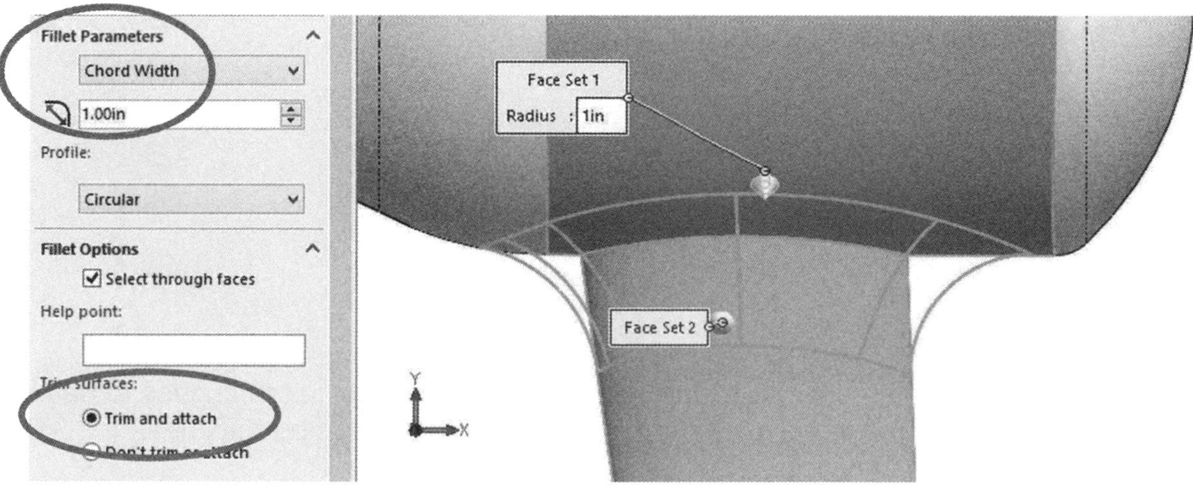

21.16. – To see the effect of thickening the surface we need to hide the *"Surface-Fill1"* body. Select the surface in the graphics area or the FeatureManager and click in the "Hide" icon.

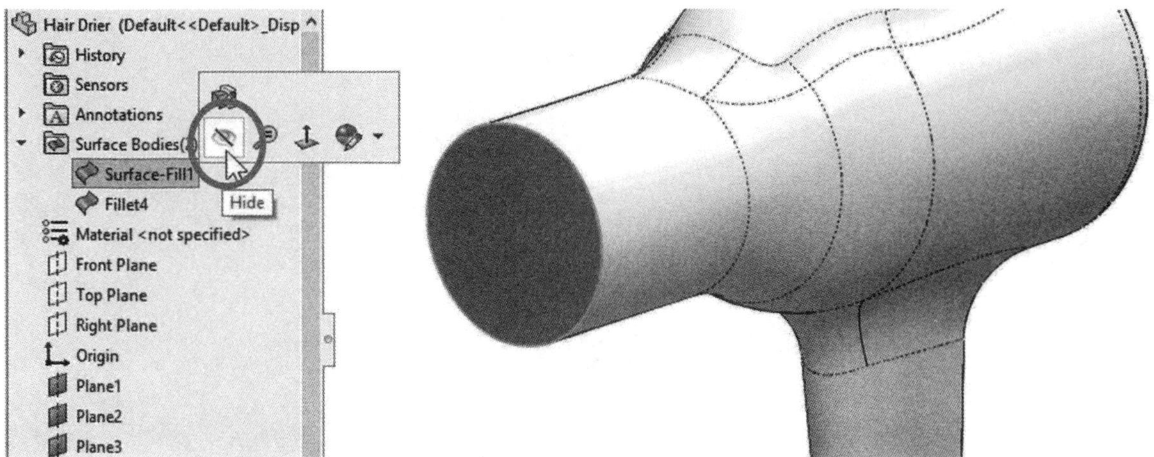

21.17. - The first method to turn the surface into a solid is to thicken it. Select the **"Thicken"** command in the Surfaces tab, or the menu "**Insert, Boss/Base, Thicken**." In "Surfaces to Thicken," select the *"Fillet4"* surface.

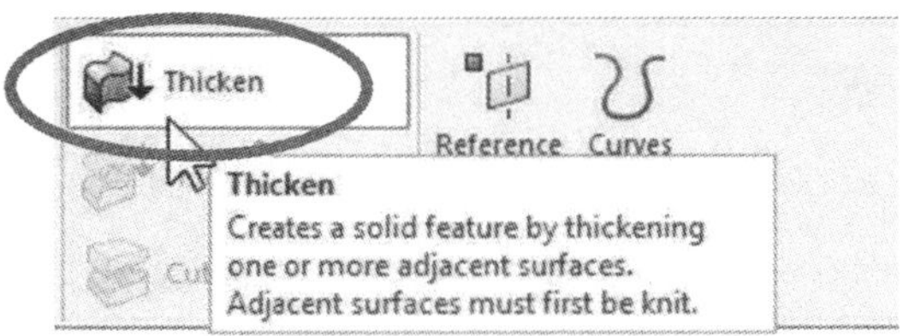

When we thicken a surface we can add material to one side, the other side, or 'mid plane' to both sides of the surface. For our example, we'll make the surface 0.1" thick inside to keep our model's dimensions as outside dimensions. Notice the preview when we change the side to thicken. Click OK to finish.

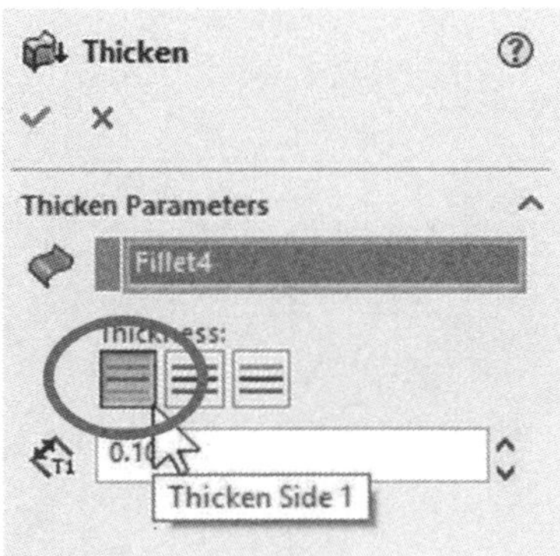

Using the "Thicken Both Sides" option adds the specified thickness to each side, making the end result twice as thick as the specified thickness.

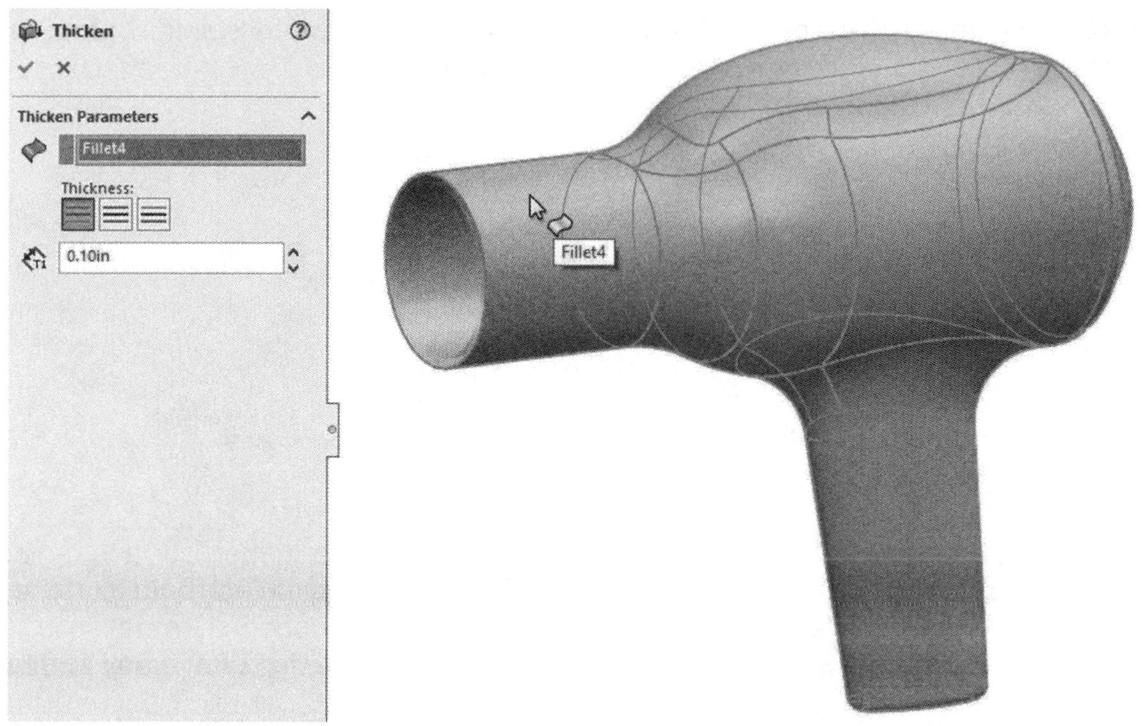

Now we have a solid body and a (hidden) surface body.

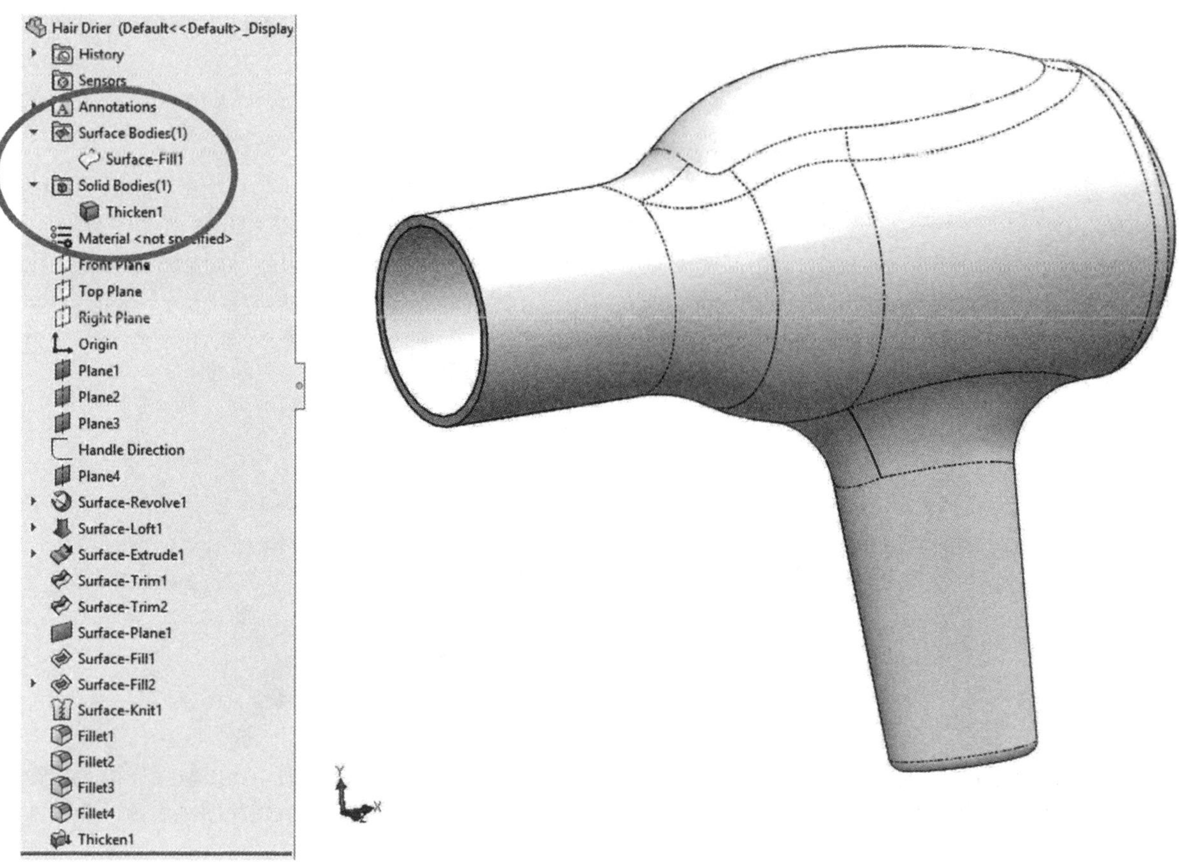

21.18. - To learn the second method to convert surfaces to solids, delete the *"Thicken1"* feature, and show the hidden surface.

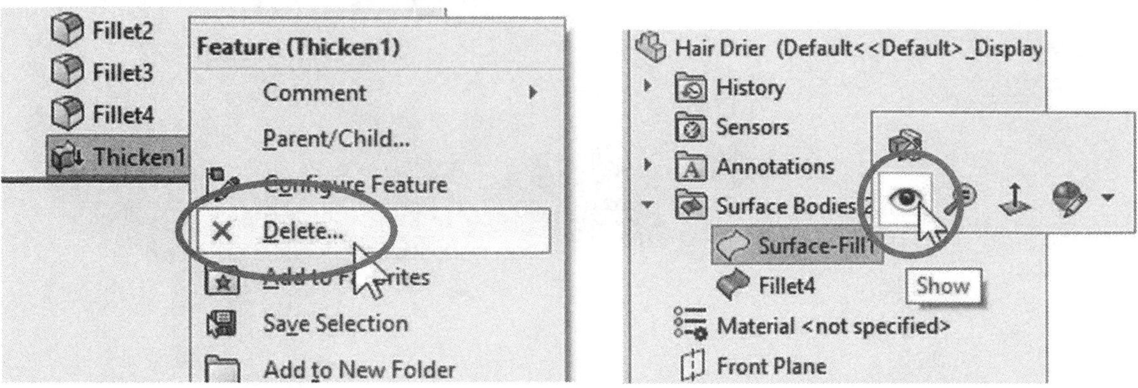

21.19. - After deleting the thickened body and showing the hidden surface, select the "**Knit Surface**" command from the Surfaces tab and select both surfaces. When the knitted surfaces form a closed volume, the option "Create solid" is enabled. Turn on this option and click OK to finish. Now we don't have any surface bodies and our model is a solid, non-hollow body.

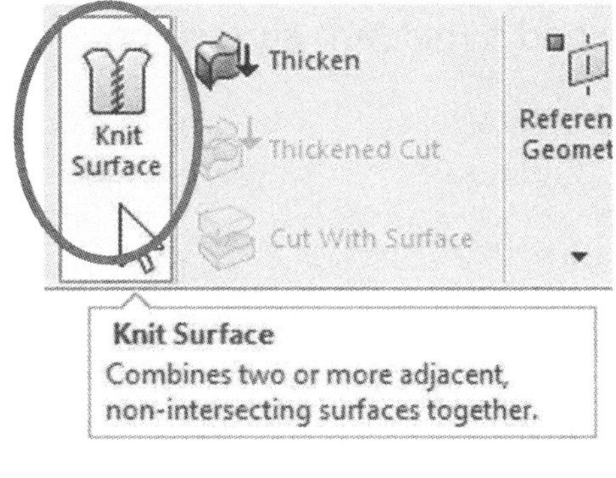

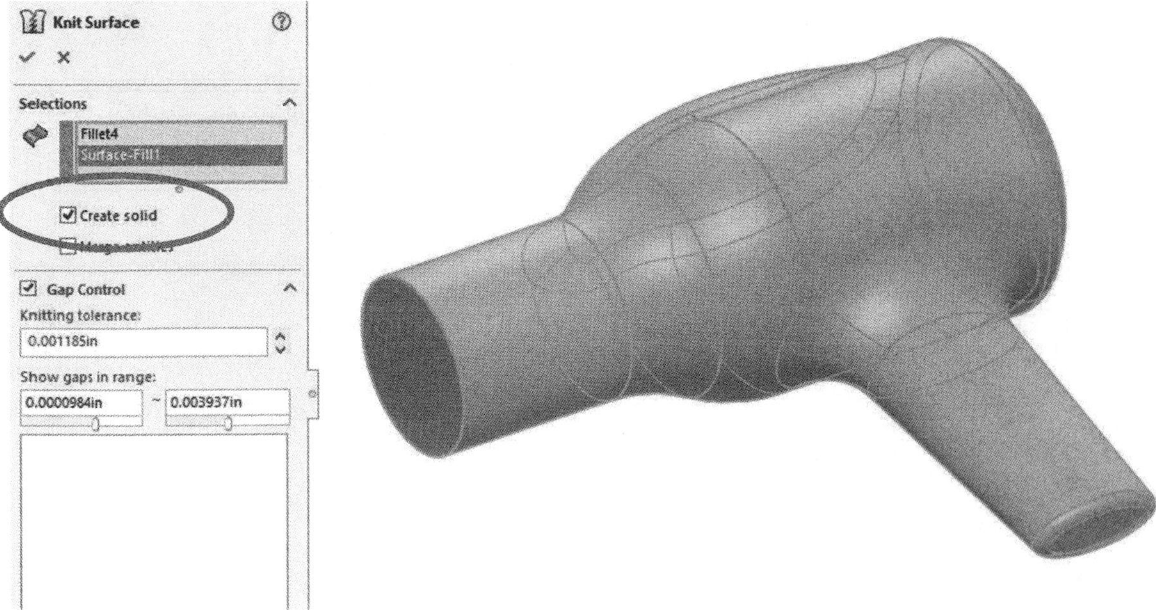

21.20. - From this point on we can treat our model as any regular solid body; we don't have any surface bodies left in the model and a single solid body. The last feature to complete the model will be a "**Shell**" feature. Make it 0.1″ thick and remove the front face of the hair drier. Save the model to finish.

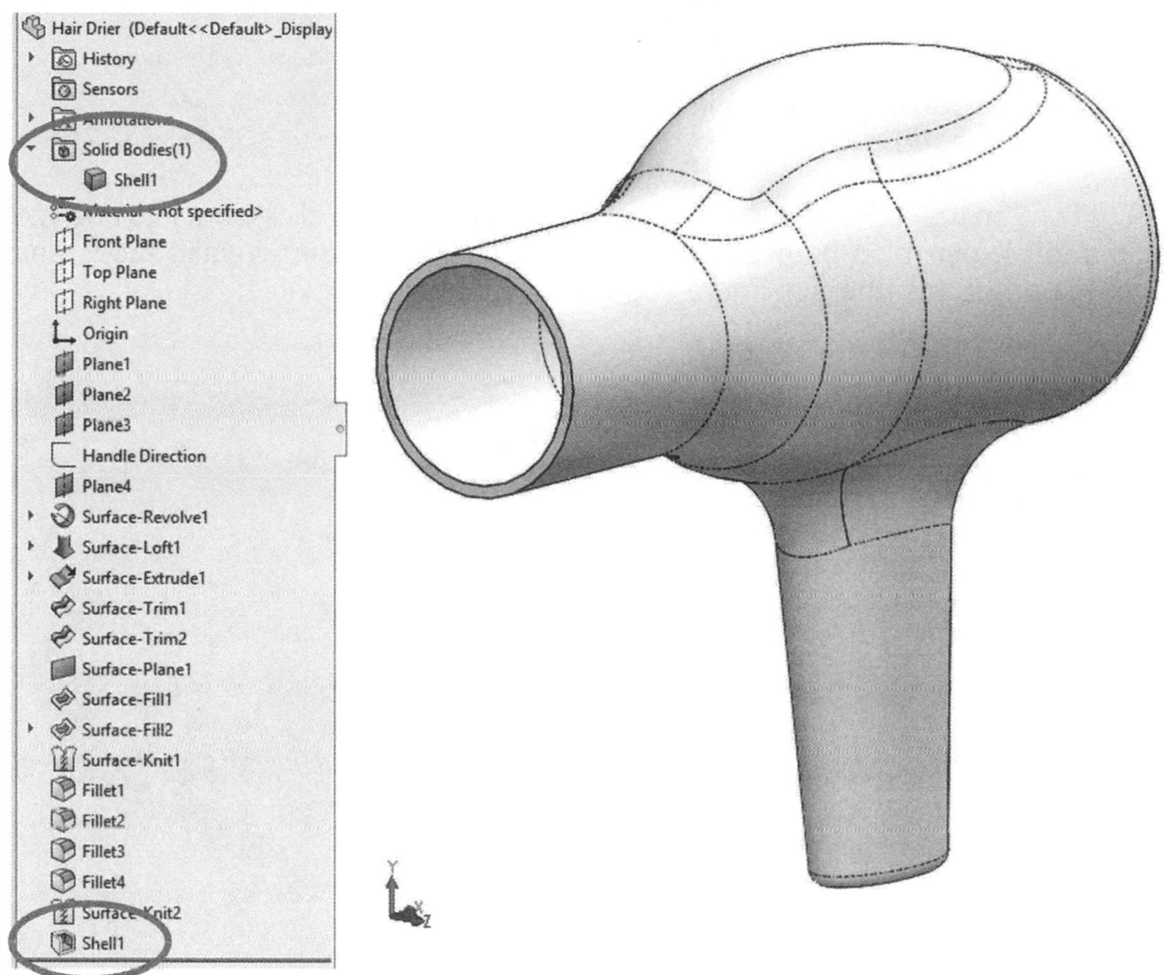

Choosing an option to convert surfaces to solid depends on different factors; sometimes one method fails and we need to use a different approach.

A common error message when working with surfaces or shelled parts is when a face overlaps with another after thickening, or a face offsets into another after shelling the part. For example, if we change the fillet at the bottom of the handle to a 0.5″ radius, the "**Thicken**" command fails. In this case, we would have to add the fillet *after* thickening the surfaces.

There are no 'hard rules' to select one approach over the other, and just as both methods may work, both may fail. If this is the case, we may have to evaluate different ways to accomplish our objective, but a good starting point is to run a model check and look for the smallest radius in the model. To evaluate the model, use the "Check" command from the Evaluate tab in the CommandManager.

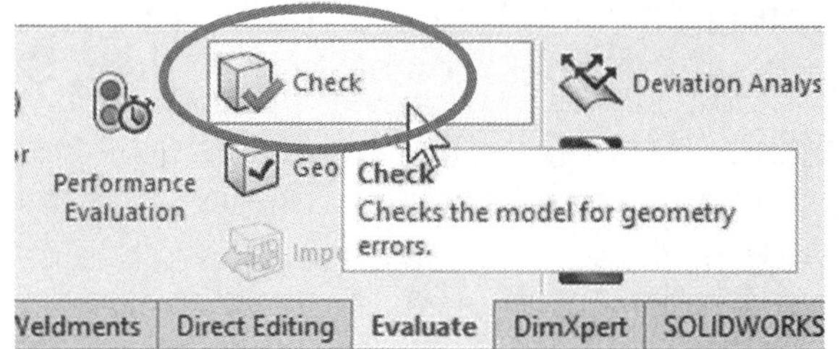

Turn on the "Minimum radius of curvature" option and click Check to analyze the model. It can also help us find geometry errors including invalid faces and edges, short edges, the maximum edge and vertex gap.

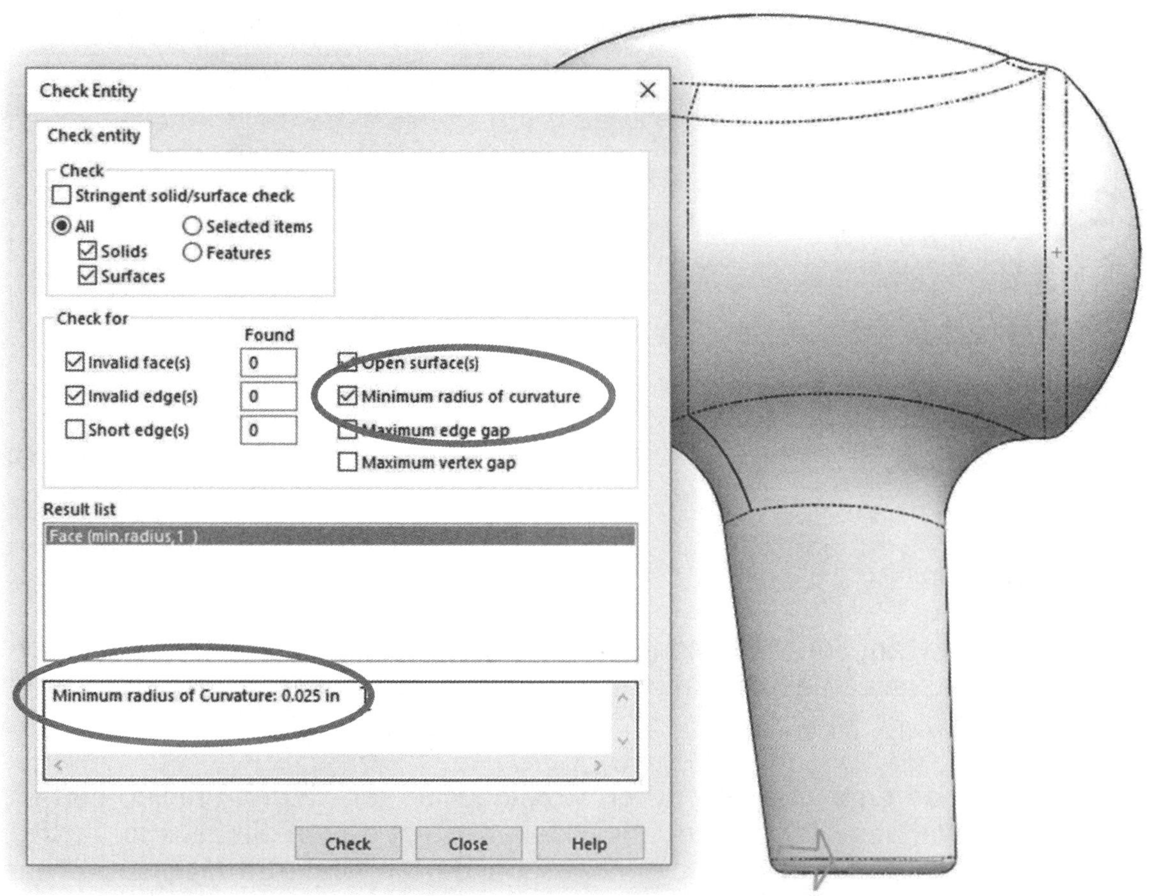

Master Model

When we design a product made of multiple parts that need to maintain a specific shape when assembled, like our hair drier, we use a technique called *Master Model.* Using this technique, we start our design with a part that has the overall shape of the finished assembly (the '*Hair Drier*' part), then it's split into a multi-body part, and each split body is saved to an individual part and finished with the features specific to each one. Finally, the completed components are assembled to form the finished product design. This is the general process:

<table>
<tr><td>Master Model
(1 Part)</td><td>Split Master Model
(1 Part, 3 Solid Bodies)</td><td>Save Split Part's Bodies individually (3 single-body parts)</td></tr>
<tr><td>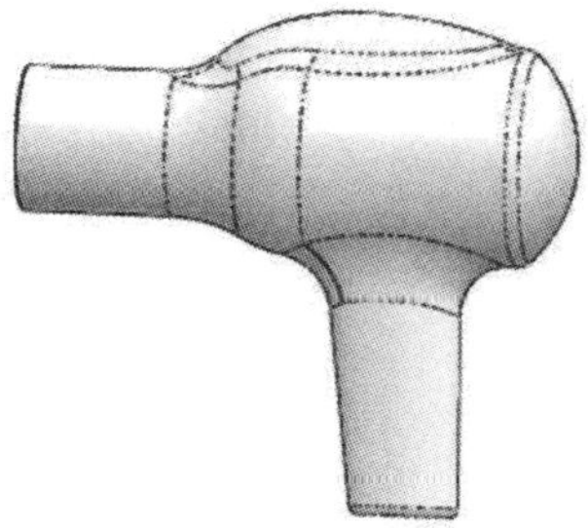</td><td>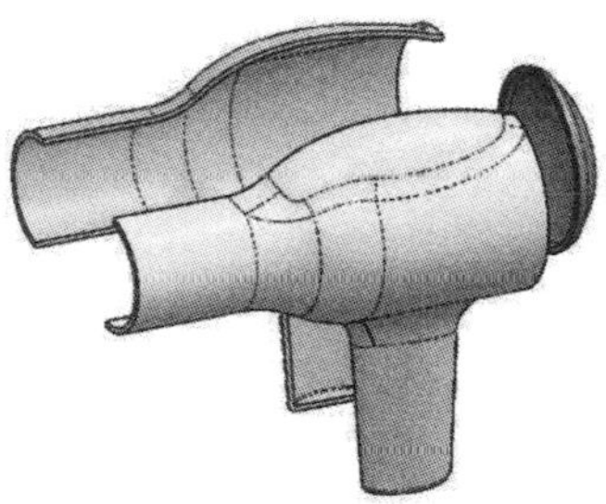</td><td>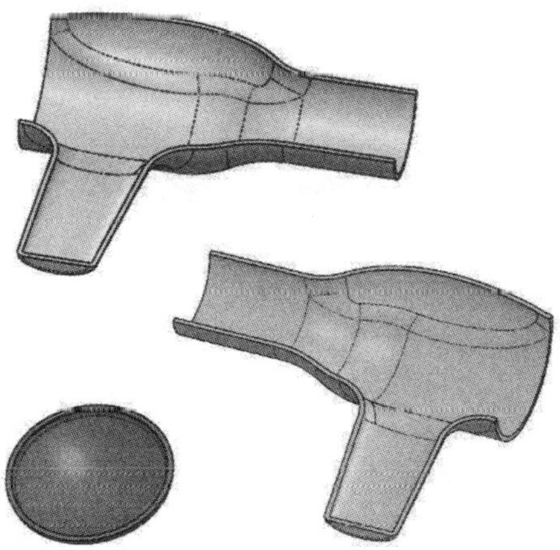</td></tr>
<tr><td colspan="2">Add Missing Features to Individual Parts
(3 parts)</td><td>Assemble finished components
(1 Assembly)</td></tr>
<tr><td colspan="2">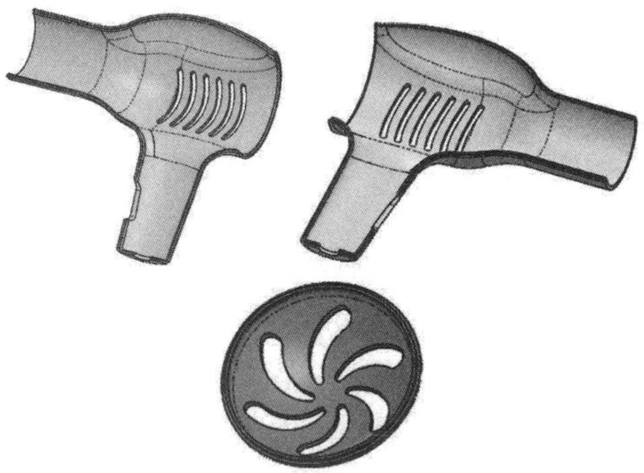</td><td></td></tr>
</table>

22.1. - The first body we are going to split from the hair drier's master model will become the back cover. To split a solid into multiple bodies we can use a plane, a surface (Curved or flat) that crosses the entire model or a sketch (open or closed).

Add a sketch in the *"Front Plane"* and draw a line 0.125" to the left of the origin. When using an open sketch to split a model, the sketch *must* cross the model; the split will be made by projecting the sketch normal to the sketch plane, as if we made an extruded surface in both directions and cut our model with it.

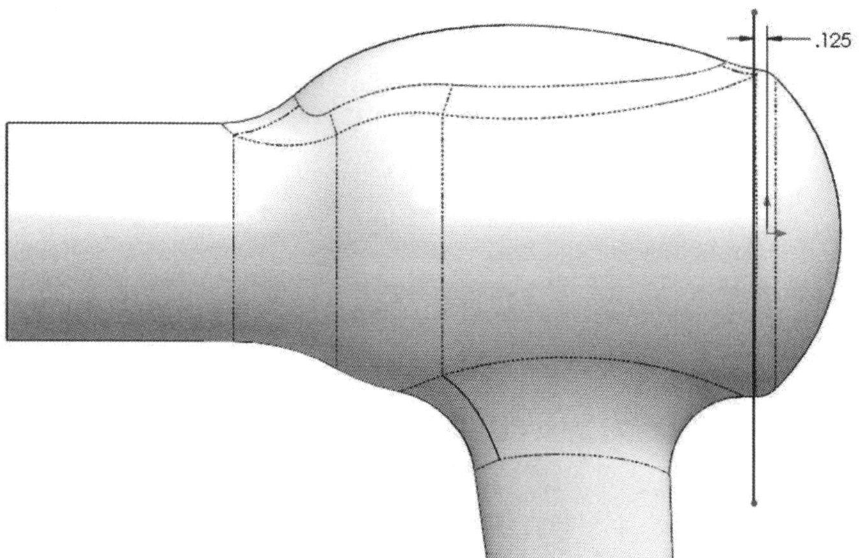

22.2. - While still editing the sketch select the "**Split**" command from the menu "**Insert, Features, Split.**" If asked, select part and assembly templates. In the Split command the current sketch will be added to the "**Trim Tools**" selection box. Selecting the "**Cut Part**" button will split the solid body in two parts.

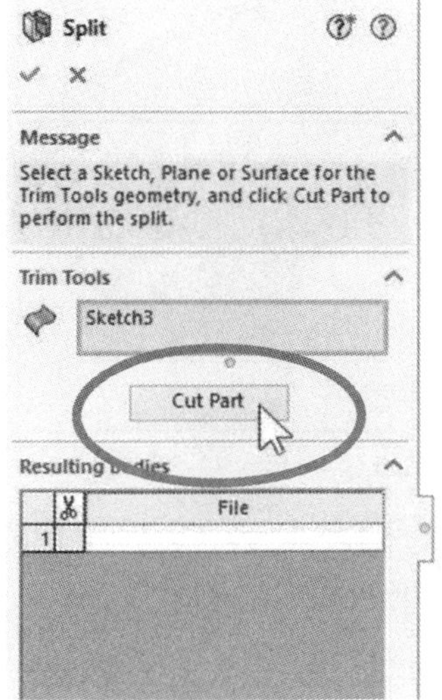

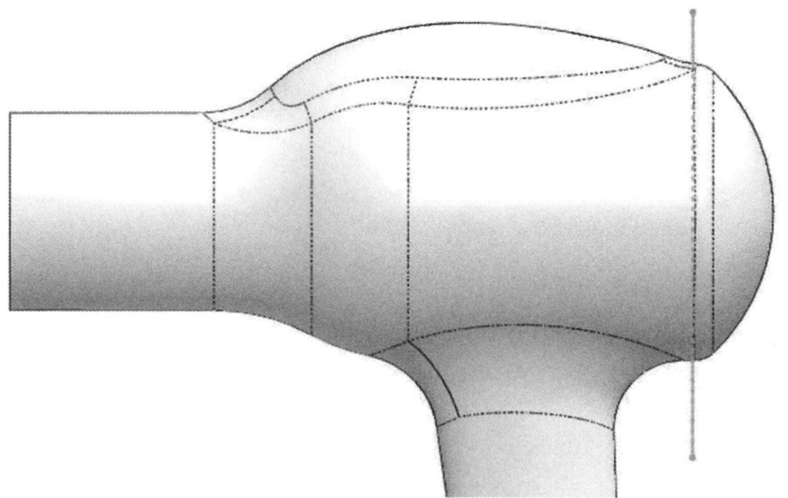

After cutting the part we have two bodies listed in the "Resulting Bodies" list. In this step we'll separate the back cover only. Select the body for the back cover and click in the *<None>* tag to give the new part a name. Save this body as *'H-D Cover'*.

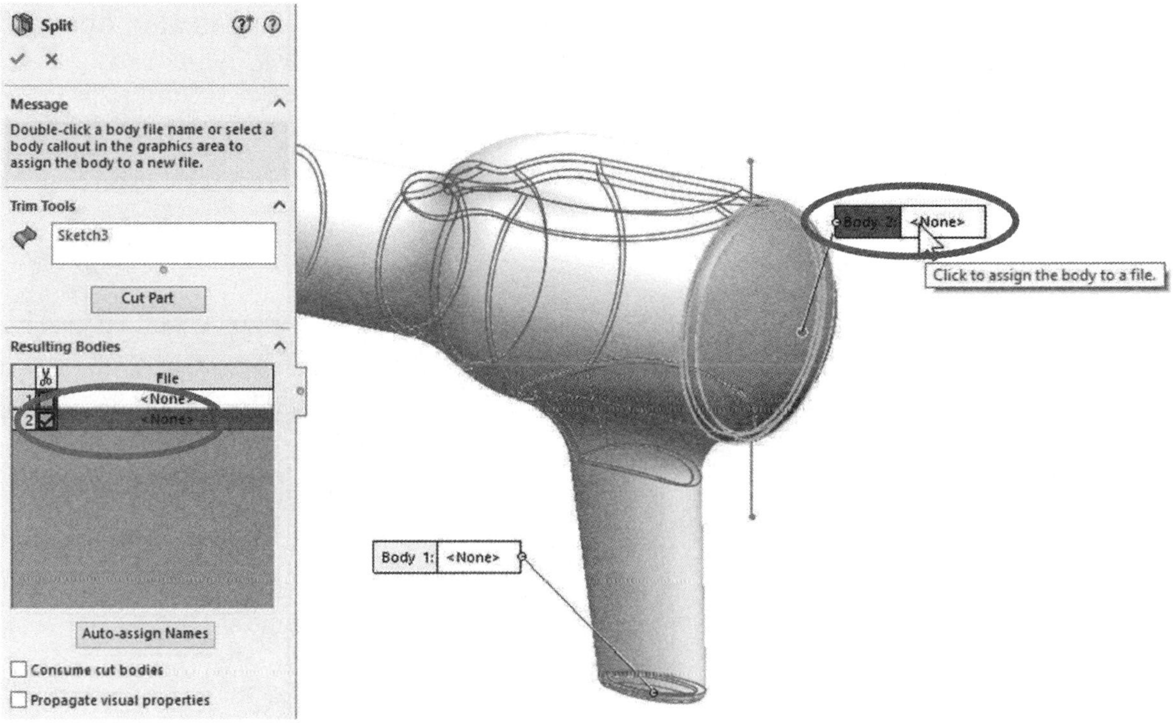

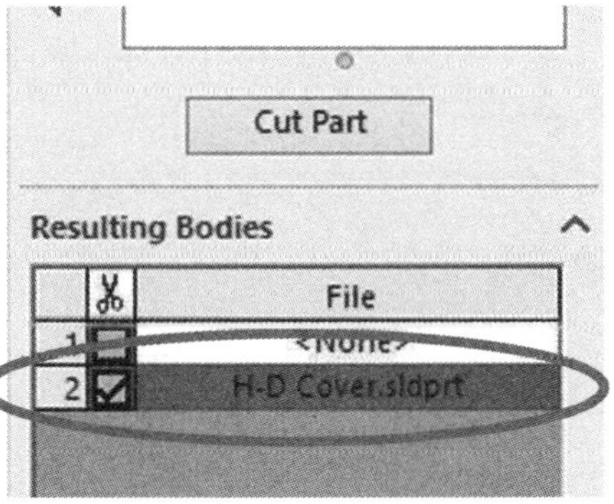

If a system error is displayed after naming the split body saying, "The above file name is invalid" you need to enable write access to the SOLIDWORKS' install directory.

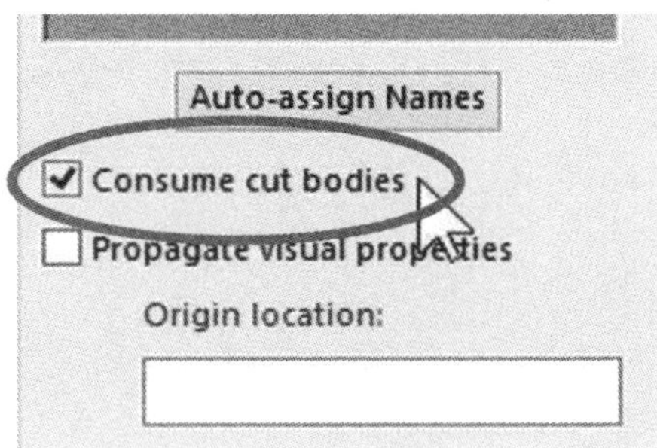

Before completing the "**Split**" command, turn on the "Consume cut bodies" option. By doing this the body we are saving to an external file will be deleted from the master model. Click OK to split the part and finish.

After completing the command, a *"Split1"* feature is added to the FeatureManager of the master model and the back cover is now missing from it. The body we cut away is added to a new part with a single feature called ***"Stock-**Parent_File"*; it has an external reference to the *'Hair Drier'* part and its state is "In Context." Split parts follow the same external reference rules as top down design parts covered earlier in the book. When this new part is created it is also opened. Select the menu "**Window**" and switch to the new part to view it.

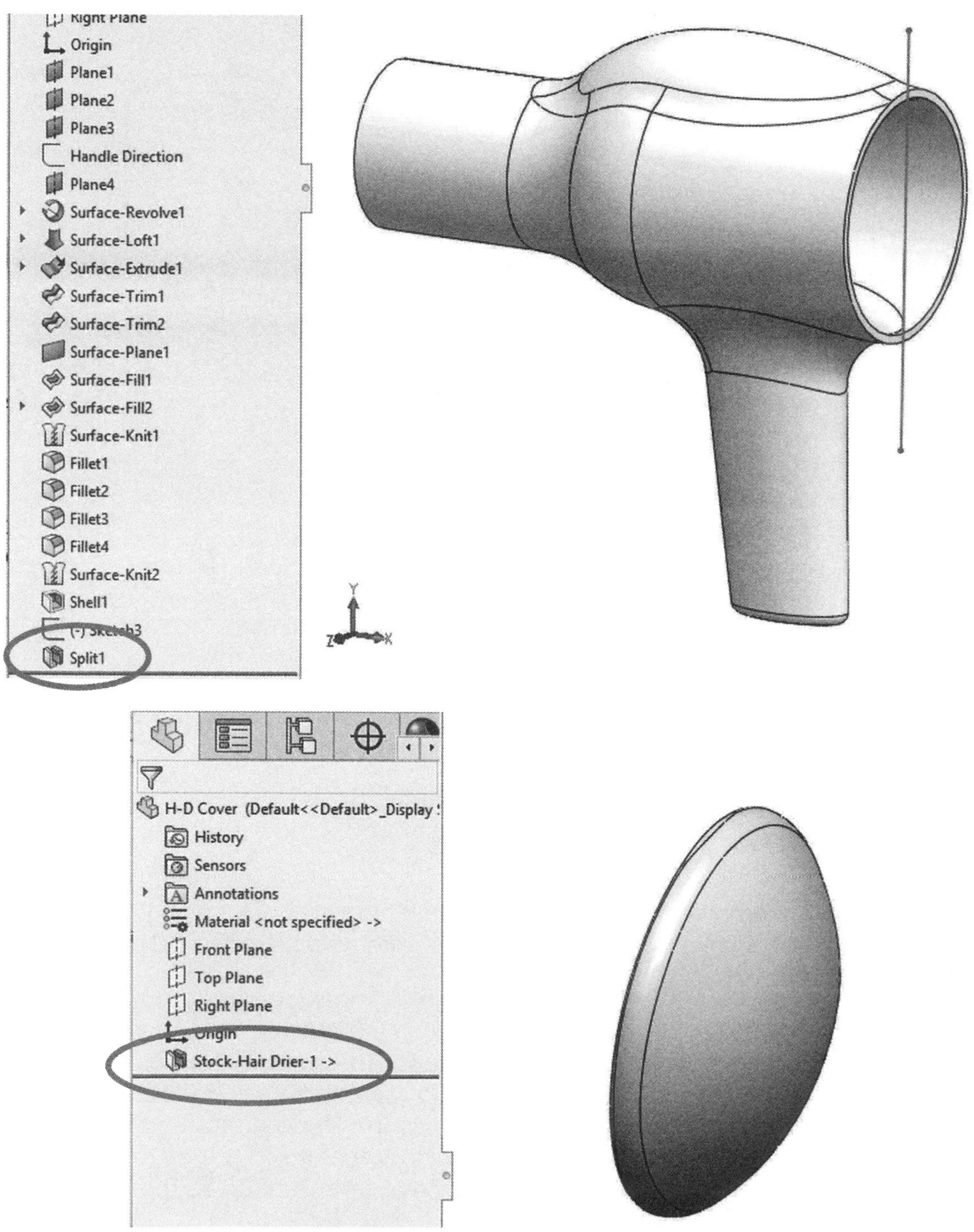

22.3. - The next step will be to add a second split feature to divide the remaining body in two parts, using the *"Front Plane"* as the cutting tool. Go back to the hair drier model, select the "**Split**" command again, add the *"Front Plane"* to the "Trim Tools" selection box and cut the part. Save both bodies and name them *'H-D Left'* and *'H-D Right'*. Turn on the "Consume cut bodies" option and click OK to finish.

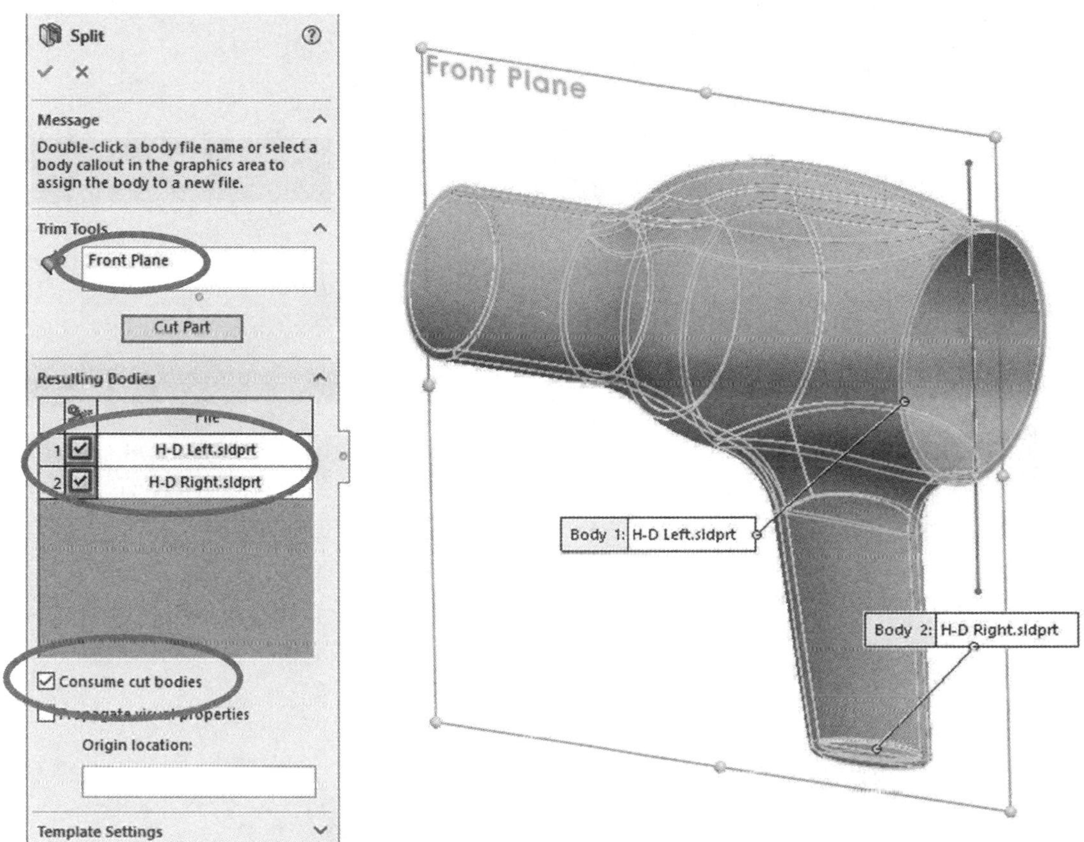

After completing the second "**Split**" command, there are no more bodies in our master part and the two bodies have been added to the named files.

The "Split" command can use multiple surfaces, planes, or sketches at the same time to cut a part in multiple directions.

It is not necessary to use the "Consume bodies" option; the split bodies can remain in the master model, but be aware that features added *after* the "Split" features will not propagate to the split bodies' parts. We used the option in this exercise to show the reader its effect. Likewise, we don't have to save to an external part; the split bodies can remain in the master file and we can keep adding features to each body as local operations. The problem with this approach is that the FeatureManager can get very long, keeping track of features will be difficult, rebuild times can be long slowing down the system, and eventually we'll have to save each body to a part anyway. In any case, it's to our advantage to do it at a point where all the common features have been added to the master file, and keep the Master Model and the split files manageable.

From this point on we can individually add the features specific to each part, make an assembly with them and, if needed, add any features in the context of the assembly to finish our design.

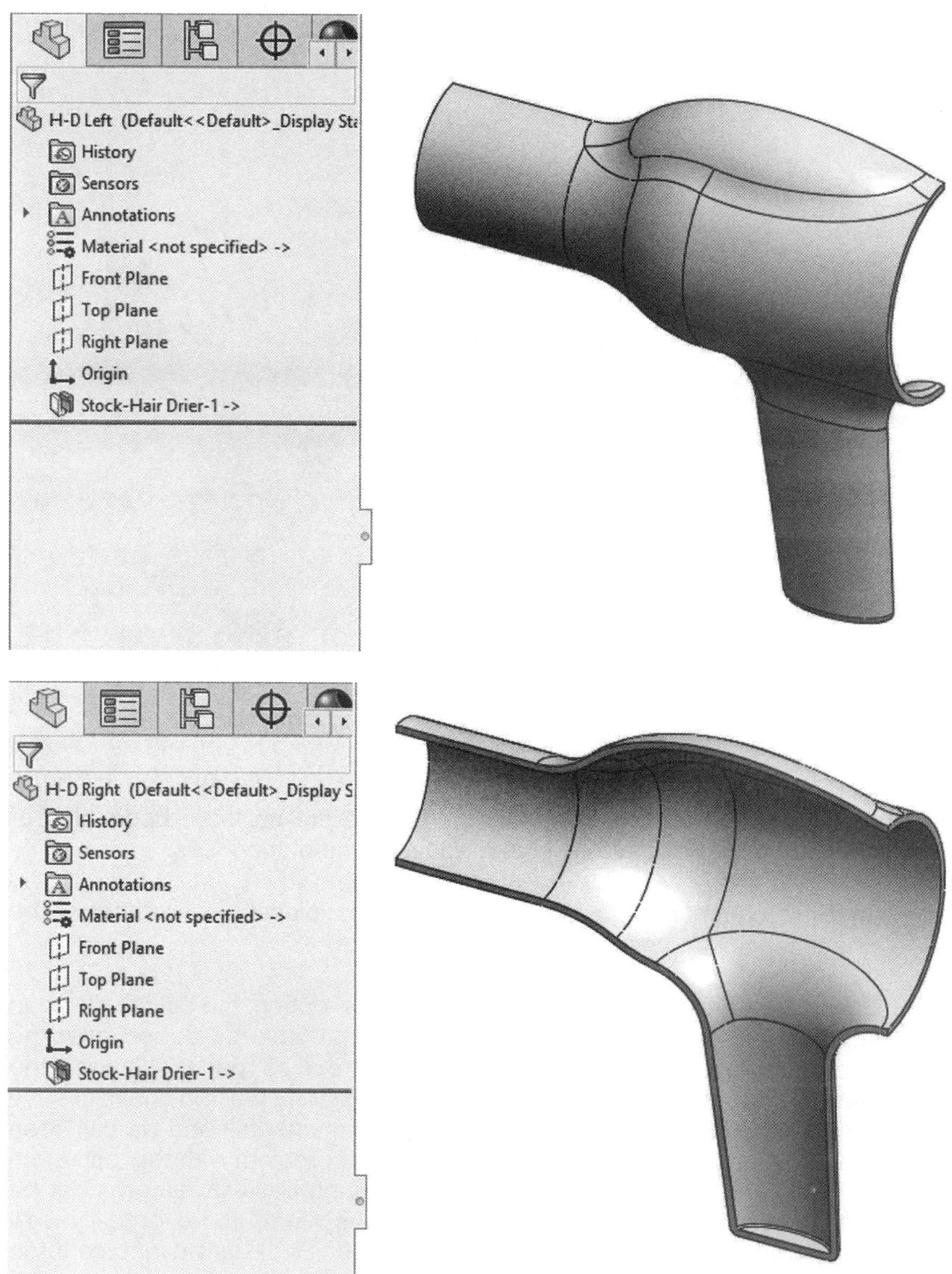

22.4. - Make a new assembly with the three components and insert them at the assembly's origin. When adding components from a split part to an assembly, if added at the assembly's origin they will be fixed in the same place they were in reference to the part's origin before they were split.

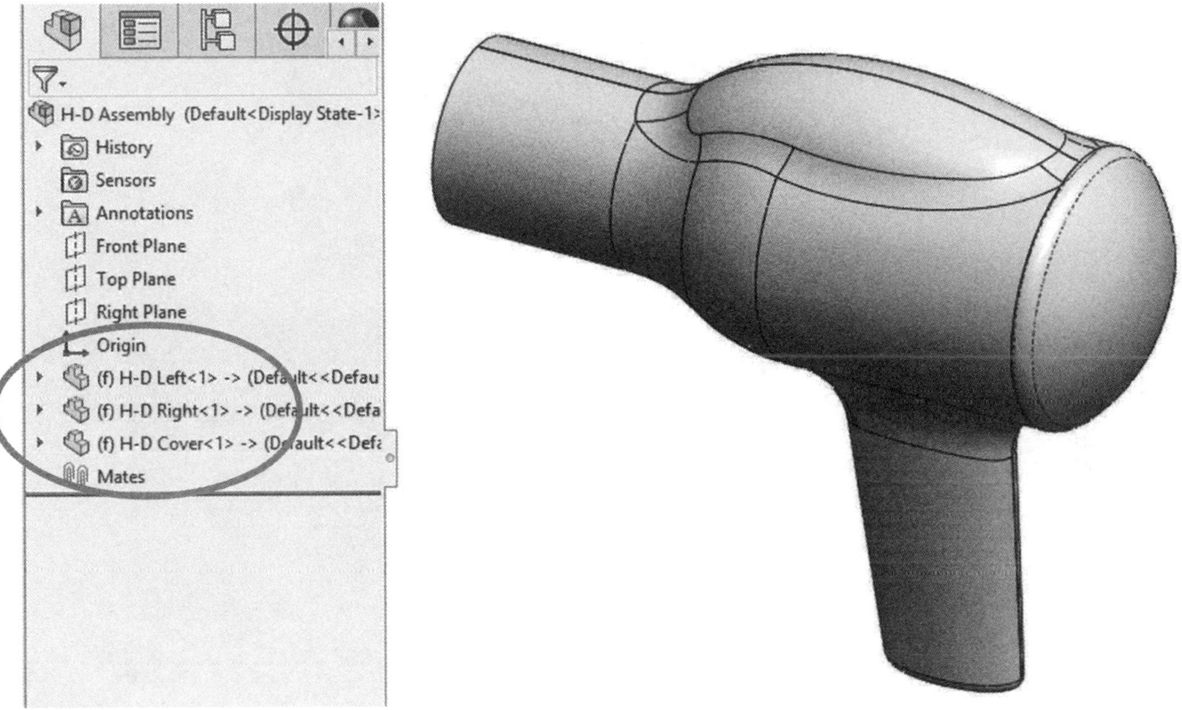

Notes:

Fastening Features

When designing plastic components, we usually have to fit parts together using a variety of methods, including screws, hooks and grooves, mounting bosses, etc. SOLIDWORKS includes several tools to automate this process when adding these features to our components.

Fastening Feature	Description
Mounting Boss Creates a parameterized mounting boss typically used in plastic design.	A mounting boss creates a boss with a hole or a pin with support fins to mate to each other or to use screws.
Snap Hook Creates a parameterized snap hook typically used in plastic design.	Adds a snap hook to latch on a matching groove to assemble plastic components.
Snap Hook Groove Creates a groove to mate with a selected snap hook feature.	Adds the matching groove to catch a snap hook. Must be made after the snap hook.
Vent Uses sketch elements to create a vent for airflow in both a plastic or sheet metal design.	A vent is a feature that allows us to make a vent opening using a grill sketch, defining the ribs, spars and boundaries of the vent.
Lip/Groove Creates a lip, groove, or both lip and groove typically used in plastic design.	Creates a lip and groove to (usually) matching plastic components to assemble them. A lip to one part, a groove to the other.

23.1. - To learn about fastening features, open the file "*Plastic Box*" from the accompanying files. It has already been split in two bodies making a matching box and cover.

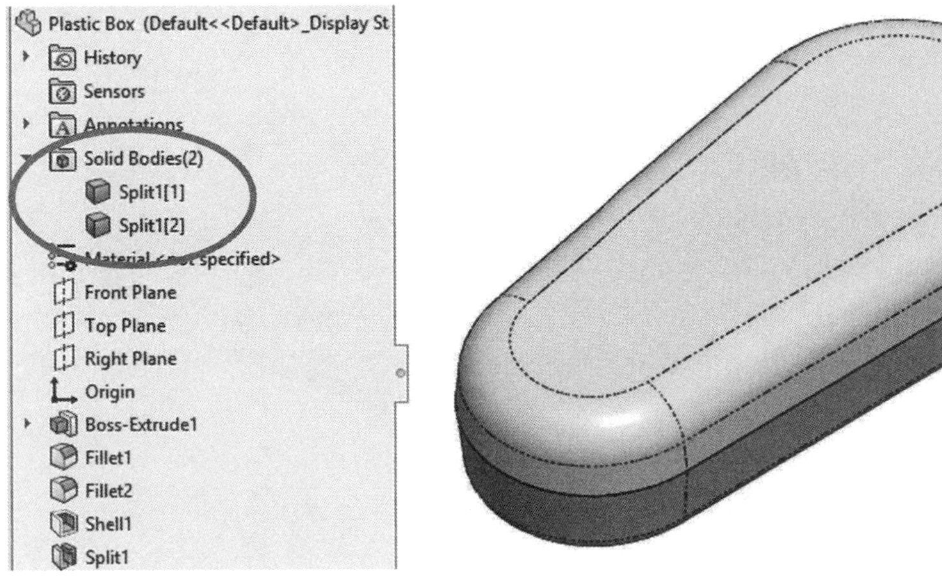

The first fastening feature will be a lip and groove, which are primarily used to align and assemble components adding a groove to one part and a lip to the other. Select the "**Lip/Groove**" from the menu "**Insert, Fastening Feature, Lip/Groove,**" or show the Fastening Feature toolbar.

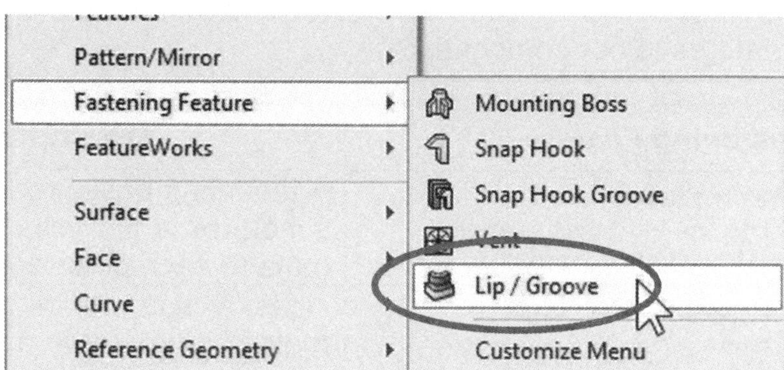

The lip will be added to one body and the groove to the other body; we can add both at the same time, or they can be added separately, either to a body in a multi body part, or a part. In our case the groove will be added to the bottom body, and the lip to the top body. First, we need to select the body that will have the groove. During this command the body will be colored green.

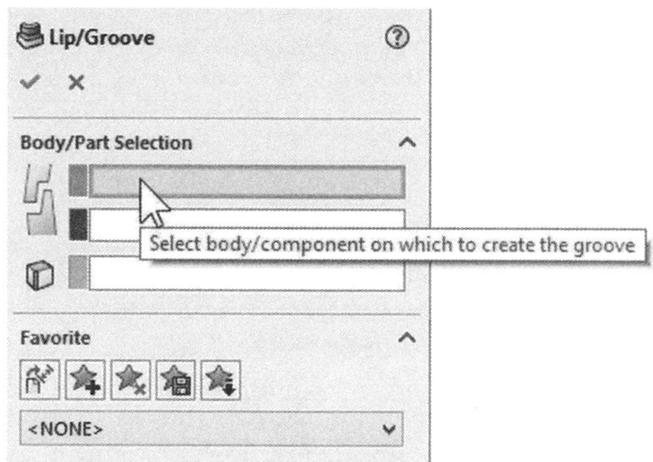

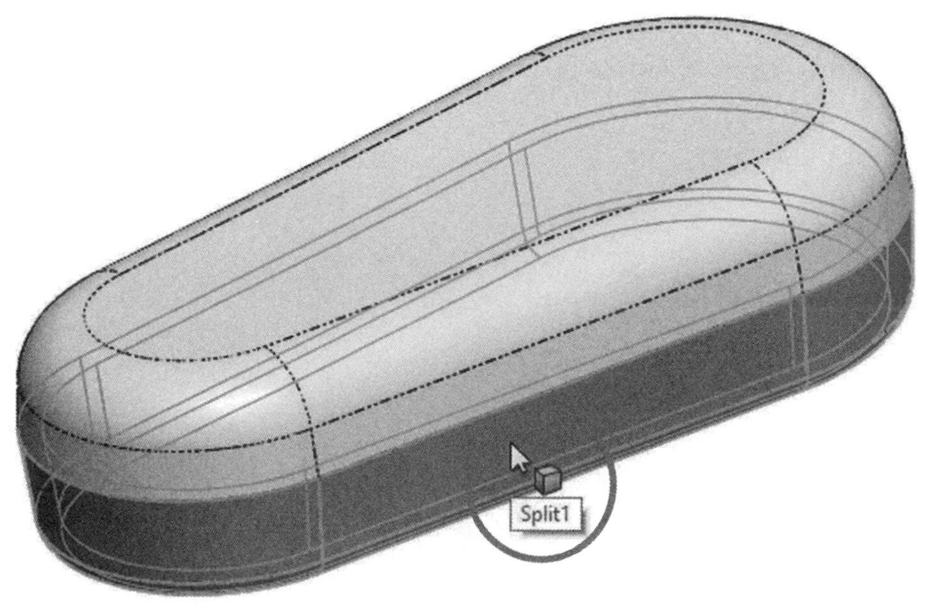

Immediately after selecting the groove body, the “Groove Selection” options are displayed and we are asked to select the body to which the Lip feature will be added. After adding the top body to the lip selection box the “Lip Selection” options are also displayed. The lip body will be colored purple.

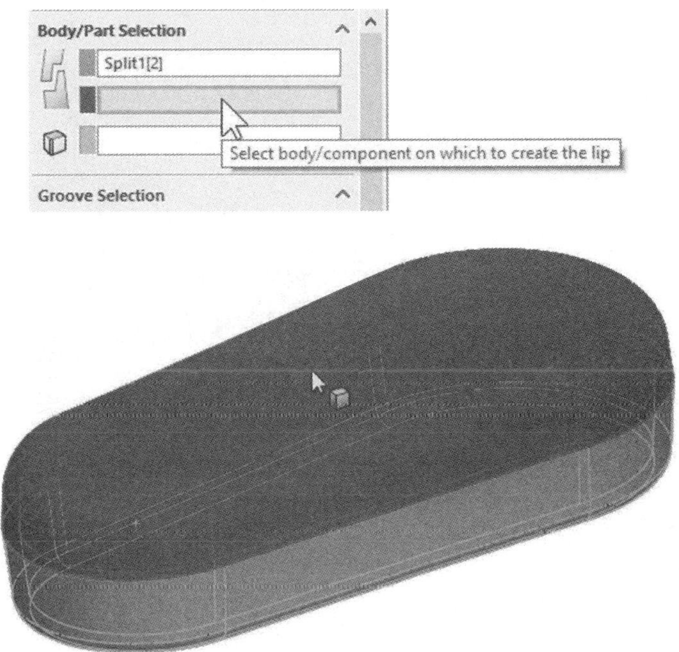

23.2. - After selecting the Lip body, we have to define the faces to which a groove/lip will be added, and the edge where it will be located. Activate the face selection box in the “Groove Selection” area. Automatically the Groove body is the only one visible, and the face matching the Lip body is already pre-selected. This is the only face that needs to be selected. If the groove is to be added to more faces, they would need to be selected in this step.

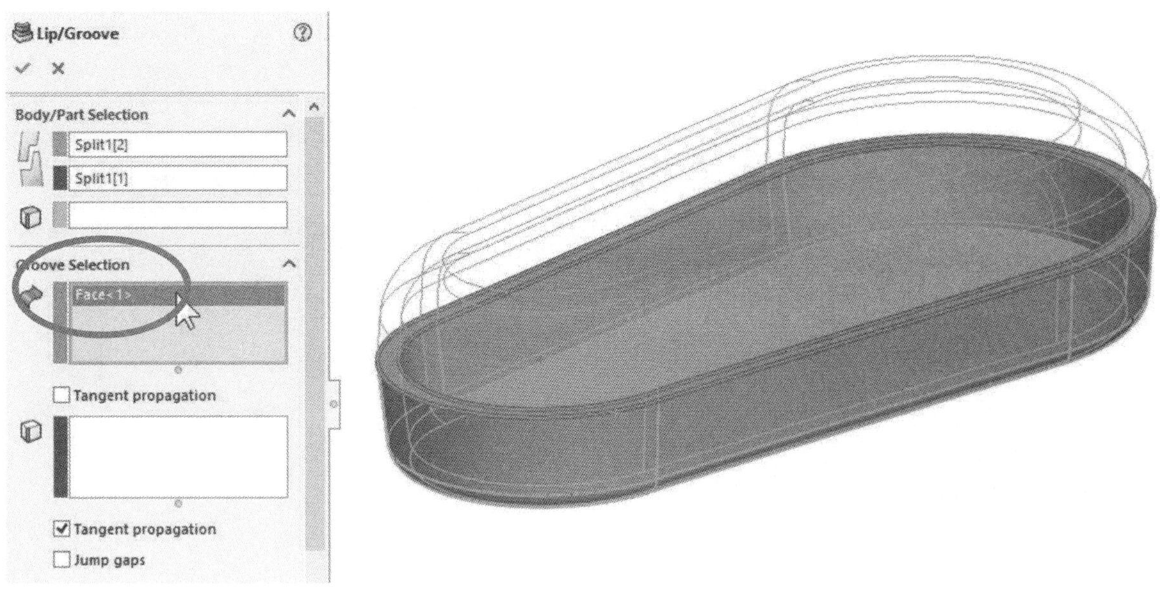

The next step is to select the edge where the groove will be located. Activate the edge selection box; for our example, we'll use the inside edge. Make sure the "Tangent Propagation" option below is turned on; this way the groove will be added all around the body. After selecting the edge, we see a preview of the groove.

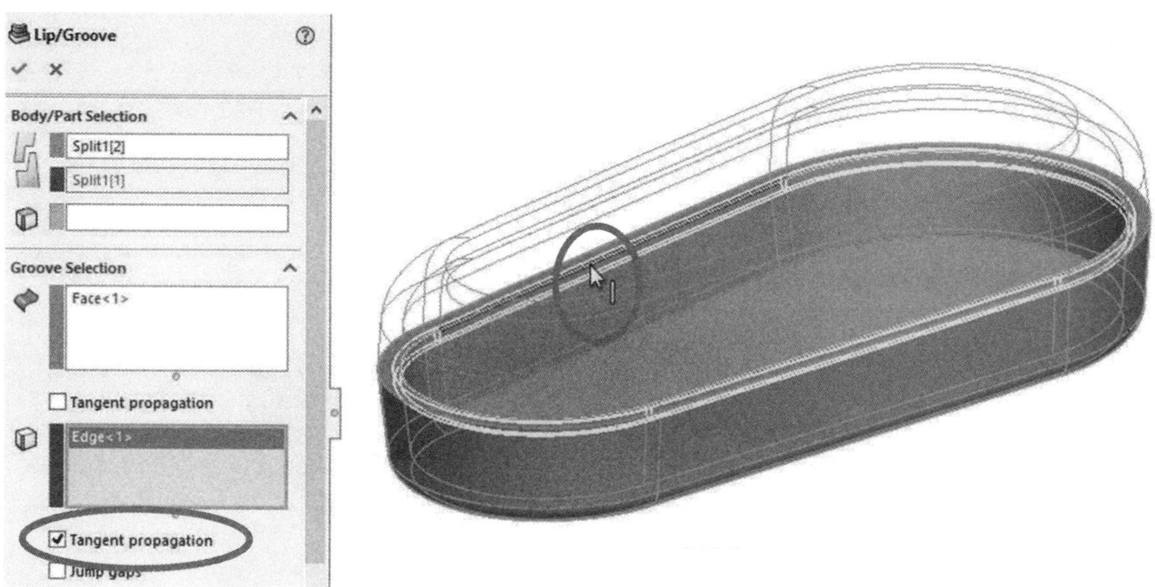

23.3. - After defining the groove location, activate the Lip faces selection box. The groove body is hidden and the Lip body is shown. The same matching face with the groove body is already pre-selected; this is the only face needed. Rotate the model for a better view and activate the edge selection box. Select the inner edge to add the lip to it. Be sure the "Tangent Propagation" checkbox is selected to add the lip around the entire body.

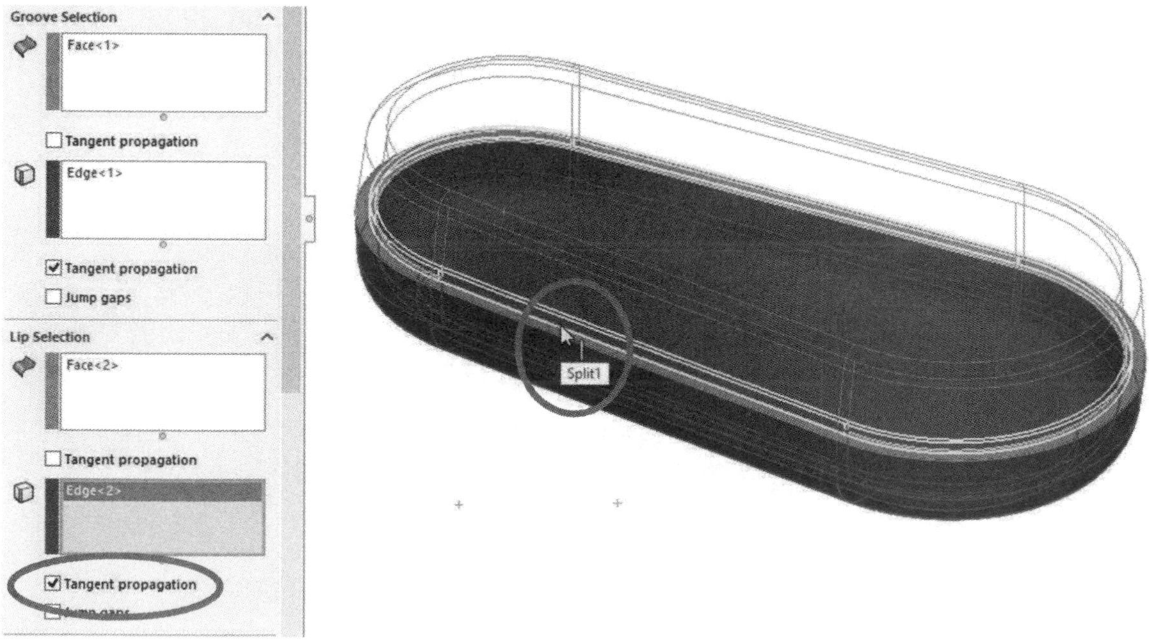

23.4. - After selecting the edges to locate the lip and groove, scroll down to the parameters section, enter the lip and groove dimensions, and click OK to finish.

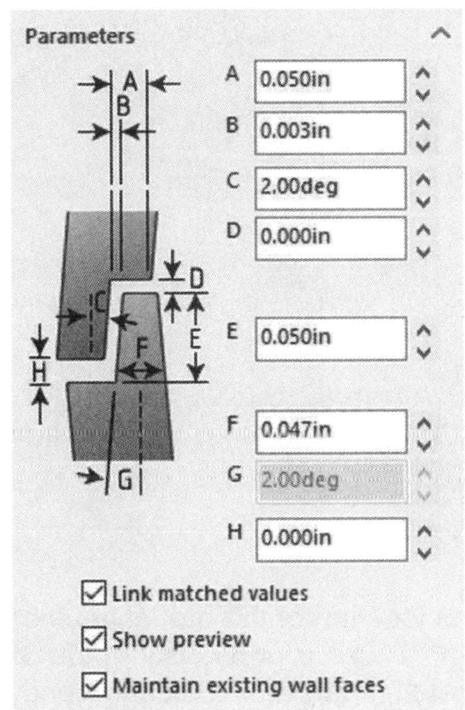

A: Groove width
B: Spacing between groove and lip
C: Groove draft angle
D: Upper gap between lip and groove
E: Lip height
F: Lip width
G: Lip draft angle (matched to groove draft angle)
H: Gap between lip and groove

The "*Lip&Groove1-Groove*" and "*Lip&Groove1-Lip*" are added to the FeatureManager as separate features. Activate the section view command using the "*Front Plane*" to see a cross section to see the lip and groove just made.

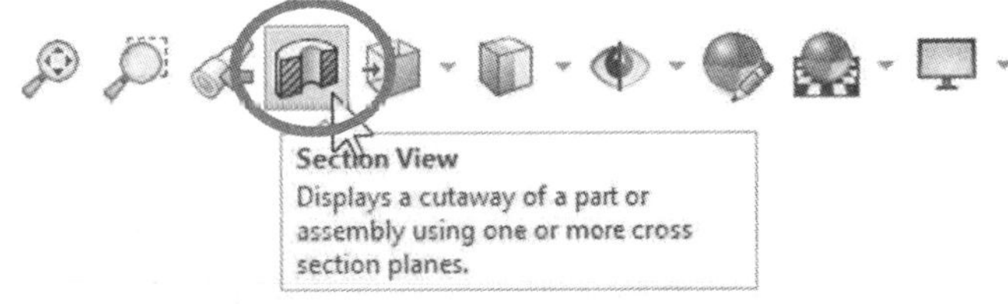

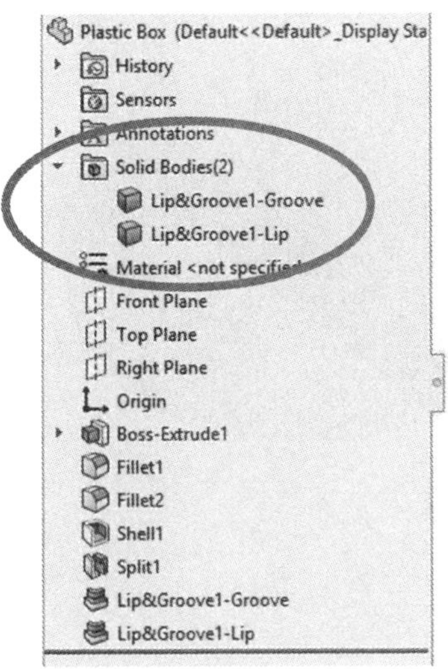

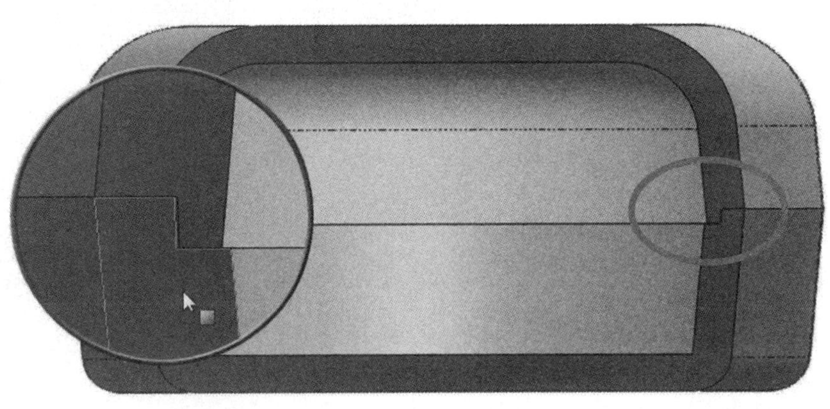

23.5. - After adding the Lip and Groove, the next step is to add a mounting boss in the bottom body. The mounting boss feature allows us to add bosses to add screws, mounting bosses for electronic boards, etc. Hide the top body and select the "**Mounting Boss**" command from the menu "Insert, **Fastening Feature, Mounting Boss**."

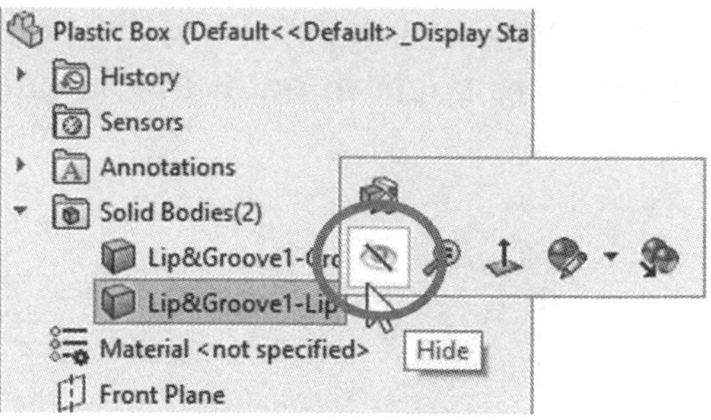

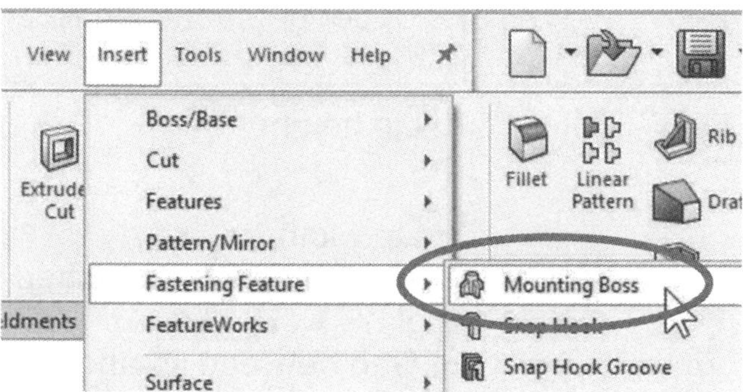

The first thing we need to do is to select the location of the mounting boss; we can either click to select a location in the part, or use a previously made 3D sketch point. Since we don't have a pre-made sketch point, click inside the body, near the center of the wider side. After selecting the location, a preview of the mounting boss is displayed. The boss will be precisely located after the boss is added by editing the automatically created 3D sketch.

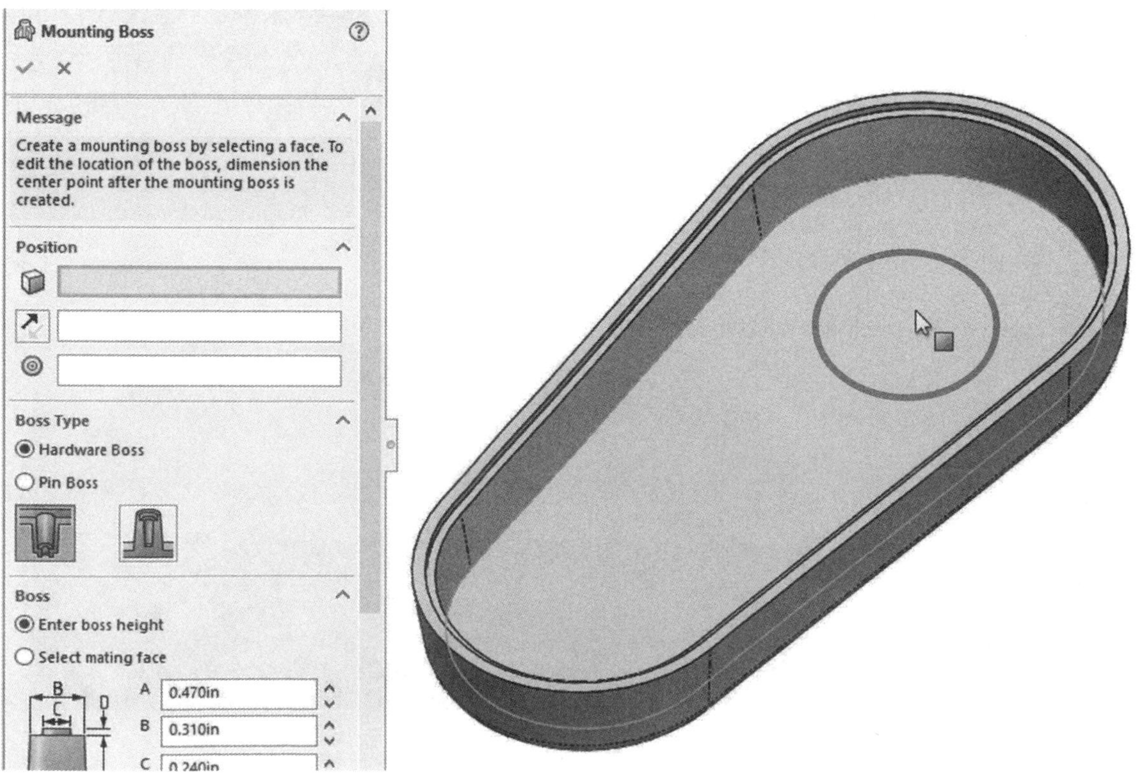

In the lower body, we'll add a Hardware boss; this boss allows us to put a screw from the back of the part, and secure it to the other half of the box. The boss can optionally have fins for reinforcement.

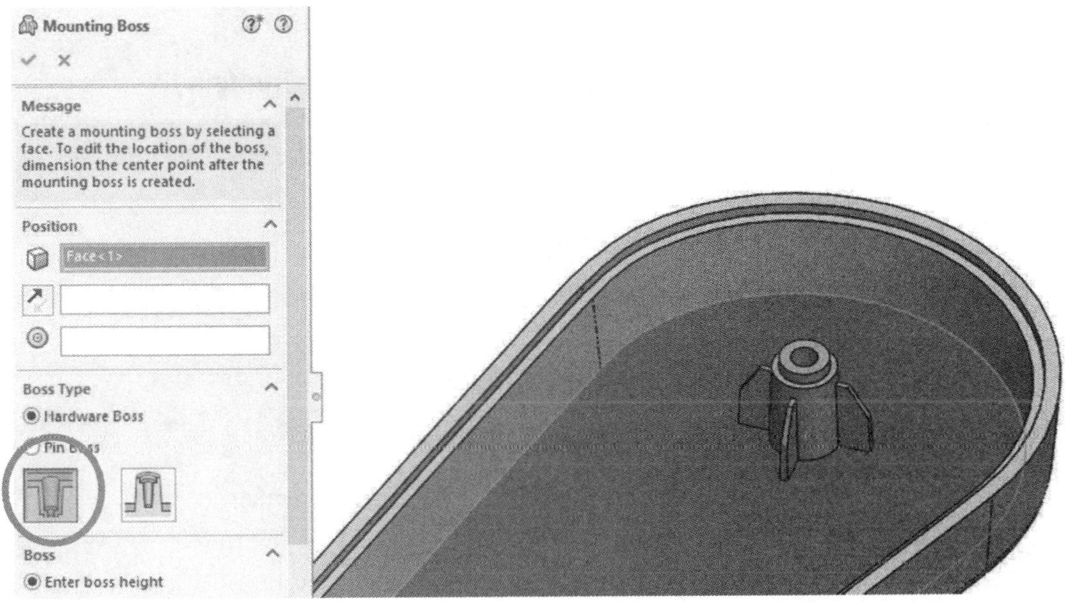

After selecting the location, enter the following dimensions for the hardware boss.

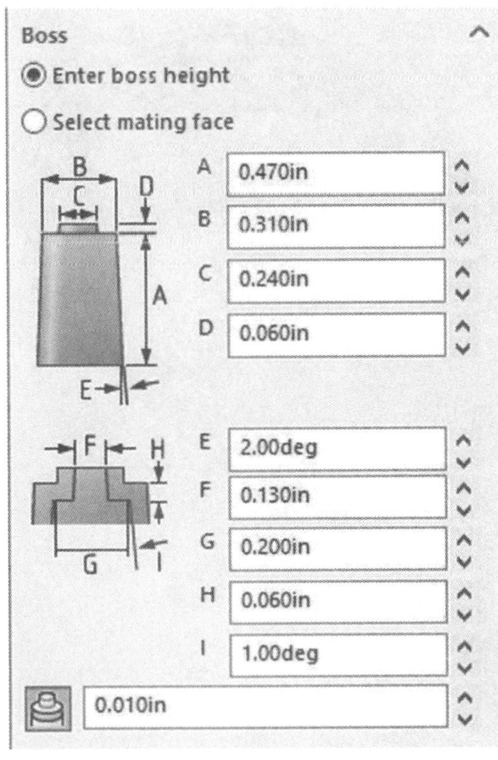

Enter the boss' height

Define boss' height using a model face

A= Boss height

B= Boss diameter

C= Diameter of the boss step

D= Height of the boss step

E= Draft angle of the boss

F= Diameter of the inside hole

G= Diameter of the inside counter bore

H= Depth inside the counter bore

I= Draft angle inside the hole

Clearance value for the boss height

In the Fins section select the "*Right Plane*" as the direction to align the fins, enter 4 for the number of fins per mounting boss and enter the following values for the fin's dimensions. Click OK to add the mounting boss.

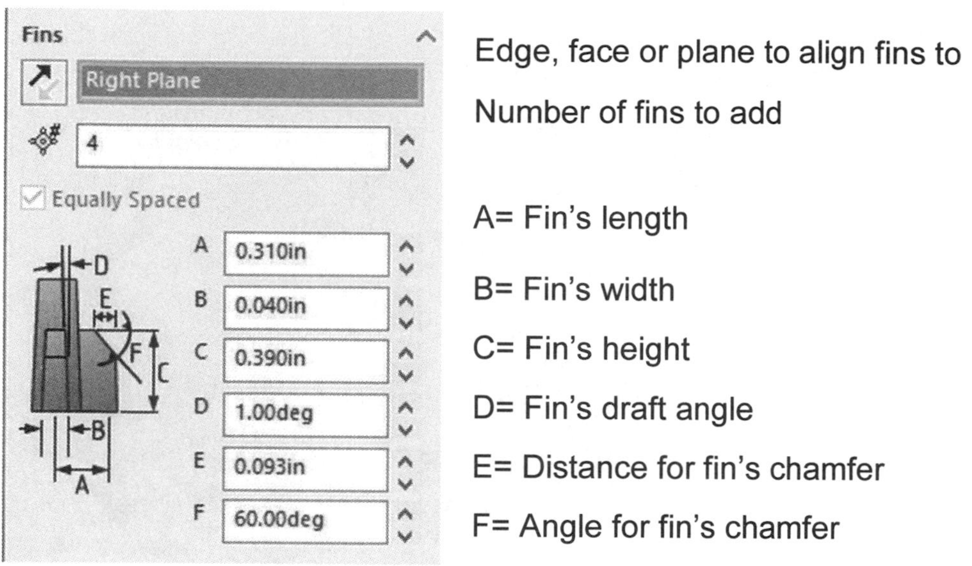

23.6. - After adding the first mounting boss, edit the 3D sketch in the "*Mounting Boss1*" feature to locate it.

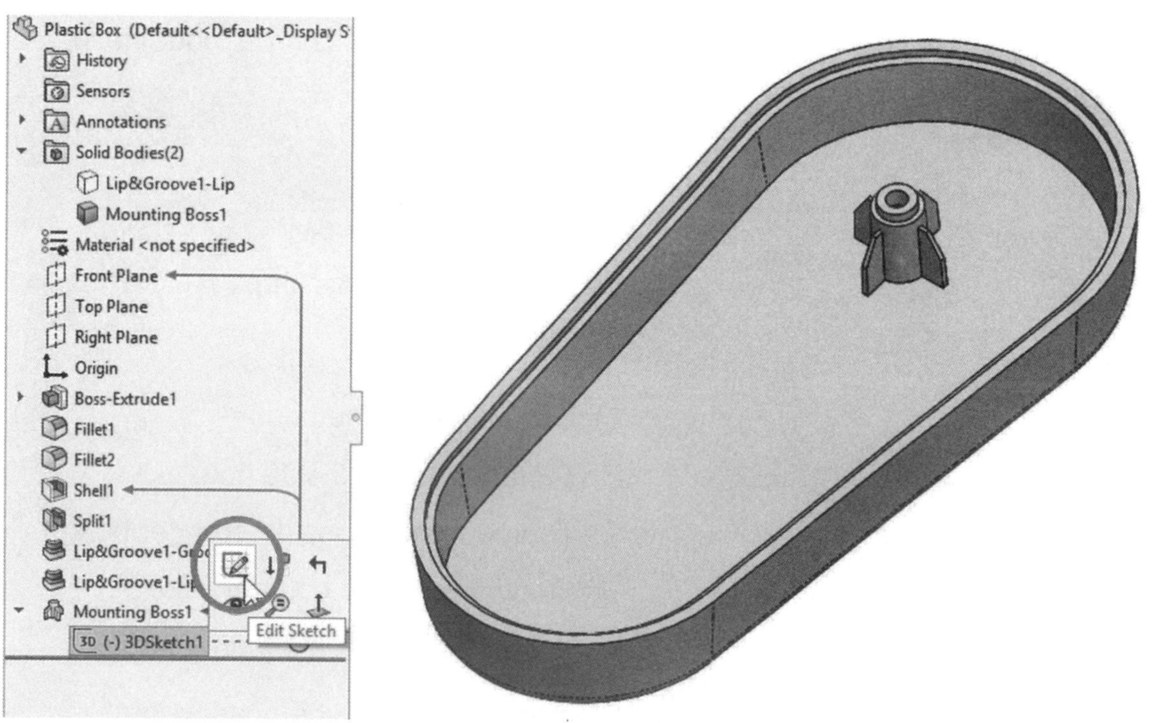

Select the 3D sketch point and make it concentric to the round edge at the inside of the shell and exit the sketch when done.

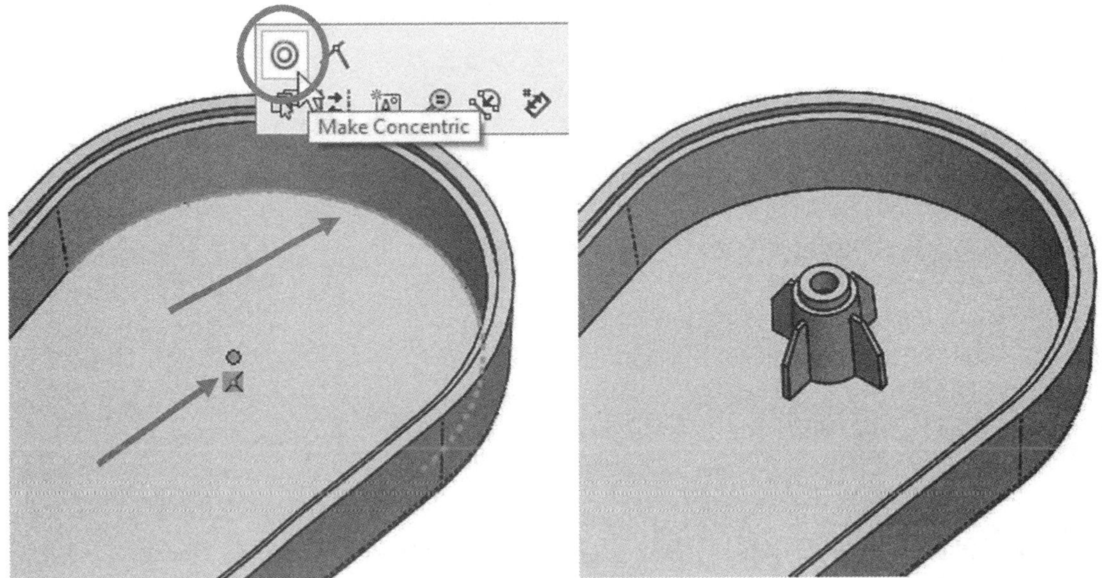

23.7. - As is the case with most plastic components, multiple mounting bosses are usually required, and their location is typically driven by the electronics board inside. To add two more identical mounting bosses, we'll use a sketch driven pattern. Draw the following 2D sketch inside the bottom body; each Point in the sketch will define the location of a new pattern instance. Exit the sketch when finished to continue.

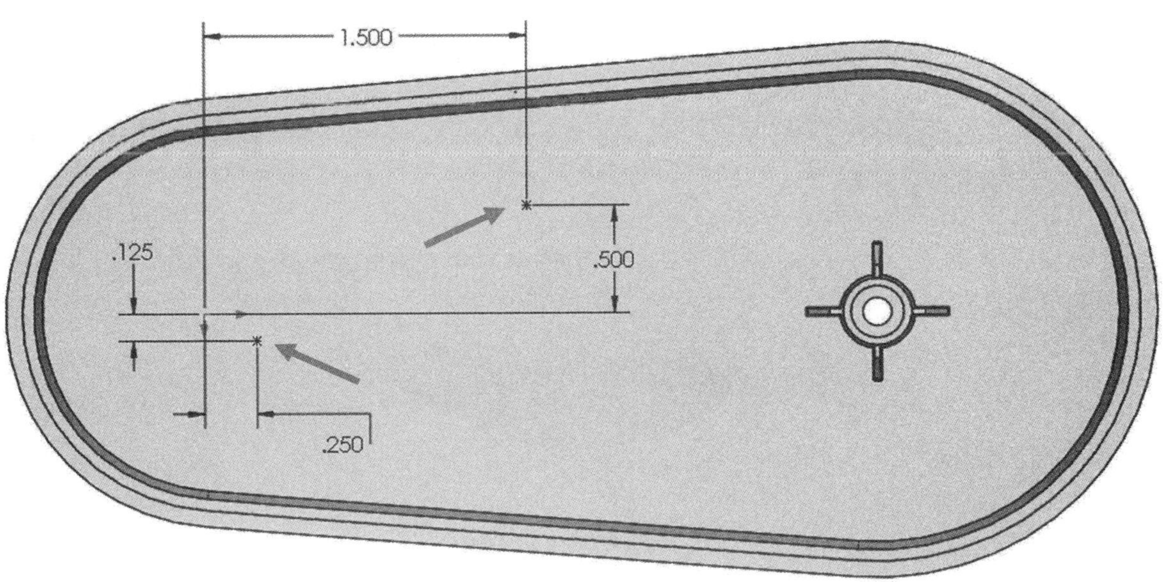

23.8. – After adding the sketch with the location of the new copies, select the "**Sketch Driven Pattern**" from the "Linear Pattern" drop down menu, or the menu "**Insert, Pattern/Mirror, Sketch Driven Pattern.**"

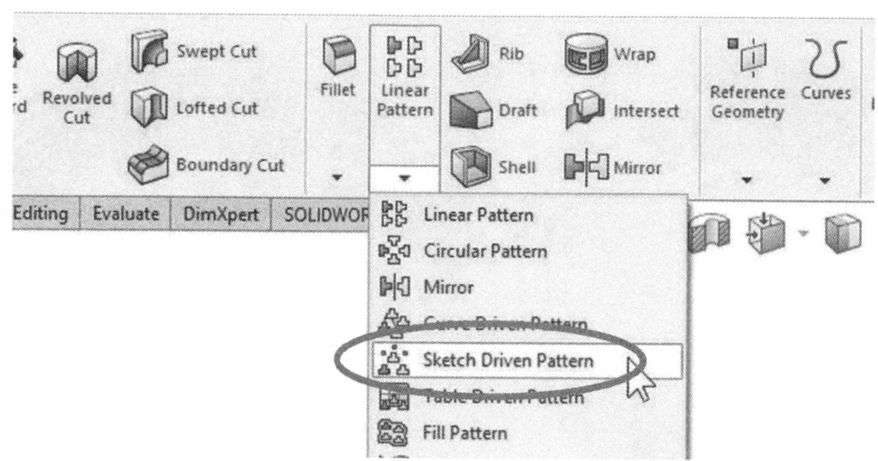

In the "Reference Sketch" selection box, select any point in the previously made sketch, and in the "Features to Pattern" selection box select any face of the mounting boss. The Sketch Driven Pattern adds a copy of the mounting boss at each point in the sketch, making it easy to create an oddly distributed pattern. Click OK to add the pattern.

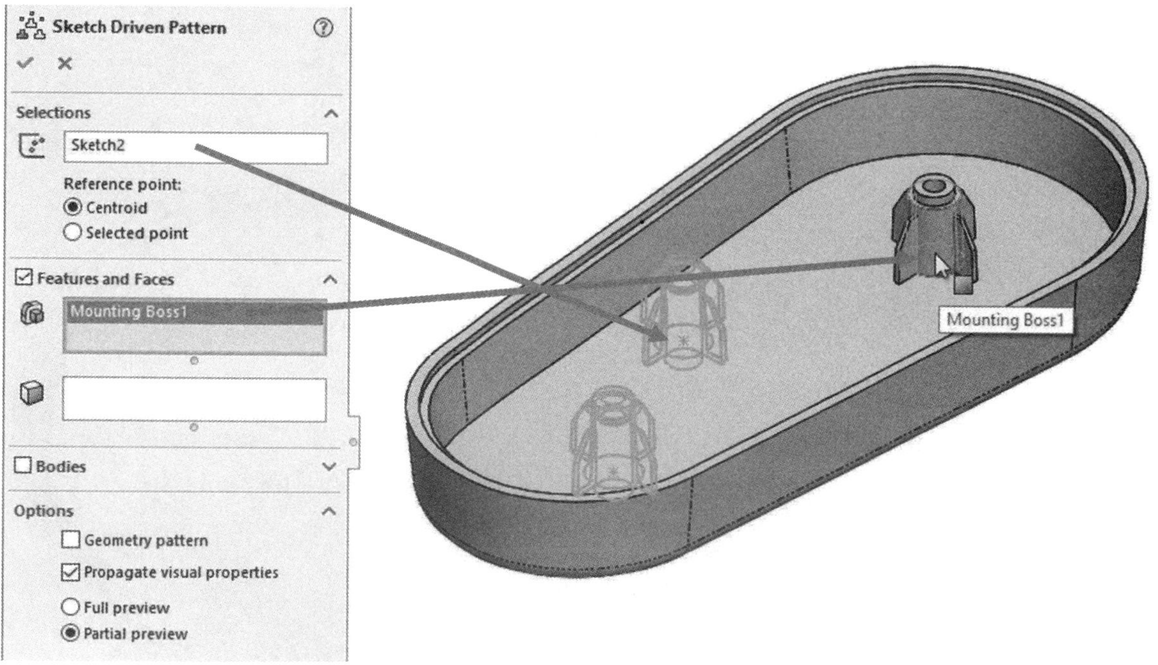

23.9. - The next step is to add the mounting boss opposite of the previous boss in the upper half body; this boss will have a hole for the screw coming from the other side of the bottom half. To add the new mounting boss, we must show the upper half body to continue.

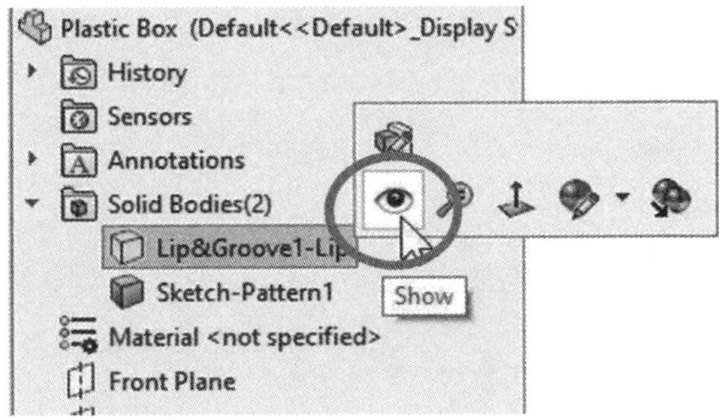

23.10. - Select the "**Mounting Boss**" command from the menu "**Insert, Fastening Feature, Mounting Boss**," in the position selection box, select the inside face of the top body using the "Select Other" command from the right-mouse-button menu. The outer face is automatically hidden, and the inside face that we need to select is immediately available. Left click to select it.

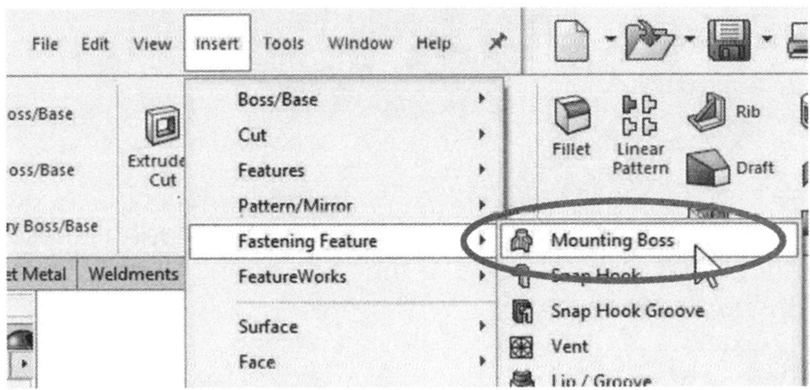

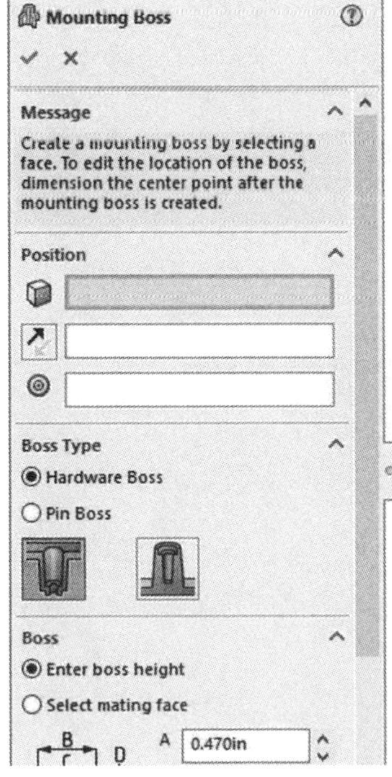

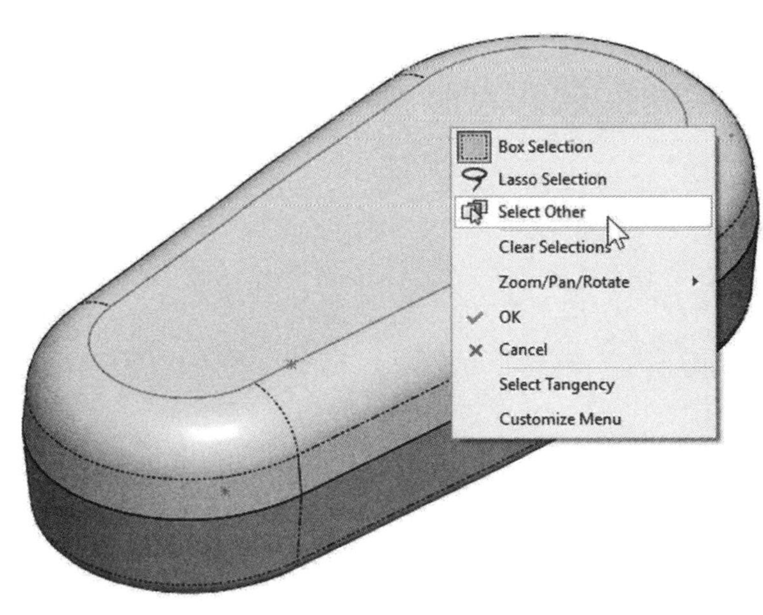

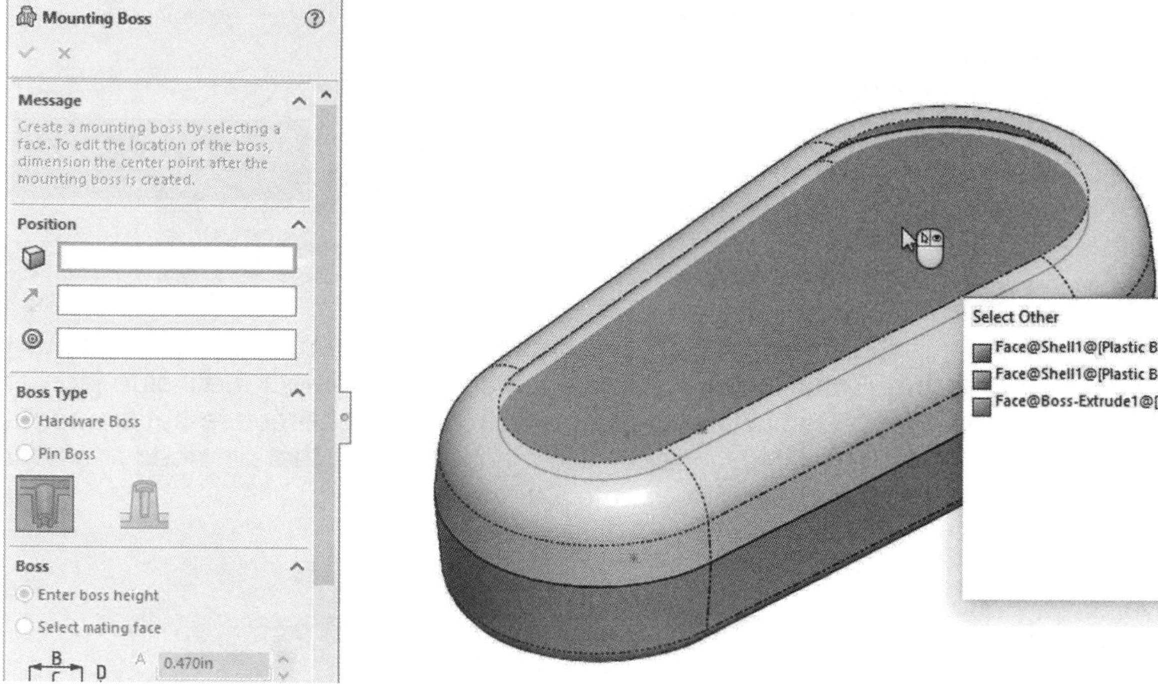

After the inside face is selected, activate the "Select circular edge to position the mounting boss" selection box, change the view to "Hidden Lines Visible" mode and select the topmost circular edge of the first mounting boss; this way the new boss will match the first one.

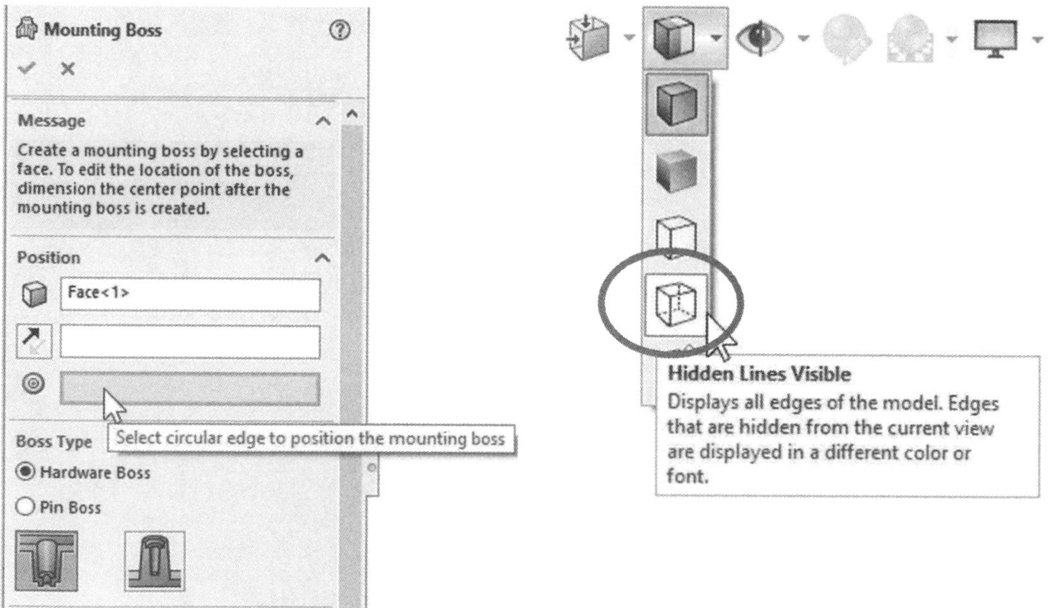

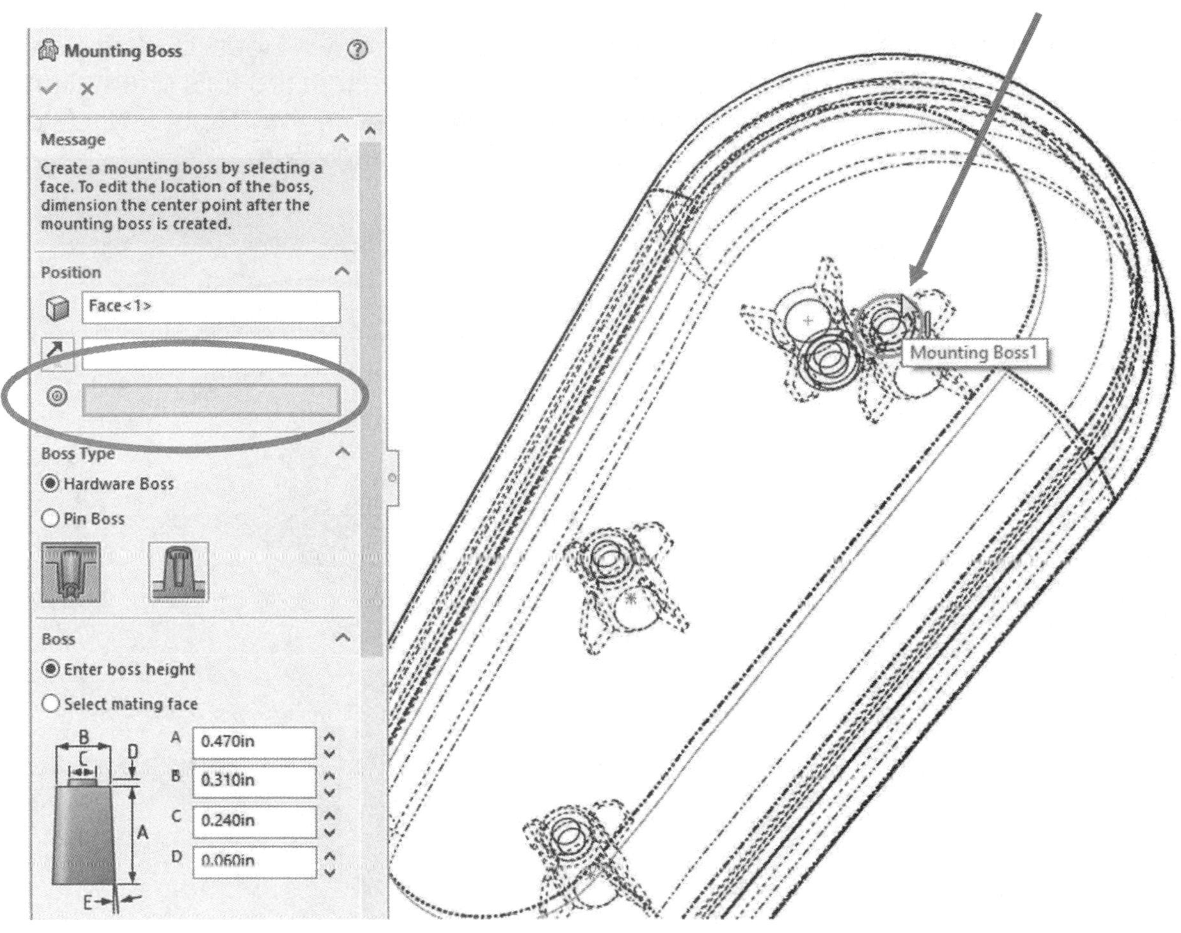

After selecting the edge to locate the new boss, select the "Thread" option in "Boss Type." The preview now shows the new boss matching the previous one.

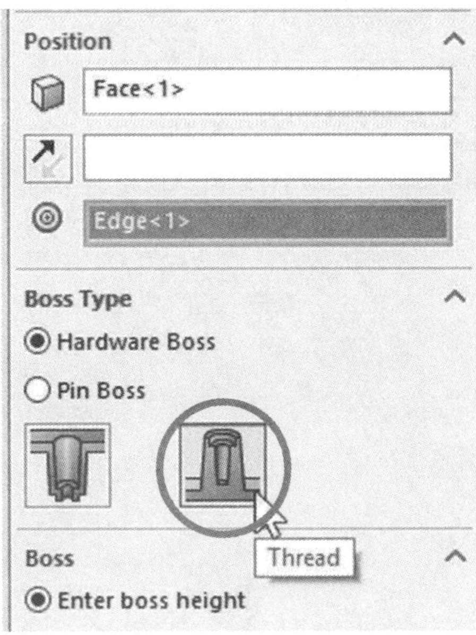

23.11. - Now that the threaded mounting boss is located, we need to define its dimensions. Enter the values listed for the boss and the fins. In the fin's orientation selection box, select an edge from the first mounting boss or the "Right Plane" to align it. Click OK when done to finish the "Thread" mounting boss.

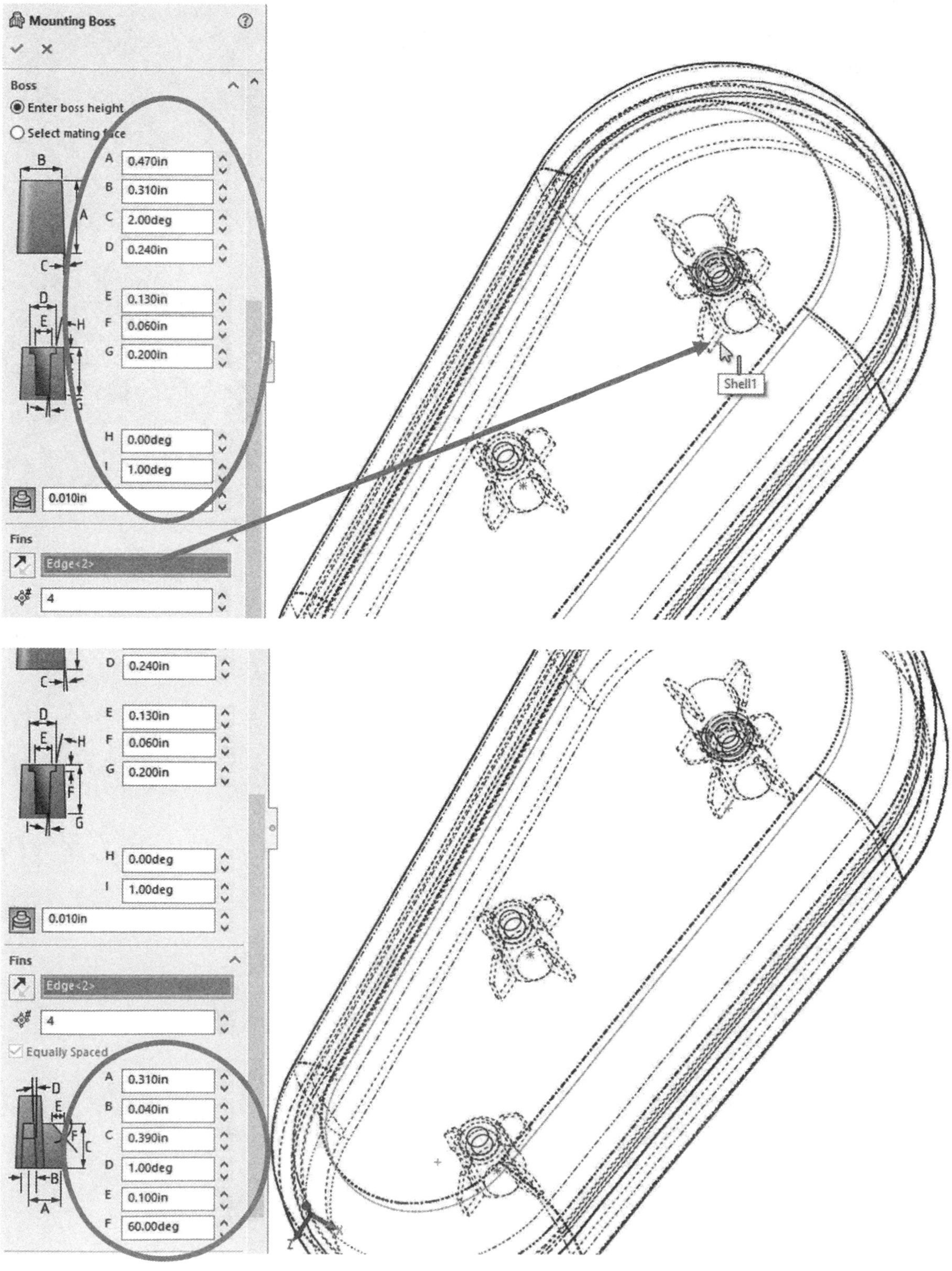

23.12. - Make the top body transparent and change the view to "Shaded With Edges" mode using the "Display Pane" (Default shortcut F8) to see the result.

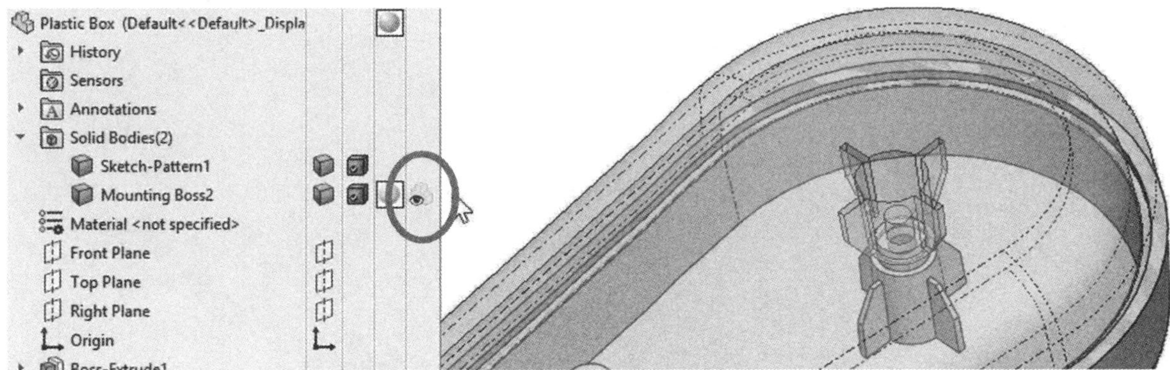

23.13. - To add the other two threaded mounting bosses, add a new "**Sketch Driven Pattern**" using the new mounting boss and the same sketch used for the first pattern. Click OK to finish, and hide the sketch used in the pattern.

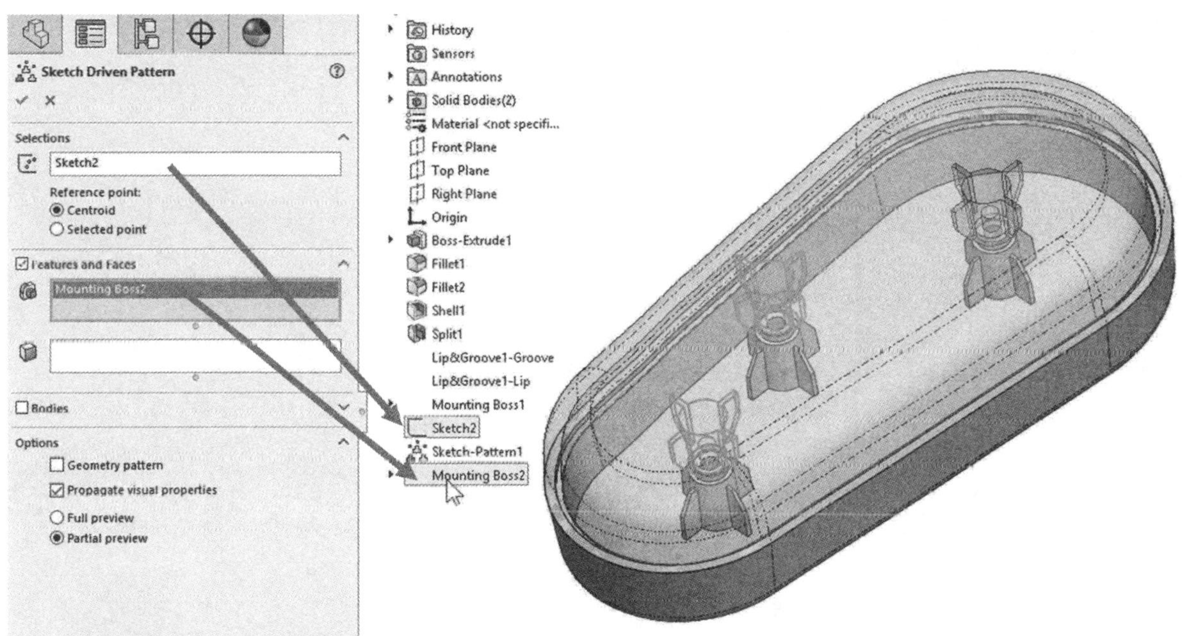

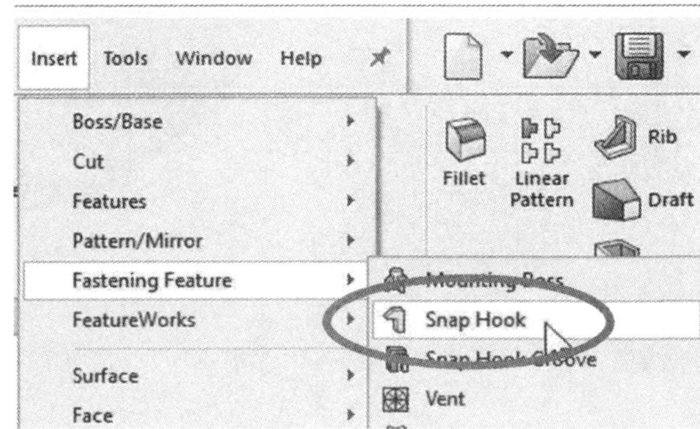

23.14. - The next step is to add the Snap Hooks. Snap hooks are commonly used to hold plastic components together without using fasteners or adhesives. Select the "**Snap Hook**" command from the menu "**Insert, Fastening Features, Snap Hook.**" The snap hooks will be added to the top body, and the grooves to catch the hooks in the lower body. For the first step (hooks) make the top body opaque and hide the bottom body. Rotate the view to see inside the part.

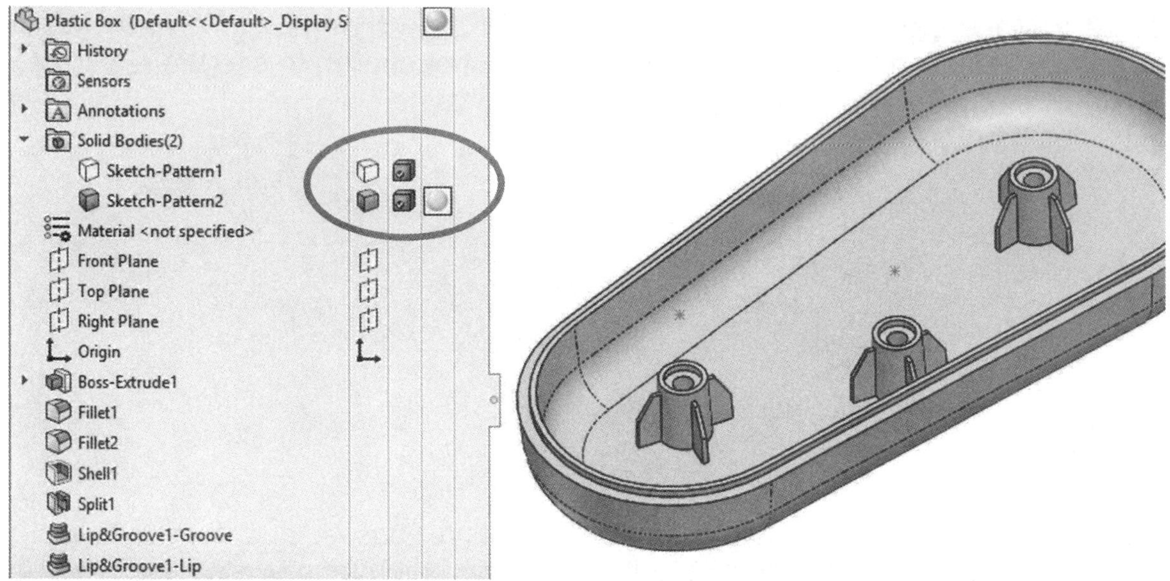

To locate the snap hook feature, select the inside edge of the lip feature near the left side of the part. A 3D Sketch point will be added at this location, which will be later edited to accurately locate the snap hook.

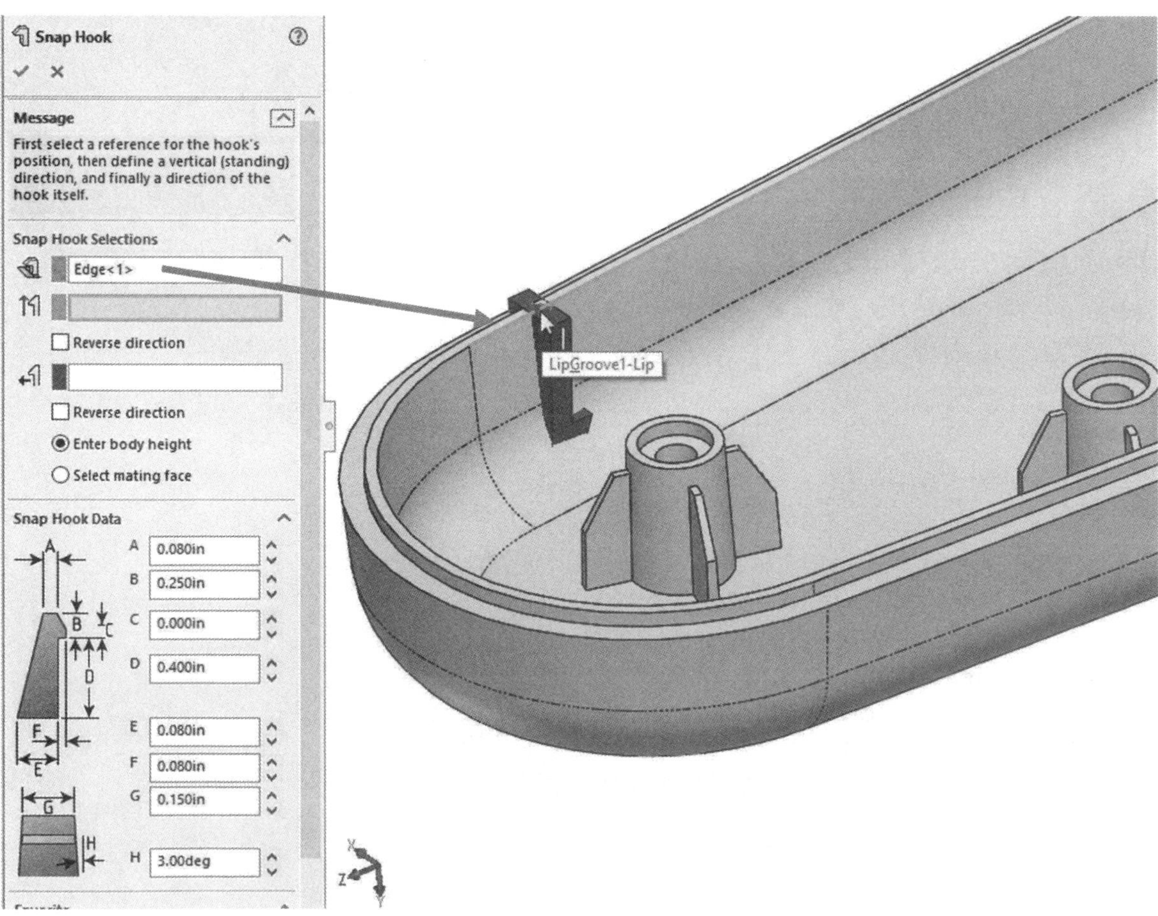

The next step is to define the vertical direction of the snap hook. We can select an edge along the desired direction, or a plane perpendicular to the snap hook. In this case, we'll select the top face of the lip. As soon as the face is selected, the hook is automatically rotated.

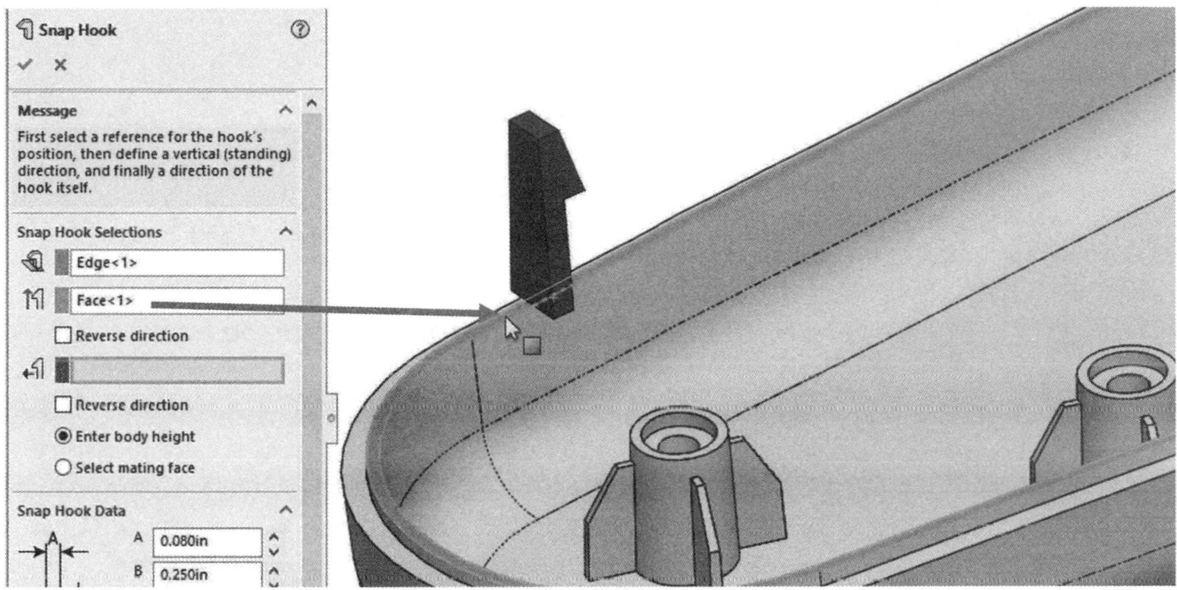

The next selection is to define the snap hook's orientation. For our example select the face inside. If needed, turn on the "Reverse direction" option to make the snap hook face outward.

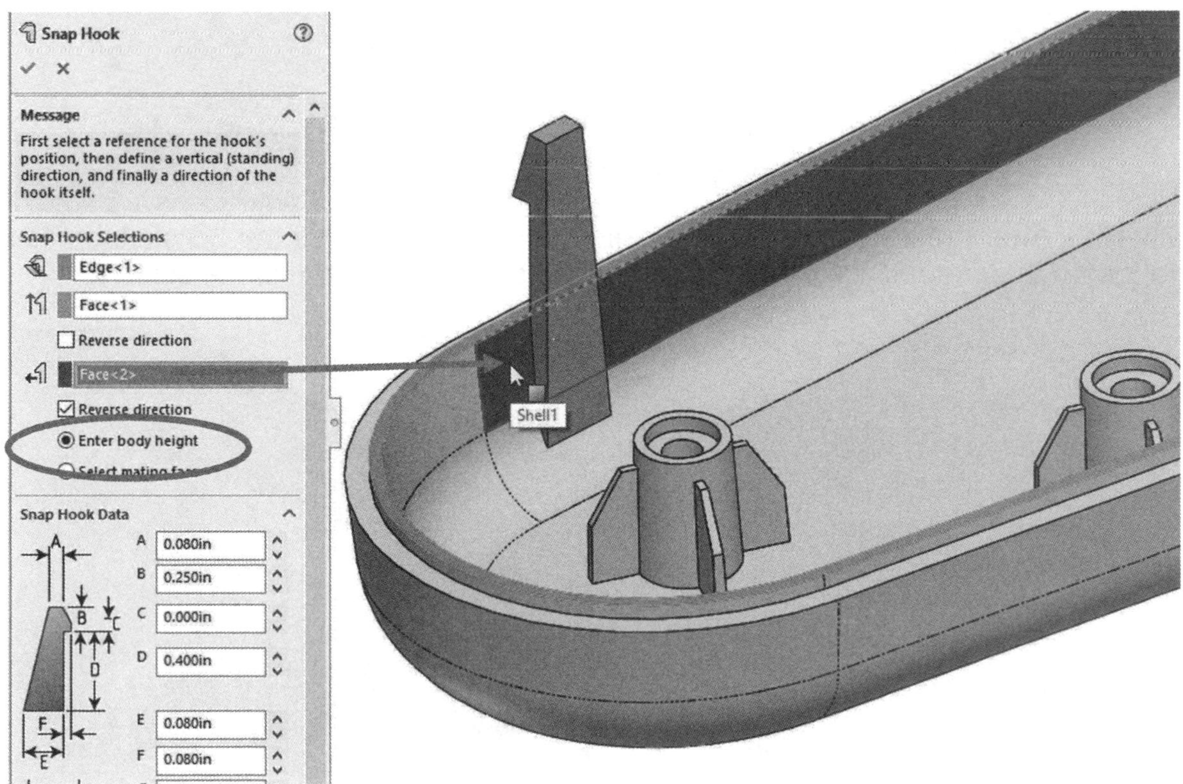

23.15. - Enter the following dimensions to define the size of the snap hook, and click OK to finish. To the right, we listed the description of each dimension.

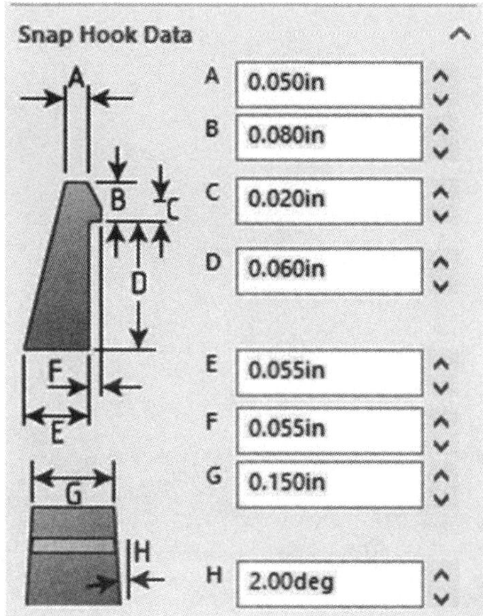

Depth at the top of the hook

Hook height

Hook lip height

Body height

Hook overhang

Depth at the base of the hook

Total width

Top draft angle

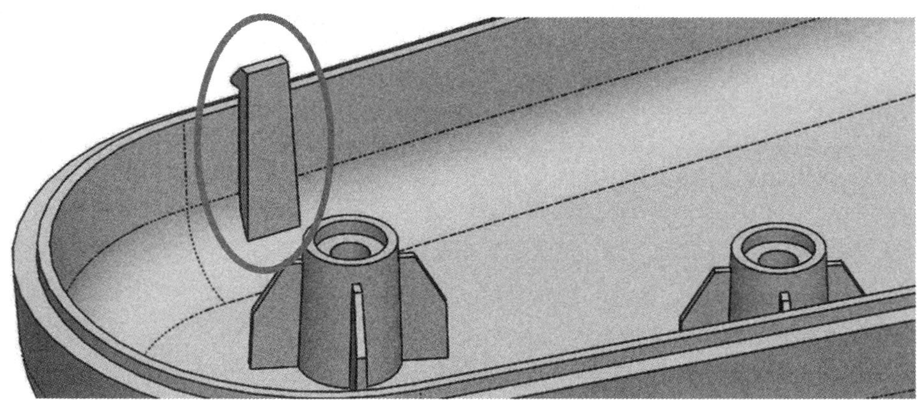

23.16. - After the snap hook is completed, expand it and edit the 3D sketch used to locate it; dimension the point to locate the snap hook and exit the sketch.

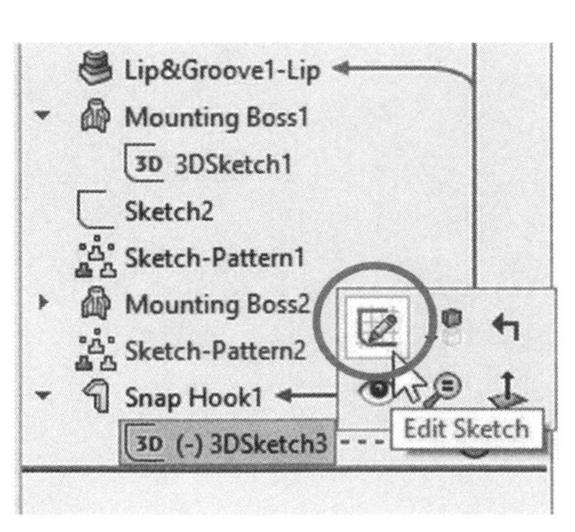

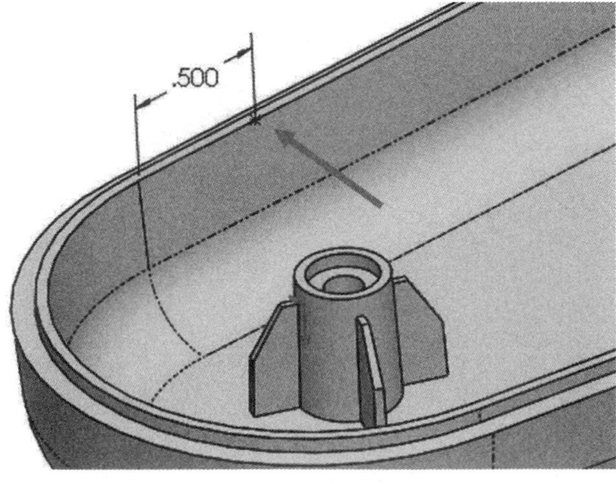

23.17. - Add a linear pattern to add a total of 3 snap hooks along the same edge, spaced 1" between them.

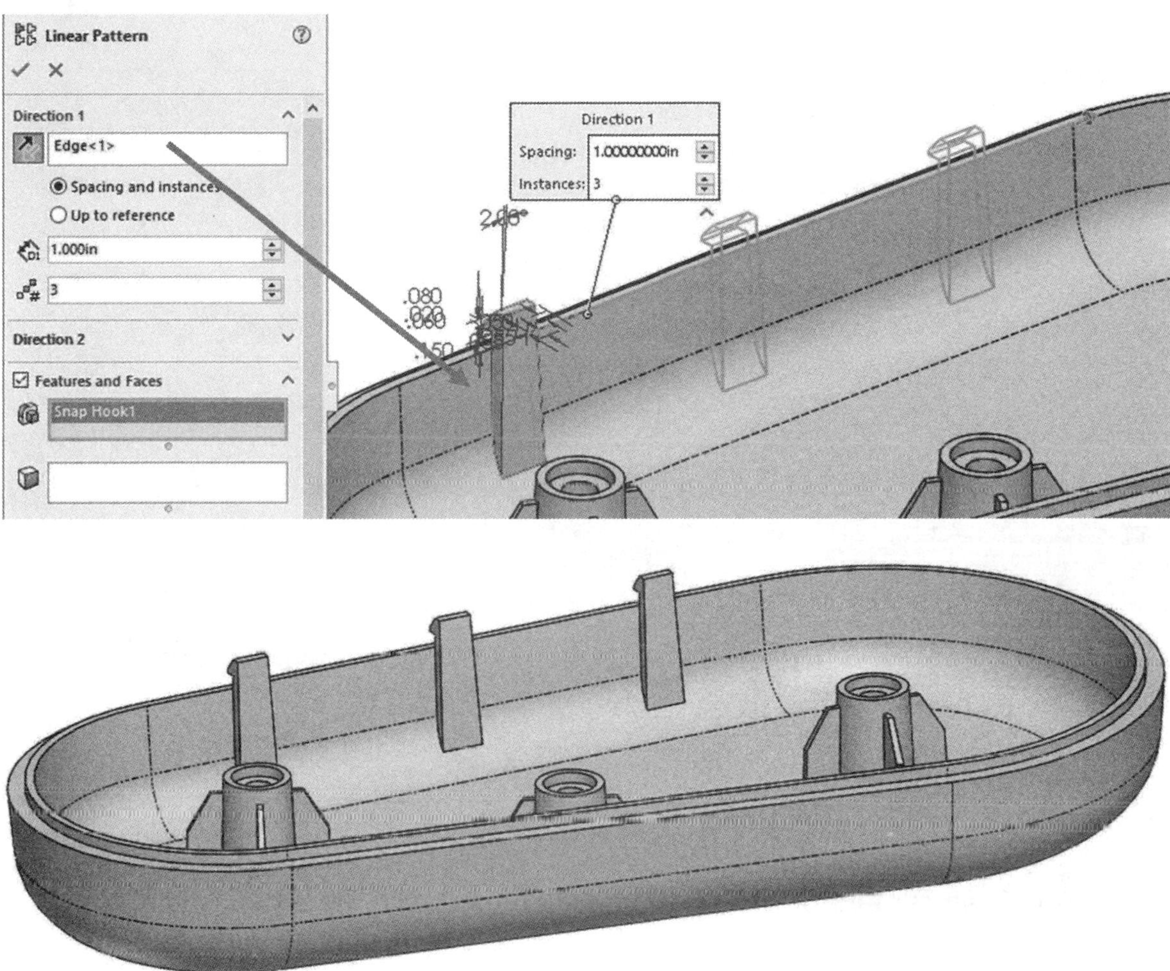

23.18. - Add a mirror feature to copy the three snap hooks to the other side.

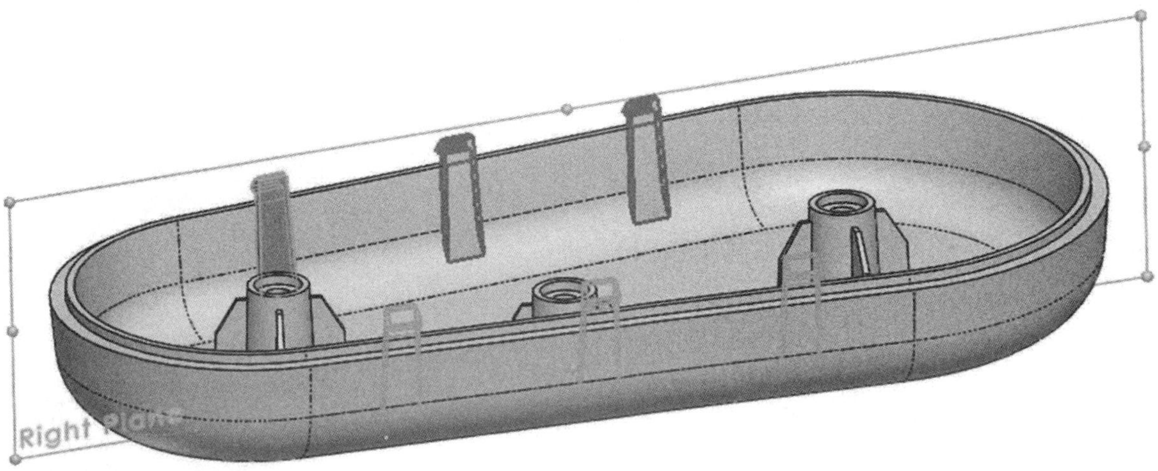

23.19. - The next step is to add the matching grooves in the other body to catch the snap hooks and hold the two pieces together. To improve visibility, make the bottom body visible, and the body with the hooks (top) transparent using the view palette (Shortcut F8). Select the "**Snap Hook Groove**" from the menu "**Insert, Fastening Features, Snap Hook Groove.**"

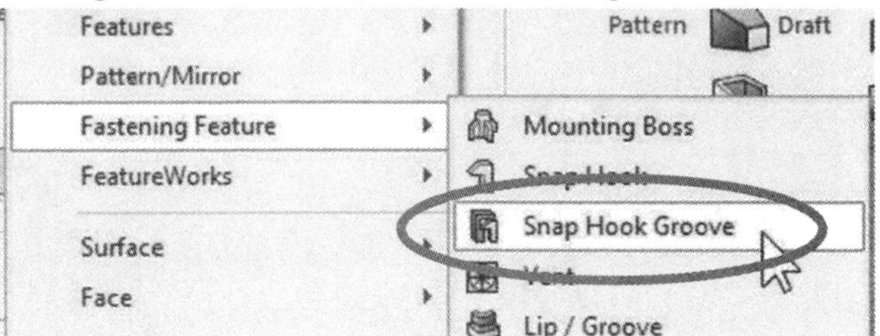

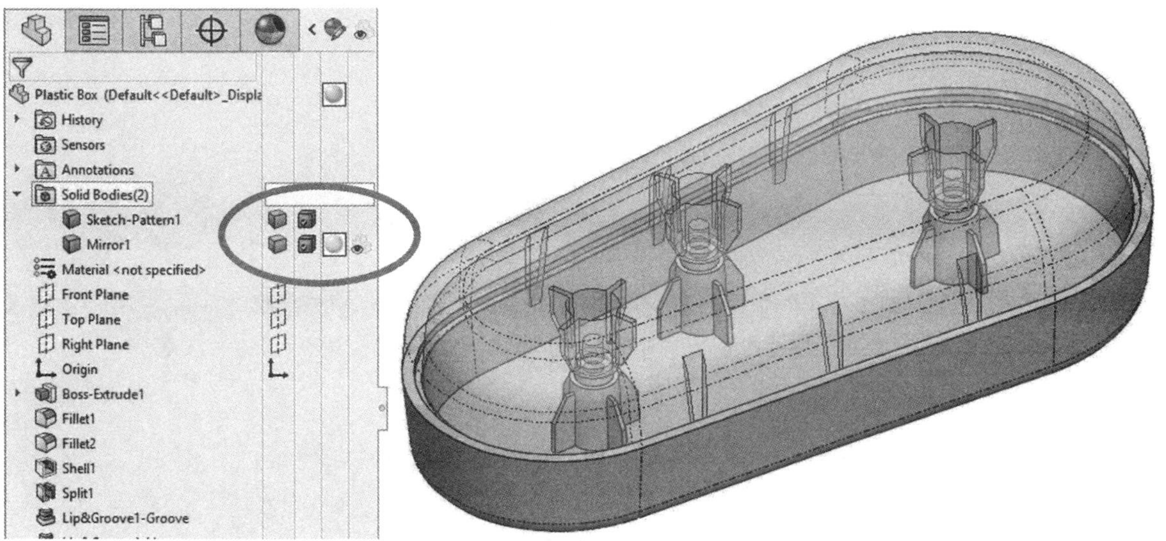

First, we need to select a snap hook feature to match the groove to it. Open the fly-out FeatureManager and select the "Snap Hook1" feature.

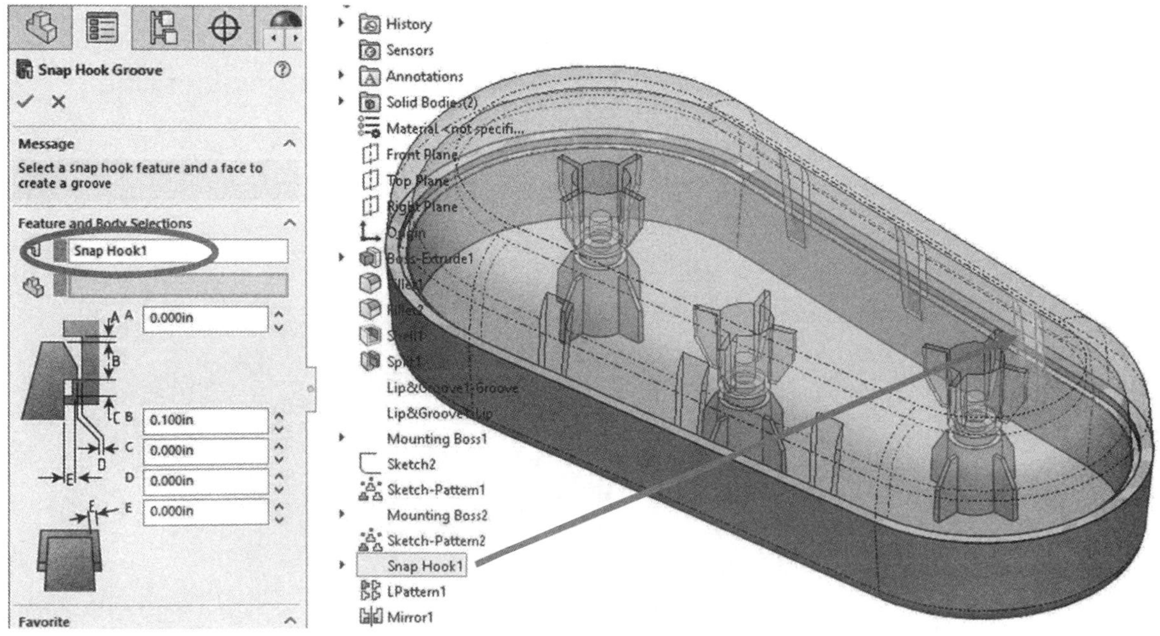

The second selection is the body where the **snap hook groove** will be added. Select the lower body and set the groove's dimensions as indicated. These dimensions provide a clearance between the snap hook and the groove to make assembly easy. After the lower body is selected, the top body is hidden and a preview of the groove is displayed. Click OK to finish and add the groove.

23.20. – Hide the top body, and just like we did with the snap hooks, add the same linear pattern and a mirror to add a matching groove to each of the six snap hooks. The top body was hidden and a section view added to see the hook and groove fit. Save and close the file.

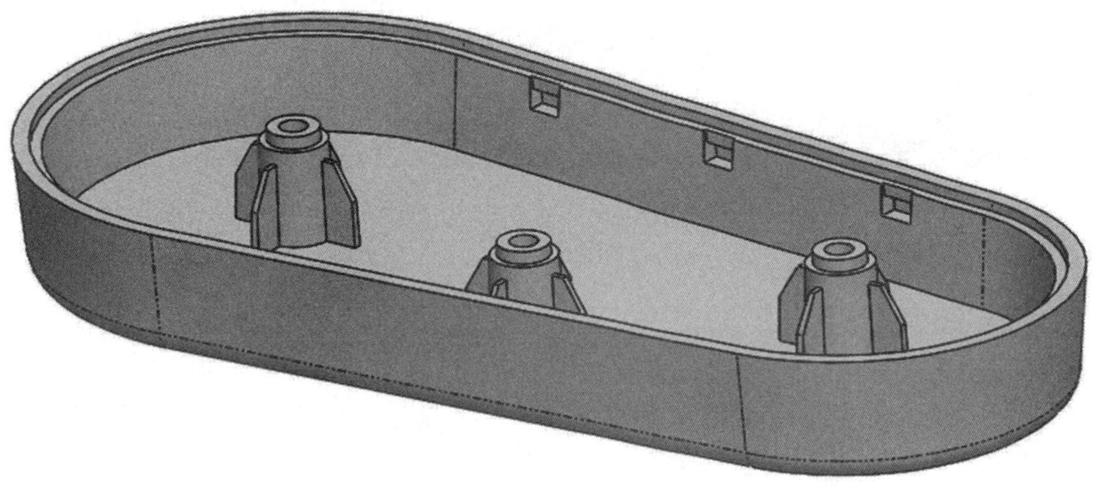

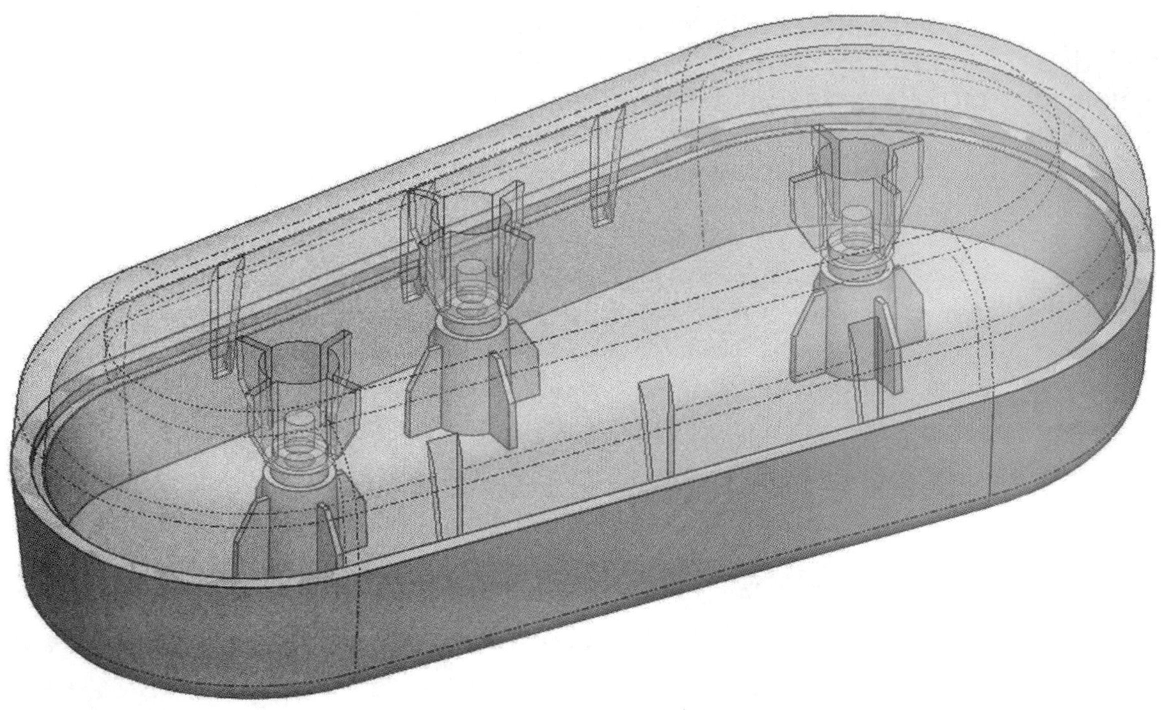

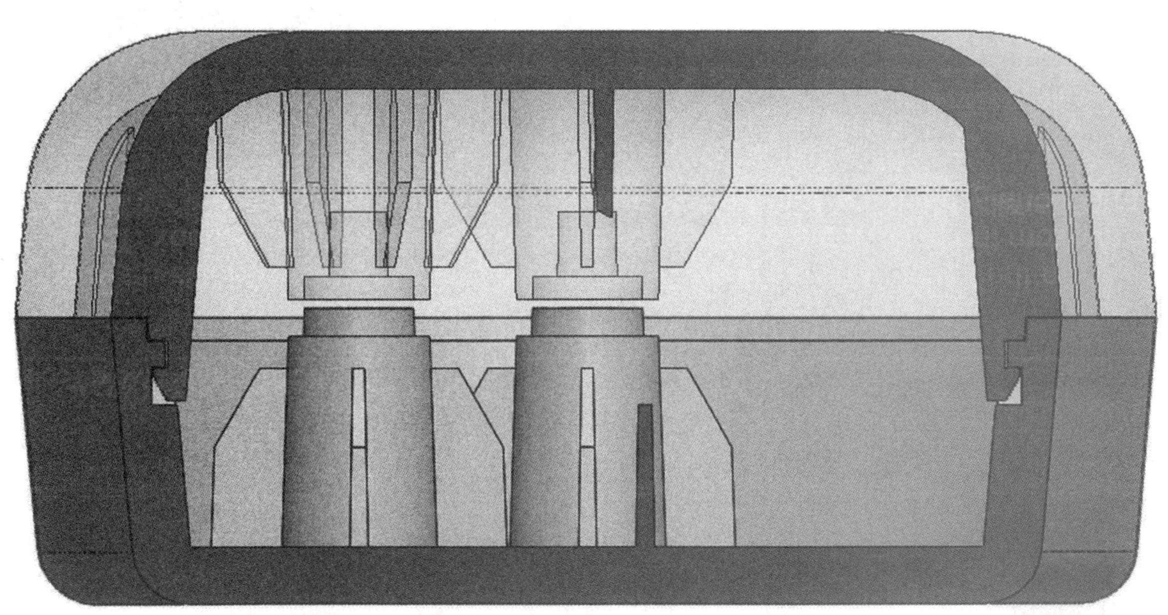

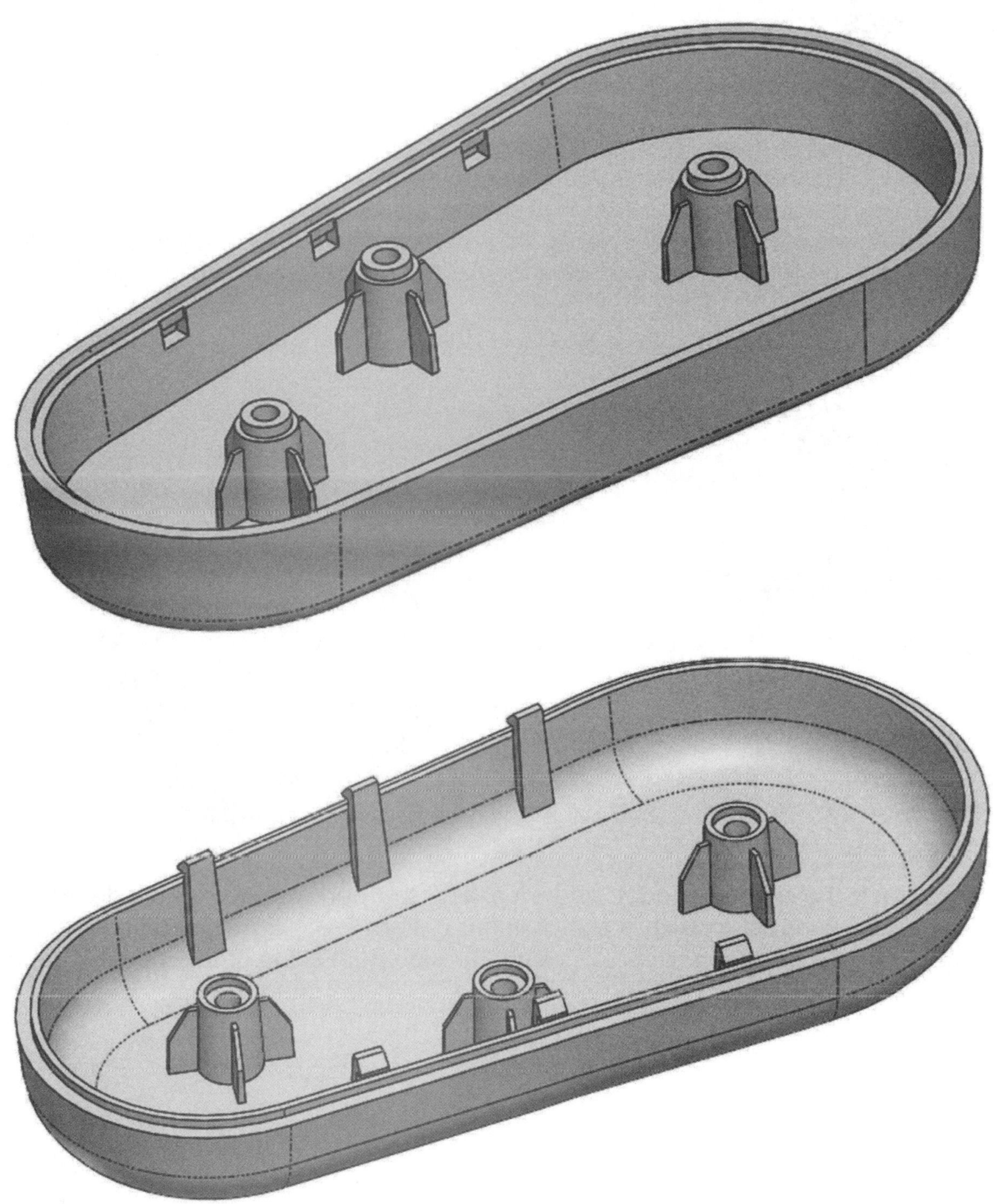

EXERCISES:

Using the hair drier files made in this lesson, add the following features, and select the best place to add them (in the master file before splitting the part *or* in the split body after saving them to a part). Remember, to add features before the split feature we must roll back the FeatureManager.

Hint: Features present in more than one part can be added in the master file; features unique to each part can be added in the master file or the split part.

- Add vents to the sides and back cover (this is your chance to be creative ☺).

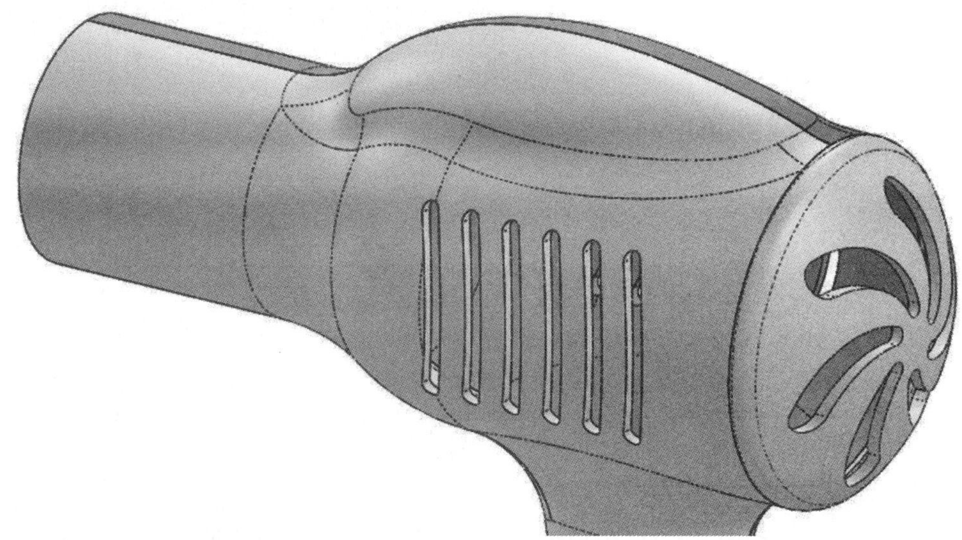

- Rollback the master model before splitting the part, and add the side vents. Here is a suggestion with a sketch linear pattern using an offset on both sides capped with an arc, and making a cut through both sides of the model. (Menu "View, Display, Tangent Edges Removed" for these images.)

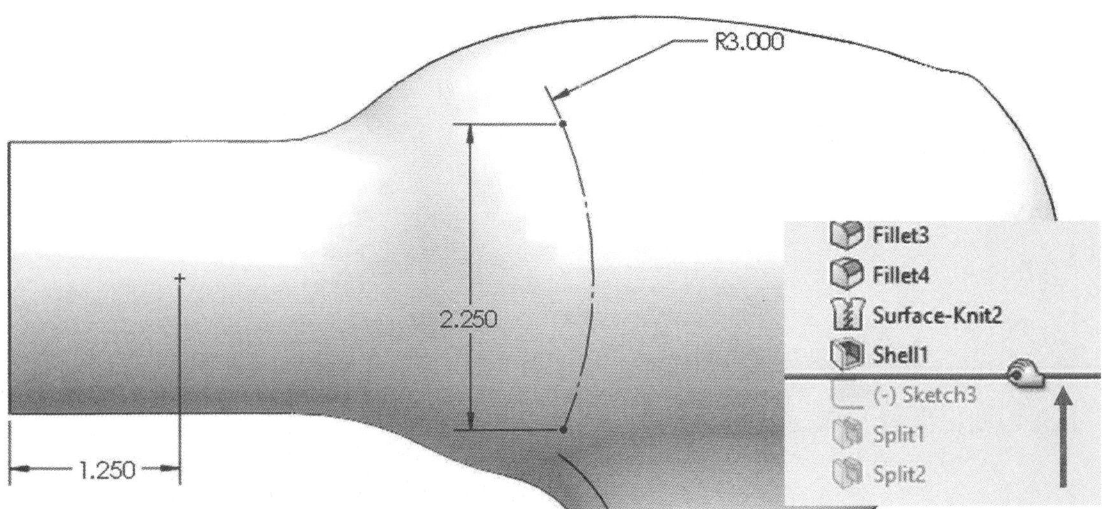

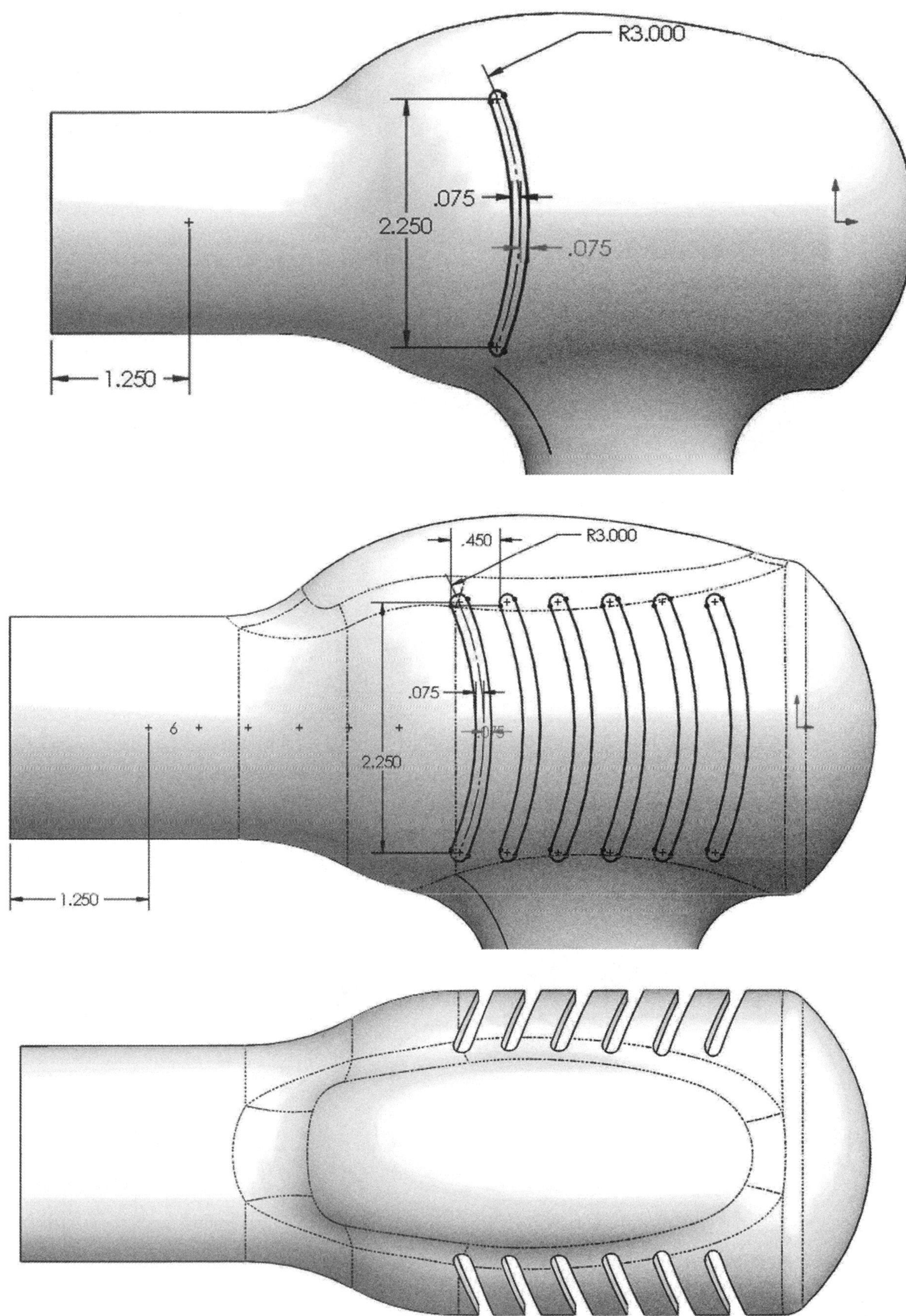
R3.000
.075
2.250
.075
1.250
.450
R3.000
.075
6
2.250
1.250

- Add a new cut for the switch and the power cord, then roll forward past the split features.

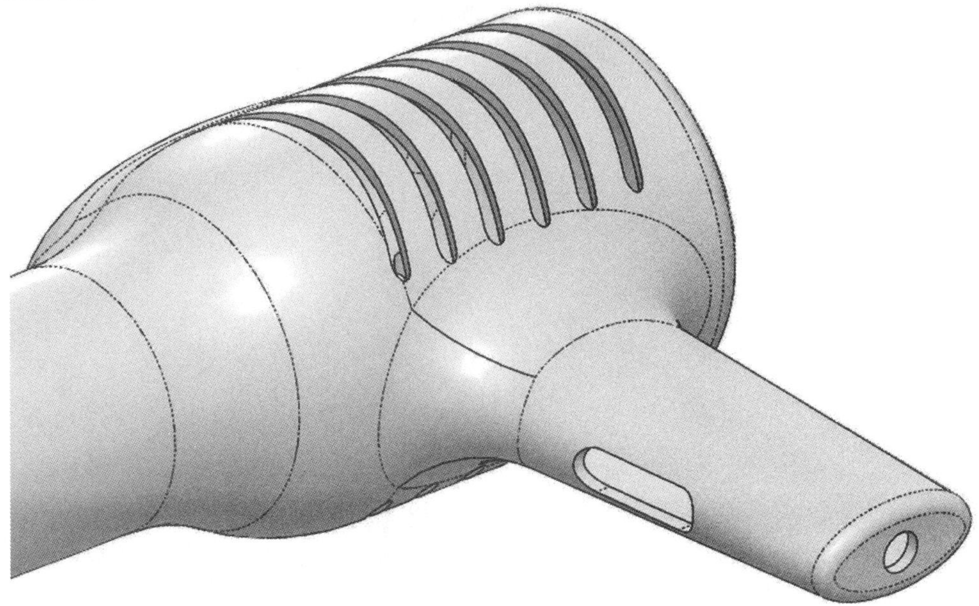

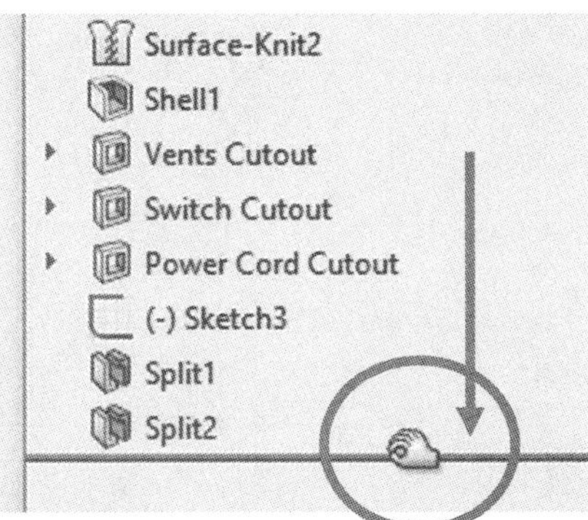

- Open the cover, add a vent and a circular pattern.

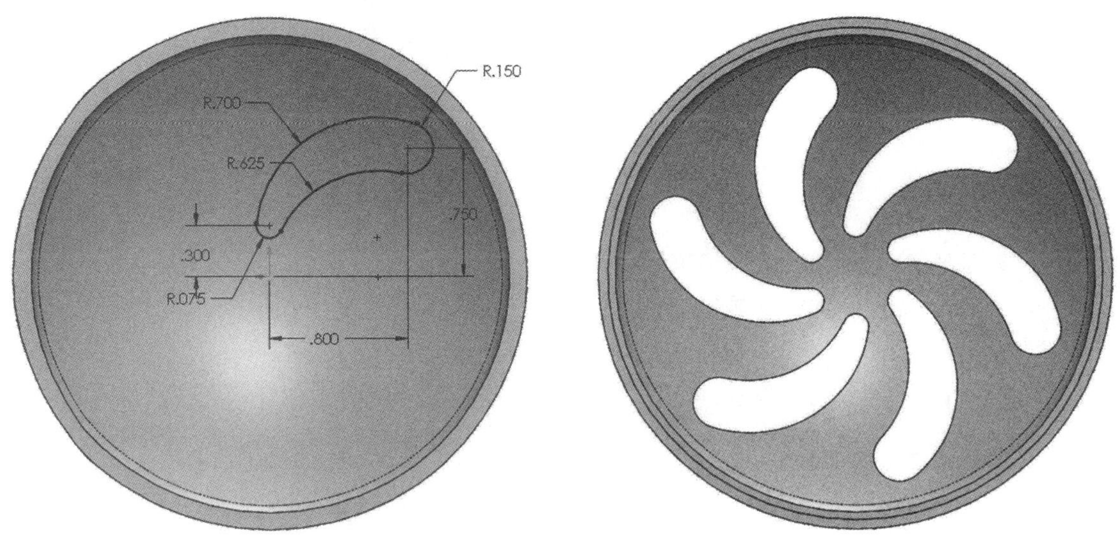

- Add a Lip feature to assemble with the other 2 parts.

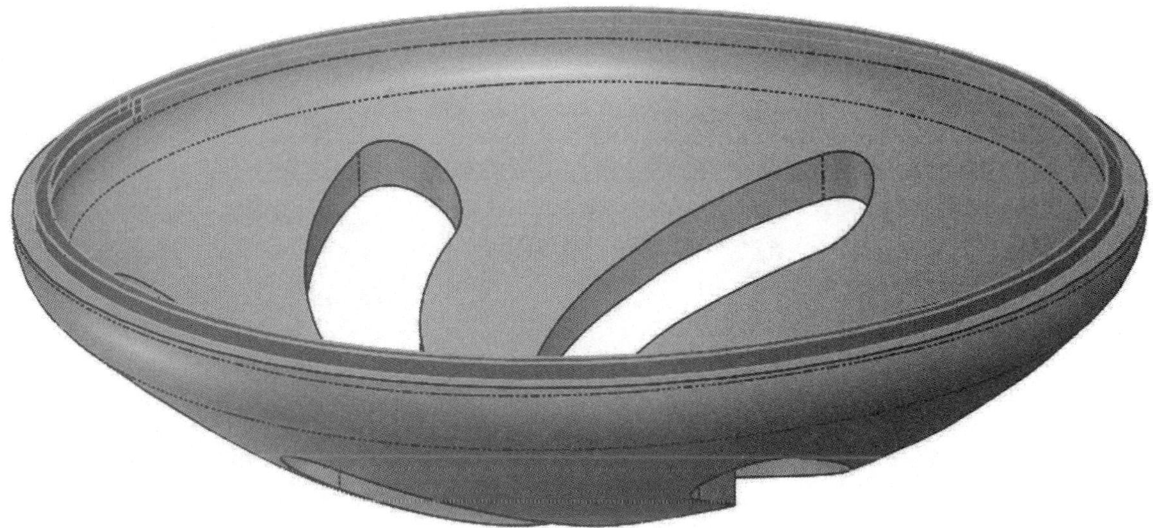

- Add the first Groove feature to the left side to assemble with the right side.

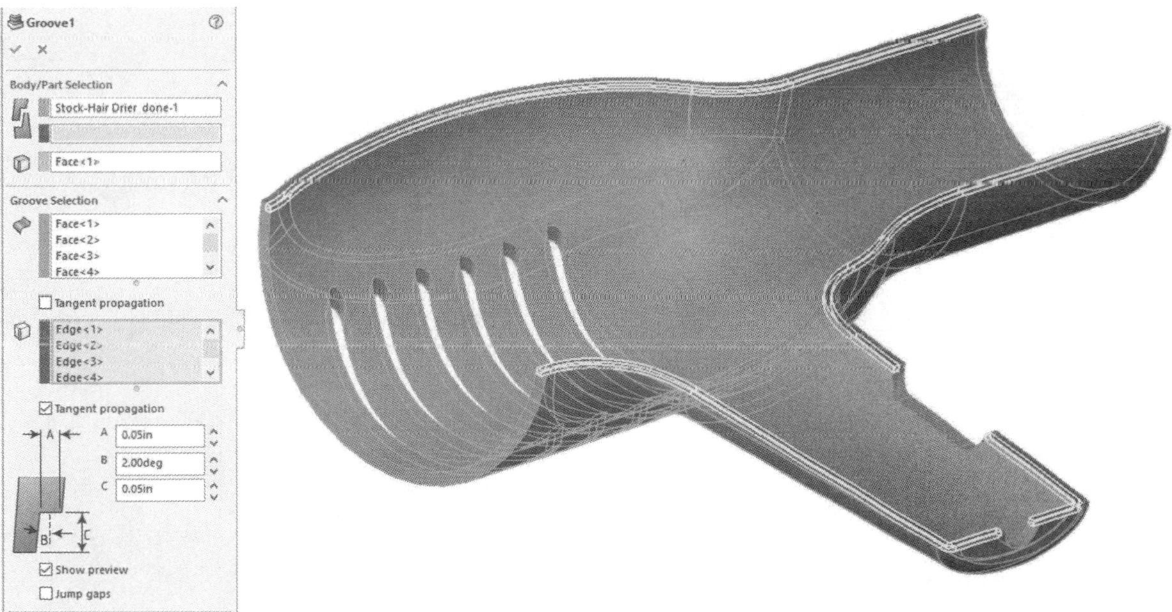

- Add the second Groove feature to assemble with the cover.

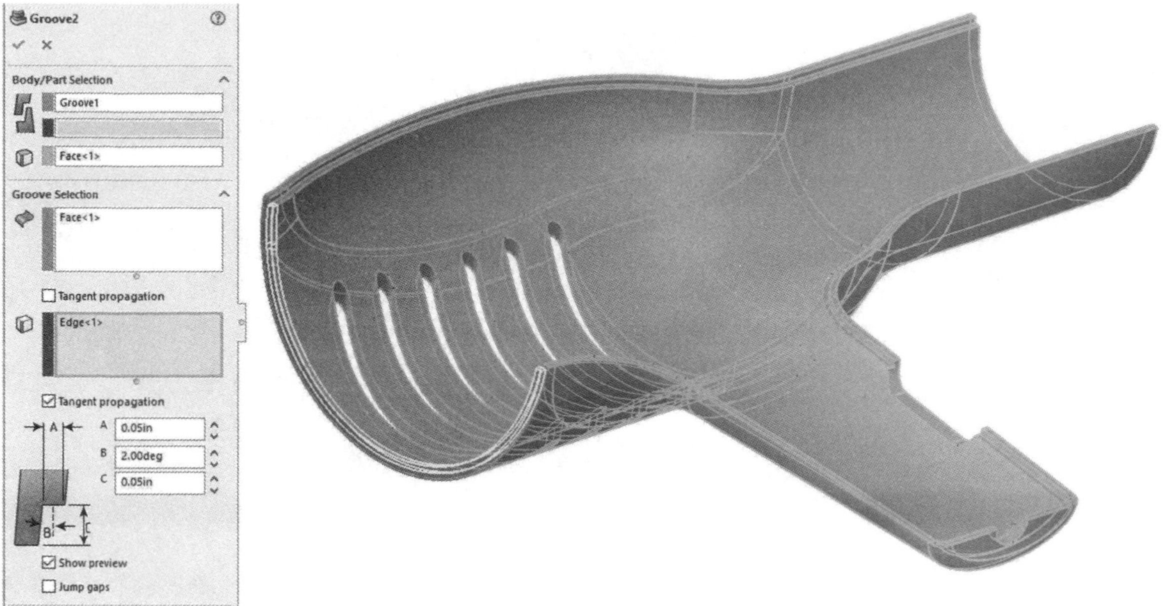

- Add the same Groove feature to the back of the Right side of the hair dryer.

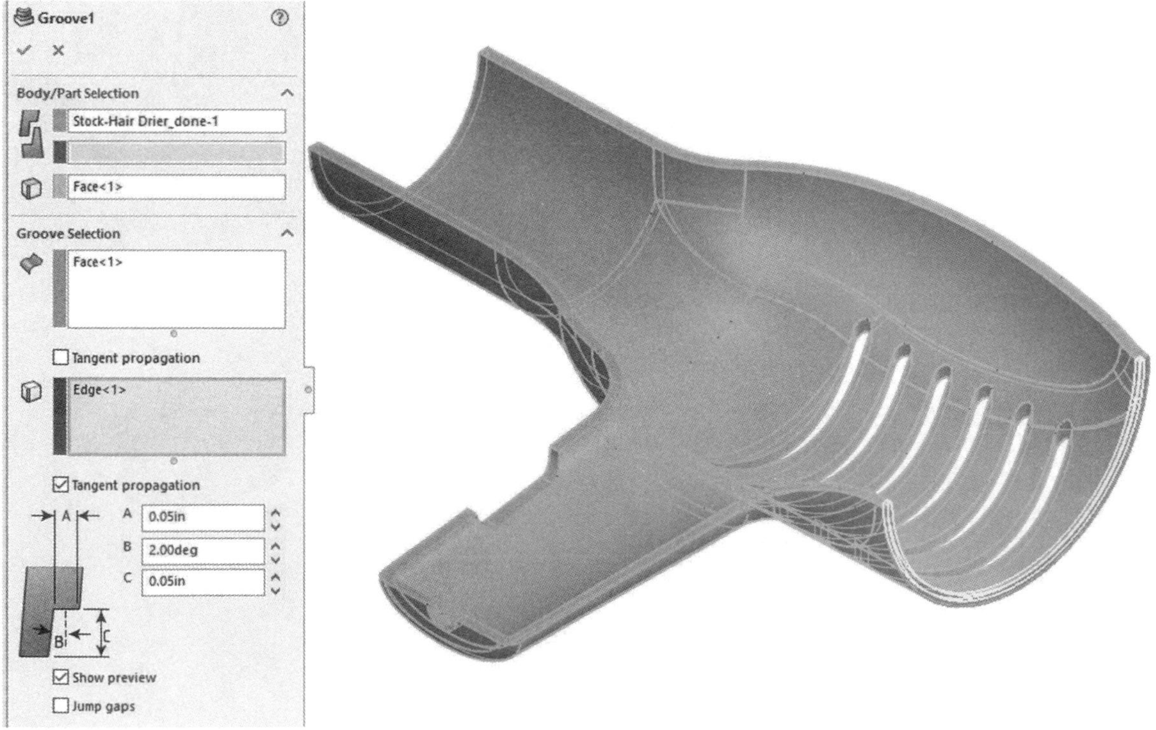

- And a Lip feature to match the groove in the Left side.

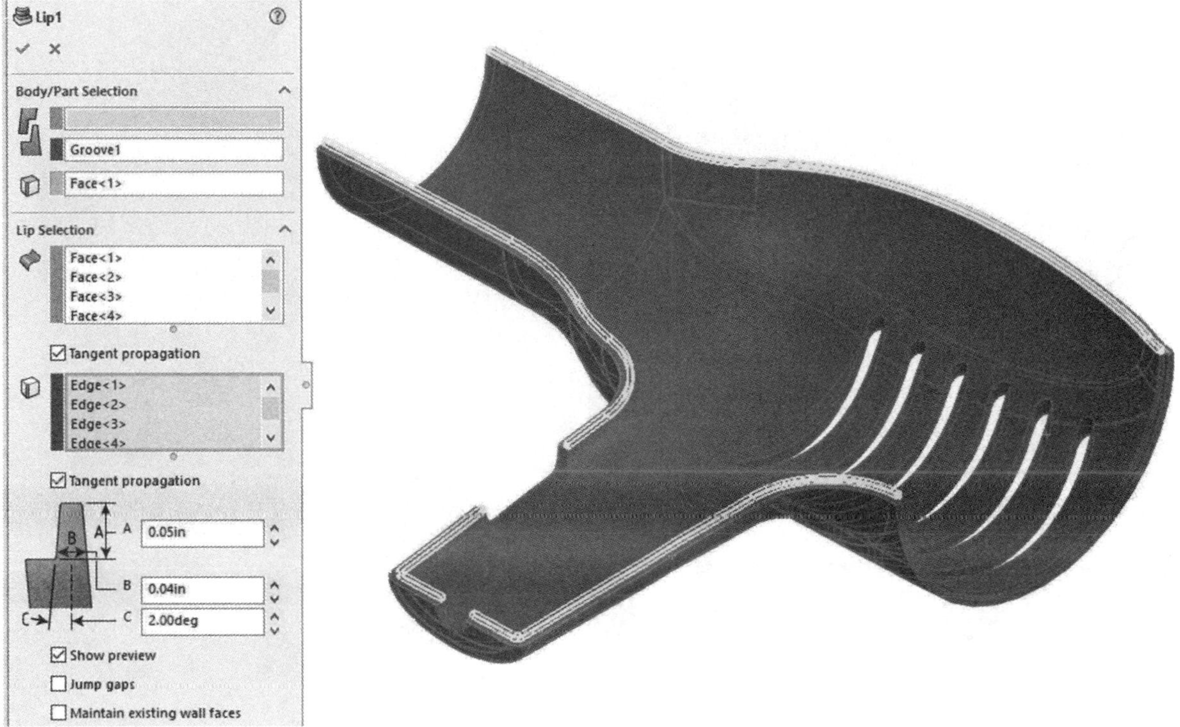

- Make sure there are no interferences in the assembly.

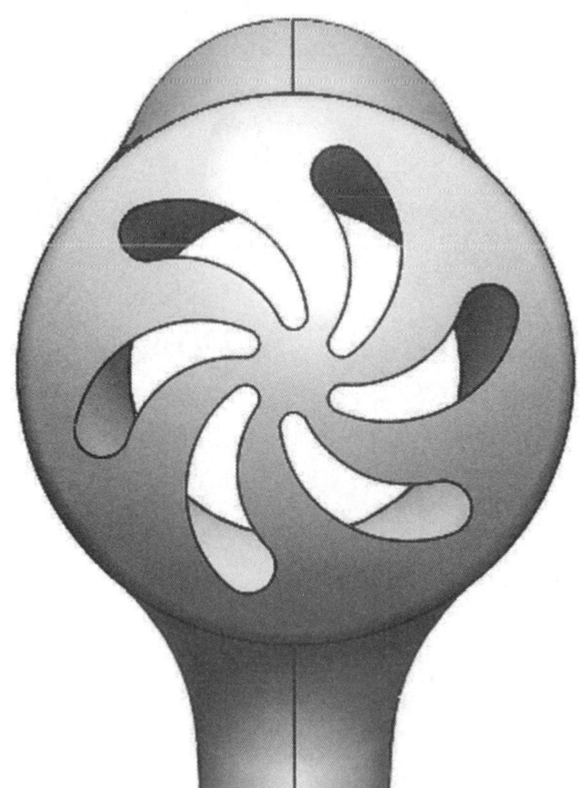

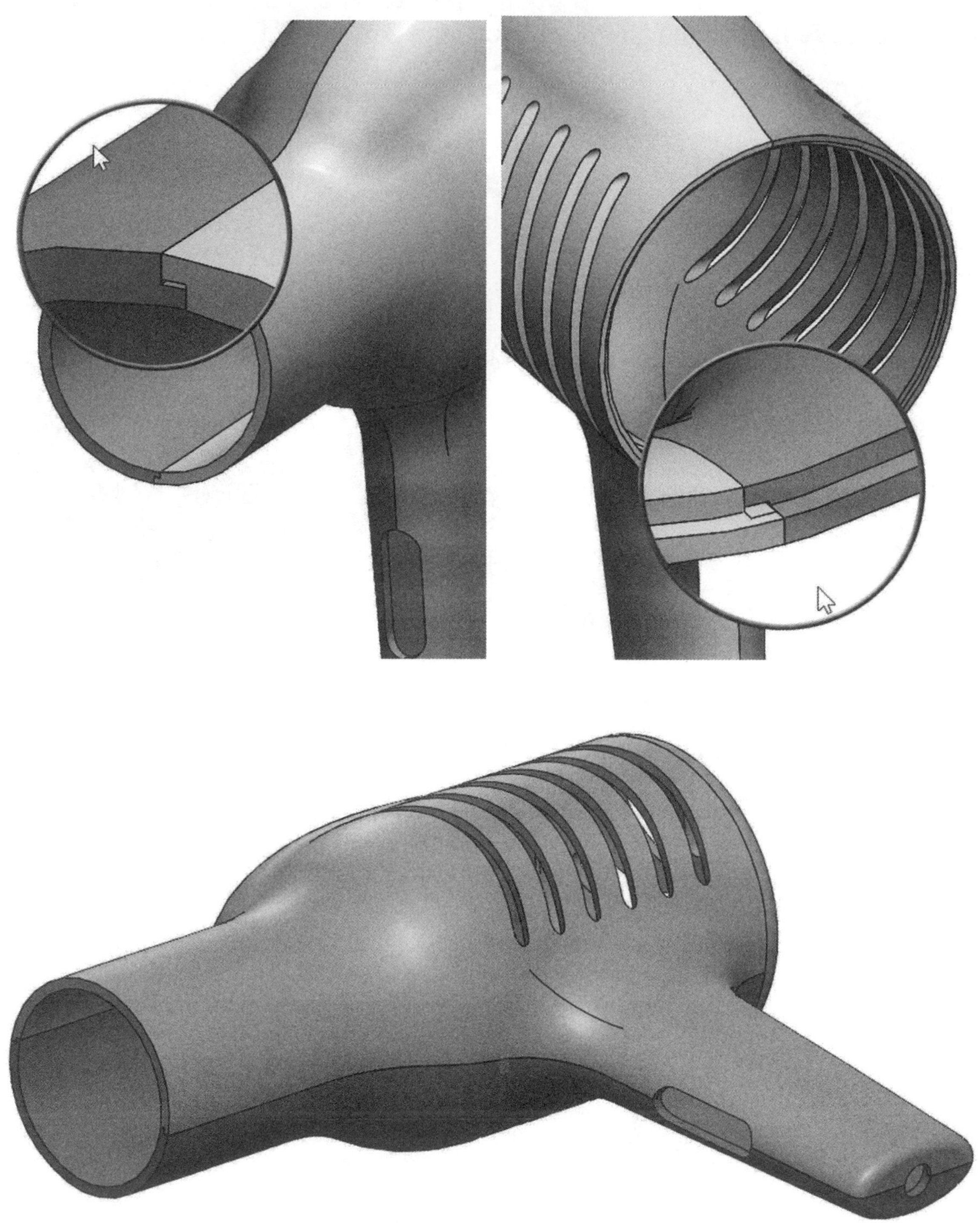

Bucket With Surfaces

For our next example, we'll design a bucket using surfaces and, at the end, convert them into a solid model. The process we'll follow to build it is purposely not the most efficient way, but it will show the reader how to learn different surfacing tools using a simple model.

24.1. – The first surface in our model will be a revolved surface. Make a new part in inches and add the following sketch in the "*Front Plane.*" Don't forget the vertical centerline at the origin.

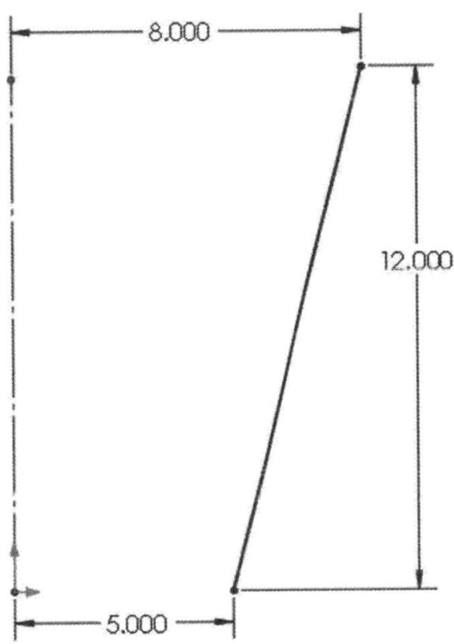

24.2. - Select the "**Revolved Surface**" command from the Surfaces tab, or the menu "**Insert, Surface, Revolved.**" Make the revolved surface 360 degrees. The first surface body will be revolved about the sketch centerline.

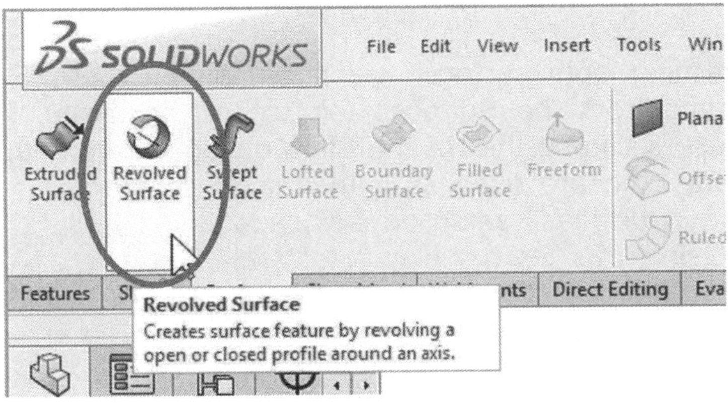

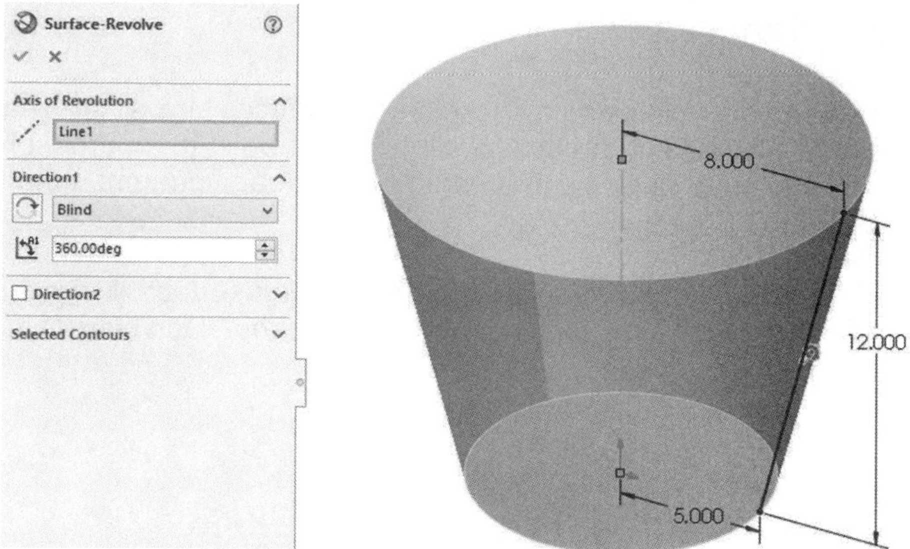

24.3. – Add a new sketch in the *"Front Plane"* and make a second revolved surface; this will form the bucket's spout. Make the top endpoint horizontal to the top edge of the first surface.

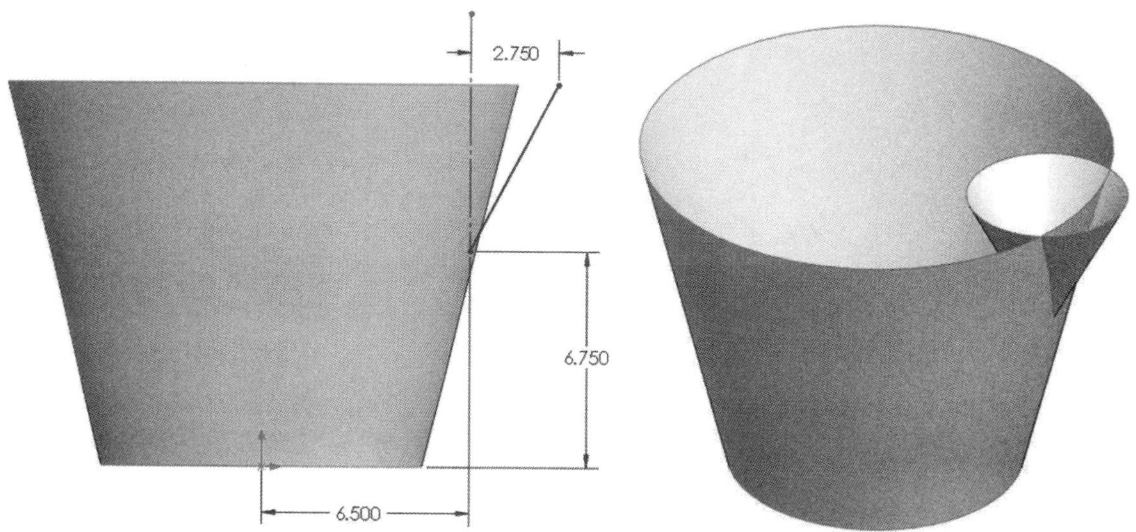

24.4. - After creating both surfaces we need to trim them. Select the "**Trim Surface**" command from the Surfaces toolbar, or the menu "**Insert, Surface, Trim.**" In this step, we'll use the "Mutual" trim option to trim both surfaces at the same time, where both are used as a trimming boundary. In the "Surfaces" selection box, add both surfaces. Using the "Keep selections" and "Show excluded surfaces" options, select the main body of the bucket and the outside face of the spout surface. After the selection, the surfaces to be removed are the only ones visible. Click OK to finish.

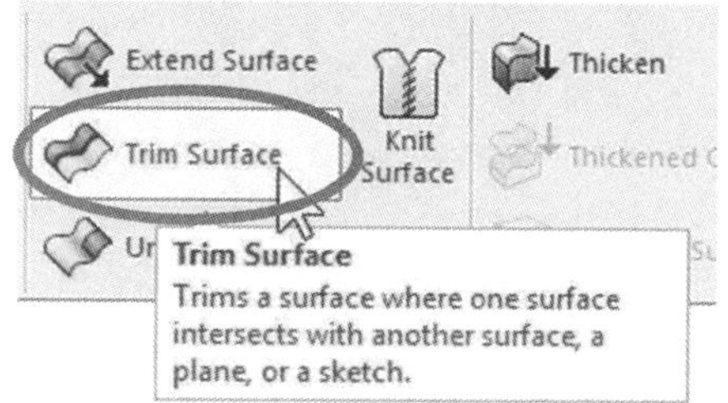

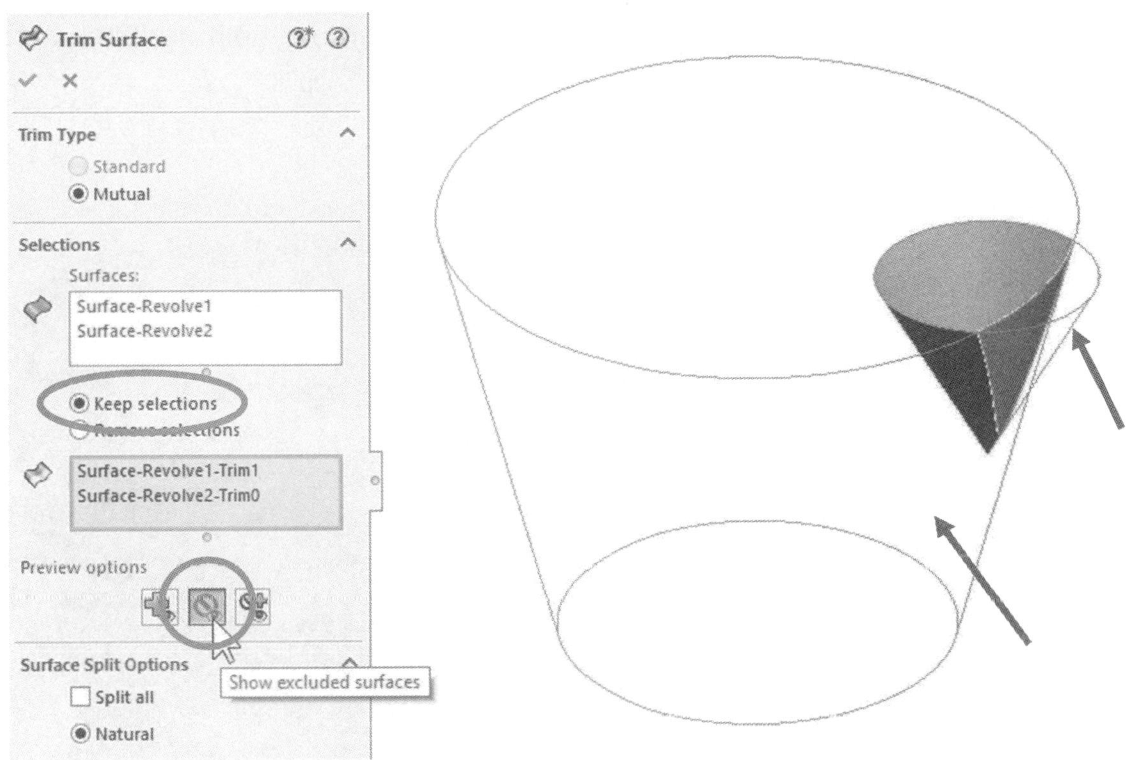

When multiple surfaces are mutually trimmed, they are merged into a single surface body. By default, open surface edges are shown using a light blue color. In our surface, the upper and lower surface edges are open (in blue) and the edge where the trimmed surfaces meet is black.

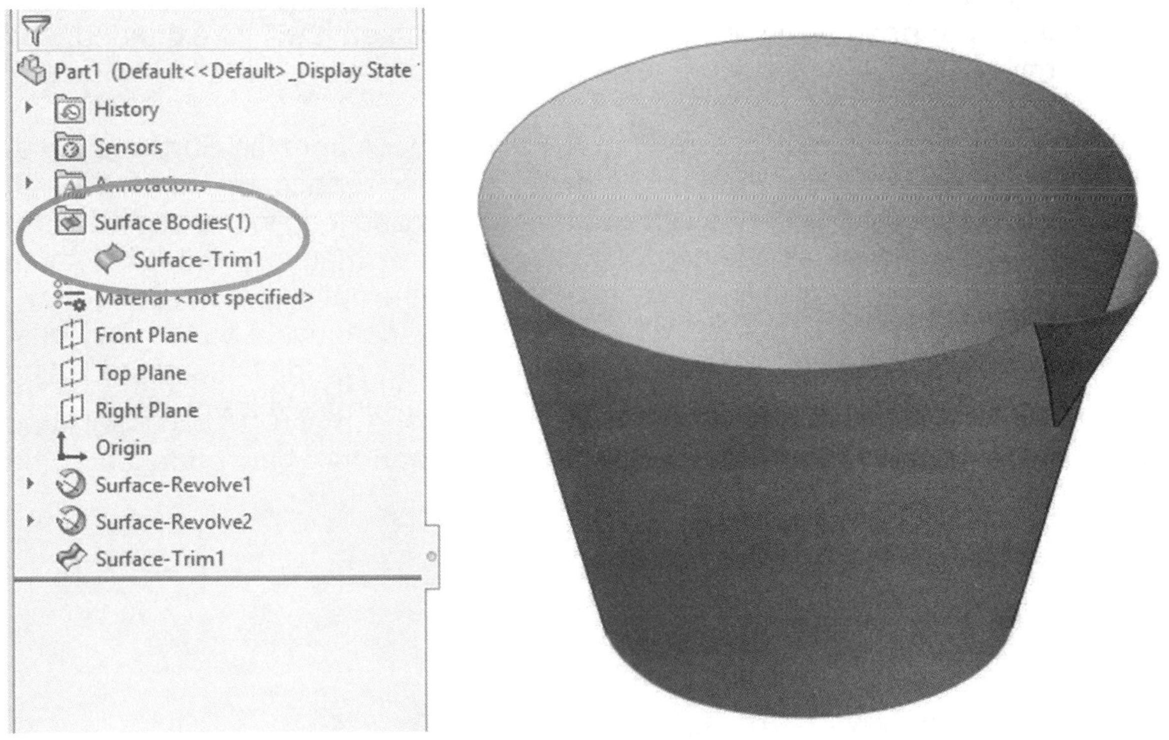

24.5. – To round the spout, add a fillet in the edge where both surfaces meet. Select the "**Fillet**" command, and add a 1″ radius fillet.

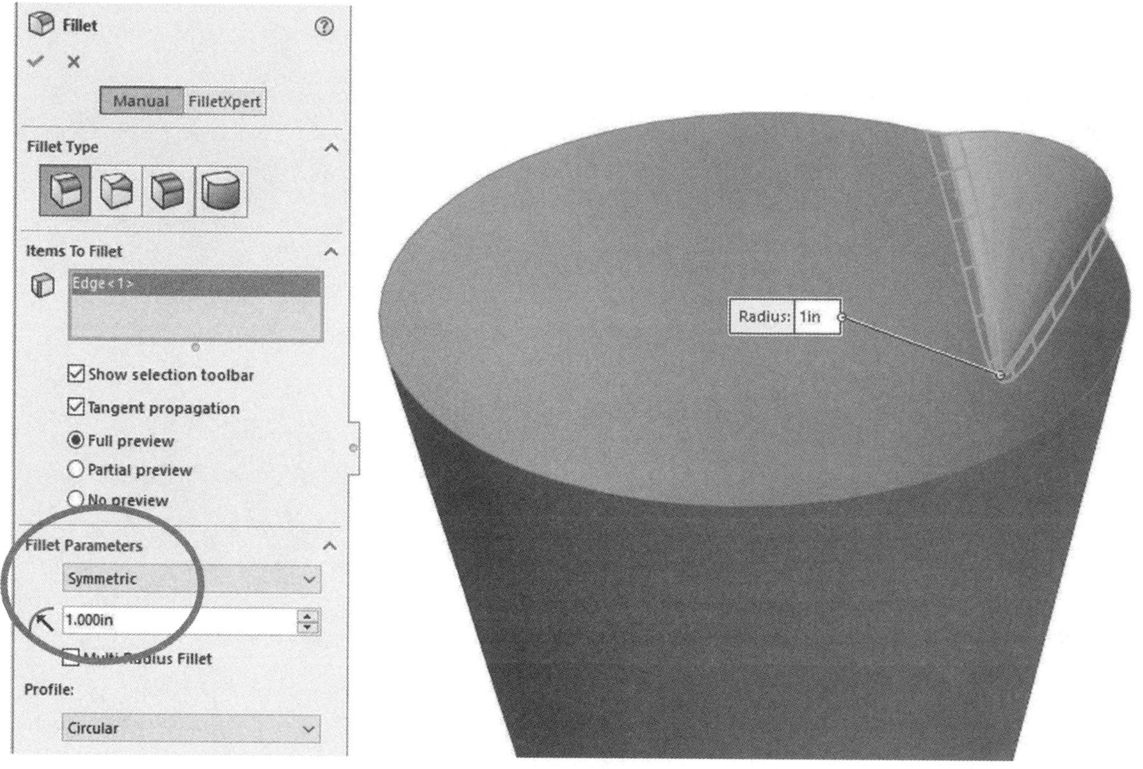

24.6. - We need to add a lip to the top of the bucket, but instead of creating a plane at the top of the bucket and making a sketch to make the surface needed, first we'll make a planar surface using the existing open edge, then it will be extended, and finally trimmed.

For the first step, select the "**Planar Surface**" icon from the Surfaces tab or the menu "**Insert, Surface, Planar**." To make a planar surface, we need to select either a 2D sketch that defines a closed area, or a closed loop formed by edges in a 2D plane. To select all the top edges, instead of manually selecting them individually, right mouse click in one edge and select the option "Select Tangency." This option automatically selects all tangent edges connected to the first one, which in our case are all the edges needed. The preview of the flat surface will be automatically shown covering the bucket. Click OK to complete the surface.

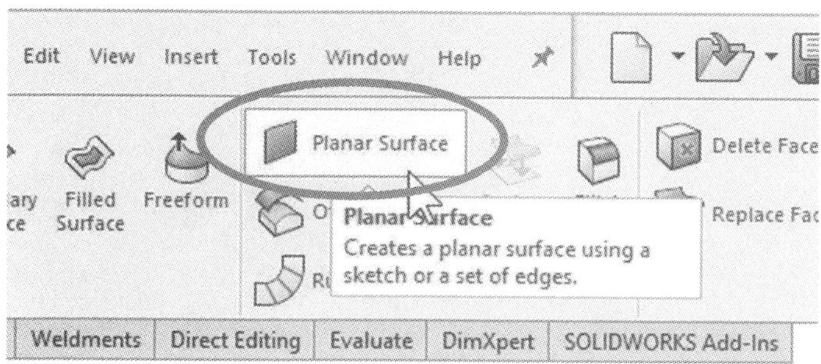

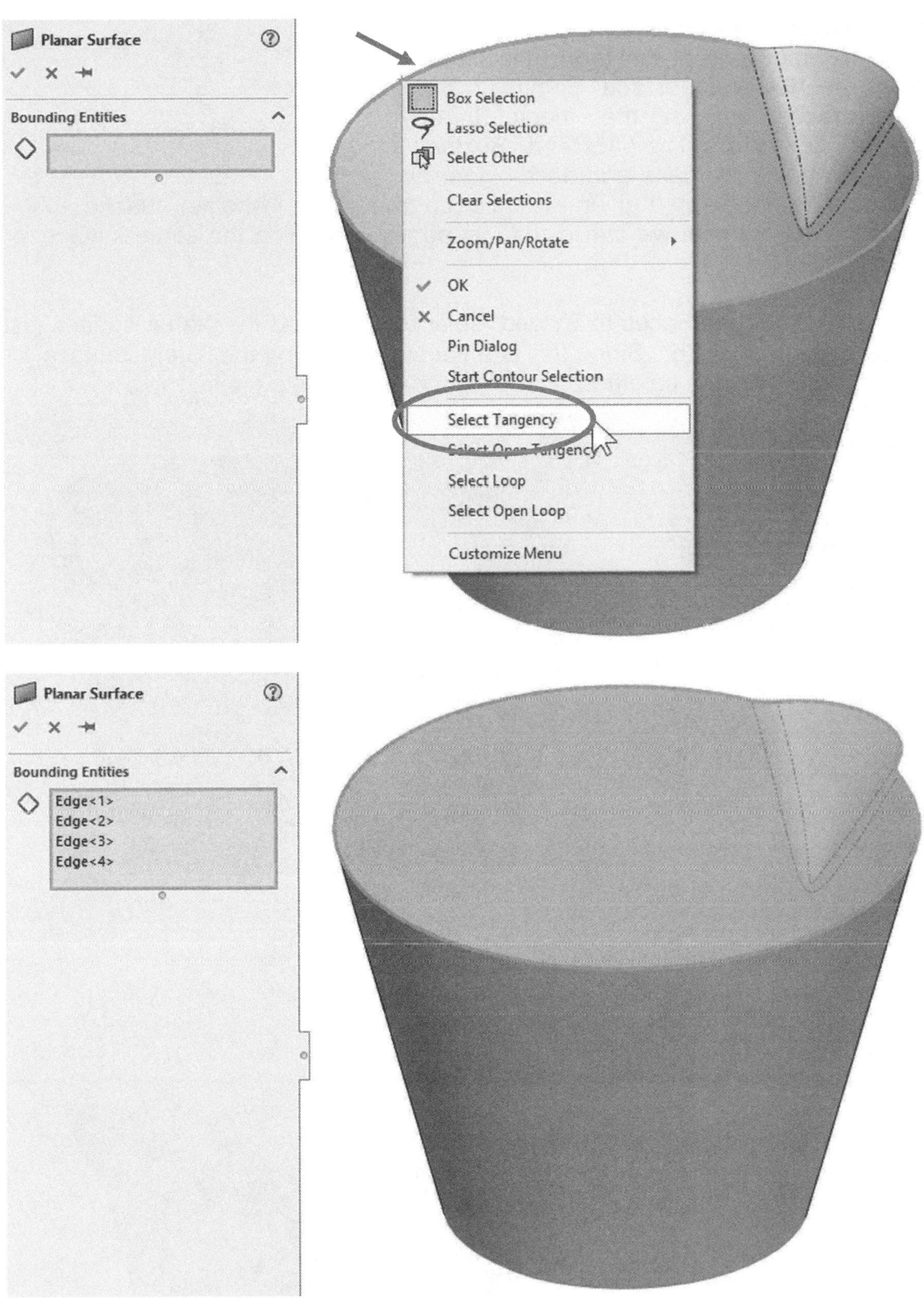
Planar Surface
Bounding Entities
Box Selection
Lasso Selection
Select Other
Clear Selections
Zoom/Pan/Rotate
OK
Cancel
Pin Dialog
Start Contour Selection
Select Tangency
Select Open Tangency
Select Loop
Select Open Loop
Customize Menu
Planar Surface
Bounding Entities
Edge<1>
Edge<2>
Edge<3>
Edge<4>

24.7. – The second step is to make this surface bigger to build a lip at the top of the bucket. Select the "**Extend Surface**" command from the Surfaces tab, or the menu "**Insert, Surface, Extend.**" The "**Extend Surface**" command helps us make a surface bigger by extending its edges by a given distance, up to a point or another surface. In the case of a 3D surface, we can extend its edges continuing the same surface, or linearly.

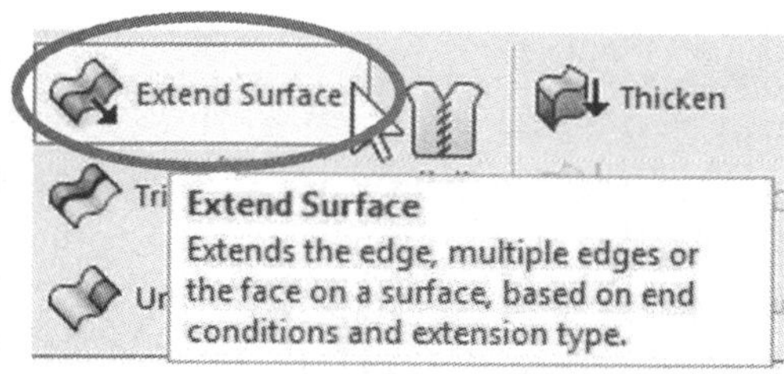
Extend Surface
Thicken
Extend Surface
Extends the edge, multiple edges or the face on a surface, based on end conditions and extension type.

In the "Edges/Faces to Extend" selection box, add the planar surface just made, and enter 0.875". Since this is a flat surface, using the "Same surface" or "Linear" option makes no difference. Click OK to complete the surface.

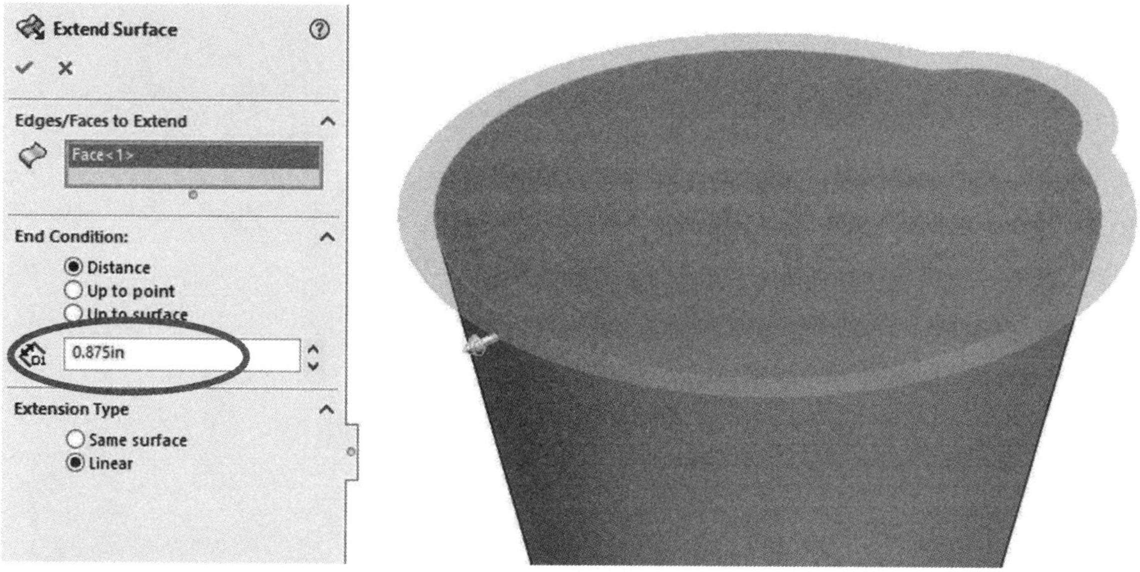

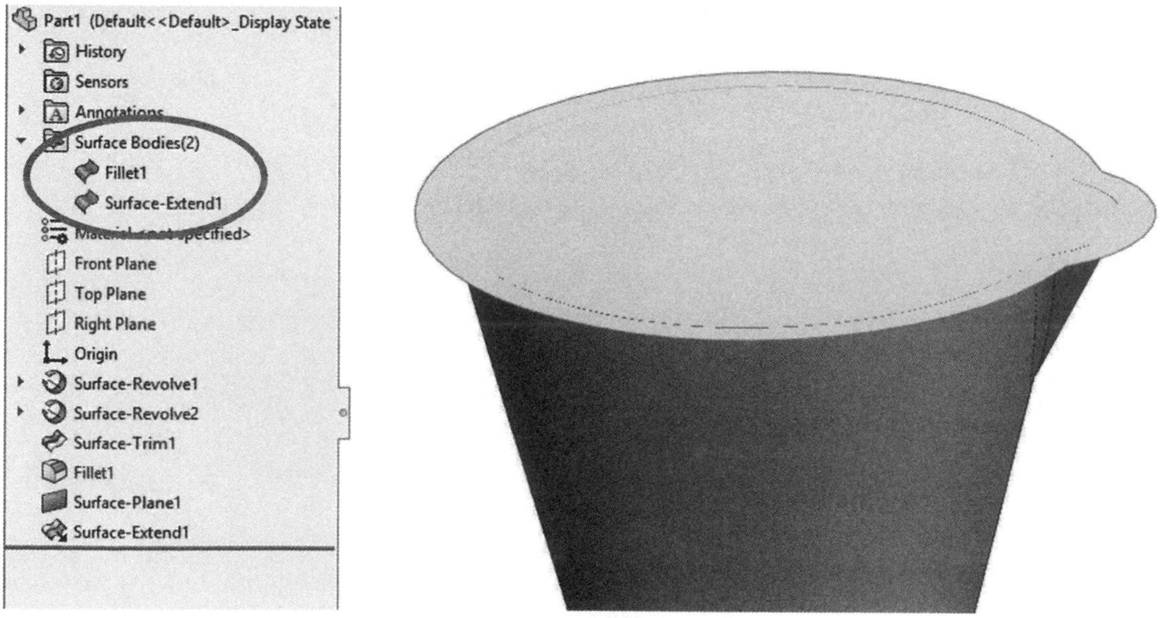

24.8. - The last step is to trim and round this surface to form the bucket's lip. Open the "**Fillet**" command and select the "Face fillet" type. Enter a 0.375″ radius and select the (now extended) planar face in the "Face set 1" selection box and the bucket's body in the "Face set 2" selection box. If a fillet preview is not visible, click in the "Reverse Face Normal" buttons next to the selection boxes until the desired fillet preview is visible. Under "Fillet Options" make sure the "**Trim and attach**" option is selected. By using this option, the selected faces will be trimmed by the fillet, the trimmed unused faces will be removed, and the fillet will be merged with the selected faces into a single surface body. Click OK to complete.

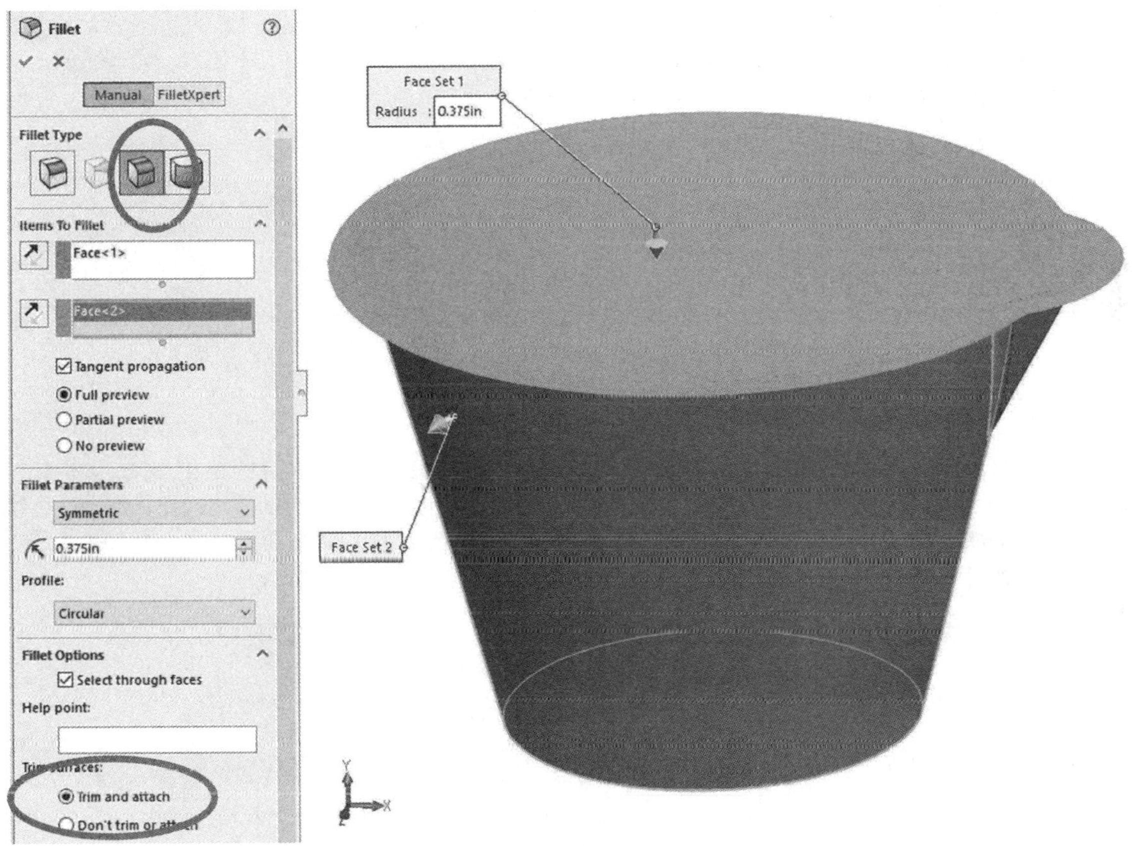

After completing the fillet, the planar face is trimmed, the fillet is blended along the perimeter of the bucket, and all surfaces are merged into a single body.

24.9. – To complete the faces needed to form the bucket, add a new planar surface to format the bottom of the bucket using the "**Planar Surface**" command. Select the round edge at the bottom and click OK to continue. After adding this surface, we'll have two surface bodies again.

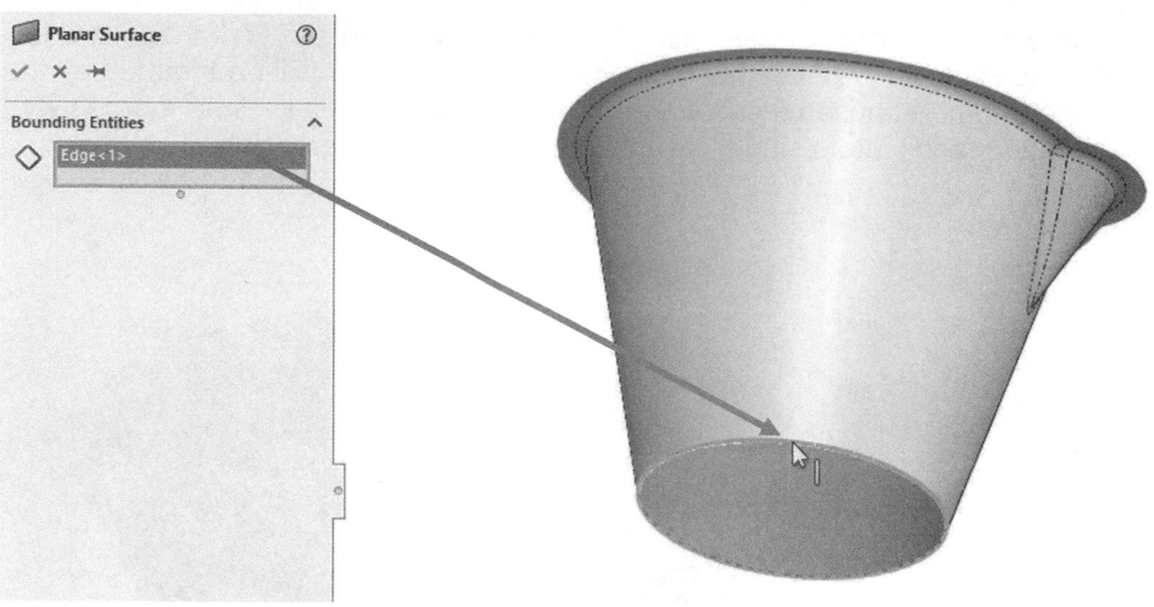

24.10. - Add a 0.625″ face fillet between the bucket and the last surface using the same "Trim and attach" setting to trim and merge the filleted surfaces. Reverse the normal direction arrows as needed to obtain the correct fillet.

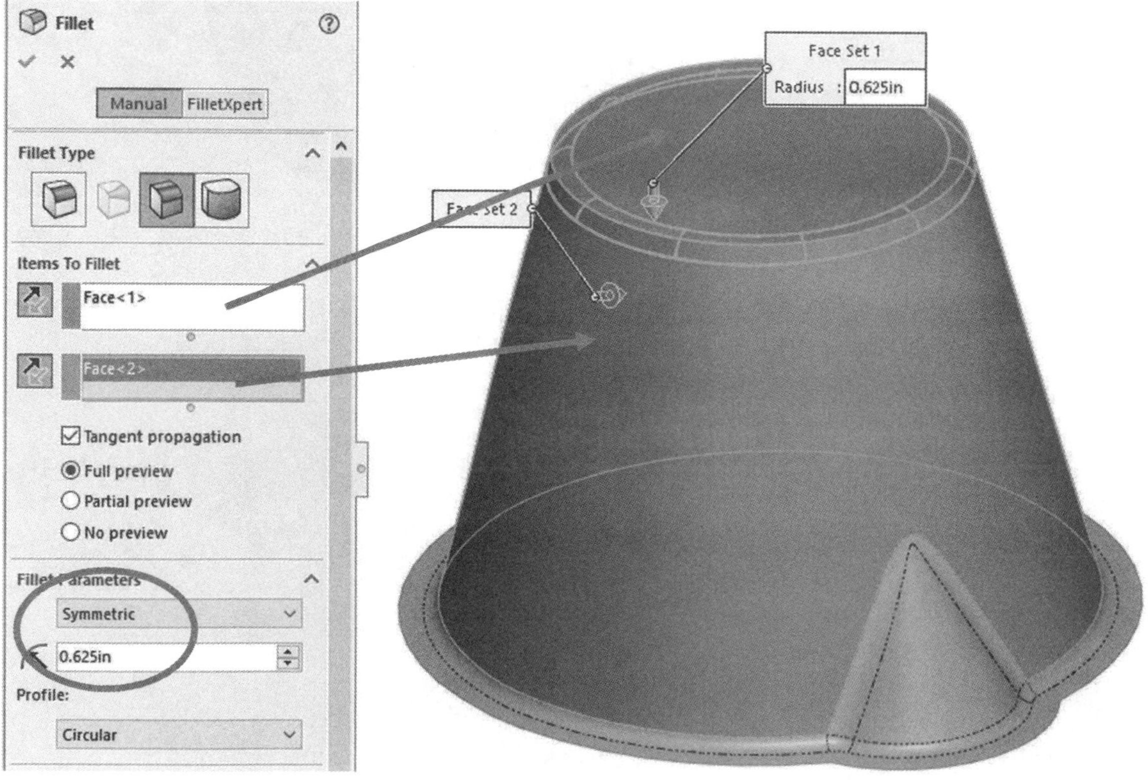

Now the bucket's surface is complete.

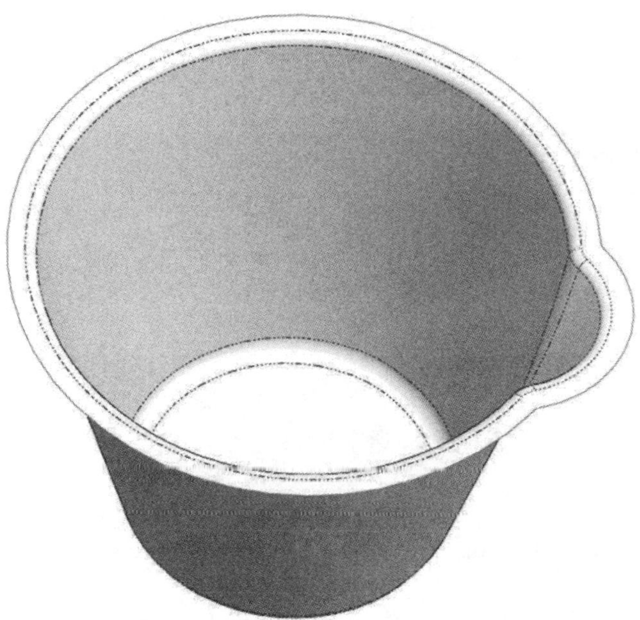

24.11. - To convert the surface to a solid, select the "**Thicken**" command from the Surfaces tab or the menu "**Insert, Boss/Base, Thicken**." Make the surface 0.125″ thick, adding the material inside the bucket and click OK to finish.

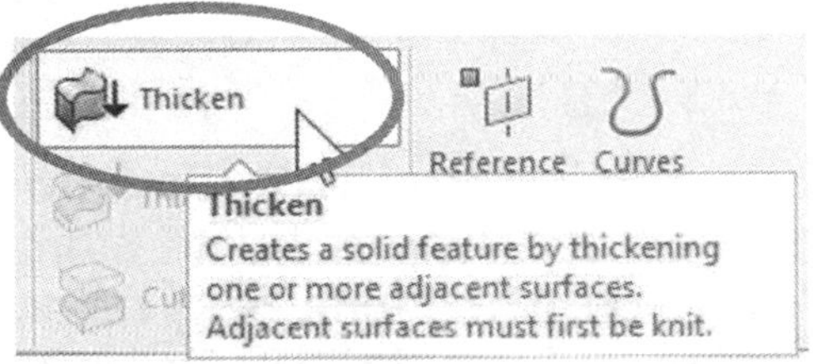

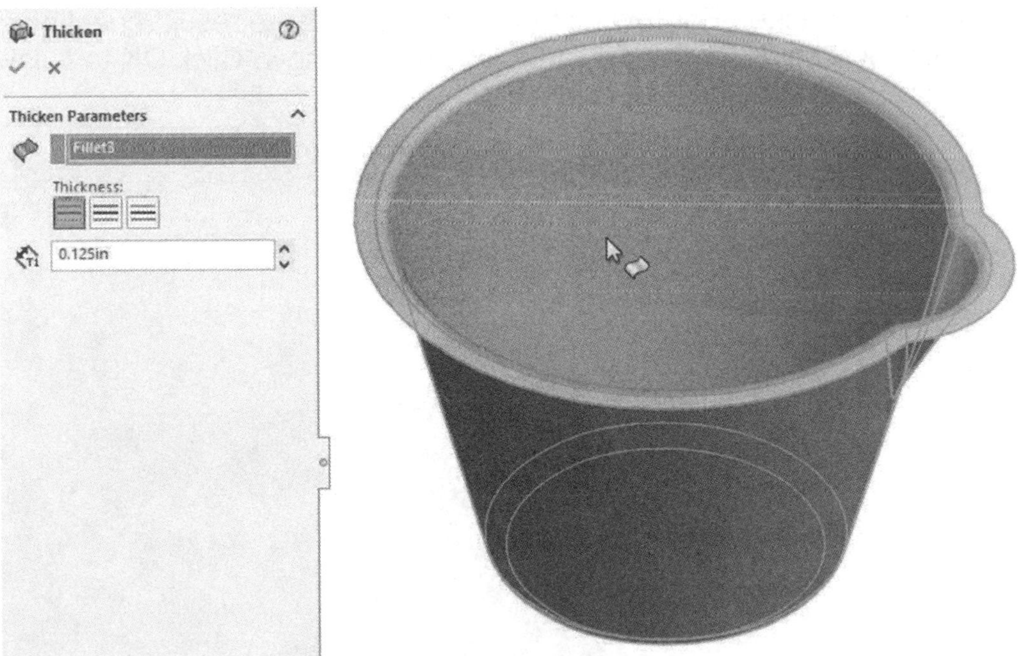

Now our part is a single solid body (no "Solid Bodies" or "Surface Bodies" folder), and the surface body was absorbed by the "**Thicken**" command.

24.12. - To complete the bucket's design, we'll add a couple of attachment points for a handle. To make the first extrusion we'll use the "**Extrude From**" option. Switch to a front view and make the following sketch in the *"Front Plane."* Note the sketch is in the middle of the part, add a Vertical relation between the circle's center and the origin.

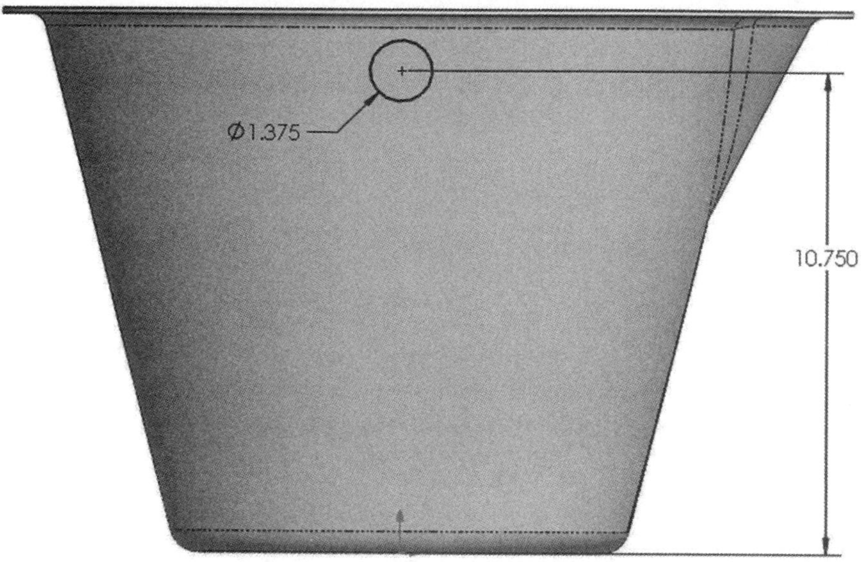

Select the "**Extruded Boss**" command. In the "From" selection box, use "Offset" and type 8.25″ to start the sketch away from the sketch plane. In the "Direction 1" options box, reverse the direction if needed to point towards the bucket and use the "Up To Next" end condition to make the extrusion conform to the outside of the bucket. For this feature, clear the "Merge result" checkbox; we'll continue modeling the bucket as a multi body part for now. Click OK to finish.

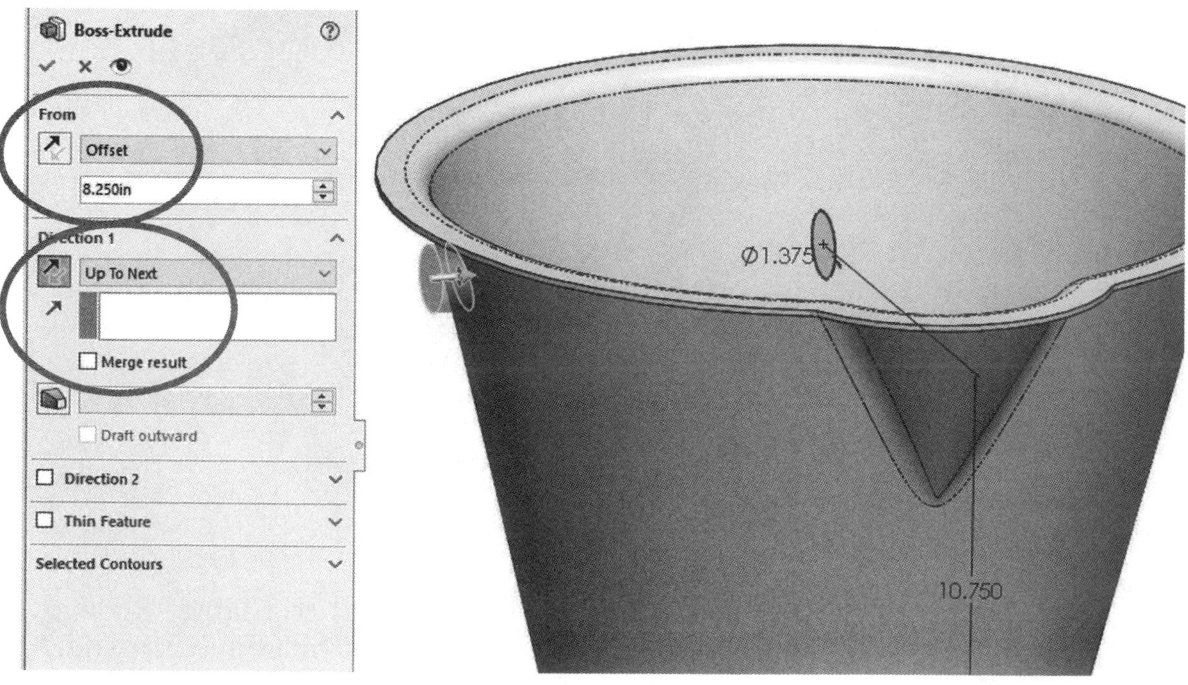

Now we have two solid bodies listed under the "Solid Bodies" folder in the FeatureManager.

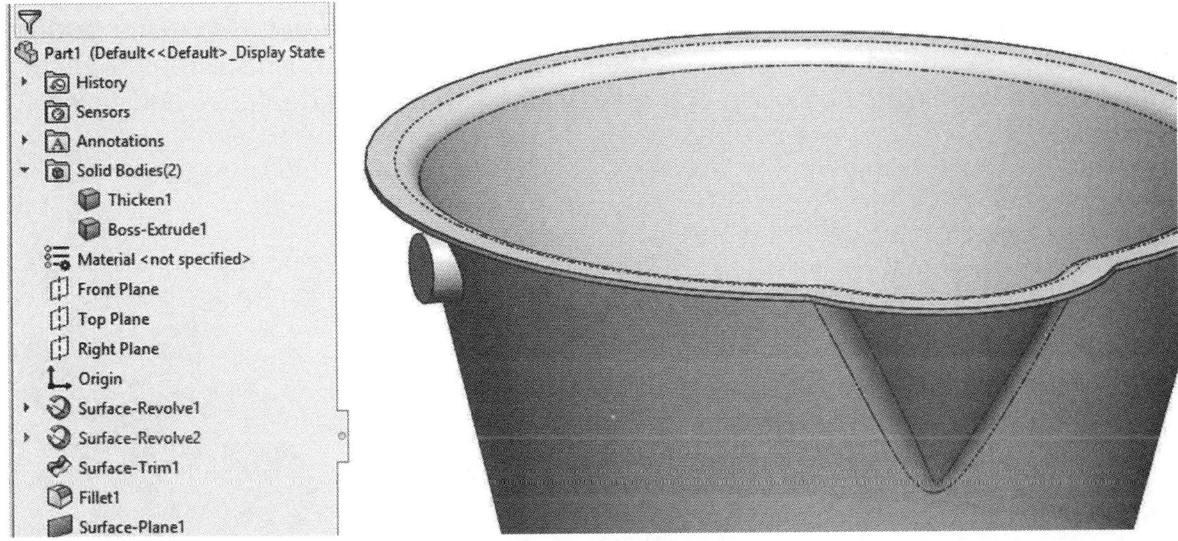

24.13. - Add a new sketch in the front face of the boss we just made with a concentric circle 0.625″ in diameter. Make an extruded cut with the "Through All" end condition. Before finishing the cut extrude, under the "Feature Scope" options, clear the "Auto-select" checkbox and select the body made in the previous step. This way we'll only affect this solid body with the cut and not the entire bucket. We could have added the hole in the extrusion, but chose to show how to do it this way instead to learn more about multi body operations. Click OK to finish.

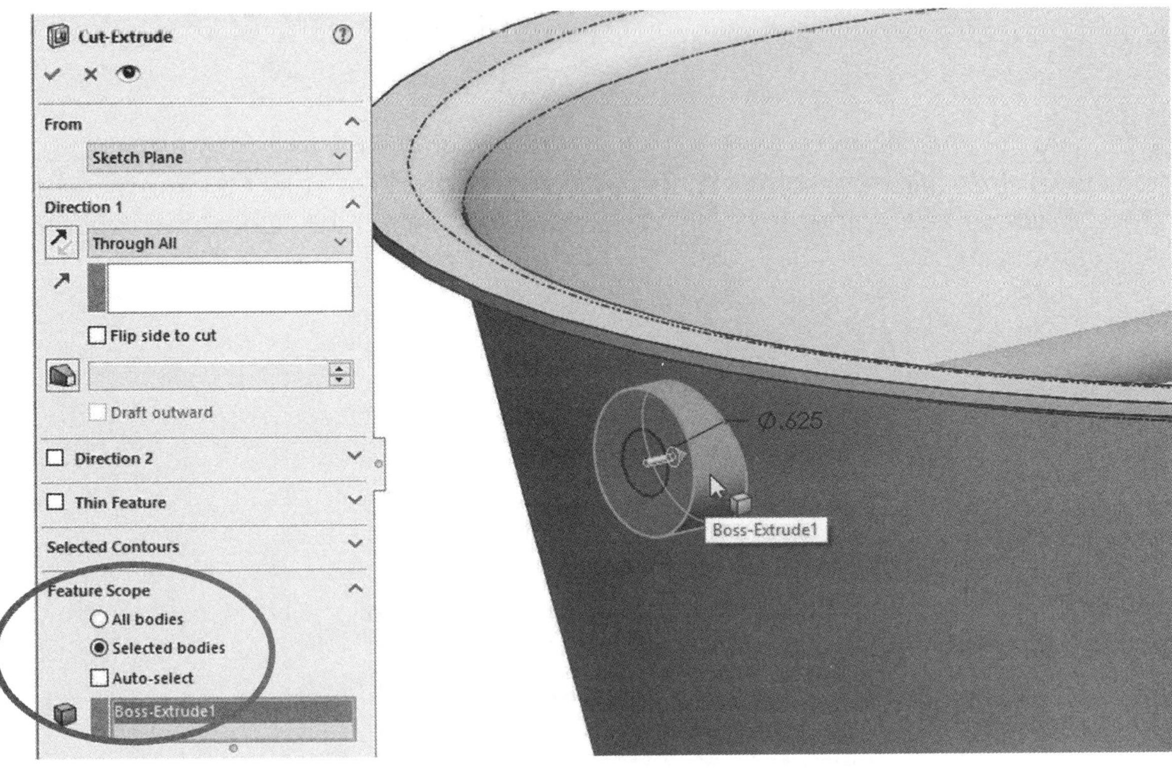

24.14. - In the next step, we'll make a mirror of the body to have two attachment points in opposite sides for the bucket's handle. Select the "**Mirror**" command using the *"Front Plane"* as the mirror plane, activate the "Bodies to Mirror" selection box and select the *"Cut-Extrude1"* body to be mirrored. Do not merge the bodies at this time; we'll do that in the next step. Click OK to finish. Notice that selecting the "Bodies to Mirror" selection box automatically hides the features and faces to mirror boxes.

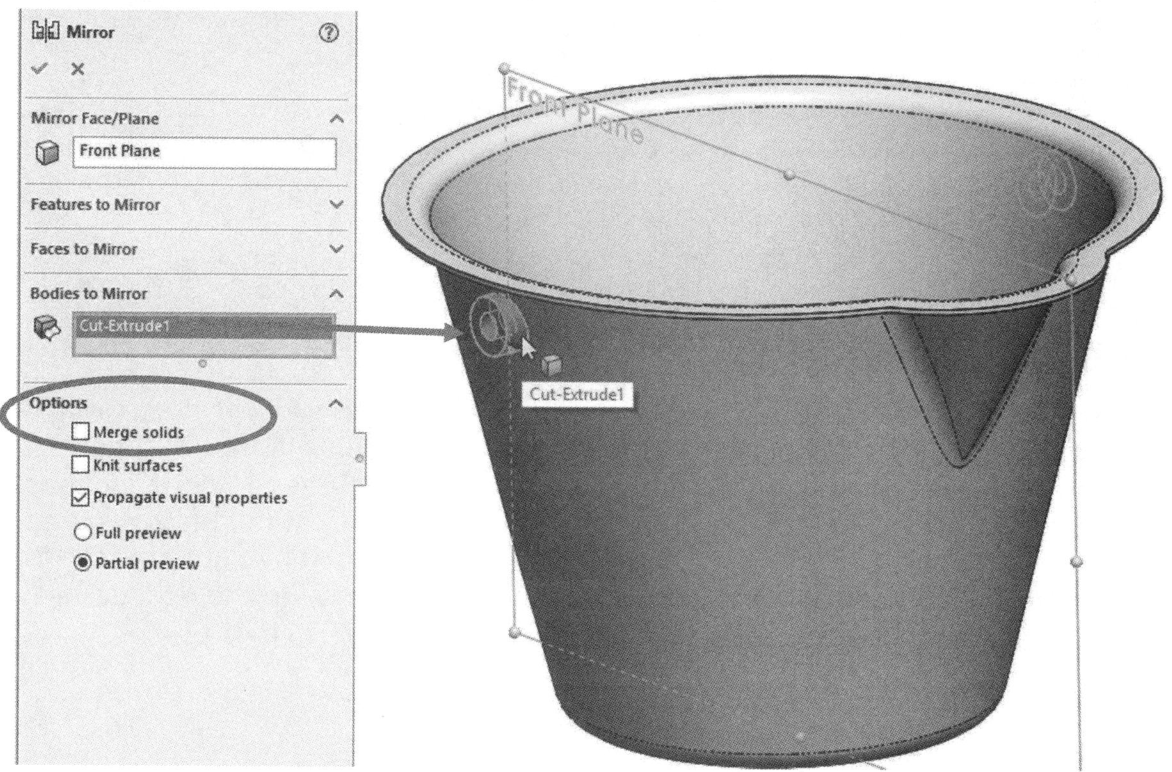

24.15. - At this point we have 3 solid bodies, and now we need to combine them into one. Select the menu "**Insert, Features, Combine**" or select all three bodies in the *"Solid Bodies"* folder and right mouse click to select the "**Combine**" option.

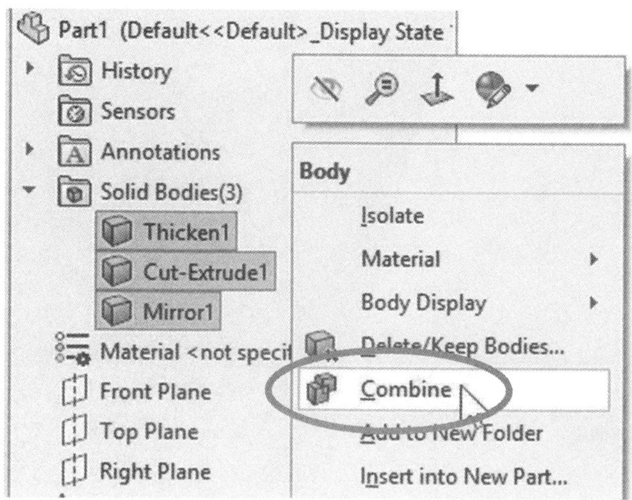

Select "Add" under the "Operation Type" options and click OK to complete. This will merge all three bodies into one.

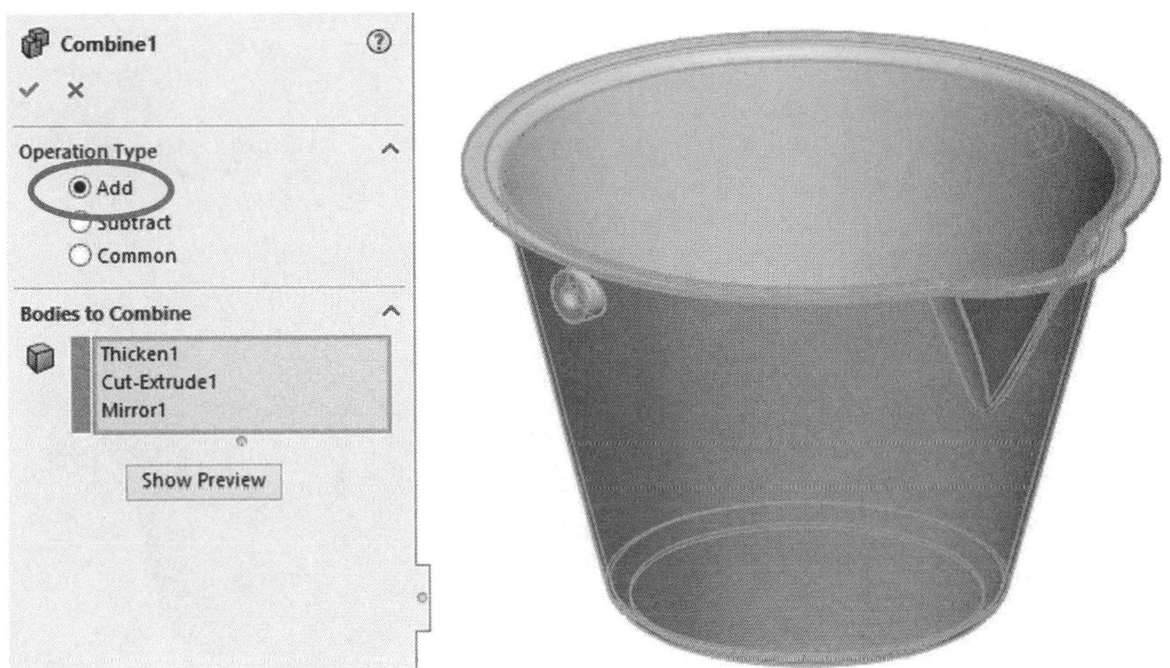

24.16. - The last step is to add a 0.25″ fillet to the bosses. We can select the outer face of the bucket or the two edges.

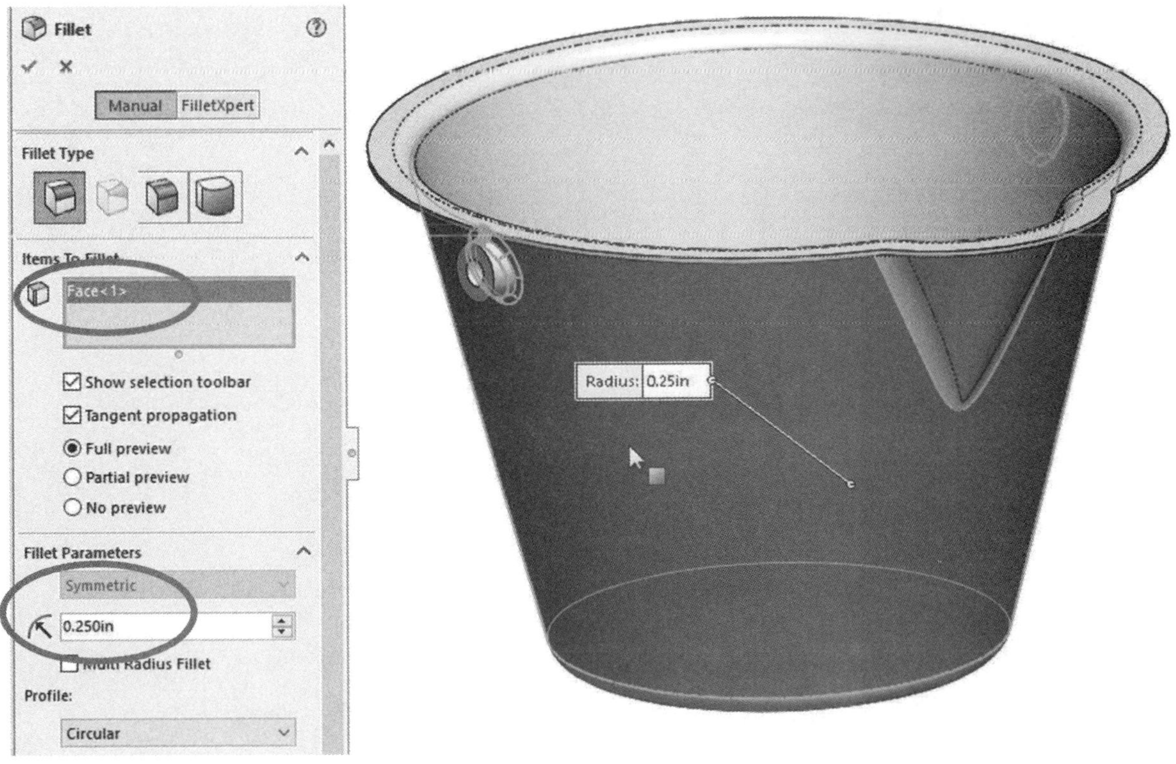

24.17. - Save the part as *'Bucket'* and close it; we'll use it later to design a mold to make it. Our finished part now looks like this:

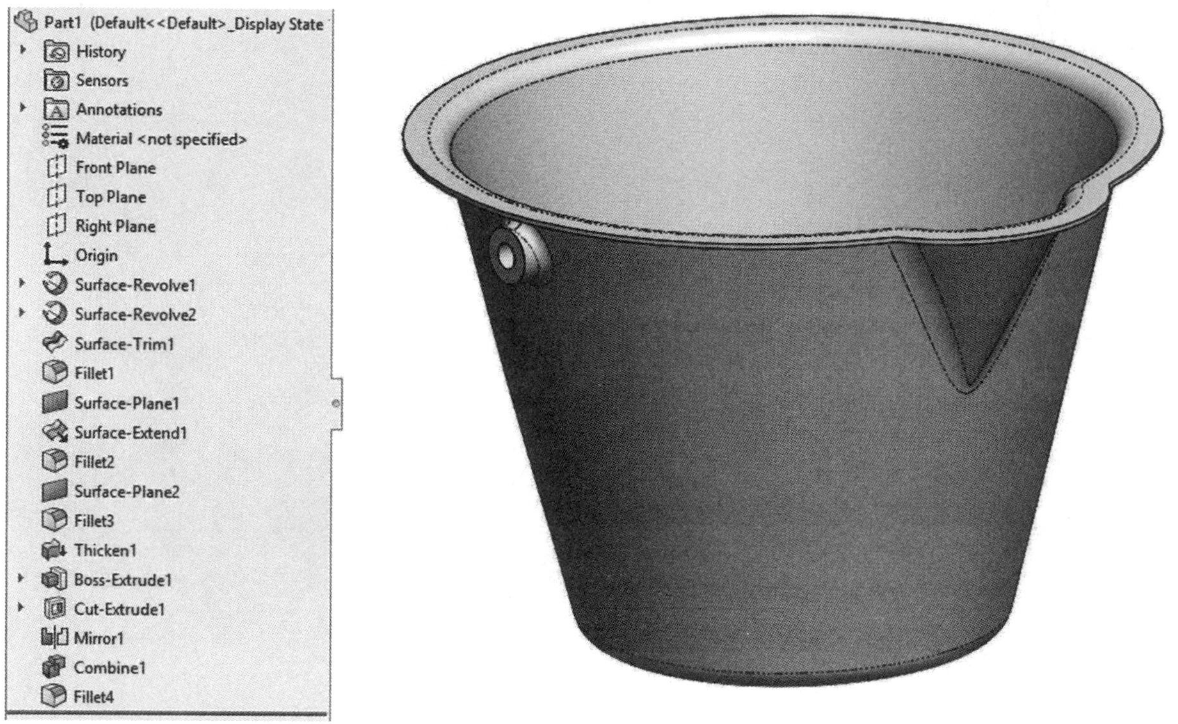

More about Surfaces

When modeling more complex shapes, we may have to resort to surfaces in order to create auxiliary geometry as support for our models in the form of trim surfaces, start or end conditions for extrusions and cuts, to generate guide curves, etc. In the next model, we'll use two surfaces to generate a curve that will be used for a sweep. Look at the following image of a sword's grip.

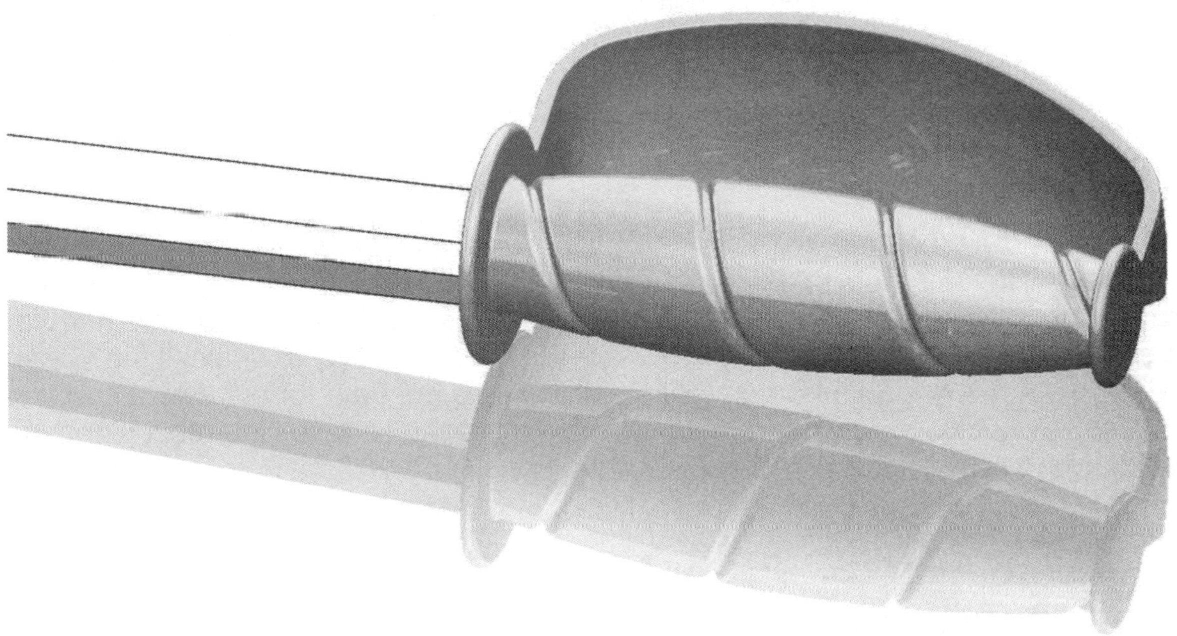

At first glance it seems to be a simple enough model, but if we look closer, we notice the groove in the grip is going in a spiral around an uneven surface, which makes for an interesting challenge which we are going to show how to create in the next few steps.

25.1. - The first thing we are going to do is to define the profile of the grip. Start a new part in inches and make the following sketch in the *"Front Plane."* The horizontal line is a centerline (construction geometry) and the top is a 3 point arc.

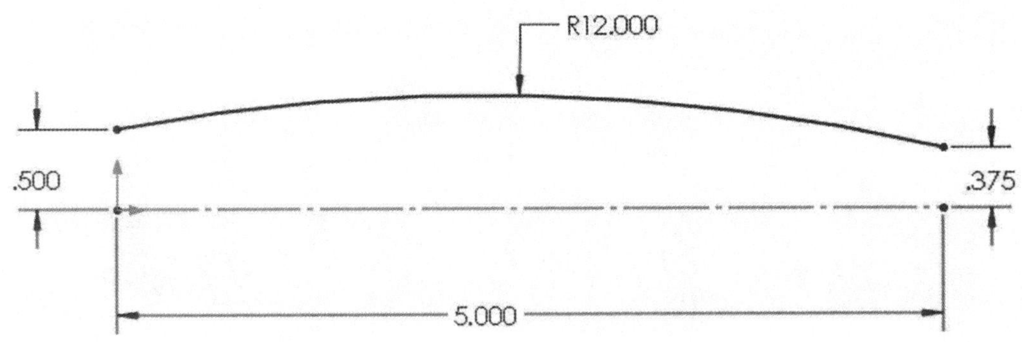

Select the "**Revolved Surface**" command and make a 360 deg. revolved surface.

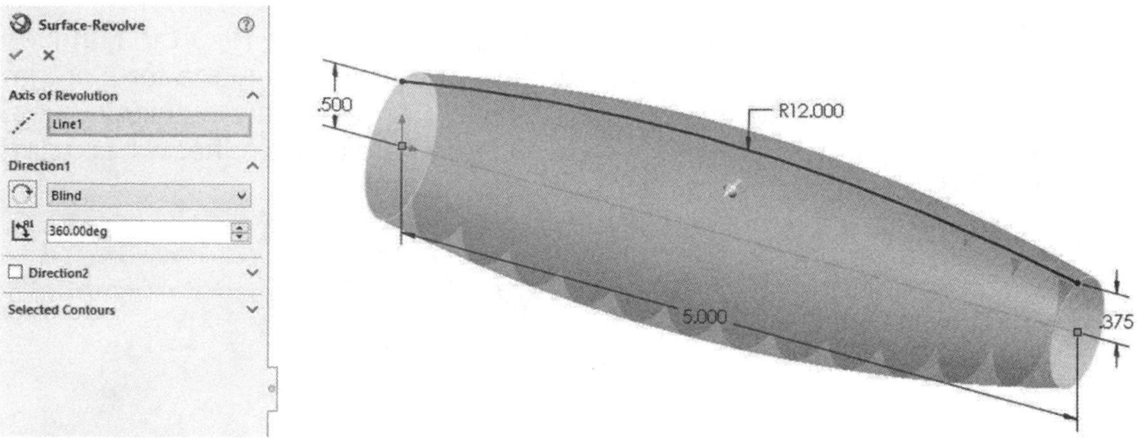

25.2. - For the next step, we'll make a swept surface that will twist along a path like a screw. The purpose of this surface will be to create an intersection with the revolved surface; we'll trim the surface and the resulting edge will be the path for a sweep cut. Make the following new sketch in the *"Front Plane"* and exit the sketch when done. It's a single horizontal line; this will be the path for our swept surface. Make the sketch longer than the part to make sure the surfaces intersect along the full length of the grip. Exit the sketch when done.

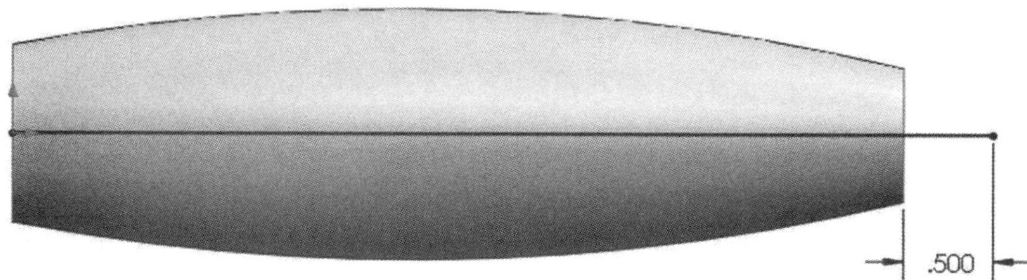

25.3. - Make a second sketch also in the *"Front Plane"*; this time it will be a vertical line as shown. Be sure to make this line higher than the revolved surface. Exit the sketch when done.

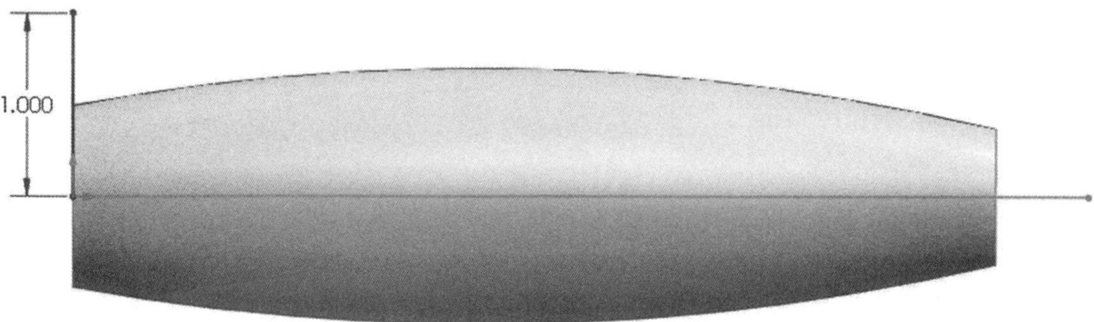

25.4. - Select the "**Swept Surface**" command from the Surfaces toolbar or the menu "**Insert, Surface, Sweep**." Use the vertical sketch line as the "Profile" and the horizontal sketch line as the "Path."

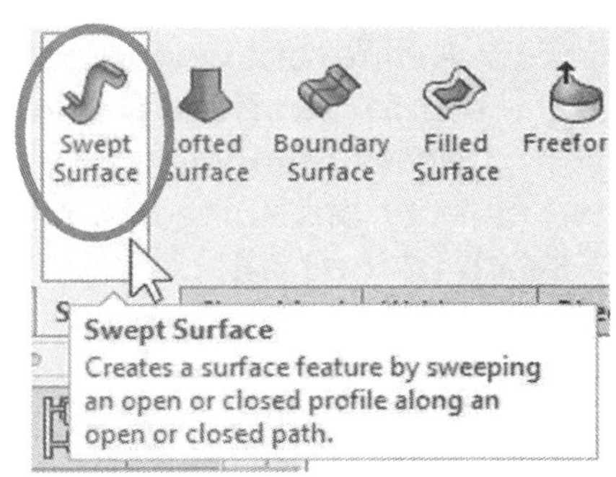

Expand the "Options" box; under "Profile Twist" select "Specify Twist Value," and in "Twist control" select "Revolutions" and enter a value of 4. With these settings the "Profile" sketch will twist along the length of the "Path" sketch 4 turns. Click OK to finish.

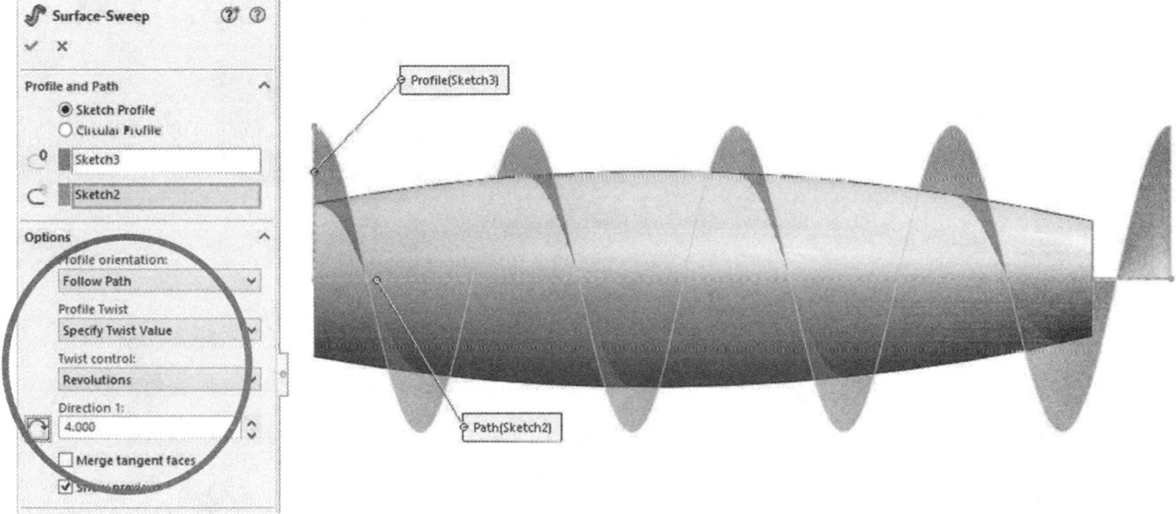

25.5. - What we are interested in is the curve along the intersection of both surfaces, so we'll trim the swept surface with the revolved surface, and use the resulting edge as a path for a swept cut. Select the "**Trim Surface**" command; using the "Standard" trim type, select the revolved surface in the "**Trim tool**" selection box, activate the "Keep selections" option, and select the outside of the swept surface to keep it. Click OK to finish.

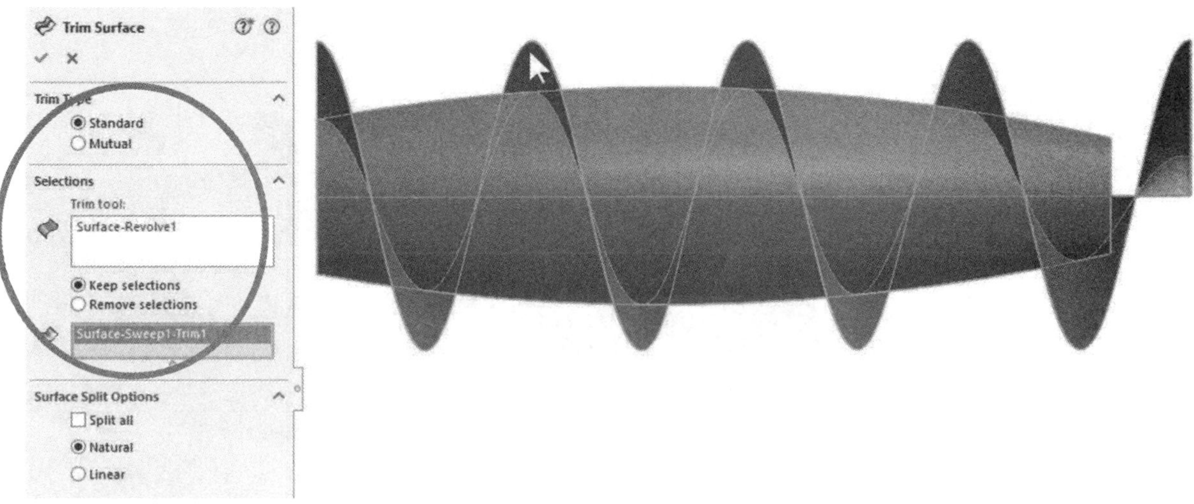

A different way to obtain this curve is by using the "**Intersection Curve**" command from the drop down "Convert Entities" icon. Using this option we select both surfaces, and the curve is added in a 3D Sketch at their intersection.

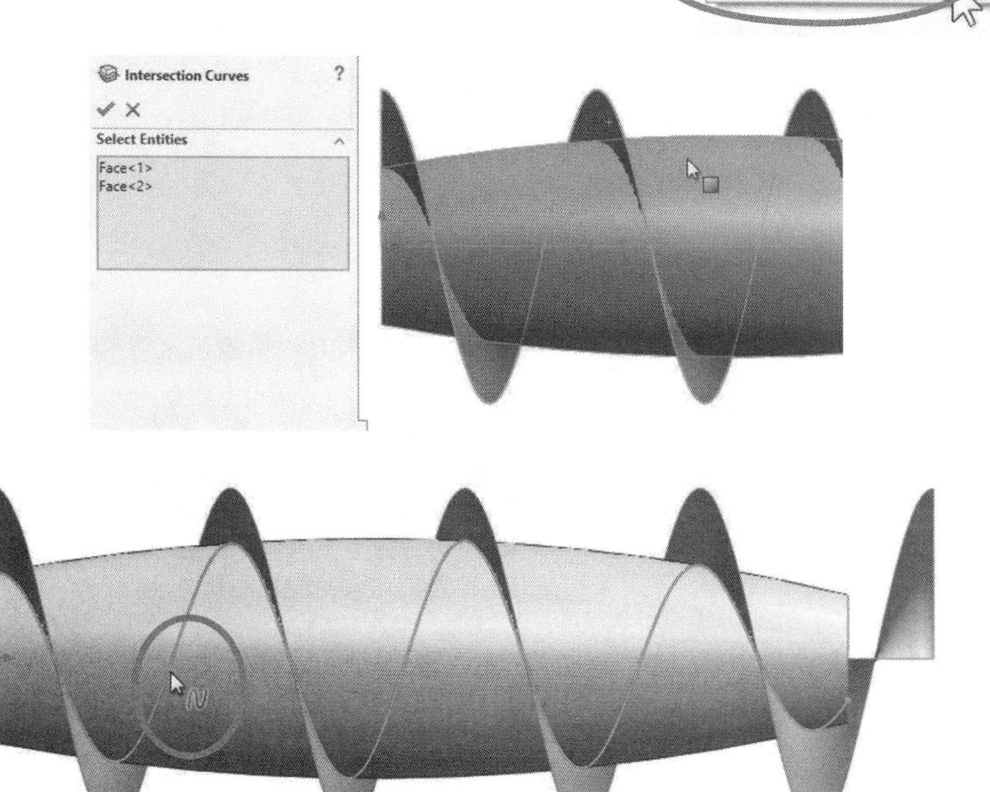

25.6. – After the surface is trimmed, hide the revolved surface.

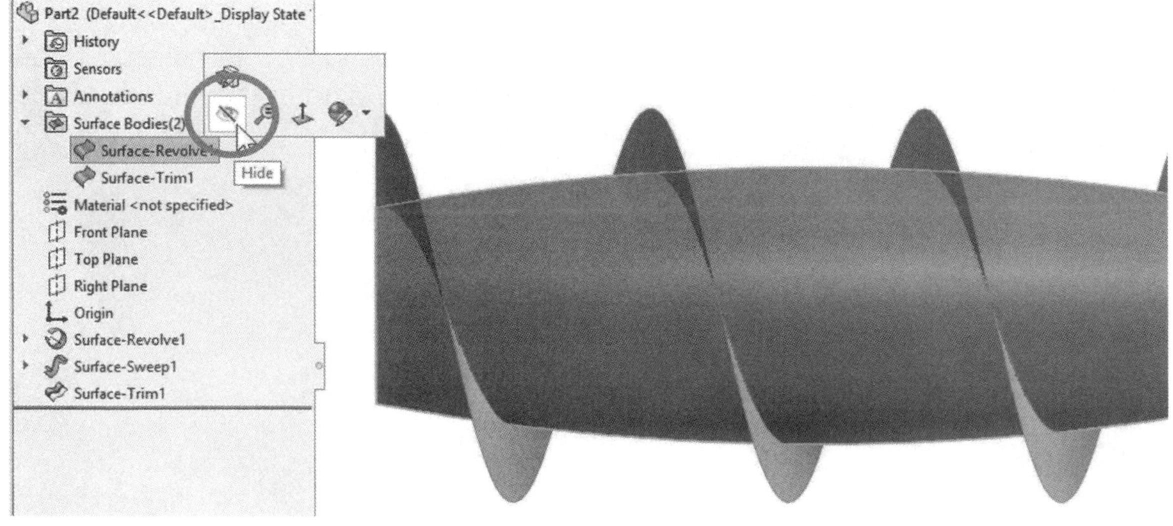

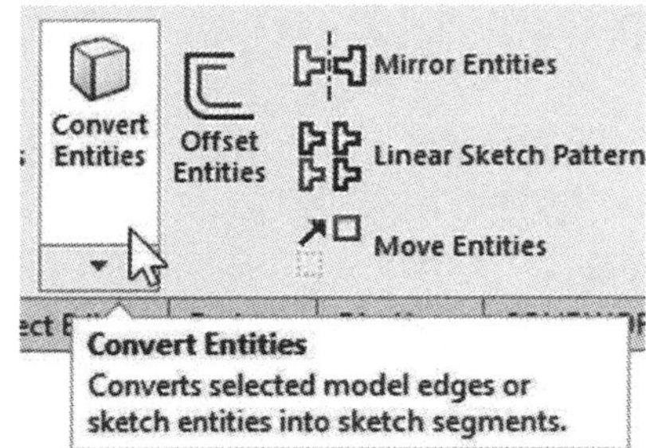

25.7. - Now we need to make a solid body for the grip. Show the sketch in the *"Surface-Revolve1"* and start a new sketch in the *"Front Plane."* Select the top line in the revolved surface sketch, use "**Convert Entities**" to project it into the new sketch and complete the sketch to form a closed contour.

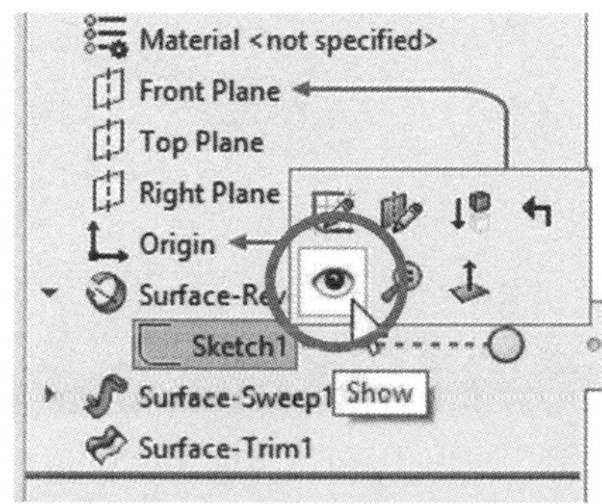

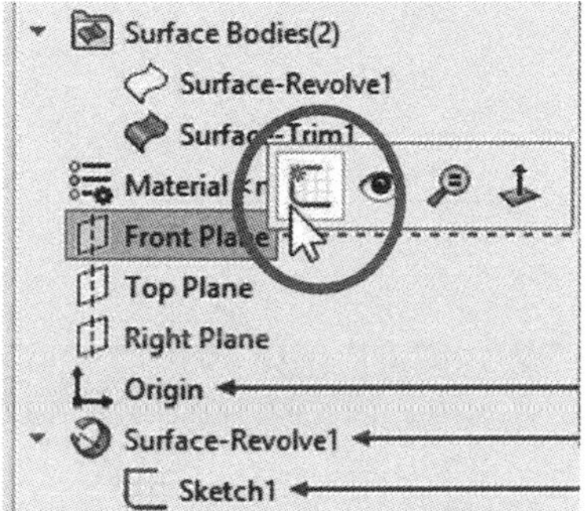

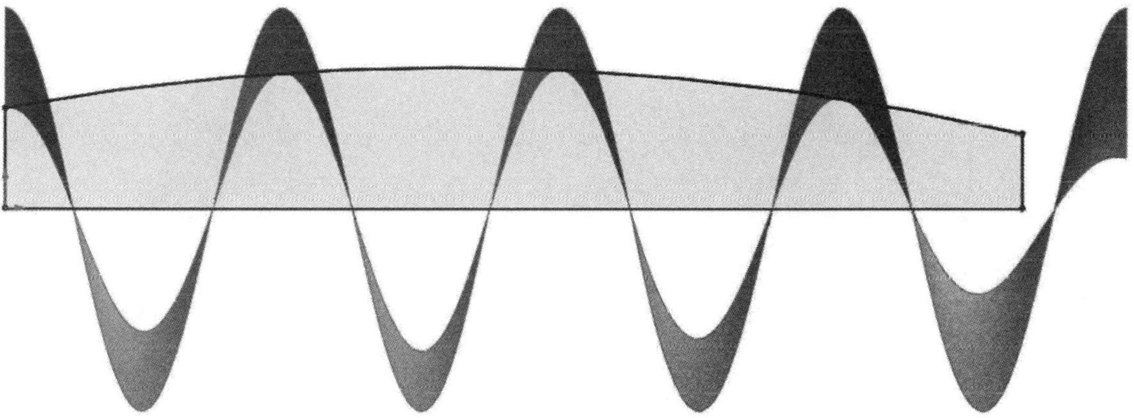

Select the "**Revolved Boss/Base**" command from the Features toolbar (this will be a solid feature), and select the horizontal line to use as the Axis of Revolution to make the revolved feature around it. Click OK to complete it. Now we have a solid body and two surface bodies in our part. Hide the revolved surface sketch after we are done.

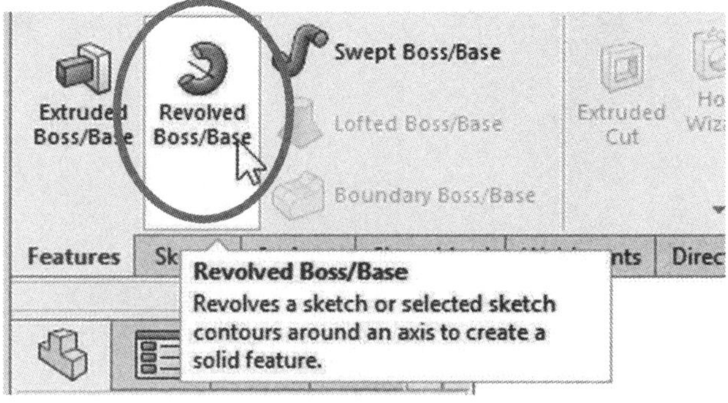

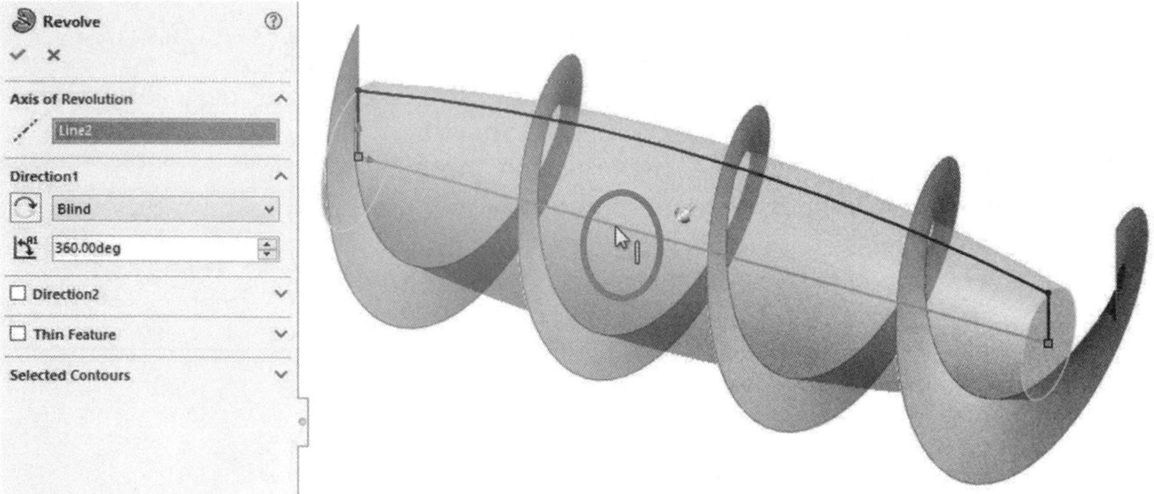

25.8. - After completing the solid we need to make the profile for a Sweep Cut. Add a new sketch in the *"Front Plane,"* and draw it as shown close to the left of the swept surface. Draw an equilateral triangle (hint: make all three lines equal) and add the construction line and a sketch point as indicated, then add a "Pierce" relation between the point and the swept surface's edge. Exit the sketch to finish.

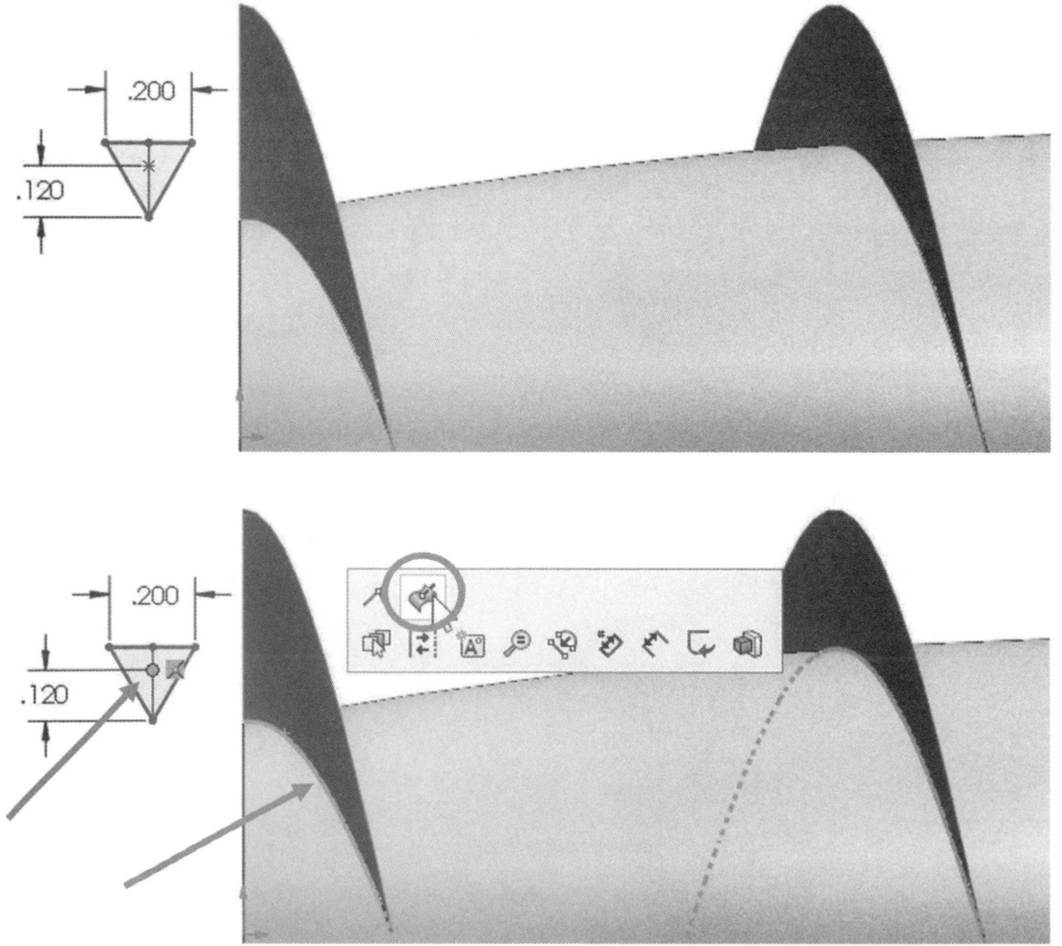

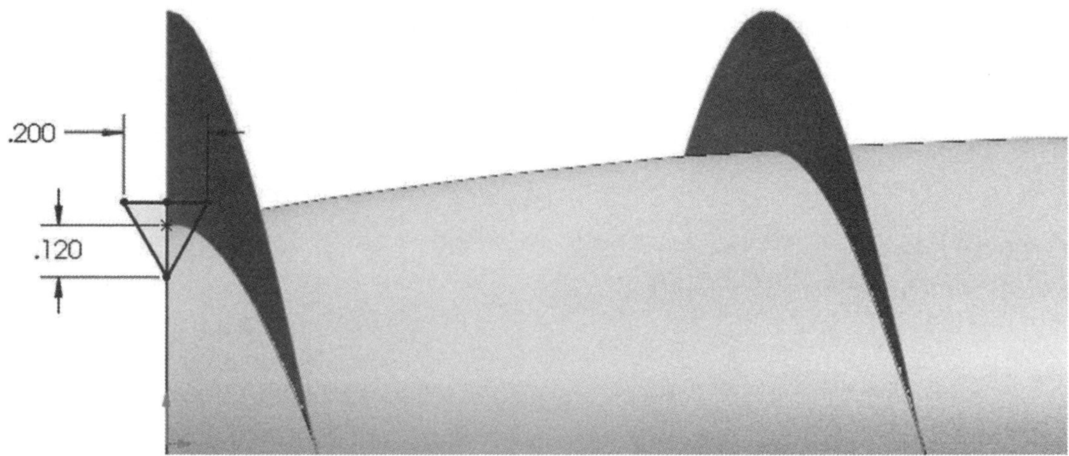

25.9. - Select the "**Swept Cut**" command from the Features tab or the menu "**Insert, Cut, Sweep**," and select the triangular sketch as the "Profile" and the inside edge of the trimmed surface as the "Path."

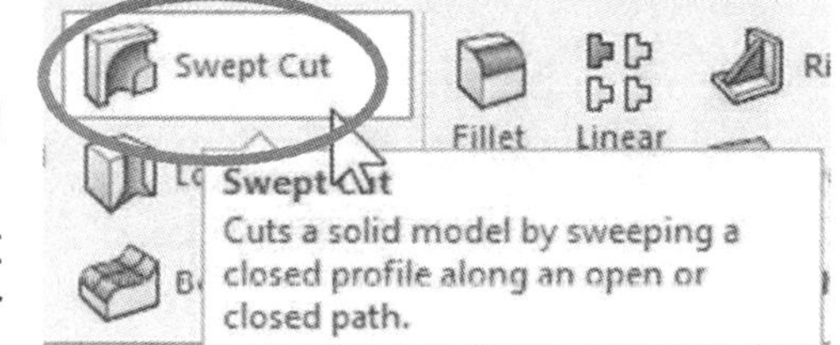

Expand the "Guide Curves" section and select the outside edge of the swept surface; adding this edge as a guide curve will prevent the profile from twisting, resulting in a smoother surface and a better looking result.

In the "Options" section select "Follow Path" in "Profile Orientation" and "Follow Path and First Guide Curve" in "Profile Twist." Click OK to finish.

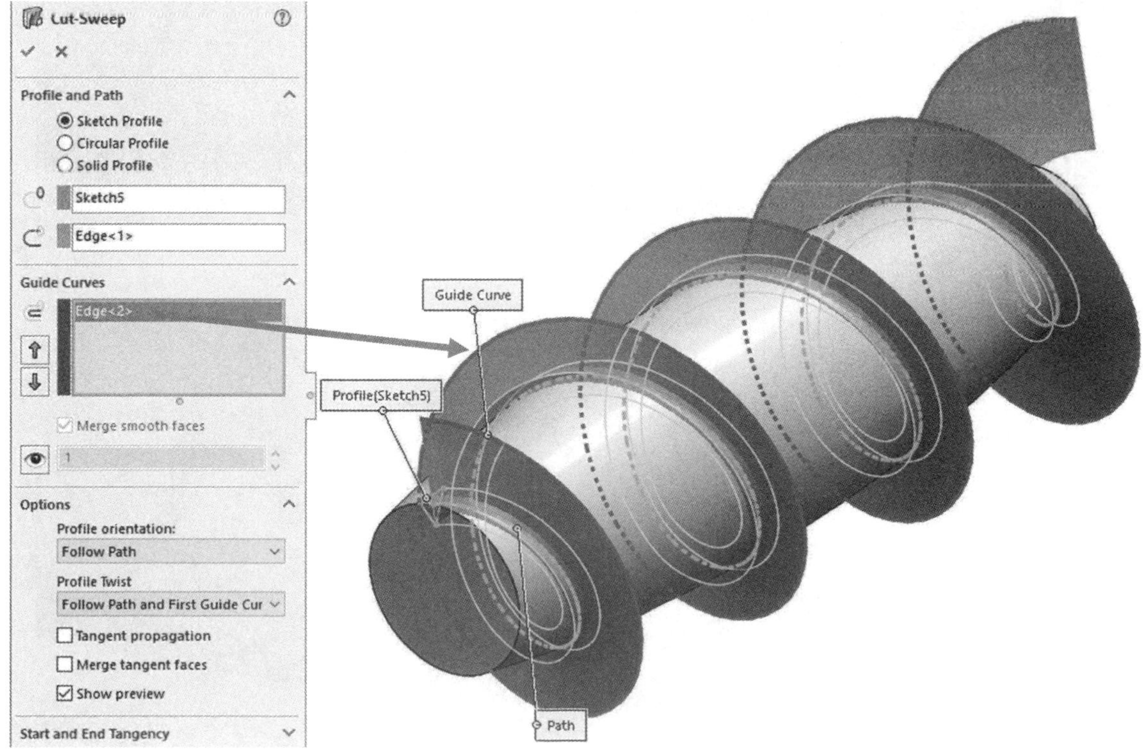

When using guide curves to control a sweep or loft, it is better to add the guide curves before making the profile; this way we can add pierce relations between the profile and the guide curve. In the previous case, the guide curve works as expected because the profile is coincident to the edge of the swept surface.

25.10. - Hide the trimmed surface to view the resulting cut in the solid body.

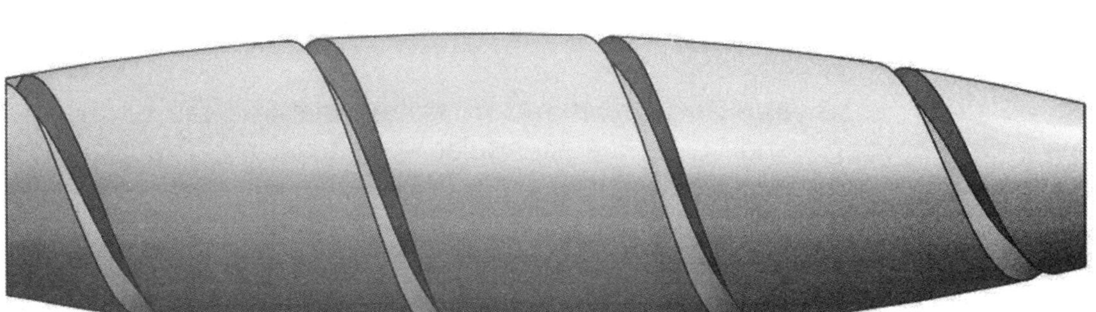

25.11. - Add a 0.035″ radius fillet at the bottom edge of the swept cut.

25.12. - Add a 0.060″ radius fillet to the two outer edges of the swept cut.

25.13. - As a final touch, add a guard and pommel to the grip. Add a new sketch in the *"Front Plane"* and make a revolved boss. The Revolved Boss is overlapping the ends of the spiral cut to cover the ends.

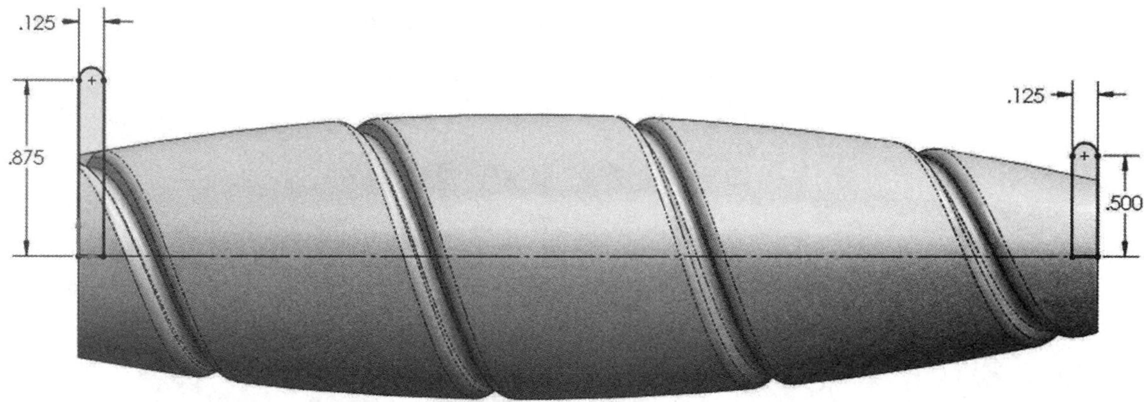

25.14. - Save the file as *'Sword Grip'* and close it.

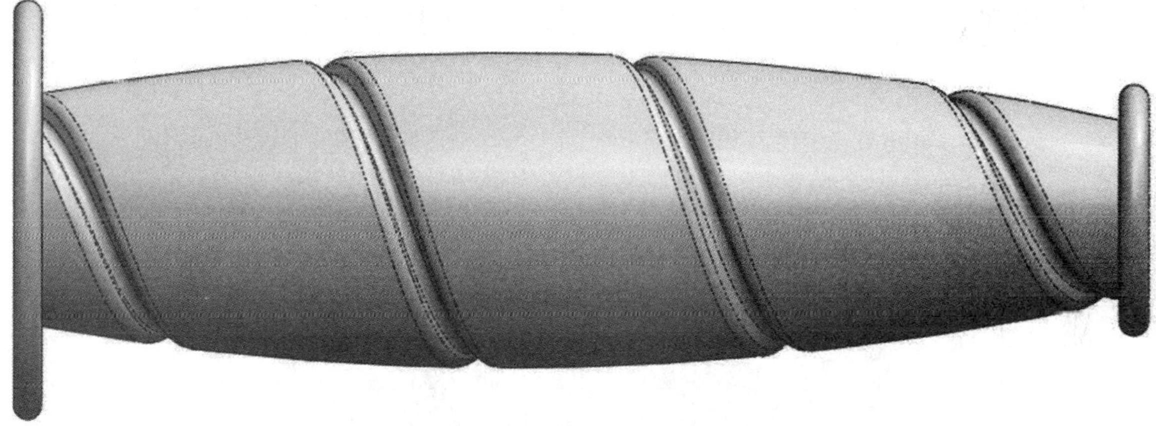

EXERCISE:

Add a hand guard and a blade to the sword grip to finish your sword. Below is a suggestion. Use the "Display Pane" to change features, bodies, or parts' appearance to your liking. These images were made using RealView graphics.

TIPS: Add an open sketch in the Front plane, and make a Thin revolved feature.

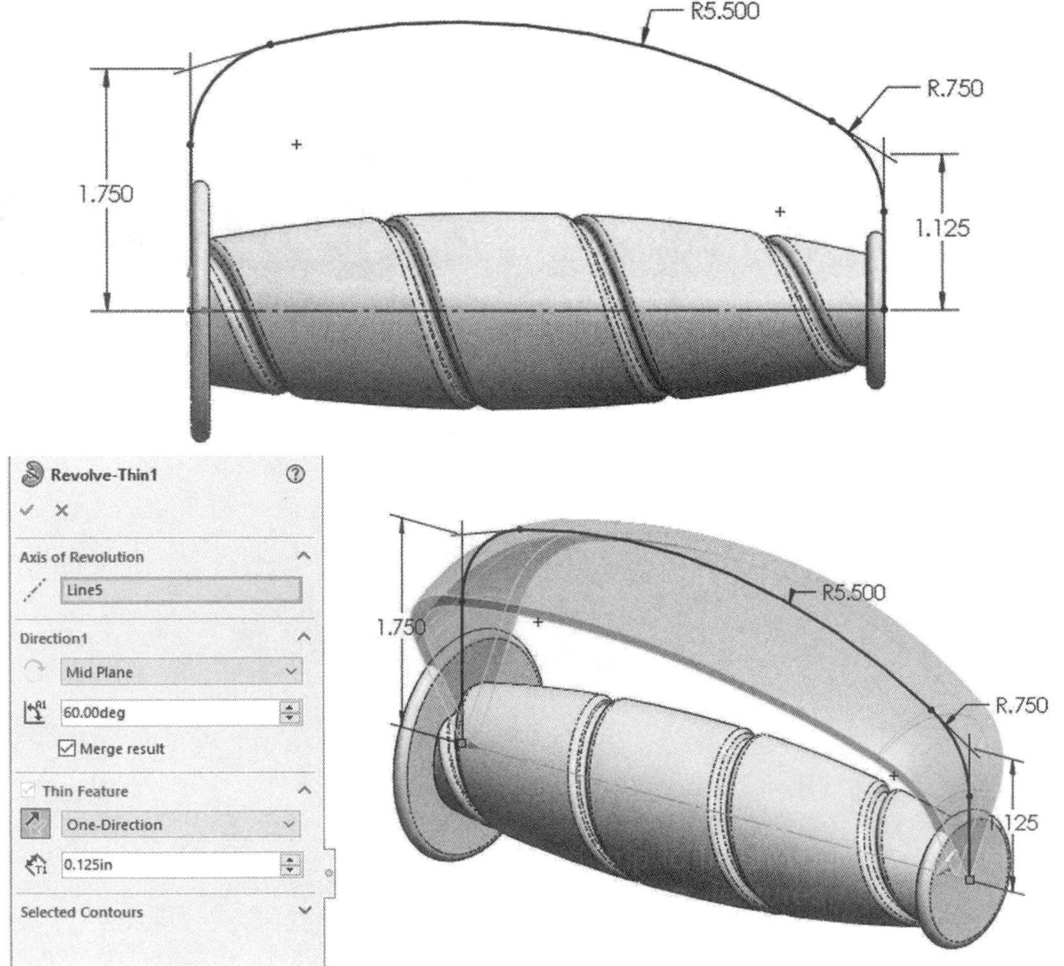

- Add Fillets

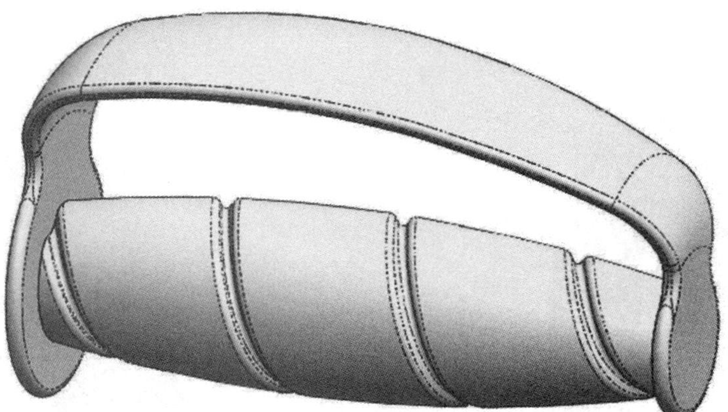

- Add a Blade

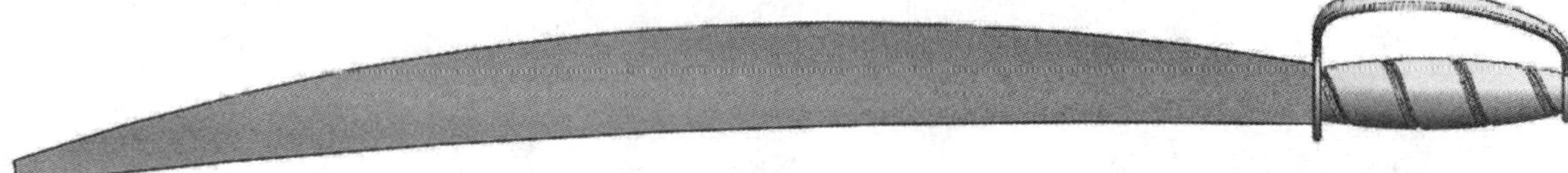

- Add a sharp edge with a couple of chamfers

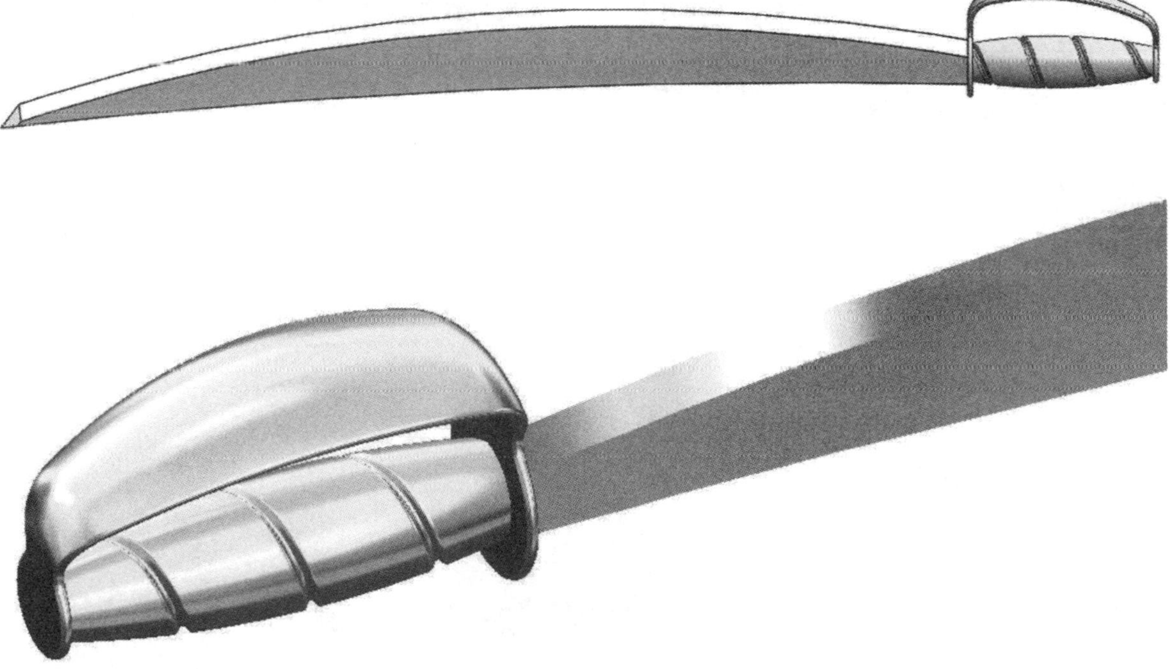

Exercise: Using the provided file *'Chair Surfaces.sldprt'* create the next model using the following instructions.

Sequence of operations:

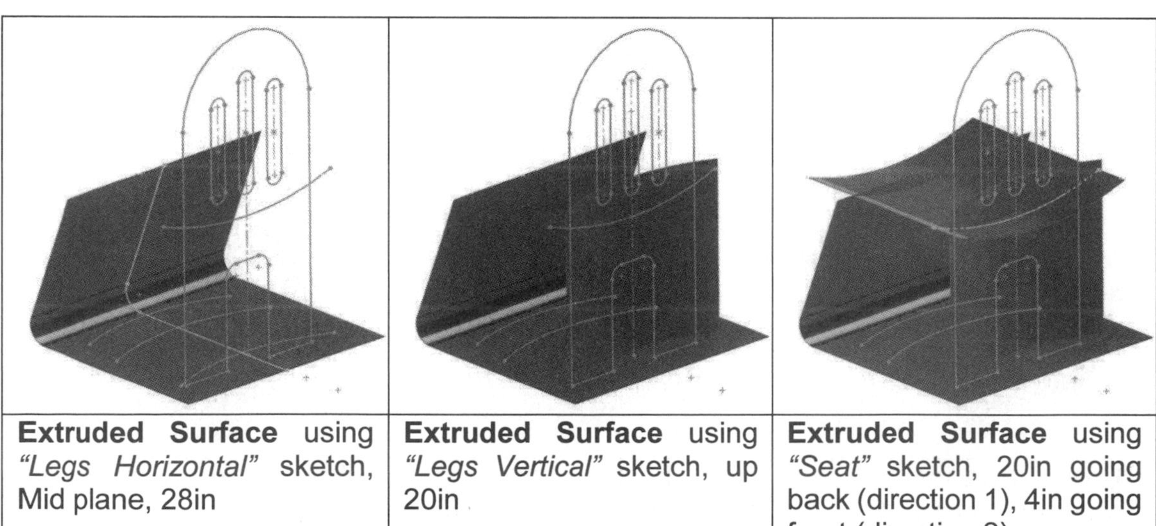

Extruded Surface using *"Legs Horizontal"* sketch, Mid plane, 28in	**Extruded Surface** using *"Legs Vertical"* sketch, up 20in	**Extruded Surface** using *"Seat"* sketch, 20in going back (direction 1), 4in going front (direction 2)

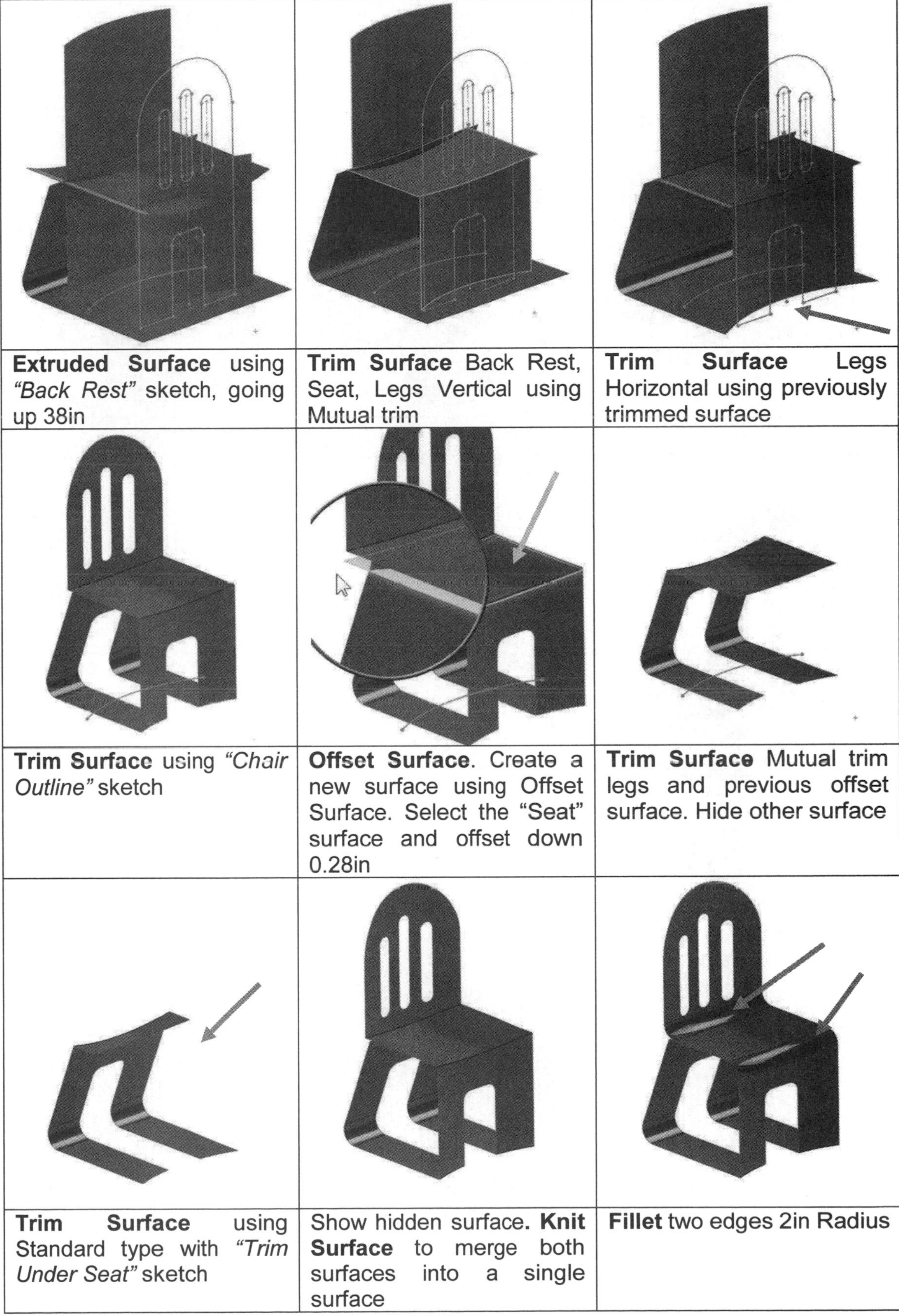

Extruded Surface using *"Back Rest"* sketch, going up 38in	**Trim Surface** Back Rest, Seat, Legs Vertical using Mutual trim	**Trim Surface** Legs Horizontal using previously trimmed surface
Trim Surface using *"Chair Outline"* sketch	**Offset Surface**. Create a new surface using Offset Surface. Select the "Seat" surface and offset down 0.28in	**Trim Surface** Mutual trim legs and previous offset surface. Hide other surface
Trim Surface using Standard type with *"Trim Under Seat"* sketch	Show hidden surface**. Knit Surface** to merge both surfaces into a single surface	**Fillet** two edges 2in Radius

<table>
<tr>
<td>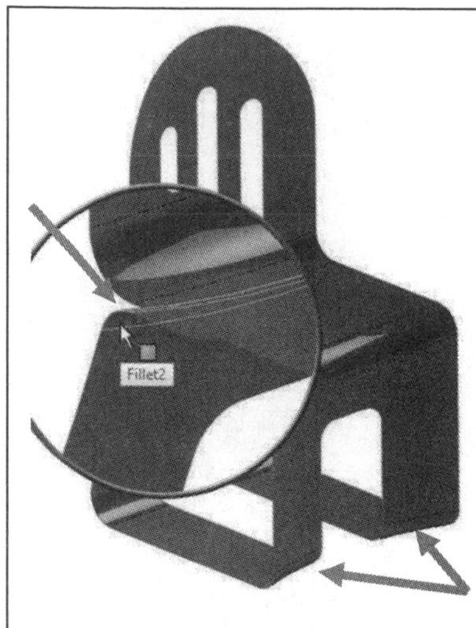
</td>
<td></td>
<td>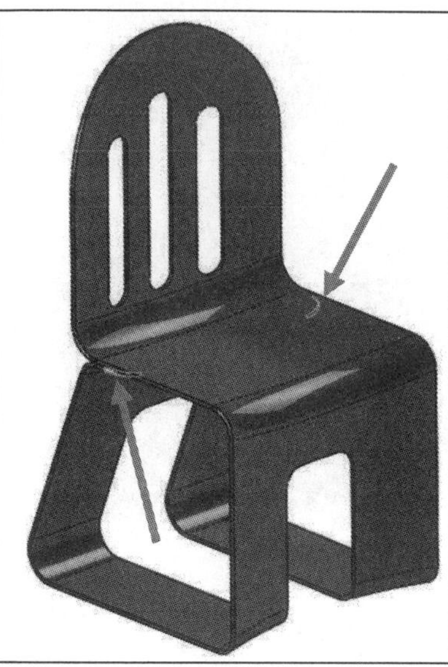</td>
</tr>
<tr>
<td>Fillet edge under seat in back, front legs at bottom (3 edges total) 1.5in radius</td>
<td>Thicken surface 0.25in thick, material going inside</td>
<td>Fillet corners under seat 1.5in radius</td>
</tr>
<tr>
<td></td>
<td></td>
<td></td>
</tr>
<tr>
<td>Fillet Select all tangent faces on both sides of the chair, fillet 0.1in radius</td>
<td></td>
<td></td>
</tr>
</table>

Save the finished part and close it.

Review and Questions:

a) What is the thickness of a surface?

b) When two surfaces intersect each other, they merge. True or False.

c) The option to extrude along a specified direction is available **only** when extruding a surface. True or False.

d) Name two types of elements that can be used to trim a surface.

e) When we trim two surfaces using the "Mutual Trim" option, how many surface bodies do we end up with?

f) Name two types of geometry that can be used to create a planar surface.

g) Which of the following is a requirement to knit two or more surfaces:
A - Surfaces must be touching at the edges.
B - Surfaces must be open.
C - Surfaces must be closed.

h) When thickening a surface, the option "Try to form a solid" can be used if:
A - The surface is open.
B - The surface is closed.
C - The surface is a loft.

i) Name two elements that can be used to split a part into multiple bodies.

Answers:
a - Zero, it has no thickness.
b - False. When two surfaces intersect they need to be trimmed and knitted to be merged.
c - False. This option is available with any extrusion or cut feature, either solid or surface.
d - Another surface, a Plane or a sketch that will project onto the surface.
e - One, the mutually trimmed surfaces are merged automatically.
f - A closed 2D sketch or a group of closed edges on a plane.
g - A. Surfaces cannot be knitted if they are not touching at the edges.
h - B. When thickening a closed surface the option "Try to form a solid" will be enabled.
i - A Plane, a surface or sketch that extends past the solid body.

Notes:

Mold Tools

After designing plastic or cast metal parts, it is necessary to design a mold if the part is going to be made by plastic injection or forge. Mold making is a manufacturing specialty where experience plays as big a role as preparation and study, and it's not an easy trade to master.

When designing a mold, many factors must be taken into consideration, including the shape and size of the part, the material to be used (plastic, resin, metal, etc.), the process, the tooling necessary to make the mold, etc. In other words, it's a complex process.

The purpose of this chapter is not to teach the mold making trade, as that by itself is enough to fill several books and there would still be much more to learn. Our intention is to show the user the tools available in SOLIDWORKS for mold design, learn and understand how and why to use them, and briefly cover design considerations that can affect manufacturing of a molded part, like draft and parting line selection.

Be aware that a complete mold design includes multiple components, starting with a mold base; from there we have to design cooling lines, runners, gates, add hardware components like springs, nuts, bolts, ejection pins, dowels, O-rings, slides, lifters, etc. Our focus in this chapter will be limited to showing how to create the core, cavity, side cores and inserts, as most of the time these are the most difficult tasks; adding the rest of the components and other design elements for a complete, manufacturing ready mold design is the complex process we were just referring to. Many of those factors affect the finished part and even if the part can be properly molded.

In this chapter, we'll learn how to make molds for a few of the parts we have previously made and a couple others, where we'll learn different techniques to make a core and a cavity. These exercises will give the reader a good idea of what needs to be done and, more importantly, how these techniques can be used in different situations. More complex parts will require additional steps, multi part molds, inserts, sliders and such, but that's precisely where the skill and knowledge of the designer comes into play.

We are only showing how to use the most common tools available for that purpose; the rest of the design is completed with mold making knowledge, experience, and skill. This is the stage of designing a mold when you want to involve a toolmaker in the process, and plan the design together.

Notes:

Card Holder Mold

26.1. - We'll start by making a simple two-part mold for a business card holder. A two-part mold is one that, as the name implies, is made of just two parts, a core and a cavity. Open the *'Card Holder.sldprt'* from the included files.

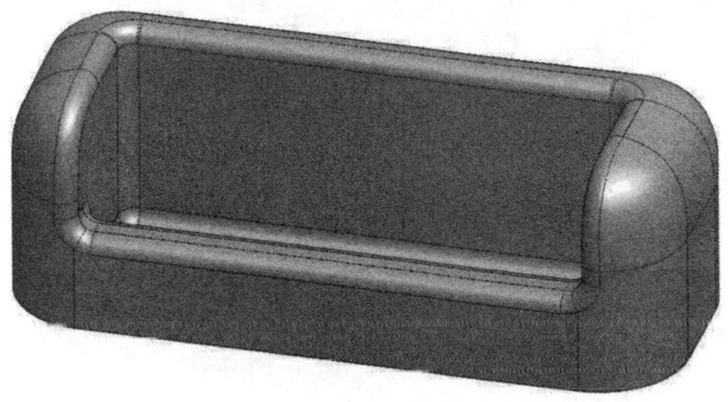

One of the most critical and important details when designing plastic parts is to know if we can get the part out of the mold. The first thing we need to do is to make a "**Draft Analysis**." A draft analysis will tell us if we have the required draft in the 'vertical' faces to eject the part from the mold. By 'vertical' we are referring to the faces parallel to the direction the part will separate from the mold, or 'direction of pull'. Picture it this way: when we make a cake or a pie, the mold's vertical walls have an inclination, or "**Draft**" angle, to facilitate releasing the cake from the mold. If the walls were truly vertical, it would be very difficult to get the cake out of the mold in one piece. Even worse would be to have the walls inclined inside; this is a condition known as a negative draft, and a part would never come out of the mold without breaking the part or the mold.

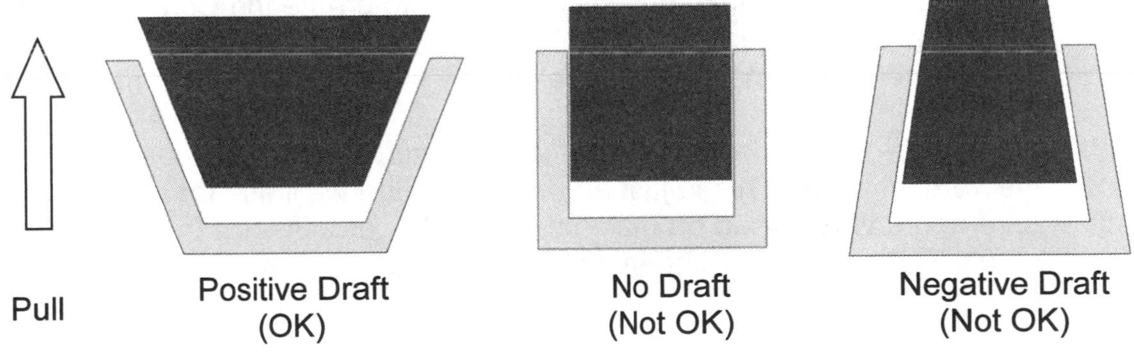

Pull | Positive Draft (OK) | No Draft (Not OK) | Negative Draft (Not OK)

The required draft to easily separate the part from the mold will vary depending on the part's size, the material used, whether the part has a texture or not, the process used, etc. What we know for sure is that we <u>always</u> want to have a positive draft and as much as possible. If our part's design forces us to have a negative draft in one or more faces, we must use a different approach, like a multi part mold, side cores or a combination of both. Later in this lesson we'll cover how to deal with this situation.

26.2. - To start, make sure the "**Mold Tools**" toolbar is open in the Command-Manager. (Right mouse click in a tab and select "**Mold Tools**" or turn on the **Mold Tools** toolbar.)

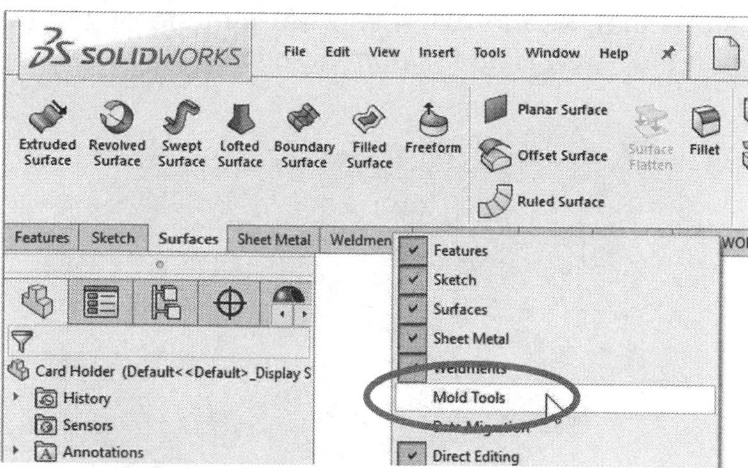

In the Mold Tools toolbar, we have some commands available in other toolbars, for example, from Features, Surfaces, Evaluate tools and a few mold specific tools.

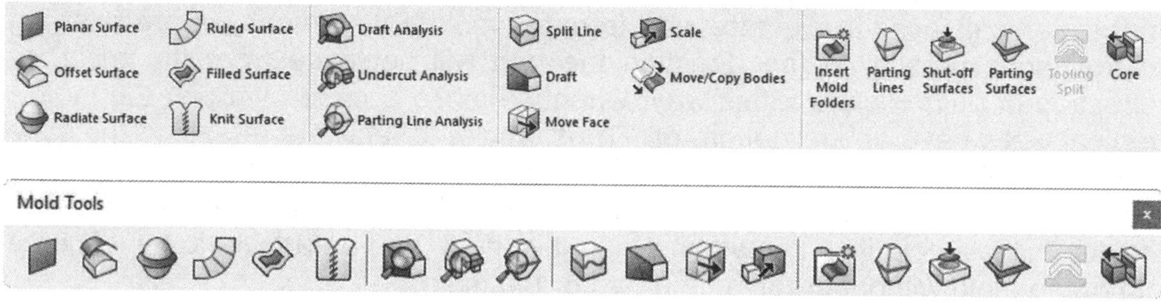

When we design a mold, surfaces are used to create the Core and Cavity. A Core is defined as the 'male' half of the mold, and the Cavity as the 'female' half of the mold. Some molds only have either a core or a cavity.

26.3. - The first step is to analyze our part to make sure we have at least 3 degrees draft in the vertical walls. This is just an arbitrary value we'll use for our example; draft requirements vary based on material, geometry, etc. Select "**Draft Analysis**" from the Mold Tools toolbar or the menu "**View, Display, Draft Analysis**."

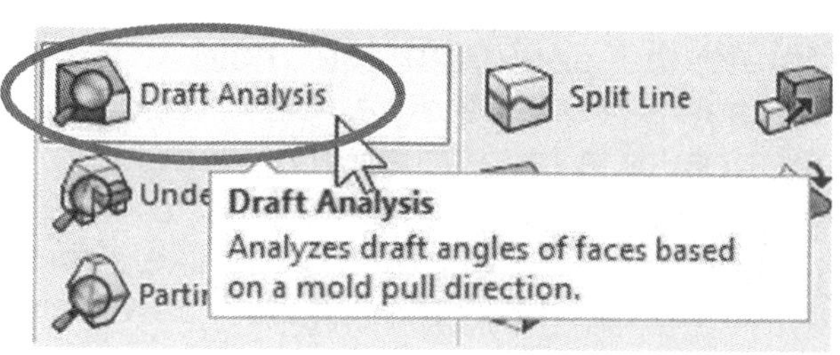

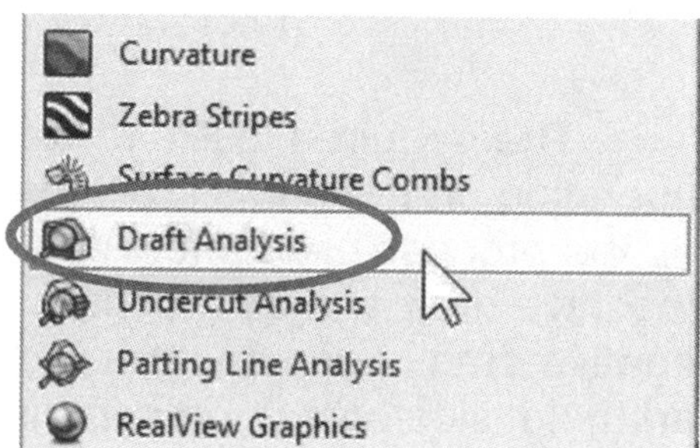

For the "**Draft Analysis**" tool we must select a face or plane <u>perpendicular</u> to the desired direction of pull, or an edge along the direction of pull. In our case, we'll use the bottom of the *'Card Holder'* for our direction of pull (we can also use the *"Top Plane"*). Enter 3 degrees in the "Draft Angle" value box. Immediately after selecting the direction of pull, the part is color coded to let us know which faces have the required draft and which ones don't. Our part's faces are colored with:

Green	Positive draft, indicated by the arrow direction
Yellow	Faces with less than the required draft angle
Red	Faces with negative draft

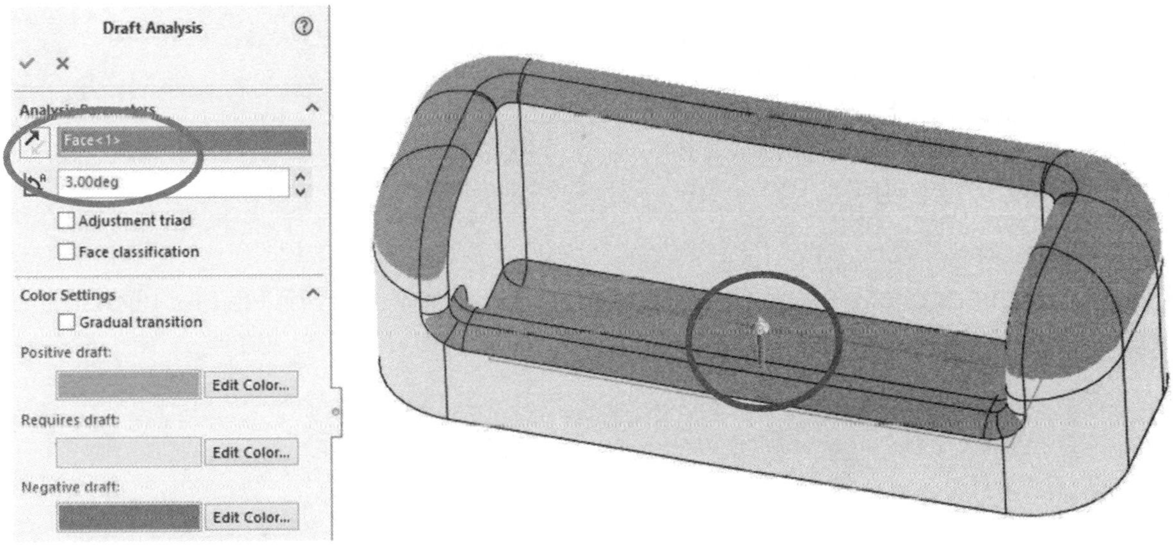

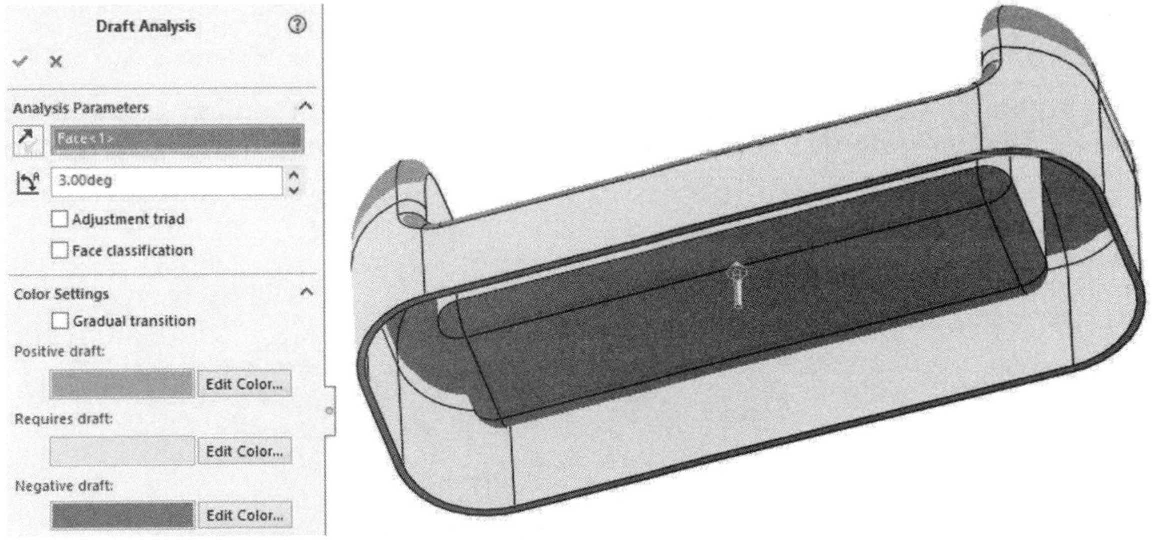

Note that as we move the mouse around the part, the draft angle of the face is displayed next to the mouse pointer.

"**Positive**" and "**Negative**" are only telling us which side of the mold a face will be made with. This direction can be reversed using the "Reverse Direction" icon next to the "Direction of Pull" selection box. When we have a Core and Cavity mold, "Positive" faces will be on one side, and "Negative" faces will be on the other. In our case the green faces will be made in the Cavity and the red faces in the Core. What we are mostly interested in are the yellow faces that need to be given a draft to properly release the part from the mold, and these are the ones that need to be modified to be either "Positive" (green) or "Negative" (red).

26.4. - After identifying the faces that need to be given a draft, cancel the "**Draft Analysis**" tool. If we click OK, the draft analysis colors remain visible and we can see which faces need to be modified as we edit the part. The first feature to be modified will be the *"Boss-Extrude1."* Select it in the FeatureManager or the graphics area and edit it.

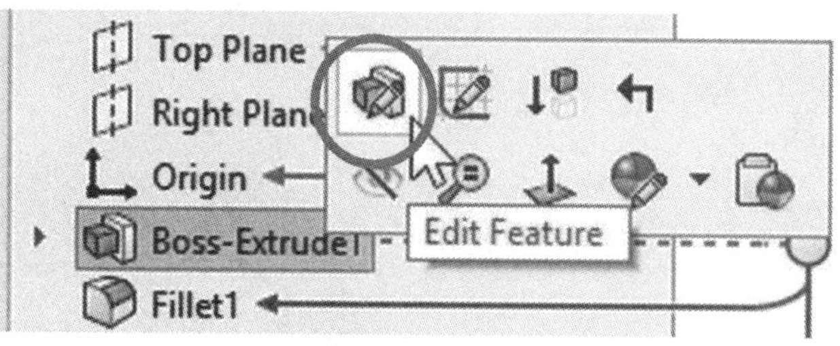

If the user clicks OK and the Draft Analysis view remains visible, click in the "**Draft Analysis**" icon again to turn it off; it works as a toggle.

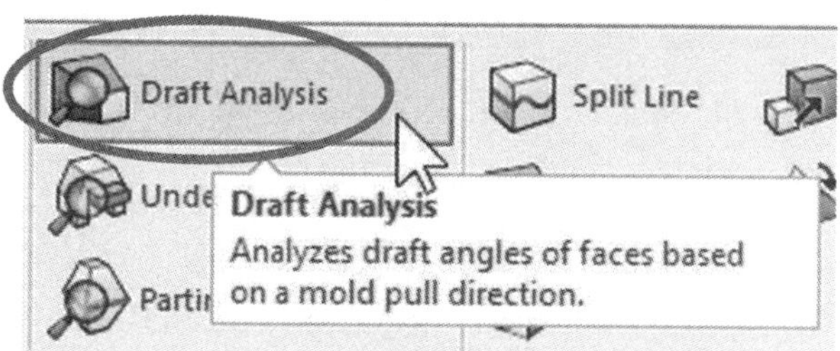

In the "Direction 1" settings, activate the "Draft On/Off" option and enter 3 degrees. You will see the faces in the preview 'leaning' in the part. Click OK to continue.

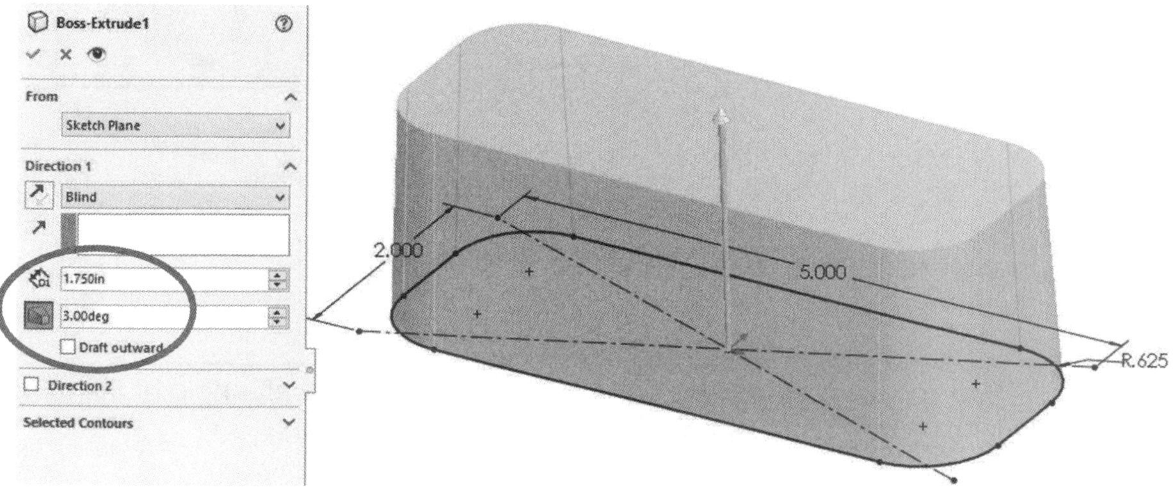

DRAFT ANALYSIS VIEW: After completing this change make a Draft Analysis again. We can see that only a few faces still require draft and are still yellow: some in the top (Cavity side) and some in the bottom (Core side).

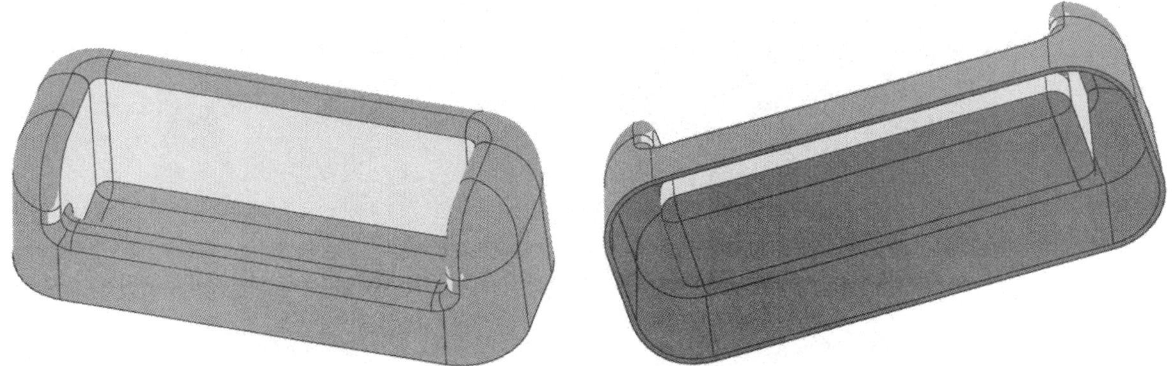

26.5. - Since not all faces can be given a draft at the time a feature is created, we may have to add it as a secondary operation. To add a draft to the remaining faces that need it, we'll use the "**Draft**" command. Faces to be drafted must not be connected to a fillet or the draft will fail. Analyzing our model, we can see that we have a few operations, then fillets and a shell at the end to make the part hollow.

What we need to do is to 'rollback' the model just before the *"Fillet2"* feature to be able to add the necessary draft. Since the part is shelled at the end, fixing the faces in the top (Cavity side) will also take care of the corresponding faces in the bottom (Core side). Select the *"Fillet2"* feature, and from the pop-up menu select "**Rollback**." By doing this we go *'back in time'* in the FeatureManager, to the moment right before the fillets and the shell were created. By doing this, we can add a new feature at this position in the FeatureManager and when we are finished we can "**Roll Forward**" to finish building the part, essentially un-suppressing the remaining features. Notice the Rollback bar is now located just below the *"Cut-Extrude2"* feature and the remaining features are grayed out below it ("Suppressed").

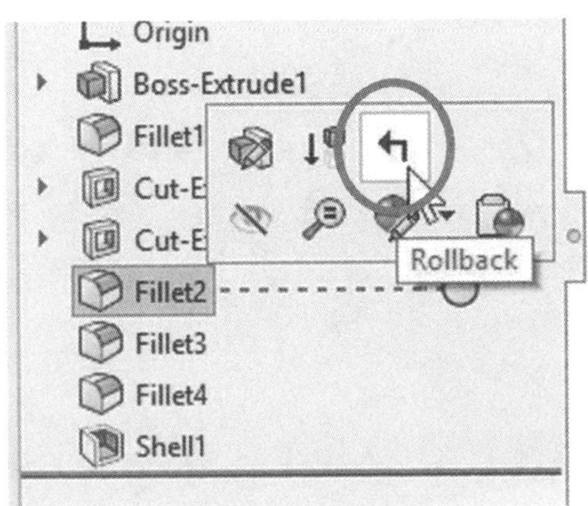

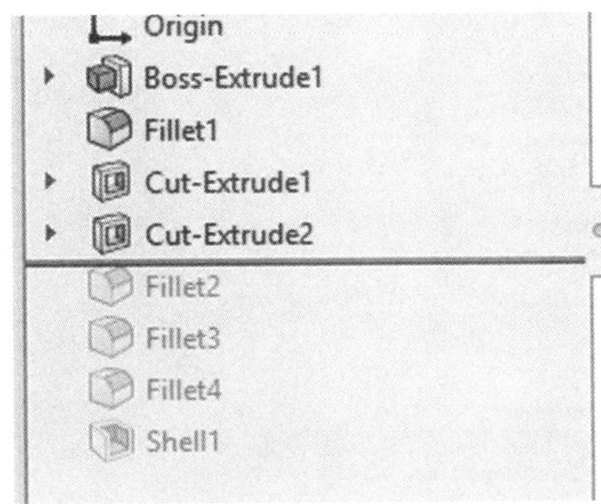

26.6. - Select the "**Draft**" command from the CommandManager or the menu "**Insert, Features, Draft**." The "**Draft**" command is available in the Features and the Mold Tools toolbars. (To make selection easier, the "**Draft Analysis**" view is off.)

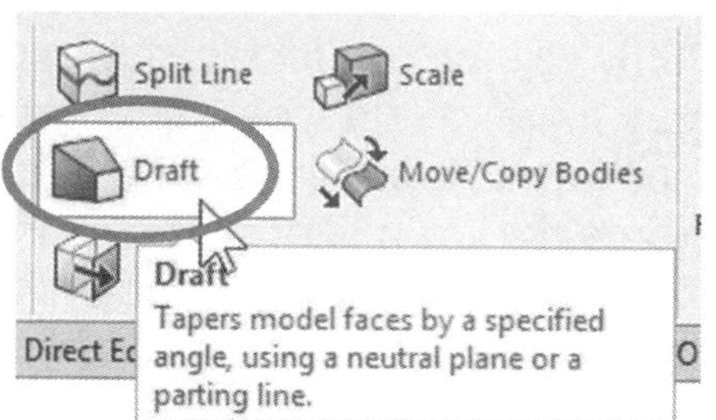

The faces we need to draft are the four inside faces, where business cards are placed. Select "Neutral Plane" in the "Type of Draft" selection box. The "Neutral Plane" is a reference flat face or plane that will be used to start measuring the draft angle. This is also where the faces to be drafted will start to *'incline'*; think of it as a hinge where the face starts to move. Select the bottom face of the cavity as the "Neutral Plane" and enter 3 degrees in the "Draft Angle" value box. Notice the arrow in the corner of the selected face is pointing up; this is the "**Direction of Pull**" we want.

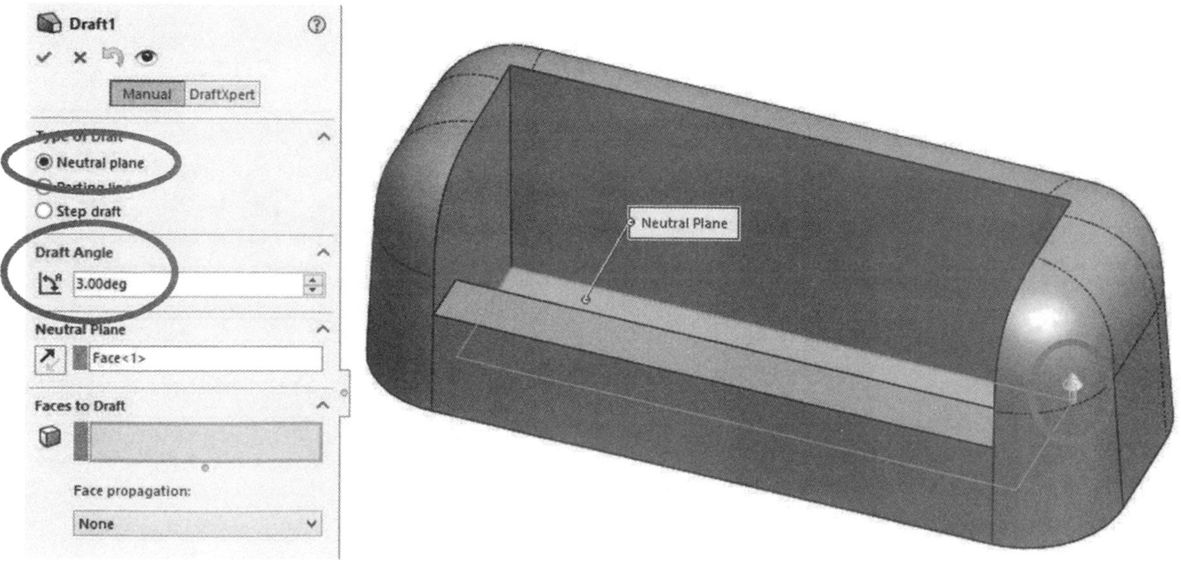

More information about the "**Neutral Plane**":

The plane we select as the "Neutral Plane" defines how the selected faces will be drafted. Look at the following images and the neutral plane selections. In all cases the "Direction of Pull" is pointing up. If the "Direction of Pull" is reversed the draft is added in the other direction. The difference between the different "Neutral Plane" selections is that the faces to be drafted are projected to the "Neutral Plane" and at this point is where they will be 'hinged' and start the draft.

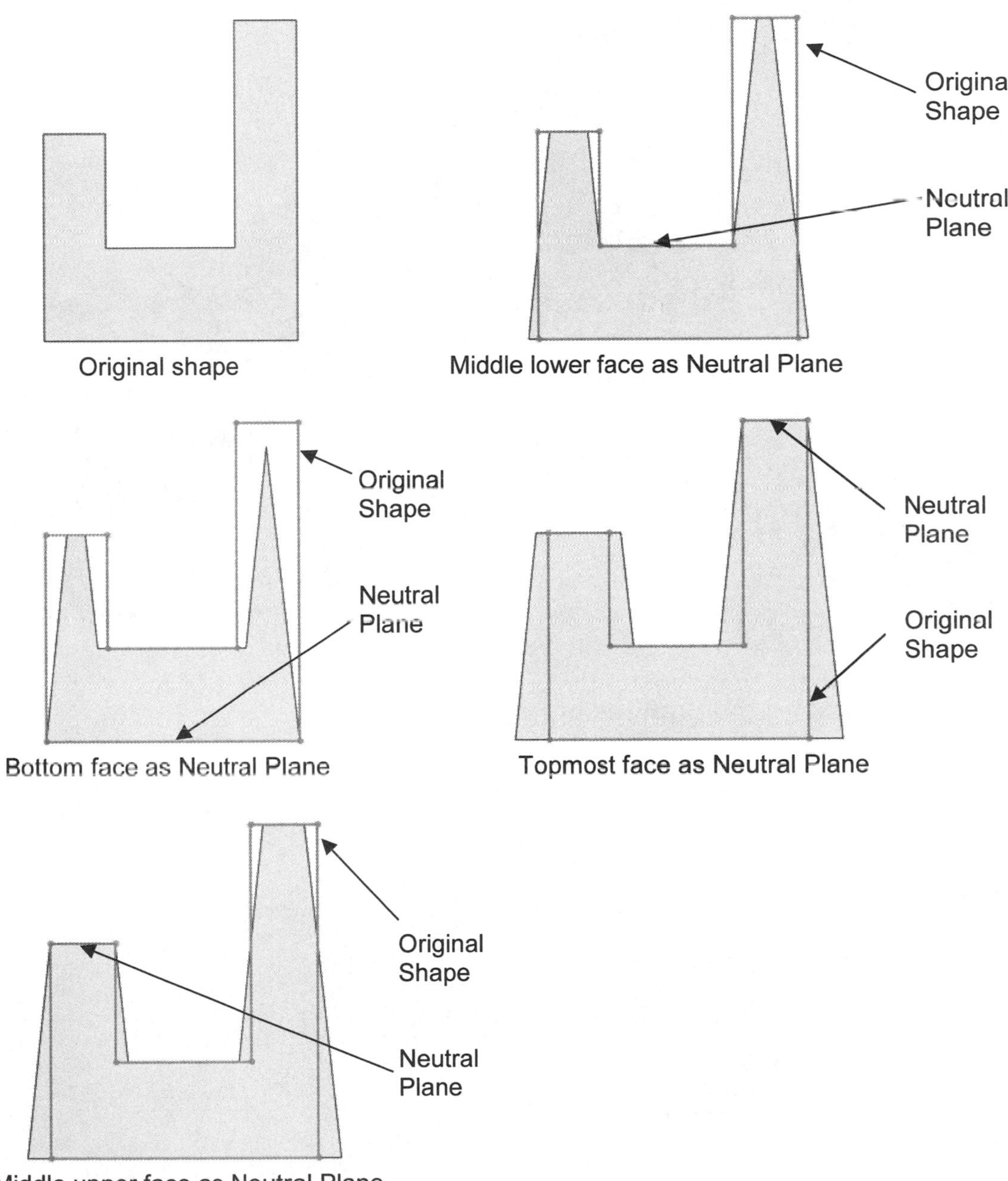

Depending on the desired result, we may have to add multiple "Draft" features using different "Neutral Plane" and "Faces to Draft" selections.

The draft "Direction of Pull" can be reversed using the "Reverse Direction" icon next to the "Neutral Plane" selection box. Reversing the direction of pull inverts the direction of the draft.

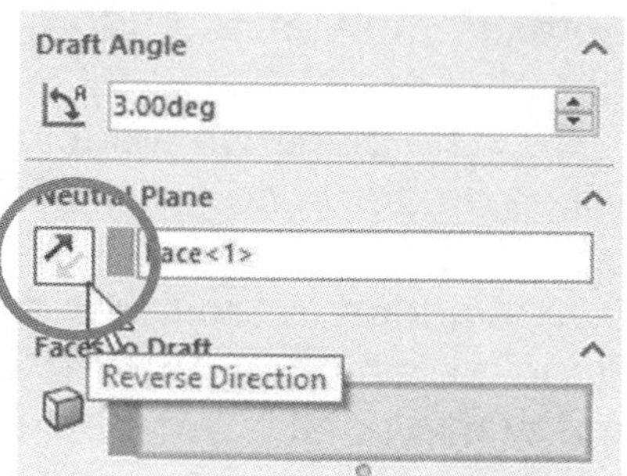

26.7. - Add the four internal vertical faces to the "Faces to Draft" selection box and click OK to finish. Notice the faces moving after the draft is completed. Make sure the arrow indicating the draft direction is pointing up.

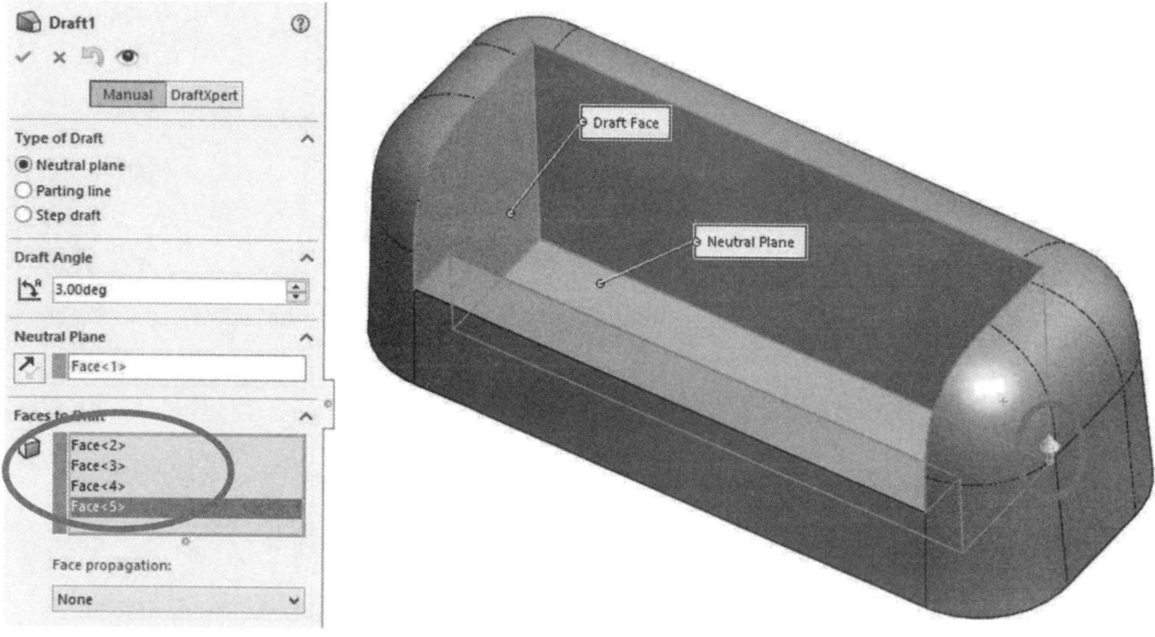

Top view before adding the draft feature…

Top view after adding the draft.

26.8. – After adding the draft, select the “Rollback bar” and drag it all the way to the bottom to rebuild the rest of the features.

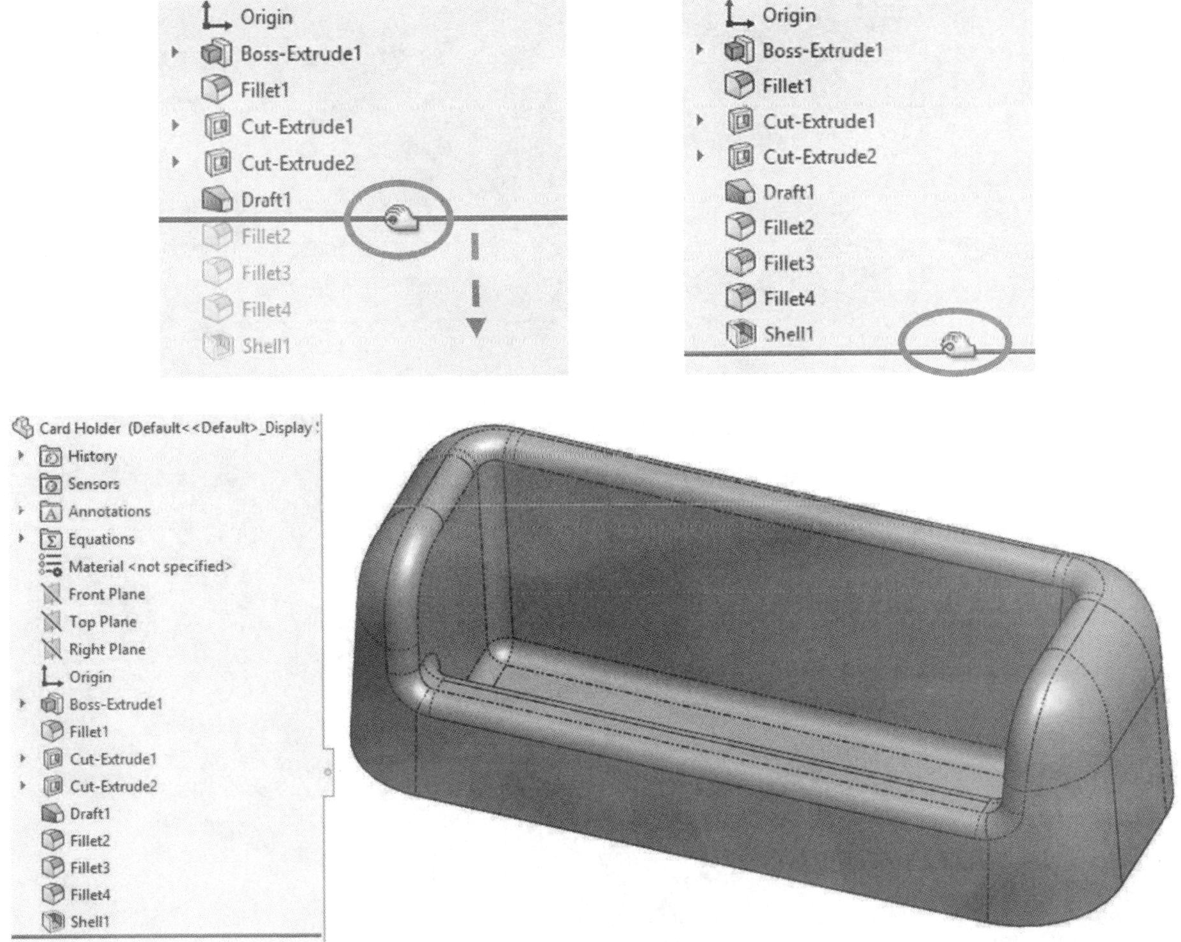

After adding features while in rollback mode, errors may occur after rolling the model forward due to geometric changes made to the model.

26.9. - Use the "**Draft Analysis**" view again and verify that all faces have the required 3 degrees of draft to continue. One side is now completely green ("Positive" draft), the other side is completely red ("Negative" draft) and there are no yellow faces in the model. Click Cancel to continue (if we press OK we'll keep the "Draft Analysis" view colors).

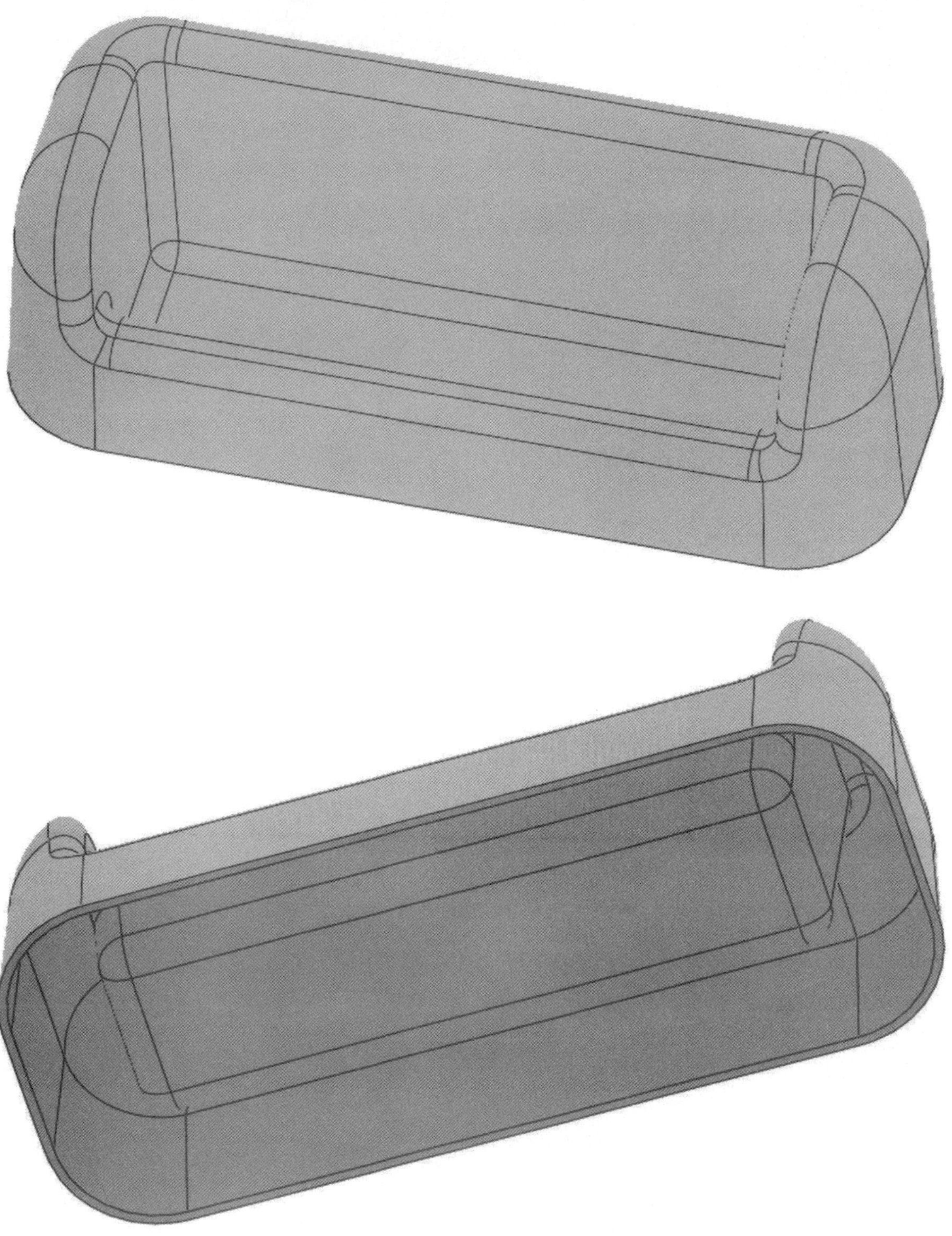

26.10. - When we make a mold for a part, the designed part is usually scaled up to compensate for part shrinkage in the mold when the molten plastic or metal solidifies. To scale the part (up or down), select the "**Scale**" command in the Mold Tools tab or the menu "**Insert, Features, Scale.**"

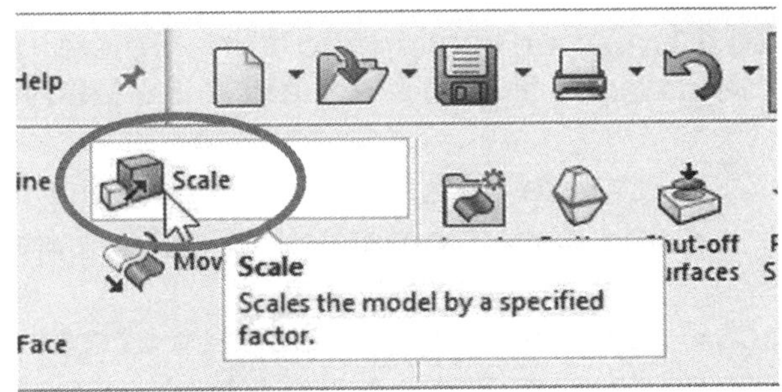

Scale the part about its "Centroid" using a scaling factor of 1.03; this will make the part 3% bigger to compensate for mold shrinkage. Click OK to complete. Notice the part is slightly bigger when finished and a "Scale" feature is added to the FeatureManager.

The shrinkage values used to scale a part before creating a mold vary widely depending on the material used and the part's geometry. For example, if a part is elongated, it may shrink differently along one axis than the other, and in a case like this a non "Uniform scaling" option is used. As far as the material, the shrinkage value is a physical property of the material used in the molding process and is usually provided by the material's manufacturer.

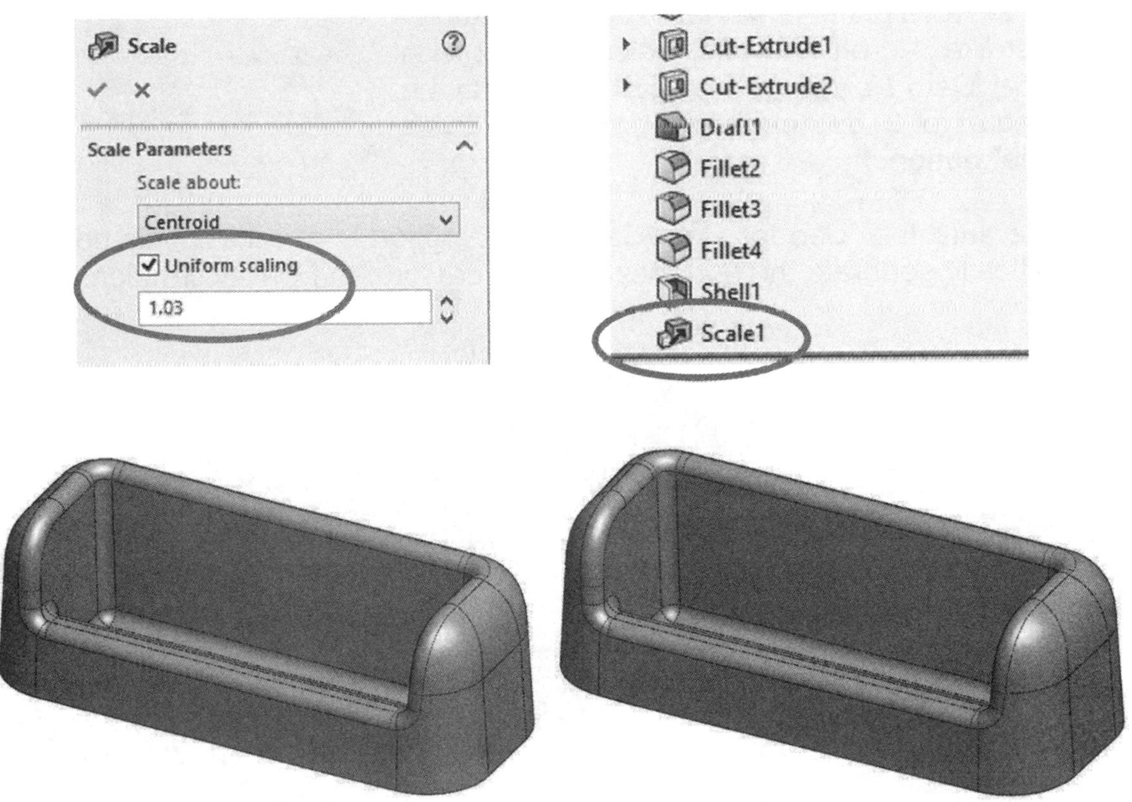

See the difference? Roll back and forward the *"Scale1"* feature to see the change; subtle, but visible.

26.11. - After scaling the part, we are ready to make the mold. In general terms, the process to make a mold in SOLIDWORKS is:

- Define a Parting Line where the mold halves meet.
- If needed, create Shut-off Surfaces to close holes in the part connecting the core and the cavity.
- Create a Parting Surface to split the mold at the parting line.
- Add a Tooling Split feature to generate the Core and Cavity using the Parting Surface made in the previous step.
- If needed, split the core and/or cavity model to create side cores and inserts.

A **parting line** is formed by the model edges where the two halves of the mold meet. An easy way to identify the parting line is using a "Draft Analysis." In simple terms, the parting line will be the edges where the faces with "Positive" draft (green) meet the faces with "Negative" draft (red). In a simple two-piece mold (single core and cavity) like this, the parting line is usually automatically found and selected.

Select the **"Parting Lines"** command from the Mold Tools tab or the menu "**Insert, Molds, Parting Line**." The command is like "**Draft Analysis**" with additional options. In the "Mold Parameters" options, select a plane or flat face for a "Direction of Pull" just like a draft analysis. Select the flat face inside the *'Card Holder'* or the *"Top Plane,"* set the draft angle to 3 degrees and press the "**Draft Analysis**" button.

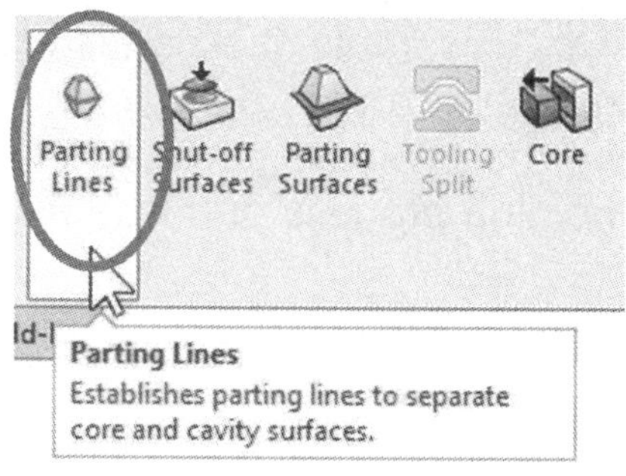

Make sure the "Use for Core/Cavity Split" option is checked. This option will automatically generate the surfaces needed for Core and Cavity creation.

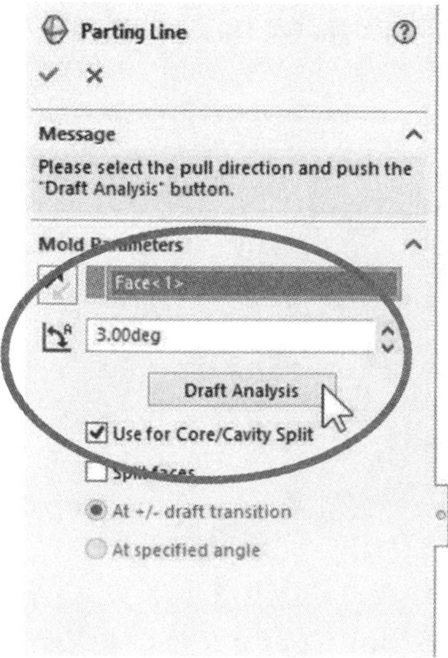

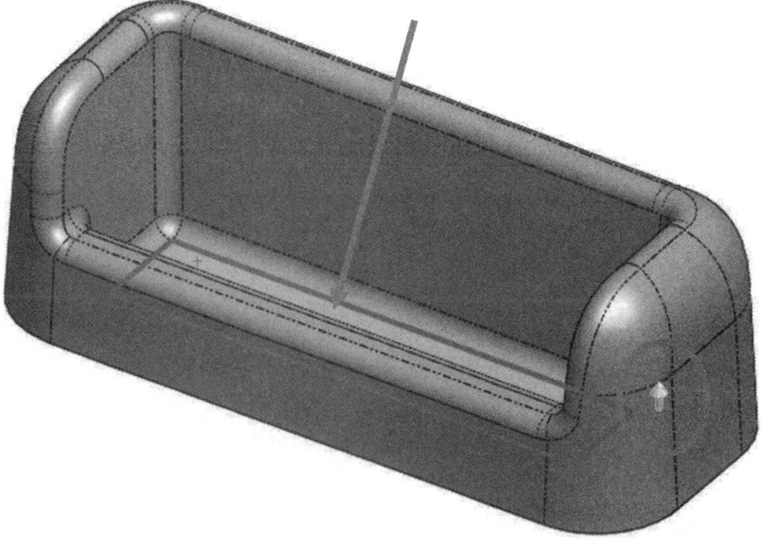

After "Draft Analysis" is selected, the edges needed for the parting line are automatically added to the "Parting Lines" selection box, and the message "The parting line is complete" is highlighted in green at the top, letting us know that the mold can be separated into core and cavity. Click OK to generate the parting line and continue.

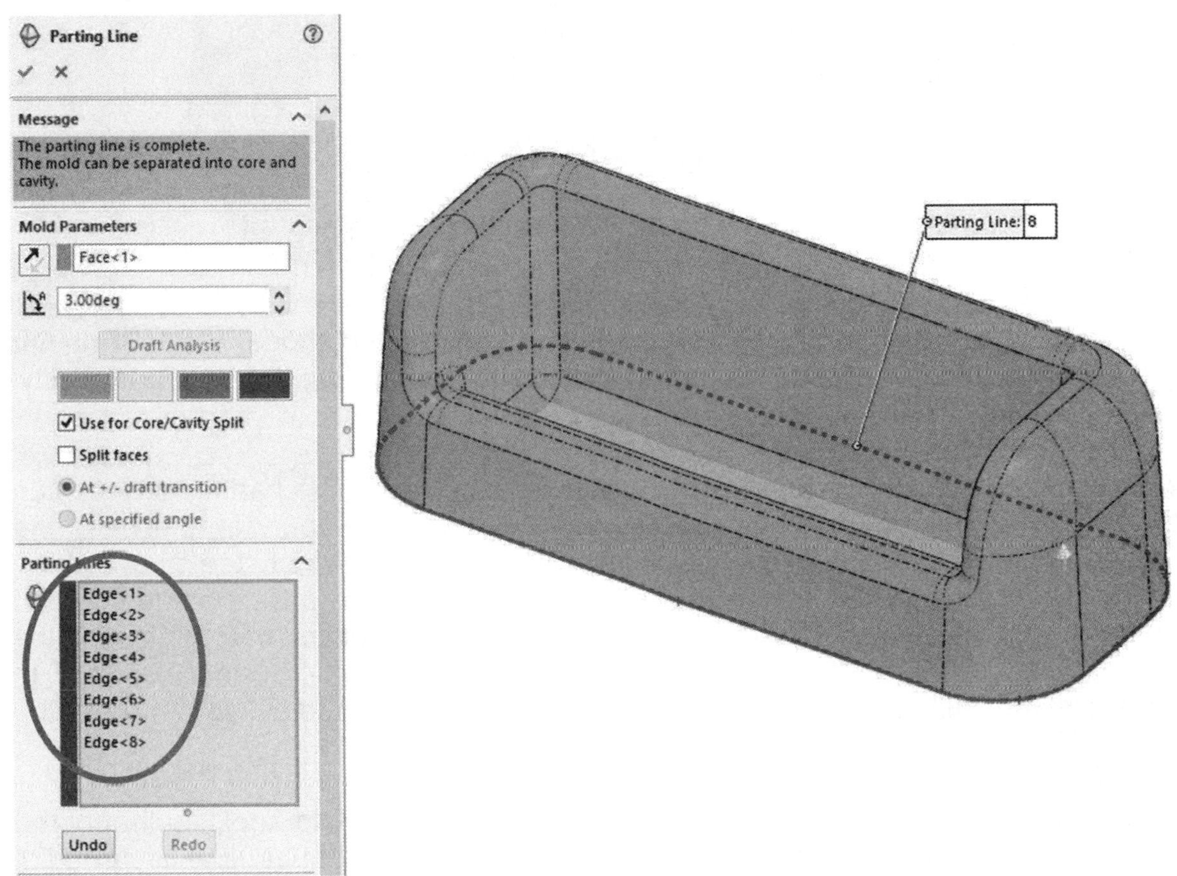

26.12. - After completing the parting line, the *"Parting Line1"* feature is added, and in the *"Surface Bodies"* folder two sub-folders are created, one with a surface for the Core and the other for the Cavity. These surfaces are automatically created with the "**Parting Line**" command. In this example, the *"Cavity Surface Bodies"* are the faces with a "Positive" draft (green) and the *"Core Surface Bodies"* are the faces with a "Negative" draft (red). The parting line remains visible on the screen and can be hidden like other features, if needed.

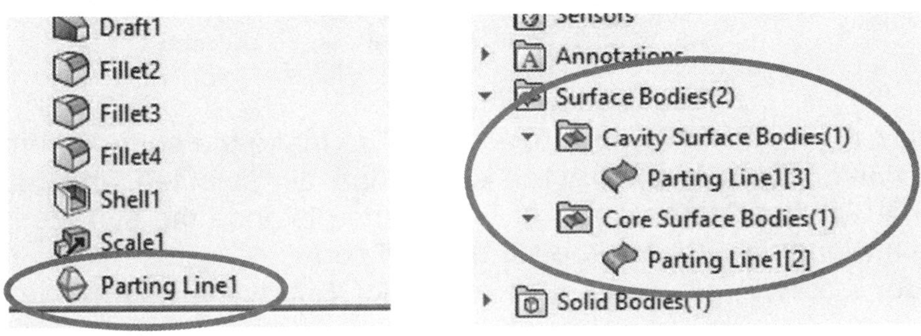

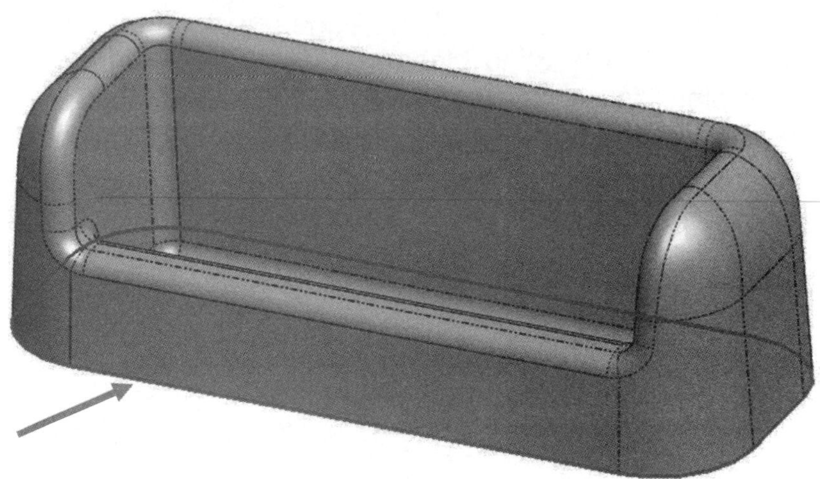

The next step would be to close any holes in the model with "**Shut Off Surfaces**," but since our part has no holes connecting the core and cavity, this step is not needed.

26.13. - The next step is to generate the parting surface. The parting surface can be generated manually or automatically and serves as a boundary between the Core and Cavity. Think of it as a surface used to split a block of steel in two parts, the core and the cavity. The parting surface is connected to the parting line and (generally) radiates away from it perpendicular to the "Direction of Pull" used in the parting line. The parting surface is used with the "Cavity Surface Bodies" to generate the mold's cavity, and with the "Core Surface Bodies" to generate the mold's core.

Select the "**Parting Surfaces**" command from the Mold Tools toolbar or the menu "**Insert, Molds, Parting Surfaces**." In this part the parting line lies in a flat surface, and it's easy to generate.

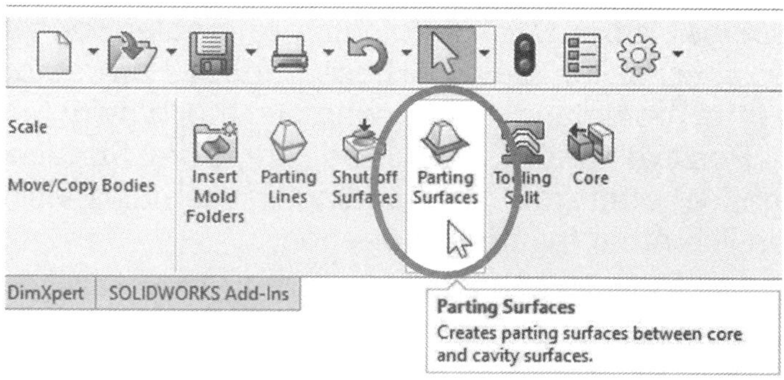

Select the option "Perpendicular to pull" to make the surface perpendicular to the direction of pull. The parting line is automatically selected, and enter a value of 1″ for the "Parting Surface" value. This is the distance the surface is radiated away from the parting line. The option "Knit all surfaces" automatically merges all the faces generated into a single surface. Click OK to finish when done.

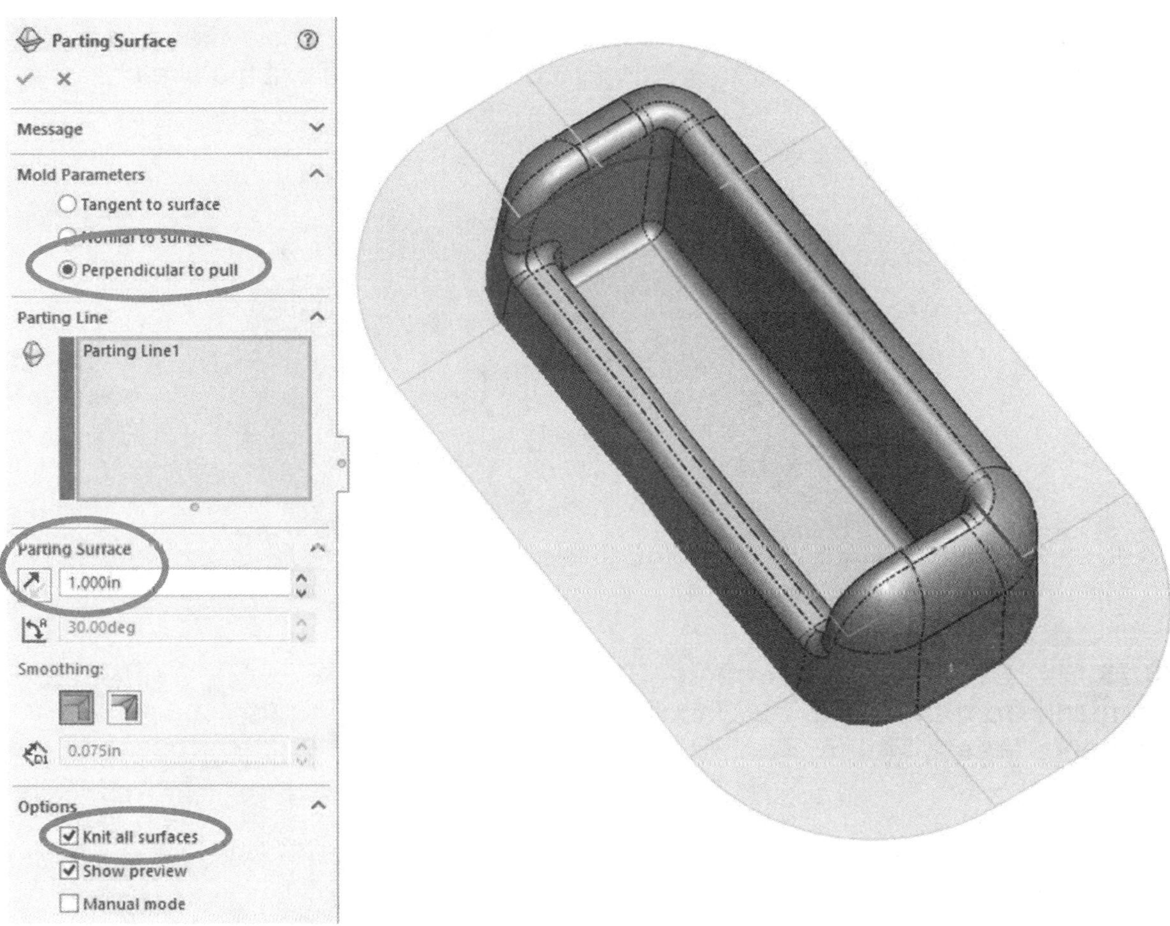

After the parting surface is made, a new folder is added to the *"Surface Bodies"* folder called *"Parting Surface Bodies"* and the new surface is listed here.

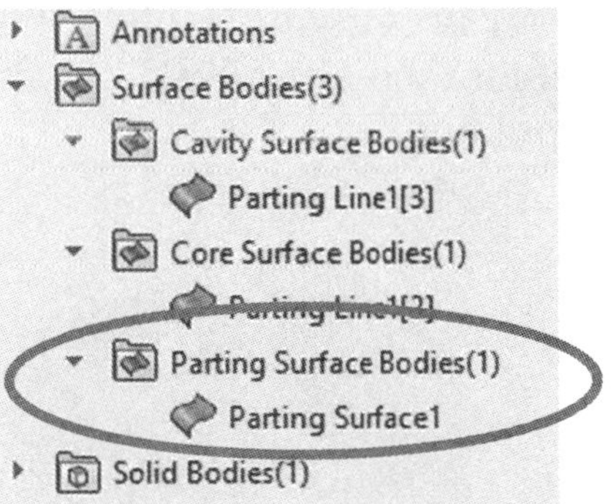

26.14. - The next step is to generate the tooling split. In this step, we make the solid bodies for the core and cavity. To make these bodies, we need to make a sketch perpendicular to the "Direction of Pull" that fits inside the parting surface; the reason is that the parting, core and cavity surfaces will be used to split this body into core and cavity. We can make the sketch first and then make the tooling split, or start the "**Tooling Split**" command and then make the sketch at that time. In this example, we'll make the sketch first.

Since the "*Parting Surface1*" is flat, select it and add a new sketch in it. Use "**Convert Entities**" to project the parting surface edges and exit the sketch.

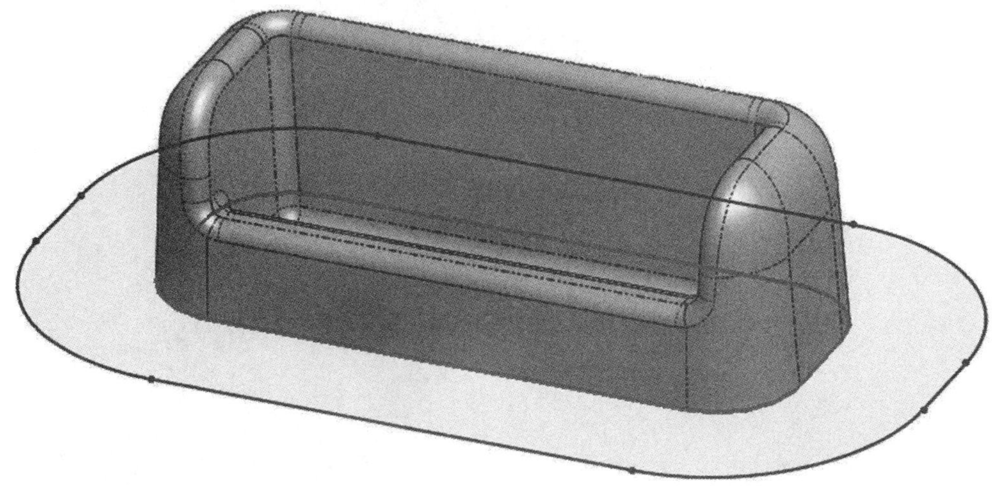

26.15. - Select the "**Tooling Split**" command from the Mold Tools toolbar or the menu "**Insert, Molds, Tooling Split**."

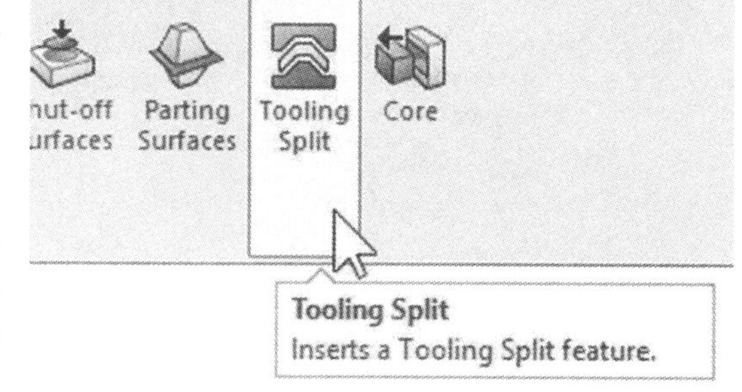

The "**Tooling Split**" command is enabled only after a parting line feature has been added to the model.

When we use the "**Tooling Split**" command, we are first asked for a plane or flat face to add a sketch, or an existing sketch (if a sketch is pre-selected this step is skipped). If asked, select the sketch we just made in the graphics area or in the FeatureManager to continue.

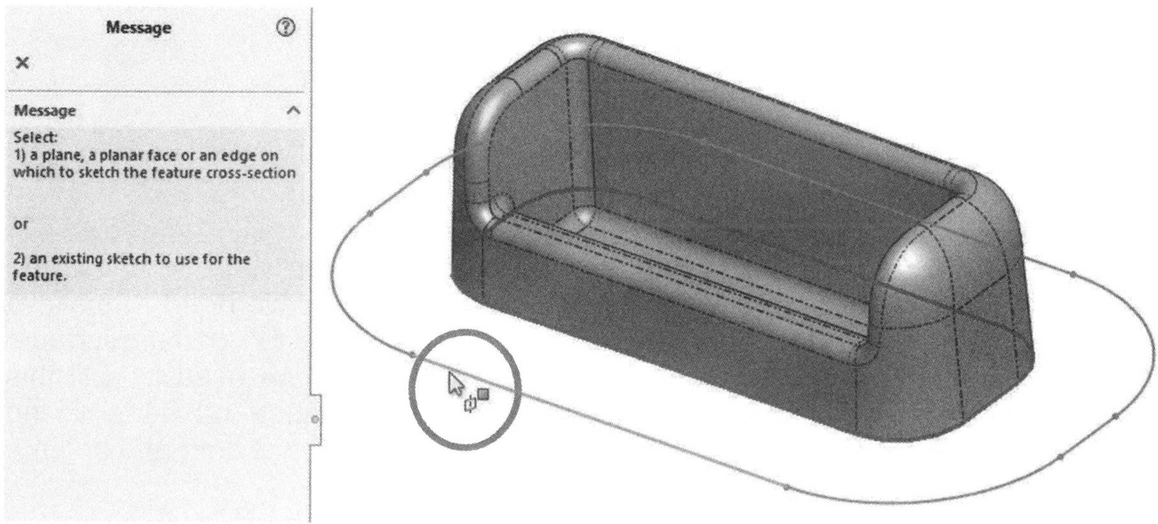

After selecting the sketch, we must enter the "Block Size" for the core and cavity measured from the sketch plane. In the "Core," "Cavity" and "Parting Surface" selection boxes the corresponding surface bodies are automatically selected from the corresponding *"Surface Bodies"* sub-folders. Alternatively, we can manually select the surfaces if needed.

Enter a block size big enough to completely enclose the *'Card Holder'* part, in our case 2.25″ up and 0.5″ down will be enough. Click OK to finish.

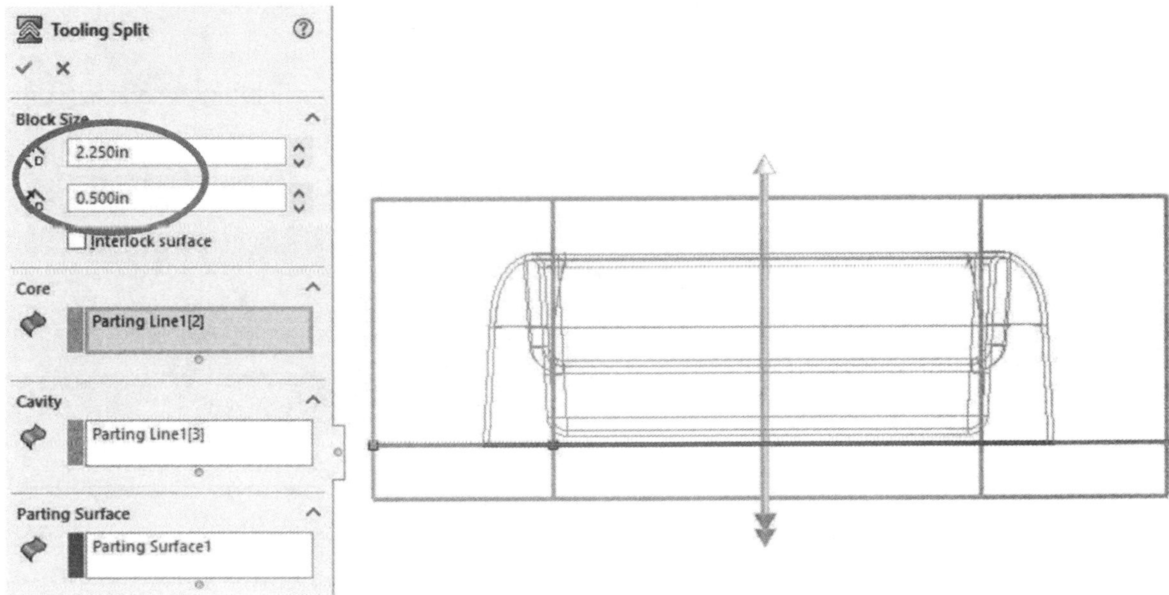

Knowing the size of the mold base that will be used, or the sizes available, will help when deciding how big to make the core and cavity. This way they will be big enough to fit in the mold base and not be bigger than needed.

26.16. - Now the core and cavity are finished. Hide the *"Parting Line1"* feature by selecting it and clicking in the "Hide" command.

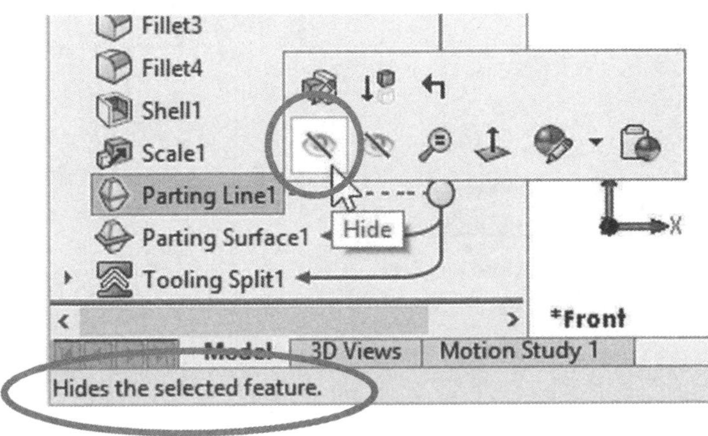

Now in the *"Solid Bodies"* folder we have three bodies: one is the part, one is the core and the third is the cavity. At this point, the core and cavity bodies can be saved into a new part file to continue designing the mold and adding the rest of the mold specific features, like cooling lines, injection ports, ejector pins, etc., using the same approach we used when working with a master model.

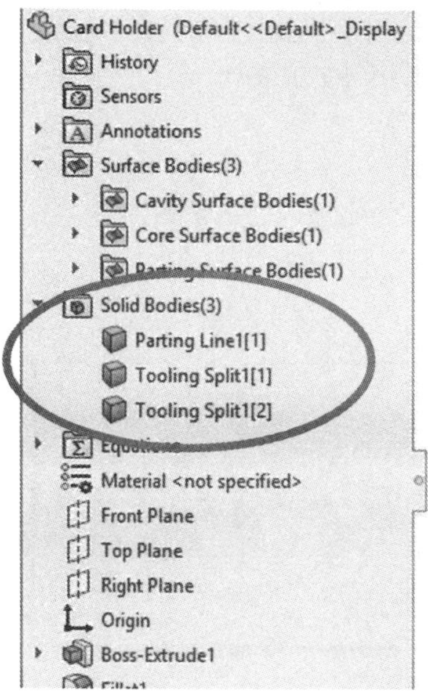

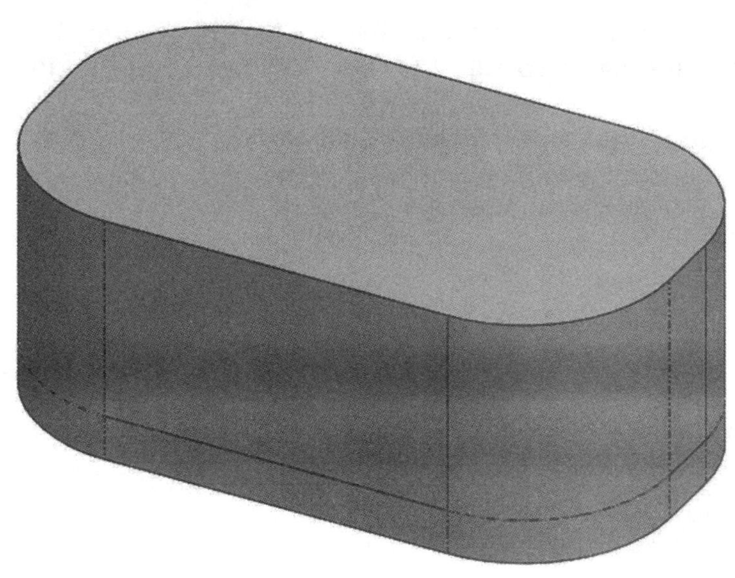

To finish our design, save the core and cavity bodies to a new part using the "Insert into New Part" command from the right-mouse-button menu. Name them *'Card Holder Core'* and *'Card Holder Cavity'* to continue. When finished save and close all the parts.

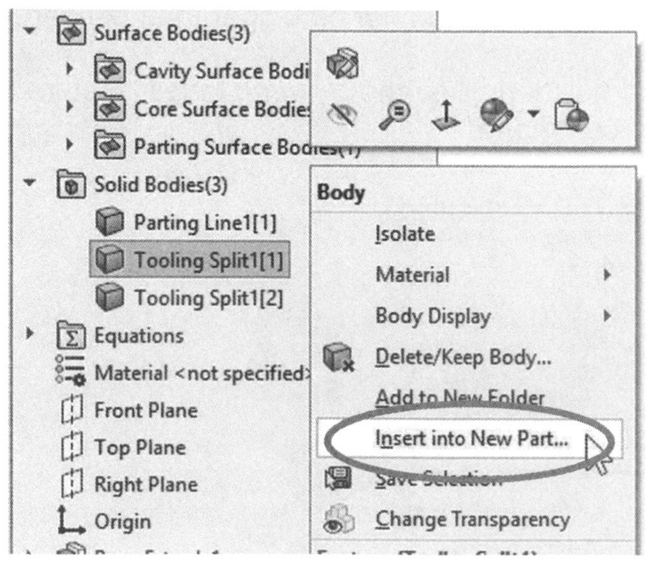

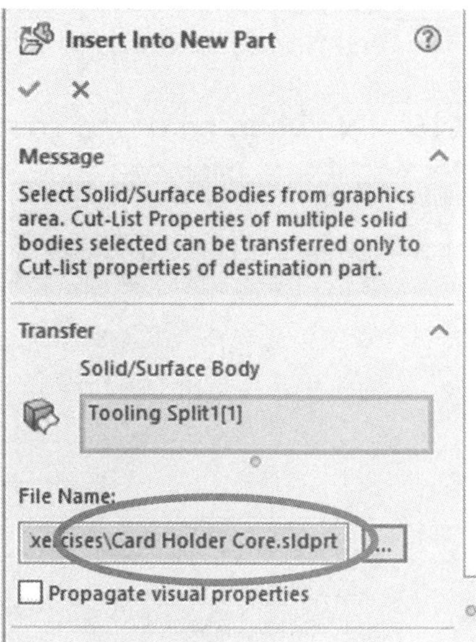

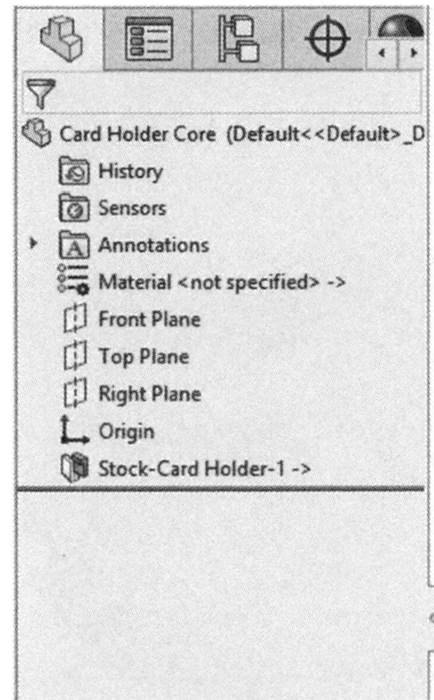

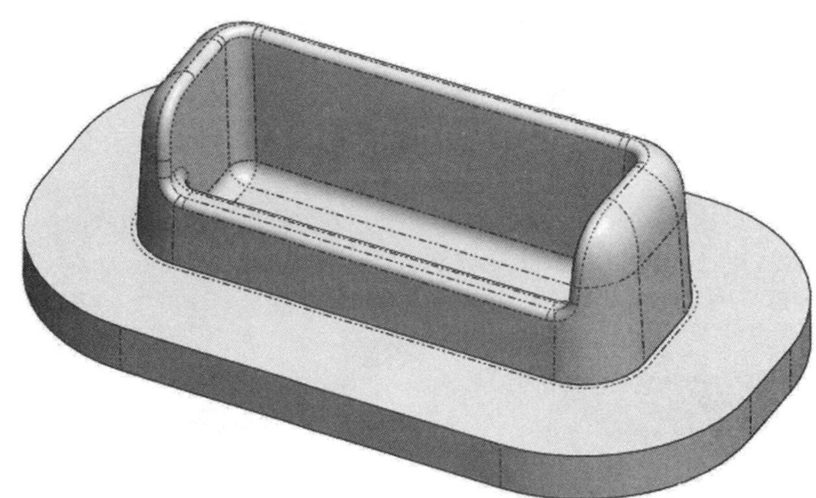

Card holder mold's Core part

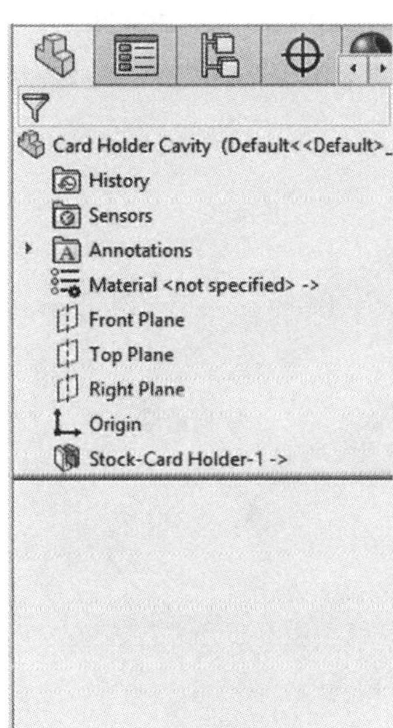

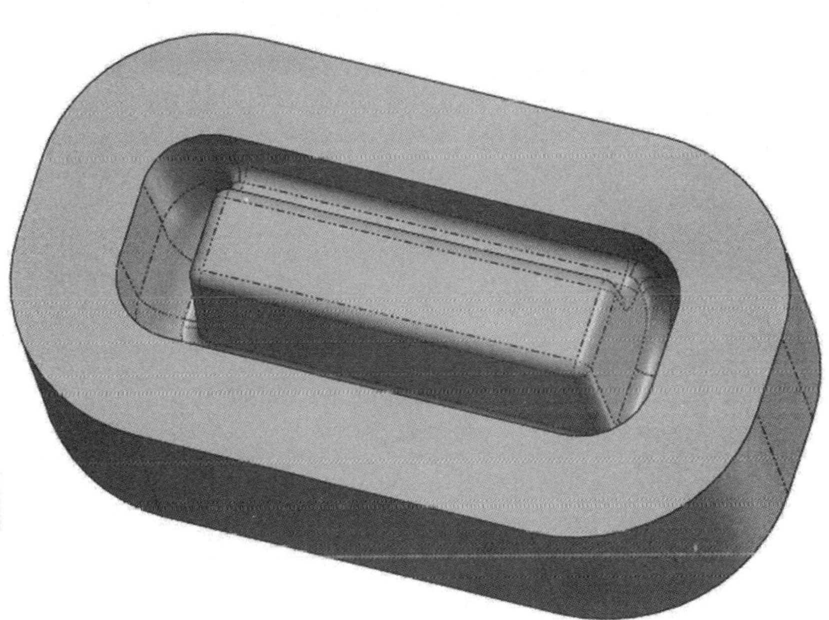

Card holder mold's Cavity part

In the original part, the solid bodies and surface bodies can be hidden to see the result before saving the bodies to a new part file.

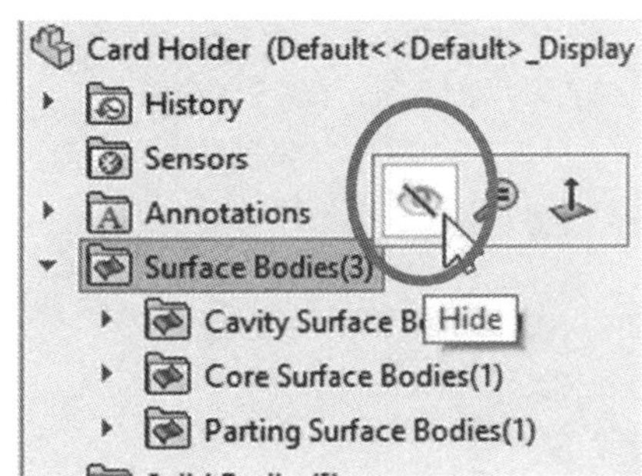

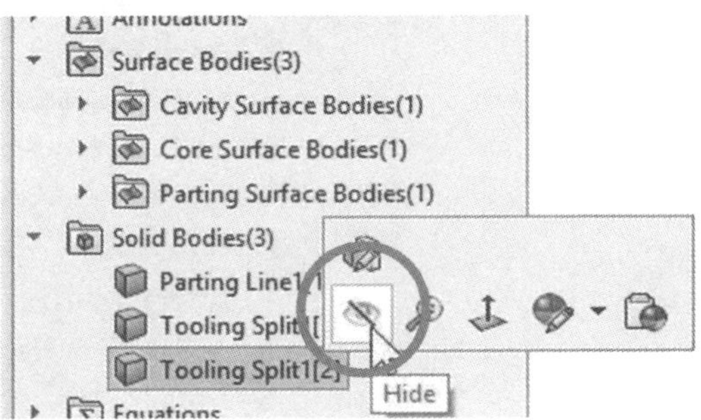

- Original part and cavity bodies hidden. Core's body color changed for visibility.

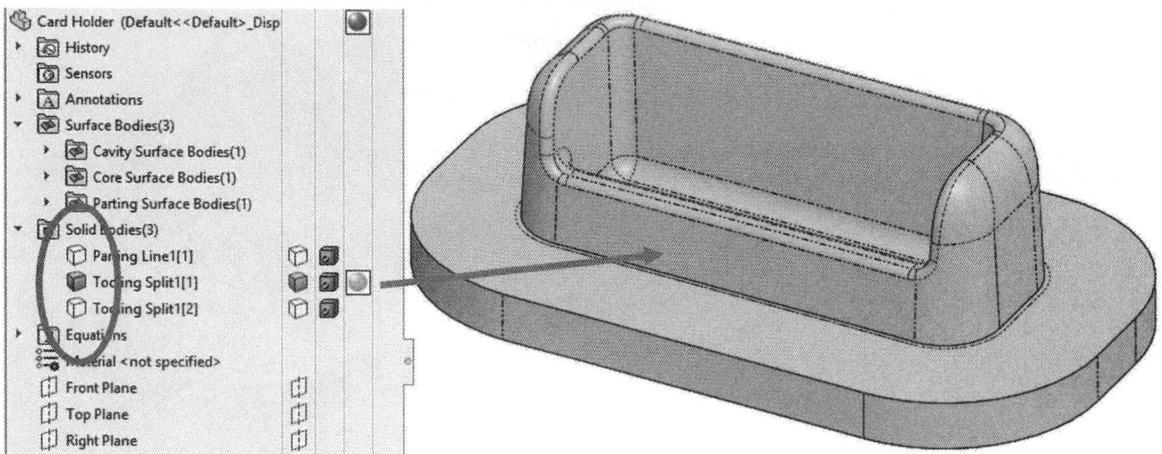

- Original part and core bodies hidden. Cavity's color changed for visibility.

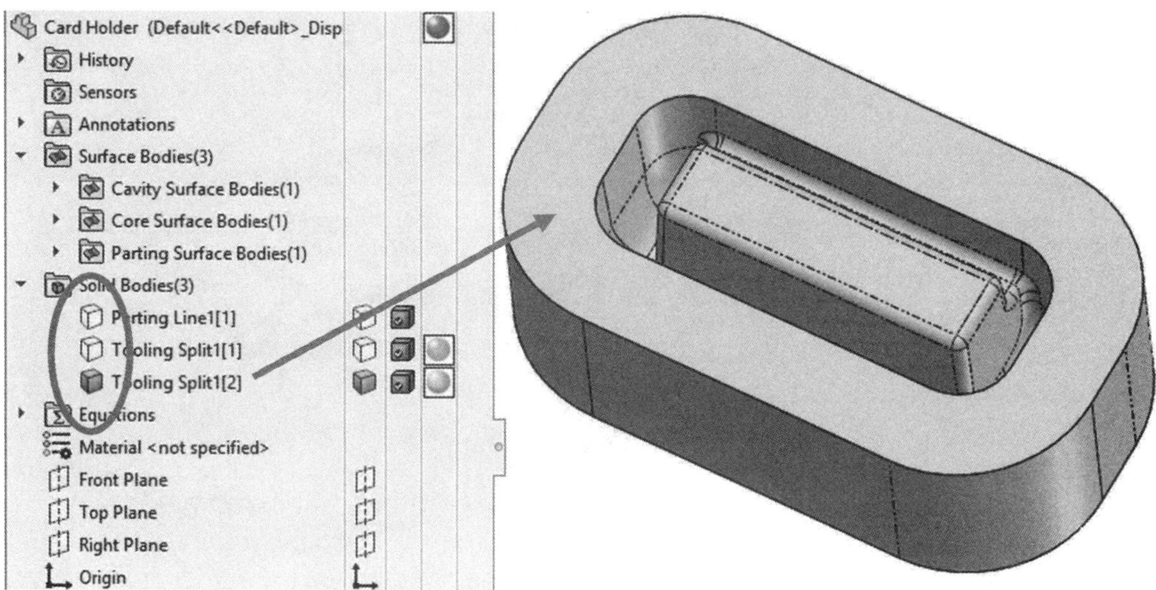

Solid and surface bodies' appearance can be changed in the Display Pane (shortcut F8).

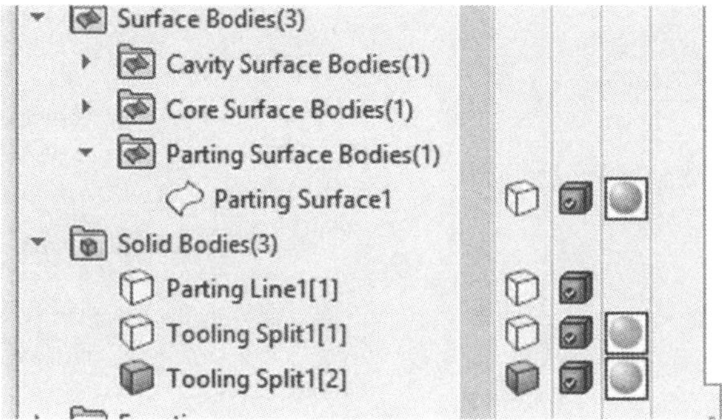

Hair Drier Cover Mold

In the next example, we'll make a mold for the back cover of the hair drier. To make the mold for this part, we must:

- Make a draft analysis to verify if the part has the necessary draft.
- Add draft to the faces that don't meet the minimum draft requirement.
- Add a parting line, a parting surface and shut off surfaces.
- Split the mold into core and cavity.

27.1. – Locate and open the *'Hair Drier Cover-Mold.sldprt'* part from the included files.

This part's external references have been broken to avoid potential conflicts with user generated files.

27.2. - This part is similar to the card holder, in the sense that it can also be made with a simple two-part mold; the difference is this part has holes going through it (vents) that the previous part did not. With this part, we'll learn how to work with a part that has holes in it. The first thing we need to do is to verify if our part's faces have at least three degrees of draft. (Remember this value is arbitrary for our exercises and is not representative of any material.)

Select the "**Draft Analysis**" command from the Mold Tools tab and select the *"Right Plane"* for "Direction of Pull."

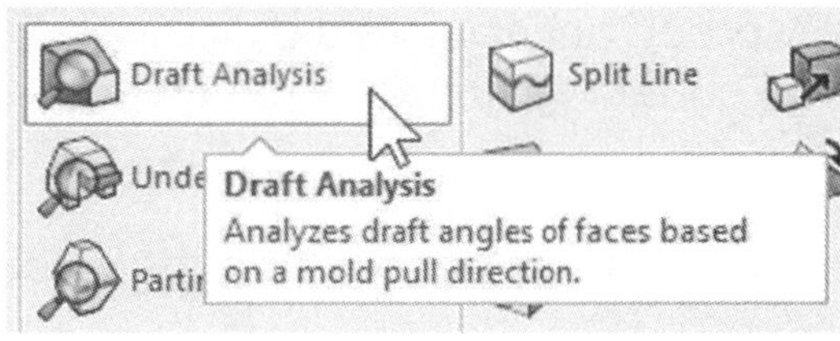

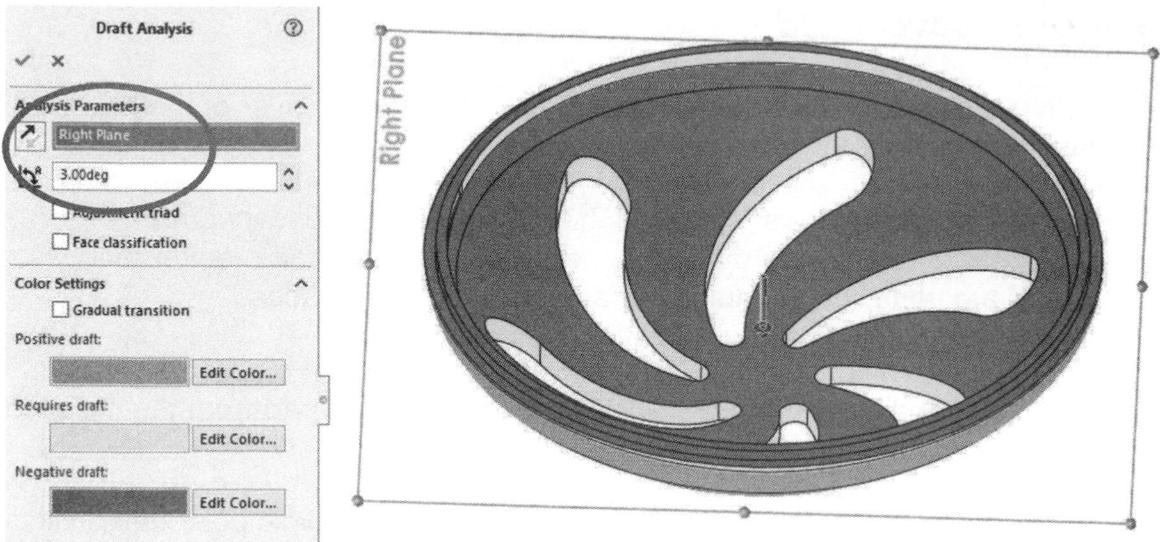

From the analysis, we can see that the ventilation holes and the lip built to assemble the cover don't have the required three degrees of draft, and therefore, we need to correct this before we can make a mold. Click Cancel to continue. (If we click OK, the Draft Analysis colors will remain visible.)

27.3. - To add a draft to the inside face of the "*Lip1*" feature, we must make two steps before. In the first step, we'll add a face fillet; the reason to do this is because the inside face is broken into multiple regions, and generating a single face will allow us to more easily fix this problem. In the second step, we'll add a split line to the previous face fillet, allowing us to add the required draft.

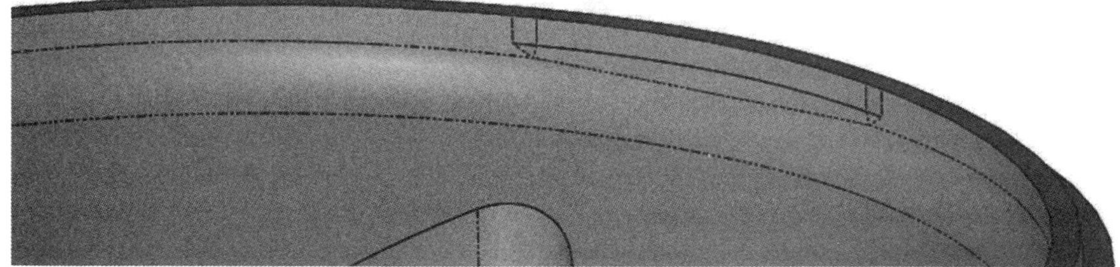

For the first step select the "**Fillet**" command; under "Fillet Type" select the "Face Fillet" option and add the indicated faces to the "Face Set 1" and "Face Set 2" selection boxes. In the "Fillet Parameters" section use the option "Hold Line," and select the inner edge at the top of "*Lip1*." With this option the selected edge becomes the fillet's boundary, and its radius is defined by the faces to fillet and the hold line. An advantage of using a face fillet is that the intermediate faces between the two selections are absorbed and replaced by a single fillet surface, eliminating the previously sub-divided surface giving us a clean face to work. Click OK to add the fillet and continue.

The edge in the hold line must be touching the selected face for the fillet to work.

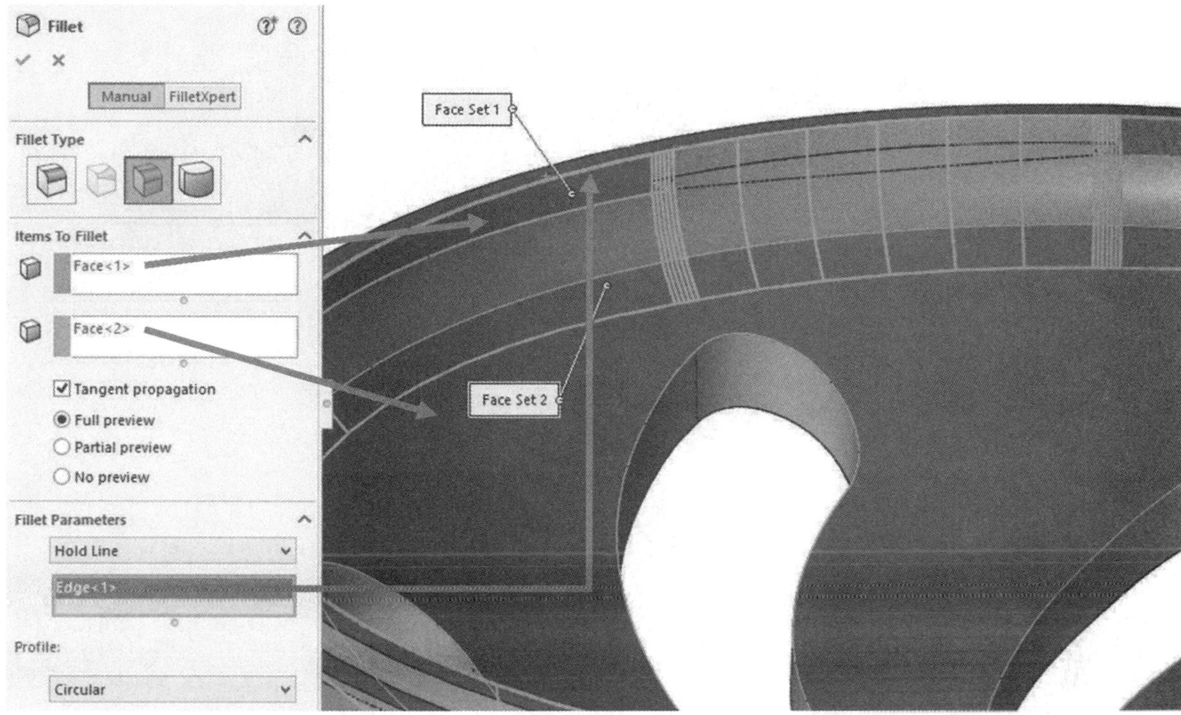

After adding the face fillet, the small faces inside are absorbed by the new fillet.

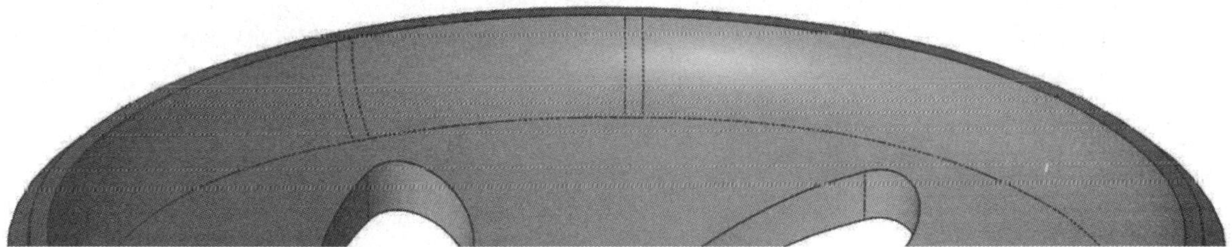

27.4. - In the second step we'll add a parting line to split the new fillet, and then we'll be able to draft the top faces. Add a new sketch in the "*Top Plane*" and project the indicated outside edge using the "**Convert Entities**" command.

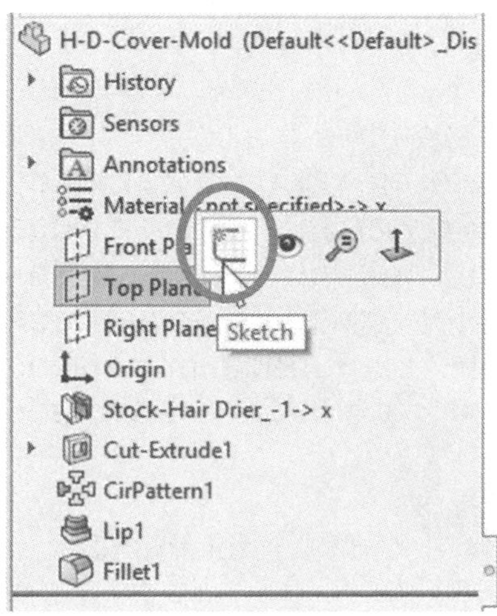

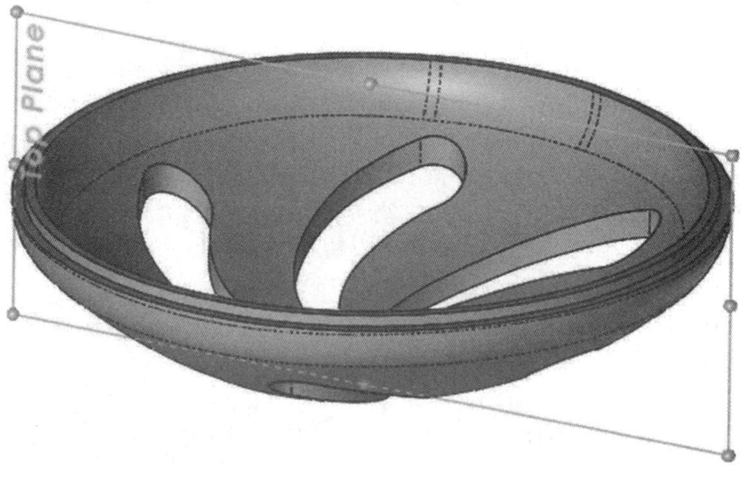

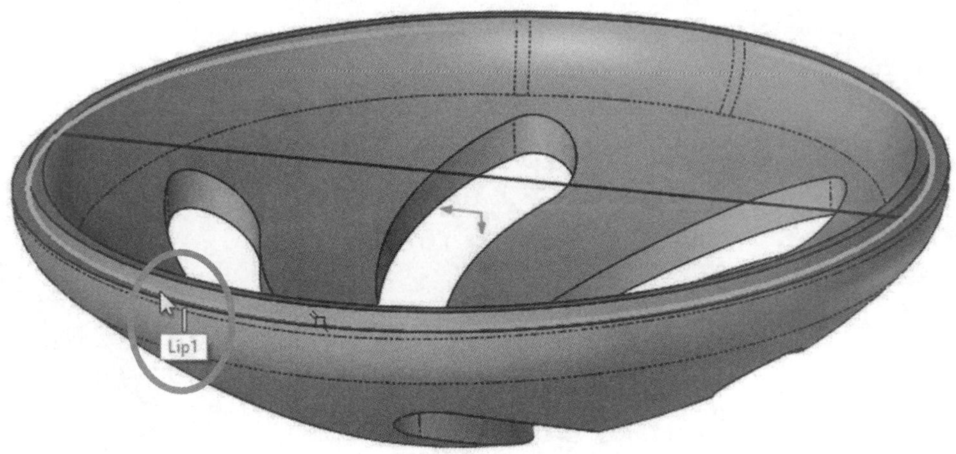

After projecting the edge in the sketch, select the menu "**Insert, Curve, Split Line**." The sketch we are working on will be pre-selected; using the "Projection" type, add all the faces of the new fillet in the "Faces to Split" selection box. Click OK to split the faces and continue.

The resulting fillet's split faces look like this:

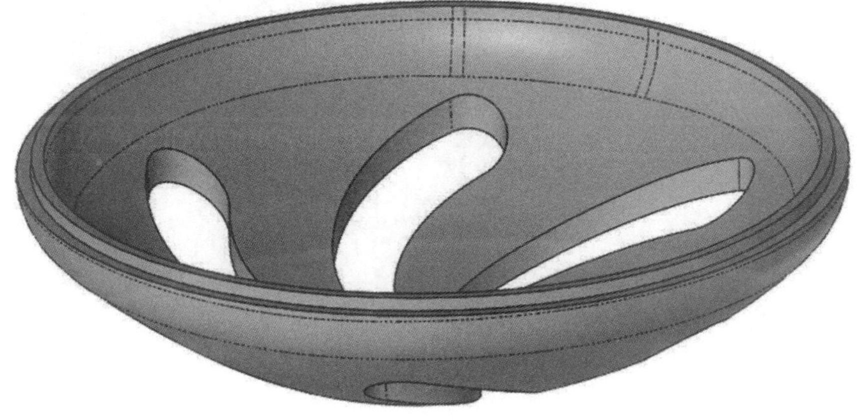

27.5. - After splitting the fillet's face, select the "**Draft**" command from the Mold Tools tab in the CommandManager. Select the "Parting line" type, and set the draft angle to 4 degrees. In the "Direction of Pull" select the "*Right Plane*" making sure the direction of pull arrow is pointing away from the part.

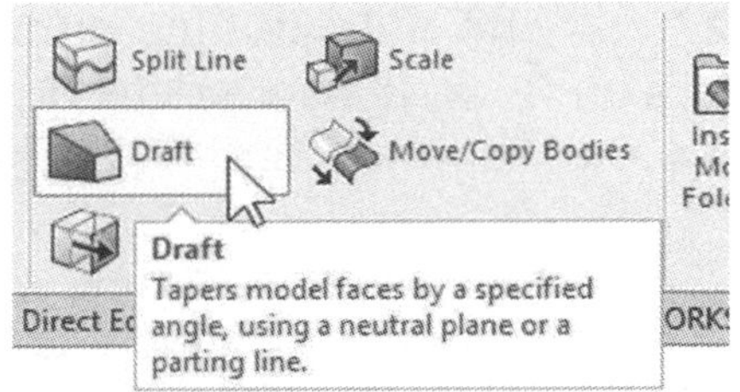

In the "Parting Lines" selection box, select the four edges previously created by the split line command; after selecting an edge make sure the orange direction arrow is pointing towards the face we want to draft. If it is not, click in the "Other Face" button to reverse it and continue selecting the rest of the parting line edges. When you finish selecting the edges, click OK to add the draft.

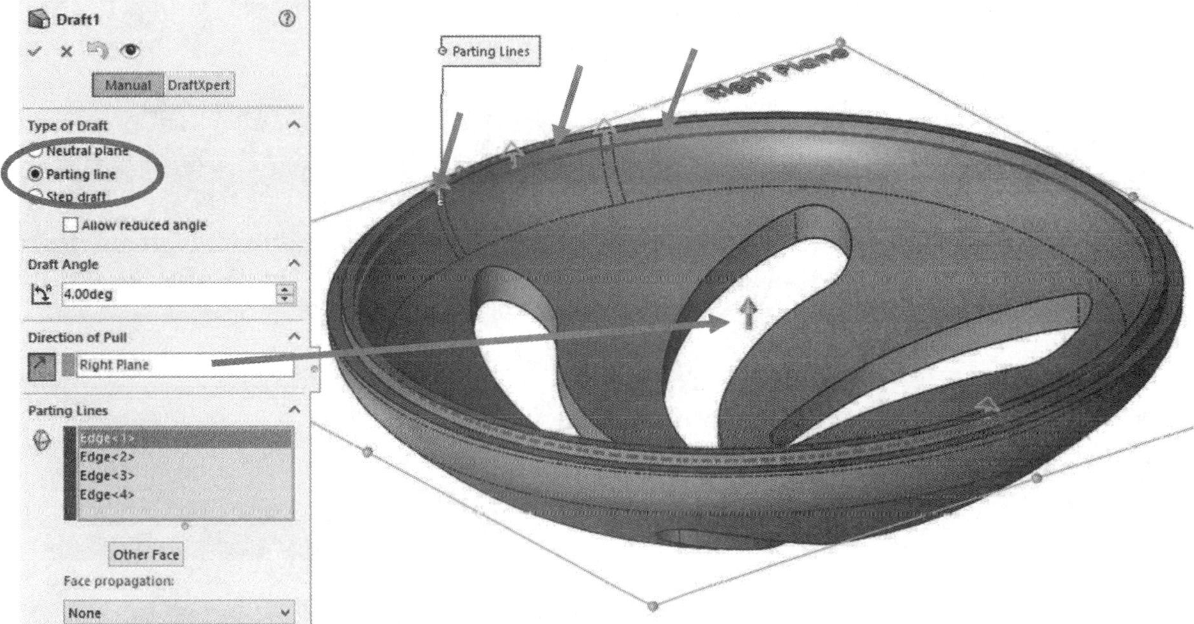

27.6. - Making a "**Draft Analysis**" using the "*Right Plane*" for "Direction of pull" we see the inside faces of the lip have the required draft, and the outside face shows a small cylindrical face that needs to be drafted, as well as the ventilation holes.

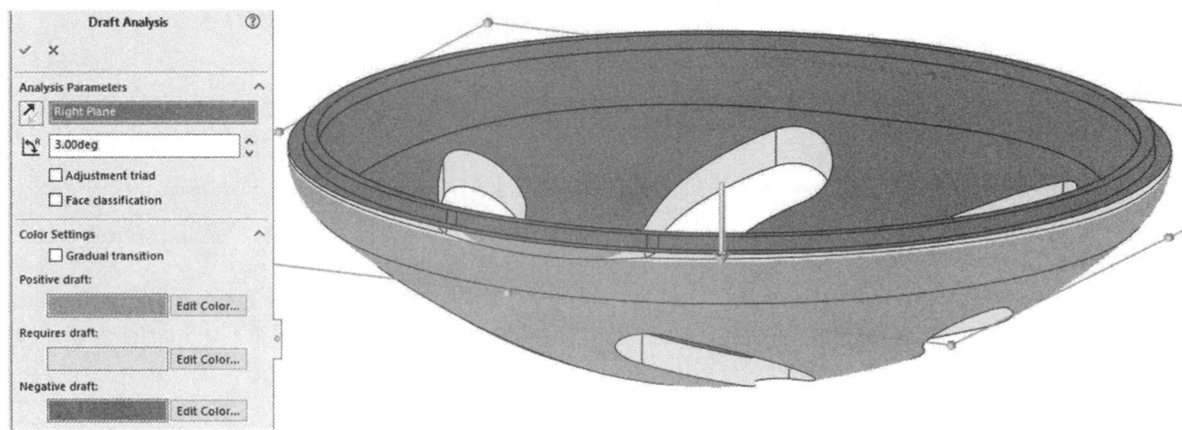

27.7. - The outside face cannot be fixed as the previous one using a face fillet. In this case, we'll use a different approach: we'll delete the faces, add a new one with the correct draft, and finally knit the faces into a solid.

Just as we can delete surface body faces, we can delete solid body *faces*. In a solid, when we delete a face, the rest of the model is automatically converted into a surface body.

Select the "**Delete Face**" command from the Surfaces tab in the CommandManager, or the menu "**Insert, Face, Delete.**" Using the "Delete" option, locate and select the four outside faces that don't have the required draft and the fillet's face. Click OK to continue. After deleting the faces, we have a single surface body.

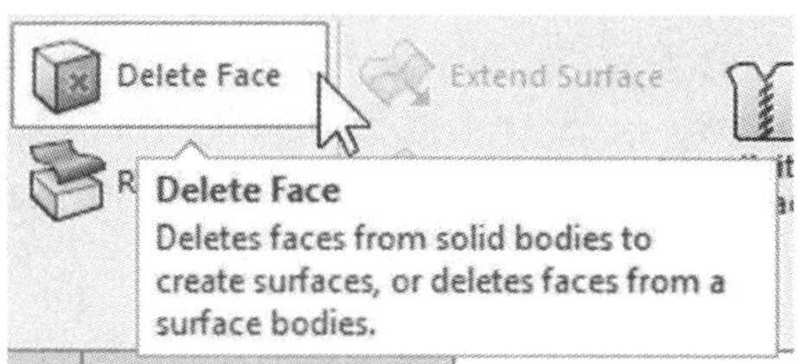

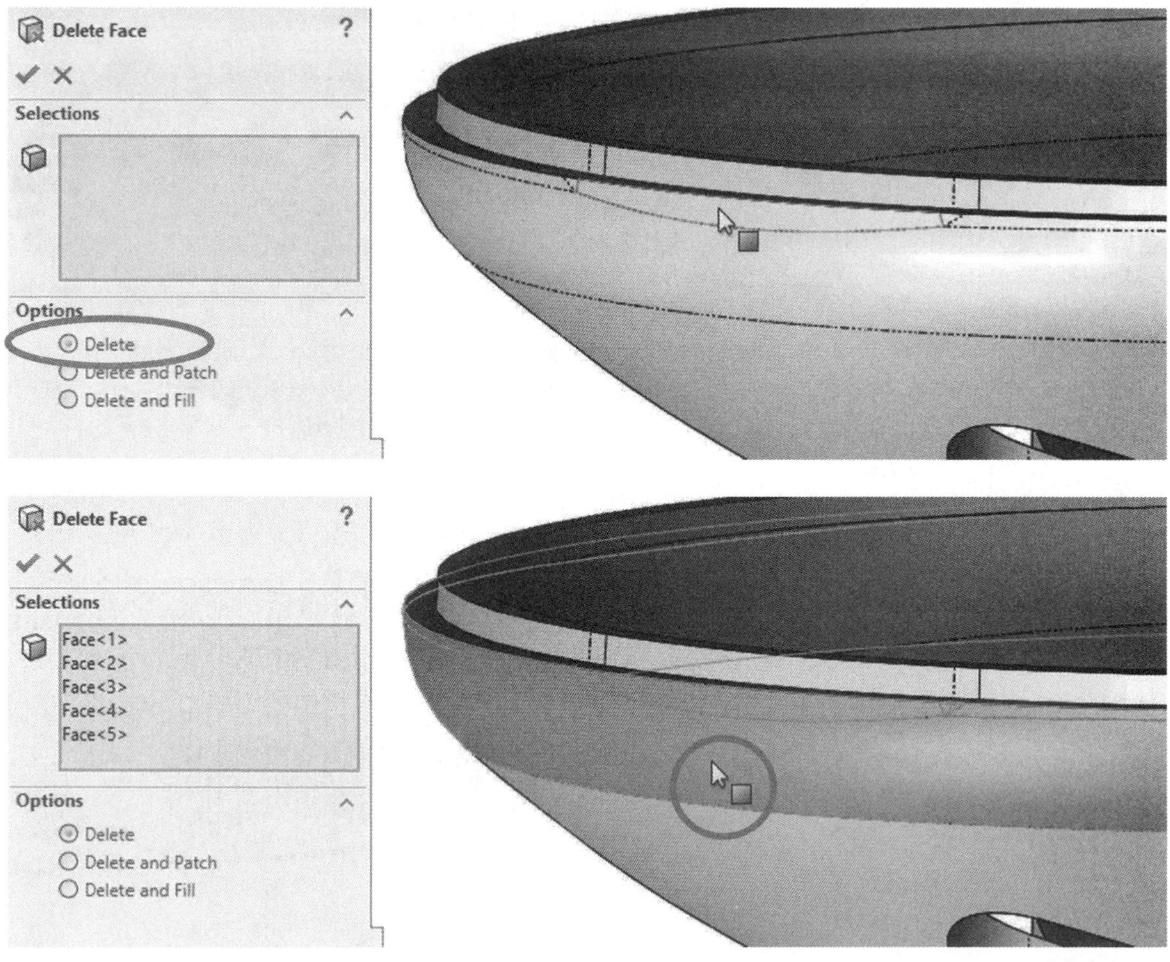

The difference between "Delete and Patch" and "Delete and Fill" option is that "Patch" will add faces and trim them to fill the gap. "Fill" will add a single surface to fill the gap. In our example, neither option produces the desired drafted face.

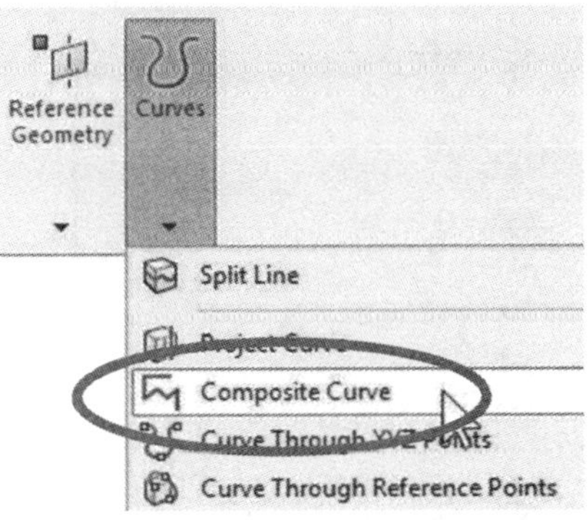

27.8. - The gap created by deleting the faces will be filled with a swept surface that will go around the part connecting both sides of the opening. When the previous faces were deleted the resulting edge is a combination of edge segments, and a swept surface needs a continuous edge or curve for either a path or a guide curve. Select the "**Composite Curve**" from the "**Curves**" command or the menu "**Insert, Curve, Composite**." To select all the edges around the surface to create the continuous curve we need, right-mouse-click in one edge of the upper surface, and select "Select Tangency." All of the edges tangent to the selected edge will create a single, continuous curve for our swept surface. Click OK to finish the curve and continue.

The other open edge is a single, continuous edge, and therefore we don't need to make a composite curve to use it as a path or guide for a swept surface.

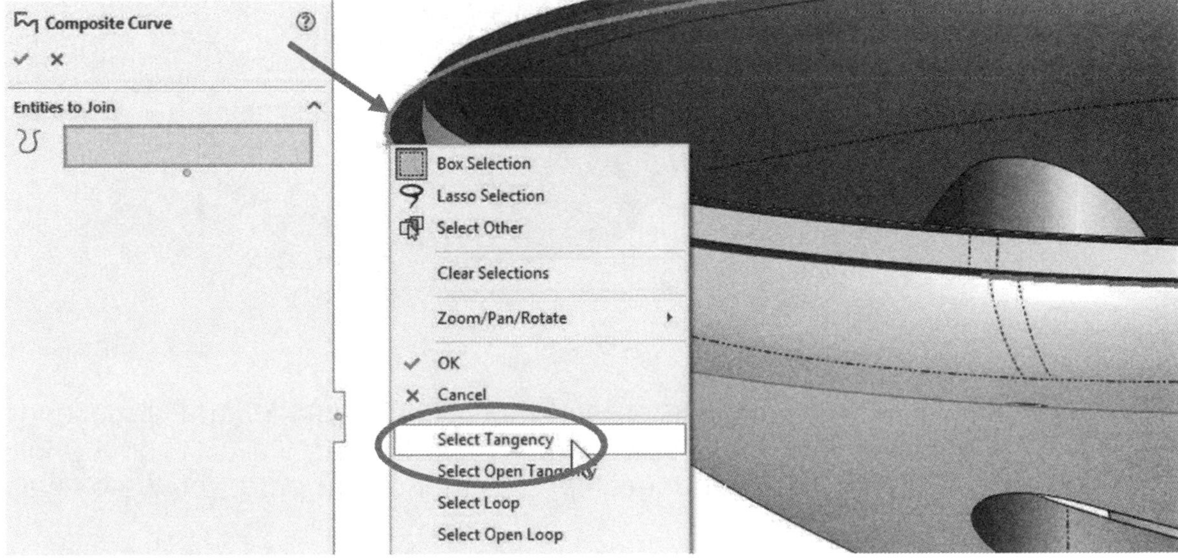

27.9. - Switch to a Right View, and press "Shift + down arrow key." Holding the Shift key while pressing the arrow key rotates the screen in 90 degree increments. This will give us the orientation we need to make the swept surface profile.

Add a sketch in the "*Top Plane*" and draw the following sketch close to the left side of the surface. The straight line and the arc are tangent to each other, and the construction line is added to add the angular dimension to the line, which will be the draft we need in the part.

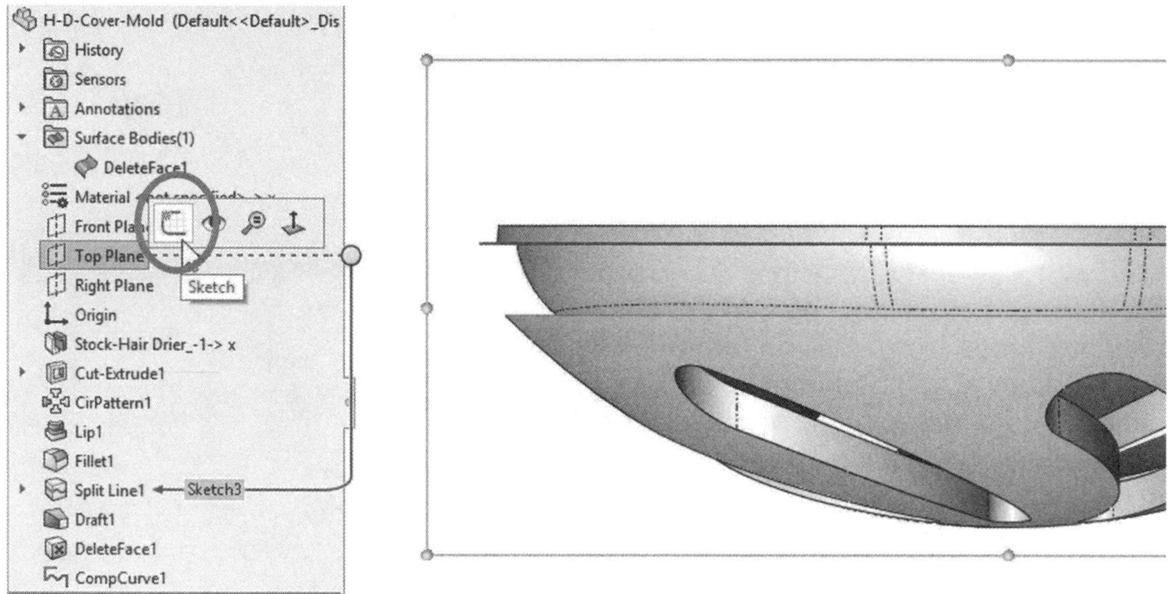

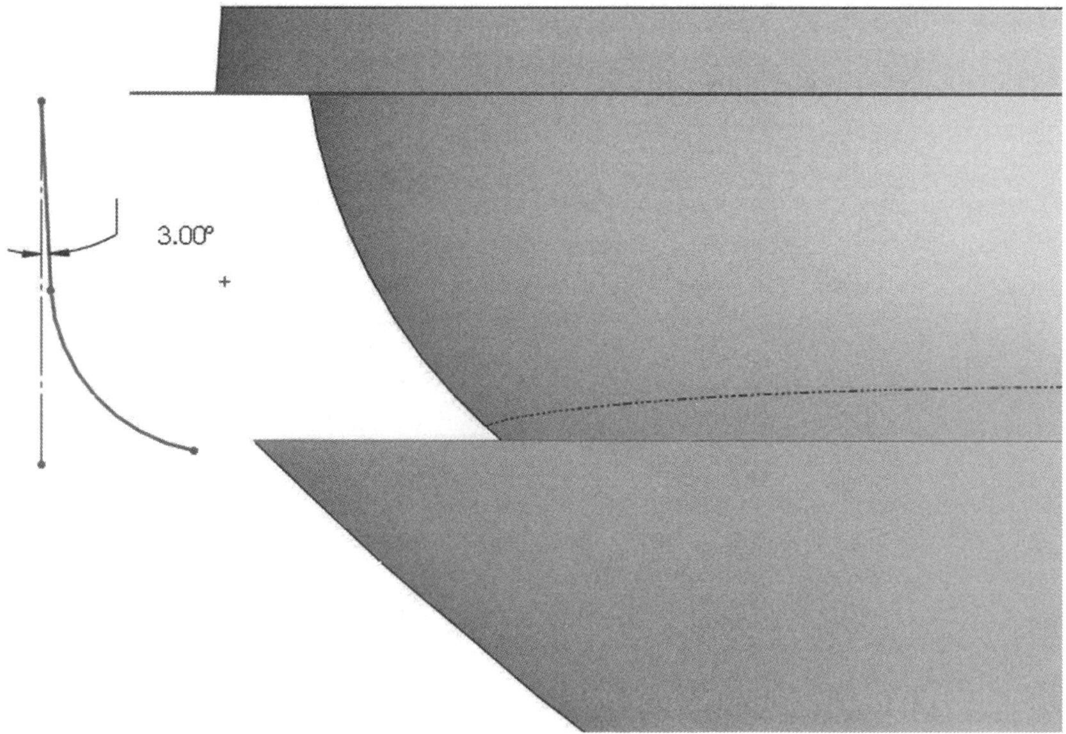

27.10. - To make the sketch arc tangent to the lower curved surface, we need to project the surface's silhouette onto the sketch. To do this we need to use the "**Intersection Curve**" command from the drop-down menu in the "Convert Entities" command. The intersection curve will add sketch entities at the intersection of the selected face and the current sketch plane, giving us the curve we need.

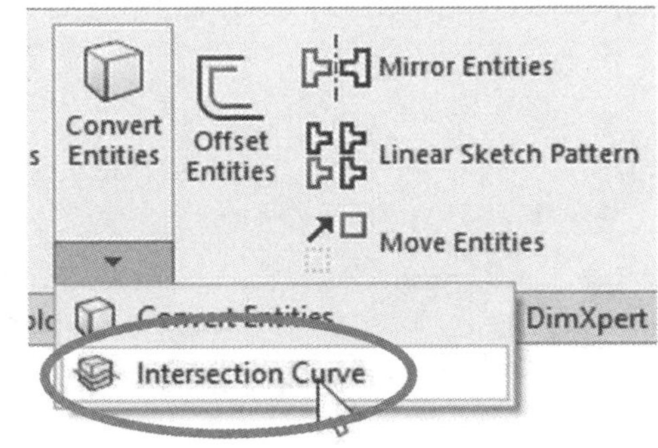

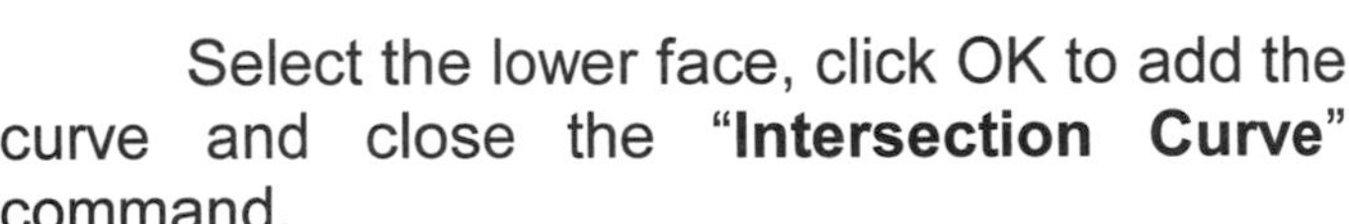

Select the lower face, click OK to add the curve and close the "**Intersection Curve**" command.

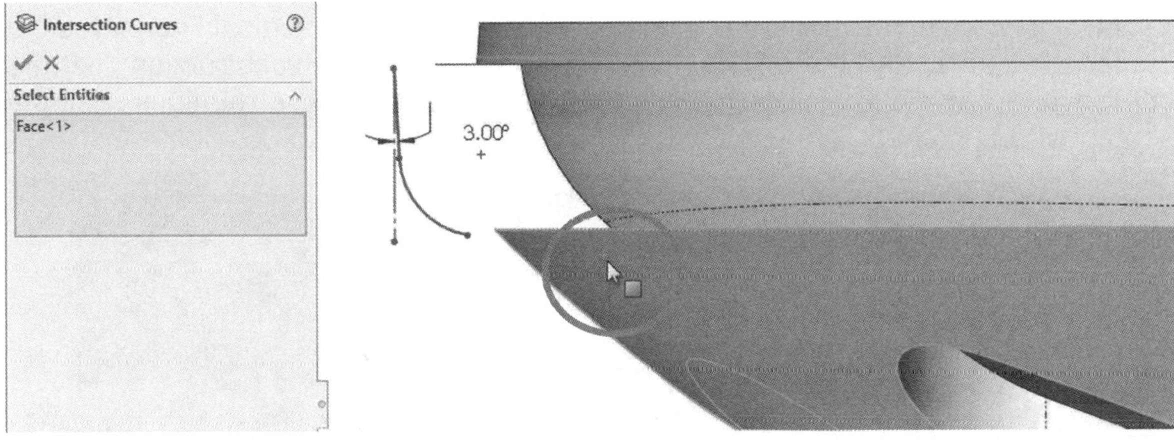

The resulting curves are added at all places where the selected surface crosses the sketch plane. Convert the segment closer to our sketch to construction geometry, and delete the other two segments, since they will not be needed.

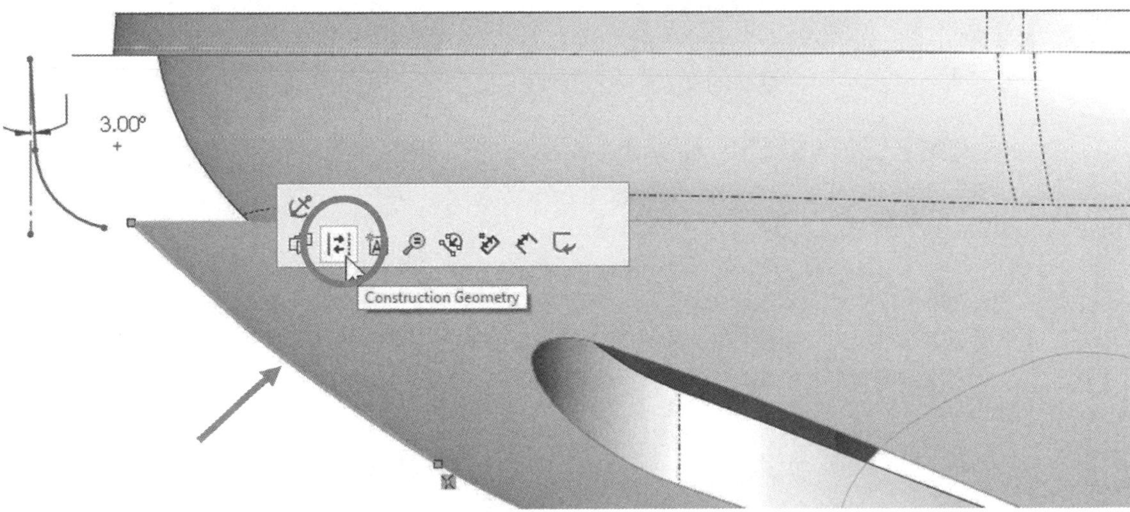

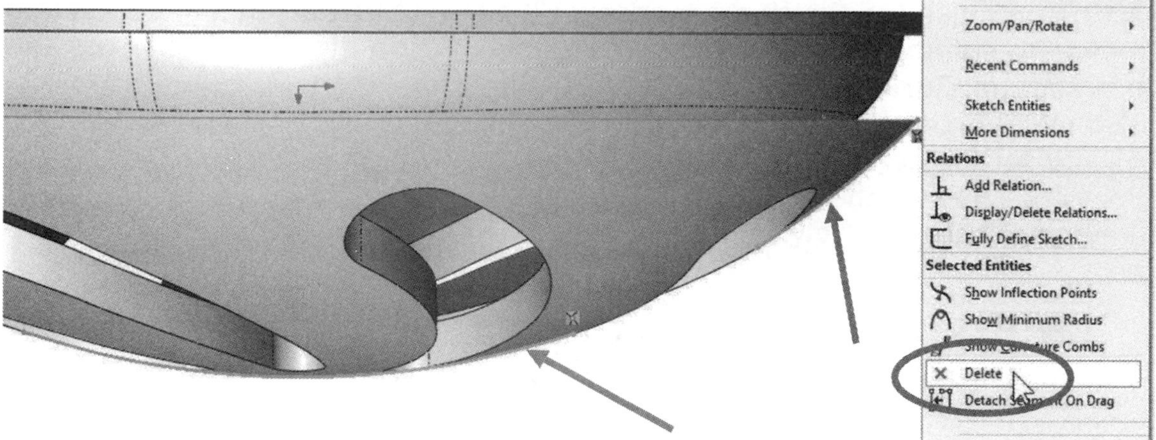

27.11. – To locate the sketch, make the top endpoint of the vertical line pierce the composite curve, and the arc coincident and tangent to the construction line (the intersecting curve). The sketch will be fully defined with these relations. Exit the sketch when finished.

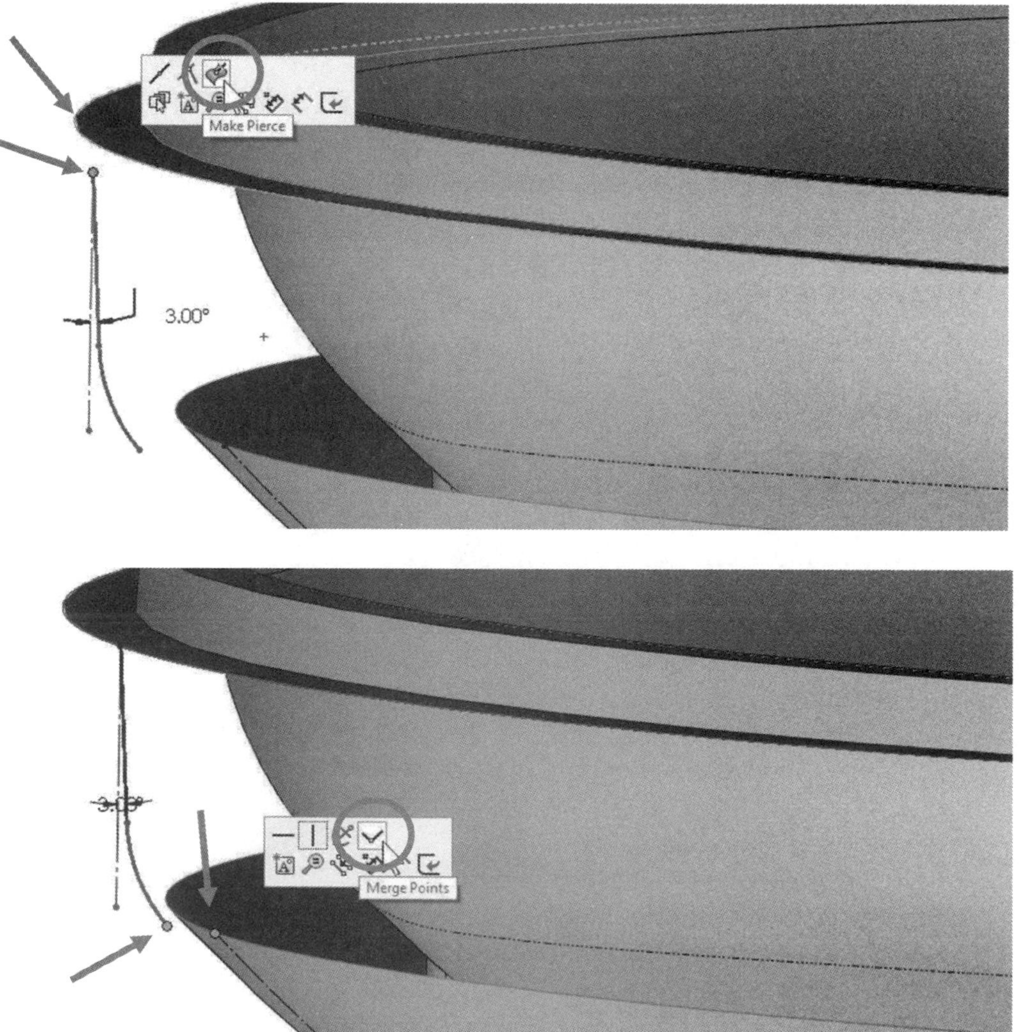

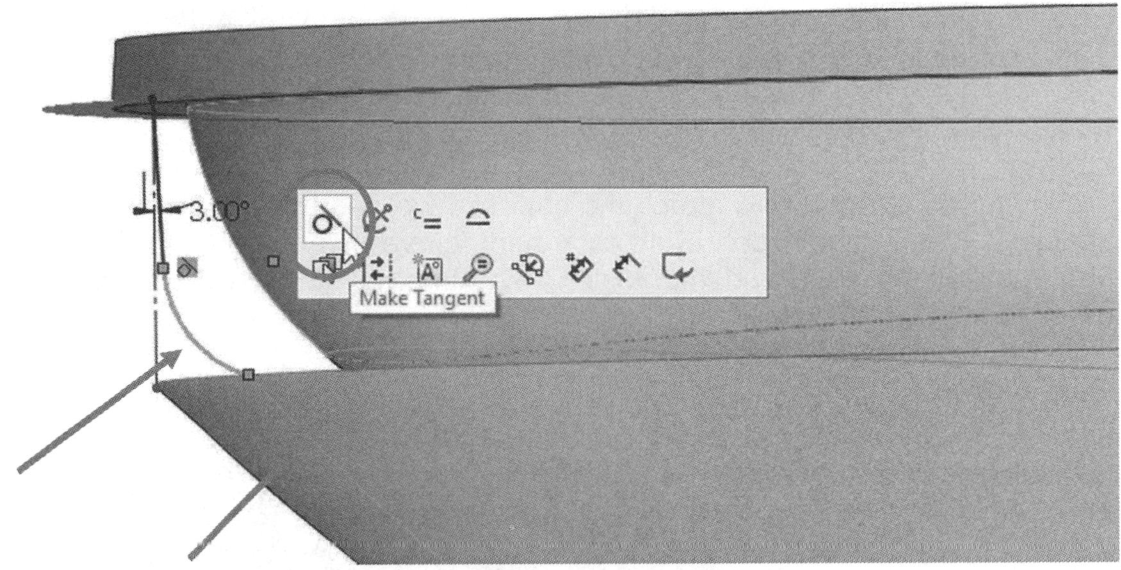

27.12. – After completing the profile sketch, select the "**Swept Surface**" command from the Surfaces tab, or the menu "**Insert, Surface, Sweep…**" using the previous sketch as a profile and the Composite curve as a path.

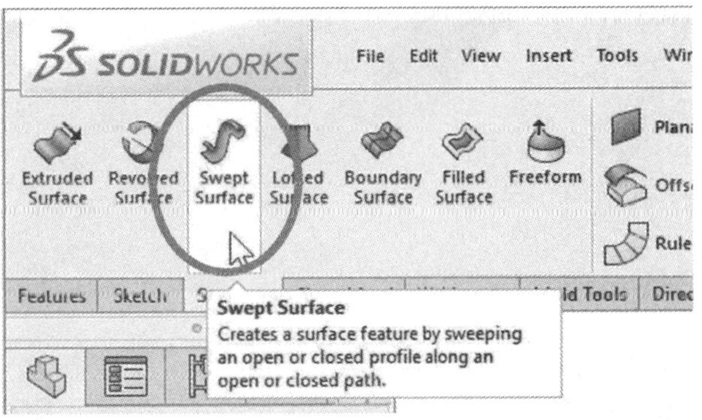

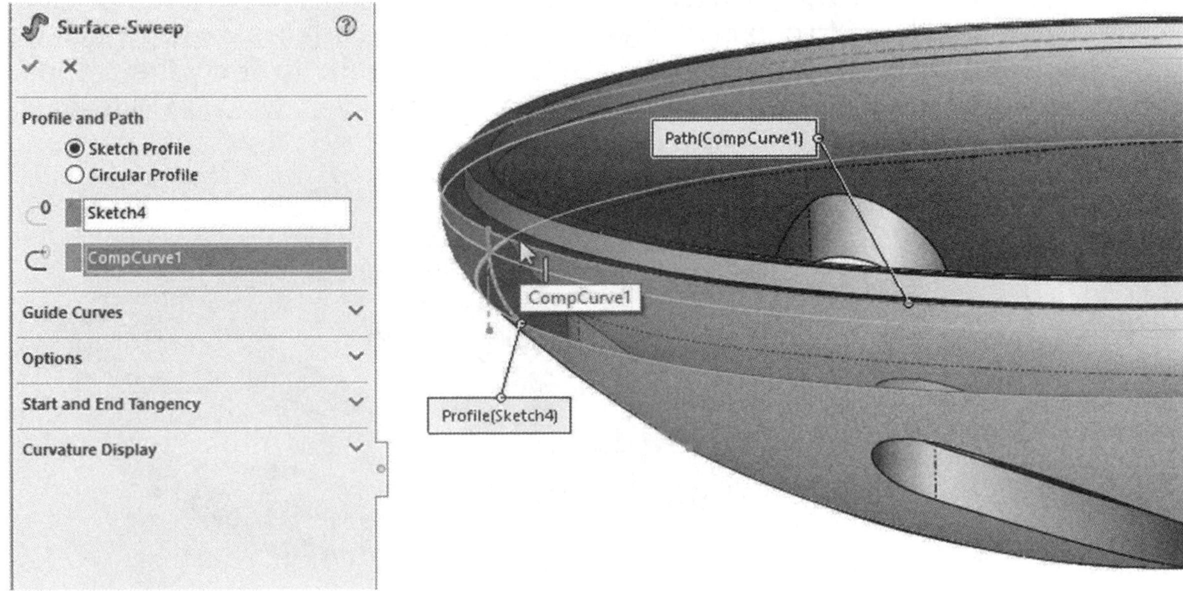

In this case, using the lower edge as a path instead of creating the composite curve first would give the same result, but instead we chose to use this opportunity to practice how to create and use a composite curve to take advantage of the learning opportunity.

In this example, to match the surfaces we need to use the "Guide Curve" option. Add the lower open edge as a guide curve and change the "Profile Twist" option to "Follow Path and 1st Guide Curve" to obtain the correct result. Click OK to finish the swept surface.

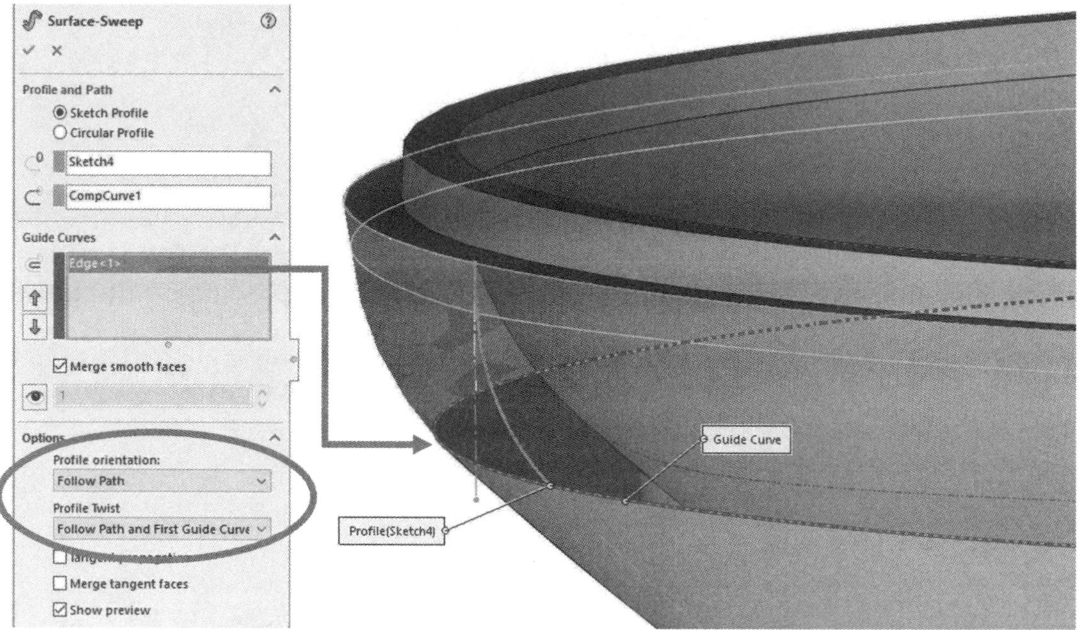

27.13. - After adding the swept surface, we have two surface bodies. Use the "**Knit Surface**" command to merge both surfaces into a single body. When the knitted surfaces form a closed volume, the option "Create solid" is enabled. Turn this option On and click OK to continue. After knitting the two surfaces we have a single solid body and no surface bodies. Make a "**Draft Analysis**" to verify the part has the correct draft in every face except for the ventilation holes, which we'll fix next.

27.14. – The next step in preparing this part to make a mold for it, is to add the necessary draft to the ventilation holes. We need to decide if they will be made in the Core or the Cavity, which will define the "Direction of Pull" for the draft. In this case, to make the mold easier to build, we'll add the holes to the Core side (male side) and make the Cavity part completely round inside.

If we edit the *"Cut-Extrude1"* feature and turn the "Draft" option On, the holes would become smaller than what we had originally intended because the cut feature starts in the face where the part was originally split, and adding a 3 deg. draft would shrink the holes too much as it moves farther away from the sketch plane. Cancel this change as this is not the result we are looking for.

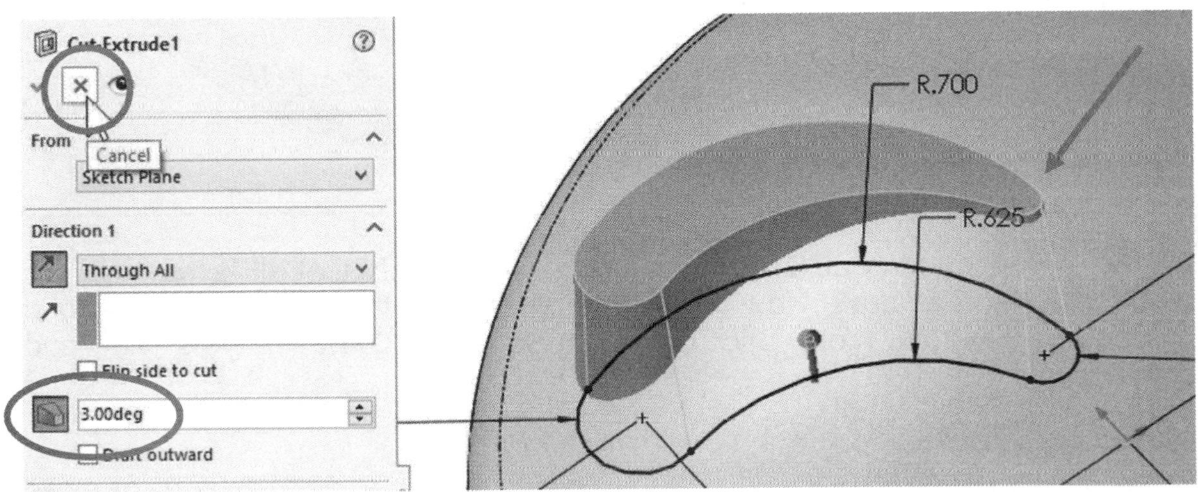

27.15. - To add a draft to the vent cuts, select the "**Draft**" command. Under "Type of Draft" select the "Parting Line" option. Adding the draft using this option will allow us to draft the faces starting at a selected parting line, which in our case will be the top edge of each hole. For the "Direction of Pull" select the "*Right Plane*" making sure the arrow is pointing down, as we intend to make the holes with the core side of the mold; this will be easier to visualize when the mold is finished. In the "Parting Lines" selection box add the top edges of each hole using the "Select Tangency" option.

After selecting all the edges make sure the arrows are pointing to the correct faces, otherwise the draft will be added in the wrong direction. If an arrow is pointing in the wrong direction, select the corresponding edge in the "Parting Lines" list and press the "Other Face" button to reverse it. Click OK to add the draft and continue. By adding the draft this way, the holes are wider inside and narrower outside, allowing us to make the holes in the Core side as we intended.

NOTE: As of the writing of this book, adding the draft feature to the first hole and using the pattern to copy both the hole and the draft features, or making a pattern using model faces fails, or produces the wrong result; that is the reason why we will add a draft to all the holes at the same time.

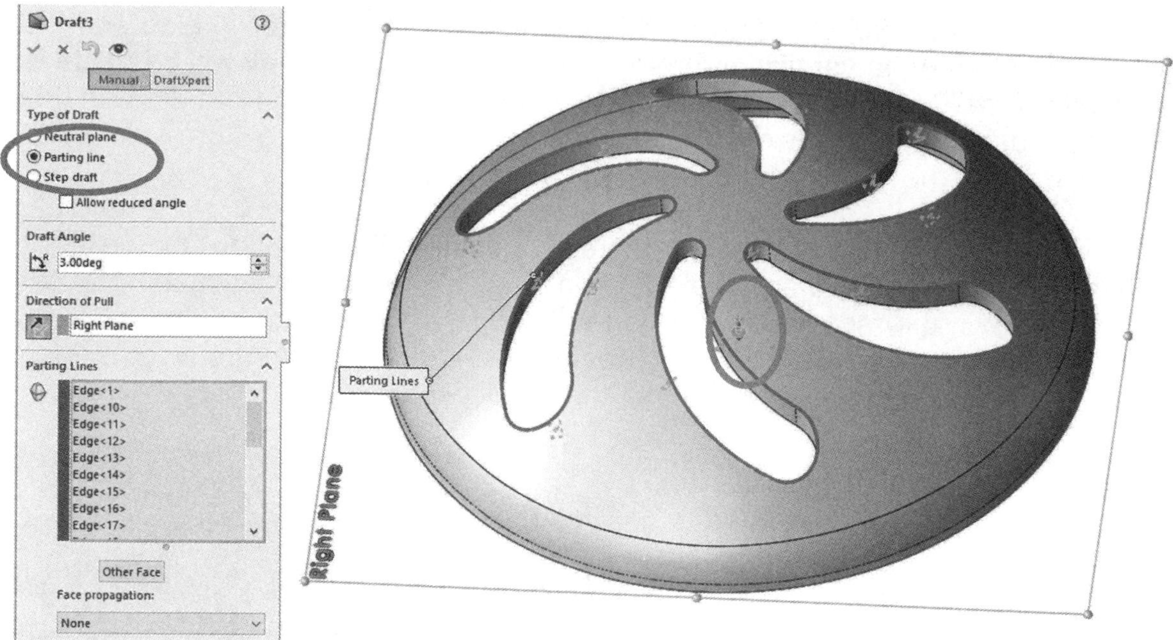

27.16. – After running a "**Draft Analysis**" we can verify that all faces have the required 3 degrees draft. Now that all the faces have the required draft we can proceed to make the mold. Click Cancel to close the "**Draft Analysis**" view and continue.

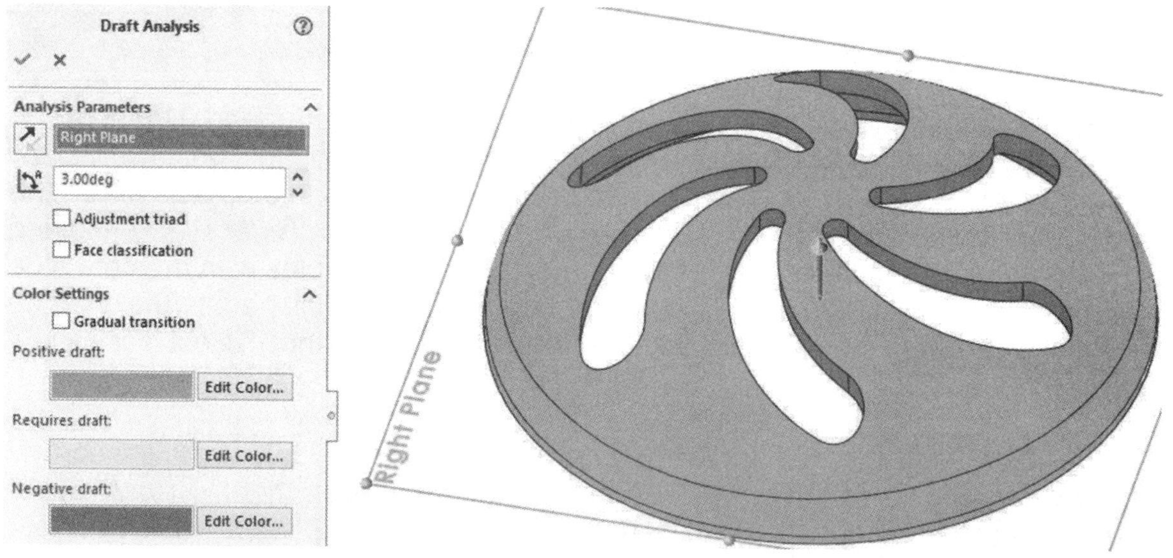

Sometimes it's difficult to identify which faces need draft. In these cases, we can use the "Face Classification" option of the Draft Analysis tool. Using this option allows us to hide or show faces by color, and gives us as a count for each case. Using this option a new color indicating the "Straddle faces" is added. Straddle faces are defined as faces that have both positive and negative draft, typically cylindrical or curved faces located on the positive and negative draft sides.

For the next image, some faces were not drafted to show how the "Face classification" works. To turn a group of faces On or Off, click in the Show/Hide icon next to each color. By turning off the positive and negative draft faces we can clearly see the faces that don't meet the required draft.

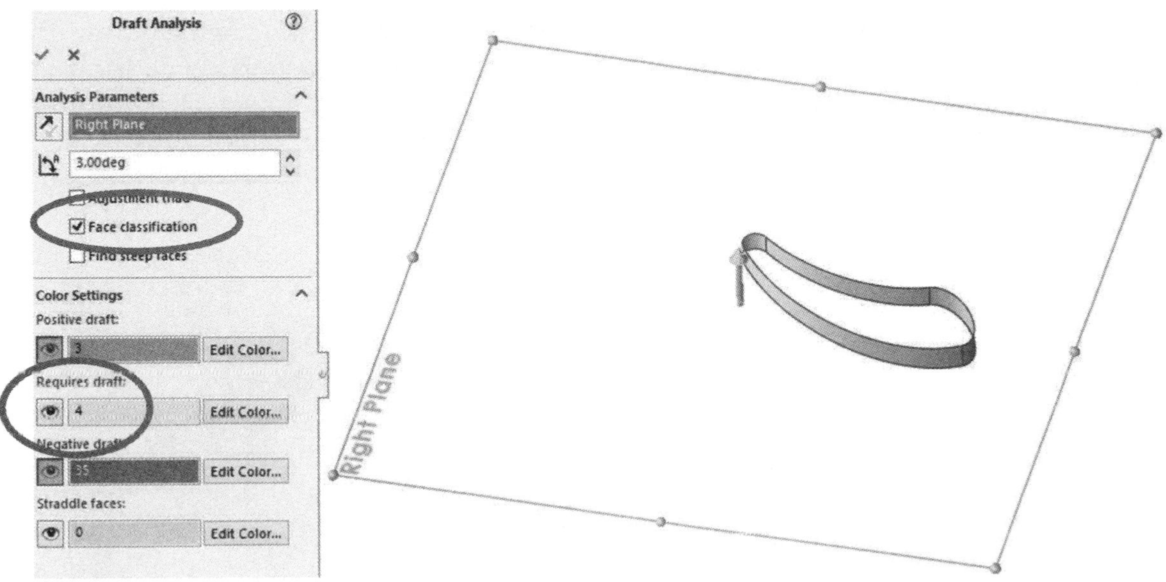

27.17. – After making sure all faces have the required draft, the next step is to generate the parting line.

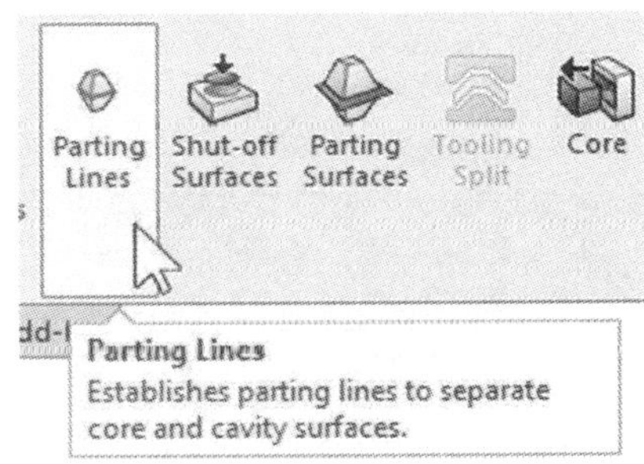

Select the "**Parting Line**" command from the Mold Tools toolbar or the menu "**Insert, Molds, Parting Line**." Select the *"Right Plane"* or a flat face parallel to it for the "Direction of Pull," enter 3 degrees as before, and click the "**Draft Analysis**" button.

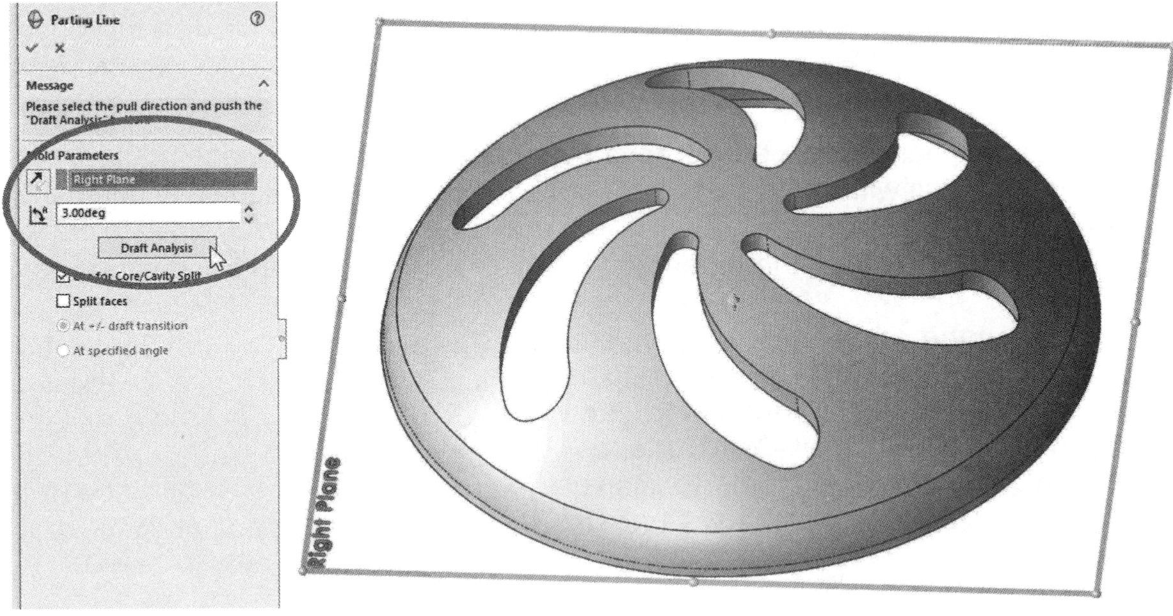

If the parting line edges are not automatically selected, we must select them manually. When a part has faces that need draft along the parting line, the edges may not be automatically selected because there may not be a clear choice as to which side of the mold a face should go in.

If the parting line edges are not automatically selected, manually add the edges where the red faces meet the green faces along the outside of the part. Notice the perimeter is made of multiple edges; be sure to select all of them *or* right mouse click on one edge and use the "Select Tangency" option.

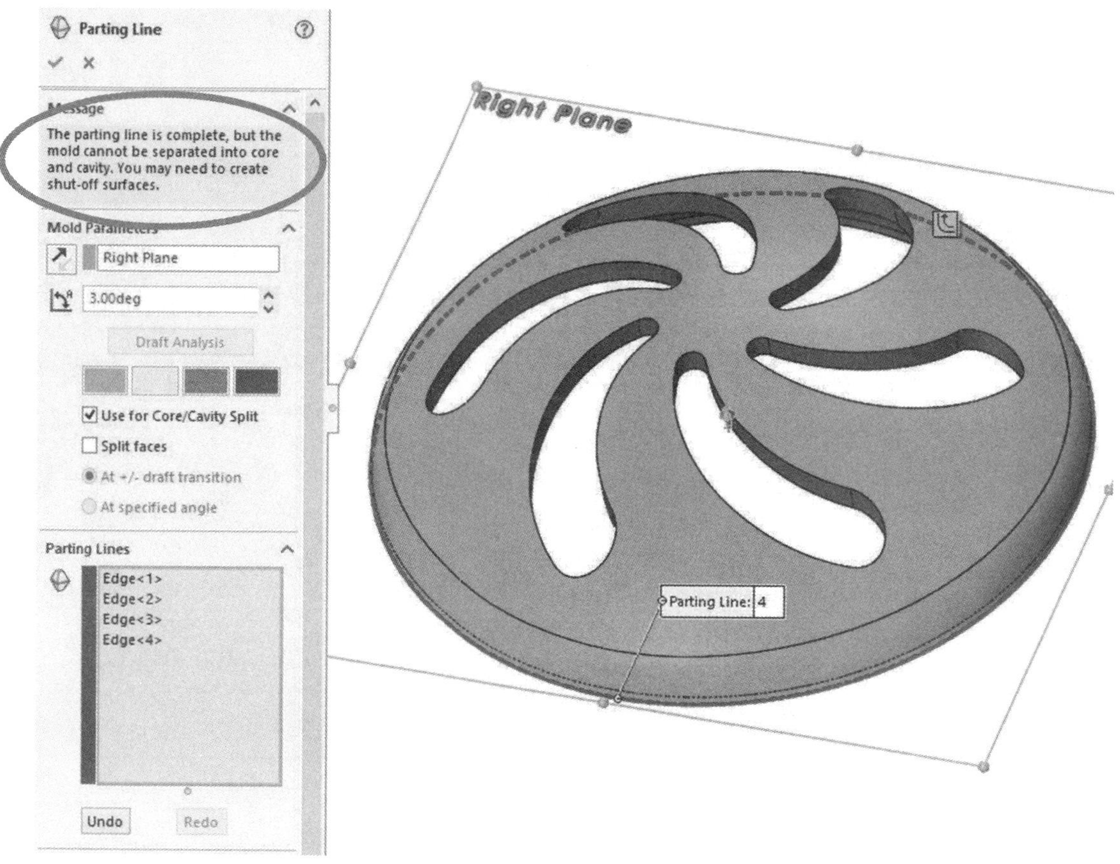

After the parting line edges have been selected, the message at the top of the command (still with yellow background) says that the parting line is complete, but the mold cannot be separated into Core and Cavity, and we may need to create shut-off surfaces to close the holes in the part. Click OK to finish the parting line.

27.18. - Once the parting line is complete, we need to close the holes in the part to be able to split the tooling into Core and Cavity. Select the "**Shut-Off Surfaces**" command from the Mold Tools toolbar or the menu "**Insert, Molds, Shut-Off Surfaces**."

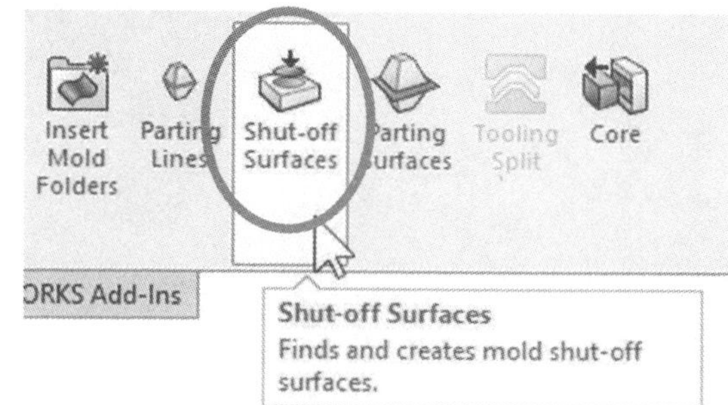

After selecting the "**Shut-off Surfaces**" command, the open loops are automatically selected based on the "Direction of Pull" in the parting line settings. Be sure to leave the "Knit" option checked to have the new faces automatically merged with the rest of the Core and Cavity faces when we finish this command.

By turning the "Show Preview" option on, we can see the resulting surfaces. The message at the top of the command now has a green background letting us know the mold can be separated into Core and Cavity. If a hole had not been automatically selected, we'd have to manually select its edges, also where the Core faces (Red) meet the Cavity faces (Green). Click OK to complete the command.

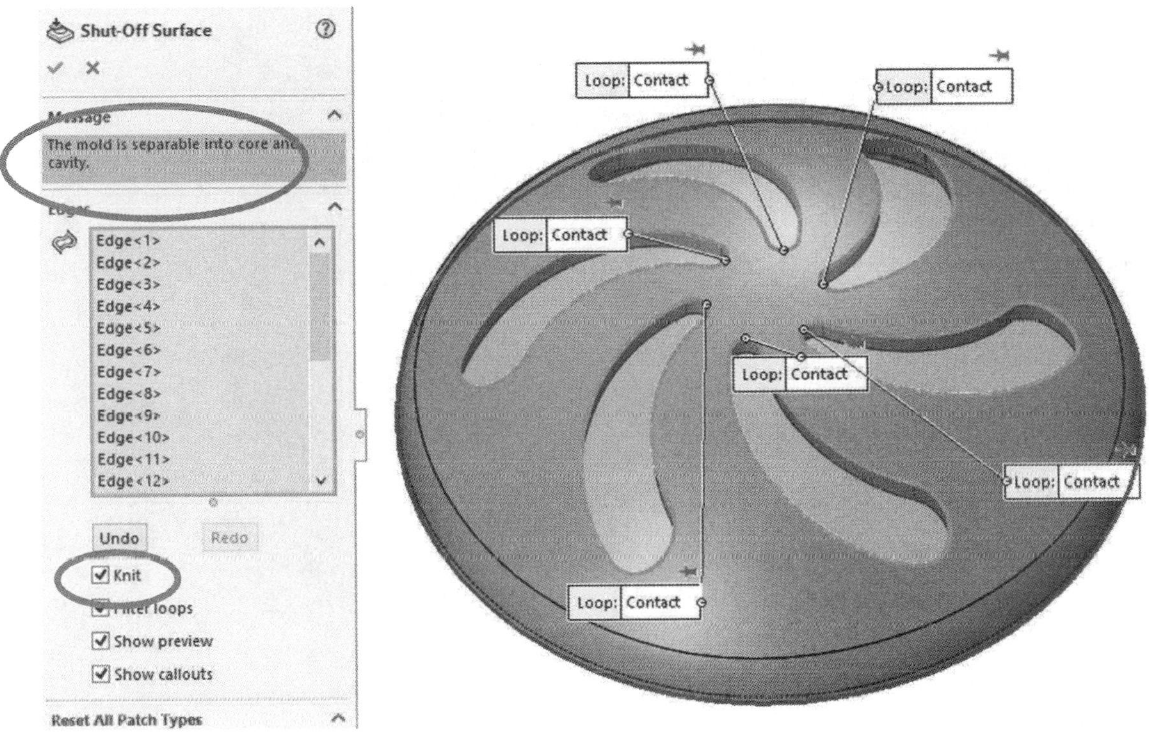

After the shut off surfaces are completed, the vents are closed and ready to split the model into a core and cavity.

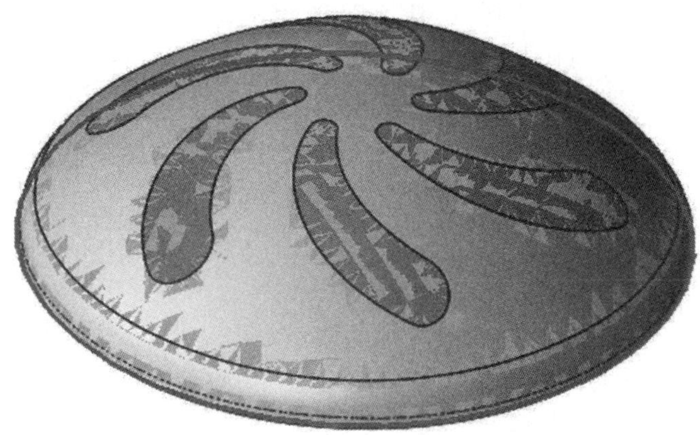

27.19. - The next step is to generate the parting surface to separate the mold into core and cavity. Select the "**Parting Surfaces**" command from the Mold Tools toolbar or the menu "**Insert, Molds, Parting Surfaces**." The "Perpendicular to pull" option is automatically selected, as well as *"Parting Line1."* Enter a value of 1″ for the "Parting Surface" distance and be sure the "Knit all surfaces" option is checked. Click OK to finish.

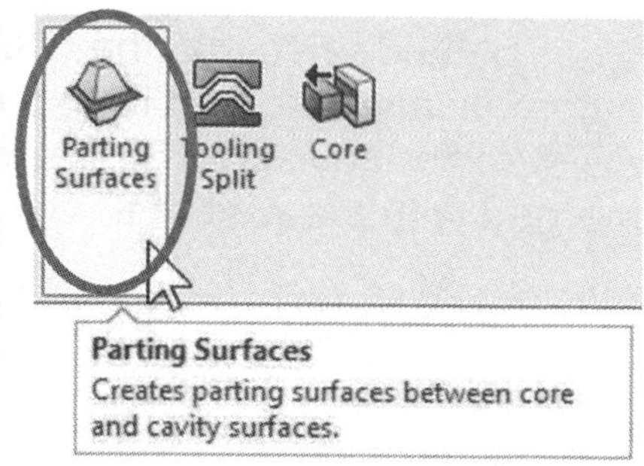

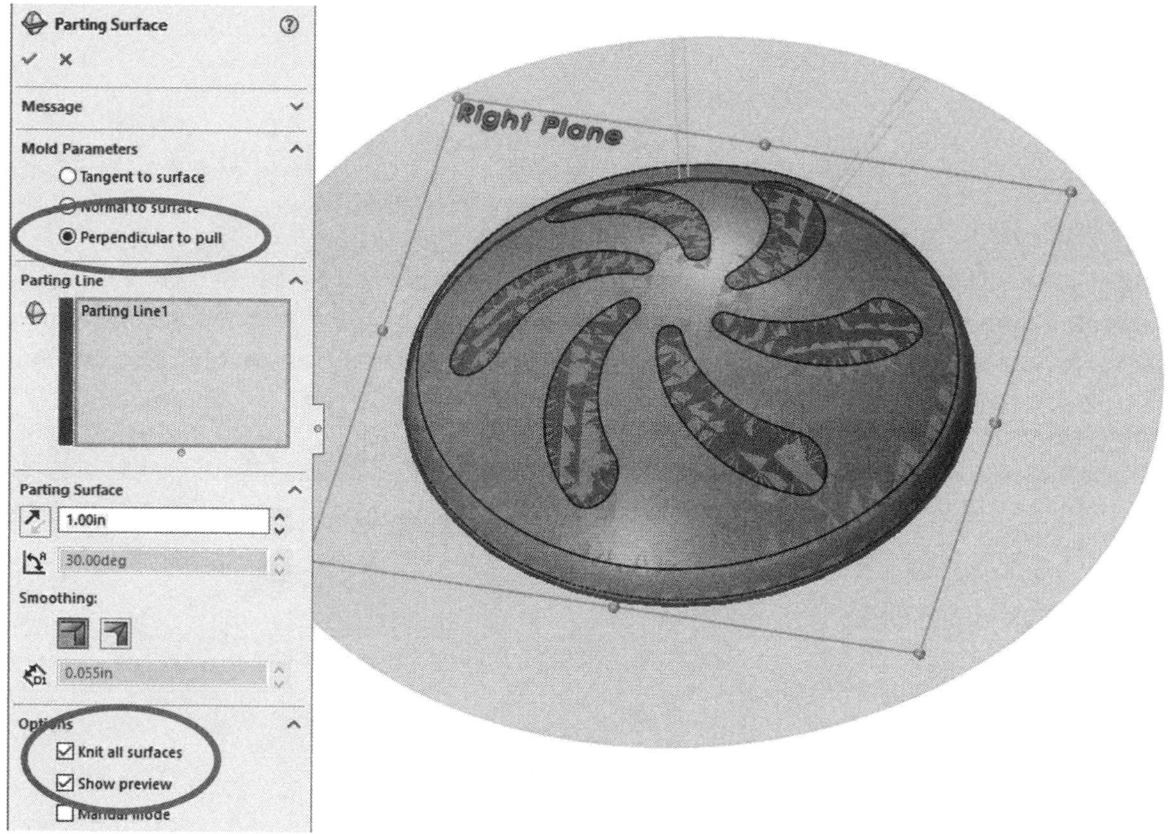

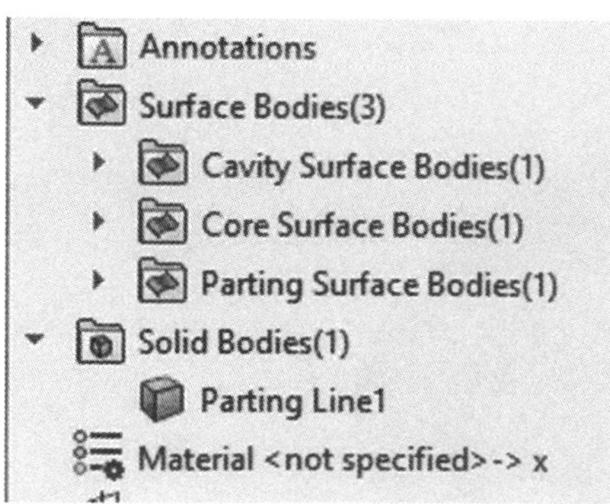

Expanding the *"Surface Bodies"* folder will make all surface bodies visible. The Cavity surfaces are green and the Core surfaces are red. The green (cavity) and red (core) are overlapping because the shut-off surfaces are knitted to both the Core and Cavity surfaces, the cavity surface (green) and the part overlap in the Cavity side, and the core surface (red) and the part overlap in the Core side. Hide the surfaces if wanted, but remember the parting surface needs to be visible to make the tooling split.

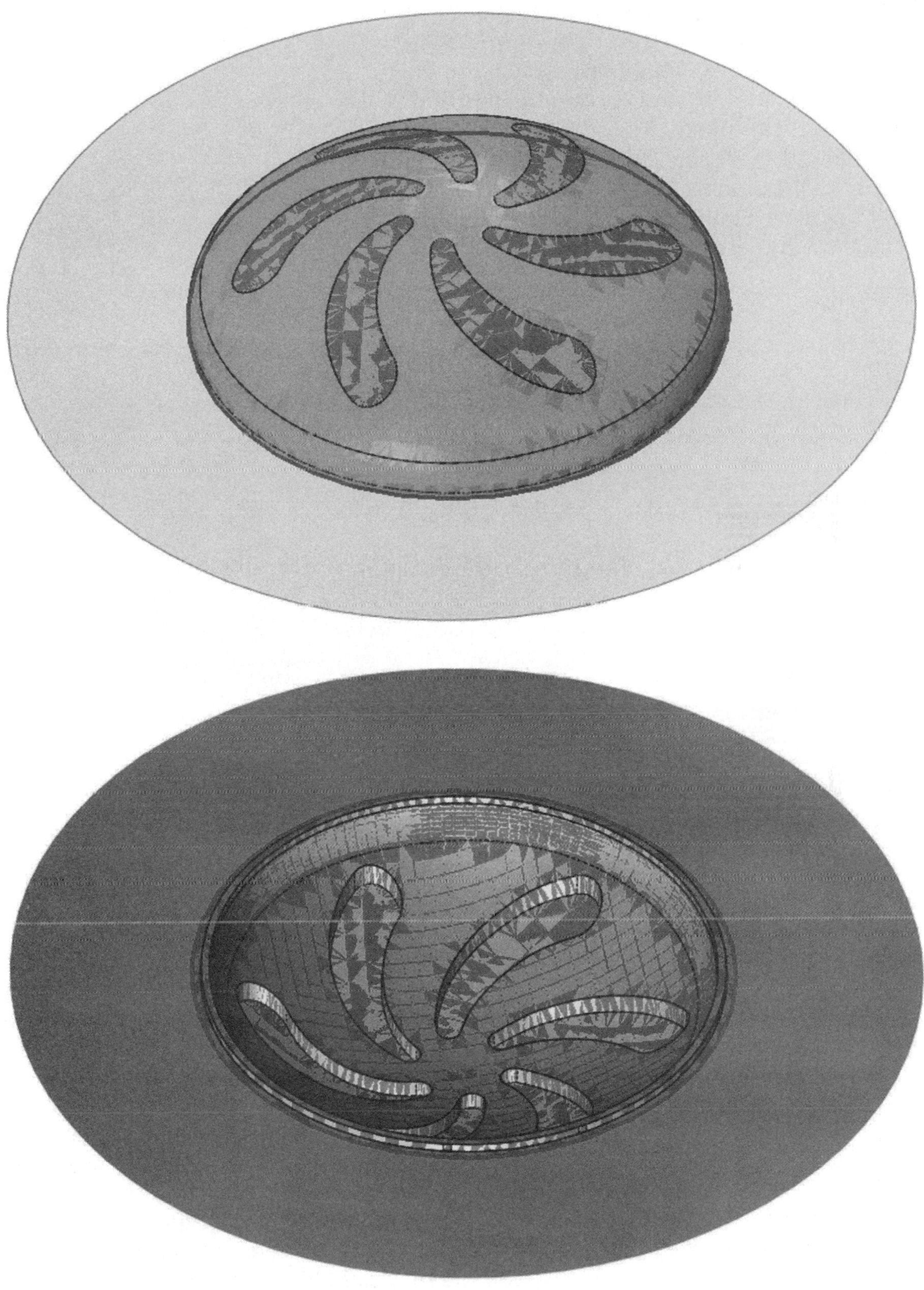

27.20. - Now we are ready to make the core and cavity. Select the "**Tooling Split**" command from the Mold Tools tab. In this example, we'll make the tooling sketch at the same time we make the tooling split. After selecting the "Tooling Split" command, select the parting surface to add a new sketch in it; use "**Convert Entities**" to project the parting surface's edge and exit the sketch to continue.

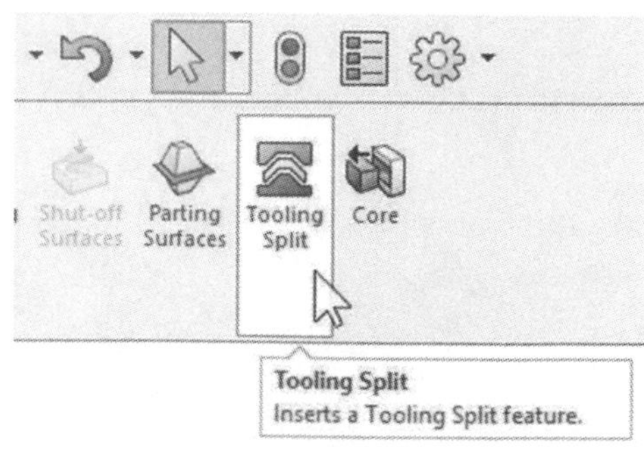

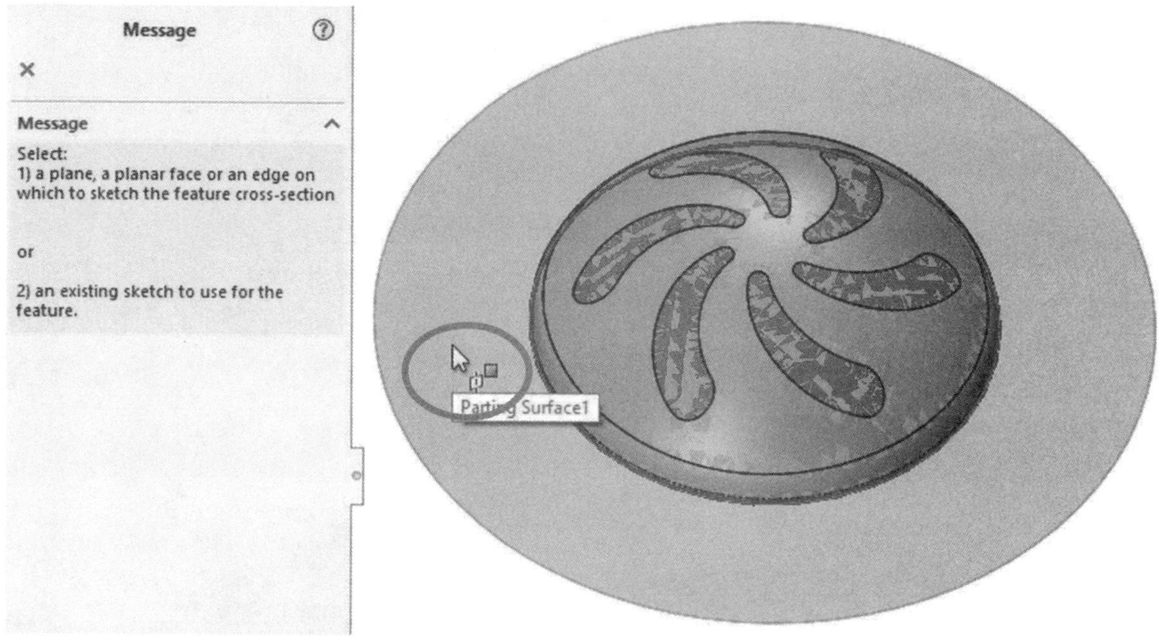

27.21. - After using "**Convert Entities**" to project the Parting Surface the sketch is complete. Exit the sketch to return to the "**Tooling Split**" command.

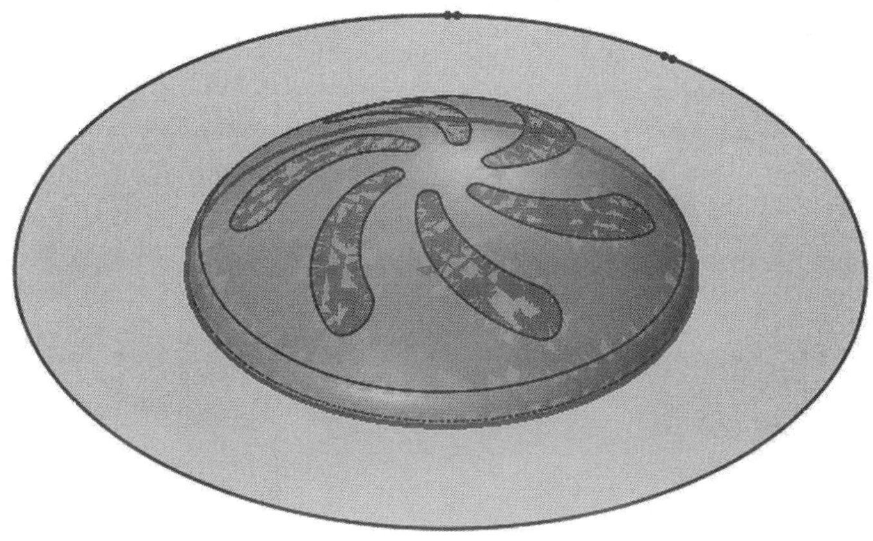

The necessary surfaces are automatically added to the "Core," "Cavity" and "Parting Surface" selection boxes. Make the "Block Size" 1″ up and 0.5″ down to completely enclose the part and click OK to complete the tooling split.

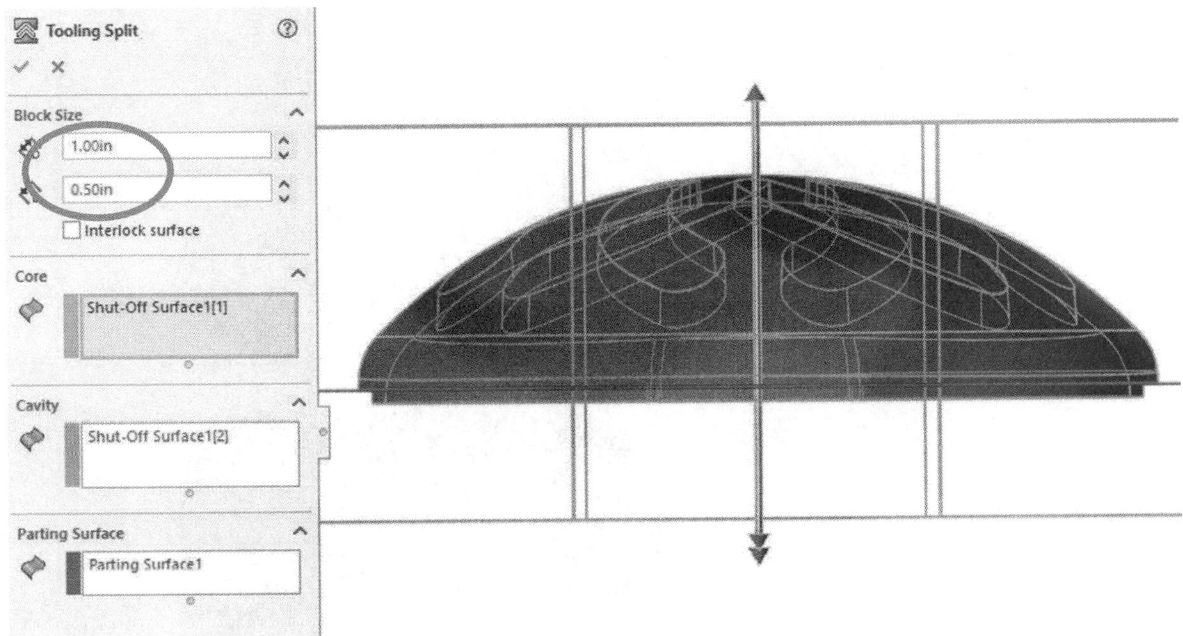

27.22. - To view the core, cavity and part bodies as an open mold in the part, select the menu "**Insert, Features, Move/Copy….**" This command allows us to move and/or copy surfaces and solid bodies in the part as if we were moving parts in an assembly.

We'll separate the core and cavity bodies away from the design part's body only to show how this feature works and see how the bodies look. There is no good reason to move the bodies in this step of the process, other than visually inspecting the resulting bodies.

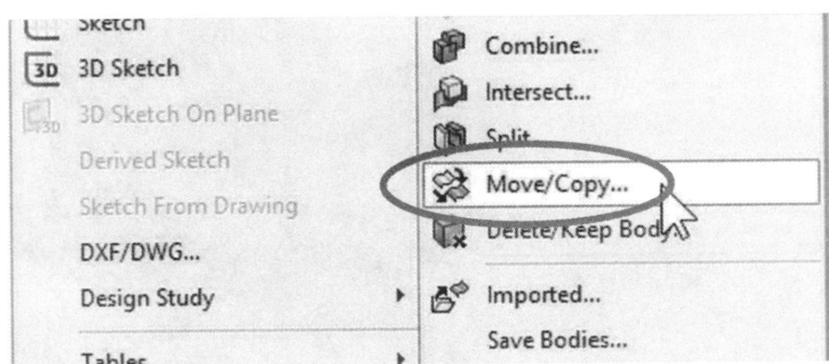

Select the top body (Cavity) and either enter a value to move it along the "X" direction, or drag the direction arrow in the graphics area, like exploding an assembly.

If we see the "Mate Settings" option to move the bodies like in an assembly, select the "Translate/Rotate" button in the options to continue.

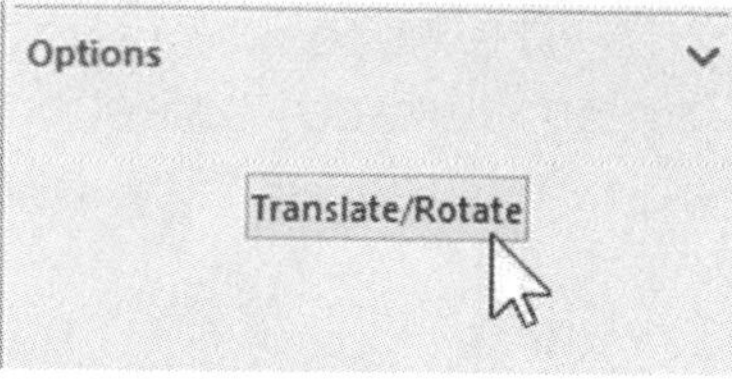

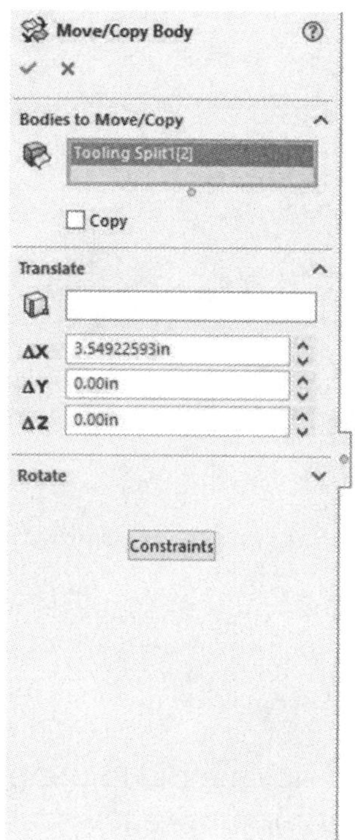

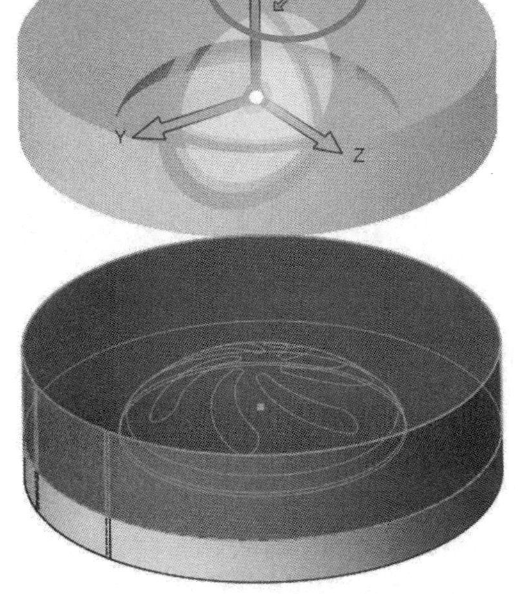

Repeat the **Move/Copy** command to move the lower body (Core) down. Make sure the "Copy" option is not checked.

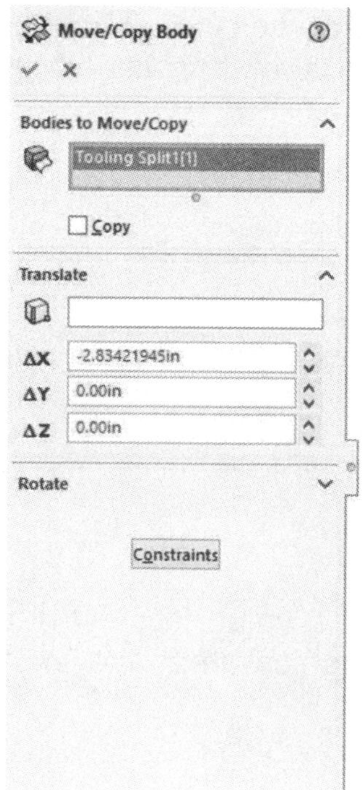

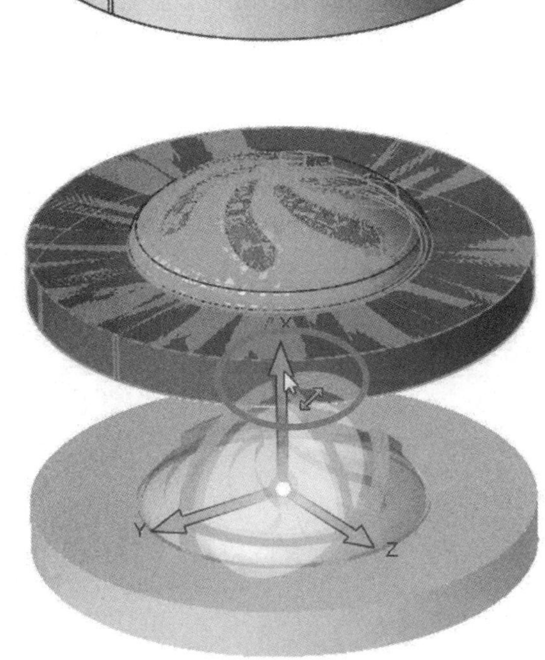

If needed, hide all the surfaces at the same time by selecting the *"Surface Bodies"* folder and click "Hide" as well as the *"Parting Line1."*

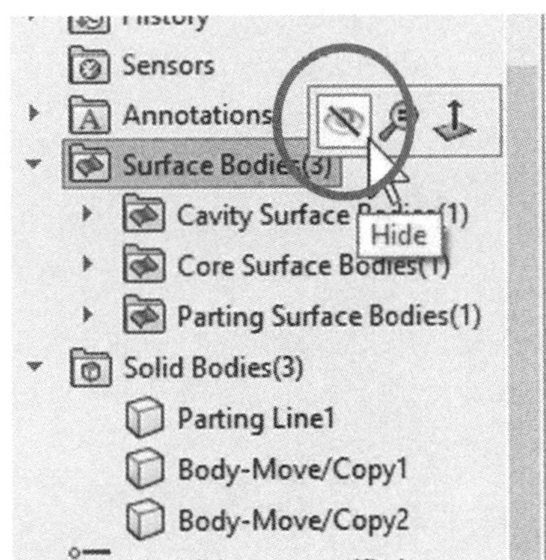

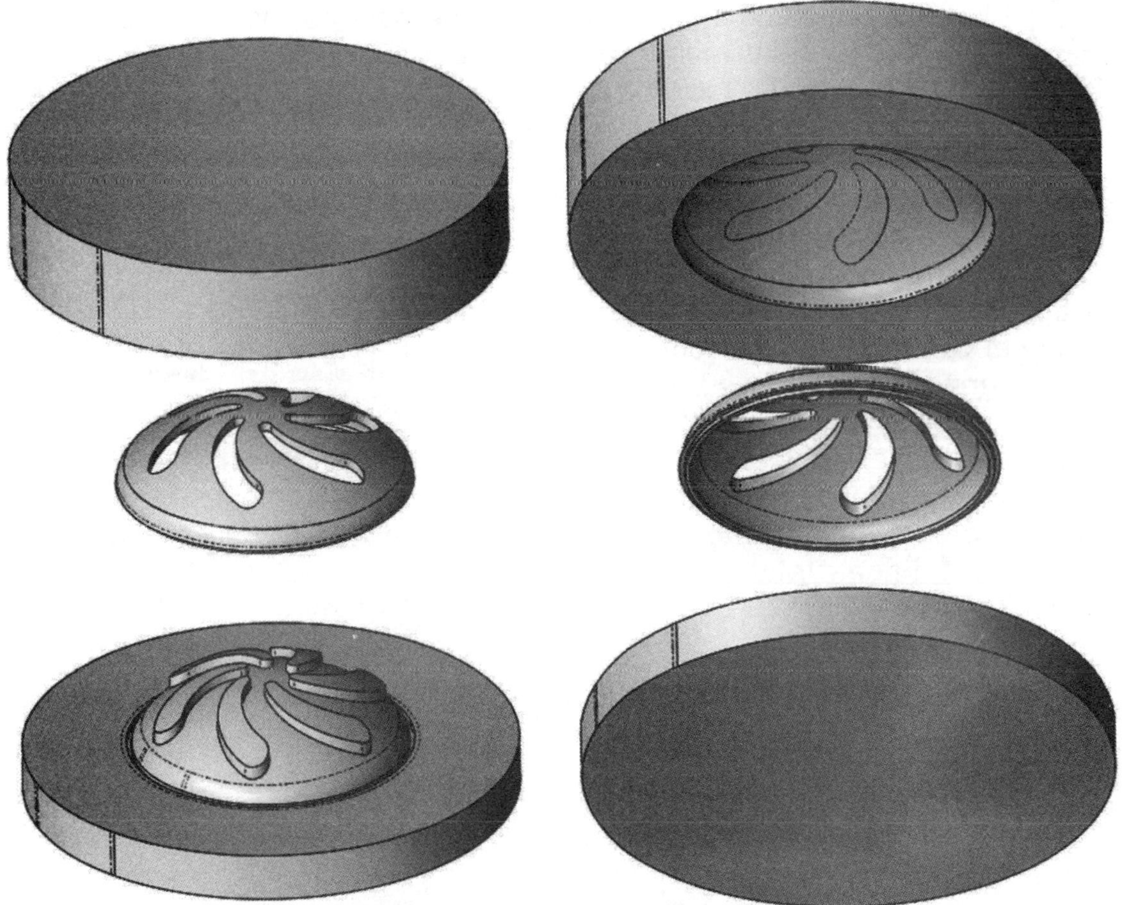

When making multi body operations like combining bodies where one or more bodies are absorbed, it's often useful to copy a body that will be absorbed in the same locations as the originals, so when a body is consumed we have a copy of the original body.

27.23. - The next step is to save the core and cavity bodies to separate files as we did with the previous part. Select the core and cavity bodies and save them to new parts as *'Hair Drier Cover-Core'* and *'Hair Drier Cover-Cavity'*.

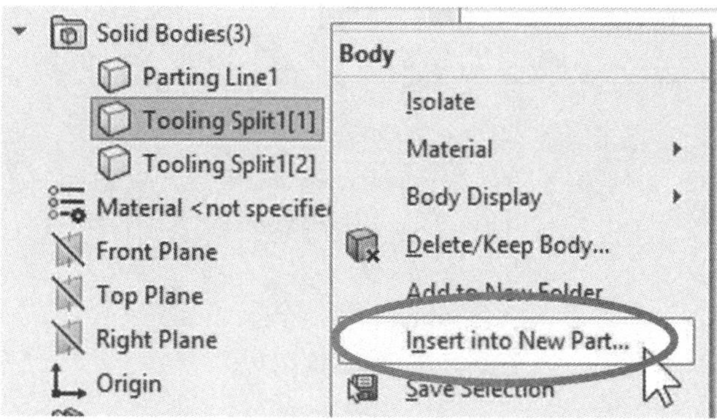

If we save a core/cavity/body to a new part after moving them with the "Move/Copy" command, the body in the new part will be located in the same position relative to the origin it had before saving it to a new part. To retain the original body location in reference to the origin, suppress the "Move/Copy" commands *before* saving the bodies to a new part.

27.24. - After saving the bodies to a new part, open the cavity. Notice, the holes of the cover are outlined in the cavity. A part like this is usually made with a Computer Numerical Control (CNC) machine, and for manufacturing purposes, it's better to have a single surface than a surface split in multiple areas like in this case. One way we can fix this is by deleting the faces from the solid body, and patching them with a new one. Select the "**Delete Face**" command from the Surfaces toolbar or the menu "**Insert, Face, Delete.**"

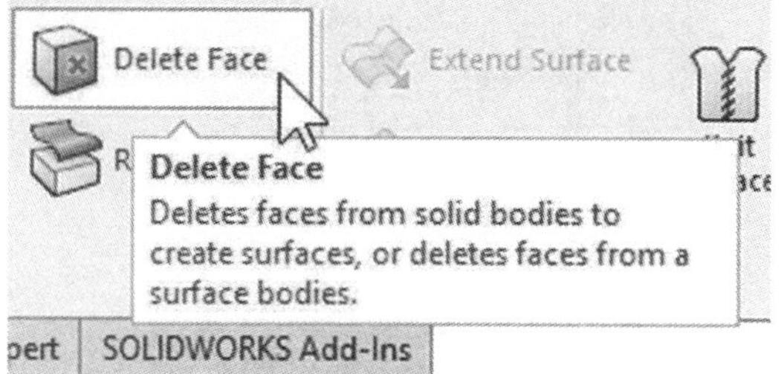

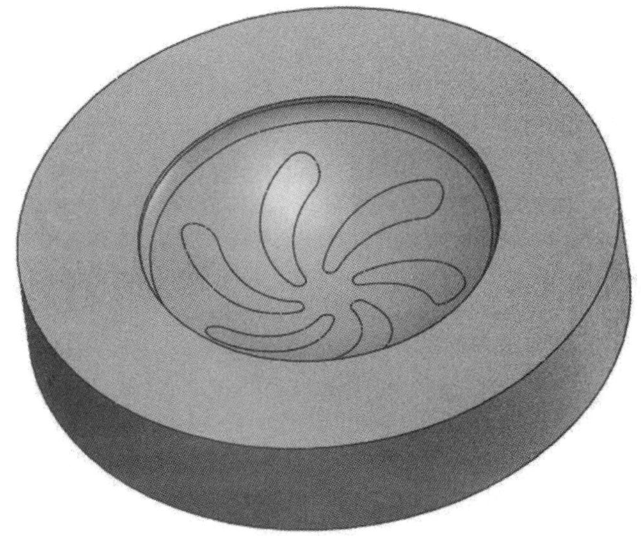

Select all the faces inside the Cavity that outline a hole. Use the "Delete and Patch" option and click OK to finish. This way we'll remove the outlines and patch the surface merging it with the cavity to produce a single smooth surface.

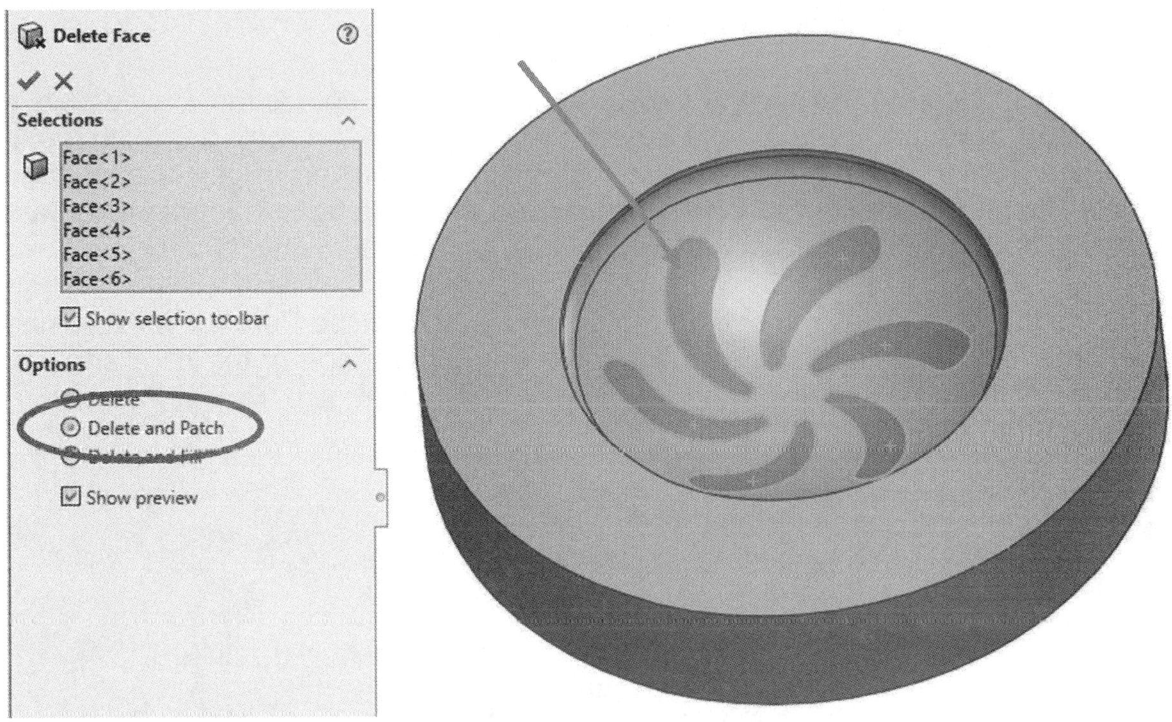

And now we have a single surface in the Cavity that can be easily machined with a CNC. Save and close the Cavity.

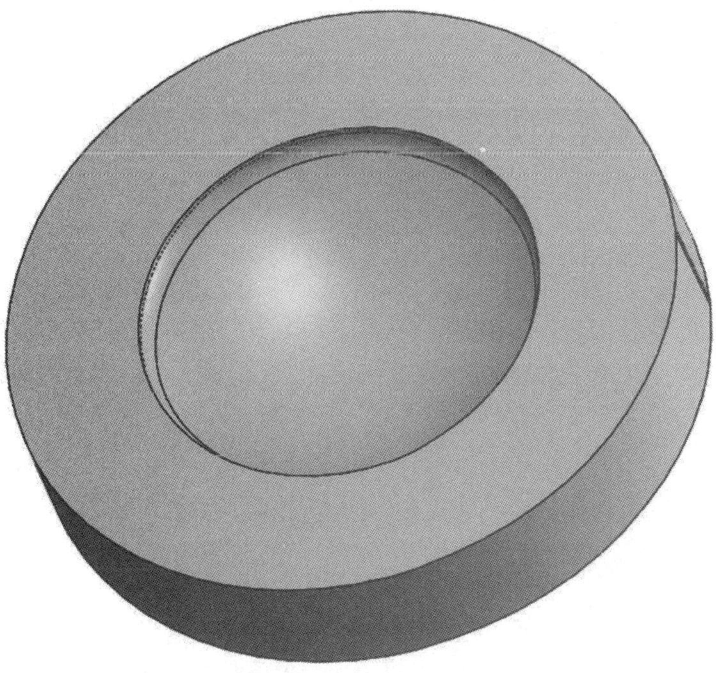

27.25. - A common operation in mold making is to add inserts to a mold. Inserts are usually made in areas of the mold that may experience wear and are cheaper to replace than an entire mold. In the Core part the protrusions that make the holes in the cover will be made using a replaceable mold insert; the inserts are made by splitting one or more bodies from the core or cavity.

The reason to use inserts instead of a single part core is that the second is usually more difficult to make, and if these protrusions wear out, it's harder to repair and expensive to replace, whereas an insert can be replaced easier, cheaper and faster. This is a decision that can be answered with help from the tool maker and depends on the volume of parts to be molded, cost of using inserts or not, etc.

In this step, we'll practice how to cut an insert from the Core (or cavity) using the "**Split**" command. Add a new sketch in the bottom face of the core and use "**Convert Entities**" to project the lower edge of one insert; use "**Select Tangency**" in the insert's lower edge.

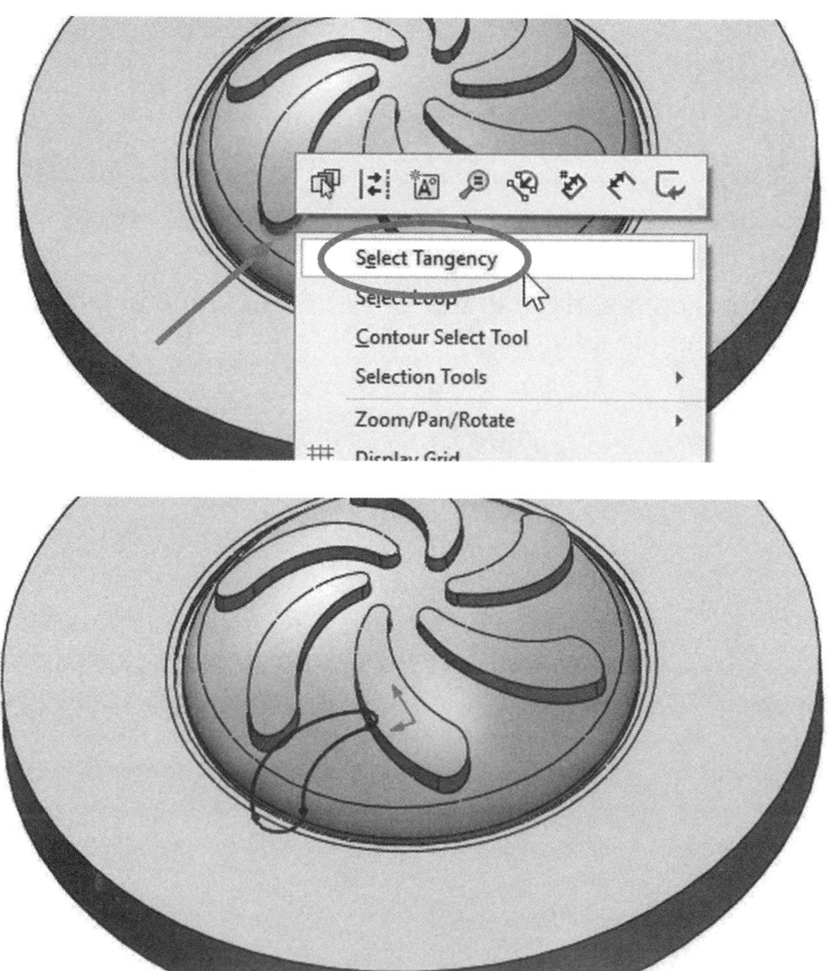

If we select the top edge to "**Convert Entities**" the resulting insert will be wrong. The top edge of the insert is smaller than the lower edge because of the draft added to the holes to release it from the mold.

27.26. – Now we need to separate the insert from the Core. We can use the sketch as the split tool just like we did with the main body of the *'Hair Drier'*, but we'll show a different way using a surface body. While still editing the sketch, from the Surfaces tab select "**Extruded Surface**" or the menu "**Insert, Surface, Extrude**," extrude the surface past the top of the Core and click OK to finish.

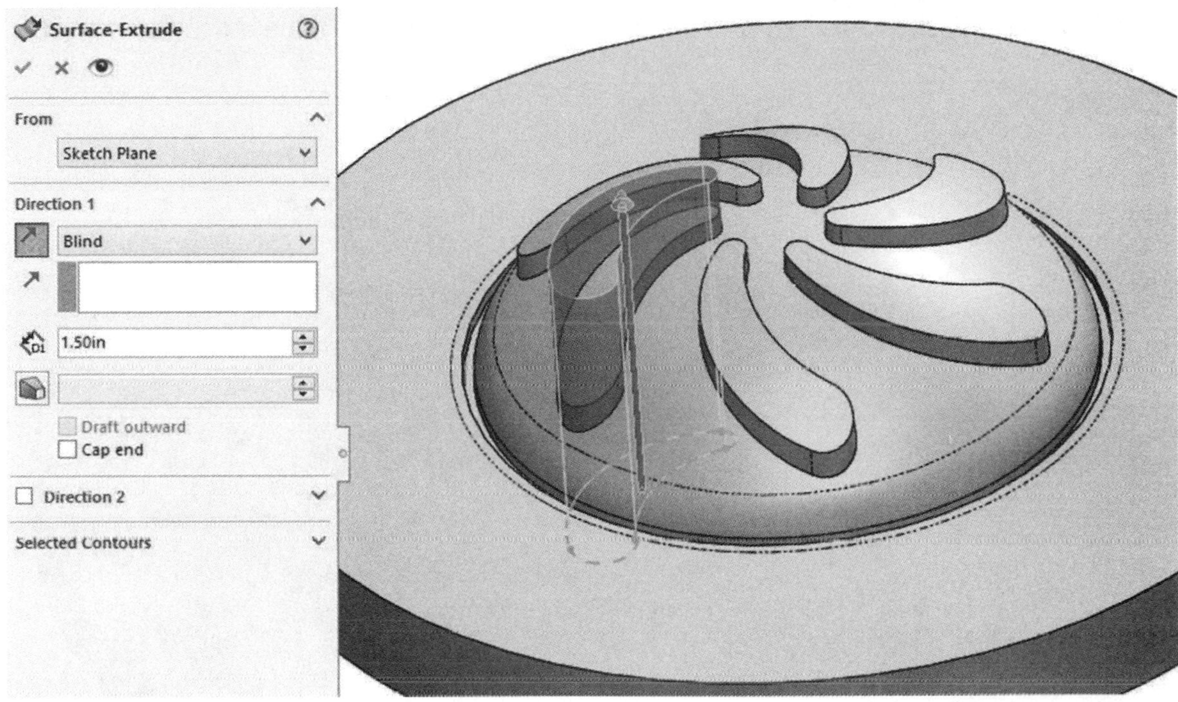

27.27. – Now we can split the insert from the Core using the previously extruded surface. Select the menu "**Insert, Features, Split**," select the extruded surface in the "Trim Tools" selection box and press "Cut Bodies" to divide the Core's body.

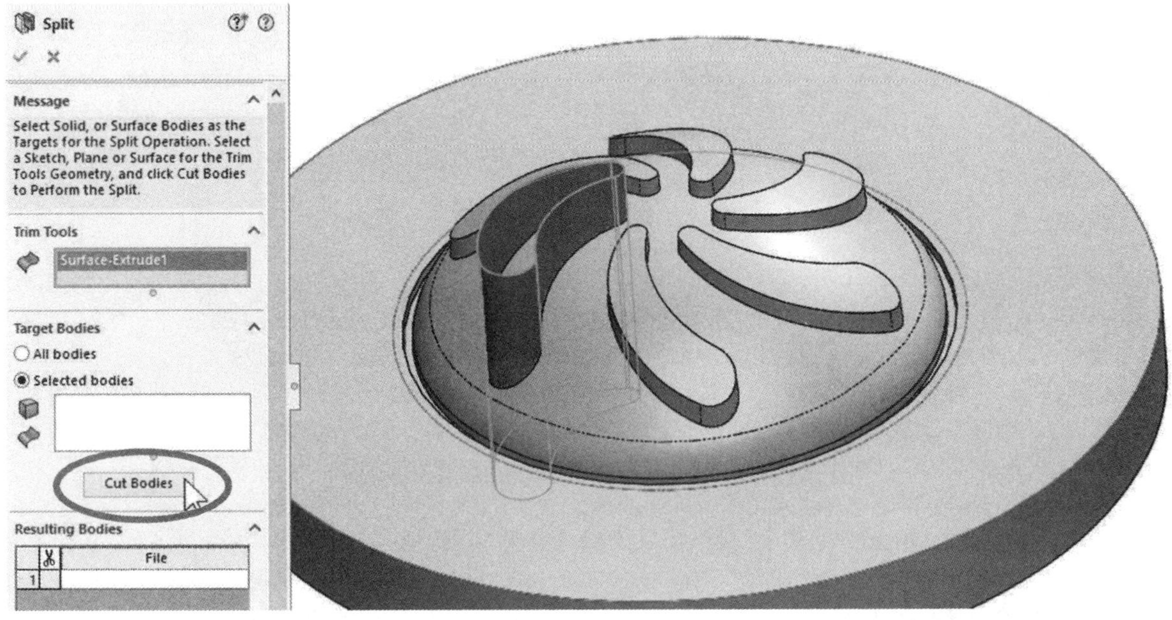

In the "Target bodies" selection box we can define which bodies to cut. If we use the "All bodies" option, every body (solid and surface) intersected by the trim tool will be split; if we use the "Selected bodies" option we can select which body to split. In our case, we only have one body, and it is automatically selected. Double click the insert to give it a name and save it as *'Hair Drier Cover-Core Insert'*. Use the "Consume cut bodies" option to remove it from the Core, click OK to split and save the insert to a new part. When finished, hide the extruded surface.

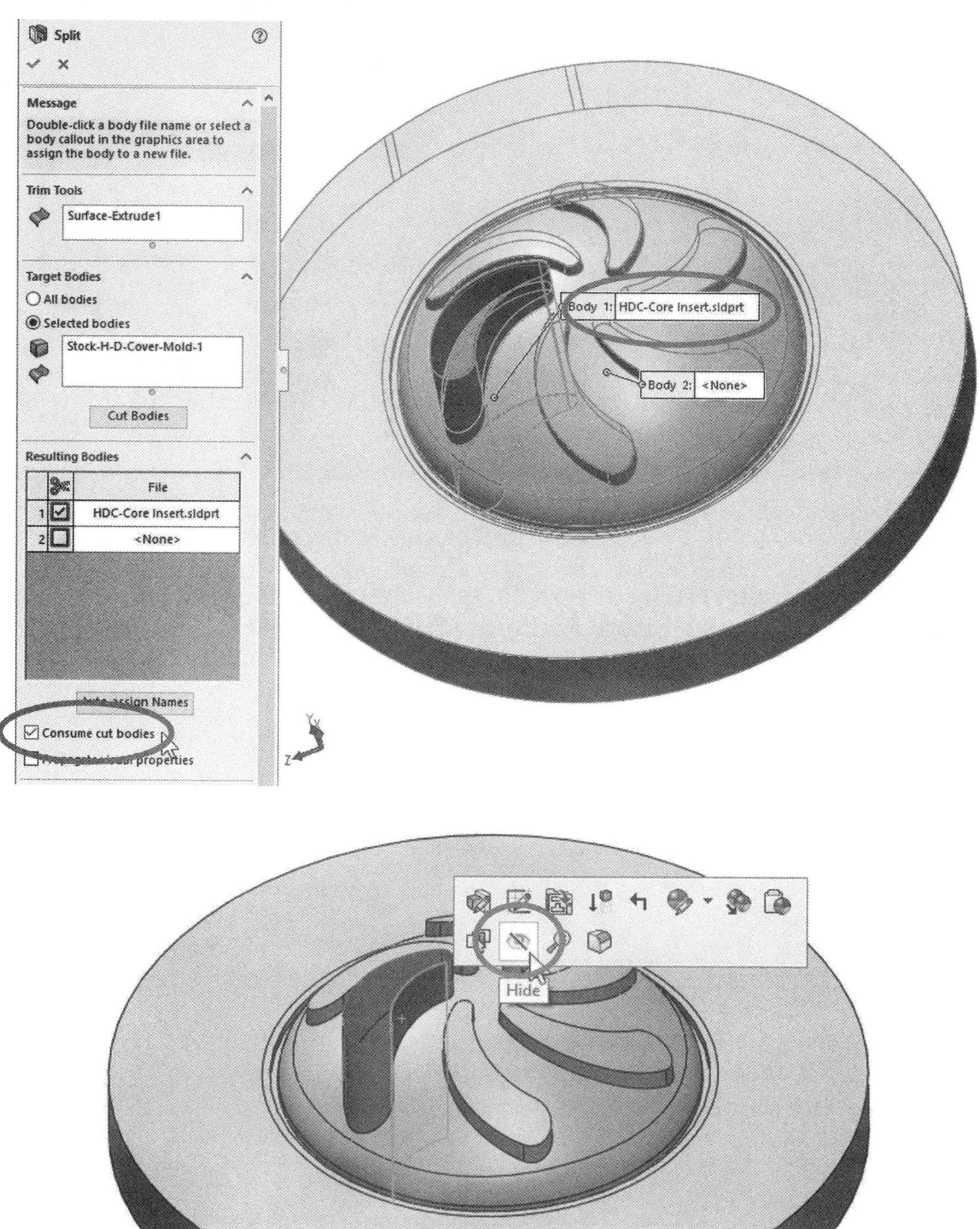

27.28. - To make the rest of the inserts, we'd follow the same procedure, but since all the inserts are equal in this part we can make one insert and use a cut extrude feature to cut out the rest of the holes.

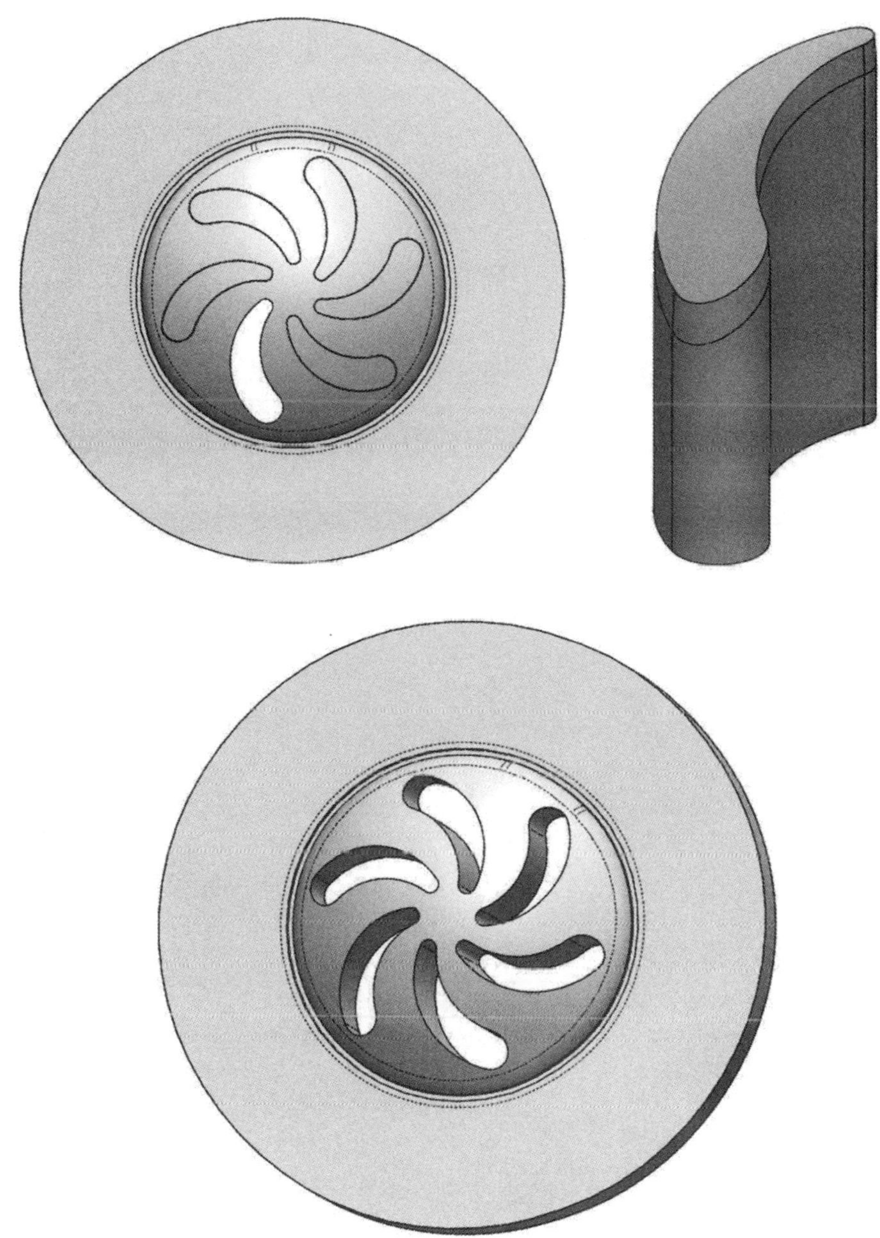

When designing molded parts keep in mind how they are going to be molded, assembled, and if at all possible, involve the toolmaker in the design process to get his input before it's too late. He'll be glad you did, and you will be too. ☺

Save and close all parts for the Hair Drier Cover's mold.

Notes:

Hair Drier Body's Mold

28.1. – For the next exercise, we are going to use a part that will not generate the correct parting line and we'll have to manually select it. Locate the *'H-D Right Mold'* from the included files and open it.

*This part is finished including vents, power cord and switch cutouts; drafts have been added to most faces and external relations have been broken to prevent possible conflicts with user generated files.

28.2. – The first step is to run a draft analysis to identify if the faces have the required draft angle. Select the "**Draft Analysis**" command using the *"Front Plane"* for the "Direction of Pull" and enter 3 degrees. Turn on the "Face classification" checkbox to identify the model faces as "Positive Draft," "Negative Draft," "Requires draft" or "Straddle faces."

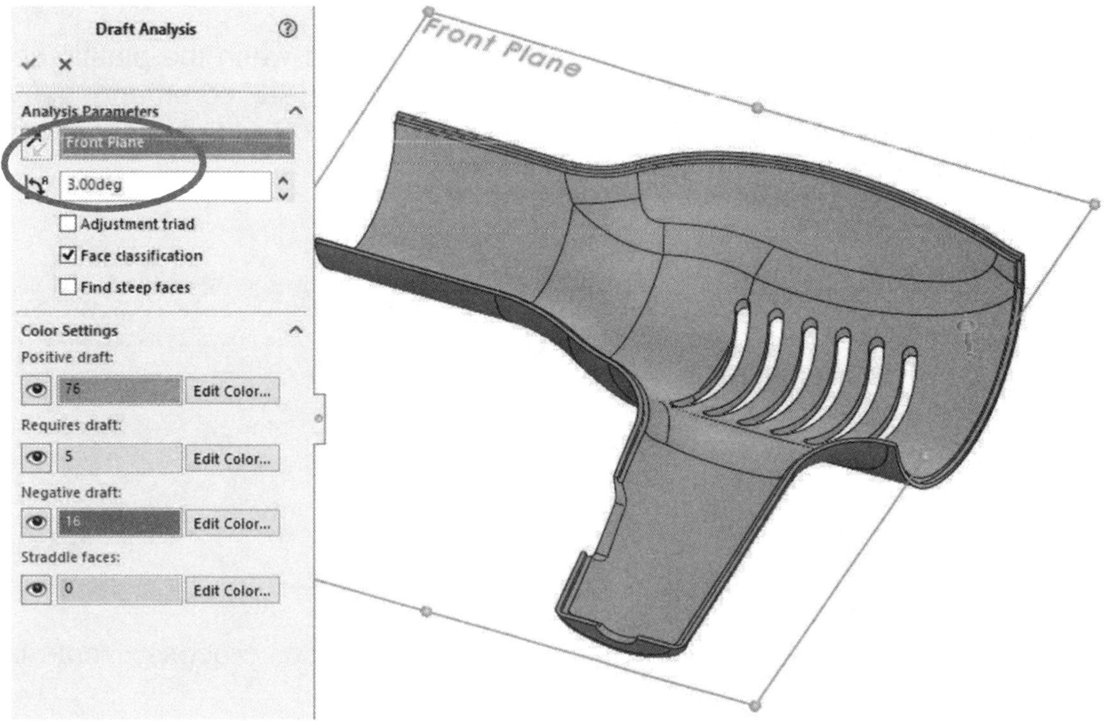

Click in the Hide/Show icons next to the green and red boxes to turn off both "Positive" and "Negative" draft faces, and isolate the faces that require draft. Here we can see that a few faces do not have the required three degrees draft. At this time we'll proceed to make the core and cavity ignoring these faces. The reason we are not adding draft to these faces is because adding draft would generate the correct parting line automatically, which in general is good; however, we are overlooking this detail to force an incorrect solution to learn additional functionality. Cancel the "**Draft Analysis**" to continue.

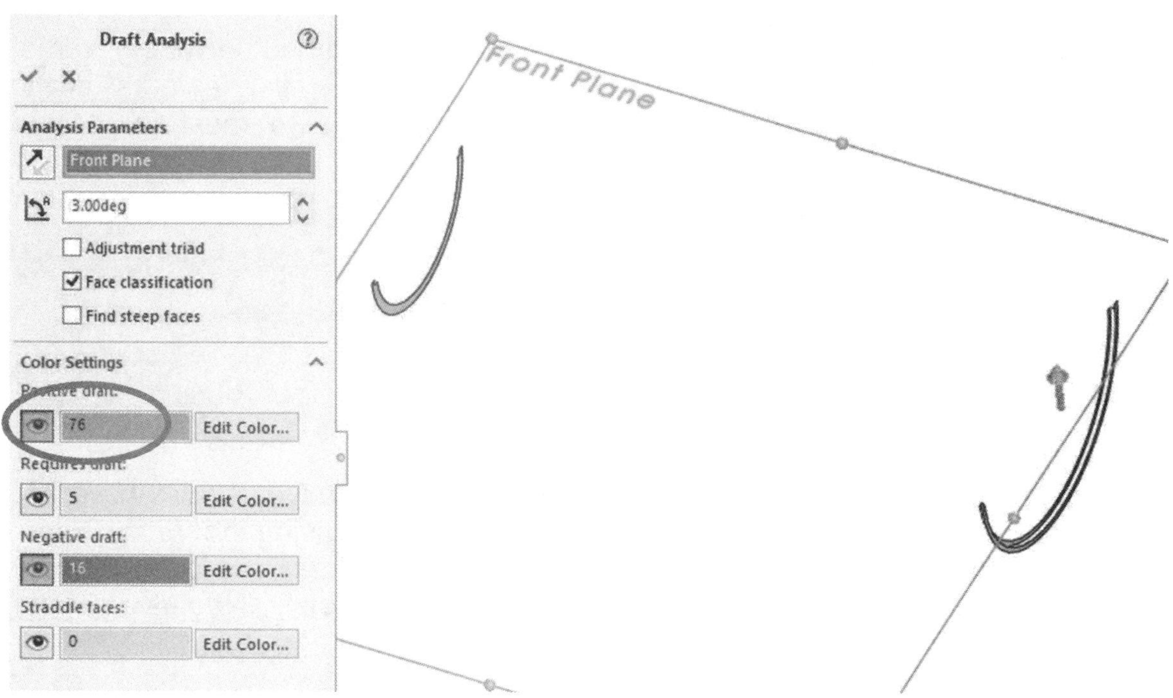

28.3. - To compensate for material shrinkage in the mold when the plastic cools down, scale the part, and make it 2% bigger. Select the "**Scale**" command, use the option "Uniform Scaling" about the "Centroid" and enter 1.02 as the scaling factor. Click OK to scale the part and continue.

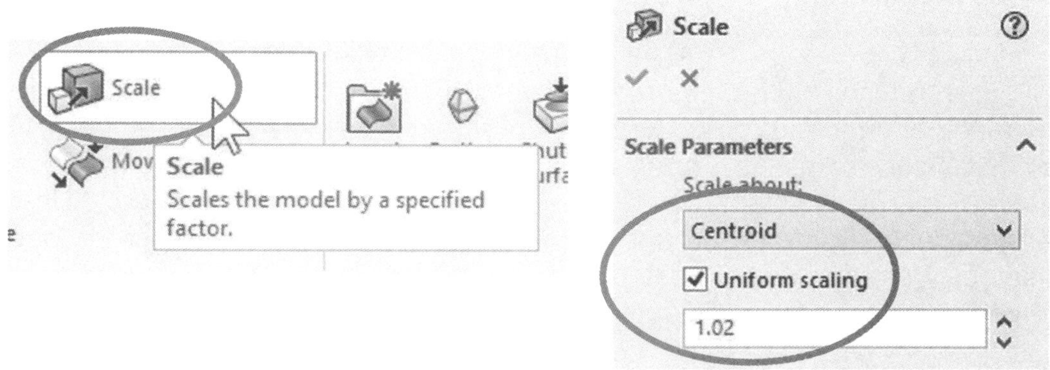

Remember that material shrinkage depends on process, materials, geometry, etc. The value used is just an example.

28.4. - Select the "**Parting Lines**" command; use the *"Front Plane"* for the "Direction of Pull," enter 3 degrees for the draft angle, and press the "Draft Analysis" button to calculate it. A parting line is automatically selected, but since we left a few faces with no draft, we need to review and make sure the parting line is where we want it. Be aware that experience in mold design is helpful when making these decisions.

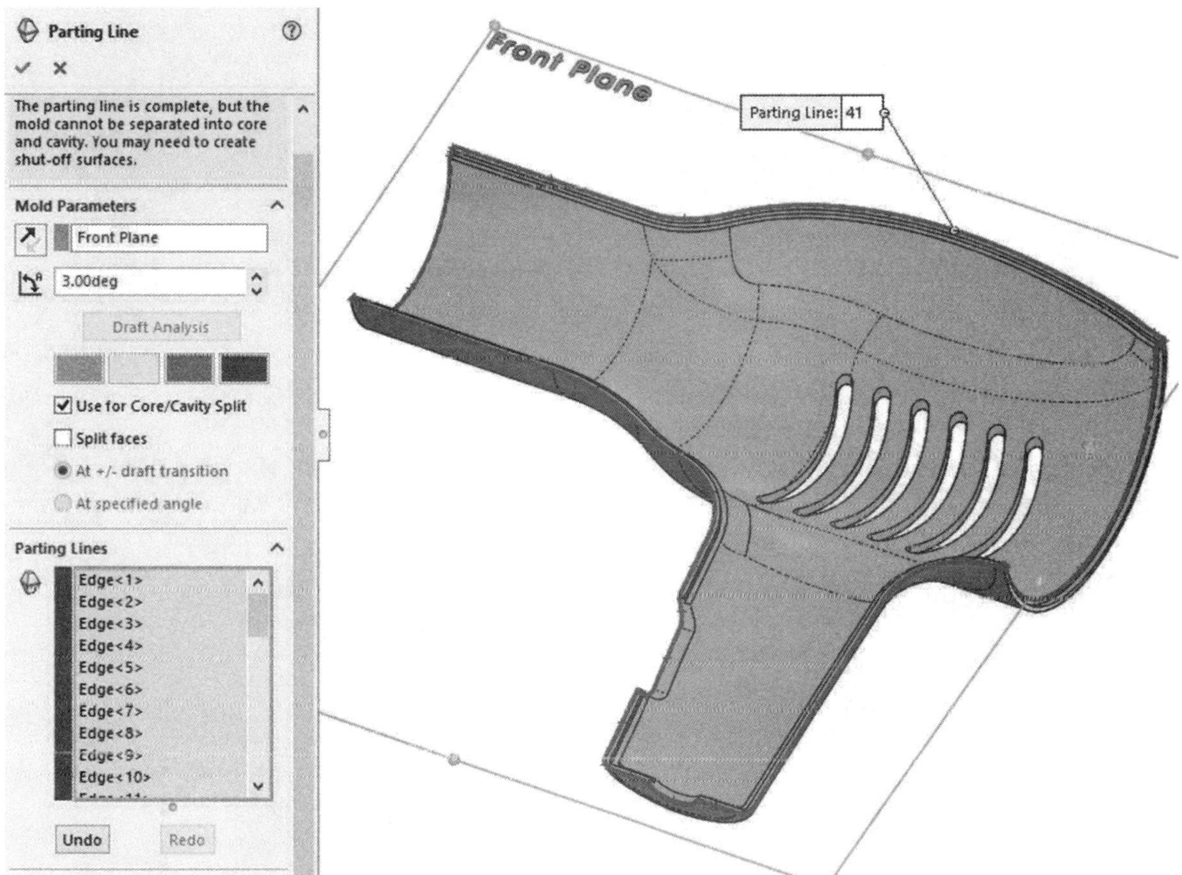

After reviewing the automatically selected parting line, we can see the edges selected in the front and back going across the faces that need draft. Since it may be easier to make the faces that need draft (yellow) in the Core (green side), we have to manually select the edges we want the parting line to follow. Unselect the edges between the green and yellow faces in the front and back of the part, and select the edge between the red and yellow faces. The edge labels are displayed when an edge is selected in the "Parting Lines" selection box.

If needed, re-run the "Draft Analysis" after re-selecting the edges. Change the draft angle to reset the analysis button.

- One side in the back of the *'Hair Drier'* (where the back cover fits)...

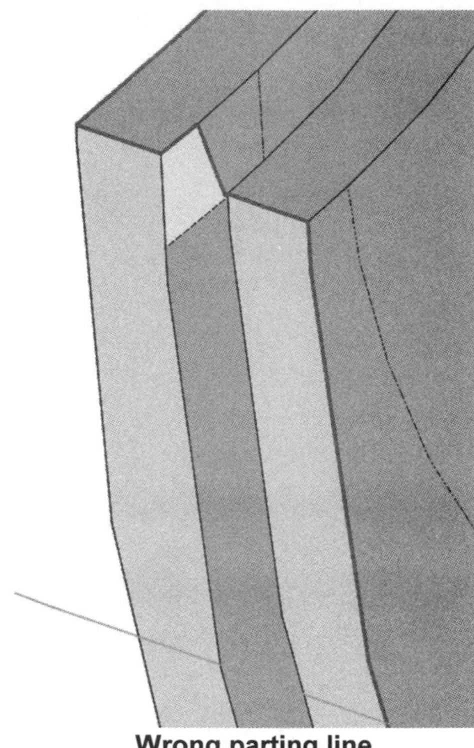

Wrong parting line

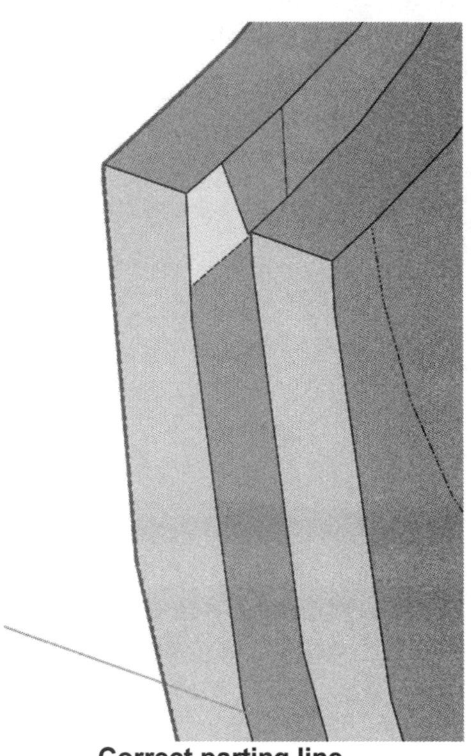

Correct parting line

- The other side in the back of the *'Hair Drier'*.

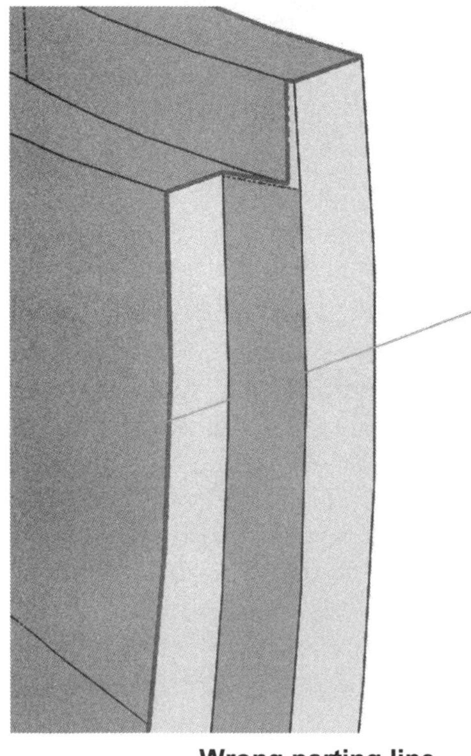

Wrong parting line

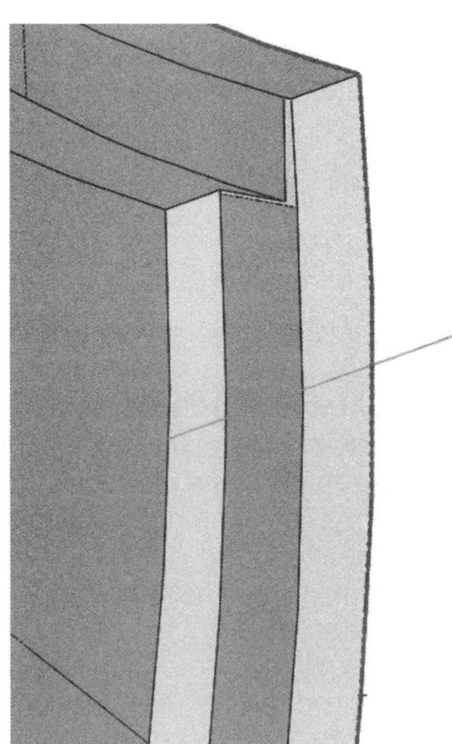

Correct parting line

- One side in the front of the *'Hair Drier'*...

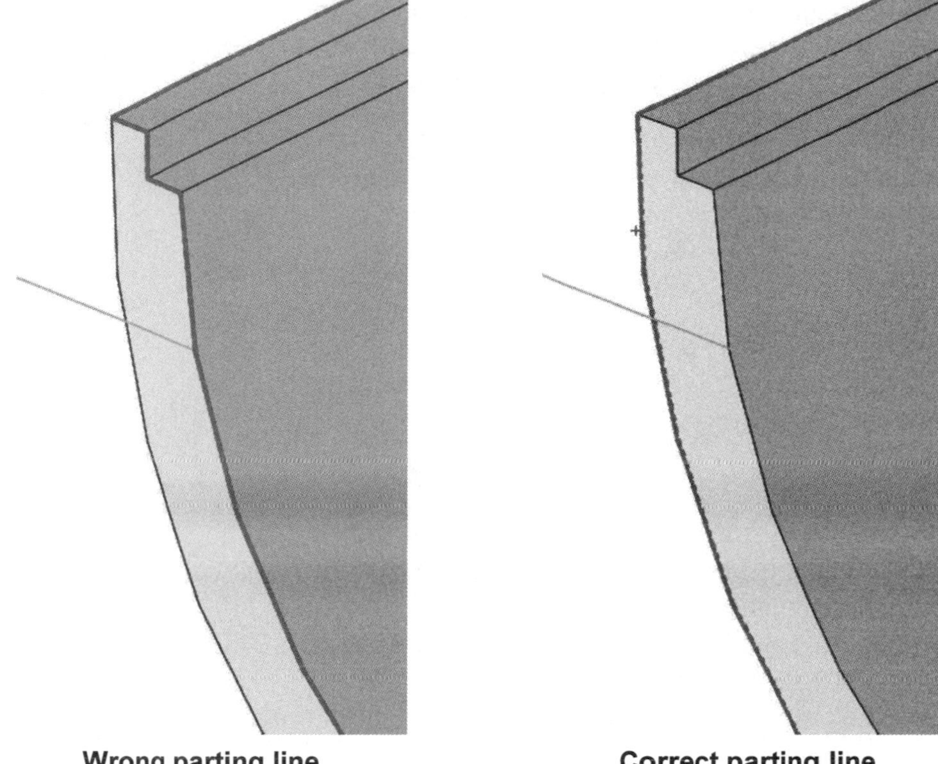

Wrong parting line **Correct parting line**

- The other side in the front of the *'Hair Drier'*.

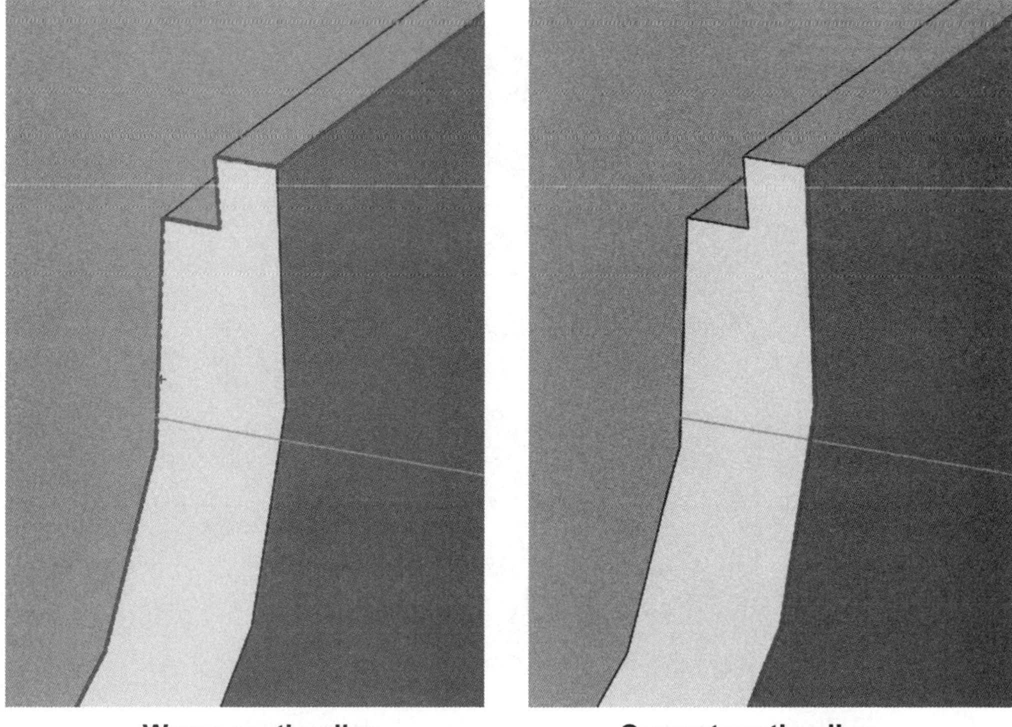

Wrong parting line **Correct parting line**

After the correct edges are selected, the message at the top is letting us know that the parting line is complete but we still need to create shut-off surfaces before we can split the mold into Core and Cavity. Click OK to add the parting line and continue.

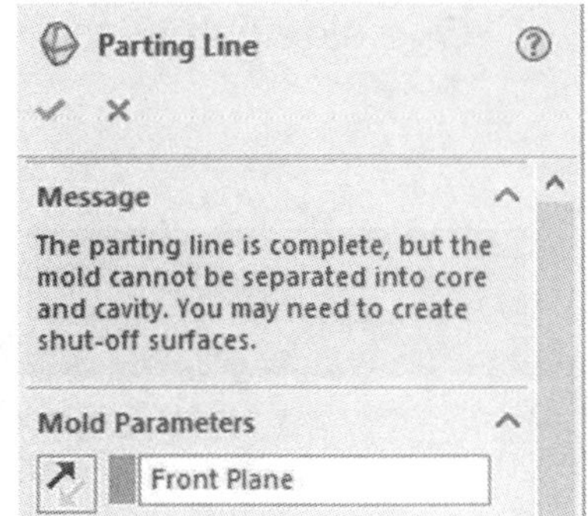

28.5. - Select the "**Shut-off Surfaces**" command from the Mold Tools toolbar or the menu "**Insert, Molds, Shut-off Surfaces**."

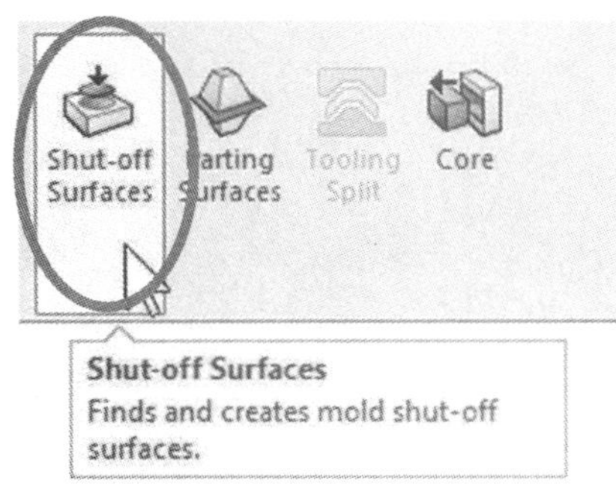

After selecting the command, all the ventilation holes are automatically selected. If any redundant edges are selected, remove them from the selection list by unselecting them, or deleting them from the "Edges" selection list. We need to have the edges outside the hair dryer. After they are closed; our mold can now be separated into core and cavity. Click OK to continue.

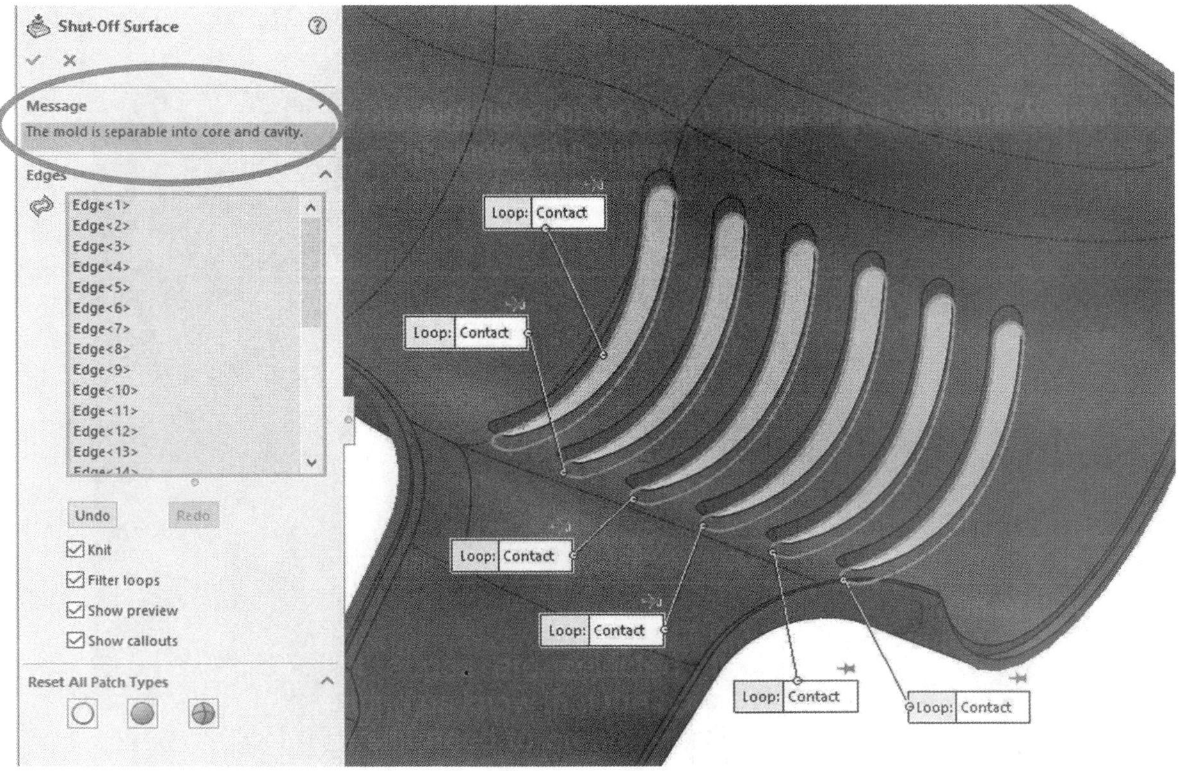

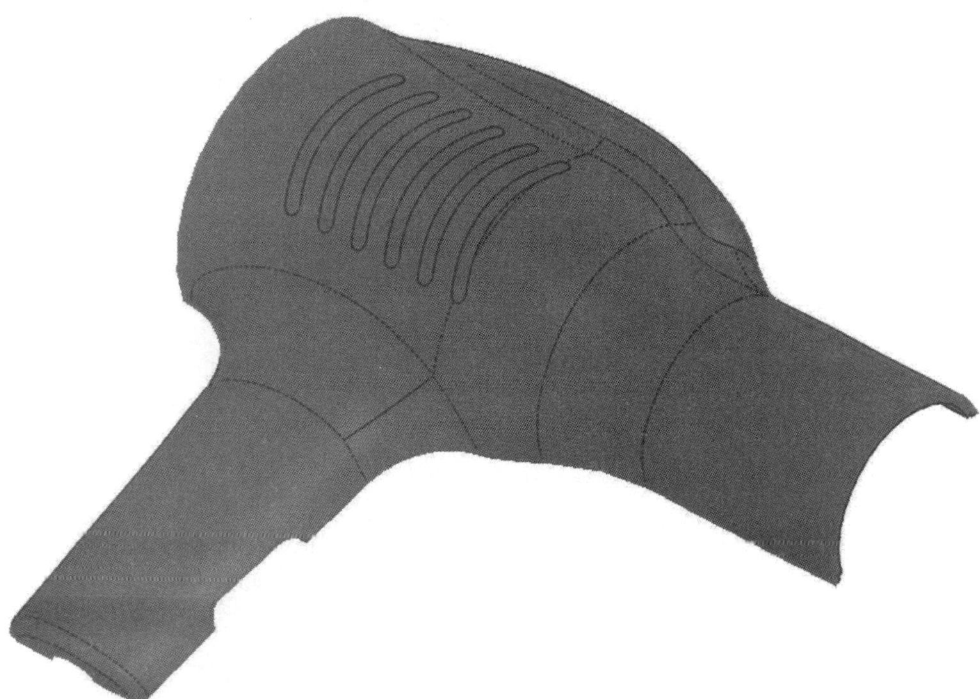

28.6. - The last step before creating the core and cavity is to create a parting surface. In the previous examples the parting surface was generated automatically saving us time. In this case an acceptable parting surface cannot be automatically generated with any combination of parameters and options, and therefore, we have to manually modify and complete it.

Select "**Parting Surfaces**" from the Mold Tools tab or the menu "**Insert, Molds, Parting Surface**." Under Mold Parameters select "Perpendicular to pull," enter a surface distance of 0.75″ and turn on the options "Knit all surfaces" and "Manual mode." Change to a Front view for visibility. Enabling the "Manual mode" allows us to manipulate the resulting parting surface.

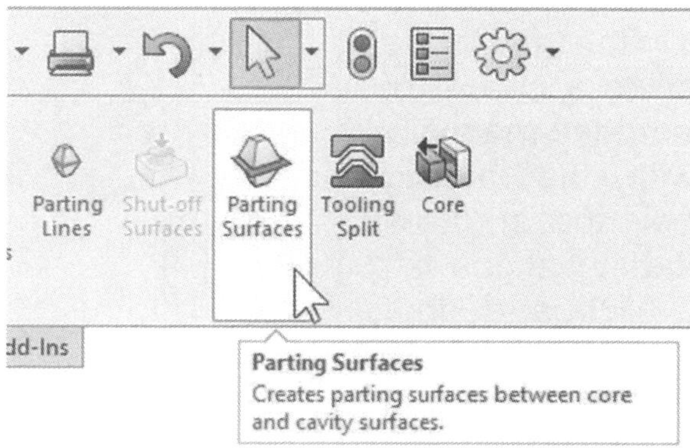

Feel free to explore the different combinations of parameters to see the (unacceptable) resulting surfaces before continuing, and the reason to manually modify it.

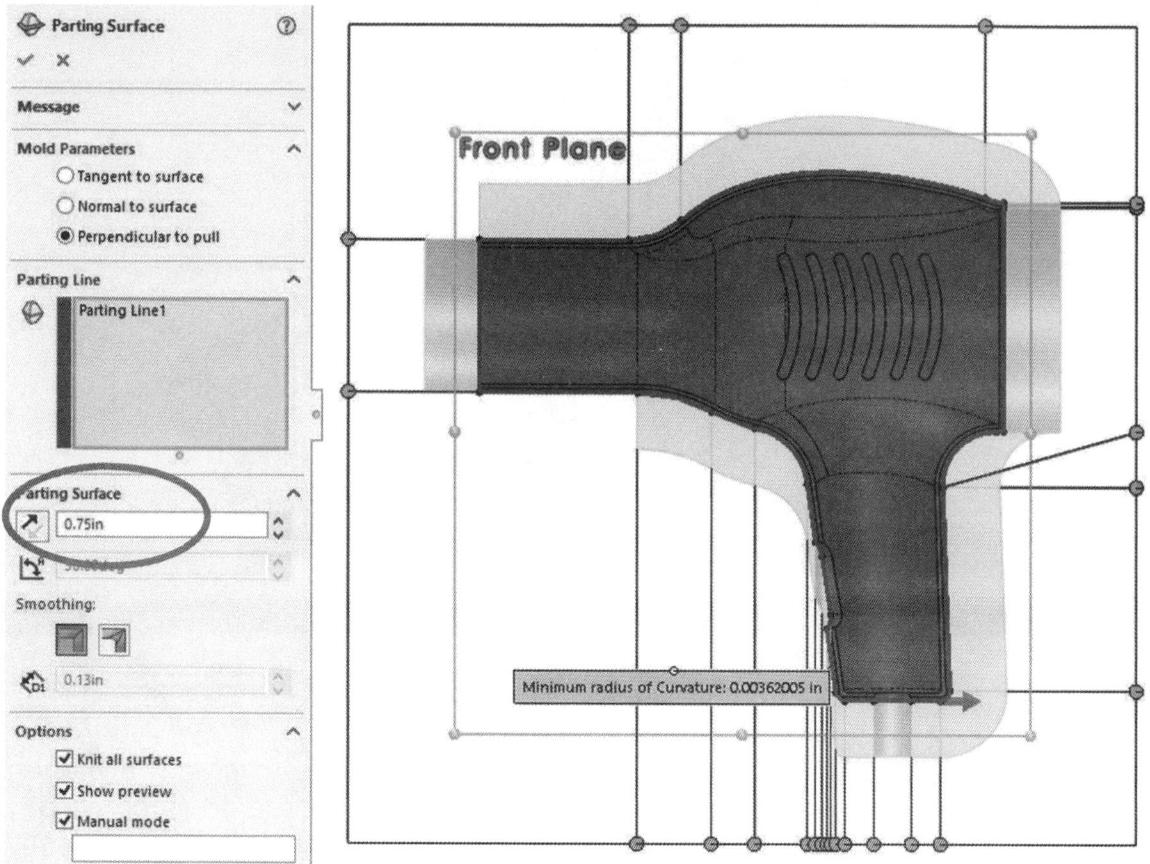

When the "Manual mode" is activated, every vertex in the parting line edges are connected to a handle in an outside frame, where we can click and drag the handles to define the direction of the surface at each vertex. In the area around the switch opening we can see all handles are attached to the bottom frame, and in the front of the hair drier we are missing a couple of surfaces; these will be added manually and then knitted with the parting surface later. To repair the surfaces around the switch, starting at the left most switch handle, click and drag each handle along the outside frame to the left, and then up; when we get to the projection of the vertex in the vertical frame the handle will turn black to let us know we are aligned with the vertex. A preview will let us know which vertex we are working with.

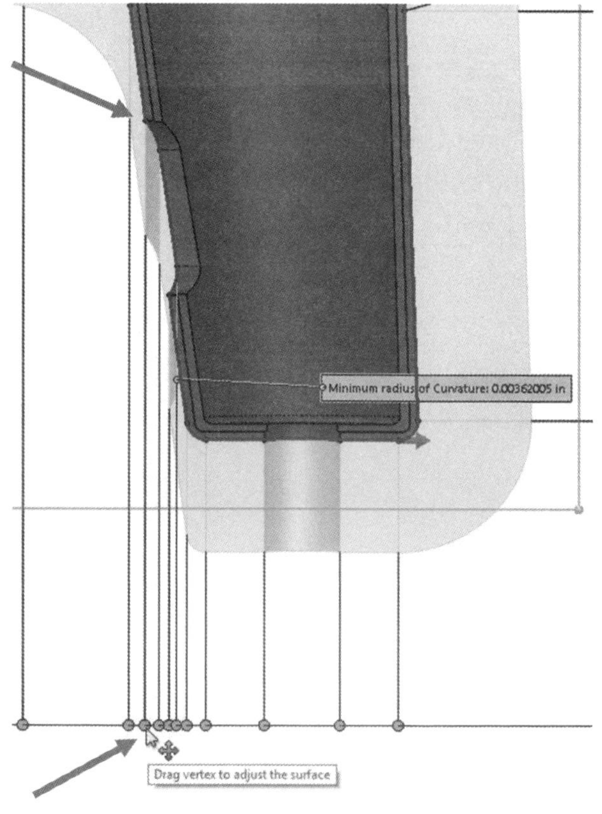

Move the rest of the handles to the vertical frame on the left to fix the surface.

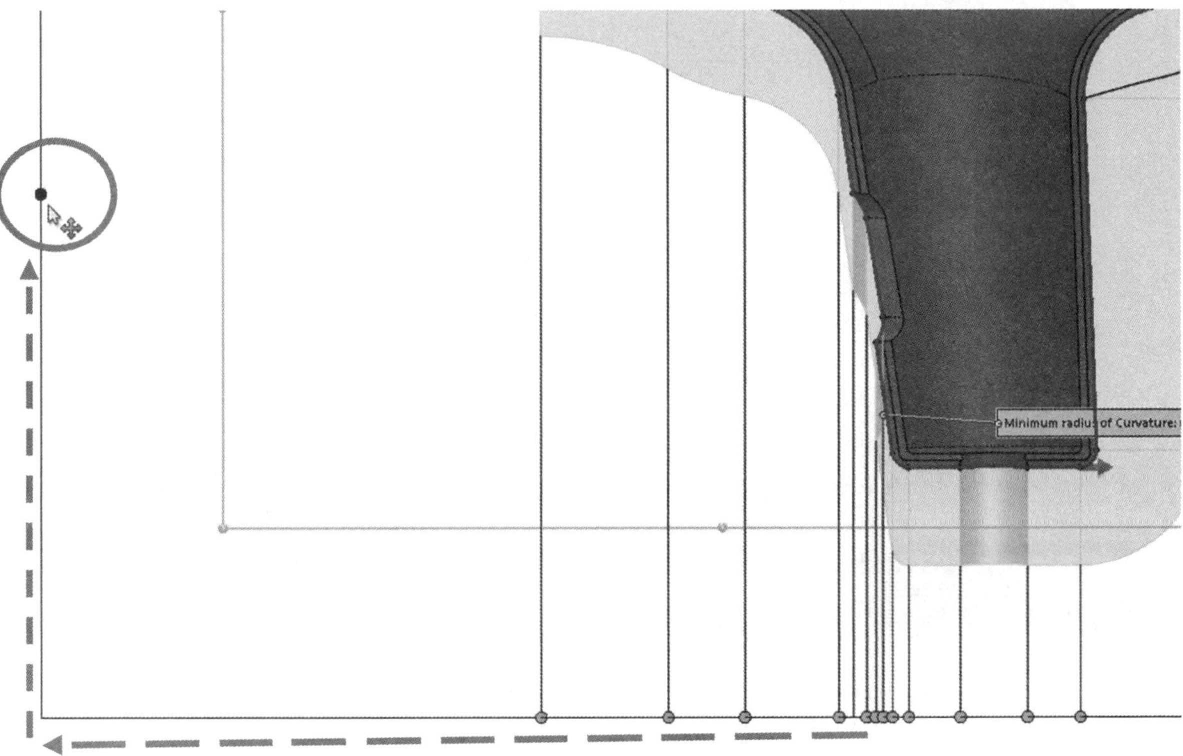

After correctly locating the switch cutout surface handles on the left side frame, click OK to generate the parting surface. Next, we'll add the missing surfaces.

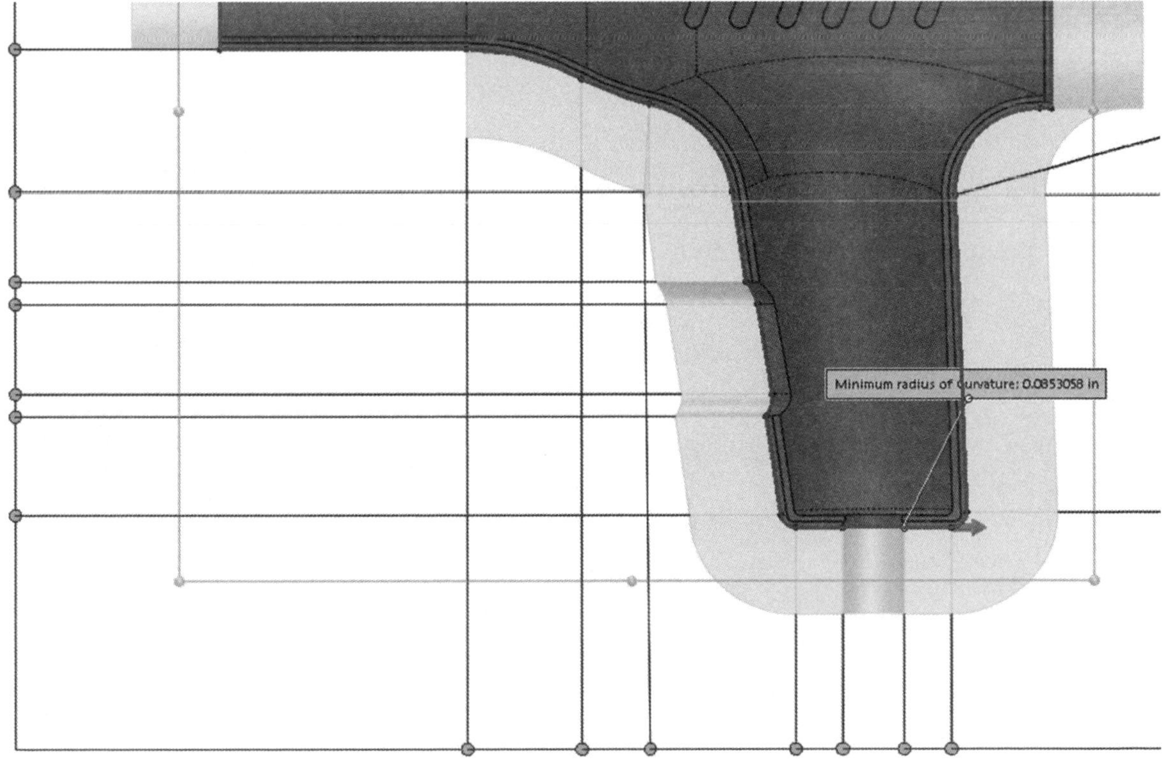

28.7. - To complete the upper part of the surface, add a sketch at the top of the "*Parting Surface1*", and add a tangent arc as shown. This sketch will be used as a guide for a loft surface. Be sure to capture relations to the vertices and Exit the sketch. The center of the arc is located in the corner at the intersection of the faces.

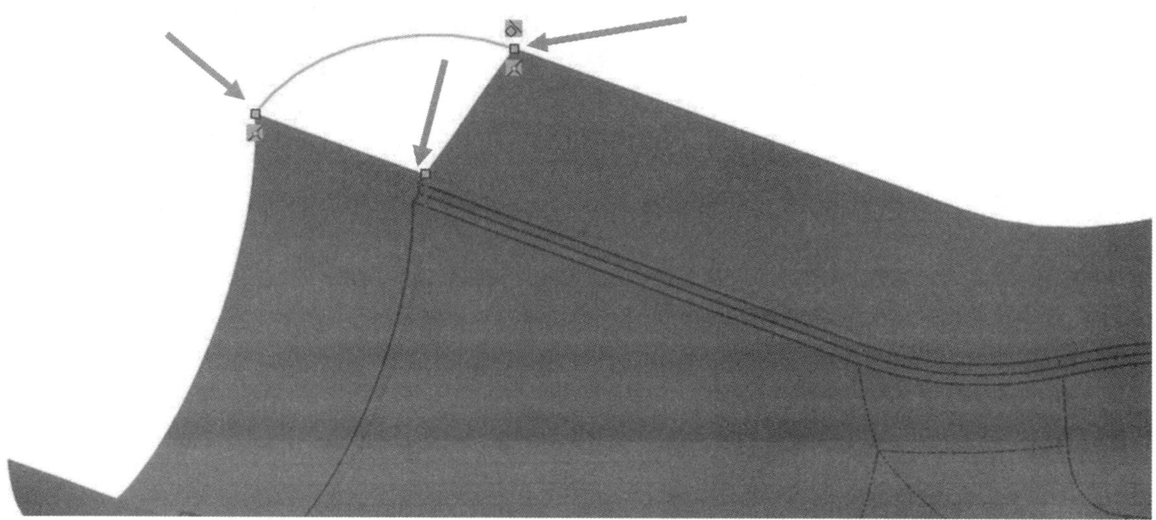

28.8. - Select the "**Lofted Surface**" command to complete the next surface. In the "Profiles" selection box, select the two surface edges indicated near the outside of the surface, and the previous sketch in the "Guide Curves" selection box.

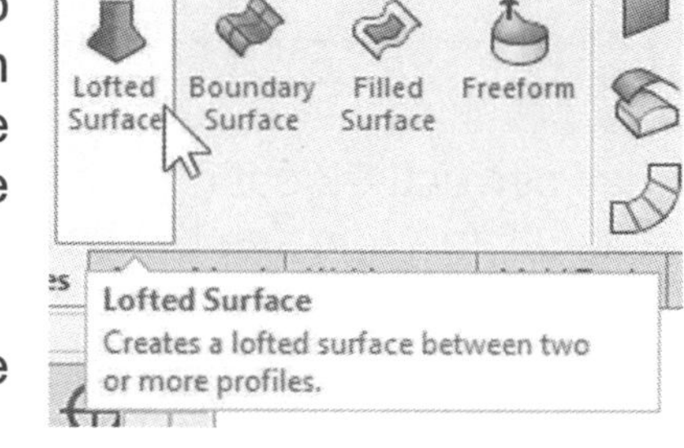

Click OK to generate the new lofted surface and finish.

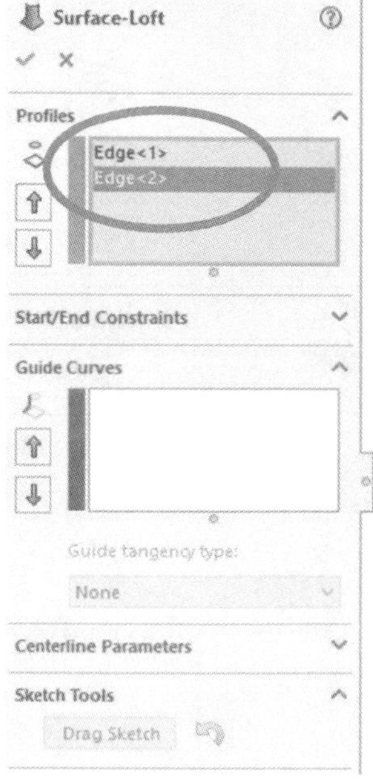

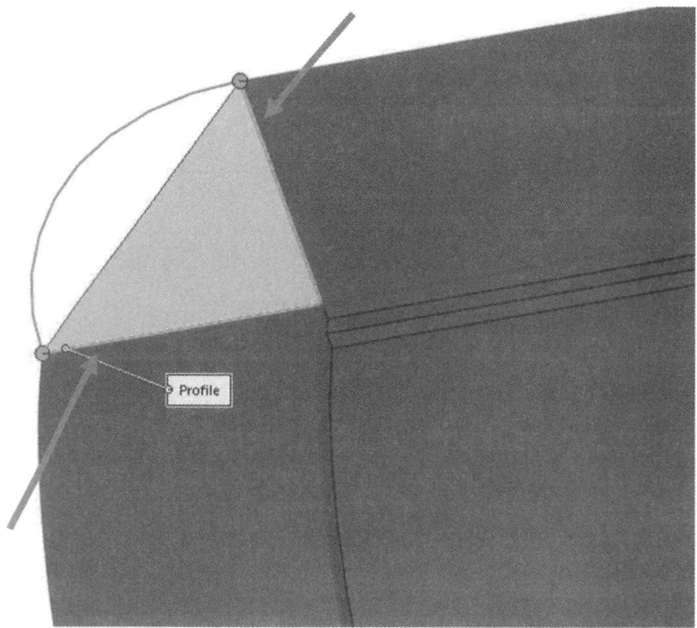

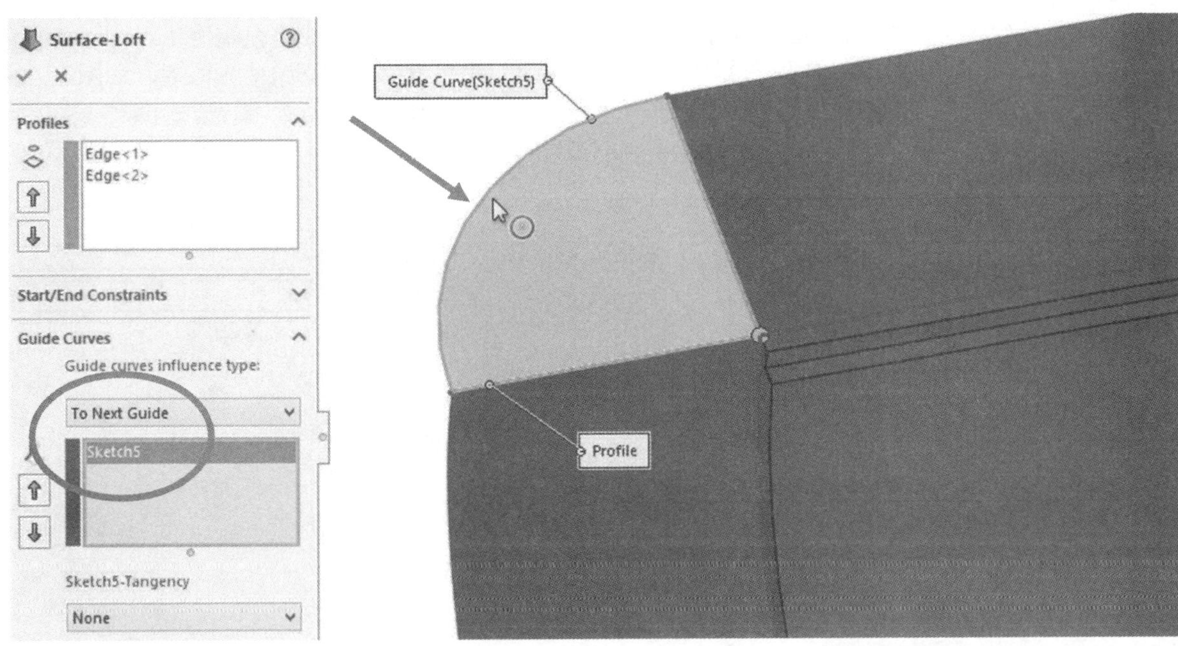

28.9. - To make the next surface add a new sketch in the parting surface just as in the previous step, and add the line and arc indicated. Make the arc's center point coincident to the horizontal edge of the front surface to fully define the sketch. This sketch will also be used as a guide curve. Exit the sketch when done.

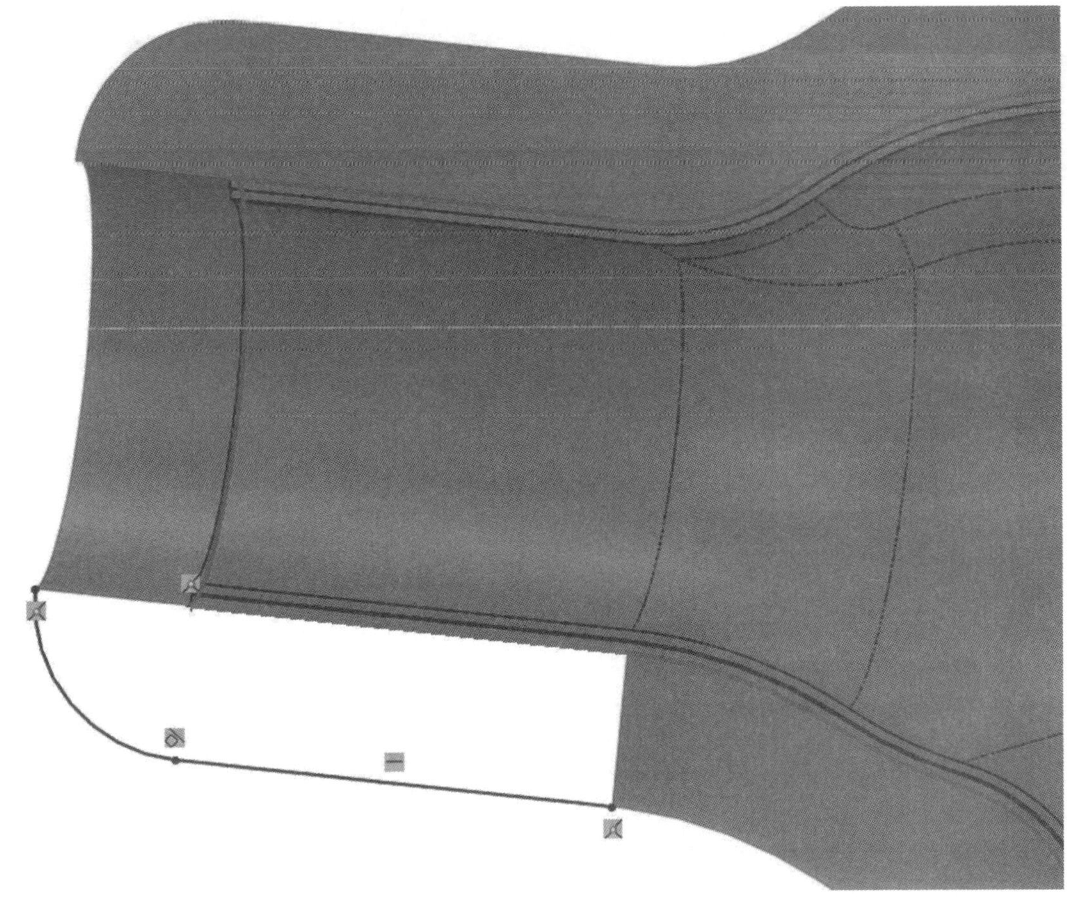

28.10. - Select the "**Lofted Surface**" command again and select the two existing surface edges in the "Profiles" selection box and the previous sketch and the opposite model's edge in the "Guide Curves" selection box, to use two guide curves to control the lofted surface and obtain a better result. Click OK to add the surface and continue.

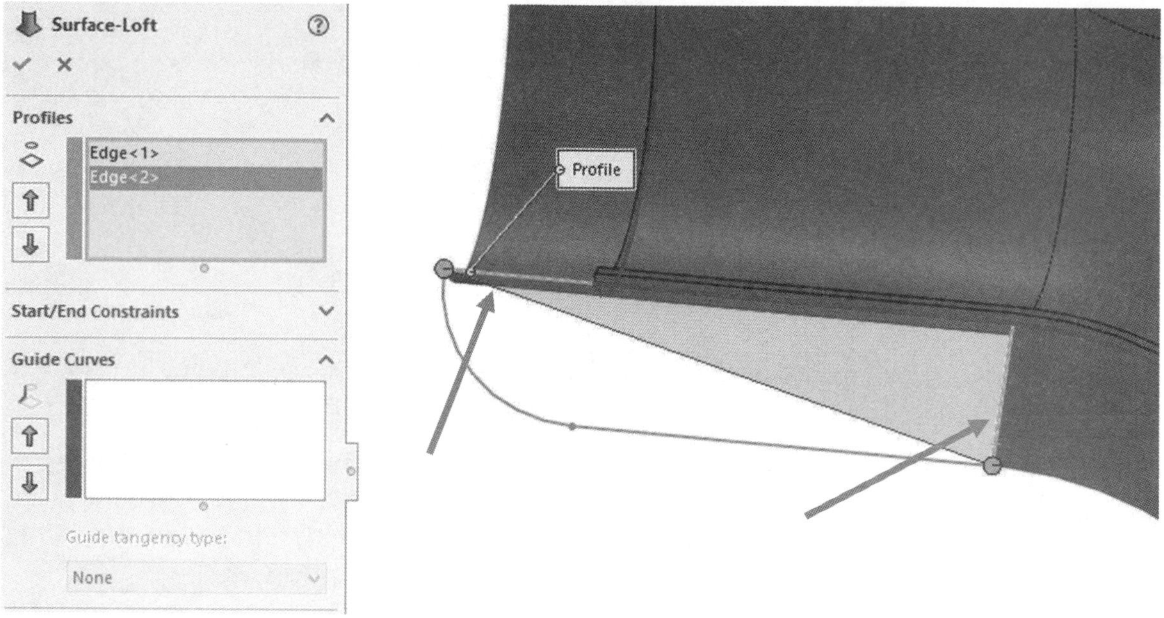

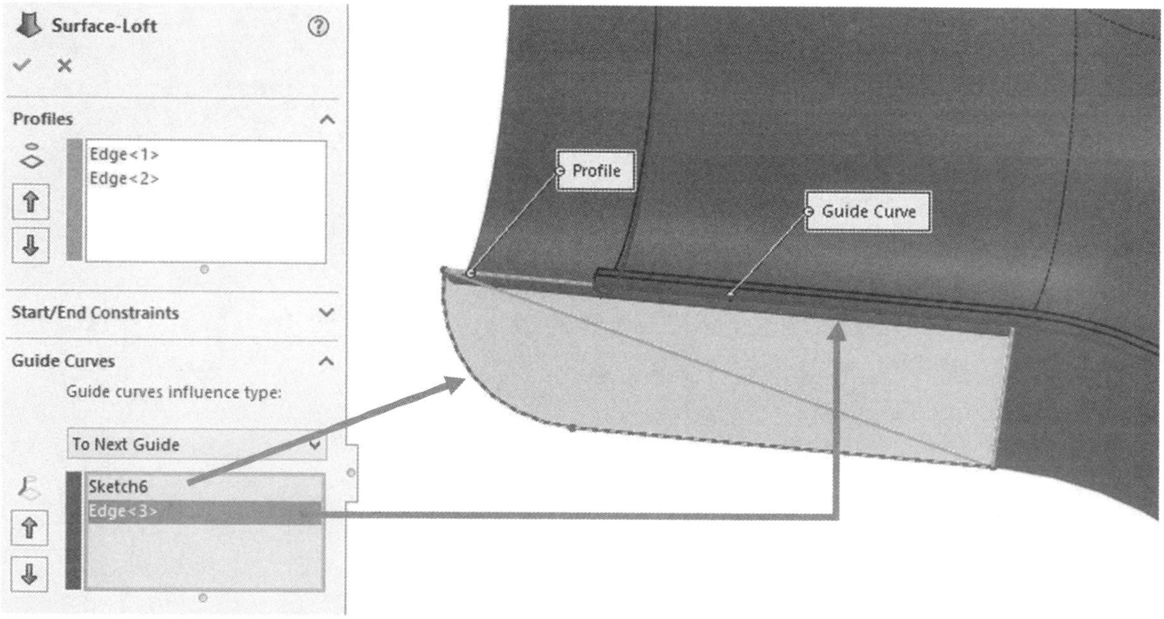

If after selecting the surface edges in the "Profiles" selection box we cannot see a preview, it's probably because the edges were selected in opposite ends causing the loft to twist. To correct it drag the green dot on one of the edges to the other vertex.

After adding the two lofted surfaces we have a total of six surface bodies. Keep in mind that after expanding the "Surface Bodies" folder, all surfaces within it will become visible, and overlapping surfaces and model faces will be visible at the same time.

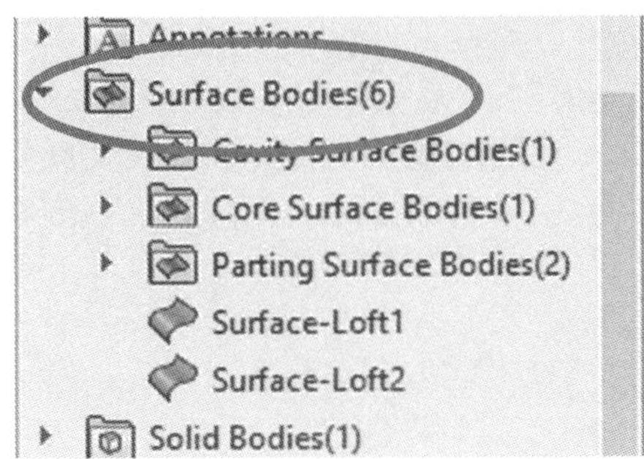

28.11. - To use the lofted surfaces for a tooling split we have a couple of options:

- The surfaces can to be added to the *"Parting Surface Bodies"* folder.
- They can be selected at the time of making the tooling split.
- We can knit them to the existing *"Parting Surface1"* to have a single surface and then add it to the *"Parting Surface"* folder.

In our example, we are going to use the last option to show additional functionality and make it easier to create the tooling split sketch.

Select the "**Knit Surface**" command and knit the two lofted surfaces created with the two *"Parting Surface1"* bodies.

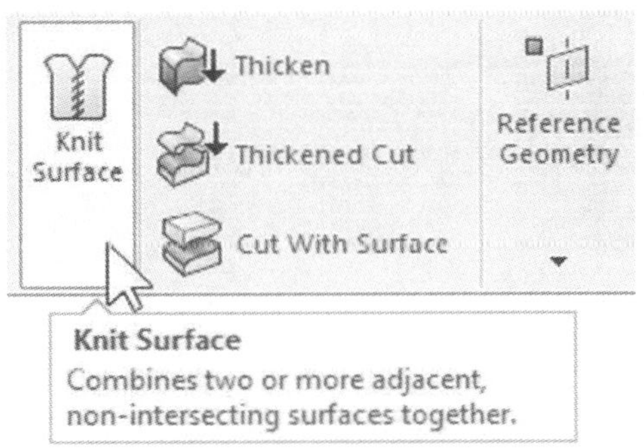

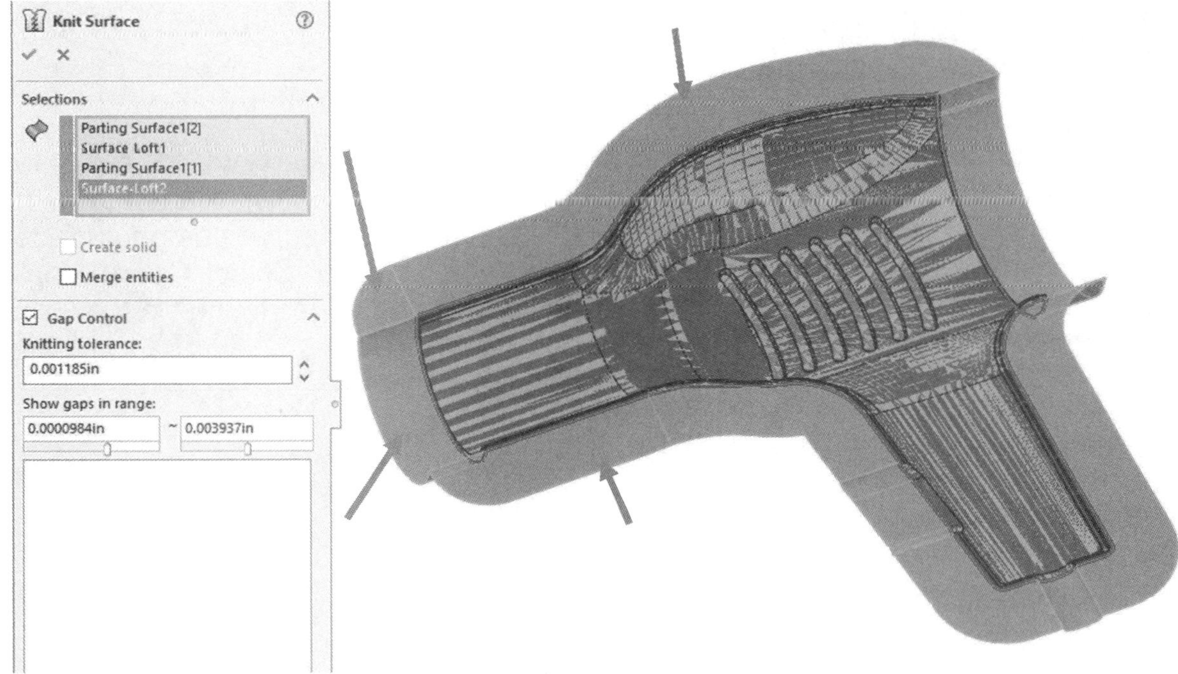

28.12. - After knitting the surfaces, the lofted surfaces become part of the parting surface. Now we need to add the new knitted surface to the *"Parting Surfaces"* folder. Click and drag the *"Surface-Knit1"* surface into the folder.

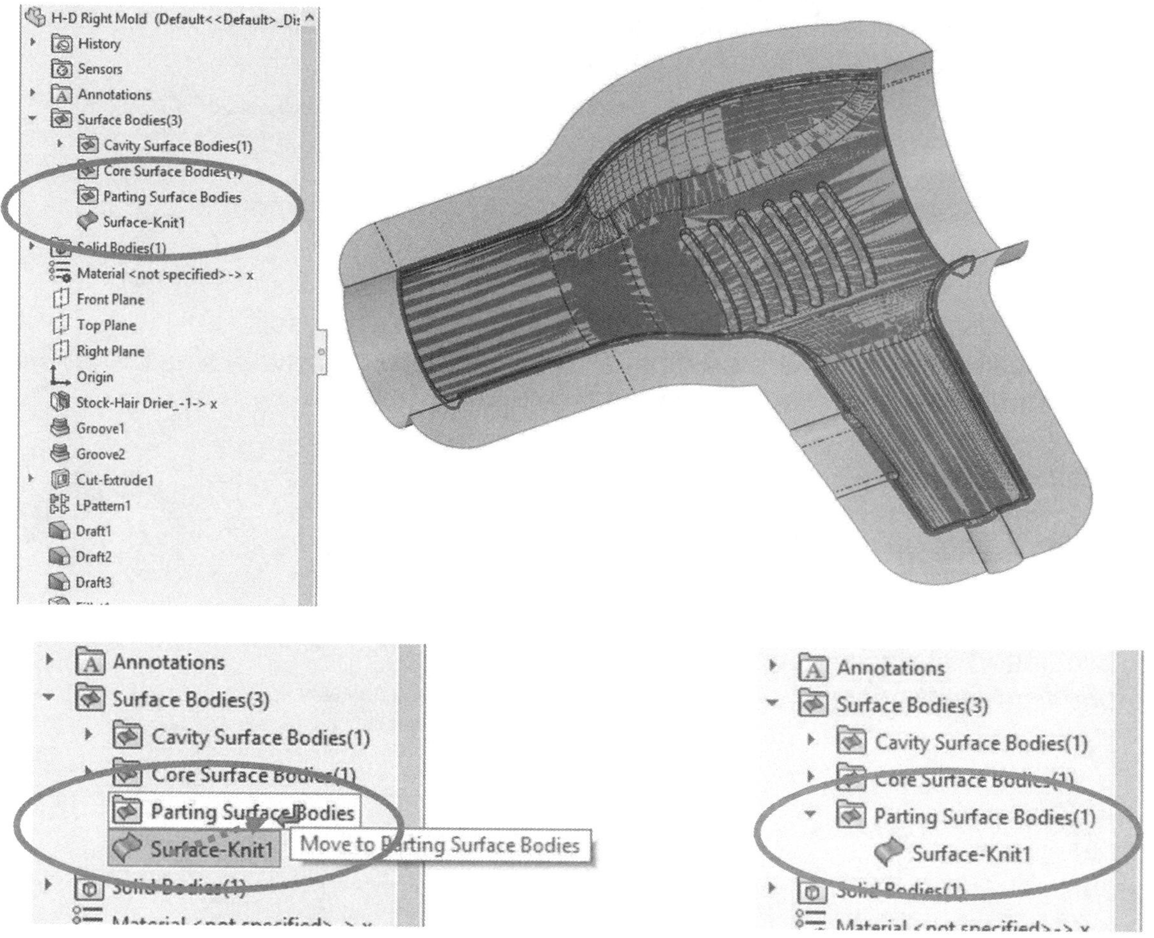

28.13. – After assigning the surface bodies to the correct folders, we are ready to make a tooling split. Select the "**Tooling Split**" command, and select a flat area of the parting surface to add a new sketch on it. To select the entire parting surface outline, right-mouse-click in an external edge and use the "Select Open Loop" command. Use "**Convert Entities**" to project all selected edges to the sketch. Exit the sketch when done to continue the "**Tooling Split**."

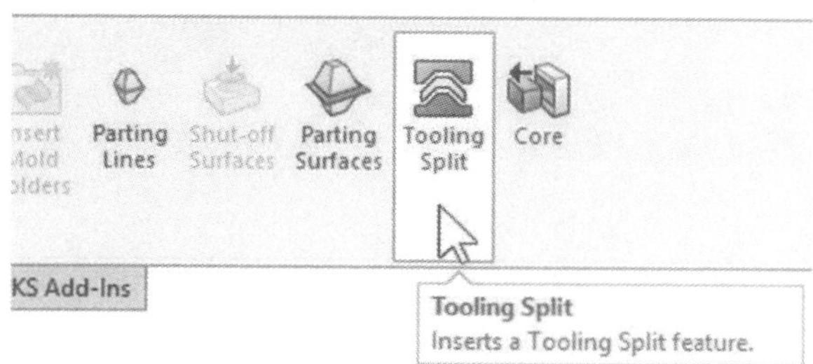

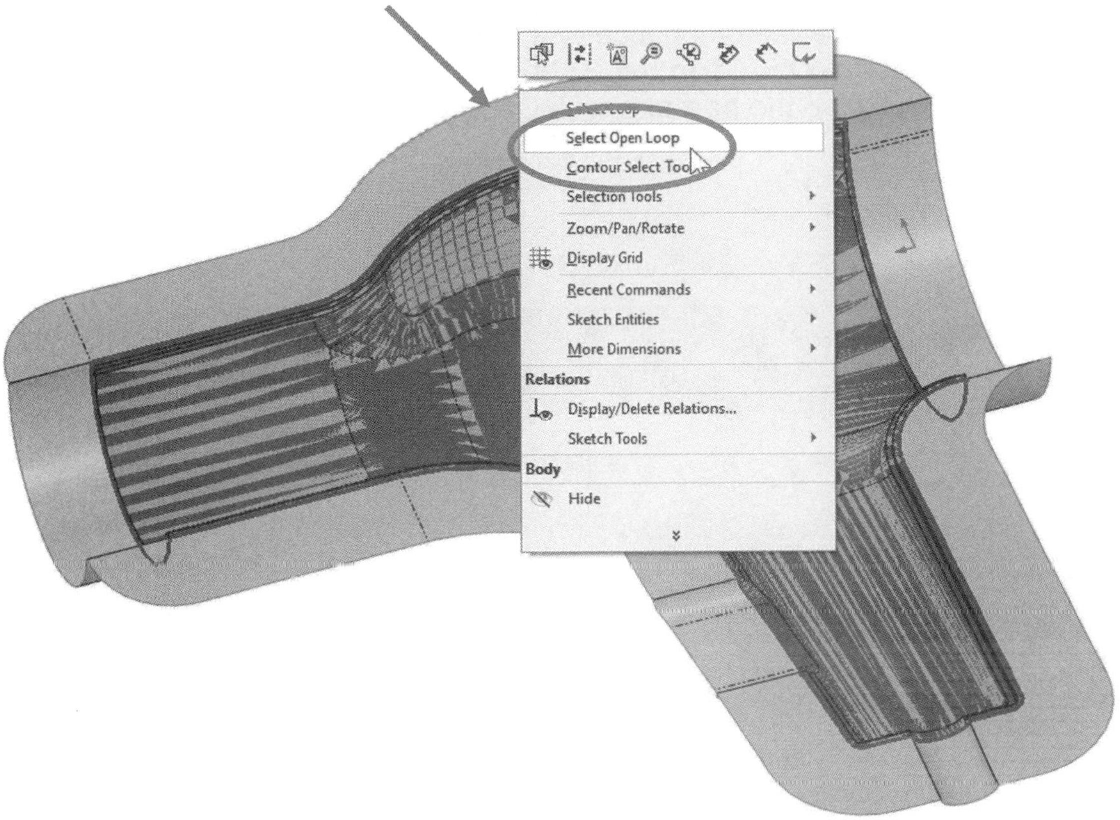

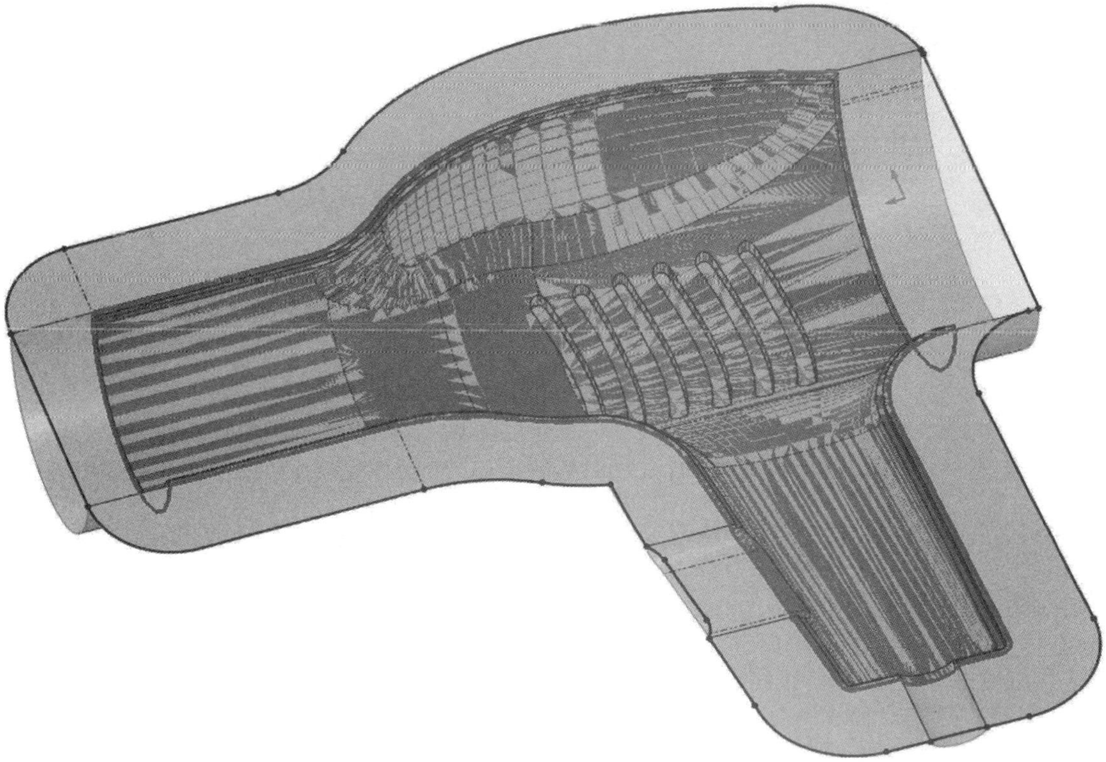

The tooling sketch must be bigger than the parting line and as big as or smaller than the parting surface.

After exiting the sketch change to a Top or Bottom view to preview the size of the tooling block. In the "**Tooling Split**" command, make the "Block Size" big enough to fully enclose the part and click OK to continue.

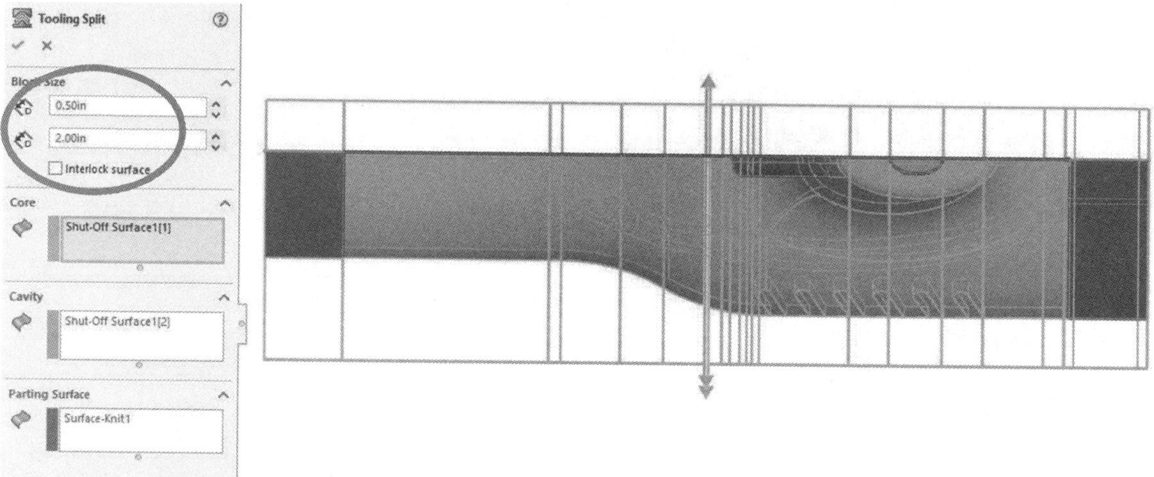

28.14. - Hide the *"Parting Line1"* feature and all *"Surface Bodies"*...

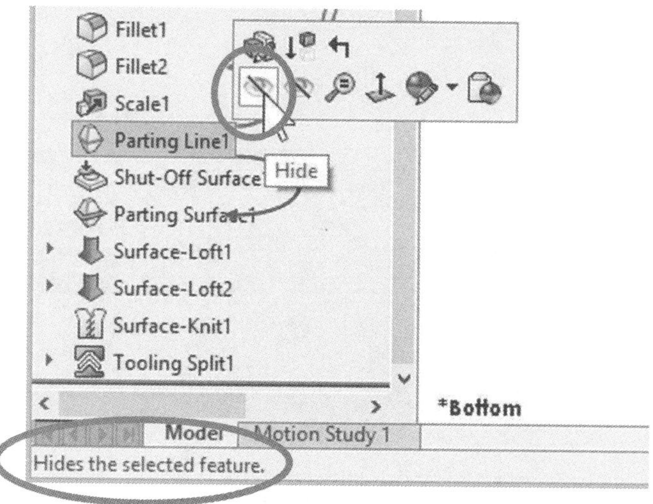

In the pop-up menu we have two Hide/Show icons, the first one is to control the feature's visibility, and the second is to control the solid body's visibility.

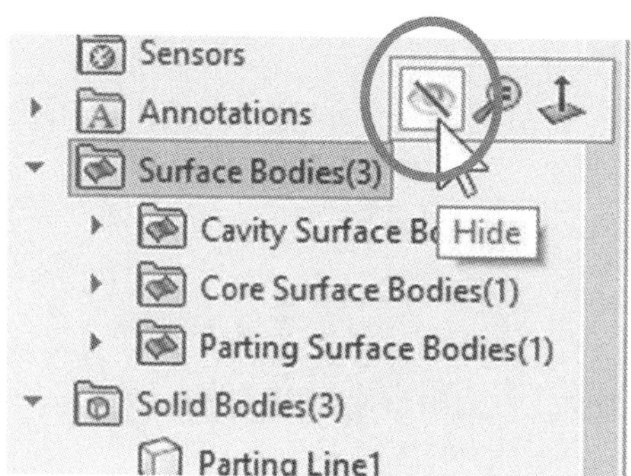

Hide the body at the top of the mold to see the cavity.

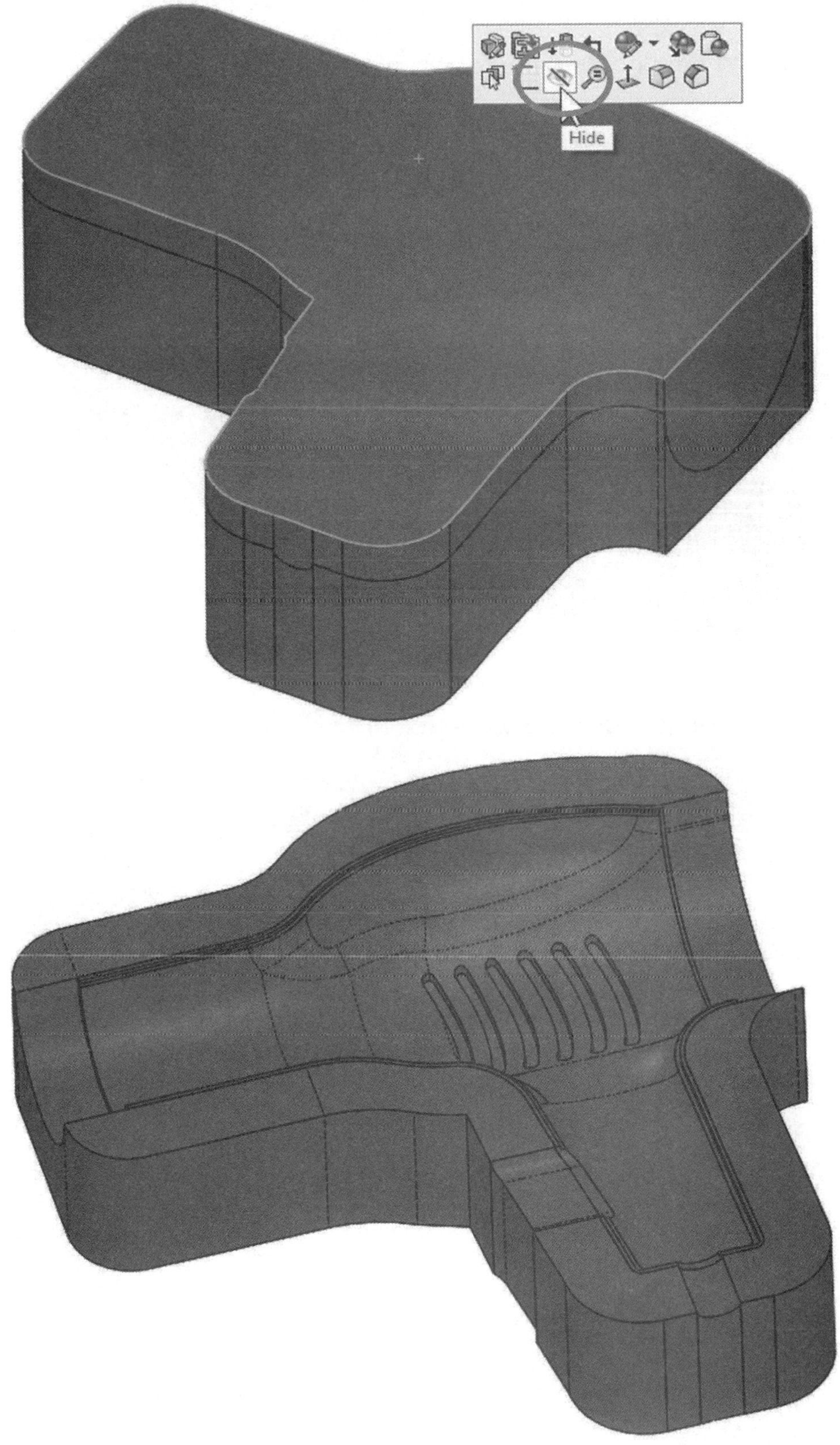

28.15. - Optionally we can change the color of the molded part's body (just the body's color) to make it easier to identify. Select the molded part, and from the pop-up menu select the "**Appearance**" icon. In the drop-down menu we can choose to change a face, feature, body or the entire part's appearance. Select the "Body" option and change its color for visualization purposes.

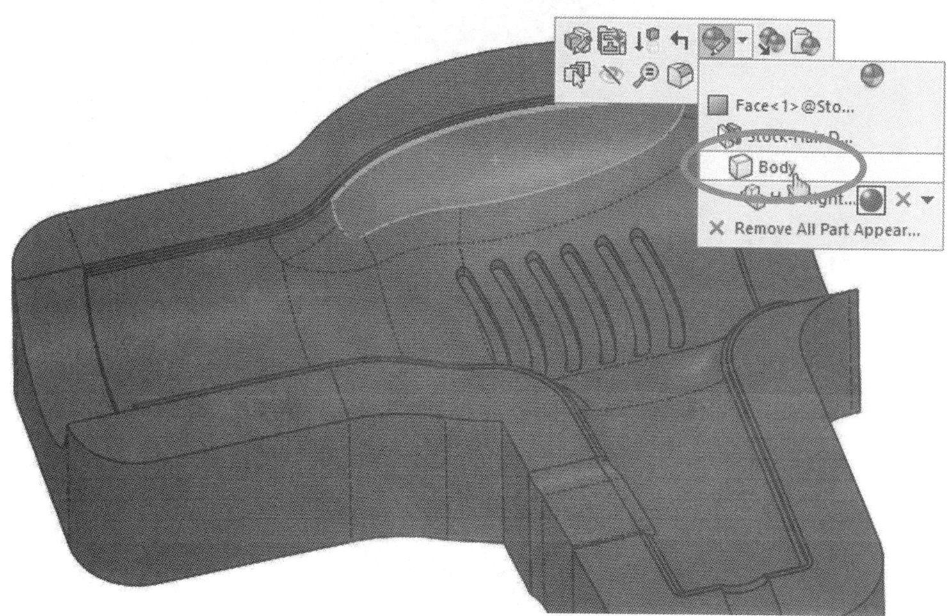

Or select the body in the *"Solid Bodies"* folder and change its color this way.

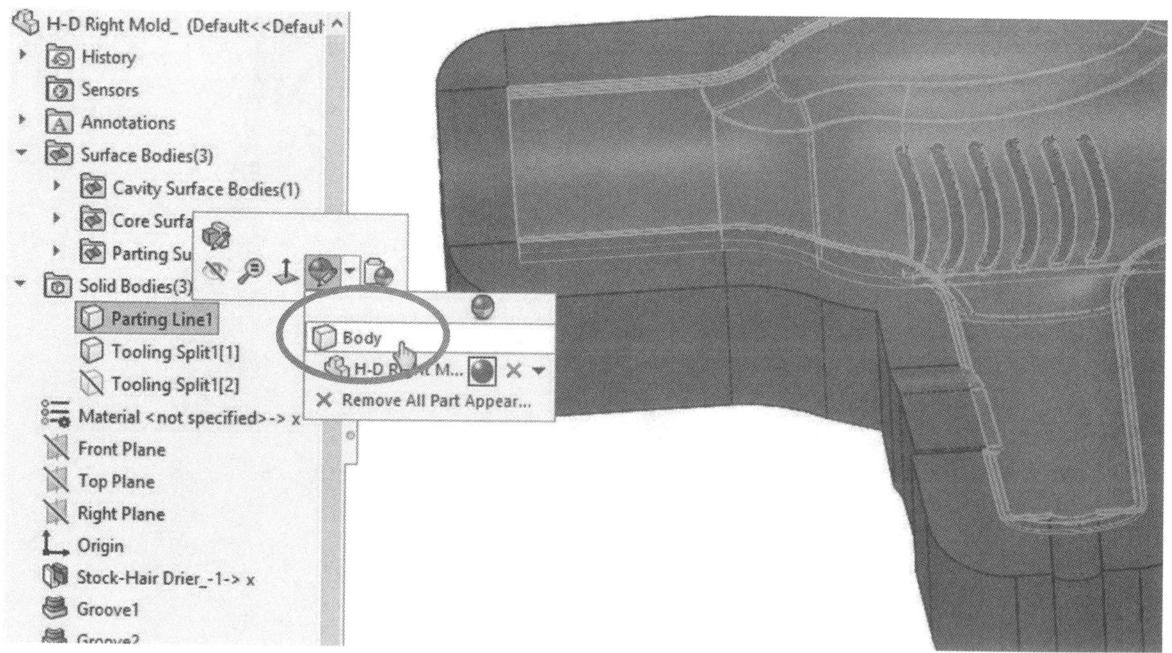

Optionally we can use the "Display Pane" (shortcut F8).

The finished Core and Cavity with the molded part inside. Save all files and close.

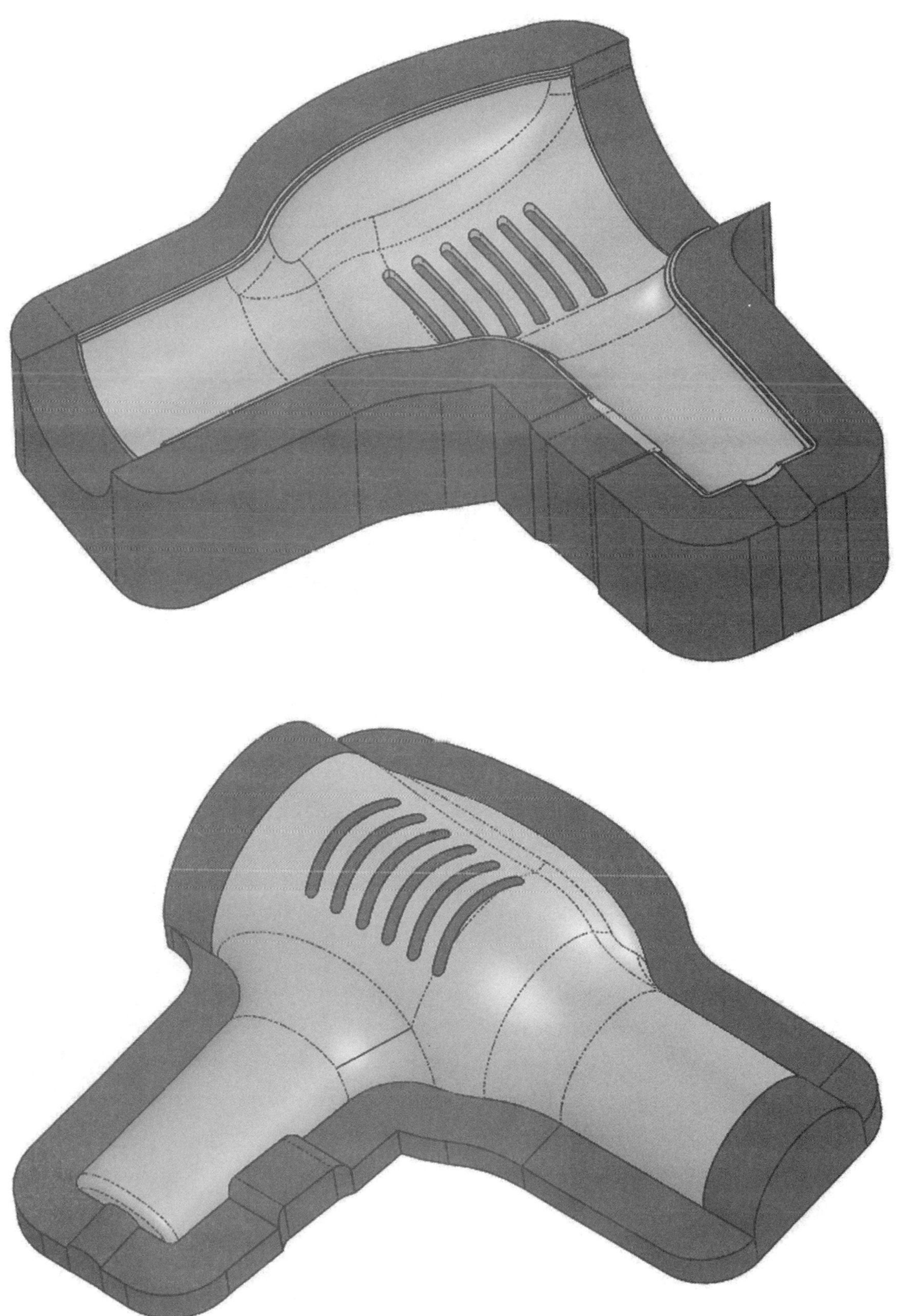

Exercise: Create new part files from the Core and Cavity bodies.

- In the Cavity part, eliminate the vent hole markings to make a smooth surface (like we did with the back cover) using the "Delete Face" command.

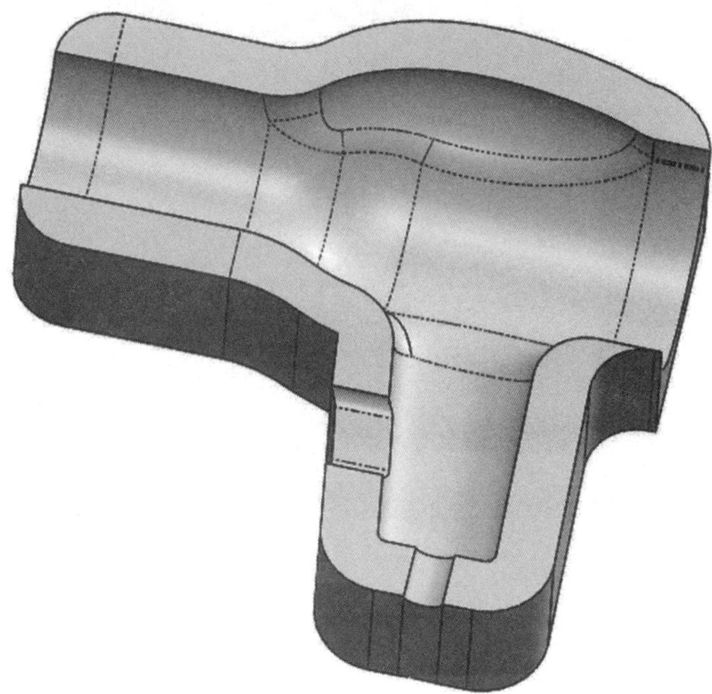

- In the Core part, cut an insert for one of the vent holes.

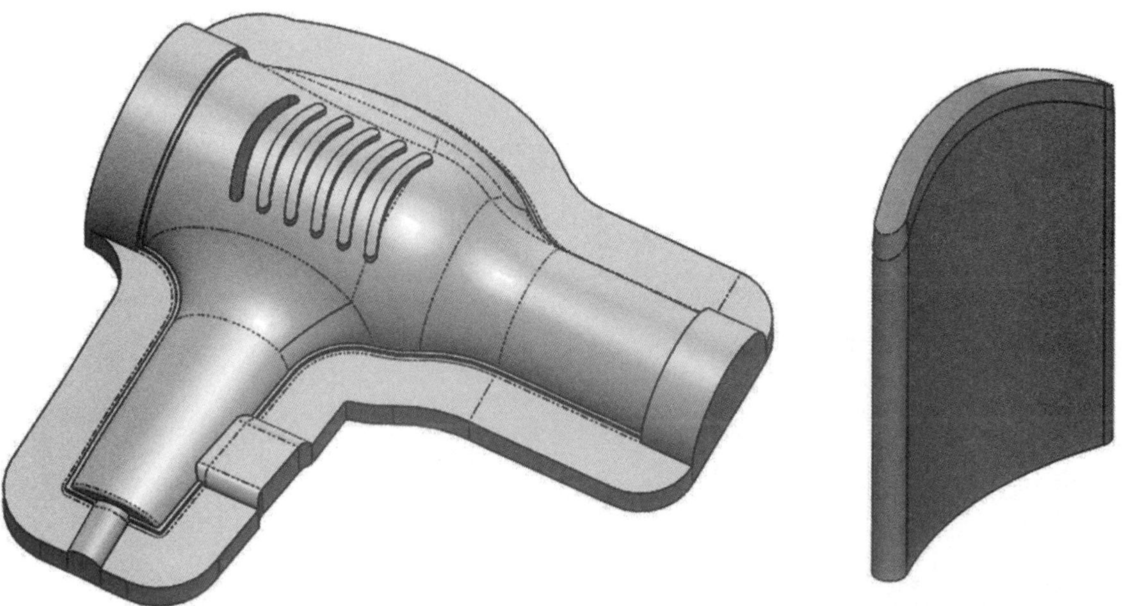

Ribs Feature

29.1. - When designing plastic parts, it's common practice to add reinforcement ribs to make parts stronger. Reinforcement ribs can be created using regular modeling tools like extrusions, cuts, drafts, etc.; however, using this approach can be very time consuming. The "**Rib**" feature automates and simplifies the creation of reinforcement ribs. To add a couple of ribs to our part, go back to the '*H-D Right Mold'* part, and move the rollback bar before the *"Scale"* feature to add the ribs before the mold related features.

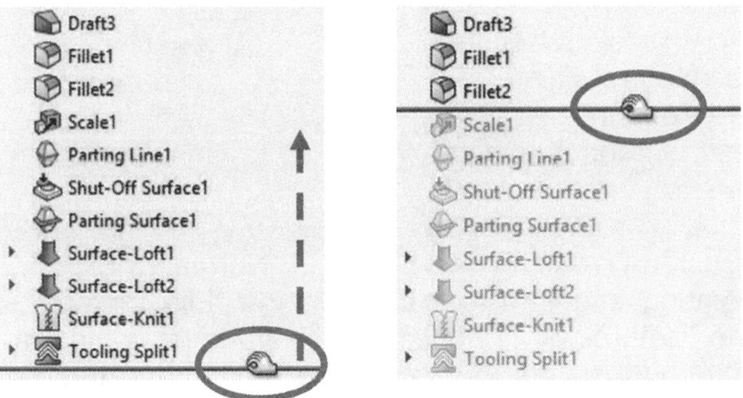

Using this approach we can add the ribs and other features, then roll forward to the end and have the mold's core and cavity update.

29.2. - Add a new sketch in the flat face of the switch cutout; draw the next three lines and dimension them. Notice the lines do not go all the way to the edges of the part. The rib feature will extend until it reaches the sides of the part.

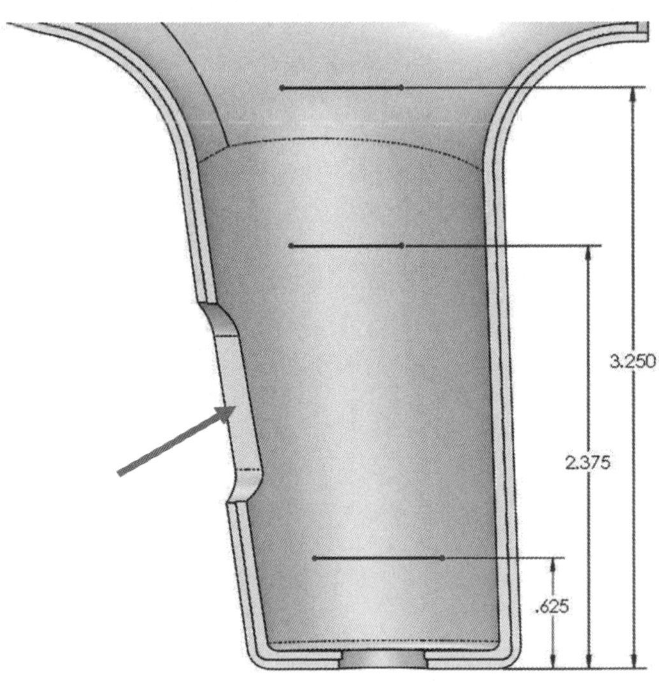

29.3. – The Rib feature uses a sketch with single or multiple open and/or closed profiles. To create the ribs we have to define their direction, thickness and draft angle.

Select the "**Rib**" command from the Features toolbar or the menu "**Insert, Features, Rib**." Set the "Thickness" to 0.075". Using the mid plane option, activate the "Draft" option and make it 2 degrees. After the "Draft" option is activated we'll see the option to select where to start the draft. We'll use the default setting "At sketch plane," meaning the sketch plane will also be the "Neutral Plane" for the draft. (The option "At wall interface" makes the neutral plane at the bottom of the rib.)

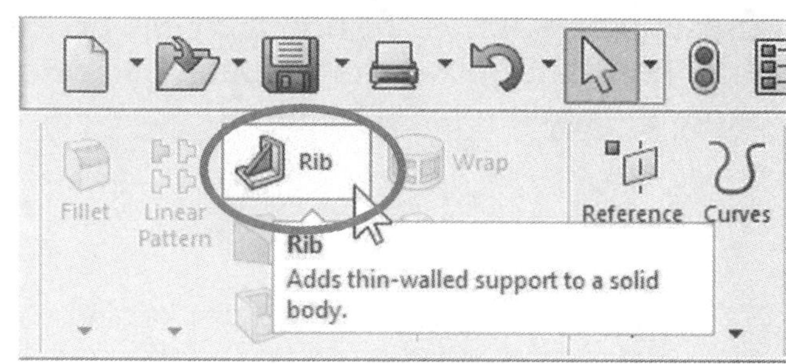

The "Extrusion Direction:" will be set to "Normal to Sketch." If the direction arrow is not pointing into the part, turn On the "Flip material side" checkbox, otherwise the rib feature will fail because the material could not be added. The "Draft Outward" option makes the rib feature grow in the direction of the arrow. Click OK to complete the first rib feature.

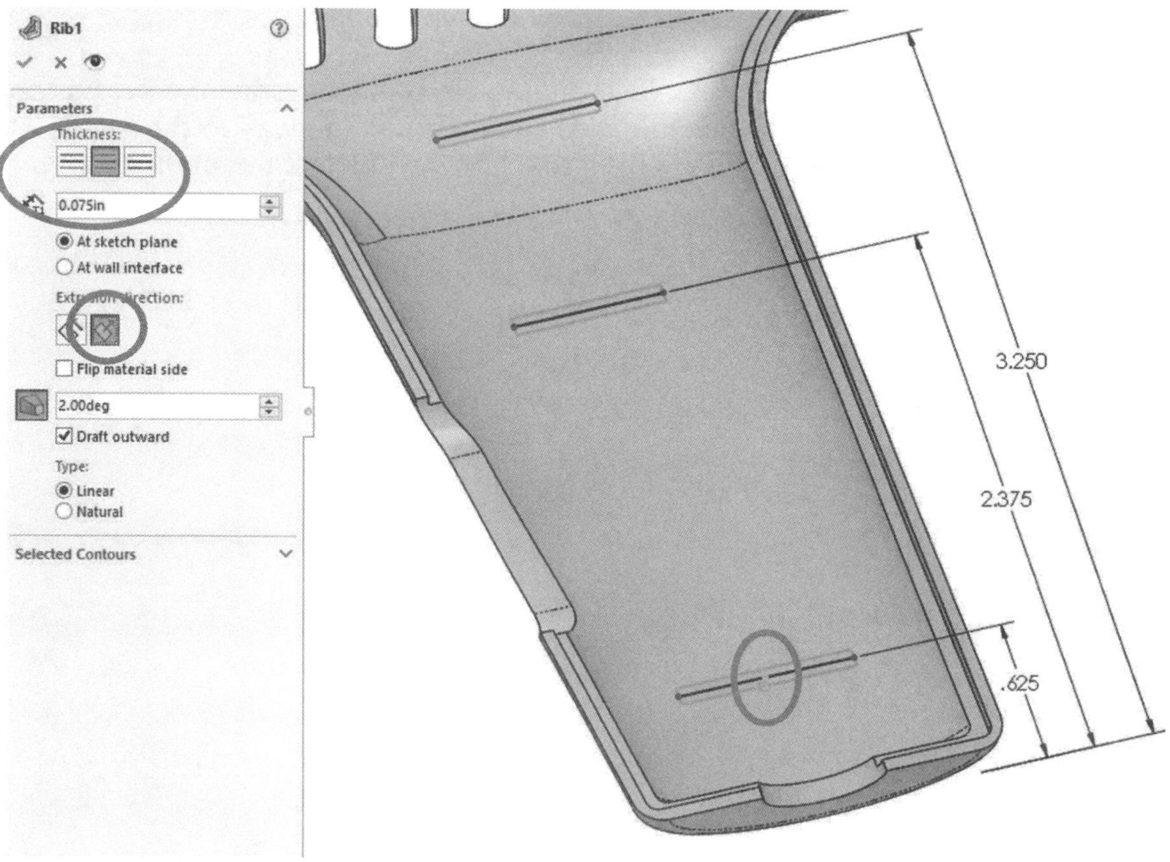

The finished rib will add material using an *'up to next'* end condition in the direction of the extrusion and extending the sketch lines until they reach either a part's wall or another rib, making it a powerful feature because it can match the geometry under the rib feature. An important thing to know about the rib feature is that if we don't have a stopping face for the rib in either direction, the feature will fail.

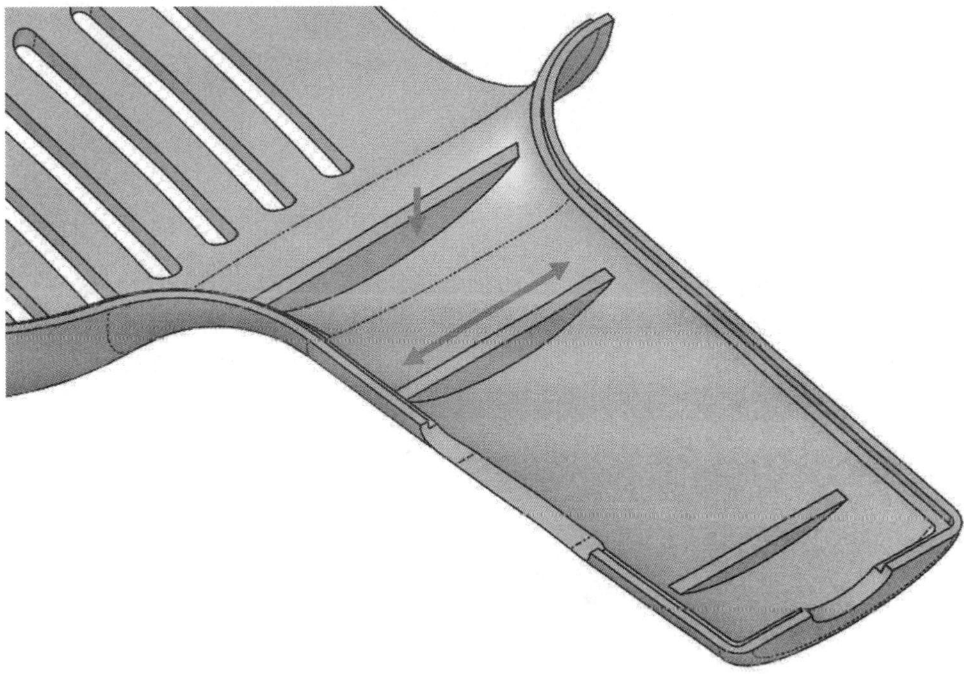

29.4. - For the second rib, add a plane parallel to the *"Right Plane"* 6 inches to the left and add the following sketch in it. Again, the sketch is not touching the edges. The arc's center point is located at the sketch origin.

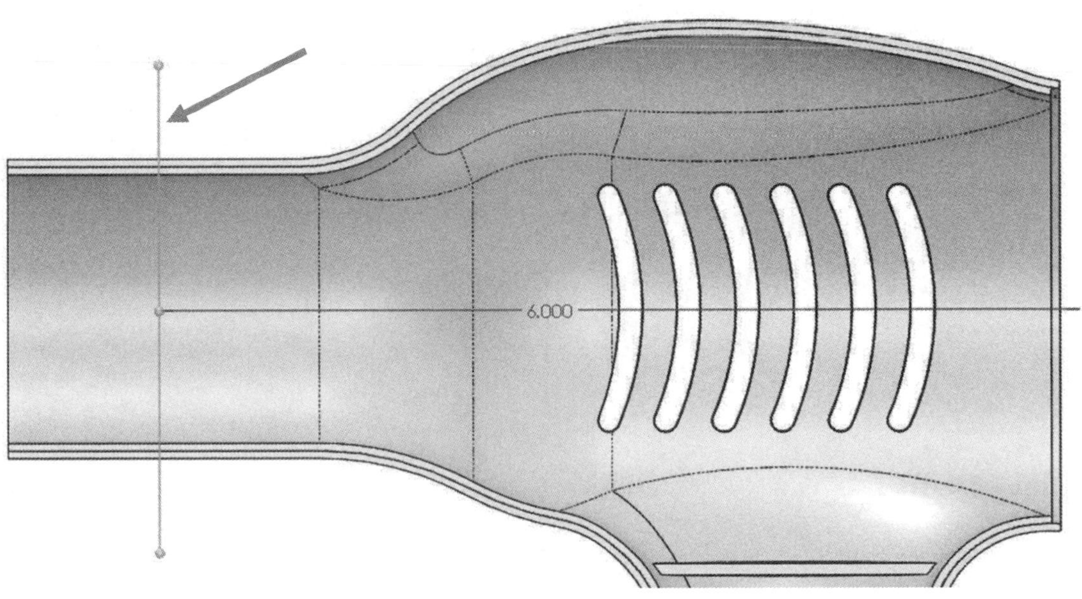

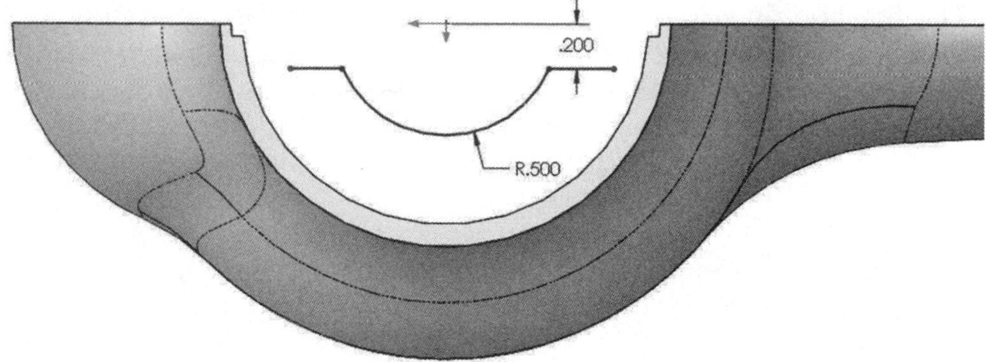

Select the "**Rib**" command and use the same settings as the previous rib, but instead we'll use the option "Parallel to plane" which is automatically pre-selected. After activating the "Draft" button, a new button labeled "Next Reference" allows us to select the sketch segment that will be used to start measuring the draft. Click on it to change the reference to one of the straight lines if needed.

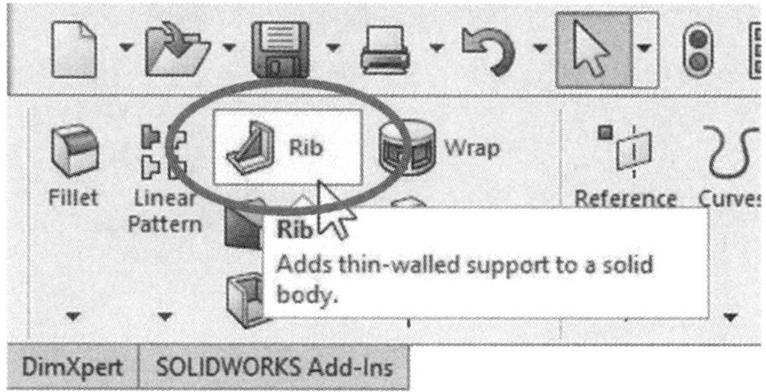

Make sure the material direction arrow is pointing into the part, otherwise check the "Flip material side" option and click OK to continue.

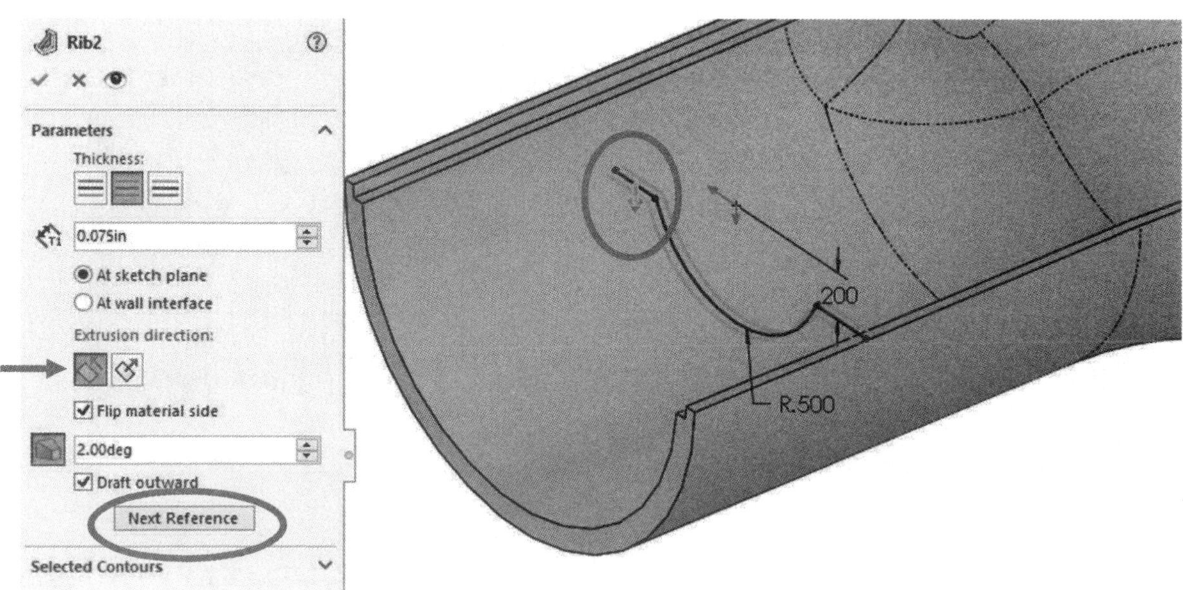

The second rib is complete.

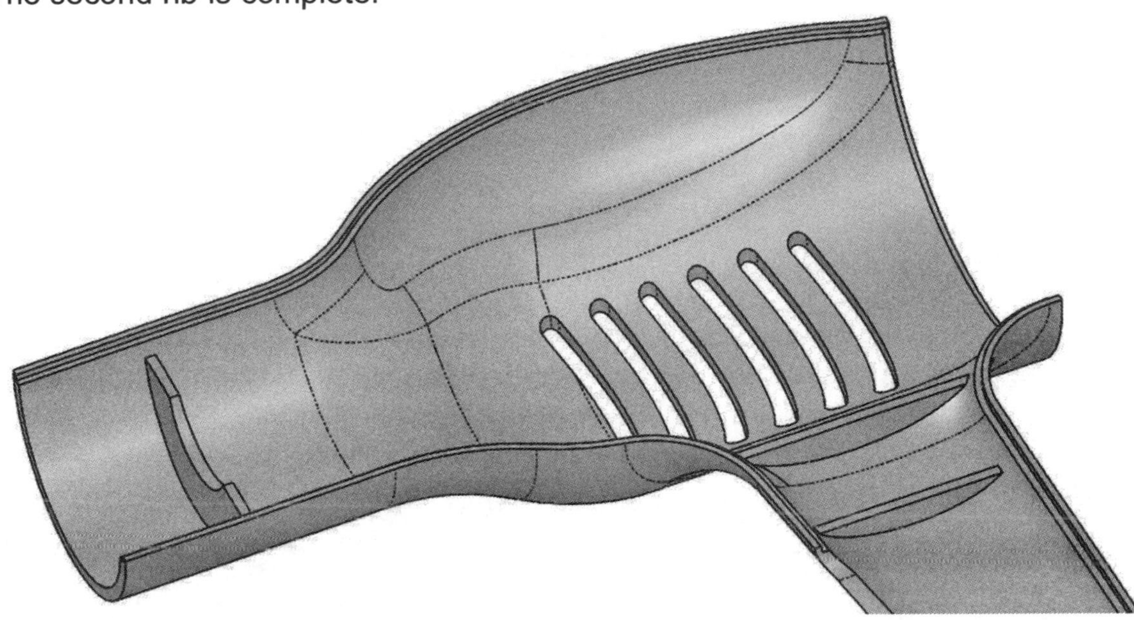

29.5. - For the next step, we'll add a pattern using the second rib, using an option to skip pattern instances. Make a linear pattern of the second rib with four instances and a spacing of 0.625 inches. Before finishing the pattern, expand the "Instances to Skip" selection box. Here we can *'turn off'* pattern instances by selecting the unwanted instance in the screen; click in the third instance to turn it off, and notice its preview is gone. Click OK to finish the pattern. As expected, the third instance is removed and the other rib instances conform to the varying part's geometry.

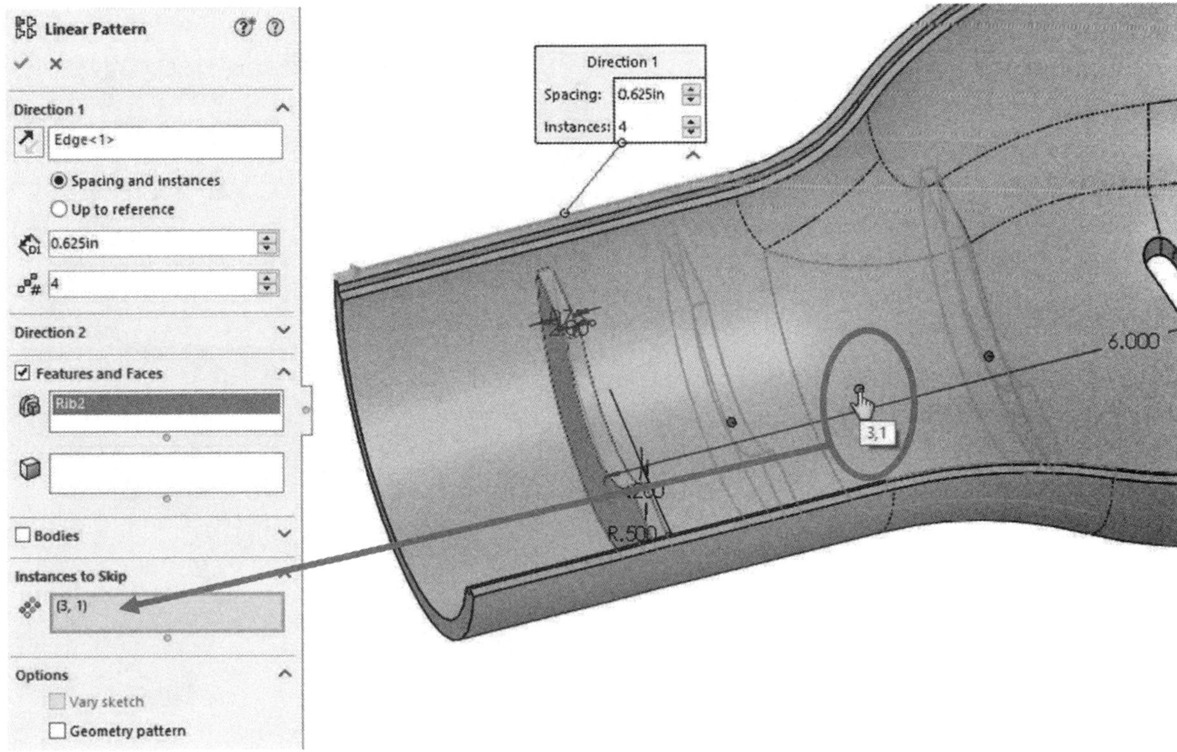

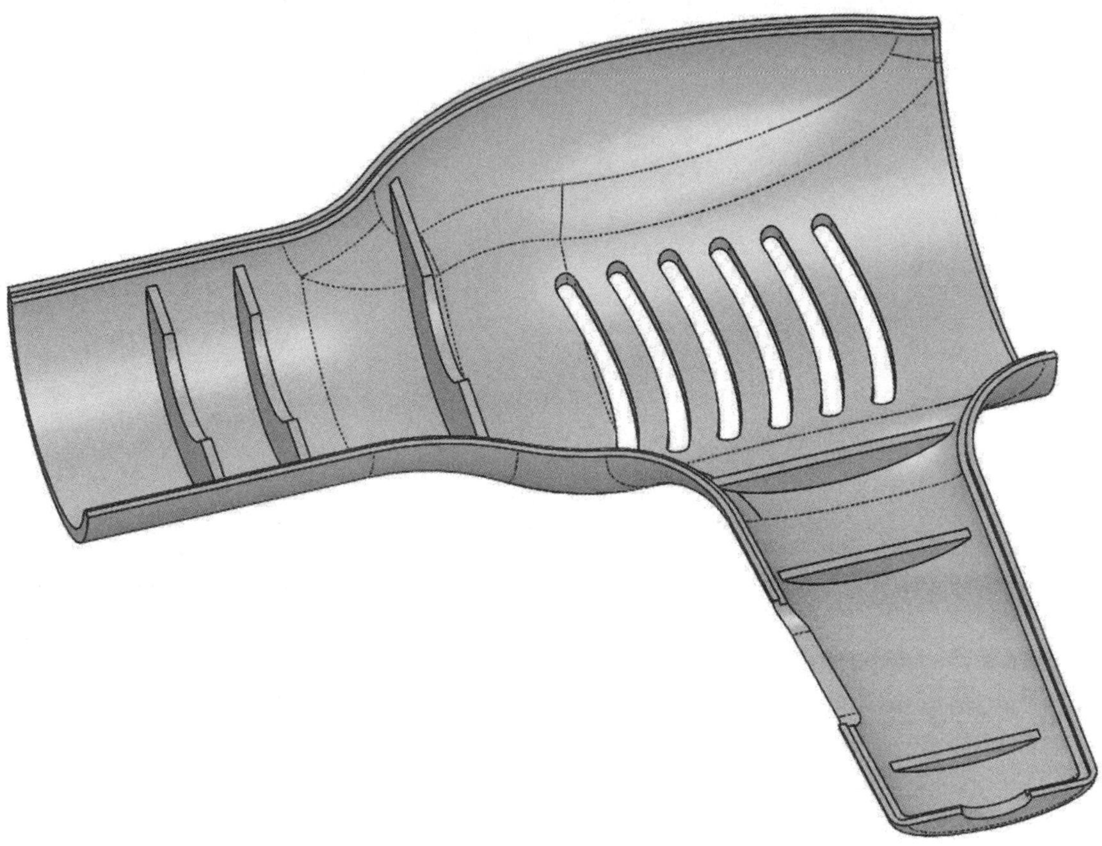

29.6. - As a final touch add a 0.020″ fillet to both rib features and the pattern.

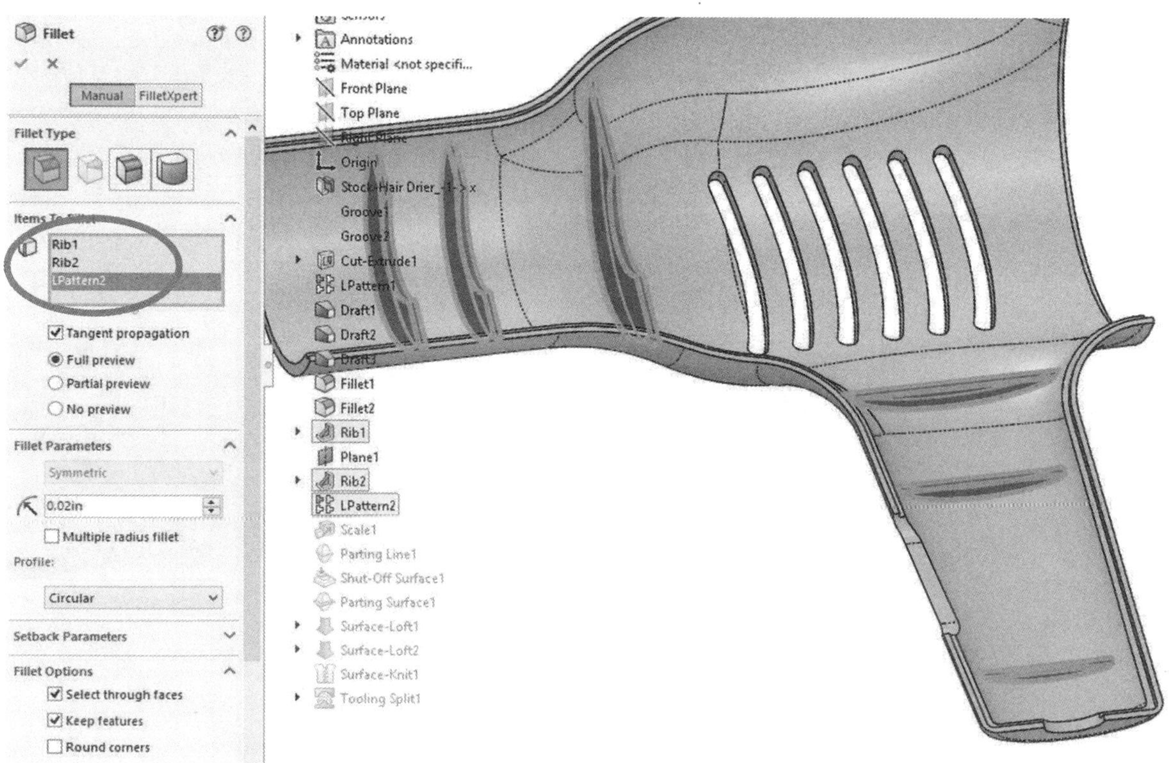

29.7. – After the ribs are complete we must rebuild all of the mold features. To do this we need to move the “Rollback” bar to the bottom. After rebuilding the rest of the features, the mold rebuilds correctly as expected because none of the edges along the parting line or the parting surface were modified. If we had modified any of the edges or faces used in the mold features, there is a possibility we would have to fix errors in subsequent features.

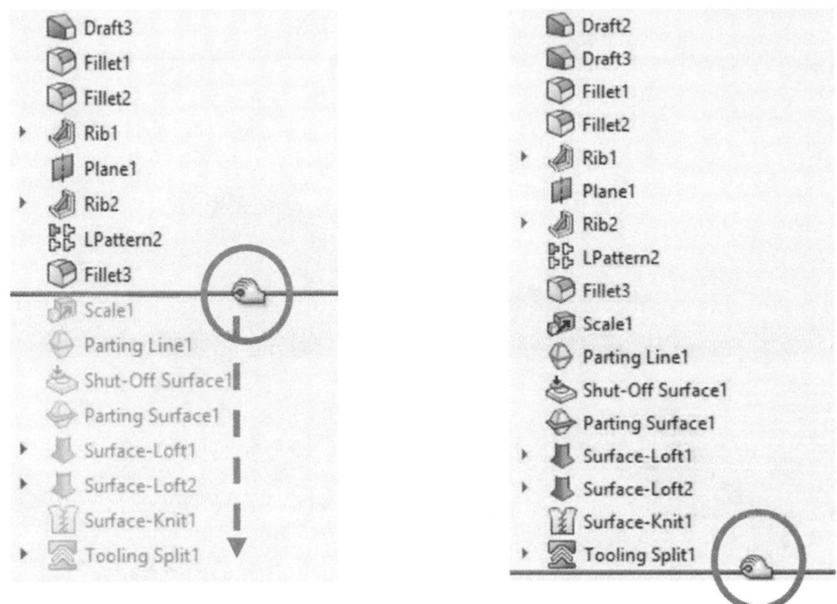

29.8. - With the finished part open to propagate changes to the Core and Cavity parts, open the mold’s Core to propagate the changes added in this lesson. The Cavity would be the same as only the core side was changed.

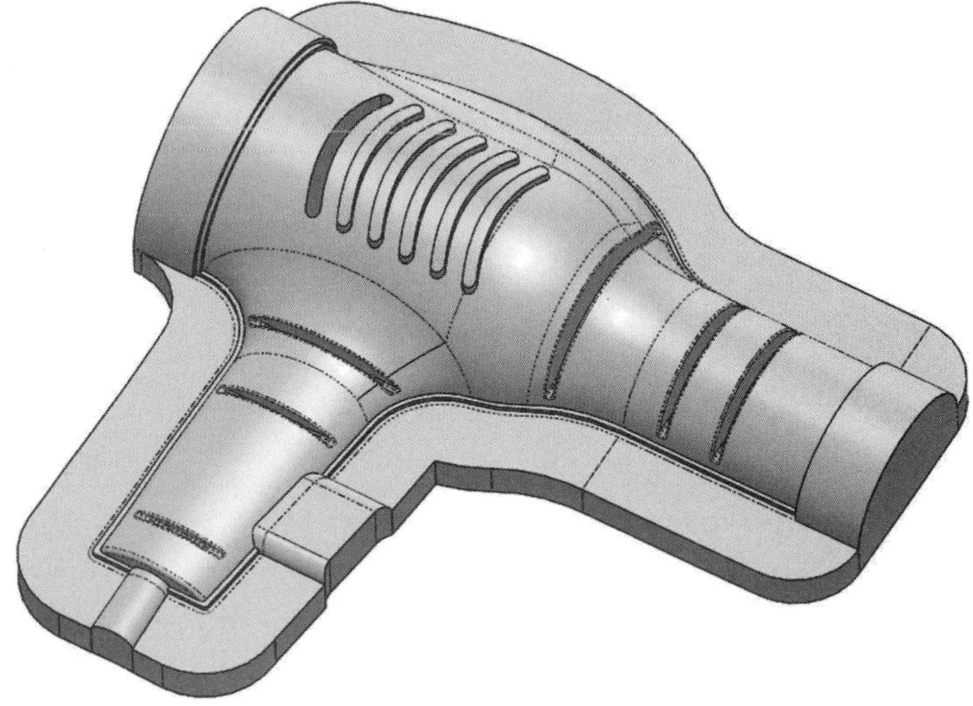

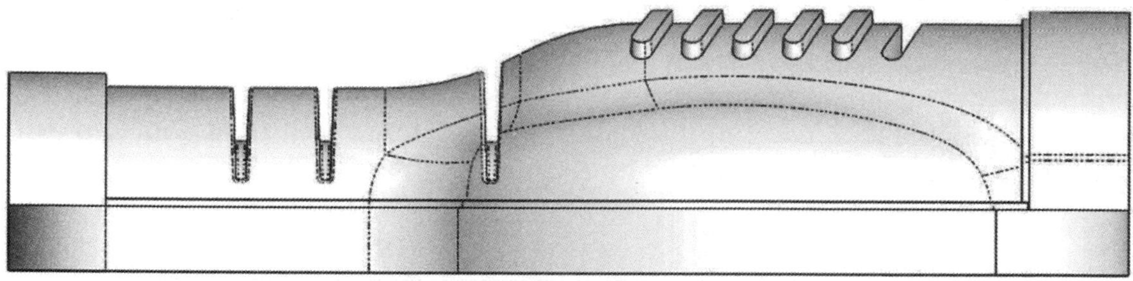

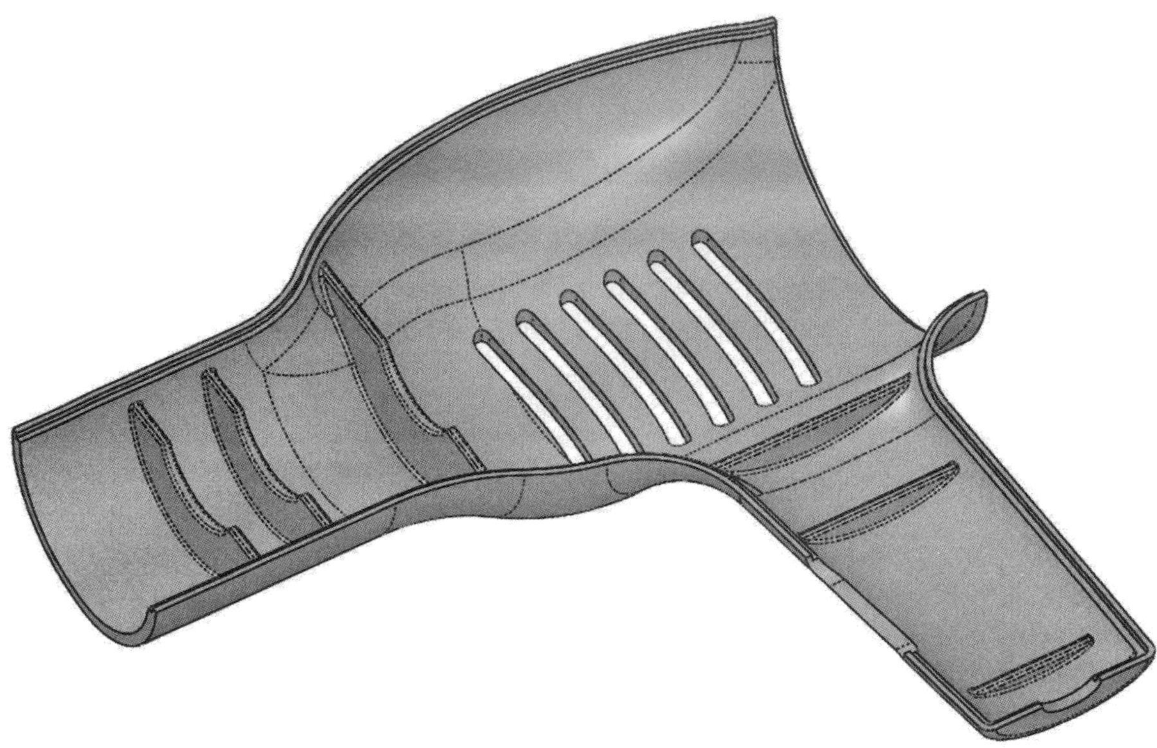

Bucket's Mold

30.1. - In the last mold exercise, we'll learn how to make a "**Side Core**." Think of a side core as a part split from a core or cavity to help us release the molded part from the mold, like a hole perpendicular to the "Direction of Pull," an undercut, or a negative draft face in the positive draft side of the part.

Open the part *'Bucket Mold'* from the included files (or use the bucket model made in the Surfacing lesson) and run a "**Draft Analysis**" with 3 degrees using the *"Top Plane"* as "Direction of Pull." Use the "Face classification" option to show the "Straddle faces" in and around the attachment holes for the handle. The blue faces will not allow the part to release from a two-part mold, making this part a good example of when we need to make a side core.

After molding a part with a side core, the side cores are pulled away from the mold assembly first in order to release the faces that would otherwise be trapped, then the mold is opened and the molded part is released. Cancel the *"Draft Analysis"* to continue; it was done to identify faces that need a draft and the straddle faces that need to be made using a side core.

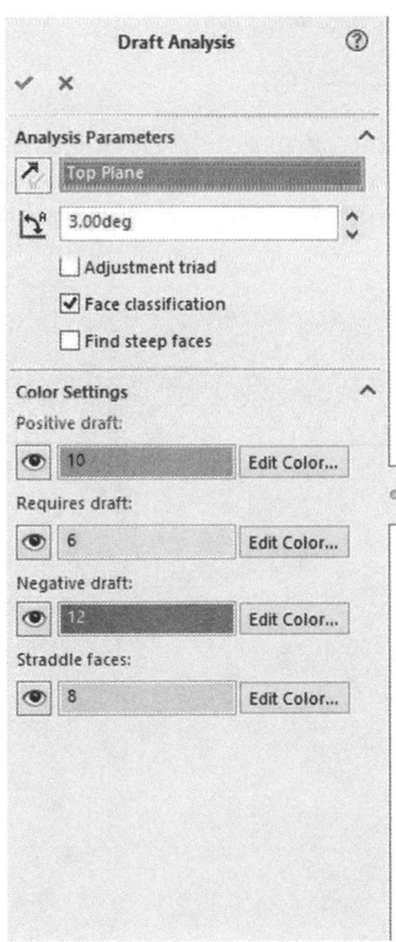

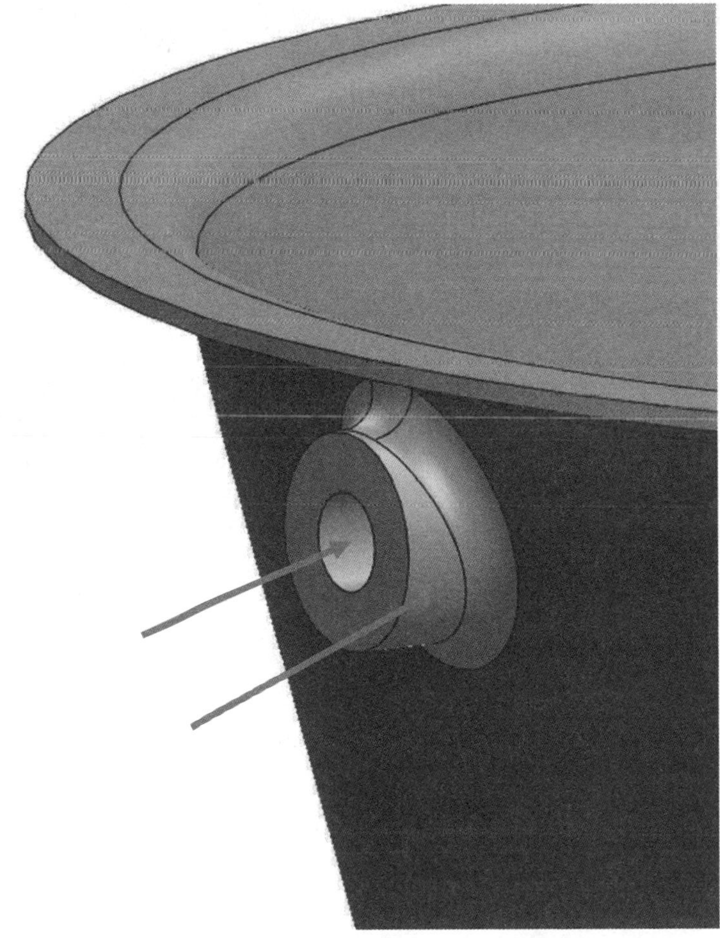

30.2. - From the draft analysis we see the lip at the top of the bucket requires a draft. Add a 3 degree draft to the outside face of the lip at the top of the bucket, in the negative draft direction (down) to mold the lip with the Cavity. Use the "Neutral Plane" and "Along Tangent" face propagation options to automatically select the rest of the faces that need a draft. Click OK to complete it.

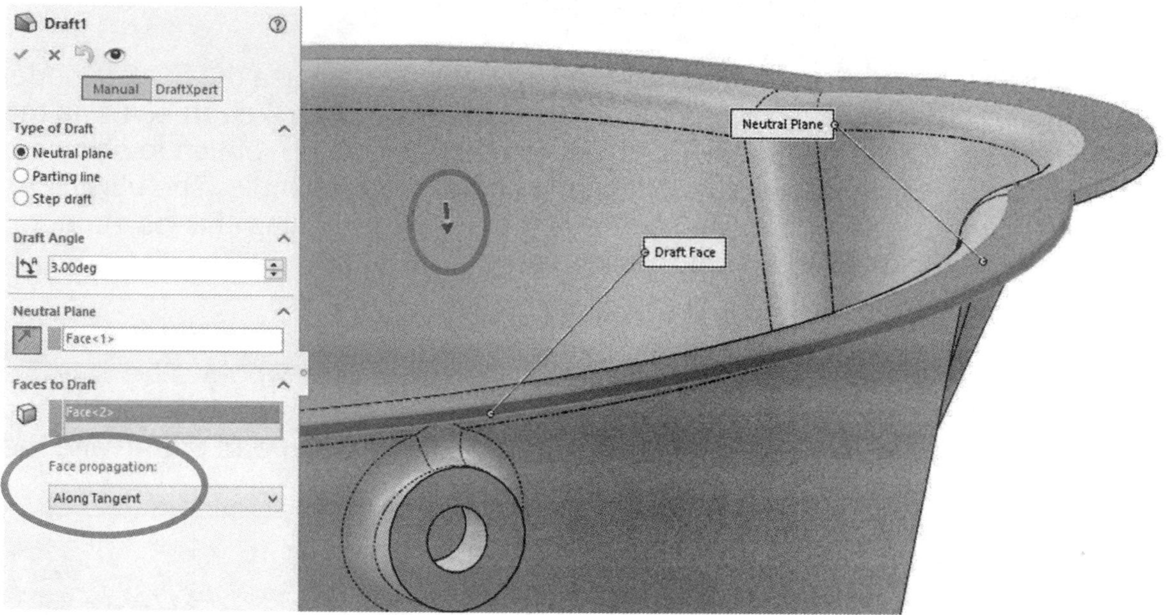

30.3. - Run a "**Draft Analysis**" using the indicated face as direction of pull; this analysis will help us identify the faces that need a draft in the side cores. The yellow faces in the handle's attachment point need to be drafted, otherwise the side core will not be able to be released.

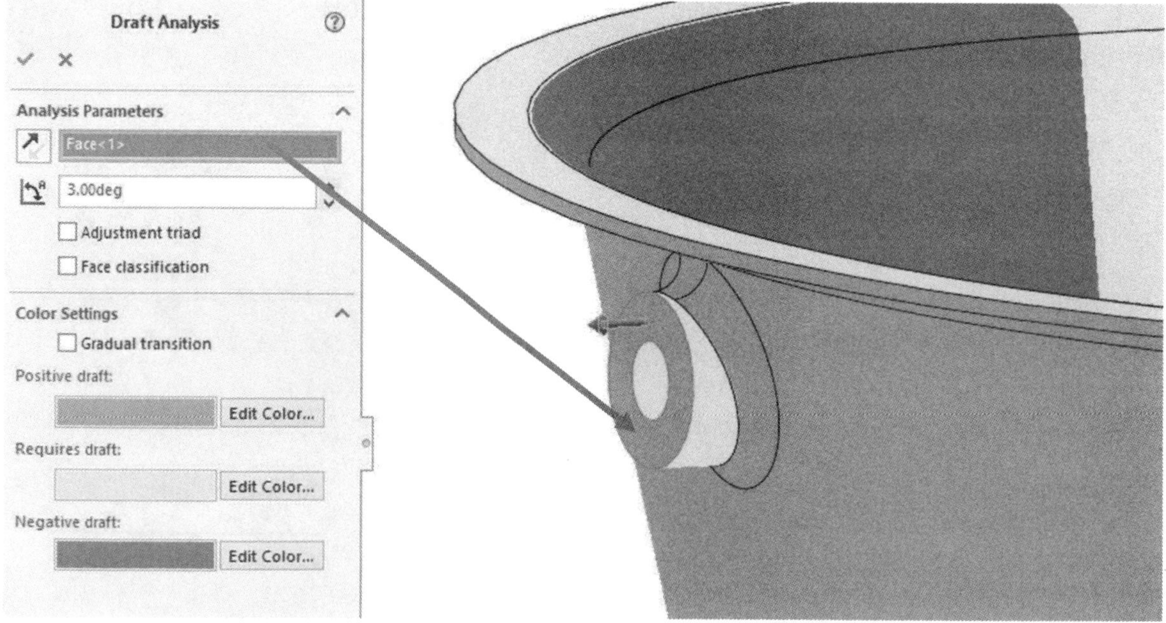

30.4. – Now that the faces that need a draft have been identified, we need to modify the extruded boss to give it a draft. Edit the *"Boss-Extrude1"* feature and activate the "Draft" option, and enter 3 degrees and turn on the "Draft outward" option. This way the boss will 'grow' as it is extruded towards the bucket. Click OK to continue.

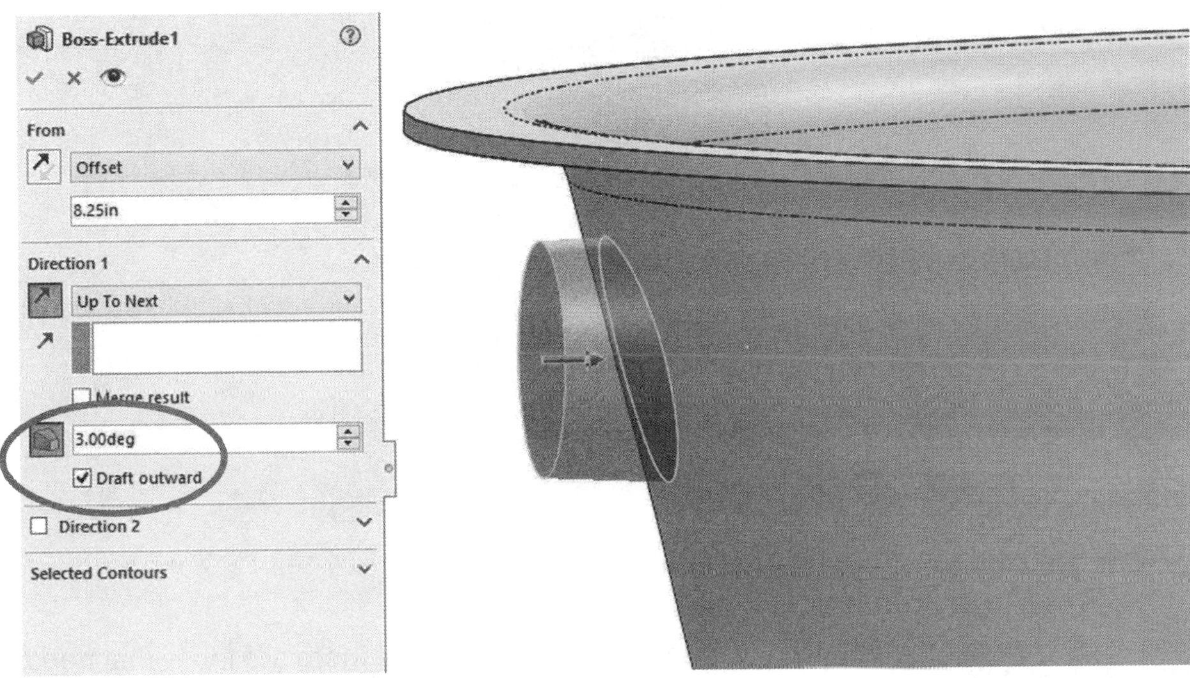

30.5. - Now edit the *"Cut-Extrude1"* feature and also activate the "Draft" option but in this case with a 2 degrees draft. Leave the "Draft Outward" option off.

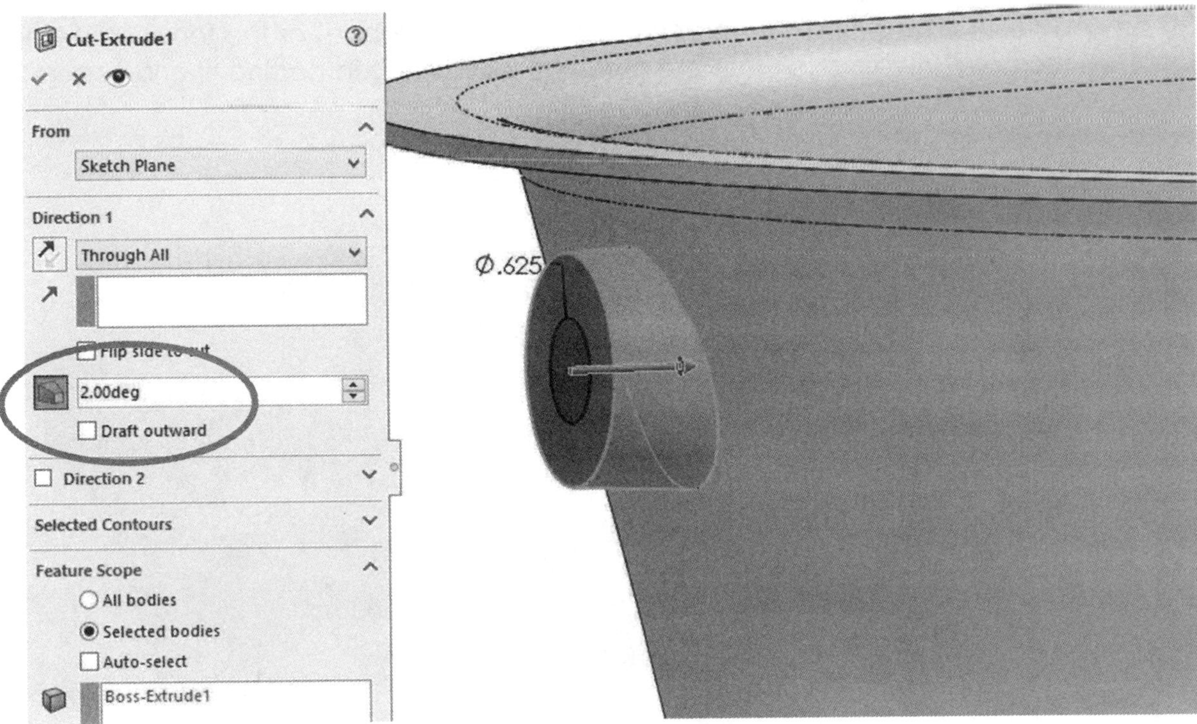

30.6. - Run a new "**Draft Analysis**" using the indicated face for "Direction of Pull" with 2 degrees. Now we can see the boss and the hole in the attachment point have the required draft to properly release the side core. Since the boss on the other side is a mirror copy it is also fixed. Cancel the "**Draft Analysis**" to continue.

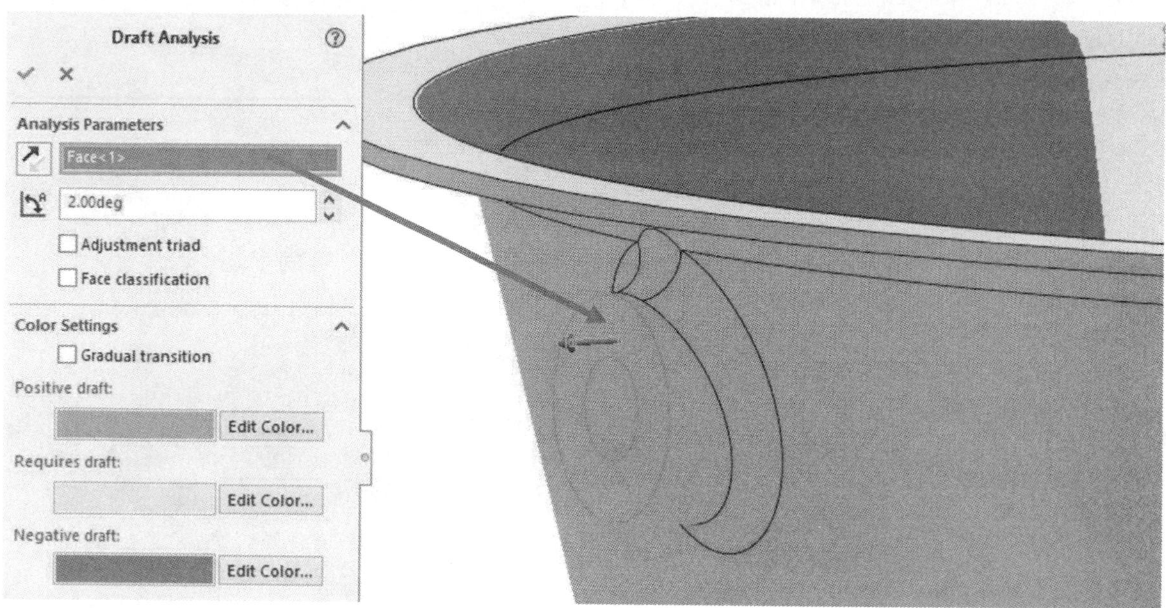

Using this face for the direction of pull will tell us that there are faces that need draft, but since we are only interested in the faces that belong to the boss, using this face for direction of pull we are sure the faces created with the side core have the required draft and will work as expected.

30.7. - Add a "**Parting Line**" command using the *"Top Plane"* or the top most face for "Direction of Pull." In this case the automatically selected parting line is correct. Click OK to continue.

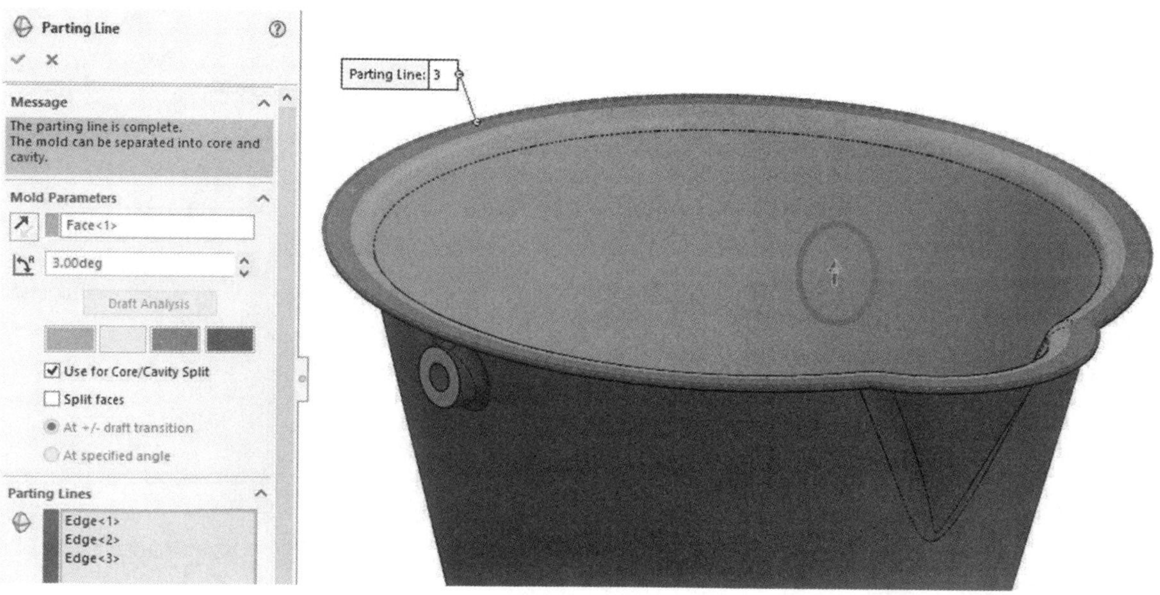

30.8. – Since the part doesn't have any holes connecting the Core and Cavity, there is no need to add shut-off surfaces and we can continue to the "**Parting Surface**." Select the "**Parting Surface**" command using the "Perpendicular to pull" option, and make the distance 2 inches. Turn on the "Smooth" option to make the transition between adjacent surfaces smoother, giving us a better result.

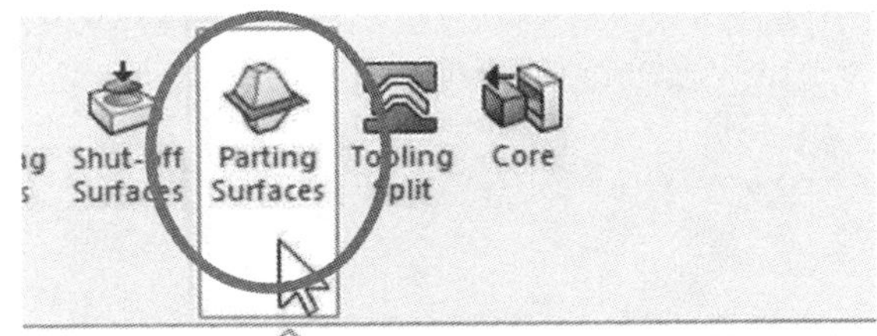

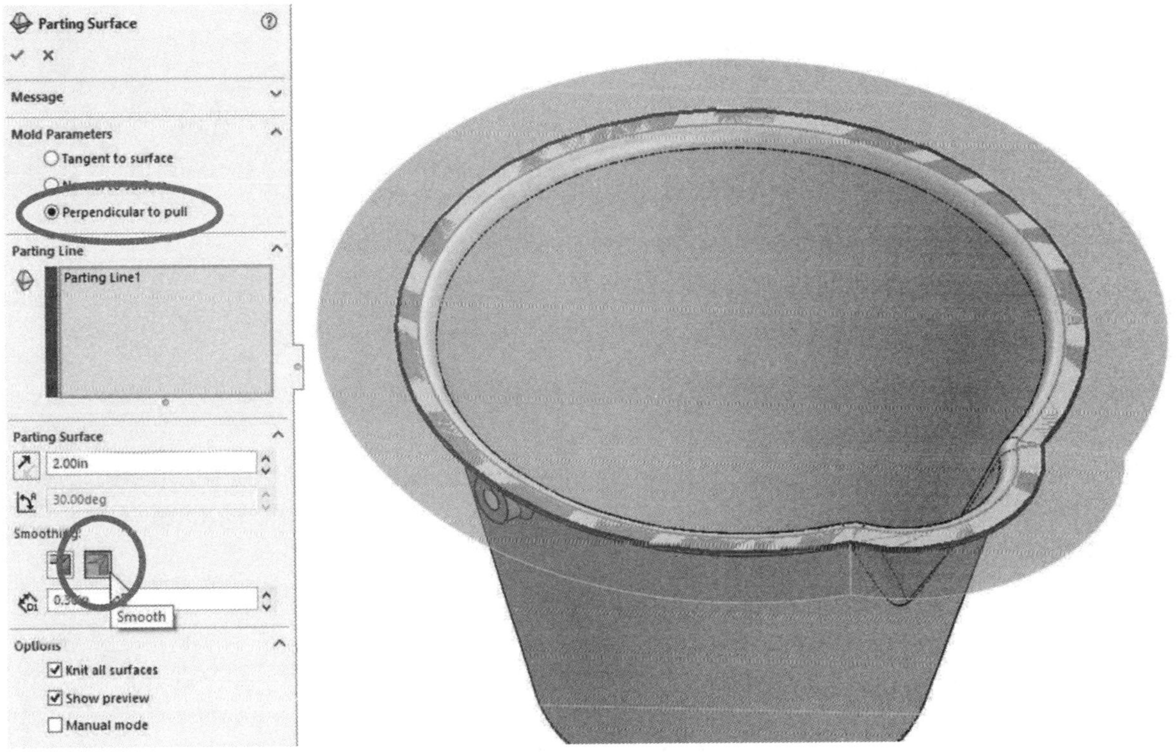

30.9. - Using the "**Tooling Split**" command, add the sketch in the parting surface, select the parting surface and project its edges into the sketch using the "**Convert Entities**" command.

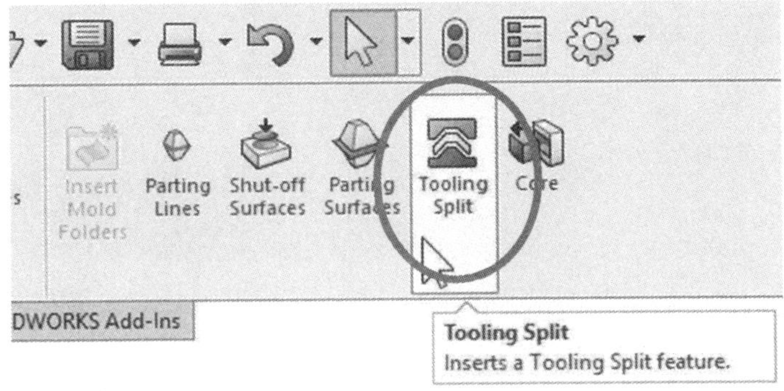

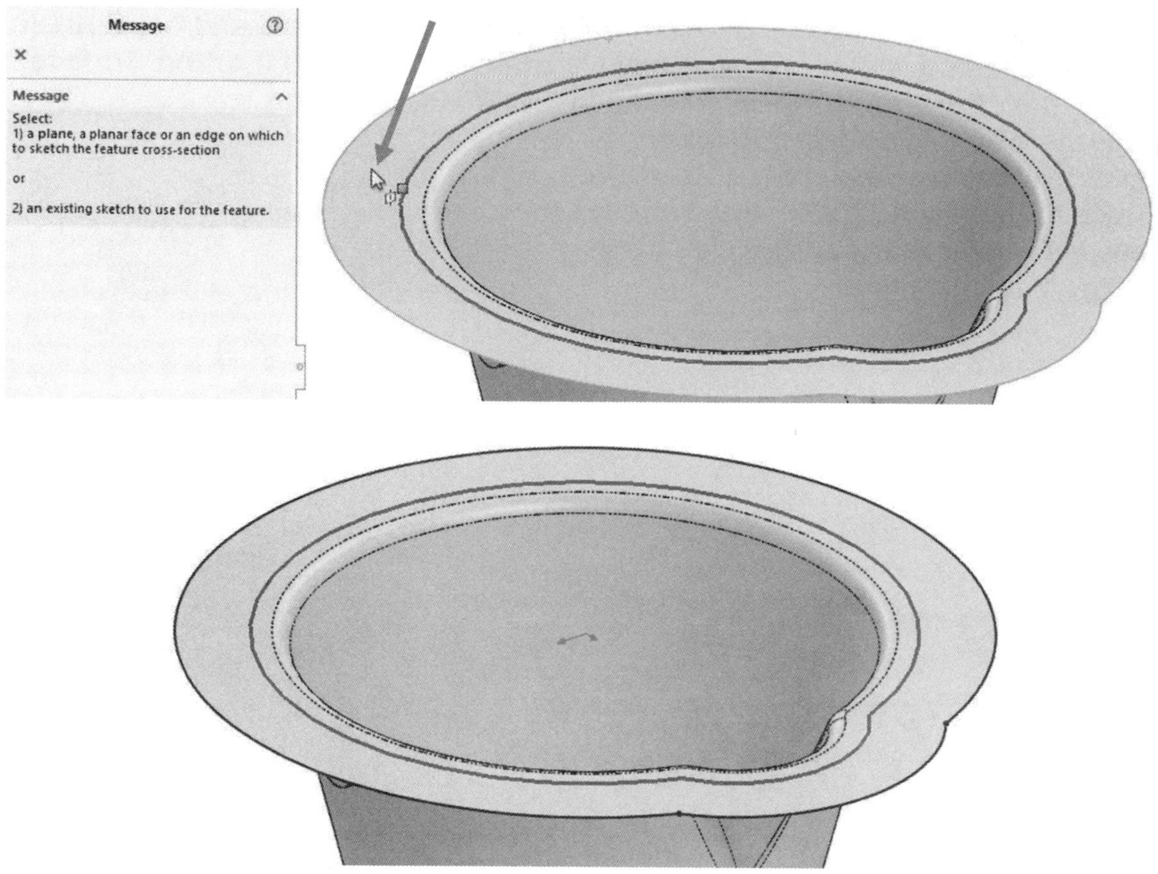

After the parting surface's edges are projected, exit the sketch and make the block size 1″ up and 13″ down as shown to fully enclose the bucket. Click OK to finish and hide the *"Parting Line1"* feature.

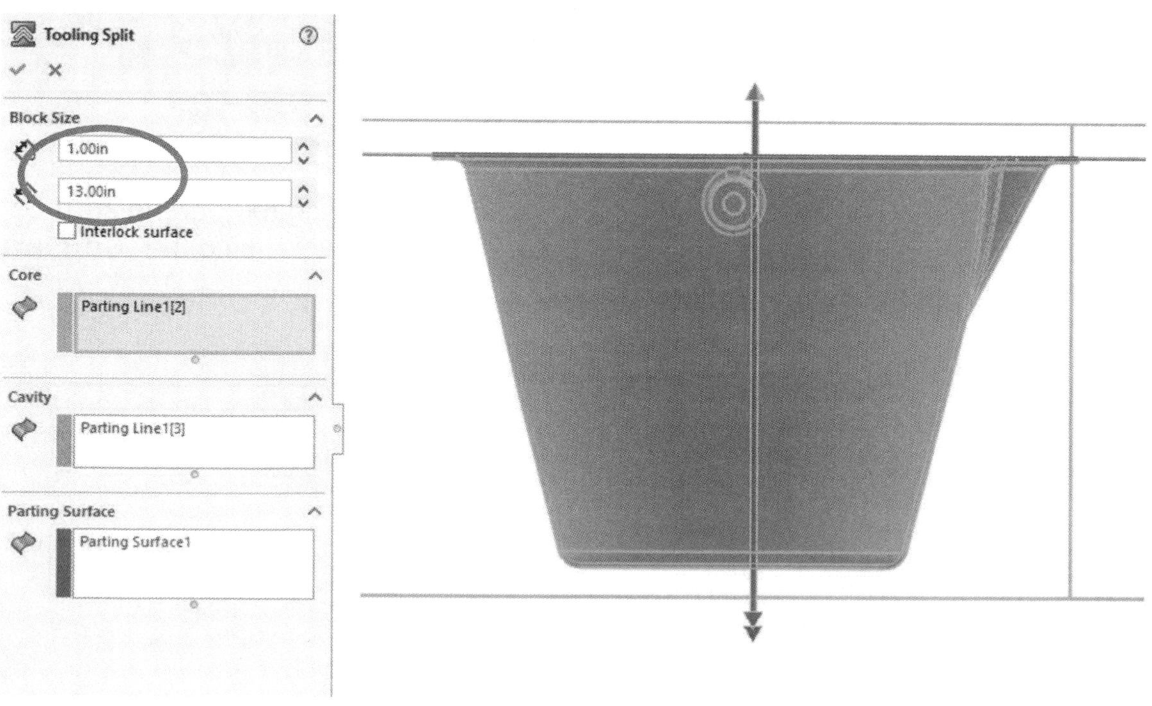

30.10. - After the tooling split is made we have the Core and Cavity bodies. The side cores are made by splitting a body from these bodies, in this case from the cavity. Switch to a Front view with hidden lines visible mode, and zoom in the attachment point; this is the area from which we'll split the first side core.

Add the following sketch in the "*Front Plane.*" The sketch is a closed profile and starts at the parting line. The objective is to split a body (the side core) from the cavity body using this sketch as a cutting tool. Exit the sketch to continue.

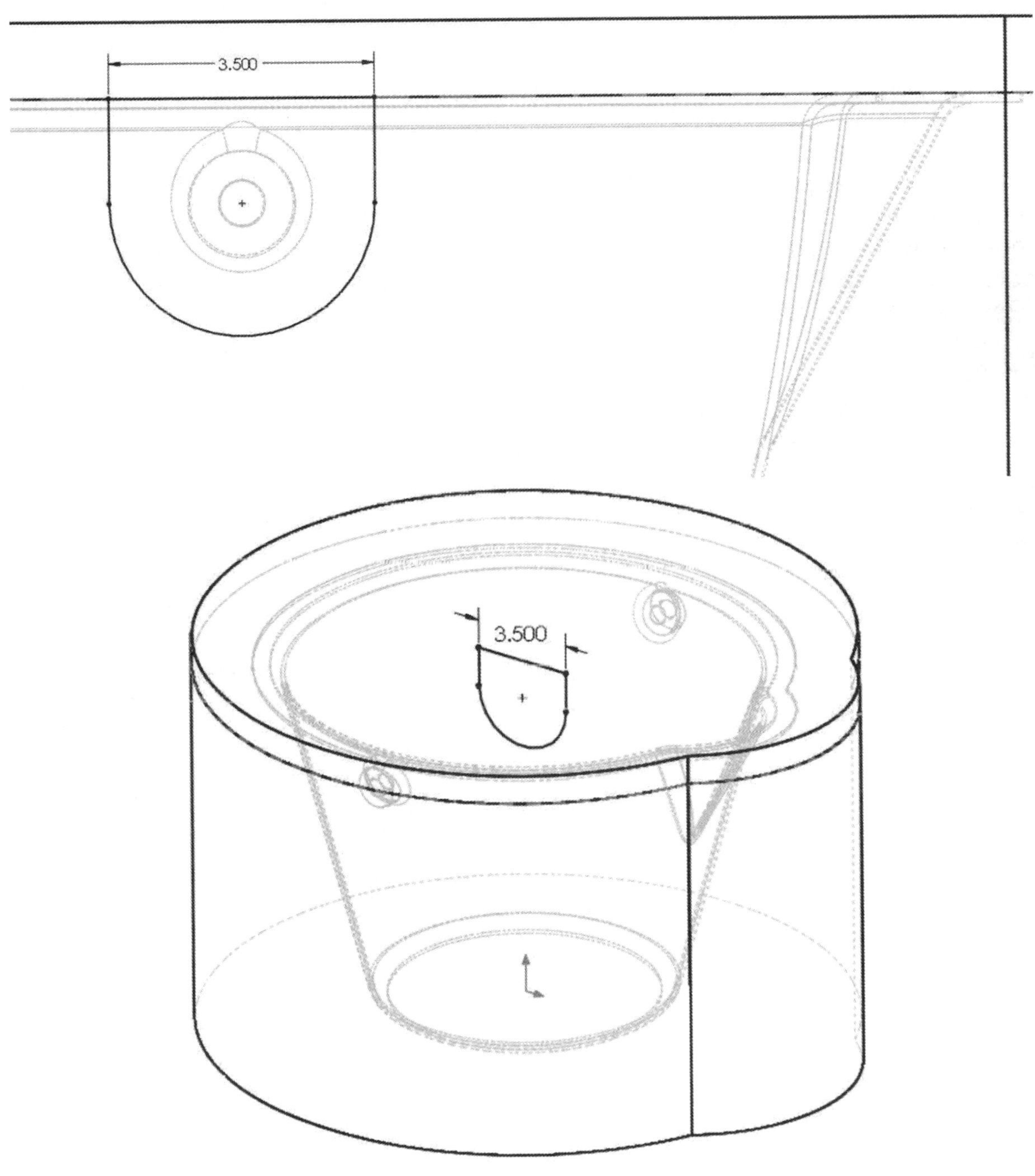

30.11. - Select the "**Core**" command from the Mold Tools tab or the menu "**Insert, Molds, Core**." When asked to select a planar face or an existing sketch, select the previous sketch to continue. It works the same as the "**Tooling Split**"; we can make the sketch before or after selecting the command.

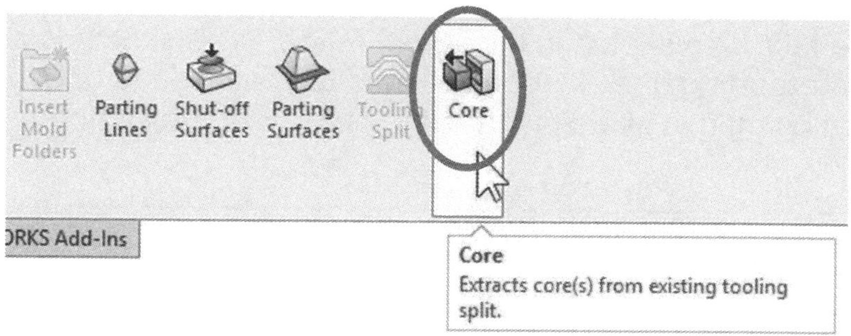

The "**Core**" command allows us to extract the side cores from either the core or the cavity body; in essence, this command splits a solid body and offers a few more options than the "**Split**" command. In the "Core/Cavity body" selection box select the cavity's body, which is the one we want to split the core from, and in the "Parameters" options use the "Through All" end condition to cut the cavity. The preview will show the Core body that will be split from the cavity body. Enter a draft angle of 1 degree; keep in mind that this draft will be added to the outside faces of the core to allow it to easily separate from the mold assembly after the part is molded, and is different from the part's draft. Click OK to split the first side core.

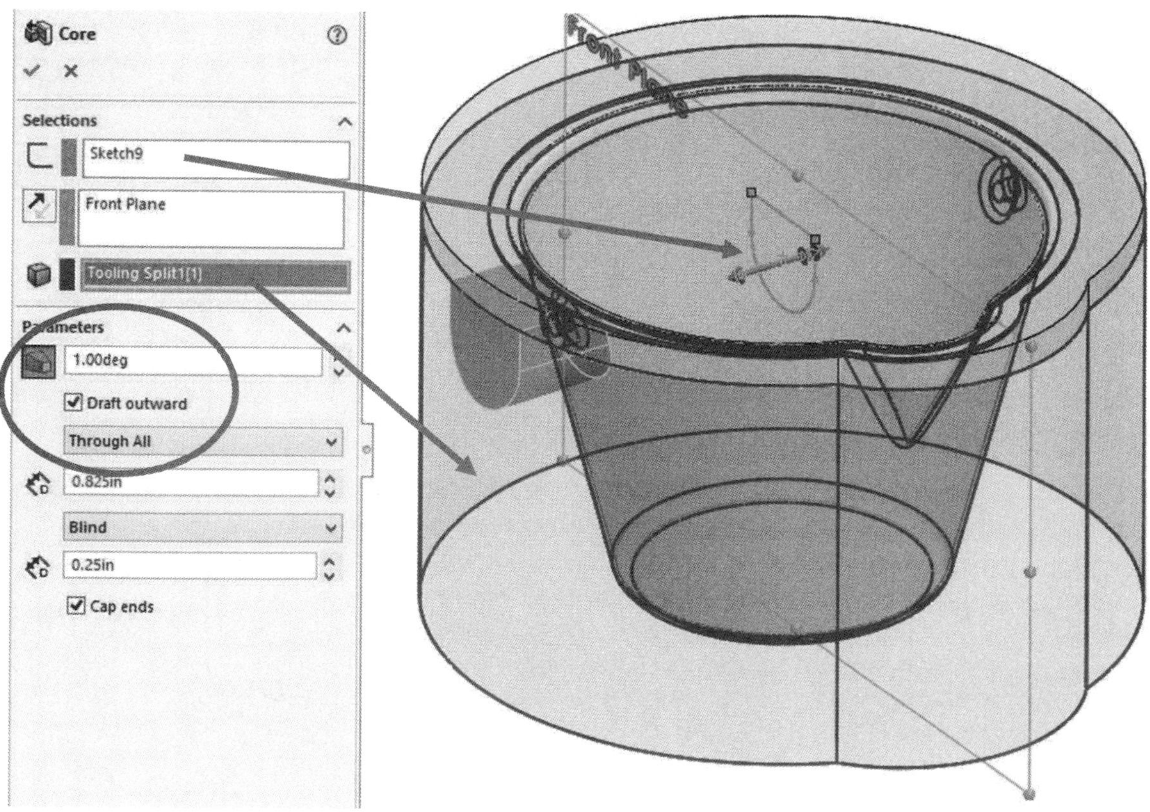

After the core is split from the cavity body it is added to a *"Core bodies"* folder in the *"Solid Bodies"* folder.

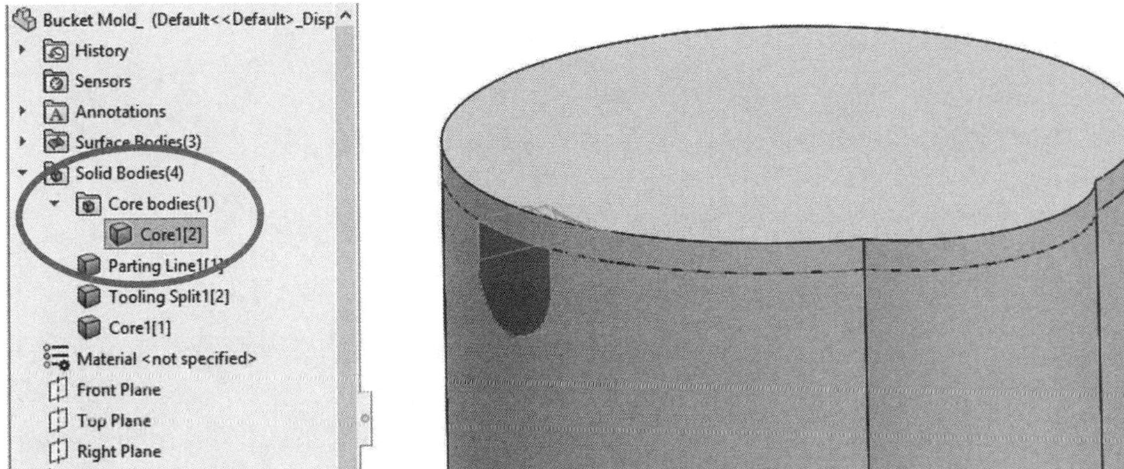

30.12. – For the second side core, we'll add a derived sketch of the previous core also in the *"Front Plane."* Pre-select the core's sketch and the "*Front Plane*" and select the menu "**Insert, Derived Sketch**." Add the necessary relations to fully define it just like the first core. Exit the sketch and select the "**Core**" command from the Mold Tools tab. Select the derived sketch and make a second core using the same settings as before, but this time going in the opposite direction. Click OK to finish the second core. The reason to add a derived sketch is because once a sketch is used in a Core feature, it cannot be reused in another Core.

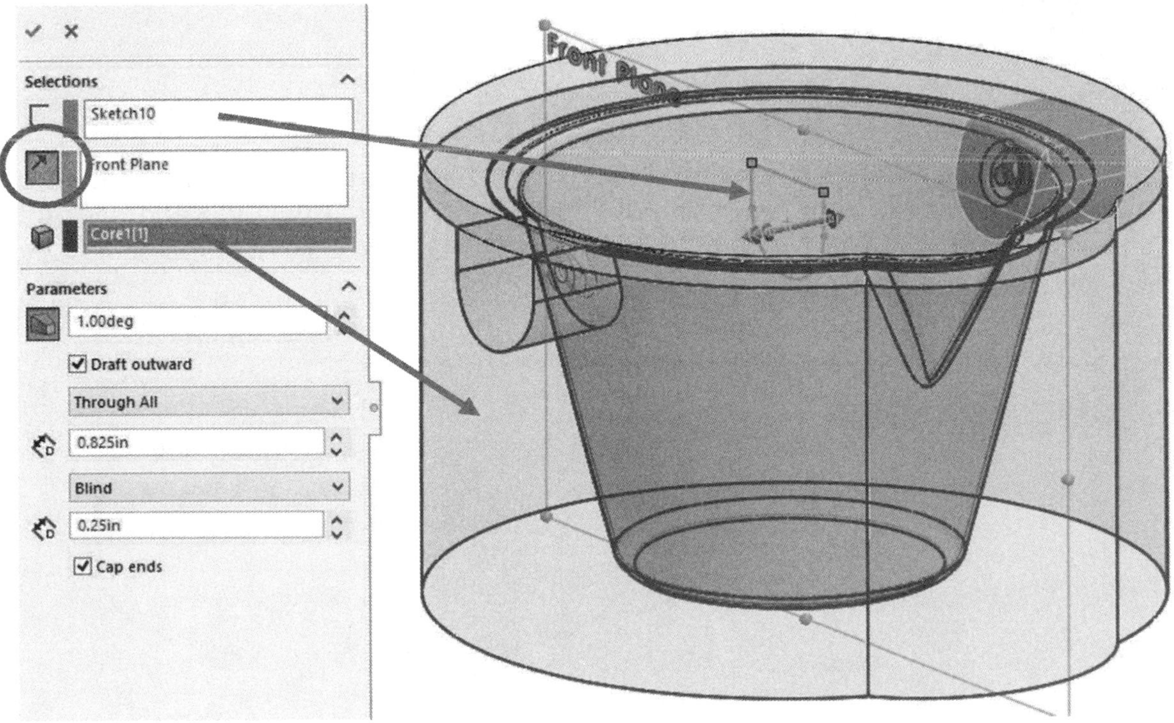

The two side cores are highlighted after splitting them from the cavity body. The bucket, all surface bodies and the Core's body have been hidden for visibility.

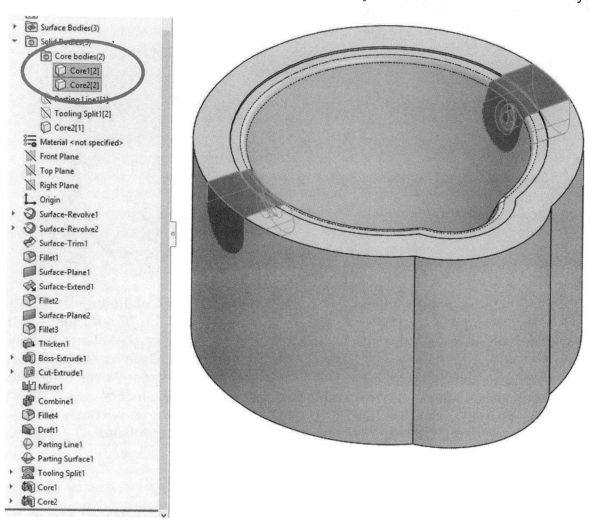

30.13. - Hide all the surface bodies, show the solid bodies and use the "**Move/Copy**" command (menu "**Insert, Features, Move/Copy**") to separate and see the different bodies of the mold including the side cores. The bucket's body color was changed for visibility.

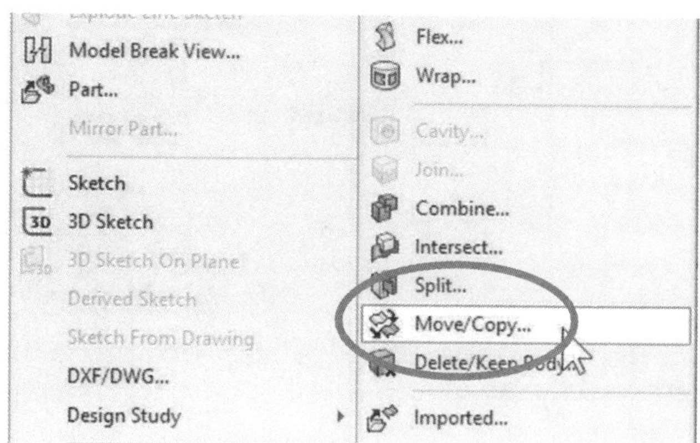

The "**Move/Copy Bodies**" command and other solid and **surface editing** tools are available in the Direct Editing tab in the CommandManager.

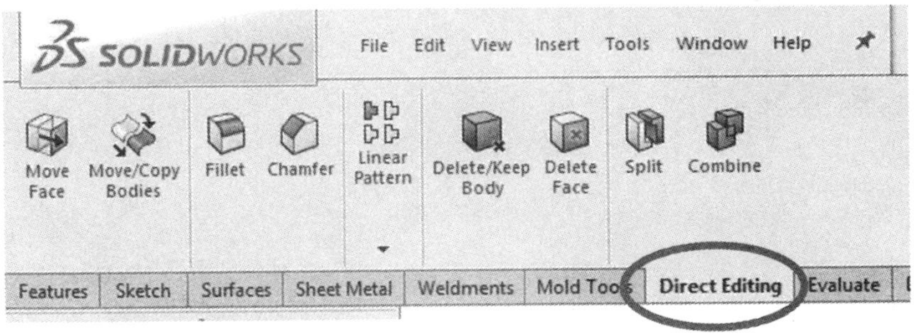

At this point, the different bodies can be inserted into a new part to add the remaining mold specific features to each one. Save the part and close.

Exercise 1: Save the core, cavity and side cores of the bucket's mold to external files and make an assembly with all the components. Add an exploded view for presentation purposes.

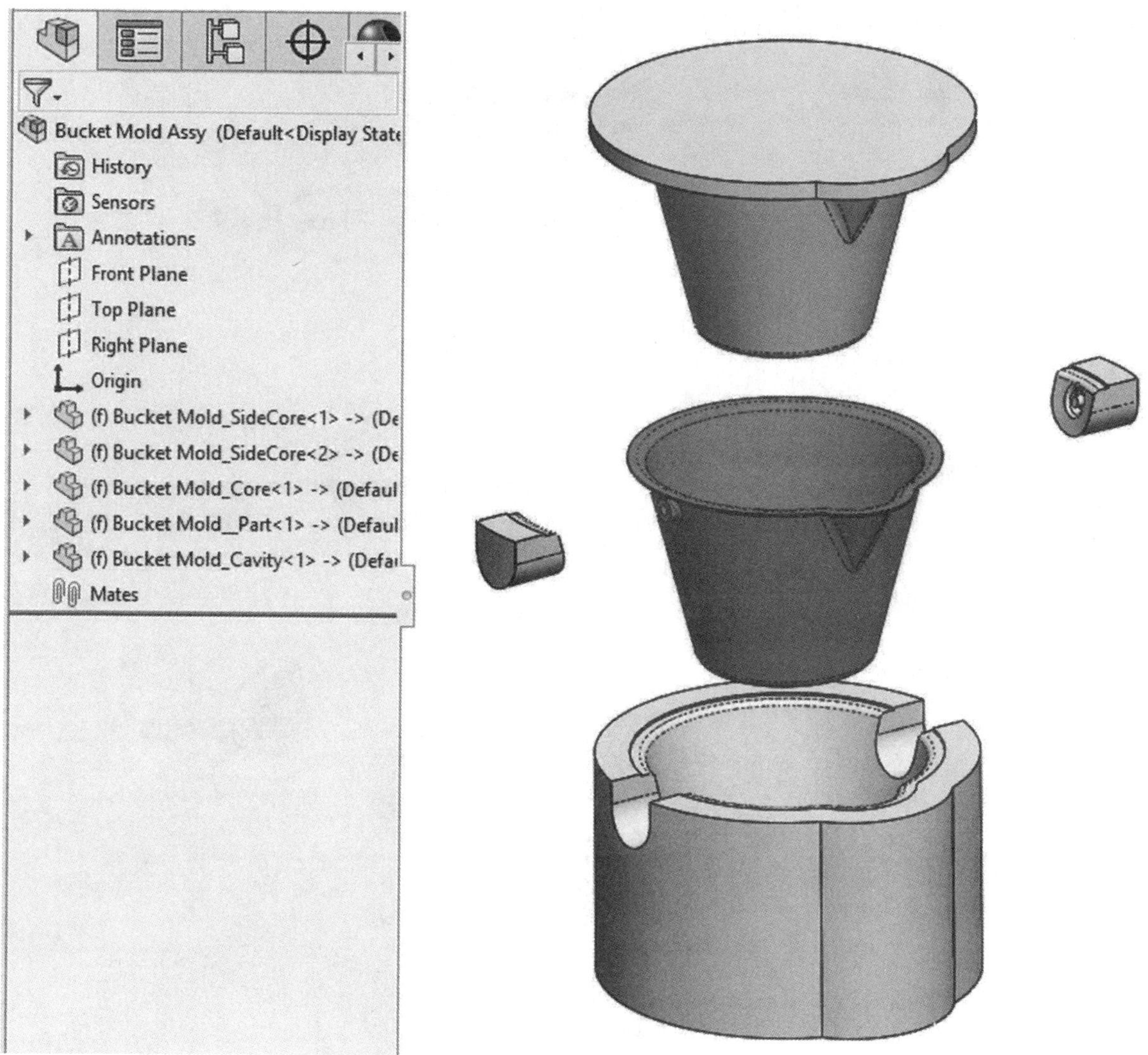

Exercise 2: Open the provided part *'Remote Master'*. Save the top body and the bottom body each to a new part, make a tooling split for each half of the remote control, and create side cores for the trapped areas of the mold.

'Remote Master' (2 solid bodies)

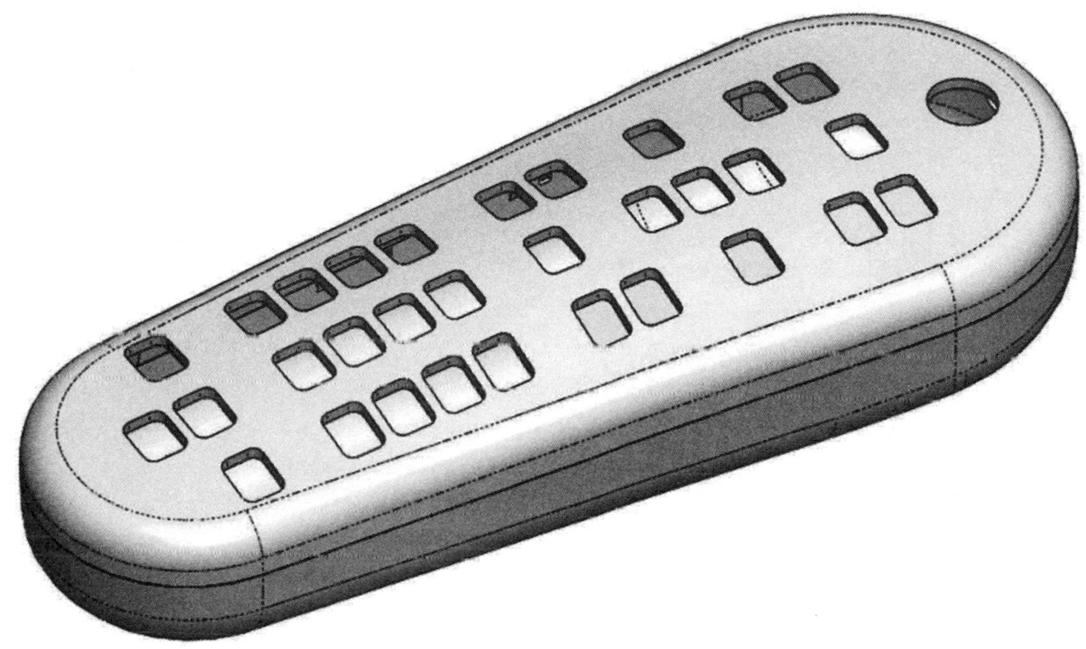

Completed parts and molds

Save the 'Remote Bottom' to a file, add mounting bosses and ribs.

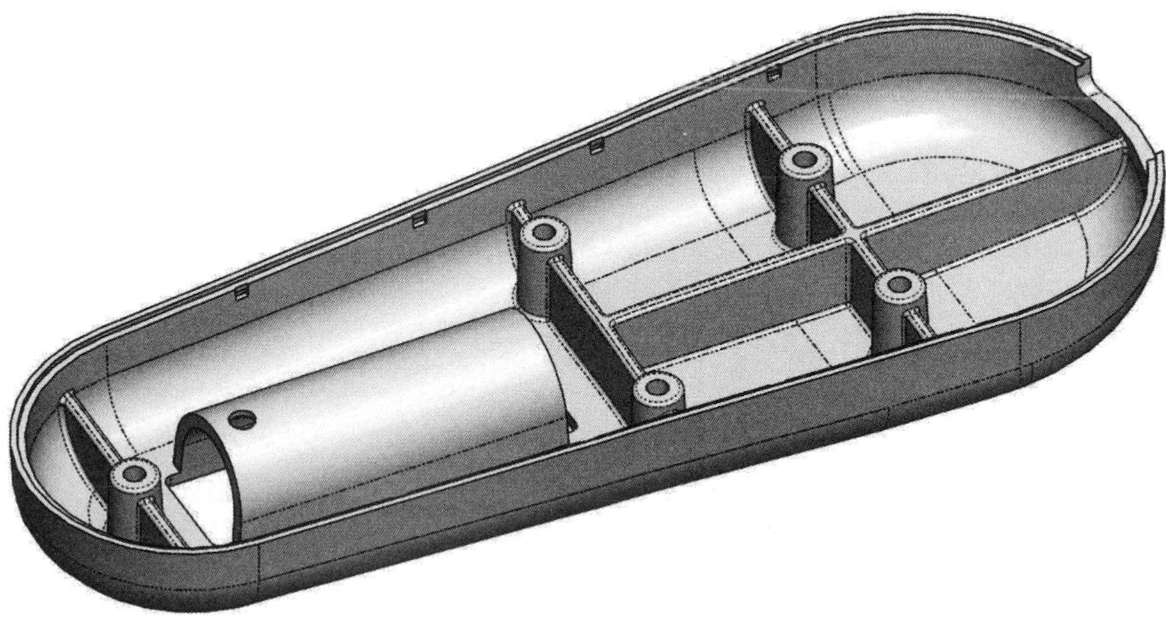

Suggestion:

Add a plane 0.150” below the parting line, add the sketch and extrude the bosses using “Up to Next” with a 2° draft.

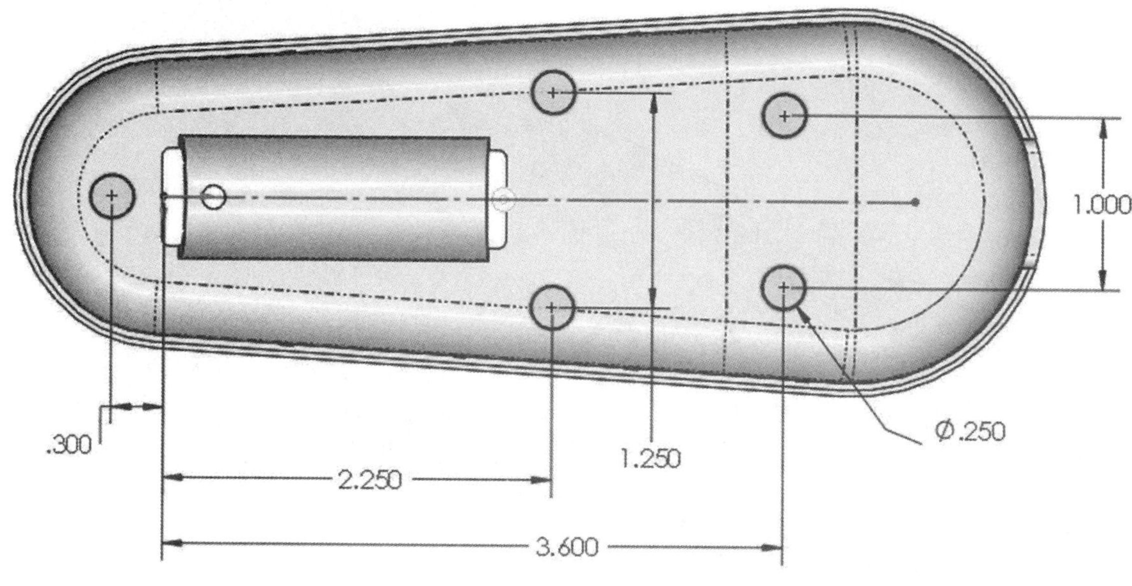

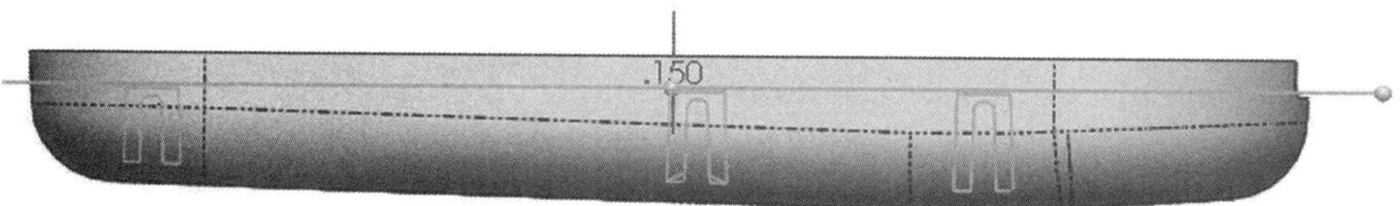

Add the ribs using these settings, and add a fillet to bosses and ribs at the end.

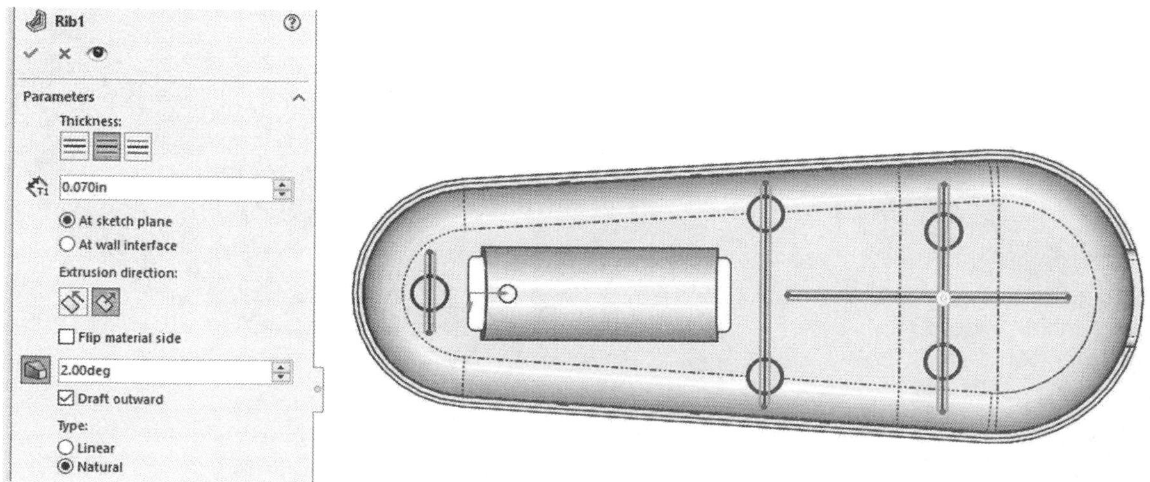

Add a Parting line, Shut-Off Surface to close the openings, Parting Surface and Tooling Split.

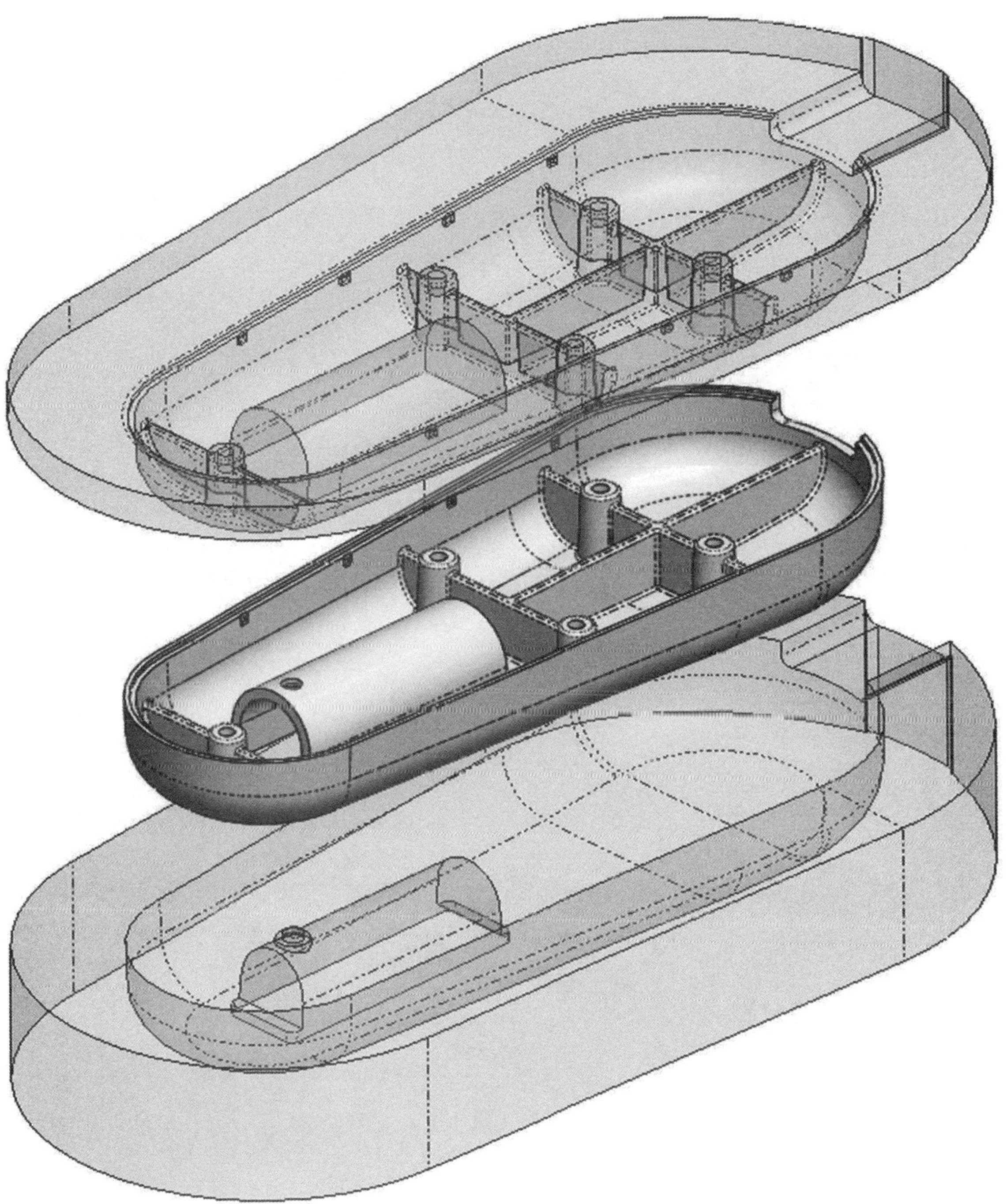

After making the tooling split of the 'Remote Bottom', save the 'Remote Bottom-Cavity'.

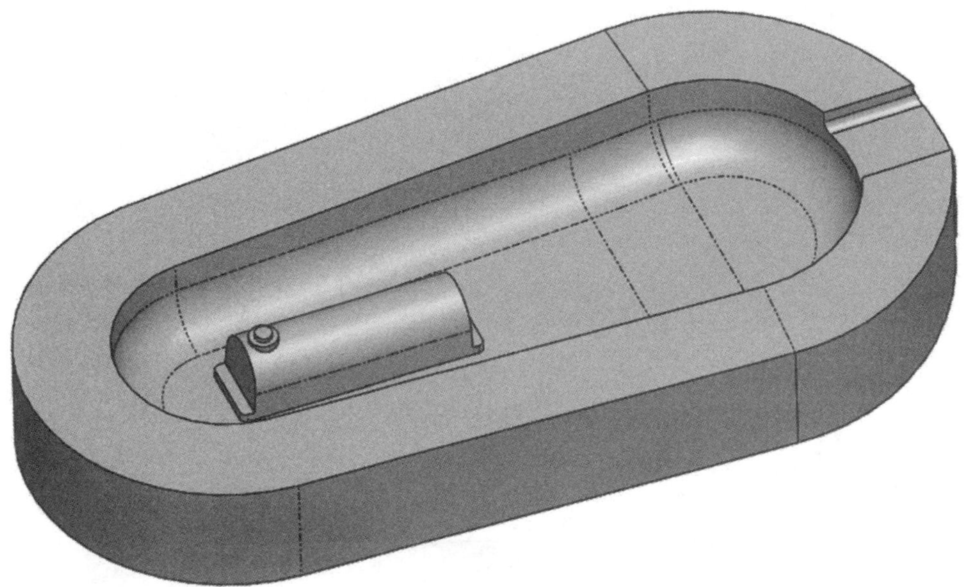

Save the 'Remote Bottom-Core' and create a side core with a multiple sketch to cut all cores at the same time. Save the 'Remote Bottom–Core-Insert' to a separate part.

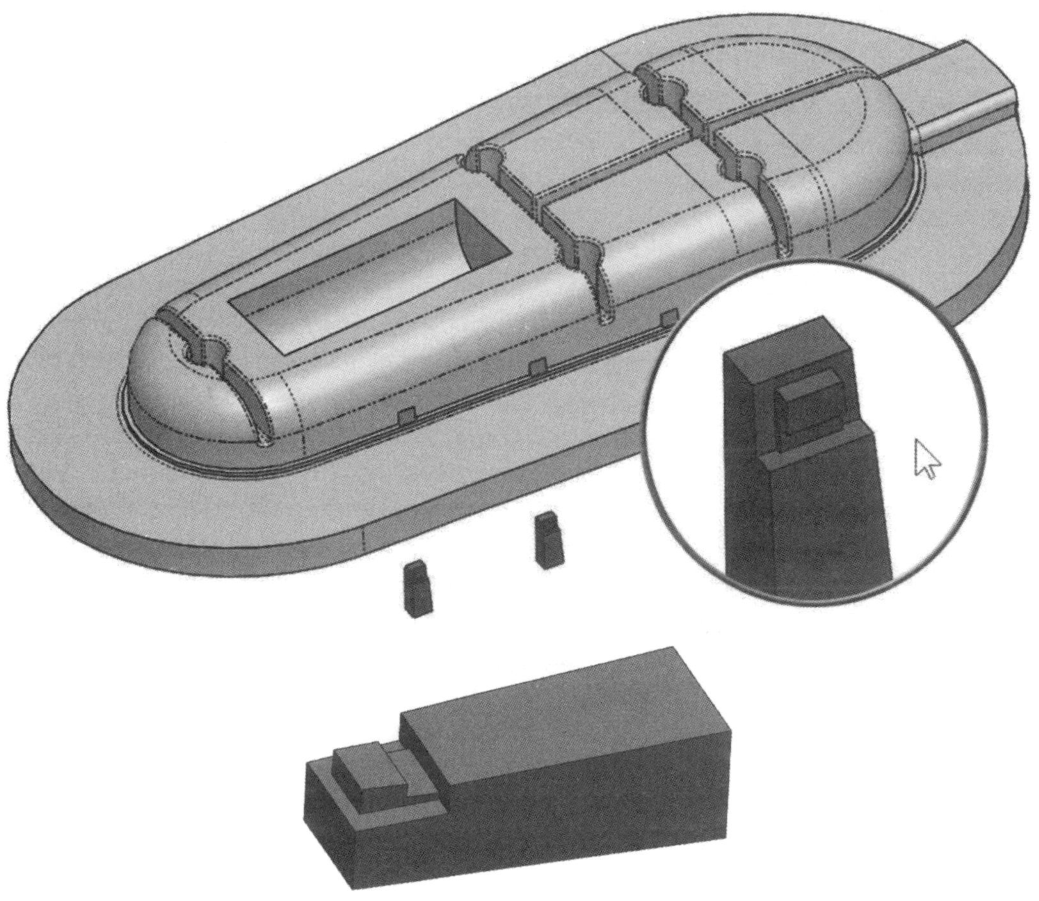

Save the 'Remote Top' to a part file, add the necessary ribs, mounting boss and fillets.

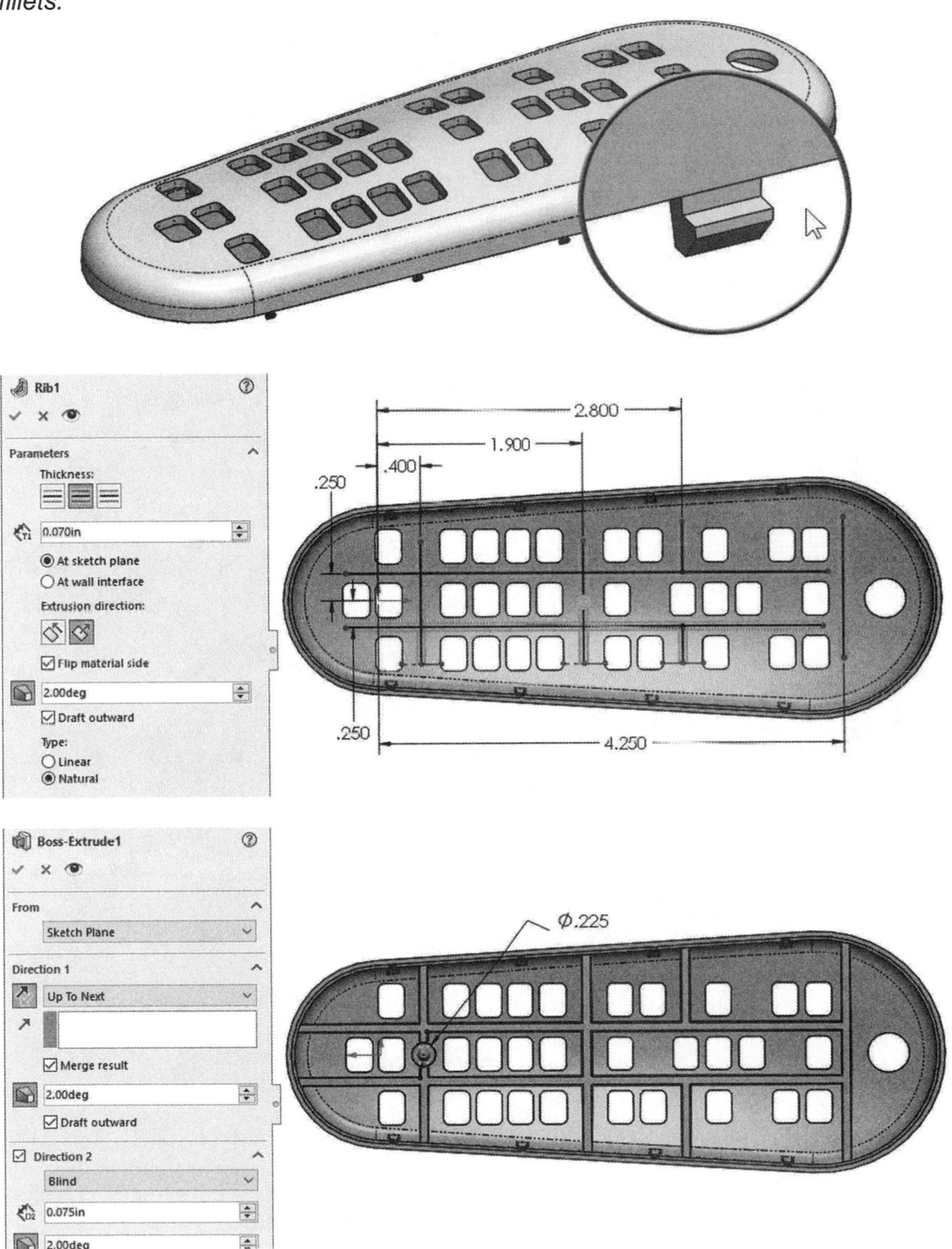

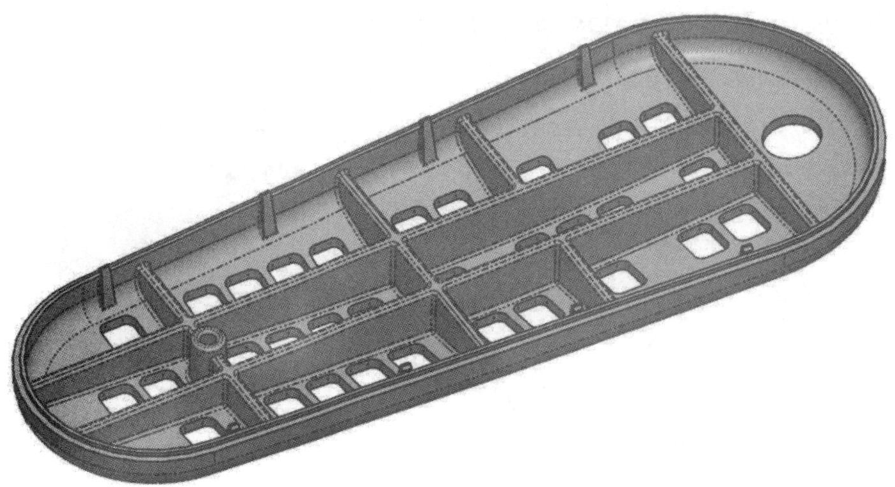

Add a Parting Line, Shut-Off Surfaces, Parting Surface and Tooling Split.

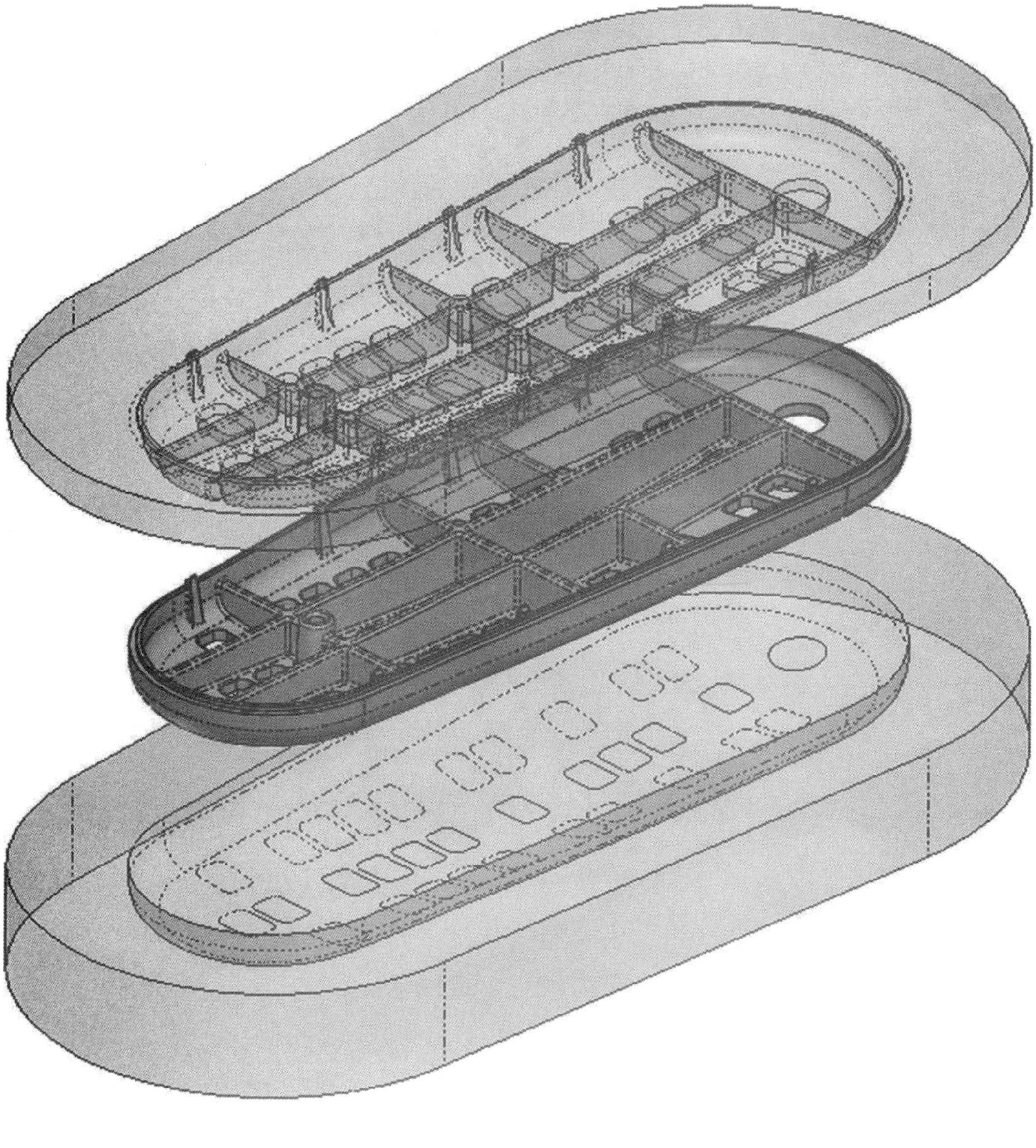

Save the 'Remote Top Cavity' and eliminate the split surface inside using the "Delete Face" command.

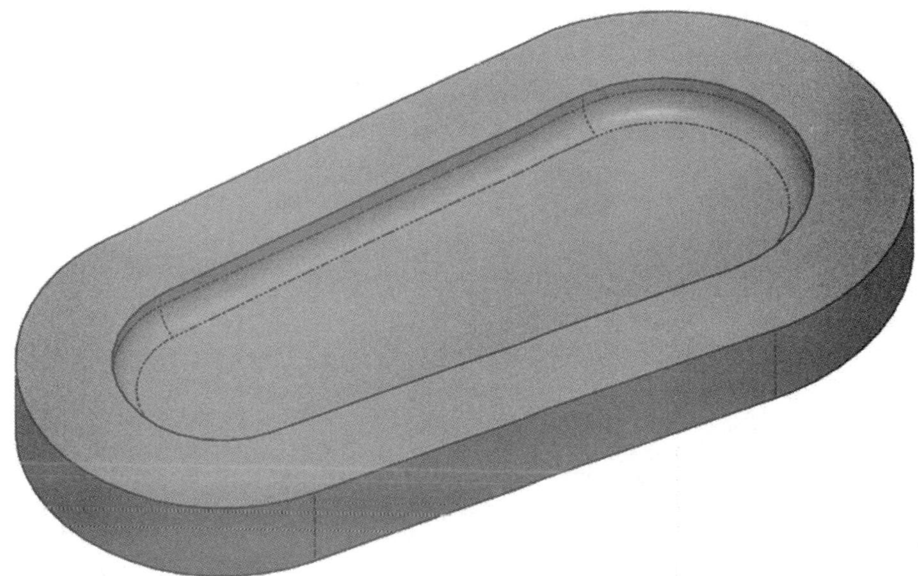

Save the 'Remote Top-Core' and split all side cores from one side at the same time. Save the 'Remote Top–Core-Insert' to a part.

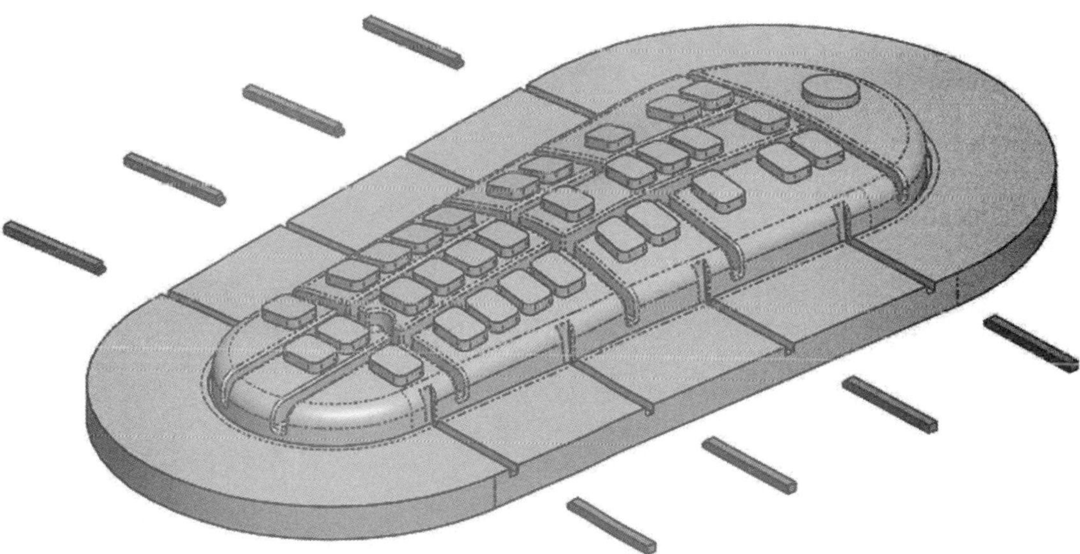

'Remote Top Core-Insert'

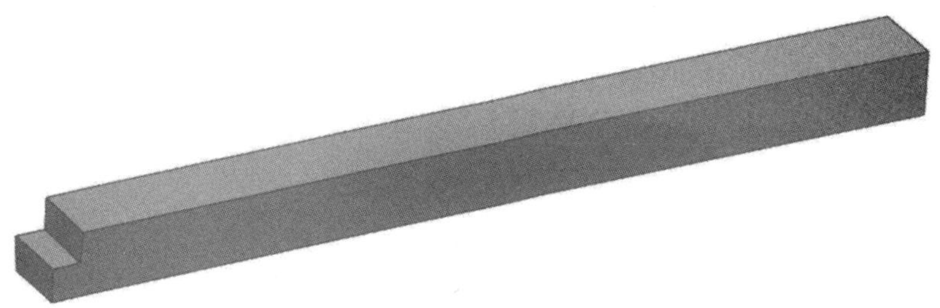

Notes:

Final Comments

After going through the exercises in the book, the reader will have acquired a good understanding of several topics including sheet metal, welded structures, 3D Sketching, Top Down design, editing tools, Multi Body parts, Surfacing, Mold tools and a handful of tips and tricks that can help a designer work faster and more efficiently.

Even though most of the topics included in this book are usually considered advanced, they are presented using examples that break down the concepts and make them easier to understand. Topics covered explore the most commonly used options to enable a reader to use them efficiently in different design tasks.

The Sheet Metal, Mold tools and Welded structures topics fall within the scope of each individual manufacturing trade, each of which has very specific tools, techniques and *fine details* that are (usually) learned after studying and by experience; for example, mold making uses most of the tools available in SOLIDWORKS, going from the most basic features all the way up to advanced surfacing tools.

The amount of trade specific training that can be covered in a CAD oriented book is limited, since the objective of the book is to teach how to use SOLIDWORKS and not a trade. If a reader has experience in one of these fields, he/she will quickly understand the purpose of these specific tools, but we also tried our best to explain in simple terms the terminology and processes, so that a reader unfamiliar with a trade can understand the purpose of a tool and the reason for these operations.

We welcome your suggestions and ideas. As this book evolves, new functionality and examples will be added, but rest assured we'll always try our best to make sure the models shown maximize the functionality covered in as few pages as possible, helping you to understand and master each topic. If you have any comments about this book, please let me know. I promptly reply to all email, and am always happy to hear from readers like you.

Sincerely,

Alejandro Reyes,
Certified SOLIDWORKS Expert and Instructor.

alejandro@mechanicad.com
Mechanicad Inc.

Notes:

Index

Notes: